CDMA Systems
Engineering Handbook

For a complete listing of the *Artech House Mobile Communications Library*, turn to the back of this book.

CDMA Systems Engineering Handbook

Jhong Sam Lee
Leonard E. Miller

Artech House
Boston • London

Library of Congress Cataloging-in-Publication Data
Lee, Jhong S.
    CDMA systems engineering handbook /Jhong S. Lee, Leonard E. Miller.
        p.   cm. — (Artech House mobile communications library)
    Includes bibliographical references and index.
    ISBN 0-89006-990-5 (alk. paper)
    1. Code division multiple access.   2. Mobile communication systems.
    I. Miller, Leonard E.   II. Title.   III. Series.
TK5103.45.L44   1998
621.382—dc21   98-33846
                                                                          CIP

British Library Cataloguing in Publication Data
Lee, Jhong S.
    CDMA systems engineering handbook.— (Artech House mobile communications
library)
    1. Code division multiple access—Handbooks, manuals, etc.
    I. Title   II. Miller, Leonard E.
    621.3'84'56

    ISBN 0-89006-990-5

Cover design by Lynda Fishbourne

International Standard Book Number: 0-89006-990-5
Library of Congress Catalog Card Number: 98-33846

10 9 8 7 6 5 4

To our wives, Helen Lee and Fran Miller

Table of Contents

Preface

The first North American digital cellular standard, based on time-division multiple-access (TDMA) technology, was adopted in 1989. Immediately thereafter, in 1990, Qualcomm, Inc. proposed a spread-spectrum digital cellular system based on code-division multiple-access (CDMA) technology, which in 1993 became the second North American digital cellular standard, known as the IS-95 system. This book is written for those who are interested in learning all about the technical basis of the design and the operational principles of the IS-95 CDMA cellular system and of related personal communication services (PCS) systems, such as the one specified as the standard J-STD-008. These CDMA communications systems are spread-spectrum systems and make use of most of the modern communication and information-theoretic techniques that have so far been discovered by so many scientists and engineers. The primary objective of this book is to explain *in a tutorial manner* the technical elements of these remarkable wireless communication systems *from the ground level up*. The book also provides in the beginning chapter all the *tools*, in the form of systems analysis basics, for those who need them to understand the main flow of the text in the succeeding chapters. In that sense, the book is self-contained. We have generated many *problems, each with a solution,* to aid the reader in gaining clear and complete understanding of the subjects presented. In order to keep the reader's attention focused on the main flow of the discussion, where necessary, involved mathematical derivations are put into an appendix in each chapter.

This book is based on materials used for extensive technical courses conducted by the first author of this book in Korea and the United States since 1993, under variations on the generic title of *Elements of the Technical Basis for CDMA Cellular System Design*. In Korea alone over 1100 hours of lectures on these topics were given at various organizations: the Electronics and Telecommunications Research Institute (ETRI); the Central Research Center of Korea Mobile Telecommunications Coporation (now SK Telecom); Shinsegi Telecommunications, Inc. (STI); Seoul Communication Technology Co. (SCT); Hyundai Electronics Industries Co., Ltd. (HEI); Hansol PCS; and Korea Radio Tower, Inc. (KRT). A tutorial on this subject was also given at the 2nd CDMA International Conference (CIC) held in Seoul in October, 1997. The style and manner of presentation of the technical issues were motivated from the experience of teaching practicing engineers at these industrial facilities. The book is organized into eleven chapters:

Chapter 1, *Introduction and Review of System Analysis Basics*, begins by reviewing the fundamental concepts of SS and CDMA systems. The rest of the chapter contains tutorial coverage of systems analysis basics relevant to succeeding chapters, including sampling theory, waveshaping for spectrum control, and the use of probability functions in systems analysis.

Chapter 2, *Mobile Radio Propagation Considerations*, provides an overview of radio propagation loss models and mobile radio channel models, including the use of these models in cellular design. The material is given in some detail in order to enable the engineer to use such models as cell-design tools and to discern which models are useful in particular situations.

Chapter 3, *Basic Cellular Systems Engineering*, explains the fundamentals of telephone traffic theory, conventional cellular system architecture, and cellular engineering tradeoffs. Since it is important for engineers to know the reasoning behind any equation or table they use, each topic that is presented is given in complete form rather than as an unexplained "cookbook" ingredient. We have striven to make these topics clear so that engineers can "think cellular" on their own.

Chapter 4, *Overview of the IS-95 Standard*, gives a systematic summary of the main features of the CDMA common air interface standard, with more detailed discussions of selected aspects in the chapter's appendices. This chapter is the distillation of several hundred pages of documentation into a coherent summary that indicates not only what the system design is, but also the design philosophy behind it, including detailed explanations not available in the IS-95 document, such as the theory behind the interleaving and the hash functions used in the system.

Chapter 5, *Walsh Functions and CRC Codes*, sets forth the theory and application of the Walsh functions used for orthogonal multiplexing on the IS-95 forward link and for orthogonal modulation on the reverse link. The relations between Walsh, Hadamard, and Rademacher functions are shown, and several ways of generating and demodulating Walsh functions are given. The material presented on Walsh functions in this chapter is much more than what is required in understanding their application to IS-95 CDMA systems. The CRC codes used by the system to detect frame errors are explained in a tutorial manner to the extent needed to understand their use in IS-95.

Chapter 6, *Theory and Application of Pseudonoise Sequences*, gives the mathematical background of the PN sequence generators used in IS-95. It is shown how to derive the "mask" vectors that are used extensively in the system to assign different PN code offsets to base stations and different PN code starting positions to mobile users. The correlation and spectral properties of PN code waveforms are also treated, with a view toward their application to code tracking in the following chapter.

Chapter 7, *Modulation and Demodulation of IS-95 Spread Spectrum Signals*, gives performance analyses of forward and reverse link modulations. The advantages of the QPSK spread-spectrum modulation used in IS-95 over BPSK spread-sprectrum modulation are thoroughly analyzed. The principles of PN code acquisition and tracking are explained. The effect of using shaped and unshaped PN code references are also quantitatively analyzed.

Chapter 8, *Convolutional Codes and Their Use in IS-95*, treats the theory and practice of convolutional codes in a tutorial manner, leading up to their use in the CDMA cellular system. The convolutional codes used in IS-95 are evaluated based on the theory and the formulas developed in the chapter.

Chapter 9, *Diversity Techniques and Rake Processing*, begins with the treatment of generic diversity techniques, such as selection diversity, equal gain diversity, and maximal ratio combining. The original Rake concept is explained. The chapter concludes with a description of the application of the Rake concept in the IS-95 system.

Chapter 10, *CDMA Cellular System Design and Erlang Capacity*, begins by examining the CDMA cellular system forward and reverse link power budgets in great detail, providing a systematic treatment of the signal and interference parameters, such as loading, that directly influence the operation of the system. Methods for characterizing CDMA cell size and for balancing forward and reverse link coverage areas are described, and the effect of CDMA soft handoff on link reliability is summarized. The merit of the CDMA system is assessed in terms of Erlang capacity. Methods are shown for analyzing the area coverage of the system.

Chapter 11, *CDMA Optimization Issues*, deals with the control of parameters related to system optimization in terms of the selection of base station PN code offsets and the allocation of power to forward link channels. The fade margins achievable on the forward link are derived as functions of the system parameters. A new concept of forward and reverse link "capacity balancing" is introduced. Finally, the implementation of dynamic forward power allocation to the signaling (pilot, sync, and paging) channels is discussed in conjunction with closed-loop forward traffic-channel power control.

Acknowledgments

The authors are grateful for the support of their families during the process of writing this book. We are also pleased to acknowledge the contributions of William A. Jesser, Jr. and Dr. Soon Young Kwon (former employees of J. S. Lee Associates, Inc.) for the research work on Walsh functions, interleaving techniques, convolutional codes, hash functions, and random number generators.

We also wish to acknowledge the support given to the first author of this book, during the period of giving CDMA lectures in Korea, by Dr. Jung Uck Seo (now President of SK Telecom), who was appointed by the Minister of Communications and Information to supervise the entire Korea CDMA Development Program; Dr. Seungtaik Yang, then President of Electronics and Telecommunications Research Institute (ETRI) (now President of the University of Information and Communications); Dr. Hang Gu Bahk, then Vice President of ETRI (now Vice President of Hyundai Electronics Industries Co., Ltd. (HEI)); Dr. Hyuck Cho Kwon, founding President of Shinsegi Telecom, Inc. (STI); Mr. Tai-Ki Chung, President of STI; Mr. Byung Joon Chang, Vice President of Telecommunications Systems Division of HEI; and Mr. Limond Grindstaff of AirTouch Communications, who served as fonding Technical Director of STI. These men in their respective roles did much to help Korea to become the world's first and most successful deployment of CDMA cellular technology, which as of this writing brought about nearly ten million CDMA system subscribers since the beginning of 1996. In addition, we want to recognize the leaders and engineers of the following organizations whose efforts made CDMA a practical reality in Korea and beyond: ETRI; Hansol PCS; HEI; Korea Telecom Freetel, Inc.; LG Information and Communications, Ltd.; LG Telecom, Ltd.; and Samsung Electronics Co., Ltd.

Jhong Sam Lee
Leonard E. Miller
Rockville, Maryland USA
September, 1998

1

Introduction and Review of Systems Analysis Basics

1.1   Introduction

The purpose of this book is to provide a clear understanding of code-division multiple access (CDMA) technology and build a solid understanding of the technical details and engineering principles behind the robust new IS-95 digital cellular system standard. The book is intended to help practicing cellular engineers better understand the technical elements associated with CDMA systems and how they are applied to the IS-95 standard, which was developed in response to the requirement for the design of a second-generation cellular telephone system. The CDMA cellular system uses state-of-the-art digital communications hardware and techniques and is built on some of the more sophisticated aspects of modern statistical communications theory. The book is designed to be self-contained in that it includes in this chapter all the systems analysis basics and statistical tools that are pertinent to the technical discussions in the later chapters.

The "second-generation" means digital, as opposed to the "first-generation" analog system. The current U.S. analog cellular system, known as the Advanced Mobile Phone System (AMPS), operates in a full-duplex fashion using frequency-division duplexing (FDD), with a 25-MHz bandwidth in each direction over the following frequency allocations:

- From mobile to base station: 824–849 MHz;

- From base station to mobile: 869–894 MHz.

The Federal Communications Commission (FCC) further divided the 25-MHz bandwidth equally between two service providers, known as the "A" (wire) and the "B" (nonwire) carriers, each with 12.5 MHz of spectrum allocated for each direction.

1

In AMPS, each channel occupies 30 kHz of bandwidth in a frequency-division multiple access (FDMA) system, using analog frequency modulation (FM) waveforms. The frequencies that are used in one cell area are reused in another cell area at a distance such that mutual interference gives a carrier-to-interference power ratio of no less than 18 dB. Given this performance requirement and the fact that in the mobile radio environment the attenuation of carrier power usually is proportional to the fourth power of the distance from the emitter to a receiver, the analog cellular system utilizes seven-cell clusters, implying a frequency reuse factor of seven. The resulting capacity is then just one call per 7×30 kHz $= 210$ kHz of spectrum in each cell, and in the total of 12.5 MHz allocated there can be no more than 60 calls per cell.[1]

In 1988, the Cellular Telecommunications Industry Association (CTIA) released cellular service requirements for the next-generation (second-generation) digital cellular system technology, known as a users' performance requirements (UPR) document. The key requirements included a tenfold increase in call capacity over that of AMPS, a means for call privacy, and compatibility with the existing analog system. The compatibility requirement arose from the fact that the FCC did not allocate a separate band for the digital system, so the second-generation system must operate in the same band as AMPS.

In 1989, a committee of the Telecommunications Industry Association (TIA) formulated an interim standard for a second-generation cellular system that was published in 1992 as IS-54 [1]. In that standard, which became the first U.S. digital cellular standard, the committee adopted a time-division multiple access (TDMA) technology approach to the common air interface (CAI) for the digital radio channel transmissions. The IS-54 TDMA digital cellular system employs digital voice produced at 10 kbps (8 kbps plus overhead) and transmitted with $\pi/4$ differentially encoded quadrature phase-shift keying ($\pi/4$ DQPSK) modulation. The design envisioned noncoherent demodulation, such as by using a limiter-discriminator or a class of differential phase detectors. Because the IS-54 system permits 30 kHz/10 kbps $= 3$ callers per 30-kHz channel spacing, the increase of capacity over AMPS is only a factor of three (180 calls per cell), and the TDMA digital cellular system so far falls short of meeting the capacity objective of the UPR.

Immediately following the emergence of the IS-54 digital cellular standard, Qualcomm, Inc., in 1990 proposed a digital cellular telephone system based on CDMA technology, which in July 1993 was adopted as a second

---

[1] As is shown in Section 3.2.5, the actual capacity is lower than 60 calls per cell because of the allocation of some channels to signaling traffic.

U.S. digital cellular standard, designated IS-95 [2]. Using spread-spectrum signal techniques, the IS-95 system provides a very high capacity, as will be convincingly shown in this book, and is designed to provide compatibility with the existing AMPS, in compliance with the specifications of the UPR document.

1.1.1 Multiple Access Techniques

The first cellular generation's AMPS and the second generation's IS-54 and IS-95 are generic examples of the three basic categories of multiple access (MA) techniques:

- FDMA;
- TDMA;
- CDMA.

The three basic techniques can be combined to generate such hybrids as combined frequency division and time division (FD/TDMA), combined frequency division and code division (FD/CDMA), and others [3]. The IS-95 system employs FD/CDMA techniques, whereas the IS-54 system uses an FD/TDMA method. All these multiple access strategies are competing techniques, each aiming at distributing signal energy per access within the constrained time-frequency plane resource available.

In an FDMA system, the time-frequency plane is divided into, say, M discrete frequency channels, contiguous along the frequency axis as depicted in Figure 1.1. During any particular time, a user transmits signal energy in one of these frequency channels with a 100% duty cycle. In a TDMA system, the time-frequency plane is divided into M discrete timeslots, contiguous along the time axis as shown in Figure 1.2. During any particular time, a user transmits signal energy in one of these timeslots with low duty cycle. In a CDMA system, the signal energy is continuously distributed throughout the entire time-frequency plane. In this scheme, the frequency-time plane is not divided among subscribers, as done in the FDMA and TDMA systems. Instead, each subscriber employs a wideband coded signaling waveform [3] as illustrated in Figure 1.3.

Having defined the three MA techniques employed by AMPS, IS-54, and IS-95 systems, one may wonder why the capacities of these systems differ from one another! Is it the inherent property of the MA technique that sets

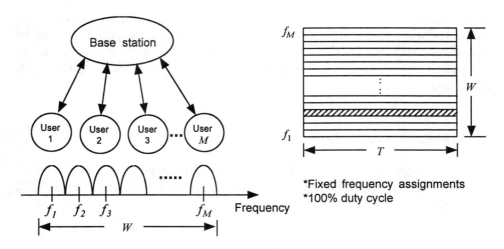

Figure 1.1 Frequency-Division Multiple Access.

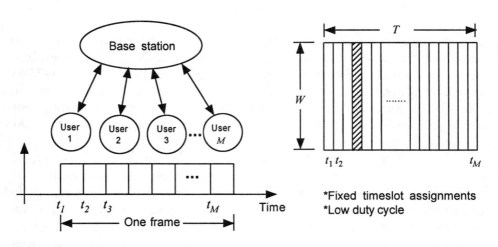

Figure 1.2 Time-Division Multiple Access.

one apart from another? Before focusing on the reasons for the differences, let us consider an ideal, hypothetical situation as follows: Suppose that each of the three MA systems has bandwidth W MHz and each user employs an uncoded bit rate $R_b = 1/T_b$, where T_b is the bit duration. Let us also assume that each MA system employs orthogonal signaling waveforms, as suggested in Figures 1.1 to 1.3. Then the maximal number of users is given by

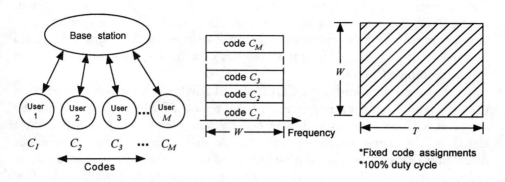

Figure 1.3 Code-Division Multiple Access.

$$M = \text{capacity} \leq \frac{W}{R_b} = WT_b \tag{1.1a}$$

Let us now assume that the received power of each user in any MA system is S_r. Then the total received power P_r is

$$P_r = MS_r \tag{1.1b}$$

The required signal-to-noise power ratio (SNR) or $E_b/\mathcal{N}_0$ (bit energy-to-noise spectral density ratio) is assumed to be equal to the actual value, giving

$$\left(\frac{E_b}{\mathcal{N}_0}\right)_{\text{req}} = \left(\frac{E_b}{\mathcal{N}_0}\right)_{\text{actual}} = \frac{S_r/R_b}{\mathcal{N}_0} = \frac{P_r/M}{\mathcal{N}_0 R_b} \tag{1.1c}$$

from which we obtain

$$M = \frac{(P_r/\mathcal{N}_0)}{R_b \cdot (E_b/\mathcal{N}_0)_{\text{req}}} \tag{1.1d}$$

Therefore, in an ideal situation, in principle each MA technique can deliver equal capacity, namely

$$M_{FDMA} = M_{TDMA} = M_{CDMA} = \frac{(P_r/\mathcal{N}_0)}{R_b \cdot (E_b/\mathcal{N}_0)_{\text{req}}} \tag{1.1e}$$

Yet, in reality each MA system as used in the cellular telephone industry does not deliver the same capacity. In this introduction, we can offer some answers as to why a certain system does not deliver the full capacity it could be capable of providing, but not all the answers until we penetrate deep into the succeeding chapters.

As stated previously, the FDMA-based AMPS falls far short of meeting the orthogonal signaling requirements of the ideal FDMA system, in that the signal at a given frequency of one cell cannot tolerate interference from a signal at the same frequency in another cell unless a mutual distance separation is such that the relative power ratio (carrier-to-interference power ratio) is above 18 dB. This is one of the mitigating realities for the low capacity of the signal design of the first-generation AMPS. The IS-54 system enhances AMPS capacity by about three times, far short of meeting the UPR objectives. The three types of MA techniques used in the FDMA, TDMA, and CDMA cellular communications systems may be compared with respect to their use of time and frequency resources as shown in Figure 1.4.

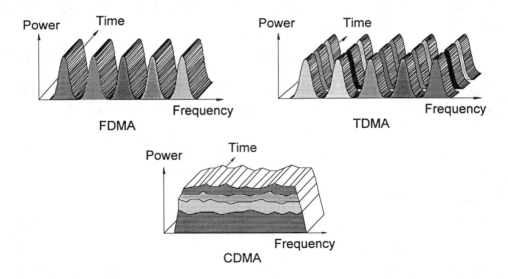

Figure 1.4 Multiple access techniques of FDMA (AMPS), TDMA (IS-54), and CDMA (IS-95).

The question then is: does CDMA meet the UPR's requirement in terms of capacity being at least ten times the capacity of AMPS? The authors believe strongly that the answer is a positive "yes"! This book is written not only to provide explanations of the basic technology involved in the IS-95 CDMA system design, but also to prove by analysis the reasons why the CDMA system can meet such high-capacity requirements. Chapter 11 offers ways of meeting optimality requirements of an IS-95 system, in terms of an optimal forward link power control scheme that will provide the high capacity the CDMA system is capable of delivering.

The reason for the high capacity of the IS-95 system is not merely the near-orthogonality of the signals in any user channels, but also the system's exploitation of the fractional duty cycle of human speech voice activity, as well as the employment of three or more directional sector antennas that increase *cell capacity* directly through full frequency-time plane reuse, features unique to the CDMA system and not in common with either FDMA or TDMA systems. In the IS-95 CDMA system, each user is given one out of a set of orthogonal codes with which the data is spread, yielding code orthogonality. The orthogonality property allows the multiple users to be distinguished from one another.[2] Although users operate on the same frequency at the same time, the spreading of the baseband signal spectrum allows interference from other users to be suppressed, which increases the capacity of the CDMA technique.

The IS-95 system, conceived and promoted by Qualcomm, Inc., is an elegant example of a commercial application of a spread-spectrum system, which has opened a new era of spread-spectrum wireless communications in nonmilitary applications. It seems appropriate to say that, if one ever wanted to see a communications system that is designed and built using most of the modern communications and information theoretic disciplines, the IS-95 CDMA system could be a good example of it [4].

1.1.2 Spread-Spectrum Techniques

Spread-spectrum techniques involve the transmission of a signal in a radio frequency bandwidth substantially greater than the information bandwidth to achieve a particular operational advantage. Once only of interest to the

---

[2] In a mobile environment, multipath receptions may contribute to the interference power for each mobile station. This subject will be dealt with in detail in Chapter 10 in which CDMA systems engineering issues are discussed.

military, spread-spectrum techniques have now been adapted for commercial applications, using nearly all the operational advantages that rendered the techniques important to the military, as suggested in Table 1.1.

The manner in which spread-spectrum signals can be processed to yield gains for interference rejection can be understood by calculating the jamming margin for a spread-spectrum system. Let the following parameters be defined:

S = received power for the desired signal in W.

J = received power for undesired signals in W (jamming, other multiple access users, multipath, etc.).

$R = 1/T_b$ = Data rate (data signal bandwidth in Hz).

W = spread bandwidth in Hz.

E_b = received energy per bit for the desired signal in W-sec.

$\mathcal{N}_0$ = equivalent noise spectral power density in W/Hz.

Then the ratio of the equivalent "noise" power J to S is

$$\frac{J}{S} = \frac{\mathcal{N}_0 W}{E_b/T_b} = \frac{WT_b}{E_b/\mathcal{N}_0} = \frac{W/R}{E_b/\mathcal{N}_0} \tag{1.2a}$$

When the value of $E_b/\mathcal{N}_0$ is set to that required for acceptable performance of the communications system, then the ratio J/S bears the interpretation of a jamming margin:

Table 1.1 Applications of spread spectrum systems

Purposes	Military	Commercial
Antijamming	✓	✓
Multiple access	✓	✓
Low detectability	✓	
Message privacy	✓	✓
Selective calling	✓	✓
Identification	✓	✓
Navigation	✓	✓
Multipath protection	✓	✓
Low radiated flux density	✓	✓

$$\frac{J}{S} = \text{tolerable excess of interference over desired signal power}$$

$$= \frac{W/R}{(E_b/\mathcal{N}_0)_{\text{req}}} \tag{1.2b}$$

or

$$\text{Margin (dB)} = \frac{W}{R} \text{ (dB)} - \left(\frac{E_b}{\mathcal{N}_0}\right)_{\text{req}} \text{(dB)} \tag{1.2c}$$

The quantity W/R is called the spread-spectrum processing gain. Note that, for a system that is not spread in bandwidth (i.e., $W = R$), the value of $E_b/\mathcal{N}_0$ is numerically equal to the signal-to-noise power ratio (SNR).

Example 1.1 Let the information bandwidth be $R = 9{,}600\,\text{Hz}$, corresponding to digital voice, and let the transmission bandwidth be $W = 1.2288$ MHz. If the required SNR is 6 dB, what is the antijam margin?
    Solution: Applying (1.2c), we have

$$\text{Margin} = 10\log_{10}\left(\frac{1.2288 \times 10^6}{9.6 \times 10^3}\right) - 6 = 21.1 - 6 = 15.1\,\text{dB} \tag{1.2d}$$

The implication of this example is that the communicator can achieve its objective (by attaining 6 dB SNR) even in the face of interference (jamming) power in excess of $10^{1.5} = 32$ times (due to jamming margin) the communicator's SNR requirement, due to the processing gain of $W/R = 128$. In other words, 15.1 dB is a *leftover* (margin), and the question is how one can use this jamming margin. If there is a second communicator in the spread-spectrum bandwidth, it can "use up" the entire margin of 15.1 dB if it communicates with the base station 15.1 dB closer, relative to the first user. Thus the idea is to expend the jamming margin of 15.1 dB by accommodating the maximal number of communicators in an MA communications system. In a CDMA communications system, the cochannel communicators, occupying the frequency-time plane simultaneously, account for the interference (jamming) power. This implies that, if every user in the spread-spectrum bandwidth supplies the identical amount of signal power to the base station antenna through a *perfect* power control scheme, regardless of location, then $10^{1.5} = 32$ other MA users can be accommodated.
    The example considered above explains the essence of a CDMA system such as IS-95. The digital voice data at the rate of $9{,}600$ bps is coded for channel error protection and then spread-spectrum modulated, resulting in a

wideband signal at 1.2288×10^6 chips per sec[3] and is transmitted to a receiving party in the presence of other users. The basic operation that needs to be performed to convert the information data signal to a wideband spread-spectrum signal and to retrieve it at the receiver is accomplished simply by a multiplication process, as indicated in Figure 1.5. We devote all of Chapter 6 to the theory and practice of the spreading code sequences and explain their use in the IS-95 system.

1.1.3 IS-95 System Capacity Issues

The capacity of a CDMA system is proportional to the processing gain of the system, which is the ratio of the spread bandwidth to the data rate. This fact may be shown as follows: assuming first that the system in question is isolated from all other forms of outside interference (i.e., a single cell), the carrier power $C \equiv S = E_b/T_b = R E_b$ as in (1.2a). Similarly, analogous to the jamming power in (1.2a), the interference power at the base station receiver may be defined as

$$I = W \cdot \mathcal{N}_0 \qquad (1.3)$$

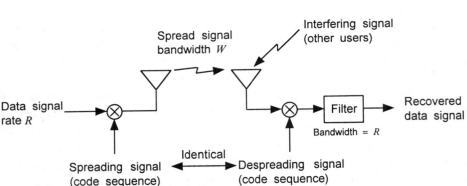

- Data signal is spread at transmitter (first multiplication).

- Data signal is recovered at receiver (second multiplication).

- Interfering signal is spread at receiver (one multiplication).

Figure 1.5 Basic spread-spectrum operations.

---

[3] The PN sequence code rate is described as the "chip rate," while the information sequence is described as "bits."

where W is the transmission bandwidth and N_0 is the interference power spectral density. Thus a general expression for the carrier-to-interference power ratio for a particular mobile user at the base station is given by

$$\frac{C}{I} = \frac{R \cdot E_b}{W \cdot N_0} = \frac{E_b/N_0}{W/R} \tag{1.4}$$

The quantity E_b/N_0 is the bit energy to noise power spectral density ratio, and W/R is the processing gain of the system. Let M denote the number mobile users. If power control is used to ensure that every mobile has the same received power at the base station, then (neglecting thermal noise) the interference power caused by the $M - 1$ interferers is

$$I = C \cdot (M - 1) \tag{1.5}$$

Substituting for I, the carrier-to-interference ratio can now be expressed as

$$\frac{C}{I} = \frac{C}{C \cdot (M - 1)} = \frac{1}{M - 1} \tag{1.6}$$

substituting C/I from equation (1.6) into (1.4), the capacity for a CDMA system is found to be

$$M \approx M - 1 = \frac{W}{R} \cdot \frac{1}{E_b/N_0} \tag{1.7}$$

Thus, the capacity of a CDMA system is proportional to the processing gain.

The capacity of a CDMA system is limited by the interference caused by other users simultaneously occupying the same frequency band; this interference is reduced by the processing gain of the system. Because techniques that decrease the amount of interference received by the base station translate into equivalent gains in capacity, the CDMA system is capable of increasing capacity by virtue of its ability to reduce interference by the amount of the spread-spectrum processing gain. This processing gain is based on the fact that in the CDMA receiver the interfering users' signals continue to have the bandwidth W, while the selected user's signal is despread by the removal of the spreading code. The receiver then can perform filtering at the selected user's despread (baseband) bandwidth, which admits only a small fraction of the energy received from the interfering user signals.

Studies have shown that the average duty cycle of a full-duplex voice conversation is approximately 35%. It is therefore possible in a digital system to reduce the transmission rate or to use intermittent transmissions during these pauses in speech. Because transmission is not eliminated entirely but reduced during such pauses, the effective duty cycle of the digital waveform is closer to 40% or 50%. If the duty cycle of the speech traffic channels in the CDMA system is denoted by the variable α, then the capacity equation becomes

$$M \approx \frac{W}{\alpha R} \cdot \frac{1}{E_b/\mathcal{N}_0} = \frac{W}{R} \cdot \frac{1}{E_b/\mathcal{N}_0} \cdot \frac{1}{\alpha} \qquad (1.8)$$

The capacity equations shown thus far do not consider the directivity, if any, of the base station receiving antennas. If the base station employs directional antennas that divide the cell into sectors, each antenna will only receive a fraction of the interference from within the cell. In practice, the receiving antennas have overlapping coverage areas of approximately 15%. Standard implementations divide the cell into three sectors, which provides an effective capacity increase of $G = 3 \cdot 0.85 = 2.55$, and the corresponding equation for capacity is

$$M \approx \frac{W}{R} \cdot \frac{1}{E_b/\mathcal{N}_0} \cdot \frac{1}{\alpha} \cdot G \qquad (1.9)$$

Up to this point, the capacity equation assumes a single, isolated CDMA cell; the number M in (1.9) therefore is comparable with the number of frequencies available to users in a sector of a single-cell AMPS deployment. In a multicell system, the interference from signals originating in other cells must be taken into account when determining the capacity of a particular "home" cell; such interference is, of course, diminished by the attenuation incurred by the interferers in propagating to the home cell—for all multiple access schemes. In conventional cellular systems using FDMA, the base station receiver must have a C/I of at least 18 dB to ensure acceptable analog voice quality. Under typical propagation conditions, to achieve this C/I, adjacent cells cannot use the same frequency, and the same frequency can be used in only one out of seven cells. Thus, the capacity is of an FDMA multicell system is reduced by a factor of seven from that of a single-cell FDMA system.

In a CDMA multicell system, the majority of interference emanates from mobiles within the same cell (while for FDMA there is no interference from mobiles in the same cell). Therefore, the total amount of interference for a multicell CDMA system is not much greater than for a single-cell CDMA system. Simulations performed by Qualcomm have shown that interference from other cells accounts for only 35% of that received at the base station [5]. Based on this information, the equation for CDMA capacity may be modified to include a reuse efficiency[4] for other-cell interference, F_e:

$$M \approx \frac{W}{R} \cdot \frac{1}{E_b/\mathcal{N}_0} \cdot \frac{1}{\alpha} \cdot G \cdot F_e \tag{1.10}$$

An example using realistic parameters for the IS-95 system is included below:

$$W/R = 128 \qquad E_b/\mathcal{N}_0 = 7\,\mathrm{dB} = 5$$
$$\alpha = 0.5 \qquad G = 2.55 \tag{1.11a}$$
$$F_e = 0.65.$$

These parameters yield the following value of capacity:

$$M = 128 \cdot \tfrac{1}{5} \cdot \tfrac{1}{0.5} \cdot 2.55 \cdot 0.65 = 85 \tag{1.11b}$$

The current AMPS can accommodate six users per cell in the same bandwidth[5] (1.25 MHz), yielding a capacity gain of over 14 for the CDMA system in this example. (1.10) normalized by the number of AMPS users per sector is plotted as a function of α and of F_e, respectively, in Figures 1.6 and 1.7, with $E_b/\mathcal{N}_0$ as a parameter.

The capacity (1.10) as plotted in Figures 1.6 and 1.7 is a theoretical limit that suggests realizability for assumed conditions. A real-world capacity estimate must be made, however, under realistic assumptions. We deal with this capacity issue in detail in Chapter 10 in terms of the "Erlang capacity" of a CDMA cellular system. We find there that a real-world CDMA system with the IS-95 design does deliver the capacities expected under near-optimal design parameter choices.

---

[4] A detailed discussion of reuse efficiency is included in Chapter 10.
[5] AMPS has 30-kHz FDMA channels and a frequency reuse factor of seven. Thus, 1.25 MHz can accommodate $1.25/7/.03 \approx 6$ users per cell.

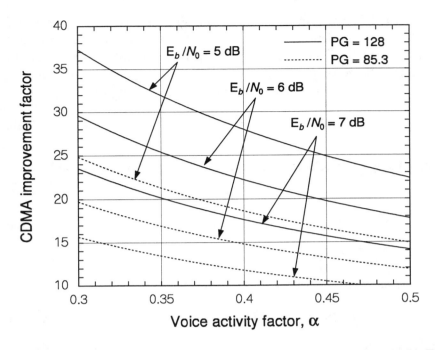

Figure 1.6 CDMA capacity improvement over FDMA versus voice duty factor α, with $F_e = 0.65$ and $G = 2.55$. PG = 85.3 corresponds to a data rate of 14.4 kbps.

1.1.4 Categories of Spread-Spectrum Systems

There are two major techniques for obtaining a spread-spectrum signal: frequency hopping (FH) and direct sequence (DS) pseudorandom noise (PN) spreading. Both types of spread-spectrum signal are employed in current wireless systems, including cellular systems. The second-generation U.S. CDMA cellular system uses DS spread-spectrum techniques.

A typical FH waveform pattern is depicted in Figure 1.8 in terms of bandwidth occupancy as a function of time. Within each hop period T_h, there can be more than one modulation symbol, with period T_s. If T_h is a multiple of T_s, with multiple symbols per hop, as is typical for systems employing continuous phase modulation, the system is said to be "slow hopping." If T_h is a submultiple of T_s, with multiple hops per symbol, as is typical for systems employing noncoherent frequency-shift keying, the system is said to be "fast hopping" [6–12].

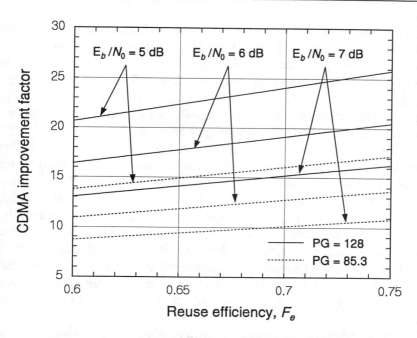

Figure 1.7 CDMA capacity improvement over FDMA versus reuse efficiency F_e, with $\alpha = 0.5$ and $G = 2.55$. PG = 85.3 corresponds to a data rate of 14.4 kbps.

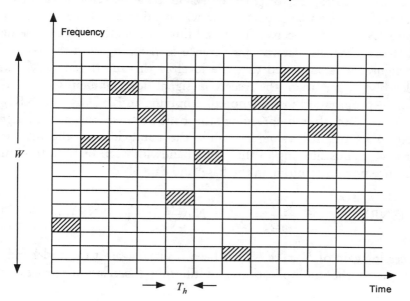

Figure 1.8 Typical frequency-hopping waveform pattern.

To implement an FH system, a baseband or IF (intermediate frequency) communications waveform is shifted pseudorandomly in frequency on each hop using a digitally programmable frequency synthesizer, and it is necessary to have an identical, synchronized pseudorandom code generator[6] and synthesizer in the intended receiver. When frequency hopping is used as a CDMA technique, users broadcast FH signals simultaneously and care is taken to avoid or minimize collisions of hops from different users.

In DS spread-spectrum systems [6, 13, 14], typically the symbol rate $(1/T_s)$ of a $\pm$ binary baseband data waveform is increased by multiplying it with a pseudorandom $\pm$ binary waveform whose "chip" rate $(1/T_c)$ is much faster than the symbol rate $(T_s = NT_c)$. As illustrated conceptually in Figure 1.9, the effect of this operation is to spread the instantaneous bandwidth of the waveform by the factor N, which for the same signal power causes the spectral density of the waveform to be quite low and "noise-like." In the figure, which shows a single-sided RF spectrum[7], the signal power is indicated to be $P_S = A_1 W = A_0 B$, implying that the spectral density of the spread signal is reduced by the factor $A_1/A_0 = W/B = 1/N$ from its level without spreading.

At the receiver, "despreading" (multiplication by the same $\pm$ binary spreading waveform as at the transmitter) and removal of the carrier modulation restores the original baseband data waveform, allowing the receiver to filter out a large part of the wideband interference. Assuming, as in Figure 1.9, that the receiver front end filter admits the W Hz-wide desired signal, the receiver also admits interference or jamming with this bandwidth. In the figure, it is suggested that the level of this interference, $\mathcal{N}_0$, can be considerably larger than the received signal level (taken to be A_1 for simplicity—propagation loss is ignored), implying an RF SNR of $(SNR)_{RF} = A_1/\mathcal{N}_0 < 1$. But after despreading, the bandwidth of the desired signal is reduced to its original value, B, while the bandwidth of the interference remains W. Thus filtering to the signal bandwidth can be used to eliminate interference power, resulting in the baseband data SNR

$$(SNR)_{baseband} = \frac{A_0 B}{\mathcal{N}_0 B} = \frac{A_0}{\mathcal{N}_0} = N \cdot \frac{A_1}{\mathcal{N}_0} = \frac{W}{B} \cdot (SNR)_{RF} \qquad (1.12)$$

A processing gain of $N = W/B$ has been realized against the wideband interference using the multiplication and filtering (correlation) process. If a

---

[6] A thorough treatment of pseudorandom code generators is given in Chapter 6.
[7] The basics of spectra and waveform analysis are summarized later in this chapter.

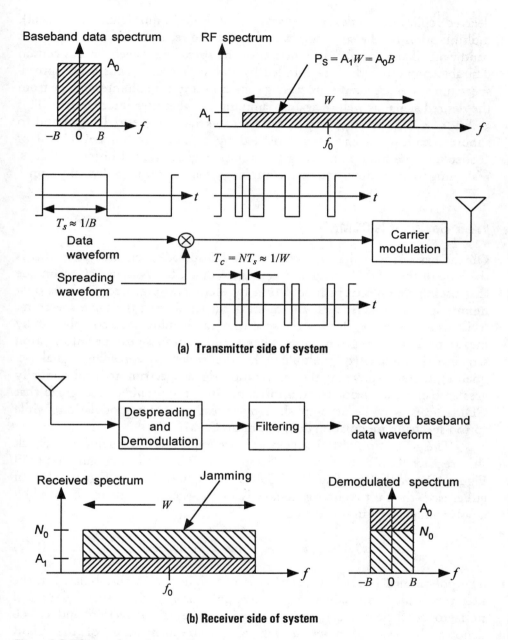

(a) Transmitter side of system

(b) Receiver side of system

Figure 1.9 DS spread-spectrum system.

delayed replica of the desired signal is received (i.e., a multipath component), multiplication by the spreading waveform at the receiver does not reduce its bandwidth if the correlation function of the spreading waveform has certain desirable properties that are fulfilled by PN sequences. Thus a DS spread-spectrum system realizes a processing gain against multipath interference from the desired signal as well as against jamming or other-user interference. This ability of a DS spread spectrum system to extract the desired signal and to suppress multipaths has been exploited by a "rake" technique [15, 16] of "collecting" the multipaths using PN sequence generators at different delays, realigning them in time, and then combining them to realize a diversity gain.[8]

1.1.5 So, What Is CDMA?

On the basis of what has been said of the DS spread-spectrum system that is the core of the CDMA system, we can state that CDMA is an MA technique that uses spread-spectrum modulation by each accessing party with its own unique spreading code, with all accessing parties sharing the same spectrum. It is also clear now that spread-spectrum modulation is accomplished by means of PN codes. The narrowband information signal or information sequence is modulated (multiplied) by the wideband spreading signal (sequence), thereby spreading the information signal spectrum to a substantially greater bandwidth prior to transmission. It is important to recognize that CDMA can only be accomplished by spread-spectrum modulation, while spread-spectrum modulation does not mean CDMA.

The generation of PN sequences is accomplished using a linear feedback shift register (LFSR), in either the "simple" or "Fibonacci" configuration as in Figure 1.10 or the "modular" or "Galois" configuration as in Figure 1.11. In either case, the shift register generator is a finite-state machine mechanized by a polynomial given in the form of

$$f(x) = x^n + c_{n-1}x^{n-1} + \cdots + c_2x^2 + c_1x + 1 \qquad (1.13)$$

The polynomial (1.13) is a special type of polynomial, well tabulated in the literature, called an irreducible primitive polynomial, which specifies a set of nonzero coefficients $\{c_i\}$, where $c_i = 1$ denotes a connection and $c_i = 0$ denotes the lack of a connection in the mechanization of the LFSR

---

[8] Rake and other diversity techniques are discussed in Chapter 9.

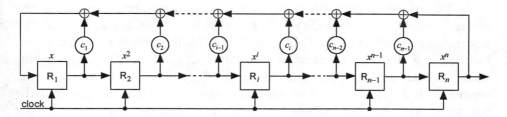

Figure 1.10 Fibonacci or simple shift register generator (SSRG) configuration of
$f(x) = 1 + c_1 x + c_2 x^2 + \cdots + c_i x^i + \cdots + c_{n-1} x^{n-1} + x^n.$

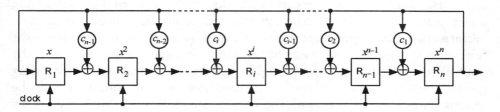

Figure 1.11 Galois or modular shift register generator (MSRG) configuration of
$f(x) = 1 + c_1 x + c_2 x^2 + \cdots + c_i x^i + \cdots + c_{n-1} x^{n-1} + x^n.$

configuration. As is thoroughly discussed in Chapter 6, the sequences generated by such an LFSR with an initial loading of nonzero n-tuples in the n stages, are periodic sequences with length $P = 2^n - 1$, and there are P different sequences of length P that are shifted versions of the given initial sequence of length P. The sequences generated in this way are the ones used. What is so special about these sequences that makes it possible to realize such CDMA systems as IS-95? There are three most important properties associated with a PN sequence, aside from the basic property that it has the maximal length of $2^n - 1$, where n is the number of stages of the LFSR. Two of the three remaining properties have to do with the randomness of the sequence, which we have the occasion to observe in a later chapter, but the one we wish to mention here is the correlation property. What it means is that if a complete sequence of length $2^n - 1$ is compared, bit by bit, with any shift of itself (one of $2^n - 2$ remaining sequences), the number of agreements differs from the number of disagreements by at most 1. This means that when two identical sequences are compared, bit by bit, the number of agreements minus the number of disagreements is equal to the number of agreements, which is $2^n - 1$.

The property just described, namely the correlation property, is the reason behind the possibility of accomplishing the extraction of the information signal from the "MA noise" or "MA interference" environment. Further, it is the reason behind the possibility of the Rake diversity scheme, which is such an important part of the IS-95 system. Moreover, it is the reason behind the possibility of rejecting other users from coming into the particular user's baseband channel. This is the mechanism responsible for the processing gain we spoke of earlier. This is the mechanism responsible for multipath rejection when the multipath is not of use for signal processing purposes. Consider Figure 1.12. A signal received by way of a reflected path can cause destructive interference with the signal received by way of the direct path. In a conventional system, this multipath signal can degrade performance. In a spread-spectrum system, however, the multipath can be rejected if not useful, to the extent the processing gain can provide (i.e., the multipath signal strength can be suppressed by the factor of the processing gain). There is a condition for this fortune, however: the multipath must be separated in time compared with the arrival time of the direct path[9] by at least one PN sequence chip duration. Such a multipath is called a resolvable multipath. In the mobile communications environment, there are many multipaths, and the receiver must select a few good resolvable multipaths that are strong enough to collect and process. This is the principle of Rake diversity reception, which we discuss in Chapter 9. The delay of the reflected signal (multipath), relative to the signal received over the direct path, is

$$\Delta T = (R_1 + R_2 - D)/c \qquad (1.14)$$

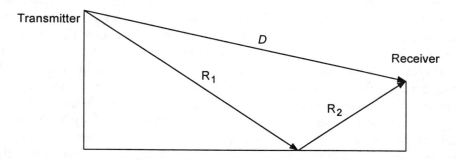

Figure 1.12 Two-ray multipath model.

---

[9] The "direct path" itself could be a reflected path signal. It is a matter of an arrival-time difference between the two paths.

where c is the speed of light. If $\Delta T > T_c$, where T_c is the PN code chip duration or $T_c = 1/R_{code}$ where R_{code} is the PN code sequence rate, then the reflected signal is decorrelated (rejected) with the direct path signal, and this is the mechanism of multipath suppression by processing gain. Therefore, for sufficiently high PN code rates, the multipath signal can be effectively suppressed and only the good resolvable multipaths are processed in the IS-95 system. The base station uses a maximum of four multipaths (four "fingers"), whereas the mobile station processes a maximum of three fingers in the Rake receiver.

1.1.6 Battle of Jamming Power Versus Processing Gain

Spread-spectrum modulations have long been used by the military to combat intentional jamming by a hostile transmitter. As indicated in Table 1.1, spread-spectrum radio provides antijam (AJ) capabilities through a processing gain (PG) that results from using a wideband (large bandwidth) signal. As for commercial applications of spread-spectrum systems, the seemingly inefficient use of the radio spectrum was thought to be impractical in the past [17]. In commercial spread-spectrum systems, however, interference (jamming) comes from other similar users in the band, and these interferences, unlike hostile jammers, can be controlled, coordinated, and managed for the overall users' benefit in a CDMA digital cellular system. Though each commercial user has the same PN code, the coordination permits users to be distinguished by code phase in the application of the autocorrelation property of PN sequences that was described earlier. Not having the luxury of such coordination, each user in a military CDMA system generally has a distinct PN code generator to ensure a strong correlation with only one signal. Therefore, in addition to jamming, a military user is subject to MA inteference that is due to "cross-correlation" with different sequences, which is considerably larger than that of a single-PN-generator based commercial system.

The miltary advantage that the spread-spectrum system offers can also be illustrated in terms of a communication range extension capability over a conventional non-spread-spectrum communication system. Consider a situation where a communicator, who requires a 10-dB SNR, employs a spread-spectrum modulation that provides a PG of 30 dB. Also assume that a hostile jammer at a 200-unit distance away uses jamming power equal to the communicator's transmitter power. Our assumptions are as follows:

- The communications transmitter with power P_t is located at position $(0, 0)$.

- The jammer with power P_t is located at position $(200, 0)$.

- Assume that a communications receiver is located at position (x, y).

- Let J be the jamming power received by the communications receiver at position (x, y).

- Let S be the signal power received by the communications receiver at position (x, y).

The received signal-to-jammer power ratio, S/J, is then given by

$$\left(\frac{S}{J}\right)_{rec} = \left(\frac{R_{JR}}{R_{TR}}\right)^k \tag{1.15a}$$

where k denotes the kth power propagation law and

$R_{TR} = $ range from transmitter at $(0, 0)$ to receiver at (x, y)

$R_{JR} = $ range from jammer at $(200, 0)$ to receiver at (x, y)

For the special case of free-space propagation, the power law is $k = 2$; in this case we have

$$\left(\frac{S}{J}\right)_{rec} = \left(\frac{R_{JR}}{R_{TR}}\right)^2 = \frac{(x - 200)^2 + y^2}{x^2 + y^2} \tag{1.15b}$$

Manipulation of the equation for S/J yields

$$x^2 + \frac{400x}{\left(\frac{S}{J}\right)_{rec} - 1} + y^2 = \frac{(200)^2}{\left(\frac{S}{J}\right)_{rec} - 1}$$

which results in

$$\left[x + \frac{200}{\left(\frac{S}{J}\right)_{rec} - 1}\right]^2 + y^2 = \left[\frac{200\sqrt{(S/J)_{rec}}}{\left(\frac{S}{J}\right)_{rec} - 1}\right]^2 \tag{1.15c}$$

which is the equation for a circle with

$$\text{Center position} = \left(-\frac{200}{\left(\frac{S}{J}\right)_{\text{rec}} - 1}, 0\right) \tag{1.15d}$$

and

$$\text{Radius} = \frac{200\sqrt{(S/J)_{\text{rec}}}}{\left|\left(\frac{S}{J}\right)_{\text{rec}} - 1\right|} \tag{1.15e}$$

For example, for $(S/J)_{\text{rec}} = 0.1 = -10\,\text{dB}$, the center and radius of the circle of positions are $(200/0.9, 0) = (222, 0)$ and $200\sqrt{0.1}/0.9 = 70.3$, respectively. For $(S/J)_{\text{rec}} = 2 = 3\,\text{dB}$, the center and radius of the circle of positions are $(-200, 0)$ and $200\sqrt{2}/1 = 283$, respectively. These example cases of (1.15c) are shown in Figure 1.13, among other cases.

The implications of Figure 1.13 are very interesting in that the communication can be "heard" at a 182-unit distance away if it uses a spread-spectrum modulation that provides PG of 30 dB in the presence of an equal-power jammer. Without the spread-spectrum modulation, he could only be heard at a 48-unit distance away on the x-axis. This example is a good illustration of the range extension capability of the digital spread-spectrum cordless telephones now available in the commercial market.

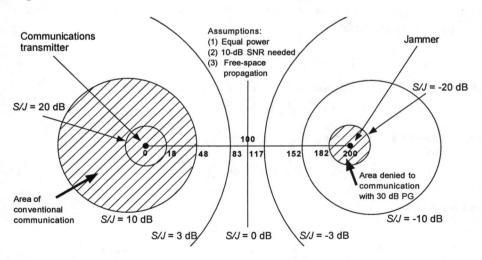

Figure 1.13 Illustration of jamming power versus processing gain "battle" [18], assuming a spread-spectrum system with 30-dB processing gain.

1.2 Review of Linear Systems Analysis Fundamentals

As mentioned in the introduction, it is our intention that this book provide the reader with the background needed to understand CDMA cellular systems. In fulfilling that intention, beginning with this section we include a carefully selected collection of systems analysis topics, each of which is essential for analyzing CDMA cellular systems and is put to use later in the book.

1.2.1 Linear Systems

Consider a time-invariant system G shown in Figure 1.14, where the input is denoted by $x(t)$ and the output, by $y(t)$. Let $h(t)$ be the response of the system G when the impulse $\delta(t)$ is applied at the input. That is,

$$h(t) = G\{\delta(t)\} \tag{1.16a}$$

$$= \text{Response of } G \text{ at time } t \text{ to a unit impulse}$$
$$\text{applied at time } t = 0$$

$$\triangleq \text{Impulse response of the filter } G$$

The response of the filter (system) G at time t due to unit impulse applied at time $t = \tau$ is then denoted

$$h(t, \tau) = G\{\delta(t - \tau)\} \tag{1.16b}$$

$$= \text{Response of } G \text{ at time } t \text{ to a unit impulse}$$
$$\text{applied at time } t = \tau$$

For a time-invariant system, we have

$$h(t, \tau) = G\{\delta(t - \tau)\} = h(t - \tau) \tag{1.16c}$$

and for a causal system

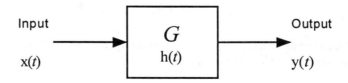

Figure 1.14 Linear time-invariant system G.

$$h(t, \tau) = 0 \quad \text{for} \quad t < \tau \tag{1.16d}$$

By definition, a causal system can have no response (output) before the input is applied to the system.

Now if we assume that the input waveform $x(t)$ and the impulse response $h(t)$ are resolved into unit impulses, we write

$$x(t) = \int_{-\infty}^{\infty} x(\lambda)\, \delta(t - \lambda)\, d\lambda \tag{1.17a}$$

and

$$h(t) = \int_{-\infty}^{\infty} h(\lambda)\, \delta(t - \lambda)\, d\lambda \tag{1.17b}$$

The response $y(t)$ of the system G is then given by

$$y(t) = G\{x(t)\} = G\left\{ \int_{-\infty}^{\infty} x(\lambda)\, \delta(t - \lambda)\, d\lambda \right\}$$

$$= \int_{-\infty}^{\infty} x(\lambda)\, G\{\delta(t - \lambda)\}\, d\lambda = \int_{-\infty}^{\infty} x(\lambda)\, h(t, \lambda)\, d\lambda \tag{1.17c}$$

The integral shown in (1.17c),

$$y(t) = \int_{-\infty}^{\infty} x(\lambda)\, h(t, \lambda)\, d\lambda$$

is known as the *superposition integral* and is a definition of a linear system (i.e., if a system is linear, then the superposition integral holds). If the system is time-invariant, $h(t, \tau) = h(t - \tau)$ and (1.17c) becomes

$$y(t) = \int_{-\infty}^{\infty} x(\lambda)\, h(t - \lambda)\, d\lambda \tag{1.17d}$$

The integral (1.17d) is known as the *convolution integral* and is the most fundamental integral in linear system analysis. In a way, it is a prediction integral in that if one knows the impulse response of the linear, time-invariant filter (system), one can predict or calculate the response of the filter due to the input $x(t)$. The integral is often denoted by an asterisk ($*$) and written as

$$y(t) = x(t) * h(t) \triangleq \int_{-\infty}^{\infty} x(\lambda) \, h(t - \lambda) \, d\lambda \qquad (1.17e)$$

Note the symmetry property of

$$y(t) = x(t) * h(t) = \int_{-\infty}^{\infty} x(\lambda) \, h(t - \lambda) \, d\lambda$$

$$= \int_{-\infty}^{\infty} x(t - \lambda) \, h(\lambda) \, d\lambda = h(t) * x(t) \qquad (1.17f)$$

For a *linear*, *causal*, and *time-invariant* system, the convolution integral becomes

$$y(t) = x(t) * h(t) = \int_{-\infty}^{\infty} x(\lambda) \, h(t, \lambda) \, d\lambda \qquad \leftarrow \text{linear}$$

$$= \int_{-\infty}^{t} x(\lambda) \, h(t, \lambda) \, d\lambda \qquad \leftarrow \text{causal}$$

$$= \int_{-\infty}^{t} x(\lambda) \, h(t - \lambda) \, d\lambda \qquad \leftarrow \text{time-invariant}$$

$$= \int_{0}^{t} x(\lambda) \, h(t - \lambda) \, d\lambda \qquad \leftarrow \text{no input for } t < 0$$

$$= \int_{0}^{t} x(t - \lambda) \, h(\lambda) \, d\lambda \qquad \leftarrow \text{symmetry}$$

It is obvious that for $t = T$ (the observation time T), we have

$$y(t) = \int_{0}^{T} x(\lambda) \, h(T - \lambda) \, d\lambda \qquad (1.17g)$$

If the input waveform $x(t)$ is to be viewed in the frequency domain, we need to obtain the Fourier transform of $x(t)$, denoted $\mathcal{F}\{x(t)\}$, and given by

$$\mathcal{F}\{x(t)\} \triangleq X(f) = \int_{-\infty}^{\infty} x(t) \, e^{-j2\pi ft} \, dt \qquad (1.18a)$$

with the inverse Fourier transform, denoted $\mathcal{F}^{-1}\{X(f)\}$, being given by

$$\mathcal{F}^{-1}\{X(f)\} \triangleq x(t) = \int_{-\infty}^{\infty} X(f)\, e^{j2\pi ft}\, df \qquad (1.18b)$$

The integrals (1.18a) and (1.18b) form a Fourier transform pair, denoted by

$$x(t) \leftrightarrow X(f) \qquad (1.18c)$$

If we choose the angular frequency notation $\omega = 2\pi f$, we have the alternative Fourier transform pair expressions

$$\mathcal{F}\{x(t)\} \triangleq X(\omega) = \int_{-\infty}^{\infty} x(t)\, e^{-j\omega t}\, dt, \qquad (1.18d)$$

and

$$\mathcal{F}^{-1}\{X(\omega)\} \triangleq x(t) = \frac{1}{2\pi} \int_{-\infty}^{\infty} X(\omega)\, e^{j\omega t}\, df \qquad (1.18e)$$

The Fourier transform of the impulse response of a linear filter is the *filter transfer function* or amplitude frequency characteristic of the filter. That is,

$$h(t) \leftrightarrow H(f)$$

where

$$\mathcal{F}\{h(t)\} \triangleq H(f) = \int_{-\infty}^{\infty} h(t)\, e^{-j2\pi ft}\, dt \qquad (1.18f)$$

The filter characteristic $H(f)$ of a linear filter characterized by an impulse response $h(t)$ shows a particular frequency response characteristic in amplitude as well as phase for the impulse response waveform. We can now redraw the linear filter shown in Figure 1.14 as shown in Figure 1.15.

Example 1.2 *Fourier transform pairs.* Fourier transform pairs for the Dirac delta function and the sinusoidal waveforms are shown in Table 1.2 below.

Example 1.3 Let the time functions $u(t)$ and $v(t)$ be given over the time interval $(-\infty, \infty)$. If $u(t) \leftrightarrow U(f)$ and $v(t) \leftrightarrow V(f)$, then

$$\int_{-\infty}^{\infty} u(t)\, v^*(t)\, dt = \int_{-\infty}^{\infty} U(f)\, V^*(f)\, df \qquad (1.18g)$$

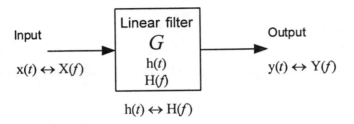

Figure 1.15 Characterization of a linear filter G in both time and frequency domains.

Table 1.2 Selected Fourier transform pairs

$x(t)$	$X(f)$	$X(\omega)$
$\delta(t)$	1	1
$\delta(t - t_0)$	$e^{-j2\pi f t_0}$	$e^{-j\omega t_0}$
$\delta(t + t_0)$	$e^{j2\pi f t_0}$	$e^{j\omega t_0}$
1	$\delta(f)$	$2\pi\delta(\omega)$
$e^{j2\pi f_0 t} = e^{j\omega_0 t}$	$\delta(f - f_0)$	$2\pi\delta(\omega - \omega_0)$
$e^{-j2\pi f_0 t} = e^{-j\omega_0 t}$	$\delta(f + f_0)$	$2\pi\delta(\omega + \omega_0)$
$\cos 2\pi f_0 t = \cos\omega_0 t$ $= \frac{1}{2}\left(e^{j2\pi f_0 t} + e^{-j2\pi f_0 t}\right)$	$\frac{1}{2}[\delta(f - f_0) + \delta(f + f_0)]$	$\pi[\delta(\omega - \omega_0) + \delta(\omega + \omega_0)]$
$\sin 2\pi f_0 t = \sin\omega_0 t$ $= \frac{1}{2j}\left(e^{j2\pi f_0 t} - e^{-j2\pi f_0 t}\right)$	$\frac{1}{2j}[\delta(f - f_0) + \delta(f + f_0)]$	$\frac{\pi}{j}[\delta(\omega - \omega_0) + \delta(\omega + \omega_0)]$

where the asterisk as a superscript (*) denotes complex conjugate. This relationship is known as *Parseval's theorem*, and its use is diverse in linear system theory. Note that if $u(t) = v(t)$, then we have

$$\int_{-\infty}^{\infty} |u(t)|^2 \, dt = \int_{-\infty}^{\infty} |U(f)|^2 \, df \qquad (1.18h)$$

(1.18h) implies that the total energy in the time domain delivered by the "voltage" $u(t)$ is also found (accounted for) in the frequency domain, where $|U(f)|^2$ is the energy spectral density. The proof of (1.18g) starts from the observations that

$$\int_{-\infty}^{\infty} u(t)\, v(t) \, dt = \int_{-\infty}^{\infty} U(f)\, V(-f) \, df \qquad (1.18i)$$

and

$$\mathcal{F}\{u^*(t)\} = U^*(-f) \tag{1.18j}$$

The convolution theorems. The calculation of the Fourier transform or inverse Fourier transform of a function (waveform) that is given as a product of two functions (waveforms) is facilitated by using the so-called *convolution theorems.* There are two convolution theorems, for the time domain and for the frequency domain.

(1) Time domain convolution theorem. Let $v(t) = x(t) * y(t)$, with $x(t) \leftrightarrow X(f)$ and $y(t) \leftrightarrow Y(f)$. Then

$$V(f) = \mathcal{F}\{x(t) * y(t)\} = X(f)\,Y(f) \tag{1.19a}$$

(2) Frequency domain convolution theorem. Let $z(t) = x(t) \times y(t)$, with $x(t) \leftrightarrow X(f)$ and $y(t) \leftrightarrow Y(f)$. Then

$$Z(f) = \mathcal{F}\{x(t) \times y(t)\} = X(f) * Y(f) \tag{1.19b}$$

The proofs of the results shown in (1.19a) and (1.19b) are very easily obtained through a straightforward application of the integrals defining the Fourier transforms and the convolutions.

Example 1.4 The Fourier transforms of rectangular waveforms are very important in the analysis of linear systems. To facilitate the analysis in a succint and simple manner, we use some simplifying notations introduced by Woodward [19]. Let the rectangular pulse function be defined as

$$\text{Rect}(t) \triangleq \begin{cases} 1, & |t| \le \tfrac{1}{2} \\ 0, & |t| > \tfrac{1}{2} \end{cases} \tag{1.20a}$$

The Fourier transform of the *Rect* function is the *sinc* function:

$$\text{Rect}(t) \leftrightarrow \text{sinc}(f) \triangleq \frac{\sin(\pi f)}{\pi f} \tag{1.20b}$$

The derivation of this transform relationship is easily shown:

$$\mathcal{F}\{\text{Rect}(t)\} = \int_{-\infty}^{\infty} \text{Rect}(t)\, e^{-j2\pi ft} dt = \int_{-1/2}^{1/2} e^{-j2\pi ft} dt = \frac{\sin(\pi f)}{\pi f}$$

Graphs of this Fourier transform pair are shown in Figure 1.16. It is readily seen that $\int_{-\infty}^{\infty} \text{sinc}x \, dx = 1$ because, from the Fourier transform pair, we have

$$\text{Rect}(t) = \int_{-\infty}^{\infty} \text{sinc}(f) \, e^{j2\pi ft} df \Rightarrow \text{Rect}(0) = 1 = \int_{-\infty}^{\infty} \text{sinc}(f) \, df$$

By a straightforward extension of the definition for the rectangular function, we find the following Fourier transform pair for a unit-amplitude rectangular pulse with a T_c-second "chip" width, as illustrated in Figure 1.17:

$$\text{Rect}(t/T_c) \leftrightarrow T_c \, \text{sinc}(T_c f) \tag{1.20c}$$

Example 1.5 Let us assume that a linear filter, designated "ideal filter #1," has a rectangular amplitude frequency characteristic given by $H(f) = \text{Rect}(f/R_c)$, where R_c is the cutoff frequency. The impulse response of this filter is obtained by taking the inverse Fourier transform of $H(f)$:

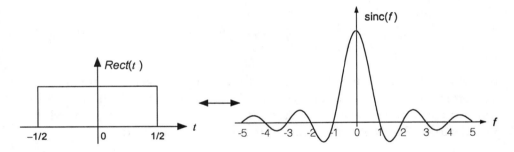

Figure 1.16 Fourier transform pair $\text{Rect}(t) \leftrightarrow \text{sinc}(f)$.

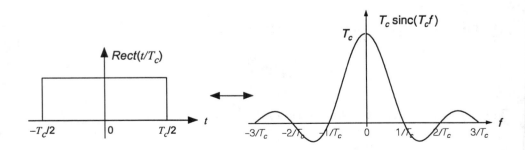

Figure 1.17 Fourier transform pair $\text{Rect}(t/T_c) \leftrightarrow T_c\text{sinc}(T_c f)$.

$$\mathcal{F}^{-1}\{H(f)\} \triangleq h(t) = \int_{-\infty}^{\infty} \text{Rect}\left(\frac{f}{R_c}\right) e^{j2\pi ft} df$$

$$= R_c \frac{\sin \pi R_c t}{\pi R_c t} = R_c \text{sinc}(R_c t) \qquad (1.20d)$$

This transform pair is illustrated in Figure 1.18. Note that ideal filter #1 is has an impulse response that is neither finite (in time extent) nor causal.

Example 1.6 We will show that shifted sinc functions are mutually orthogonal, namely that

$$\int_{-\infty}^{\infty} \text{sinc}(t - m) \, \text{sinc}(t - n) \, dt = \begin{cases} 1, & m = n \\ 0, & m \neq n \end{cases} \qquad (1.20e)$$

where m and n are integers. This is most easily verified by applying Parseval's theorem that was discussed in an example above. Because

$$\mathcal{F}\{\text{sinc}(t - m)\} = \text{Rect}(f) \, e^{-j2\pi fm}$$

and $$\qquad \mathcal{F}\{\text{sinc}(t - n)\} = \text{Rect}(f) \, e^{-j2\pi fn}$$

from Parseval's theorem, $\int_{-\infty}^{\infty} u(t) \, v(t) \, dt = \int_{-\infty}^{\infty} U(f) \, V(-f) \, df$, we have

$$\int_{-\infty}^{\infty} \text{sinc}(t - m) \, \text{sinc}(t - n) dt = \int_{-\infty}^{\infty} [\text{Rect}(f)]^2 \, e^{-j2\pi f(m-n)} df$$

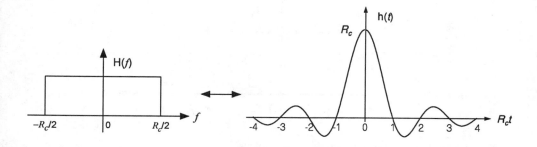

Figure 1.18 Fourier transform pair for ideal filter #1.

$$= \int_{-1/2}^{1/2} e^{-j2\pi f(m-n)} df = \int_{-1/2}^{1/2} \cos[2\pi f(m-n)] \, df$$

$$= \begin{cases} 1, & m = n \\ 0, & m \neq n \end{cases} = \delta_{mn}$$

where δ_{mn} is the Kronecker delta.

Example 1.7 Let us consider another linear filter, designated "ideal filter #2," whose frequency characteristic is shown in Figure 1.19 and is given by

$$\text{H}(f) = \begin{cases} 1, & |f| < f_1 \\ \frac{f_2-|f|}{f_2-f_1}, & f_1 \leq |f| \leq f_2 \\ 0, & |f| > f_2 \end{cases} \tag{1.20f}$$

The impulse response for this filter is given by

$$\text{h}(t) = \mathcal{F}^{-1}\{\text{H}(f)\} = \int_{-\infty}^{\infty} \text{H}(f) \, e^{j2\pi ft} \, df$$

$$= (f_1 + f_2) \text{sinc}[(f_1 - f_2)t] \, \text{sinc}[(f_1 + f_2)t] \tag{1.20g}$$

The derivation of (1.20g) is included in Appendix 1A.

In a later section of this chapter, we make an important application of (1.20g) in the discussion of the waveshaping filters that are used in the IS-95 system.

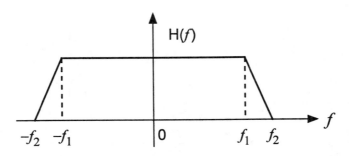

Figure 1.19 Amplitude characteristic of ideal filter #2.

1.2.2 Finite Impulse Response Filter

The filter impulse response examples given above all have infinitely persisting time functions, and also are not causal. If the impulse response were finite in duration, even if noncausal, in principle the filter could be synthesized if one is willing to accept a finite delay, as suggested in Figure 1.20 and as will be shown to be the case for the IS-95 baseband filters.

To obtain a finite impulse response (FIR), the frequency transfer characteristic cannot be "ideal" in the sense of having a zero value in the region outside of the cutoff frequencies, known as the "stop band." An FIR can be thought of as the result of "windowing" an infinite impulse response function. For example, consider the following windowing operation depicted in Figure 1.21.

$$h_1(t) \triangleq \text{finite impulse response} = \overbrace{(1/T_c)\text{sinc}(t/T_c)}^{\text{Infinite response}} \cdot \overbrace{\text{Rect}(t/NT_c)}^{\substack{\text{Rectangular window} \\ |t| \le NT_c/2}}$$

$$= h(t) \times w(t) \tag{1.21a}$$

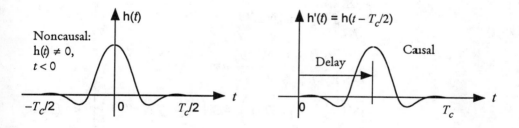

Figure 1.20 Noncausal and causal filter impulse responses.

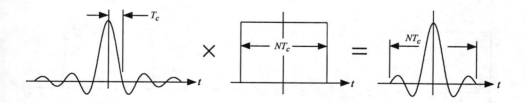

Figure 1.21 Synthesis of FIR waveform.

To obtain the amplitude frequency response characteristic (filter response characteristic), we need to calculate the inverse Fourier transform of $h_1(t)$, which is the product of two separate functions. Thus the resulting frequency transfer characteristic is the convolution of the ideal rectangular (frequency) characteristic $H(f) = \mathcal{F}\{h(t)\}$ and the Fourier transform of the window function. That is, by invoking the frequency domain convolution theorem, we have

$$H_1(f) = \mathcal{F}\{h_1(t)\} = \mathcal{F}\left\{\frac{1}{T_c}\operatorname{sinc}\left(\frac{t}{T_c}\right) \times \operatorname{Rect}\left(\frac{t}{NT_c}\right)\right\}$$

$$= \underbrace{\operatorname{Rect}(T_c f)}_{\text{Ideal } H(f)} * \underbrace{NT_c \operatorname{sinc}(NT_c f)}_{\text{Transform of window}} \tag{1.21b}$$

The windowing has the effect of adding "ripples" to the frequency characteristic, as depicted in Figure 1.22. When the convolution indicated in (1.21b) is carried out, the result in given by

$$H_1(f) = H(f) * W(f) = \int_{-\infty}^{\infty} H(\nu)\,W(f-\nu)\,d\nu$$

$$= \int_{-1/2T_c}^{1/2T_c} NT_c \operatorname{sinc}[NT_c(f-\nu)]\,d\nu$$

$$= \frac{1}{\pi}\int_{-1/2T_c-f}^{1/2T_c-f} \frac{\sin(\pi NT_c\nu)}{\nu}\,d\nu$$

$$= \frac{1}{\pi}\left[\operatorname{Si}\left(\frac{N\pi}{2} - N\pi T_c f\right) + \operatorname{Si}\left(\frac{N\pi}{2} + N\pi T_c f\right)\right] \tag{1.21c}$$

where $\operatorname{Si}(\,\cdot\,)$ is the sine integral, which is plotted in Figure 1.23 and defined by [20]:

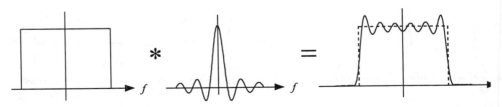

Figure 1.22 Ripple effects from windowing.

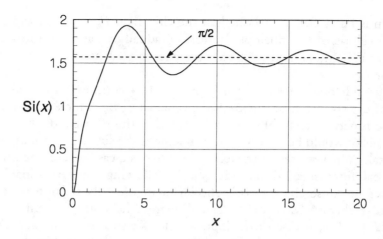

Figure 1.23 The sine integral.

$$\text{Si}(z) \triangleq \int_0^z \frac{\sin x}{x} \, dx, \text{ with } \text{Si}(0) = 0, \text{Si}(\infty) = \frac{\pi}{2}, \text{Si}(-z) = -\text{Si}(z)$$

The square of the magnitude of the transform, $|H_1(f)|^2$, which is the energy spectrum, is shown plotted in Figure 1.24 in dB units relative to the value at the center of the band, for the case of $N = 12$. The oscillations or ripples are clearly seen as deep nulls in the sidelobes of this spectrum. Note that the

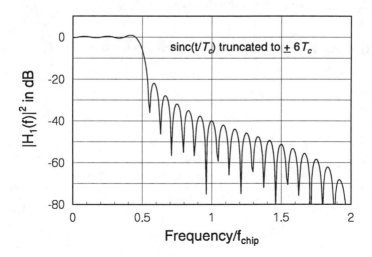

Figure 1.24 Energy spectrum for finite sinc function response.

spectrum is fairly flat out to one-half the chip rate, $R_c = 1/T_c$, and then the upper envelope of the sidelobe decreases gradually. Later we show that the particular FIR filters used in IS-95 were chosen to achieve a specific sidelobe behavior.

The sidelobe behavior seen in Figure 1.24 contrasts with the "infinite slope" of the spectrum of the ideal filter characteristic, which is the transform of a sinc function, as it falls to a zero value in the stop band. Note that an infinite delay would be necessary to implement the sinc function, because it is noncausal. What we see in the example above is a case where one can obtain a nonideal filter spectral characteristic by allowing some reasonable delay. The filter response characteristic shown in Figure 1.24 is far from the ideal rectangular shape. Most of the signal energy, however, is confined to the band $|fT_c| < 0.6$, with a small ripple effect. We shall see that this example is very close to the practical filter design that is implemented in the IS-95 system. We need to review a few more fundamentals before we take up the subject of the IS-95 FIR filter, however.

1.2.3 Fourier Series

1.2.3.1 Trigonometric and Exponential Fourier Series

Trigonometric Fourier series. A function $f(t)$ can be represented in terms of $\cos n\omega_0 t$ and $\sin n\omega_0 t$ over any interval, say, $(t_0, t_0 + T)$:

$$f(t) = A_0 + \sum_{n=1}^{\infty} (A_n \cos n\omega_0 t + B_n \sin n\omega_0 t), \quad t_0 < t < t_0 + T \quad (1.22a)$$

where

$$\omega_0 = 2\pi f_0 = 2\pi/T$$

and

$$A_n = \frac{\displaystyle\int_{t_0}^{t_0+T} f(t)\, \cos n\omega_0 t \, dt}{\displaystyle\int_{t_0}^{t_0+T} \cos^2 n\omega_0 t \, dt}, \qquad A_0 = \frac{1}{T}\int_{t_0}^{t_0+T} f(t)\, dt$$

$$B_n = \frac{\displaystyle\int_{t_0}^{t_0+T} f(t)\,\sin n\omega_0 t\,dt}{\displaystyle\int_{t_0}^{t_0+T} \sin^2 n\omega_0 t\,dt}$$

Exponential (complex) Fourier series. The function $f(t)$ can also be represented over the interval $(t_0, t_0 + T)$ as follows:

$$f(t) = c_0 + c_1 e^{j\omega_0 t} + c_2 e^{j2\omega_0 t} + \cdots + c_n e^{jn\omega_0 t} + \cdots$$
$$+ c_{-1} e^{-j\omega_0 t} + c_{-2} e^{-j2\omega_0 t} + \cdots + c_{-n} e^{-jn\omega_0 t} + \cdots$$
$$= \sum_{n=-\infty}^{\infty} c_n e^{jn\omega_0 t}, \qquad t_0 < t < t_0 + T \tag{1.22b}$$

where

$$c_n = \frac{\displaystyle\int_{t_0}^{t_0+T} f(t)\left(e^{jn\omega_0 t}\right)^* dt}{\displaystyle\int_{t_0}^{t_0+T} e^{jn\omega_0 t}\left(e^{jn\omega_0 t}\right)^* dt} = \frac{1}{T}\int_{t_0}^{t_0+T} f(t)\,e^{-jn\omega_0 t}\,dt$$

1.2.3.2 Fourier Transform of a Periodic Function

Let $u(t)$ be a periodic function (waveform) of period T as shown in Figure 1.25. Let us now define $y(t)$ as the aperiodic function, illustrated in Figure 1.26, that is nonzero over the interval $(-T/2, T/2)$. That is, $y(t)$ is one period of $u(t)$:

$$y(t) \triangleq \begin{cases} u(t), & |t| \leq T/2 \\ 0, & \text{elsewhere} \end{cases}$$

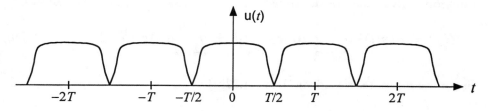

Figure 1.25 A periodic function (waveform).

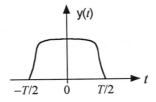

Figure 1.26 An aperiodic function.

We can then represent the periodic function $u(t)$ in terms of the aperiodic $y(t)$ as follows:

$$u(t) = \cdots + y(t + 2T) + y(t + T) + y(t) + y(t - T) + y(t - 2T) + \cdots$$

$$= \sum_{n=-\infty}^{\infty} y(t - nT) = y(t) \text{ repeated in every } T\text{-second interval.}$$

Woodward [19] introduced another notation for a periodic (repeated) function and called it the "repetition function," denoted $\text{Rep}_T(\cdot)$ and defined as

$$u(t) = \text{Rep}_T\{y(t)\} \triangleq \sum_{n=-\infty}^{\infty} y(t - nT) \qquad (1.23a)$$

Note that the summation in (1.23a) is an orthodox notation that we all are familiar with, whereas $u(t) = \text{Rep}_T\{y(t)\}$ is a Woodward notation. As we see in what follows, the Woodward notation is more than a notation in that its Fourier transform pair is most conveniently represented in yet another notation in such a way as to provide an intuitive picture of the physical processes involved.

We proceed now to find the Fourier transform of the $\text{Rep}_T(\cdot)$ function, which in effect is to find the Fourier transform of a periodic function. Given the Fourier transform pair $y(t) \leftrightarrow Y(f)$, then by definition

$$Y(f) = \mathcal{F}\{y(t)\} = \int_{-\infty}^{\infty} y(t)\, e^{-j2\pi ft}\, dt = \int_{-T/2}^{T/2} u(t)\, e^{-j2\pi ft}\, dt \qquad (1.23b)$$

From the Fourier series for $u(t)$ [see (1.22b)]:

$$u(t) = \sum_{n=-\infty}^{\infty} c_n\, e^{j2\pi n f_0 t}, \qquad f_0 = \frac{1}{T}$$

where, using (1.23b):

$$c_n = \frac{1}{T} \int_{-T/2}^{T/2} u(t)\, e^{j2\pi n f_0 t}\, dt = \frac{1}{T} Y(n f_0) = \frac{1}{T} Y\left(\frac{n}{T}\right) \tag{1.23c}$$

Now, the Fourier transform[10] of $u(t)$ can be developed as follows:

$$U(f) = \mathcal{F}\{u(t)\} = \mathcal{F}\left\{ \sum_{n=-\infty}^{\infty} c_n\, e^{j2\pi n f_0 t} \right\}$$

$$= \sum_{n=-\infty}^{\infty} c_n\, \mathcal{F}\{e^{j2\pi n f_0 t}\}$$

$$= \sum_{n=-\infty}^{\infty} c_n\, \delta(f - n f_0) = \sum_{n=-\infty}^{\infty} c_n\, \delta\left(f - \frac{n}{T}\right) \tag{1.23d}$$

Substituting the Fourier coefficients $\{c_n\}$ given in (1.23c), we have

$$U(f) = \sum_{n=-\infty}^{\infty} \frac{1}{T} Y(n f_0)\, \delta(f - n f_0) = \sum_{n=-\infty}^{\infty} \frac{1}{T} Y\left(\frac{n}{T}\right) \delta\left(f - \frac{n}{T}\right) \tag{1.23e}$$

Woodward defined a function called the "comb function" as follows:

$$\mathrm{Comb}_F\{z(\alpha)\} \triangleq \sum_{n=-\infty}^{\infty} z(nF)\, \delta(\alpha - nF) \tag{1.23f}$$

Using this notation, we can express the Fourier transform of $u(t)$ as

$$U(f) = \sum_{n=-\infty}^{\infty} \frac{1}{T} Y\left(\frac{n}{T}\right) \delta\left(f - \frac{n}{T}\right) = \frac{1}{T} \mathrm{Comb}_{1/T}\{Y(f)\} \tag{1.23g}$$

Therefore, we have the Fourier transform pair:

$$\mathrm{Rep}_T\{y(t)\} \leftrightarrow \frac{1}{T} \mathrm{Comb}_{1/T}\{Y(f)\} \tag{1.23h}$$

The $\mathrm{Comb}_{1/T}\{\,\cdot\,\}$ function is illustrated in Figure 1.27, where we see that the Fourier transform of the periodic function is a line spectrum with lines at intervals of $1/T$ and the envelope $Y(f)$, the Fourier transform of one period of the periodic function.

---

[10] Strictly speaking, one cannot take the Fourier transform of a periodic signal because periodic signals have infinite energy.

The Fourier transform pair in (1.23h) has validity if time and frequency are exchanged. This is shown by finding the transform of the Comb function. Let $u(t)$ be the sampled version of the waveform $y(t)$:

$$u(t) \triangleq \text{Comb}_T\{y(t)\} = \sum_{n=-\infty}^{\infty} y(nT)\,\delta(t-nT)$$

Then the Fourier transform of the Comb function is

$$\mathcal{F}\{u(t)\} \triangleq U(f) = \mathcal{F}\{\text{Comb}_T[y(t)]\}$$

$$= \mathcal{F}\left\{ \sum_{n=-\infty}^{\infty} y(nT)\,\delta(t-nT) \right\} = \mathcal{F}\left\{ y(t) \sum_{n=-\infty}^{\infty} \delta(t-nT) \right\}$$

$$= \mathcal{F}\{y(t)\,\text{Rep}_T[\delta(t)]\} = Y(f) * \tfrac{1}{T}\,\text{Comb}_{1/T}\{1\}$$

by the frequency domain convolution theorem. Continuing, we find that

$$U(f) = Y(f) * \frac{1}{T} \sum_{n=-\infty}^{\infty} \delta\left(f - \frac{n}{T}\right)$$

$$= \frac{1}{T} \sum_{n=-\infty}^{\infty} \int_{-\infty}^{\infty} Y(f-\nu)\,\delta\left(\nu - \frac{n}{T}\right) d\nu$$

$$= \frac{1}{T} \sum_{n=-\infty}^{\infty} Y\left(f - \frac{n}{T}\right) = \frac{1}{T}\,\text{Rep}_{1/T}\{Y(f)\}$$

Therefore, we have the transform pair

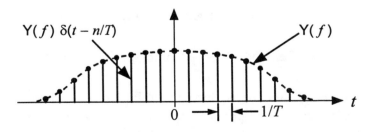

Figure 1.27 The function $\text{Comb}_{1/T}\{Y(f)\}$, the spectrum of $\text{Rep}_T\{y(t)\}$.

$$\text{Comb}_T\{y(t)\} \leftrightarrow \frac{1}{T}\text{Rep}_{1/T}\{Y(f)\} \tag{1.23i}$$

1.3 Sampling Theorems

1.3.1 Sampling Theorem in the Frequency Domain

Consider a time-limited (finite energy) signal $u(t)$ in the interval $|t| \leq T/2$ that is not necessarily even about $t = 0$, as illustrated in Figure 1.28. Now, let us construct a periodic signal $v(t)$ as the repetition of $u(t)$. That is, let $v(t)$ be $v(t) = \text{Rep}_T\{u(t)\}$, as shown in Figure 1.29. Then, in terms of $v(t)$ we may write

$$u(t) = \text{Rect}(t/T) \times v(t) = \text{Rect}(t/T) \times \text{Rep}_T\{u(t)\}$$

and

$$U(f) = T\operatorname{sinc}(Tf) * \frac{1}{T}\text{Comb}_{1/T}\{U(f)\}$$

$$= \sum_{n=-\infty}^{\infty} U\left(\frac{n}{T}\right)\delta\left(f - \frac{n}{T}\right) * \operatorname{sinc}(Tf)$$

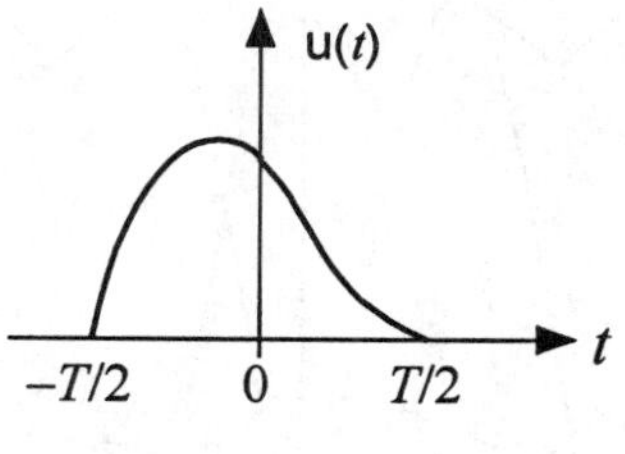

Figure 1.28 Finite energy signal.

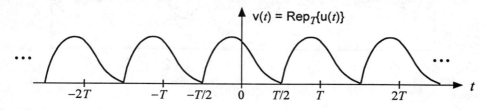

Figure 1.29 A periodic waveform constructed from an aperiodic waveform.

$$= \sum_{n=-\infty}^{\infty} U\left(\frac{n}{T}\right) \text{sinc}\left[T\left(f - \frac{n}{T}\right)\right] \tag{1.24}$$

Thus, the frequency-domain sampling theorem states that, for any real time waveform that is zero outside the interval $(-T/2, T/2)$, the spectrum is completely determined by its values at the frequencies nf_0, where n is an integer and $f_0 \triangleq 1/T$. The value of the spectrum between the sample values is found by interpolation; in (1.24) the interpolation is performed perfectly by the sinc function, as illustrated conceptually in Figure 1.30.

Note that the sampling theorem can be written in terms of convolution:

$$U(f) = \sum_{n=-\infty}^{\infty} U\left(\frac{n}{T}\right) \text{sinc}\left[T\left(f - \frac{n}{T}\right)\right] = \text{Comb}_{1/T}\{U(f)\} * \text{sinc}(Tf)$$

Therefore, the spectrum can be viewed as the result of passing the spectrum samples through an interpolation filter, as suggested by Figure 1.31.

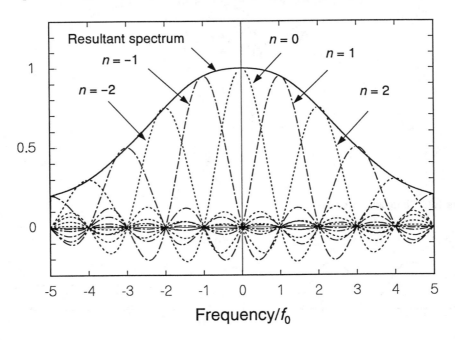

Figure 1.30 Interpolation of frequency-domain samples.

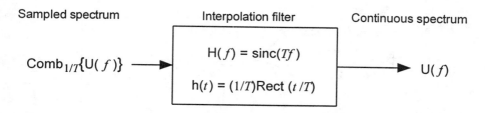

Figure 1.31 Interpolation filter concept for frequency-domain sampling theorem.

1.3.2 Sampling Theorem in the Time Domain

Similarly, if $u(t)$ is a bandlimited signal confined to the band $(-W, W)$, the signal spectrum can be viewed as the product of its periodic extension (repetition) $\text{Rep}_{2W}\{U(f)\}$ and the rectangular window $\text{Rect}(f/2W)$:

$$U(f) = \text{Rect}(f/2W) \times \text{Rep}_{2W}\{U(f)\}$$

Taking the inverse Fourier transform of both sides results in

$$u(t) = 2W \text{sinc}(2Wt) * \frac{1}{2W} \text{Comb}_{1/2W}\{u(t)\}$$

$$= \sum_{n=-\infty}^{\infty} u\left(\frac{n}{2W}\right) \delta\left(t - \frac{n}{2W}\right) * \text{sinc}(2Wt)$$

$$= \sum_{n=-\infty}^{\infty} u\left(\frac{n}{2W}\right) \text{sinc}\left[2W\left(t - \frac{n}{2W}\right)\right] \tag{1.25}$$

Thus, the time-domain sampling theorem states that if $u(t)$ is low-pass limited to the band $(-W, W)$, the waveform is completely determined by its values taken at the interval $1/2W$. This interval is called the "Nyquist interval" and the sampling rate, $2W$ samples per second, is called the "Nyquist rate" [21].

The continuous function $u(t)$, according to (1.21), can be reconstructed from samples by using a sinc interpolating function, as illustrated in Figure 1.32. Note that this process can be written in terms of a convolution:

$$u(t) = \text{Comb}_{1/2W}\{u(t)\} * \text{sinc}(2Wt)$$

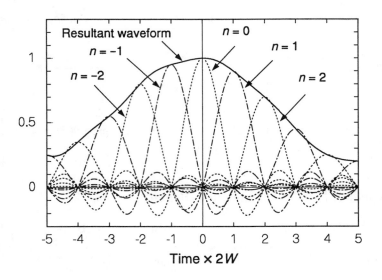

Figure 1.32 Interpolation of time-domain samples.

This convolution relationship implies that the interpolation can be performed using a linear filter with the impulse response, $h(t) = \text{sinc}(2Wt)$, which is the ideal rectangular lowpass filter. This concept is illustrated in Figure 1.33.

Example 1.8 *Representation of continuous waveforms.* Let $x(t)$ and $y(t)$ be waveforms, both limited to the band $(W, -W)$. Then, from the lowpass waveform sampling theorem, we have

$$x(t) = \overbrace{\text{Comb}_{1/2W}\{x(t)\}}^{\text{Sampling}} * \overbrace{\text{sinc}(2Wt)}^{\text{Interpolation}} = \sum_{n=-\infty}^{\infty} x\left(\frac{n}{2W}\right) \text{sinc}(2Wt - n)$$

and

Sampled signal   Interpolation filter   Continuous signal

$$\text{Comb}_{1/2W}\{u(t)\} \longrightarrow \boxed{\begin{array}{c} h(t) = \text{sinc}(2Wt) \\[2mm] H(f) = (1/2W)\text{Rect}\,(f/2W) \end{array}} \longrightarrow u(t)$$

Figure 1.33 Interpolation filter concept for time-domain sampling theorem.

$$y(t) = \text{Comb}_{1/2W}\{y(t)\} * \text{sinc}(2Wt) = \sum_{n=-\infty}^{\infty} y\left(\frac{n}{2W}\right) \text{sinc}(2Wt - n)$$

When two functions are multiplied and the product is integrated, the process is a measure of the "correlation" between the two functions. Using the sampled expressions for $x(t)$ and $y(t)$, their correlation can be written

$$\int_{-\infty}^{\infty} x(t)\, y(t)\, dt = \sum_{m=-\infty}^{\infty} \sum_{n=-\infty}^{\infty} x\left(\frac{m}{2W}\right) y\left(\frac{n}{2W}\right)$$

$$\times \int_{-\infty}^{\infty} \text{sinc}(2Wt - n) \,\text{sinc}(2Wt - n)$$

$$= \frac{1}{2W} \sum_{m=-\infty}^{\infty} \sum_{n=-\infty}^{\infty} x\left(\frac{m}{2W}\right) y\left(\frac{n}{2W}\right) \delta_{mn}$$

so that

$$2W \int_{-\infty}^{\infty} x(t)\, y(t)\, dt = \sum_{n=-\infty}^{\infty} x\left(\frac{n}{2W}\right) y\left(\frac{n}{2W}\right) \tag{1.26a}$$

A special case of (1.26a) for $y(t) = x(t)$ is

$$2W \int_{-\infty}^{\infty} x^2(t)\, dt = \sum_{n=-\infty}^{\infty} x^2\left(\frac{n}{2W}\right) \tag{1.26b}$$

Because the samples $\{x(n/2W),\ n = \ldots, -1, 0, 1, \ldots\}$ taken at intervals of $1/2W$ constitute a vector, the meaning of (1.26b) is that the square of the length of the vector from the origin to the signal point in the signal space is proportional to the total energy of the waveform. That is, let

$$\mathbf{x} = (x_1, x_2, \ldots, x_n, \ldots), \qquad x_n \triangleq \frac{1}{\sqrt{2W}}\, x(n/2W)$$

and

$$\mathbf{y} = (y_1, y_2, \ldots, y_n, \ldots), \qquad y_n \triangleq \frac{1}{\sqrt{2W}}\, y(n/2W)$$

Then

$$\int_{-\infty}^{\infty} x(t)\, y(t)\, dt = \langle \mathbf{x}, \mathbf{y} \rangle = \sum_n x_n y_n$$

so that

$$\int_{-\infty}^{\infty} x^2(t)\, dt = \langle \mathbf{x}, \mathbf{x} \rangle = |\mathbf{x}|^2 = \sum_n x_n^2 \tag{1.26c}$$

The geometric interpretation of (1.26b), and hence (1.26c), was employed by Shannon in 1949 in his famous seminal paper, "Communications in the Presence of Noise" [22].

If the waveforms $x(t)$ and $y(t)$ are assumed to be time-limited to $(0, T)$ while still (approximately) bandlimited to $(-W, W)$[11], and samples are taken at the Nyquist interval $1/2W$, one obtains a total of $2WT$ samples in the time T seconds. Then, from (1.25), the sampling theorem, we have

$$x(t) \cong \sum_{n=0}^{2WT-1} x\left(\frac{n}{2W}\right) \mathrm{sinc}(2Wt - n) \qquad (1.27a)$$

$$y(t) \cong \sum_{n=0}^{2WT-1} y\left(\frac{n}{2W}\right) \mathrm{sinc}(2Wt - n) \qquad (1.27b)$$

and

$$\int_0^T x(t)\,dt \cong \int_0^T dt \sum_{n=0}^{2WT-1} x\left(\frac{n}{2W}\right) \mathrm{sinc}(2Wt - n)$$

$$= \sum_{n=0}^{2WT-1} x\left(\frac{n}{2W}\right) \int_0^T \mathrm{sinc}(2Wt - n)\,dt$$

or

$$\int_0^T x(t)\,dt \approx \frac{1}{2W} \sum_{n=0}^{2WT-1} x\left(\frac{n}{2W}\right) \qquad (1.27c)$$

since the integral of the sinc function is approximately equal to $1/2W$, as shown in Appendix 1B.

The correlation between $x(t)$ and $y(t)$ is now given by

$$\int_0^T x(t)\,y(t)\,dt \approx \frac{1}{2W} \sum_{n=0}^{2WT-1} x\left(\frac{n}{2W}\right) y\left(\frac{n}{2W}\right) \qquad (1.27d)$$

so that, for $x(t) = y(t)$:

$$\int_0^T x^2(t)\,dt \approx \frac{1}{2W} \sum_{n=0}^{2WT-1} x^2\left(\frac{n}{2W}\right) \qquad (1.27e)$$

---

[11] It is, of course, impossible for a waveform or signal to be both time-limited and bandlimited. Hence, such an assumption can only be made in an approximate sense.

The implication of (1.27e) is that one can obtain the signal energy by means of sampling the waveform at a rate greater than or equal to the Nyquist rate, as suggested by the scheme diagrammed in Figure 1.34.

1.3.3 Sampling Theorem for Bandpass Waveforms

Consider a bandpass waveform $x(t)$ whose spectrum is confined to the bands

$$f_0 - \tfrac{1}{2}W < |f| < f_0 + \tfrac{1}{2}W, \quad f_0 \geq \tfrac{1}{2}W$$

where f_0 is the carrier or center frequency and W is the bandwidth over which the signal energy is confined, as depicted in Figure 1.35. In view of the time-domain sampling theorem developed in Section 1.3.2 for a lowpass waveform, we might be tempted to apply the formula given by (1.25) by taking $f_0 + W/2$ as the highest frequency in place of W in (1.25). But this is not possible because, in the first place, the bandpass signal contains no frequency components from 0 to $\pm(f_0 - W/2)$ Hz, while the representation of the signal by (1.25) contains frequency components in this interval.

If we consider the spectrum on the positive frequency axis only, then the resultant time-domain expression, being complex, will consist of a "real part" and an "imaginary part," and we can extract the signal as the real part. This is the basis of the development for the sampling theorem for high-frequency waveforms. Consider the upper single sideband of the spectrum $X(f)$, as illustrated in Figure 1.36, and call it $Z(f)$. Then we have

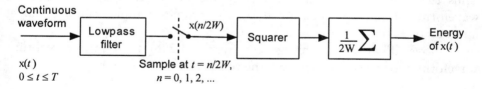

Figure 1.34 Use of samples to measure the energy of a waveform.

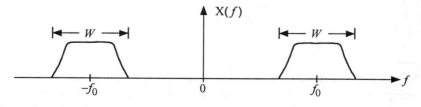

Figure 1.35 Spectrum of a bandpass waveform.

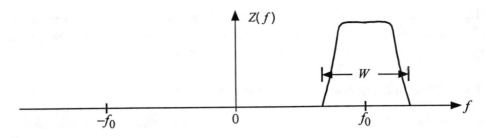

Figure 1.36 The one-sided spectrum of the complex signal $z(t)$.

$$Z(f) = \text{Rep}_W\{Z(f)\} \times \text{Rect}[(f - f_0)/W] \qquad (1.28a)$$

We know that only a *complex waveform*[12] $z(t)$ will give a spectrum that appears only on the positive frequency axis, and thus our objective is to obtain the real part of the inverse Fourier transform of (1.28a), which will give the required sampling theorem.

Now, by applying the convolution theorem to (1.28a), we get

$$\mathcal{F}^{-1}\{Z(f)\} \triangleq z(t) = \left[\frac{1}{W}\,\text{Comb}_{1/W}\{z(t)\}\right] * \left[(W\,\text{sinc}\,Wt)\,e^{j2\pi f_0 t}\right]$$

$$= \sum_{n=-\infty}^{\infty} z\!\left(\frac{n}{W}\right)\text{sinc}\left[W\!\left(t - \frac{n}{W}\right)\right]e^{j2\pi f_0(t-n/W)} \qquad (1.28b)$$

Thus the sampling theorem for high-frequency waveforms states that the waveform is uniquely determined by its complex sample values taken at intervals of $1/W$.

Now, if we apply this general result (1.28b) to a purely real-time waveform, we have only to extract the real part of (1.28b). Let

$$z(t) = x(t) + j\,y(t) \qquad (1.28c)$$

where

$$x(t) = \text{Re}\{z(t)\}$$

12　The Fourier transform of a real-time signal always produces a spectrum that is symmetric with respect to the $f = 0$ axis, whereas that of the *complex representation* of the signal always produces a spectrum which appears only on the positive frequency axis.

A straightforward expansion of (1.28b) results in the real-part expression given by

$$x(t) = \text{Re}\{(\text{Comb}_{1/W}\{z(t)\}) * (\text{sinc}Wt\, e^{j2\pi f_0 t})\}$$

$$= \text{Re}\{(\text{Comb}_{1/W}\{x(t) + j\,y(t)\})$$

$$* (\text{sinc}Wt \cdot [\cos 2\pi f_0 t + j\sin 2\pi f_0 t])\}$$

$$= \sum_{n=-\infty}^{\infty} R\left(\frac{n}{W}\right) \text{sinc}\left[W\left(t - \frac{n}{W}\right)\right] \cdot \cos\left[2\pi f_0\left(t - \frac{n}{W}\right) + \phi\left(\frac{n}{W}\right)\right]$$

where
$$\tag{1.29a}$$

$$R(t) \triangleq \sqrt{x^2(t) + y^2(t)} \quad \text{and} \quad \phi(t) \triangleq \tan^{-1}\left[\frac{y(t)}{x(t)}\right] \tag{1.29b}$$

An interpretation of this result is that an amplitude $R(n/W)$ and an instantaneous carrier phase angle $\phi(n/W)$ need to be specified at each sampling instant taken at intervals of $1/W$. This scheme is depicted in Figure 1.37.

1.3.4 Discrete Time Filtering

In the analysis of discrete time systems, the unit impulse function (also known as the Kronecker delta function) plays an important role; it is defined by

$$\delta(n - k) = \delta_{nk} \triangleq \begin{cases} 0, & n \neq k \\ 1, & n = k \end{cases} \tag{1.30a}$$

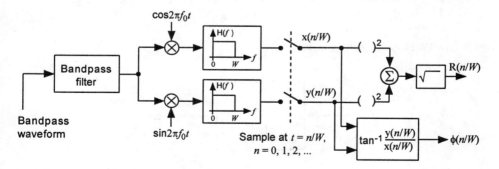

Figure 1.37 Samples needed to reconstruct a bandpass waveform.

and it is illustrated in Figure 1.38. The resolution of an arbitrary discrete-time function x(n) into unit impulse functions, as a counterpart of (1.17a) for the case of a continuous waveform, is

$$x(n) = \sum_{k=-\infty}^{\infty} x(k)\, \delta(n-k) \tag{1.30b}$$

As we have done for the case of a continuous-time system, if G is assumed to be a discrete-time system as depicted in Figure 1.39, we obtain the output response y(n) as

$$
\begin{aligned}
y(n) = G\{x(n)\} &= G\left\{ \sum_{k=-\infty}^{\infty} x(k)\, \delta(n-k) \right\} \\
&= \sum_{k=-\infty}^{\infty} x(k)\, G\{\delta(n-k)\} \\
&= \sum_{k=-\infty}^{\infty} x(k)\, h(n,k) \tag{1.31a}
\end{aligned}
$$

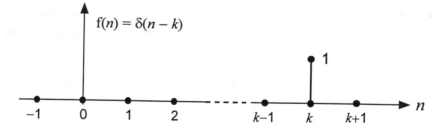

Figure 1.38 Unit impulse (delta) function.

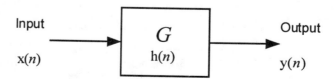

Figure 1.39 Discrete-time system.

This result is the discrete-time system counterpart of the *superposition integral* defined in (1.17c) and is called a *superposition summation*. We can state the analogies with the continuous-time system as follows:

$$y(n) = \sum_{k=-\infty}^{\infty} x(k)\, h(n, k) \qquad \longleftarrow \text{ linear}$$

$$= \sum_{k=-\infty}^{n} x(k)\, h(n, k) \qquad \longleftarrow \text{ causal}$$

$$= \sum_{k=-\infty}^{n} x(k)\, h(n - k) \qquad \longleftarrow \text{ time-invariant}$$

$$= \sum_{k=0}^{n} x(k)\, h(n - k) \qquad \longleftarrow \text{ no input for } n < 0$$

$$= \sum_{k=0}^{n} x(n - k)\, h(k) \qquad \longleftarrow \text{ symmetry}$$

These summations are called *convolution* summations and can be denoted as

$$y(n) = x(n) * h(n) = \sum_{k=-\infty}^{\infty} x(k)\, h(n - k) \qquad (1.31b)$$

Example 1.9 Suppose that the response of a discrete-time system to a unit impulse function is as shown on the left in Figure 1.40. Find the response of this system to the input $x(n)$ shown on the right in Figure 1.40.

Solution: The response, sketched in Figure 1.41, is given by

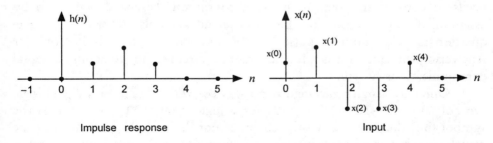

Figure 1.40 Impulse response and input for an example discrete-time system.

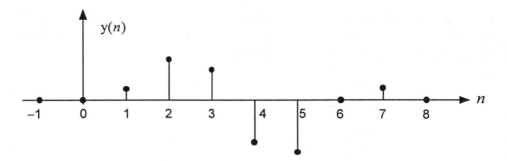

Figure 1.41 Example discrete-time system response.

$$y(n) = \sum_{k=-\infty}^{\infty} x(k)\, h(n-k)$$

$$= x(0)\, h(n) + x(1)\, h(n-1) + x(2)\, h(n-2)$$

$$+ x(3)\, h(n-3) + x(4)\, h(n-4)$$

1.4 Baseband Pulse Shaping for Bandlimited Transmission

In modern digital communications systems, the digital data symbols must be transmitted by means of a continuous pulse shape of some appropriate form to accomplish propagation over the channel. For example, when a string of binary data such as "1 0 1 1 0 0 1 0 1" is to be transmitted to a distant receiver, in addition to translating the data signal to a suitable carrier frequency, one needs to convert this data string to a continuous baseband waveform by assigning different pulses for the basic digital symbols "0" and "1." The question is, what kind of pulse for "0" and what kind of pulse for "1"? Answering this question is a fundamental requirement for designing a digital communications system.

One may assign a rectangular pulse, $\text{Rect}(t/T - 1/2)$, for a digital logic symbol "0" and $-\text{Rect}(t/T - 1/2)$ for a logic symbol "1," where T is the symbol (bit) duration. But this scheme is not desirable because, to pass the waveform energy without introducing distortion and intersymbol interference (ISI), the system bandwidth must be significantly in excess of $1/T$, the inverse of the symbol rate. Then what is the practical approach for designing an impulse response waveform for a digital system to meet the requirements of (a) confinement of the spectrum to a required bandwidth and (b) minimum possible ISI? This question is the central issue of this section; we review succinctly a few examples of approaches to the stated waveform requirements and describe the techniques used in the IS-95 system.

1.4.1 Bandlimited Waveforms for Digital Applications

Let us consider the basic model of a digital transmission system that is shown in Figure 1.42. For simplicity, the use of a high frequency carrier is not considered. We first assume that the digital data stream is represented by a series of impulses whose polarities are based on the 0s and 1s of the source data. The transmitter filter is characterized by an impulse response $g(t) = \text{Rect}(t/T)$. The baseband signal representing the digital data stream at the output of this filter can thus be written as

$$D(t) = \sum_{k=-\infty}^{\infty} d_k\, g(t - kT) = \sum_{k=-\infty}^{\infty} d_k\, \text{Rect}\left(\frac{t}{T} - k\right) \qquad (1.32)$$

where the $\{d_k\}$ are $+1$ or -1 values, corresponding respectively to logical 0s or 1s, generated at T-second intervals. The transmitter filter characteristic is illustrated in Figure 1.43.

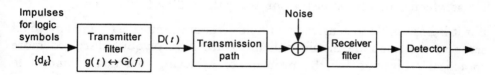

Figure 1.42 Basic model of a data transmission system.

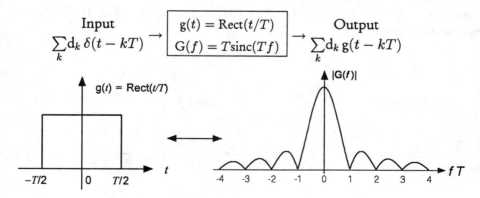

Figure 1.43 Transmitter filter characteristic.

As stated at the outset of this section, the desire to conserve the bandwidth, while minimizing or eliminating ISI (if possible), is the central concern of baseband waveform design philosophy. Nyquist [21] considered the problem of designing the pulse shape g(t) such that no ISI occurs at the sampling instants. He showed that the minimum bandwidth needed to transmit R symbols per second without ISI is $R/2$ Hz. This occurs when the filter has the amplitude frequency characteristic

$$G(f) = \text{Rect}(f/R) \tag{1.33a}$$

and the impulse response

$$g(t) = \mathcal{F}^{-1}\{G(f)\} = R\,\text{sinc}(Rt) = \frac{1}{T}\,\text{sinc}\left(\frac{t}{T}\right) \tag{1.33b}$$

which are illustrated in Figure 1.44. Note that if the information symbol occurs at the rate of one in every T seconds, and the receiver samples without error at the time instants of $t = iT$, $i = 0, 1, 2, \ldots$, then there is no ISI, as seen on the right-hand side of Figure 1.44. One then can state the Nyquist result in connection with this figure as follows:

> Given a channel with bandwidth W Hz, the maximal theoretical rate at which symbols can be transmitted without ISI is $2W$ symbols/sec.

But it is impossible to achieve this result without infinite delay, and any small error in the sampling instant will introduce ISI. Nyquist discussed more "practical" solutions for the problem of achieving maximal symbol rate without introducting ISI, by proposing a different filter frequency response shape, G(f). To maintain a symbol rate of R symbols/sec, the practical filters must require more bandwidth than the Nyquist bandwidth of $R/2$ Hz. One "practical shape" is the raised cosine (RC) filter, which has 100% "excess bandwidth," given by

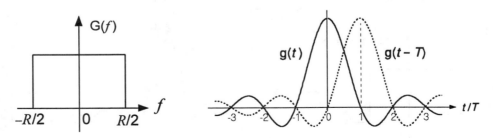

Figure 1.44 Nyquist filter and impulse response waveform.

$$G(f) = \cos^2\left(\frac{\pi f}{2R}\right) = \frac{1}{2}\left[1 + \cos\left(\frac{\pi f}{R}\right)\right], \quad |f| \le R \qquad (1.34a)$$

and this filter shape in the frequency domain is depicted in Figure 1.45. By 100% excess bandwidth, it is meant that the cutoff frequency of $R/2$ is doubled to become R Hz, as shown in the figure. Thus the situation is now that R symbols/sec in R Hz is the maximal symbol transmission rate in relation to bandwidth. In other words, one can pack only one symbol/sec per Hz of bandwidth.

Note that the transition of the RC filter characteristic in Figure 1.45 from full value to zero is smooth and begins at $f = 0$, with the amplitude reduced to $1/2$ at $f = R/2$; thus the transition takes place over the entire range of frequencies that are passed by the filter. In practical communications systems, there is a whole family of RC filters based on the width of the transition region, each a RC filter reduced to half amplitude at $f = R/2$ but starting the transition at the frequency $(1 - \alpha)R/2$ and ending it at $(1 + \alpha)R/2$, as illustrated in Figure 1.46. The parameter α is equal to the fraction of excess bandwidth, and is called the "rolloff factor." The transfer function for the RC filter with rolloff α is given by

$$G(f; \alpha) = \begin{cases} 1, & |fT| \le \frac{1}{2}(1-\alpha) \\ \cos^2\left\{\frac{\pi}{2\alpha}\left[fT - \frac{1}{2}(1-\alpha)\right]\right\}, & \frac{1}{2}(1-\alpha) < fT \le \frac{1}{2}(1+\alpha) \quad (1.34b) \\ 0, & |fT| > \frac{1}{2}(1+\alpha) \end{cases}$$

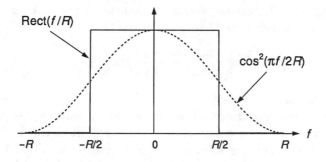

Figure 1.45 Raised cosine (RC) filter with 100% excess bandwidth.

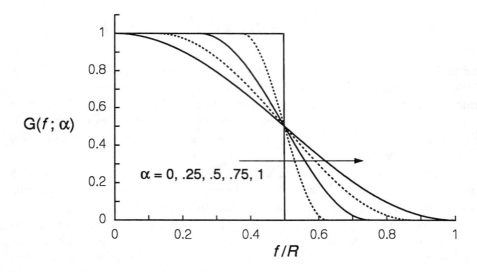

Figure 1.46 RC filter with rolloff factor α.

where an alternative expression in the transition region is

$$\cos^2\left\{\tfrac{\pi}{2\alpha}\left[fT-\tfrac{1}{2}(1-\alpha)\right]\right\} = \tfrac{1}{2}\left[1+\cos\left\{\tfrac{\pi}{\alpha}\left[fT-\tfrac{1}{2}(1-\alpha)\right]\right\}\right]$$
$$= \tfrac{1}{2}\left[1-\sin\left\{\tfrac{\pi}{\alpha}\left(fT-\tfrac{1}{2}\right)\right\}\right]$$

If we take the inverse transform of $T \cdot G(f; \alpha)$ in (1.34b), as shown in Appendix 1C, we obtain the impulse response of the filter as the Nyquist pulse shape with rolloff factor α:

$$g(t; \alpha) = \frac{\cos(\pi\alpha t/T)}{1-(2\alpha t/T)^2}\,\operatorname{sinc}\left(\frac{t}{T}\right) \tag{1.34c}$$

which is plotted in Figure 1.47.

In practical applications of waveshaping techniques, the Nyquist RC filter is used with an appropriate choice of the rolloff factor α. For example, in the IS-54 digital cellular telephone system based on TDMA [1], the RC filter is used with a rolloff factor value of $\alpha = 0.35$. To achieve the desired pulse shape, filters employing the square-root of $G(f; \alpha)$ are used in both transmitter and receiver. The impulse response of the $\sqrt{\text{RC}}$ filter with rolloff factor α is shown in [23] to be

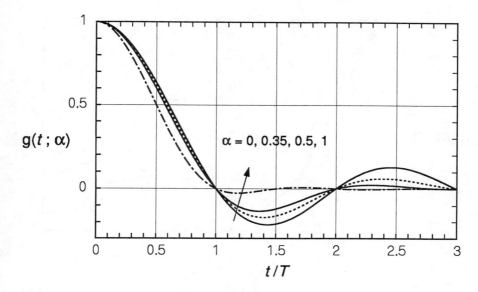

Figure 1.47 Nyquist pulse shapes for different values of α.

$$g_1(t) = \frac{\sin[(1-\alpha)\pi t/T] + 4\alpha(t/T)\cos[(1+\alpha)\pi t/T]}{(\pi t/T)[1 - (4\alpha t/T)^2]} \qquad (1.34d)$$

which has the particular values $g_1(0) = 1 - \alpha + 4\alpha/\pi$ and $g_1(T/4\alpha) = \left(\alpha/\sqrt{2}\right)\left[(1 + \frac{2}{\pi})\sin(\pi/4\alpha) + (1 - \frac{2}{\pi})\cos(\pi/4\alpha)\right]$. This pulse shape is shown normalized by $g_1(0)$ in Figure 1.48. Note how $g_1(t)$ is in general a narrower pulse than $g(t)$, but broadens into $g(t)$ when it is convolved with itself. Note also that the pulse shape $g_1(t)$ causes ISI because it is not generally equal to zero at the symbol times, nT; thus, a system using this square-root pulse shape actually transmits a signal that has ISI, but the identical impulse response of the receiver filter *equalizes* this condition and removes the ISI.

It is important to note that, when the transmission waveform is "shaped" in the form of Nyquist pulses with rolloff factor α as shown in Figure 1.47, the signal energy is confined to the bandwidth indicated in the spectral characteristic given in Figure 1.46, and no ISI is introduced when sampling times are error free. In practice, however, timing errors do occur and filter designs do depart from the theoretical specifications, thereby inducing some performance degradation.

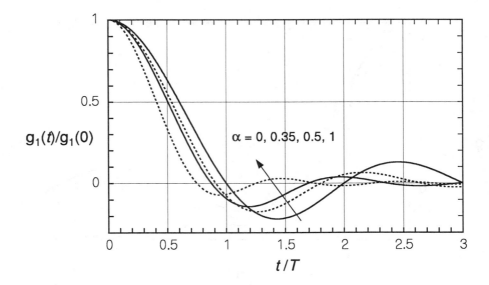

Figure 1.48 Impulse response of square-root RC filter.

With the notion of baseband waveform shaping and its purposes, we can now examine how the IS-95 system implements its waveshaping filter with the background that we have reviewed up to this point.

1.4.2 FIR Pulse Shaping in IS-95

The IS-95 system uses baseband waveshaping filters designed to meet a bandwidth constraint while minimizing ISI. The filters are specified to have a frequency response $H(f)$ that satisfies the limits shown in Figure 1.49, with a less than 1.5-dB ripple in the passband ($0 \leq f \leq f_p = 590\,\text{kHz}$) and a 40-dB attenuation in the stop band ($f > f_s = 740\,\text{kHz}$), with a nominal one-sided bandwidth of $f_c = 614.4\,\text{kHz}$. In addition to these frequency domain limits, IS-95 specifies that the impulse response of the filter be a close fit to that of a certain 48-tap FIR filter with response $h_0(n)$.

A FIR filter impulse response assumes a sample period T_s, and the discrete time output $y(n)$ of the filter for an input $x(n)$ is of the form

$$y(n) \equiv y(nT_s) = \sum_{k=0}^{K-1} x(n-k)\, h_0(k) \tag{1.35a}$$

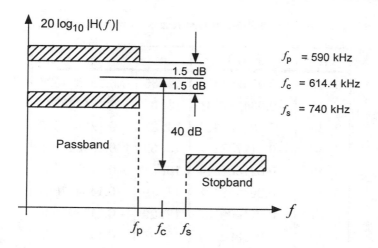

Figure 1.49 Filter specifications.

so that, if the input $x(n)$ is a sequence of data impulse values given by

$$x(n) = \sum_m d_m \delta(n - n_m) \qquad (1.35b)$$

then the discrete-time output of the filter with response $h_0(n)$ is

$$y(n) = \sum_k x(k)\, h_0(n - k) = \sum_k \sum_m d_m \delta(k - n_m)\, h_0(n - k)$$
$$= \sum_m d_m\, h_0(n - n_m) \qquad (1.36a)$$

The notation in (1.35b) and (1.36a) allows for the input samples to occur at multiples of the filter sample time, T_s; for example, the values can be $\{n_m\} = \{4m = 0, 4, 8, \dots\}$ as in IS-95 to allow the discrete filter to make a smooth transition between data symbols. The continuous-time output (following digital-to-analog conversion and smoothing) is the superposition of shifts of the pulse shape $h_0(t)$:

$$y(t) = \sum_m d_m\, h_0(t - n_m T_s) \qquad (1.36b)$$

The IS-95 FIR impulse response coefficients $\{h_0(k),\ k = 0, 1, \dots, 47\}$ are listed in Table 1.3, and the required fit is described by

Table 1.3 IS-95 FIR impulse response coefficients [2]

n	$h_0(n)$	n	$h_0(n)$
0, 47	-0.025288315	12, 35	0.007874526
1, 46	-0.034167931	13, 34	0.084368728
2, 45	-0.035752323	14, 33	0.126869306
3, 44	-0.016733702	15, 32	0.094528345
4, 43	0.021602514	16, 31	-0.012839661
5, 42	0.064938487	17, 30	-0.143477028
6, 41	0.091002137	18, 29	-0.211829088
7, 40	0.081894974	19, 28	-0.140513128
8, 39	0.037071157	20, 27	0.094601918
9, 38	-0.021998074	21, 26	0.441387140
10, 37	-0.060716277	22, 25	0.785875640
11, 36	-0.051178658	23, 24	1.000000000

$$\text{Mean squared error} = \sum_{k=0}^{\infty} \beta \left[\text{h}(kT_s - \tau) - \text{h}_0(k) \right]^2 \leq 0.03 \qquad (1.37a)$$

where the tap (sample) spacing is $T_s = T_c/4$, one-fourth the duration of a PN sequence chip, and β and τ may be selected to minimize the mean square error. The total duration of the impulse response, from the first sample to the last, is $48\,T_s = 12\,T_c$, and the (unsmoothed) response is symmetric about the delay $\tau = 6T_c$ as shown in Figure 1.50 in comparison with a Nyquist (sinc) pulse shape that was delayed and scaled to fit the mainlobe of the impulse response. Note that the delay is necessary to achieve the symmetric pulse shape with a causal filter. It is clear from observing this comparison that the IS-95 pulse shape is not a truncated sinc pulse; the smoothed IS-95 pulse shape, although it is very close to being zero-valued at $\pm T_c$ from the peak, will cause a small amount of ISI because it is generally nonzero at the sample times $\pm nT_c$ from the peak.

In [24–27], methods for designing FIR filters are discussed in which the Nyquist pulse shape is multiplied by a finite-duration window function that is selected to achieve desirable passband and stopband spectral characteristics. We saw previously (c.f., Figure 1.22) that truncation of the Nyquist pulse shape resulted in ripples in the frequency transfer function; the magnitude of

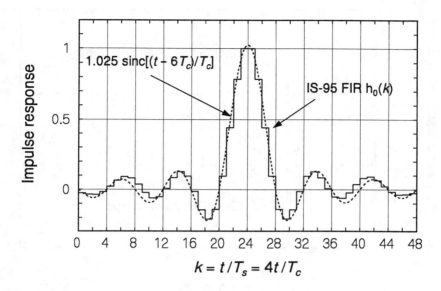

Figure 1.50 Comparison of IS-95 FIR pulse shape with that of ideal lowpass filter.

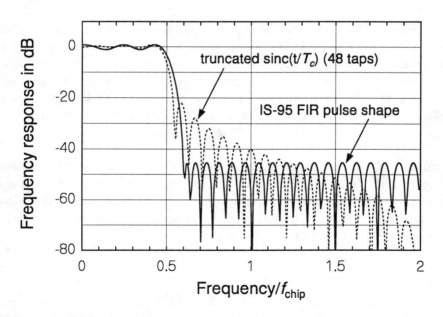

Figure 1.51 Comparison of transforms of IS-95 FIR pulse shape and truncated sinc pulse.

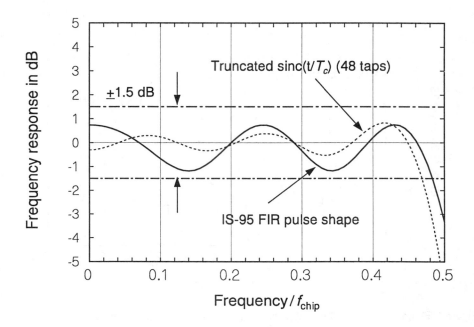

Figure 1.52 Comparison of frequency response ripples in the passband.

the ripples in the passband and the stopband depends on the particular window function [24] that is used for truncation. As demonstrated by the squared magnitude of its Fourier transform that is shown in Figures 1.51 and 1.52, the shape of the IS-95 pulse may be viewed as a windowed sinc pulse based on the spectrum design criterion of IS-95. Thus the stopband performance of the IS-95 pulse shape, being greater than the required 40 dB down, was the dominant factor in the selection of this shape.

Example 1.10 A simulated portion of an IS-95 PN code waveform[13] is shown in Figure 1.53. In the top part of the figure, a rectangular pulse shape is used, while in the bottom part, the actual pulse shape $h_0(n)$ is used. The binary PN chip sequence has a period of 32767 chips, with the reference period (ending in 15 zeros) given by

---

[13] The PN code used is the in-phase (I) quadrature channel "short" code, without the insertion of an extra zero at the end of the sequence that is discussed in Chapters 4 and 6.

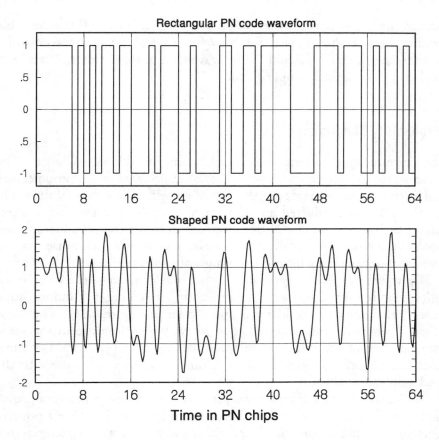

Figure 1.53 Comparison of rectangular and shaped PN waveforms.

$$1010100100111010001101111100110001 \cdots 0000000000$$

This chip data sequence, using the mapping logic $0 \rightarrow +1$ and logic $1 \rightarrow -1$, is clearly seen in the top part of Figure 1.53, artificially delayed by six chips to match the filter delay. Because of the delay, the plot begins with six of the zeros at the end of the previous sequence period. As suggested by Figure 1.50, the filter output is represented as a smoothed, continuous waveform by plotting the sample values at the midpoints of the sample intervals. That is, the figure displays the filter output as

$$y(t) = \sum_m d_m \, p(t - mT_c), \text{ with } p\big[(k + \tfrac{1}{2})T_s\big] = h_0(k) \qquad (1.37b)$$

Note from the time-domain presentation of Figure 1.53 that (1) the desirable bandwidth characteristics of the IS-95 waveform are obtained at the cost of a nonconstant signal envelope, and (2) the waveform is not necessarily at a local peak value at the nominal sampling time.

1.5 Probability Functions

Every communication system is disturbed by unavoidable interference. The most fundamental interference that must be considered in communication systems is the noise generated by the receiver itself, called *thermal noise*. In 1928, Johnson [28] conducted an experiment that showed that noise arises because of the *random motion* of free electrons in the electric conductors, and his work is the basis of the quantitative measure of the thermal noise that we take into account in the analysis of communications systems.

The waveform that is generated within a system, say, in a receiver or amplifier that is turned on with no input, will resemble a "random noise waveform" that might look like the one shown in Figure 1.54. The noise waveform "behaves" randomly and is in no way deterministic in the sense that we can predict the value of the waveform at any given instant of time. Yet it is affecting the desired signal by adding a random component to the received amplitude values.

The behavior of a random waveform can only be described in a probabilistic sense. The noise value that affects the signal value in either a positive or negative way can be described probabilistically once we characterize the noise *random variable (RV)* in statistical terms. So, what is a random variable? When we turn on a receiver (without any input connected) or a noise generator, we observe a noise waveform. If we turn off the system and then turn it on again, the waveform we observe is a random amplitude waveform such as shown in Figure 1.54, but not the same as the one observed earlier.

Figure 1.54 Sample function x(t) of a typical random process.

Now suppose that we have 100 or more identical noise generators and that we turned them on at the same time and observed their noise output waveforms. The situation might be described as in Figure 1.55, where $x(i; t)$ denotes the waveform function for noise generator i. The experiment depicted in Figure 1.55 is intended to define the concepts of random process and random variable. The collection of all those noise waveforms is an *ensemble*,

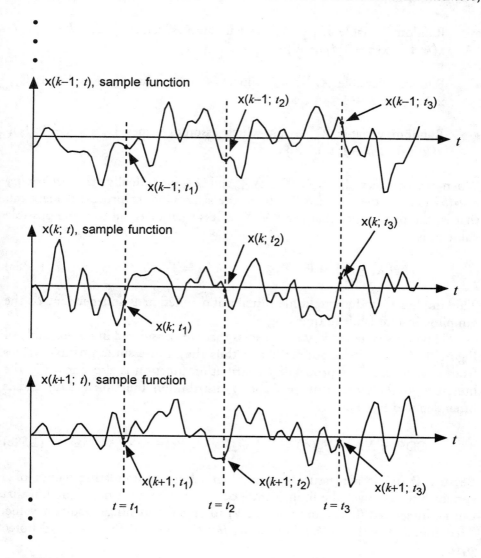

Figure 1.55 Concepts of random process and random variable.

and together we call them a *process*; that is, a random process is a collection or ensemble of member functions called sample functions. Therefore, a random process is an ensemble of sample functions. We now define random variables by sampling the random process as indicated in Figure 1.55. For example, that figure shows three sampling instants t_1, t_2, and t_3, and we generate three RVs as follows:

- Random variable $X_{t_1} \equiv X_1$, which assumes values $\{\ldots, \mathrm{x}(k-1; t_1), \mathrm{x}(k; t_1), \mathrm{x}(k+1; t_1), \ldots\}$;

- Random variable $X_{t_2} \equiv X_2$, which assumes values $\{\ldots, \mathrm{x}(k-1; t_2), \mathrm{x}(k; t_2), \mathrm{x}(k+1; t_2), \ldots\}$;

- Random variable $X_{t_3} \equiv X_3$, which assumes values $\{\ldots, \mathrm{x}(k-1; t_3), \mathrm{x}(k; t_3), \mathrm{x}(k+1; t_3), \ldots\}$.

When we consider the RVs X_1, X_2, and X_3 individually, the *probability distribution function* of each RV from the same random process is identical; that is, the probability that the RV X_i is less than or equal to some possible value α, is

$$\Pr\{X_i \leq \alpha\} = \Pr\{X \leq \alpha\} \quad \text{for} \quad i = 1, 2, 3 \tag{1.38a}$$

This implies that the amplitude distribution function is independent of the sampling instant (shift of time origin).

The RVs generated by the process that we considered in the example of Figure 1.55 are continuous RVs, and thus they can assume possible values from $-\infty$ to ∞. The probability distribution function is also known in the literature as the *cumulative* probability distribution function (CDF) and is often denoted $F_X(\alpha)$:

$$\text{CDF of } X: \quad F_x(\alpha) \triangleq \Pr\{X \leq \alpha\} \equiv \Pr\{-\infty < X \leq \alpha\} \tag{1.38b}$$

Because $F_x(\alpha)$ is a measure that is obtained by accumulating values of a certain function over the limit from $-\infty$ to α, there must be a function that can be integrated (for a continuous RV) to give the identical resultant value. That function is the *probability density function (pdf)* of the RV X, denoted $p_x(\alpha)$:

$$F_x(\alpha) = \Pr\{-\infty < X \le \alpha\} = \int_{-\infty}^{\alpha} p_x(\lambda)\, d\lambda \qquad (1.39a)$$

and thus

$$p_x(\alpha) \triangleq \frac{dF_x(\alpha)}{d\alpha} = \frac{d}{d\alpha}\Pr\{-\infty < X \le \alpha\} \qquad (1.39b)$$

Note that any pdf must be such that $p_x(\alpha) \ge 0$ (because a CDF must only increase as α increases) and $\int_{-\infty}^{\infty} p_x(\alpha)\, d\alpha = 1$ (because this is the probability that X takes any value at all). As will be demonstrated shortly, (1.39b) is a most convenient formula to obtain the pdf of a new RV that is a transformation of an old RV, such as the input-output relationship of a communications system.

Before leaving the subject of one-dimensional probability distribution functions, we need to observe the following basic properties of CDFs:

$$0 \le F_x(\alpha) \le 1$$

$$F_x(-\infty) \triangleq \Pr\{-\infty < X < -\infty\} \equiv 0$$

$$F_x(\infty) = \Pr\{-\infty < X < \infty\} = 1 \qquad (1.40)$$

$$F_x(\alpha_1) \le F_x(\alpha_2) \quad \text{for} \quad \alpha_1 \le \alpha_2$$

Now, returning to the joint sampling of the random process as depicted in Figure 1.55, we can define a *joint probability density function*, denoted $p_{x_1, x_2, \ldots, x_M}(\alpha_1, \alpha_2, \ldots, \alpha_M)$, where α_i is a possible value that the random variable X_i can assume, $i = 1, 2, \ldots, M$. If the joint pdf $p_{x_1, x_2, \ldots, x_M}(\bullet)$ is equal to the product of the individual pdfs of the $\{X_i\}$; that is, if for all M

$$p_{x_1, x_2, \ldots, x_M}(\alpha_1, \alpha_2, \ldots, \alpha_M) = p_{x_1}(\alpha_1)\, p_{x_2}(\alpha_2) \cdots p_{x_M}(\alpha_M) \qquad (1.41)$$

then the RVs $X_1, X_2, \ldots, X_M$ are *statistically independent*.

We now review pdfs that are well known and pertinent to the analysis of communication systems. Readers are referred to the excellent references given in [29–32] for further insights and in-depth discussions on probability theory and its applications to communications systems.

1.5.1 Probabilities

Consider a situation where we conduct an experiment that can produce three possible outcomes: A, B, C. If we ask for the probability that the outcome A occurs, we cannot answer the question unless we know that probability from previously conducted experimental "relative frequencies of occurrence" data. The way we determine the probability of the outcomes is as follows: we repeat the experiment N times and record the number of times each event occurs. Let us denote these by $N(A)$, $N(B)$, and $N(C)$, respectively. The probability of occurrence of the events, written $P(A)$, $P(B)$, and $P(C)$, can be assessed in the limit as N approaches infinity. That is,

$$P(A) = \lim_{N \to \infty} \frac{N(A)}{N}$$

$$P(B) = \lim_{N \to \infty} \frac{N(B)}{N} \qquad (1.42)$$

$$P(C) = \lim_{N \to \infty} \frac{N(C)}{N}$$

In the example we discuss here, all possible outcomes A, B, or C constitute the *sample space* of the experiment, and each of the possible events is a *sample point*. These are analogous to the ensemble or process and to the sample functions that we discussed earlier.

Consider another experiment in which events A and B can both happen, and the possibilities are A alone, B alone, A and B together, or neither A nor B. Then we can speak of the *conditional event* that, say, B occurs, given that A occurs, denoted $B \mid A$, and the *conditional probability* of event B given event A, written $P(B \mid A)$. This probability is defined as

$$P(B \mid A) = P(A, B)/P(A), \quad P(A) \neq 0 \qquad (1.43a)$$

where $P(A, B)$ is the probability that the events A and B occur jointly, and likewise, we have

$$P(A \mid B) = P(A, B)/P(B), \quad P(B) \neq 0 \qquad (1.43b)$$

If $P(A \mid B) = P(A)$, when the occurrence of the event A is independent of the occurrence of the event B, it follows from (1.43b) that

$$P(A, B) = P(A) \cdot P(B), \quad A \text{ and } B \text{ independent} \tag{1.43c}$$

If the occurrence of the event A does not depend in any way on the occurrence of the event B, the two events are statistically independent, and (1.43c) can be used as the definition of the *statistical independence* of A and B.

The event "A or B" is sometimes written $A \cup B$ ("A union B") in probability theory, and the event "A and B" is written $A \cap B$ ("A intersection B"). In general, the probability of $A \cup B$ is

$$P(A \cup B) = P(A) + P(B) - P(A \cap B) \leq P(A) + P(B) \tag{1.43d}$$

which indicates that the probability of the union of events is bounded above by the sum of the event probabilities. If events A and B are mutually exclusive (cannot occur at the same time), then the event $A \cap B$ does not occur and $P(A \cup B) = P(A) + P(B)$. It follows that if A, B, and C exhaust the possibilities of the experiment (the sample space contains only A, B, and C), $P(A \cup B \cup C) = 1$ and if these three possibilities are mutually exclusive, then (1.39d) leads to the expression

$$P(A \cup B \cup C) = 1 = P(A) + P(B) + P(C) \tag{1.43e}$$

Consider a continuous RV X that assumes values over the range $(-\infty, \infty)$. We define an event A, which is the event that X is less than or equal to a value α. If we assume that X has a pdf $p_x(\alpha)$, we can then write the probability that the event A occurs as follows:

$$P(A) = \Pr\{-\infty < X \leq \alpha\} = F_x(\alpha) = \int_{-\infty}^{\alpha} p_x(\beta)\, d\beta \tag{1.44a}$$

In (1.44a) we used the notation $A \triangleq \{-\infty < X \leq \alpha\}$ to describe the event under consideration. This may appear somewhat inconsistent in that we used two different expressions for the same quantity:

$$P(A) = \Pr\{A\} \tag{1.44b}$$

The right-hand side brackets { } are used to denote the *description* of an event, whereas the left-hand side parentheses ( ) are used to denote an event that is already defined. In any case, these expressions should be understood in their context without attaching a rigorous meaning to the way the notation is used.

1.5.2 Probability Distribution Functions

From the defining equations for the CDF and pdf given in (1.38) and (1.39), it is straightforward to state that the probability that the RV X falls within the interval (α, β) is given by

$$\Pr\{\alpha < X \leq \beta\} = F_x(\beta) - F_x(\alpha) = \int_\alpha^\beta p_x(\lambda)\, d\lambda \qquad (1.45a)$$

Suppose that we have two random processes $\{x(t)\}$ and $\{y(t)\}$, where $x(t)$ and $y(t)$ denote sample functions of the respective random processes. Assume now that we have obtained M RVs from each of the two random processes, $(X_1, X_2, \ldots, X_M)$ and $(Y_1, Y_2, \ldots, Y_M)$. An example that fits this situation is the input and output processes of an arbitrary system. The two random processes are said to be statistically independent if

$$p_{x_1,x_2,\ldots,x_M,y_1,y_2,\ldots,y_M}(\alpha_1, \alpha_2, \ldots, \alpha_M, \beta_1, \beta_2, \ldots, \beta_M)$$
$$= p_{x_1,x_2,\ldots,x_M}(\alpha_1, \alpha_2, \ldots, \alpha_M) \times p_{y_1,y_2,\ldots,y_M}(\beta_1, \beta_2, \ldots, \beta_M) \qquad (1.45b)$$

The RVs $(X_1, X_2, \ldots, X_M)$ are said to be statistically independent if and only if

$$p_{x_1,x_2,\ldots,x_M}(\alpha_1, \alpha_2, \ldots, \alpha_M) = p_{x_1}(\alpha_1)\, p_{x_2}(\alpha_2) \cdots p_{x_M}(\alpha_M) \qquad (1.45c)$$

The same can be said of the RVs generated from the random process $\{y(t)\}$:

$$p_{y_1,y_2,\ldots,y_M}(\beta_1, \beta_2, \ldots, \beta_M) = p_{y_1}(\beta_1)\, p_{y_2}(\beta_2) \cdots p_{y_M}(\beta_M) \qquad (1.45d)$$

is the condition for statistical independence of the RVs $(Y_1, Y_2, \ldots, Y_M)$.

Let us now consider two RVs X_1 and X_2 generated from a single random process $\{x(t)\}$. Their *joint probability density function*, denoted $p_{x_1,x_2}(\alpha_1, \alpha_2)$, is defined as

$$p_{x_1,x_2}(\alpha_1, \alpha_2) = \frac{\partial^2}{\partial\alpha_1 \partial\alpha_2} \Pr\{X_1 \leq \alpha_1, X_2 \leq \alpha_2\}$$

$$= \frac{\partial^2}{\partial\alpha_1 \partial\alpha_2} F_{x_1,x_2}(\alpha_1, \alpha_2) \qquad (1.46a)$$

where $F_{x_1,x_2}(\alpha_1, \alpha_2)$ is the joint cumulative probability distribution function of RVs X_1 and X_2. From (1.45c) we then have the following results:

$$F_{x_1,x_2}(\alpha_1, \alpha_2) = \int_{-\infty}^{\alpha_2} d\beta \int_{-\infty}^{\alpha_1} d\alpha\, p_{x_1,x_2}(\alpha, \beta) = \Pr\{X_1 \leq \alpha_1, X_2 \leq \alpha_2\}$$

(1.46b)

$$F_{x_1}(\alpha_1) = \int_{-\infty}^{\infty} d\beta \int_{-\infty}^{\alpha_1} d\alpha\, p_{x_1,x_2}(\alpha, \beta) = \Pr\{X_1 \leq \alpha_1\} \qquad (1.46c)$$

$$F_{x_2}(\alpha_2) = \int_{-\infty}^{\infty} d\alpha \int_{-\infty}^{\alpha_2} d\beta\, p_{x_1,x_2}(\alpha, \beta) = \Pr\{X_2 \leq \alpha_2\} \qquad (1.46d)$$

from which we can easily show that

$$p_{x_1}(\alpha) = \int_{-\infty}^{\infty} p_{x_1,x_2}(\alpha, \beta)\, d\beta \qquad (1.46e)$$

$$p_{x_2}(\beta) = \int_{-\infty}^{\infty} p_{x_1,x_2}(\alpha, \beta)\, d\alpha \qquad (1.46f)$$

The pdf obtained by integrating the joint pdf over one of the RVs, as in (1.46e) and (1.46f), is called a *marginal probability density function*. Analogous to (1.43a) and (1.43b), we can define the conditional pdf of X_1, given that $X_2 = \lambda$, as

$$p_{x_1}(\alpha \mid X_2 = \lambda) = \frac{p_{x_1,x_2}(\alpha, \lambda)}{p_{x_2}(\lambda)} \qquad (1.47a)$$

and thus

$$p_{x_1,x_2}(\alpha, \lambda) = p_{x_1}(\alpha \mid X_2 = \lambda)\, p_{x_2}(\lambda) \qquad (1.47b)$$

The uses of pdfs are diverse, as will be shown in many places in this chapter. We have already shown in (1.44a) that the probability of an event can be calculated by integrating the pertinent pdf. The conditional pdf is also used in calculating the conditional probability of an event, given another event in a given sample space. Suppose that we define a sample space in which two events A and B are the only outcomes. The conditional probability of A, given B, is defined as

$$P(A \mid B) = P(A, B)/P(B), \quad P(B) \neq 0 \qquad (1.48a)$$

Let us assume that the events A and B are associated with two random variables, X_1 and X_2, respectively, on the sample space, and are defined as

$$\text{Event } A \triangleq \{\alpha_1 < X_1 \leq \alpha_2\}$$
$$\text{Event } B \triangleq \{\beta_1 < X_2 \leq \beta_2\} \tag{1.48b}$$

Then (1.48a) can be written, in view of (1.47a), as

$$P(A \mid B) = \frac{\displaystyle\int_{\beta_1}^{\beta_2} \int_{\alpha_1}^{\alpha_2} \mathrm{p}_{x_1, x_2}(\alpha, \beta) \, d\alpha \, d\beta}{\displaystyle\int_{\beta_1}^{\beta_2} \mathrm{p}_{x_2}(\beta) \, d\beta} \tag{1.48c}$$

where we assumed that

$$\int_{\beta_1}^{\beta_2} \mathrm{p}_{x_2}(\beta) \, d\beta \neq 0$$

If, however, we want to compute the conditional probability of the event A, given that $X_2 = \lambda$, we need to calculate

$$P(A \mid X_2 = \lambda) = \int_{\alpha_1}^{\alpha_2} \mathrm{p}_{x_1}(\alpha \mid X_2 = \lambda) \, d\alpha \tag{1.48d}$$

The conditional probability of the event A, given that the event B occurred, is written as

$$P(A \mid B) = \int_{\alpha_1}^{\alpha_2} \mathrm{p}_{x_1}(\alpha \mid B) \, d\alpha \tag{1.48e}$$

where $\mathrm{p}_{x_1}(\alpha \mid B)$ is the conditional pdf of X_1, given the event B. We can also write the probability of the joint event $AB \equiv A \cap B$ in terms of the joint pdf of the RV X_1 and the event B:

$$P(AB) = \int_{\alpha_1}^{\alpha_2} \mathrm{p}_{x_1, B}(\alpha, B) \, d\alpha \tag{1.48f}$$

Note that

$$P(AB) = P(B) \, P(A \mid B)$$
$$= \int_{\alpha_1}^{\alpha_2} P(B) \, \mathrm{p}_{x_1}(\alpha \mid B) \, d\alpha \tag{1.48g}$$

We therefore have the result

$$p_{x_1,B}(\alpha, B) = P(B)\, p_{x_1}(\alpha \mid B) \tag{1.48h}$$

From (1.48d) we also have

$$P(A \mid X_2 = \lambda) = \int_{\alpha_1}^{\alpha_2} p_{x_1}(\alpha \mid X_2 = \lambda)\, d\alpha$$

$$= \int_{\alpha_1}^{\alpha_2} \frac{p_{x_1,x_2}(\alpha, \lambda)}{p_{x_2}(\lambda)}\, d\alpha = \frac{p_{A,x_2}(A, \lambda)}{p_{x_2}(\lambda)} \tag{1.48i}$$

so that

$$P(A \mid X_2 = \lambda)\, p_{x_2}(\lambda) = p_{A,x_2}(A, \lambda) \tag{1.48j}$$

where $p_{A,x_2}(A, \lambda)$ is called the joint pdf of the RV X_2 and the event A.

Transformation of random variables. Consider a random variable X with pdf $p_X(\alpha)$, which is the input to a system G that produces the corresponding output random variable Y as a transformation of the input RV X. The situation is depicted in Figure 1.56.

Before we go further, we need to make some important remarks concerning the notation and its meaning. When we say that the RV X is the input to the system, we really mean that the input is a sample function x(t) of the random process {x(t)} and that X is obtained by sampling the process (the ensemble of sample functions) at a particular time t_i. That is, $X \triangleq$ x(t_i), or X_i in the notation of the beginning of the section. If we understand this fact, it is all right to say that the input to the system G is a sample function x(t) and the output is a sample function y(t) of the output process {y(t)}. Further, we can say that the input and output of the system G are the random processes {x(t)} and {y(t)}, respectively, although it is not possible to apply an ensemble to the input of a system or to obtain an ensemble at its output. The sample functions x(t) and y(t) do, however, represent the underlying ensembles of sample functions (i.e., processes), and the RVs X and Y take their distributional properties from those ensembles, so that, unless otherwise

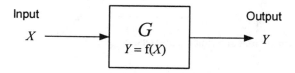

Figure 1.56 Transformation of a random variable.

indicated, the notation $x(t)$ refers to both a sample function and its underlying process. With this reasoning understood, we can also represent the situation depicted in Figure 1.56 as follows: Consider a system G as shown in Figure 1.57, whose input is a random process $x(t)$ with pdf $p_X(\alpha)$ and whose output process is $y(t)$ with the transformation $y(t) = f[x(t)]$. We normally suppress the notation of time when a single time-sampling instant is involved.

Example 1.11 Consider the case of the transformation

$$y(t) = x(t) + m \quad \text{or} \quad Y = X + m$$

where $x(t) \triangleq X$ is an input RV with a known pdf and m is a constant. Find the pdf of the output RV $y(t) \triangleq Y$.

    Solution. A straightforward way of obtaining the pdf of the output RV is to obtain the cumulative probability distribution function of Y, namely $F_y(\alpha)$, and then obtain $p_Y(\alpha)$ by differentiating $F_y(\alpha)$ as shown in (1.39b). Thus

$$F_y(\alpha) = \Pr\{Y \le \alpha\} = \Pr\{X + m \le \alpha\}$$
$$= \Pr\{X \le \alpha - m\} = F_x(\alpha - m)$$

Differentiating with respect to α, we obtain

$$p_y(\alpha) = \frac{d}{d\alpha} F_y(\alpha) = \frac{d}{d\alpha} F_x(\alpha - m) = p_x(\alpha - m) \tag{1.49a}$$

If we specify $p_x(\alpha)$, the pdf of Y is easily written down as $p_x(\alpha)$ translated by m units to the right. For example, if X is a Gaussian RV with zero mean and

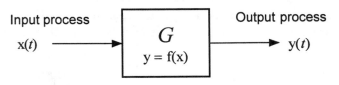

Figure 1.57 Transformation of a random process.

variance σ^2, which we denote[14] as $X = G(0, \sigma^2)$ or $X = N(0, \sigma^2)$, the pdf is given by

$$p_x(\alpha) = \frac{1}{\sqrt{2\pi}\sigma} e^{-\alpha^2/2\sigma^2} \tag{1.49b}$$

and the transformation $Y = X + m$ gives the pdf of Y to be, from (1.49a):

$$p_y(\alpha) = p_x(\alpha - m) = \frac{1}{\sqrt{2\pi}\sigma} e^{-(\alpha-m)^2/2\sigma^2} \tag{1.49c}$$

The Gaussian pdf is the one used for describing thermal noise, which Johnson discussed [28], and as such, the pdf given in (1.49b) characterizes the front-end noise of communication receivers. The CDF of $X = G(0, \sigma^2)$ is given by

$$F_x(\alpha) = \Pr\{X \leq \alpha\} = \int_{-\infty}^{\alpha} \frac{1}{\sqrt{2\pi}\sigma} e^{-\lambda^2/2\sigma^2} d\lambda$$

$$= 1 - Q\left(\frac{\alpha}{\sigma}\right) \tag{1.49d}$$

in which we define[15] $Q(\alpha)$ as the *complementary cumulative probability distribution function (CCDF)* of a "standard" zero-mean, unit-variance Gaussian RV, $G(0, 1)$:

$$Q(\alpha) \triangleq \int_{\alpha}^{\infty} \frac{1}{\sqrt{2\pi}} e^{-\lambda^2/2} d\lambda \tag{1.49e}$$

The pdf, CDF and CCDF of the standard Gaussian distribution are shown plotted in Figure 1.58, which indicates some values of $p_X(\alpha)$ and $Q(\alpha)$ for special values of the argument. Often it is desired to find the argument of $Q(\alpha)$ for which the function equals some special value, such as $Q(\alpha) = 0.1$; Table 1.4 lists values of the argument of $Q(\alpha)$ for several special values of the

---

[14] In a section to follow, we shall give definitions of moments such as the mean (average) and variance of a RV. The notation $X = G(m, \sigma^2)$ or $X = N(m, \sigma^2)$ is often used to indicate that the RV X is a Gaussian or normal RV with mean m and variance σ^2.

[15] The notation $P(\alpha) = 1 - Q(\alpha)$ is often employed [29], as are the *error function* $\mathrm{erf}(\alpha) = 1 - 2Q(\alpha\sqrt{2})$ and *complementary error function* $\mathrm{erfc}(\alpha) = 1 - \mathrm{erf}(\alpha) = 2Q(\alpha\sqrt{2})$. A useful relation is $Q(\alpha) = \frac{1}{2}\left[1 - \mathrm{erf}(\alpha/\sqrt{2})\right] = \frac{1}{2}\mathrm{erfc}(\alpha/\sqrt{2})$.

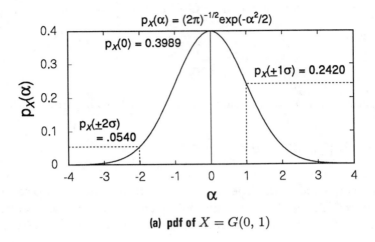

(a) pdf of $X = G(0, 1)$

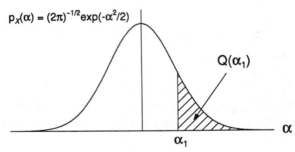

(b) Definition of the $Q(\alpha)$ function

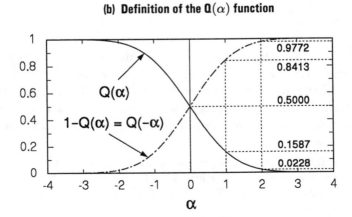

(c) Cumulative distribution function of $X = G(0, 1)$

Figure 1.58 Probability functions of the Gaussian distribution.

Table 1.4 Special values of the argument of $Q(\alpha)$

$Q(\alpha)$	α	$Q(\alpha)$	α
0.5	0.00000	10^{-3}	3.09023
0.4	0.25335	10^{-4}	3.71902
0.3	0.52440	10^{-5}	4.26489
0.2	0.84162	10^{-6}	4.75342
0.1	1.28155	10^{-7}	5.19934
0.01	2.32635	10^{-8}	5.61200

function. Accurate functional approximations for computing the function are given in [20].

The popularity for the use of $Q(\alpha)$ in communications systems analysis involving the Gaussian noise channel cannot be overemphasized. As seen above, $Q(\alpha)$ must be obtained from a table or by numerical integration. For that reason, an approximation for $Q(\alpha)$ is often used for analysis purposes, such as the upper bounds [29]

$$Q(\alpha) < \frac{1}{\sqrt{2\pi\alpha}} e^{-\alpha^2/2} < \tfrac{1}{2} e^{-\alpha^2/2} \qquad (1.49f)$$

These bounds are plotted in Figure 1.59, showing their relative tightness.

Example 1.12 Let us consider another example of determining the pdf of a function of a random variable. Consider the square-law transformation

$$y = x^2 \qquad (1.50a)$$

which is plotted in Figure 1.60. This transformation is also known as the full-wave quadratic rectifier transformation or the full-wave square-law detector. We again use the formulation of the CDF to obtain the pdf of the RV Y as a transformation of the RV X. Clearly, from observation of Figure 1.60,

$$F_y(\alpha) = \Pr\{Y \le \alpha\} = \Pr\{X^2 \le \alpha\} = \Pr\{-\sqrt{\alpha} < X \le \sqrt{\alpha}\}$$
$$= F_x(\sqrt{\alpha}) - F_x(-\sqrt{\alpha}), \quad \alpha \ge 0$$

We therefore have

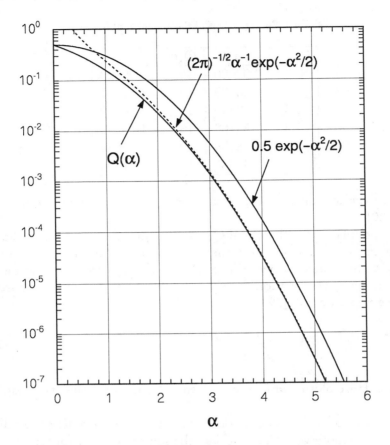

Figure 1.59 Comparison of $Q(\alpha)$ with two upper bounds.

$$p_y(\alpha) = \frac{d}{d\alpha} F_y(\alpha) = \frac{d}{d\alpha}\left[F_x(\sqrt{\alpha}) - F_x(-\sqrt{\alpha})\right]$$

$$= \frac{1}{2\sqrt{\alpha}}\left[p_x(\sqrt{\alpha}) + p_x(-\sqrt{\alpha})\right], \ \alpha \geq 0$$

That is,

$$p_y(\alpha) = \begin{cases} 0, & \alpha < 0 \\ \dfrac{1}{2\sqrt{\alpha}}\left[p_x(\sqrt{\alpha}) + p_x(-\sqrt{\alpha})\right], & \alpha \geq 0 \end{cases} \tag{1.50b}$$

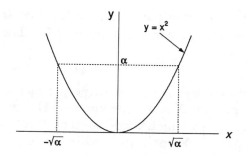

Figure 1.60 Square-law transformation.

Let us now consider a specific form for $p_X(\alpha)$, say, the Gaussian pdf of (1.49b), which after substitution in (1.50b) yields

$$p_y(\alpha) = \begin{cases} 0, & \alpha < 0 \\ \dfrac{1}{\sqrt{2\pi\alpha}}\, e^{-\alpha/2}, & \alpha \geq 0 \end{cases} \qquad (1.50c)$$

which is known as the *chi-squared* pdf with one degree of freedom (more is said about chi-squared RVs in a later section).

If we assume a different specific form of $p_x(\alpha)$, the pdf known as the Rayleigh pdf:

$$p_x(\alpha) = \begin{cases} 0, & \alpha < 0 \\ \dfrac{2\alpha}{b}\, e^{-\alpha^2/b}, & \alpha \geq 0 \end{cases} \qquad (1.50d)$$

where b is a positive constant, the pdf of Y is then obtained as

$$p_y(\alpha) = \frac{1}{2\sqrt{\alpha}}\left[p_x\left(\sqrt{\alpha}\right) + p_x\left(-\sqrt{\alpha}\right)\right], \quad \alpha \geq 0$$

$$= \frac{1}{2\sqrt{\alpha}}\left[\frac{2\sqrt{\alpha}}{b}\, e^{-\alpha/b} + 0\right], \quad \alpha \geq 0$$

$$= \begin{cases} 0, & \alpha < 0 \\ \dfrac{1}{b}\, e^{-\alpha/b}, & \alpha \geq 0 \end{cases} \qquad (1.50e)$$

which is an exponential pdf (Y is an exponentially distributed RV).

Example 1.13 Determine the pdf of a new RV that is the result of the transformation of two RVs as indicated in Figure 1.61. The transformation is

$$z = x + y$$

Solution: Our objective is to obtain the pdf of the RV Z, under the assumption that the pdfs for X and Y are known. This problem can easily be solved by applying the conditional pdf concept that we discussed above. First, assume that the RV Y took up a value β from its allowed domain. Then we have a transformation of the form

$$z = x + \beta \tag{1.51a}$$

where β is a constant. This situation is then identical to that considered in an example above where, from (1.45a) with a change of notation, we had

$$p_z(u) = p_x(u - \beta) \tag{1.51b}$$

Therefore, for the current situation, we can write the pdf of RV Z, given that $Y = \beta$,

$$p_z(u \mid Y = \beta) = p_x(u - \beta \mid Y = \beta) \tag{1.51c}$$

From (1.47a) and (1.47b), the joint pdf of Z and Y is obtained by multiplying both sides of (1.51c) by $p_y(\beta)$:

$$
\begin{aligned}
p_{z,y}(u, \beta) &= p_z(u \mid Y = \beta)\, p_y(\beta) \\
&= p_x(u - \beta \mid Y = \beta)\, p_y(\beta) \\
&= p_{x,y}(u - \beta, \beta)
\end{aligned}
\tag{1.51d}
$$

and the marginal pdf of Z is obtained by integrating out the RV Y:

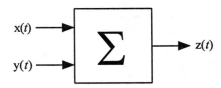

Figure 1.61 Transformation that gives the sum of two RVs.

$$p_z(u) = \int_{-\infty}^{\infty} p_{z,y}(u, \beta)\, d\beta = \int_{-\infty}^{\infty} p_{x,y}(u - \beta, \beta)\, d\beta \qquad (1.51e)$$

where $Z = X + Y$. If the RVs X and Y are statistically independent, then

$$p_{x,y}(\alpha, \beta) = p_x(\alpha)\, p_y(\beta)$$

and (1.51e) becomes

$$p_z(u) = \int_{-\infty}^{\infty} p_x(u - \beta)\, p_y(\beta)\, d\beta = p_x(u) * p_y(u) \qquad (1.51f)$$

which is the convolution of the pdfs of X and Y.

Averages. Consider a discrete RV X that assumes M discrete values x_1, x_2, ..., x_M with probabilities $P(x_i)$, $i = 1, 2, \ldots, M$. We define the statistical average (the expectation or ensemble average) of X by[16]

$$E\{X\} = \overline{X} = \sum_{i=1}^{M} x_i P(x_i) \qquad (1.52a)$$

which is the weighted sum of the values that the RV can assume, with the weight being the probability of occurrence of the value.

For a continuous RV X, which assumes any value over the interval $(-\infty, \infty)$, the statistical average (the expectation or ensemble average) of X is defined by

$$E\{X\} = \overline{X} = \int_{-\infty}^{\infty} \alpha\, p_x(\alpha)\, d\alpha \qquad (1.52b)$$

In some cases, we need to calculate the statistical average of a RV Y that is a transformation of a RV X:

$$Y = f(X) \qquad (1.52a)$$

---

[16] The notation for the statistical average (expectation) of a RV X can be either $E\{X\}$ or $\overline{X}$, depending on choice and convenience.

in which f($\bullet$) denotes the relationship connecting the "old" RV X with the "new" RV Y. The statistical average of the new RV can be obtained with the knowledge of the probability distribution of the old RV, as follows:

$$E\{Y\} = \overline{Y} = E\{f(X)\} = \sum_{i=1}^{M} f(x_i)\, P(x_i) \qquad (1.53b)$$

for a discrete RV X, and

$$E\{Y\} = \overline{Y} = E\{f(X)\} = \int_{-\infty}^{\infty} f(\alpha)\, p_x(\alpha)\, d\alpha \qquad (1.53c)$$

for a continuous RV X. These results are known as the *theorem of expectation* [31].

Moments of a random variable. The expected value of the nth power of the discrete or continuous RV X is called the nth *moment* of the RV X, defined by

$$E\{X^n\} = \overline{X^n} = \sum_{i=1}^{M} x_i^n\, P(x_i) \qquad (1.54a)$$

and

$$E\{X^n\} = \overline{X^n} = \int_{-\infty}^{\infty} \alpha^n\, p_x(\alpha)\, d\alpha \qquad (1.54b)$$

for discrete and continuous RVs, respectively. Therefore, the first moment is mean, expectation, or statistical average, $E\{X\} = m_x$. The nth *central moment* is defined as

$$E\{(X - \overline{X})^n\} = \overline{(X - m_x)^n} = \int_{-\infty}^{\infty} (\alpha - m_x)^n p_x(\alpha)\, d\alpha \qquad (1.54c)$$

Note that for $n = 0$, $E\{(X - m_x)^0\} = E\{1\} = 1$, and for $n = 1$, $E\{(X - m_x)^1\} = E\{X - m_x\} = 0$.

The second central moment is called the *variance*, denoted σ_x^2 and defined by

$$\sigma_x^2 \triangleq E\{(X - m_x)^2\} = E\{(X - \overline{X})^2\} = E\{X^2\} - 2\overline{X}\,E\{X\} + \overline{X}^2$$

$$= \overline{X^2} - \overline{X}^2 = E\{X^2\} - m_x^2 \qquad (1.54d)$$

The square root of the variance, σ_x, is called the *standard deviation*.

Joint moments of two random variables. Consider two continuous RVs X and Y with their joint pdf $p_{x,yY}(\alpha, \beta)$. The $(n + k)$th order *joint moment* of the RVs X and Y is defined by

$$E\{X^n Y^k\} = \overline{X^n Y^k} = \int_{-\infty}^{\infty} \int_{-\infty}^{\infty} \alpha^n \beta^k p_{x,y}(\alpha, \beta) \, d\alpha \, d\beta \qquad (1.55a)$$

and the corresponding *joint central moment* is given by

$$E\left\{ (X - \overline{X})^n (Y - \overline{Y})^k \right\} = \overline{(X - \overline{X})^n (Y - \overline{Y})^k}$$

$$= \int_{-\infty}^{\infty} \int_{-\infty}^{\infty} (\alpha - m_x)^n (\beta - m_y)^k p_{x,y}(\alpha, \beta) \, d\alpha \, d\beta \qquad (1.55b)$$

where $m_x \triangleq \overline{X}$ and $m_y \triangleq \overline{Y}$. When $n = 1$ and $k = 1$, the joint central moment has a special significance:

$$\sigma_{xy} \triangleq E\{(X - m_x)(Y - m_y)\} \qquad (1.56a)$$

which is called the *covariance* of the RVs X and Y. The *normalized covariance* or *correlation coefficient* of X and Y is defined by

$$\rho_{xy} \triangleq \frac{E\{(X - m_x)(Y - m_y)\}}{\sqrt{E\{(X - m_x)^2\} E\{(Y - m_y)^2\}}} = \frac{\sigma_{xy}}{\sigma_x \sigma_y} \qquad (1.56b)$$

in which σ_x and σ_y are the standard deviations of X and Y, respectively.

Example 1.14 The mean and variance of a sum of independent RVs will be found. Let Y be the weighted sum of N RVs:

$$Y = \sum_{i=1}^{N} a_i X_i = a_1 X_1 + a_2 X_2 + \cdots + a_N X_N \qquad (1.57a)$$

The mean of Y is

$$E\{Y\} = m_y = E\left\{ \sum_{i=1}^{N} a_i X_i \right\} = \sum_{i=1}^{N} a_i E\{X_i\} = \sum_{i=1}^{N} a_i m_{x_i} \qquad (1.57b)$$

in which we have used the fact that $E\{a_i X_i\} = a_i E\{X_i\}$. Thus, the mean of a weighted sum of independent RVs is the weighted sum of their means.[17] The mean square of Y is

$$
\begin{aligned}
E\{Y^2\} &= E\left\{\sum_{i=1}^{N}\sum_{j=1}^{N} a_i a_j X_i X_j\right\} = \sum_{i=1}^{N}\sum_{j=1}^{N} a_i a_j E\{X_i X_j\} \\
&= \sum_{i=1}^{N} a_i^2 E\{X_i^2\} + \sum_{\substack{i=1 \\ i \neq j}}^{N}\sum_{j=1}^{N} a_i a_j m_{x_i} m_{x_j} \\
&= \sum_{i=1}^{N} a_i^2 E\{X_i^2\} + \left(\sum_{i=1}^{N} a_i m_{x_i}\right)^2 - \sum_{i=1}^{N} a_i^2 m_{x_i}^2 \\
&= \sum_{i=1}^{N} a_i^2 \left[E\{X_i^2\} - m_{x_i}^2\right] + m_y^2 \\
&= \sum_{i=1}^{N} a_i^2 \sigma_{x_i}^2 + m_y^2
\end{aligned}
\tag{1.57c}
$$

in which we used the fact that $E\{X_i X_j\} = m_{x_i} m_{x_j}$ for $i \neq j$. It follows directly that the variance of a weighted sum of independent RVs is the weighted sum of their variances:

$$
\sigma_y^2 = E\{Y^2\} - m_y^2 = \sum_{i=1}^{N} a_i^2 \sigma_{x_i}^2
\tag{1.57d}
$$

1.5.3 Characteristic Function

In the analysis of communications systems, techniques for determining the pdf of the relevant RVs are important. For example, the relevant RVs are often decision variables for deciding which information symbol has been sent over the channel. We have considered some methods above for this purpose, such as the CDF formulation method and the application of conditional pdfs. There is another method, known as the *characteristic function* method, which

---

[17] Note that the independence of the $\{X_i\}$ was not used to derive (1.53a); it is a general relation for sums of RVs, independent or dependent.

is very useful in some cases for determining the pdf of a sum of independent RVs. The characteristic function (CHF) is also useful in the art of determining the moments of a RV as well, as we see later.

The characteristic function $C_x(j\nu)$ of the probability distribution of a real RV X is defined as the statistical average of $\exp\{j\nu X\}$:

$$C_x(j\nu) = \mathrm{E}\{e^{j\nu X}\} = \int_{-\infty}^{\infty} e^{j\nu\alpha} p_x(\alpha)\, d\alpha \tag{1.58a}$$

where ν is real. Note that the CHF has the form of the Fourier transform of the pdf $p_x(\alpha)$. Because the pdf is nonnegative and $e^{j\nu X}$ has unit magnitude:

$$\left| \int_{-\infty}^{\infty} e^{j\nu\alpha} p_x(\alpha)\, d\alpha \right| \leq \int_{-\infty}^{\infty} |e^{j\nu\alpha}|\, |p_x(\alpha)|\, da = \int_{-\infty}^{\infty} p_x(\alpha)\, d\alpha = 1 \tag{1.58b}$$

hence the CHF as defined by the integral in (1.58a) always exists. Thus, the CHF $C_x(j\nu)$ and the pdf $p_x(\alpha)$ form a Fourier transform pair, and the pdf can be obtained from the CHF:

$$p_x(\alpha) = \frac{1}{2\pi} \int_{-\infty}^{\infty} e^{-j\nu\alpha} C_x(j\nu)\, d\nu \tag{1.58c}$$

If X is a discrete RV, (1.58a) becomes

$$C_x(j\nu) = \sum_i e^{j\nu x_i} P(x_i) \tag{1.58d}$$

Example 1.15   The CHF for a Gaussian RV $Y = G(m_y, \sigma_y^2)$ is found as follows:

$$C_y(j\nu) = \mathrm{E}\{e^{j\nu Y}\} = \overline{e^{j\nu Y}} = \int_{-\infty}^{\infty} e^{j\nu\alpha} p_y(\alpha)\, d\alpha$$

$$= \int_{-\infty}^{\infty} e^{j\nu\alpha} \frac{1}{\sqrt{2\pi}\sigma_y} \exp\left\{ -\frac{(\alpha - m_y)^2}{2\sigma_y^2} \right\} d\alpha$$

$$= \exp\{j\nu m_y - \tfrac{1}{2}\nu^2\sigma_y^2\} \tag{1.58e}$$

For the standardized Gaussian RV $Z = G(0, 1)$ the CHF has the simple form

$$C_z(j\nu) = e^{-\nu^2/2} \tag{1.58f}$$

Example 1.16 The CHF for a sum of N independent Gaussian random variables will be found. Consider $Y = \sum_{i=1}^{N} a_i X_i$, where $X_i = G(m_i, \sigma_i^2)$. From (1.57b) and (1.57d) we have

$$m_y = \sum_{i=1}^{N} a_i m_i \quad \text{and} \quad \sigma_y^2 = \sum_{i=1}^{N} a_i^2 \sigma_i^2 \tag{1.59a}$$

The CHF of Y is given by

$$C_y(j\nu) = E\{e^{j\nu Y}\} = E\left\{\exp\left(j\nu \sum_{i=1}^{N} a_i X_i\right)\right\} = E\left\{\prod_{i=1}^{N} e^{j\nu a_i X_i}\right\} \tag{1.59b}$$

Since the RVs $\{X_i\}$ are independent, their joint pdf factors into the product of their individual pdfs:

$$p_{x_1, x_2, \ldots, x_N}(\alpha_1, \alpha_2, \ldots, \alpha_n) = p_{x_1}(\alpha_1) p_{x_2}(\alpha_2) \cdots p_{x_N}(\alpha_N) \tag{1.59c}$$

and calculation of the CHF of Y is found to be

$$C_y(j\nu) = \prod_{i=1}^{N} \int_{-\infty}^{\infty} e^{j\nu a_i X_i} p_{x_i}(\alpha_i) = \prod_{i=1}^{N} C_{x_i}(j\nu a_i) \tag{1.59d}$$

the product of the individual CHFs of the weighted variables $\{a_i X_i\}$. Substituting (1.58e) appropriately for the individual CHFs, we obtain

$$C_y(j\nu) = \prod_{i=1}^{N} \exp\{j\nu a_i m_i - \tfrac{1}{2}\nu^2 a_i^2 \sigma_i^2\} = \exp\left\{\sum_{i=1}^{N} \left(j\nu a_i m_i - \tfrac{1}{2}\nu^2 a_i^2 \sigma_i^2\right)\right\}$$

$$= \exp\left\{j\nu \left(\sum_{i=1}^{N} a_i m_i\right) - \tfrac{1}{2}\nu^2 \left(\sum_{i=1}^{N} a_i^2 \sigma_i^2\right)\right\}$$

$$= \exp\{j\nu m_y - \tfrac{1}{2}\nu^2 \sigma_y^2\} \tag{1.59e}$$

From this result, we learn the important fact that the weighted sum of independent Gaussian RVs is itself a Gaussian RV. It is also true (not shown here) that a sum of correlated Gaussian RVs is a Gaussian RV.

1.5.4 Moment Generating Function

Another ensemble average related to the distribution of a RV X that is very closely related to the characteristic function $C_x(j\nu)$ is the *moment generating function*, $M_x(\nu)$. The moment generating function (MGF) for X is the statistical average of $\exp(\nu X)$:

$$M_x(\nu) = \mathrm{E}\{e^{\nu X}\} = \overline{e^{\nu X}} = \int_{-\infty}^{\infty} e^{\nu\alpha} \mathrm{p}_x(\alpha)\, d\alpha \qquad (1.60a)$$

If we differentiate $M_x(\nu)$ with respect to its argument ν, we obtain

$$\frac{dM_x(\nu)}{d\nu} = \frac{d}{d\nu} \int_{-\infty}^{\infty} e^{\nu\alpha} \mathrm{p}_x(\alpha)\, d\alpha = \int_{-\infty}^{\infty} \alpha e^{\nu\alpha} \mathrm{p}_x(\alpha)\, d\alpha$$

If we evaluate both sides of this equation at $\nu = 0$, the result is the first moment of the RV X:

$$m_x = \int_{-\infty}^{\infty} \alpha\, \mathrm{p}_x(\alpha)\, d\alpha = \left. \frac{dM_x(\nu)}{d\nu} \right|_{\nu=0}$$

That is,

$$M_x'(0) \triangleq \left. \frac{dM_x(\nu)}{d\nu} \right|_{\nu=0} = m_x = \mathrm{E}\{X\} \qquad (1.60b)$$

If an MGF $M_x(\nu)$ exists, then we may take its successive derivatives by successively differentiating inside the integral in (1.56a). Then we obtain

$$M_x'(\nu) \triangleq \frac{dM_x(\nu)}{d\nu} = \int_{-\infty}^{\infty} \alpha e^{\nu\alpha} \mathrm{p}_x(\alpha)\, d\alpha = \mathrm{E}\{X e^{\nu X}\}$$

$$M_x''(\nu) \triangleq \frac{d^2 M_x(\nu)}{d\nu^2} = \int_{-\infty}^{\infty} \alpha^2 e^{\nu\alpha} \mathrm{p}_x(\alpha)\, d\alpha = \mathrm{E}\{X^2 e^{\nu X}\} \qquad (1.60c)$$

$$\vdots$$

$$M_x^{(n)}(\nu) \triangleq \frac{d^n M_x(\nu)}{d\nu^n} = \int_{-\infty}^{\infty} \alpha^n e^{\nu\alpha} \mathrm{p}_x(\alpha)\, d\alpha = \mathrm{E}\{X^n e^{\nu X}\}$$

By evaluating these expressions at $\nu = 0$, we obtain

$$E\{X\} = M_x'(0) = \frac{d}{d\nu} M_x(\nu)\bigg|_{\nu=0}$$

$$E\{X^2\} = M_x''(0) = \frac{d^2}{d\nu^2} M_x(\nu)\bigg|_{\nu=0} \qquad (1.60d)$$

$$\vdots$$

$$E\{X^n\} = M_x^{(n)}(0) = \frac{d^n}{d\nu^n} M_x(\nu)\bigg|_{\nu=0}$$

In the case of X being a discrete RV which assumes value x_i with probability $P(x_i)$, $i = 1, 2, \ldots, K$, the MGF is given by

$$M_x(\nu) = \sum_{i=1}^{K} e^{\nu x_i} P(x_i) \qquad (1.60e)$$

If the MGF $M_x(\nu)$ of a RV X exists, then it possesses a Maclaurin series expansion:

$$M_x(\nu) = 1 + E\{X\}\nu + E\{X^2\}\frac{\nu^2}{2!} + \cdots + E\{X^n\}\frac{\nu^n}{n!} + \cdots \qquad (1.61a)$$

which is easily proved if we recognize the series

$$e^{\nu X} = 1 + X\nu + X^2\frac{\nu^2}{2!} + \cdots + X^n\frac{\nu^n}{n!} + \cdots \qquad (1.61b)$$

and substituting the series in the defining equation (1.60a). An advantage of this fact is that we can easily obtain the nth moment $E\{X^n\}$ if we can readily obtain the power series expansion of $M_x(\nu)$, because $E\{X^n\}$ is the coefficient of the term $\nu^n/n!$ in the series expansion of $M_x(\nu)$.

Example 1.17   The Bernoulli random variable.   Any experiment that produces the two possible outcomes, "success" and "failure," is called a *Bernoulli trial*. If we denote the event of success by 1 and failure by 0, then the discrete Bernoulli RV X assumes two values: $X = 1$ with probability $P(X = 1) = p$ and $X = 0$ with probability $P(X = 0) = q = 1 - p$. The MGF, for $-\infty < \nu < \infty$, is

$$M_x(\nu) = \sum_{i=0}^{1} e^{\nu x_i} P(X = x_i) = e^{\nu \cdot 0} P(X = 0) + e^{\nu \cdot 1} P(X = 1)$$

$$= q + p e^{\nu} = 1 - p + p e^{\nu} \qquad (1.62a)$$

Now:

$$M_x'(\nu) = p e^{\nu} \quad \rightarrow \quad E\{X\} = M_x'(0) = p$$

$$M_x''(\nu) = p e^{\nu} \quad \rightarrow \quad E\{X^2\} = M_x''(0) = p$$

and the variance σ_x^2 of the Bernoulli RV is given by

$$\sigma_x^2 = E\left\{ (X - \overline{X})^2 \right\} = E\{X^2\} - E^2\{X\} = p - p^2$$

$$= p(1 - p) = pq \qquad (1.62b)$$

Example 1.18   The binomial random variable.   When we perform n *independent, repeated* Bernoulli trials with the probability p of success at each trial, the experiment is called a *binomial experiment*.   In the binomial experiment, the probability law gives the *number of successes* in the n independent, repeated Bernoulli trials.   Assume for example that $0\,1\,1\,0\,1\,0\,1$ denotes the outcome of seven such trials, where "1" denotes a success and "0" a failure.   This is a case of four successes out of seven trials, and the probability of this event is $p^4(1 - p)^{7-4} = p^4 q^3$.   But this is one of $\binom{7}{4}$ possible success patterns that can occur with a total of four successes.   Thus, the binomial probability law with parameters n and p, where $n = 1, 2, \ldots$ and $0 \le p \le 1$, is given by

$$P(x) = \begin{cases} \binom{n}{x} p^x (1 - p)^{n-x}, & x = 0, 1, 2, \ldots, n \\ 0, & \text{otherwise} \end{cases} \qquad (1.63a)$$

Now, the MGF is

$$M_x(\nu) = \sum_{x=0}^{n} e^{\nu x} P(x) = \sum_{k=0}^{n} e^{\nu k} \binom{n}{k} p^k (1 - p)^{n-k}$$

$$= \sum_{k=0}^{n} \binom{n}{k} (p e^{\nu})^k q^{n-k} = (p e^{\nu} + q)^n \qquad (1.63b)$$

The first two moments are derived as

$$M_x'(\nu) = np\,e^\nu(p\,e^\nu + q)^{n-1} \quad \rightarrow \quad \mathrm{E}\{X\} = M_x'(0) = np$$

$$M_x''(\nu) = np\,e^\nu(p\,e^\nu + q)^{n-1} + n(n-1)p^2 e^{2\nu}(p\,e^\nu + q)^{n-2}$$

$$\rightarrow \quad \mathrm{E}\{X^2\} = M_x''(0) = npq + n^2 p^2$$

The variance σ_x^2 is found from

$$\sigma_x^2 = \mathrm{E}\{X^2\} - \mathrm{E}^2\{X\} = npq + n^2 p^2 - n^2 p^2 = npq \qquad (1.63c)$$

Example 1.19 The Poisson random variable. The Poisson probability law has a wide application in system analysis involving telephone traffic theory. We make good use of the results related to the Poisson distribution to be dealt with in this example when we analyze the Erlang capacity of CDMA cellular telephone systems in a later chapter.

The Poisson probability law with parameter λ, where $\lambda > 0$, is that the (Poisson) RV assumes the integer value k, where $k > 0$, with probability

$$P(x) = \begin{cases} e^{-\lambda}\dfrac{\lambda^x}{x!}, & x = 0, 1, 2, \ldots \\ 0, & \text{otherwise.} \end{cases} \qquad (1.64a)$$

The MGF is

$$M_x(\nu) = \sum_{x=0}^{\infty} e^{\nu x} P(x) = \sum_{k=0}^{\infty} e^{\nu k} \cdot e^{-\lambda}\frac{\lambda^k}{k!} = e^{-\lambda}\sum_{k=0}^{\infty}\frac{(\lambda e^\nu)^k}{k!}$$

$$= e^{-\lambda}e^{\lambda e^\nu} = e^{\lambda(e^\nu - 1)} \qquad (1.64b)$$

The first two derivatives and moments are

$$M_x'(\nu) = e^{\lambda(e^\nu-1)}\lambda e^\nu \quad \rightarrow \quad \mathrm{E}\{X\} = M_x'(0) = \lambda$$

$$M_x''(\nu) = e^{\lambda(e^\nu-1)}\lambda^2 e^{2\nu} + \lambda e^\nu e^{\lambda(e^\nu-1)}$$

$$\rightarrow \quad \mathrm{E}\{X^2\} = M_x''(0) = \lambda^2 + \lambda$$

The variance σ_x^2 is therefore given by

$$\sigma_x^2 = \mathrm{E}\{X^2\} - \mathrm{E}^2\{X\} = \lambda^2 + \lambda - \lambda^2 = \lambda \qquad (1.64c)$$

The Poisson RV gives the *mean* and the *variance* the identical value of λ, which will be nicely applied in the analysis of Erlang capacity of CDMA systems in a later chapter.

Example 1.20 The exponential random variable. The exponential probability law is also an important one and is used in telephone switching theory as the distribution law for call duration. The well known Erlang B formula is also based on this distribution, as discussed in Chapter 3.

The exponential probability law with parameter $\mu > 0$ is specified by the pdf

$$p_x(\alpha) = \begin{cases} \mu e^{-\mu\alpha}, & \alpha \geq 0 \\ 0, & \text{otherwise} \end{cases} \tag{1.65a}$$

The MGF is

$$M_x(\nu) = \int_0^\infty e^{\nu\alpha} \cdot \mu e^{-\mu\alpha} d\alpha = \mu \int_0^\infty e^{(\nu-\mu)\alpha} d\alpha$$

$$= \frac{\mu}{\mu - \nu} = \frac{1}{1 - \nu/\mu} \tag{1.65b}$$

The first two derivatives and moments are

$$M_x'(\nu) = \frac{\mu}{(\mu - \nu)^2} \quad \rightarrow \quad E\{X\} = M_x'(0) = \frac{1}{\mu}$$

$$M_x''(\nu) = \frac{2\mu}{(\mu - \nu)^3} \quad \rightarrow \quad E\{X^2\} = M_x''(0) = \frac{2}{\mu^2}$$

The variance σ_x^2 is found to be

$$\sigma_x^2 = E\{X^2\} - E^2\{X\} = \frac{2}{\mu^2} - \frac{1}{\mu^2} = \frac{1}{\mu^2} \tag{1.65c}$$

so that the standard deviation σ_x is given by $1/\mu$, which is equal to the mean, $m_x = 1/\mu$.

Example 1.21 The uniform random variable. Another important continuous RV is the uniformly distributed random variable. The pdf of the uniform RV X is given by

$$p_x(\alpha) = \begin{cases} \dfrac{1}{2b}\text{Rect}\left(\dfrac{\alpha}{2b}\right), & |\alpha| \leq b \\ 0, & \text{otherwise} \end{cases} \tag{1.66a}$$

Because the calculations for the mean m_x and the variance σ_x^2 for this case are simple, it is more expedient to find them directly than to use the MGF:

$$E\{X\} = m_x = \int_{-\infty}^{\infty} \alpha \, p_x(\alpha) \, d\alpha = \int_{-b}^{b} \frac{\alpha}{2b} \, d\alpha = 0$$

$$E\{X^2\} = \int_{-\infty}^{\infty} \alpha^2 p_x(\alpha) \, d\alpha = \int_{-b}^{b} \frac{\alpha^2}{2b} \, d\alpha = \frac{b^2}{3}$$

and the variance σ_x^2 is given by

$$\sigma_x^2 = E\{X^2\} - E^2\{X\} = \frac{b^2}{3} \tag{1.66b}$$

Example 1.22 The Gaussian (normal) random variable. The Gaussian RV X with mean m_x and variance σ_x^2, denoted $X = G(m_x, \sigma_x^2)$, has the pdf given by

$$p_x(\alpha) = \frac{1}{\sqrt{2\pi}\sigma_x} e^{-(\alpha - m_x)^2 / 2\sigma_x^2}, \ |\alpha| < \infty \tag{1.67a}$$

The MGF is

$$M_x(\nu) = \int_{-\infty}^{\infty} e^{\nu\alpha} \frac{1}{\sqrt{2\pi}\sigma_x} e^{-(\alpha - m_x)^2 / 2\sigma_x^2} \, d\alpha$$

$$= e^{\nu m_x} \frac{1}{\sqrt{2\pi}} \int_{-\infty}^{\infty} e^{\nu\sigma_x\alpha} e^{-\alpha^2/2}$$

$$= e^{\nu m_x + \sigma_x^2\nu^2/2} \overbrace{\int_{-\infty}^{\infty} \frac{1}{\sqrt{2\pi}} e^{-(\alpha - \sigma_x\nu)^2/2}}^{= 1}$$

$$= e^{\nu m_x + \sigma_x^2\nu^2/2} \tag{1.67b}$$

The first two derivatives and moments are

$$M_x'(\nu) = (m_x + \sigma_x^2\nu)e^{\nu m_x + \sigma_x^2\nu^2/2} \ \rightarrow \ E\{X\} = M_x'(0) = m_x$$

$$M_x''(\nu) = \sigma_x^2 e^{\nu m_x + \sigma_x^2\nu^2/2} + (m_x + \sigma_x^2\nu)^2 e^{\nu m_x + \sigma_x^2\nu^2/2}$$

$$\rightarrow \ E\{X^2\} = M_x''(0) = \sigma_x^2 + m_x^2$$

and the variance σ_x^2 is obtained, as expected, as

$$\sigma_x^2 = E\{X^2\} - E^2\{X\} = \sigma_x^2 + m_x^2 - m_x^2 = \sigma_x^2 \tag{1.67c}$$

Example 1.23 The multivariate Gaussian distribution. Let $\boldsymbol{X}$ be a K-dimensional vector of jointly Gaussian RVs: $\boldsymbol{X} = (X_1, X_2, \ldots, X_K)$. The *mean vector* $\boldsymbol{m}_x$ is

$$\boldsymbol{m}_x = \mathrm{E}\{\boldsymbol{X}\} = (\mathrm{E}\{X_1\}, \mathrm{E}\{X_2\}, \ldots, \mathrm{E}\{X_K\})$$

$$= (m_{x1}, m_{x2}, \ldots, m_{xK}) \tag{1.67d}$$

The second-order central moments are given by [33]

$$\sigma_{ij} \triangleq \mathrm{E}\{(X_i - m_{xi})(X_j - m_{xj})\} = \rho_{ij}\sigma_i\sigma_j \tag{1.67e}$$

where ρ_{ij} is the normalized correlation between X_i and X_j, form the $K \times K$ symmetric *covariance matrix* $\boldsymbol{\Sigma}_x$ given by

$$\boldsymbol{\Sigma}_x = \| \sigma_{ij} \| = \begin{bmatrix} \sigma_1^2 & \rho_{12}\sigma_1\sigma_2 & \cdots & \rho_{1K}\sigma_1\sigma_K \\ \rho_{12}\sigma_2\sigma_1 & \sigma_2^2 & \cdots & \rho_{2K}\sigma_2\sigma_K \\ \vdots & & & \vdots \\ \rho_{1K}\sigma_K\sigma_1 & \rho_{2K}\sigma_K\sigma_2 & \cdots & \sigma_K^2 \end{bmatrix} \tag{1.67f}$$

Note that the covariance matrix $\boldsymbol{\Sigma}_x$ is the expected value of the outer product $(\boldsymbol{X} - \boldsymbol{m}_x)^{\mathrm{T}}(\boldsymbol{X} - \boldsymbol{m}_x)$. In terms of the mean vector and covariance matrix, the joint pdf of the RVs in the vector $\boldsymbol{X}$ is

$$\mathrm{p}_K(\boldsymbol{\alpha}; \boldsymbol{m}_x, \boldsymbol{\Sigma}_x) = \mathrm{p}_K(\alpha_1, \alpha_2, \ldots, \alpha_K; \boldsymbol{m}_x, \boldsymbol{\Sigma}_x)$$

$$= \frac{1}{(2\pi)^{K/2}\sqrt{|\boldsymbol{\Sigma}_x|}} \exp\left\{-\tfrac{1}{2}(\boldsymbol{\alpha} - \boldsymbol{m}_x)\boldsymbol{\Sigma}_x^{-1}(\boldsymbol{\alpha} - \boldsymbol{m}_x)^{\mathrm{T}}\right\} \tag{1.67g}$$

in which $|\boldsymbol{\Sigma}_x|$ is the determinant of $\boldsymbol{\Sigma}_x$ and $\boldsymbol{\Sigma}_x^{-1}$ is its inverse. The argument of the exponential, a scalar quantity known as a *quadratic form*, is of the form $\boldsymbol{a}\mathbf{A}\boldsymbol{a}^{\mathrm{T}}$, where the vector $\boldsymbol{a}$ is $1 \times K$ the matrix $\mathbf{A}$ is $K \times K$. The moment generating function corresponding to the multivariate Gaussian pdf is a function of K variables forming a vector $\boldsymbol{\nu} = (\nu_1, \nu_2, \ldots, \nu_K)$ and is found as

$$M_x(\boldsymbol{\nu}) = \mathrm{E}\{\exp(\nu_1 X_1 + \nu_2 X_2 + \cdots + \nu_K X_K)\}$$

$$= \mathrm{E}\{\exp(\boldsymbol{\nu}\boldsymbol{X}^{\mathrm{T}})\} = \mathrm{E}\{\exp(\tfrac{1}{2}\boldsymbol{\nu}\boldsymbol{X}^{\mathrm{T}} + \tfrac{1}{2}\boldsymbol{X}\boldsymbol{\nu}^{\mathrm{T}})\}$$

$$= \frac{1}{(2\pi)^{K/2}\sqrt{|\boldsymbol{\Sigma}_x|}} \int_{-\infty}^{\infty} \cdots \int_{-\infty}^{\infty} \exp\{-\tfrac{1}{2}Q\} d\alpha_1 \cdots d\alpha_K$$

where the quadratic form Q is

$$Q = (\boldsymbol{\alpha} - \boldsymbol{m}_x)\Sigma_x^{-1}(\boldsymbol{\alpha} - \boldsymbol{m}_x)^{\mathrm{T}} - \boldsymbol{\nu}\boldsymbol{\alpha}^{\mathrm{T}} - \boldsymbol{\alpha}\boldsymbol{\nu}^{\mathrm{T}}$$
$$= (\boldsymbol{\alpha} - \boldsymbol{m}_x - \Sigma_x\boldsymbol{\nu})\Sigma_x^{-1}(\boldsymbol{\alpha} - \boldsymbol{m}_x - \Sigma_x\boldsymbol{\nu})^{\mathrm{T}} - \boldsymbol{\nu}\Sigma_x\boldsymbol{\nu}^{\mathrm{T}} - \boldsymbol{\nu}\boldsymbol{m}_x^{\mathrm{T}} - \boldsymbol{m}_x\boldsymbol{\nu}^{\mathrm{T}}$$
$$= (\boldsymbol{\alpha} - \boldsymbol{m}_x')\Sigma_x^{-1}(\boldsymbol{\alpha} - \boldsymbol{m}_x')^{\mathrm{T}} - \boldsymbol{\nu}\Sigma_x\boldsymbol{\nu}^{\mathrm{T}} - 2\boldsymbol{\nu}\boldsymbol{m}_x^{\mathrm{T}}$$

Therefore, the integral is the expectation of a constant for a multivariate Gaussian distribution with the mean vector $\boldsymbol{m}_x'$, leading to

$$M_x(\boldsymbol{\nu}) = \exp\left\{\boldsymbol{\nu}\boldsymbol{m}_x^{\mathrm{T}} + \tfrac{1}{2}\boldsymbol{\nu}\Sigma_x\boldsymbol{\nu}^{\mathrm{T}}\right\} \qquad (1.67\mathrm{h})$$

1.5.5 Correlation Functions and Power Spectra

Suppose that X_1 and X_2 are RVs obtained from a random process by taking samples at the instants $t = t_1$ and $t = t_2$, that the joint probability density function for X_1 and X_2 is $\mathrm{p}_{x_1,x_2}(\alpha, \beta)$, and that the probability density function for each random variable X_i, $i = 1, 2$ is given by $\mathrm{p}_{x_1}(\alpha)$ and $\mathrm{p}_{x_2}(\alpha)$. If $\mathrm{p}_{x_1}(\alpha)$ and $\mathrm{p}_{x_2}(\alpha)$ are the same, that is, the choice of sampling instant t_i ($i = 1, 2, \ldots$) did not make the pdf different, then we say that the random process is *strictly stationary*. If the pdf $\mathrm{p}_{x_i}(\alpha)$ is different in that the choice of sampling instant itself produces a different pdf, then the process is *nonstationary*.

  If we multiply the two RVs X_1 and X_2 and take the ensemble average of the product with respect to the joint pdf $\mathrm{p}_{x_1,x_2}(\alpha, \beta)$, it is called the autocorrelation function of the random process $\mathrm{x}(t)$. That is, we define the autocorrelation function $\mathrm{R}_x(t_1, t_2)$ of a real random process $\{\mathrm{x}(t)\}$ as

$$\mathrm{R}_x(t_1, t_2) = \mathrm{E}\{X_1 X_2\} = \int\int \alpha\beta\, \mathrm{p}_{x_1,x_2}(\alpha, \beta)\, d\alpha\, d\beta \qquad (1.68)$$

where the notation $\mathrm{R}_x(t_1, t_2)$ implies that the random process $\mathrm{x}(t)$ was sampled at time instants $t = t_1$ and $t = t_2$ to produce RVs X_1 and X_2. The prefix *auto-* in the word autocorrelation simply means that the RVs are generated from the *same* random process $\mathrm{x}(t)$. The integration limits in (1.68) are the range of the RVs as they are defined. If we assume that the time instant $t = t_1$ is τ sec later than the time instant $t = t_2$, that is, $\tau = t_1 - t_2$, we have $\mathrm{R}_x(t_1, t_2) = \mathrm{R}_x(t_1, t_1 - \tau)$. If the random process is stationary, the joint probability distribution for X_1 and X_2 depends only on the time difference τ, not on the particular values of t_1 and t_2 [31]. The autocorrelation function is then a function only of the time difference τ and we write

$$R_x(t_1, t_1 - \tau) = R_x(t, t - \tau) = R_x(\tau) \qquad (1.69a)$$

The covariance function. In addition to the correlation function, we also define a function $C_x(t_1, t_2)$, called the *covariance function* of $x(t)$, defined by

$$C_x(t_1, t_2) \triangleq E\{[x(t_1) - m_x(t_1)][x(t_2) - m_x(t_2)]\} \qquad (1.69b)$$

where $m_x(t_1)$ and $m_x(t_2)$ are the mean functions of the RVs $X_1 = x(t_1)$ and $X_2 = x(t_2)$ of the random process $\{x(t)\}$. The covariance function and the correlation function are related by

$$
\begin{aligned}
C_x(t_1, t_2) &= \overline{[x(t_1) - m_x(t_1)][x(t_2) - m_x(t_2)]} \\
&= \overline{x(t_1)\,x(t_2)} - \overline{x(t_1)}\,m_x(t_2) - m_x(t_1)\,\overline{x(t_2)} + m_x(t_1)\,m_x(t_2) \\
&= \overline{X_1 X_2} - \overline{X_1} m_2 - m_1 \overline{X_2} + m_1 m_2 \\
&= R_x(t_1, t_2) - m_x(t_1) m_x(t_2) \qquad (1.69c)
\end{aligned}
$$

For a *wide sense stationary* random process—one for which the correlation function depends only on the time difference $t_1 - t_2 = \tau$ and the mean is constant—the covariance function is of the form

$$C_x(\tau) = R_x(\tau) - m_x^2 \qquad (1.69d)$$

Analogous to the normalized covariance of (1.56b), we define the *normalized covariance function* of the random process $\{x(t)\}$ by

$$
\begin{aligned}
\rho_x(t_1, t_2) &= \frac{E\{(X_1 - m_1)(X_2 - m_2)\}}{\sigma_1 \sigma_2} = \frac{R_x(t_1, t_2) - m_x(t_1)\,m_x(t_2)}{\sigma_x(t_1)\,\sigma_x(t_2)} \\
&= \frac{R_x(t_1, t_2) - m_1 m_2}{\sigma_1 \sigma_2} \qquad (1.69e)
\end{aligned}
$$

where $m_i = m_x(t_i) = E\{x(t_i)\}$ and $\sigma_i^2 = \sigma_x^2(t_i) = E\{[x(t_i) - m_x(t_i)]^2\}$. If the process is stationary and $t_1 - t_2 = \tau$, we have

$$\rho_x(\tau) = \frac{R_x(\tau) - m_x^2}{\sigma_x^2} \tag{1.69f}$$

Note that, for a wide sense stationary process, because $R_x(t, t - \tau) = \overline{x(t)\,x(t - \tau)}$, we have

$$R_x(0) = \overline{x^2(t)} = \text{mean square value} = \text{average power of } x(t) = \sigma_x^2 \tag{1.69g}$$

The reader is referred to [29–32] for further reading on this subject, including discussions on complex random processes.

Power spectrum. The power spectrum or power spectral density (PSD) of a wide sense stationary random process $x(t)$ is defined as the Fourier transform[18] of the autocorrelation function:

$$\mathcal{F}\{R_x(\tau)\} = \mathcal{S}_x(f) = \text{PSD} \tag{1.70a}$$

or

$$\mathcal{S}_x(f) = \int_{-\infty}^{\infty} R_x(\tau)\, e^{-j2\pi f \tau}\, d\tau \tag{1.70b}$$

and

$$R_x(\tau) = \mathcal{F}^{-1}\{\mathcal{S}_x(f)\} = \int_{-\infty}^{\infty} \mathcal{S}_x(f)\, e^{+j2\pi f \tau}\, df \tag{1.70c}$$

Note that

$$|R_x(\tau)| = \left| \int_{-\infty}^{\infty} \mathcal{S}_x(f)\, e^{j2\pi f \tau}\, df \right|$$

$$\leq \int_{-\infty}^{\infty} |\mathcal{S}_x(f)|\, |e^{j2\pi f \tau}|\, df = \int_{-\infty}^{\infty} \mathcal{S}_x(f)\, df = R_x(0)$$

that is:

$$R_x(\tau) \leq R_x(0) \tag{1.70d}$$

Example 1.24 Let us asssume that we have a sinewave process $\{x(t) = A\cos(\omega_0 t + \phi)\}$, where ω_0 and A are constants and ϕ is an RV uniformly distributed over the interval $(0, 2\pi)$:

---

[18] This Fourier transform pair relationship between the autocorrelation function and the PSD is also known as the *Wiener-Kinchine theorem*.

$$p_\phi(\alpha) = \begin{cases} \frac{1}{2\pi}, & 0 \le \alpha \le 2\pi \\ 0, & \text{otherwise} \end{cases} \tag{1.71a}$$

We will obtain the covariance function of this process.

$$m_x(t) = E\{A\cos(\omega_0 t + \phi)\} = \int_{-\infty}^{\infty} A\cos(\omega_0 t + \alpha)\, p_\phi(\alpha)\, d\alpha$$

$$= \frac{1}{2\pi} \int_0^{2\pi} A\cos(\omega_0 t + \alpha)\, d\alpha = 0 \tag{1.71b}$$

Thus

$$C_x(t, t-\tau) = E\{[x(t) - m_x(t)][x(t-\tau) - m_x(t-\tau)]\}$$

$$= E\{x(t)\,x(t-\tau)\} = R_x(t, t-\tau) \tag{1.71c}$$

$$R_x(t, t-\tau) = E\{A\cos(\omega_0 t + \phi) \cdot A\cos(\omega_0 t - \omega_0\tau + \phi)\}$$

$$= \tfrac{1}{2}A^2 E\{\cos\omega_0\tau\} + \tfrac{1}{2}A^2 E\{\cos(2\omega_0 t - \omega_0\tau + 2\phi)\}$$

But

$$E\{\cos(2\omega_0 t - \omega_0\tau + 2\phi)\} = \frac{1}{2\pi}\int_0^{2\pi}\cos(2\omega_0 t - \omega_0\tau + 2\phi)\,d\phi = 0$$

$$\tag{1.71d}$$

so that

$$C_x(t, t-\tau) = C_x(\tau) = R_x(\tau) = \tfrac{1}{2}A^2\cos\omega_0\tau \tag{1.71e}$$

Observe from (1.69g) and (1.71e) that

$$\text{Average power of } x(t) = \sigma_x^2 = E\{x^2(t)\}$$

$$= R_x(0) = \int_{-\infty}^{\infty} S_x(f)\, df = \tfrac{1}{2}A^2 \tag{1.71f}$$

and thus $S_x(f)$ is clearly the power density that is distributed in the frequency domain, whose total collection is equal to the total average power $x^2(t) = R_x(0)$. This justifies the reasonableness of the definition of the power spectral density as the Fourier transform of the correlation function.

Input and output random processes of a linear system. Consider a linear time invariant system with impulse response $h(t)$ as illustrated in Figure 1.62. We assume that a sample function $x(t)$ from a random process $\{x(t)\}$

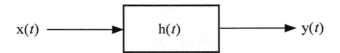

Figure 1.62 Linear system (filter) with impulse response h(t).

is applied to the system and its output is a sample function $y(t)$ of random process $\{y(t)\}$. Then, we have the output $y(t)$ given by

$$y(t) = x(t) * h(t) = \int_{-\infty}^{\infty} x(\alpha)\, h(t - \alpha)\, d\alpha \tag{1.72a}$$

The mean function $m_y(t)$ is then obtained from

$$m_y(t) = E\{y(t)\} = E\left\{ \int_{-\infty}^{\infty} x(\alpha)\, h(t - \alpha)\, d\alpha \right\}$$

$$= \int_{-\infty}^{\infty} \overline{x(\alpha)}\, h(t - \alpha)\, d\alpha = \int_{-\infty}^{\infty} m_x(\alpha)\, h(t - \alpha)\, d\alpha \tag{1.72b}$$

The autocorrelation function of $y(t)$ is found to be

$$R_y(t_1,\, t_2) \triangleq \overline{y(t_1)\, y(t_2)}$$

$$= E\left\{ \int_{-\infty}^{\infty} x(\alpha)\, h(t_1 - \alpha)\, d\alpha \int_{-\infty}^{\infty} x(\beta)\, h(t_2 - \beta)\, d\beta \right\}$$

$$= \int_{-\infty}^{\infty} \int_{-\infty}^{\infty} \overline{x(\alpha)\, x(\beta)}\, h(t_1 - \alpha)\, h(t_2 - \beta)\, d\alpha\, d\beta$$

$$= \int_{-\infty}^{\infty} \int_{-\infty}^{\infty} R_x(\alpha,\, \beta)\, h(t_1 - \alpha)\, h(t_2 - \beta)\, d\alpha\, d\beta \tag{1.72c}$$

Assuming that the random process $\{x(t)\}$ is wide sense stationary, that is, $R_x(t_1,\, t_2) = R_x(\tau)$ with $\tau = t_1 - t_2$, we have

$$R_y(t_1,\, t_2) = \int_{-\infty}^{\infty} \int_{-\infty}^{\infty} R_x(\alpha - \beta)\, h(t_1 - \alpha)\, h(t_2 - \beta)\, d\alpha\, d\beta$$

By making the change of variables $t_1 - \alpha = \mu$ and $t_2 - \beta = \gamma$, we obtain

$$R_y(t_1, t_2) = \int_{-\infty}^{\infty} \int_{-\infty}^{\infty} R_x(t_1 - t_2 + \gamma - \mu)\, h(\mu)\, h(\gamma)\, d\mu\, d\gamma \qquad (1.72d)$$

From (1.72d), we observe that $R_y(t_1, t_2) = R_y(\tau)$, and thus we have

$$R_y(\tau) = \int_{-\infty}^{\infty} \int_{-\infty}^{\infty} R_x(\tau + \gamma - \mu)\, h(\mu)\, h(\gamma)\, d\mu\, d\gamma \qquad (1.72e)$$

Applying the Wiener-Kinchine theorem on $R_y(\tau)$ and $R_x(\tau)$, that is:

$$\mathcal{S}_y(f) = \int_{-\infty}^{\infty} R_y(\tau)\, e^{-j2\pi f\tau} \quad \leftrightarrow \quad R_y(\tau) = \int_{-\infty}^{\infty} \mathcal{S}_y(f)\, e^{j2\pi f\tau} \qquad (1.72f)$$

$$\mathcal{S}_x(f) = \int_{-\infty}^{\infty} R_x(\tau)\, e^{-j2\pi f\tau} \quad \leftrightarrow \quad R_x(\tau) = \int_{-\infty}^{\infty} \mathcal{S}_x(f)\, e^{j2\pi f\tau} \qquad (1.72g)$$

the output autocorrelation function given in (1.72e) becomes

$$R_y(\tau) = \int_{-\infty}^{\infty} \int_{-\infty}^{\infty} \int_{-\infty}^{\infty} \mathcal{S}_x(f)\, e^{j2\pi f(\tau + \gamma - \mu)}\, h(\mu)\, h(\gamma)\, df\, d\mu\, d\gamma$$

$$= \int_{-\infty}^{\infty} \mathcal{S}_x(f)\, e^{j2\pi f\tau} df \overbrace{\int_{-\infty}^{\infty} h(\mu)\, e^{-j2\pi f\mu} d\mu}^{H(f)} \overbrace{\int_{-\infty}^{\infty} h(\gamma)\, e^{j2\pi f\gamma} d\gamma}^{H^*(f)}$$

where $H(f)$ and $H^*(f)$ are the filter's transfer function and its complex conjugate, respectively. Thus

$$R_y(\tau) = \int_{-\infty}^{\infty} \mathcal{S}_x(f)\, |H(f)|^2 e^{j2\pi f\tau} df \qquad (1.72h)$$

By comparing (1.72h) with (1.72f), we see the identity

$$\mathcal{S}_y(f) = \mathcal{S}_x(f)\, |H(f)|^2 \qquad (1.72i)$$

Analogous to (1.70d), we have

$$|R_y(t_1, t_2)| = |R_y(t_1 - t_2)| = |R_y(\tau)| \le R_y(0) = \overline{y^2(t)} \qquad (1.72j)$$

Crosscorrelation function. Consider a situation in which an input process $\{x(t)\}$ is applied to two linear filters characterized by impulse response functions $h_y(t)$ and $h_z(t)$ with transfer functions $H_y(f)$ and $H_z(f)$, respectively, as shown in Figure 1.63. We define the crosscorrelation function of the two output processes $\{y(t)\}$ and $\{z(t)\}$ as follows.

$$
\begin{aligned}
R_{yz}(t_1, t_2) &= \overline{y(t_1)\,z(t_2)} \\
&= \mathrm{E}\left\{ \int_{-\infty}^{\infty} x(t_1 - \alpha)\,h_y(\alpha)\,d\alpha \int_{-\infty}^{\infty} x(t_2 - \beta)\,h_z(\beta)\,d\beta \right\} \\
&= \int_{-\infty}^{\infty}\int_{-\infty}^{\infty} \overline{x(t_1 - \alpha)\,x(t_2 - \beta)}\,h_y(\alpha)\,h_z(\beta)\,d\alpha\,d\beta \\
&= \int_{-\infty}^{\infty}\int_{-\infty}^{\infty} R_x(t_1 - \alpha,\, t_2 - \beta)\,h_y(\alpha)\,h_z(\beta)\,d\alpha\,d\beta. \qquad (1.73a)
\end{aligned}
$$

For a stationary process $\{x(t)\}$, we have

$$
\begin{aligned}
R_{yz}(t_1, t_2) &= \int_{-\infty}^{\infty}\int_{-\infty}^{\infty} R_x(t_1 - t_2 + \beta - \alpha)\,h_y(\alpha)\,h_z(\beta)\,d\alpha\,d\beta \\
&= \int_{-\infty}^{\infty} S_x(f)\,e^{j2\pi f(t_1 - t_2)}\,df \int_{-\infty}^{\infty} h_y(\alpha)\,e^{-j2\pi f\alpha}\,d\alpha \\
&\qquad\qquad \times \int_{-\infty}^{\infty} h_z(\beta)\,e^{j2\pi f\beta}\,d\beta \qquad (1.73b)
\end{aligned}
$$

But

$$
H_y(f) = \int_{-\infty}^{\infty} h_y(\alpha)\,e^{-j2\pi f\alpha}\,d\alpha, \quad H_z^*(f) = \int_{-\infty}^{\infty} h_z(\beta)\,e^{j2\pi f\beta}\,d\beta
$$

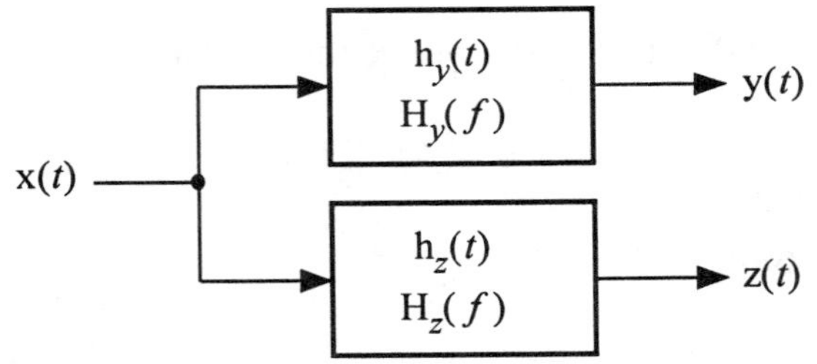

Figure 1.63 Two linear filters in parallel.

and thus (1.73b) becomes

$$R_{yz}(t_1 - t_2) = R_{yz}(\tau) = \int_{-\infty}^{\infty} S_x(f)\, H_y(f)\, H_z^*(f)\, e^{j2\pi f\tau} df \qquad (1.73c)$$

Gaussian random process. Consider a random process $\{x(t)\}$. If we obtain N random variables $X_1 = x(t_1)$, $X_2 = x(t_2)$, $\ldots$, $X_N = x(t_N)$ by sampling the process at N different time instants $t_1, t_2, \ldots, t_N$, and find that the N-dimensional joint pdf $p_{x_1,x_2,\ldots,x_N}(\alpha_1, \alpha_2, \ldots, \alpha_N)$ is identical to an N-dimensional joint Gaussian pdf, we say that the process $\{x(t)\}$ is a Gaussian random process [29]. That is, the "name" of a particular process, such as Gaussian, Rayleigh, uniform, and so forth, is so called through the characterization of the pdf that is identified with the particular process.

Suppose we have two RVs X_1 and X_2, with $\overline{X_i} = 0$, $\sigma_i^2 = \overline{X_i^2}$, $i = 1, 2$ and $\rho_{12} = \overline{X_1 X_2}/\sigma_1\sigma_2$. The two-dimensional pdf $p_{x_1,x_2}(\alpha, \beta)$ is given by

$$p_{x_1,x_2}(\alpha, \beta) = \frac{1}{2\pi\sigma_1\sigma_2\sqrt{1 - \rho_{12}^2}}$$

$$\times \exp\left\{-\frac{1}{2(1 - \rho_{12}^2)}\left[\frac{\alpha^2}{\sigma_1^2} - 2\rho_{12}\frac{\alpha\beta}{\sigma_1\sigma_2} + \frac{\beta^2}{\sigma_2^2}\right]\right\} \qquad (1.74a)$$

Random variables X_1 and X_2 are jointly Gaussian if and only if their joint pdf $p_{x_1,x_2}(\alpha, \beta)$ has the form given in (1.70a). If $\overline{X_i^2} = \sigma^2$, $i = 1, 2$ so that $\rho_{12} = \rho$, we have

$$p_{x_1,x_2}(\alpha, \beta) = \frac{1}{2\pi\sigma^2\sqrt{1 - \rho^2}} \exp\left\{-\frac{\alpha^2 - 2\rho\alpha\beta + \beta^2}{2\sigma^2(1 - \rho^2)}\right\} \qquad (1.74b)$$

One can easily show by integration that the marginal pdfs $p_{x_1}(\alpha)$ and $p_{x_2}(\alpha)$ are the same:

$$p_{x_1}(\alpha) = p_{x_2}(\alpha) = \int_{-\infty}^{\infty} p_{x_1,x_2}(\alpha, \beta)\, d\beta = \frac{1}{\sqrt{2\pi}\sigma}\, e^{-\alpha^2/2\sigma^2} \qquad (1.74c)$$

If X_1 and X_2 are statistically independent:

$$p_{x_1,x_2}(\alpha, \beta) = p_{x_1}(\alpha)\, p_{x_2}(\beta) = \frac{1}{2\pi\sigma^2} \exp\left\{-\frac{\alpha^2 + \beta^2}{2\sigma^2}\right\} \qquad (1.74d)$$

This result for the two-dimensional case can be extended to the arbitrary N-dimensional case, and we summarize the result as follows [29]:

1. The pdf $p_{x_1, x_2, \ldots, x_N}(\alpha_1, \alpha_2, \ldots, \alpha_N)$ of jointly Gaussian RVs depends only on the means $\overline{X_i}$, $i = 1, 2, \ldots, N$ and the covariances (central moments):

    $$\lambda_{ij} = \lambda_{ji} \triangleq \overline{(X_i - \overline{X_i})(X_j - \overline{X_j})}, \ i, j = 1, 2, \ldots, N$$

2. A set of N zero-mean Gaussian RVs $(X_1, X_2, \ldots, X_N)$ are statistically independent if and only if the covariances $\lambda_{ij} = 0$ for all $i \neq j$, and in this case:

    $$p_{x_1, x_2, \ldots, x_N}(\alpha_1, \alpha_2, \ldots, \alpha_N) = \prod_{i=1}^{N} p_{x_i}(\alpha_i)$$

    $$= \frac{1}{(2\pi)^{N/2} \sigma_1 \sigma_2 \cdots \sigma_N} \exp\left\{-\frac{1}{2} \sum_{i=1}^{N} \frac{\alpha_i^2}{\sigma_i^2}\right\} \qquad (1.74e)$$

    where $\lambda_{ii} = \sigma_i^2$ since $\overline{X_i} = 0$.

3. Any linear transformation of a set of N jointly Gaussian RVs produces new RVs that are also jointly Gaussian.

White Gaussian noise. When the PSD of a zero-mean noise process $\{n(t)\}$ is flat for all frequencies, we call such a process *white noise*. When the process is Gaussian, it is called *white Gaussian noise* with the power spectrum

$$S_n(f) \triangleq \tfrac{1}{2} \mathcal{N}_0, \quad -\infty < f < \infty \qquad (1.75a)$$

where $\tfrac{1}{2}\mathcal{N}_0$ is called the "two-sided" PSD. White noise is a fictitious entity because its total average power becomes

$$\overline{n^2(t)} = R_n(0) = \int_{-\infty}^{\infty} S_n(f)\,df = \infty \qquad (1.75b)$$

In practice, the power spectral density that is flat over the bandwidth of interest can be assumed to be white noise. Let us assume that the transfer function of a linear filter is $H(f)$, and the input is white noise with two-sided PSD $\tfrac{1}{2}\mathcal{N}_0$ W/Hz. Then from (1.72i) with $x(t) = n(t)$, we have

$$S_y(f) = \tfrac{1}{2}\mathcal{N}_0|H(f)|^2 \tag{1.75c}$$

and $\qquad \overline{y^2(t)} = R_y(0) = \int_{-\infty}^{\infty} S_y(f)\,df = \tfrac{1}{2}\mathcal{N}_0 \int_{-\infty}^{\infty} |H(f)|^2\,df \tag{1.75d}$

The autocorrelation function of white noise is

$$R_n(\tau) = \mathcal{F}^{-1}\{S_n(f)\} = \mathcal{F}^{-1}\{\tfrac{1}{2}\mathcal{N}_0\} = \tfrac{1}{2}\mathcal{N}_0\,\delta(\tau) \tag{1.75e}$$

which is also a fictitious but useful result in that its application can explain many important physical phenomena in communication theory.

Example 1.25 Consider a linear filter with an ideal lowpass transfer function given by

$$H(f) = \text{Rect}(f/2W) \tag{1.76a}$$

Assume that the input to this filter is white Gaussian noise, $n(t)$, with $S_n(f) = \tfrac{1}{2}\mathcal{N}_0$. The situation is depicted in Figure 1.64. Since zero-mean white Gaussian noise is the input to the filter, the filter output is also a zero-mean random process, called *bandlimited white Gaussian noise*:

$$m_{n_o}(t) = m_n(t) * h(t) = \int_{-\infty}^{\infty} \overline{n(\alpha)}\,h(t-\alpha)\,d\alpha = 0$$

Since $m_{n_o}(t) = 0$, the covariance and correlation functions of the filter output process are

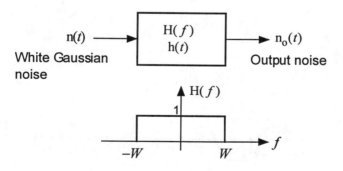

Figure 1.64 Ideal lowpass filter with white Gaussian noise as input.

$$C_{n_o}(\tau) = R_{n_o}(\tau) = \mathcal{F}^{-1}\{S_{n_o}(f)\} = \mathcal{F}^{-1}\{S_n(f)|H(f)|^2\}$$

$$= \mathcal{F}^{-1}\{\tfrac{1}{2}\mathcal{N}_0|\mathrm{Rect}(f/2W)|^2\}$$

$$= \tfrac{1}{2}\mathcal{N}_0 \cdot 2W\,\mathrm{sinc}(2W\tau) = \mathcal{N}_0 W\,\mathrm{sinc}(2W\tau) \qquad (1.76b)$$

The autocorrelation function of the output noise is plotted in Figure 1.65. Note that $C_{n_o}(0) = R_{n_o}(0) = \overline{n_o^2(t)} = \mathcal{N}_0 W$. Suppose we take N samples $n_o(t_1), n_o(t_2), \ldots, n_o(t_N)$ from the output process $\{n_o(t)\}$ at the time instants $t_1, t_2, \ldots, t_N$, where

$$t_i = \frac{i}{2W} + t_0, \quad i = 1, 2, \ldots, N \qquad (1.76c)$$

and t_0 is any constant. Let us now observe the following:

(1) $\qquad \overline{n_o(t_i)} = m_{n_o}(t_i) = 0$

(2) $\qquad \lambda_{ij} = \overline{n_o(t_i)\,n_o(t_j)} = C_{n_o}(t_i - t_j) = C_{n_o}\left(\dfrac{i-j}{2W}\right)$

$$= \mathcal{N}_0 W\,\mathrm{sinc}\left[2W\left(\frac{i-j}{2W}\right)\right] = \mathcal{N}_0 W\,\mathrm{sinc}(i-j)$$

$$= \begin{cases} \mathcal{N}_0 W, & i = j \\ 0, & i \neq j \end{cases} \qquad (1.76d)$$

which means that the samples $\{n_o(t_i)\}$ are statistically independent RVs, and thus we have

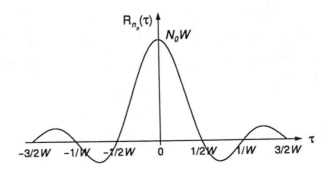

Figure 1.65 Autocorrelation function of bandlimited white Gaussian noise.

$$P_{n_o(t_1),n_o(t_2),\ldots,n_o(t_N)}(\alpha_1, \alpha_2, \ldots, \alpha_N) = \frac{1}{(2\pi\mathcal{N}_0 W)^{N/2}} \exp\left\{-\tfrac{1}{2}\sum_{i=1}^{N}\frac{\alpha_i^2}{\sigma_i^2}\right\}$$

The implication of this result is that if we take samples of bandlimited white Gaussian noise at Nyquist intervals, the samples are statistically independent.

Note that if we take samples $n(t_1)$, $n(t_2)$, $\ldots$, $n(t_N)$ from the white Gaussian noise process $\{n(t)\}$ and examine their correlations, we obtain from (1.75e)

$$\lambda_{ij} = R_n(t_i - t_j) = \overline{n(t_i)\,n(t_j)} = \tfrac{1}{2}\mathcal{N}_0\,\delta(t_i - t_j)$$

$$= \begin{cases} \tfrac{1}{2}\mathcal{N}_0, & i = j \\ 0, & i \neq j \end{cases}$$

which indicates that the samples are statistically independent, no matter how closely the sample intervals are taken. Note also that $R_n(0) = \overline{n^2(t)} = \tfrac{1}{2}\mathcal{N}_0\,\delta(0) = \tfrac{1}{2}\mathcal{N}_0$, which indicates that white-Gaussian-noise-generated RVs are zero-mean Gaussian RVs with variance $\tfrac{1}{2}\mathcal{N}_0$; that is, $n(t) = G\left(0, \tfrac{1}{2}\mathcal{N}_0\right)$.

1.5.6 Central Limit Theorem

In many communications systems, we encounter situations in which we need to compute the probability measure of the event that the sum of M different signal-plus-noise sources exceeds a certain threshold value.

For example, in a cellular mobile communications system such as IS-95, the signals from M different mobile stations are received at the base station, and each signal transmitted by the respective mobile terminal has experienced many traumatic propagation path anomalies such as multipath, shadowing, and attenuation before reaching the base station receiver, where additional Gaussian thermal noise is added. Thus, if we need to compute the probability that the combined waveform exceeds a certain threshold value to obtain a performance measure, we need to know the probability distribution of such a combined waveform random variable. If we denote the signal from mobile station k by x_k, we can express the combination of M such signals as

$$y(t) \triangleq x_1(t) + x_2(t) + \cdots + x_M(t) \tag{1.77a}$$

We can obtain all kinds of moments of the RV Y obtained as a sample of $y(t)$ as well as the probability measure (cumulative or complementary cumulative distribution function) if we know the pdf of Y.

First, let us assume that we know the pdf of each RV X_i contributing to the sum Y, $i = 1, 2, \ldots, M$. Assuming statistical independence of these RVs, one can theoretically compute the pdf of Y, $p_y(\alpha)$, through iterated convolutions (see (1.47e)). But the labor that is involved in obtaining $p_y(\alpha)$ for $M > 2$ is considerable! Second, suppose that we do *not* know the characteristic of each RV X_i, $i = 1, 2, \ldots, M$, but that we only know their moments. Yet we need to obtain a complementary CDF such as

$$\Pr\{Y > y_T\} = 1 - F_y(y_T) \tag{1.77b}$$

We resort to an approximation technique in computing $F_y(y_T)$ in such adverse conditions, and this is where the *Central Limit Theorem (CLT)* [29–30] helps us out.

The essence of the CLT is that, for large M, we can compute $\Pr\{Y > y_T\}$ as if the RV Y were a Gaussian RV; that is, as if

$$Y = G\left(m_y,\, \sigma_y^2\right) \tag{1.77c}$$

where m_y and σ_y^2 are obtained from the expressions

$$m_y = \mathrm{E}\left\{\sum_{i=1}^{M} X_i\right\},\quad \sigma_y^2 = \overline{Y^2} - \overline{Y}^2 = \mathrm{E}\left\{\left(\sum_{i=1}^{M} X_i\right)^2\right\} - \left(\sum_{i=1}^{M} \overline{X_i}\right)^2 \tag{1.77e}$$

The CLT does not imply that $p_y(\alpha)$ itself approaches the Gaussian pdf, but that the integral

$$F_y(y_T) = \int_{-\infty}^{y_T} p_y(\alpha)\, d\alpha \tag{1.77f}$$

approaches a value that is obtained by the integral of the Gaussian pdf. In other words, one should not say that $p_y(\alpha)$ has the Gaussian shape, but a correct result is obtained for the calculation of $F_y(y_T)$ if we pretend that $p_y(\alpha)$ has the Gaussian form.

More formally stated, let $\{X_i\}$ be a set of statistically independent, zero-mean RVs, each with the same pdf $p_{x_i}(\alpha) = p_x(\alpha)$ and finite variance $\sigma_{x_i}^2 = \sigma_x^2$. Let Z be the sample mean $\bar{x}$ normalized by its standard deviation:[19]

$$Z \triangleq \frac{\bar{x}}{\sigma(\bar{x})} = \frac{\frac{1}{M}\sum_{i=1}^{M}X_i}{\frac{\sigma_x}{\sqrt{M}}} = \frac{1}{\sqrt{M}}\sum_{i=1}^{M}\frac{X_i}{\sigma_x} \tag{1.78a}$$

Then the CLT says that [27]

$$\lim_{M \to \infty} F_z(\alpha) = \lim_{M \to \infty} \Pr\{Z \leq \alpha\} = \lim_{M \to \infty} \Pr\left\{\sum_{i=1}^{M}\frac{X_i}{\sigma_x\sqrt{M}}\right\}$$

$$= \int_{-\infty}^{\alpha}\frac{1}{\sqrt{2\pi}}e^{-\lambda^2/2}d\lambda \tag{1.78b}$$

Consequently, if Y is a sum of M RVs and the mean and standard deviation of Y are m_y and σ_y, respectively, then as M becomes large:

$$\Pr\{Y > \alpha\} = \Pr\left\{\frac{Y - m_y}{\sigma_y} > \frac{\alpha - m_y}{\sigma_y}\right\} \to Q\left(\frac{\alpha - m_y}{\sigma_y}\right) \tag{1.78c}$$

and for any two integration limits α_1 and α_2,

$$\lim_{M \to \infty}\int_{\alpha_1}^{\alpha_2}p_y(\alpha)\,d\alpha = \int_{\alpha_1}^{\alpha_2}\frac{1}{\sqrt{2\pi}\sigma_y}e^{-(\alpha - m_y)^2/2\sigma_y^2}\,d\alpha \tag{1.78d}$$

We make use of the CLT in several places in this book, and in particular, when we evaluate the Erlang capacity of CDMA systems.

1.5.7 Chernoff Bounds

The CLT that we have considered in the previous section is a form of approximation in the calculation of certain probability measures and is often used in the analysis of communications systems. Another limit theorem that

---

[19] We know from (1.53c) that if the variances of the $\{X_i\}$ are all equal, then the variance of their sum is $M\sigma_x^2$. Therefore, the variance of the sample mean is σ_x^2/M.

falls within the domain of often-practiced approximation in probability theory is a bounding technique known as the *Chernoff bound (CB)* [34,35]. The CB has to do with an approximation technique for estimating the first moment (statistical average) of a RV that is a sum of M independent RVs, for $M = 2, 3, \ldots$, where the sum RV is subjected to a binary decision relative to a threshold value, say, zero.

Consider that we have a RV Y that is a sum of M independent RVs $\{X_i\}$, each with mean $\overline{X_i} = m_x$ and variance $\sigma_{x_i}^2 = \sigma_x^2$, $i = 1, 2, \ldots, M$. The form of Y is then

$$Y = \sum_{i=1}^{M} X_i \tag{1.79a}$$

The specific characterization of the RVs $\{X_i\}$ depends on the system or situation under consideration. For example, X_i could represent a RV that is the difference of two squared Gaussian RVs that fits the description of a diversity reception of a noncoherent binary communication system [8]. For the purpose and the reasoning here, we do not need to specify the actual characterization of the random variables at this time.

We often need to obtain the probability that $Y > 0$ in the analysis of system performance:

$$\Pr\{Y > 0\} = \Pr\left\{\sum_{i=1}^{M} X_i > 0\right\} \tag{1.79b}$$

But to calculate this measure, we resort to an approximation or bounding technique. The technique is to form a unit step function, Z, with Y as the variable. That is,

$$Z = \mathrm{U}_{-1}\left(\sum_{i=1}^{M} X_i\right) = \begin{cases} 1, & Y = \sum\limits_{i=1}^{M} X_i \geq 0 \\ \\ 0, & Y = \sum\limits_{i=1}^{M} X_i < 0 \end{cases} \tag{1.79c}$$

as shown in Figure 1.66, where $\mathrm{U}_{-1}(\bullet)$ denotes the unit step function. Since the RV $Z = \mathrm{U}_{-1}(Y)$ can take only the two values 0 and 1, if we calculate the mean of Z, we have

$$E\{Z\} = \overline{Z} = 0 \cdot \Pr\{Z = 1\} + 1 \cdot \Pr\{Z = 1\}$$

$$= \Pr\{Z = 1\} = \Pr\left\{\sum_{i=1}^{M} X_i > 0\right\} \tag{1.79d}$$

Thus, the desired probability (1.79b) equals the ensemble average (expected value) of Z, and we have

$$\Pr\left\{\sum_{i=1}^{M} X_i > 0\right\} = E\{Z\} = E\left\{U_{-1}\left(\sum_{i=1}^{M} X_i\right)\right\} \tag{1.79e}$$

The difficulty is that there is no easy way of evaluating the right-hand side of (1.79e). We therefore resort to the CB. The CB is based on recognizing that the unit step function $U_{-1}(Y)$ is overbounded by an exponential function $e^{\lambda Y}$, as shown in Figure 1.66, and thus the bound is given by

$$\Pr\left\{\sum_{i=1}^{M} X_i > 0\right\} \le E\left\{\exp\left(\lambda\sum_{i=1}^{M} X_i\right)\right\} \tag{1.79f}$$

where λ is the bound-tightening parameter, which can be chosen to minimize (tighten) the upper bound. The CB is an "exponentially tight" bound [29], meaning that the bound based on (1.79f), when put in the exponential form $e^{-\xi}$, has an exponent ξ that is the largest possible value.

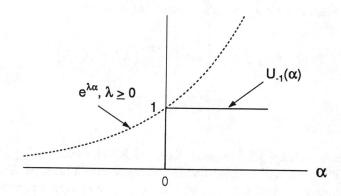

Figure 1.66 Unit step function for the Chernoff bound derivation.

Refined Chernoff bound. Jacobs [35] has shown that the right-hand side of (1.79f) can always be multiplied by one-half without destroying the validity of the relationship of the CB. Thus, we may have a refined Chernoff bound (RCB) as

$$\Pr\left\{\sum_{i=1}^{M} X_i > 0\right\} \leq \frac{1}{2} E\left\{\exp\left(\lambda \sum_{i=1}^{M} X_i\right)\right\} \qquad (1.79\text{g})$$

where the bound-tightening parameter λ has the same meaning as defined earlier. The "factor of one-half" therefore shall always be a part of the CB expression, and the reader is referred to [35] for the reasoning behind this remarkable improvement (refinement) of the original CB.

Example 1.26 A CB example. Let us consider the RV Y given by

$$Y = \sum_{i=1}^{L}(X_{i1}^2 - X_{i2}^2) \qquad (1.80\text{a})$$

where X_{i1} and X_{i2} are statistically independent, zero-mean Gaussian RVs with variances σ_1^2 and σ_2^2, respectively, with $\sigma_2^2 > \sigma_1^2$. That is:

$$X_{i1} = G(0, \sigma_1^2), \ X_{i2} = G(0, \sigma_2^2), \ i = 1, 2, \ldots, L \qquad (1.80\text{b})$$

We wish to calculate $\Pr\{Y > 0\}$. The exact solution is difficult to obtain, and so we wish to find a CB. The bound is given by

$$\Pr\{Y > 0\} \leq \frac{1}{2} E\left\{\exp\left(\lambda \sum_{i=1}^{L}(X_{i1}^2 - X_{i2}^2)\right)\right\}$$

$$= \frac{1}{2} \prod_{i=1}^{L} \overline{e^{\lambda X_{i1}^2} \cdot e^{-\lambda X_{i2}^2}} \qquad (1.80\text{c})$$

To calculate $E\{\exp(\lambda X_{i1}^2)\}$ and $E\{\exp(\lambda X_{i2}^2)\}$, where X_{i1} and X_{i2} are Gaussian RVs, we will invoke a well-known formula [29]. If X is a Gaussian RV with mean m_x and variance σ_x^2, and w is any complex constant such that $2\sigma_x^2 \operatorname{Re}\{w\} < 1$, then

$$\overline{\exp\{wX^2\}} = \frac{\exp\left\{\dfrac{wm_x}{1 - 2w\sigma_x^2}\right\}}{\sqrt{1 - 2w\sigma_x^2}} \tag{1.80d}$$

Thus, by using the formula (1.80d), we have for $i = 1, 2, \ldots, L$,

$$\overline{e^{\lambda X_{i1}^2}} = \frac{1}{\sqrt{1 - 2\lambda\sigma_1^2}}; \qquad \lambda < \frac{1}{2\sigma_1^2}$$

$$\overline{e^{-\lambda X_{i2}^2}} = \frac{1}{\sqrt{1 + 2\lambda\sigma_2^2}}; \qquad \lambda > \frac{-1}{2\sigma_2^2} \tag{1.80e}$$

The CB expression then becomes

$$\Pr\{Y > 0\} \leq \frac{1}{2} \prod_{i=1}^{L} \left(1 - 2\lambda\sigma_1^2\right)^{-\frac{1}{2}} \left(1 + 2\lambda\sigma_2^2\right)^{-\frac{1}{2}}$$

$$= \frac{1}{2} \left[\frac{1}{1 + 2\lambda(\sigma_2^2 - \sigma_1^2) - 4\lambda^2\sigma_1^2\sigma_2^2}\right]^{L/2} \tag{1.80f}$$

We now need to choose λ in such a way that the bound is as tight as possible. The value of λ that minimizes the right-hand side of (1.80f) is the same as the value of λ that maximizes the denominator inside the brackets, and is determined by differentiation as follows:

$$\frac{d}{d\lambda}\left[1 + 2\lambda(\sigma_2^2 - \sigma_1^2) - 4\lambda^2\sigma_1^2\sigma_2^2\right] = 0 \quad \rightarrow \quad \lambda = \frac{\sigma_2^2 - \sigma_1^2}{4\sigma_1^2\sigma_2^2} \tag{1.80g}$$

By substituting (1.80g) in (1.80f), we get for the CB

$$\Pr\{Y > 0\} = \Pr\left\{\sum_{i=1}^{L}(X_{i1}^2 - X_{i2}^2) > 0\right\} \leq \frac{1}{2}\left(\frac{2\sigma_1\sigma_2}{\sigma_1^2 + \sigma_2^2}\right)^L \tag{1.80h}$$

or

$$\Pr\{Y > 0\} = \Pr\left\{\sum_{i=1}^{L}(X_{i1}^2 - X_{i2}^2) > 0\right\} \leq \frac{1}{2}\left(\frac{2\sqrt{\mu}}{1 + \mu}\right)^L \tag{1.80i}$$

using

$$\mu \triangleq \sigma_1^2/\sigma_2^2 \tag{1.80j}$$

Example 1.27 A comparison of the CLT and the CB. Suppose we calculate $\Pr\{Y > 0\}$ by using the CLT, with Y defined in (1.80a), and compare the result with the CB (1.80j). Because

$$Y = \sum_{i=1}^{L} (X_{i1}^2 - X_{i2}^2) \tag{1.81a}$$

in order to apply the CLT we need to calculate the mean and variance of Y, which is to be looked on as if it is a Gaussian RV, $Y = G(m_y, \sigma_y^2)$. Now:

$$m_y = \overline{Y} = \overline{\sum_{i=1}^{L} (X_{i1}^2 - X_{i2}^2)} = \sum_{i=1}^{L} \left(\overline{X_{i1}^2} - \overline{X_{i2}^2} \right)$$

$$= \sum_{i=1}^{L} (\sigma_1^2 - \sigma_2^2) = L(\sigma_1^2 - \sigma_2^2) \tag{1.81b}$$

and

$$\sigma_y^2 = \overline{(Y - \overline{Y})^2} = \overline{Y^2} - \overline{Y}^2$$

where

$$\overline{Y^2} = \sum_{i=1}^{L} \sum_{j=1}^{L} \overline{(X_{i1}^2 - X_{i2}^2)(X_{j1}^2 - X_{j2}^2)}$$

$$= \sum_{i=1}^{L} \overline{(X_{i1}^2 - X_{i2}^2)^2} + \sum_{\substack{i,j=1 \\ i \neq j}}^{L} \overline{(X_{i1}^2 - X_{i2}^2)} \cdot \overline{(X_{j1}^2 - X_{j2}^2)}$$

$$= L\left(\overline{X_{i1}^4} - 2\sigma_1^2\sigma_2^2 + \overline{X_{i2}^4} \right) + L(L-1)(\sigma_1^2 - \sigma_2^2)^2 \tag{1.81c}$$

The fourth moments needed in (1.77c) are found using the Gaussian MGF:

$$\overline{X_{ik}^4} = \frac{d^4}{d\nu^4} e^{\sigma_k^2 \nu^2 / 2} \bigg|_{\nu=0} = \frac{d^4}{d\nu^4} \left[1 + \frac{\sigma_k^2 \nu^2}{2} + \frac{\sigma_k^4 \nu^4}{8} + \cdots \right] \bigg|_{\nu=0}$$

$$= 3\sigma_k^4, \qquad k = 1, 2 \tag{1.81d}$$

Thus $\overline{Y^2} = 2L(\sigma_1^4 + \sigma_2^4) + L^2(\sigma_1^2 - \sigma_2^2)^2 = 2L(\sigma_1^4 + \sigma_2^4) + \overline{Y}^2$, and

$$\sigma_y^2 = 2L(\sigma_1^4 + \sigma_2^4) \tag{1.81e}$$

Substituting m_y and σ_y in the CLT formula (1.78d) yields the following expressions involving the Gaussian complementary CDF:

$$\Pr\{Y > 0\} = \Pr\left\{\sum_{i=1}^{L}(X_{i1}^2 - X_{i2}^2) > 0\right\} \approx 1 - Q\left(\frac{m_y}{\sigma_y}\right)$$

$$= 1 - Q\left(\frac{L(\sigma_1^2 - \sigma_2^2)}{\sqrt{2L(\sigma_1^4 + \sigma_2^4)}}\right) = 1 - Q\left(\frac{\sqrt{L}(\mu - 1)}{\sqrt{2(\mu^2 + 1)}}\right) \qquad (1.81f)$$

using $\mu \triangleq \sigma_1^2/\sigma_2^2$ as in the CB expression (1.76i). A comparison of the CLT expression (1.81f) with the CB expression (1.76i) is shown in Figure 1.67. Since for large L, the CLT gives a very close approximation, we learn from Figure 1.57 for the case of $L = 20$ that the CB, although it is exponentially tight, is rather loose compared with the CLT approximation. The main advantage of using the CB is the ease of calculation compared with that of the CLT, as demonstrated in this example. Both the CB and CLT approach the exact probability asymptotically with L. These approximations will be compared with the exact probability in Section 1.5.9.2.

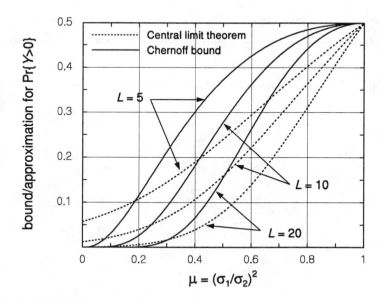

Figure 1.67 Comparison of CB and CLT approximation.

1.5.8 The Narrowband Gaussian Random Process

A random process $\{x(t)\}$ is called a *narrowband random process* [31] if its spectral density $S_x(f)$ is such that the significant region of the spectrum is small compared with the center frequency of that region. A narrowband spectral density is depicted in Figure 1.68.

If one observes a sample function $x(t)$ of a narrowband process on an oscilloscope, one may see a display like that illustrated in Figure 1.69, which has the appearance of a sinusoidal waveform with a slowly varying amplitude, whose absolute value is called the envelope, and slowly varying phase. We may express the sample function waveform in the form

$$x(t) = R(t) \cos[\omega_0 t + \phi(t)] \tag{1.82a}$$

where $\omega_0 = 2\pi f_0$ is the mean angular center frequency of the spectral band and the *envelope* $R(t)$ and the *phase* $\phi(t)$ are slowly varying functions of time. The statistical analysis pertaining to narrowband Gaussian random processes

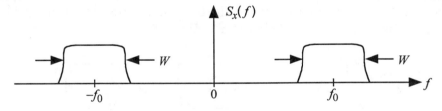

Figure 1.68 Spectral density of a narrowband random process ($W \ll f_0$).

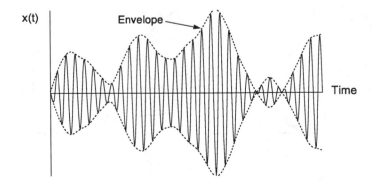

Figure 1.69 Typical time domain observation of a sample function of a narrowband random process.

is mostly due to the work of Rice [36–38], and our objective here is to summarize the major results of his work to the extent needed in the systems analysis in this book. The details of Rice's work are treated in many well-known books, such as [29–31].

1.5.8.1 Rayleigh Distributions

If we expand $x(t)$ given in (1.82a), we obtain

$$x(t) = x_c(t) \cos\omega_0 t - x_s(t) \sin\omega_0 t \tag{1.82b}$$

where

$$x_c(t) = R(t) \cos\phi(t) \tag{1.82c}$$

and

$$x_s(t) = R(t) \sin\phi(t) \tag{1.82d}$$

are called the *in-phase and cross-quadrature components* of $x(t)$, respectively, or sometimes more briefly, the in-phase and quadrature components of the *Rician decomposition* of the process. Hence, we have the identities

$$R(t) = \sqrt{x_c^2(t) + x_s^2(t)} \tag{1.82e}$$

and

$$\phi(t) = \tan^{-1}\left[\frac{x_s(t)}{x_c(t)}\right] \tag{1.82f}$$

where $R(t) \geq 0$ and $0 \leq \phi(t) \leq 2\pi$. It is known that the RVs $x_c(t) \triangleq X_c$ and $x_s(t) \triangleq X_s$, sampled at the same instant, are statistically independent Gaussian RVs with

$$\overline{x_c(t)} = \overline{x_s(t)} = 0 \tag{1.82g}$$

$$\sigma_{x_c}^2 = \sigma_{x_s}^2 = \sigma_x^2 = \sigma^2 \tag{1.82h}$$

and the joint pdf of X_c and X_s is given by

$$p_{x_c,x_s}(\beta, \gamma) = \frac{1}{2\pi\sigma^2} \exp\left\{-\frac{\beta^2 + \gamma^2}{2\sigma^2}\right\} \tag{1.83a}$$

We now need to obtain the joint pdf of the envelope and phase RVs $R(t)$ and $\phi(t)$, respectively, from the joint pdf $p_{x_c,x_s}(\beta, \gamma)$ given in (1.83a). The situation at hand is a case of transformation from the joint RVs

$[x_c(t), x_s(t)]$ to the new joint RVs $[R(t), \phi(t)]$, and the objective is to obtain $p_{R,\phi}(\alpha, \beta)$. The solution is best found using the *Jacobian of the transformation method*, and the reader is referred to [29–31] for further background on this subject. Here, the method is simply stated, as follows: Assume that a set of RVs $(X_1, X_2, \ldots, X_N)$ with the joint pdf $p_{x_1, x_2, \ldots, x_N}(\alpha_1, \alpha_2, \ldots, \alpha_N)$ is transformed to a new set of RVs $(Y_1, Y_2, \ldots, Y_N)$. The transformation $(X_1, X_2, \ldots, X_N) \rightarrow (Y_1, Y_2, \ldots, Y_N)$ is reversible in the sense that the inverse transformation $(Y_1, Y_2, \ldots, Y_N) \rightarrow (X_1, X_2, \ldots, X_N)$ also exists. That is, the desired transformations are given by

$$
\begin{aligned}
Y_1 &= f_1(X_1, X_2, \ldots, X_N) \\
Y_2 &= f_2(X_1, X_2, \ldots, X_N) \\
&\vdots \\
Y_N &= f_N(X_1, X_2, \ldots, X_N)
\end{aligned}
\tag{1.83b}
$$

whereas the inverse transformations are

$$
\begin{aligned}
X_1 &= g_1(Y_1, Y_2, \ldots, Y_N) \\
X_2 &= g_2(Y_1, Y_2, \ldots, Y_N) \\
&\vdots \\
X_N &= g_N(Y_1, Y_2, \ldots, Y_N)
\end{aligned}
\tag{1.83c}
$$

Then the desired pdf is given by

$$
\begin{aligned}
p_{y_1, \ldots, y_N}(\beta_1, \ldots, \beta_N) = |J| \\
\times \; p_{x_1, \ldots, x_N}(\alpha_1 = g_1(\beta_1, \ldots, \beta_N), \ldots, \alpha_N = g_N(\beta_1, \ldots, \beta_N))
\end{aligned}
\tag{1.83d}
$$

where the Jacobian J is given by

$$
J = \left| \frac{\partial(\alpha_1, \alpha_2, \ldots, \alpha_N)}{\partial(\beta_1, \beta_2, \ldots, \beta_N)} \right| \triangleq
\begin{vmatrix}
\frac{\partial g_1}{\partial \beta_1} & \cdots & \frac{\partial g_N}{\partial \beta_1} \\
\vdots & & \vdots \\
\frac{\partial g_1}{\partial \beta_N} & \cdots & \frac{\partial g_N}{\partial \beta_N}
\end{vmatrix}
\tag{1.83e}
$$

That is, the Jacobian is a determinant whose elements are the partial derivatives of "old" RVs with respect to each of the "new" RVs.

In the present case, the transformation $(X_1, X_2, \ldots, X_N) \rightarrow (Y_1, Y_2, \ldots, Y_N)$ pertains to the transformation of (X_c, X_s) into $R = R(t)$

and $\Phi = \phi(t)$ according to (1.82e) and (1.82f), with the reverse transformation being given by (1.82c) and (1.82d). The Jacobian is therefore given by

$$J = \left| \frac{\partial(X_c, X_s)}{\partial(R, \Phi)} \right| = \begin{vmatrix} \frac{\partial X_c}{\partial R} & \frac{\partial X_s}{\partial R} \\ \frac{\partial X_c}{\partial \Phi} & \frac{\partial X_s}{\partial \Phi} \end{vmatrix} = \begin{vmatrix} \cos\Phi & \sin\Phi \\ -R\sin\Phi & R\cos\Phi \end{vmatrix}$$

$$= R\cos^2\Phi + R\sin^2\Phi = R \tag{1.83f}$$

It follows that the joint pdf of R and Φ is

$$p_{R,\phi}(\alpha, \beta) = |J|\, p_{x_c, x_s}(\alpha\cos\beta, \alpha\sin\beta)$$

$$= \frac{\alpha}{2\pi\sigma^2} \exp\left\{ -\frac{\alpha^2\cos^2\beta + \alpha^2\sin^2\beta}{2\sigma^2} \right\}$$

$$= \begin{cases} \dfrac{\alpha}{2\pi\sigma^2}\, e^{-\alpha^2/2\sigma^2}, & \alpha \geq 0,\ 0 \leq \beta \leq 2\pi \\ 0, & \alpha < 0 \end{cases} \tag{1.84a}$$

The pdf of R, $p_R(\alpha)$, can be obtained by integrating over the values of Φ from 0 to 2π, and hence the marginal pdf is found to be

$$p_R(\alpha) = \begin{cases} \dfrac{\alpha}{\sigma^2}\, e^{-\alpha^2/2\sigma^2}, & \alpha \geq 0 \\ 0, & \alpha < 0 \end{cases} \tag{1.84b}$$

This is the Rayleigh pdf and is plotted in Figure 1.70. If we integrate the

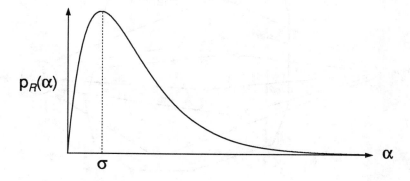

Figure 1.70 The Rayleigh pdf.

joint pdf in (1.84a) over the RV R from 0 to ∞, we obtain

$$p_\phi(\beta) = \frac{1}{2\pi}, \quad 0 \leq \beta < 2\pi \qquad (1.84c)$$

which means that Φ is a uniformly distributed random phase that is independent of the envelope RV, R, so that

$$p_{R,\phi}(\alpha, \beta) = p_r(\alpha)\, p_\phi(\beta) \qquad (1.84d)$$

We have just seen that the envelope of a narrowband Gaussian noise process is Rayleigh distributed. The Rayleigh distribution is an important topic in the discussion of communication, radar, and, in particular, mobile cellular communications channels.

1.5.8.2 Rayleigh Fading

Consider the situation depicted in Figure 1.71, where a transmitter emits a clean sinusoidal wave of the form $A\cos\omega_0 t$, where A and ω_0 are the constant amplitude and angular frequency, respectively. Out of the scattering channel one obtains a waveform $x(t)$, which is composed of in-phase and quadrature components that are the sums of many delayed replicas of the transmitted waveform. That is:

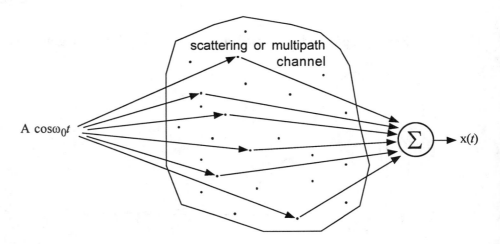

Figure 1.71 Scattering or multipath channel.

$$x(t) = \sum_i \left[A_i \cos \omega_0 t + B_i \cos (\omega_0 t + \pi/2) \right]$$

$$= \sum_i \left[A_i \cos \omega_0 t - B_i \sin \omega_0 t \right]$$

$$= x_c(t) \cos \omega_0 t - x_s(t) \sin \omega_0 t \tag{1.85a}$$

where

$$x_c(t) \triangleq \sum_i A_i, \quad x_s(t) \triangleq \sum_i B_i \tag{1.85b}$$

The idea is that the time-varying nature of the scattering channel (multipath channel) makes the sums $\sum_i A_i$ and $\sum_i B_i$ be time-varying amplitudes for the in-phase and quadrature components of the received waveform when there is relative movement between the transmitter and the receiver. The waveform $x(t)$ is clearly a sample function of a random process $\{x(t)\}$, and further, if we invoke the CLT, we can regard $x_c(t)$ and $x_s(t)$ as if they are Gaussian RVs. If $x(t)$ of (1.85a) is expressed in terms of the envelope and phase, we have

$$x(t) = R(t) \cos[\omega_0 t + \phi(t)] \tag{1.85c}$$

Thus, we have the same expression for a narrowband Gaussian random process as for the sample function out of the scattering or multipath channel when the input (the transmitted signal) is a clean sinusoidal waveform. It follows that the sum of multipath waveforms, such as experienced in the mobile communications channel, has an envelope with a Rayleigh distribution. Fading channel characteristics are discussed in detail in Chapter 2, which deals with mobile radio propagation considerations. The root mean square value of the Rayleigh envelope is

$$R_{rms} = \sqrt{E\{R^2\}} = \left[\int_0^\infty \alpha^2 \cdot \frac{\alpha}{\sigma^2} e^{-\alpha^2/2\sigma^2} d\alpha \right]^{1/2} = \sqrt{2}\,\sigma \tag{1.85d}$$

and when the signal envelope R falls below the rms envelope value R_{rms}, we say that the signal is experiencing a "fade." This is the reason that the so-called "Rayleigh fading" channel is usually assumed in models of multipath mobile communications channels. What should be noted here is that the Rayleigh distribution of the amplitude RV is due not only to the presence of narrowband Gaussian noise, such as receiver-generated thermal noise, but also to the composite waveform of the scattered or multipath signals.

1.5.8.3 Sinewave Plus Narrowband Gaussian Random Process

We now determine the pdfs of the envelope and phase of the sum of a sinewave process and a narrowband Gaussian random process. The sinewave process is given by

$$\{A \cos(\omega_0 t + \theta)\} \tag{1.86a}$$

where A and ω_0 are constant amplitude and constant angular frequency, respectively, and θ is a phase random variable uniformly distributed over the interval $(0, 2\pi)$. A typical sample function $z(t)$ of a sinewave plus narrowband Gaussian random process $\{x(t)\}$ is given by

$$z(t) = A \cos(\omega_0 t + \theta) + x(t) \tag{1.86b}$$

where the sample function $x(t)$ is given by (1.82b).

Expanding the right-hand side of (1.86b) and substituting (1.82b) for $x(t)$, we have

$$z(t) = z_c(t) \cos\omega_0 t - z_s(t) \sin\omega_0 t \tag{1.87a}$$

where

$$z_c(t) \triangleq A \cos\theta + x_c(t) \tag{1.87b}$$

$$z_s(t) \triangleq A \sin\theta + x_s(t) \tag{1.87c}$$

If we express $z(t)$ in terms of an envelope and phase, we have

$$z(t) = R(t) \cos[\omega_0 t + \phi(t)] \tag{1.87d}$$

where

$$R(t) \triangleq \sqrt{z_c^2(t) + z_s^2(t)}, \quad R(t) \geq 0 \tag{1.87e}$$

and

$$\phi(t) \triangleq \tan^{-1}\left[\frac{z_s(t)}{z_c(t)}\right], \quad 0 \leq \phi(t) \leq 2\pi \tag{1.87f}$$

From the identities of (1.87d) and (1.87a), it is clear that

$$z_c(t) = R(t) \cos \phi(t) \tag{1.87g}$$

$$z_s(t) = R(t) \sin \phi(t) \tag{1.87h}$$

which are in the same form as (1.82c) and (1.82d). Our objective is to obtain the joint pdf of $R(t)$ and $\phi(t)$ from the knowledge of the joint pdf of $z_c(t)$ and $z_s(t)$, given a phase value θ of the sinewave sample function.

Now, because $x_c(t) = G(0, \sigma^2)$ and $x_s(t) = G(0, \sigma^2)$ are independent Gaussian RVs, $z_c(t)$ and $z_s(t)$ are also independent Gaussian RVs, given the phase θ. That is:

$$z_c(t) = G(A \cos \theta, \sigma^2) \tag{1.88a}$$

$$z_s(t) = G(A \sin \theta, \sigma^2) \tag{1.88b}$$

Thus, the joint pdf of $z_c(t)$ and $z_s(t)$, given θ, is

$$p_{z_c,z_s}(\alpha, \beta \mid \theta = \gamma) = \frac{1}{2\pi\sigma^2} \exp\left\{ -\frac{(\alpha - A\cos\gamma)^2 + (\beta - A\sin\gamma)^2}{2\sigma^2} \right\} \tag{1.88c}$$

The joint pdf of $R(t)$ and $\phi(t)$ can now be determined by the Jacobian method used in the previous subsection. From (1.87g) and (1.87h), the Jacobian of the transformation is

$$J = \left| \frac{\partial(z_c, z_s)}{\partial(R, \phi)} \right| = \begin{vmatrix} \frac{\partial z_c}{\partial R} & \frac{\partial z_s}{\partial R} \\ \frac{\partial z_c}{\partial \phi} & \frac{\partial z_s}{\partial \phi} \end{vmatrix} = \begin{vmatrix} \cos\phi & \sin\phi \\ -R\sin\phi & R\cos\phi \end{vmatrix} = R$$

and therefore we have

$$p_{R,\phi}(\alpha, \beta \mid \theta = \gamma) = |J| \, p_{z_c,z_s}(\alpha \cos\beta, \alpha \sin\beta \mid \theta = \gamma)$$

$$= \frac{\alpha}{2\pi\sigma^2} \exp\left\{ -\frac{(\alpha\cos\beta - A\cos\gamma)^2 + (\alpha\sin\beta - A\sin\gamma)^2}{2\sigma^2} \right\}$$

$$= \frac{\alpha}{2\pi\sigma^2} \exp\left\{ -\frac{1}{2\sigma^2}[\alpha^2 + A^2 - 2A\alpha\cos(\gamma - \beta)] \right\} \tag{1.88d}$$

Integrating over the phase ϕ, we have

$$p_R(\alpha \mid \theta = \gamma) = \int_0^{2\pi} p_{R,\phi}(\alpha, \beta \mid \theta = \gamma)\, d\beta$$

$$= \frac{\alpha}{2\pi\sigma^2}\, e^{-(\alpha^2 + A^2)/2\sigma^2} \int_0^{2\pi} \exp\left\{ \frac{A\alpha}{\sigma^2} \cos(\gamma - \theta) \right\} d\beta \qquad (1.88e)$$

The integral over the phase ϕ is given by [38]

$$\int_0^{2\pi} \exp\left\{ \frac{A\alpha}{\sigma^2} \cos(\gamma - \theta) \right\} d\beta = 2\pi\, I_0\left(\frac{A\alpha}{\sigma^2} \right) \qquad (1.88f)$$

where $I_0(x)$ is the modified Bessel function of order zero [20] given by

$$I_0(x) = \sum_{n=0}^{\infty} \frac{x^{2n}}{2^{2n}(n!)^2} \qquad (1.88g)$$

$$\approx \begin{cases} 1 + \frac{1}{4}x^2 + \frac{1}{64}x^4, & x < 2.5 \\ e^{1.1x}/1.9x, & 2.5 < x < 10 \end{cases} \qquad (1.88h)$$

The function $I_0(x)$ is shown plotted in Figure 1.72. The conditional pdf of the envelope R given that the sinewave phase is $\theta = \gamma$ is independent of the phase. Thus, the pdf of the envelope of a sinewave plus a narrowband Gaussian random noise process is given by

$$p_R(\alpha) = \begin{cases} \dfrac{\alpha}{\sigma^2}\, e^{-(\alpha^2 + A^2)/2\sigma^2} I_0\left(\dfrac{A\alpha}{\sigma^2} \right), & \alpha \geq 0 \\ 0, & \text{otherwise} \end{cases} \qquad (1.89a)$$

This pdf is known as the Rician pdf, and its application in the field of radar and communications is considerable. We shall discuss this pdf a bit more: let us define a normalized envelope $r = R/\sigma$ and a normalized sinewave amplitude $a = A/\sigma$. The Rician pdf is then expressed as

$$p_r(\alpha) = \begin{cases} \alpha\, e^{-(\alpha^2 + a^2)/2} I_0(a\alpha), & \alpha \geq 0 \\ 0, & \text{otherwise} \end{cases} \qquad (1.89b)$$

which is plotted in Figure 1.73 for different values of a. Note that $a^2/2 = A^2/2\sigma^2 = \text{SNR}$. It should be observed that as $A \to 0$, the Rician pdf approaches the Rayleigh pdf, as expected.

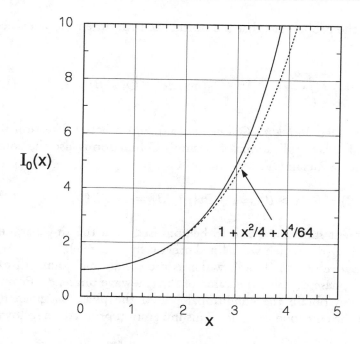

Figure 1.72 Modified Bessel function and approximation for it.

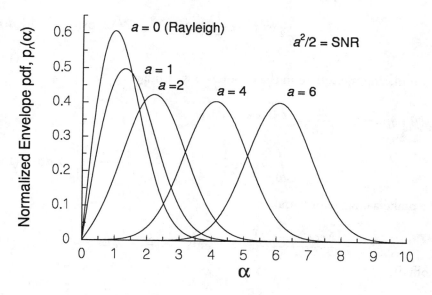

Figure 1.73 Rician pdf for different values of $a = \sqrt{2 \cdot \text{SNR}}$.

If we compute the probability that the normalized Rician envelope exceeds a value β, we have

$$\Pr\{r > \beta\} = \int_{\beta}^{\infty} \alpha \, e^{-(\alpha^2+a^2)/2} I_0(a\alpha) \, d\alpha \triangleq Q(a, \beta) \qquad (1.90a)$$

This integral is called *Marcum's Q-function* [39], and it is the detection probability in typical radar applications. Marcum's Q-function is also the complementary CDF for a Rician RV:

$$Q(a, \beta) = \Pr\{r > \beta\} = 1 - \Pr\{r \le \beta\} = 1 - F_r(\beta) \qquad (1.90b)$$

This two-argument function is not to be confused with the single-argument Gaussian CCDF $Q(\alpha)$, which we defined previously in (1.49e).

Let us now return to (1.88d), which is the conditional joint pdf of the envelope R and phase ϕ, given the value of the sinewave phase, θ. From this joint pdf, we can obtain the conditional pdf of the phase ϕ, given θ, by integrating out the envelope variable. This integration proceeds as follows:

$$
\begin{aligned}
p_\phi(\beta \mid \theta = \gamma) &= \int_0^\infty p_{R,\phi}(\alpha, \beta \mid \theta = \gamma) \, d\alpha \\
&= \int_0^\infty \frac{\alpha}{2\pi\sigma^2} \exp\left\{ -\frac{1}{2\sigma^2} [\alpha^2 + A^2 - 2A\alpha\cos(\gamma - \beta)] \right\} d\alpha
\end{aligned}
$$

$$(1.91a)$$

By completing the square in the exponent, we obtain

$$
p_\phi(\beta \mid \theta = \gamma) = \frac{1}{2\pi\sigma^2} e^{-A^2[1-\cos^2(\gamma-\beta)]/2\sigma^2}
$$
$$
\times \int_0^\infty \alpha \exp\left\{ -\frac{1}{2} \left[\frac{\alpha - A\cos(\gamma - \beta)}{\sigma} \right]^2 \right\} d\alpha \qquad (1.91b)
$$

If we make a change of variable:

$$\frac{A\cos(\gamma - \beta)}{\sigma} = \lambda \quad \text{and} \quad \frac{\alpha}{\sigma} - \lambda = \mu \qquad (1.91c)$$

we obtain

$$p_\phi(\beta \mid \theta = \gamma) = \frac{1}{2\pi} e^{-A^2 \sin^2(\gamma-\beta)/2\sigma^2} \int_{-\lambda}^{\infty} (\mu + \lambda) e^{-\mu^2/2} \, d\mu$$

$$= \frac{1}{2\pi} e^{-A^2 \sin^2(\gamma-\beta)/2\sigma^2} \left[e^{-\lambda^2/2} + \lambda \sqrt{2\pi} \int_{-\lambda}^{\infty} \frac{e^{-\mu^2/2}}{\sqrt{2\pi}} \, d\mu \right]$$

$$= \frac{1}{2\pi} \left[e^{-A^2/2\sigma^2} + \lambda \sqrt{2\pi} \, e^{-A^2 \sin^2(\gamma-\beta)/2\sigma^2} Q(-\lambda) \right] \qquad (1.91\text{d})$$

After recognizing that $Q(-\lambda) = 1 - Q(\lambda)$ and using the normalized amplitude $a = A/\sigma$, we have the result

$$p_\phi(\beta \mid \theta = \gamma) = \frac{1}{2\pi} e^{-\frac{1}{2}a^2} + \frac{a\cos(\gamma-\beta)}{\sqrt{2\pi}} e^{-\frac{1}{2}a^2 \sin^2(\gamma-\beta)} \{1 - Q[a\cos(\gamma-\beta)]\}$$

$$(1.91\text{e})$$

The pdf (1.91e) is plotted in Figure 1.74 with a as a parameter and with the case $\theta = \pi$ radians selected for convenience. Note that as $a \to 0$ (no signal present), the pdf $p_\phi(\beta \mid \theta = \gamma)$ becomes the uniform distribution's pdf ($= 1/2\pi$) as expected.

It is worth mentioning that when a sinewave signal term (specular component) is received in the presence of a Rayleigh component, the channel

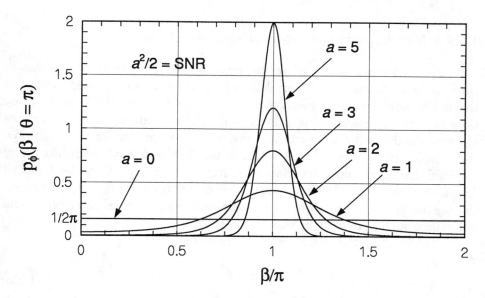

Figure 1.74 The pdf of the phase of a sinewave plus narrowband Gaussian noise.

inducing such a case is referred to as a Rician channel. An example situation is that of a satellite emitting a sinewave signal that is received along with multipath components by a terrestrial receiver.

Approximation to Marcum's Q-function. The probability integral (1.90a) has extensive application in radar and communications systems analysis, and it has been tabulated because it is difficult to compute accurately for the interesting cases of $\beta \gg a$ and small values of the probability. But the table is not widely available. Therefore, it is useful to have an easily computed yet close approximation to Marcum's Q-function. It can be shown that this function is the CCDF for a noncentral chi-squared RV (to be discussed in Section 1.5.9), and can be approximated by [40]

$$Q(a, \beta) \approx Q\left(\sqrt{\beta^2 - \tfrac{1}{2}} - \sqrt{a^2 + \tfrac{1}{2}}\right), \qquad \beta^2 > \tfrac{1}{2} \tag{1.92a}$$

and

$$Q(a, \beta) \approx Q(\beta - a) \quad \text{for} \quad a \gg 1,\ \beta \gg 1 \tag{1.92b}$$

This compact, closed-form approximation to Marcum's Q-function is effectively used in obtaining a very simple expression for the probability of error for a $\pi/4$ DQPSK modulation system in [41].

1.5.8.4 Modeling and Simulation of Bandpass Noise

We have seen that samples of white noise, no matter how close together in time, are uncorrelated. This correlation property is directly related to the fact that white noise has a flat spectrum. Although, strictly speaking, white noise does not exist in nature, it is a useful concept for modeling communications systems. In this section, we consider the correlation properties of bandlimited white noise; that is, white noise with spectral level η that has been filtered by an ideal bandpass filter, so that its spectrum is confined to the band of frequencies that is B Hz wide about some center frequency, $\pm f_c$. Then we discuss the properties of white noise bandlimited by practical filter types, and show how such noise may be simulated.

The bandlimited white noise spectrum is given by

$$S_x(f) = \begin{cases} \eta, & f \in (f_c \pm \tfrac{1}{2}B) \text{ and } f \in (-f_c \pm \tfrac{1}{2}B) \\ 0, & \text{otherwise} \end{cases} \tag{1.93a}$$

The autocorrelation function is the inverse Fourier transform of the spectral density function:

$$R_x(\tau) = \int_{-\infty}^{\infty} e^{j2\pi f\tau} \, \mathcal{S}_x(f) \, df = 2 \int_{0}^{\infty} \cos(2\pi f\tau) \, \mathcal{S}_x(f) \, df, \quad \mathcal{S}_x(f) \text{ even}$$

$$= 2\eta \int_{f_c-B/2}^{f_c+B/2} \cos(2\pi f\tau) \, df = \frac{\eta}{\pi\tau} [\sin(2\pi f\tau)]_{f=f_c-B/2}^{f=f_c+B/2} \tag{1.93b}$$

$$= 2\eta B \cdot \text{sinc}(B\tau) \cdot \cos(2\pi f_c\tau) \tag{1.93c}$$

It is customary when dealing with bandpass signals and noise to use the notation $\eta = \frac{1}{2}\mathcal{N}_0$ for the "two-sided" noise spectral density level. With this notation, the autocorrelation function is

$$R_x(\tau) = \mathcal{N}_0 B \cdot \text{sinc}(B\tau) \cdot \cos(2\pi f_c\tau) \tag{1.93d}$$

as illustrated in Figure 1.75. The noise power is $\sigma^2 = R_x(0) = \mathcal{N}_0 B = 2\eta B$. By contrast, when dealing with a lowpass waveform contained in the interval $|f| \leq B/2$, (1.93b) becomes

$$R_x(\tau) = 2\eta \int_{0}^{B/2} \cos(2\pi f\tau) \, df = \frac{\eta}{\pi\tau} [\sin(2\pi f\tau)]_{f=0}^{f=B/2} \tag{1.94a}$$

$$= \eta B \cdot \text{sinc}(B\tau) = \mathcal{N}_0 B \, \text{sinc}(B\tau) \tag{1.94b}$$

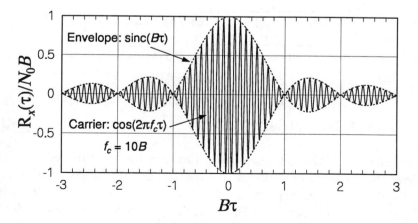

Figure 1.75 Autocorrelation function for bandlimited white noise.

Thus, the spectral density for *lowpass* bandlimited white noise is $\eta = \mathcal{N}_0$ in the passband.

The form of the autocorrelation function suggests that the bandlimited white noise can be modeled as baseband noise modulating the carrier f_c. Indeed, the Rician decomposition of the bandlimited noise waveform $x(t)$ is given by (1.109b) with $\omega_c = 2\pi f_c$ replacing ω_0 in that expression:

$$x(t) = x_c(t) \cos(2\pi f_c t) - x_s(t) \sin(2\pi f_c t) \tag{1.95a}$$

where $x_c(t)$ and $x_s(t)$ are independent baseband (lowpass) noise quadrature terms. It is instructive to calculate the autocorrelation function of the Ricean decomposition form of the bandlimited white noise, and compare the result with the autocorrelation function already found. Let $x_c(t)$ and $x_s(t)$ be independent noise processes, each with the spectrum $S_0(f) = \mathcal{N}_0$, $|f| \leq B/2$. Thus the autocorrelation functions of $x_c(t)$ and $x_s(t)$ are the same:

$$R_{x_c}(\tau) = R_{x_s}(\tau) = \mathcal{N}_0 \int_{-B/2}^{B/2} e^{j2\pi f \tau} \, df = \mathcal{N}_0 B \operatorname{sinc}(B\tau) \tag{1.95b}$$

The autocorrelation function of the Rician decomposition form of $x(t)$ is

$$
\begin{aligned}
R_x(\tau) &= \mathrm{E}\{x(t)\,x(t-\tau)\} \\
&= \mathrm{E}\{x_c(t)\,x_c(t-\tau)\} \cos(2\pi f_c t)\cos(2\pi f_c t - 2\pi f_c \tau) \\
&\quad + \mathrm{E}\{x_s(t)\,x_s(t-\tau)\} \sin(2\pi f_c t)\sin(2\pi f_c t - 2\pi f_c \tau) \\
&\quad \overbrace{- \mathrm{E}\{x_c(t)\,x_s(t-\tau)\}}^{=\,0} \cos(2\pi f_c t)\sin(2\pi f_c t - 2\pi f_c \tau) \\
&\quad \overbrace{- \mathrm{E}\{x_s(t)\,x_c(t-\tau)\}}^{=\,0} \sin(2\pi f_c t)\cos(2\pi f_c t - 2\pi f_c \tau) \\
&= \mathcal{N}_0 B \operatorname{sinc}(B\tau) \left[\cos(2\pi f_c t)\cos(2\pi f_c t - 2\pi f_c \tau) \right. \\
&\quad \left. + \sin(2\pi f_c t)\sin(2\pi f_c t - 2\pi f_c \tau) \right] \\
&= \mathcal{N}_0 B \operatorname{sinc}(B\tau) \cos(2\pi f_c \tau) \tag{1.95c}
\end{aligned}
$$

This calculation of the autocorrelation function for bandlimited white noise confirms the accuracy of modeling the spectral density functions of $x_c(t)$ and $x_s(t)$ by the rectangular spectrum $S_0(f)$. Note that the noise powers in the baseband noise waveforms are $R_{x_c}(0) = R_{x_s}(0) = \mathcal{N}_0 B$ W; this is the same amount of power as in the Rician decomposition, $x(t)$, because

$$R_x(0) = E\{x^2(t)\} = E\{[x_c(t)\cos(2\pi f_c t) - x_s(t)\sin(2\pi f_c t)]^2\}$$

$$= \overbrace{E\{x_c^2(t)\}}^{\mathcal{N}_0 B}\cos^2(2\pi f_c t) - 2\overbrace{E\{x_c(t)\,x_s(t)\}}^{0}\sin(2\pi f_c t)\cos(2\pi f_c t)$$

$$+ \overbrace{E\{x_s^2(t)\}}^{\mathcal{N}_0 B}\sin^2(2\pi f_c t)$$

$$= \mathcal{N}_0 B \tag{1.95d}$$

As discussed in Section 1.3, to represent the bandlimited white noise process $x(t)$ by samples, it is sufficient to take samples of $x_c(t)$ and $x_s(t)$ at the Nyquist rate of B pairs of samples per second. The correlation between samples of either of these waveforms is

$$R_{x_c}(kT_s) = \mathcal{N}_0 B\,\text{sinc}(B\cdot kT_s) = \mathcal{N}_0 B\,\text{sinc}(k) = 0, \quad \text{for } T_s = 1/B \tag{1.95e}$$

That is, the samples of the two baseband noise waveforms taken at the Nyquist rate are uncorrelated. The joint pdf of N pairs of samples $\{x_{ci}, x_{si}\}$ of the in-phase and quadrature components of $x(t)$ therefore is

$$p_x(x_c, x_s) = \prod_{i=1}^{N}\frac{1}{2\pi\sigma^2}\exp\left\{-\frac{x_{ci}^2 + x_{si}^2}{2\sigma^2}\right\}, \quad \text{with } \sigma^2 = \mathcal{N}_0 B \tag{1.96}$$

Bandpass (colored) Gaussian noise models. In practice, the receiver filter in a communications system is not ideal in the sense of having a transfer function that is flat across the passband, with the result that the received noise spectrum is no longer "white" (flat) in the passband, but it is "colored," as illustrated in Figure 1.76, in which one definition of bandwidth is used—the "3-dB bandwidth" B_3. After considering several bandpass filter characteristics and their corresponding autocorrelation functions, we consider the effect of filter shape on the correlations between noise samples.

The impulse response $h(t)$ of a bandpass filter that is symmetric about its center frequency f_c can be modeled by the expression

$$h(t) = \begin{cases} 2h_0(t)\cos(2\pi f_c t), & t \geq 0 \\ 0, & \text{otherwise} \end{cases} \tag{1.97a}$$

where $h_0(t)$ is an equivalent baseband filter impulse response. The Fourier transform of $h(t)$ is

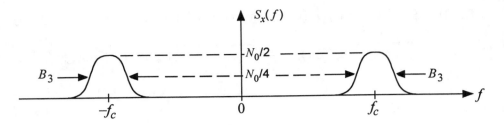

Figure 1.76 Spectrum of colored bandpass noise.

$$H(f) = \int_0^\infty e^{-j2\pi ft} \cdot 2\,h_0(t) \cos\left(2\pi f_c t\right) dt$$

$$= \int_0^\infty e^{-j2\pi(f-f_c)t} h_0(t)\, dt + \int_0^\infty e^{-j2\pi(f+f_c)t} h_0(t)\, dt$$

$$= H_0(f - f_c) + H_0(f + f_c) \tag{1.97b}$$

When the input to the filter is white noise, the noise spectrum at the output of the bandpass filter is

$$\mathcal{S}(f) = \tfrac{1}{2}\mathcal{N}_0 \cdot |H(f)|^2 = \tfrac{1}{2}\mathcal{N}_0 \left|H_0(f - f_c) + H_0(f + f_c)\right|^2$$

$$= \tfrac{1}{2}\mathcal{N}_0\big\{|H_0(f - f_c)|^2 + |H_0(f + f_c)|^2$$

$$+ H_0(f - f_c)\,H_0^*(f + f_c) + H_0(f + f_c)\,H_0^*(f - f_c)\big\}$$

$$\approx \tfrac{1}{2}\mathcal{N}_0 \left\{|H_0(f - f_c)|^2 + |H_0(f + f_c)|^2\right\} \tag{1.97c}$$

The total received noise power is

$$\sigma^2 = \int_{-\infty}^\infty \mathcal{S}(f)\, df = \tfrac{1}{2}\mathcal{N}_0\left\{\int_{-\infty}^\infty |H_0(f - f_c)|^2 df + \int_{-\infty}^\infty |H_0(f + f_c)|^2 df\right\}$$

$$= \mathcal{N}_0 \int_{-\infty}^\infty |H_0(f)|^2\, df \tag{1.97d}$$

The *3-dB bandwidth* of the filter, denoted B_3, is found from the equation

$$\left|H_0\left(\tfrac{1}{2}B_3\right)\right|^2 \overset{\text{set}}{=} \tfrac{1}{2}|H_0(0)|^2 \tag{1.98}$$

The *noise bandwidth* or *equivalent rectangular bandwidth* of the bandpass filter, denoted B_N, is defined as

$$B_N \triangleq \frac{1}{|H_0(0)|^2} \int_{-\infty}^{\infty} |H_0(f)|^2 \, df \qquad (1.99)$$

Selection of filter models. For analysis purposes, it is often convenient to assume that the filters are "ideal" (rectangular) filters, which do not distort the signal waveform. When the bandwidth of the ideal filter is matched to the noise bandwidth of the actual filter, nearly the same systems analysis results are obtained. When it is desired to model specifically the effects of filtering on the signal, it is convenient for analysis to assume a "realistic" filter shape that is mathematically tractable, such as the Gaussian-shaped filter.

For simulation purposes, the filter model does not have to be mathematically tractable. A class of filters such as the Butterworth filters is sometimes convenient for simulation because the shape of the filter is easily modified by changing a single parameter. Table 1.5 gives the transfer function, 3-dB bandwidth, and noise bandwidth for several well-known filter types.

Analysis of autocorrelation for Butterworth baseband filters. The Butterworth family of filters is preferred in many analog applications because its characteristic approaches that for an ideal bandpass filter, one that does not change the spectral characteristics of the signal, as illustrated in Figure 1.77. Next we calculate the baseband autocorrelation function $R_c(\tau)$ for the Butterworth family of filters.

It can shown that the baseband autocorrelation function for bandpass noise whose spectrum has been shaped by a Butterworth filter is

Table 1.5 Example bandpass filter parameters

Equivalent lowpass filter type	$\|H_0(f)\|^2$	B_3	B_N
Butterworth (K-pole) (maximally flat $\lim\limits_{K \to \infty}$ → ideal BPF)	$\dfrac{1}{1 + (f/B)^{2K}}$	$2B$	$2B \cdot \dfrac{(\pi/2K)}{\sin(\pi/2K)}$ $= 1.57B_3, \ K = 1$ $= 1.11B_3, \ K = 2$ $= 1.05B_3, \ K = 3$
Gaussian-shaped ("realistic" BPF)	$e^{-\pi(f/2B)^2}$	$4B\sqrt{\dfrac{\ln 2}{\pi}}$	$2B = 1.0645B_3$

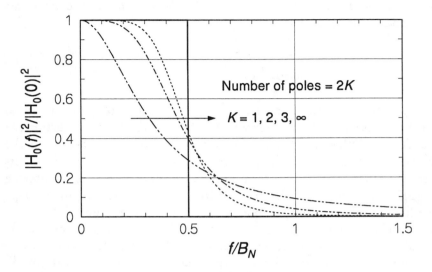

Figure 1.77 Normalized transfer function for Butterworth filters.

$$R_c(\tau) = \int_{-\infty}^{\infty} e^{+j2\pi f\tau} \mathcal{N}_0 |H_0(f)|^2 df = 2\mathcal{N}_0 \int_0^{\infty} \frac{\cos(2\pi f\tau)}{1 + (2f/B_3)^{2K}} df$$

$$= \mathcal{N}_0 B_3 \int_0^{\infty} \frac{\cos(\pi B_3 \tau x)}{1 + x^{2K}} dx$$

$$= \begin{cases} \mathcal{N}_0 B_N \, e^{-2B_N|\tau|}, & K = 1 \\ \mathcal{N}_0 B_N \, e^{-2B_N|\tau|} \sqrt{2} \sin\left(\frac{\pi}{4} + 2B_N|\tau|\right), & K = 2 \\ \mathcal{N}_0 B_N \left\{ \frac{1}{2} e^{-3B_N|\tau|} + e^{-\frac{3}{2}B_N|\tau|} \sin\left(\frac{\pi}{6} + \frac{3\sqrt{3}}{2} B_N|\tau|\right) \right\}, & K = 3 \end{cases} \quad (1.100)$$

The autocorrelation function $R_c(\tau)$ in (1.100) is plotted in Figure 1.78 for $K = 1, 2$, and 3, corresponding respectively to 2, 4, and 6 poles. Recall that samples of the baseband noise terms x_c and x_s are uncorrelated when the bandpass filter is ideal (flat) and the samples are taken at the rate $1/B = 1/B_N$. In general, for other filters, there is some correlation between samples. As the number of poles, $2K$, becomes large, the Butterworth filter becomes the ideal bandpass filter. We see from Figure 1.78 that for as few as $2K = 6$ poles, the baseband autocorrelation function is very close to ideal, and the correlations between samples at the rate $1/B_N$ can be ignored.

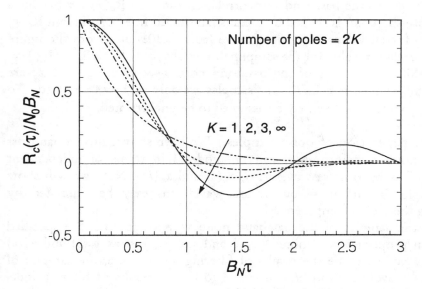

Figure 1.78 Baseband autocorrelation function for bandpass noise with spectrum shaped by K th order Butterworth filter.

Analysis of autocorrelation for Gaussian-shaped baseband filters. The Gaussian-shaped baseband filter may be used to represent the class of "realistic" filters with nonflat passbands. For digital communications, the best receiver filter is not necessarily flat in the passband, but instead has a frequency transfer characteristic matching the spectrum of the signal. The baseband autocorrelation function for bandpass noise whose spectrum has been shaped by a Gaussian-shaped filter is

$$
R_c(\tau) = \int_{-\infty}^{\infty} e^{j2\pi f\tau}\, \mathcal{N}_0 |H_0(f)|^2 df = \mathcal{N}_0 \int_{-\infty}^{\infty} e^{j2\pi f\tau}\, e^{-\pi(f/B_N)^2} df
$$

$$
= \mathcal{N}_0 \int_{-\infty}^{\infty} \exp\left\{ -\frac{\pi}{B_N^2}\left[(f - jB_N^2\,\tau)^2 + B_N^4\,\tau^2 \right] \right\} df
$$

$$
\overset{\sigma\sqrt{2\pi}\,\times\,\text{Gaussian pdf, with } \sigma = B_N/\sqrt{2\pi}}{= \mathcal{N}_0\, e^{-\pi(B_N\tau)^2} \int_{-\infty}^{\infty} \overbrace{\exp\left\{ -\frac{\pi}{B_N^2}\left(f - jB_N^2\,\tau \right)^2 \right\}}}
$$

$$
= \mathcal{N}_0 B_N\, e^{-\pi(B_N\tau)^2} \tag{1.101}
$$

Figure 1.79 shows the baseband autocorrelation function $R_c(\tau)$ for bandpass noise whose spectrum is shaped by a Gaussian-shaped filter. Although $R_c(\tau)$ is nonzero for all values of τ, we observe from (1.101) and from the figure that the correlation value for the sampling interval $\tau = 1/B_N$ is $R_c(1/B_N) = e^{-\pi} = 0.0432$. If samples of the baseband noise waveforms x_c and x_s are taken at the Nyquist rate $1/B_N$, the samples are only slightly correlated. To a close approximation they may be assumed to be uncorrelated.

Computer generation of noise samples. We have shown how a bandpass Gaussian noise waveform $x(t)$ can be modeled in terms of independent quadrature baseband noise waveforms, $x_c(t)$ and $x_s(t)$. Now we will show how samples of the noise quadrature waveforms may be simulated by generating them on a computer.

Let the samples of the baseband noise waveforms with the specified correlation properties be denoted $\{x_{cn}\}$ and $\{x_{sn}\}$. Just as $x_c(t)$ and $x_s(t)$ can be considered to be the results of filtering white noise, the samples of these noise waveforms can be considered to be the results of filtering independent noise samples $\{\xi_{cn}\}$ and $\{\xi_{sn}\}$, respectively, using a discrete filter, such as a FIR discrete filter. To generate samples of both quadrature noise components, conceptually two Gaussian random number generators and two FIRs are needed. The "whiteness" of the input noise is simulated by the fact that (in non-real time) samples of the input noise are independent, no matter how frequently the numbers are generated. The FIRs (actually a subprogram

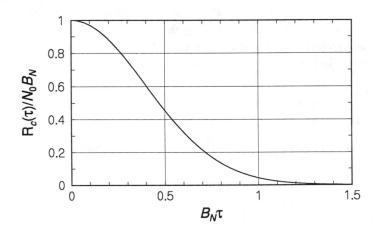

Figure 1.79 Baseband autocorrelation function for bandpass noise with spectrum shaped by Gaussian-shaped filter.

that simulates them) can be discrete-time approximations to the desired filter impulse response, with a finite number of points, say N_{FIR} points.

Most programming languages provide a uniform random number generator, a function that returns a pseudorandomly selected real number between 0 and 1. Let a uniform random variable be denoted U. Two independent zero-mean Gaussian random variables with variance σ^2 may be generated from two independent uniform random variables U_1 and U_2 in the following manner:

1. Take the natural logarithm of one of the uniform variables (yielding a negative number), multiply the result by $-2\sigma^2$, and then take the square root. As we will show, these operations give an "envelope" random variable, $R = \sqrt{-2\sigma^2 \cdot \ln U_1}$; that is, Rayleigh distributed.

2. Multiply the other uniform variable by 2π. This gives a "phase" variable, $\theta = 2\pi U_2$; that is, uniformly distributed between 0 and 2π radians.

3. Obtain the two independent Gaussian random variables as $x_c = R \cos\theta$ and $x_s = R \sin\theta$.

This process is diagrammed in Figure 1.80. The mathematical justification for this procedure can be stated as follows: Let U_1 and U_2 be two independent uniform RVs. Their joint pdf is a constant:

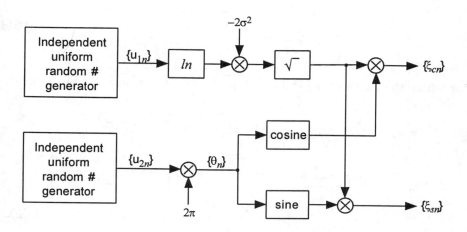

Figure 1.80 Diagram of the generation of two independent Gaussian random variables from two independent uniform random variables.

$$p_{u_1,u_2}(\alpha, \beta) = p_{u_1}(\alpha)\, p_{u_2}(\beta) = \begin{cases} 1, & 0 \le \alpha, \beta \le 1 \\ 0, & \text{otherwise} \end{cases} \qquad (1.102a)$$

Noting that $ln\, U_1 \le 0$ because $U_1 \le 1$, we define the transformation

or

$$R = \sqrt{-2\sigma^2 \cdot ln\, U_1} \qquad\qquad \theta = 2\pi \cdot U_2$$
$$U_1 = e^{-R^2/2\sigma^2} \qquad\qquad U_2 = \theta/2\pi \qquad (1.102b)$$

for which the Jacobian is $J = \left(R\, e^{-R^2/2\sigma^2}\right)/2\pi\sigma^2$. The joint pdf of R and θ is found as

$$p_{R,\theta}(r, \vartheta) = |J| \cdot p_{u_1,u_2}\left(e^{-r^2/2\sigma^2}, \frac{\vartheta}{2\pi}\right)$$

$$= \begin{cases} \dfrac{r}{2\pi\sigma^2}\, e^{-r^2/2\sigma^2}, & 0 \le \vartheta \le 2\pi,\ r \ge 0 \\ 0, & \text{otherwise} \end{cases}$$

$$= p_R(r)\, p_\theta(\vartheta) \qquad (1.102c)$$

the product of the pdfs of two RVs, a Rayleigh RV with power σ^2 and 2π times a uniform RV.

Now we define another transformation of variables as

or

$$x_c = r\cos\vartheta, \qquad\qquad x_s = r\sin\vartheta$$
$$r = \sqrt{x_c^2 + x_s^2}, \qquad\qquad \vartheta = \tan^{-1}(x_s/x_c) \qquad (1.102d)$$

for which the Jacobian is $J = 1/\sqrt{x_c^2 + x_s^2}$. This Jacobian is used to write

$$p_{x_c,x_s}(\alpha, \beta) = \frac{1}{\sqrt{\alpha^2 + \beta^2}} \cdot p_{R,\theta}\left(\sqrt{\alpha^2 + \beta^2},\ \tan^{-1}(\beta/\alpha)\right)$$

$$= \frac{1}{\sqrt{\alpha^2 + \beta^2}} \cdot \frac{\sqrt{\alpha^2 + \beta^2}}{2\pi\sigma^2} \cdot \exp\left\{-\frac{\left(\sqrt{\alpha^2 + \beta^2}\right)^2}{2\sigma^2}\right\}$$

$$= \frac{1}{2\pi\sigma^2} \cdot \exp\left\{-\frac{\alpha^2 + \beta_s^2}{2\sigma^2}\right\} \qquad (1.102e)$$

which is the joint pdf of two independent Gaussian random variables (x_c, x_s), each with zero mean and variance σ^2. Thus, we have shown that the two independent Gaussian random variables representing quadrature noise samples can be generated on a computer in the manner presented above.

Using $\sigma^2 = 1$ in the method described above to generate independent noise samples, the typical white noise waveform shown in Figure 1.81 was generated as the samples $\{x_n = x(nT_s) = x(nT_d/4)\}$, where $1/T_s$ is the sampling rate. The sampling rate was chosen to be four times the data symbol rate $1/T_d = B$ to be considered below. Because the variance is unity, the amplitudes in Figure 1.81 rarely exceed $\pm 2\sigma = \pm 2$.

A total of $N = 512$ samples are displayed in Figure 1.81. The discrete Fourier transform[20] for this data is the set of coefficients $\{S_{xk}\}$, $k = 0, 1, \ldots, N - 1$ given by

$$S_{xk} = S_x(k\Delta f) = \frac{1}{\sqrt{N}} \sum_{n=0}^{N-1} x_n e^{-j2\pi nk/N}, \quad \Delta f = \frac{1}{NT_s} = \frac{4}{NT_d} \quad (1.103a)$$

The power spectrum corresponding to (1.103a) may be written

$$P_{xk} = |S_{xk}|^2 = \frac{1}{N} \sum_{n=0}^{N-1} \sum_{m=0}^{N-1} x_n x_m e^{j2\pi(m-n)k/N}$$

$$= \frac{1}{N} \left\{ \sum_{n=0}^{N-1} x_n^2 + 2 \sum_{m>n} x_n x_m \cos\left[\frac{2\pi(m-n)k}{N}\right] \right\} \quad (1.103b)$$

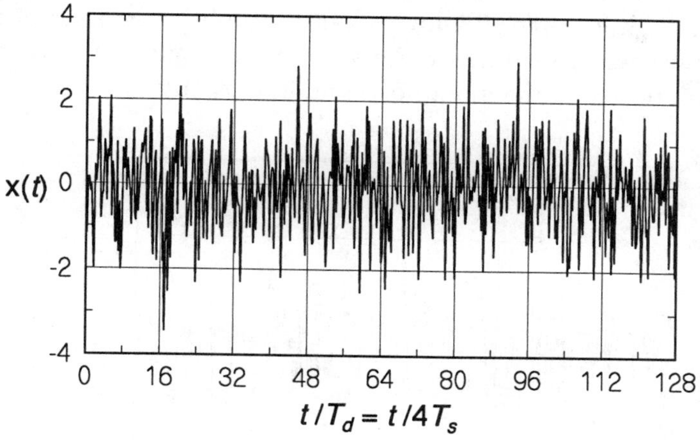

Figure 1.81 Simulated white noise waveform.

[20] Here we use the normalization $1/\sqrt{N}$ for the DFT. In other applications, the normalization $1/N$ is sometimes used.

which, because the zero-mean RVs x_n and x_m are independent for $n \neq m$, has the average value

$$\overline{P_{xk}} = \frac{1}{N} \sum_{n=0}^{N-1} \overline{x_n^2} = \frac{N\sigma^2}{N} = \sigma^2 \qquad (1.103c)$$

Note that the average power is not a function of frequency, so that (1.103c) has the interpretation of the (lowpass) white noise power spectral density level $\mathcal{N}_0 = \sigma^2 = 1$. The power spectrum for the example data of Figure 1.81 is shown in Figure 1.82, and its average agrees well with the prediction of $\sigma^2 = 1$ given by (1.103c). Note that the frequency range in Figure 1.82 is $\frac{1}{2}N\Delta f = 2/T_s = 2B$, the factor of $\frac{1}{2}$ being due to the fact that only half the spectrum is displayed: the transform is even about $f = 0$, and it is sufficient to plot the power for positive frequencies only.

The departure of Figure 1.82 from the theoretical constant spectrum is due to two factors. First, the finiteness of the sample causes the randomness of the spectrum to be significant; it can be shown that the variance of the power spectrum P_{xk} is proportional to $1/N$. Second, the random number generators of U_1 and U_2 do not produce either perfectly uniform or perfectly independent samples.

Example 1.28 Filtering to obtain lowpass noise from white noise using the IS-95 FIR filter. The simulated white noise samples displayed in Figure 1.81 were passed through a 48-tap FIR lowpass filter to obtain samples of a baseband, bandlimited white Gaussian noise waveform. Thus, the output noise samples $\{y_m\}$ are related to the input noise samples $\{x_n\}$ by

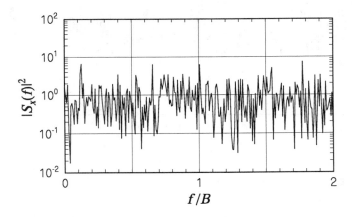

Figure 1.82 Power spectrum for example noise simulation.

$$y_k = \sum_{n=0}^{47} h_n \cdot x_{k-n} \qquad (1.104a)$$

where the $\{h_n\}$ are the filter coefficients given above in Table 1.3 for the pulse-shaping filter employed in the IS-95 digital cellular system to control the bandwidth of the transmissions. For that filter, the variance of the output noise samples theoretically is, because the variance of a sum of independent RVs is the sum of their variances:

$$\mathrm{Var}\{y_m\} = \sum_{n=0}^{47} \mathrm{Var}\{h_n x_{m-n}\} = \sum_{n=0}^{47} h_n^2 \, \mathrm{Var}\{x_{m-n}\} = \sum_{n=0}^{47} h_n^2 \, \sigma^2$$

$$= \sigma^2 \sum_{n=0}^{47} = 3.9403\sigma^2 = \sigma_y^2 \qquad (1.104b)$$

yielding the value of $\sigma_y = 1.9850\sigma$. After the filtering, the noise waveform has the appearance shown in Figure 1.83. This filter output noise waveform $y(t)$ clearly has a lower bandwidth than the filter input noise waveform $x(t)$, and has a larger dispersion than the input, because $\sigma_y \approx 2\sigma$. The power spectrum of this sample output is shown in Figure 1.84.

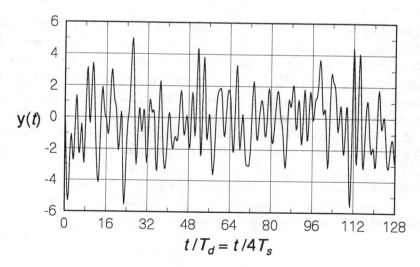

Figure 1.83 Noise samples at the output of the IS-95 FIR filter when the input samples are simulated independent (white noise) samples.

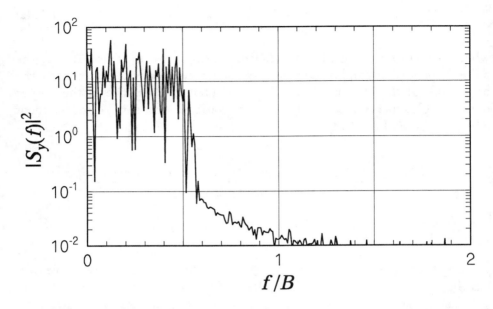

Figure 1.84 Power spectrum of noise at the IS-95 FIR filter output when the input samples are simulated white noise samples.

Figure 1.84 is consistent with the theoretical value of $\sigma_y^2 \approx 4$ when we recall that the power in $y(t)$ is mainly contained in the passband, $f \in (-B/2, B/2)$. The average spectral level in the passband is therefore approximately equal to $\sigma_y^2/B = 4/1 = 4$ according to the scale chosen for this numerical example. Except for a constant factor, ideally the shape of the spectrum of y(t) when the input x(t) is white noise should be that of the IS-95 filter that was plotted in Figure 1.51. Although the example spectrum in Figure 1.84 does feature a sharp cutoff that resembles the transition from passband to stopband in Figure 1.51, the sidebands in the spectrum of the numerical example are somewhat larger than the theoretical and are not flat as expected. These departures from the theoretical can be attributed to the sample size and the imperfection of the random number generators.

1.5.9 Chi-Squared Distributions

In the analysis of communications systems, such as performance measure calculations for certain signal reception schemes, we often encounter a situation where the incoming signal plus noise or noise-only waveforms are *squared* (a square-law transformation) and then subjected to further signal-processing schemes, such as being compared with a predetermined threshold value for a detection decision. In digital communications systems, we employ a sampling technique for these signal-processing schemes.

As we discussed in regard to the sampling theorems, when we sample at the Nyquist rate, the samples are statistically independent. When we sample a Gaussian random process, bandlimited to the $(-W, W)$, we obtain $2WT$ statistically independent samples over a time interval $(0, T)$ if the samples are taken at the Nyquist sampling rate of $2W$ samples per second.

Now, if the signal-processing scheme requires that one takes $N = 2WT$ samples of the Gaussian random process in the time interval of T seconds and then pass each of the N samples through a square-law transformation device prior to summing the N samples, we have a sum of N squared Gaussian RVs. This is the kind of decision variable that is used in many signal-processing schemes in the communications and radar fields. When an RV (decision variable) is formed as the sum of N squared statistically independent, identically distributed Gaussian RVs, we call that RV a *chi-squared* (χ^2) RV. As is made clear in what follows, we also categorize the sum as either a *central chi-squared RV* or a *noncentral chi-squared RV*, depending on whether the Gaussian RVs are zero-mean or nonzero-mean, respectively.

1.5.9.1 Central Chi-Squared Distribution

Let $X_1, X_2, \ldots, X_N$ be statistically independent Gaussian RVs with zero means and unit variances; that is, $X_i = G(0, 1)$ for $i = 1, 2, \ldots, N$. The sum Y of the squares of these RVs:

$$Y = X_1^2 + X_2^2 + \cdots + X_N^2 \tag{1.105a}$$

is called a central[21] chi-squared (χ^2) distributed RV with N degrees of freedom [42]. We wish to derive the pdf for the chi-squared RV Y by first obtaining its characteristic function:

$$C_y(j\nu) = E\{e^{j\nu Y}\} = E\{e^{j\nu(X_1^2+X_2^2+\cdots+X_N)}\}$$

$$= E\left\{\prod_{i=1}^{N} e^{j\nu X_i^2}\right\} = \prod_{i=1}^{N} \overline{e^{j\nu X_i^2}} \qquad (1.105\text{b})$$

The idea that we wish to exploit here in deriving the pdf of Y is to first obtain the characteristic function $C_{x_i^2}(j\nu) \triangleq \overline{e^{j\nu X_i^2}}$, which is the Fourier transform of the pdf of the RV X_i^2, which was obtained in (1.50c) to be

$$p_{x_i^2}(\alpha) = \begin{cases} \dfrac{1}{\sqrt{2\pi\alpha}} e^{-\alpha/2}, & \alpha \geq 0 \\ 0, & \text{otherwise} \end{cases} \qquad (1.105\text{c})$$

It can be shown that

$$C_{x_i^2}(j\nu) = \mathcal{F}\{p_{x_i^2}(\alpha)\} = \int_{-\infty}^{\infty} \frac{1}{\sqrt{2\pi\alpha}} e^{-\alpha/2} e^{j\nu\alpha} \, d\alpha$$

$$= \frac{1}{(1-j2\nu)^{1/2}} \qquad (1.105\text{d})$$

Thus the CHF of $Y = \sum_{i=1}^{N} X_i^2$, expressed in (1.105b), is given by

$$C_y(j\nu) = \prod_{i=1}^{N} \overline{e^{j\nu X_i^2}} = \frac{1}{(1-j2\nu)^{N/2}} \qquad (1.105\text{e})$$

The required pdf of Y is then obtained by the inverse Fourier transform of (1.105e), and the result is given by [42, 43]

$$p_y(\alpha) = p_{\chi^2}(\alpha; N) \triangleq \begin{cases} \dfrac{1}{2^{N/2}\Gamma(N/2)} \alpha^{(N-2)/2} e^{-\alpha/2}, & \alpha \geq 0 \\ 0, & \text{otherwise} \end{cases} \qquad (1.105\text{f})$$

---

[21] The term *central* indicates that each Gaussian RV has zero mean. The nonzero-mean Gaussian case will be termed *noncentral*.

where $\Gamma(x)$ is the Gamma function. This is the pdf for the χ^2 (chi-squared) distributed RV Y with N degrees of freedom.

Scaled central chi-squared RV. Consider an RV Z defined by

$$Z = Y_1^2 + Y_2^2 + \cdots + Y_N^2 \tag{1.106a}$$

where the $\{Y_i\}$ are statistically independent, zero-mean Gaussian RVs: $Y_i = G(0, \sigma^2)$ for $i = 1, 2, \ldots, N$. The pdf of Z can easily be obtained as follows: First we note that $Z = \sigma^2(X_1^2 + X_2^2 + \cdots + X_N^2) = \sigma^2 Y$ when we compare (1.105a) with (1.106a). That is, Z is a *scaled* chi-squared RV with N degrees of freedom, as opposed to a *standard* chi-squared RV, which is defined as the sum of squares of zero-mean, *unit-variance* Gaussian RVs. Now

$$F_z(\alpha) = \Pr\{Z \le \alpha\} = \Pr\{\sigma^2 Y \le \alpha\} = \Pr\{Y \le \alpha/\sigma^2\}$$
$$= F_y(\alpha/\sigma^2) \tag{1.106b}$$

Thus

$$p_z(\alpha) = \frac{d}{d\alpha} F_z(\alpha) = \frac{d}{d\alpha} F_y\left(\frac{\alpha}{\sigma^2}\right) = \frac{1}{\sigma^2} p_y\left(\frac{\alpha}{\sigma^2}\right)$$
$$= \frac{1}{\sigma^2} p_{\chi^2}\left(\frac{\alpha}{\sigma^2}; N\right) \tag{1.106c}$$

so that

$$p_z(\alpha) = \begin{cases} \dfrac{1}{(\sigma^2)^{N/2} 2^{N/2} \Gamma(N/2)} \, \alpha^{(N-2)/2} e^{-\alpha/2\sigma^2}, & \alpha \ge 0 \\[2ex] 0, & \text{otherwise} \end{cases} \tag{1.106d}$$

Note that we summed N squared Gaussian RVs, each with *identical* variances σ^2. The pdf (1.106c) is also known as that of the *gamma distribution*, which necessarily implies that each RV in the sum has identical variance. If we consider the case of $N = 2$, (1.106c) becomes

$$p_z(\alpha) = \frac{1}{2\sigma^2} e^{-\alpha/2\sigma^2}, \qquad \alpha \ge 0 \tag{1.106d}$$

which is the pdf of the exponentially distributed RV, and we obtained this pdf in (1.50e) by considering a *square-law* transformation of a Gaussian RV.

The central χ^2 distributed RV with N degrees of freedom pertains to a situation where N independent samples of a zero-mean Gaussian random process (or sample function) are passed through a square-law device (transformation) and then summed, as suggested in Figure 1.85.

1.5.9.2 Noncentral Chi-Squared Distribution

Consider an RV U defined by

$$U = Z_1^2 + Z_2^2 + \cdots + Z_N^2 \tag{1.107a}$$

where the $\{Z_i\}$ are statistically independent, nonzero-mean, unit-variance Gaussian RVs: $Z_i = G(m_i, 1)$. The RV U is called a *noncentral* chi-squared RV with N degrees of freedom. Our objective is to obtain the pdf of U. To this end, let us first consider the RV $Z_i = m_i + X_i$, where $X_i = G(0, 1)$. The pdf of this RV is of course

$$p_{z_i}(\alpha) = \frac{1}{\sqrt{2\pi}} \exp\left\{ -\tfrac{1}{2}(\alpha - m_i)^2 \right\} \tag{1.107b}$$

From this pdf we can easily obtain the pdf of $Z_i^2 = (m_i + X_i)^2$ by recognizing that

$$F_{z_i^2}(\alpha) = \Pr\{Z_i^2 \le \alpha\} = F_{z_i}\left(\sqrt{\alpha}\right) - F_{z_i}\left(-\sqrt{\alpha}\right)$$

$$p_{z_i^2}(\alpha) = \frac{d}{d\alpha} F_{z_i^2}(\alpha) = \frac{1}{2\sqrt{\alpha}} p_{z_i}\left(\sqrt{\alpha}\right) + \frac{1}{2\sqrt{\alpha}} p_{z_i}\left(-\sqrt{\alpha}\right)$$

$$= \frac{1}{\sqrt{8\pi\alpha}} \left[\exp\left\{ -\tfrac{1}{2}\left(\sqrt{\alpha} - m_i\right)^2 \right\} + \exp\left\{ -\tfrac{1}{2}\left(\sqrt{\alpha} + m_i\right)^2 \right\} \right] \tag{1.107c}$$

Carrying out the indicated squaring operations in the exponents in (1.107c) and recognizing that $e^x + e^{-x} = 2\cosh x$, we obtain

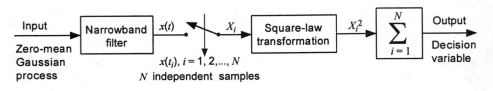

Figure 1.85 Generation of χ^2 distributed RV.

$$P_{z_i^2}(\alpha) = \frac{1}{\sqrt{2\pi\alpha}} e^{-(\alpha+m_i^2)/2} \cosh\left(m_i\sqrt{\alpha}\right) \qquad (1.107\text{d})$$

By taking the Fourier transform of (1.107d), we obtain the characteristic function [42]

$$C_{z_i^2}(j\nu) = \int_{-\infty}^{\infty} P_{z_i^2}(\alpha) e^{j\nu\alpha} \, d\alpha = \int_{-\infty}^{\infty} \frac{e^{-(\alpha+m_i^2)/2}}{\sqrt{2\pi\alpha}} \cosh\left(m_i\sqrt{\alpha}\right) d\alpha$$

$$= \frac{1}{\sqrt{1-j2\nu}} \cdot \exp\left\{ -\frac{m_i^2}{2} + \frac{m_i^2/2}{1-j2\nu} \right\} \qquad (1.107\text{e})$$

The characteristic function for U is therefore given by

$$C_u(j\nu) = \mathrm{E}\left\{ e^{j\nu(Z_1^2+Z_2^2+\cdots+Z_N^2)} \right\} = \prod_{i=1}^{N} \overline{e^{j\nu Z_i^2}}$$

$$= \left(\frac{1}{1-j2\nu}\right)^{N/2} \exp\left\{ \sum_{i=1}^{N} \left(-\frac{m_i^2}{2} + \frac{m_i^2/2}{1-j2\nu} \right) \right\}$$

$$= \left(\frac{1}{1-j2\nu}\right)^{N/2} \exp\left\{ -\frac{\lambda}{2} + \frac{\lambda/2}{1-j2\nu} \right\} \qquad (1.107\text{f})$$

where

$$\lambda = \sum_{i=1}^{N} m_i^2 \qquad (1.107\text{g})$$

is defined as the *noncentrality parameter*. The pdf of U is found as the inverse Fourier transform of $C_u(j\nu)$ [42]:

$$p_u(\alpha) = p_{\chi^2}(\alpha; N, \lambda) \triangleq \begin{cases} \frac{1}{2}\left(\frac{\alpha}{\lambda}\right)^{(N-2)/4} e^{-(\alpha+\lambda)/2} I_{\frac{N}{2}-1}\left(\sqrt{\lambda\alpha}\right), & \alpha \geq 0 \\ \\ 0, & \text{otherwise} \end{cases}$$

$$(1.107\text{g})$$

where $I_{\frac{N}{2}-1}(\bullet)$ is the modified Bessel function of the first kind and order $\frac{N}{2}-1$. Note that we have derived the pdf of a noncentral chi-squared RV with N degrees of freedom by considering the independent Gaussian RVs to be of unit variance ($\sigma^2 = 1$).

Scaled noncentral chi-squared RV. Consider an RV V defined by

$$V = V_1^2 + V_2^2 + \cdots + V_N^2 \tag{1.108a}$$

where the $\{V_i\}$ are statistically independent Gaussian RVs with $V_i = G(A_i, \sigma^2) = A_i + Y_i$ for $i = 1, 2, \ldots, N$ and where $Y_i = G(0, \sigma^2)$. Let us note that the RVs $\{V_i\}$ are *not* unit-variance Gaussian RVs, and therefore we need to normalize them to conform with the definition of a noncentral chi-squared RV. First we note that $V = \sigma^2(Z_1^2 + Z_2^2 + \cdots + Z_N^2)$ when we compare (1.135a) with (1.134a). That is, V is a *scaled* noncentral chi-squared RV with N degrees of freedom, as opposed to a *standard* noncentral chi-squared RV, which is defined as the sum of squares of nonzero mean, unit-variance Gaussian RVs.

$$V = \sigma^2 \sum_{i=1}^{N} \left(\frac{A_i}{\sigma} + \frac{Y_i}{\sigma} \right)^2 = \sigma^2 \sum_{i=1}^{N} (m_i + X_i)^2 = \sigma^2 U \tag{1.108b}$$

where $m_i \triangleq A_i/\sigma$, since $X_i = Y_i/\sigma = G(0, 1)$. Now

$$F_v(\alpha) = \Pr\{V \le \alpha\} = \Pr\{\sigma^2 U \le \alpha\} = F_u(\alpha/\sigma^2) \tag{1.108c}$$

Thus

$$p_v(\alpha) = \frac{d}{d\alpha} F_v(\alpha) = \frac{d}{d\alpha} F_u\left(\frac{\alpha}{\sigma^2} \right) = \frac{1}{\sigma^2} p_u\left(\frac{\alpha}{\sigma^2} \right)$$

$$= \frac{1}{\sigma^2} p_{\chi^2}\left(\frac{\alpha}{\sigma^2}; N, \lambda \right) \tag{1.108d}$$

so that

$$p_v(\alpha) = \begin{cases} \dfrac{(\alpha/\sigma^2 \lambda)^{(N-2)/4}}{2\sigma^2} \exp\left\{ -\tfrac{1}{2}\left(\frac{\alpha}{\sigma^2} + \lambda \right) \right\} I_{\frac{N}{2}-1}\left(\sqrt{\frac{\lambda \alpha}{\sigma^2}} \right), & \alpha \ge 0 \\ \\ 0, & \text{otherwise} \end{cases}$$

$$\tag{1.108e}$$

where the *noncentrality parameter* is given by

$$\lambda = \sum_{i=1}^{N} m_i^2 = \sum_{i=1}^{N} \frac{A_i^2}{\sigma^2} \tag{1.108f}$$

Notation for chi-squared RVs. It is often convenient to use the notation $\chi^2(N; \lambda)$ to refer to a noncentral chi-squared RV with N degrees of freedom and noncentrality parameter λ. A central chi-squared RV can be written $\chi^2(N; 0) \equiv \chi^2(N)$. Using this notation, a scaled noncentral chi-squared RV can be denoted as $\sigma^2 \chi^2(N; \lambda)$ and a scaled central chi-squared RV as $\sigma^2 \chi^2(N)$, where σ^2 denotes the variance of the Gaussian RVs.

Example 1.29 Let us revisit (1.80a), considered in Example 1.26, where the RV Y was given by

$$Y = \sum_{i=1}^{L} (X_{i1}^2 - X_{i2}^2) \tag{1.109a}$$

and where X_{i1} and X_{i2} are statistically independent, zero-mean Gaussian RVs with variances σ_1^2 and σ_2^2, respectively, with $\sigma_2^2 > \sigma_1^2$. In the examples of (1.80a) and (1.81a), $\Pr\{Y > 0\}$ was computed by using the CB and the CLT, which are both approximation methods. In this example, we compute the probability exactly using the pdfs of the scaled central chi-square RVs with L degrees of freedom, X_1 and X_2, given by the sums

$$X_1 \triangleq \sum_{i=1}^{L} X_{i1}^2 \qquad \text{and} \qquad X_2 \triangleq \sum_{i=1}^{L} X_{i2}^2 \tag{1.109b}$$

The pdfs $p_{x_1}(\alpha)$ and $p_{x_2}(\alpha)$ for these RVs are given by

$$p_{x_i}(\alpha) = \begin{cases} \dfrac{1}{(\sigma_i^2)^{L/2} 2^{L/2} \Gamma(L/2)} \alpha^{(L-2)/2} e^{-\alpha/2\sigma_i^2}, & \alpha \geq 0 \\ 0, & \text{otherwise} \end{cases} \quad i = 1, 2 \tag{1.109c}$$

In computing the probability measure $\Pr\{Y > 0\}$, our inclination is first to obtain the pdf of the RV Y and then to integrate it over the interval $(0, \infty)$. To this end, one can obtain the pdf of $Y = X_1 - X_2$ by using the conditional pdf method, and the result can be shown to be the evaluation of the integrals indicated below:

$$p_y(\gamma) = \int_{-\infty}^{\infty} p_{y,x_1}(\gamma, \alpha)\, d\alpha = \int_{-\infty}^{\infty} p_{x_2}(\alpha - \gamma \mid X_1 = \alpha)\, p_{x_1}(\alpha)\, d\alpha$$

$$= \int_{-\infty}^{\infty} p_{x_1,x_2}(\alpha, \alpha - \gamma)\, d\alpha = \int_{-\infty}^{\infty} p_{x_1}(\alpha)\, p_{x_2}(\alpha - \gamma)\, d\alpha$$

$$
= \begin{cases}
\int_0^\infty p_{x_1}(\alpha)\, p_{x_2}(\alpha - \gamma)\, d\alpha, & \gamma < 0 \\[2em]
\int_\gamma^\infty p_{x_1}(\alpha)\, p_{x_2}(\alpha - \gamma)\, d\alpha, & \gamma \geq 0
\end{cases}
\tag{1.109d}
$$

The calculations required to execute the integrals in (1.109d) analytically are no mean task when the pdfs, $p_{x_i}(\alpha)$, $i = 1, 2$ from (1.109c), are used. The reader may try and find out for himself or herself how laborious it is to accomplish the required result. The ultimate objective is to obtain the probability measure

$$
\Pr\{Y > 0\} = \int_0^\infty p_y(\alpha)\, d\alpha
\tag{1.109e}
$$

which may still present some difficulty, depending on the form of the pdf $p_y(\alpha)$. Instead, we resort to another approach to calculate the required probability measure. We obtain the result without the use of the pdf of Y by noting that

$$
\Pr\{Y > 0\} = \Pr\{X_1 > X_2\} = \Pr\{\sigma_1^2 \chi_1^2(L) > \sigma_2^2 \chi_2^2(L)\}
$$
$$
= \Pr\left\{\chi_2^2(L) < \frac{\sigma_1^2}{\sigma_2^2} \chi_1^2(L)\right\}
\tag{1.110a}
$$

where X_1 and X_2 are scaled central chi-squared RVs with L degrees of freedom as defined in (1.109b), and $\chi_1^2(L)$ and $\chi_2^2(L)$ are standard chi-squared RVs with L degrees of freedom. Note that we have standardized the chi-squared RVs X_1 and X_2 by normalizing as follows:

$$
\chi_1^2(L) \triangleq X_1/\sigma_1^2 \qquad \text{and} \qquad \chi_2^2(L) \triangleq X_2/\sigma_2^2
\tag{1.110b}
$$

The final result for (1.110a) is derived in Appendix 1D and is given by

$$
\Pr\{Y > 0\} = \begin{cases}
\dfrac{\mu^{L/2}\, \Gamma(L)}{(1+\mu)^L \Gamma(\frac{L}{2}) \Gamma(\frac{L}{2}+1)}\; {}_2F_1\left(1, L; \tfrac{L}{2}+1; \dfrac{\mu}{1+\mu}\right), & L \text{ odd} \\[2em]
\left(\dfrac{\mu}{1+\mu}\right)^{L/2} \displaystyle\sum_{k=0}^{L/2-1} \binom{\frac{L}{2}+k-1}{k} \left(\dfrac{1}{1+\mu}\right)^k, & L \text{ even}
\end{cases}
\tag{1.110c}
$$

where $_2F_1(\bullet)$ is the *gaussian hypergeometric function* and $\mu \triangleq \sigma_1^2/\sigma_2^2$ as defined earlier in (1.80j).

One of the reasons for including this particular example is to calculate the exact probability measure $\Pr\{Y > 0\}$ as a function of μ with L as a parameter and to compare the results with the approximations we made for this measure using the Chernoff bound and the CLT. In Figure 1.67, these approximations were compared with each other, and in Figures 1.86 and 1.87 they are compared with the exact probability.

The comparisons shown in Figures 1.86 and 1.87 reveal that, in this case, the CLT approximation is generally a better approximation than the CB, and becomes a closer approximation as L increases. Thus, we demonstrated the usefulness of the CLT to obtain an estimate of the probability without the labor of finding the exact probability. The CB is shown to be an upper bound, while the CLT result is neither an upper nor a lower bound, and the CB provides a closer approximation for small values of the probability and small values of μ, for which $\sigma_1 \ll \sigma_2$ in this particular example.

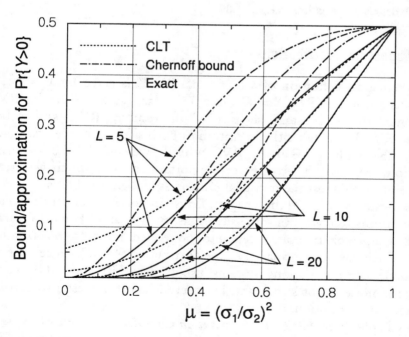

Figure 1.86 Comparison of exact and approximate probabilities.

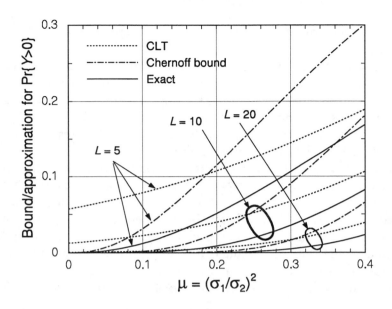

Figure 1.87 Details of comparison in Figure 1.86.

1.5.10 Lognormal Distributions

The mathematical definition of a lognormal RV Y is given by $Y = e^X$, in which X is a Gaussian RV, $X = G(m, \sigma^2)$. In fact, when a Gaussian RV is the exponent of any constant, such as "e," the resulting RV is said to be a *lognormal* RV. If one takes the logarithm of both sides of $Y = e^X$ to obtain $\ln Y = X$, the result is a Gaussian RV. Why do we consider the lognormal distribution, anyway? There are very important reasons, which we briefly describe and then discuss the statistical properties of lognormal RVs.

As we stated earlier in this chapter, Gaussian noise is ever present in any communications receiver. In some situations, such as at VLF and LF carrier frequencies, atmospheric noise external to the receiver must also be taken into account, which has a lognormal envelope distribution [44–48]. In cellular and PCS mobile communications systems using the VHF and UHF bands, atmospheric noise is not significant, but there is another reason to consider the lognormal distribution.

In Chapter 2, *Mobile Radio Propagation Considerations*, one of the semi-empirical computer propagation models discussed, known as the *Longley-Rice model*, characterizes the propagation loss measured at the receiver in logarithmic (dB) units as a Gaussian RV:

$$L(\text{dB}) = L_0(\text{dB}) + \sigma_c(\text{dB}) \times G(0,1) = G(L_0, \sigma_c^2)\,(\text{dB}) \qquad (1.111a)$$

where $L_0(\text{dB})$ is the median value of the distribution of the propagation loss and $\sigma_c(\text{dB})$ is its standard deviation, which can vary over the range of 8 to 10 dB. This empirical model continues to be supported as reality in such real channels as the mobile wireless channel. Because $L(\text{dB}) = L_0(\text{dB}) + \sigma_c(\text{dB}) \times G(0,1)$ and can be written $L(\text{dB}) \equiv L_0(\text{dB}) + \sigma_c(\text{dB}) \times X$, where X is a zero-mean, unit-variance normal RV, then in absolute numbers, the loss is the RV L_a, where

$$L_a = 10^{L(\text{dB})/10} = 10^{[L_0(\text{dB})+\sigma_c(\text{dB})X]/10}$$

$$= L_{0a} \cdot 10^{\sigma_c X/10} = L_{0a} \cdot V \qquad (1.111b)$$

where $V \triangleq 10^{\sigma_c X/10}$ is a lognormal RV and $L_{0a} \triangleq 10^{L_0(\text{dB})/10}$. The propagation loss is thus understood to be a lognormal RV. The variation in L_a is due to many factors, including the presence or absence of obstacles in the line of sight that can "shadow" the receiver. The term "shadowing" refers to the fact that a hill or other obstruction can block the radio signal, much like it does the light from the sun, as depicted in Figure 1.88. The effect of shadowing on link reliability is discussed in Chapter 10.

The zero-mean, unit-variance Gaussian random variable X can vary between $-\infty$ and $+\infty$; however, over 99% of its variation is within the range $-3 < X < 3$. Thus over 99% of the variation in the propagation loss is within $\pm 3\sigma_c(\text{dB})$, or

$$-3\sigma_c + L_0 < L < L_0 + 3\sigma_c \text{ with probability} > 0.99 \qquad (1.111c)$$

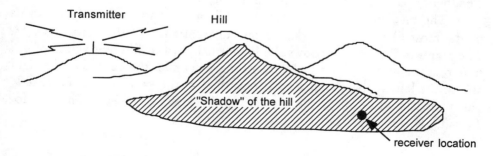

Figure 1.88 Lognormal shadow zone.

1.5.10.1 Probability Density Function of a Lognormal RV

The pdf of the zero-mean, unit-variance RV X has the function form

$$p_x(w) = \frac{1}{\sqrt{2\pi}}\, e^{-w^2/2}, \quad -\infty < w < \infty \tag{1.112a}$$

and the median value of X is $x_m = 0$. Let us define the lognormal variable V:

$$V = 10^{\sigma_c X/10} = e^{\beta \sigma_c X} \tag{1.112b}$$

where

$$\beta = (ln10)/10 = 0.23 \tag{1.112c}$$

Then the cumulative distribution function for V is

$$F_v(\alpha) = \Pr\{V \le \alpha\} = \Pr\{10^{\sigma_c X/10} \le \alpha\} = \Pr\left\{X \le \tfrac{1}{\beta \sigma_c}\cdot ln\,\alpha\right\}$$

$$= \int_{-\infty}^{(ln\,\alpha)/\beta\sigma_c} dw\, p_x(w) = \frac{1}{\sqrt{2\pi}}\int_{-\infty}^{(ln\,\alpha)/\beta\sigma_c} dw\, e^{-w^2/2} \tag{1.112d}$$

$$\Rightarrow p_v(\alpha) = \frac{\partial}{\partial\alpha} F_v(\alpha) = \left|\frac{\partial}{\partial\alpha}\frac{1}{\beta\sigma_c}\, ln\,\alpha\right|\cdot p_x\left(\frac{1}{\beta\sigma_c}\, ln\,\alpha\right)$$

$$= \begin{cases} \dfrac{1}{\beta\sigma_c\alpha}\cdot\dfrac{1}{\sqrt{2\pi}}\,\exp\left\{-\dfrac{(ln\,\alpha)^2}{2\beta^2\sigma_c^2}\right\}, & \alpha \ge 0 \\[2ex] 0, & \text{otherwise} \end{cases} \tag{1.112e}$$

The lognormal pdf is shown plotted in Figure 1.89.

The median of V is $v_m = 10^{x_m} = 10^0 = 1$ for any value of σ_c and its mode (most likely value) is found from differentiation of the pdf to be $e^{-\sigma_c^2}$. The graph in Figure 1.89 shows the lognormal pdf for different values of σ_c; note that the median, for which 50% of the area under the pdf lies to the right and to the left, is always equal to 1, regardless of the value of σ_c. Because V has a median value of 1, the median of the loss L_a is L_{0a}. The pdf for $L_a = L_{0a}V$ is found to be

$$p_L(u) = \frac{1}{L_{0a}}\, p_v\left(\frac{u}{L_{0a}}\right) = \frac{1}{\beta\sigma_c\sqrt{2\pi}\, u}\,\exp\left\{-\frac{[\,ln\,(u/L_{0a})]^2}{2\beta^2\sigma_c^2}\right\} \tag{1.112f}$$

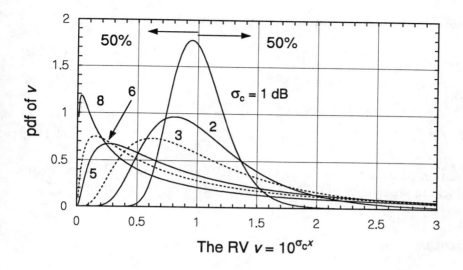

Figure 1.89 Lognormal pdf.

1.5.10.2 Moments of Lognormal RVs

The moments of the propagation loss RV are most conveniently found by writing the loss as

$$L_a = L_{0a}V = L_{0a}\,e^{\beta\sigma_c X} = e^{\beta(L_0 + \sigma_c X)} \tag{1.113a}$$

where L_0 is the median loss in dB and X is a zero-mean, unit-variance Gaussian RV. Then the kth moment of the loss is

$$E\{L_a^k\} = E\{e^{k\beta(L_0+\sigma_c X)}\} = e^{k\beta L_0}\,E\{e^{k\beta\sigma_c X}\} = L_{0a}^k E\{e^{(k\beta\sigma_c)X}\} \tag{1.113b}$$

$$= L_{0a}^k \cdot M_x(k\beta\sigma_c) \tag{1.113c}$$

in which $M_x(\,\cdot\,) = E\{e^{k\beta\sigma_c X}\}$ is the moment generating function for the RV X. In this light, the moments of the propagation loss are

$$E\{L_a^k\} = L_{0a}^k\,e^{k^2\beta^2\sigma_c^2/2} = e^{k\beta L_0 + k^2\beta^2\sigma_c^2/2} \tag{1.113d}$$

For example, the mean, mean square, and variance of the loss are

$$E\{L_a\} = L_{0a}\, e^{\beta^2 \sigma_c^2/2} \tag{1.113e}$$

$$E\{L_a^2\} = L_{0a}^2\, e^{2\beta^2 \sigma_c^2} \tag{1.113f}$$

$$Var\{L_a\} = E\{L_a^2\} - [E\{L_a\}]^2$$

$$= L_{0a}^2\, e^{\beta^2 \sigma_c^2}\left(e^{\beta^2 \sigma_c^2} - 1\right) \tag{1.113g}$$

Application of these results for lognormal RVs will be made when we discuss the blocking probability of CDMA cellular systems in a later chapter.

References

[1] "Dual-Mode Subscriber Equipment-Network Equipment Compatibility Specifications," TIA/EIA Interim Standard 54 (IS-54), Washington, DC: Telecommunications Industry Association, 1989.

[2] "Mobile Station-Base Station Compatibility Standard for Dual-Mode Wideband Spread Spectrum Cellular System," TIA/EIA Interim Standard 95 (IS-95), Washington, DC: Telecommunications Industry Association, July 1993 (amended as IS-95-A in May 1995).

[3] Wittman, J. H., "Categorization of Multiple-Access/Random Access Modulation Techniques," *IEEE Trans. on Communication Technology*, Vol. COM-15, No. 5, pp. 724–725, Oct. 1967.

[4] Lee, J. S., "Overview of the Technical Basis of Qualcomm's CDMA Cellular Telephone System Design—A View of North American TIA/EIA IS-95" (invited paper), *Proc. 1994 IEEE International Conference on Communication Systems (ICCS'94)*, pp. 353–358, Nov. 14–18, 1994, Singapore.

[5] Gilhousen, K. S., I. M. Jacobs, R. Padovani, A. J. Viterbi, L. A. Weaver, Jr., and C. E. Wheatley III, "On the Capacity of a Cellular CDMA System," *IEEE Trans. on Vehicular Technology*, Vol. 40, pp. 303–312, May 1991.

[6]   Simon, M. K., J. K. Omura, R. A. Scholtz, and B. K. Levitt, *Spread Spectrum Communications Handbook* (revised edition), New York: McGraw-Hill, 1994.

[7]   Torrieri, D. J., *Principles of Secure Communication Systems* (2nd edition), Boston: Artech House, 1992.

[8]   Miller, L. E., J. S. Lee, and D. J. Torrieri, "Analysis of Transients Affecting the Selection of Hopping Rate for FH/CPFSK Systems," *Proc. IEEE 1991 Military Commun. Conf.*, pp. 26.1–26.5, Nov. 4–7, 1991, Washington.

[9]   Dixon, R. C., *Spread Spectrum Systems*, New York: Wiley, 1976.

[10] Torrieri, D. J., "Frequency Hopping with Multiple Frequency-Shift Keying and Hard Decisions," *IEEE Trans. on Commun.*, Vol. COM-32, pp. 574–582, May 1984.

[11] Lee, J. S., L. E. Miller, and R. H. French, "The Analyses of Uncoded Performances for Certain ECCM Receiver Design Strategies for Multi-hops/Symbol FH/MFSK Waveforms," *IEEE J. Selected Areas in Commun.*, Vol. SAC-3, pp. 611-621, Sept. 1985.

[12] Lee, J. S., L. E. Miller, and R. H. French, "Error-Correcting Codes and Nonlinear Diversity Combining Against Worst-Case Partial-Band Jamming of Frequency-Hopping MFSK Systems," *IEEE Trans. on Commun.*, Vol. COM-36, pp. 471–478, Apr. 1988.

[13] Holmes, J. K., *Coherent Spread Spectrum Systems*, New York: Wiley, 1982.

[14] Cooper, G. R., and C. D. McGillem, *Modern Communications and Spread Spectrum*, McGraw-Hill, New York, 1986.

[15] Price, R., and P. E. Green, "A Communication Technique for Multipath Channels," *Proc. IEEE*, Vol. 46, pp. 555–570 (March 1958).

[16] Stuber, G. L., *Principles of Mobile Communication*, Boston: Kluwer Academic Publishers, 1996.

[17] Viterbi, A. J., "When Not to Spread Spectrum—A Sequel," IEEE *Communications Magazine*, Vol. 23, pp. 12–17, Apr. 1985.

[18] Kahn, C. R., "Spread Spectrum Applications and State-of-the-Art Equipments," Paper No. 5, AGARD-NATO Lecture Series No. 58 on

"Spread Spectrum Communications," May 28–June 6, 1973, Torrance, CA: Magnavox, Nov. 1972.

[19] Woodward, P. M., *Probability and Information Theory, with Applications to Radar* (2nd ed.), New York: Pergamon Press, 1964.

[20] Abramowitz, M., and I. A. Stegun (eds.), *Handbook of Mathematical Functions*, Washington, DC: National Bureau of Standards Applied Mathematics Series 55, Government Printing Office, 1964 (ninth printing, 1970).

[21] Nyquist, H., "Certain Topics in Telegraph Transmission Theory," *Trans. AIEE*, Vol. 47, pp. 617–644, 1929.

[22] Shannon, C. E., "Communication in the Presence of Noise," *Bell System Technical Journal*, Vol. 27, 1948. Also in *C. Elwood Shannon: Collected Papers*, edited by N. J. A. Sloane and A. D. Wyner, New York: IEEE Press, 1993.

[23] Chennakeshu, S., and G. J. Saulnier, "Differential Detection of $\pi/4$-Shifted DQPSK for Digital Cellular Radio," *IEEE Trans. on Vehicular Technology*, Vol. 42, pp. 46–57, Feb. 1993.

[24] Frerking, M. E., *Digital Signal Processing in Communication Systems*, New York: Van Nostrand Reinhold, 1994.

[25] McClellan, J. H., T. W. Parks, and L. R. Rabiner, "A Computer Program for Designing Optimum FIR Linear Phase Digital Filters," *IEEE Trans. on Audio and Electroacoustics*, Vol. AU-21, No. 6, Dec. 1973.

[26] McClellan, J. H., "FIR Filter Design and Synthesis," Ch. 5 of *Programs for Digital Signal Processing*, New York: IEEE Press, 1979.

[27] Harris, F. J., "On the Use of Windows for Harmonic Analysis With Discrete Fourier Transform," *Proc. IEEE*, Vol. 66, No. 1, pp. 51–83, Jan. 1978.

[28] Johnson, J. B., "Thermal Agitation of Electricity in Conductors," *Physics Reviews*, Vol. 32, pp. 97–109, July 1928.

[29] Wozencraft, J. M., and I. M. Jacobs, *Principles of Communications Engineering*, New York: Wiley, 1965.

[30] Whalen, A. D., *Detection of Signals in Noise*, San Diego, CA: Academic Press, 1971.

[31]   Davenport, Jr., W. B., and W. L. Root, *An Introduction to the Theory of Random Signals and Noise*, New York: McGraw-Hill, 1958. (Reprinted by IEEE Press in 1996.)

[32]   Hogg, R. V., and A. T. Craig, *Introduction to Mathematical Statistics* (2nd edition), New York: Macmillan, 1965.

[33]   Anderson, T. W., *An Introduction to Multivariate Statistical Analysis*, New York: Wiley, 1958.

[34]   Chernoff, H., "A Measure of Asymptotic Efficiency for Tests of a Hypothesis Based on a Sum of Observations," *Ann. Math. Statistics*, Vol. 23, pp. 493–507, 1952.

[35]   Jacobs, I. M., "Probability-of-Error Bounds for Binary Transmission on the Slowly Fading Rician Channel," *IEEE Trans. on Information Theory*, Vol. IT-12, pp. 431–441, Oct. 1966.

[36]   Rice, S. O., "Mathematical Analysis of Random Noise," *Bell System Technical Journal*, Vol. 23, pp. 283–332, 1944; also Vol. 24, pp. 46–156, 1945.

[37]   Wax, N., *Selected Papers on Noise and Stochastic Processes*, New York: Dover, 1954.

[38]   Rice, S. O., "Statistical Properties of a Sine-wave Plus Random Noise," *Bell System Technical Journal*, Vol. 27, pp. 109–157, Jan. 1948.

[39]   Marcum, J. I., and P. Swerling, "Studies of Target Detection by Pulsed Radar," *IRE Trans. on Information Theory*, Vol. IT-6, Apr. 1960.

[40]   Miller, L. E., J. S. Lee, and D. J. Torrieri, "Frequency-Hopping Signal Detection Using Partial Band Coverage," *IEEE Trans. on Aerospace and Elec. Syst.*, Vol. 29, pp. 540–553, Apr. 1993.

[41]   Miller, L. E., and J. S. Lee, "Bit Error Rate Expressions for Differentially Detected $\pi/4$ DQPSK Modulation," *IEEE Trans. on Communications*, Vol. 46, pp. 71–81, Jan. 1998.

[42]   Papoulis, A., *Probability, Random Variables, and Stochastic Processes*, New York: McGraw-Hill, 1965.

[43]   Campbell, G. A., and R. M. Foster, *Fourier Integrals for Practical Applications*, Princeton, NJ: Van Nostrand, 1954.

[44]   Furutsu, K., and T. Ishida, "On the Theory of Amplitude Distribution of Impulse Noise," *J. Applied Physics*, Vol. 32, pp. 1206–1221, July 1961.

[45] Kneuer, J. G., "A Simplified Physical Model for Amplitude Distribution of Impulsive Noise," *IEEE Trans. on Communications*, Vol. COM-12, p. 220, Dec. 1964.

[46] Omura, J. K., "Statistical Analysis of LF/VLF Communications Modems," Research report DASA-2324, Menlo Park, CA: Stanford Research Institute, Aug. 1969.

[47] Hildebrand, F. B., *Methods of Applied Mathematics*, New York: Prentice-Hall, 1952.

[48] Friedland, B., *Principles and Techniques of Applied Mathematics*, New York: Wiley, 1956.

Appendix 1A Impulse Response of Ideal Filter #2

The frequency domain characteristic of ideal filter #2 is given in the text as

$$H(f) = \begin{cases} 1, & |f| < f_1 \\ \frac{f_2 - |f|}{f_2 - f_1}, & f_1 \le |f| \le f_2 \\ 0, & |f| > f_2 \end{cases} \tag{1A.1}$$

The impulse response is found as the inverse Fourier transform of $H(f)$:

$$h(t) = \int_{-\infty}^{\infty} e^{j2\pi ft} H(f)\, df = 2\int_{0}^{\infty} H(f)\cos 2\pi ft\, df \qquad \text{H is even}$$

$$= 2\int_{0}^{f_1} \cos 2\pi ft\, df + 2\int_{f_1}^{f_2} \underbrace{\frac{f_2 - f}{f_2 - f_1}}_{u(f)}\, \underbrace{\cos 2\pi ft}_{v'(f)}\, df \qquad \text{integ. by parts}$$

$$= 2\frac{\sin 2\pi f_1 t}{2\pi t} + 2\left[\underbrace{\frac{f_2 - f}{f_2 - f_1}}_{u(f)} \cdot \underbrace{\frac{\sin 2\pi ft}{2\pi t}}_{v(f)}\right]_{f_1}^{f_2} - 2\int_{f_1}^{f_2} df \underbrace{\frac{-1}{f_2 - f_1}}_{u'(f)}\underbrace{\frac{\sin 2\pi ft}{2\pi t}}_{v(f)}$$

$$= \frac{\sin 2\pi f_1 t}{\pi t} + 0 - \frac{\sin 2\pi f_1 t}{\pi t} + \frac{1}{f_2 - f_1} \cdot \frac{2}{(2\pi t)^2}\underbrace{[\cos 2\pi f_1 t - \cos 2\pi f_2 t]}_{\text{form: } \cos(A\text{-}B)\, -\, \cos(A\text{+}B)}$$

$$= \frac{1}{f_2 - f_1} \cdot \frac{1}{(\pi t)^2} \sin[(f_1 + f_2)\pi t] \sin[(f_2 - f_1)\pi t]$$

$$= (f_1 + f_2) \operatorname{sinc}[(f_1 + f_2)\pi t] \operatorname{sinc}[(f_2 - f_1)\pi t] \tag{1A.2}$$

Appendix 1B Integral of sinc Function

A general formula for the integral of a sinc function is

$$\int_a^b \operatorname{sinc}(cx)\, dx = \int_a^b \frac{\sin(\pi cx)}{\pi cx}\, dx = \frac{1}{\pi c} \int_{\pi ac}^{\pi bc} \frac{\sin x}{x}\, dx$$

$$= \frac{1}{\pi c}[\operatorname{Si}(\pi bc) - \operatorname{Si}(\pi ac)] \tag{1B.1}$$

where $\operatorname{Si}(\,\cdot\,)$ is the sine integral [20]. Therefore, the integral in Section 1.3.2 is

$$I_n = \int_0^T \operatorname{sinc}(2Wt - n)\, dt = \int_{-n/2W}^{T-n/2W} \operatorname{sinc}(2Wt)\, dt$$

$$= \frac{1}{2\pi W}[\operatorname{Si}(2\pi WT - n\pi) - \operatorname{Si}(-n\pi)]$$

$$= \frac{1}{2\pi W}[\operatorname{Si}(2\pi WT - n\pi) + \operatorname{Si}(n\pi)] \tag{1B.2}$$

because the sine integral is an odd function. Note in Table 1B.1 that the value of $\operatorname{Si}(k\pi)$ is very close to $\pi/2$ for $k > 0$. Thus, assuming that $2WT$ is a large integer, the value of the integral is approximately

$$I_n \approx \frac{1}{2\pi W} \times \begin{cases} \frac{\pi}{2} + 0 = \frac{\pi}{2}, & n = 0 \\ \frac{\pi}{2} + \frac{\pi}{2} = \pi, & 1 \le n < 2WT \\ 0 + \frac{\pi}{2} = \frac{\pi}{2}, & n = 2WT \end{cases}$$

$$= \frac{1}{2W} \times \begin{cases} 1/2, & n = 0,\ 2WT \\ 1, & 1 \le n < 2WT \end{cases} \tag{1B.3}$$

Overlooking the value of the first point ($n = 0$), the value of the integral is well approximated by $1/2W$.

Table 1B.1 Values of Si$(k\pi)$

n	1	2	3	4	5	6	7	8	9	10
$2\,\text{Si}(n\pi)/\pi$	1.18	0.90	1.07	0.95	1.04	0.97	1.03	0.97	1.02	0.98

Appendix 1C Impulse Response of RC Filter

The Nyquist frequency characteristic, as a function of the excess bandwidth factor α, is (after multiplying by T for normalization purposes)

$$G(f;\alpha) = \begin{cases} T, & |fT| \le \frac{1}{2}(1-\alpha) \\ T\cos^2\{\frac{\pi}{2\alpha}[fT-\frac{1}{2}(1-\alpha)]\}, & \frac{1}{2}(1-\alpha) < fT \le \frac{1}{2}(1+\alpha) \\ 0, & |fT| > \frac{1}{2}(1+\alpha) \end{cases}$$

$$(1C.1)$$

The corresponding pulse shape, parametric in α, is obtained as the inverse Fourier transform of $G(f;\alpha)$. Because $G(f;\alpha)$ is an even function of frequency, its inverse transform is

$$g(t;\alpha) = 2T \int_0^{(1-\alpha)/2T} \cos(2\pi ft)\, df$$

$$+ 2T \int_{(1-\alpha)/2T}^{(1+\alpha)/2T} \cos^2\{\tfrac{\pi}{2\alpha}[fT-\tfrac{1-\alpha}{2}]\}\cos(2\pi ft)\, df$$

$$= \frac{2T}{2\pi t}\sin\left[(1-\alpha)\frac{\pi t}{T}\right] + f(t)$$

where

$$f(t) = 2T \int_0^{\alpha/T} \cos^2\left[\tfrac{\pi}{2\alpha}\, fT\right]\cos\left[2\pi t\left(f + \tfrac{1-\alpha}{2T}\right)\right]df$$

$$= T \int_0^{\alpha/T} [1 + \cos(\pi fT/\alpha)]\cos\left[2\pi t\left(f + \tfrac{1-\alpha}{2T}\right)\right]df$$

$$= T \int_0^{\alpha/T} \cos\left[2\pi t\left(f + \tfrac{1-\alpha}{2T}\right)\right]df$$

$$+ T \int_0^{\alpha/T} \underbrace{\cos(\pi fT/\alpha)\cos\left[2\pi t\left(f + \tfrac{1-\alpha}{2T}\right)\right]}_{\frac{1}{2}(\text{sum} + \text{difference})}df$$

$$= \frac{T}{2\pi t} \sin\left[2\pi t\left(f + \frac{1-\alpha}{2T}\right)\right]\Big|_0^{\alpha/T}$$

$$+ \frac{T}{4\pi(t + T/2\alpha)} \sin\left[2\pi f\left(t + \frac{T}{2\alpha}\right) + \frac{\pi t(1-\alpha)}{T}\right]\Big|_0^{\alpha/T}$$

$$+ \frac{T}{4\pi(t - T/2\alpha)} \sin\left[2\pi f\left(t - \frac{T}{2\alpha}\right) + \frac{\pi t(1-\alpha)}{T}\right]\Big|_0^{\alpha/T}$$

$$= \frac{T}{2\pi t}\left\{\sin\left[\frac{\pi t(1+\alpha)}{T}\right] - \sin\left[\frac{\pi t(1-\alpha)}{T}\right]\right\}$$

$$+ \frac{T}{4\pi\left(t + \frac{T}{2\alpha}\right)}\left\{-\sin\left[\frac{\pi t(1+\alpha)}{T}\right] - \sin\left[\frac{\pi t(1-\alpha)}{T}\right]\right\}$$

$$+ \frac{T}{4\pi\left(t - \frac{T}{2\alpha}\right)}\left\{-\sin\left[\frac{\pi t(1+\alpha)}{T}\right] - \sin\left[\frac{\pi t(1-\alpha)}{T}\right]\right\}$$

$$= \frac{T}{2\pi t}\left\{\sin\left[\frac{\pi t(1+\alpha)}{T}\right] - \sin\left[\frac{\pi t(1-\alpha)}{T}\right]\right\}$$

$$- \frac{tT}{2\pi\left[t^2 - \left(\frac{T}{2\alpha}\right)^2\right]}\left\{\sin\left[\frac{\pi t(1+\alpha)}{T}\right] + \sin\left[\frac{\pi t(1-\alpha)}{T}\right]\right\}$$

Thus

$$g(t) = \frac{T}{2\pi t}\left[1 - \frac{t^2}{t^2 - \left(\frac{T}{2\alpha}\right)^2}\right]\left\{\sin\left[\frac{\pi t(1+\alpha)}{T}\right] + \sin\left[\frac{\pi t(1-\alpha)}{T}\right]\right\}$$

$$= \frac{T}{2\pi t}\left[-\frac{\left(\frac{T}{2\alpha}\right)^2}{t^2 - \left(\frac{T}{2\alpha}\right)^2}\right]\left\{2\sin\left(\frac{\pi t}{T}\right)\cos\left(\frac{\pi t\alpha}{T}\right)\right\}$$

$$= \mathrm{sinc}\left(\frac{t}{T}\right) \cdot \frac{\cos(\pi\alpha t/T)}{1 - (2\alpha t/T)^2} \tag{1C.2}$$

Appendix 1D Probability for a Difference of Chi-Squared RVs

From the text, the probability to be found is $\Pr\{Y > 0\}$, where the RV Y is the difference of two independent scaled (central) chi-squared RVs, each with L degrees of freedom:

$$Y = \sigma_1^2 \chi_1^2(L) - \sigma_2^2 \chi_2^2(L) \tag{1D.1}$$

with $\sigma_1 < \sigma_2$ and using the notation $\mu = \sigma_1^2/\sigma_2^2$. The formulation of the probability proceeds as follows:

$$\Pr\{Y > 0\} = \Pr\{\sigma_1^2\chi_1^2(L) > \sigma_2^2\chi_2^2(L)\}$$

$$= \Pr\left\{\chi_2^2(L) < \frac{\sigma_1^2}{\sigma_2^2}\,\chi_1^2(L)\right\} = \Pr\{\chi_2^2(L) < \mu\chi_1^2(L)\} \quad \text{(1D.2a)}$$

To obtain the probability distribution function indicated in (1D.2a), we first assume that the RV χ_1^2 assumes a value, say α. We then obtain the conditional distribution function of $F_{\chi_2^2}(\mu\alpha \mid \chi_1^2 = \alpha)$. Then the conditional distribution, given that $\chi_1^2 = \alpha$, is

$$F_{\chi_2^2}(\mu\alpha \mid \chi_1^2 = \alpha) = \Pr\{\chi_2^2(L) < \mu\chi_1^2(L) \mid \chi_1^2 = \alpha\}$$

$$= \int_0^{\mu\alpha} p_{\chi_2^2}(\beta; L \mid \chi_1^2 = \alpha)\,d\beta \quad \text{(1D.2b)}$$

and the unconditional probability is

$$\Pr\{Y > 0\} = \int_0^{\infty} p_{\chi_1^2}(\alpha; L)\,F_{\chi_2^2}(\mu\alpha; L \mid \chi_1^2 = \alpha)\,d\alpha \quad \text{(1D.2c)}$$

$$= \int_0^{\infty} p_{\chi_1^2}(\alpha; L) \int_0^{\mu\alpha} p_{\chi_2^2}(\beta; L)\,d\beta\,d\alpha \quad \text{(1D.2d)}$$

in which we used the independence of the two variables in going from (1D.2c) to (1D.2d) and

$$p_{\chi_i^2}(\alpha; L) = \frac{\alpha^{(L-2)/2}}{2^{L/2}\,\Gamma(L/2)}\,e^{-\alpha/2}, \quad \alpha \geq 0, \quad i = 1, 2 \quad \text{(1D.3)}$$

Note that (1D.2a) can also be written as follows, to change the order of integration:

$$\Pr\{Y > 0\} = \Pr\left\{\chi_1^2(L) > \frac{1}{\mu}\chi_2^2(L)\right\}$$

$$= \int_0^{\infty} p_{\chi_2^2}(\alpha; L) \int_{\alpha/\mu}^{\infty} p_{\chi_1^2}(\beta; L)\,d\beta\,d\alpha \quad \text{(1D.4)}$$

For even-valued L. When $L/2 = \nu$, an integer, the second form of the probability (1D.4) is more convenient. The inner integral in (1D.4) is

$$\int_{\alpha/\mu}^{\infty} p_{\chi_1^2}(\beta; L) \, d\beta = \frac{1}{2^\nu \, \Gamma(\nu)} \int_{\alpha/\mu}^{\infty} \beta^{\nu-1} \, e^{-\beta/2} \, d\beta$$

$$= \frac{1}{\Gamma(\nu)} \int_{\alpha/2\mu}^{\infty} x^{\nu-1} \, e^{-x} \, dx$$

$$= \frac{\Gamma(\nu; \alpha/2\mu)}{\Gamma(\nu)} = e^{-\alpha/2\mu} \sum_{k=0}^{\nu-1} \frac{(\alpha/2\mu)^k}{k!} \tag{1D.5}$$

in which $\Gamma(\nu; \alpha/2\mu)$ is an *incomplete gamma function* [20]. Substituting (1D.5) in (1D.4) yields

$$\Pr\{Y > 0\} = \frac{1}{2^\nu \Gamma(\nu)} \sum_{k=0}^{\nu-1} \frac{(1/2\mu)^k}{k!} \int_0^\infty \alpha^{k+\nu-1} \, e^{-(1+1/\mu)\alpha/2} \, d\alpha$$

$$= \frac{1}{\Gamma(\nu)} \sum_{k=0}^{\nu-1} \frac{(1/\mu)^k}{k!} \int_0^\infty x^{k+\nu-1} \, e^{-(1+1/\mu)x} \, dx$$

$$= \frac{1}{\Gamma(\nu)} \sum_{k=0}^{\nu-1} \frac{(1/\mu)^k}{k!} \left(\frac{\mu}{1+\mu}\right)^{k+\nu} \Gamma(k+\nu)$$

$$= \left(\frac{\mu}{1+\mu}\right)^{L/2} \sum_{k=0}^{\frac{L}{2}-1} \binom{\frac{L}{2}+k-1}{k} \left(\frac{1}{1+\mu}\right)^k \tag{1D.6}$$

For odd-valued L. When $L/2 = \nu + \frac{1}{2}$, where ν is an integer, the first form of the probability (1D.2d) is more convenient. The inner integral in (1D.2d) is

$$\int_0^{\mu\alpha} p_{\chi_2^2}(\beta; L) \, d\beta = \frac{1}{2^{\nu+\frac{1}{2}} \, \Gamma(\nu + \frac{1}{2})} \int_0^{\mu\alpha} \beta^{\nu-\frac{1}{2}} \, e^{-\beta/2} \, d\beta$$

$$= \frac{1}{\Gamma(\nu + \frac{1}{2})} \int_0^{\mu\alpha/2} x^{\nu-\frac{1}{2}} \, e^{-x} \, dx = \frac{\gamma(\nu + \frac{1}{2}; \mu\alpha/2)}{\Gamma(\nu + \frac{1}{2})}$$

$$= \frac{1}{\Gamma(\nu + \frac{3}{2})} \left(\frac{\mu\alpha}{2}\right)^{\nu+\frac{1}{2}} e^{-\mu\alpha/2} \, _1F_1\left(1; \nu + \frac{3}{2}; \frac{\mu\alpha}{2}\right) \tag{1D.7}$$

where $\gamma(\nu + \frac{1}{2}; \mu\alpha/2)$ is another *incomplete gamma function* [20] and the *confluent hypergeometric function* $_1F_1(\bullet)$ is [20]

$$_1F_1(a; b; x) = \frac{\Gamma(b)}{\Gamma(a)} \sum_{n=0}^{\infty} \frac{x^n}{n!} \frac{\Gamma(n+a)}{\Gamma(n+b)} = 1 + \frac{ax}{b} + \frac{a(a+1)x^2}{2b(b+1)} + \cdots \quad (1D.8)$$

Substituting (1D.7) and (1D.8) into (1D.2) yields

$$\begin{aligned}
\Pr\{Y > 0\} &= \frac{(\mu/2)^{\nu+\frac{1}{2}}}{2^{\nu+\frac{1}{2}}\Gamma(\nu+\frac{1}{2})} \sum_{n=0}^{\infty} \frac{(\mu/2)^n}{\Gamma(n+\nu+\frac{3}{2})} \int_0^{\infty} \alpha^{n+2\nu} e^{-(1+\mu)\alpha/2} d\alpha \\
&= \frac{\mu^{\nu+\frac{1}{2}}}{\Gamma(\nu+\frac{1}{2})} \sum_{n=0}^{\infty} \frac{\mu^n}{\Gamma(n+\nu+\frac{3}{2})} \int_0^{\infty} x^{n+2\nu} e^{-(1+\mu)x} dx \\
&= \frac{\mu^{\nu+\frac{1}{2}}}{\Gamma(\nu+\frac{1}{2})} \sum_{n=0}^{\infty} \frac{\mu^n}{\Gamma(n+\nu+\frac{3}{2})} \left(\frac{1}{1+\mu}\right)^{n+2\nu+1} \Gamma(n+2\nu+1) \\
&= \frac{\mu^{\nu+\frac{1}{2}}}{(1+\mu)^{2\nu+1}} \frac{\Gamma(2\nu+1)}{\Gamma(\nu+\frac{1}{2})\Gamma(\nu+\frac{3}{2})} \, _2F_1\left(1, 2\nu+1; \nu+\frac{3}{2}; \frac{\mu}{1+\mu}\right) \\
&= \frac{\mu^{L/2}}{(1+\mu)^L} \frac{\Gamma(L)}{\Gamma(\frac{L}{2})\Gamma(\frac{L}{2}+1)} \, _2F_1\left(1, L; \frac{L}{2}+1; \frac{\mu}{1+\mu}\right) \quad (1D.9)
\end{aligned}$$

in which the *gaussian hypergeometric function* $_2F_1(\bullet)$ is [20]

$$_2F_1(a, b; c; x) = \frac{\Gamma(c)}{\Gamma(a)\Gamma(b)} \sum_{n=0}^{\infty} \frac{x^n}{n!} \frac{\Gamma(n+a)\Gamma(n+b)}{\Gamma(n+c)}$$

$$= 1 + \frac{abx}{c} + \frac{a(a+1)b(b+1)x^2}{2c(c+1)} + \cdots \quad (1D.10)$$

2

Mobile Radio Propagation Considerations

This chapter provides background material on the propagation of radio waves in the mobile environment that is needed for the consideration of the design and performance of analog and digital cellular systems. After a brief overview of the more general propagation theory and models, we give the mobile radio environment particular attention, and the chapter concludes with discussions on the application of propagation models to link budgets and to cellular system design.

2.1   Overview of Propagation Theory and Models

In this section, we present an overview on selected topics in propagation theory that are important for mobile radio communications scenarios at UHF frequencies. The propagation of radio waves in free space is first discussed, then we show how we can use the free-space propagation as a reference and show how to account for the differences between free space and the Earth's atmosphere and terrain. A more detailed presentation of this subject can be found in [1, 2].

2.1.1   Free-Space Propagation

A sinusoidal waveform emitted by a point source would propagate radially in a vacuum (free space) and for that reason we can call the source an *isotropic radiator*. If the emitted power is P_{rad} W, at a distance d meters from the source, the magnitude of the wave's Poynting vector (power per unit area) is

$$P_{fs} = \frac{P_{rad}}{4\pi d^2} \quad \text{in W/m}^2 \tag{2.1a}$$

In (2.1), fs denotes free space. For an antenna (radiator) that is not isotropic, at a distance large compared with the size of the antenna, the radiated power P_{rad} may be replaced by $P_t G_t$, where

$$P_t = \text{power delivered to the transmitter antenna} \tag{2.1b}$$

$$G_t = \text{transmitter antenna gain} \tag{2.1c}$$

For derivation of the effects of Earth on radio wave propagation, it is sometimes more convenient to speak in terms of the root-mean-square electric field intensity (in volts per meter) given by

$$E_{fs} = \sqrt{Z_{fs} P_{fs}} \tag{2.2a}$$

where Z_{fs}, the impedance of free space, is given by [3]

$$Z_{fs} = \sqrt{\mu_{fs}/\epsilon_{fs}} \approx 120\pi \doteq 377 \, \Omega \tag{2.2b}$$

and $\mu_{fs} = 4\pi \times 10^{-7} \, \text{H/m} = 4\pi \times 10^{-7} \, \text{V-s/A-m}$ and $\epsilon_{fs} = (10^{-9}/36\pi) \, \text{F/m}$ $= (10^{-9}/36\pi) \, \text{A-sec/V-m}$ are the permeability and the permittivity (dielectric constant), respectively, of free space. At a relatively large distance from a nonisotropic radiator, then, the electric field intensity in free space is

$$E_{fs} = \sqrt{120\pi \cdot \frac{P_t G_t}{4\pi d^2}} = \frac{\sqrt{30 P_t G_t}}{d} \, \text{V/m} \tag{2.3}$$

Whether or not there is free-space propagation, if the electric field intensity and power at a receiving antenna are E_{rec} and P_{rec}, respectively, the maximal useful power that could be intercepted by a matched receiver using an isotropic antenna is given by [4]

$$P_r = \frac{\lambda^2}{4\pi} \cdot P_{rec} \tag{2.4a}$$

For a nonisotropic antenna (with antenna power gain G_r) and a matched receiver, the received power is, using $E_{rec} = \sqrt{Z_{fs} P_{rec}}$:

$$P_r = \frac{\lambda^2}{4\pi} \cdot P_{rec} G_r = \frac{\lambda^2}{4\pi} \cdot \frac{E_{rec}^2}{Z_{fs}} \cdot G_r = \left(\frac{E_{rec}\lambda}{2\pi}\right)^2 \cdot \frac{G_r}{120} \tag{2.4b}$$

If (2.3) is solved for transmitter power, the ratio of received power to transmitted power is found to be

$$\frac{P_r}{P_t} = \left(\frac{E_{rec}\lambda}{2\pi}\right)^2 \frac{G_r}{120} \div \frac{(E_{fs}d)^2}{30\,G_t} = \begin{cases} \left(\dfrac{\lambda}{4\pi d}\right)^2 G_t G_r \left(\dfrac{E_{rec}}{E_{fs}}\right)^2, \text{ general} \\[2mm] \left(\dfrac{\lambda}{4\pi d}\right)^2 G_t G_r, \text{ free space} \end{cases} \quad (2.5)$$

The general relation in (2.5) shows how to assess the effects of propagation in terms of the received electric field intensity relative to the value in free space. That is, in dB the propagation loss is

$$\text{Path loss} = -10\log_{10}\left(\frac{P_r}{P_t} \cdot \frac{1}{G_t G_r}\right)$$

$$= 20\log_{10}\left(\frac{4\pi d}{\lambda}\right) + 20\log_{10}\left(\frac{E_{fs}}{E_{rec}}\right) \quad (2.6a)$$

$$= L_{fs} + L_{nfs} \quad (2.6b)$$

where the subscript nfs denotes "non-free-space" and the free-space loss is

$$L_{fs} = 20\log_{10}\left(\frac{4\pi d}{\lambda}\right) = 20\log_{10}\left(\frac{4\pi \cdot 1,000\,d_{km}}{299.8/f_{MHz}}\right)$$

$$= 32.45\,\text{dB} + 20\log_{10}(d_{km}\,f_{MHz}) \quad (2.7a)$$

$$= 36.58\,\text{dB} + 20\log_{10}(d_{mi}\,f_{MHz}) \quad (2.7b)$$

using the notations $d_{km} \equiv d$ in kilometers, $d_{mi} \equiv d$ in miles and $f_{MHz} \equiv f$ in megahertz. Figure 2.1 shows the free-space path loss for frequencies and distances that are relevant to mobile communications.

2.1.2 Radio Horizon and Propagation Modes

In this section, we show how the refractive properties of the atmosphere and Earth's terrain affect not only the received signal power but the curvature of the path take by the radio waves. The length of the path relative to the *radio horizon*, or line-of-sight distance, is shown as a basis for modeling the propagation according to different propagation modes.

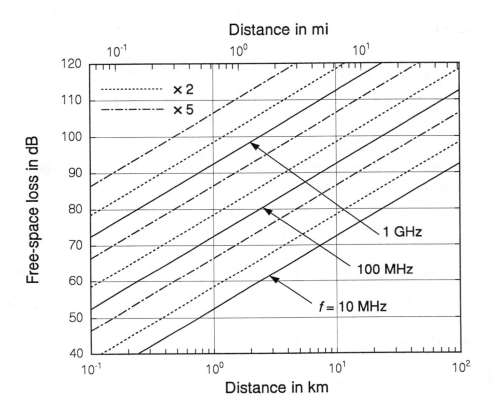

Figure 2.1 Free-space path loss.

2.1.2.1 Effect of the Atmosphere

For ground-to-ground communications at VHF and UHF, the departure of the value of the received electric field intensity E_{rec} from the value that would occur in free space for the same distance arises chiefly from effects of the interaction of the radio wave with Earth. There are, however, significant effects due to the non-free-space properties of the medium of propagation itself—the air in the troposphere, that portion of the atmosphere extending to about 10 mi above Earth.

Because of the presence of various gases in the atmosphere, including water vapor, the dielectric constant ϵ_r of the air in the troposphere is slightly greater than unity, the value of the dielectric constant in free space; because the density of these gases generally decreases with height, ϵ_r and the refractive index of the air, $n = \sqrt{\epsilon_r}$, decrease with height. This variation of refractive index gives rise to propagation phenomena such as refraction, reflection, scattering, duct transmission, and fading of signals.

The path of a radio wave ray traveling in the atmosphere can be described in terms of the variation with height of the refractive index of air, n, as a function of h, the height above ground. The effects of interest are related to the derivative of the index of refraction, which typically has a negative value in the troposphere, with a value for the "standard atmosphere" such that the slope of n is about $-.012/\text{ft} = -.039/\text{m}$. Because of the decrease of the index of refraction with increase in height, the velocity of propagation increases with height in such a way that the ray paths are very nearly arcs of circles over the distances involved in ground-to-ground mobile communications, with the radius of curvature $r_w \approx 4r_e$ in the standard atmosphere, where

$$r_e = \text{radius of Earth (6,370 km)} \qquad (2.8a)$$

A horizontally emitted ray is curved, bending toward Earth instead of continuing in a straight line, but not intersecting Earth (assumed to be a sphere), as illustrated in Figure 2.2.

Because radio wave paths are curved, the distance to the horizon for radio signals is larger than the straight-line distance (optical path). The geometry of the situation in which an inclined curved ray path from or to antenna i with effective height h_{ei} is tangent to the earth is illustrated in Figure 2.3 for a "smooth Earth" with no terrain features. The path distance to the tangent point is the smooth-Earth radio horizon distance for that antenna, denoted d_{Lsi} in the diagram on the left side of Figure 2.3.

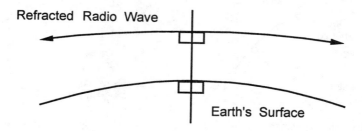

Figure 2.2 Curvature of radio wave paths in air.

Because for analysis of communications paths it is more convenient to deal with straight-line ray paths, it is desirable to change the geometrical coordinate system so that the refracted rays appear to be straight lines. For this purpose, a modified tangential geometry involving a fictitious Earth, having a radius $a = kr_e > r_e$, is postulated as shown in the diagram on the right side of Figure 2.3 with the same distance d_{Lsi}. For equivalence of results, the solution for k is constrained by the requirement that the distance of the ray path above Earth, h, remain the same for either model for any distance along the tangential ray path. Under this constraint, it can be shown [2] that the fictitious Earth's radius should be taken to be $a = 4r_e/3 = 8,493$ km. The use of the effective Earth radius formula $a = 4r_e/3$ is said to allow ray paths to be considered as being straight so long as they lie within the first kilometer above Earth's surface [5].

The radio horizon distance for a smooth Earth, as a function of antenna height, is calculated using the geometry of the right side of Figure 2.3. Recognizing that d_{Lsi} is one side of a right triangle, with the other side being the fictitious Earth radius a and the hypotenuse being $a + h_{ei}$, the solution for d_{Lsi} is

$$d_{Lsi} = \sqrt{(a + h_{ei})^2 - a^2} \approx \sqrt{2ah_{ei}} \qquad (2.8b)$$

when Earth's radius and the antenna height are expressed in the same units. For typical units:

$$d_{Lsi}(\text{km}) \approx \sqrt{0.002\, a(\text{km})\, h_{ei}(\text{m})} = \sqrt{17\, h_{ei}(\text{m})} \qquad (2.8c)$$

$$d_{Lsi}(\text{mi}) \approx \sqrt{2\, h_{ei}(\text{ft})} \qquad (2.8d)$$

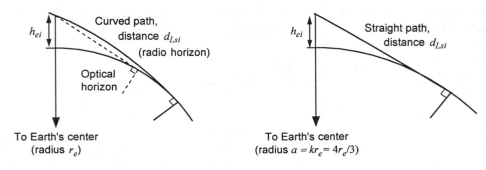

Figure 2.3 Geometry of radio waves tangent to Earth.

A transmitter and receiver are said to be within radio line-of-sight (LOS) when the link distance d is such that

$$d_{Ls} \triangleq d_{Lst} + d_{Lsr} > d \tag{2.9}$$

The actual radio horizon distance, as opposed to the idealized value given by (2.8b), is typically smaller, because of the effect of terrain and buildings.

2.1.2.2 Characterization of Terrain and Its Effects

Obviously, if a hill or building is "in the way" of the tangential path depicted in the smooth-Earth model of Figure 2.3, the actual horizon distance is the distance to a tangent with that object. Therefore, the more hilly or populous the area is, the more likely it is that the radio horizon is closer than predicted by the smooth-Earth model of (2.8). The presence of terrain or buildings also affects the LOS distance through the effective antenna height: if the antenna is put on top of a hill or a building, the radio horizon is much farther away.

Typically, the altitude above sea level of the land in the vicinity of the transmitting or receiving antenna, and along the great circle surface path between them, varies to some degree above and below a "reference plane." That is, the land altitude at a distance x along the path from transmitter to receiver can be expressed as

$$h_a(x) = h_{ref}(x) + h_s(x), \quad x = 0 \text{ (transmitter) to } d \text{ (receiver)} \tag{2.10}$$

in which $h_{ref}(x)$ denotes the reference or average terrain height along the path and $h_s(x)$ denotes the variations in surface height. For the purposes of modeling the effects of terrain on horizon distance and other factors affecting propagation, the reference plane can be a straight line fit to the points of $h_a(x)$ if these data are available. The deviations of the surface height described as samples of $h_s(x)$ are likely to have a roughly symmetrical distribution, with a zero-valued median and mean. The degree of variation of the terrain can be parameterized by the "terrain irregularity parameter" Δh, the interdecile range—the difference between the value of $h_s(x)$ that is exceeded for 10% of the samples and the value that is not exceeded for 10% of the samples, as illustrated in Figure 2.4.

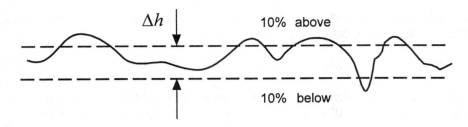

Figure 2.4 Terrain irregularity parameter.

From studies of many terrain profiles (reported in [5] and [6]), it has been determined that the median interdecile height as a function of the length of the path d is given by

$$\Delta h(d) = \Delta h \left(1 - 0.8\, e^{-d_{km}/50}\right) \tag{2.11}$$

When an interdecile range based on analysis of a particular terrain profile is given, it is treated as the value of $\Delta h(d)$ for that distance, and the asymptotic value Δh may be estimated by solving (2.11). When detailed path terrain profiles are not available, the analyst can specify a value for Δh that is chosen to fit one of the terrain descriptions in Table 2.1 [5].

Let the structural height of antenna i (either $i = t$ for transmitter or $i = r$ for receiver) above the local ground be denoted h_{gi}; then the effective antenna height for that antenna can be formulated as

$$h_{ei} \triangleq \max[h_{gi}\,,\, h_{gi} + h_a(x_i) - h_{ref}(x_i)], \quad i = t,\, r \tag{2.12}$$

Table 2.1 Terrain irregularity parameter values

Terrain description	Δh in meters
Water or very smooth plains	0–5
Smooth plains	5–20
Slightly rolling plains	20–40
Rolling plains	40–80
Hills	80–150
Mountains	150–300
Rugged mountains	300–700
Extremely rugged mountains	> 700

Let h_s denote the height of the terrain surface in a given antenna location, relative to the local average terrain height. If the antenna is on a hill ($h_s > 0$), (2.12) indicates that the terrain height is added to the structural height to obtain the effective antenna height; otherwise, the effective height is the actual, structural height. This definition of effective antenna height is illustrated in Figure 2.5.

For prediction purposes in the absence of specific terrain data, the parameter Δh can be related to a statistical treatment of the location of antenna sites. If an antenna site is randomly selected, as tends to be the case for a mobile radio unit, then it is reasonable to assume that, on the average, the altitude $h_a(x)$ at the antenna site equals the reference value h_{ref}. It is to be expected that a more carefully located antenna would be placed on a hill. In [5], based on the analysis of many path profiles, the following empirical formula for effective antenna height was developed for computer propagation calculations when the structural height of the antenna is 10m or less:

$$h_{ei} = \begin{cases} h_{gi}, & \text{random siting} \\ h_{gi} + \left[1 + c \cdot \max\left\{1, \sin\left(\frac{\pi h_{gi}}{10\,\text{m}}\right)\right\}\right] e^{-2h_{gi}/\Delta h}, & \text{selected siting} \end{cases} \quad (2.13a)$$

where

$$c = \begin{cases} 4, & \text{careful siting} \\ 9, & \text{very careful siting} \end{cases} \quad (2.13b)$$

Note that the siting practices make no difference when $\Delta h = 0$; for $\Delta h \neq 0$, as h_{gi} increases, it rises quickly but then approaches the value $h_{ei} = h_{gi}$ asymptotically, as shown in Figure 2.6.

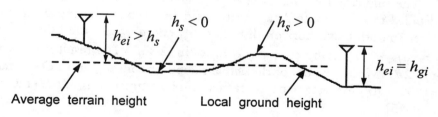

h_s = Local ground height – average terrain height

h_{ei} = effective antenna height $\qquad h_{gi}$ = structural antenna height

Figure 2.5 Effective antenna height.

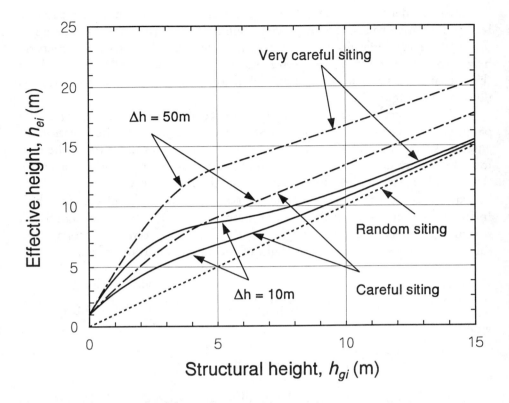

Figure 2.6 Effect of terrain and siting practice on effective antenna height [2, 5].

The empirical formula for effective antenna height (2.13), in the absence of detailed terrain profile data, can be used to estimate the transmitter and receiver smooth-Earth horizon distances d_{Lst} and d_{Lsr} and the LOS distance $d_{Ls} = d_{Lst} + d_{Lsr}$ using $d_{Lsi} = \sqrt{2ah_{ei}}$ as in (2.8b). The result is shown in Figure 2.7. Median values of the actual radio horizons d_{Li} and LOS distance $d_L = d_{L1} + d_{L2}$ for irregular terrain then can be estimated from the empirical formula [5]

$$d_{Li} = d_{Lsi} \times e^{-0.07\sqrt{\Delta h/\max(h_{ei},\, 5\,\mathrm{m})}} \qquad (2.14)$$

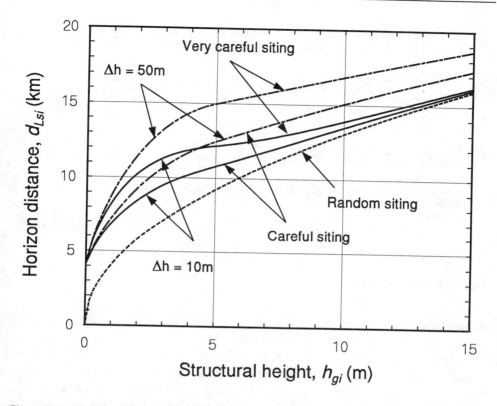

Figure 2.7 Effect of terrain and siting practice on horizon distance [2, 5].

2.1.2.3 Propagation Modes

The effects of the presence of Earth on the ground-to-ground propagation of radio waves at VHF and UHF depend on path length. For shorter, LOS paths, the main influence of Earth is to cause a reflected ray to arrive at the receiver in addition to the direct ray, causing destructive or constructive interference. For beyond-line-of-sight (BLOS) or transhorizon paths, the propagation of the signal over the horizon is possible because of either diffraction of the wave over the horizon, scattering of the signal from the troposphere, or both. As illustrated in Figure 2.8, generally there is a gradual transition between one mode of propagation and another.

As Figure 2.8 suggests, the dominant mode of propagation is determined by the link distance d. If d is less than the combined horizon

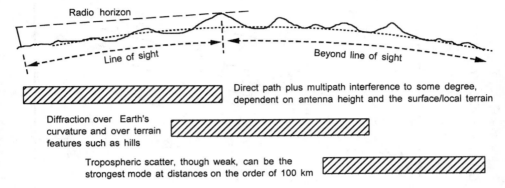

Figure 2.8 Propagation modes.

distances of the transmitter and receiver antennas, which is called the *LOS distance*, the dominant propagation mode is LOS.

From the "random siting" curve of Figure 2.7 it is evident that a mobile unit in a cellular system, with an antenna height of about $h_{gm} = h_{em} = 2\,\mathrm{m}$, has a smooth-Earth radio horizon of about $d_{Lsm} = 5\,\mathrm{km}$. In a rural setting, the likelihood that there is a terrain obstacle reduces this distance according to the empirical formula given in (2.14). For example, if the terrain can be characterized as "hills" with a terrain irregularity parameter value of $\Delta h = 100\mathrm{m}$, the actual horizon distance for the mobile is about

$$d_{Lm} = 5 \cdot e^{-0.07\sqrt{100/\max(2,\,5)}} = 5 \cdot e^{-0.07\sqrt{20}} = 3.7\,\mathrm{km} \qquad (2.15)$$

A base station in the same cellular system will typically have an antenna structural height of about $h_{gb} = 100\text{--}200$ ft or 30–60 m; for definiteness, let us assume that $h_{gb} = 50\mathrm{m}$. On a smooth Earth, using (2.8b) we can see that this antenna height corresponds to a horizon distance of about $d_{Lsb} = \sqrt{16.99\,h_{gb}} = 29.1\,\mathrm{km}$. In the case of actual terrain, the base station antenna would be placed on a hill and the horizon distance without obstacles is even greater— according to the empirical "careful siting" formula given in (2.13), the effective antenna height of an elevated 50m base station antenna is about

$$h_{eb} = h_{gb} + \left[1 + 4 \cdot \max\left\{1, \sin\left(\frac{\pi h_{gb}}{10}\right)\right\}\right] e^{-2h_{gb}/\Delta h}$$

$$= 50 + [1 + 4 \cdot \max\{1, \sin(5\pi)\}]e^{-2\cdot 50/100} = 51.8\,\text{m} \tag{2.16}$$

This effective antenna height gives a smooth-Earth horizon distance of $d_{Lsb} = 29.7\,\text{km}$, which is reduced to a likely actual horizon distance of

$$d_{Lb} = 29.7 \times e^{-0.07\sqrt{100/51.8}} = 27.0\,\text{km} \tag{2.17}$$

Thus, the total LOS distance for a base-to-mobile link or mobile-to-base link is likely to be on the order of $d_L = d_{Lb} + d_{Lm} = 27.0 + 3.7 = 30.7\,\text{km}$.

For cellular communications, the distances involved dictate that the propagation mode that is most often involved is LOS. However, in an urban setting, the buildings often act as man-made terrain obstacles that reflect and block the direct LOS path between the antennas, so that the mode of propagation is almost always a complex combination of reflected paths that make their way around buildings and diffracted paths that are bent as they interact with the tops of buildings. For that reason it is quite difficult to predict propagation loss in urban settings using a theoretical model, and a number of empirical formulas are used for prediction.

2.1.3 LOS and Diffraction Propagation Modes

Although in most cellular radio situations the propagation mechanisms are rather complex because of the interaction with the radio waves with buildings, it is useful to understand the basic effects of multipath reception in the LOS mode and the diffraction of signals over natural and man-made obstacles.

2.1.3.1 Propagation in the LOS Region

For an LOS situation, the geometry is as shown in the left side of Figure 2.9; for a given receiver antenna location, there is both a direct ray with path length r_1 and a reflected ray with path length r_2. Because the orientation of the electric field is reversed when reflected (giving an apparent 180° phase shift for horizontally polarized waves), the reflected ray tends to act as destructive interference when $r_1 \approx r_2$, which occurs for lower antenna heights and longer distances. One effect of Earth's curvature relative to the

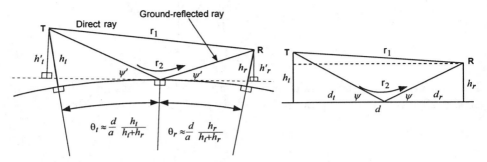

Reflection from a smooth spherical Earth   Reflection from a plane smooth Earth

Figure 2.9 LOS ray path geometries.

geometry for a flat or plane Earth shown in the right side of Figure 2.9 can be seen as the apparent antenna heights h'_i $(i = t, r)$ are shorter than the actual antenna heights, h_i, resulting in a smaller difference between r_1 and r_2 (for the same distance along the surface), tending toward more destructive interference. Another effect is divergence of the reflected wavefront, tending to reduce the strength of the reflected ray. These two effects counter each other, so that to a sufficient degree of accuracy the propagation of LOS paths can be analyzed using the plane Earth model in the right side of Figure 2.9.

From Figure 2.9, the difference in path length for the direct and reflected rays is seen as

$$
\begin{aligned}
r_2 - r_1 &= \sqrt{(h_t + h_r)^2 + d^2} - \sqrt{(h_t - h_r)^2 + d^2} \\
&\approx d\left[1 + \frac{1}{2}\left(\frac{h_t + h_r}{d}\right)^2\right] - d\left[1 + \frac{1}{2}\left(\frac{h_t - h_r}{d}\right)^2\right] \\
&= \frac{2h_t h_r}{d}
\end{aligned}
\tag{2.18a}
$$

This path-length difference gives rise to the phase difference

$$
\Delta = \frac{4\pi h_t h_r}{\lambda d} = 1.3343\pi \times 10^{-5}\, \frac{f_{\mathrm{MHz}} h_t h_r}{d_{\mathrm{km}}}
\tag{2.18b}
$$

when the antenna heights are expressed in meters. The grazing angle or angle of reflection, ψ, is given by

$$\tan \psi = \frac{h_t}{d_t} = h_t \div \frac{h_t \, d}{h_t + h_r} = \frac{h_t + h_r}{d} \qquad (2.18c)$$

Ignoring a component of the signal that reaches the receiver by groundwave propagation from the point of reflection that is negligible at VHF and UHF [7], the squared magnitude of the ratio between received and free-space electric field intensities is

$$\left| \frac{E_{rec}}{E_{fs}} \right|^2 \approx \left| 1 + \mathrm{R} \, e^{j\Delta} \right|^2 = 1 + |\mathrm{R}|^2 + 2|\mathrm{R}| \cos(\Delta + \xi) \qquad (2.19a)$$

where the complex reflection coefficient R is related to the grazing angle and the ground impedance z by

$$\mathrm{R} = \frac{\sin\psi - z}{\sin\psi + z} \equiv |\mathrm{R}| \, e^{j\xi} \qquad (2.19b)$$

The ground impedance z is related to ϵ_g, the complex dielectric constant of the partially conducting Earth [8, 9] and is given by

$$z = \begin{cases} \sqrt{\epsilon_g - \cos^2\psi} \approx \sqrt{\epsilon_g - 1}, & \text{horizontal polarization} \\ \sqrt{\epsilon_g - \cos^2\psi}/\epsilon_g \approx \sqrt{\epsilon_g - 1}/\epsilon_g, & \text{vertical polarization} \end{cases} \qquad (2.20a)$$

The value of ϵ_g is related to the free-space permittivity ϵ_{fs} as well as to ϵ and σ, the relative permittivity and conductivity of the ground, respectively, by

$$\epsilon_g = \epsilon - j \frac{\sigma}{2\pi f \epsilon_{fs}} = \epsilon - 1.796 \times 10^4 \, \frac{j\sigma}{f_{\mathrm{MHz}}} \qquad (2.20b)$$

Typical values of Earth parameters are $\sigma = 0.005$ and $\epsilon = 15$, giving $\epsilon_g = 15 - j90/f_{\mathrm{MHz}}$; for $f = 100\,\mathrm{MHz}$, the resulting value of z is $3.75\angle-1.84°$ for horizontal polarization and $0.25\angle1.59°$ for vertical polarization. For $f = 1{,}000\,\mathrm{MHz}$, z is $3.74\angle-0.18°$ for horizontal polarization and $0.25\angle0.16°$ for vertical polarization. Thus, for UHF cellular frequencies and very low grazing angles, $\mathrm{R} \approx -1$ and (2.19a) becomes

$$\left|\frac{E_{rec}}{E_{fs}}\right|^2 \approx 2(1 - \cos\Delta) = 4\sin^2(\Delta/2) = 4\sin^2\left(\frac{2\pi h_t h_r}{\lambda d}\right) \qquad (2.21)$$

The relation shown in (2.21) indicates that LOS propagation over a smooth Earth results in a 6-dB signal power gain over free space when $\Delta/2$ is an odd multiple of $\pi/2$, when $\sin(\Delta/2) = \pm 1$, and results in signal cancellation when $\Delta/2$ is a multiple of π, when $\sin(\Delta/2) = 0$. A variation in this angle can be due to a variation in antenna heights, link distance, or both.

For example, consider a base station antenna height $h_t = 50$ m and a mobile antenna height of $h_r = 2$ m, as depicted in Figure 2.10. If the frequency is 850 MHz, then (2.18b) gives $\Delta = 1.134\pi/d_{km}$. For $d < 1.134$ km, the angle $\Delta/2$ in (2.21) is greater than $\pi/2$, and the gain described by (2.21) oscillates as the mobile moves toward the base station, as shown in the graph in Figure 2.11. For $d > 1.134$ km, the angle $\Delta/2$ is always less than $\pi/2$, and there are no oscillations in the attenuation as the mobile moves farther away from the base station. Note in Figure 2.11 that the oscillations are smaller if the magnitude of the ground reflection is less than unity; as discussed in Appendix 2A.2, the roughness of Earth generally causes $|R|$ to be less than 1.

The distance at which $\Delta = n\pi$ is given by

$$d_n = 4h_t h_r/n\lambda \quad \Leftrightarrow \quad \Delta = n\pi \qquad (2.22)$$

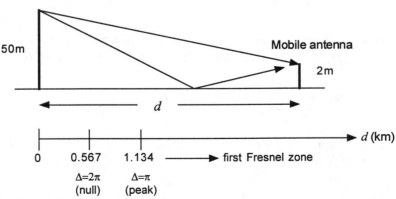

Figure 2.10 Example mobile two-ray multipath situation.

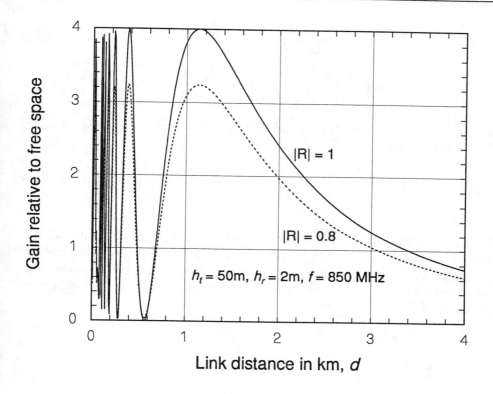

Figure 2.11 Attenuation relative to free space for the example of Figure 2.10.

Note in particular that if $d > d_1 = 4h_t h_r/\lambda$, then $\Delta/2 < \pi/2$ and no oscillations of (2.21) occur in the sense that the received signal power always decreases as d increases.

The set of positions for which the condition $\Delta < n\pi$ holds is referred to as the nth Fresnel zone; the first Fresnel zone corresponds to $d > d_1$. For $d > d' \triangleq 12h_t h_r/\lambda$, then $\Delta/2 < \pi/6$, making $\sin(\Delta/2) < 0.5$, and the received LOS power is always less than that for free space, according to (2.21). Also, the sine function in (2.21) can be replaced by its argument, giving

$$\left| \frac{E_{rec}}{E_{fs}} \right|^2 \approx \left(\frac{4\pi h_t h_r}{\lambda d} \right)^2, \quad d > d' \tag{2.23a}$$

The net gain for the path is (2.21) multiplied by the free-space loss:

$$\frac{P_r}{P_t} = 4\sin^2\left(\frac{2\pi h_t h_r}{\lambda d}\right) \times \left(\frac{\lambda}{4\pi d}\right)^2 \tag{2.23b}$$

$$\approx 4\left(\frac{2\pi h_t h_r}{\lambda d}\right)^2\left(\frac{\lambda}{4\pi d}\right)^2 = \frac{(h_t h_r)^2}{d^4}, \quad d > d' \tag{2.23c}$$

which shows a fourth-power dependence on distance for $d > d'$. In dB units, this LOS propagation loss is

$$L_{LOS} = 10\log_{10}\left[\frac{(h_t h_r)^2}{d^4}\right] = 120 + 40\log_{10}d_{km} - 20\log_{10}(h_{tm}h_{rm}) \tag{2.23d}$$

For the same example parameters as in Figure 2.11, the net path gain (2.23b) is plotted in Figure 2.12, along with the asymptotic fourth-power law expression (2.23c). The boundaries of the first and second Fresnel zones are also noted in this figure, and the influence of the magnitude of the ground reflection coefficient, $|R|$, is shown.

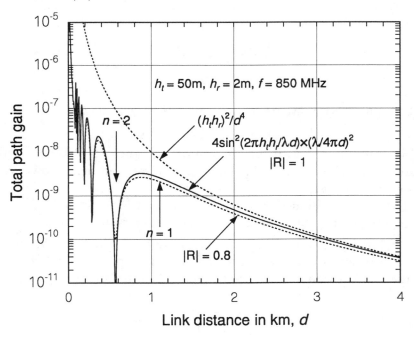

Figure 2.12 Net path gain for the example of Figure 2.10.

Using an example, it has been shown that the ground reflection points are within the first Fresnel zone when the mobile is relatively far from the base station. When the terrain is not flat or when there are buildings in the vicinity of the direct path, however, other reflection points are possible that can cause destructive multipath interference. In general, the first Fresnel zone is defined as the elliptical volume containing reflection points for which the difference in path lengths of direct and reflected rays is less than half a wavelength [10]. As illustrated in Figure 2.13, the volume is elliptical because an ellipse is the locus of points for which the combined distance from two focal points (here, the length of the reflected path from transmitter to receiver) is a constant. Note that in Figure 2.13, the terrain heights along the path from transmitter to receiver have been adjusted to account for Earth's curvature, so that the direct path can be plotted as a straight line.

The radius h_0 of the first Fresnel zone at the some point along a path of length d that is distance d_t from the transmitter and distance d_r from the receiver is found by solving the following equation for h_0:

$$\lambda/2 = \text{length of reflected path} - d$$

$$= \text{distance from transmitter to reflection point}$$
$$+ \text{distance from reflection point to receiver} - d$$

$$= \sqrt{d_t^2 + h_0^2} + \sqrt{d_r^2 + h_0^2} - d$$

$$= d_t \left[1 + \frac{h_0^2}{2d_t^2} + \cdots \right] + d_r \left[1 + \frac{h_0^2}{d_r^2} + \cdots \right] - (d_t + d_r)$$

$$\approx \frac{h_0^2}{2} \cdot \frac{d_t + d_r}{d_t d_r} \Rightarrow h_0(\text{m}) = \sqrt{\frac{\lambda d_t d_r}{d}} = 548 \sqrt{\frac{d_{t\text{km}} d_{r\text{km}}}{d_{\text{km}} f_{\text{MHz}}}} \qquad (2.24)$$

$$h_0 = \text{Radius of first Fresnel zone}$$

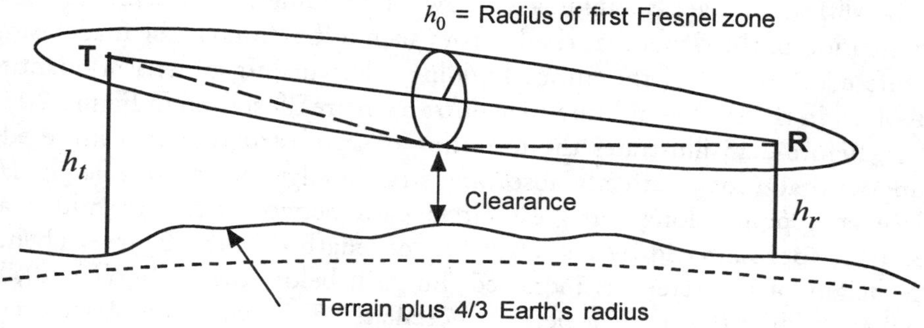

Figure 2.13 First Fresnel zone and path clearance.

Thus, for example, midway in a $d = 5\,\text{km}$ path, the radius of the first Fresnel zone is $h_0 = 21\text{m}$ for $f = 850\,\text{MHz}$ and $h_0 = 14\text{m}$ for $f = 1,900\,\text{MHz}$. If the direct path clears the terrain and any buildings by this amount, the reflected ray will constructively interfere with the direct ray; otherwise, there is the possibility of destructive multipath interference that increases with frequency. For this reason, it is desirable to mount the base station antenna high off the ground when possible, such as on top of a building; in an urban area, it usually is not possible to locate the antenna at a height that guarantees first Fresnel zone clearance for all potential mobile locations.

2.1.3.2 Diffraction Over Terrain and Buildings

When the link distance d is greater than the LOS distance d_{Ls}, or when a natural or man-made obstacle blocks the direct path, there still may be a useful amount of signal power at the receiver from diffraction.

Radio waves may be transmitted beyond the LOS into a shadow zone by the phenomenon of *diffraction*. Diffraction is a fundamental property of wave motion, and in optics it is the correction to be applied to geometrical optics (ray theory) to obtain an accurate description of the way light waves bend around obstructions. "The order of magnitude of the diffraction at radio frequencies may be obtained by recalling that a 1,000-MHz radio wave has about the same wavelength as a 1,000-Hz sound wave in air so that these two types of waves may be expected to bend around absorbing obstacles with approximately equal facility" [7].

Figure 2.14 illustrates the concept of a shadow zone in which the signal is received, with additional loss, from the diffraction phenomenon. In an urban setting, instead of a being diffracted over a hill, the signal is more likely to be diffracted over a building or row of buildings. An estimate of the attenuation of the signal that results from such diffraction can be made using a "knife edge" model of the hill or building. The main geometrical quantities involved in the theory of knife-edge diffraction are illustrated in Figure 2.15.

Figure 2.15 illustrates the modeling of an obstruction as a knife edge. Suppose that a low, perfectly absorbing straight-edge screen with height H is place at a point along the great circle path between the transmitter and receiver, approximated by a straight line of length d in the figure. Then, as the height of the screen is increased (but still below the path—$H < 0$), the field strength at the receiver begins to oscillate. The amplitude of the oscillation is maximum when the top of the screen is in line with the direct path ($H = 0$); for this case the attenuation is 6 dB. In the theory of the diffraction

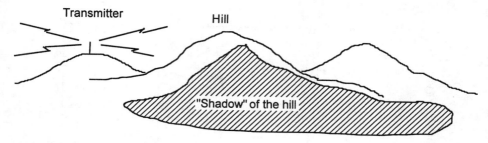

Figure 2.14 Shadow zone.

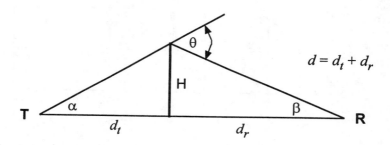

Figure 2.15 Knife-edge geometry.

of electric fields over a knife edge [3, 11], the loss over the knife edge is related, through classical optical diffraction theory, to the Fresnel sine and cosine integrals. For $H > 0$, the loss in dB may be approximated by [5, 6]

$$A(v) \approx \begin{cases} 6.02 + 9.11v - 1.27v^2, & v \leq 2.4 \\ 12.953 + 20\log_{10}v, & v > 2.4 \end{cases} \tag{2.25a}$$

where

$$v = \sqrt{\frac{2d}{\lambda}}\tan\alpha\tan\beta = H\sqrt{\frac{2d}{d_t d_r \lambda}} \tag{2.25b}$$

and the angles α and β are defined in Figure 2.15. The functional approximation $A(v)$ is plotted in Figure 2.16.

When the path from transmitter to receiver is subject to more than one obstacle, or more than one edge, the total loss from diffraction may be estimated by adding the losses in dB of the several edges, each with its own

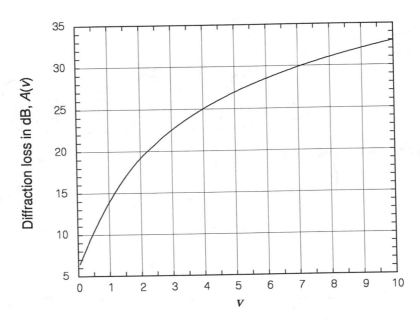

Figure 2.16 Knife-edge diffraction loss.

value of the quantity v that is determined by the geometry of the situation [7]. In Appendix 2A.3, this method is used to compute the diffraction loss from different terrain or building obstacles near to both the transmitter and the receiver.

2.1.4   Empirical Propagation Formulas

The practicing cellular radio or systems engineer is not likely to be involved in detailed modeling of propagation paths but will use existing models and commerically available computer programs for the prediction of propagation losses and the determination of coverage areas.

There are many mathematical models for radio propagation loss ranging from purely empirical ones, based on various measurements of loss, to semi-empirical models that use theoretical considerations to predict propagation effects based on measurements of physical parameters other than propagation loss itself. The utility of a particular model is proportional to the confidence that can be put in its applicability to the physical situations in which the radio system will actually operate. For example, a 1988 study of existing propagation models found fault with nearly all of them in some respect for the purpose of predicting cellular coverage in the 800/900-MHz frequency range [1]. Modeling propagation is field in which active research is still ongoing [12].

In general, a model is more useful if it formulates the propagation loss in terms of several parameters whose values can be made specific to the situation. The complexity of existing propagation models varies according to whether the calculation involves estimation of relevant physical parameters from terrain data or from field strength measurements. In this subsection, several well-known empirical formulas are discussed; in Section 2.1.5, some more complex computer models are discussed.

2.1.4.1 Hata and CCIR Formulas

Following extensive measurements of urban and suburban radio propagation losses, Okumura et al. [13] published many empirical curves useful for cellular systems planning. These empirical curves were subsequently reduced to a convenient set of formulas by Hata [14], known as the *Hata model*, that are widely used in the industry. The basic formula for the median propagation loss given by Hata is

$$L \text{ (dB)} = 69.55 + 26.16 \log_{10} f_{MHz} - 13.82 \log_{10} h_1 - a(h_2)$$
$$+ (44.9 - 6.55 \log_{10} h_1) \log_{10} d_{km} - K \qquad (2.26a)$$

where h_1 and h_2 are base station and mobile antenna heights in meters, respectively, d_{km} is the link distance in kilometers, f_{MHz} is the center frequency in megahertz, and the term $a(h_2)$ is an antenna height-gain correction factor that depends upon the environment:

$$a(h_2) = \begin{cases} (1.1 \log_{10} f_{MHz} - 0.7)h_2 - (1.56 \log_{10} f_{MHz} - 0.8) & \text{medium–small city} \\ 8.29 (\log_{10} 1.54 \, h_2)^2 - 1.1 & \text{large city, } f_{MHz} < 200 \\ 3.2 (\log_{10} 11.75 \, h_2)^2 - 4.97 & \text{large city, } f_{MHz} > 400 \end{cases}$$
$$(2.26b)$$

The factor K in (2.26a) is used to correct the small city formula for suburban and open areas:

$$K = \begin{cases} 2 \left[\log_{10}(f_{MHz}/28)\right]^2 + 5.4 & \text{suburban area} \\ 4.78(\log_{10}f_{MHz})^2 - 18.33 \log_{10}f_{MHz} + 40.94 & \text{open area} \end{cases} \quad (2.26c)$$

An empirical formula for the combined effects of free-space path loss and terrain-induced path loss was published by the CCIR (Comité Consultatif International des Radio-Communication, now ITU-R) and is given by [15]

$$L \ (dB) = 69.55 + 26.16 \log_{10}f_{MHz} - 13.82 \log_{10}h_1 - a(h_2)$$

$$+ (44.9 - 6.55 \log_{10}h_1) \log_{10}d_{km} - B \quad (2.27a)$$

where h_1 and h_2 are base station and mobile antenna heights in meters, respectively, d_{km} is the link distance in kilometers, f_{MHz} is the center frequency in megahertz, and

$$a(h_2) = (1.1 \log_{10}f_{MHz} - 0.7)h_2 - (1.56 \log_{10}f_{MHz} - 0.8) \quad (2.27b)$$

$$B = 30 - 25 \log_{10}(\% \text{ of area covered by buildings}) \quad (2.27c)$$

This formula is recognizable as the Hata model for medium–small city propagation conditions, supplemented with a correction factor, B. The term B is such that the correction $B = 0$ is applied for an urban area, one that is about 15% covered by buildings; for example, if 20% of the area is covered by buildings, then $B = 30 - 25 \log_{10}20 = -2.5 \, dB$.

The CCIR formula yields the following path-loss equation for the typical cellular parameter values $f = 850 \, MHz$ and $h_2 = 2m$:

$$L(dB) = 144.9 - 13.82 \log_{10}h_1 + (44.9 - 6.55 \log_{10}h_1) \log_{10}d_{km} - B$$

$$= \alpha + \beta \log_{10}d_{km} - B \quad (2.28)$$

in which α and β describe a "power law" propagation loss curve (in absolute, not dB, units) of the form $L = a \times d^{\gamma}$. Table 2.2 gives example values of α, β, and γ for different values of the base station antenna height.

Table 2.2 shows that the CCIR formula gives a power law between 3.5 and 4 for cellular link parameters, compared with a free-space power law of $\gamma = 2$. As shown in Section 2.1.3.1, a power law of $\gamma = 4$ is typical of a two-ray multipath situation well within the first Fresnel zone, for mobile positions that are relatively far from the base station.

Table 2.2 CCIR/Hata effective power law

h_1	α	β	γ
10m	131.09	38.35	3.84
20m	126.93	36.38	3.64
30m	124.50	35.22	3.52

As shown in the example presented in Section 2.1.3.1, when the mobile is close to the base station, the link is subject to severe fluctuation in received signal strength because of destructive multipath interference. This fluctuation tends to stop when the distance of the mobile from the base station exceeds the distance corresponding to the boundary of the first Fresnel zone, however. In the actual mobile environment, the physical situation is seldom as simple as that depicted in the example, and moreover is highly dynamic. However, there are usually many opportunities for reflecting paths to occur (buildings, vehicles, etc.), and the basic geometry remains the same.

A graphical comparison of the Hata and CCIR propagation loss formulas in shown in Figure 2.17. It is evident from this comparison that the simple correction factor B in the CCIR formula achieves the same effect as

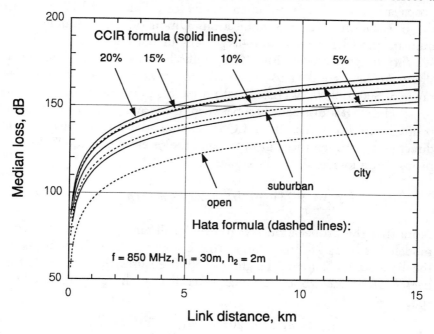

Figure 2.17 Comparison of CCIR and Hata propagation loss formulas.

the more complicated correction factor K in the Hata model, which is to produce a dependence of the the formula on the density of buildings, with the loss increasing with the density of buildings in accordance with experimental observations.

A number of comparisons of the Okumura/Hata/CCIR propagation loss predictions have been made with measurements and with other models [1], [16–19]. The consensus has been that the Japanese urban data gives propagation loss values that are about 10 dB lower than those measured in urban situations in England, Canada, and the United States, but suburban results agree with measurements in those locations.

2.1.4.2 Walfisch-Ikegami Formula

In Europe, a result of research under the Cooperation in the Field of Scientific and Technical Research (COST) program has been the development of improved empirical and semideterministic models for mobile radio propagation [12]. In particular, Project 231 (COST 231), entitled "Evolution of Land Mobile Radio Communications," has resulted in the adoption of propagation modeling recommendations for cellular and PCS applications by the International Telecommunications Union (ITU), including a semideterministic model for medium-to-large cells in built-up areas that is called the *Walfisch-Ikegami model* [20]. The Walfisch-Ikegami model (WIM) has been shown to be a good fit to measured propagation data for frequencies in the range of 800 to 2,000 MHz and path distances in the range of 0.02 to 5 km.

The WIM distinguishes between LOS and non-line-of-sight (NLOS) propagation situations. In a LOS situation, there is no obstruction in the direct path between the transmitter and the receiver, and the WIM models the propagation loss in dB by the equation

$$L_{LOS} = 42.64 + 26 \log_{10} d_{km} + 20 \log_{10} f_{MHz}, \quad d_{km} \geq 0.02 \qquad (2.29)$$

Note that the propagation law (power of distance) for the LOS situation is modeled as being $26/10 = 2.6$, so that $L_{LOS} \propto d^{2.6}$. This model assumes that the base station antenna height ($\geq 30m$) ensures that the path has a high degree of Fresnel zone clearance. Recall that the propagation loss in free space is given by

$$L_{fs} = 32.45 + 20 \log_{10} d_{km} + 20 \log_{10} f_{MHz} \qquad (2.30a)$$

the LOS propagation loss (2.29) can be written as

$$L_{LOS} = L_{fs} + 10.19 + 6\log_{10} d_{km} = L_{fs} + 6\log_{10}(50\, d_{km})$$
$$= L_{fs} + 6\log_{10}(d_m/20) \tag{2.30b}$$

where d_m is the distance in meters. Thus, as shown graphically in Figure 2.18, the WIM model for LOS propagation equals the free-space loss at 20 m, and increases with distance 6 dB per decade faster than the free-space loss for longer distances.

For NLOS path situations, the WIM gives an expression for the path loss that uses the parameters illustrated in Figure 2.19. The quantities referred to in Figure 2.19 are the following:

h_b = Base antenna height over street level, in meters (4 to 50m)

h_m = Mobile station antenna height in meters (1 to 3m)

h_B = Nominal height of building roofs in meters

$\Delta h_b = h_b - h_B$ = Height of base antenna above rooftops in meters

$\Delta h_m = h_B - h_m$ = Height of mobile antenna below rooftops in meters

b = Building separation in meters (20 to 50m recommended if no data)

w = Width of street ($b/2$ recommended if no data)

ϕ = Angle of incident wave with respect to street (use 90° if no data)

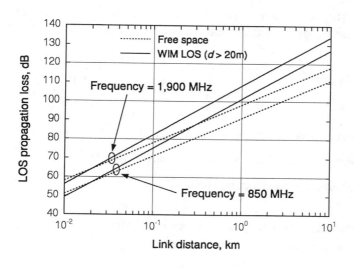

Figure 2.18 Comparison of WIM LOS formula with free-space loss.

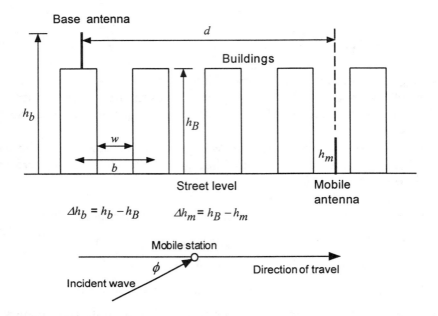

Figure 2.19 WIM NLOS parameters.

In the absence of data, building height in meters may be estimated by three times the number of floors, plus 3 m if the roof is pitched instead of flat. The model works best for base antennas well above the roof height.

Using the parameters listed above, for NLOS propagation paths the WIM gives the following expression for the path loss in dB:

$$L_{NLOS} = \begin{cases} L_{fs} + L_{rts} + L_{mds}, & L_{rts} + L_{mds} \geq 0 \\ L_{fs}, & L_{rts} + L_{mds} < 0 \end{cases} \qquad (2.31a)$$

where

$$L_{fs} = \text{Free-space loss} = 32.45 + 20\log_{10} d_{km} + 20\log_{10} f_{MHz} \qquad (2.31b)$$

$$L_{rts} = \text{Roof-to-street diffraction and scatter loss, and} \qquad (2.31c)$$

$$L_{msd} = \text{Multiscreen diffraction loss} \qquad (2.31d)$$

The loss terms L_{rts} and L_{msd} are functions of the NLOS parameters shown previously. The formula given for L_{rts} is

$$L_{rts} = -16.9 - 10\log_{10} w + 10\log_{10} f_{MHz} + 20\log_{10} \Delta h_m + L_{ori} \qquad (2.32a)$$

where

$$L_{ori} = \begin{cases} -10 + 0.354\,\phi, & 0 \leq \phi \leq 35° \\ 2.5 + 0.075(\phi - 35°), & 35° \leq \phi \leq 55° \\ 4.0 - 0.114(\phi - 55°), & 55° \leq \phi \leq 90° \end{cases} \qquad (2.32\text{b})$$

is an orientation loss. From the formula, it is evident that L_{rts} decreases for wider streets and increases for taller buildings. The effect of the angle of incidence ϕ is seen from the graph in Figure 2.20 to be

- To subtract 10 dB if the signal is arriving in line with the street ($\phi = 0°$);

- To add as much as 4 dB if the angle of arrival is oblique; and

- To subtract 1 dB if the signal arrives perpendicular to the street ($\phi = 90°$).

The formula given for the multiscreen diffraction loss term L_{msd} is

$$L_{msd} = L_{bsh} + k_a + k_d \log_{10} d_{km} + k_f \log_{10} f_{MHz} - 9 \log_{10} b \qquad (2.33\text{a})$$

In this expression, L_{bsh} is shadowing gain (negative loss) that occurs when the base station antenna is higher than the rooftops:

$$L_{bsh} = \begin{cases} -18 \log_{10}(1 + \Delta h_b), & \Delta h_b > 0 \\ 0, & \Delta h_b \leq 0 \end{cases} \qquad (2.33\text{b})$$

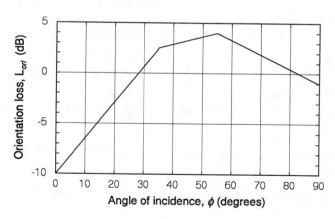

Figure 2.20 Orientation loss used in the WIM.

From the formula, it is evident that L_{msd} decreases for wider building separation (b). The quantities k_a, k_d, and k_f (to be described below) determine the dependence of the loss on the distance (d_{km}) and the frequency (f_{MHz}). The term k_a in the formula for the multiscreen diffraction loss is given by

$$k_a = \begin{cases} 54, & \Delta h_b > 0 \\ 54 + 0.8\,|\Delta h_b|, & \Delta h_b \leq 0 \text{ and } d_{km} \geq 0.5 \\ 54 + 0.8\,|\Delta h_b|(d_{km}/0.5), & \Delta h_b \leq 0 \text{ and } d_{km} < 0.5 \end{cases} \qquad (2.33c)$$

This relation results in a 54-dB loss term if the base station antenna is above the rooftops ($\Delta h_b > 0$), but more than 54 dB if it is below the rooftops. The increase from 54 dB is less if the link distance is rather small (less than 500m).

The factor k_d in the formula for L_{msd} is given by

$$k_d = \begin{cases} 18, & \Delta h_b > 0 \\ 18 + 15\,(|\Delta h_b|/h_B), & \Delta h_b \leq 0 \end{cases} \qquad (2.33d)$$

This relation causes L_{msd} to increase with distance at 18 dB per decade if the base antenna is above the rooftops ($\Delta h_b > 0$). But if the base antenna is below the rooftops, the increase with distance is higher (*e.g.*, 30 dB per decade of distance when the base antenna is only 20% as high as the buildings ($\Delta h_b/h_B = 0.8$). The factor k_f in the formula for the multiscreen diffraction loss is given by

$$k_f = -4 + \begin{cases} 0.7\left(\frac{f_{MHz}}{925} - 1\right), & \text{medium city and suburban} \\ 1.5\left(\frac{f_{MHz}}{925} - 1\right), & \text{metropolitan area} \end{cases} \qquad (2.33e)$$

Recall that the free-space and diffraction/scatter loss terms L_{fs} and L_{rts} together give an increase of 30 dB per decade of frequency. The expression for k_f indicates that this dependence on frequency should be adjusted downward for

- Frequencies < 6.21 GHz for medium city and suburban environments.

- Frequencies < 2.29 GHz for a metropolitan area.

For a typical cellular frequency of 850 MHz, the value of k_f is

$$k_f = \begin{cases} -4.06, & \text{medium city and suburban} \\ -4.12, & \text{metropolitan area} \end{cases} \qquad (2.33\text{f})$$

Thus the total dependence on frequency for the 800-MHz cellular band is about 26 dB per decade of frequency.

Combining the expressions for the different NLOS propagation loss terms comprising the WIM results in the loss formulas shown below:

$\Delta h_b > 0$:

$$L_{NLOS} = 69.55 + 38 \log_{10} d_{km} + 26 \log_{10} f_{MHz} - 10 \log_{10} w - 9 \log_{10} b$$
$$+ 20 \log_{10} \Delta h_m - 18 \log_{10}(1 + \Delta h_b) + L_{ori} \qquad (2.34\text{a})$$

$\Delta h_b \leq 0, \ d_{km} \geq 0.5$:

$$L_{NLOS} = 69.55 + (38 + 15|\Delta h_b|/h_B) \log_{10} d_{km} + 26 \log_{10} f_{MHz}$$
$$- 10 \log_{10} w - 9 \log_{10} b$$
$$+ 20 \log_{10} \Delta h_m + 0.8|\Delta h_b| + L_{ori} \qquad (2.34\text{b})$$

$\Delta h_b \leq 0, \ d_{km} < 0.5$:

$$L_{NLOS} = 69.55 + (38 + 15|\Delta h_b|/h_B) \log_{10} d_{km} + 26 \log_{10} f_{MHz}$$
$$- 10 \log_{10} w - 9 \log_{10} b + 20 \log_{10} \Delta h_m$$
$$+ 0.8|\Delta h_b|(d_{km}/0.5) + L_{ori} \qquad (2.34\text{c})$$

At 850 MHz, the WIM NLOS loss formulas become

$\Delta h_b > 0$:

$$L_{NLOS} = 145.7 + 38 \log_{10} d_{km} - 10 \log_{10} w - 9 \log_{10} b$$
$$+ 20 \log_{10} \Delta h_m - 18 \log_{10}(1 + \Delta h_b) + L_{ori} \qquad (2.35\text{a})$$

$\Delta h_b \leq 0, \ d_{km} \geq 0.5$:

$$L_{NLOS} = 145.7 + (38 + 15|\Delta h_b|/h_B) \log_{10} d_{km} - 10 \log_{10} w - 9 \log_{10} b$$
$$+ 20 \log_{10} \Delta h_m + 0.8|\Delta h_b| + L_{ori} \qquad (2.35\text{b})$$

$\Delta h_b \leq 0, \ d_{km} < 0.5$:

$$L_{NLOS} = 145.7 + (38 + 15|\Delta h_b|/h_B)\log_{10}d_{km} - 10\log_{10}w - 9\log_{10}b$$
$$+ 20\log_{10}\Delta h_m + 0.8\,|\Delta h_b|(d_{km}/0.5) + L_{ori} \qquad (2.35c)$$

Example calculations of the WIM, to be shown below, were performed to examine the effects of the building height parameter h_B, the building separation parameter b, the base station antenna height h_b, and the mobile antenna height h_m. In what follows, we plot path losses versus link distance for the WIM for various parameter values. A key to understanding the significance of the features of the propagation loss curves is given pictorially in Figure 2.21. For example, graphs are plotted on a log-log scale, so that the slope of any curve indicates the propagation power law. The meaning of "Slope = $10 \times$ power law (γ)" in Figure 2.21 can be explained with an example: suppose that the slope of a propagation curve is 40 dB per decade of distance; then the power law γ is found from

$$\text{Slope} = 40 = 10 \times \text{power law } (\gamma) \qquad (2.35d)$$

from which we determine that the power law is $\gamma = 4$.

Calculations of WIM formulas (2.35a) through (2.35c) are presented in Figures 2.22 to 2.25.

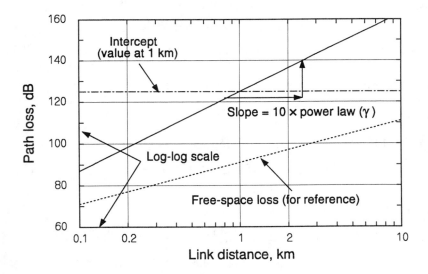

Figure 2.21 Features of propagation loss curves.

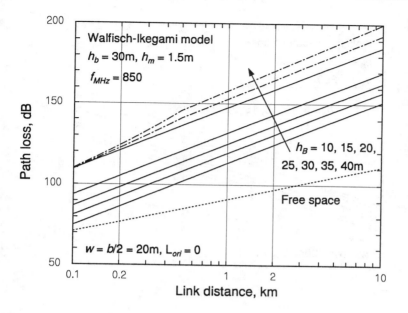

Figure 2.22 Effect of building height.

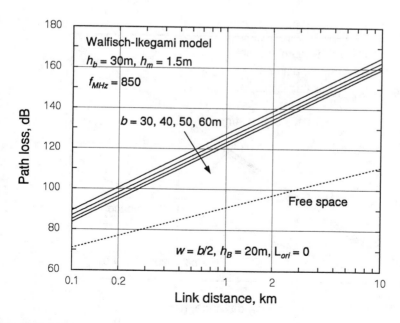

Figure 2.23 Effect of building separation.

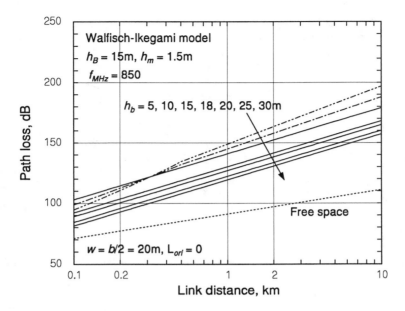

Figure 2.24 Effect of base station antenna height.

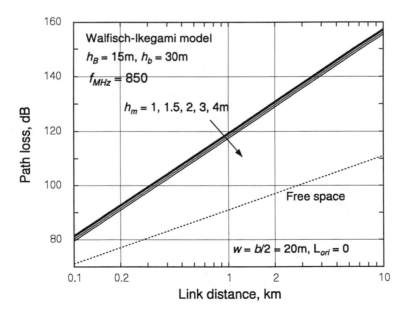

Figure 2.25 Effect of mobile antenna height.

The effects observed in Figures 2.22 through 2.25 may be summarized as follows: (1) The propagation loss is very sensitive to the base station antenna height and to the building height, through the parameters Δh_b and Δh_m, especially at distances greater than 500m. For buildings lower than the base station antenna height ($\Delta h_b > 0$), the power law for the loss as a function of distance is fixed at $\gamma = 3.8$. For buildings higher than the base station antenna height ($\Delta h_b < 0$), the power law is between $\gamma = 3.8$ (at $h_b = h_B$) and $\gamma = 5.3$ (at $h_b = 0$).

(2) The propagation loss is relatively insensitive to the building separation b.

(3) The propagation loss is very insensitive to the mobile station antenna height h_m for realistic values of this parameter.

2.1.5 Computer Propagation Loss Models

In addition to the formulas for mobile radio propagation path loss that have been discussed, there are several well-known computer models that use somewhat more sophisticated techniques for estimating model parameters from terrain and other input data and automating the selection of empirical parameter values that apply in a given mobile situation. Two of these computer models are the Longley-Rice model [2, 5, 6]—more properly known as the Institute for Telecommunications Sciences Irregular Terrain Model (ITSITM)—and the Terrain Integrated Rough Earth Model (TIREM) [21, 22]. In addition, there are several proprietary models offered by cellular engineering consulting firms.

2.1.5.1 The Longley-Rice and TIREM Models

Both the Longley-Rice and TIREM models consider the parameters of the terrain along the great circle path from the transmitter to the receiver, and develop a customized formula for the propagation loss in excess of the free-space loss L_{fs} as a function of the link distance along that path. For example, the output of the Longley-Rice model is a reference median dB value of the excess propagation loss (attenuation), based on the formula

$$A_{ref}(d) = \begin{cases} A_{el} + k_1 d + k_2 \log_{10} d, & d_{min} < d < d_{Ls} \\ A_{ed} + m_d d, & d_{Ls} < d < d_x \\ A_{es} + m_s d, & d > d_x \end{cases} \tag{2.36}$$

where the three distance regions correspond to those dominated by LOS, diffraction, and scattering propagation modes. Each distance region in (2.36) is determined from the program input and features the intercept, slope, and power-law parameters shown. The details of the methodology and calculations of the ITSITM are summarized in [2, 23, 24], and a listing of a BASIC implementation of it is included in [2].

The values of these parameters of (2.36) are chosen in such a way as to make the overall curve piecewise continuous as a function of distance. Figure 2.26 shows an example of the total propagation loss in dB calculated by the Longley-Rice program for an 850-MHz mobile-to-mobile communications scenario. For the example shown in Figure 2.26, the LOS distance d_{Ls} is 29.4 km for a smooth Earth. The expression for the free-space loss in dB for the example is $FSL = 91.04 + 20 \log_{10} d_{km}$. The values of the LOS and diffraction region parameters in (2.36) that pertain to the curves in Figure 2.26 are listed in Table 2.3. Note that the total propagation loss for the example of Figure 2.26 in absolute units (not dB) can be written

$$L = 10^{9.104} d_{km}^2 \times 10^{(A_{el} + k_1 d_{km})/10} d_{km}^{k_2/10}$$

$$= 10^{9.104 + A_{el}/10} \times e^{\beta k_1 d_{km}} \times d_{km}^{2 + k_2/10}, \quad \beta = \ln 10/10 \qquad (2.37)$$

By inspection of (2.37), the power law of the propagation loss as a function of distance in the LOS region is $\gamma = 2 + k_2/10$. In Table 2.3, and in the lower part of Figure 2.26, observe that this power law is $\gamma = 3.3$ for a very smooth Earth ($\Delta h = 1$ m) and decreases to $\gamma = 2$ as the terrain irregularity parameter becomes large. This dependency results from reflected rays that are more severely attenuated than direct rays, or even are blocked by terrain, for rough terrain.

In addition to the reference attenuation in dB and a version of it adjusted for climate and other factors, the Longley-Rice program outputs an

Table 2.3 Longley-Rice parameters

Δh	A_{el}	k_1	k_2	A_{ed}	m_d
1	-4.19	0.500	13.0	8.40	0.722
10	-0.442	0.495	9.47	10.4	0.598
100	14.4	0.623	0	20.2	0.428
200	27.0	0.547	0	32.3	0.370

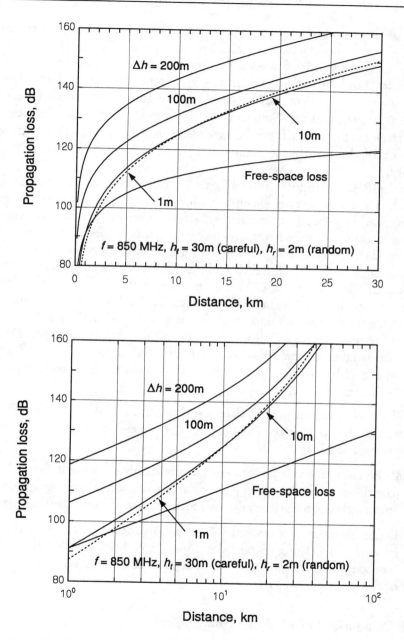

Figure 2.26 Example Longley-Rice calculations.

estimate of the variability of the actual propagation loss in dB, expressed in terms of a standard deviation for the loss σ_c. Typically, σ_c is about 8 dB. Thus, the program can be considered as modeling the propagation loss in dB as a Gaussian random variable with the given values for the median (also the mean in the case of a Gaussian variable) and standard deviation. Using the values of A_{ref}, σ_c, and the Gaussian distribution, the program calculates the value of attenuation that will not be exceeded for some specified percentage of the time or of the possible locations or situations for which the input data may apply.

The TIREM computer model bases its calculation of the propagation loss on data for the terrain elevations along the transmitter-receiver path. Using digitized map data, the model evaluates the terrain profile between the two antenna sites and, depending on the geometrical properties of the profile, selects a mode of propagation for calculating a median transmission loss. The main difference between TIREM and the Longley-Rice model is in the calculation of the loss in the LOS region, those link distances for which an unobstructed path extends between transmitter and receiver. While the Longley-Rice model uses a mixture of two-ray multipath and spherical Earth diffraction to calculate LOS loss in excess of the free-space loss, TIREM chooses the smaller of the smooth Earth diffraction loss and a reflection-region loss found by calculating the clearance of the the direct path above the terrain:

- If the clearance is greater than 1.5 times the radius of the first Fresnel zone, the program outputs the free-space loss as the effective propagation loss.

- If the clearance is less that one-half the radius of the first Fresnel zone (the direct path comes very near Earth's surface), the program uses an empirical rough Earth multipath formulation to compute the loss.

- If the clearance is between 0.5 and 1.5 times the radius of the first Fresnel zone, the program develops a weighted combination of the free-space and rough Earth multipath loss formulations as the estimate of the propagation loss.

2.1.5.2 Comparison of WIM and Longley-Rice

It is interesting to compare the propagation curves developed by the ITS (Longley-Rice) model with those developed by the WIM. For the purpose of comparing the ITS model and the WIM, let a typical cellular scenario be defined by the following common parameter values: $f_{MHz} = 850$, $h_m = 1.5$, $h_b = 30$.

For the WIM, additional assumptions about the height and separation of buildings are needed. Let the building height be $h_B = 15m$ and the building separation be $b = 40m$.

For the ITS model, additional assumptions about the environment and siting practices are needed. Let the standard earth parameter assumptions be used, and let the value of Δh be 50m, which corresponds to rolling plains. This number implies that 80% of the elevations along a given propagation path are within $\pm 25m$ of the average terrain height.

The propagation curves under these assumptions are shown in Figure 2.27. From the figure, it is obvious that the ITS model, for any of the siting practices (random, careful, or very careful), corresponds very closely to the WIM LOS propagation model for the assumed parameters. This fact highlights the different propagation modes that are included as options in the WIM: either LOS or NLOS. The ITS model does not specifically account for buildings and other obstructions, but simply assumes that a path distance less than the radio horizon is a LOS path.

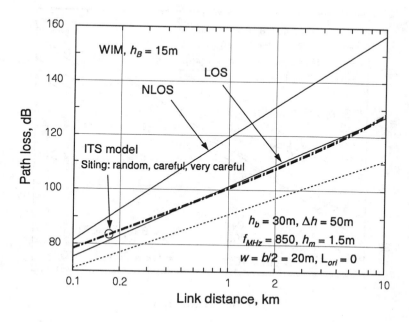

Figure 2.27 Comparison of WIM and ITS models in the LOS region.

It may be possible, however, to manipulate the parameters of the ITS model to make it predict the propagation loss correctly in an urban setting. Two possible approaches for doing this are:

- Input a smaller base station antenna height, because the ITS model calculates an effective antenna height by adding site location elevation to the antenna height.

- Input a larger value of the terrain irregularity parameter Δh to simulate the diffraction effects of buildings with that of hills.

Figure 2.28 shows the result of trying such a matchup. The conclusion to be drawn from the figure is that it is not practical to try making the ITS model produce urban propagation predictions (other than LOS) comparable with the WIM, by means of manipulating the terrain parameter. Although the amount of propagation loss can be made similar in a certain range of link distances, the propagation power law is clearly not the same, being about 2 instead of the value of 3.8 used by the WIM.

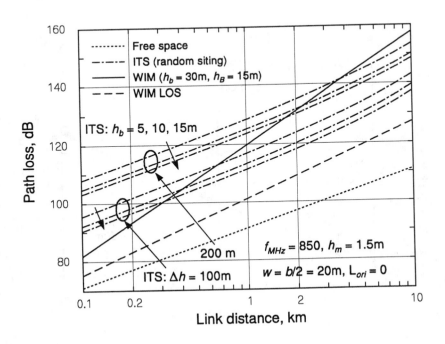

Figure 2.28 Comparison of WIM and Longley-Rice in NLOS situations.

2.1.6   The Use of Propagation Models in Cellular Design

The best radio propagation models provide a good prediction of the median value of the propagation loss as a function of distance, frequency, and environmental parameters. The models therefore are useful for developing preliminary designs of cellular systems by enabling an analysis of the required number of cells and the best locations for cell sites in the intended service area. Because there is a great deal of variation in the actual propagation loss due to shadowing and other effects of terrain and buildings, the predictions of the propagation loss must be confirmed by field measurements.

A rough idea of the desirable cell size can be derived from the maximum value of propagation loss, L_{max}, that can be tolerated by the forward and reverse link budgets. The value of L_{max} may vary with location because the expected number of cellular users (potential interferers in CDMA cellular systems) can vary with location as well as with the time of day. Typically, the cell size needs to be smaller where a higher concentration of traffic is anticipated because there is a limit to the capacity of a cell. The size of the cell is also limited by the amount of allowance for propagation loss that is left over in the link budget after the interference and link margin requirements are taken into account.

In the absence of detailed data on factors affecting propagation, the value of L_{max} determines the radius of a circular area (the cell) inside of which the loss is less than L_{max}. The initial concept of the cellular system's coverage area therefore is the intersection of circles of various sizes, which leads to a hexagonal cell geometry, as discussed in Chapter 3. A more refined cell design takes into account terrain, building density, and other factors that influence propagation loss. For example, suppose the sketch in Figure 2.29 describes the propagation environment for a proposed cell site in terms of building density. Given this depiction of the environment surrounding the proposed cell site, the propagation loss in different locations can be analyzed in several ways. One method is to consider the propagation loss along various radials (subsectors) proceeding from the position of the base station, as suggested in Figure 2.30.

Another method for analyzing the propagation loss in the area surrounding the proposed cell site is to overlay a grid coordinate system on the area and determine the values of the environmental factors affecting propagation loss in each element of the grid. Figure 2.31 illustrates this concept for the example area.

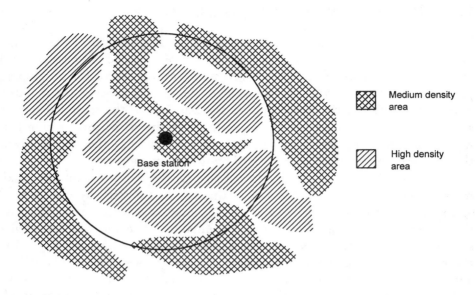

Figure 2.29 Building density of area surrounding base station.

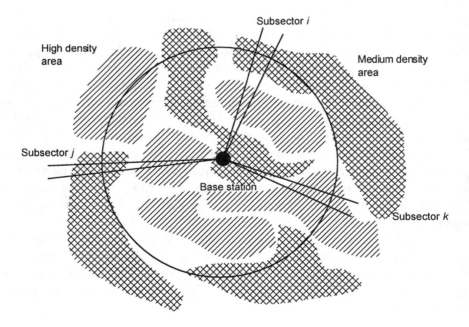

Figure 2.30 Radial subsectors.

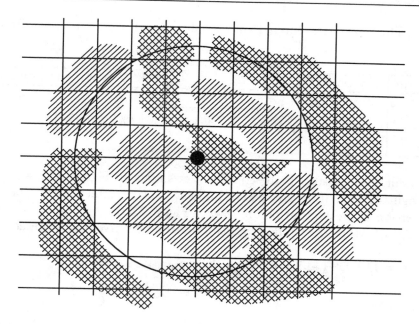

Figure 2.31 Grid coordinates overlaid on base station coverage area.

2.1.6.1 Numerical Example of a Propagation Loss Contour

Using the grid method sketched above, a plotting program was used to produce contours of constant propagation loss by following these steps:

- The grid was assumed to cover a 4-km by 4-km area, with the candidate base station location in the center.

- In anticipation of using the CCIR propagation model, the areas identified by shading as being of high building density were assigned a density value of 40%; those of medium density were assigned a value of 20%, and the remainder of the area was assumed to have a building density of 10%.

- The distances of points spaced at 100m intervals on the grid were calculated; using the distance and the percentage of building density, the propagation loss for each point was calculated.

- The data were entered into a table (spreadsheet) in the following form:

x in km	y in km	d_{km}	%	propagation loss
$\vdots$	$\vdots$	$\vdots$	$\vdots$	$\vdots$
x_i	y_i	$d_i = \sqrt{(x_i - 2)^2 + (y_i - 2)^2}$	$(\%)_i$	$L_i = L[d_i; (\%)_i]$
$\vdots$	$\vdots$	$\vdots$	$\vdots$	$\vdots$

The result of these steps is the contour plot shown in Figure 2.32, alongside the original sketch of the area and its building density pattern (a higher resolution version of the contour plot follows in Figure 2.33).

At 850 MHz, the CCIR propagation loss model gives the formula

$$L = 119.95 + (44.9 - 6.55 \log_{10} h_b) \log_{10} d_{km} - 2.52\, h_m$$
$$- 13.82 \log_{10} h_b + 25 \log_{10}(\%) \tag{2.38a}$$

where h_b is the base station antenna height (e.g, 30m), h_m is the mobile station antenna height (e.g., 1.5m), d_{km} is the link distance in km, and % denotes the percentage of land covered by buildings. Substituting for the antenna heights gives the formula

$$L = 95.76 + 35.225 \log_{10} d_{km} + 25 \log_{10}(\%) \tag{2.38b}$$

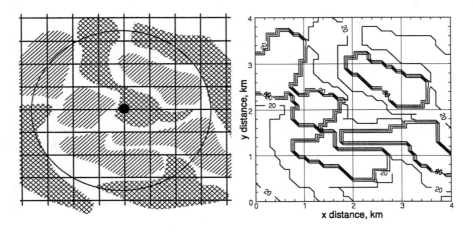

Figure 2.32 Contour plot of discretized building density in percentage.

Contour plots of the propagation loss are shown in Figure 2.33 for a 5-dB resolution and in Figure 2.34 for a 2-dB resolution.

If the building density were the same over the entire area, the propagation loss contours would be concentric circles. The tendency toward concentric circles can be seen in the contour plots shown.

Note that the propagation loss can be less in a low-density area farther away from the base station than a high-density area, giving rise to "pockets" or small areas inside the larger area in which propagation is either very poor in comparison with the surrounding area, or very good. Although the existence of such pockets cannot always be eliminated entirely, a candidate cell site position that results in many pockets needs some change in the design to improve the situation. It may be that a relatively small change in position will give better results. An increase in base station antenna height may improve performance enough to make the propagation loss in the pockets tolerable. The dependence of the coverage area on the maximum tolerable propagation loss term in the link budget is illustrated next.

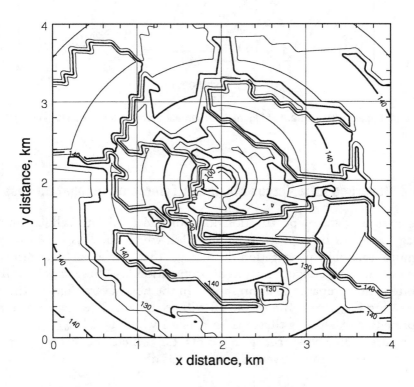

Figure 2.33 Contour plot of propagation loss using 5-dB resolution.

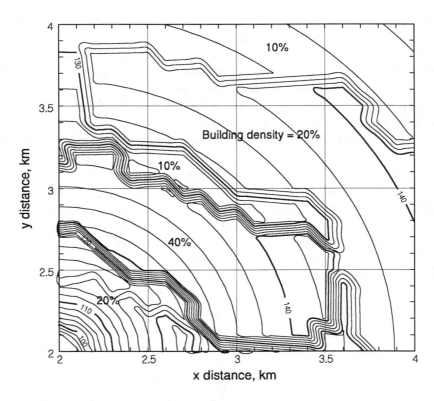

Figure 2.34 Upper right quadrant of Figure 2.33: contour plot of propagation loss using 2-dB resolution.

2.1.6.2 Coverage Area Versus Maximum Tolerable Propagation Loss

From contour data of propagation loss for the candidate cellular site, the coverage area can be seen as those areas for which the loss is less than the maximum tolerable propagation loss L_{max} that has been determined by analysis of the link budget. Instead of propagation loss, the received SNR can be plotted and compared to a threshold. In the next several figures, the effect of increasing L_{max} from 135 to 145 dB will be shown. The coverage area will be represented by shading those grid points for which $L < L_{max}$.

Figure 2.35 shows the grid points for which the propagation loss, according to the formula given, is less than 135 dB.

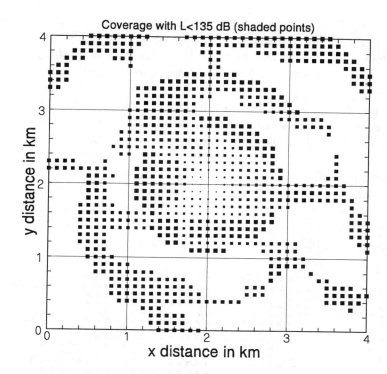

Figure 2.35 Areas covered with loss less than 135 dB.

Comments on the example of $L_{max} = 135\ dB$: As seen in the figure, if $L_{max} = 135\ dB$, the cell site under consideration can deliver a reliable signal level only to an area within about $0.9\ km = 900m$ of the site. This would be the radius of coverage if the site were selected, and there would need to be other sites spaced at roughly 1.7 times this distance, or about 1.5 km.

The existence of rather large pockets of coverage outside the core reception area could be a problem, if the site is used, because mobiles in those pockets (affiliated with an adjacent base station) would be likely to hand off to the center base station when it is not the nearest base station. A solution to this problem is to find another site that does not have these large pockets of out-of-cell coverage or to increase the coverage area by increasing the value of L_{max} by some system improvement that frees additional margin, such as coding or additional transmitter power. An alternative measure is to raise the antenna at the base station by mounting the antenna on a high building to reduce the propagation loss to the unshaded areas, or to lower the antenna to increase the loss, thereby eliminating the pockets.

In Figure 2.36, increasing the maximum tolerable propagation loss by 5 dB increases the radius of the core service area from about 0.9 km to about 1.3 km. Coverage is sporadic well beyond this distance in some directions. The coverage could be viewed as having a 2-km radius, with some pockets of insufficient signal strength—the centers of the subareas that have high building density. Possible remedies for this situation include installing repeater antennas to boost the cell site signal in the high-density areas or installing microcells in the high-density areas, which probably will have a high concentration of call traffic that warrants having separate cell sites of their own within the larger cell.

In Figure 2.37, the coverage is shown for $L_{max} = 145$ dB: Raising the maximum tolerable propagation loss to 145 dB has resulted in a solid coverage area with a radius of about 1.7 km. The two high-density areas with insufficient signal can be covered by an adjacent cell site.

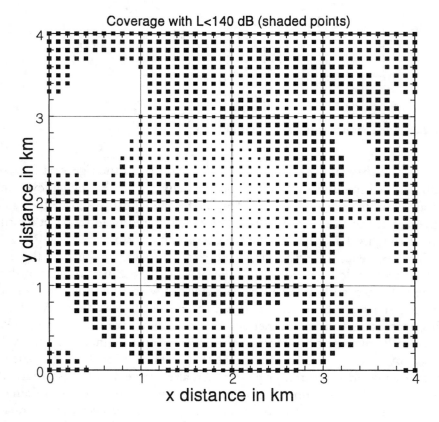

Figure 2.36 Area coverage with loss less than 140 dB.

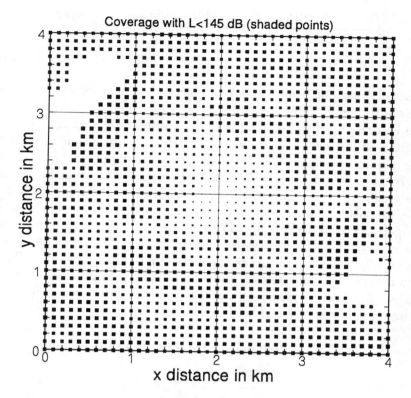

Figure 2.37 Areas covered with loss less than 145 dB.

Although a good-sized, solid-core coverage area has been created by raising the tolerable loss (most likely by increasing power), interference is now a problem from this candidate cell site to areas in adjacent cells if a regular pattern of cells is employed. We may use sector antennas that can be tilted downward to control the extent of the coverage area in a particular direction, or, if the call traffic distribution allows, implement the cell as an oddly-shaped cell with a noncircular coverage area. For the parameters used in this example, the CCIR propagation loss formula as a function of base station antenna height, distance, and building percentage is

$$L(dB) = 119.95 + (44.9 - 6.55 \log_{10} h_b) \log_{10} d_{km} - 2.52\, h_m$$
$$- 13.82 \log_{10} h_b + 25 \log_{10}(\%)$$

$$= \underbrace{116.17 - 13.82 \log_{10} h_b}_{A} + \underbrace{(44.9 - 6.55 \log_{10} h_b)}_{B} \log_{10} d_{km}$$

$$+ 25 \log_{10}(\%) \quad (2.39)$$

The values of the coefficients A and B as functions of the base station antenna height are as shown in Table 2.4.

In Figure 2.38, the effect on the $L_{max} = 140$ dB coverage area of raising the antenna to $h_b = 50$m is shown. As the graphical depictions of the coverage areas for $L_{max} = 140$ dB show, the raising of the base station antenna from $h_b = 30$m to $h_b = 50$m has the effect of increasing the radius of the core area of the transmitter's coverage from about 1.3 km to about 1.7 km. Thus, raising the antenna by 10m has produced the same result as increasing L_{max} from 140 dB to 145 dB while keeping the antenna height constant at $h_b = 30$m. From a systems viewpoint, it is much easier to raise an antenna by 10m than to increase the tolerable propagation loss by 5 dB. As discussed previously in connection with the $L_{max} = 145$ dB graph, if the 1.7-km radius core area is taken to be the cell, the interference from this transmitter to adjacent cells will need to be dealt with in some manner.

Table 2.4 Coefficients A and B for the example

h_b	A	B
30m	95.76	35.22
40m	94.03	34.41
50m	92.69	33.77

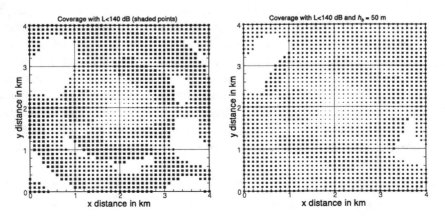

Figure 2.38 Comparison of coverage areas when $h_b = 30$m and $h_b = 50$ m.

2.2   The Mobile Radio Environment

We have just covered the subject of propagation modeling in cellular design with an emphasis on cell site area coverage, using specific examples to illustrate the approach. We now characterize the mobile radio environment in terms of the effect of the communications channel on the radio signal and discuss such topics as delay spread function, which characterizes a channel model by showing the distribution of signal multipath replicas in time. In addition to this time-domain description of the channel, we also discuss frequency-domain characterizations, including the channel frequency transfer function and the Doppler-spread function. The mobile channel almost always involves time dispersion and, as we have shown in Chapter 1, this phenomenon results in fading due to the combination of multiple paths at the receiver. The characterization of fading channels is given in this section, and several fading channel types are discussed in terms of time- and frequency-selective dispersion, slow and fast fading, and flat and nonflat fading. We define these terms precisely. We also consider the relation between fade rate and vehicular speed, as a function of fade depth.

2.2.1   Channel Models

Between the antenna of the transmitter and the antenna of the receiver, the transmitted signal undergoes a variety of effects. The term "channel" refers to these effects on the signal and can be thought of as a filter acting on the transmitted signal. Included in these effects are attenuation due to path loss and various fading phenomena. In an attempt to describe the channel mathematically, many channel models have been developed [25–28]. Some of these models simply describe the loss due to the propagation distance while others describe the fading phenomena. Because the channel can be thought of as a filter, the channel models can be modeled as the transfer function of a filter.

Most channel models that describe the effect of the channel on the signal vary with time. Thus, these models are given mathematically as a function of time, which can also be described in the frequency domain. An important component of the channel is the arrival of multiple replicas of the signal interfering with the desired signal. Another important effect is the shift in frequency of the signal due to the relative motion between the transmitter and receiver.

2.2.1.1 Delay-Spread Function

One effect of the channel is to create multiple replicas of the signal arriving with various amplitudes and phase. This effect can be measured and expressed mathematically. Consider a channel that is presented with a complex envelope input signal $z(t)$, where

$$z(t) = I_z(t) + jQ_z(t) \qquad (2.40a)$$

The channel will affect the input signal in such a manner as to produce the complex envelope signal $w(t)$ at the output, where

$$w(t) = I_w(t) + jQ_w(t) \qquad (2.40b)$$

This situation is illustrated in Figure 2.39. The input and output can be expressed as a function of frequency by taking the Fourier transform, yielding

$$Z(f) = \mathcal{F}\{z(t)\} = \int_{-\infty}^{\infty} z(t)\, e^{-j2\pi ft}\, dt \qquad (2.40c)$$

and

$$W(f) = \mathcal{F}\{w(t)\} = \int_{-\infty}^{\infty} w(t)\, e^{-j2\pi ft}\, dt \qquad (2.40d)$$

In the time domain, the complex envelope of the output of the channel can be expressed as the convolution of the complex envelope of the input $z(t)$ with the channel function $g(\tau; t)$:

$$w(t) = \int_{-\infty}^{\infty} z(t - \tau)\, g(\tau; t)\, d\tau \qquad (2.41)$$

where the channel function $g(\tau; t)$ is the *delay-spread function* of the channel and is generally a complex variable. Note the similarity of (2.41) to (1.17c),

$$z(t) \;\rightarrow\; \boxed{\text{Channel}(t)} \;\rightarrow\; w(t)$$

Figure 2.39 The effect of the channel on the input signal $z(t)$.

which is the superposition integral for a linear system characterized by an impulse response function $h(\lambda; t)$. Thus, $g(\tau; t)$ is equivalent to an impulse response of the channel. If the input to the channel were an impulse at a time $t = \tau_0$, then using $z(t) = \delta(t - \tau_0)$ in (2.41) gives

$$w(t) = \int_{-\infty}^{\infty} \delta(t - \tau_0 - \tau)\, g(\tau; t)\, d\tau = g(t - \tau_0; t) \tag{2.42}$$

This example shows that the delay-spread function $g(\tau; t)$ is the channel response at time t to an impulse at a time $t - \tau$. Thus, this impulse was transmitted at a time τ seconds earlier than the response at time t. The example also shows that the delay-spread function of a channel can be found by sending an impulse and measuring the received impulses.

The channel can also be characterized in terms of discrete intervals $(i \cdot \Delta\tau)$ of time delay as the sum

$$w(t) \approx \sum_i z(t - i\Delta\tau)\, g(i\Delta\tau; t)\, \Delta\tau \tag{2.43}$$

Thus, the channel can also be thought of as a tapped delay line with each tap weighted by an amplitude coefficient. The output is the sum of delayed versions of the input signal, each weighted by the amplitude coefficient. A diagram of this representation is given in Figure 2.40.

For a causal system such as a propagation channel:

$$g(\tau; t) = 0 \qquad \text{for} \qquad \tau < 0 \tag{2.44}$$

In addition, if the system is non-time-varying, then the time dependency can be dropped, yielding

$$g(\tau; t) \equiv g(\tau) \tag{2.45a}$$

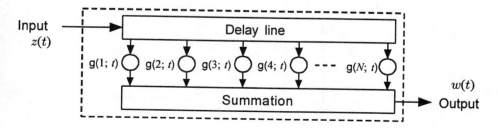

Figure 2.40 The channel produces replicas of the input signal.

Therefore, the channel acts like an ordinary filter whose output is the convolution of the input and the impulse response.

$$w(t) = z(t) * g(t) = \int_{-\infty}^{\infty} z(t - \tau) g(\tau) \, d\tau \qquad (2.45b)$$

A typical multipath channel is used to demonstrate the delay-spread function: Suppose that the input to the system is given by $z(t) = s(t)$ (real-valued) and the output of the channel is given by

$$w(t) = \alpha_1 s(t - \tau_1) + \alpha_2 s(t - \tau_2) + \cdots + \alpha_N s(t - \tau_N) \qquad (2.46a)$$

$$= \sum_{k=1}^{N} \alpha_k \, s(t - \tau_k) \qquad (2.46b)$$

$$= \int_{-\infty}^{\infty} s(t - \tau) \cdot \sum_{k=1}^{N} \alpha_k \delta(\tau - \tau_k) \, d\tau \qquad (2.46c)$$

A comparison of (2.46c) with (2.41) yields an delay-spread function that is the superposition of delayed impulses:

$$g(\tau; t) = \sum_{k=1}^{N} \alpha_k \delta(t - \tau_k) \qquad (2.47)$$

This form of the delay-spread function can be depicteded graphically as in Figure 2.41.

The delay spread, denoted Δ_τ, indicates the degree of time spreading in the channel. The delay spread of the channel can be found from various

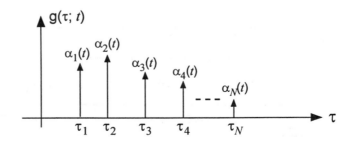

Figure 2.41 The delay-spread function of a typical mobile channel.

moments of the delay-spread function, $g(\tau; t)$. The delay spread is the square root of the difference between the mean of the delay squared and the square of the mean delay, which is expressed as

$$\Delta_\tau(t) = \sqrt{\overline{\tau^2} - \overline{\tau}^2} = \left\{ \frac{\int_{-\infty}^{\infty} \tau^2\, g(\tau; t)\, d\tau}{\int_{-\infty}^{\infty} g(\tau; t)\, d\tau} - \left[\frac{\int_{-\infty}^{\infty} \tau\, g(\tau; t)\, d\tau}{\int_{-\infty}^{\infty} g(\tau; t)\, d\tau} \right]^2 \right\}^{1/2} \qquad (2.48)$$

The relationship between some of the parameters in (2.48) and the delay-spread function of a typical multipath channel is shown in Figure 2.42.

The delay spread of a channel depends in part on the proximity of scattering objects to the transmitter and receiver. Another factor is the amount of scattering agents in the environment. Table 2.5 gives some typical examples of delay spreads in different environments [9].

The minimum delay τ_0 is the time required for the signal to propagate directly from the transmitter to the receiver. Most of the multipath signals

Table 2.5 Typical values of delay spread [9]

Environment type	Delay spread
In-building	$< 0.1\ \mu s$
Open area	$< 0.2\ \mu s$
Suburban area	$0.5\ \mu s$
Urban area	$3\ \mu s$

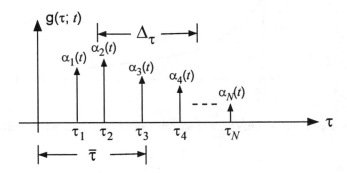

Figure 2.42 The delay spread can be calculated using various moments of the delay-spread function.

are grouped just after the earliest received signal. Large delays relative to τ_0 are seldom received. Thus, it is common to model the distribution of the delay spread as the exponential distribution shifted by τ_0, as illustrated in Figure 2.43. If this distribution is normalized by its delay spread, then it is given by the expression

$$\frac{g(\tau; t)}{\int_{-\infty}^{\infty} g(\tau; t)\, d\tau} = \frac{1}{\Delta_\tau} \cdot e^{-(\tau - \tau_0)/\Delta_\tau}, \qquad \tau \geq \tau_0 \qquad (2.49)$$

The mean of the normalized exponential distribution is

$$\bar{\tau} = \frac{1}{\Delta_\tau} \int_{\tau_0}^{\infty} \tau\, e^{-(\tau - \tau_0)/\Delta_\tau}\, d\tau = \Delta_\tau + \tau_0 \qquad (2.50a)$$

The mean square of the distribution is given by

$$\overline{\tau^2} = \frac{1}{\Delta_\tau} \int_{\tau_0}^{\infty} \tau^2 e^{-(\tau - \tau_0)/\Delta_\tau}\, d\tau = \Delta_\tau^2 + (\Delta_\tau + \tau_0)^2 \qquad (2.50b)$$

Thus the standard deviation for a delay with an exponential distribution is

$$\sigma_\tau \triangleq \sqrt{\overline{\tau^2} - \bar{\tau}^2} = \Delta_\tau \qquad (2.50c)$$

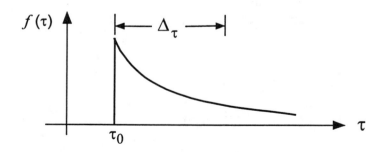

Figure 2.43 The distribution of delays can be approximated by the exponential function.

2.2.1.2 Frequency Transfer Function

The delay-spread function and the associated delay-spread parameter Δ_τ have been described in the time domain. The delay-spread function was described as different multipath components with varying delays and amplitudes. The effects of the delay-spread function can also be described in the frequency domain. In the frequency domain, the effects are characterized as the transfer function of the channel.

The transfer function of a system is defined as the output divided by the input. This can be expressed in the frequency domain as

$$T(f; t) \triangleq \frac{W(f)}{Z(f)} \tag{2.51a}$$

and

$$T(f; t) = \int_{-\infty}^{\infty} g(\tau; t)\, e^{-j2\pi f \tau}\, d\tau = \mathcal{F}\{g(\tau; t)\} \tag{2.51b}$$

In general, $T(f; t)$ is a complex-valued, time-varying function.

If the input to a linear time-invariant system is $e^{j2\pi f_d t}$, whose Fourier transform is $\delta(f - f_d)$, then the output is given by

$$\mathcal{F}^{-1}\{H(f)\, \delta(f - f_d)\} = \int_{-\infty}^{\infty} H(f)\, \delta(f - f_d)\, e^{j2\pi f t}\, df = H(f_d)\, e^{j2\pi f_d t} \tag{2.52}$$

The quantity $H(f_d)$ by definition describes the gain of the system to the input $e^{j2\pi f_d t}$. If the system is time varying, then the transfer function will have a similar relationship.

Assume that the real-valued bandpass signal is

$$x(t) = \cos\left[2\pi(f_0 + f_d)t\right] = \cos\left(2\pi f_0 t + 2\pi f_d t\right)$$
$$= R(t)\cos\left[2\pi f_0 t + \phi(t)\right] \tag{2.53}$$

observe that the value of the envelope is $R(t) = 1$ and the phase function is $\phi(t) = 2\pi f_d t$. The complex envelope is then given by

$$x(t) = R(t)\, e^{j\phi(t)} = 1 \cdot e^{j2\pi f_d t} = \cos(2\pi f_d t) + j\sin(2\pi f_d t) \tag{2.54}$$

When the complex envelope of the time-varying channel input is $z = e^{j2\pi f_d t}$, then the complex envelope of the channel output is, by definition,

$$\int_{-\infty}^{\infty} g(\tau; t)\, e^{j2\pi f_d(t-\tau)}\, d\tau = e^{j2\pi f_d t} \int_{-\infty}^{\infty} g(\tau; t)\, e^{-j2\pi f_d \tau}\, d\tau \tag{2.55}$$

The integral on the right-hand side of the equation is the Fourier transform of the delay-spread function with respect to the delay variable τ. Thus, the envelope of the output is

$$e^{j2\pi f_d t} \int_{-\infty}^{\infty} g(\tau; t)\, e^{-j2\pi f_d \tau}\, d\tau = e^{j2\pi f_d t} \cdot T(f_d; t) \tag{2.56}$$

Therefore, the time-varying frequency transfer function embodies the very same concept of input-output relationship that the non-time-varying transfer function does.

The channel transfer function now will be used to explain the concept of a *coherence bandwidth* for the channel about frequency f_0 and at time t. Suppose that two components of the signal, z_1 and z_2, have frequencies $f_0 + f_d$ and $f_0 - f_d$. That is, the signal contains the sinusoids

$$z_1(t) = A_1 \cos[2\pi(f_0 + f_d)t + \phi_1] = \text{Re}\{A_1 e^{j2\pi f_d t + j\phi_1} \cdot e^{j2\pi f_0 t}\} \tag{2.57a}$$

and

$$z_2(t) = A_2 \cos[2\pi(f_0 - f_d)t + \phi_2] = \text{Re}\{A_2 e^{-j2\pi f_d t + j\phi_2} \cdot e^{j2\pi f_0 t}\} \tag{2.57b}$$

with the respective complex envelopes

$$z_1(t) = A_1 e^{j2\pi f_d t + j\phi_1} \quad \text{and} \quad z_2(t) = A_2 e^{-j2\pi f_d t + j\phi_2} \tag{2.58}$$

The complex envelopes at these frequencies at the output of the channel are

$$w_1(t) = z_1 \cdot T(f_d; t) = A_1 e^{j2\pi f_d t + j\phi_1} \cdot T(f_d; t) \tag{2.59a}$$

and

$$w_2(t) = z_2 \cdot T(-f_d; t) = A_2 e^{-j2\pi f_d t + j\phi_2} \cdot T(-f_d; t) \tag{2.59b}$$

in which the center frequency f_0 is implicit in the definition of $T(f; t)$ because the operation of the channel on the complex envelope of the signal is under consideration. A measure of the coherence or agreement between the signal components in the frequency domain, after being affected by the channel, is the correlation or coherence function

$$E\{w_1 \cdot w_2\} = E\{z_1 \cdot z_2\}E\{T(f_d; t) \cdot T(-f_d; t)\}$$

$$= E\{z_1 \cdot z_2\}E\{|T(f_d; t)|^2\} \triangleq C(f_d; t) \qquad (2.60)$$

in which $E\{\cdot\}$ denotes expectation (statistical averaging). (2.60) uses the fact that $T(-f_d; t)$ is the complex conjugate of $T(f_d; t)$, a fact that follows from the definition of the frequency transfer function in (2.56).

In concept, the coherence bandwidth of the received process is the frequency separation beyond which frequency components of the received complex envelope are statistically independent, under the assumption that the scattering effects of the channel are adequately modeled by a random delay-spread function $g(\tau; t)$ that is a Gaussian RV, for which statistical independence is equivalent to a zero correlation between the components. As a practical matter, instead of zero correlation we may define the coherence bandwidth B_c as the frequency separation $f_d = B_c$ such that

$$C(B_c; t) = 0.5 \qquad (2.61)$$

The frequency separation at which the coherence between signal components falls to $1/2$ is twice the coherence bandwidth or $2B_c$. Assuming that the correlation between the original signal components, $E\{z_1 \cdot z_2\}$, is relatively independent of frequency, the coherence then is proportional to the average value of $|T(f_d; t)|^2$, the squared magnitude of the channel transfer function. In any event, the coherence bandwidth of the channel can be defined as the value of $f_d = B_c$ that causes the value of $|T(f_d; t)|^2$ to fall to one-half its value for $f_d = 0$. That is, B_c can be thought of as the 3-dB bandwidth of the channel about the assumed center frequency f_0, as illustrated in Figure 2.44.

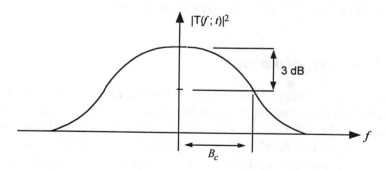

Figure 2.44 Definition of coherence bandwidth of the channel in terms of the frequency transfer function $|T(f_d; t)|^2$.

Example 2.1 Suppose that the delay-spread function is non-time-varying and is given by

$$g(\tau; t) = g_e(\tau) = \kappa\, e^{-(\tau-\tau_0)/\Delta_\tau} \tag{2.62}$$

where $\tau \geq \tau_0$ and κ = a constant. Find the transfer function for the channel.

   Solution: The transfer function, also non-time-varying, is

$$T(f; t) = T_e(f) = \mathcal{F}\{g_e(\tau)\} = \int_{\tau_0}^{\infty} \kappa\, e^{-(\tau-\tau_0)/\Delta_\tau}\, e^{-j2\pi f\tau}\, d\tau \tag{2.63a}$$

$$= e^{-j2\pi f\tau_0} \int_{\tau_0}^{\infty} \kappa\, e^{-(j2\pi f\Delta_\tau+1)(\tau-\tau_0)/\Delta_\tau}\, d\tau \tag{2.63b}$$

$$= \frac{\kappa\, e^{-j2\pi f\tau_0}}{1+j2\pi f\Delta_\tau} \int_0^{\infty} e^{-x/\Delta_\tau}\, dx = \frac{\kappa\Delta_\tau}{1+j2\pi f\Delta_\tau}\, e^{-j2\pi f\tau_0} \tag{2.63c}$$

The coherence measure for this example is

$$\frac{|T_e(f_d)|^2}{|T_e(0)|^2} = \frac{1}{|1+j2\pi f_d\Delta_\tau|^2} = \frac{1}{1+(2\pi f_d\Delta_\tau)^2} \tag{2.64}$$

and the coherence bandwidth is found from

$$\frac{1}{1+(2\pi f_d\Delta_\tau)^2}\bigg|_{f_d=B_c} = \frac{1}{2} \quad\Rightarrow\quad B_c = \frac{1}{2\pi\Delta_\tau} \tag{2.65}$$

In the absence of specific information about the delay-spread function or the channel frequency transfer function, it is customary to use (2.65) to define the coherence bandwidth B_c in terms of the delay spread Δ_τ.

2.2.1.3 Doppler-Spread Function

In the frequency domain, the Fourier transform of the output complex envelope, $W(f)$, with respect to the delay parameter τ can be modeled as the product of the channel frequency transfer function $T(f; t)$ and the transform of the input complex envelope $Z(f)$; that is, $W(f) = Z(f) \cdot T(f; t)$. The transform of the output complex envelope can also be thought of as the convolution of the transform of the input complex envelope and a function called the *Doppler-spread function*, denoted $H(\nu; f)$:

$$W(f) = \int_{-\infty}^{\infty} Z(f - \nu) \, H(\nu; f) \, d\nu = [Z(\alpha) * H(\alpha; f)]_{\alpha = f} \qquad (2.66)$$

Suppose that the Doppler-spread function is an impulse: $H(\nu; f) = K \, \delta(\nu - f_d)$; then the output of the channel in the frequency domain is

$$W(f) = \int_{-\infty}^{\infty} Z(f - \nu) \, K \, \delta(\nu - f_d) \, d\nu = K \, Z(f - f_d) \qquad (2.67)$$

The value of $H(\nu; f)$ at ν is the portion of the output spectrum at frequency f that is due to shifting the input spectrum up by ν Hz. Reflections from moving objects can cause a Doppler shift in the frequency domain. The shift in frequency is proportional to the relative motion of the transmitter, scatterers, and receiver. In a multipath situation involving motion, therefore, the channel output spectrum can be the superposition of several shifted versions of the input spectrum. The Doppler-spread function is used to model this phenomenon.

The channel can also be characterized in terms of discrete intervals $(i \cdot \Delta\nu)$ as the sum

$$W(f) \approx \sum_i Z(f - i\Delta\nu) \times H(i\Delta\nu; f) \, \Delta\nu \qquad (2.68a)$$

$$= \int_{-\infty}^{\infty} Z(f - \nu) \, H(\nu; f) \, d\nu \qquad (2.68b)$$

(2.68b) shows that the channel output spectrum is a weighted sum of shifted versions of the input spectrum. This concept is illustrated in Figure 2.45. The degree of Doppler spreading is characterized by the Doppler spread F_s, which is defined to be the root-mean-square second moment of $H(\nu; f)$, as depicted in Figure 2.46:

$$(F_s)^2 \triangleq \int_{-\infty}^{\infty} \nu^2 \, H(\nu; f) \, d\nu \qquad (2.69)$$

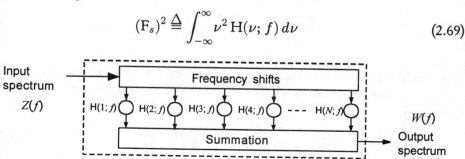

Figure 2.45 The output spectrum is a weighted sum of shifts of the input spectrum.

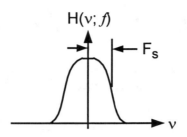

Figure 2.46 Relation of the Doppler spread to the distribution of the frequency shift ν.

2.2.1.4 Combined Delay Spread and Doppler Spread

The delay-spread function has been shown to model the propagation channel as shifts in the time domain. Delays are caused by the signal reflecting off of multiple objects. The delay-spread channel is modeled by convolving the delay-spread function with the input function. Similarly, the Doppler-spread function models the channel as shifts in the frequency domain. The shifts in frequency are due to the motion of the scattering objects. The Doppler-spread channel is modeled by convolving the Doppler-spread function with the channel input in the frequency domain. These models involve a single convolution and/or Fourier transforms. Next, the two channel models will be combined to produce a model that includes both a convolution and Fourier transform.

The channel output may be written

$$w(t) = \int_{-\infty}^{\infty} z(t-\tau)\, g(\tau; t)\, d\tau \tag{2.70a}$$

$$= \int_{-\infty}^{\infty} z(t-\tau) \left[\int_{-\infty}^{\infty} U(\tau; \nu)\, e^{j2\pi\nu t}\, d\nu \right] d\tau \tag{2.70b}$$

in which $U(\tau; \nu)$ is introduced as the Fourier transform of $g(\tau; t)$ with respect to the time variation of the channel (the second argument of the function):

$$U(\tau; \nu) \triangleq \int_{-\infty}^{\infty} g(\tau; t)\, e^{-j2\pi\nu t}\, dt \tag{2.71}$$

$U(\tau; \nu)$ is a delay-Doppler-spreading function that models both time-delay spreading and frequency-shift spreading. From the defining equation for $U(\tau; \nu)$, the channel output is seen to be modeled as the weighted sum of delayed and frequency-shifted versions of the input:

$$w(t) \approx \sum_i \sum_k \Delta\tau\,\Delta\nu\, z(t - i\Delta\tau) \times U(i\Delta\tau; k\Delta\nu) \times e^{j2\pi k\Delta\nu t} \tag{2.72a}$$

$$= \int_{-\infty}^{\infty} z(t - \tau) \int_{-\infty}^{\infty} U(\tau; \nu)\, e^{j2\pi\nu t}\, d\nu\, d\tau \tag{2.72b}$$

Example 2.2 Consider the propagation environment depicted in Figure 2.47. The received signal has three components: the direct path d_0, the path reflected from a fixed reflector d_1, and the path reflected from a moving reflector d_2. The delays for each of these paths are listed below:

Direct path:
$$\tau_0 = \frac{d_0}{c} = \text{constant}$$

Reflection 1:
$$\tau_1 = \frac{d_{11} + d_{12}}{c} = \text{constant}$$

Reflection 2:
$$\tau_2 = \frac{d_{21}(t) + d_{22}(t)}{c} = \tau_2(t) \doteq \tau_{20} + \kappa t$$

Find the delay-spreading function and combined delay- and Doppler-spreading function.

Solution: The delay-spreading function for this example is

$$g(\tau; t) = \alpha_0 \delta(\tau - \tau_0) + \alpha_1 \delta(\tau - \tau_1) + \alpha_2 \delta(\tau - \tau_{20} - \kappa t) \tag{2.73}$$

The combined delay and Doppler-spreading function, then, is

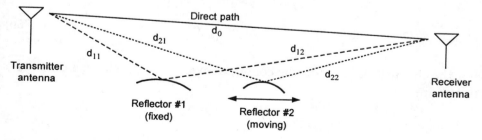

Figure 2.47 Example propagation environment.

$$U(\tau; \nu) = \int_{-\infty}^{\infty} g(\tau; t)\, e^{-j2\pi\nu t}\, dt \tag{2.74a}$$

$$= \alpha_0 \delta(\tau - \tau_0)\, \mathcal{F}\{1\} + \alpha_1 \delta(\tau - \tau_1) \mathcal{F}\{1\}$$
$$+ \alpha_2 \mathcal{F}\{\delta(\tau - \tau_{20} - \kappa t)\} \tag{2.74b}$$

$$= \alpha_0 \delta(\tau - \tau_0)\, \delta(\nu) + \alpha_1 \delta(\tau - \tau_1)\, \delta(\nu)$$
$$+ \alpha_2 \mathcal{F}\{\delta(\tau - \tau_{20} - \kappa t)\} \tag{2.74c}$$

The first two terms in (2.74) include a delay with no Doppler. The third term includes both delay and Doppler. The transform needed to complete the calculation for the example is

$$\mathcal{F}\{\delta(\tau - \tau_{20} - \kappa t)\} = \int_{-\infty}^{\infty} \delta(\tau - \tau_{20} - \kappa t)\, e^{-j2\pi\nu t}\, dt \tag{2.75a}$$

Let

$$\alpha = \tau - \tau_{20} - \kappa t \quad \text{and} \quad dt = -d\alpha/\kappa \tag{2.75b}$$

then

$$\mathcal{F}\{\delta(\tau - \tau_{20} - \kappa t)\} = \frac{1}{\kappa} \int_{-\infty}^{\infty} \delta(\alpha)\, e^{-j2\pi\nu(\tau - \tau_{20} - \alpha)/\kappa}\, d\alpha \tag{2.75c}$$

$$= \frac{1}{\kappa} \cdot e^{-j2\pi\nu(\tau - \tau_{20})/\kappa} \tag{2.75d}$$

Using the result of the calculation, the combined delay-spreading and Doppler-spreading function for the example is found to be

$$U(\tau; \nu) = \int_{-\infty}^{\infty} g(\tau; t)\, e^{-j2\pi\nu t}\, dt \tag{2.76a}$$

$$= \alpha_0 \delta(\tau - \tau_0)\, \delta(\nu) + \alpha_1 \delta(\tau - \tau_1)\, \delta(\nu)$$
$$+ \frac{\alpha_2}{\kappa}\, e^{-j2\pi\nu(\tau - \tau_{20})/\kappa} \tag{2.76b}$$

For the channel input $e^{j2\pi f_d t}$, the complex envelope of the channel output for this example is

$$w(t) = \int_{-\infty}^{\infty} z(t - \tau)\, g(\tau; t)\, d\tau \tag{2.77a}$$

$$= \alpha_0\, e^{j2\pi f_d(t - \tau_0)} + \alpha_1\, e^{j2\pi f_d(t - \tau_1)} + \frac{\alpha_2}{\kappa}\, e^{j2\pi f_d[t(1-\kappa) - \tau_{20}]} \tag{2.77b}$$

Note that the first two terms in (2.77b) have only a delay component. The third term has both a delay component and a frequency shift (Doppler) component. The delay components are a result of the delay in the propagation distance between the receiver and transmitter. The Doppler component is a product of the motion of the second scattering object.

2.2.2 Fading and Fade Rate

2.2.2.1 Characterization of the Random Fading Channel

It has been shown previously that the effects of a time- and/or frequency-dispersive channel can be modeled by writing the channel output $w(t)$ as the superposition of delayed replicas of the channel input $z(t)$:

$$w(t) = \sum_n \alpha_n(t)\, z[t - \tau_n(t)] \tag{2.78}$$

with the effect on the respective complex envelopes with respect to the center frequency f_0 being modeled by

$$w(t) = \sum_n \alpha'_n(t)\, z[t - \tau_n(t)], \quad \alpha'_n(t) = \alpha_n(t)\, e^{-j2\pi f_0 \tau_n(t)} \tag{2.79a}$$

The time-varying delay-spreading function that corresponds to this model is

$$g(\tau; t) = \sum_n \alpha'_n(t)\, \delta[\tau - \tau_n(t)] \tag{2.79b}$$

The time-varying attentuation factors $\{\alpha'_n(t)\}$ are complex numbers that represent the propagation losses and center-frequency phase shifts on the individual paths, which are indexed by n. A kind of "fading" of the received signal occurs when either the receiver or the transmitter moves, causing variation in these propagation losses as, for example, discrete paths alternately interfere with each constructively and destructively.

The time delays $\{\tau_n(t)\}$ are real numbers that represent the path delays. The rate at which these delays vary, due perhaps to receiver or transmitter motion, strongly influences the multipath channel's effect on the signal.

Suppose that the transmitted signal is a unit-amplitude sinusoid at frequency f_0 and that there is no relative motion between transmitter and receiver. Then the complex envelope of the signal and its Fourier transform are

$$z(t) = 1 \quad \text{and} \quad Z(f) = \delta(f) \tag{2.80a}$$

The complex envelope of the channel output in the time domain is

$$w(t) = \int_{-\infty}^{\infty} 1 \cdot g(\tau; t)\, d\tau$$

$$= \sum_n \alpha_n'(t) = \sum_n \alpha_n(t)\, e^{-j2\pi f_0 \tau_n(t)} = \sum_n \alpha_n(t)\, e^{-j\theta_n(t)} \tag{2.80b}$$

The time variation in the attenuation factors $\{\alpha_n(t)\}$ is usually insignificant except for large dynamic changes in the medium. The phase angles $\{\theta_n(t)\}$ can rapidly change by 2π, however, whenever a delay τ_n varies by as little as $\pm 1/f_0$.

The time delays can vary randomly and at different rates. The net effect of this variation is to produce a channel output complex envelope that can be modeled in the time domain as a complex Gaussian random process. In this case, the delay-spread function $g(\tau; t)$ can be viewed as a complex Gaussian random process in the t variable. The variation in the output complex envelope is known as *fading*. The magnitude of a zero-mean complex Gaussian process $|g(\tau; t)|$ is a Rayleigh distributed RV. Therefore, the phenomena created by variations in the time delays is termed *Rayleigh fading*.

Given that the delay-spread function $g(\tau; t)$ is a zero-mean, complex Gaussian random variable, it can be described by the correlation function

$$R_g(\tau; \Delta t) \triangleq \tfrac{1}{2}\, E\{g^*(\tau; t)\, g(\tau; t + \Delta t)\} \tag{2.81}$$

in which the asterisk (*) denote complex conjugation. In the special non-time-varying case, the correlation function is called the *multipath intensity profile*, $R_g(\tau; t) \equiv \phi_g(\tau)$ [29]. This function can take the form

$$\phi_g(\tau) = K\, e^{-(\tau - \tau_0)/T_m}, \quad \tau \geq \tau_0 \tag{2.82}$$

The value T_m is an alternative measure of delay spread that applies to random channels, whereas previously Δ_τ was used.

Recall that $g(\tau; t)$ and $T(f; t)$ are a Fourier transform pair. Thus, if $g(\tau; t)$ is a zero-mean RV, then $T(f; t)$ is also a zero-mean RV. Given that $T(f; t)$ is a zero-mean, complex Gaussian random process, it can be described by the correlation function

$$R_T(\Delta f; \Delta t) \triangleq \tfrac{1}{2} E\{T^*(f; t)\, T(f + \Delta f; t + \Delta t)\} \tag{2.83}$$

This function is sometimes called the *spaced-frequency spaced-time correlation function* of the channel [26]. For the non-time-varying case, $R_T(\Delta f; 0) \equiv \phi_T(\Delta f)$ is the Fourier transform of $\phi_g(\tau)$:

$$\phi_T(\Delta f) = \int_{-\infty}^{\infty} e^{-j2\pi\Delta f \tau}\, \phi_g(\tau)\, d\tau \tag{2.84}$$

Using the exponential model shown in (2.82), the value of $\phi_T(\Delta f)$ is

$$\phi_T(\Delta f) = \int_{\tau_0}^{\infty} e^{-j2\pi\Delta f \tau} \cdot K\, e^{-(\tau-\tau_0)/T_m}\, d\tau \tag{2.85a}$$

$$= \frac{K\, T_m\, e^{-j2\pi\Delta f \tau_0}}{1 + j\, 2\pi(\Delta f) T_m} \tag{2.85b}$$

Note that $|\phi_T(\Delta f)|$, the magnitude of the frequency-domain correlation function, is maximum at $\Delta f = 0$ and falls to one-half its value for $2\pi(\Delta f) T_m$ equal to 1. Thus, $\beta_c \triangleq 1/T_m$ can be defined as a measure of the coherence bandwidth of the random channel.

When $\beta_c = 1/T_m$ is small in comparison with the signal bandwidth W_s, the frequency content of the signal is distorted by the channel, and the channel fading then is said to be *frequency-selective*, which has the same meaning as *time-dispersive* since a small β_c is equivalent to a large T_m:

$$\frac{W_s}{\beta_c} = W_s T_m \gg 1 \rightarrow \text{time-dispersive or frequency-selective} \tag{2.86}$$

When $\beta_c = 1/T_m$ is large in comparison with the signal bandwidth, the frequency content of the signal is not distorted by the channel, and the channel fading then is said to be *frequency-flat*.

$$\frac{W_s}{\beta_c} = W_s T_m \ll 1 \quad \rightarrow \quad \text{non-time-dispersive or frequency-flat} \quad (2.87)$$

It has been shown above that a time-dispersive channel yields a frequency-selective fading channel. It is possible, however, that a channel can be frequency-selective and yet not be a fading channel. The frequency selectivity is a result of multiple transmission paths, giving rise to the delay spread. The fading that may or may not accompany the frequency selectivity is a result of the time variation of the delays.

When the bandwidth of the signal greatly exceeds the coherence bandwidth of the channel, it follows that the symbol duration is much smaller than the delay spread, allowing for possible resolution of multipath components:

$$W_s \gg \beta_c \qquad \rightarrow \qquad T_s = \frac{1}{W_s} \ll T_m = \frac{1}{\beta_c} \qquad (2.88)$$

When the bandwidth of the signal is much less than the coherence bandwidth of the channel, it follows that the symbol duration is much greater than the delay spread. This reduces intersymbol interference due to the multiple paths, but the paths themselves are not resolvable:

$$W_s \ll \beta_c \qquad \rightarrow \qquad T_s = \frac{1}{W_s} \gg T_m = \frac{1}{\beta_c} \qquad (2.89)$$

Similar to the spaced-frequency correlation function $R_T(\Delta f; 0) \equiv \phi_T(\Delta f)$, a spaced-time correlation function [27] can be defined as $R_T(0; \Delta t) = \psi_T(\Delta t)$.

Note that the Fourier transform can be taken of the spaced-frequency, spaced-time correlation function with respect to Δt, the time spacing. The transform with respect to a time parameter reflects the "frequency (Doppler shift) content" of the time variations in the channel, parametric in the "frequency (selectivity) content" of the delay spreading:

$$S_T(\Delta f; \lambda) \triangleq \mathcal{F}\{R_T(\Delta f; \Delta t)\} \qquad (2.90a)$$

$$= \int_{-\infty}^{\infty} R_T(\Delta f; \Delta t)\, e^{-j2\pi\lambda(\Delta t)}\, d(\Delta t) \qquad (2.90b)$$

In particular, for $\Delta f = 0$, we have

$$S(\lambda) \equiv S_T(0; \lambda) = \int_{-\infty}^{\infty} \psi_T(\Delta t)\, e^{-j2\pi\lambda(\Delta t)}\, d(\Delta t) \tag{2.91}$$

which is called the *Doppler power spectrum* of the random channel in [26]. The "bandwidth" of $S(\lambda)$ is known as the Doppler spread of the channel and may be denoted β_d. Because $S(\lambda)$ and $\psi_T(\Delta t)$ are a Fourier transform pair, we understand that a large β_d corresponds to a short correlation time (*coherence time* T_c) in $T(f; t)$, or *fast fading*. Similarly, a small β_d corresponds to a long correlation time (*coherence time* T_c) in $T(f; t)$, or *slow fading*.

A summary of the relationship between the spaced-frequency correlation function and the multipath intensity profile is illustrated in upper part of Figure 2.48 [28]. The relationship between the spaced-time correlation function and the Doppler power spectrum can be found in the lower part of Figure 2.48.

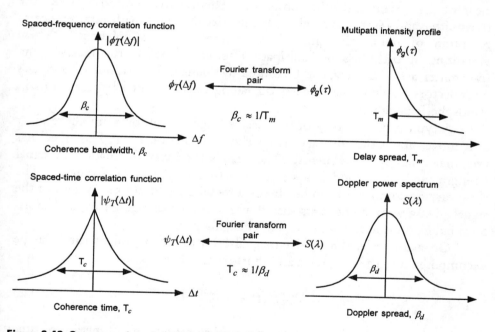

Figure 2.48 Summary of random channel characteristics.

2.2.2.2 Commonly Used Fading Terms

Several commonly used terms describe the fading channel. Among these are *long-term* and *short-term fading*, which relate to the local average and instantaneous signal levels. Frequency-selective fading, as the name implies, affects only certain frequencies. Time-selective fading is analogous to frequency-selective fading except that it is in the time domain. A channel that is both time- and frequency-selective is described as *doubly dispersive*. In a flat-fading channel, there are no dominant fading mechanisms, but fading occurs from the random channel fluctuations.

Fluctuation in a received signal level experienced by mobile radio users, known as *fading*, is due to various effects. These include the variation in propagation path loss as a receiver or transmitter moves from position to position. The attenuation on the path changes as it becomes more or less clear of the terrain and other obstacles. This relative motion between the transmitter and receiver also changes the effects of multipath interference as reception of multipath signals become more or less coherent. These variations in path loss or multipath interference while the receiver and transmitter are stationary, can be caused by changes in positions of obstacles or reflectors. Additional impacts on fading can be caused by variations in atmospheric parameters.

A typical time history of the received signal level for a mobile receiver as it moves about the coverage area is shown in Figure 2.49. This signal has two major fading components. The first is the rapid fluctuation in signal strength, and the second is a slower variation in the signal strength. The slow-varying component can be obtained by taking a local time average of the signal. Long-term fading is associated with the slow-changing part, and short-term fading is associated with the fast-changing part.

On a dB scale, the received signal level as a function of time can be decomposed into the long-term and short-term fading parts:

$$V_{dB}(t) = V_{dB}^{(\text{long})}(t) + V_{dB}^{(\text{short})}(t) \tag{2.92a}$$

On an absolute scale, therefore, the received signal level as a function of time is the product of long-term and short-term variations:

$$V(t) = V^{(\text{long})}(t) \times V^{(\text{short})}(t) \tag{2.92b}$$

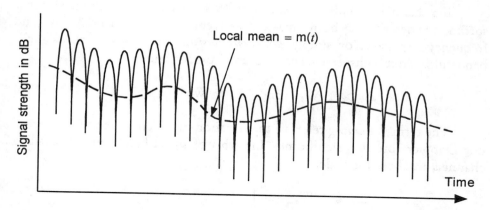

Figure 2.49 Short- and long-term fading.

In general, the long-term fading is due to propagation factors, while short-term fading is due to multipath factors (Rayleigh fading).

When the channel is time-dispersive, that is, stretches pulses in time because of multipath propagation, the combination of time delays has a frequency-dependent effect on the signal. The terms *frequency-selective* and *time-dispersive* are equivalent. A channel is considered frequency-selective when the delay spread is significantly larger than the inverse of the signal bandwidth, which can be expressed as

$$\Delta_\tau \gg 1/W_s \tag{2.93}$$

If the channel has a significant delay spread, but an insignificant Doppler spread, it can be described as a *time-dispersive-flat* channel.

When the channel is frequency-dispersive, that is, stretches the signal bandwidth in the frequency domain due to Doppler effects, the combination of Doppler shifts has a time-dependent effect on the signal. The terms *time-selective* and *frequency-dispersive* are equivalent. A channel is considered time-selective when the Doppler spread is significantly larger than the signal bandwidth:

$$F_s \gg W_s \tag{2.94}$$

If the channel has a significant Doppler spread but an insignificant delay spread, it can be described as a *frequency-dispersive-flat* channel.

If a channel has both a significant delay spread and a significant Doppler spread, it is described as both time- and frequency-selective, or both time- and frequency-dispersive, or simply *doubly-dispersive*. In terms of the signal's bandwidth, doubly-dispersive channels are those for which

$$\Delta_\tau \gg 1/W_s \quad \text{and} \quad F_s \gg W_s \tag{2.95}$$

The term *flat fading* refers to the situation in which neither delay spread nor Doppler spread is significant. In terms of signal bandwidth, flat-fading channels are those for which

$$\Delta_\tau \ll 1/W_s \quad \text{and} \quad F_s \ll W_s \tag{2.96}$$

The changes in reception that make up the fading do take place in time. For flat fading, however, they are associated with the randomness of the channel rather than with motion or other sources of time variation.

A summary of fading channel types is given in Figure 2.50.

$W_s T_s = 1$	
$W_s \ll \beta_c$ and $T_s \ll T_c$ $\Downarrow$ $W_s T_s = 1 \ll \beta_c T_c, \beta_d T_m = \frac{1}{\beta_c T_c} \ll 1$ • Flat, slow fading • One apparent path • Channel constant during symbol	$W_s \ll \beta_c$ and $T_s \gg T_c$ • Flat, fast fading • One apparent path • Channel varies during symbol
$W_s \gg \beta_c$ and $T_s \ll T_c$ • Frequency selective, slow fading • Resolvable multiple paths • Channel constant during symbol	$W_s \gg \beta_c$ and $T_s \gg T_c$ $\Downarrow$ $W_s T_s = 1 \gg \beta_c T_c, \beta_d T_m = \frac{1}{\beta_c T_c} \gg 1$ • Frequency selective, fast fading • Resolvable multiple paths • Channel varies during symbol

Figure 2.50 Summary of fading channel types.

2.2.2.3 Fade Rate and Vehicular Speed

In mobile radio systems, the received signal level varies with time. In part, the variation in time is due to the change in the propagation loss that occurs as the vehicle travels from one position to another, giving rise to relatively slow-changing shadowing effects as the propagation path encounters natural and man-made obstacles. In addition, there may be a relatively fast-changing fading effect due to the interaction of multiple signal components arriving over slightly different paths that interfere with one another, sometimes constructively and sometimes destructively. In many cases, the amplitudes of the multiple signal components that contribute to fast fading are subject to random variation that can be modeled by a Rayleigh distribution. The rate of the variations in the fast-fading effects can be related to the vehicle's speed, through the Doppler effect. In this subsection, background is provided on the Doppler effect and its influence on the rate and duration of fast fades.

Consider the transmission of a signal consisting of a pure tone at frequency f_0 between two antennas, separated by distance d, that may be moving. The Doppler effect of the relative motion between transmitter and receiver antennas is seen at the receiver as a shift in frequency to f_1:

$$f_1 = f_0 \left(1 + \frac{\dot{d}}{c}\right) = f_0 + \Delta f \qquad (2.97)$$

where c is the speed of light and $\dot{d}$ is the time derivative of the distance between the antennas, as illustrated in Figure 2.51. Let the transmitter be at the position $(x_0, y_0, z_0) = (0, 0, 0)$ and the receiver at the position (x_1, y_1, z_1). Then the distance from the transmitter to the receiver in a rectangular coordinate system is

$$d = \sqrt{(x_1 - x_0)^2 + (y_1 - y_0)^2 + (z_1 - z_0)^2} = \sqrt{x_1^2 + y_1^2 + z_1^2} \qquad (2.98)$$

The position of the receiver, relative to that of the transmitter, can be described in terms of the distance d as the length of the vector

$$\boldsymbol{r} = (x_1 - x_0)\,\boldsymbol{u}_x + (y_1 - y_0)\,\boldsymbol{u}_y + (z_1 - z_0)\,\boldsymbol{u}_z \qquad (2.99a)$$

$$= d \times (\cos\theta_x \cdot \boldsymbol{u}_x + \cos\theta_y \cdot \boldsymbol{u}_y + \cos\theta_x \cdot \boldsymbol{u}_z) \qquad (2.99b)$$

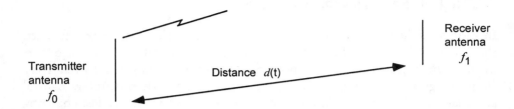

Figure 2.51 Situation involving the Doppler effect.

where u_x, u_y, and u_z are unit vectors in the x, y, and z directions, respectively, and θ_x, θ_y, and θ_z are the respective direction angles of r with the axes in the coordinate system depicted in Figure 2.52.

The time derivative of the receiver's relative position is

$$\dot{d} = \frac{(x_1 - x_0)(\dot{x}_1 - \dot{x}_0) + (y_1 - y_0)(\dot{y}_1 - \dot{y}_0) + (z_1 - z_0)(\dot{z}_1 - \dot{z}_0)}{\sqrt{(x_1 - x_0)^2 + (y_1 - y_0)^2 + (z_1 - z_0)^2}}$$

$$\text{(2.100a)}$$

$$= (\dot{x}_1 - \dot{x}_0)\cos\theta_x + (\dot{y}_1 - \dot{y}_0)\cos\theta_y + (\dot{z}_1 - \dot{z}_0)\cos\theta_z \quad \text{(2.100b)}$$

$$= \left\langle \dot{r}, r/|r| \right\rangle = \text{Inner product of } \dot{r} \text{ and } r/d \quad \text{(2.100c)}$$

Defining v as the magnitude of the time derivative of the distance vector, the inner product can be expressed as

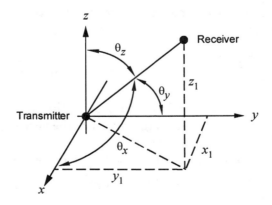

Figure 2.52 Geometry of the receiver position relative to the transmitter.

$$\dot{d} = v \times \left\langle \frac{\dot{r}}{|\dot{r}|}, \frac{r}{|r|} \right\rangle = v \cos \alpha \qquad (2.101)$$

using α to denote the angle between the distance vector and its derivative; that is, the angle the relative motion of the receiver makes with the direction from the transmitter to the receiver. Thus, the shifted frequency at the receiver is

$$f_1 = f_1(\alpha) = f_0 + \Delta f(\alpha) \qquad (2.102a)$$

where $$\Delta f(\alpha) = \frac{\dot{d}}{c} f_0 = \frac{v \cos \alpha}{c} f_0 = f_m \cos \alpha \qquad (2.102b)$$

using f_m to denote the maximum shift in frequency. The quantity f_m is given by

$$f_m(\text{Hz}) = \frac{v}{c} f_0 = v(\text{m/s}) \times \frac{f_0(\text{Hz})}{c(\text{m/s})} = v(\text{m/s}) \times \frac{f_0(\text{MHz})}{300} \qquad (2.103)$$

In this expression, $v(\text{m/s})$ denotes the numerical value of the speed of the receiver relative to the transmitter when expressed in units of m/s, and $f_0(\text{MHz})$ denotes the numerical value of the frequency when expressed in units of MHz.

Example 2.3 Obtain formulas for the maximum Doppler shift for speed expressed in units of m/s, km/hr, ft/s, and mi/hr, and plot the value of maximum Doppler shift as a function of speed in km/hr and mi/hr for center frequencies of 500 MHz, 850 MHz, 1 GHz, and 2 GHz.

Solution: For the different units of speed, the following formulas for the maximum Doppler shift f_m may be used:

special case of $f_0 = 850$ MHz

$$f_m(\text{Hz}) = \frac{f_0(\text{MHz})}{300} \cdot v(\text{m/s}) \qquad = 2.83 \times v(\text{m/s}) \qquad (2.104a)$$

$$= \frac{f_0(\text{MHz})}{1080} \cdot v(\text{km/hr}) \qquad = 0.787 \times v(\text{km/hr}) \qquad (2.104b)$$

$$= \frac{f_0(\text{MHz})}{984.3} \cdot v(\text{ft/s}) \qquad = 0.864 \times v(\text{ft/s}) \qquad (2.104c)$$

$$= \frac{f_0(\text{MHz})}{671.1} \cdot v(\text{mi/hr}) \qquad = 1.27 \times v(\text{mi/hr}) \qquad (2.104d)$$

The value of f_m in Hz is shown as a function of v in km/hr in Figure 2.53 and as a function of v in mi/hr in Figure 2.54.

To discuss the rate of fading, it is necessary to consider the distribution of fading signal amplitudes. A signal that is undergoing Rayleigh fading has the following pdf for its envelope (see Section 1.5.8.2) R:

$$p_R(\alpha) = \frac{\alpha}{\sigma^2} e^{-\alpha^2/2\sigma^2}, \alpha \geq 0 \tag{2.105}$$

where σ^2 denotes the average value of $\frac{1}{2}R^2$, the signal power; (2.105) is plotted in Figure 2.55. The mean square value of R was shown in Chapter 1 to be $2\sigma^2$, with the root-mean-square value $R_{rms} = \sigma\sqrt{2}$. The mode of the Rayleigh distribution, that is, the most likely value, is easily shown by differentiation of the pdf to be

$$R_{mode} = \sigma = R_{rms}/\sqrt{2} \tag{2.106}$$

The cumulative distribution function (CDF) for the Rayleigh distribution is the probability that the random variable is less than or equal to some given value:

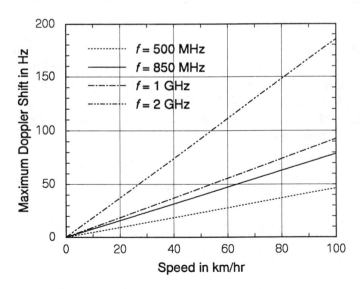

Figure 2.53 Maximum Doppler shift versus speed in km/hr.

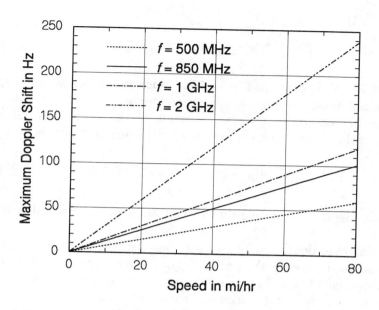

Figure 2.54 Maximum Doppler shift versus speed in mi/hr.

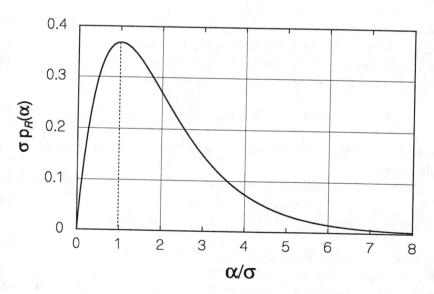

Figure 2.55 Rayleigh pdf.

$$F_R(\alpha) \triangleq \Pr\{R \le \alpha\} = \int_0^\alpha \frac{x}{\sigma^2} e^{-x^2/2\sigma^2} dx \qquad (2.107a)$$

$$= 1 - e^{-\alpha^2/2\sigma^2} = 1 - e^{-\alpha^2/R_{rms}^2} \qquad (2.107b)$$

A plot of $F_R(\alpha)$ is given in Figure 2.56. Viewed on logarithmic paper, the Rayleigh cdf has a very linear appearance for most of the distribution, as shown in Figure 2.57. An example value of the cumulative distribution function: at the envelope rms value (R_{rms}), the value of the cdf is

$$F_R(R_{rms}) = F_R\left(\sigma\sqrt{2}\right) = 1 - e^{-1} = 0.63 \qquad (2.107c)$$

That is, the signal envelope is below its rms value with a probability of 63%.

Measure of fade depth, ρ. The relative value of the signal envelope R at a given time can be greater than or less than the rms envelope value, R_{rms}. If the envelope falls below its rms value to some degree, the signal is said to be experiencing a "fade." It is convenient to use the parameter

$$\rho \triangleq R/R_{rms} \qquad (2.108a)$$

to indicate the degree of fading. In terms of this parameter, the cdf for the envelope is $1 - e^{-\rho^2}$. As we have seen, the probability that ρ is less than 1 is 0.63; that is, there is a 63% chance of a fade of some degree. The *fade depth* or degree of fading of signal power is usually expressed as

$$\text{Fade depth in dB} = -20 \log_{10} \rho = -20 \log_{10}(R/R_{rms}) \qquad (2.108b)$$

Example 2.4 Find the probabilities of a 20-dB fade and a 10-dB fade for Rayleigh fading.
    Solution: A 20-dB fade means $\rho = 10^{-20/20} = 0.1$, with the probability $F_R(0.01) = 1 - e^{-0.01} = 0.01$. A 10-dB fade means $\rho = 10^{-10/20} = \sqrt{0.1} = 0.316$, with the probability $F_R(0.1) = 1 - e^{-0.1} = 0.095$.

    The fade rate of Rayleigh fading mobile radio signals has been modeled as the average level-crossing rate of the envelope of a narrowband Gaussian process with a Doppler-shifted spectrum [30]. An expression for this average is

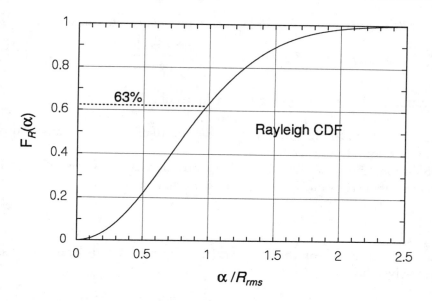

Figure 2.56 Rayleigh cumulative distribution function.

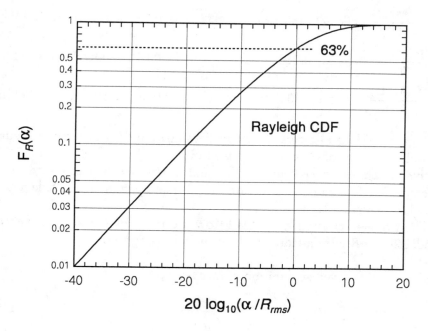

Figure 2.57 Rayleigh cdf plotted on a logarithmic scale.

N_R = Average number of level crossings per second (fade rate)

$$= f_m \rho \sqrt{2\pi}\, e^{-\rho^2}, \qquad \rho = \frac{\text{Envelope level}}{\text{Envelope rms value}} \qquad (2.109)$$

A derivation of this expression is given in Appendix 2B.

Example 2.5 Find the formulas for the rate of 10-dB fades as a function of speed in m/s, km/hr, ft/s, and mi/hr.

Solution: For a 10-dB fade in signal power relative to rms power, the fade depth parameter value is $\rho = 0.316$, resulting in a factor of

$$\rho \sqrt{2\pi}\, e^{-\rho^2} = 0.7172$$

In that case, formulas for the fade rate (crossing rate) for different units of speed are given below:

special case of $f_0 = 850\,\text{MHz}$

$$N_R = f_m \cdot 0.7172 = \frac{f_0(\text{MHz})}{300} \cdot v(\text{m/s}) \cdot 0.7172 = 2.03 \times v(\text{m/s}) \qquad (2.110\text{a})$$

$$= \frac{f_0(\text{MHz})}{300} \cdot v(\text{km/hr}) \cdot 0.1992 \qquad\qquad = 0.565 \times v(\text{km/hr}) \quad (2.110\text{b})$$

$$= \frac{f_0(\text{MHz})}{300} \cdot v(\text{ft/s}) \cdot 0.2186 \qquad\qquad = 0.619 \times v(\text{ft/s}) \qquad (2.110\text{c})$$

$$= \frac{f_0(\text{MHz})}{300} \cdot v(\text{mi/hr}) \cdot 0.3206 \qquad\qquad = 0.909 \times v(\text{mi/hr}) \quad (2.110\text{d})$$

A graph of fade rate versus the fade-depth parameter ρ is given in Figure 2.58, and the rate of 10-dB fades is plotted in Figures 2.59 and 2.60, respectively, against speed in km/hr and mi/hr with frequency as a parameter. The fade rate as a function of speed and parametric in fade depth is plotted in Figure 2.61.

The average duration of a fade below a certain level is also derived in Appendix 2B for Rayleigh fading and has the following value:

$$\bar{\tau} = \text{Average duration of fade level } \rho$$

$$= \left(1 - e^{-\rho^2}\right) \cdot \frac{1}{N_R} = \frac{e^{\rho^2} - 1}{f_m \rho \sqrt{2\pi}} \qquad (2.111)$$

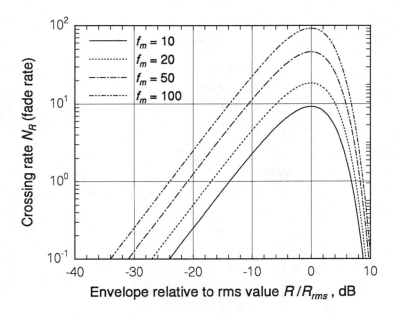

Figure 2.58 Fade rate as a function of fade depth.

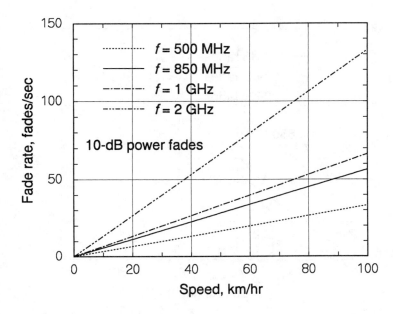

Figure 2.59 Fade rate versus speed in km/hr.

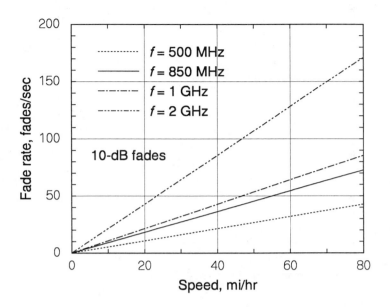

Figure 2.60 Fade rate versus speed in mi/hr.

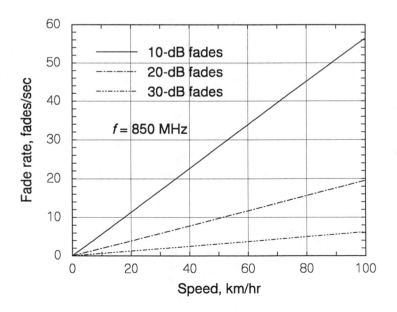

Figure 2.61 Fade rate versus speed in km/hr, fade depth varied.

Example 2.6 Find the average duration of 10-dB fades as a function of speed in m/s, km/hr, ft/s, and mi/hr.

Solution: For a 10-dB fade in signal power relative to rms power, $\rho = 10^{-.5} = 0.316$, giving $1 - e^{-\rho^2} = 0.0952$. In that case, formulas for the fade duration for different units of speed are given below:

special case of $f_0 = 850\,\text{MHz}$

$$\bar{\tau} = \left[\frac{f_0(\text{MHz})}{300} \cdot v(\text{m/s}) \cdot 0.7172 \right]^{-1} \cdot 0.0952 \qquad = 0.047/v(\text{m/s}) \quad (2.112a)$$

$$= \left[\frac{f_0(\text{MHz})}{300} \cdot v(\text{km/hr}) \cdot 0.1992 \right]^{-1} \cdot 0.0952 \qquad = 0.169/v(\text{km/hr}) \,(2.112b)$$

$$= \left[\frac{f_0(\text{MHz})}{300} \cdot v(\text{ft/s}) \cdot 0.2186 \right]^{-1} \cdot 0.0952 \qquad = 0.154/v(\text{ft/s}) \quad (2.112c)$$

$$= \left[\frac{f_0(\text{MHz})}{300} \cdot v(\text{mi/hr}) \cdot 0.3206 \right]^{-1} \cdot 0.0952 \qquad = 0.105/v(\text{mi/hr}) \,(2.112d)$$

2.2.3 Lognormal Shadowing

In addition to the reference value of attenuation, the Longley-Rice propagation loss model calculates an estimated standard deviation $\sigma_c(\text{dB})$ for the propagation loss in dB. The implication is that

$$L(\text{dB}) = \underbrace{A_{ref}(\text{dB}) + L_{fs}(\text{dB})}_{L_{med}(\text{dB})} + \sigma_c(\text{dB}) \times G(0, 1) \qquad (2.113)$$

where $G(0, 1)$ denotes a zero-mean Gaussian RV with unit variance, L_{fs} is the free-space loss, and A_{ref} is the reference (median) excess attenuation or loss due to terrain. A typical value of $\sigma_c(\text{dB})$ is from 8 to 10 dB. As discussed in Section 1.5.10, the propagation loss is thus understood to be a lognormal RV. The variation in L_{med} is due to many factors, including the presence or absence of obstacles in the LOS that can shadow the receiver. The term *shadowing* refers to the fact that a hill or other obstruction can block the radio signal, much like it does the light from the sun, as illustrated previously in Figure 2.14. In Section 1.5.10, the distribution functions and moments of lognormal random variables are given. Also, in Chapter 10, we use lognormal statistics in the evaluation of CDMA link reliability and Erlang capacity calculations.

References

[1]   IEEE Vehicular Technology Society Committee on Radio Propagation, "Coverage Prediction for Mobile Radio Systems Operating in the 800/900 MHz Frequency Range," *IEEE Trans. on Vehicular Technology*, special issue on mobile radio propagation, Vol. 37, No. 1, pp.3–71, Feb. 1988.

[2]   Miller, L. E., "Propagation Model Sensitivity Study," J. S. Lee Associates, Inc. report under contract DAAL02-89-C-0040, July 1992, DTIC accession number AD-B166479.

[3]   Jordan, E. C., and K. G. Balmain, *Electromagnetic Waves and Radiating Systems*, Englewood Cliffs, NJ: Prentice-Hall, 1968 (2nd ed.).

[4]   Balanis, C. A., *Antenna Theory: Analysis and Design*, New York: Harper & Row, 1982.

[5]   Longley, A. G., and P. L. Rice, *Prediction of Tropospheric Radio Transmission Loss Over Irregular Terrain: A Computer Method—1968*, U.S. Dept. of Commerce, Environmental Science Services Admin. (ESSA) technical report ERL 79-ITS 67, July 1968 (DTIC accession number AD-676874).

[6]   Hufford, G. A., A. G. Longley, and W. A. Kissick, *A Guide to the Use of the ITS Irregular Terrain Model in the Area Prediction Mode*, Natl. Telecom. and Info. Admin. (NTIA) report 82-100, Apr. 1982. Available from NTIS as document number PB82-217977.

[7]   Bullington, K., "Radio Propagation for Vehicular Communications," *IEEE Trans. on Vehicular Technology*, Vol. VT-26, pp. 295–308, Nov. 1977, reprinted in [31].

[8]   Barton, D. K., *Modern Radar Systems Analysis*, Norwood, MA: Artech House, 1988.

[9]   Lee, W. C. Y., *Mobile Communications Design Fundamentals*, 2nd ed., New York: Wiley, 1993.

[10]   Parsons, J. D., "Radio Wave Propagation," ch. 2 in *Land Mobile Radio Systems* (R. J. Holbeche, ed.), London: Peter Peregrinus, Ltd., 1985.

[11] Norton, K. A., P. L. Rice, and L. E. Vogler, "The Use of Angular Distance in Estimating Transmission Loss and Fading Range for Propagation through a Turbulent Atmosphere Over Irregular Terrain," *Proc. IRE*, Vol. 43, No. 10, pp. 1488–1526, Oct. 1955.

[12] Fleury, B. H., and P. E. Leuthold, "Radiowave Propagation in Mobile Communications: An Overview of European Research," *IEEE Communications Magazine*, pp. 70–81, Feb. 1996.

[13] Okumura, Y., *et al.*, "Field Strength and Its Variability in VHF and UHF Land-Mobile Radio Service," *Reviews of the Electrical Communications Laboratory* (Japan), Vol. 16, pp. 825–873, Sept./Oct. 1968, reprinted in [31].

[14] Hata, M., "Empirical Formula for Propagation Loss in Land Mobile Radio Services," *IEEE Trans. on Vehicular Technology*, Vol. VT-29, pp. 317–325, Aug. 1980.

[15] Boucher, N., *Cellular Radio Handbook* (2nd ed.), Mill Valley, CA: Quantum Publishing, 1992.

[16] Steele, R., *Mobile Radio Communications*, London: Pentech Press, 1992.

[17] Delisle, G. Y., *et al.*, "Propagation Loss Prediction: A Comparative Study with Application to the Mobile Channel," *IEEE Trans. on Vehicular Technology*, Vol. VT-34, No. 2, pp. 86–96, May 1985.

[18] Aurand, J. F., and R. E. Post, "A Comparison of Prediction Methods for 800-MHz Mobile Radio Propagation," *IEEE Trans. on Vehicular Technology*, Vol. VT-34, No. 4, pp. 149–153, Nov. 1985.

[19] Allesbrook, K., and J. D. Parsons, "Mobile Radio Propagation in British Cities at Frequencies in the VHF and UHF Bands," *IEEE Trans. on Vehicular Technology*, Vol. VT-26, No. 4, pp. 313–323, Nov. 1977.

[20] Stuber, G. L., *Principles of Mobile Communication*, Boston: Kluwer Academic Publishers, 1996.

[21] Weissberger, M., *et al.*, "Radio Wave Propagation: A Handbook of Practical Techniques for Computing Basic Transmission Loss and Field Strength," IIT Research Institute report to Electronic Compatibility Analysis Center, Annapolis, MD, Sept. 1982 (DTIC accession number AD A 122 090).

[22] Sciandra, R. M., "TIREM/SEM Programmer's Reference Manual," Electromagnetic Compatibility Analysis Center report ECAC-CR-90-039, Annapolis, MD, July 1990.

[23] Durkin, J., "Computer Prediction of Service Areas for VHF and UHF Land Mobile Radio Services," *IEEE Trans. on Vehicular Technology*, Vol. VT-26, pp. 323–327, Nov. 1977, reprinted in [31].

[24] Weiner, M. W., "Use of the Longley-Rice and Johnson-Gierhart Tropospheric Radio Propagation Programs: 0.02–20 GHz," *IEEE J. Selected Areas in Communications*, Vol. SAC-4, pp. 297–307, Mar. 1986.

[25] Goldberg, B., ed., *Communications Channels: Characterization and Behavior*, New York: IEEE Press, 1976.

[26] Bello, P. A., "Characterization of Randomly Time-Variant Linear Channels," *IEEE Trans. on Communications Systems*, Vol. CS-11, pp. 360–393, Dec. 1963, reprinted in [25].

[27] Brayer, K., ed., *Data Communications via Fading Channels*, New York: IEEE Press, 1975.

[28] Proakis, J., *Digital Communications* (2nd ed.), New York: McGraw-Hill, 1989.

[29] Cox, D. C., and R. P. Leck, "Correlation Bandwidth and Delay Spread Multipath Propagation Statistics for 910-MHz Urban Mobile Radio Channels," *IEEE Trans. on Communications*, Vol. COM-23, pp. 1271–1280, Nov. 1975, reprinted in [31].

[30] Jakes, W. C., ed., *Microwave Mobile Communications*, New York: IEEE Press, 1994.

[31] Bodson, D., G. F. McClure, and S. R. McConoughley, eds., *Land-Mobile Communications Engineering*, publication number PCO1685, New York: IEEE Press, 1984.

[32] Reudink, D. O., "Properties of Mobile Radio Propagation Above 400 MHz," *IEEE Trans. on Vehicular Technology*, Vol. VT-23, pp. 143–159, Nov. 1974, reprinted in [31].

[33] Gradshteyn, I. S., and I. M. Rhyzhik, *Table of Integrals, Series, and Products* (4th ed.), New York: Academic Press, 1965.

Appendix 2A Details of Propagation Loss for Irregular Terrain

2A.1 Angles of Elevation

Most propagation calculations use the angles of elevation of the radio horizons. If Earth were a smooth sphere, as illustrated in Figure 2A.1, then the horizon distances would be d_{Lst} and d_{Lsr}, and the angles of elevation θ_{et} and θ_{er} above the local horizontals would be the negatives of the Earth angles (angular distances) θ_t and θ_r, respectively. That is, in radians:

$$\theta_{et} = -\left[\frac{\pi}{2} - \left(\frac{\pi}{2} - \theta_t\right)\right] = -\theta_t = -\frac{d_{Lst}}{a}$$

and
$$\theta_{er} = -\theta_r = -\frac{d_{Lsr}}{a} \tag{2A.1a}$$

Therefore, from Figure 2.61 and the definitions of the angles in radians, in the smooth-Earth case and for many LOS cases the great circle distance (d), angular distance (θ), and elevation angles are related by

$$\frac{d}{a} = \theta - \theta_{et} - \theta_{er} \tag{2A.1b}$$

which by transposition gives

$$\theta = \frac{d - d_{Lst} - d_{Lsr}}{a} \tag{2A.1c}$$

showing that θ is the angular distance in excess of LOS. Median values of the horizon elevation angles for irregular terrain may be estimated from the following empirical formula [5]:

$$\theta_{ei} = \text{smooth-Earth value} + \text{increase for } \Delta h > 0 \tag{2A.2a}$$

$$= -\frac{d_{Lsi}}{a} + 0.65\,\Delta h \left(\frac{1}{d_{Li}} - \frac{1}{d_{Lsi}}\right) \tag{2A.2b}$$

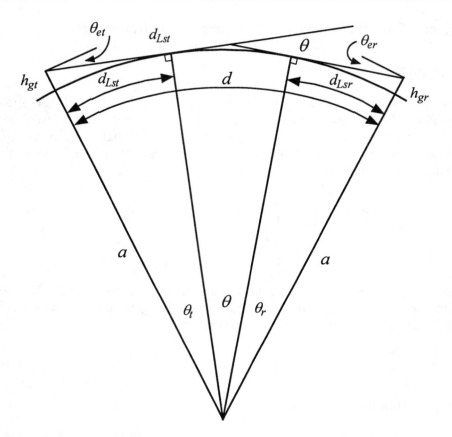

Figure 2A.1 Smooth-Earth angular geometry.

in which Δh is the terrain irregularity parameter and d_{Li} $(i = t, r)$ is the rough-Earth horizon distance, as discussed in the text.

For calculating propagation loss parameters for transhorizon paths, the angular distance θ in excess of the LOS distance for a great circle distance d is then estimated by [5]

$$\theta = \theta_e + \frac{d}{a} \qquad (2A.3a)$$

where

$$\theta_e \triangleq \max(\theta_{e1} + \theta_{e2}, -d_L/a) \qquad (2A.3b)$$

in the light of (2A.2b), this expression for θ has the interpretation

$$\theta = \frac{d - d_L}{a} + (d_{Ls1} - d_{L1}) \max \left\{ \frac{0.65\,\Delta h}{d_{Ls1}d_{L1}}, \frac{1}{a} \right\}$$

$$+ (d_{Ls2} - d_{L2}) \max \left\{ \frac{0.65\,\Delta h}{d_{Ls2}d_{L2}}, \frac{1}{a} \right\} \tag{2A.3c}$$

$$= \text{smooth-Earth value} + \text{terms positive for } \Delta h > 0 \tag{2A.3d}$$

In [5], it is observed that values of θ calculated using estimated horizon distances tend to be considerably less accurate than those using actual horizon distances from terrain profiles.

2A.2 LOS Path Loss

Another parameter describing the terrain is the *terrain roughness factor* σ_h, defined in [5] as "the root-mean-square deviation of terrain and terrain clutter within the limits of the first Fresnel zone in the dominant reflecting plane." In [32], σ_h is defined as the standard deviation of surface irregularities relative to the mean height of the surface.

The roughness factor pertains to LOS situations with little or no path clearance and affects the value of the ground reflection coefficient used in calculation of the propagation loss over a smooth Earth. The empirical formula given in [5] for σ_h in meters is

$$\sigma_h(d) = \begin{cases} 0.78\Delta h(d)\,e^{-0.5[\Delta h(d)]^{1/4}}, & \Delta h(d) > 4\text{m} \\ 0.39\Delta h(d), & \Delta h(d) \leq 4\text{m} \end{cases} \tag{2A.4}$$

Because of the terrain, generally the maximal great circle distance d for which LOS propagation can occur is less than it is for a smooth Earth, since $d_L < d_{Ls}$ for $\Delta h > 0$. When $d < d_L$, however, LOS propagation can occur and the transmission path loss is less than that for a smooth Earth because the terrain surface roughness tends to weaken the reflected ray [9]. The effective magnitude of the ground reflection coefficient in [5] is estimated by the empirical formula

$$R_e = \sqrt{\frac{G_{r1}G_{r2}}{G_{o1}G_{o2}}} \begin{cases} |R|\,e^{-2\pi\sigma_h\sin\psi/\lambda}, & |R|\,e^{-2\pi\sigma_h\sin\psi/\lambda} > \max\left(\sqrt{\sin\psi}, 0.5\right) \\ \sqrt{\sin\psi}, & \text{otherwise} \end{cases}$$

$$\tag{2A.5a}$$

where

G_{ri} = antenna gain of radio i in the direction of the reflecting point (2A.5b)

G_{oi} = antenna gain of radio i in the direction of the other radio (2A.5c)

Unless the antenna beams are very narrow or are directed away from Earth's surface to minimize reflection from the surface, $G_{ri} = G_{oi}$.

The attenuation factor $2\pi\sigma_h\sin\psi/\lambda$ in (2.118a) is related to the Rayleigh criterion, given in [32] as the quantity $C = 2\pi\sigma_h\psi/\lambda$; the surface is smooth enough for specular reflection if $C < 0.1$, and a surface is considered "rough" if $C > 10$.

With a reflection coefficient $R = R_e e^{j(\pi-c)}$ instead of $R = -1$, the LOS path loss in excess of free space becomes

$$L_{LOS}(d) = -10\log_{10}\left[1 + R_e^2 - 2R_e\cos\left(\frac{4\pi h_{et}h_{er}}{\lambda d} - c\right)\right] \qquad (2A.6a)$$

$$= -10\log_{10}\left(1 + R_e^2\right) - 10\log_{10}\left[1 - \frac{2R_e}{1 + R_e^2}\cos\left(\frac{4\pi h_{et}h_{er}}{\lambda d} - c\right)\right]$$

$$\approx -10\log_{10}\left(1 + R_e^2\right) + \frac{8.686R_e}{1 + R_e^2}\cos\left(\frac{4\pi h_{et}h_{er}}{\lambda d} - c\right), R_e \ll 1 \qquad (2A.6b)$$

As Δh increases, R_e decreases from the idealized value of 1, and the amount of interference from the reflected ray decreases, tending to cause the total LOS path loss to be close to that of free space ($L_{LOS} \rightarrow 0$). Another effect of the terrain irregularity, however, is that as Δh increases, the diffraction mode of propagation becomes more important, even at LOS distances.

2A.3 Diffraction Loss

For irregular terrain, there are two diffraction theories used for prediction of BLOS path loss, corresponding to (1) a nonspherical but smooth Earth and (2) a highly irregular Earth such that the radio horizons are determined by *knife-edge* obstacles. In addition, an empirical adjustment called a *clutter factor* in [5] is used to fit the models to measured path losses. For cellular applications, where distances are well within smooth-Earth horizon distances, and terrain and building obstacles are the likely causes of diffraction, the contribution of the first term is negligible. These considerations give rise to two nonnegligible diffraction loss terms—denoted here by L_{d2} and A_{fo}.

The knife-edge model of diffraction, giving rise to an estimate of diffraction loss, L_{d2}, is based on the assumption that the horizon distances d_{L1} and d_{L2} are determined by rather sharp terrain obstacles, and the geometrical models shown in Figure 2A.2 are used for analysis. The concept of applying the knife-edge theory to BLOS diffraction is developed as suggested in the top part of Figure 2A.2, in which two triangles similar to the one at the top of the figure are superimposed on a path profile, with the horizons acting as the knife edges. In [5] and [6], the total loss over the two edges is calculated using the geometry illustrated at the bottom part of Figure 2A.2:

$$L_{d2} = A(v_1) + A(v_2) \tag{2A.7}$$

Using the approximate geometry shown and the fact that the angles involved are so small that tangents and sines can be replaced by the angles, the arguments v_1 and v_2 may be derived from

$$v_i^2 \approx \frac{2(d - d_L + d_{Li})}{\lambda} \alpha_i \beta_i, \quad i = 1, 2 \tag{2A.8a}$$

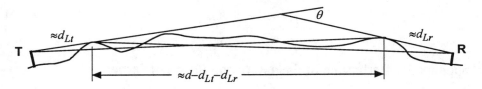

Concept of the horizons as knife edges

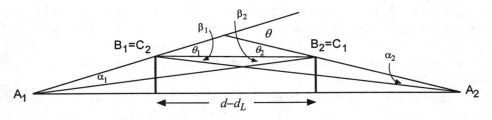

Approximate knife-edge geometry

Figure 2A.2 Geometry for diffraction calculations.

using

$$\frac{\sin \alpha_i}{\sin \beta_i} = \frac{d - d_L}{d_{Li}} \approx \frac{\alpha_i}{\beta_i} \tag{2A.8b}$$

and

$$\alpha_i + \beta_i = \theta_i \approx \theta/2 \tag{2A.8c}$$

it is found that

$$v_i^2 \approx \frac{2(d - d_L + d_{Li})}{\lambda} \left[\frac{d - d_L}{d - d_L + d_{Li}} \cdot \frac{\theta}{2} \right] \left[\frac{d_{Li}}{d - d_L + d_{Li}} \cdot \frac{\theta}{2} \right]$$

$$= \frac{(d - d_L) \theta^2}{2\lambda} \cdot \frac{d_{Li}}{d - d_L + d_{Li}} \tag{2A.9}$$

In addition to L_{d2}, an empirical third component of diffraction loss called the "clutter factor" is given by [5]

$$A_{fo} = \min \left\{ 5 \log_{10} \left[1 + 10^{-5} \cdot h_{g1} \, h_{g2} \, f_{\text{MHz}} \, \sigma_h(d_{Ls}) \right], 15 \, \text{dB} \right\} \tag{2A.10}$$

A total diffraction path loss estimate, then, is

$$L_{diff} = L_{d2} + A_{fo} \tag{2A.11}$$

Appendix 2B Derivation of Fade Rate and Duration Formulas

The mobile radio signal is subject to fading of different kinds: Rayleigh fading due to unresolvable multiple paths that interfere with one another, and lognormal shadowing due to terrain masking. Both kinds of fading are highly dependent on the vehicle's position relative to terrain and buildings that can either block or reflect the signal. Even if the vehicle is not moving, these objects may be moving, so that in general, the fading manifests itself as a time-varying signal amplitude.

The effect of vehicular motion on the spectrum of the received signal can be analyzed as follows: The maximum Doppler shift of the signal's frequency (f_c) is $f_m = \frac{V}{c} \cdot f_c = \frac{V}{\lambda_c}$, where V is the vehicle's speed and λ_c is the signal wavelength. Because the angle α between the direction of arrival of the signal and the direction of the vehicle's travel varies, the amount of shift varies also.

Let the received frequency be $f(\alpha) = f_c + f_m \cos \alpha$. Then, if the angle of arrival is a uniformly distributed RV, the pdf for the received frequency is

$$p_f(x) = \left\{ \left| \frac{\partial \alpha}{\partial f} \right| p_\alpha[\alpha(f)] \right\}\bigg|_{f=x} = \left| \frac{\partial}{\partial f} \cos^{-1}\left(\frac{f - f_c}{f_m} \right) \right|_{f=x} \cdot \frac{1}{2\pi}$$

$$= \frac{1}{\sqrt{f_m^2 - (x - f_c)^2}} \cdot \frac{1}{2\pi} \tag{2B.1}$$

which gives a rough indication of the spectral distribution of the signal from the probabilistic point of view. Alternatively, we can postulate many multipath arrivals, so that there is an arrival at every direction. Assuming these arrivals have equal energy, the total received waveform due to the original transmission of the single frequency f_c has the power σ^2 and the spectrum

$$\mathcal{S}(f) = \frac{\sigma^2}{2\pi} \cdot \frac{1}{\sqrt{f_m^2 - (f \mp f_c)^2}}, \quad |f \mp f_c| < f_m \tag{2B.2}$$

The form of this spectrum is plotted in Figure 2B.1.

The autocorrelation function that corresponds to the Doppler spectrum shown in (2B.2) is found as the inverse Fourier transform of $\mathcal{S}(f)$:

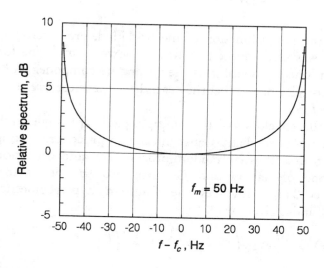

Figure 2B.1 Doppler-induced power spectrum.

$$R(\tau) = \int_{-\infty}^{\infty} \mathcal{S}(f)\, e^{j2\pi f\tau}\, df = 2\int_{0}^{\infty} \mathcal{S}(f)\cos(2\pi f\tau)\, df, \quad \mathcal{S}(f) \text{ even}$$

$$= \frac{\sigma^2}{\pi} \int_{-f_m+f_c}^{f_m+f_c} \frac{\cos(2\pi f\tau)}{\sqrt{f_m^2 - (f - f_c)^2}}\, df \tag{2B.3a}$$

where

$$\cos(2\pi f\tau) = \cos[2\pi f_c\tau + 2\pi(f - f_c)\tau]$$

$$= \cos(2\pi f_c\tau)\cos[2\pi(f - f_c)\tau]$$

$$- \sin(2\pi f_c\tau)\sin[2\pi(f - f_c)\tau] \tag{2B.3b}$$

Because $\sin[2\pi(f - f_c)\tau]$ is an odd function in the interval $|f - f_c| < f_m$, only the cosine term is significant, and after a change of variable ($\nu = f - f_c$), the correlation function is found to be

$$R(\tau) = \frac{2\sigma^2}{\pi} \cos(2\pi f_c\tau) \int_{0}^{f_m} \frac{\cos(2\pi \nu\tau)}{\sqrt{f_m^2 - \nu^2}}\, d\nu$$

$$= \frac{2\sigma^2}{\pi} \cos(2\pi f_c\tau) \int_{0}^{1} \frac{\cos(2\pi f_m\tau x)}{\sqrt{1 - x^2}}\, dx$$

$$= \sigma^2 J_0(2\pi f_m\tau)\cos(2\pi f_c\tau) \tag{2B.3c}$$

where $J_0(\cdot)$ is the Bessel function of the first kind, order zero, and integral 3.753.2 of [33] was used. Note that $R(0) = \text{power} = \sigma^2$. The form of this autocorrelation function reveals that the quadrature components of the fading signal are not correlated, and that they each have the autocorrelation function $g(\tau) = \sigma^2 J_0(2\pi f_m\tau)$.

Because the envelopes of the multiple paths are Rayleigh-distributed, the multipaths are narrowband Gaussian random processes, making the total received waveform a Gaussian random process. The autocorrelation function of the envelope of narrowband Gaussian process with autocorrelation function $g(\tau)\cos 2\pi f_c\tau$ involves the Gaussian hypergeometric function $_2F_1(a, b; c; x)$ and is given by [30]

$$R_R(\tau) = \frac{\pi\sigma^2}{2}\, _2F_1\left(-\frac{1}{2}, -\frac{1}{2}; 1; \frac{g^2(\tau)}{\sigma^4}\right)$$

$$
= \frac{\pi\sigma^2}{2} \sum_{n=0}^{\infty} \frac{1}{n!} \left[\frac{g^2(\tau)}{\sigma^4} \right]^n \frac{\left[(-\frac{1}{2})(\frac{1}{2}) \cdots (-\frac{3}{2}+n) \right]^2}{n!}
$$

$$
= \frac{\pi\sigma^2}{2} \left[1 + \frac{g^2(\tau)}{4\sigma^4} + \frac{g^4(\tau)}{64\sigma^8} + \cdots \right]
$$

$$
\approx \frac{\pi\sigma^2}{2} \left[1 + \frac{g^2(\tau)}{4\sigma^4} \right] = \frac{\pi\sigma^2}{2} \left[1 + \frac{1}{4} J_0^2(2\pi f_m \tau) \right] \qquad (2B.4)
$$

The (baseband) spectrum of the envelope, needed to compute certain statistics later, is found as the Fourier transform of $R_R(\tau)$:

$$
\mathcal{S}_R(f) = \frac{\pi\sigma^2}{2} \delta(f) + \underbrace{\frac{\pi\sigma^2}{8} \int_0^{\infty} J_0^2(2\pi f_m \tau) \cos(2\pi f \tau) \, d\tau}_{(\pi/8\sigma^2)\mathcal{F}\{g^2(\tau)\}} \qquad (2B.5)
$$

The Fourier transform of $g^2(\tau)$ can be calculated as the convolution of the Fourier transform of $g(\tau)$ with itself:

$$
\mathcal{F}\{g^2(\tau)\} = \mathcal{F}\{g(\tau) \times g(\tau)\} = G(f) * G(f) \qquad (2B.6a)
$$

where

$$
G(f) = \mathcal{F}\{g(\tau)\} \qquad (2B.6b)
$$

Recall that $g(\tau)$ was found as the inverse Fourier transform of $\mathcal{S}(f)|_{f_c=0}$, the baseband version of the Doppler spectrum. Therefore

$$
G(f) = \mathcal{S}(f - f_c) = \frac{\sigma^2}{\pi} \cdot \frac{1}{\sqrt{f_m^2 - f^2}}, \qquad |f| < f_m \qquad (2B.7)
$$

and the Fourier transform of $g^2(\tau)$ is

$$
\mathcal{F}\{g^2(\tau)\} = \left(\frac{\sigma^2}{\pi} \right)^2 \int_{-f_m}^{f_m - f} \frac{1}{\sqrt{f_m^2 - x^2}} \cdot \frac{1}{\sqrt{f_m^2 - (x + f)^2}} \, dx, \, 0 < f < 2f_m
$$

$$
= \left(\frac{\sigma^2}{\pi} \right)^2 \frac{1}{f_m} \int_{-1}^{1 - f/f_m} \frac{1}{\sqrt{1 - x^2}} \cdot \frac{1}{\sqrt{1 - (x + f/f_m)^2}} \, dx \qquad (2B.8)
$$

The integral in (2.132) can be manipulated as follows:

$$I \triangleq \int_{-1}^{1-f/f_m} \frac{1}{\sqrt{1-x^2}} \cdot \frac{1}{\sqrt{1-(x+f/f_m)^2}} \, dx$$

$$= \int_{-(1-f/2f_m)}^{1-f/2f_m} \frac{1}{\sqrt{1-(x-f/2f_m)^2}} \cdot \frac{1}{\sqrt{1-(x+f/2f_m)^2}} \, dx$$

(translation)

$$= 2\int_0^{1-a} \frac{1}{\sqrt{1-(x-a)^2}} \cdot \frac{1}{\sqrt{1-(x+a)^2}} \, dx \,, \; a = \frac{f}{2f_m} \quad \text{(evenness)}$$

$$= 2\int_0^{1-a} \frac{1}{\sqrt{(1-x-a)(1-x+a)(1-x-a)(1+x+a)}} \, dx$$

$$= 2\int_0^{1-a} \frac{1}{\sqrt{(1-a)^2-x^2}\sqrt{(1+a)^2-x^2}} \, dx$$

$$= \frac{2}{(1-a)(1+a)}\int_0^{1-a} \frac{1}{\sqrt{1-\left(\frac{x}{1-a}\right)^2} \cdot \sqrt{1-\left(\frac{x}{1+a}\right)^2}} \, dx$$

$$= \frac{2}{1+a}\int_0^{1} \frac{1}{\sqrt{1-y^2} \cdot \sqrt{1-\left(\frac{1-a}{1+a}\right)^2 y^2}} \, dy$$

(transformation: $y = \frac{x}{1-a}$)

$$= \frac{2}{1+a} K\left(\frac{1-a}{1+a}\right) \tag{2B.9a}$$

where $K(\cdot)$ is the complete elliptic integral of the first kind, a tabulated function. By a further transformation relationship for this function, it can be shown also that

$$I = K\left(\sqrt{1-a^2}\right) \tag{2B.9b}$$

Thus, the baseband spectrum of the signal envelope that is subject to fading can be written

$$S_R(f) = \frac{\pi\sigma^2}{2}\delta(f) + \frac{\pi}{8\sigma^2}\mathcal{F}\{g^2(\tau)\} = \frac{\pi\sigma^2}{2}\delta(f) + \frac{\pi}{8\sigma^2}\left(\frac{\sigma^2}{\pi}\right)^2\frac{1}{f_m}\cdot I$$

$$= \frac{\pi\sigma^2}{2}\delta(f) + \frac{\sigma^2}{8\pi f_m}K\left(\sqrt{1-\left(\frac{f}{2f_m}\right)^2}\right) \tag{2B.10}$$

This spectrum describes the frequency content of the fluctuations in the fading signal envelope.

The fluctuating nature of the fading process, as described by its spectral characteristics, determines that deep fades of the signal envelope below some critical level will occur at a certain rate, on the average, and have a certain average duration. The fade rate is measured in terms of $N_R(r)$, the *level crossing rate* for the envelope process, which is defined as the expected positive value of the envelope time derivative, conditioned on the value of the envelope (r) and weighted by the pdf for the envelope:

$$N_R(r) = p_R(r) \times \int_0^\infty \dot{r}\, p_{\dot{R}}(\dot{r}\mid r)\, d\dot{r} = \frac{r}{\sigma^2}e^{-r/2\sigma^2}\int_0^\infty \dot{r}\, p_{\dot{R}}(\dot{r}\mid r)\, d\dot{r} \tag{2B.11}$$

The time derivative of the envelope for a narrowband Gaussian random process can be shown to be independent of the envelope, and to have a zero-mean Gaussian distribution with variance equal to $\sigma_1^2 = 2\sigma^2\pi^2 f_m^2$, resulting in

$$N_R(r) = \frac{r}{\sigma^2}e^{-r/2\sigma^2}\int_0^\infty \dot{r}\,\frac{1}{\sigma_1\sqrt{2\pi}}e^{-\dot{r}^2/2\sigma_1^2}\, d\dot{r} \tag{2B.12}$$

Performing the integration, the level crossing rate is found to be

$$N_R(r) = \frac{r}{\sigma^2}e^{-r/2\sigma^2}\int_0^\infty \dot{r}\,\frac{1}{\sigma_1\sqrt{2\pi}}e^{-\dot{r}^2/2\sigma_1^2}\, d\dot{r}$$

$$= \frac{r}{\sigma^2}e^{-r/2\sigma^2} \times \frac{\sigma_1}{\sqrt{2\pi}}\int_0^\infty e^{-u}\, du = f_m\sqrt{\pi}\cdot\frac{r}{\sigma}e^{-r^2/2\sigma^2} \tag{2B.13a}$$

$$= f_m\sqrt{2\pi}\cdot\rho\, e^{-\rho^2}, \quad \text{using} \quad \rho = \frac{r}{\sigma\sqrt{2}} = \frac{r}{R_{rms}} \tag{2B.13b}$$

which is the formula shown for the fade rate in (2.109) in the text.

Note that the fade rate is maximum for $\rho = 1$, giving

$$\max_{\rho}\{N_R\} = f_m \sqrt{2\pi} \cdot e^{-1} = 0.92 f_m \qquad (2B.13c)$$

A graph of the level crossing rate, parametric in ρ and f_m, is shown above in the text of this chapter.

The level crossing rate N_R, having been expressed in terms of Doppler frequency and relative envelope level, indicates the rate of fading fluctuations. For example, the rate of so-called "deep fades" can be estimated by setting the envelope level at a small value relative to the envelope's rms value. For example, when $\rho = 0.1$, the level crossing rate is $N_R = f_m \sqrt{2\pi}(0.1)e^{-0.01} \approx f_m/4$.

Derivation for average fade duration. The duration of any deep fades is significant. How long will the envelope be in a deep fade? Or, how long an integration time can be used to estimate power for a power control algorithm? One simple measure of fade duration is the inverse of the level crossing rate, interpreted as an average period for the fading fluctuations:

$$D_R = (N_R)^{-1} = \frac{1}{f_m \rho \sqrt{2\pi}} e^{\rho^2} \qquad (2B.14)$$

Another way to analyze fade duration is to weight the quantity D_R, the average period of fading fluctuations, by the probability that the fade is below a certain level. Using this approach, an average duration for fades below the threshold $R = \rho R_{rms}$ can be formulated as

$$\bar{\tau} = D_R \times \Pr\{R \le \rho R_{rms}\} = D_R \times \left(1 - e^{-\rho^2}\right)$$

$$= \frac{1}{f_m \rho \sqrt{2\pi}} e^{\rho^2} \times \left(1 - e^{-\rho^2}\right) = \frac{1}{f_m \rho \sqrt{2\pi}} \left(e^{\rho^2} - 1\right) \qquad (2B.15)$$

which is the formula used in (2.111) in the text.

Since the Doppler frequency f_m is directly proportional to vehicle speed, there is a relationship between the average fade duration and vehicle speed, through the variable f_m. Substituting for f_m in the expression for $\bar{\tau}$ gives the relationship between speed and fade duration:

$$\bar{\tau} = \frac{1}{f_m \rho \sqrt{2\pi}} \left(e^{\rho^2} - 1 \right) = \frac{1{,}080 \left(e^{\rho^2} - 1 \right)}{V_{kph} \, f_{MHz} \, \rho \sqrt{2\pi}} \tag{2B.16a}$$

For $V_{kph} = 80$, $f_{MHz} = 850$, and $\rho = 1$, the fade duration is

$$\bar{\tau} = \frac{1{,}080 \left(e^{\rho^2} - 1 \right)}{V_{kph} \, f_{MHz} \, \rho \sqrt{2\pi}} = \frac{1{,}080 \left(e^1 - 1 \right)}{80 \cdot 850 \cdot 1 \cdot \sqrt{2\pi}}$$

$$= 0.0109 \text{ sec} = 11 \text{ ms} \tag{2B.16b}$$

For $f_{MHz} = 850$, $\rho = 0.1$, and $\bar{\tau} = 2 \text{ ms}$, the speed is

$$V_{kph} = \frac{1{,}080 \left(e^{\rho^2} - 1 \right)}{\bar{\tau} \, f_{MHz} \, \rho \sqrt{2\pi}} = \frac{1{,}080 \left(e^{0.01} - 1 \right)}{.002 \cdot 850 \cdot 0.1 \cdot \sqrt{2\pi}}$$

$$= 25.5 \text{ km/hr} \tag{2B.16c}$$

3

Basic Cellular Systems Engineering

In this chapter, we introduce the basic systems engineering aspects of a cellular system by discussing them as they apply to conventional frequency-division multiple-access (FDMA) cellular systems. These systems engineering aspects include the conventions of telephone traffic theory, the conceptual basis of cellular systems, and the tradeoffs that determine the parameters of cellular systems.

3.1   Review of Telephone Traffic Theory

3.1.1   Telephone Connectivity

The placement of a telephone call requires that a connection be made between the caller's telephone and that of the called party. The connection provided for a particular telephone call is typically a combination of wire and radio links. As illustrated in Figure 3.1, the telephone system provides the connection by switching the needed links over to the use of the call for its duration.

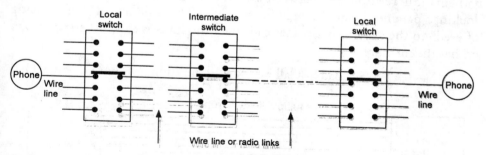

Figure 3.1 Telephone connectivity.

The switches in the telephone system are connected by groups of wire and radio links that are called trunks. The links in a trunk carry both call traffic and the signaling among switches of information that is necessary for call routing and other system control purposes, as suggested in Figure 3.2. For example, when the caller picks up the phone, an "off-hook" signal is detected at that phone's nearest switch, indicating a request for services from the telephone system. If an outgoing link is free, the caller hears a "dial tone"; otherwise, he or she hears a "busy signal."

In a given location, for economic reasons there are usually much fewer links available for calls than the number of telephones or potential callers. There is a finite probability that someone desiring to make a call will find that all the available links are in use—the call then is said to be "blocked" or "time congested" [1, 2]. The more links available, the less the system's probability of blocking is, providing better service to the users. The cost of the telephone system is related to the number of links, so there is a tradeoff between the cost of telephone service and its quality.

The telephone system carrier determines the number of links to make available on the basis of a target value for the probability of blocking. That value, along with other measures of the quality of the system, constitute the grade of service that is being delivered to the user.

3.1.2 Traffic Load and Trunk Size

Suppose that a trunk has N links. After defining measures for traffic, the mathematical relationship between the traffic load and the probability of blocking, parametric in N, will be shown. The amount of traffic being "offered" to the switch that routes calls over the links is called the *traffic load* and has the measure

$$\text{Load} = \text{Call rate} \times \text{Call duration} \tag{3.1}$$

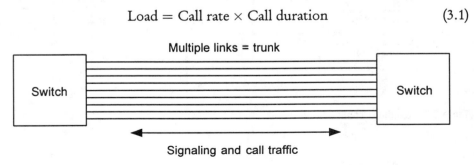

Figure 3.2 Telecommunications trunk.

Because call rate is measured in calls/second and call duration is in seconds/call, the traffic load is a dimensionless quantity that, however, is given in units of *Erlangs*. For example, if the call rate in a small office building is 50 calls/hr and the average call length is 5 minutes, then the traffic load for the trunk carrying the calls out of the building is 50 calls/hr × (5/60) hr/call = 250/60 = 4.2 E. This is the traffic load in Erlangs, usually denoted by A.

The theoretically maximal traffic load that can be carried by a trunk with N links is N Erlangs. In this context, an Erlang corresponds to a fully used link. It is possible for the "offered load" to be greater than the "carried load," and the relationship between these quantities depends on how the specific system handles the calls when the N links are fully occupied [1, 2]. For example, calls that arrive when all N links are in use may be rejected immediately ("dropped"), briefly delayed ("held") on the chance a link will soon become available, and so on. Because calls start and stop randomly and are not coordinated, and because it takes some time to switch a link from one user to another, the N links will always carry fewer than N Erlangs of traffic.

The number of links needed to keep the blocking probability below a certain value is primarily a function of the amount of offered traffic. The maximum amount of traffic fluctuates with the time of day and the day of the week or year. For systems planning purposes, the traffic offered during a "busy hour" of the day is often used. Typically, a user is likely to be on the telephone at any given time during the busy hour with a probability of 0.02 to 0.10; that is, each subscriber is considered to offer $A_0 = 0.02$–0.10 E of traffic during the busy hour. The number of subscribers that can be supported by the switch during the busy hour—for a specified blocking probability—is then

$$M = A/A_0 \tag{3.2}$$

3.1.3 Erlang B Statistics

In the conventional modeling of telephone traffic and the servicing of calls by the telephone switching system, it is assumed that N lines are available at the local switch for handling calls from a much larger number of telephone subscribers, each of whom uses the telephone only a small fraction of the time. The uncoordinated initiation of calls by a large number of subscribers ($M \gg N$) gives rise to a randomly spaced sequence of call requests, termed "call arrivals" that, to a good degree of approximation, does not depend on how many of those subscribers are engaged in active calls at a given time. The arrival of calls from different users at the switch under these conditions is assumed to be a Poisson random process, with the average rate of arriving calls λ calls/sec. The duration (holding time) of a call is assumed to have an exponential probability distribution, with an average call length of T seconds. This distribution is valid for a call departure or completion process that is Poisson, with average departure rate of $1/T = \mu$ calls/sec [2].

In a time interval Δt, the Poisson call arrival process with average arrival rate λ has the following discrete probability function for N_A, the number of calls that arrive in the interval:

$$\Pr\{N_A = n\} = \frac{(\lambda \, \Delta t)^n}{n!} \, e^{-\lambda \Delta t}, \qquad n = 0, 1, 2, \ldots \tag{3.3}$$

For example, the probability of exactly one arrival in Δt is $\lambda \Delta t \cdot e^{-\lambda \Delta t}$. The average number of arrivals in Δt is $\overline{N}_A = \lambda \Delta t$. Similarly, in a time interval Δt, the Poisson call departure process with average departure rate μ has the following discrete probability function for N_D, the number of calls that depart in the interval:

$$\Pr\{N_D = n\} = \frac{(\mu \, \Delta t)^n}{n!} \, e^{-\mu \Delta t}, \qquad n = 0, 1, 2, \ldots \tag{3.4}$$

The average number of departures in Δt is $\overline{N}_D = \mu \Delta t$.

As the random call traffic arrives and departs, the number of lines occupied by ongoing calls can vary from 0 to N. In the short interval between the times t and $t + dt$, the probability that exactly one call will arrive is $\lambda \, dt \, e^{-\lambda \, dt} \approx \lambda \, dt$ and the probability that one call will depart from an occupied line is $\mu \, dt \, e^{-\mu \, dt} \approx \mu \, dt$. Thus, given that k of the N lines are occupied at time t, in the interval $(t, t + dt)$ there are the probabilities

$$\lambda \, dt = \Pr\{1 \text{ call will arrive in the interval}\} \tag{3.5a}$$

and
$$\binom{k}{1} (\mu \, dt)^1 (1 - \mu \, dt)^{k-1} \approx k \mu \, dt$$

$$= \Pr\{1 \text{ call will depart in the interval}\} \tag{3.5b}$$

The probability that neither an arrival nor a departure occurs in the interval is approximated by

$$(1 - \lambda \, dt)(1 - k\mu \, dt) = 1 - \lambda \, dt - k\mu \, dt + (\lambda \, dt)(k\mu \, dt)$$

$$\approx 1 - \lambda \, dt - k\mu \, dt \tag{3.5c}$$

With this understanding of the probabilities for the events that can occur in the interval $(t, t + dt)$, the transition of the probabilities of the various numbers of lines occupied can be expressed by

$$
\underset{\substack{\text{# occupied} \\ \text{after } dt \\ \downarrow}}{P(k;} \quad t + dt) = \underset{\substack{\text{prob. of no change} \\ \overbrace{\hspace{3cm}}}}{(1 - \lambda\, dt - k\mu\, dt)} \cdot P(k;\ t) + \underset{\substack{\text{same} \\ \text{#} \\ \downarrow}}{\overset{\substack{\text{prob. of} \\ \text{increase}}}{\widetilde{\lambda\, dt}}} \cdot \underset{\substack{\text{smaller #} \\ \downarrow}}{P(k-1;\ t)}
$$

$$
+ \underset{\substack{\overbrace{\hspace{2.5cm}} \\ \text{prob. of} \\ \text{decrease}}}{(k+1)\mu\, dt} \cdot \underset{\substack{\uparrow \\ \text{larger #}}}{P(k+1;t)}, \qquad 0 < k < N \qquad (3.6a)
$$

using $P(k; t)$ to denote the probability that k lines are occupied at time t. The special cases of $k = 0$ and $k = N$ are

$$
P(0; t + dt) = (1 - \lambda\, dt) \cdot P(0; t) + \mu\, dt \cdot P(1; t) \qquad (3.6b)
$$

and

$$
P(N; t + dt) = (1 - \lambda\, dt - N\mu\, dt) \cdot P(N; t) + \lambda\, dt \cdot P(N-1; t) \qquad (3.6c)
$$

At steady state, it is assumed that the probabilities are not functions of time:

$$
P(k; t + dt) = P(k; t) = P_k, \qquad k = 0, 1, 2, \ldots, N \qquad (3.7)
$$

Therefore, the steady-state transition equations become

$$
P_k = (1 - \lambda\, dt - k\mu\, dt) \cdot P_k + \lambda\, dt \cdot P_{k-1} + (k + 1)\mu\, dt \cdot P_{k+1}
$$
$$
\Rightarrow \quad 0 = [(-\lambda - k\mu)\, P_k + \lambda\, P_{k-1} + (k + 1)\mu\, P_{k+1}] \cdot dt
$$
$$
\Rightarrow \quad (\lambda + k\mu)P_k = \lambda\, P_{k-1} + (k+1)\mu\, P_{k+1}, \quad 0 < k < N \qquad (3.8a)
$$

Similarly, the equations for the special cases of $k = 0$ and $k = N$ become

$$
\lambda\, P_0 = \mu\, P_1, \ k = 0 \qquad (3.8b)
$$

and

$$
(\lambda + N\mu)P_N = \lambda\, P_{N-1}, \ k = N \qquad (3.8c)
$$

Because the number of lines occupied is restricted to $0 \le k \le N$, the probabilities $\{P_k\}$ also must satisfy the requirement that

$$
P_0 + P_1 + \cdots + P_N = \sum_{k=0}^{N} P_k = 1 \qquad (3.9a)
$$

The form of the probability P_k that satisfies all the conditions is shown in Appendix 3A to be

$$P_k = \frac{(\lambda/\mu)^k/k!}{\sum\limits_{k=0}^{N}(\lambda/\mu)^k/k!}, \quad \text{all } k \tag{3.9b}$$

The state of having all lines occupied may be momentary, so it is referred to as "time congestion." If calls are rejected when all N lines are occupied, then P_k for the case of $k = N$ is the probability that a call is rejected or "blocked":

$$B = P_N = \frac{(\lambda/\mu)^N/N!}{\sum\limits_{k=0}^{N}(\lambda/\mu)^k/k!} = \frac{A^N/N!}{\sum\limits_{k=0}^{N}A^k/k!}, \quad A = \frac{\lambda}{\mu} \tag{3.10}$$

This expression for the blocking probability is known as the *Erlang B formula*.

Note that blocking (time congestion) is associated with the occupancy of the lines or channels through which the callers access the system. Without queueing the arrivals—putting them on a waiting list instead of rejecting them—the system is temporarily blocked to new calls during the short time when all N lines are occupied. Note also that the blocking probability is a function of the dimensionless quantity

$$\frac{\lambda \text{ (calls/sec)}}{\mu \text{ (calls/sec)}} = A \text{ (traffic "load" in Erlangs)} \tag{3.11}$$

For a finite number of users (M), the load in Erlangs can also be expressed in terms of ρ, the fraction of time that each user occupies a telephone line: $A = M\rho$.

It is shown in Appendix 3B that the mean of the number of the N lines that are occupied is $A(1 - B) \approx A$; thus the offered load A can be interpreted as the average number of lines occupied when the blocking probability is small. In terms of A, the blocking probability increases with the load A and decreases with N, the number of lines at the switch, as shown graphically in Figure 3.3. Note from Figure 3.3 that, for the same value of blocking probability B, the normalized load A/N is proportional to the number of lines (trunk size) N. This phenomenon is referred to as "trunking efficiency": the value of the load A that can be accommodated for a given value of blocking increases by more than the increase in the number of lines, when

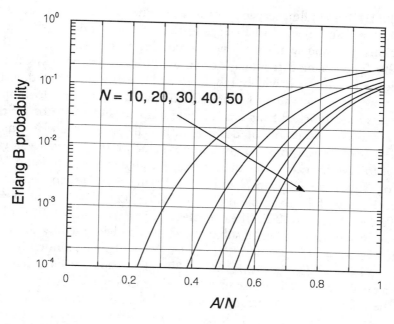

Figure 3.3 Erlang B probability.

N is increased. For example, the values of A and A/N shown in Table 3.1 hold for the case of $B = 0.01$. Doubling N from 10 to 20 causes A to increase by the factor $12.0/4.5 = 2.67$. Clearly, there is a very significant gain in efficiency at the 1% blocking level when a switch with $N = 10$ lines is upgraded to 20 or 30 lines, but relatively little gain to be had by increasing N once the switch has 40 or 50 lines.

Another way to interpret the behavior of the curves in Figure 3.3 is to observe that, for the same value of normalized load, B is inversely proportional to N. From this point of view, trunking efficiency is due to the fact that the average number of calls being completed in a short interval is directly proportional to N; for a higher value of N, there is a greater chance that a line is free the moment that a new call arrives.

Table 3.1 Example: Proportionality of A/N to N for $B = 0.01$

N	10	20	30	40	50	60	70	80	90	100
A	4.5	12.0	20.3	29.0	37.9	46.9	56.1	65.4	74.7	84.1
A/N	0.45	0.60	0.68	0.73	0.76	0.78	0.80	0.82	0.83	0.84

The traffic-handling capacity of a telephone switch can be assessed by finding the maximum load in Erlangs A that can be serviced while keeping the blocking probability at or below a certain level, for example $B = 0.01$ or $B = 0.02$. The Erlang B formula is often used for this purpose as a matter of convenience because it is widely tabulated and is easily interpreted, although it tends to give a low value of blocking compared with other formulas; the formula assumes blocked calls are not reattempted and may not reflect the system's particular method of dealing with calls when there is time congestion.[1] Because the formula gives a low value of B for given values of A and N, it follows that the value of A is high for given values of B and N.

For estimating A, given B and N, an Erlang B table such as that shown in Table 3.2 is useful. Example use of the table: Let $N = 20$ and $B = 0.02$. From Table 3.2, $A = 13.2$ E. If each subscriber offered $A_0 = 0.03$ E of traffic on average, then the number of subscribers that can be served by the switch is estimated as $M = A/A_0 = 440$. Now, let $N = 30$; from the table we find that $A = 21.9$ and estimate that $M = 730$. The capacity has increased by 66% as a result of a 50% increase in N, due to trunking efficiency.

The effects of blocked calls returning, because of repeated attempts to place a particular call after it is blocked, can be estimated by considering the blocking probability to be a function of the effective load A', where

$$\underset{\underset{A' = A}{\downarrow}}{\text{1st attempt}} \quad \underset{\underset{+\, AB}{\downarrow}}{\text{2nd}} + \underset{\underset{(AB)B}{\downarrow}}{\text{3rd}} + \underset{\underset{(AB^2)B}{\downarrow}}{\text{4th}} + \cdots = \frac{A}{1-B} \tag{3.12}$$

The Erlang B table can be used to find A', then A can be estimated as $A = (1 - B)A'$. For example, for $B = 0.02$ and $N = 30$, Table 3.2 gives $A = 21.9$, which is taken to be the value of A'; the value of A for blocked calls returning is then estimated as $A = (1 - 0.02)21.9 = 0.98 \cdot 21.9 = 21.5$. The difference seems small, but for a per-subscriber load of $A_0 = 0.03$, the difference leads to $M = A'/A_0 = 730$ subscribers for blocked calls not returning and $M = A/A_0 = 715$ subscribers for calls returning.

---

[1] For example, blocking can be reduced by "holding" incoming calls in a queue briefly until a line becomes free; this technique is modeled by the *Erlang C* formula. A summary and comparison of other blocking probability formulations is given in Appendix 3C.

Table 3.2 Erlang B: offered load, given B, N

N		B			N		B		
	0.02	0.01	0.005	0.001		0.02	0.01	0.005	0.001
1	0.02	0.01	0.005	0.001	11	5.8	5.2	4.6	3.6
2	0.22	0.15	0.105	0.046	12	6.6	5.9	5.3	4.2
3	0.6	0.46	0.35	0.19	13	7.4	6.6	6.0	4.8
4	1.1	0.9	0.7	0.44	14	8.2	7.4	6.7	5.4
5	1.7	1.4	1.1	0.8	15	9.0	8.1	7.4	6.1
6	2.3	1.9	1.6	1.1	16	9.8	8.9	8.1	6.7
7	2.9	2.5	2.2	1.6	17	10.7	9.6	8.8	7.4
8	3.6	3.1	2.7	2.1	18	11.5	10.4	9.6	8.0
9	4.3	3.8	3.3	2.6	19	12.3	11.2	10.3	8.7
10	5.1	4.5	4.0	3.1	20	13.2	12.0	11.1	9.4

N		B			N		B		
	0.02	0.01	0.005	0.001		0.02	0.01	0.005	0.001
21	14.0	12.8	11.9	10.1	31	22.8	21.2	19.9	17.4
22	14.9	13.7	12.6	10.8	32	23.7	22.0	20.7	18.2
23	15.8	14.5	13.4	11.5	33	24.6	22.9	21.5	19.0
24	16.6	15.3	14.2	12.2	34	25.5	23.8	22.3	19.7
25	17.5	16.1	15.0	13.0	35	26.4	24.6	23.2	20.5
26	18.4	17.0	15.8	13.7	36	27.3	25.5	24.0	21.3
27	19.3	17.8	16.6	14.4	37	28.3	26.4	24.8	22.1
28	20.2	18.6	17.4	15.2	38	29.2	27.3	25.7	22.9
29	21.0	19.5	18.2	15.9	39	30.1	28.1	26.5	23.7
30	21.9	20.3	19.0	16.7	40	31.0	29.0	27.4	24.4

N		B			N		B		
	0.02	0.01	0.005	0.001		0.02	0.01	0.005	0.001
41	31.9	29.9	28.2	25.2	51	41.2	38.8	36.9	33.3
42	32.8	30.8	29.1	26.0	52	42.1	39.7	37.7	34.2
43	33.8	31.7	29.9	26.8	53	43.1	40.6	38.6	35.0
44	34.7	32.5	30.8	27.6	54	44.0	41.5	39.5	35.8
45	35.6	33.4	31.7	28.4	55	44.9	42.4	40.4	36.6
46	36.5	34.3	32.5	29.3	56	45.9	43.3	41.2	37.5
47	37.5	35.2	33.4	30.1	57	46.8	44.2	42.1	38.3
48	38.4	36.1	34.2	30.9	58	47.8	45.1	43.0	39.1
49	39.3	37.0	35.1	31.7	59	48.7	46.0	43.9	40.0
50	40.3	37.9	36.0	32.5	60	49.6	46.9	44.8	40.8

Table 3.2 (continued) Erlang B: offered load, given B, N

N	\multicolumn{4}{c}{B}	N	\multicolumn{4}{c}{B}
	0.02	0.01	0.005	0.001		0.02	0.01	0.005	0.001
61	50.6	47.9	45.6	41.6	71	60.1	57.0	54.6	50.1
62	51.5	48.8	46.5	42.5	72	61.0	58.0	55.5	50.9
63	52.5	49.7	47.4	43.3	73	62.0	58.9	56.4	51.8
64	53.4	50.6	48.3	44.2	74	62.9	59.8	57.3	52.7
65	54.4	51.5	49.2	45.0	75	63.9	60.7	58.2	53.5
66	55.3	52.4	50.1	45.8	76	64.9	61.7	59.1	54.4
67	56.3	53.4	51.0	46.7	77	65.8	62.6	60.0	55.2
68	57.2	54.3	51.9	47.5	78	66.8	63.5	60.9	56.1
69	58.2	55.2	52.8	48.4	79	67.7	64.4	61.8	57.0
70	59.1	56.1	53.7	49.2	80	68.7	65.4	62.7	57.8

N	\multicolumn{4}{c}{B}	N	\multicolumn{4}{c}{B}
	0.02	0.01	0.005	0.001		0.02	0.01	0.005	0.001
81	69.6	66.3	63.6	58.7	91	79.3	75.6	72.7	67.4
82	70.6	67.2	64.5	59.5	92	80.2	76.6	73.6	68.2
83	71.6	68.2	65.4	60.4	93	81.2	77.5	74.5	69.1
84	72.5	69.1	66.3	61.3	94	82.2	78.4	75.4	70.0
85	73.5	70.0	67.2	62.1	95	83.1	79.4	76.3	70.9
86	74.5	70.9	68.1	63.0	96	84.1	80.3	77.2	71.7
87	75.5	71.9	69.0	63.9	97	85.1	71.2	78.2	72.6
88	76.4	72.8	69.9	64.7	98	86.0	82.2	79.1	73.5
89	77.3	73.7	70.8	65.6	99	87.0	83.1	80.0	74.4
90	78.3	74.7	71.8	66.5	100	88.0	84.1	80.9	75.2

3.2 The Cellular Concept

Early civilian mobile radio voice communications were developed primarily for police and taxi dispatching and used frequencies in the VHF band [3]. Gradually, the demand for private mobile telephone applications grew and additional portions of the radio spectrum were allocated at higher frequencies in the UHF band. This process continued until it was realized that development was required of more efficient methods for using radio spectrum for voice communications.

The need to accommodate large numbers of mobile telephone subscribers using limited spectral resources led to the concept of cellular radio

[4]. The principal features of the cellular concept are (1) the reuse of the spectra in more than one geographical section (cell) of a larger service area through careful control of transmitter powers and (2) provision for expansion of capacity through the division of cells into smaller cells (cell splitting). In this section, the parameters affecting frequency reuse in cellular systems will be discussed.

3.2.1 Expansion of Mobile System Capacity Through Frequency Reuse

Before the late 1970s, a limited number of mobile radio-telephones were available in a given service area, proportional to the number of FDMA channels supported by the spectral allocation. In effect, the population of mobile subscribers—say, M subscribers—all contended for the use of a fixed number of channels N over which to make calls to a central switching center, which then would connect the call to the public switched telephone network (PSTN). The situation was analogous to that of an office building with many (M) telephones, much more than the actual number of telephone lines (N) connecting the building's local switch to the PSTN. In fact, the frequency channels of the mobile phone system interfaced with a dedicated switch, the mobile telephone switching office (MTSO), as depicted in Figure 3.4. As suggested by Figure 3.4, by its nature the local switch for the wire subscribers services a relatively small physical area or neighborhood, while the MTSO services wireless subscribers who are in general located over a wide area

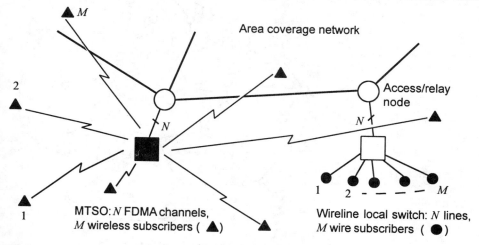

Figure 3.4 Comparison of fixed and mobile telephone switches.

through radio transmissions to and from a base station associated with the MTSO.

The effective service area of a base station is a function of its power, antenna pattern, and the characteristics of mobile radio signal propagation in the service area. For the early mobile telephone systems, when the number of mobile subscribers in a service area was relatively small, it was important to locate the base station transmitter appropriately and to use enough power. Figure 3.5 depicts a hypothetical population center that is serviced by both the PTSN and by a mobile telephone system. The figure shows many local switches for wireline subscribers, and the density of the locations of these switches is roughly equivalent to the density of the wireline subscribers. There is one central MTSO/base station in this hypothetical example, with an auxiliary MTSO/repeater station that has been added to enable coverage of the entire service area, because mobile subscribers can be anywhere in the area. Note that the subscriber capacity of the wireline system is proportional to the number of local switches; when needed, additional switches are installed, and the throughput of the network is upgraded to handle the increased traffic.

Active wireline users in different neighborhoods do not interfere with one another because their means of access to the network (wires) are separated in space. The means of access to the network for mobile subscribers are radio

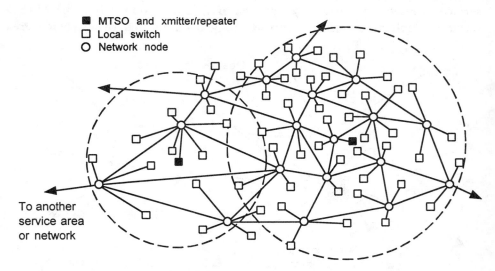

Figure 3.5 Network of fixed and mobile telephone switching offices.

channels instead of wires. Under the cellular concept, capacity is increased, not by adding wires and local switches, but by frequency reuse and adding MTSOs. The frequency reuse is achieved by reducing transmitter power so that many smaller coverage areas (cells) replace the few larger coverage areas.

3.2.2 Cell Geometry

Conceptually, an omnidirectional base station transmitter has a circular coverage area, defined as the area for which the received base station signal power exceeds a certain threshold. As illustrated in Figure 3.6, a large geographical area can be divided into overlapping circular areas. If the circles completely cover the area (there are no "holes" in coverage) and are all the same size, they support the concept of hexagonal "cells," each defined as the locations affiliated with the nearest base station, as suggested in Figure 3.7. Note that the "size" of a hexagonal cell can be given as R_c, the radius of the coverage area, or as $R = R_c \cos 30° = (\sqrt{3}/2)R_c = 0.866\,R_c$.

In actual practice, however, the coverage area for a particular base station is not circular because the propagation loss is affected by natural and manmade terrain. For example, consider the built-up area surrounding a base station that is depicted in Figure 3.8. The propagation loss as a function of distance in any particular subsector (i j, or k in the figure), is unique because the variation in building density with distance from the base station is different in each subsector.

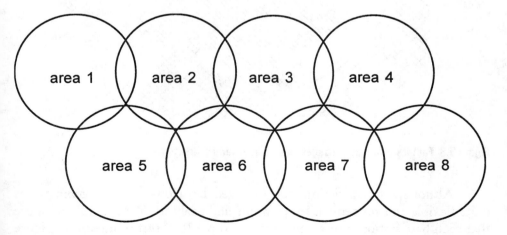

Figure 3.6 Idealized (circular) coverage areas.

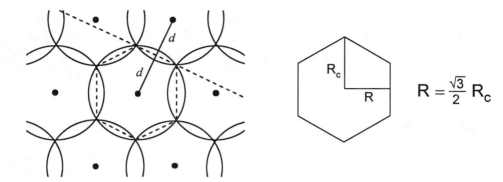

Figure 3.7 Hexagonal shape associated with circular coverage areas.

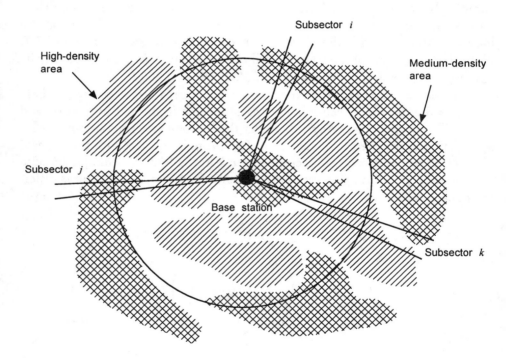

Figure 3.8 Factors affecting propagation loss in different directions.

Although it is a fiction, the idealized hexagonal cellular geometry is quite flexible and convenient for planning and analysis. For example, suppose that each base station transmitter employs three 120° sector antennas. Figure 3.9 shows that the transmitter locations can be considered to be either at the

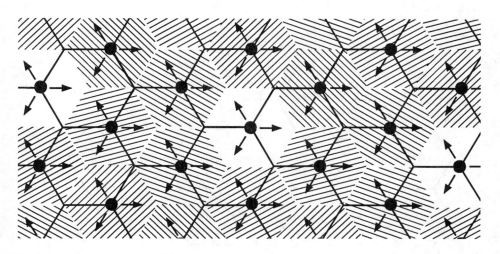

Figure 3.9 Hexagonal cell patterns.

centers of the cells (as indicated by the pattern of shaded cells) or at the corners of the cells (as indicated by the pattern of outlined cells).

3.2.2.1 Cellular Coordinate Systems

Figure 3.10 shows a layout of hexagonal cells in a nonorthogonal coordinate system [4, 5]. Using the cell position coordinates (u, v) in this system, the cell centers are located at the positions

$$(u, v) = (2Ri, 2Rj) = \left(R_c \sqrt{3}i, R_c \sqrt{3}j \right) \tag{3.13a}$$

In terms of the (u, v) coordinate system, an arbitrary position in a rectangular (x, y) coordinate system with the same origin is

$$x = u \cos 30° = \tfrac{1}{2} u \sqrt{3}, \quad y = u \sin 30° + v = \tfrac{1}{2} u + v \tag{3.13b}$$

The square of the distance between the centers of cell a, whose center is located at (u_a, v_a), and cell b, whose center is located at (u_b, v_b), therefore can be computed as

$$d_{ab}^2 = (x_a - x_b)^2 + (y_a - y_b)^2 = \tfrac{3}{4}(u_a - u_b)^2 + \left(\tfrac{1}{2} u_a + v_a - \tfrac{1}{2} u_b - v_b \right)^2$$

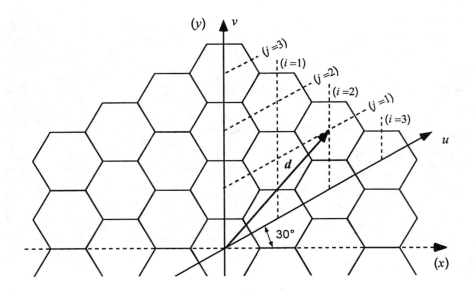

Figure 3.10 Nonorthogonal cellular coordinate system.

$$= (u_a - u_b)^2 + (v_a - v_b)^2 + (u_a - u_b)(v_a - v_b) \qquad (3.14a)$$

$$= (2R)^2 \left[(i_a - i_b)^2 + (j_a - j_b)^2 + (i_a - i_b)(j_a - j_b) \right] \qquad (3.14b)$$

$$= (3R_c)^2 \left[(i_a - i_b)^2 + (j_a - j_b)^2 + (i_a - i_b)(j_a - j_b) \right] \qquad (3.14c)$$

In particular, the distance of a cell's center from the origin in this coordinate system is

$$d = 2R\sqrt{i^2 + j^2 + ij} = 3R_c\sqrt{i^2 + j^2 + ij} \qquad (3.15)$$

Another cellular geometry emphasizes the rotational symmetry of the hexagonal grid system by using the notion of hexagonal "rings" of cells around a center cell [6], as shown in the diagram in Figure 3.11. The diagram consists of the center cell and one of the six 60° sectors around the origin. The coordinates of a cell in the sector are (n, i), where n is the "ring" number and $i = 1, 2, \ldots, n$ indexes the cells in the sector that are in ring n. The squared distance of the ith cell in the nth ring is

$$d^2(n, i) = (2Rn)^2 + (2Ri)^2 - 2(2Rn)(2Ri)\cos 60°$$
$$= 4R^2(n^2 + i^2 - ni) \qquad (3.16a)$$

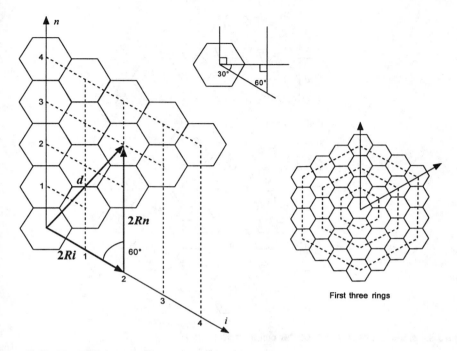

Figure 3.11 Ring cellular coordinate system.

which gives the distance formula

$$d(n, i) = 2R\sqrt{n^2 + i^2 - ni} = R_c\sqrt{3}\,\sqrt{n^2 + i^2 - ni} \qquad (3.16b)$$

The distances of cells in different rings from the center cell are listed in Table 3.3 and illustrated in Figure 3.12 as factors of the cell size parameters R and R_c. Note than there are $6n$ cells in each ring, but the cells in each ring do not all have the same distance from the center cell.

Table 3.3 Cellular rings (n = ring number, i = position)

n	i	Distance	Number	n	i	Distance	Number
1	1	$2R = R_c\sqrt{3}$	6	4	1	$2R\sqrt{13} = R_c\sqrt{39}$	6
2	1	$2R\sqrt{3} = 3R_c$	6		2	$4R\sqrt{3} = 6R_c$	6
	2	$4R = 2R_c\sqrt{3}$	6		3	$2R\sqrt{13} = R_c\sqrt{39}$	6
3	1	$2R\sqrt{7} = R_c\sqrt{21}$	6		4	$8R = 4R_c\sqrt{3}$	6
	2	$2R\sqrt{7} = R_c\sqrt{21}$	6	$\vdots$		$\vdots$	$\vdots$
	3	$6R = 3R_c\sqrt{3}$	6	n	i	$2R\sqrt{n^2 + i^2 - ni}$	6

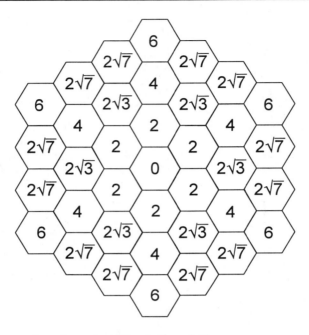

Figure 3.12 Distances from the center cell as factors of R.

3.2.2.2 Clusters of Hexagonal Cells

For frequency reuse purposes, as is discussed below, it is advantageous to treat a "cluster" of adjacent cells as an entity. The number of cells in the cluster is called the *cluster size*. Because of the restrictions of the hexagonal geometry, only certain cluster sizes can be used and still cover the service area without any gaps. One of the constraints of the hexagonal geometry for cellular radio applications is that each cluster is surrounded by six similar clusters with the same orientation. It is argued in [4] that, under this constraint, each cluster has a total area equivalent to what can be called a *superhexagon*. Therefore, as we show below, the number of cells in a cluster that satisfies the constraint is the ratio of the area in the superhexagon to that in one hexagon.

The distance from any cell in a particular position in one cluster to a cell in an adjacent cluster in the same position in its cluster is a particular case of the intercellular distance (3.15):

$$d(i, j) = 2R\sqrt{i^2 + j^2 + ij} \qquad (3.17a)$$

Without loss of generality, then, let the "center" cell in the cluster (which may not be precisely in the physical center of the cluster) be the reference cell, so that its center is the center of the superhexagon, as illustrated in Figure 3.13 for the case of $i = j = 2$. Because a single hexagon is made up of six equilateral triangles with base R_c and height R, the area of a hexagonal cell is given by

$$a_{\text{cell}} = 6 \times \tfrac{1}{2} \, (\text{base} \times \text{height}) = 3R_c R = \tfrac{3\sqrt{3}}{2} R_c^2 = 2\sqrt{3}\, R^2 \quad (3.17b)$$

The radius of the superhexagon in Figure 3.13 is $R' = 3R_c = 2R\sqrt{3}$; in general, the radius that is equivalent to R is

$$R' = \tfrac{1}{2} \, d(i,j) = R\sqrt{i^2 + j^2 + ij} \qquad (3.18)$$

Let the cluster size (i.e., the number of cells in the cluster) be denoted K. Thus, the equivalent cluster size for the superhexagon is

$$K = \frac{\text{Area of superhexagon}}{\text{Area of one hexagon}} = \frac{2\sqrt{3}\,(R')^2}{a_{\text{cell}}}$$

$$= \left(\frac{R'}{R}\right)^2 = i^2 + j^2 + ij \qquad (3.19)$$

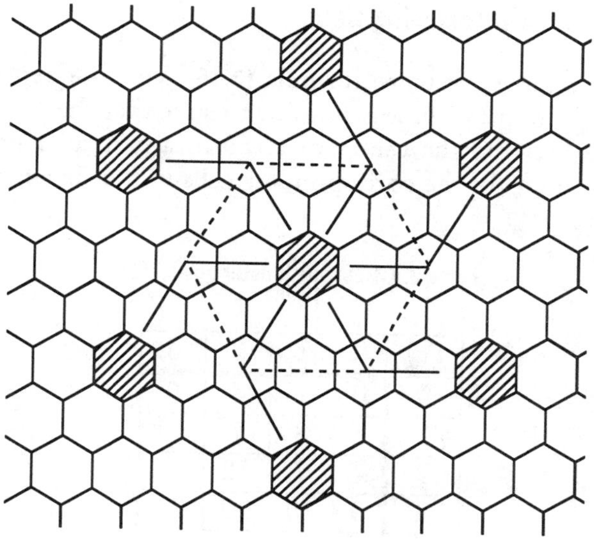

Figure 3.13 Superhexagon concept for $i = j = 2$.

From this equation it is apparent that the cluster sizes that can be used to construct a superhexagon include the cases of $K = 1$ $(i = 1, j = 0), K = 3$ $(i = j = 1)$, and $K = 7$ $(i = 2, j = 1)$. Table 3.4 lists the possible cluster sizes for i and j less than four.

Having determined the cluster sizes, a further constraint is needed to actually construct the clusters, because there is more than one possible cluster configuration for some values of K. For example, when $K = 3$, the three cells can be lined up in a straight row or in a triangular configuration. In addition to the constraints that have already been stated, it is desirable to select the cluster configurations that are more "compact" in some sense. For this purpose, a procedure such as *Dijkstra's algorithm* [7–9] is applicable.

Dijkstra's algorithm is often employed to determine least-cost routing in networks. It works iteratively by adding, to a set of connections that is being built up, another connection that increases some cost function by the least amount. To use this algorithm to construct cellular clusters, the cost of adding the next cell in the cluster may be defined as the distance of that cell from the center of the cluster that has been built up so far. In Figure 3.14, cells are numbered in the order in which they were added to the cluster on the basis that they were closest to the center of the cluster, up to twelve cells. The results of this cluster-growing procedure are shown in Figure 3.15 for $K = 3$, $K = 7$, and $K = 12$.

3.2.2.3 Locations of Interfering Cells

Comparing the intercell distance formula (3.15) with the expression for the cluster size K, given in (3.19), it is possible to state that the "reuse distance"— that is, distance between the center of a cell in a particular position in one cluster and the centers of the corresponding cells in the six adjacent clusters— is given by

Table 3.4 Possible cluster sizes, K

$i\backslash j$	0	1	2	3
0	0	1	4	9
1	1	3	7	13
2	4	7	12	19
3	9	13	19	27

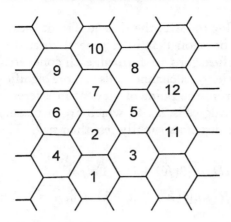

Figure 3.14 Order in which cells are accumulated to form clusters.

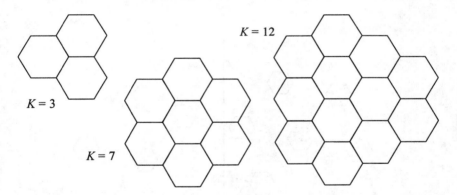

Figure 3.15 Clusters of size 3, 7, and 12 (shape based on Dijkstra's algorithm).

$$D = 2R\sqrt{i^2 + j^2 + ij}$$

$$= R_c\sqrt{3}\sqrt{i^2 + j^2 + ij} = R_c\sqrt{3K} \tag{3.20a}$$

or

$$D/R_c = \sqrt{3K} \tag{3.20b}$$

The ratio D/R_c is called the *cochannel reuse ratio* [4] because the frequency reuse concept involves dividing the available FDMA channels into K sets and operating on frequencies in a particular set in only one cell in each cluster, in a particular position in the cluster. In this manner, the same frequency is reused (and is a potential source of cochannel interference) at distances at least equal to D.

As shown in Figure 3.16, the six equidistant, nearest interfering cells at distance D form a hexagon that is in general rotated at some angle, $\theta(K)$. Progressively larger hexagons are formed, each concentric with the first ring's hexagon. The first tier of cochannel cells is more significant than the next tier of cochannel cells from the point of view of interference. For the nth ring, there are $6n$ interfering cells, six of which are at distance nD and the other $6(n-1)$ cells are at distances slightly less than nD. The locations of the six cells at distance nD are

$$x_i(n) = nD\cos[\theta(K) + (i-1)\pi/3] \tag{3.21a}$$

and $\qquad y_i(n) = nD\sin[\theta(K) + (i-1)\pi/3], \ \ i = 1, 2, \ldots, 6 \tag{3.21b}$

In vector form, the locations of the $n-1$ cells in between pairs of the six cells at distance nD for $n > 1$ can be written

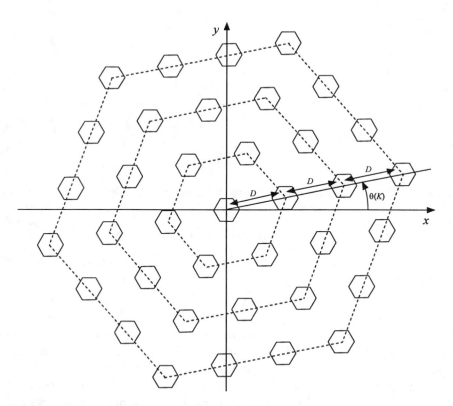

Figure 3.16 Locations of cochannel interfering cells.

$$\begin{bmatrix} x_{ij}(n) \\ y_{ij}(n) \end{bmatrix} = \left(\frac{j}{n}\right) \times \begin{bmatrix} x_i(n) \\ y_i(n) \end{bmatrix} + \left(1 - \frac{j}{n}\right) \times \begin{bmatrix} x_{i+1}(n) \\ y_{i+1}(n) \end{bmatrix} \tag{3.22}$$

$$j = 1, 2, \ldots, n-1$$

The values of D and $\theta(K)$ are given in Table 3.5 for several values of K.

3.2.3 Selection of Cluster Size

Cochannel interference, such as that illustrated in Figure 3.17 for the reverse (mobile-to-base) link, is built into a cellular system. Satisfactory system performance is achieved by controlling the cell size, frequency spacing, and other physical parameters as well as emission characteristics of base stations and mobiles.

The selection of the cluster size for frequency reuse in an FDMA cellular mobile radio system is based on a tradeoff between cochannel interference and spectral efficiency considerations. The larger the value of K, the larger the reuse distance and, consequently, the smaller the interference

Table 3.5 Location parameters for interfering cells

K	3	4	7	9	12
D/R_c	3	$2\sqrt{3}$	$\sqrt{21}$	$3\sqrt{3}$	6
$\theta(K)$	0	$\pi/6$	$\cot^{-1}\left(\sqrt{3}\right)$	$\pi/6$	0

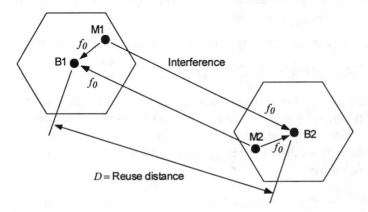

Figure 3.17 Reverse link cochannel interference (Bi = base station i, Mj = mobile j).

power. On the other hand, dividing the number of available frequency channels by K reduces the trunking efficiency for each cell and, consequently, the number of subscribers per FDMA channel that can be supported by the system.

3.2.3.1 Interference Ratio Versus Cluster Size

Consider the forward link transmissions in an FDMA cellular system for cluster size K. Assuming that each base station delivers the same power, P_t, to its antenna, the carrier-to-interference power ratio at a mobile receiver in the central cell can be formulated as

$$\frac{C}{I} = \frac{P_t\, G_c(\phi_{0m})\, \kappa\, v_0/d_{0m}^{\gamma}}{\sum_i P_t\, G_c(\phi_{im})\, \kappa\, v_i/d_{im}^{\gamma}} = \left[\sum_i \frac{G_c(\phi_{im})\, v_i}{G_c(\phi_{0m})\, v_0}\left(\frac{d_{0m}}{d_{im}}\right)^{\gamma}\right]^{-1} \quad (3.23)$$

where the center cell is designated "cell 0," the interfering cells are indexed by $i > 0$, and

$G_c(\,\cdot\,) = $ Base station antenna gain.

$\phi_{im} = $ Angle between main beam of antenna of cell i and direction to mobile.

$\kappa = $ Constant of proportionality for propagation loss.

$v_i = $ Unit-median lognormal RV representing uncertainty in the propagation loss.

$d_{im} = $ Distance from the antenna of cell i and the mobile.

$\gamma = $ Propagation loss power law.

By definition, d_{0m} is less than or equal to the cell radius, R_c, and the distances between the interfering cells and the mobile are in the neighborhood of $nD \pm R_c$. The propagation power law for ground-to-ground mobile communications is typically between $\gamma = 2.7$ and $\gamma = 4$ [10]. The antenna gain can be modeled by

$$G_c(\phi) = \text{constant}, \quad \text{omnidirectional antenna} \quad (3.24a)$$

and

$$G_c(\phi) = \begin{cases} \text{constant,} & |\phi| \leq \pi/3 \\ 0, & \text{otherwise} \end{cases} \quad 120° \text{ sector antenna} \qquad (3.24\text{b})$$

With the information on interfering cell locations given above in Section 3.2.2.3, a precise calculation of the angles and distances in (3.23) can be made in conjunction with a simulation of the lognormal RVs and perhaps variations in base station and vehicular locations to determine the sensitivity of system performance to these factors. Such a simulation was reported in [4], based on propagation measurements in several U.S. cities. A first-order estimate of the effect of the value of K can be developed analytically, however, by using the median values of the lognormal variables and by making several reasonable approximations.

Consider a mobile in the center or "home" cell that is receiving transmissions on one of the home cell base station's FDMA channels. That same channel can be used at the same time in other cells at reuse distances nD, provided that the level of cochannel interference is small due to propagation loss and possibly other factors. The worst case cochannel interference occurs when all potential interferers are using the same FDMA channel at the same time as the mobile under consideration. If the distance of the mobile from its base station is replaced with its worst case value, $d_{0m} = R_c$, and if the distances from the interfering base stations to the mobile are taken to be their median value, $d_{im} \approx nD$ for an interfering cell in the nth ring of cochannel cells, then the worst case median forward link C/I is

$$\frac{C}{I} = \left[\sum_{\substack{i \\ \text{ring 1}}} \frac{G_c(\phi_{im})}{G_c(\phi_{0m})} \left(\frac{R_c}{D}\right)^\gamma + \sum_{\substack{i \\ \text{ring 2}}} \frac{G_c(\phi_{im})}{G_c(\phi_{0m})} \left(\frac{R_c}{2D}\right)^\gamma \right.$$

$$\left. + \sum_{\substack{i \\ \text{ring 3}}} \frac{G_c(\phi_{im})}{G_c(\phi_{0m})} \left(\frac{R_c}{3D}\right)^\gamma + \cdots \right]^{-1}$$

$$= \left(\frac{D}{R_c}\right)^\gamma \left[\sum_{\substack{i \\ \text{ring 1}}} \frac{G_c(\phi_{im})}{G_c(\phi_{0m})} + \left(\frac{1}{2}\right)^\gamma \sum_{\substack{i \\ \text{ring 2}}} \frac{G_c(\phi_{im})}{G_c(\phi_{0m})} \right.$$

$$\left. + \left(\frac{1}{3}\right)^\gamma \sum_{\substack{i \\ \text{ring 3}}} \frac{G_c(\phi_{im})}{G_c(\phi_{0m})} + \cdots \right]^{-1} \qquad (3.25)$$

Substituting $D/R_c = \sqrt{3K}$ and using the model for G_c given in (3.24a), the C/I when omnidirectional antennas are used is

$$\frac{C}{I} = (3K)^{\gamma/2}\left[6 + 12\left(\frac{1}{2}\right)^{\gamma} + 18\left(\frac{1}{3}\right)^{\gamma} + \cdots\right]^{-1} \tag{3.26a}$$

$$= \frac{(3K)^{\gamma/2}}{6}\left[1 + \left(\frac{1}{2}\right)^{\gamma-1} + \left(\frac{1}{3}\right)^{\gamma-1} + \cdots\right]^{-1} \tag{3.26b}$$

$$= \frac{(3K)^{\gamma/2}}{6 \cdot \zeta(\gamma-1)} \tag{3.26c}$$

where $\zeta(\,\cdot\,)$ is the Riemann zeta function [11, §23.2], which has the values

$$\zeta(\gamma-1) \triangleq \sum_{n=1}^{\infty}\frac{1}{n^{\gamma-1}} = \begin{cases} \infty, & \gamma = 2 \\ \pi^2/6 = 1.6449, & \gamma = 3 \\ 1.2021, & \gamma = 4 \\ \pi^4/90 = 1.0823, & \gamma = 5 \end{cases} \tag{3.27}$$

Thus, for this idealized model of a cellular system, if the propagation loss is equivalent to free space ($\gamma = 2$) and omnidirectional antennas are used, the growth in the number of interferers as each ring of interferers is added makes up for the path loss and the interference grows without bound—*the cellular concept does not work for free-space propagation.* The propagation law for most mobile situations, however, is on the order of $\gamma = 4$. Thus, for omnidirectional antennas, the worst-case median forward link C/I value is estimated to be

$$\frac{C}{I} = \frac{(3K)^2}{6\,\zeta(3)} = 1.2478\,K^2 = \begin{cases} 11.23 = 10.5\,\text{dB}, & K = 3 \\ 19.97 = 13.0\,\text{dB}, & K = 4 \\ 61.14 = 17.9\,\text{dB}, & K = 7 \\ 101.1 = 20.0\,\text{dB}, & K = 9 \\ 179.7 = 22.5\,\text{dB}, & K = 12 \end{cases} \tag{3.28}$$

These estimates are conservative in that they ignore the improvement that can be obtained by exploiting base station antenna directivity in the vertical plane. It is common practice to "tilt" the base station antennas downward (using several antennas as necessary to achieve the desired horizontal pattern) to decrease the gain in the direction of a distant cell [12], as illustrated in Figure 3.18.

For sectored cells with antennas having directivity in the horizontal plane, using the model for G_c given above in (3.26b), fewer than $6n$ of the base stations at reuse distance nD will interfere with the mobile in the home cell. Assuming that all the sectors in the different cells are aligned in the same way, the numbers of interfering cells in each ring were counted and found to be as shown in Table 3.6. For example, when $K = 7$, only two of the six cells in the first ring are interfering, whereas for the second ring, four of the twelve cells are interfering, and so forth. The numbers of interfering cells in the examples given in Table 3.6 are from one-third to one-half the numbers for omnidirectional antennas. In Table 3.6, the "Average" row indicates the algebraic mean of the the number of interfering cell in the respective rings for cluster sizes from $K = 3$ to $K = 12$. For example, for the first ring the average number of interfering cells is 2.4 out of the total of 6 cells, whereas the average number of interfering cells for the second ring is 4.8 out of 12 cells, and so on. We observe that, approximately, there are $2n + 1$ interfering

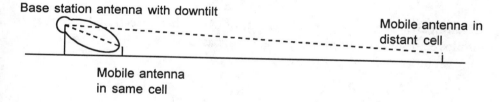

Base station antenna with downtilt

Mobile antenna in distant cell

Mobile antenna in same cell

Figure 3.18 Downtilt of base station antenna pattern.

Table 3.6 Numbers of interfering cells with directional antennas

n	1	2	3	4	5
$K = 3$	3 of 6	5 of 12	7 of 18	9 of 24	11 of 30
$K = 4$	2 of 6	5 of 12	6 of 18	9 of 24	10 of 30
$K = 7$	2 of 6	4 of 12	6 of 18	8 of 24	10 of 30
$K = 9$	2 of 6	5 of 12	6 of 18	9 of 24	11 of 30
$K = 12$	3 of 6	5 of 12	7 of 18	9 of 24	11 of 30
Average	2.4 of 6	4.8 of 12	6.4 of 5	8.8 of 24	10.6 of 30
$2n + 1$	3	5	7	9	11

cells instead of $6n$. Replacing the multiples of 6 in (3.26a) with $2n + 1$ results in

$$\frac{C}{I} = (3K)^{\gamma/2}\left[3 + 5\left(\tfrac{1}{2}\right)^{\gamma} + 7\left(\tfrac{1}{3}\right)^{\gamma} + \cdots\right]^{-1} \tag{3.29a}$$

$$= (3K)^{\gamma/2}\left[\sum_{n=1}^{\infty}\frac{2n+1}{n^{\gamma}}\right]^{-1} = \frac{(3K)^{\gamma/2}}{2\,\zeta(\gamma-1) + \zeta(\gamma)} \tag{3.29b}$$

Using $\gamma = 4$ and the zeta function values in (3.27), the worst case median forward link C/I with sector antennas becomes

$$\frac{C}{I} = \frac{(3K)^2}{2\,\zeta(3) + \zeta(4)} = 2.5814\,K^2 = \begin{cases} 23.23 = 13.7\,\text{dB}, & K = 3 \\ 41.30 = 16.2\,\text{dB}, & K = 4 \\ 126.5 = 21.0\,\text{dB}, & K = 7 \\ 209.1 = 23.2\,\text{dB}, & K = 9 \\ 371.7 = 25.7\,\text{dB}, & K = 12 \end{cases} \tag{3.30a}$$

Example 3.1 What is the improvement of C/I when directional antennas are employed instead of omnidirectional ones, based on three sectors per cell?

Solution: From (3.28), we have

$$\left(\frac{C}{I}\right)_{\text{omnidirectional}} = \frac{(3K)^2}{6\,\zeta(3)} = 1.2478\,K^2 \tag{3.30b}$$

and from (3.30a) we have

$$\left(\frac{C}{I}\right)_{\text{directional}} = \frac{(3K)^2}{2\,\zeta(3) + \zeta(4)} = 2.5814\,K^2 \tag{3.30c}$$

The improvement of directional sector antennas over omnidirectional antennas, for three sectors per cell, therefore, is given by

$$\frac{(C/I)_{\text{directional}}}{(C/I)_{\text{omnidirect}}} = \frac{2.5814\,K^2}{1.2478\,K^2} = 2.07 = 3.16\,\text{dB} \tag{3.30d}$$

For the reverse link, the worst case median value of C/I is the same as the forward link as we have analyzed here, under the same assumptions.

In [4], it was reported that subjective testing of analog voice transmissions at UHF frequencies representative of the mobile cellular band have determined that an RF SNR of 18 dB is required for "good" or "excellent" rating by listeners. The same reference reports that simulations of a cellular system, using propagation data for Philadelphia and Newark, were conducted to determine the sensitivity of system performance to factors such as cluster size and reuse distance. The analytical results for the worst case median C/I that are shown above in (3.28) and (3.30) agree well with the statement in [4] that the SNR requirement of 18 dB was found to be satisfied for at least 90% of mobile locations using $K = 12$ for omnidirectional base station antennas and for $K = 7$ using 120° sector directional antennas.

3.2.3.2 Tradeoff of Interference Ratio and Spectral Efficiency

It is evident that the median received C/I is directly proportional to the cluster size K so that there are good reasons for using relatively large cluster sizes. However, there are also good reasons for keeping the cluster size as small as possible, including the consideration of spectral efficiency, which is defined as the number of subscribers per frequency channel that can be supported by the system.

In an FDMA cellular system such as AMPS, a fixed number of frequency channels N have been allocated for cellular telephone calls. For example, in the United States, each cellular carrier is allocated 12.5 MHz of spectrum, which can be divided into $12.5/0.03 = 416$ channels, each occupying 30 kHz, as shown in Table 3.7 [13] for carriers "A" and "B." After designation of certain channels as control channels (21 for each carrier), there are $N = 395$ channels available for calls. The concept of frequency reuse that leads to multicell clustering assumes that these N channels are divided into K subsets of channels that are reused at a distance of $D = R_c\sqrt{3K}$. Therefore in each cell of the cluster, there are N/K channels available, and if each cell has three sectors, there are $N/3K$ channels available per sector.

Because the number of available channels per cell or sector is inversely proportional to the cluster size, there is a penalty for increasing K in terms of trunking efficiency. That is, for the same blocking probability, the number of subscribers that can be supported by the N frequency channels is less if the channels are divided into subsets.

For example, if all $N = 395$ AMPS channels were available to each cellular subscriber, at the 2% blocking level, the system could support $A =$

Table 3.7 Division of the cellular spectrum into 30-kHz channels

System	Bandwidth in MHz	Number of channels	Boundary channel number	Transmitter center frequency in MHz	
				Mobile	Base
(not used)		1	(990)	(824.010)	(869.010)
A″	1	33	991	824.040	869.040
			1023	825.000	870.000
A	10	333	1	825.030	870.030
			333	834.990	879.990
B	10	333	334	835.020	880.020
			666	844.980	889.980
A′	1.5	50	667	845.010	890.010
			716	846.480	891.480
B′	2.5	83	717	846.510	891.510
			799	848.970	893.970

381 E of offered traffic, according to the Erlang B model. If each subscriber offers $A_0 = 0.03$ E, then $M = A/A_0 = 381/0.03 = 12,700$ subscribers per N channels. The spectral efficiency therefore is $M/N = 12,700/395 = 32.15$ subscribers per FDMA channel.

- If the N channels are divided into $K = 3$ frequency reuse subsets, then $395/3 = 132$ channels are available in each cell, which can support $A = 119$E of offered traffic for a blocking probability of $B = 0.02$. Each cluster then can support $M = 3 \times 119/0.03 = 11,900$ subscribers with a spectral efficiency of $M/N = 11,900/395 = 30.13$ subscribers per channel.

- If each of the three cells in the cluster is divided into three sectors, then subscribers in each sector can access the cellular system through $395/3/3 = 44$ available channels, which can support $A = 34.7E$ of offered traffic for a blocking probability of $B = 0.02$. Each cluster can then support $M = 3 \times 3 \times 38.7/0.03 = 11,610$ subscribers with a spectral efficiency of $M/N = 11,610/395 = 29.39$ subscribers per channel.

Table 3.8 summarizes the tradeoff in median C/I and spectral efficiency versus cluster size for the AMPS system parameters. From the table, we may observe the following: the spectral efficiency decreases as the cluster size increases. Thus, for maximal possible spectral efficiency, one needs to select the smallest possible cluster size for the required minimum C/I. Therefore, if the required minimum C/I is 18 dB, for an omnidirectional antenna system, the cluster size must be $K = 9$ (giving $C/I = 20$ dB); whereas for a three-sector directional antenna system, the cluster size can be $K = 7$ (giving $C/I = 21$ dB).[2]

Example 3.2 Show graphically the median C/I in dB and spectral efficiency M/N as a function of cluster size, based on data given in Table 3.8.

Solution: Figure 3.19 provides the same information as in Table 3.8. Note for $C/I = 18$ dB (left vertical axis), the cluster size has to be 7 or greater

Table 3.8 Median C/I and spectral efficiency versus cluster size for $A_0 = 0.03$ E

Cluster size, K	# sect.	Median C/I	Available channels	$B = 0.01$			$B = 0.02$		
				A	M	M/N	A	M	M/N
3	1	10.5 dB	132	114	11,400	28.86	119	11,900	30.13
	3	13.7 dB	44	32.5	9,750	24.68	34.7	10,410	26.35
4	1	13.0 dB	99	83.1	11,080	28.05	87.0	11,600	29.37
	3	16.2 dB	33	22.9	9,160	23.19	24.6	9,840	24.91
7	1	17.9 dB	56	43.3	10,103	25.58	45.9	10,710	27.11
	3	21.0 dB	19	11.2	7,840	19.85	12.3	8,610	21.80
9	1	20.0 dB	44	32.5	9,750	24.68	34.7	10,410	26.35
	3	23.2 dB	15	8.1	7,290	18.46	9.0	8,100	20.51
12	1	22.5 dB	33	22.9	9,160	23.19	24.6	9,840	24.91
	3	25.7 dB	11	5.2	6,240	15.80	5.8	6,960	17.62

---

[2] The AMPS system is based on $K = 7$ for a three-sector system, based on a requirement of $C/I \geq 18$ dB.

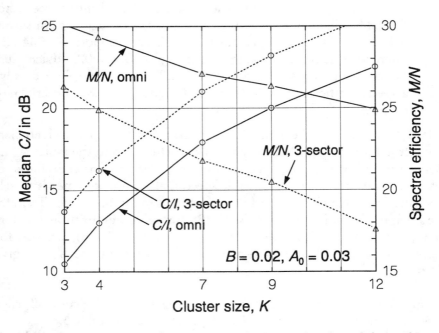

Figure 3.19 Tradeoff between C/I and spectral efficiency as functions of cluster size.

for directional antennas, and 9 or greater for omnidirectional antennas. Because spectral efficiency (right vertical axis) decreases as cluster size increases, we choose the minimum cluster sizes that give the required C/I; that is, $K = 7$ and $K = 9$, respectively.

3.2.4 Cell Splitting and Base Station Power

As shown, the use of directional base station antennas to create sectors in the cells of a cluster reduces the amount of cochannel interference. For FDMA channelization, however, the sectorization of the cell further divides the number of available channels and reduces the spectral efficiency—system subscriber capacity decreases. The subdivision of cells therefore does not necessarily expand the capacity. There is, however, a technique of cell subdivision that does expand the capacity: the second principal feature of the cellular concept as given in [4] is provision for the capacity expansion through the "splitting" of cells into smaller cells.

Methods were discussed previously for quantifying the number of subscribers M that can be accommodated by a cluster of cells, based on Erlang B telephone traffic theory and assumptions about the average amount of traffic offered by subscribers and the number of frequency channels that are allocated for cellular telephone transmissions. Although the reuse distance D is a critical parameter of cellular design, it appears in the analysis as a normalized distance D/R_c; the actual physical size of the cells and clusters has not been specified. In principle, then, the cluster and its M-subscriber capacity can be scaled to the physical dimensions required to match M with the actual number of cellular subscribers in the area covered by the cluster. The density of cells in the service area should be proportional to the density of subscribers for the same area, as suggested in Figure 3.20.

The scaling of cells is straightforward if all cells in the system are scaled by the same factor. The initial dimensioning of cells for a service area is based in part on fitting cell size to predictions of the density of potential cellular subscribers. However, the density of subscribers in general is not uniform over a large area and it is usually necessary even during the initial system design to implement different cell densities in different portions of the service area. Similarly, over the long term, the density of subscribers in general evolves unevenly. For example, the development of a business district may increase the density of subscribers in the city center, and the construction of a commuter express highway may increase the density of call traffic along the

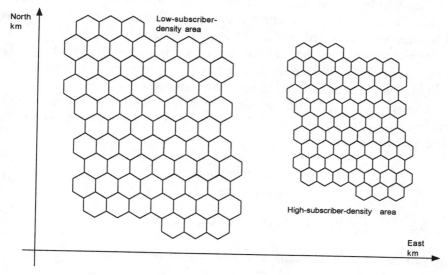

Figure 3.20 Density of cells proportional to density of subscribers.

highway and in residential areas that have highway access. Note that the potential for significant short term variation in traffic distributions is great in these examples—the subscribers, after all, are mobile!

Propagation considerations limit how small the cells can be made and still experience the cochannel interference reductions that result from clustering [14]. In urban areas especially, the propagation of interfering signals from other cells at short distances may not be inversely proportional to the fourth power of the distance because of multiple reflections of signals from buildings.

When the subscriber density in a particular area increases significantly, the cells in that area may be split (subdivided) into smaller cells. Within the region of smaller cells, the spatial frequency reuse pattern is simply a smaller version of that within the region of larger cells. There is some complexity involved in assigning FDMA channels near the borders of the two regions because the cochannel interference reuse distances of the two regions are different. It is sometimes useful to implement an "overlay" concept in the border areas—and during the cell-splitting process—in which the smaller cells are thought of as being superimposed upon the pattern of larger cells in a distinct "layer," and disjoint sets of frequencies are assigned to the two layers [4], at least near the borders of the two regions.

Just as the physical dimensions of cells are not apparent in the analysis of frequency reuse distance and clustering, neither is the actual amount of power emitted by a base station or mobile user. In deriving the forward link ratio C/I, the powers at the different base stations were assumed to be equal, and so they were canceled in the division of numerator (C) by denominator (I). It might seem that the amount of base station power is arbitrary. The correct amount of power is not arbitrary, however, and this fact is highlighted when considering different-size cells and cell clusters.

Clearly, there is a minimal amount of base station power required to provide a satisfactory SNR to a mobile at the edge of the cell in the absence of cochannel interference:

$$\text{SNR} = \frac{\text{Base station received power}}{\text{Receiver noise}} = \frac{P_t \kappa / R_c^\gamma}{N_m} > \text{SNR}_{\text{req}} \quad (3.31\text{a})$$

$$\Rightarrow P_t > P_{t0} \triangleq N_m \cdot \text{SNR}_{\text{req}} \cdot R_c^\gamma / \kappa \quad (3.31\text{b})$$

When there are multiple cells, the base station power must be increased to compensate for the cochannel interference. Let k_e be the effective number of interfering cells at the reuse distance D from the mobile; then the effective SNR for a mobile at the cell edge is

$$\text{SNR} = \frac{P_t \kappa / R_c^\gamma}{N_m + k_e P_t \kappa / D^\gamma} > \text{SNR}_{\text{req}} \tag{3.32a}$$

$$\Rightarrow P_t > \frac{N_m \cdot \text{SNR}_{\text{req}} \cdot R_c^\gamma / \kappa}{1 - \text{SNR}_{\text{req}} / [(D/R_c)^\gamma / k_e]} \tag{3.32b}$$

$$= \frac{P_{t0}}{1 - \text{SNR}_{\text{req}} / (C/I)} \tag{3.32c}$$

in which an expression for C/I in form of (3.25) has been recognized. For example, if the reuse distance is such that C/I is 3 dB more than the required SNR, then the base station power required to overcome the noise-plus-interference is twice that needed to overcome just the mobile receiver noise. These equations show that there is a minimal value of base station transmitter power, and the minimum is a function of distance. Thus, when cells are split into smaller cells, the minimum power requirement is reduced also.

The specification of a maximal amount of base station transmitter power is not as obvious as a minimal amount. Presumably, the cost of the equipment is directly proportional to power, and this is a motivation to use as little power as necessary but this factor does not rule out using an arbitrary amount of power so long as the minimal requirement is met. A consideration that leads to an upper bound on base station power is the desirability of balancing forward and reverse links: the effective radius of the base station's coverage area should not exceed the cell radius, except for some nominal amount, and the cell radius should correspond to the range of the mobile transmitters. If more base station power is used than necessary, mobiles seeking to hand off may affiliate with the base station before they are within the cell, increasing the possibility of failed handoff attempts due to the limited range of the mobile transmitters.

3.2.5 AMPS Parameters

For reference, the FDMA channel allocations for the AMPS analog cellular system are included here. The spectrum allocation for service providers A and B that was described in Table 3.7 is shown in greater detail in Table 3.9 [15]. Note that the frequency assignments to the cell sectors in each seven-cell cluster follow a simple pattern. The orientation and labeling of sectors is shown in Figure 3.21, and Table 3.10 summarizes the allocation of frequencies in terms of the numbers of channels.

Table 3.9 Channel assignments for U.S. FDMA cellular systems

Sect.	Frequency channels assigned
	Block A
1A	1, 22, 43, 64, 85, 106, 127, 148, 169, 190, 211, 232, 253, 274, 295, 667, 688, 709, 1003
2A	2, 23, 44, 65, 86, 107, 128, 149, 170, 191, 212, 233, 254, 275, 296, 668, 689, 710, 1004
3A	3, 24, 45, 66, 87, 108, 129, 150, 171, 192, 213, 234, 255, 276, 297, 669, 690, 711, 1005
4A	4, 25, 46, 67, 88, 109, 130, 151, 172, 193, 214, 235, 256, 277, 298, 670, 691, 712, 1006
5A	5, 26, 47, 68, 89, 110, 131, 152, 173, 194, 215, 236, 257, 278, 299, 671, 692, 713, 1007
6A	6, 27, 48, 69, 90, 111, 132, 153, 174, 195, 216, 237, 258, 279, 300, 672, 693, 714, 1008
7A	7, 28, 49, 70, 91, 112, 133, 154, 175, 196, 217, 238, 259, 280, 301, 673, 694, 715, 1009
1B	8, 29, 50, 71, 92, 113, 134, 155, 176, 197, 218, 239, 260, 281, 302, 674, 695, 716, 1010
2B	9, 30, 51, 72, 93, 114, 135, 156, 177, 198, 219, 240, 261, 282, 303, 675, 696, 1011
3B	10, 31, 52, 73, 94, 115, 136, 157, 178, 199, 220, 241, 262, 283, 304, 676, 697, 991, 1012
4B	11, 32, 53, 74, 95, 116, 137, 158, 179, 200, 221, 242, 263, 284, 305, 677, 698, 992, 1013
5B	12, 33, 54, 75, 96, 117, 138, 159, 180, 201, 222, 243, 264, 285, 306, 678, 699, 993, 1014
6B	13, 34, 55, 76, 97, 118, 139, 160, 181, 202, 223, 244, 265, 286, 307, 679, 700, 994, 1015
7B	14, 35, 56, 77, 98, 119, 140, 161, 182, 203, 224, 245, 266, 287, 308, 670, 701, 995, 1016
1C	15, 36, 57, 78, 99, 120, 141, 162, 183, 204, 225, 246, 267, 288, 309, 671, 702, 996, 1017
2C	16, 37, 58, 79, 100, 121, 142, 163, 184, 205, 226, 247, 268, 289, 310, 672, 703, 997, 1018
3C	17, 38, 59, 80, 101, 122, 143, 164, 185, 206, 227, 248, 269, 290, 311, 673, 704, 998, 1019
4C	18, 39, 60, 81, 102, 123, 144, 165, 186, 207, 228, 249, 270, 291, 312, 674, 705, 999, 1020
5C	19, 40, 61, 82, 103, 124, 145, 166, 187, 208, 229, 250, 271, 292, 675, 706, 1000, 1021
6C	20, 41, 62, 83, 104, 125, 146, 167, 188, 209, 230, 251, 272, 293, 676, 707, 1001, 1022
7C	21, 42, 63, 84, 105, 126, 147, 168, 189, 210, 231, 252, 273, 294, 677, 708, 1002, 1023
ctrl	313–333
	Block B
1A	355, 376, 397, 418, 439, 460, 481, 502, 523, 544, 565, 586, 607, 628, 649, 720, 741, 762, 783
2A	356, 377, 398, 419, 440, 461, 482, 503, 524, 545, 566, 587, 608, 629, 650, 721, 742, 763, 784
3A	357, 378, 399, 420, 441, 462, 483, 504, 525, 546, 567, 588, 609, 630, 651, 722, 743, 764, 785
4A	358, 379, 400, 421, 442, 463, 484, 505, 526, 547, 568, 589, 610, 631, 652, 723, 744, 765, 786
5A	359, 380, 401, 422, 443, 464, 485, 506, 527, 548, 569, 590, 611, 632, 653, 724, 745, 766, 787
6A	360, 381, 402, 423, 444, 465, 486, 507, 528, 549, 570, 591, 612, 633, 654, 725, 746, 767, 788
7A	361, 382, 403, 424, 445, 466, 487, 508, 529, 550, 571, 592, 613, 634, 655, 726, 747, 768, 789
1B	362, 383, 404, 425, 446, 467, 488, 509, 530, 551, 572, 593, 614, 635, 656, 727, 748, 769, 790
2B	363, 384, 405, 426, 447, 468, 489, 510, 531, 552, 573, 594, 615, 636, 657, 728, 749, 770, 791
3B	364, 385, 406, 427, 448, 469, 490, 511, 532, 553, 574, 595, 616, 637, 658, 729, 750, 771, 792
4B	365, 386, 407, 428, 449, 470, 491, 512, 533, 554, 575, 596, 617, 638, 659, 730, 751, 772, 793
5B	366, 387, 408, 429, 450, 471, 492, 513, 534, 555, 576, 597, 618, 639, 660, 731, 752, 773, 794
6B	367, 388, 409, 430, 451, 472, 493, 514, 535, 556, 577, 598, 619, 640, 661, 732, 753, 774, 795
7B	368, 389, 410, 431, 452, 473, 494, 515, 536, 557, 578, 599, 620, 641, 662, 733, 754, 775, 796
1C	369, 390, 411, 432, 453, 474, 495, 516, 537, 558, 579, 600, 621, 642, 663, 734, 755, 776, 797
2C	370, 391, 412, 433, 454, 475, 496, 517, 538, 559, 580, 601, 622, 643, 664, 735, 756, 777, 798
3C	371, 392, 413, 434, 455, 476, 497, 518, 539, 560, 581, 602, 623, 644, 665, 736, 757, 778, 799
4C	372, 393, 414, 435, 456, 477, 498, 519, 540, 561, 582, 603, 624, 645, 666, 737, 758, 779
5C	373, 394, 415, 436, 457, 478, 499, 520, 541, 562, 583, 604, 625, 646, 717, 738, 759, 780
6C	374, 395, 416, 437, 458, 479, 500, 521, 542, 563, 584, 605, 626, 647, 718, 739, 760, 781
7C	375, 396, 417, 438, 459, 480, 501, 522, 543, 564, 585, 606, 627, 648, 719, 740, 761, 782
ctrl	334–354

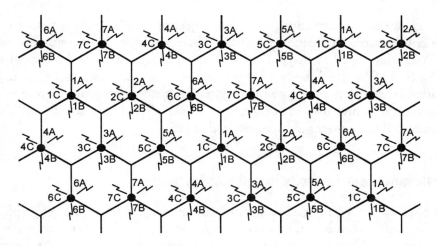

Figure 3.21 Sector frequency channel sets for seven-cell clusters.

Table 3.10 Summary of channel allocations

set	1	2	3	4	5	6	7	sum
	Frequency block A							
A	19	19	19	19	19	18	18	131
B	18	18	19	19	19	19	19	131
C	19	19	19	19	19	19	19	133
control	3	3	3	3	3	3	3	21
sum	56	56	57	57	57	56	56	416
	Frequency block B							
A	19	19	19	19	18	19	19	132
B	19	19	19	19	19	19	19	133
C	19	19	19	19	18	18	18	130
control	3	3	3	3	3	3	3	21
sum	57	57	57	57	55	56	56	416

The control channels, one for each sector, are described as "setup" channels in [4]. Mobile units, when turned on, measure the relative strengths of a standard group of these channels and determine to monitor the channel with the strongest signal, which can be assumed to originate from the closest base station transmitter; thus, one function of the setup channel signal is to provide a pilot (reference) signal. Digital data on a setup channel assist the mobile unit in its decision whether to continue monitoring that channel as the mobile travels from cell to cell, and inform it of the identification numbers of mobile units to which calls are being directed (paged).

In our discussions thus far on the subject of FDMA cellular systems, we have confined our attention to AMPS. There are other analog systems, such as TACS, developed in the U.K., NMT from Scandanavia, and Narrowband AMPS (NAMPS) developed by Motorola from the United States. Although a worldwide standard analog cellular system has not emerged, the individual systems have found use throughout the world. The characteristics of these different cellular systems are summarized in [16].

3.3 Coverage and Capacity in Cellular Systems

The coverage of a base station sector in a cellular system is the geographical area in which a mobile unit may communicate with the base station with sufficient signal level at the base station receiver to satisfy service requirements. The capacity of the same sector in the cellular system is the number of mobile users that can simultaneously access the base station with an acceptable level of mutual interference.

In this section, the tradeoff between coverage and capacity for a cellular system will be discussed as it affects the design of a cellular system.[3]

3.3.1 Coverage Limits

The coverage of individual cellular base stations determines how many base stations are required to provide service to a particular geographical area. For a single mobile user (the "coverage-limited case" [17]), the maximum range between the base station and the user is limited by the sensitivity of the base station receiver, given the amount of power that can be transmitted by the mobile unit.

The sensitivity of the base station receiver is the amount of received signal power required to achieve a certain minimal value of the SNR at the receiver's input. In practice, the stated range of a mobile transmission must consider a "margin" that allows for interference and variation in mobile signal level that may occur through fading or shadowing.

---

[3] A treatment of area coverage for CDMA cellular systems is covered in Section 10.2.3.

3.3.1.1 Generic Cellular System Link Budget

The *link power budget* or *link budget* for the reverse link in a cellular system typically has the following components:

+	Mobile transmitter power, in dBm	P_m
+	Transmitter antenna gain, in dBi	G_m
−	Median link propagation loss in dB, a function of link distance d	$L_{med}(d)$
+	Cell receiver antenna gain, in dBi	G_c
−	Cell cabling losses, in dB	L_{rc}
−	Receiver noise plus interference from other users, modeled as noise, in dBm	$(N + I)_c$
+	Receiver processing gain (if any), in dB	PG
=	Median received SNR, in dB	SNR_{med}
−	Required SNR, in dB	SNR_{req}
=	Excess of required SNR, in dB (margin)	M_{dB}

For example, an AMPS analog system has the typical link budget values that are listed in Table 3.11, assuming that the mobile unit is operating at maximum power.

Table 3.11 Typical AMPS link budget parameter values at maximum range

Parameter	Value
Mobile maximum power, P_m	27 dBm (500 mW)
Mobile transmitter antenna gain, G_m	2 dBi
Median link propagation loss, L_{med}	120 dB
Cell receiver antenna gain, G_c	14 dBi
Cell cabling losses, L_{rc}	2 dB
Receiver noise and interference, $(N + I)_c$	−121 dBm
Receiver processing gain, PG	0 dB
Median received SNR, SNR_{med}	42 dB
Required SNR, SNR_{req}	18 dB
Excess of required SNR	24 dB

3.3.1.2 Receiver Noise Calculation

The amount of receiver thermal noise is calculated using either an equivalent noise temperature or a noise figure. For the latter approach, the equation is

$\mathcal{N}_0$ = Receiver noise power spectral density level

$$= kT_0F \text{ (Boltzman's const} \times \text{Ref. noise temp.} \times \text{Noise figure)} \quad (3.33a)$$

$$= (1.38 \cdot 10^{-23} J/°K)(293°K) F = 4.043 \, F \cdot 10^{-21} \text{ W/Hz} \quad (3.33b)$$

$$= -203.9 \, \text{dBW/Hz} + \text{NF(dB)} \quad (3.33c)$$

In dBW/Hz and dBm/Hz, for a 5-dB noise figure the noise density is

$$\mathcal{N}_0 \text{ (dB)} = -204 \, \text{dBW/Hz} + 5 \, \text{dB} = -199 \, \text{dBW/Hz} \quad (3.34a)$$

$$= -174 \, \text{dBm/Hz} + 5 \, \text{dB} = -169 \, \text{dBm/Hz} \quad (3.34b)$$

In the AMPS 30-kHz (45-dBHz) receiver bandwidth, the noise power in dBm is

$$(\mathcal{N}_0 W)_c \text{ (dB)} = -169 + 45 = -124 \, \text{dBm} \quad (3.35)$$

With an equal amount of interference, the total $(N + I)_c$ equals -121 dBm.

3.3.1.3 Maximum Tolerable Propagation Loss

After a basic link budget for the system is determined, the coverage limit for the cell can be expressed in terms of the maximum tolerable propagation loss. Using this value, a preliminary plan for the location of cell sites can be developed based on propagation loss prediction algorithms and map data for the cellular service area. A survey of propagation loss prediction formulas is given in Chapter 2.

The link budget can be expressed in terms of the SNR requirement and a desired margin M_{dB} and can then be solved for the maximum propagation loss:

$$\text{SNR}_{med} = P_m + G_m - L_{med} + G_c - L_{rc} - (N + I)_c > \text{SNR}_{req} + M_{dB}$$

so that

$$L_{med} < L_{mmx} \triangleq P_m + G_m + G_c - L_{rc} - (N+I)_c - \text{SNR}_{req} - M_{dB} \quad (3.36)$$

For example, using the values in Table 3.11 gives the bound

$$L_{med} < L_{mmx} = 27 + 2 + 14 - 2 - (-121) - 18 - M_{dB}$$
$$= 144\,\text{dB} - M_{dB} \tag{3.37}$$

This tolerable amount of loss can be converted into a cell radius using one of the propagation models.

3.3.2 Coverage Versus Capacity

Sector capacity is limited by the amount of mutual interference among the mobile units that can be tolerated in order to cover a given area, while maintaining a margin against signal fading and shadowing losses. Conversely, the coverage may be viewed as being limited by capacity in that the amount of interference is determined by a given value of capacity and the degree to which this capacity is "loaded" by actual traffic in a typical situation. This results in a tradeoff between coverage and capacity.

In addition to the basic link budget parameters discussed so far, generally other factors also affect this tradeoff, including gains (if any) due to diversity and multipath processing, error control coding, soft handoff, and the like.

The tradeoff between cellular coverage and capacity is discussed in terms of the following procedure to determine the density of base stations to cover a particular service area as a function of offered traffic density:

- Determine the amount of link margin needed to maintain a specified link reliability in the face of lognormal shadowing.

- Determine any gain or loss in a multicell system due to handoff processing.

- Relate the received SNR to the traffic loading.

- Find the blocking probability as a function of traffic loading.

- Determine the increase in margin required versus the load.

- Find the cell radius as a function of loading and convert to base station density as a function of traffic density.

3.3.2.1 Link Margin for the Coverage-Limited Case

The required SNR for the cell site receiver can vary considerably as signal channel conditions vary and may need to be determined by simulation or testing. However it is determined, let the required SNR in dB be denoted SNR_{req} and let the initial consideration be given to the "coverage-limited case" in which there is just one cell—no cochannel interference.

In lognormal shadowing, as discussed in Section 2.2.3, the propagation loss in dB can be considered to be a Gaussian RV:

$$L(d) = L_{med}(d) + \sigma_{dB} X \tag{3.38}$$

where L_{med} and σ_{dB} are the median value and standard deviation of the propagation loss in dB, respectively, and $X = G(0, 1)$ is a zero-mean, unit-variance Gaussian RV. Thus, the probability density function (pdf) for the loss in dB is

$$p_L(x) = \frac{1}{\sigma_{dB}} p_G\left(\frac{x - L_{med}}{\sigma_{dB}}\right) = \frac{1}{\sqrt{2\pi}\,\sigma_{dB}} \exp\left\{-\frac{(x - L_{med})^2}{2\sigma_{dB}^2}\right\} \tag{3.39}$$

The CDF for the loss in dB is

$$P_L(x) \triangleq \Pr\{L(d) \le x\} = \Pr\{L_{med} + \sigma_{dB}G \le x\}$$

$$= \Pr\left\{G \le \frac{x - L_{med}}{\sigma_{dB}}\right\} = \int_{-\infty}^{\frac{x - L_{med}}{\sigma_{dB}}} \frac{1}{\sqrt{2\pi}}\, e^{-u^2/2}\, du$$

$$= P_G\left(\frac{x - L_{med}}{\sigma_{dB}}\right) \equiv P\left(\frac{x - L_{med}}{\sigma_{dB}}\right) \tag{3.40a}$$

The complementary CDF for the propagation loss is

$$Q_L(x) \triangleq \Pr\{L(d) > x\} = 1 - P_L(x) = 1 - P\left(\frac{x - L_{med}}{\sigma_{dB}}\right)$$

$$= \int_{\frac{x - L_{med}}{\sigma_{dB}}}^{\infty} \frac{1}{\sqrt{2\pi}}\, e^{-u^2/2}\, du = Q_G\left(\frac{x - L_{med}}{\sigma_{dB}}\right) \equiv Q\left(\frac{x - L_{med}}{\sigma_{dB}}\right) \tag{3.40b}$$

The system design includes a margin M_{dB}, intended to compensate for the variation in propagation loss. The margin is implemented by using more transmitter power than would be necessary for the same link distance if there were no variation in the propagation loss. Including a margin, the received SNR in the coverage-limited (no interference) case can be written

$$\text{SNR} = P_m + G_{\text{net}} - L_{med}(d) - \sigma_{dB}G - N_c$$

$$= \text{SNR}_{med}(d) - \sigma_{dB}G \tag{3.41a}$$

where P_m is the transmitter power in dBm, G_{net} denotes the net gain of the various fixed gains and losses in dB, N_c is the cell receiver noise in dBm, and the median value of the received SNR is

$$\text{SNR}_{med}(d) \triangleq P_m + G_{\text{net}} - L_{med}(d) - N_c \tag{3.41b}$$

The coverage area of the cell is defined by the maximal distance (radius) $d_{max} \equiv R_c$. Successful link operation in the coverage area is defined by the inequality

$$\text{SNR} = \text{SNR}_{med}(R_c) - \sigma_{dB}G > \text{SNR}_{req} \tag{3.42}$$

As the propagation loss varies randomly, what must the margin be to produce successful link operation with a specified reliability? To answer that question, we first define *reliability* as the probability of successful operation:

$$\text{Reliability} \equiv P_{rel} \triangleq \text{Pr}\{\text{SNR} > \text{SNR}_{req}\} \tag{3.43a}$$

$$= \text{Pr}\{\text{SNR}_{med}(R_c) - \sigma_{dB}G > \text{SNR}_{req}\} \tag{3.43b}$$

The link margin in dB is related to link reliability P_{rel} by

$$P_{rel} \triangleq \text{Pr}\{\text{SNR}_{med}(R_c) - \sigma_{dB}G > \text{SNR}_{req}\}$$

$$= \text{Pr}\left\{ G < \frac{\text{SNR}_{med}(R_c) - \text{SNR}_{req}}{\sigma_{dB}} \right\} \overset{\text{set}}{=} \text{Pr}\left\{ G < \frac{M_{dB}(R_c)}{\sigma_{dB}} \right\}$$

$$= P\left(\frac{M_{dB}(R_c)}{\sigma_{dB}} \right) = 1 - Q\left(\frac{M_{dB}(R_c)}{\sigma_{dB}} \right) \tag{3.44a}$$

The values of $P(x)$ and $Q(x)$ are tabulated, so that it is possible to find x_p or x_q such that $p = P(x_p)$ or $q = Q(x_q)$. Another way to express the values of these arguments giving specified probability values is $x_p = P^{-1}(p)$ or $x_q = Q^{-1}(q)$. Using that notation, the margin is found from

$$\frac{M_{dB}(R_c)}{\sigma_{dB}} = P^{-1}(P_{rel}) = Q^{-1}(1 - P_{rel}) \tag{3.44b}$$

For example, if a link reliability of $P_{rel} = 90\%$ is desired, then from a table of the Gaussian distribution we find that the link coverage-limited margin needs to be

$$M_{dB} = \sigma_{dB} \cdot P^{-1}(0.90) = 1.28155\,\sigma_{dB} \tag{3.44c}$$

This example is illustrated in Figure 3.22.

A typical value [18] of the standard deviation for the propagation loss variation is $\sigma_{dB} = 8\,\mathrm{dB}$, which becomes the standard deviation of the SNR if there is no power control. With adaptive power control, the mobile transmitter power can be adjusted to adapt to the variations in the propagation loss. A typical value [19] for the standard deviation of the SNR in that case is $\sigma_{dB} = 2.5\,\mathrm{dB}$. Thus, for 90% link reliability, the margin needs to be $1.28155 \times 8 = 10.25\,\mathrm{dB}$ without power control and $1.28155 \times 2.5 = 3.20\,\mathrm{dB}$ with adaptive power control. Table 3.12 gives the single-cell margins required for various values of the reliability and σ_{dB}.

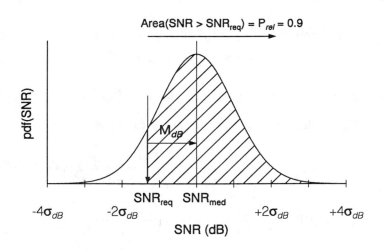

Figure 3.22 Margin required for 90% link reliability.

Table 3.12 Margins required for single cell

P_{rel}	M_{dB}	
	$\sigma_{dB} = 8\,\text{dB}$	$\sigma_{dB} = 2.5\,\text{dB}$
0.9	10.25	3.20
0.95	13.16	4.11
0.98	16.43	5.13

3.3.2.2 Determination of Multicell Margin Requirements

Having found the link margin needed for the reverse link based on an analysis of the single-user, single-cell case (i.e., no interference), we now consider how the internetworking of the base stations (and sectors of the same base station) in a cellular system affects the link budget. Specifically, we consider how the required margin in dB is modified by a "handoff gain" (which may be negative) to account for the particular handoff technique employed by the system, either:

- Hard handoff, in which the call is transferred to the adjacent cell or sector when the signal level in the current cell falls below a certain threshold.

- Soft handoff (to be discussed in Chapters 4 and 10 under CDMA handoff considerations), which allows simultaneous transmission to and from the mobile station through two base stations or cells.

Hard handoff is commonly used by FDMA cellular systems, such as AMPS, because transferring the call to another cell site involves retuning the oscillators in the mobile's transmitter and receiver to new frequencies. When the mobile is near the border of two cells, the signal levels usually fluctuate at both cell sites, resulting in a *pingpong* effect. The case of a mobile traveling a route that crosses and recrosses a cellular boundary is handled in hard handoff systems by engineering a slight amount of overlap of the cellular coverage areas. As illustrated in Figure 3.23, in order to engineer an overlap of coverage areas, the link margin must be increased to ensure reliable operation at the extended range κR_c, where R_c is the cell radius and $\kappa \geq 1$ is a cell range extension factor.

At or about the cell edge—the distance $d = R_c$ from the base station—the median propagation path loss can be modeled as being proportional to

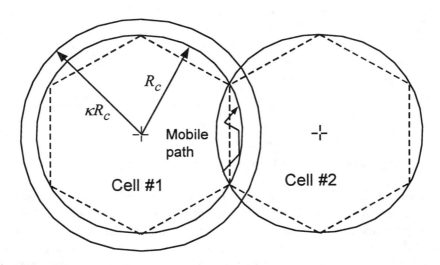

Figure 3.23 Cell overlap in terms of range extension factor.

some power (γ) of the distance. In dB, this proportionality can be expressed by

$$L_{med}(d) = K_L + 10\gamma \log_{10}d, \quad R_c - \Delta d < d < R_c + \Delta d \quad (3.45)$$

where K_L is a constant and it is said that the propagation is "γ-law" propagation. For γ-law propagation, the median SNR in dB at the distance κR_c slightly into another cell is related to the median SNR in dB at the cell edge by

$$\text{SNR}_{med}(\kappa R_c) = \text{SNR}_{med}(R_c) - 10\gamma \log_{10}\kappa \quad (3.46)$$

As shown in Figure 3.24, to make the SNR at the distance κR_c equal to the value that produces the desired reliability at the distance R_c, the link margin must be increased by $10\gamma \log_{10}\kappa$. Thus, a system using hard handoff has a negative "handoff gain" that equals $-10\gamma \log_{10}\kappa$. Example values of margin and handoff gain are given in Table 3.13 for a 90% link reliability and $\gamma = 4$.

For cellular technologies that permit simultaneous reception of the mobile at two different base stations, a form of diversity can be used in what is called a "soft" handoff.[4] In CDMA cellular systems, soft handoff is

---

[4] In this chapter, our discussion of soft handoff is somewhat hypothetical and does not include details of system implementation.

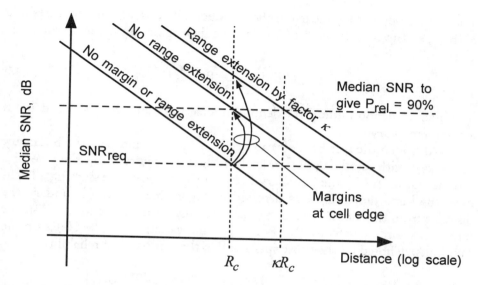

Figure 3.24 Margins with and without extended range.

Table 3.13 Margin and hard handoff gain, $P_{rel} = 0.9$

Range extension factor, κ	Required margin, $M_{dB}(R_c)$	Handoff gain, $\mu = 4$
1.00	10.25 dB	0 dB
1.05	11.10 dB	−0.85 dB
1.10	11.91 dB	−1.66 dB
1.15	12.68 dB	−2.43 dB
1.20	13.42 dB	−3.17 dB
1.25	14.13 dB	−3.88 dB

actually employed, as described in Chapter 4 and analyzed in Chapter 10—frequency retuning is not required [20–22]. When one base station's reverse link from the mobile is subject to fading, chances are the other station's reverse link is not. The result is that less margin is needed to ensure the same reverse link reliability.

The soft handoff situation involves two random events:

Event 1 (E_1): $SNR_1 \equiv SNR$ at base station 1 $> SNR_{req}$ (3.47a)

Event 2 (E_2): $SNR_2 \equiv SNR$ at base station 2 $> SNR_{req}$ (3.47b)

The link is operating satisfactorily if either SNR is above the threshold SNR_{req}, so the overall event of satisfactory link operation is the union of E_1 and E_2. That is,

$$P_{rel} = \Pr\{E_1 \cup E_2\} = 1 - \Pr\{\overline{E}_1 \cap \overline{E}_2\} \tag{3.48}$$

in which $\cup$ denotes union, $\cap$ denotes intersection, and the overbar denotes a complementary event.

Although the soft handoff can occur at any position near the boundary of the cell with another cell, the analysis here concentrates on the case of a mobile transmitter at the boundary between the two cells. Assuming all cells have the same dimensions, as depicted in Figure 3.25, at the boundary the distance from the mobile to the two base stations is the same and that distance (r) varies from $0.866R_c$ to R_c. For lognormal variations in the SNRs on the two links, denoted by the subscripts 1 and 2, the soft handoff reliability is

$$P_{rel} = \Pr\{(SNR_1 > SNR_{req}) \cup (SNR_2 > SNR_{req})\}$$

$$= \Pr\{(SNR_{med}(r) - \sigma_{dB}G_1 > SNR_{req})$$
$$\cup (SNR_{med}(r) - \sigma_{dB}G_2 > SNR_{req})\}$$

$$= \Pr\left\{\left(G_1 < \frac{M_{dB}(r)}{\sigma_{dB}}\right) \cup \left(G_2 < \frac{M_{dB}(r)}{\sigma_{dB}}\right)\right\} \tag{3.49a}$$

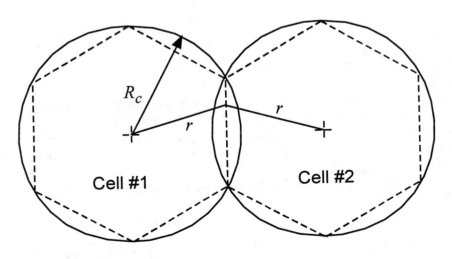

Figure 3.25 Soft handoff scenario.

which is given by

$$P_{rel} = 1 - \Pr\left\{\left(G_1 > \frac{M_{dB}(r)}{\sigma_{dB}}\right) \cup \left(G_2 > \frac{M_{dB}(r)}{\sigma_{dB}}\right)\right\} \qquad (3.49b)$$

where G_1 and G_2 are correlated zero-mean, unit-variance Gaussian variables. The propagation paths from the mobile to the two base stations have at least one thing in common—the position of the mobile: if the mobile moves to a high position, both losses tend to *decrease*; if the mobile moves to a low position, both losses tend to *increase*. Thus the possibility of correlated propagation losses must be considered.

The propagation losses in dB have been treated as Gaussian RVs, resulting in SNRs in dB that are Gaussian RVs. The correlation coefficient between two RVs X and Y is defined as

$$\rho \triangleq \frac{E\{(X - \overline{X})(Y - \overline{Y})\}}{\sigma_X \sigma_Y} = \frac{E\{X \cdot Y\} - \overline{X} \cdot \overline{Y}}{\sigma_X \sigma_Y} \qquad (3.50)$$

If the propagation losses are correlated, so are the SNRs at the two base stations. The correlation coefficient for the two SNRs is found by substituting into (3.50) the following parameters:

$$X = \text{SNR}_1 = \text{SNR}_{med}(r) - \sigma_{dB}G_1, \quad \overline{X} = \text{SNR}_{med}(r) \qquad (3.51a)$$

$$Y = \text{SNR}_2 = \text{SNR}_{med}(r) - \sigma_{dB}G_2, \quad \overline{Y} = \text{SNR}_{med}(r) \qquad (3.51b)$$

$$\sigma_X = \sigma_Y = \sigma_{dB} \qquad (3.51c)$$

The resulting correlation coefficient for the SNRs is

$$\rho = \frac{E\{(X - \overline{X})(Y - \overline{Y})\}}{\sigma_X \sigma_Y} = \frac{E\{(-\sigma_{dB}G_1)(-\sigma_{dB}G_2)\}}{\sigma_{dB}^2}$$

$$= E\{G_1 G_2\} \qquad (3.52)$$

That is, the correlation coefficient for the SNRs is the same as that for the two underlying zero-mean, unit-variance Gaussian variables G_1 and G_2.

Given the correlation coefficient ρ for the two propagation losses, the reliability expression (3.49b) is

$$P_{rel} = 1 - \int_{M_{dB}(r)/\sigma_{dB}}^{\infty} dx \int_{M_{dB}(r)/\sigma_{dB}}^{\infty} p_{G_1,G_2}(x,\, y;\, \rho)\, dy \qquad (3.53a)$$

$$= 1 - L\left(\frac{M_{dB}(r)}{\sigma_{dB}},\, \frac{M_{dB}(r)}{\sigma_{dB}},\, \rho\right) \qquad (3.53b)$$

where

$$p_{G_1,G_2}(x,\, y;\, \rho) = \frac{1}{2\pi(1-\rho^2)}\exp\left\{-\frac{x^2 - 2\rho xy + y^2}{2(1-\rho^2)}\right\} \qquad (3.53c)$$

is the joint pdf for G_1 and G_2, and the domain of the integration is represented by the *unshaded* area in Figure 3.26.

In (3.53b) the function $L(a, b, \rho)$ is the bivariate Gaussian probability integral, a function that is tabulated for certain special cases [11, §26]; it is the area under the bivariate pdf that is illustrated by the *shaded* area in Figure 3.27. For equal first and second arguments, the following special relationship holds:

$$L(a,\, a,\, \rho) = 2\,L\left(a,\, 0,\, -\sqrt{\frac{1-\rho}{2}}\right) \qquad (3.54)$$

and the function $L(\cdot,\, 0,\, \cdot)$ is tabulated and graphed in [11]. For 90% reliability at the cell boundary, the worst case (largest margin) is found by combining (3.53b) and (3.54) to obtain

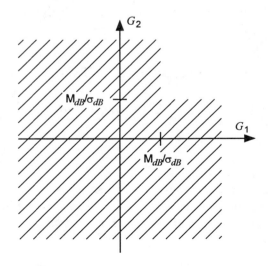

Figure 3.26 Domain of integration for joint probability.

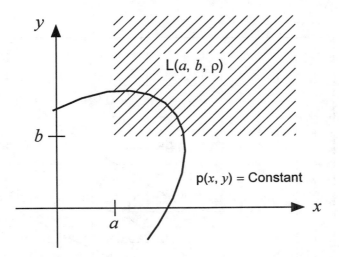

Figure 3.27 Bivariate probability integral.

$$L\left(\frac{M_{dB}(R_c)}{\sigma_{dB}}, 0, -\sqrt{\frac{1-\rho}{2}}\right) = \frac{1 - P_{rel}}{2} = 0.05 \qquad (3.55)$$

Graphically solving (3.55) from the curves in [11] for the margin as a function of ρ results in the data plotted in Figure 3.28 for different values of the reliability. For example, for zero correlation ($\rho = 0$), the solution of (3.55) for 90% reliability gives $M_{dB}(R_c) = 0.48\sigma_{dB}$; for a correlation coefficient of $\rho = 0.5$, the same conditions yield $M_{dB}(R_c) = 0.77\sigma_{dB}$. Table 3.14 gives values of the soft handoff margin and handoff gain for several situations.

Note that the required margin for soft handoff is low for negative correlation, when it is likely that the other link is good when one of them is bad, and the required margin is high for positive correlation, when the links tend to be either good or bad at the same time. Because the two links have the mobile location in common, it is to be expected that the correlation is positive.

The combination of negative handoff gains for cellular systems with hard handoff, such as FDMA, and of positive handoff gains for those with soft handoff, such as CDMA, gives the latter significant advantages in the reverse link power budget. The net handoff gain advantage of soft handoff over hard handoff is shown in Table 3.15 for $\rho = 0.5$, $\mu = 4$, $\kappa = 1.05$, and for $\sigma_{dB} = 8$ dB (without power control) and $\sigma_{dB} = 2.5$ dB (with power control). The table shows that the advantage is about 5 dB without power

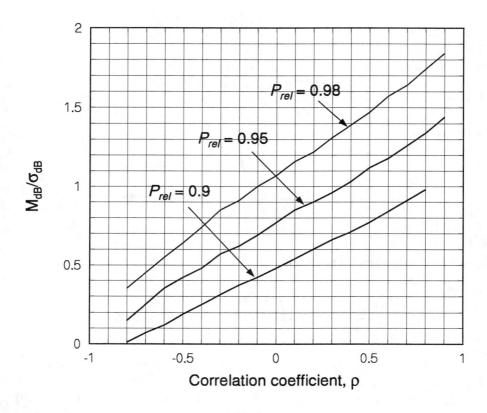

Figure 3.28 Relative margin in dB versus correlation.

Table 3.14 Soft handoff margin and gain

P_{rel}	ρ	$\dfrac{M_{dB}(R_c)}{\sigma_{dB}}$	$\sigma_{dB} = 8\,\text{dB}$		$\sigma_{dB} = 2.5\,\text{dB}$	
			$M_{dB}(R_c)$	Handoff gain	$M_{dB}(R_c)$	Handoff gain
0.9	−0.5	0.19	1.52	8.73	0.48	2.72
	0.0	0.48	3.84	6.41	1.20	2.00
	0.5	0.77	6.16	4.09	1.93	1.27
0.95	−0.5	0.42	3.36	9.80	1.05	3.06
	0.0	0.77	6.16	7.00	1.93	2.18
	0.5	1.12	8.96	4.20	2.80	1.31
0.98	−0.5	0.64	5.12	11.31	1.60	3.53
	0.0	1.07	8.56	7.87	2.68	2.45
	0.5	1.47	11.76	4.67	3.68	1.45

Table 3.15 Handoff gain advantage of soft handoff over hard

P_{rel}	$M_{dB}(\text{Hard}) - M_{dB}(\text{Soft})$	
	$\sigma_{dB} = 8\,\text{dB}$	$\sigma_{dB} = 2.5\,\text{dB}$
0.90	4.94	2.12
0.95	5.05	2.16
0.98	5.52	2.30

control and about 2 dB with power control. The soft handoff system can use this advantage to reduce mobile transmitter power, which lowers the drain on the battery and also increases system capacity because of the reduction in mutual interference. Or, for the same power, a larger cell size can be used [19].

3.3.2.3 Reverse Link C/I and C/N as Functions of System Loading

In the discussions in this section so far, the background presented has established the hypothetical comparison shown in Table 3.16 of the link power budgets of analog cellular systems with hard and soft handoff in terms of the received signal levels required on the reverse link for 90% link reliability and a standard deviation for the SNR in dB of $\sigma_{dB} = 8\,\text{dB}$. Next we show how cellular size and coverage considerations can be associated with capacity considerations. In doing so, the received C/I ratio is related to the reverse link traffic and to the blocking probability for the mobile telephone calls. Then the required signal level is found as a function of the reverse link traffic level in Erlangs. These results will then be put together to develop the trade-off between cellular system coverage and traffic loads.

Table 3.16 Comparison of reliable signal levels

	Hard handoff	Soft handoff
Required SNR in dB	18 dB	
+ N_0 for 5-dB noise figure	−169 dBm/Hz	
+ RF bandwidth in dB	44.8 dBHz	
Minimum required signal	−106.2 dBm	
+ 1-cell margin for 90% reliability	10.3 dB	
− Handoff gain	−0.8 dB	4.1 dB
Reliable signal level	−95.1 dBm	−100.0 dBm

When the cellular system is loaded with calls to its capacity, by definition the received C/I is just sufficient to meet the link requirements, but will not meet the requirements if another user is added to the link. In the case of an FDMA system, these considerations have led to the selection of three-sector, corner-fed cells in seven-cell clusters with a capacity of 18 or 19 channels per sector [4]. The analog FDMA cellular system requires a C/I value of 18 dB for satisfactory operation. By design, the C/I can be assumed to be equal to the required 18 dB for a fully loaded system; that is:

$$\frac{C}{I_{max}} = 18\,\text{dB} \tag{3.56}$$

It follows that if $I = kI_{max}$, where k is the fractional loading, then a general expression for C/I is

$$\frac{C}{I} = \frac{C}{kI_{max}} = \frac{C}{I_{max}} \cdot \frac{1}{k} = 18\,\text{dB} - 10\log_{10}k. \tag{3.57}$$

This relationship between the C/I in dB and the fractional traffic load is shown in Figure 3.29.

The fractional loading of the sector k can be related to the quantity of traffic in Erlangs. An Erlang corresponds to a continuously occupied channel. Thus, when the fraction k of the sector's channels are in use (systemwide), then it can be said that the traffic load is kN_{ch} Erlangs, where N_{ch} is the number of channels per sector—18 or 19 in the AMPS system, as discussed in Section 3.2. This correspondence is approximate because it is unlikely that every sector in the system will have the same loading as its inter-fering sectors at any given time.

Thus related to Erlangs, the C/I for AMPS can also be related to a value of blocking probability (B) using some formula for blocking as a function of traffic load in Erlangs (A) and the number of channels (N). A widely used formula for blocking probability is the Erlang B formula, which gives the values for blocking as a function of load in Erlangs shown in Table 3.17 (up to 3% blocking) for the numbers of channels per sector in an AMPS cellular system. These values of blocking probability are plotted in Figure 3.30 along with the C/I, now a function of load in Erlangs. Note that the sectors with 19 channels have a significant advantage over those with 18 channels. Note also that the C/I is about 20 dB for a 2% blocking probability, a commonly accepted grade of service for cellular telephones.

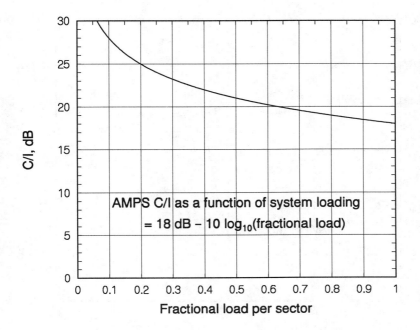

Figure 3.29 C/I as a function of loading for an FDMA cellular system.

Table 3.17 Blocking probabilities for an AMPS system

Load in Erlangs A	Blocking for $N_{ch} = 18$	Blocking for $N_{ch} = 19$
8.0	.000945	.000398
8.5	.00171	.000763
9.0	.0029	.00137
9.5	.00466	.00233
10.0	.00714	.00375
10.5	.0105	.00575
11.0	.0148	.00848
11.5	.0201	.012
12.0	.0265	.0165
12.5	.031	.0219
13.0	—	.0284
13.5	—	.0358

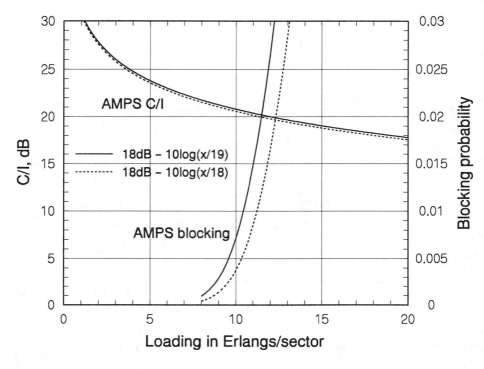

Figure 3.30 AMPS C/I and blocking probability versus load in Erlangs.

The blocking mechanism in AMPS is the unavailability of a discrete channel assignment at the moment that a call is desired. Because the callers' activities are uncoordinated, low blocking probabilities are achieved only when the average number of active users (represented by the load in Erlangs) is considerably less than the number of channels. For example, it can be seen in Figure 3.30 that a 2% blocking probability is exceeded when the sector loading is 12 users (11.5E) for the sectors with 18 FDMA channels and 13 users (12.3E) for the sectors with 19 FDMA channels.

The required C/N, and therefore the required received signal power C, can be found as

$$\frac{C}{N} = \frac{C}{(N+I)-I} = \left[\frac{N+I}{C} - \frac{I}{C}\right]^{-1} = \left[\left(\frac{C}{N+I}\right)_{req}^{-1} - \left(\frac{C}{I}\right)^{-1}\right]^{-1} \tag{3.58}$$

Substituting 18 dB for the required value of $C/(N+I)$ and $18\,\text{dB} - 10\log_{10}k$ for C/I, we find that

$$\frac{C}{N} = \frac{1}{10^{-1.8} - k \cdot 10^{-1.8}} = \frac{10^{1.8}}{1 - k} \tag{3.59a}$$

or, in dB,

$$\frac{C}{N} = 18\,\text{dB} - \overbrace{10\log_{10}(1 - k)}^{\leq 0} $$

which implies that

$$C(\text{dBm}) = C(\text{dBm})|_{\text{no interference}} - 10\log_{10}(1 - k) \tag{3.59b}$$

Thus, the received signal level in dBm must increase by the amount $-10\log_{10}(1 - k)$, where k is the fractional loading of the reverse link.

Previously we found in the case of no interference (see Table 3.16) that the required signal level at the AMPS base station receiver must be $-106\,\text{dBm}$ without a margin and $-95.1\,\text{dBm}$ including a margin that ensures a 90% link reliability with a hard-handoff extension of the radius of the cell of 5% to reduce the pingpong effect. Substituting the required signal level with no interference in (3.59b) results in

$$C(\text{dBm}) = -95.1\,\text{dBm} - 10\log_{10}(1 - M/N_{ch}) \tag{3.59c}$$

A graph of required signal level in dBm with and without a reliability margin is given in Figure 3.31, in which the blocking probability is also plotted, both as a function of the sector loading in Erlangs. We note from this figure that at a 2% blocking probability, the required signal level is about $-91\,\text{dBm}$ with a reliability margin and about $-102\,\text{dBm}$ without such a margin. Figure 3.32 shows the required signal level as a function of traffic, assuming a hard-handoff reliability margin is used, for different target values of the reliability.

3.3.2.4 System Coverage Versus Traffic Load

Having found the values of received signal strength needed by the AMPS cellular system as a function of user traffic, the signal strength is next converted into a cell radius through the use of a propagation loss model, thereby relating the cell size to the traffic load. This relation is then used to show the area density of base stations needed by the system as functions of traffic.

Along with these operations, we also bring into the methodology specific consideration of additional link budget parameters such as cable losses

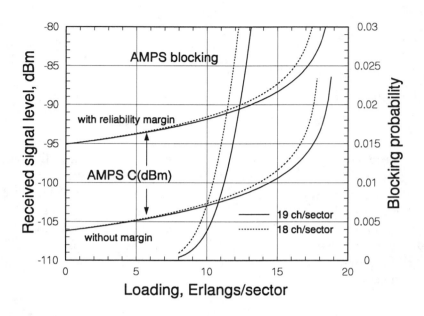

Figure 3.31 AMPS required signal level versus traffic loading.

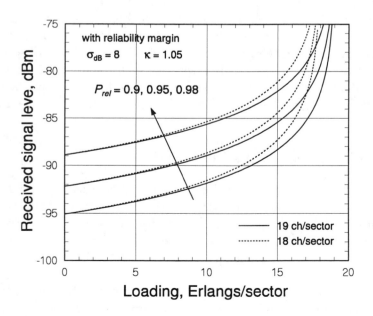

Figure 3.32 AMPS required signal level, reliability varied.

and antenna gains, as well as an allowance for the additional signal attenuation that may occur when the mobile user is inside a building.

The CCIR empirical formula for the median propagation path loss, discussed in Chapter 2, is given by

$$L(d) = 69.55 + 26.16 \log_{10} f_{MHz} - 13.82 \log_{10} h_1 - a(h_2)$$
$$+ (44.9 - 6.55 \log_{10} h_1) \log_{10} d_{km} - B \qquad (3.60)$$

where
$$d_{km} = \text{Link distance in kilometers}$$
$$f_{MHz} = \text{Center frequency in megahertz}$$
$$h_1 = \text{Base station antenna height in meters}$$
$$h_2 = \text{Mobile receiver antenna height in meters}$$
$$a(h_2) = \text{Antenna height-gain function}$$
$$= (1.1 \log_{10} f_{MHz} - 0.7) h_2 - (1.56 \log_{10} f_{MHz} - 0.8)$$
$$B = 30 - 25 \log_{10}(\%)$$

where (%) is the percentage of area covered by buildings. The cell radius will be obtained by finding the distance at which the propagation loss, when used in the link power budget, results in the received signal level being equal to its required value, as a function of the traffic load. Substituting the typical cellular parameter values $f_{MHz} = 850$, $h_1 = 30$, and $h_2 = 1.5$ into the CCIR formula results in

$$L(d) = 95.76 + 35.22 \log_{10} d_{km} + 25 \log_{10}(\%) \qquad (3.61a)$$
$$= 120.76 + 35.22 \log_{10} d_{km}, \qquad 10\% \text{ buildings} \quad (3.61b)$$
$$= 113.23 + 35.22 \log_{10} d_{km}, \qquad 5\% \text{ buildings} \quad (3.61c)$$

Table 3.18 gives a summary of certain link budget parameters that are needed for the development that follows [17].

Relating the link budget for AMPS at the edge of the cell, where the distance equals the cell radius R_c, results in

$$C(\text{dBm}) = \overbrace{P_m - L_{tm} + G_m - L_p - L_b + G_c - L_{rc}}^{29.2\,\text{dBm}} - L(R_c) \qquad (3.62a)$$
$$= -95.1\,\text{dBm} - 10 \log_{10}(1 - M/N_{ch}) \qquad (3.62b)$$

Table 3.18 Detailed link budget parameters

P_m	mobile transmitter power	28 dBm
L_{tm}	mobile cable losses	0 dBm
G_m	mobile antenna gain	2.1 dBm
L_p	mobile antenna orientation loss	3 dB
L_b	allowance for building penetration	10 dB
G_c	cell site antenna gain	14.1 dBi
L_{rc}	cell site cable losses	2 dB
$P_m - L_{tm} + G_m - L_p - L_b + G_c - L_{rc}$		29.2 dBm

which by eliminating C and rearranging leads to an expression for $L(R_c)$ as a function of traffic. This can be compared with the CCIR formula:

$$L(R_c) = 124.3\,\text{dB} + 10\log_{10}(1 - \overbrace{M/N_{ch}}^{k}) \tag{3.62c}$$

$$= 95.8\,\text{dB} + 35.22\log_{10}R_{km} + 25\log_{10}(\%) \tag{3.62d}$$

Solving these two expressions for $L(R_c)$ for the cell radius gives

$$\log_{10}R_{km} = \frac{28.5 + 10\log_{10}(1-k) - 25\log_{10}(\%)}{35.22}$$

$$= 0.809 + 0.284\log_{10}(1-k) - 0.710\log_{10}(\%) \tag{3.63a}$$

or

$$R_{km} = 10^{0.809}(1-k)^{0.284}(\%)^{-0.71}$$

$$= 6.442\,(1-k)^{0.284}(\%)^{-0.71} \tag{3.63b}$$

Figure 3.33 shows the cell radius of AMPS as a function of k for two values for the percentage of buildings, 5% and 10%.

The area coverage of a cellular system can be expressed in terms of the number of base stations required per unit area, or for a nominal area such as $1,000\,\text{km}^2$. Having found an expressions for the cell radius R_c, parametric in the fractional user load (k) and the percentage of buildings in the service area, the area of the sector of a hexagonal cell with three sectors is easily formulated as

$$a_{\text{sector}} \triangleq \frac{\sqrt{3}}{2}R_c^2 = 0.866\left[c(1-k)^{0.284}\right]^2\text{km}^2 \tag{3.64}$$

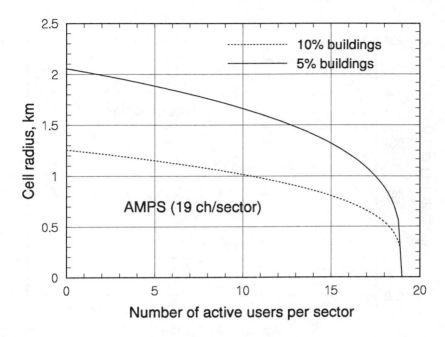

Figure 3.33 Cell radius versus number of active users per sector.

where the parameter $c = 2.055$ for 5% buildings and $c = 1.256$ for 10% buildings when $P_{rel} = 0.9$ and $\sigma_{dB} = 8$ dB.

Let a_0 be defined as a unit area expressed in square kilometers, and let the base station density be denoted y. The relation for y when $a_0 = 1,000$ km^2 is found from the ratio

$$\frac{1 \text{ base station}}{3\,a_{\text{sector}}} = \frac{y}{a_0} \quad \Rightarrow \quad y = \frac{a_0}{3\,a_{\text{sector}}} = \frac{1,000}{3\,a_{\text{sector}}} \tag{3.65}$$

where a_{sector} is defined in (3.64). In Figure 3.34, the base station density y is shown as a function of the fractional loading when $P_{rel} = 0.9$ and $\sigma_{dB} = 8$ dB. The base station density increases with the loading because the radius of the cells decreases with the loading in order to maintain the required SNR under higher interference conditions.

Another view of AMPS cellular system coverage can be made in terms of the amount of user traffic per unit area that can be carried. The number of active users (Erlangs of traffic) being carried per sector is kN_{ch}, where the values $N_{ch} = 19$ may be taken. Per unit area, the user traffic density is

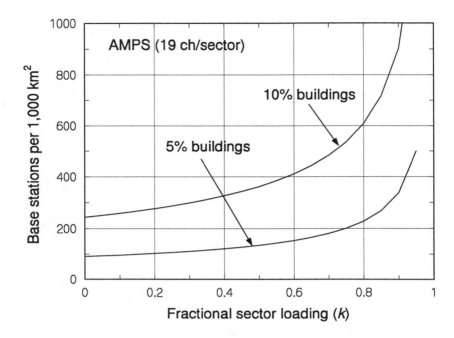

Figure 3.34 Base station density versus fractional sector loading.

$$x = \text{traffic density} = M/a_{\text{sector}} = kN_{ch}/a_{\text{sector}} \qquad (3.66)$$

As shown above in (3.64), the sector area a_{sector} itself is a function of the fractional loading, so that (3.66) gives the user traffic density x as the ratio of two functions of k.

Figure 3.35 shows the user traffic density of an AMPS cellular system when $P_{rel} = 0.9$ and $\sigma_{dB} = 8\,\text{dB}$, for 5% and 10% buildings. The figure shows that, at 50% loading of the system and 10% of the service area covered by buildings, the AMPS system can service about 10E/km^2.

As a final step in showing the tradeoff between area coverage and traffic capabilities for a cellular system, it is possible to plot the base station density $y = y(k)$ as given in (3.65) against the user traffic density $x = x(k)$ as given in (3.66). Such a plot eliminates the parametric variable k. In Figure 3.36, the base station density is plotted against the active user density when $P_{rel} = 0.8$ and $\sigma_{dB} = 8\,\text{dB}$. For example, if there are, on the average, five active users per square kilometer and 10% buildings, the AMPS system will require about 300 base stations per $1,000\text{ km}^2$.

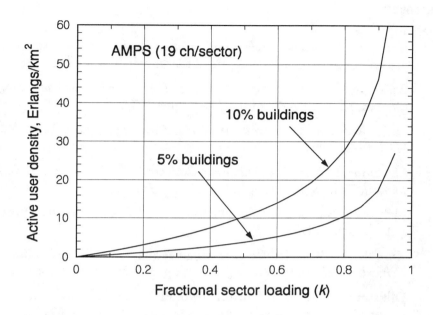

Figure 3.35 User traffic density versus fractional loading.

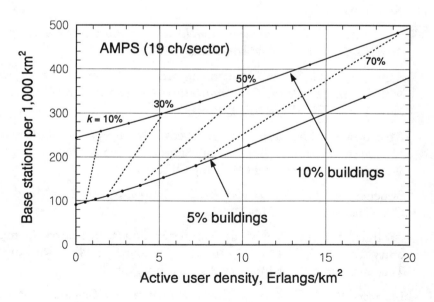

Figure 3.36 Base station density versus user traffic density.

References

[1]   Bear, D., *Principles of Telecommunication Traffic Engineering*, revised 3rd ed., London: Peter Peregrinus Ltd. on behalf of the IEE, 1988.

[2]   Beckmann, P., *Elementary Queuing Theory and Telephone Traffic*, Geneva, IL: Lee's ABC of the Telephone, 1977.

[3]   Jakes, W. C., ed., *Microwave Mobile Communications*, New York: IEEE Press, 1994.

[4]   MacDonald, V. H., "The Cellular Concept," *The Bell System Technical Journal*, Vol. 58, No. 1, pp. 15–41, Jan. 1979.

[5]   Mehrotra, A., *Cellular Radio: Analog and Digital Systems*, Boston: Artech House, 1994.

[6]   Kim, K. I., "CDMA Cellular Engineering Issues," *IEEE Trans. on Vehicular Technology*, Vol. 42, No. 3, pp. 345–350, Feb. 1995.

[7]   Dijkstra, E. W., "A Note on Two Problems in Connection with Graphs," *Numerical Mathematics*, Vol. 1, pp. 269–271, 1959.

[8]   Claiborne, J. D., *Mathematical Preliminaries for Computer Networking*, New York: Wiley, 1990.

[9]   Schwartz, M., *Telecommunication Networks: Protocols, Modeling and Analysis*, New York: Addison-Wesley, 1987.

[10]   Rappaport, T. S., R. Muhamed, and V. Kapoor, "Propagation Models," in *The Mobile Communications Handbook*, J. R. Gibson, ed., New York: CRC Press/IEEE Press, 1996.

[11]   Abramowitz, M., and I. Stegun, eds., *Handbook of Mathematical Functions*, National Bureau of Standards Math Series No. 55, Washington: Government Printing Office, 1970.

[12]   Boucher, N. J., *The Cellular Radio Handbook*, 2nd ed., Mendicino, CA: Quantum Publishing, 1992.

[13]   "Cellular System Dual-Mode Mobile Station-Base Station Compatibility Standard," EIA/TIA Interim Standard IS-54B, Washington: Telecommunications Industry Association, Apr. 1992.

[14]   Macario, R. C. V., *Cellular Radio Principles and Design*, New York: McGraw-Hill, 1993.

[15]   Lee, W. C. Y., *Mobile Cellular Telecommunications Systems*, New York: McGraw-Hill, 1989.

[16]   Walker, J., *Mobile Information Systems*, Boston: Artech House, 1990.

[17]   Wheatley, C. E., "Trading Coverage for Capacity in Cellular Systems: A System Perspective," *Microwave Journal*, pp. 62–79, July 1995.

[18]   Miller, L. E., "Propagation Model Sensitivity Study," J. S. Lee Associates, Inc., report JC-2092-1-FF under contract DAAL02-89-C-0040, July 1992. (DTIC accession number AD-B166479.)

[19]   Viterbi, A. J., A. M. Viterbi, and E. Zehavi, "Performance of Power-Controlled Wideband Terrestrial Digital Communication," *IEEE Trans. on Communications*, Vol. 41, No. 4, pp. 559–569, Apr. 1993.

[20]   Lee, J. S., "Overview of the Technical Basis of Qualcomm's CDMA Cellular Telephone System Design—A View of North American TIA/EIA IS-95" (invited paper), *Proc. IEEE 1994 Singapore Internat'l Conf. on Communicatios Systems*, pp. 353–358, Nov. 14–18, 1994.

[21]   Lee, J. S., L. E. Miller, and W. A. Jesser, Jr., "Comparative Views of Modulation Schemes for the North American TDMA (IS-54) and CDMA (IS-95) Mobile Cellular Systems," *Proc. 1994 Internat'l Conf. on Personal, Mobile Radio and Spread Spectrum Communications*, pp. 4.3.1–4, Beijing, Oct. 12–14, 1994.

[22]   Viterbi, A. J., A. M. Viterbi, K. S. Gilhousen, and E. Zehavi, "Soft Handoff Extends CDMA Cell Coverage and Increases Reverse Link Capacity," *IEEE J. on Selected Areas in Communications*, Vol. 12, No. 8, pp. 1281–1288, Oct. 1994.

Appendix 3A Demonstration That the Form of P_k Satisfies the Equations

The general equation to be satisfied is

$$(\lambda + k\mu)P_k = \lambda\, P_{k-1} + (k+1)\mu\, P_{k+1}, \quad 0 < k < N \qquad (3A.1)$$

Substituting for P_k on the left-hand side gives

$$(\lambda + k\mu) \cdot P_k = \lambda \cdot P_k + k\mu \cdot P_k$$

$$= \frac{(\lambda^{k+1}/\mu^k)/k!}{\sum\limits_{k=0}^{N}(\lambda/\mu)^k/k!} + \frac{(\lambda^k/\mu^{k-1})/(k-1)!}{\sum\limits_{k=0}^{N}(\lambda/\mu)^k/k!}$$

$$= (k+1)\mu \cdot \frac{(\lambda/\mu)^{k+1}/(k+1)!}{\sum\limits_{k=0}^{N}(\lambda/\mu)^k/k!} + \lambda \cdot \frac{(\lambda/\mu)^{k-1}/(k-1)!}{\sum\limits_{k=0}^{N}(\lambda/\mu)^k/k!}$$

$$= (k+1)\mu\, P_{k+1} + \lambda\, P_{k-1} \qquad \text{QED} \qquad (3\text{A.2})$$

Appendix 3B Moments for the Erlang B Distribution

Given the values of $\{P_k, \; k = 0, 1, \dots, N\}$, the average number of lines occupied at any given time is

$$E\{k\} = \sum_{k=0}^{N} k P_k = \frac{\sum\limits_{k=0}^{N} k \cdot \frac{A^k}{k!}}{\sum\limits_{k=0}^{N} \frac{A^k}{k!}} = \frac{A \cdot \sum\limits_{k=0}^{N-1} \frac{A^k}{k!}}{\sum\limits_{k=0}^{N} \frac{A^k}{k!}} = A(1 - B) \qquad (3\text{B.1})$$

The mean square of the number of occupied lines similarly is found to be

$$E\{k^2\} = \sum_{k=0}^{N} k^2 P_k = \frac{\sum\limits_{k=0}^{N} k^2 \cdot \frac{A^k}{k!}}{\sum\limits_{k=0}^{N} \frac{A^k}{k!}} = \frac{\sum\limits_{k=1}^{N} \frac{kA^k}{(k-1)!}}{\sum\limits_{k=0}^{N} \frac{A^k}{k!}} = \frac{A \cdot \sum\limits_{k=0}^{N-1} \frac{(k+1)A^k}{k!}}{\sum\limits_{k=0}^{N} \frac{A^k}{k!}}$$

$$= \frac{A \cdot \left[A \cdot \sum\limits_{k=0}^{N-2} \frac{A^k}{k!} + \sum\limits_{k=0}^{N-1} \frac{A^k}{k!} \right]}{\sum\limits_{k=0}^{N} \frac{A^k}{k!}} = A(1 - B - NB) + A^2(1 - B) \qquad (3\text{B.2})$$

Therefore, the variance of the number of lines being occupied is

$$\text{Var}\{k\} = E\{k^2\} - [E\{k\}]^2 = A(1 - NB - B^2) \qquad (3\text{B.3})$$

Note that, for small values of B, which holds in general for $N > A$, the mean and variance are both approximately equal to A.

Appendix 3C Summary of Blocking Formulas

The Erlang B formula for blocking probability is a convenient method for relating the number of channels (N) and the amount of offered traffic (A) for a given probability of blocking. The formula strictly applies, however, only to the case of a very large number of traffic sources—so that the rates of arrival and departure do not depend on the number of calls in progress—and to the case of "lost calls cleared." When a call is blocked (lost) because the channels are all occupied, the Erlang B formula results from assuming that the call is not tried again.

Table 3C.1 gives a summary of several blocking formulas and the assumptions on which they are based. A comparison of the formulas for a particular example ($M = 88$, $N = 30$) is shown in Figure 3C.1. The several

Table 3C.1 Summary of blocking formulas

Blocking analysis	Treatment of lost calls	Blocking formula
Formulas for large (infinite) numbers of traffic sources		
Erlang B	Calls "cleared," do not recur	$B_1 = P_N / \sum\limits_{k=0}^{N} P_k$, with $P_k = A^k/k!$
Lost calls return [3]	Calls reenter until served	Given B, effective load is $A' = A(1 - B)$
Erlang C [2]	Lost calls "held" in infinite queue, eventually served	$B = B_1 / [1 - A(1 - B_1)/N]$
Molina [1], [3]	Same as Erlang C	$B = 1 - e^{-A} \sum\limits_{k=0}^{N-1} P_k = e^{-A} \sum\limits_{k=N}^{\infty} P_k$
Formulas for finite number of traffic sources, M		
Engset [1], [2], [3]	Lost calls are cleared	$B_2(\rho) = p_N / \sum\limits_{k=0}^{N} p_k$, $p_k = \binom{M}{k} \rho^k$
	(compute B, A parametric)	and $A(\rho) \approx M\rho/[1 + \rho B_2(\rho)]$
Bernoulli [3]	Lost calls held	$B = \sum\limits_{k=N}^{M} \binom{M}{k} \left(\frac{A}{M}\right)^k \left(1 - \frac{A}{M}\right)^{M-k}$

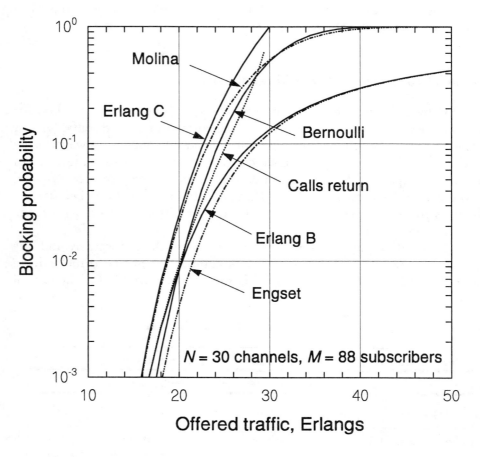

Figure 3C.1 Comparison of blocking probability formulas.

formulas are due to the fact that, in practice, telephone systems attempt to increase efficiency and customer satisfaction by "holding" or delaying calls that cannot be serviced immediately because the N lines are all in use. The calls are held in a queue for a short time on the chance that a line will become available soon. The event of "blocking" in this type of system is not that the lines are all in use when the call arrives, but that the call is delayed.

4

Overview of the IS-95 Standard

This book is designed to explain the principles underlying CDMA digital cellular systems, and this chapter, by providing an overview of the IS-95 CDMA standard [1], functions as a "gateway" or point of departure for the contents in the rest of the book. The overview is intended for the reader to gain an overall familiarity with the system.

The full title of the IS-95 standard is "Mobile Station-Base Station Compatibility Standard for Dual-Mode Wideband Spread Spectrum Cellular System," which indicates the fact that the document is a *common air interface (CAI)*—it does not specify completely how a system is to be implemented, only the characteristics and limitations to be imposed on the signaling protocols and data structures. Different manufacturers may use different methods and hardware approaches, but they all must produce the waveforms and data sequences specified in IS-95.

Just as little is said in IS-95 about the implementation of a particular requirement, the theory motivating and justifying the various requirements is not given in any detail. Because in many cases the significance of a particular requirement is not obvious, there is a need for the type of background information provided in this book. Once the principles behind the requirements are understood, the operational significance of the requirements is often enhanced, and the implementation alternatives are apparent.

The IS-95 standard refers to a "dual-mode" system, one that is capable of both analog and digital operation to ease the transition between current analog cellular systems and digital systems. Therefore, there are sections of IS-95 for both analog and digital cellular systems. Although attention is paid in Chapter 3 to analog cellular systems engineering issues, in general this book concentrates on the CDMA cellular system, and in this chapter only the sections of IS-95 dealing with the digital system are summarized.

The IS-95 system, like all cellular systems, interfaces with the PSTN through an MTSO, as suggested in Figure 4.1. In that figure, the mobile stations are shown to communicate with base stations over "forward"

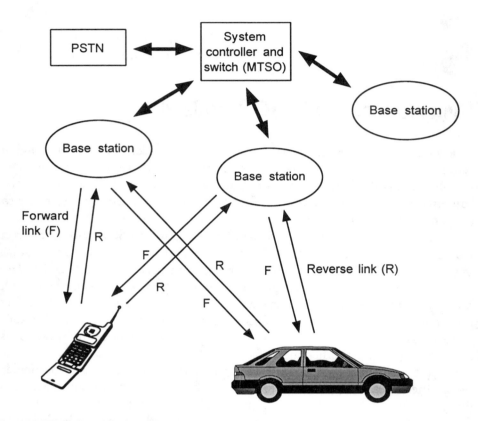

Figure 4.1 Cellular system architecture.

(base-to-mobile) and "reverse" (mobile-to-base) radio links, also sometimes called "downlink" and "uplink," respectively.

The radio communications over the forward and reverse links of the digital communications system that are specified by IS-95 are organized into "channels." Figure 4.2 illustrates the different channel types that are designated in IS-95: pilot, synchronization, paging, and traffic channels for the forward link; and access and traffic channels for the reverse link.

As will be shown, the modulation techniques and even the multiple access techniques are different on the forward and reverse links. Both links of the IS-95 system, however, depend on the conformity of all transmissions to strict frequency and timing requirements. Therefore, before proceeding to describe the forward and reverse link channels in Sections 4.2 and 4.3, respectively, we discuss the use of frequency and time in the system.

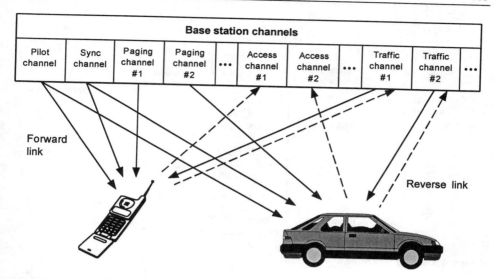

Figure 4.2 IS-95 forward and reverse link channels.

4.1 Coordination of Frequency and Time

The IS-95 standard concerns communications over the radio links between base station transceivers in one or more geographically dispersed fixed locations and subscriber transceivers in geographically dispersed mobile locations, as depicted previously in Figure 4.1. A number of base station sites are connected with and controlled by a base station controller that is associated with an MTSO, which interfaces with the PSTN through a mobile switching center (MSC). The several MTSOs in a region are in turn under the control of an operations management center (OMC). One of the critical functions of an MTSO in a CDMA system (also called a base station controller—BSC) is the maintenance of frequency and time standards. As suggested in Figure 4.3, each CDMA MTSO has a time and frequency coordination function, in addition to the normal cellular functions of the MTSO, that provides required timing and frequency signals such as a frequency reference, time of day, synchronization reference, and system clocks, based on reference signals broadcast by satellites in the Global Positioning System (GPS).

In this section, we review the aspects of the IS-95 standard that relate to the coordination of frequency and time resources and the maintenance of frequency and time standards.

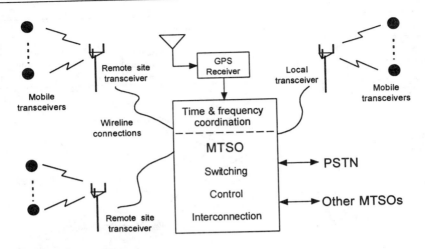

Figure 4.3 MTSO functions.

4.1.1 Cellular Frequency Bands and Channels

In North America, the first-generation analog cellular system, AMPS [2], uses FDMA to divide the allocated spectrum into 30-kHz analog FM voice channels. The frequency allocations for these voice channels are diagrammed in Figure 4.4. The "A" and "B" designations of the frequency bands refers to the Government's policy of licensing two cellular providers in a given geographical area: a carrier (A) whose business is primarily wireless systems, and a carrier (B) whose business traditionally has been wireline telephone communications. For historical, rather than technical, reasons the bands assigned to a particular provider are not contiguous—originally a contiguous band of 10 MHz was assigned to each provider; this allotment was later increased to 12.5 MHz by adding the smaller bands indicated in Figure 4.4. The dual-mode compatibility requirement dictates that the analog portion of equipment based on IS-95 be capable of operating in these bands and using these channels.

The correspondence between channel number N and center frequency is given for mobile and base station by

$$f_{\text{mobile}} = \begin{cases} 0.030\,N + 825.000\,\text{MHz}, & 1 \leq N \leq 799 \\ 0.030(N - 1023) + 825.000\,\text{MHz}, & 990 \leq N \leq 1023 \end{cases} \quad (4.1a)$$

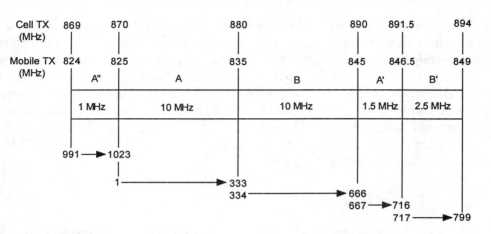

Figure 4.4 Cellular frequency allocations.

and

$$
f_{\text{base}} = \begin{cases} 0.030\, N + 870.000\,\text{MHz}, & 1 \le N \le 799 \\ 0.030(N - 1023) + 870.000\,\text{MHz}, & 990 \le N \le 1023 \end{cases} \tag{4.1b}
$$

For example, for channel 689, the forward link center frequency is $f_{\text{base}} = 890.670\,\text{MHz}$ and the reverse link center frequency is $f_{\text{mobile}} = 845.670\,\text{MHz}$.

Unlike an FDMA cellular system, a CDMA cellular system does not require the use of "clusters" of cells to enforce a minimum reuse distance between cells using the same frequency channels in order to control the amount of cochannel interference. Instead, adjacent CDMA cells reuse the identical spectrum and utilize spread-spectrum processing gain to overcome interference.

The nominal bandwidth of the IS-95 CDMA waveform is 1.25 MHz. Figure 4.4 shows that the smallest contiguous segment of bandwidth is 1.5 MHz. The IS-95 choice of chip rate was dictated in part by a desire to operate a CDMA system in the 1.5-MHz A' band. In concept, the transition from analog to digital cellular in a given service area can be based on overlaying the CDMA waveform on a subset of the analog frequencies (simultaneous operation), or by arranging that the analog system does not use those frequencies (disjoint operation).

Because of the relatively wide bandwidth of the IS-95 digital waveform, its center frequencies are restricted to those cellular channels shown within the rectangles drawn in Figure 4.5. For example, the lowest mobile center frequency is 824.700 MHz (channel 1013), which is about half the CDMA signal bandwidth away from the lower edge of the A″ band. This restriction allows for up to five simultaneous CDMA waveforms at different center frequencies (frequency assignments or FAs) for operator A, and up to six for operator B, depending upon the size of the guard band used to separate CDMA waveforms in frequency.

4.1.2   System Time

While the maintenance of a systemwide time standard is not essential to the analog cellular system, it is a critical component of the CDMA digital cellular system design. Each base station of the IS-95 system is required to maintain a clock that is synchronized to GPS time signals. The known beginning of the GPS time count (time 00:00:00 on Jan. 6, 1980) is traceable to Universal Coordinated Time (UTC) and is required to be aligned in a particular fixed way with the PN codes used by every base station in the CDMA system—two "short" PN codes having 26.66-ms periods, and a "long" PN code that has a period that is over 41 days long, as we discuss in greater detail in what follows.

The particular alignment of base station PN codes with the beginning of GPS time is discussed in Section 6.3.4.2.

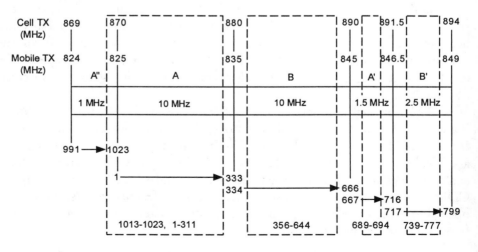

Figure 4.5 Allowed CDMA center frequency regions for A and B operators.

The synchronization of time standards for the several CDMA base stations in a cellular service area is necessary because each base station transmits on the same center frequency and uses the same two short PN codes to spread its (forward link) waveform, the different base station signals being distinguished at a mobile receiver only by their unique short PN code starting positions (phase offsets), as illustrated in Figure 4.6.

Although the base stations are synchronized to each other because each is synchronized to GPS, in general each mobile unit receives the base station signals with a different combination of propagation delays, because the mobile is usually at a different distance from each base station. (Traveling at the speed of light, signals are delayed by 3.336 ns/m or 1.017 ns/ft.) The system time reference for a particular mobile is established when it acquires a certain base station signal, usually that from the nearest base station, affiliates with that base station, and reads the sychronization message broadcast by that base station. The message contains information that enables the mobile unit to synchronize its long PN code and time reference with those of that particular base station—but slightly delayed, due to the distance of the mobile from the selected base station. Using this delayed version of system time, the mobile transmits on the reverse link, and its signal is received with additional delay at its affiliated base station.

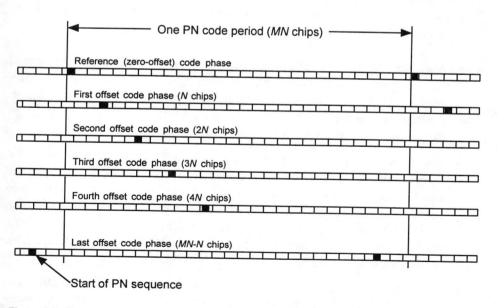

Figure 4.6 Short PN code offsets are assigned to different base stations.

4.2 Description of Forward Link Operations

The forward link channel structure consists of the transmission of up to 64 simultaneous, distinct channels with varying functions that are orthogonally multiplexed onto the same RF carrier. One of these channels is a high-power pilot signal that is transmitted continuously as a coherent phase reference for reception of carriers modulated by information. Another of these channels is a continuously transmitted synchronization channel that is used to convey system information to all users in the cell. Up to seven paging channels are used to signal incoming calls to mobiles in the cell and to convey channel assignments and other signaling messages to individual mobiles. The remainder of the channels are designated as traffic channels, each transmitting voice and data to an individual mobile user.

4.2.1 Forward Link CAI Summary

The IS-95 document [1] specifies a CAI for the CDMA cellular system that different manufacturers of equipment for the system can use to ensure that their design implementations (possibly different in some respects) work together. Before going into details of the various aspects of the IS-95 forward link design, we first provide a summary of the main features of the forward link CAI.

- *Multiplexing:* the forward link channelization is based on an orthogonal code-division multiplexing scheme using an orthogonal set of "sub-carrier" digital waveforms known as Walsh functions. In this primary, multiple-access sense, the "C" in CDMA on the forward link refers to Walsh function multiplexing.

- *Interference rejection:* the forward link waveform is modulated by direct-sequence PN code spread-spectrum techniques to isolate the signals received from a particular base station and to discriminate against signals received from other base stations. In the sense that the mobile receiver uses PN code phase to distinguish base station signals (but not individual channels), the "C" in CDMA on the forward link refers to the PN code phase as well as the Walsh functions.

- *Modulation:* the forward link waveform features modulation of I (cosine) and Q (sine) RF carriers by different PN-coded bipolar ($\pm$) baseband digital data streams, thereby generating a form of quaternary phase-shift keying (QPSK).

- *Pulse shaping:* the shape of the baseband digital pulses in the I and Q output channels is determined by a FIR filter that is designed to control the spectrum of the radiated power for minimal adjacent-frequency interference.

- *PN chip rate:* the PN code chip rate, which is 1.2288 Mcps, is 128 times the maximal source data rate of 9.6 kbps.

- *Effective bandwidth:* for the PN chip rate and spectrum control specified, the energy of the IS-95 forward link signal is almost entirely contained in a 1.25-MHz bandwidth.

- *Voice coding:* a variable-rate vocoder is specified, with data rates 1,200, 2,400, 4,800, and 9,600 bps depending on voice activity.

- *Error-control coding:* the forward link uses rate $\frac{1}{2}$ convolutional coding, with Viterbi decoding.

- *Interleaving:* to protect against burst error patterns (a distinct possibility on the fading mobile communications channel), the forward link interleaves code symbols before transmission, using a 20-ms span.

4.2.2 Orthogonal Multiplexing Scheme

On the forward link, each channel is distinguished by a distinct orthogonal Walsh sequence modulated by the encoded data, much like an individual telemetry subcarrier that is modulated by one of several data sources. Walsh sequences are the rows of Hadamard matrices whose dimensions are powers of 2 and are orthogonal when correlated over their period. Data on the paging and traffic channels are scrambled using an assigned phase offset of a long PN code that provides a degree of privacy, but is not used to distinguish the channels. The channel assignments are shown in Figure 4.7 with their corresponding Walsh sequence "covers," H_i, where $i = 0, 1, \ldots, 63$. Hadamard and Walsh sequences are discussed in detail in Chapter 5.

The baseband data rate from each of the channels being multiplexed varies, as discussed below, with the highest rate being 19.2 kilosymbols per second (ksps). Each channel's baseband data stream is combined with an assigned 64-chip Walsh sequence that repeats at the 19.2-ksps rate. Thus, the orthogonally multiplexed combination of forward link channels forms a baseband data stream with a rate of 64×19.2 ksps $= 1.2288$ Mcps.

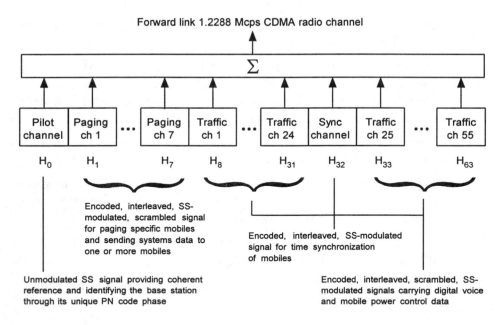

Figure 4.7 Forward link channel assignments.

As indicated in Figure 4.8, the multiplexed data stream for a particular channel is combined separately with two different short PN codes that are identified with I- and Q-quadrature carrier components.   The I- and Q-channel PN codes are denoted by $PN_I(t, \theta_i)$ and $PN_Q(t, \theta_i)$, respectively, which are generated by fifteen ($n = 15$) stage linear feedback shift registers (LFSR).   The notation θ_i denotes the *PN code offset* phase assigned to a particular base station, which is selected from 512 possible values. We discuss the details of how 512 values come about shortly. Thus, unlike conventional QPSK, which assigns alternate baseband symbols to the I- and Q-quadratures, the IS-95 system assigns the same data to each quadrature channel. There is a good reason for the IS-95 system to adopt this particular scheme—an analytical explanation of the rationale is given in Section 7.3. It is common to speak of the operations depicted in Figure 4.8 as "quadrature spreading" [3]. The two quadrature digital baseband waveforms are *shaped* using FIR filters, as shown Figure 4.8, in order to control the shape of the emitted spectrum, which we discussed in Section 1.4.2. The shaped I- and Q-channel signals are modulated by in-phase carrier ($\cos 2\pi f_c t$) and quadrature-phase carrier (sin $2\pi f_c t$) and are then combined and transmitted.

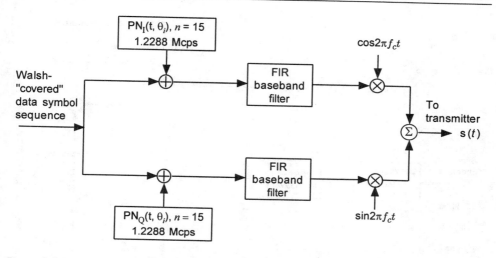

Figure 4.8 Quadrature spreading for the forward link.

4.2.3 Forward Link Channels

The orthogonal multiplexing operations on the forward link are shown in Figure 4.9. The forward link channels consist of pilot channel, sync channel, paging channels, and traffic channels as indicated in the figure. Each channel is modulated by a channel-specific Walsh sequence, denoted H_i, $i = 0, 1, \ldots,$ 63. The IS-95 standard assigns H_0 for the pilot channel, H_{32} for the synchronization channel, H_1 to H_7 for the paging channels, and the remainder of the H_i to the traffic channels. Note that we use the notation H_i to designate Walsh function i, which is also denoted as W_i. In this book, we use H_i and W_i synonymously. The reasons behind this notation are fully explained in Chapter 5. Note that the I-channel PN sequence and the Q-channel PN sequence simultaneously quadrature-modulate each of the forward link channels. The quadrature spreading circuit for each channel is as shown in Figure 4.8.

For modularity and the ability to give each channel a different gain and RF power, each channel is separately filtered for RF modulation (Chapter 11 covers this aspect in detail). Each channel is spread-sprectrum modulated by quadrature PN codes. We now discuss the channel types.

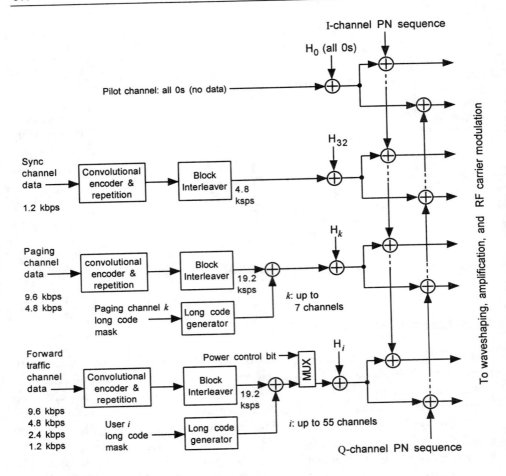

Figure 4.9 Forward link multiplexing operations.

4.2.3.1 Pilot Channel and Quadrature PN Codes

The *pilot channel* is used primarily as a coherent phase reference for demodulating the other channels; for this reason, IS-95 requires that the chip timing and carrier phase of each forward link channel be in very close agreement. The pilot channel is easily acquired by the mobile receiver because it has no data modulation, only the quadrature PN codes shown above in Figure 4.8 and Figure 4.9. Because the pilot channel is necessary for crucial timing information, it is transmitted at a higher power level than the other channels.

The engineering principles involved in choosing the relative powers of the forward link channels for optimal CDMA systems operation are fully discussed in Chapter 11.

The pilot baseband "data" (which are a constant logical 0) are modulated at a Walsh chip rate of 1.2288 Mcps by H_0, the 0th row of the 64×64 Hadamard matrix, which is the Walsh sequence consisting of 64 zeros—thus, in effect, it is not modulated at all. The multiplexed baseband data input from the pilot channel to the forward link quadrature spreading modulator of Figure 4.8 is a constant logical 0.[1]

The two distinct short PN codes in Figures 4.8 and 4.9 are maximal-length sequences generated by 15-stage shift registers and lengthened by the insertion of one chip per period in a specific location in the PN sequence. Thus, these modified short PN codes have periods equal to the normal sequence length of $2^{15} - 1 = 32,767$ plus one chip, or $32,768$ chips. As indicated previously in Figures 4.6 and 4.8, each base station is distinguished by a different phase offset θ_i of the in-phase (I) and quadrature-phase (Q) PN sequences. Each offset is a multiple of 64 PN chips, which yields $32,768/64 = 512$ possible 64-chip offsets.

At a rate of 1.2288 Mcps, the I- and Q-sequences repeat every 26.66 ms, or 75 times every 2 sec. The characteristic polynomials of the in-phase and quadrature sequence are given by[2]

$$f_I(x) = 1 + x^2 + x^6 + x^7 + x^8 + x^{10} + x^{15} \qquad (4.2)$$

$$f_Q(x) = 1 + x^3 + x^4 + x^5 + x^9 + x^{10} + x^{11} + x^{12} + x^{15} \qquad (4.3)$$

which can be generated using the modular shift register generator (MSRG) shown in Figures 4.10 and 4.11. The figures show how a "mask" vector can be used to select a shift of the sequence that is different for each base station.

---

[1] Note that a logical 0 is equivalent to a positive voltage in a ± 1 binary system, and a logical 1 is equivalent to negative voltage, so that the modulo-2 addition operation in logic (also known as exclusive OR) is equivalent to multiplication. Therefore the modulo-2 combination of baseband data and a Walsh function in $(0, 1)$ logic can be implemented by a product of the two in $(+1, -1)$ logic.

[2] The characteristic polynomials $P(x)$ given in IS-95 are the *reciprocal* polynomials for the *sequences* specified. In our treatment of this subject, we consistently refer to the characteristic polynomial $f(x) = x^n P(x^{-1})$ as that whose inverse $1/f(x)$ generates the sequence algebraically. See Chapter 6.

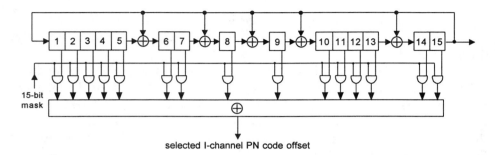

Figure 4.10 In-phase PN generator and offset-selecting mask.

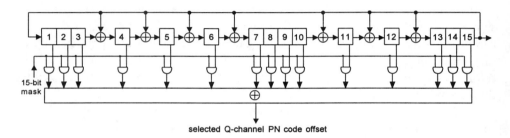

Figure 4.11 Quadrature-phase PN generator and mask.

The theory and practice of generating PN codes is discussed at length in Chapter 6, including the use of masks to control PN code phase and the significance of the insertion of an extra zero in a PN sequence.

In the absence of baseband filtering, the quadrature PN modulated data streams are binary-valued. If the logic values are mapped from (0, 1) to (+1, −1), the PN modulation produces the two-dimensional constellation with the transitions shown in Figure 4.12. Ideally, because the unfiltered data modulation has only two values (±1), the resulting RF waveform has a constant envelope, and the transitions in Figure 4.12 would be constrained by that fact. In practice, however, the binary data streams are filtered to produce a baseband waveshape whose spectrum is more concentrated than the ideal baseband data, as discussed in Chapter 1, resulting in an RF signal envelope that is not constant. We give detailed explanations on QPSK and offset QPSK (OQPSK) modulations in Section 7.2.1.

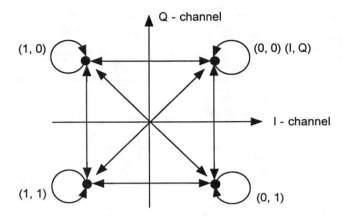

Figure 4.12 Signal constellation for the forward link quadrature modulation.

4.2.3.2 Synchronization Channel

The *synchronization channel* is demodulated by all mobiles and contains important system information conveyed by the *sync channel message*, which is broadcast repeatedly. This signal identifies the particular transmitting base station and conveys long PN code synchronization information, at a rate of 1.2 kbps, or 32 sync channel data bits per $80/3 = 26.66$-ms sync channel frame and 96 bits per 80-ms sync channel "superframe." Note that the sync channel frame length is equal to one period of the short PN codes, and the sync channel frames are in fact aligned with those periods so that, once the mobile has acquired a particular base station's pilot signal, the mobile automatically knows the basic timing structure of the sync channel. As shown in Figure 4.13, this information is provided error-control coding protection by being convolutionally encoded by a rate 1/2 encoder; this operation results in a coded binary symbol rate of $2 \times 1.2\,\text{kbps} = 2.4\,\text{kilosymbols/sec}$ (ksps), or 64 code symbols per frame; the state of the convolutional encoder is not reset after each frame, so that no encoder tail is used. The theory and practice of convolutional coding are reviewed in Chapter 8, where special attention is paid to the particular convolutional codes used in IS-95.

For further protection against possible bursts of errors, such as can occur during signal fading on the mobile channel, a combination of time diversity and interleaving is used on the sync channel. By rearranging the

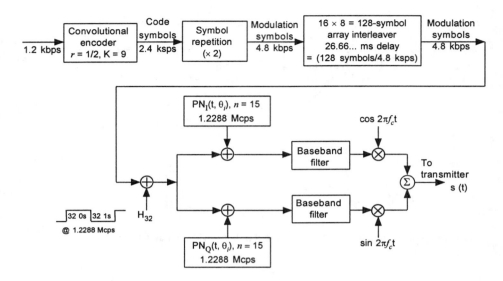

Figure 4.13 Sync channel modulation.

order of the binary symbols before transmission (interleaving them) and then restoring the correct order at the receiver (deinterleaving them), patterns of consecutive, dependent bit errors can be broken into isolated, independent errors for more reliable error detection and correction. On the sync channel, each symbol is repeated once, giving 128 coded symbols per sync channel frame. The block of 128 symbols per frame then is interleaved using a special technique designed to provide as much separation as possible between adjacent bits received in error when they are deinterleaved at the receiver. The interleaving process introduces a delay of one frame duration, or 26.66 ms. Principles of the IS-95 interleaving are explained in Section 4.4.

The interleaved sync channel data are then modulated by H_{32}, the 32nd row of the Hadamard matrix, which is the Walsh sequence consisting of 32 zeros followed by 32 ones—a "square wave"—with the chip rate of 1.2288 Mcps. The use of this particularly simple Walsh sequence facilitates acquisition of the sync channel by the mobile. The period of this square wave (64 chips) is $64/1.2288 = 52.083\ \mu s$, which is one-fourth the period of one data symbol; the combining of the sync data and this Walsh sequence can be viewed as the data's modulating the polarity of a square wave, four periods at a time.

The Walsh-multiplexed sync channel data are spread in bandwidth by the quadrature PN code modulator, as shown above in Figure 4.13. The effective spreading factor (processing gain) for the coding, diversity, multiplexing, and PN modulation is the ratio 1.2288 Mcps/1,200 bps = 1,024. A summary of sync channel modulation parameters is given in Table 4.1.

The sync channel message provides the information that is necessary for the mobile to obtain synchronization to the system timing. After this synchronization has been accomplished, the mobile can receive the paging channel and transmit on the access channel. Because all channels except the pilot and sync channels are scrambled using the long PN code, the sync message is necessary to correctly demodulate any other channel. As mentioned previously, the sync channel is divided into 26.66-ms frames, conveniently made identical to the period of the short PN codes, of which three form an 80-ms superframe. Every sync channel frame and hence sync channel superframe begins at the same instant as the short PN sequence. Thus, when the pilot channel is obtained, the beginning of the PN sequence is known as is the beginning of the sync channel frame. The sync channel message begins at the start of a sync channel superframe. The system time and long code state given in the message are valid at the start of the fourth superframe after the last superframe containing the message. The message is padded to be contained in an integer number of superframes. The interleaver, symbol repetition, and convolutional encoding used by the sync channel are also synchronized to the beginning of the frame. The valid system time is referenced to the zero-offset superframe, as shown in Figure 4.14. In other words, the offset of the base station is subtracted from the received frame time to yield the zero-offset time. The base station's offset is explicitly transmitted as part of the sync channel message to identify the base station.

Table 4.1 Sync channel modulation parameters

Parameter	Value	Units
Data rate	1,200	bps
PN chip rate	1.2288	Mcps
Code rate	1/2	bits/code symbol
Code repetition	2	mod symbols/code symbols
Modulation symbol rate	4,800	sps
PN chips/mod symbol	256	
PN chips/bit	1,024	

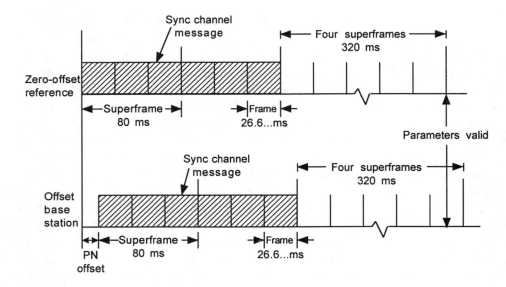

Figure 4.14 Time at which sync channel parameters become valid.

After receiving the sync channel message, the initial loading of the long code register is known. These data are loaded by the mobile into its long-code shift register, which is started at the precise moment the parameters become valid. With the long-code generator running, it can be accessed when needed. It is also useful to note that the beginning of the paging and traffic channel frames coincide with the start of a sync channel superframe.

4.2.3.3 Paging Channels

The *paging channels* are used to alert the mobile to incoming calls, to convey channel assignments, and to transmit system overhead information. Paging information is generated at either 9.6 kbps or 4.8 kbps, as shown in Figure 4.15. The information is convolutionally encoded, repeated, and interleaved. Note that the repetition of the code symbols is adapted to the data rate to fix the rate of symbols being interleaved at 19.2 ksps. Other than for the sync channel, the data frames for the IS-95 channels are 20 ms in length, and the interleaving is performed on a frame-by-frame basis. The 19.2-ksps rate requires the interleaver to process 384 coded symbols. It does so by forming the incoming data into 16 columns of 24 consecutive symbols, then reading them out in such a way as to maximize the distance between symbols that

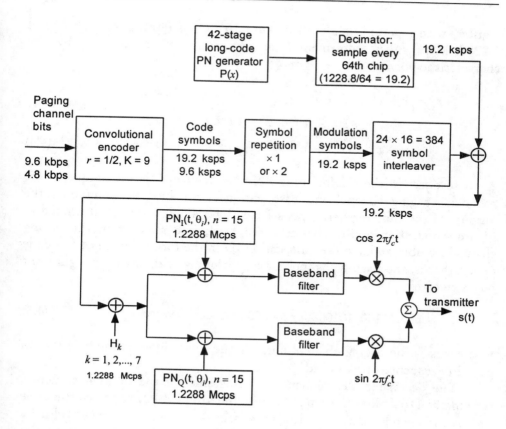

Figure 4.15 Paging channel modulation.

formerly were adjacent, as explained in detail in Section 4.4. All the paging channel modulation symbols are transmitted at the same power and baseband data rate for a given CDMA system, in contrast to the frame-by-frame variation of voice data rate and reduced power for lower rate symbols in the forward traffic channel, as is discussed below.

As with the sync channel, the convolutional encoder for a paging channel is not reset after each frame; thus, for both types of channel, symbols at the end of one interleaver block affect the symbols at the beginning of the next block. Unlike the sync channel, the encoded and interleaved paging channel symbols are scrambled with a 42-stage long-code PN sequence running at 1.2288 Mcps that is decimated to a 19.2-ksps rate by sampling every 64th PN code chip. The long PN code is generated by a 42-stage shift register, with a period of $2^{42} - 1 \approx 4.4 \times 10^{12}$ chips (lasting over 41 days at 1.2288 Mcps), that implements the recursion that is specified by the characteristic polynomial

$$
\begin{aligned}
f(x) = 1 &+ x^7 + x^9 + x^{11} + x^{15} + x^{16} + x^{17} + x^{20} + x^{21} + x^{23} \\
&+ x^{24} + x^{25} + x^{26} + x^{32} + x^{35} + x^{36} + x^{37} + x^{39} \\
&+ x^{40} + x^{41} + x^{42}
\end{aligned}
\tag{4.4}
$$

A phase offset of the original long PN code sequence that is unique to the particular paging channel and base station is obtained by combining the outputs of the shift register stages selected by a 42-bit mask, as illustrated in Figure 4.16. For paging channels, the mask is derived from the paging channel number and and the base station ID in the form of its pilot PN code offset; that is, the paging channel mask (beginning with the most significant bit) is given by

$$
11000110011010000\text{xxx}000000000000\text{yyyyyyyyy} \tag{4.5}
$$

where xxx is the paging channel number and yyyyyyyy is the base station pilot PN sequence offset index.

The decimated PN sequence is some shift of the original sequence, running at 1/64 times the rate. Because the sampled PN code has the same rate as the encoded data, the data waveform is not spread in bandwidth by combining with this particular PN code. The factors affecting the selection of the long-code masks used on the forward and reverse links are discussed in Chapter 6, as is the effect of decimating a PN sequence.

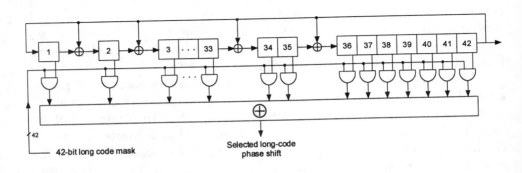

Figure 4.16 42-Stage long-code PN generator with user-distinct mask.

Each paging channel symbol data stream is spread by its designated Walsh sequence, corresponding to one of the first through seventh rows of the Hadamard matrix H_1–H_7, then spread in quadrature by the short PN codes. Paging channel modulation parameters are summarized in Table 4.2.

A paging message directed to a particular mobile unit could be broadcast periodically on one of the paging channels, requiring mobile n to scan the active paging channels for a possible message addressed to mobile n. Or, to relieve the mobile of the burden of scanning the different paging channels, the paging message could be broadcast periodically on all paging channels; however, this procedure is wasteful of paging channel capacity. In IS-95, each active paging channel has a number of periodically recurring message slots (e.g., 2,048 slots) available for transmitting pages and other base-to-mobile messages. When a message is queued up for a particular mobile, using a *hash function*, the base station pseudorandomly selects one of the paging channels and pseudorandomly selects one of the message slots in that paging channel for transmission to the particular mobile. The mobile knows exactly which paging channel and message slot to monitor for possible messages because the pseudorandom selection is based on its own identification number and known system parameters. The purpose of the hash function is to distribute the message traffic evenly among the paging channels and message slots. The IS-95 hash functions are discussed in Appendix 4B.

4.2.3.4 Traffic Channels

There are nominally up to 55 forward *traffic channels* that carry the digital voice or data to the mobile user. It is possible to use one or more of the seven Walsh sequences normally assigned to paging channels for traffic channels, for a total of up to 62 traffic channels, if the system becomes heavily loaded.

Table 4.2 Paging channel modulation parameters

Parameter	Value		Units
Data rate	9600	4800	bps
PN chip rate	1.2288	1.2288	Mcps
Code rate	1/2	1/2	bits/code symbol
Code repetition	1	2	mod symbols/code symbols
Modulation symbol rate	19,200	19,200	symbols/sec (sps)
PN chips/mod symbol	64	64	
PN chips/bit	128	256	

It is usually the case, however, that mutual interference on the reverse link limits the number of simultaneous calls to fewer than 55 calls, so the forward link's capacity is more than adequate for the traffic that can be supported by the system. Walsh sequences designated for use on the traffic channels are H_8–H_{31} and H_{33}–H_{63}. A block diagram for the forward traffic channel modulation is given in Figure 4.17. As shown in the diagram, voice data for the mth user is encoded on a frame-by-frame basis using a variable-rate voice coder, which generates data at 8.6, 4.0, 2.0, or 0.8 kbps depending on voice activity, corresponding respectively to 172, 80, 40, or 16 bits per 20-ms frame. A cyclic redundancy check (CRC) error-detecting code calculation is made at the two highest rates, adding 12 bits per frame for the highest rate and 8 bits per frame at the second highest rate. At the mobile receiver, which voice

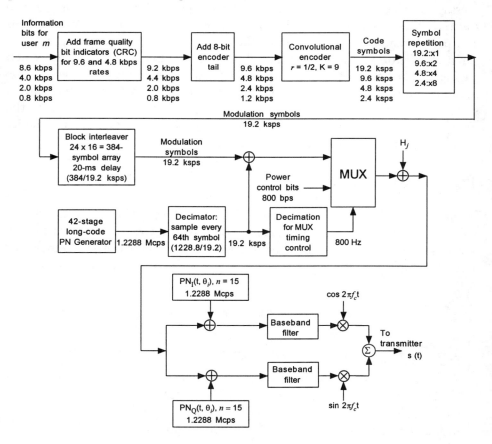

Figure 4.17 Traffic channel modulation.

data rate is being received is determined in part from performing similar CRC calculations, which also provide frame error reception statistics for forward link power control purposes. The theory and use of CRC codes are discussed in Chapter 5.

In anticipation of convolutional coding on a block basis (code symbols in one frame not affecting those in adjacent frames), a convolutional encoder "tail" of 8 bits is added to each block of data to yield blocks of 192, 96, 48, or 24 bits per frame, corresponding to the data rates of 9.6, 4.8, 2.4, and 1.2 kbps going into the encoder, which are defined as full, one-half, one-quarter, and one-eighth rates, respectively. Convolutional encoding is performed using a rate 1/2, constraint length 9 code, resulting in coded symbol rates of 19.2, 9.6, 4.8, and 2.4 ksps.

Coded symbols are repeated as necessary to give a constant number of coded symbols per frame, giving a constant symbol data rate of 19.2 ksps (i.e., 19.2 ksps × 1, 9.6 ksps × 2, 4.8 ksps × 4, 2.4 ksps × 8). The 19.2 ksps × 20 ms = 384 symbols within the same 20-ms frame are interleaved to combat burst errors due to fading, using the same interleaving scheme as on the paging channels.

Each traffic channel's encoded voice or data symbols are scrambled to provide voice privacy by a different phase offset of the long PN code, decimated to yield a code rate of 19.2 kcps. Note that different mobile users are distinguished on the forward link by the orthogonal Walsh sequence associated with the particular traffic channel, not by the user-specific long-code phase offset.

The scrambled data are punctured (overwritten) at an average rate of 800 bps by symbols that are used to control the power of the mobile station; the details of this "power control subchannel" are discussed in Section 4.4.

One of 64 possible Walsh-Hadamard periodic sequences is modulo-2 added to the data stream at 1.2288 Mcps, thus increasing the rate by a factor of 64 chips/modulation symbol (scrambled code symbol). Each symbol for a given traffic channel is represented by the same assigned 64-chip Walsh sequence for a data symbol value of 0 and the sequence's complement for a data symbol value of 1. Walsh-Hadamard sequences of order 64 have the property that all 64 of the sequences are mutually orthogonal. A unique sequence is assigned to each traffic channel so that upon reception at their respective mobile stations, the traffic channels can be distinguished (demultiplexed) based on the orthogonality of the assigned sequences. This orthogonally spread data stream is passed to the quadrature modulator for PN spreading and RF transmission.

Note from Figure 4.17 that, regardless of the data rate, the modulated channel symbol rate must be 19.2 ksps. This is accomplished by means of code symbol repetition for rates less than the 9.6-ksps data rate. The relationship between code symbol energy and the information bit rate is specified in Table 4.3 for different data rates.

4.3 Description of Reverse Link Operations

The IS-95 reverse link channel structure consists of two types of channels: access and traffic channels. To reduce interference and save mobile power, a pilot channel is not transmitted on the reverse link. A mobile transmits on either an access or a traffic channel but never both at the same time. Thus, as far as the mobile is concerned, the reverse link "channel" is an operating mode.

4.3.1 Reverse Link CAI Summary

Before going into details of the various aspects of the IS-95 reverse link design, we first provide a summary of the main features of the reverse link CAI.

- *Multiple access:* the reverse link channelization is based on a conventional spread-spectrum PN code-division multiple-access scheme in which different mobile users are distinguished by distinct phase offsets of the 42-stage long PN code, which serve as user addresses. Thus, the "C" in CDMA on the reverse link refers to spread-spectrum multiple access by means of PN codes.

Table 4.3 Forward traffic channel modulation parameters

Parameter	Value				Units
Parameter	9,600	4,800	2,400	1,200	bps
PN chip rate	1.2288	1.2288	1.2288	1.2288	Mcps
Code rate	1/2	1/2	1/2	1/2	bits/code symbol
Code repetition	1	2	4	8	mod sym/code sym
Mod symbol rate	19,200	19,200	19,200	19,200	sps
Code sym energy	$E_b/2$	$E_b/4$	$E_b/8$	$E_b/16$	
PN chips/symbol	64	64	64	64	
PN chips/bit	128	256	512	1,024	

- *Quadrature spreading:* in addition to the long PN code, the reverse link data stream is direct-sequence modulated in quadrature by the same two short PN codes as on the forward link; each mobile station in each cell uses the reference or zero-offset phases of these two codes.

- *Modulation:* the reverse link waveform features 64-ary orthogonal modulation using sequences of 64 chips to represent six binary data symbols. The quadrature modulation of I (cosine) and Q (sine) RF carriers by the two different PN-coded bipolar ($\pm$) baseband digital data streams, with the Q-quadrature stream delayed by half a PN chip, generates a form of offset quaternary phase-shift keying (OQPSK).

- *Pulse shaping:* the shape of the baseband digital pulses in the I and Q output channels is determined by a FIR filter that is designed to control the spectrum of the radiated power for minimal adjacent-frequency interference.

- *PN chip rate:* the PN code chip rate, which is 1.2288 Mcps, is 128 times the maximal source data rate of 9.6 kbps.

- *Acquisition:* the base station's acquisition and tracking of mobile signals is aided by the mobile's transmission of a preamble containing no data.

- *Voice coding:* a variable-rate vocoder is specified, with data rates 1,200, 2,400, 4,800, and 9,600 bps depending on voice activity in a particular 20-ms frame. The transmission duty cycle of the reverse link signal during a call is proportional to the data rate.

- *Error-control coding:* the reverse link uses rate 1/3 convolutional coding, with Viterbi decoding.

- *Interleaving:* to protect against possible burst-error patterns, the reverse link interleaves code symbols before transmission, using a 20-ms span.

4.3.2 Multiple Access Scheme

There is at least one reverse link access channel for every paging channel on the forward link, with a maximum of 32 access channels per paging channel. The access channels are used for the mobile to initiate a call or respond to a page or information request from the base station.

The number of traffic channels on the reverse link, transmitting voice and data to the base station, is equal to the number of traffic channels on the forward link. When no paging channels are used on the forward link, the maximum number of traffic channels is 62. In practice, the number of channels is limited by interference from users on the reverse link.

Each reverse link channel is distinguished by a user-distinct phase offset of the same 42-stage long-code PN sequence used on the forward link. Therefore, the "C" in CDMA on the reverse link is the user-distinct PN code, while on the forward link, it is the set of mutually orthogonal Walsh sequences. The channels as received at the base station are shown in Figure 4.18, where n (the number of paging channels) and m (the number of traffic channels) are limited by interference.

The reverse link transmitter consists of a convolutional encoder and modulator, and a quadrature modulator and RF circuitry, shown conceptually in Figure 4.19. Each of these components is used during mobile transmissions on both reverse link channels. Note that the quadrature modulator for the reverse link is different from that used on the forward link in that a half-chip delay is inserted in the quadrature-phase (Q) channel to achieve a form of OQPSK modulation. The effect of this delay is to constrain the I and Q chip data transitions to one quadrature at a time, rather than allowing them to be simultaneous as on the forward link. The one-half chip offset

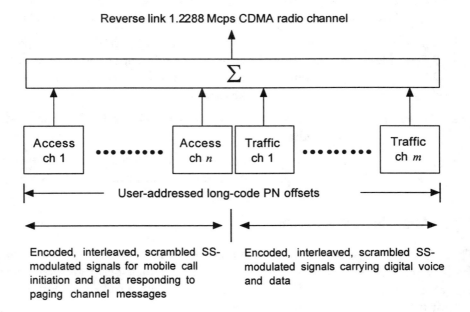

Figure 4.18 Reverse link channel assignments at the base station.

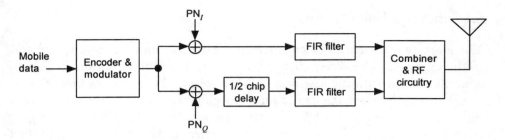

Figure 4.19 Block diagram of the reverse link transmitter.

eliminates phase transitions through the origin to provide a modulation scheme that gives a relatively constant envelope. The binary I and Q signals are mapped into phase producing the signal constellation shown in Figure 4.20. The transitions shown in the figure do not reflect the influence of the waveshaping filters that constrain the phase trajectory through their effect on the quadrature waveforms. Note, however, that there is no transition through the origin. We have more to say about OQPSK signaling in Section 7.2.

The transmission of the same data by means of a two-quadrature modulation scheme is a form of diversity. As explained in Section 7.3, its analytical justification shows that the QPSK CDMA system has a 3-dB advantage over BPSK CDMA system in terms of intersymbol interference performance and also has cochannel interference advantages.

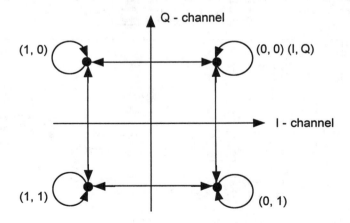

Figure 4.20 Phase transitions for OQPSK.

4.3.3 Reverse Link Channels

There are two categories of channels in the reverse link: access channel and traffic channel. The access channel is used by the mobile station to initiate communication with the base station and to respond to messages that are sent to the mobile by the base station on the paging channels. As for the forward link, the reverse link also has variable data rates for traffic channels, ranging from 1.2 to 9.6 kbps. The transmission of different data rates is accomplished by the *data burst randomizer*, as is discussed in detail below.

4.3.3.1 Access Channel

Access channel data are generated at a rate of 88 bits per 20-ms frame, to which eight encoder tail bits are appended because the encoder state is reset after each frame. The rate into the encoder then is $(88 + 8)/.02 = 4{,}800\,\text{bps}$, as illustrated in Figure 4.21. These data are encoded with a rate 1/3, constraint length 9 convolutional encoder. The rate from the encoder is $3 \times 4.8 = 14.4\,\text{ksps}$.

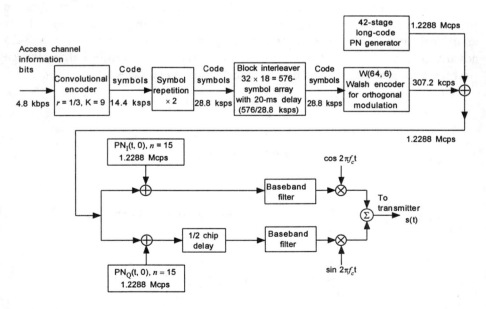

Figure 4.21 Access channel modulation.

To accommodate the same block-interleaving scheme on the access and reverse traffic channels, the access channel code symbols are repeated, making the rate into the interleaver $2 \times 14.4 = 3 \times 9.6 = 28.8$ ksps, the same as the rate into the interleaver for the highest reverse traffic channel data rate, 9,600 bps. The interleaving scheme, described more fully in Section 4.4, reads a frame's worth of data ($28.8 \times 20 = 576$ code symbols) by column into an array of 18 columns of 32 consecutive symbols, which are read out by rows in a certain order to separate the data symbols in time.

The interleaved code symbols are passed to a (64, 6) Walsh encoder. The Walsh encoding consists of using each group of six encoded symbols (c_0, c_1, ..., c_5) to select one of $2^6 = 64$ Walsh sequences H_i of order 64. The selection is performed by computing the index i according to the rule

$$i = c_0 + 2c_1 + 4c_2 + 8c_3 + 16c_4 + 32c_5 \tag{4.6}$$

where i is the row of the 64×64 Hadamard matrix, and the $\{c_j\}$ are encoded binary (0, 1) symbols. The symbol rate is therefore increased by 64/6, from 28,800 sps to 307, 200 cps, in units of "Walsh chips" per second. This step can be thought of as the encoding sequence of an ($n = 64$, $k = 6$) error-correcting code. It also is readily interpreted as a form of 64-ary orthogonal modulation using binary channel symbols. An analysis of the reverse link's 64-ary orthogonal modulation scheme is presented in Section 7.2.2.5.

The access channel signal is further spread by a factor of four using a particular phase offset of the 42-stage long PN code clocked at the rate of 1.2288 Mcps. A phase offset of the original sequence is obtained by forming the inner product of the vector consisting of the output of each shift register stage with a user-distinct 42-bit mask sequence. For access channels, the mask is constructed using pseudorandomly calculated access channel and associated paging channel numbers, and by base-station identification parameters.

The long PN code-spread baseband data stream is combined separately with I- and Q-quadrature short PN code sequences prior to waveshaping and transmission, with the Q channel combination delayed by one-half chip to implement OQPSK modulation and quadrature diversity. Note that the signal is not further spread by this operation, because the short PN codes are clocked at the same rate, 1.2288 Mcps. Note also that Figure 4.21 indicates that the zero-offset code phases of the short PN codes are used for all mobiles in all cells; the different user signals are distinguished only by their unique long PN code phases. Table 4.4 summarizes the modulation parameters of the access channel.

Table 4.4 Access channel modulation parameters

Parameter	Value	Units
Data rate	4800	bps
PN chip rate	1.2288	Mcps
Code rate	1/3	Bits/code symbol
Code symbol repetition	2	Symbols/code symbol
Transmit duty cycle	100	%
Code symbol rate	28,800	sps
Modulation	6	Code sym/mod sym
Modulation rate	4800	sps
Walsh chip rate	307.2	kcps
Mod symbol duration	208.33	μs
PN chips/code symbol	42.67	
PN chips/mod symbol	256	
PN chips/Walsh chip	4	

Access channel transmissions by particular mobiles are permitted during assigned intervals called *access channel slots*, which are an integer number of 20-ms frames in length. Each transmission within an access channel slot begins with a short random delay to distribute the transmission start times of the various mobiles that may be transmitting in the same slot on different channels, and with a preamble of 96 data zeros to aid the base station in acquiring the signal. The first time that a mobile uses an access channel, its transmissions are confined to "probe" messages formatted according to a certain procedure until the proper power level for that particular mobile has been determined.

4.3.3.2 Reverse Traffic Channel

Up to 62 traffic channels receive voice or data from the mobile at the base station. A block diagram of traffic channel processing is given in Figure 4.22. A variable rate vocoder is used to generate a digital voice signal at rate varying from 0.8 to 8.6 kbps in a given 20-ms traffic channel frame. Depending on the data rate, the data frame is encoded with a CRC block code to enable the base station receiver to determine if the frame has been received with error. An 8-bit encoder tail is added to the frame to ensure that the convolutional encoder, which follows, is reset to the all-zero state at the end of the frame. These operations result in data rates of 9,600 (full rate), 4,800 (half rate),

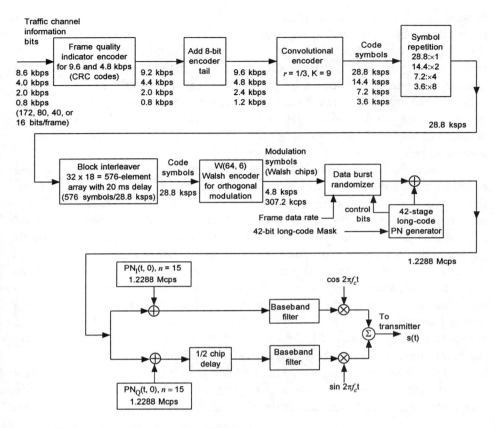

Figure 4.22 Reverse traffic channel modulation.

2,400 (1/4 rate), or 1,200 (1/8 rate) bps with, respectively, 192, 96, 48, or 24 bits per frame. The frame is then convolutionally encoded at a 1/3 rate, resulting in $3 \times 192 = 576$ code symbols per frame at full rate, or 28.8 ksps. For other voice data rates, the code symbols are repeated as necessary to cause each rate to input the same number of code symbols to the interleaver in a frame. The interleaver reads the data consecutively by column into an array of 32 rows and 18 columns; this procedure causes symbol repetitions to be located in different rows. As described more fully in Section 4.4, the rows of the interleaver array, each containing 18 code symbols, are read out in an order that depends on how many times the code symbols are repeated.

For example, if the symbols are repeated twice (i.e., the voice data rate is 4,800 bps), then every other symbol in a column of 32 symbols is a repeated symbol, making every other row in the array a repeated row. In this case, IS-

95 specifies that the rows be read out as follows: the first two odd-numbered rows, then their repeated rows; the second two odd-numbered rows, then their repetitions; and so forth. This procedure results in a sequence of two rows' worth of code symbols (36 symbols), their 36 repeated symbols, another 36 symbols followed by their 36 repeated symbols, and so on. In other words, the data repetitions are rearranged to occur by groups of 36 symbols. The purpose of this regrouping of the repeated symbols is to enable deletion of repeated symbols by groups. The motivation for such deletion is explained below.

Every six consecutive encoded symbols out of the interleaver are used to select a 64-chip Walsh sequence for orthogonal modulation according to the rule given in (4.6), with a chip rate of $28.8 \times 64/6 = 307.2$ kcps. Thus, each row of 18 symbols out of the interleaver generates three Walsh-encoded orthogonal modulation symbols, and each frame of 576 code symbols generates 96 orthogonal modulation symbols; because of the way that the symbols are read out from the interleaver array, these modulation symbols occur in alternating groups of six modulation symbols and $6(n - 1)$ repeated modulation symbols, where n is the order of repetition. Altogether in a frame interval there are $96/6 = 16$ groups of six orthogonal modulation symbols, each composed of $6 \times 64 = 384$ Walsh chips. Table 4.5 summarizes the grouping of these symbols in terms of their repetitions.

To reduce the average amount of reverse link interference and thereby increase user capacity, on reverse link transmissions, repeated symbols are gated off. A "data burst randomizer" is used to select in a pseudorandom

Table 4.5 Grouping of orthogonal modulation symbols

Voice rate	Repetition	Modulation symbol grouping
9,600 bps	× 1	$\underbrace{\text{6 symbols}}_{96/6 = 16 \text{ groups of } 6}$; 6 symbols; etc. (no repetitions)
4,800 bps	× 2	$\underbrace{\text{6 symbols, 6 repeats}}_{96/12 = 8 \text{ groups of } 12}$; 6 symbols, 6 repeats; etc.
2,400 bps	× 4	$\underbrace{\text{6 symbols, 3 groups of 6 repeats}}_{96/24 = 4 \text{ groups of } 24}$; etc.
1,200 bps	× 8	$\underbrace{\text{6 symbols, 7 groups of 6 repeats}}_{96/48 = 2 \text{ groups of } 48}$; etc.

manner which of the groups of six symbols are transmitted, based on "control bits" that are values of long PN code sequence bits at a certain time. Figure 4.23 illustrates the concept of "filling" six-symbol slots of the frame with pseudorandomly selected transmissions.

The user-distinct offset of the 42-stage long PN code is then used to further spread the signal and ensure that the channels can be distinguished. The offset is implemented using a mask that depends on the electronic serial number (ESN) of the mobile. The masks and offsets on both forward and reverse traffic channels are identical for a given mobile user.

The modulation parameters of the reverse traffic channel are summarized in Table 4.6.

Transmissions on the reverse traffic channel begin with a preamble of all-zero data frames to aid the base station in acquiring the signal. Signaling messages from the mobile to the base station may be sent on the reverse traffic channel as well as the access channel. When a message is to be sent, it can be sent in a "dim and burst" mode during periods of active speech, in which a portion of the voice data in a frame is overwritten by the message data, or in a "blank and burst" mode during periods of speech inactivity, in which all the data in the frame are message data.

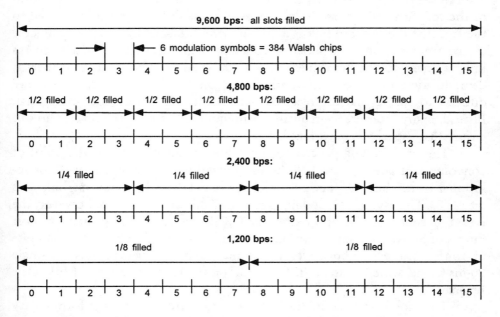

Figure 4.23 Data-burst randomizer operation.

Table 4.6 Reverse traffic channel modulation parameters

Parameter	Value				Units
Data rate	9,600	4,800	2,400	1,200	bps
PN chip rate	1.2288	1.2288	1.2288	1.2288	Mcps
Code rate	1/3	1/3	1/3	1/3	$\frac{bits}{code\ symbol}$
Transmit duty cycle	100	50	25	12.5	%
Code symbol rate	28,800	28,800	28,800	28,800	sps
Modulation rate	6	6	6	6	$\frac{code\ sym}{mod\ sym}$
Mod symbol rate	4,800	4,800	4,800	4,800	sps
Walsh chip rate	307.2	307.2	307.2	307.2	kcps
Mod symbol duration	208.33	208.33	208.33	208.33	μs
PN chips/code symbol	42.67	42.67	42.67	42.67	
PN chips/mod symbol	256	256	256	256	
PN chips/Walsh chip	4	4	4	4	

4.3.4 Comparison of Forward and Reverse Links

We can see that many similarities and differences exist between the operation of the forward and reverse links. Most of these differences can be traced to the fact that the forward link is a one-to-many transmission that is conducive to the transmission of a pilot for coherent demodulation. On the reverse link, a many-to-one transmission, a pilot is impractical and noncoherent demodulation must be employed. Thus, on the reverse link, it becomes difficult to ensure that the received signals from all mobiles are orthogonal. Instead, many interference-reducing techniques are used on the reverse link to ensure that capacity is maximized.

The voice-coding process for both links is identical up to the point of convolutional coding. Both links use powerful constraint-length 9 convolutional codes, but the forward link uses rate 1/2 and the reverse link uses rate 1/3. The asymptotic coding gain is 4.77 dB for both codes; however, the rate 1/3 code exhibits approximately 0.3 dB more coding gain over most ranges of SNR. Symbol repetition on both links is used to bring the data rate up to that of the 9.6-kbps code. Block interleaving is performed over a single 20-ms frame, which contains 384 symbols on the forward link and 576 on the reverse.

The mobile-to-base link uses Walsh-Hadamard sequences to form 64-ary modulation symbols. On the base-to-mobile link, the Walsh sequences are assigned by the base station based on the usage of the channel. Thus, on the reverse link, the Walsh sequences are chosen by the encoded data symbols, and on the forward link, they are chosen by the use of the channel.

The forward link uses the Walsh sequences to distinguish among different users, with one period of a Walsh sequence per scrambled code symbol. The reverse link employs a distinct phase offset of the long PN code to distinguish among users. The long PN code on the forward channel is decimated to scramble the data for voice privacy, because the scrambling has to be done at the modulation symbol rate to preserve Walsh sequence orthogonality among users. The reverse long code chip rate can be greater than the Walsh chip rate because Walsh symbol orthogonality is not used to distinguish users on the reverse link.

All repeated symbols on a forward traffic channel are transmitted, while repeated symbols on a reverse traffic channel are gated off using a randomization process. The forward channels employ QPSK modulation, but the reverse channels use OQPSK modulation. The comparison of the two links can be summarized as shown in Table 4.7.

4.4 Special Features of the IS-95 System

In this subsection, we detail some of the modulation features mentioned briefly in the summaries of the IS-95 forward and reverse links that are given above, including details of the interleaving and power control schemes. Also, we draw attention to special features of the system operation not mentioned previously, including the various uses of diversity made by the system and the handoff procedures employed.

Table 4.7 Comparison between forward and reverse traffic channels

	Forward link	Reverse link
Coding	Rate 1/2, constraint length 9	Rate 1/3, constraint length 9
Interleaving	1 frame, 384 symbols	1 frame, 576 symbols
Walsh modulation	Distinguishes a particular user	Provides 64-ary modulation
Long PN code	Scrambles data	Distinguishes users
	Decimated to symbol rate	No need to change rate
Repeated symbols	All symbols transmitted	Repeated symbols gated off
Modulation	QPSK	OQPSK

4.4.1 Power Control

One of the goals in a multiple-access system, such as the IS-95 CDMA digital cellular telephone system, is to maximize the number of simultaneous users it can accommodate. If each mobile unit's transmitter is adjusted so that the signal-to-interference ratio received at the base station is at the minimal acceptable level, the capacity of the system will be maximized. Any increase in mobile power raises the interference in the system, and capacity is compromised. A highly complex and dynamic scheme must be used to control the received power and thus maximize the channel capacity.

In a cellular system, one mobile may be close to the base station, while another may be several miles away. The distance between different mobiles and the base station can vary by a factor of 100. Because propagation path loss is governed approximately by an inverse fourth-power law, variations of up to 80 dB can be found between different users. A method of power control with a high dynamic range must be used.

Another characteristic of the cellular channel is fading caused by the signal's being reflected off various objects, creating a signal whose multipath components mutually interfere in a manner that is very sensitive to the position of the antenna. The average rate of the fades as a mobile travels in an area characterized by multipath fading is a function of mobile speed, being approximately one fade per second per mile per hour of mobile speed in the 850-MHz frequency range. These fades can cause the signal to be attenuated by more than 30 dB in extreme cases. The power control method must be able to track the majority of these fades. Figure 4.24 illustrates the effect of fast fading on the received mobile signal power. The principles underlying propagation path loss, fading, and their mathematical modeling were the subject of Chapter 2.

The IS-95 system uses a combination of open-loop and closed-loop power control. In the open-loop scheme, the measured received signal strength from the base station is used to determine the transmit power for the mobile; a decrease in the average received base station signal power is a real-time indication of a degradation in the mobile channel that can be caused by variations in the characteristics of the signal path, such as terrain and manmade structures that introduce "shadowing" of the signal. Assuming that the reverse link is subject to the same changes in average path loss as the forward link, under open-loop power control the mobile's average transmit power is adjusted accordingly. This open-loop method of power control provides a quick response to changes in signal conditions. Because the

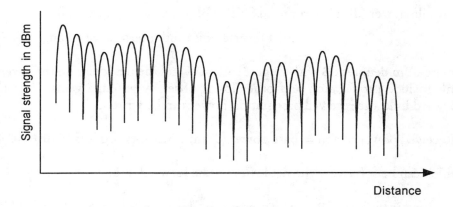

Figure 4.24 Effect of fast fading on signal strength.

forward and reverse links are separated in frequency (45 MHz for cellular and 80 MHz for PCS), however, they tend to fade independently. The correct setting of mobile transmitter power cannot be derived exactly using only the base station's average signal strength as measured at the mobile; feedback from the base station on its measurement of received mobile power must be also used.

In the IS-95 system, the base station measures the mobile signal strength and then transmits a power control command for the mobile to raise or lower its transmit power, thus providing a closed-loop refinement to the individual mobile's setting for its transmit power level.

4.4.1.1 Open-Loop Power Control

As mentioned above, each mobile station measures the received signal strength of the pilot signal. From this measurement and from information on the link power budget that is transmitted during initial synchronization, the forward link path loss is estimated. Assuming a similar path loss for the reverse link, the mobile uses this information to determine its transmitter power. A simplified link budget equation for the reverse link can be written

Received SNR (dB) = mobile power (dBm) − net reverse losses (dB)

$$- \text{ total reverse link noise and interference (dBm)} \qquad (4.7a)$$

so that the mobile power to be transmitted is determined by

Mobile power (dBm) = received SNR (dB) + net reverse losses (dB)

$$+ \text{ total reverse noise and interference (dBm)} \quad (4.7b)$$

where dBm denotes dB with respect to 1 mW and the net losses on the reverse link include propagation and other losses offset by antenna gains. For the forward link, the received base station power can be written

Received power (dBm) = base power (dBm) − net forward losses (dB)   (4.8a)

which can be solved for the net losses on the forward link:

Net forward losses (dB) = base power (dBm) − received power (dBm)   (4.8b)

This equation neglects the fact that the mobile measurement of received base station power is corrupted by forward link noise and interference. Substituting the net forward loss of (4.8b) in (4.7b) as an estimate of the net reverse loss and substituting a target value of the reverse link received SNR result in the equation

Mobile power (dBm) = target SNR (dB) + base station power (dBm)

$$+ \text{ total reverse noise and interference (dBm)}$$

$$- \text{ received power (dBm)} \quad (4.9a)$$

$$= \text{ constant (dB)} - \text{ received power (dBm)} \quad (4.9b)$$

In IS-95, the nominal value of the constant in (4.9b) is specified to be −73 dB. This value can be attributed to the nominal values of −13 dB for the target SNR, −100 dBm for the reverse link noise and interference, and 40 dBm (10 W) for the base station power. The actual values of these parameters may be different, however—the point is that the required amount of mobile transmitter power is inversely proportional to the amount of received forward link power, as expressed by the relationship in (4.9b). Data for calibrating the constant in (4.9b) are broadcast to the mobiles on the sync channel.

By design, the temporal response of the mobile to this information is nonlinear. If the pilot is suddenly received with a high signal strength, the mobile transmitter power is reduced immediately, within several microseconds, on the principle that a higher received value of forward link power is a better estimate of *average* link loss, apart from fading. But if the measured signal strength drops, the mobile power is increased slowly, on the order of a millisecond. The reason for this procedure is that if the power is not decreased quickly when a improvement in the path is encountered, the mobile will cause an increase in interference for the other users until the problem is corrected. Similarly, if the power is increased rapidly, the path loss may in fact be less than inferred from the forward link, perhaps because the base station signal faded and the mobile signal did not, and again the mobile would cause interference for the other users. Therefore, the system accepts a degradation in a single user's signal to prevent increased interference for all users.

Figure 4.25 shows the mobile transmit power corresponding to the channel in Figure 4.24 as controlled only by the open-loop nonlinear method. The dashed line in Figure 4.25 represents the transmit power if the received signal were traced without the nonlinear filter, and the solid line represents the transmit power with the filter.

4.4.1.2 Closed-Loop Power Control

Because the forward and reverse links may fade independently, closed-loop power control is employed in IS-95. The calculation of forward link path loss through the measurement of the base station received signal strength can be used as a rough estimate of the path loss on the reverse link. The true value, however, must be measured at the base station upon reception of the mobile's signal.

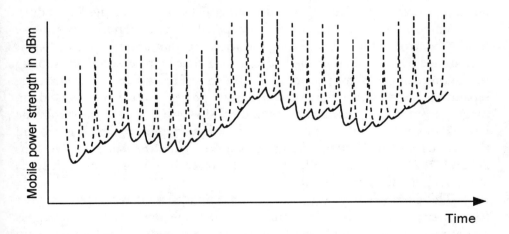

Figure 4.25 Mobile transmitter power as governed by the open-loop power control only.

At the base station, the measured signal strength is compared with the desired strength, and a power adjustment command is generated. The signal strength is estimated from intermediate outputs in the reverse link 64-ary orthogonal demodulator and is averaged over a "power control group" period of 1.25 ms (six Walsh 64-ary symbols). This power estimate is compared with the desired mobile received power level. If the average power level is greater than the threshold, the power command generator generates a "1" to instruct the mobile to decrease power. If the average power is less than the desired level, a "0" is generated to instruct the mobile to increase power. These commands instruct the mobile to adjust transmitter power by a predetermined amount, usually 1 dB.

On the forward link, in each 1.25-ms interval there are $19.2 \times 1.25 = 24$ modulation symbols (i.e., scrambled and interleaved code symbols), as discussed previously in connection with the forward link traffic channel modulation diagram in Figure 4.17. The power control command is inserted into the forward traffic channel data stream following the interleaving and the scrambling by "puncturing" (overwriting) two consecutive modulation symbols with the power control bit. Because the code rate is 1/2, a power control symbol has an energy equivalent to one information bit. The position of the power control command in the data stream is pseudo-randomly selected using values of the long-code chips in the preceding 1.25-ms interval.

Because the power control commands should affect the mobile transmitter power level as soon as possible, they are added to the data stream after the encoding and interleaving in order to eliminate the more than 20-ms delay associated with deinterleaving and decoding. The first 16 of the 24 modulation symbols in a 1.25-ms interval are possible starting positions for the power control bit. The starting position of the power control bit is determined by the values of the last four long-code chips used for scrambling in the preceding 1.25-ms period, as illustrated in Figure 4.26. Therefore, the separation between the start of consecutive power control bits is random, ranging from 9 to 39 modulation symbols. The 24th PN code chip from the last 1.25-ms period is the most significant bit in determining the position, and the 21st chip is the least significant bit. In the figure, the values of the chips 24, 23, 22, 21 are $1\ 0\ 1\ 1 = 11$; thus, in this case, the power control bit starts at the 11th modulation symbol.

The closed-loop power control also employs what is termed an *outer loop power control*. This mechanism ensures that the power control strategy is operating correctly. The frame error rate at the base station is measured and compared with the desired error rate. If the difference between error

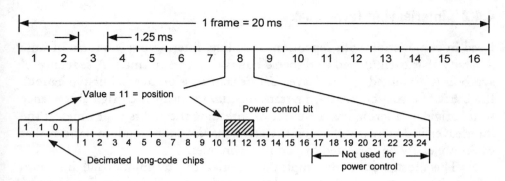

Figure 4.26 Example of power control bit position.

rates is large, then the power command threshold is adjusted to yield the desired FER.

4.4.1.3 Forward Link Power Control

In some mobile locations, the forward link may be subject to poor reception conditions, as evidenced at the mobile by an unacceptably high frame error rate. The mobile in such a situation can request an increase in the power transmitted to it by the base station.

When forward link power control is enabled, the base station periodically reduces the power transmitted to an individual mobile unit. This process continues until the mobile senses an increase in the forward link frame error rate. The mobile reports the number of frame errors to the base. With this information the base station can decide whether to increase power by a small amount, nominally 0.5 dB. Before the base station complies with the request, it must consider other requests, loading, and the current transmitted power.

The rate of forward link power adjustment is slower than that of reverse link power control, increments being made on the order of the frame interval, 20 ms. Also, the dynamic range of forward link power adjustments is limited to about ± 6 dB about the nominal power.

The effect of forward link power control is studied in Chapter 11 in connection with the optimal allocation of forward link power.

4.4.2 Interleaving Techniques

Used in conjunction with repetition or coding, interleaving is a form of time diversity that is employed to disperse bursts of errors in time. A sequence of symbols is permuted or interleaved before transmission over a bursty channel. If a burst of errors occurs during transmission, restoring the original sequence to its original ordering has the effect of spreading the errors over time. If the interleaver is designed well, then the errors have a more random pattern that can more easily be corrected by coding techniques.

The most commonly employed interleaving techniques fall into two classes. The more common type is *block interleaving*. This form is used when the data are divided into blocks or frames, such as in the IS-95 system. *Convolutional interleaving*, on the other hand, is more practical for a continuous data stream. Block interleaving is known for its ease of implementation, whereas convolutional interleaving has performance advantages. Continuous operation allows the original overhead associated with convolutional interleaving to become insignificant. IS-95 uses other forms of interleaving that are based on block-like techniques and are described further. A brief review of the general subject of interleaving is given in Appendix 4A of this chapter.

Several parameters describe interleaver performance. One of the most important is the minimum separation, S, which is the minimum distance by which consecutive errors in a burst are dispersed. Naturally this parameter depends on the length of the burst, decreasing as burst length increases. As an extreme, consider the case where the burst length is the same as the entire sequence. The minimum separation will be one because no matter how the data are permuted, an error always resides next to another error. Because interleaving involves storing some symbols in memory elements while reading others from memory elements, a delay is experienced. In general, this same delay is also experienced upon deinterleaving. The delay D is expressed as the number of additional read/write operations required to perform both interleaving and deinterleaving. As just mentioned, the process requires a number of memory elements for implementation denoted by the parameter M. To achieve good interleaver performance, the minimum separation should be as great as possible, while the delay and memory requirement should be as small as possible. This fact often leads to performance described by ratios of minimum separation-to-delay S/D and minimum separation-to-memory S/M.

An (I, J) block interleaver can be viewed as an array of storage locations which contains I columns and J rows. The data are written into the array by columns and read out by rows, as demonstrated in Figure 4.27. The first symbol written into the array is written into the top left corner, but the first symbol read out is from the bottom left corner. Continuous processing of data requires two arrays: one has data written into it, the other has data being read from it. The deinterleaving process simply requires duplicate arrays in order to reverse the interleaving procedure.

Properties of the block interleaver are readily apparent by observing the array. Let the length of a burst of errors be given by B. The minimum separation of any two errors in the burst of length B is given by

$$S = \begin{cases} J & B \le I \\ 1 & B > I \end{cases} \tag{4.10}$$

The interleaver delay is IJ at the transmitter and IJ at the receiver. Thus, the total delay is

$$D = 2IJ \tag{4.11}$$

For continuous operation, two arrays are required, giving a memory requirement of

$$M = 2IJ \tag{4.12}$$

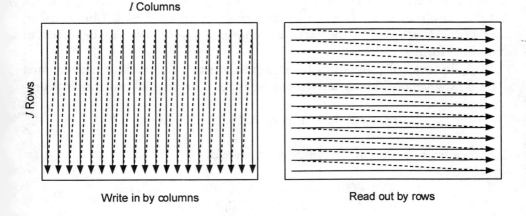

Figure 4.27 Reading to and writing from an (I, J) block interleaver.

The minimum separation of the interleaver can be changed by reading the rows out in a different order. The delay and memory requirement are essentially unchanged by this action. The largest minimum separation for $B \leq I$ is achieved by reading the rows in the order described above. However, this method yields $S = 1$ for $B > I$. Other reading strategies can reduce the minimum separation for $B \leq I$ while increasing it for $B > I$. The IS-95 system employs such techniques. Unless the reading strategy is determined carefully, the minimum separation is generally reduced.

The IS-95 system interleaves data from single data frames, which have durations of 20 ms for all channels except the sync channel, which has a frame duration of 26.66 ms. Thus, all the IS-95 interleavers operate on blocks of symbols. Strictly speaking, however, block interleaving is not employed, but the type of interleaving depends on the channel and original data rate. For example, the reverse link gates the power amplifier on and off to remove repeated symbols, and the interleaver is designed to facilitate this action. The forward link reads the rows from the array in a nonconventional order to modify the minimum-separation properties.

Interleaving on the sync channel uses a nonconventional technique that is best described as *bit reversal*. As diagrammed previously in Figure 4.13, sync channel code symbols at 2.4 ksps are repeated once, resulting in a 4.8-ksps rate into the interleaver. The symbols are interleaved over a 26.667-ms period, which gives a total of 26.667 ms $\times$ 4.8 ksps = 128 symbols per frame. Let the first symbol into the interleaver be assigned position 0, and the last be 127. The read order is given by changing the position to a 7-bit number, reversing the order of the bits, and changing the binary number back to decimal form.

For example, suppose a 16-bit sequence is to be interleaved, and the first bit is assigned position 0, and the last bit is 15. The original ordering is thus

$$0, 1, 2, 3, 4, 5, 6, 7, 8, 9, 10, 11, 12, 13, 14, 15$$

Expressing this ordering in binary format yields

$$0000, 0001, 0010, 0011, 0100, 0101, 0110, 0111,$$
$$1000, 1001, 1010, 1011, 1100, 1101, 1110, 1111$$

Reversing the order of the bits in each binary number gives

$$0000, 1000, 0100, 1100, 0010, 1010, 0110, 1110,$$
$$0001, 1001, 0101, 1101, 0011, 1011, 0111, 1111$$

Converting the binary representation back to decimal form gives the order with which the symbols are read

$$0,\ 8,\ 4,\ 12,\ 2,\ 10,\ 6,\ 14,\ 1,\ 9,\ 5,\ 13,\ 3,\ 11,\ 7,\ 15$$

Through observation, the minimum spacing for symbols in a burst of length B is

$$S = \begin{cases} 4, & B = 2 \\ 2, & 3 \le B \le 4 \\ 1, & B \ge 5 \end{cases}$$

IS-95 uses the bit reversal technique over the 128 symbols in the sync channel frame. The input array is in Figure 4.28(a), and the output array in Figure 4.28(b), where the symbols are read into the input array by columns are read out of the output array by columns. As an example, the 38th symbol in the input array would occupy position 37, which is 0100101. Reversing the order gives 1010010, which is decimal 82. Note that symbol 37 in the input array shares the same position as 82 in the output array. The minimum separation can be found by observation and ranges from $S = 32$ for $B = 2$ to $S = 1$ for $B \ge 33$. The memory requirement is twice the number of symbols in a block for continuous operation: $M = 2 \times 128 = 256$. The delay is also equivalent to twice the number of symbols in the array:

$$D = 256 = \frac{256 \text{ sym}}{4{,}800 \text{ sym/s}} = 53.333 \text{ ms} \tag{4.13}$$

The paging and traffic channels use the same interleaver structure on a single 20-ms frame, which contains 384 symbols. As with the sync channel, IS-95 specifies the interleaver using input and output arrays. The paging channel can generate data at 9.6 or 4.8 kbps, but the traffic channel can operate at four rates. The full-rate 9.6-kbps interleaver is shown in Figure 4.29. The 4.8-, 2.4-, and 1.2-kbps interleavers are shown in Figures 4.30, 4.31, and 4.32, respectively. Note that all of the interleavers are identical, and the difference between the four is only the repetition of the input symbols. Thus, the true structure, without repetition, is shown by the full-rate scheme. The arrays corresponding to less than full rate show the symbol repetition. Note that adjacent symbol repetitions going into the interleaver are converted to block repetitions coming out of the interleaver.

By observing the first six symbols of the full-rate output array, we find that the symbols are separated by 64 positions. The second group of six,

0	16	32	48	64	80	96	112
1	17	33	49	65	81	97	113
2	18	34	50	66	82	98	114
3	19	35	51	67	83	99	115
4	20	36	52	68	84	100	116
5	21	37	53	69	85	101	117
6	22	38	54	70	86	102	118
7	23	39	55	71	87	103	119
8	24	40	56	72	88	104	120
9	25	41	57	73	89	105	121
10	26	42	58	74	90	106	122
11	27	43	59	75	91	107	123
12	28	44	60	76	92	108	124
13	29	45	61	77	93	109	125
14	30	46	62	78	94	110	126
15	31	47	63	79	95	111	127

(a) Interleaver input array

0	4	2	6	1	5	3	7
64	68	66	70	65	69	67	71
32	36	34	38	33	37	35	39
96	100	98	102	97	101	99	103
16	20	18	22	17	21	19	23
80	84	82	86	81	85	83	87
48	52	50	54	49	53	51	55
112	116	114	118	113	117	115	119
8	12	10	14	9	13	11	15
72	76	74	78	73	77	75	79
40	44	42	46	41	45	43	47
104	108	106	110	105	109	107	111
24	28	26	30	25	29	27	31
88	92	90	94	89	93	91	95
56	60	58	62	57	61	59	63
120	124	122	126	121	125	123	127

(b) Interleaver output array

Figure 4.28   Forward link sync channel interleaving.

```
 1   25   49   73   97  121  145  169  193  217  241  265  289  313  337  361
 2   26   50   74   98  122  146  170  194  218  242  266  290  314  338  362
 3   27   51   75   99  123  147  171  195  219  243  267  291  315  339  363
 4   28   52   76  100  124  148  172  196  220  244  268  292  316  340  364
 5   29   53   77  101  125  149  173  197  221  245  269  293  317  341  365
 6   30   54   78  102  126  150  174  198  222  246  270  294  318  342  366
 7   31   55   79  103  127  151  175  199  223  247  271  295  319  343  367
 8   32   56   80  104  128  152  176  200  224  248  272  296  320  344  368
 9   33   57   81  105  129  153  177  201  225  249  273  297  321  345  369
10   34   58   82  106  130  154  178  202  226  250  274  298  322  346  370
11   35   59   83  107  131  155  179  203  227  251  275  299  323  347  371
12   36   60   84  108  132  156  180  204  228  252  276  300  324  348  372
13   37   61   85  109  133  157  181  205  229  253  277  301  325  349  373
14   38   62   86  110  134  158  182  206  230  254  278  302  326  350  374
15   39   63   87  111  135  159  183  207  231  255  279  303  327  351  375
16   40   64   88  112  136  160  184  208  232  256  280  304  328  352  376
17   41   65   89  113  137  161  185  209  233  257  281  305  329  353  377
18   42   66   90  114  138  162  186  210  234  258  282  306  330  354  378
19   43   67   91  115  139  163  187  211  235  259  283  307  331  355  379
20   44   68   92  116  140  164  188  212  236  260  284  308  332  356  380
21   45   69   93  117  141  165  189  213  237  261  285  309  333  357  381
22   46   70   94  118  142  166  190  214  238  262  286  310  334  358  382
23   47   71   95  119  143  167  191  215  239  263  287  311  335  359  383
24   48   72   96  120  144  168  192  216  240  264  288  312  336  260  384
```

(a) Interleaver input array

```
  1    9    5   13    3   11    7   15    2   10    6   14    4   12    8   16
 65   73   69   77   67   75   71   79   66   74   70   78   68   76   72   80
129  137  133  141  131  139  135  143  130  138  134  142  132  140  136  144
193  201  197  205  195  203  199  207  194  202  198  206  196  204  200  208
257  265  261  269  259  267  263  271  258  266  262  270  260  268  264  272
321  329  325  333  323  331  327  335  322  330  326  334  324  332  328  336
 33   41   37   45   35   43   39   47   34   42   38   46   36   44   40   48
 97  105  101  109   99  107  103  111   98  106  102  110  100  108  104  112
161  169  165  173  163  171  167  175  162  170  166  174  164  172  168  176
225  233  229  237  227  235  231  239  226  234  230  238  228  236  232  240
289  297  293  301  291  299  295  303  290  298  294  302  292  300  296  304
353  361  357  365  355  363  359  367  354  362  358  366  356  364  360  368
 17   25   21   29   19   27   23   31   18   26   22   30   20   28   24   32
 81   89   85   93   83   91   87   95   82   90   86   94   84   92   88   96
145  153  149  157  147  155  151  159  146  154  150  158  148  156  152  160
209  217  213  221  211  219  215  223  210  218  214  222  212  220  216  224
273  281  277  285  275  283  279  287  274  282  278  286  276  284  280  288
337  345  341  349  339  347  343  351  338  346  342  350  340  348  344  352
 49   57   53   61   51   59   55   63   50   58   54   62   52   60   56   64
113  121  117  125  115  123  119  127  114  122  118  126  116  124  120  128
177  185  181  189  179  187  183  191  178  186  182  190  180  188  184  192
241  249  245  253  243  251  247  255  242  250  246  254  244  252  248  256
305  313  309  317  307  315  311  319  306  314  310  318  308  316  312  320
369  377  373  381  371  379  375  383  370  378  374  382  372  380  376  384
```

(b) Interleaver output array

Figure 4.29 Forward traffic and paging channels interleaving at 9,600 bps.

1	13	25	37	49	61	73	85	97	109	121	133	145	157	169	181
1	13	25	37	49	61	73	85	97	109	121	133	145	157	169	181
2	14	26	38	50	62	74	86	98	110	122	134	146	158	170	182
2	14	26	38	50	62	74	86	98	110	122	134	146	158	170	182
3	15	27	39	51	63	75	87	99	111	123	135	147	159	171	183
3	15	27	39	51	63	75	87	99	111	123	135	147	159	171	183
4	16	28	40	52	64	76	88	100	112	124	136	148	160	172	184
4	16	28	40	52	64	76	88	100	112	124	136	148	160	172	184
5	17	29	41	53	65	77	89	101	113	125	137	149	161	173	185
5	17	29	41	53	65	77	89	101	113	125	137	149	161	173	185
6	18	30	42	54	66	78	90	102	114	126	138	150	162	174	186
6	18	30	42	54	66	78	90	102	114	126	138	150	162	174	186
7	19	31	43	55	67	79	91	103	115	127	139	151	163	175	187
7	19	31	43	55	67	79	91	103	115	127	139	151	163	175	187
8	20	32	44	56	68	80	92	104	116	128	140	152	164	176	188
8	20	32	44	56	68	80	92	104	116	128	140	152	164	176	188
9	21	33	45	57	69	81	93	105	117	129	141	153	165	177	189
9	21	33	45	57	69	81	93	105	117	129	141	153	165	177	189
10	22	34	46	58	70	82	94	106	118	130	142	154	166	178	190
10	22	34	46	58	70	82	94	106	118	130	142	154	166	178	190
11	23	35	47	59	71	83	95	107	119	131	143	155	167	179	191
11	23	35	47	59	71	83	95	107	119	131	143	155	167	179	191
12	24	36	48	60	72	84	96	108	120	132	144	156	168	180	192
12	24	36	48	60	72	84	96	108	120	132	144	156	168	180	192

(a) Interleaver input array

1	5	3	7	2	6	4	8	1	5	3	7	2	6	4	8
33	37	35	39	34	38	36	40	33	37	35	39	34	38	36	40
65	69	67	71	66	70	68	72	65	69	67	71	66	70	68	72
97	101	99	103	98	102	100	104	97	101	99	103	98	102	100	104
129	133	131	135	130	134	132	136	129	133	131	135	130	134	132	136
161	165	163	167	162	166	164	168	161	165	163	167	162	166	164	168
17	21	19	23	18	22	20	24	17	21	19	23	18	22	20	24
49	53	51	55	50	54	52	56	49	53	51	55	50	54	52	56
81	85	83	87	82	86	84	88	81	85	83	87	82	86	84	88
113	117	115	119	114	118	116	120	113	117	115	119	114	118	116	120
145	149	147	151	146	150	148	152	145	149	147	151	146	150	148	152
177	181	179	183	178	182	180	184	177	181	179	183	178	182	180	184
9	13	11	15	10	14	12	16	9	13	11	15	10	14	12	16
41	45	43	47	42	46	44	48	41	45	43	47	42	46	44	48
73	77	75	79	74	78	76	80	73	77	75	79	74	78	76	80
105	109	107	111	106	110	108	112	105	109	107	111	106	110	108	112
137	141	139	143	138	142	140	144	137	141	139	143	138	142	140	144
169	173	171	175	170	174	172	176	169	173	171	175	170	174	172	176
25	29	27	31	26	30	28	32	25	29	27	31	26	30	28	32
57	61	59	63	58	62	60	64	57	61	59	63	58	62	60	64
89	93	91	95	90	94	92	96	89	93	91	95	90	94	92	96
121	125	123	127	122	126	124	128	121	125	123	127	122	126	124	128
153	157	155	159	154	158	156	160	153	157	155	159	154	158	156	160
185	189	187	191	186	190	188	192	185	189	187	191	186	190	188	192

(b) Interleaver output array

Figure 4.30 Forward traffic and paging channels interleaving at 4,800 bps.

1	7	13	19	25	31	37	43	49	55	61	67	73	79	85	91
1	7	13	19	25	31	37	43	49	55	61	67	73	79	85	91
1	7	13	19	25	31	37	43	49	55	61	67	73	79	85	91
1	7	13	19	25	31	37	43	49	55	61	67	73	79	85	91
2	8	14	20	26	32	38	44	50	56	62	68	74	80	86	92
2	8	14	20	26	32	38	44	50	56	62	68	74	80	86	92
2	8	14	20	26	32	38	44	50	56	62	68	74	80	86	92
2	8	14	20	26	32	38	44	50	56	62	68	74	80	86	92
3	9	15	21	27	33	39	45	51	57	63	69	75	81	87	93
3	9	15	21	27	33	39	45	51	57	63	69	75	81	87	93
3	9	15	21	27	33	39	45	51	57	63	69	75	81	87	93
3	9	15	21	27	33	39	45	51	57	63	69	75	81	87	93
4	10	16	22	28	34	40	46	52	58	64	70	76	82	88	94
4	10	16	22	28	34	40	46	52	58	64	70	76	82	88	94
4	10	16	22	28	34	40	46	52	58	64	70	76	82	88	94
4	10	16	22	28	34	40	46	52	58	64	70	76	82	88	94
5	11	17	23	29	35	41	47	53	59	65	71	77	83	89	95
5	11	17	23	29	35	41	47	53	59	65	71	77	83	89	95
5	11	17	23	29	35	41	47	53	59	65	71	77	83	89	95
5	11	17	23	29	35	41	47	53	59	65	71	77	83	89	95
6	12	18	24	30	36	42	48	54	60	66	72	78	84	90	96
6	12	18	24	30	36	42	48	54	60	66	72	78	84	90	96
6	12	18	24	30	36	42	48	54	60	66	72	78	84	90	96
6	12	18	24	30	36	42	48	54	60	66	72	78	84	90	96

(a) Interleaver input array

1	3	2	4	1	3	2	4	1	3	2	4	1	3	2	4
17	19	18	20	17	19	18	20	17	19	18	20	17	19	18	20
33	35	34	36	33	35	34	36	33	35	34	36	33	35	34	36
49	51	50	52	49	51	50	52	49	51	50	52	49	51	50	52
65	67	66	68	65	67	66	68	65	67	66	68	65	67	66	68
81	83	82	84	81	83	82	84	81	83	82	84	81	83	82	84
9	11	10	12	9	11	10	12	9	11	10	12	9	11	10	12
25	27	26	28	25	27	26	28	25	27	26	28	25	27	26	28
41	43	42	44	41	43	42	44	41	43	42	44	41	43	42	44
57	59	58	60	57	59	58	60	57	59	58	60	57	59	58	60
73	75	74	76	73	75	74	76	73	75	74	76	73	75	74	76
89	91	92	94	89	91	92	94	89	91	92	94	89	91	92	94
5	7	6	8	5	7	6	8	5	7	6	8	5	7	6	8
21	23	22	24	21	23	22	24	21	23	22	24	21	23	22	24
37	39	38	40	37	39	38	40	37	39	38	40	37	39	28	40
53	55	54	56	53	55	54	56	53	55	54	56	53	55	54	56
69	71	70	72	69	71	70	72	69	71	70	72	69	71	70	72
85	87	86	88	85	87	86	88	85	87	86	88	85	87	86	88
13	15	14	16	13	15	14	16	13	15	14	16	13	16	14	16
29	31	30	32	29	31	30	32	29	31	30	32	29	31	30	32
45	47	26	28	45	47	46	48	45	47	46	48	45	47	46	48
61	63	62	64	61	63	62	64	61	63	62	64	61	63	62	64
77	79	78	80	77	79	78	80	77	79	78	80	77	79	78	80
93	95	94	96	93	95	94	96	96	95	94	96	93	95	94	96

(b) Interleaver output array

Figure 4.31 Forward traffic channel interleaving at 2,400 bps.

```
1    4    7   10   13   16   19   22   25   28   31   34   37   40   43   46
1    4    7   10   13   16   19   22   25   28   31   34   37   40   43   46
1    4    7   10   13   16   19   22   25   28   31   34   37   40   43   46
1    4    7   10   13   16   19   22   25   28   31   34   37   40   43   46
1    4    7   10   13   16   19   22   25   28   31   34   37   40   43   46
1    4    7   10   13   16   19   22   25   28   31   34   37   40   43   46
1    4    7   10   13   16   19   22   25   28   31   34   37   40   43   46
1    4    7   10   13   16   19   22   25   28   31   34   37   40   43   46
2    5    8   11   14   17   20   23   26   29   32   35   38   41   44   47
2    5    8   11   14   17   20   23   26   29   32   35   38   41   44   47
2    5    8   11   14   17   20   23   26   29   32   35   38   41   44   47
2    5    8   11   14   17   20   23   26   29   32   35   38   41   44   47
2    5    8   11   14   17   20   23   26   29   32   35   38   41   44   47
2    5    8   11   14   17   20   23   26   29   32   35   38   41   44   47
2    5    8   11   14   17   20   23   26   29   32   35   38   41   44   47
2    5    8   11   14   17   20   23   26   29   32   35   38   41   44   47
3    6    9   12   15   18   21   24   27   30   33   36   39   42   45   48
3    6    9   12   15   18   21   24   27   30   33   36   39   42   45   48
3    6    9   12   15   18   21   24   28   30   33   36   39   42   45   48
3    6    9   12   15   18   21   24   28   30   33   36   39   42   45   48
3    6    9   12   15   18   21   24   28   30   33   36   39   42   45   48
3    6    9   12   15   18   21   24   28   30   33   36   39   42   45   48
3    6    9   12   15   18   21   24   28   30   33   36   39   42   45   48
3    6    9   12   15   18   21   24   28   30   33   36   39   42   45   48
```

(a) Interleaver input array

```
1    2    1    2    1    2    1    2    1    2    1    2    1    2    1    2
9   10    9   10    9   10    9   10    9   10    9   10    9   10    9   10
17   18   17   18   17   18   17   18   17   18   17   18   17   18   17   18
25   26   25   26   25   26   25   26   25   26   25   26   25   26   25   26
33   34   33   34   33   34   33   34   33   34   33   34   33   34   33   34
41   43   41   42   41   42   41   42   41   42   41   42   41   42   41   42
5    6    5    6    5    6    5    6    5    6    5    6    5    6    5    6
13   14   13   14   13   14   13   14   13   14   13   14   13   14   13   14
21   22   21   22   21   22   21   22   21   22   21   22   21   22   21   22
29   30   29   30   29   30   29   30   29   30   29   30   29   30   29   30
37   38   37   38   37   38   37   38   37   38   37   38   37   38   37   38
45   46   45   46   45   46   45   46   45   46   45   46   45   46   45   46
3    4    3    4    3    4    3    4    3    4    3    4    3    4    3    4
11   12   11   12   11   12   11   12   11   12   11   12   11   12   11   12
19   20   19   20   19   20   19   20   19   20   19   20   19   20   19   20
27   28   27   28   27   28   27   28   27   28   27   28   27   28   27   28
35   36   35   36   35   36   35   36   35   36   35   36   35   36   35   36
43   44   43   44   43   44   43   44   43   44   43   44   43   44   43   44
7    8    7    8    7    8    7    8    7    8    7    8    7    8    7    8
15   16   15   16   15   16   15   16   15   16   15   16   15   16   15   16
23   24   23   24   23   24   23   24   23   24   23   24   23   24   23   24
31   32   31   32   31   32   31   32   31   32   31   32   31   32   31   32
39   40   39   40   39   40   39   40   39   40   39   40   39   40   39   40
47   48   47   48   47   48   47   48   47   48   47   48   47   48   47   48
```

(b) Interleaver output array

Figure 4.32 Forward traffic channel interleaving at 1,200 bps.

as well as subsequent groups of six, exhibit the same property. This is a characteristic of an $(I = 6, J = 64)$ block interleaver. These observations demonstrate that the paging and traffic channels use a (6, 64) block interleaver, which is shown in Figure 4.33.

The read array for the 9,600-bps traffic and paging channels (Figure 4.29) does correspond to the (6, 64) array of Figure 4.33, but the rows are read out in a permuted order. Upon closer observation, it becomes evident that the rows are read from the (6, 64) array in bit-reversal order. The first six symbols are from row 0. The next six are from row 32. The next six are those in row 16, and so forth. Reading the rows in bit-reversal order changes the properties of the interleaver. For a conventional (6, 64) array, the minimum spacing is given by $S = 64$ for $B \leq 6$ and $S = 1$ for $B > 6$. With this permuted structure, the minimum-separation profile is found to be

$$S = \begin{cases} 64, & B \leq 5 \\ 32, & B = 6 \\ 16, & B = 7 \\ 3, & 8 \leq B \leq 48 \\ 2, & 49 \leq B \leq 96 \\ 1, & B \geq 97 \end{cases} \tag{4.14}$$

Using the bit-reversal row reading, the minimum separation for $B > I$ is increased at the expense of the separation for $B \leq I$. The memory and delay requirements are equal to those of conventional block interleavers:

$$M = 2IJ = 2 \times 6 \times 64 = 768 \tag{4.15}$$

$$D = 2IJ = 768 = \frac{768 \text{ sym}}{19,200 \text{ sym/s}} = 40 \text{ ms} \tag{4.16}$$

The encoding and interleaving process for the reverse link was shown previously in Figure 4.21 for the access channel and 4.22 for the reverse traffic channel. Similar to the forward traffic channel, the access channel reads the rows from the input array in bit-reversal order. On the reverse traffic channel, symbols at the lower data rates (1.2, 2.4, and 4.8 kbps) are repeated after convolutional encoding. The transmitter pseudorandomly gates the amplifier on and off to eliminate the repeated symbols. The traffic channel interleaver is designed to operate in conjunction with the amplifier gating.

Row 0	1	65	129	193	257	321	Row 32	33	97	161	225	289	353
Row 1	2	66	130	194	258	322	Row 33	34	98	162	226	290	354
Row 2	3	67	131	195	259	323	Row 34	35	99	163	227	291	355
Row 3	4	68	132	196	260	324	Row 35	36	100	164	228	292	356
Row 4	5	69	133	197	261	325	Row 36	37	101	165	229	293	357
Row 5	6	70	134	198	262	326	Row 37	38	102	166	230	294	358
Row 6	7	71	135	199	263	327	Row 38	39	103	167	231	295	359
Row 7	8	72	136	200	264	328	Row 39	40	104	168	232	296	360
Row 8	9	73	137	201	265	329	Row 40	41	105	169	233	297	361
Row 9	10	74	138	202	266	330	Row 41	42	106	170	234	298	362
Row 10	11	75	139	203	267	331	Row 42	43	107	171	235	299	363
Row 11	12	76	140	204	268	332	Row 43	44	108	172	236	300	364
Row 12	13	77	141	205	269	333	Row 44	45	109	173	237	301	365
Row 13	14	78	142	206	270	334	Row 45	46	110	174	238	302	366
Row 14	15	79	143	207	271	335	Row 46	47	111	175	239	303	367
Row 15	16	80	144	208	272	336	Row 47	48	112	176	240	304	368
Row 16	17	81	145	209	273	337	Row 48	49	113	177	241	305	369
Row 17	18	82	146	210	274	338	Row 49	50	114	178	242	306	370
Row 18	19	83	147	211	275	339	Row 50	51	115	179	243	307	371
Row 19	20	84	148	212	276	340	Row 51	52	116	180	244	308	372
Row 20	21	85	149	213	277	341	Row 52	53	117	181	245	309	373
Row 21	22	86	150	214	278	342	Row 53	54	118	182	246	310	374
Row 22	23	87	151	215	279	343	Row 54	55	119	183	247	311	375
Row 23	24	88	152	216	280	344	Row 55	56	120	184	248	312	376
Row 24	25	89	153	217	281	345	Row 56	57	121	185	249	313	377
Row 25	26	90	154	218	282	346	Row 57	58	122	186	250	314	378
Row 26	27	91	155	219	283	347	Row 58	59	123	187	251	315	379
Row 27	28	92	156	220	284	348	Row 59	60	124	188	252	316	380
Row 28	29	93	157	221	285	349	Row 60	61	125	189	253	317	381
Row 29	30	94	158	222	286	350	Row 61	62	126	190	254	318	382
Row 30	31	95	159	223	287	351	Row 62	63	127	191	255	319	383
Row 31	32	96	160	224	288	352	Row 63	64	128	192	256	320	384

Figure 4.33 Forward traffic and paging channel interleaving is observed to be equivalent to a (6, 64) array.

Data from the access channel are generated at 4.8 kbps, repeated once, and $R_c = 1/3$ convolutionally encoded to 28.8 ksps. Interleaving is performed within a single 20-ms frame containing $20\,\text{ms} \times 28.8\,\text{ksps} = 576$ symbols. Symbols are read from the $(I = 18, J = 32)$ array by rows in bit-reversal order. No gating is used on the access channel, and thus, both identical symbols are transmitted. The order of the rows is given by

$$0, 16, 8, 24, 4, 20, 12, 28, 2, 18, 10, 26, 6, 22, 14, 30,$$
$$1, 17, 9, 25, 5, 21, 13, 29, 3, 19, 11, 27, 7, 23, 15, 31$$

where 0 is the first row and 31 is the last row. The (18, 32) array with the symbols read out by consecutive rows is shown in Figure 4.34. The access channel minimum-separation profile can be found to be

$$S = \begin{cases} 32, & B \le 17 \\ 16, & B = 18 \\ 8, & B = 19 \\ 3, & 20 \le B \le 72 \\ 2, & 73 \le B \le 144 \\ 1, & B \ge 145 \end{cases} \tag{4.17}$$

As with conventional block interleavers, the memory requirement is twice the number of symbols in the array:

1	33	65	97	129	161	193	225	257	289	321	353	385	417	449	481	513	545
17	49	81	113	145	177	209	241	273	305	337	369	401	433	465	497	529	561
9	41	73	105	137	169	201	233	265	297	329	361	393	425	457	489	521	553
25	57	89	121	153	185	217	249	281	313	345	377	409	441	473	505	537	569
5	37	69	101	133	165	197	229	261	293	325	357	389	421	453	485	517	549
21	53	85	117	149	181	213	245	277	309	341	373	405	437	469	501	533	565
13	45	77	109	141	173	205	237	269	301	333	365	397	429	461	493	525	557
29	61	93	125	157	189	221	253	285	317	349	381	413	445	477	509	541	573
3	35	67	99	131	163	195	227	259	291	323	355	387	419	451	483	515	547
19	51	83	115	147	179	211	243	275	307	339	371	403	435	467	499	531	563
11	43	75	107	139	171	203	235	267	299	331	363	395	427	459	491	523	555
27	59	91	123	155	187	219	251	283	315	347	379	411	443	475	507	539	571
7	39	71	103	135	167	199	231	263	295	327	359	391	423	455	487	519	551
23	55	87	119	151	183	215	247	279	311	343	375	407	439	471	503	535	567
15	47	79	111	143	175	207	239	271	303	335	367	399	431	463	495	527	559
31	63	95	127	159	191	223	255	287	319	351	383	415	447	479	511	543	575
2	34	66	98	130	162	194	226	258	290	322	354	386	418	450	482	514	546
18	50	82	114	146	178	210	242	274	306	338	370	402	434	466	498	530	562
10	42	74	106	138	170	202	234	266	298	330	362	394	426	258	490	522	554
26	58	90	122	154	186	218	250	282	314	346	378	410	442	474	506	538	570
6	38	70	102	134	166	198	230	262	294	326	358	390	422	454	486	518	550
22	54	86	118	150	182	214	246	278	310	342	374	406	438	470	502	534	566
14	46	78	110	142	174	206	238	270	302	334	366	398	430	462	494	526	558
30	62	94	126	158	190	222	254	286	318	350	382	414	446	478	510	542	574
4	36	68	100	132	164	196	228	260	292	324	356	388	420	452	484	516	548
20	52	84	116	148	180	212	244	276	308	340	372	404	436	468	500	532	564
12	44	78	108	140	172	204	236	268	300	332	364	396	428	460	492	524	556
28	60	92	124	156	188	220	252	284	316	348	380	412	444	476	508	540	572
8	40	72	104	136	168	200	232	264	296	328	360	392	424	456	488	520	552
24	56	88	120	152	184	216	248	280	312	344	376	408	440	472	504	536	568
16	48	80	112	144	176	208	240	272	304	336	368	400	432	464	496	528	560
32	64	96	128	160	192	224	256	288	320	352	384	416	448	480	512	544	576

Figure 4.34 Output of the access channel interleaver.

$$M = 2IJ = 2 \times 18 \times 32 = 1152 \tag{4.18}$$

Likewise, the delay is

$$D = 2IJ = 1{,}152 = \frac{1{,}152 \text{ sym}}{28{,}800 \text{ sym/s}} = 40 \text{ ms} \tag{4.19}$$

The reverse traffic channel performs interleaving over a single 20-ms frame period. The data rate into the interleaver is 28.8 ksps, and the block size is therefore 20 ms $\times$ 28.8 ksps = 576 symbols. As with the forward traffic channel, the reverse traffic channel uses symbol repetition.

Unlike the forward link, however, the reverse link gates the amplifier to eliminate the repeated symbols. Each 20-ms frame is divided into 16 equally spaced timeslots, called "power control groups." Each timeslot has a duration of 1.25 ms or 36 symbols, as shown previously in Figure 4.23. At an original data rate of 9.6 kbps, all timeslots contain symbols, and none is gated. At 4.8 kbps, one-half of the symbols would be gated, which would be either position 0 or 1, either position 2 or 3, and so on. At 2.4 kbps, only 1/4 are filled, and at 1.2 kbps, 1/8 are filled. Which positions contain symbols is determined pseudorandomly, a process termed *data burst randomizing*.

The reverse traffic channel interleaver for all rates is an ($I = 18$, $J = 32$) array, and thus, two rows are needed to fill one power control group. The rows of the interleaver are read out in an order that ensures that groups of 36 symbols (two rows) do not contain any duplicate symbols. The reverse traffic channel interleaver write arrays showing the symbol repetition are given in Figures 4.35, 4.36, 4.37, and 4.38 for 9.6, 4.8, 2.4, and 1.2 kbps, respectively.

The symbols are read out of the interleaver by rows in the order that is given in Table 4.8.

The combination of the 576-symbol interleaver and amplifier gating forms block interleavers whose minimum separation characteristics are dependent on the original data rate. The memory requirement and delay are independent of data rate because the original array has 576 symbols. Note that the arrays are read from the top to bottom instead of bottom to top, which reduces the minimum spacing by one. The 9.6-kbps interleaver is an ($I = 18$, $J = 32$) block interleaver. The 4.8-kbps interleaver is effectively ($I = 18$, $J = 16$). The 2.4-kbps array is effectively ($I = 18$, $J = 8$), and the 1.2-kbps interleaver is effectively ($I = 18$, $J = 4$). The term *effective* implies that only the minimum-spacing parameter is affected, and not the memory or delay. The minimum separation is thus given by

1	33	65	97	129	161	193	225	257	289	321	353	385	417	449	481	513	545
2	34	66	98	130	162	194	226	258	290	322	354	386	418	450	482	514	546
3	35	67	99	131	163	195	227	259	291	323	355	387	419	451	483	515	547
4	36	68	100	132	164	196	228	260	292	324	356	388	420	452	484	516	548
5	37	69	101	133	165	197	229	261	293	325	357	389	421	453	485	517	549
6	38	70	102	134	166	198	230	262	294	326	358	390	422	454	486	518	550
7	39	71	103	135	167	199	231	263	295	327	359	391	423	455	487	519	551
8	40	72	104	136	168	200	232	264	296	328	360	392	424	456	488	520	552
9	41	73	105	137	169	201	233	265	297	329	361	393	425	457	489	521	553
10	42	74	106	138	170	202	234	266	298	330	362	394	426	458	490	522	554
11	43	75	107	139	171	203	235	267	299	331	363	395	427	459	491	523	555
12	44	76	108	140	172	204	236	268	300	332	364	396	428	460	492	524	556
13	45	77	109	141	173	205	237	269	301	333	365	397	429	461	493	525	557
14	46	78	110	142	174	206	238	270	302	334	366	398	430	462	494	526	558
15	47	79	111	143	175	207	239	271	303	335	367	399	431	463	495	527	559
16	48	80	112	144	176	208	240	272	304	336	368	400	432	464	496	528	560
17	49	81	113	145	177	209	241	273	305	337	369	401	433	465	497	529	561
18	50	82	114	146	178	210	242	274	306	338	370	402	434	466	498	530	562
19	51	83	115	147	179	211	243	275	307	339	371	403	435	467	499	531	563
20	52	84	116	148	180	212	244	276	308	340	372	404	436	468	500	532	564
21	53	85	117	149	181	213	245	277	309	341	373	405	437	469	501	533	565
22	54	86	118	150	182	214	246	278	310	342	374	406	438	470	502	534	566
23	55	87	119	151	183	215	247	279	311	343	375	407	439	471	503	535	567
24	56	88	120	152	184	216	248	280	312	344	376	408	440	472	504	536	568
25	57	89	121	153	185	217	249	281	313	345	377	409	441	473	505	537	569
26	58	90	122	154	186	218	250	282	314	346	378	410	442	474	506	538	570
27	59	91	123	155	187	219	251	283	315	347	379	411	443	475	507	539	571
28	60	92	124	156	188	220	252	284	316	348	380	412	444	476	508	540	572
29	61	93	125	157	189	221	253	285	317	349	381	413	445	477	509	541	573
30	62	94	126	158	190	222	254	286	318	350	382	414	446	478	510	542	574
31	63	95	127	159	191	223	255	287	319	351	383	415	447	479	511	543	575
32	64	96	128	160	192	224	256	288	320	352	384	416	448	480	512	544	576

Figure 4.35 Reverse traffic channel interleaving at 9,600 bps.

$$S = \begin{cases} J - 1 & B \leq I \\ 1 & B > I \end{cases} \tag{4.20}$$

where the I and J are the effective values. As with conventional block interleavers, the memory requirement is twice the number of symbols in the array

$$M = 2IJ = 2 \times 18 \times 32 = 1152 \tag{4.21}$$

Likewise, the delay is

1	17	33	49	65	81	97	113	129	145	161	177	193	209	225	241	257	273
1	17	33	49	65	81	97	113	129	145	161	177	193	209	225	241	257	273
2	18	34	50	66	82	98	114	130	146	162	178	194	210	226	242	258	274
2	18	34	50	66	82	98	114	130	146	162	178	194	210	226	242	258	274
3	19	35	51	67	83	99	115	131	147	163	179	195	211	227	243	259	275
3	19	35	51	67	83	99	115	131	147	163	179	195	211	227	243	259	275
4	20	36	52	68	84	100	116	132	148	164	180	196	212	228	244	260	276
4	20	36	52	68	84	100	116	132	148	164	180	196	212	228	244	260	276
5	21	37	53	69	85	101	117	133	149	165	181	197	213	229	245	261	277
5	21	37	53	69	85	101	117	133	149	165	181	197	213	229	245	261	277
6	22	38	54	70	86	102	118	134	150	166	182	198	214	230	246	262	278
6	22	38	54	70	86	102	118	134	150	166	182	198	214	230	246	262	278
7	23	39	55	71	87	103	119	135	151	167	183	199	215	231	247	263	279
7	23	39	55	71	87	103	119	135	151	167	183	199	215	231	247	263	279
8	24	40	56	72	88	104	120	136	152	168	184	200	216	232	248	264	280
8	24	40	56	72	88	104	120	136	152	168	184	200	216	232	248	264	280
9	25	41	57	73	89	105	121	137	153	169	185	201	217	233	249	265	281
9	25	41	57	73	89	105	121	137	153	169	185	201	217	233	249	265	281
10	26	42	58	74	90	106	122	138	154	170	186	202	218	234	250	266	282
10	26	42	58	74	90	106	122	138	154	170	186	202	218	234	250	266	282
11	27	43	59	75	91	107	123	139	155	171	187	203	219	235	251	267	283
11	27	43	59	75	91	107	123	139	155	171	187	203	219	235	251	267	283
12	28	44	60	76	92	108	124	140	156	172	188	204	220	236	252	268	284
12	28	44	60	76	92	108	124	140	156	172	188	204	220	236	252	268	284
13	29	45	61	77	93	109	125	141	157	173	189	205	221	237	253	269	285
13	29	45	61	77	93	109	125	141	157	173	189	205	221	237	253	269	285
14	30	46	62	78	94	110	126	142	158	174	190	206	222	238	254	270	286
14	30	46	62	78	94	110	126	142	158	174	190	206	222	238	254	270	286
15	31	47	63	79	95	111	127	143	159	175	191	207	223	239	255	271	287
15	31	47	63	79	95	111	127	143	159	175	191	207	223	239	255	271	287
16	32	48	64	80	96	112	128	144	160	176	192	208	224	240	256	272	288
16	32	48	64	80	96	112	128	144	160	176	192	208	224	240	256	272	288

Figure 4.36 Reverse traffic channel interleaving at 4,800 bps.

$$D = 2IJ = 1{,}152 = \frac{1{,}152 \text{ sym}}{28{,}800 \text{ sym/s}} = 40 \,\text{ms} \tag{4.22}$$

4.4.3 Diversity and Handoff

The digital spread-spectrum design of the IS-95 forward and reverse link waveforms permits the use of several forms of diversity in addition to the time diversity inherent in the repetition, encoding, and interleaving of the data symbols. These forms of diversity include multipath diversity and base station diversity, the latter being available with or without the prospect of handing off the call to a different base station.

1	9	17	25	33	41	49	57	65	73	81	89	97	105	113	121	129	137
1	9	17	25	33	41	49	57	65	73	81	89	97	105	113	121	129	137
1	9	17	25	33	41	49	57	65	73	81	89	97	105	113	121	129	137
1	9	17	25	33	41	49	57	65	73	81	89	97	105	113	121	129	137
2	10	18	26	34	42	50	58	66	74	82	90	98	106	114	122	130	138
2	10	18	26	34	42	50	58	66	74	82	90	98	106	114	122	130	138
2	10	18	26	34	42	50	58	66	74	82	90	98	106	114	122	130	138
2	10	18	26	34	42	50	58	66	74	82	90	98	106	114	122	130	138
3	11	19	27	35	43	51	59	67	75	83	91	99	107	115	123	131	139
3	11	19	27	35	43	51	59	67	75	83	91	99	107	115	123	131	139
3	11	19	27	35	43	51	59	67	75	83	91	99	107	115	123	131	139
3	11	19	27	35	43	51	59	67	75	83	91	99	107	115	123	131	139
4	12	20	28	36	44	52	60	68	76	84	92	100	108	116	124	132	140
4	12	20	28	36	44	52	60	68	76	84	92	100	108	116	124	132	140
4	12	20	28	36	44	52	60	68	76	84	92	100	108	116	124	132	140
4	12	20	28	36	44	52	60	68	76	84	92	100	108	116	124	132	140
5	13	21	29	37	45	53	61	69	77	85	93	101	109	117	125	133	141
5	13	21	29	37	45	53	61	69	77	85	93	101	109	117	125	133	141
5	13	21	29	37	45	53	61	69	77	85	93	101	109	117	125	133	141
5	13	21	29	37	45	53	61	69	77	85	93	101	109	117	125	133	141
6	14	22	30	38	46	54	62	70	78	86	94	102	110	118	126	134	142
6	14	22	30	38	46	54	62	70	78	86	94	102	110	118	126	134	142
6	14	22	30	38	46	54	62	70	78	86	94	102	110	118	126	134	142
6	14	22	30	38	46	54	62	70	78	86	94	102	110	118	126	134	142
7	15	23	31	39	47	55	63	71	79	87	95	103	111	119	127	135	143
7	15	23	31	39	47	55	63	71	79	87	95	103	111	119	127	135	143
7	15	23	31	39	47	55	63	71	79	87	95	103	111	119	127	135	143
7	15	23	31	39	47	55	63	71	79	87	95	103	111	119	127	135	143
8	16	24	32	40	48	56	64	72	80	88	96	104	112	120	128	136	144
8	16	24	32	40	48	56	64	72	80	88	96	104	112	120	128	136	144
8	16	24	32	40	48	56	64	72	80	88	96	104	112	120	128	136	144
8	16	24	62	40	48	56	64	72	80	88	96	104	112	120	128	136	144

Figure 4.37 Reverse traffic channel interleaving at 2,400 bps.

Because the forward link waveform for a particular channel is a dual-quadrature direct-sequence spread-spectrum signal, it is possible to employ spread-spectrum correlation techniques to isolate a single multipath component of that channel's signal and to discriminate not only against signals from other base stations but also against multipath components of the same channel's received signal. Using the Rake technique, in which the receiver uses several parallel receiver "fingers" to isolate multipath components, on the forward link it is possible to extract several multipath components from the total received signal and to align them for optimal combining.

In the implementation of the IS-95 system, the mobile receiver employs a "searcher" receiver and three digital data receivers that act as fingers of a rake in that they may be assigned to track and isolate particular multipath

1	5	9	13	17	21	25	29	33	37	41	45	49	53	57	61	65	69
1	5	9	13	17	21	25	29	33	37	41	45	49	53	57	61	65	69
1	5	9	13	17	21	25	29	33	37	41	45	49	53	57	61	65	69
1	5	9	13	17	21	25	29	33	37	41	45	49	53	57	61	65	69
1	5	9	13	17	21	25	29	33	37	41	45	49	53	57	61	65	69
1	5	9	13	17	21	25	29	33	37	41	45	49	53	57	61	65	69
1	5	9	13	17	21	25	29	33	37	41	45	49	53	57	61	65	69
1	5	9	13	17	21	25	29	33	37	41	45	49	53	57	61	65	69
2	6	10	14	18	22	26	30	34	38	42	46	50	54	58	62	66	70
2	6	10	14	18	22	26	30	34	38	42	46	50	54	58	62	66	70
2	6	10	14	18	22	26	30	34	38	42	46	50	54	58	62	66	70
2	6	10	14	18	22	26	30	34	38	42	46	50	54	58	62	66	70
2	6	10	14	18	22	26	30	34	38	42	46	50	54	58	62	66	70
2	6	10	14	18	22	26	30	34	38	42	46	50	54	58	62	66	70
2	6	10	14	18	22	26	30	34	38	42	46	50	54	58	62	66	70
2	6	10	14	18	22	26	30	34	38	42	46	50	54	58	62	66	70
3	7	11	15	19	23	27	31	35	39	43	47	51	55	59	63	67	71
3	7	11	15	19	23	27	31	35	39	43	47	51	55	59	63	67	71
3	7	11	15	19	23	27	31	35	39	43	47	51	55	59	63	67	71
3	7	11	15	19	23	27	31	35	39	43	47	51	55	59	63	67	71
3	7	11	15	19	23	27	31	35	39	43	47	51	55	59	63	67	71
3	7	11	15	19	23	27	31	35	39	43	47	51	55	59	63	67	71
3	7	11	15	19	23	27	31	35	39	43	47	51	55	59	63	67	71
3	7	11	15	19	23	27	31	35	39	43	47	51	55	59	63	67	71
4	8	12	16	20	24	28	32	36	40	44	48	52	56	60	64	68	72
4	8	12	16	20	24	28	32	36	40	44	48	52	56	60	64	68	72
4	8	12	16	20	24	28	32	36	40	44	48	52	56	60	64	68	72
4	8	12	16	20	24	28	32	36	40	44	48	52	56	60	64	68	72
4	8	12	16	20	24	28	32	36	40	44	48	52	56	60	64	68	72
4	8	12	16	20	24	28	32	36	40	44	48	52	56	60	64	68	72
4	8	12	16	20	24	28	32	36	40	44	48	52	56	60	64	68	72
4	8	12	16	20	24	28	32	36	40	44	48	52	56	60	64	68	72

Figure 4.38 Reverse traffic channel interleaving at 1,200 bps.

components of a base station signal; the block diagram in Figure 4.39 illustrates this parallel receiver concept. The search receiver scans the time domain about the desired signal's expected time of arrival for multipath pilot signals from the same cell site and pilot signals (and their multipath components) from other cell sites. Searching the time domain on the forward link is simplified because the pilot channel permits the coherent detection of signals. The search receiver indicates to the mobile phone's control processor where, in time, the strongest replicas of the signal can be found, and their respective signal strengths. In turn, the control processor provides timing and PN code information to the three digital data receivers, enabling each of them to track and demodulate a different signal.

Table 4.8 Order of the traffic channel read operation

Rate	Row Number
9.6 kbps	0, 1, 2, 3, 4, 5, 6, 7, 8, 9, 10, 11, 12, 13, 14, 15, 16, 17, 18, 19, 20, 21, 22, 23, 24, 25, 26, 27, 28, 29, 30, 31
4.8 kbps	0, 2, 1, 3, 4, 6, 5, 7, 8, 10, 9, 11, 12, 14, 13, 15, 16, 18, 17, 19, 20, 22, 21, 23, 24, 26, 25, 27, 28, 30, 29, 31
2.4 kbps	0, 4, 1, 5, 2, 6, 3, 7, 8, 12, 9, 13, 10, 14, 11, 15, 16, 20, 17, 21, 18, 22, 19, 23, 24, 28, 25, 29, 26, 30, 27, 31
1.2 kbps	0, 8, 1, 9, 2, 10, 3, 11, 4, 12, 5, 13, 6, 14, 7, 15, 16, 24, 17, 25, 18, 26, 19, 27, 20, 28, 21, 29, 22, 30, 23, 31

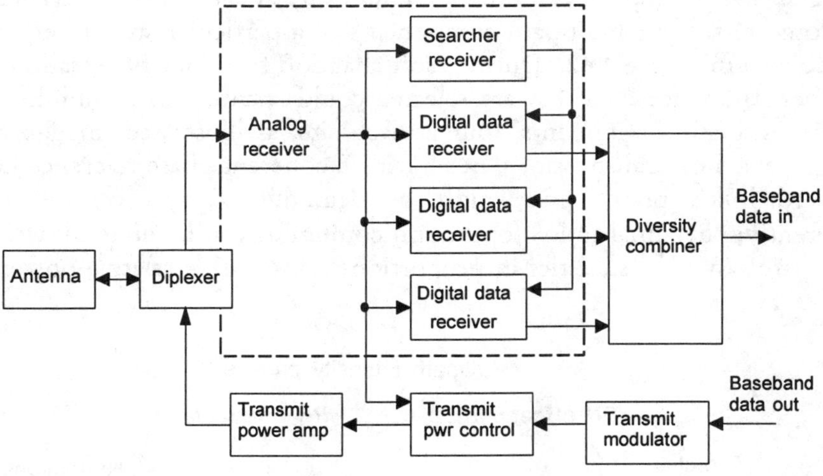

Figure 4.39 Mobile station block diagram.

The principles of diversity combining and Rake receiver processing are explained in Chapter 9.

If another cell site pilot signal becomes significantly stronger than the current pilot signal, the control processor initiates handoff procedures during which the forward links of both cell sites transmit the same call data on all their traffic channels. When both sites handle the call, additional space diversity is obtained. When handoff is not contemplated, in a cell site diversity mode the strongest paths from multiple cell sites are determined by the search receiver, and the digital data receivers are assigned to demodulate these paths. As shown conceptually in Figure 4.40, the data from all three digital receivers are combined for improved resistance to fading; in the figure, the fact that different base stations or sectors are distinguished by different short PN code offsets is highlighted by identifying the two adjacent base stations with different offsets of $64i$ PN code chips and $64j$ PN chips, respectively.

The forward link performs coherent postdetection (after demodulation) combining after ensuring that the data streams are time-aligned; performance is not compromised by using postdetection combining because the modulation technique is linear. Coherent combining, illustrated in Figure 4.41, is possible because the pilot signal from each base station provides a coherent phase reference that can be tracked by the digital data receivers.

On the reverse link, the base station receiver uses two antennas for space diversity reception, and there are four digital data receivers available for tracking up to four multipath components of a particular subscriber's signal, as depicted in Figure 4.42. During "soft" handoff from one base station site to another, the voice data that are selected could result from combining up to eight multipath components, four at each site, as illustrated in Figure 4.43. The reverse link transmissions, not having a coherent phase reference like the forward link's pilot signal, must be demodulated and combined non-coherently; "maximal ratio" (optimum) combining can be done by weighting each path's symbol statistics in proportion to the path's relative power prior

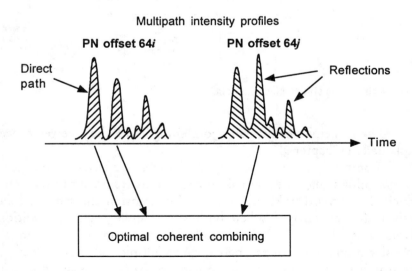

Figure 4.40 Forward link multipath/base station diversity.

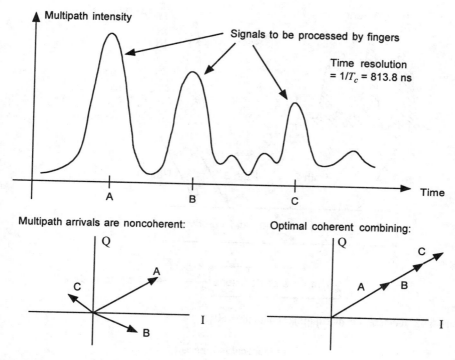

Figure 4.41 Concept of coherent combining.

to demodulation and decoding decisions. At the MTSO, the frames of data from the two base stations are selection-diversity "combined"; that is, the frame with the better quality is chosen. (These diversity combining methods are treated in Chapter 9.)

The soft handoff depicted as in progress in Figure 4.43 is termed "soft" because the connection of the mobile with one station is not broken off until one is established with the new station, rather than the "hard" handoff procedure that is necessary in analog FDMA cellular systems because the receiver must tune to a new frequency. Because the combining is done at the MTSO, however, under soft handoff, it is necessary to make hard data decisions to transmit the data frame to the MTSO. A "softer" handoff can occur in an IS-95 CDMA system when the mobile is in transition between two sectors of the same cell site. As illustrated in Figure 4.44, up to four multipath components of the mobile's transmission, as received at up to four antennas, can be combined before making a hard data decision. The figure indicates that the handoff process involves a gradually increasing preference for multipath

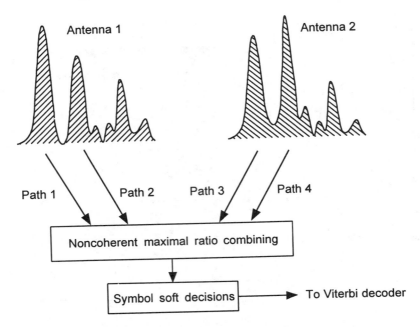

Figure 4.42 Reverse link multipath/antenna diversity.

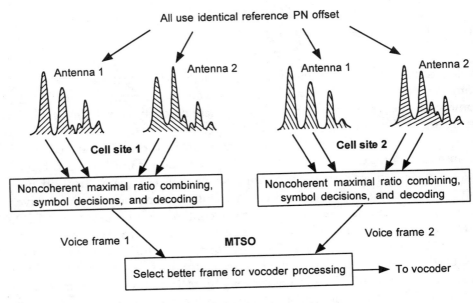

Figure 4.43 Reverse link soft handoff/multipath/base station diversity.

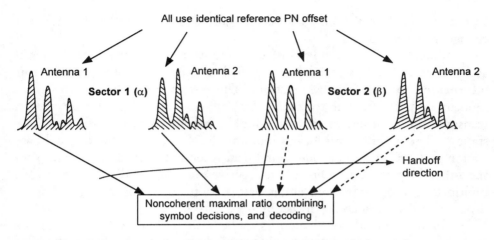

All use identical reference PN offset

Antenna 1   Antenna 2   Antenna 1   Antenna 2

Sector 1 (α)   Sector 2 (β)

Handoff
direction

Noncoherent maximal ratio combining,
symbol decisions, and decoding

Figure 4.44 Reverse link "softer" handoff between sectors of a cell site.

components received in one sector, until the mobile signal is being processed by just one sector.

Note that on reverse link transmissions, the mobile transmits using zero-offset short PN codes; during handoff, each base station must align its receiver to the unique value of propagation delay experienced by the mobile signal, using the preamble sent on reverse link access channel and traffic channel transmissions as an aid to acquisition. On the forward link, during handoff each base station or sector uses its usual short PN code offset.

As shown previously in Figures 4.8 and 4.9, the IS-95 system modulates the same data on both the in-phase and quadrature-phase channels. This modulation is different from conventional QPSK where different data are modulated on the I and Q channels. The IS-95 scheme is not conventional QPSK but a form of diversity transmission that reduces multiple access interference relative to a conventional BPSK scheme. A derivation of this fact is presented in Chapter 7.

The IS-95 CDMA system "soft handoff" greatly reduces the number of failed handoffs. As a call is initiated, the base station sends the mobile, among other parameters, a set of handoff thresholds and candidate base stations for handoff. The mobile tracks all possible pilot phase offsets in the system, with the emphasis on the candidates and the current base station. Those base stations that have a signal above the candidate threshold are kept on a periodically updated list. The list is updated when (1) a new candidate exceeds the "add threshold," (2) an old candidate remains below the "drop threshold" for a predetermined time, (3) the list becomes too large, or (4) a handoff forces a change in the candidates.

When the pilot signal level of another base station exceeds the currently active base station's signal by a specified amount, the mobile relays this information to the active base station. The system controller then assigns a modem in the new cell to handle the call and the original base station transmits the information necessary for the mobile to acquire the new base station. The mobile continues to monitor the the E_c/I_0 of the original base station and assigns at least one digital data receiver to demodulate data from the other base station. Digital data receivers are assigned to the strongest multipath components from either base station, providing a form of space diversity as discussed above in Section 4.4.2.

An example of soft handoff can be seen in Figure 4.45, where the mobile is originally communicating through base station A. When the pilot signal of base station B exceeds the add threshold, the mobile informs base station A, and base station B is put on the candidate list. Once the signal strength of base B exceeds the signal strength of base A by a specified margin, base B is put on the active list and begins to handle the call also. The signals to and from both base stations are diversity combined. As soon as the signal

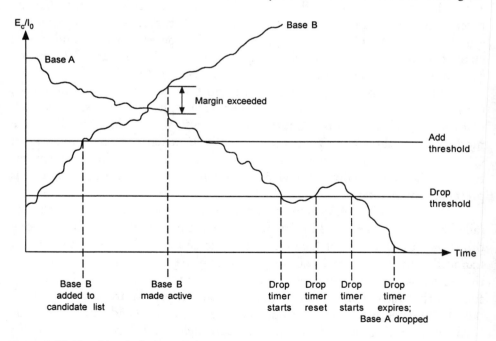

Figure 4.45 Signal levels during soft handoff.

from base station A crosses the drop threshold, the drop timer is started. The timer is reset when the signal exceeds the drop threshold. When the signal falls below the drop threshold, the drop timer is restarted. After the drop timer expires, base station A is dropped from active service.

Soft handoff methods have several advantages over conventional hard handoff methods. Contact with the new base station is made before the call is switched, which prevents the mobile from losing contact with the system if the handoff signal is not heard or incorrectly interpreted. Diversity combining is used between multiple cell sites, allowing for additional resistance to fading. Because more than one base station is used in the soft handoff region, the ping-ponging effect can be virtually eliminated. Thus, the system is relieved from unnecessary switching operations. The system is further relieved from determining signal strengths, because this processing is done at the mobile. If the new cell is loaded to capacity, handoff can still be performed if all users accept a small increase in the bit-error rate. This type of handoff can be likened to make-before-break switching. Note that in soft handoff, neither the mobile nor the base station is required to change frequency. Conventional hard handoff is performed if the mobile is required to switch to an analog system or change to another CDMA frequency. Frequency changes may be necessary if the current frequency is crowded or the mobile enters a new CDMA network.

References

[1] "Mobile Station-Base Station Compatibility Standard for Dual-Mode Wideband Spread Spectrum Cellular System," TIA/EIA Interim Standard 95 (IS-95), Washington, D.C.: Telecommunications Industry Association, July 1993 (amended as IS-95-A in May 1995).

[2] "Cellular System Dual-Mode Mobile Station–Base Station Compatibility Standard," TIA/EIA Interim Standard 54 (IS-54-B), Washington, D.C.: Telecommunications Industry Association, April 1992.

[3] Viterbi, A. J., *Principles of Spread Spectrum Multiple Access Communication*, New York: Addison-Wesley, 1995.

[4] Carlson, B. C., *Special Functions of Applied Mathematics*, New York: Academic Press, 1977.

[5]   Knuth, D. N., *Sorting and Searching*, vol. 3 of *The Art of Computer Programming*, Reading, MA: Addison-Wesley, 1973.

[6]   Lehmer, D. H. "Mathematical Methods in Large Scale Computing Units," *Annals Computation Laboratory Harvard Univ.*, Vol. 26, pp. 141–146, 1951.

Appendix 4A Theory of Interleaving

The most commonly employed interleaving techniques fall into two classes. The more common type is *block interleaving*, which is often used when the data is divided into blocks or frames, such as in the IS-95 system. *Convolutional interleaving*, on the other hand, is more practical for a continuous data stream. Block interleaving is known for its ease of implementation, whereas convolutional interleaving has performance advantages. Continuous operation allows the original overhead associated with convolutional interleaving to become insignificant.

4A.1 Block Interleaving

The fundamentals of block interleaving are covered in Section 4.4.2. This section simply gives an example demonstrating the fact that the minimum separation is affected by reading out the rows in another order. Assume that the block contains 20 symbols and is read into an $I = 5$, $J = 4$ interleaver. Let a number be assigned to each symbol that indicates its original order in the transmission sequence. After reading into the interleaver, it can be represented by the following array:

$$\begin{bmatrix} 1 & 5 & 9 & 13 & 17 \\ 2 & 6 & 10 & 14 & 18 \\ 3 & 7 & 11 & 15 & 19 \\ 4 & 8 & 12 & 16 & 20 \end{bmatrix}$$

If the symbols are read out in a conventional manner, then the output of the interleaver would be

4, 8, 12, 16, 20, 3, 7, 11, 15, 19, 2, 6, 10, 14, 18, 1, 5, 9, 13, 17

Note that the minimum separation of any burst of $B \leq I = 5$ symbols is $J = 4$, and that for $B > I = 5$, $S = 1$.

Now suppose that the symbols are read out by rows in the following order: first row, third row, second row, and then the last row. The symbols would be ordered

$$1, 5, 9, 13, 17, 3, 7, 11, 15, 19, 2, 6, 10, 14, 18, 4, 8, 12, 16, 20$$

The minimum-separation profile becomes

$$
S = \begin{cases}
4 & B \leq 4 \text{ (e.g., between 1 and 5)} \\
2 & B = 5 \text{ (e.g., between 1 and 3)} \\
1 & B > 5 \text{ (e.g., between 3 and 2)}
\end{cases}
$$

In this particular case, the minimum separation was reduced from 4 to 2 for $B = 5$. Increasing the minimum separation for $B > I$ becomes easier with larger interleaver sizes. Note also that the memory requirement and delay remain essentially unchanged.

4A.2 Convolutional Interleaving

Convolutional interleavers have a series of K parallel lines with an input commutator moving between each line much like an encoder for a convolutional code. Each of the K lines contains various numbers of storage elements, as shown in Figure 4A.1. The first line is a short from the input to output, and the second has L storage elements. Each successive line has L more units than the previous, and thus, the last contains $K(L - 1)$ storage registers. The storage elements must be loaded with some initial symbols, which are shifted out into channel and increase overhead. This loading is similar to tail bits in convolutional coding. An output commutator moves in synchronization with the input commutator to extract a symbol every time one is entered. The deinterleaver performs the complementary operations are requires a corresponding structure, depicted in Figure 4A.2.

An example, using a $K = 4$, $L = 2$ interleaver, will help illustrate the convolutional interleaving process. Let the initial loading of each register be represented by any symbol "i," and the 48 input symbols be represented by x_j, where j is the initial ordering ($j = 1, 2, \ldots, 47$). The initial conditions

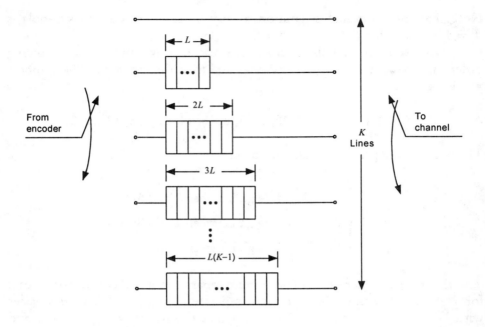

Figure 4A.1 (K, L) **Convolutional interleaver structure.**

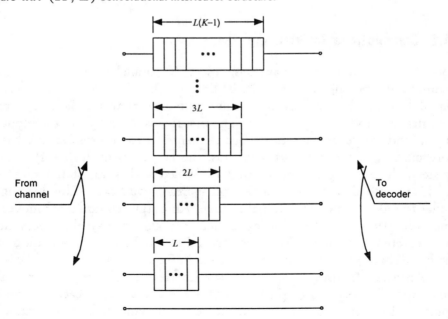

Figure 4A.2 **The deinterleaver performs the reverse operation.**

and the interleaving process for the first three commutator periods are shown in Figures 4A.3, 4A.4, 4A.5, and 4A.6.

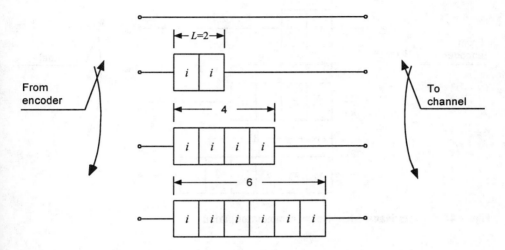

Figure 4A.3   Initial conditions of the interleaver example.

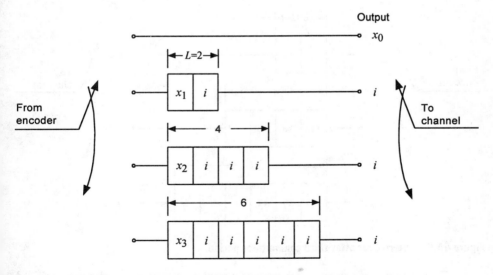

Figure 4A.4   Interleaver after first commutator period.

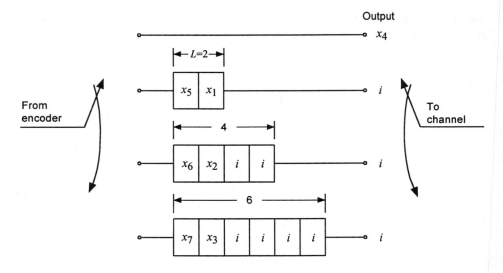

Figure 4A.5   Interleaver after second commutator period.

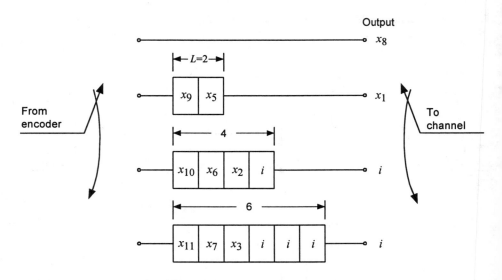

Figure 4A.6   Interleaver after third commutator period.

The interleaver input and output for the example are shown in Table 4A.1. The initial symbols are scattered throughout the desired data and are required to be gated or transmitted to the channel, which increases overhead,

Table 4A.1 Input and output of the interleaver

Period	Input	Output
1	x_0, x_1, x_2, x_3	x_0, i, i, i
2	x_4, x_5, x_6, x_7	x_4, i, i, i
3	x_8, x_9, x_{10}, x_{11}	x_8, x_1, i, i
4	$x_{12}, x_{13}, x_{14}, x_{15}$	x_{12}, x_5, i, i
5	$x_{16}, x_{17}, x_{18}, x_{19}$	x_{16}, x_9, x_2, i
6	$x_{20}, x_{21}, x_{22}, x_{23}$	x_{20}, x_{13}, x_6, i
7	$x_{24}, x_{25}, x_{26}, x_{27}$	$x_{24}, x_{17}, x_{10}, x_3$
8	$x_{28}, x_{29}, x_{30}, x_{31}$	$x_{28}, x_{21}, x_{14}, x_7$
9	$x_{32}, x_{33}, x_{34}, x_{35}$	$x_{32}, x_{25}, x_{18}, x_{11}$
10	$x_{36}, x_{37}, x_{38}, x_{39}$	$x_{36}, x_{29}, x_{22}, x_{15}$
11	$x_{40}, x_{41}, x_{42}, x_{43}$	$x_{40}, x_{33}, x_{26}, x_{19}$
12	$x_{44}, x_{45}, x_{46}, x_{47}$	$x_{44}, x_{37}, x_{30}, x_{23}$

but if continuous operation is employed, these initial symbols need only be sent once. Table 4A.2 contains the input and output of the deinterleaver corresponding to the interleaver. It is assumed that the deinterleaver is initially loaded with d symbols. Note that after the first 24 symbols, the symbol ordering is restored.

By observing the interleaver output, the minimum separation between any symbols of a burst of length B is found to be

Table 4A.2 Input and output of the deinterleaver

Period	Input	Output
1	x_0, i, i, i	d, d, d, i
2	x_4, i, i, i	d, d, d, i
3	x_8, x_1, i, i	d, d, i, i
4	x_{12}, x_5, i, i	d, d, i, i
5	x_{16}, x_9, x_2, i	d, i, i, i
6	x_{20}, x_{13}, x_6, i	d, i, i, i
7	$x_{24}, x_{17}, x_{10}, x_3$	x_0, x_1, x_2, x_3
8	$x_{28}, x_{21}, x_{14}, x_7$	x_4, x_5, x_6, x_7
9	$x_{32}, x_{25}, x_{18}, x_{11}$	x_8, x_9, x_{10}, x_{11}
10	$x_{36}, x_{29}, x_{22}, x_{15}$	$x_{12}, x_{13}, x_{14}, x_{15}$
11	$x_{40}, x_{33}, x_{26}, x_{19}$	$x_{16}, x_{17}, x_{18}, x_{19}$
12	$x_{44}, x_{37}, x_{30}, x_{23}$	$x_{20}, x_{21}, x_{22}, x_{23}$

$$S = \begin{cases} 7, & B \le 4 \\ 4, & B = 5 \\ 3, & 6 \le B \le 9 \\ 1, & B \ge 10 \end{cases}$$

The number of memory elements can be determined by direct count to be

$$M = 0 + 2 + 4 + 6 = 12$$

The delay is equal to the number of memory elements at the transmitter and the number at the receiver, or

$$D = 2 \times 12 = 24$$

These observations can be extended to any (K, L) convolutional interleaver. The minimum-separation profile is given by

$$S = \begin{cases} KL - 1, & 2 \le B \le K \\ K, & K + 1 \le B \le K(L - 1) + 1 \\ K - 1, & K(L - 1) + 2 \le B \le KL + 1 \\ 1, & B \ge KL + 2 \end{cases} \tag{4A.1}$$

The number of storage elements is given by

$$M = L \sum_{i=0}^{K-1} i = \frac{LK(K - 1)}{2} \tag{4A.2}$$

The delay is twice the number of storage elements, or

$$D = 2M = LK(K - 1) \tag{4A.3}$$

4A.3 Comparison of Block and Convolutional Interleaving

The comparison is based mainly on the S/D and S/M ratios for the two interleavers. For the block interleaver, the ratios can be found by using equations (4.10), (4.11), and (4.12):

$$\frac{S}{M}(\text{block}) = \frac{J}{2IJ} = \frac{1}{2I} \quad \text{for } B \leq I \tag{4A.4}$$

$$\frac{S}{D}(\text{block}) = \frac{J}{2IJ} = \frac{1}{2I} \quad \text{for } B \leq I \tag{4A.5}$$

For the convolutional interleaver, the ratios can be found using equations (4A.1), (4A.2), and (4A.3):

$$\frac{S}{M}(\text{convolutional}) = \frac{KL - 1}{KL(K-1)/2} \quad \text{for } B \leq K \tag{4A.6a}$$

$$\approx \frac{2KL}{KL(K-1)} = \frac{2}{K-1} \tag{4A.6b}$$

$$\frac{S}{D}(\text{convolutional}) = \frac{KL - 1}{KL(K-1)} \quad \text{for } B \leq K \tag{4A.7a}$$

$$\approx \frac{1}{K-1} \tag{4A.7b}$$

The S/M ratio is approximately four times as large with convolutional interleaving as with block interleaving. The S/D ratio for convolutional interleaving is approximately twice that for block interleaving. Convolutional interleaving also has the advantage that for $B > K$, the minimum separation is greater than unity, which is not the case for block interleaving. As noted previously, the original contents of the convolutional registers must be transmitted over the channel, which can reduce overhead. Because of its simplicity and the fact that most data are grouped in blocks, however, block interleaving is more commonly employed.

4A.4 Interleaver Design

The main purpose of interleaving is to disperse bursts of errors over time so that an error control code can correct the errors. Therefore, the longest expected burst should be dispersed over a great enough time to allow correction. The memory span N of a code is defined as the number of code symbols required to achieve the full error-correcting capability t. For block error-correcting codes, the decoder span is the block length ($N = n$). For convolutional codes, the decoder span is more difficult to define. It is generally accepted that two to five constraint lengths of channel symbols is sufficient ($N = 2nK'$ to $5nK'$, where n is the branch length and K' is the constraint length).

The following example illustrates the distribution of errors within a decoder span. Consider the (31, 16) BCH code that corrects $t = 3$ errors, and $N = n = 31$. The interleaver should ensure that no more than $t = 3$ errors occur within one memory span. To satisfy this requirement, the minimum separation for the interleaver must be greater than or equal to 11, as shown in Figure 4A.7. If the minimum separation is less than 11, then it is possible to have four errors within one memory span, as illustrated in Figure 4A.8.

The number of errors in any memory span should be less than the error-correcting capability of the code. The average number of errors in a decoder span is N/S. Because there must be t or fewer errors in a memory span:

$$t \geq \frac{N}{S} \tag{4A.8}$$

Solving for the minimum separation gives

$$S \geq \left\lceil \frac{N}{t} \right\rceil \tag{4A.9}$$

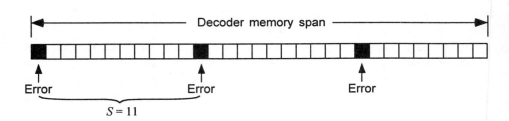

Figure 4A.7 Interleaving should ensure that at most t errors are within a memory span.

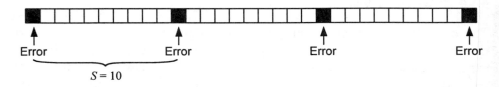

Figure 4A.8 If more than t errors occur, the code cannot correct them.

where $\lceil x \rceil$ denotes the smallest integer greater than of equal to x. For both convolutional and block interleaving, the largest minimum separation is achieved for $B \leq I$ or K. Thus, when determining interleaver structure, the longest burst expected should be less than I or K, and equation (4.31) should be satisfied.

Other practical factors affect interleaver structure. Data are usually arranged into blocks in order to be processed more manageably. Blocks can be determined by channel coding, voice coding, protocol requirements, and other framing requirements. Interleaving is usually performed within an integer number of frames. For example, in IS-95, interleaving is performed within a single frame. In IS-54 [2], interleaving is performed within two frames. Because delays of several hundred milliseconds are noticeable in speech, the total delay should be limited to this amount. Interleaver delay is a large part of the total delay, which also includes propagation, voice coding, and channel coding delays. When block codes or control messages are used, it is desirable to make the interleaver memory an integral number of code words or messages. If not, then the decoder must wait until the remainder of the code word or message is received before processing can begin.

Appendix 4B   Hash Function Used in IS-95

In IS-95, the base or mobile is often presented with a number of resources and must choose which one to use. For example, if there are seven active paging channels, the base station must decide which of these seven channels to use when paging a particular mobile subscriber. The choice should be uniform and reproducible, in the sense that it should be

Uniform. All resources should be used uniformly to prevent collisions (contention for the use of the same resource by more than one user) or overloads (exceeding the capabilities of the resource). For example, page messages should be evenly distributed over all the paging channels and all of the slot resources within the paging channels to prevent overloading a single channel.

Reproducible. It is advantageous for both the base station and the mobile to be able to independently reproduce the choice of resources from information known to both. For example, from information known by both base station and mobile, the mobile can reproduce the base's choice of paging channel and monitor only that paging channel.

One method of making a uniform choice from among a number of resources is by using a "hash function." The hash function allows this uniform distribution through a method that is reproducible at both the mobile and the base station using as inputs (a) the number of resources and (b) a parameter (hash key) that is known by both terminals. We denote the hash key as K and the number of resources as N. With the input of a hash key K, the output of the hash function $h(K)$ will be an integer value between 0 and $N - 1$, thus taking N values:

$$0 \leq h(K) \leq N - 1 \tag{4B.1}$$

Because the keys in practice may differ only slightly, such as two telephone numbers differing only in the last digit, we must be careful to find a hash function that breaks up clusters of almost identical keys to reduce the number of collisions (the same hash function outputs).

A general form for a *normalized hash function* $h'(K)$, where $0 \leq h'(K) \leq 1$, is given by

$$h'(K) = (cK) \bmod 1, \quad 0 < c < 1 \tag{4B.2}$$

Taking a number modulo 1 is the same as discarding the integer part and keeping the mantissa (the part of the number to the right of the decimal point). Examples: $10.261 \bmod 1 = 0.261$, $11 \bmod 1 = 0$. The normalized hash function can be extended to generate an integer between 0 and $N - 1$ by using the operation

$$h(K) = \lfloor N \times (cK) \bmod 1 \rfloor, \quad 0 < c < 1 \tag{4B.3}$$

where $\lfloor x \rfloor$ is the greatest integer less than x, and N is the number of resources.

One method for generating a fractional number c for use in hash functions is to take c as the ratio $c = A/w$, where $w = 2^B$, B is the word size of the processor, and $A < w$ is some positive integer that is relatively prime to w. Computers represent numbers by a certain number of binary digits; the number of digits that a processor uses is called the word size B. "Relatively prime" means that the numbers have no common divisors.

Example 4B.1 Suppose $A = 4$ and $w = 8$, where A and w are not relatively prime. Then, $c = A/w = 0.5$ and we have the values

K	1	2	3	4
$(cK) \bmod 1$	0.5	0	0.5	0

The values of $(cK) \bmod 1$ repeat for every second value of K, a highly undesirable characteristic. Now suppose that $A = 3$ and $w = 8$, where these two numbers are relatively prime. Then $c = A/w = 0.375$ and we have the values

K	1	2	3	4	5	6	7	8
$(cK) \bmod 1$	.375	.75	.125	.5	.875	.25	.625	0

When A is relatively prime to w, we have the *maximal period* of $w = 8$.

We have demonstrated that w is the maximal length of the period. Thus, we want the value of w to be large. It is also desirable to simplify the arithmetic involved in dividing A by w. Dividing a number by 2^i results in the number's being shifted i places to the right in binary notation. For example, 24 (binary 11000) divided by $4 = 2^2$ equals 6 (binary 110). Thus, we should make w the largest power of two that can be represented on the given processor, which for a B-bit processor is $w = 2^B$.

With w fixed, c is determined by the value of A, which should be chosen so that the numbers generated by the hash function have as uniform a distribution as possible. We will show that A should be chosen to be the integer relatively prime to w and closest to $0.618w$; that is, $A/w \approx 0.618 = \left(\sqrt{5} - 1\right)/2$, a fraction called the *golden ratio* in mathematics. To demonstrate that $A/w \approx 0.618$ provides a uniform distribution, a quick review of the golden ratio is covered before proceeding.

4B.1 Review of the Golden Ratio and Fibonacci Numbers

The golden ratio, denoted here by g, is defined as the ratio between the lengths of two portions of a line segment in which the ratio of the shorter portion to the longer portion equals the ratio of the longer portion to the total length:

$$g \triangleq \frac{d_2}{d_1} = \frac{d_1}{d_1 + d_2} = \frac{1}{1 + d_2/d_1} = \frac{1}{1 + g} \tag{4B.4}$$

Solving for g gives the quadratic equation $g^2 + g - 1 = 0$, for which the positive root is $g = \left(\sqrt{5} - 1\right)/2 = 0.618$.

Consider the equation $t^2 - t - 1 = 0$. The roots of this equation are

$$x = \frac{1 + \sqrt{5}}{2} = 1.618 \quad \text{and} \quad y = \frac{1 - \sqrt{5}}{2} = -0.618$$

Note that $x + y = 1$ and $xy = -1$. The *Fibonacci number* [4] F_n for integer $n = 0, 1, 2, \ldots$ is defined by

$$F_n \equiv \frac{x^n - y^n}{x - y} \quad \text{where } x \text{ and } y \text{ are the roots of } t^2 - t - 1 = 0 \qquad (4\text{B}.5)$$

Note that $F_0 = \frac{x^0 - y^0}{x - y} = \frac{1 - 1}{x - y} = 0$ and $F_1 = \frac{x^1 - y^1}{x - y} = \frac{x - y}{x - y} = 1$. Because $x^2 = x + 1$ and $y^2 = y + 1$, it follows that

$$x^{n+1} = x^n + x^{n-1} \quad \text{and} \quad y^{n+1} = y^n + y^{n-1} \qquad (4\text{B}.6)$$

giving the recursion for the Fibonacci numbers for $n > 1$:

$$F_{n+1} = \frac{x^{n+1} - y^{n+1}}{x - y} = \frac{x^n + x^{n-1} - y^n - y^{n-1}}{x - y} = F_n + F_{n-1} \qquad (4\text{B}.7)$$

Therefore, the Fibonacci numbers are

$$F_0, F_1, F_2, \ldots = 0, 1, 1, 2, 3, 5, 8, 13, 21, 34, 55, 89, 144, \ldots$$

If we define $r_n \triangleq F_{n+1}/F_n$ as the ratio of successive Fibonacci numbers, we find that

$$r_n = \frac{F_{n+1}}{F_n} = \frac{F_n + F_{n-1}}{F_n} = 1 + \frac{1}{r_{n-1}} \qquad (4\text{B}.8)$$

which leads to the continued fraction expression

$$r_n = 1 + \cfrac{1}{1 + \cfrac{1}{1 + \cdots \cfrac{1}{1 + \cfrac{1}{1 + \cfrac{1}{r_1}}}}} \tag{4B.9}$$

Because the roots x and y are such that $x > |y|$ and $\lim\limits_{n \to \infty} (y/x)^n = 0$, the ratio r_n converges to a certain asymptotic value:

$$\lim_{n \to \infty} r_n = \lim_{n \to \infty} \frac{x^{n+1} - y^{n+1}}{x^n - y^n} = \lim_{n \to \infty} \frac{x - y(y/x)^n}{1 - (y/x)^n} = x \tag{4B.10}$$

Note that the ratio r_n converges to the inverse of the golden ratio:

$$\lim_{n \to \infty} r_n = x = \frac{\sqrt{5} + 1}{2} = \frac{2}{\sqrt{5} - 1} = \frac{1}{g} \tag{4B.11}$$

showing that the golden ratio and Fibonacci numbers are closely related. For that reason, hashing that uses the golden ratio has been termed Fibonacci hashing.

4B.2 Hash Function Example

The uniformity of the hash function $h(K) = \left\lfloor N\left[\left(\frac{A}{w}K\right) \bmod 1\right]\right\rfloor$ is clearly due to its kernel, $\left(\frac{A}{w}K\right) \bmod 1$. If A/w is chosen to be the golden ratio $g = 0.618$, then the uniformity of this Fibonacci hashing can be found experimentally by plotting the various values of $(gK) \bmod 1$. These values for $K = 1, 2, \ldots, 14$ are shown in Table 4B.1. The values may be plotted as shown in Figure 4B.1, which reveals the high degree of uniformity achieved by employing the golden ratio in the hash function kernel.

Table 4B.1 Values of the hash function kernel

K	1	2	3	4	5	6	7
$(gK) \bmod 1$	0.6180	0.2360	0.8541	0.4721	0.0902	0.7082	0.3262
K	8	9	10	11	12	13	14
$(gK) \bmod 1$	0.9443	0.5623	0.1803	0.7984	0.4164	0.0344	0.6525

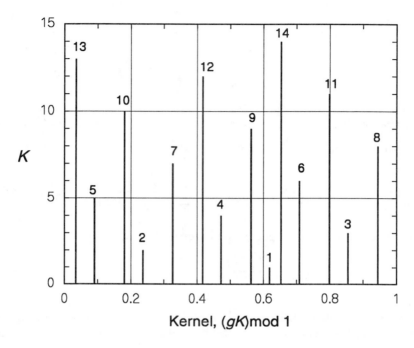

Figure 4B.1 Uniformity of the hash function kernel.

Figure 4B.1 demonstrates the truth of the following theorem, proved in [5]:

Theorem: Let θ be any irrational number. When the points $\theta \bmod 1$, $2\theta \bmod 1, \ldots, n\theta \bmod 1$ are placed in the unit interval, the next point placed, $[(n+1)\theta] \bmod 1$, falls in the largest existing segment.

In the example of Table 4B.1 and Figure 4B.1, the first point ($K = 1$) was placed in the unit interval at $g = 0.618$, creating two two segments:

$$d_1 = [0, \, g \bmod 1] \text{ with length } l_1 = 0.618$$
$$d_2 = [g \bmod 1, \, 1] \text{ with length } l_2 = 1 - 0.618 = 0.382$$

Obviously, $l_1 > l_2$, and in accordance with the theorem, the second point ($K = 2$) was placed at $2g \bmod 1 = 0.2360$ in the interval d_1, as illustrated in Figure 4B.2. After the placement of the second point, there are three segments:

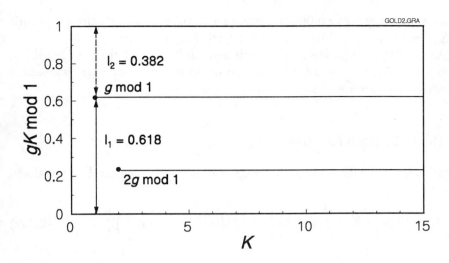

Figure 4B.2 Placement of second point.

$$d_1 = [0, \, 2g \bmod 1] \text{ with length } l_1 = 0.236$$

$$d_2 = [2g \bmod 1, \, g \bmod 1] \text{ with length } l_2 = 0.618 - 0.236 = 0.382$$

$$d_3 = [g \bmod 1, \, 1] \text{ with length } l_3 = 0.382$$

The largest of these three intervals is a tie between d_2 and d_3 because $l_2 = l_3$, and the third point ($K = 3$) is placed at $3g \bmod 1 = 0.8541$, which is in d_3, splitting this interval into two segments. Thus, d_2 becomes the longest segment and the next point ($K = 4$) should fall, according to the theorem, into d_2—it does, because $4g \bmod 1 = 0.4721$ is in d_2, as can be observed in Figure 4B.1.

The fact that there were two segments with the same length after the placement of the second point is not unusual. A corollary to the theorem stated above declares that [5] the lengths of the segments will have at most three different values. This behavior of the hash function kernel is clearly seen regarding the sixteen segments of the x-axis in Figure 4B.1. There is one interval length (= 0.0557) between the following seven pairs of points: (K_1, K_2) = (5, 13), (2, 10), (4, 12), (1, 9), (6, 14), (3, 11), and between 8 and the upper end of the (0, 1) interval. There is a second interval length (= 0.0344) between the following two pairs of points: (K_1, K_2) = (1, 14) and between 13 and the lower end of the (0, 1) interval. Finally, there is a third interval length ($= 0.0902$) between the following six pairs of points: $(K_1, K_2) = (5, 10)$, $(2, 7)$, $(7, 12)$, $(4, 9)$, $(6, 11)$, and $(3, 8)$.

According to [5], extensive experiments indicate that the two numbers, $A/w = g$ and $A/w = 1 - g$ lead to the most uniformly distributed sequences of kernel values, among all numbers θ between 0 and 1.

4B.3 The IS-95 Hash Function

The multiplicative Fibonacci hash function discussed above can be rearranged as follows:

$$h(K) = \left\lfloor N \left[\left(\frac{A}{w} K \right) \bmod 1 \right] \right\rfloor = \left\lfloor N \left(\frac{A}{w} K - \left\lfloor \frac{A}{w} K \right\rfloor \right) \right\rfloor \qquad (4B.12a)$$

because $x \bmod 1 = x - \lfloor x \rfloor$. Further, because $x \bmod y = x - y \left\lfloor \frac{x}{y} \right\rfloor$:

$$h(K) = \left\lfloor N \frac{AK - w \left\lfloor \frac{A}{w} K \right\rfloor}{w} \right\rfloor = \left\lfloor N \frac{(AK) \bmod w}{w} \right\rfloor \qquad (4B.12b)$$

In IS-95, a word size of 16 bits is assumed, giving $w = 2^{16} = 65{,}536$, and the ratio $A/w = g = \frac{\sqrt{5}-1}{2} = 0.618034$ is used, or

$$A = gw = 0.618034 \times 65{,}536 = 40{,}503.476 \qquad (4B.13)$$

The integer closest is $A = 40{,}503$, which is also relatively prime to w. Therefore, IS-95 specifies the following hash function, denoted R:

$$R \equiv h(K) = \left\lfloor N \times \frac{(40{,}503 \times K) \bmod 2^{16}}{2^{16}} \right\rfloor \qquad (4B.14)$$

where the value of N is the number of resources from which to choose, and K is the hash key.

The hash key is determined from the IS-95 parameter HASH_KEY, which is a binary number generated from either the 32-bit binary electronic serial number (ESN) or the least significant 32 bits of the binary encoded international mobile station identity (IMSI) of the mobile station; the binary coding of the IMSI is described below. These 32-bit binary numbers can be written in the general form $HASH\_KEY = L + 2^{16}H$; that is, L is the number formed by the 16 least significant bits of the number $HASH\_KEY$, and H is the number formed by the 16 most significant bits of the number $HASH\_KEY$. Using this notation, the IS-95 hash key is computed as

$$K = L \oplus H \oplus \text{DECORR} \qquad (4B.15)$$

where DECORR is a modifier intended to decorrelate the different hash function selections made by the same mobile station. Let L' denote the twelve least significant bits of $HASH\_KEY$; then Table 4B.2 shows the values of the parameters used in the IS-95 hash function.

The structures of the ESN and IMSI are shown in Figure 4B.3. Note that the ESN originates at the factory and is a fixed binary number, while the IMSI is a decimal number that is assigned to the mobile telephone unit when the country code and other information on the unit's specific use is known at the time of purchase.

As indicated in Figure 4B.3, the IMSI can be up to 15 decimal digits in length. For use in generating a hash key in IS-95, a number denoted IMSI_S is derived from the IMSI by taking its 10 least significant digits and converting it to a 34-bit binary number. If the IMSI has fewer than ten digits, zeros are appended on the most significant side of the number make it have 10 digits. For the purpose of this discussion, let us suppose that the 10-digit IMSI_S comprises a three-digit area code (d_1, d_2, d_3), a three-digit exchange number or prefix (d_4, d_5, d_6), and a four-digit local number (d_7, d_8, d_9, d_{10}). The mapping of the IMSI_S into a binary number is performed as follows:

(1) If any of the numbers d_1 to d_{10} is a zero, change it to a ten; then subtract 1 from each of the numbers. This results in the 10 numbers $(d_1', d_2', \ldots, d_{10}')$ that are rotated in value as shown in Table 4B.3.

Table 4B.2 Parameters used in IS-95 hash function

Application	N	DECORR	Return value
CDMA channel # (center frequency)	No. channels (up to 10) in list sent to mobile	0	$R + 1$
Paging channel #	No. channels (up to 7) in list sent to mobile	$2 \times L'$	$R + 1$
Paging slot #	2,048	$6 \times L'$	R
Access channel randomization	2^M, where $M \leq 512$ is in broadcast message	$14 \times L'$	R

ESN: set at factory

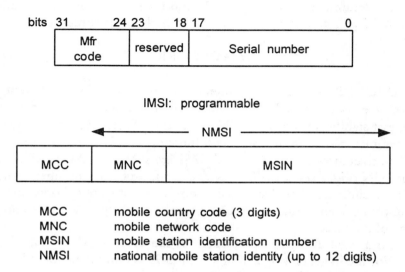

Figure 4B.3 ESN and IMSI structures.

Table 4B.3 Initial rotation of IMSI digit values

d_i	0	1	2	3	4	5	6	7	8	9
d_i'	9	0	1	2	3	4	5	6	7	8

This "rotation" procedure to determine IMSI_S digits (also used in IS-54 and AMPS to determine the mobile identification number, (MIN)) is a convention carried over from earlier cellular practice.

(2) The first three numbers (derived from the area code) are converted directly to a 10-bit binary number. For example, the area code 607 after rotation of the digits becomes 596, which in binary is $(596)_2 = 1001010100$.

(3) The next three digits (the exchange) is converted to a 10-bit binary number in the same way as the area code.

(4) The first of the last four (local number) digits is converted directly to a 4-bit binary number, except that if it is zero, it is treated as if signified the number ten and is converted to the $(10)_2 = 1010$.

(5) The last three digits are converted to a 10-bit binary number in the same way as the area code.

Example 4B.2 The 34-bit binary number corresponding to the IMSI_S given by 301-294-3463 is calculated as

$$301 \Rightarrow 290; \ (290)_2 = 0100100010$$

$$294 \Rightarrow 183; \ (183)_2 = 0010110111$$

$$(3)_2 = 0011$$

$$463 \Rightarrow 352; \ (352)_2 = 0101100000$$

$$\Rightarrow 0100100010001011011100110101100000$$

The hash key makes use of the 32 least significant bits of this number. In the notation introduced above:

$$H = 0010001000101101$$

$$L = 1100110101100000$$

$$L' = 110101100000$$

Then, the hash key to determine the paging channel number is calculated as

$$
\begin{array}{ll}
L & 1100110101100000 \\
\oplus\, H & 0010001000101101 \\
\oplus\, 2L' & \underline{0001101011000000} \\
 & 1111010110001101 \Rightarrow K = 62{,}861
\end{array}
$$

Assuming that there are $N = 7$ active paging channels, the hash function calculation (4B.14) gives

$$
R = \left\lfloor 7 \times \frac{(40{,}503 \times 62{,}861)\,\mathrm{mod}\, 2^{16}}{2^{16}} \right\rfloor
$$

$$
= \left\lfloor 7 \times \frac{2{,}546{,}059{,}083\,\mathrm{mod}\, 2^{16}}{2^{16}} \right\rfloor = \left\lfloor 7 \times \frac{51{,}019}{65{,}536} \right\rfloor
$$

$$
= \lfloor 5.4494 \rfloor = 5
$$

The IS-95 paging channel assignment in this case is $R + 1 = 6$. The reader may verify that the IMSI_S differing in only one digit, 301-294-3464, results in $K = 62{,}862$ and the selection of paging channel 3 for the same assumptions.

4B.4 IS-95 Random Number Generator

In IS-95, the mobile is sometimes presented with a choice of resources from which to choose. For example, which of the 32 access channels associated with a particular paging channel (distinguished by long PN code offsets commanded by masks) should the mobile station use when attempting an access? This choice should be uniform and random to minimize collisions and overloads on particular channels. For the pseudorandom case, as opposed to the hashing case, the receiving link is not required to reproduce the choice of resources, and thus the choice will appear random to the receiver.

A good random number generator (RNG) satisfies three basic criteria:

- The generator should produce a "nearly infinite" sequence of numbers. This means that the generator should have a "nearly infinite" period.

- The generator should produce a random sequence of numbers. In other words, the sequence should have no obvious pattern.

- The generator should be easily implementable on whatever processor it is to be used on. In other words, the arithmetic involved should not cause overflow or erroneous results.

Most RNGs produce a *periodic* sequence of numbers, each of which is based on the previous number:

$$z_{n+1} = f(z_n) \quad n = 1, 2, \ldots \tag{4B.16}$$

and is initialized by choosing an initial seed z_1. For our purposes, the form of the RNG will be limited to the linear congruent generator form

$$z_{n+1} = (az_n) \bmod m \tag{4B.17}$$

where the modulus m is a large prime integer, and the multiplier a is any integer such that $1 < a < m$. If a or $z_n = 0$, then the sequence is all zeros. Thus, $z_1 \neq 0$.

For example, let an RNG be given by $z_{n+1} = 3z_n \bmod 7$; this recursive algorithm produces the following values:

z_n	z_{n+1}
1	$3(1) \bmod 7 = 3$
3	$3(3) \bmod 7 = 2$
2	$3(2) \bmod 7 = 6$
6	$3(6) \bmod 7 = 4$
4	$3(4) \bmod 7 = 5$
5	$3(5) \bmod 7 = 1$

The sequence 1, 3, 2, 6, 4, 5, 1,... is produced. We can make two observations about this generator. First, the numbers range from 1 to $m - 1 = 6$ (0 and m are excluded); 0 is excluded because only $m \bmod m = 0$ and $az_n \neq m$, because m is prime, and m is excluded because $x \bmod m < m$, by definition of the modular operation. Second, the maximal period P before any number is repeated is $m - 1 = 6$; because 0 and m are excluded, the generator can produce all numbers from 1 to $m - 1$, which is $m - 1 = P$.

It has been suggested [6] that an appropriate choice for a modulus is $m = 2^{31} - 1$. This modulus is prime and produces a maximal period sequence of $P = m - 1 = 2^{31} - 2$, which is "nearly infinite." Because a can be any integer such that $1 < a < m$, there are $m - 1 - 1 = 2^{31} - 3$ or about 2.1×10^9 possible choices for the multiplier a. These choices can be narrowed by applying the three criteria. After applying the three criteria, only a handful of choices for a good random number generator remain.

Because we have chosen $m = 2^{31} - 1$, we must use at least a 32-bit processor to process values of m because $2^{31} - 1$ is the largest integer that can be represented by 32 bits (including one sign bit). It is desirable to implement the entire generator on a 32-bit processor, so that the generator can be compatible with a large number of computers. Because az_n can be much greater than $2^{31} - 1$, it is useful to perform the calculation in the following manner to prevent overflow:

Let $m = aq + r$ where $q = \lfloor m/a \rfloor \ll m$ (so that $aq < m$) and let $r = m \bmod a$; from the definition of the modular calculation, the generator can be expressed as

$$f(z) = az \bmod m = az - m \left\lfloor \frac{az}{m} \right\rfloor$$

Adding and then subtracting $m \lfloor z/q \rfloor = (aq + r) \lfloor z/q \rfloor$ from the equation yields

$$f(z) = az - m \left\lfloor \frac{az}{m} \right\rfloor + m \left\lfloor \frac{z}{q} \right\rfloor - (aq + r) \left\lfloor \frac{z}{q} \right\rfloor$$

$$= a\left(z - q\left\lfloor\frac{z}{q}\right\rfloor\right) \underbrace{- r\left\lfloor\frac{z}{q}\right\rfloor + m}_{} \underbrace{\left(\left\lfloor\frac{z}{q}\right\rfloor - \left\lfloor\frac{az}{m}\right\rfloor\right)}_{f_2(z)}$$

$$\underbrace{z \bmod q < q}$$

$$= \underbrace{a(z \bmod q) - r\left\lfloor\frac{z}{q}\right\rfloor}_{f_1(z)} + mf_2(z)$$

$$= f_1(z) + mf_2(z) \tag{4B.18}$$

This ordering of the calculation prevents overflow, because

$$f_1(z) = a(z \bmod q) - r\left\lfloor\frac{z}{q}\right\rfloor < a(z \bmod q) < aq < m$$

and, using ϵ to denote a positive number less than 1:

$$f_2(z) = \left\lfloor\frac{z}{q}\right\rfloor - \left\lfloor\frac{az}{m}\right\rfloor = \left\lfloor\frac{z}{q}\right\rfloor - \left\lfloor\frac{z}{q + \epsilon}\right\rfloor$$

$$< \left\lfloor\frac{z}{q}\right\rfloor - \left\lfloor\frac{z}{q + 1}\right\rfloor = \frac{z}{q(q + 1)} + \epsilon_1 - \epsilon_2$$

$$< \frac{m}{q(q + 1)} + 1 \Rightarrow mf_2(x) \gtrsim \left(\frac{m}{q}\right)^2 < a^2$$

For the IS-95 RNG, $a^2 = (16{,}807)^2 = 2.8 \times 10^8 < m = 2.15 \times 10^9$.

Even though some generators do not rank the highest on the randomness test, they are used in modern systems because they have been used in the past with great success. An example of such a generator is the one used in IS-95, given by

$$z_{n+1} = 16{,}807 z_n \bmod (2^{31} - 1)\,, \quad \text{where } 16{,}807 = 7^5 \tag{4B.19}$$

For this RNG, $m = 2^{31} - 1 = 2{,}147{,}483{,}647$ and $z_0 = $ the generator seed. The generator seed is determined from information stored in the permanent memory of the mobile and information found on the sync channel.

During mobile initiation, the mobile receives the pilot channel to obtain synchronization with the base station and obtain the sync channel. On the sync channel is found the sync channel message, which contains a 36-bit parameter SYS_TIME corresponding to the system time. The 32 least significant bits of SYS_TIME form a parameter RANDOM_TIME. The mobile computes the generator seed by using the 32-bit ESN stored in permanent memory and the 32-bit parameter RANDOM_TIME as follows:

$$z_0 = (\text{ESN} \oplus \text{RANDOM\_TIME}) \bmod m$$

If z_0 is computed to be 0, then it is replaced by $z_0 = 1$.

Example 4B.3 Find z_0, given that

ESN = 01001011101001010110100010101111

and

SYS_TIME = 101000110001011011110100110001010010.

Solution: RANDOM_TIME is the 32 least significant bits of SYSTEM_TIME:

SYS_TIME = 1010001$\underbrace{10001011011110100110001010010}$
RANDOM_TIME

$z_0 = (\text{ESN} \oplus \text{RANDOM\_TIME}) \bmod m$:

$$\begin{aligned}
\text{ESN} &= 01001011101001010110100010101111 \\
\text{RANDOM\_TIME} &= \underline{00110001011011110100110001010010} \\
\text{ESN} \oplus \text{RANDOM\_TIME} &= 01111010110010100010010011111101 \\
&= 2060068093
\end{aligned}$$

$$z_0 = (\text{ESN} \oplus \text{RANDOM\_TIME}) \bmod m$$

$$= (2060068093) \bmod 2147483647 = 2060068093$$

For those applications that require a binary fraction u_n: $0 < u_n < 1$, IS-95 specifies that the mobile station shall use the following value: $u_n = z_n/m$. For those applications that require a small integer k_n: $0 \leq k_n \leq N - 1$, the mobile station shall use the following integer value: $k_n = \lfloor N \times (z_n/m) \rfloor$.

Example 4B.4 Suppose that a mobile is attempting access to the base station to register. The link is poor, and the mobile must transmit three accesses before the base station receives and correctly acknowledges the access. Find the access channel number used for each of the accesses (IS-95 states that before each access, a new access channel number is randomly determined.)

Solution assumptions:

There are $N = 32$ access channels.

The mobile station ESN is 01001011101001010110100010101111.

The parameter SYSTEM_TIME is 1010010010111010010101101000101011111.

$$\text{SYSTEM\_TIME} = 1010\underbrace{010010111010010101101000010101111}_{\text{RANDOM\_TIME}}$$

The seed is $z_0 = (\text{ESN} \oplus \text{RANDOM\_TIME}) \bmod m$:

$$\begin{aligned} \text{ESN} &= 01001011101001010110100010101111 \\ \text{RANDOM\_TIME} &= 01001011101001010110100010101111 \\ \text{ESN} \oplus \text{RANDOM\_TIME} &= 00000000000000000000000000000000 = 0 \end{aligned}$$

$$z_0 = (\text{ESN} \oplus \text{RANDOM\_TIME}) \bmod m = 0 \bmod 2{,}147{,}483{,}647 = 0$$

because z_0 was calculated to be 0, it is replaced by $z_0 = 1$. The random number for the first access attempt is determined using

$$z_1 = a \times z_0 \bmod m = 16807 \times 1 \bmod 2{,}147{,}483{,}647 = 16{,}807$$

The access channel number for the first attempt is given by

$$k_1 = \left\lfloor \frac{N \times z_1}{m} \right\rfloor = \left\lfloor \frac{32 \times 16{,}807}{2{,}147{,}483{,}647} \right\rfloor = \lfloor 0.00025 \rfloor = 0$$

The random number for the second access attempt is determined using

$$\begin{aligned} z_2 = a \times z_1 \bmod m &= (16807 \cdot 16807) \bmod 2{,}147{,}483{,}647 \\ &= 282{,}475{,}249 \end{aligned}$$

The access channel number for the second attempt is given by

$$k_2 = \left\lfloor \frac{N \times z_2}{m} \right\rfloor = \left\lfloor \frac{32 \times 282,475,249}{2,147,483,647} \right\rfloor = \lfloor 4.209 \rfloor = 4$$

The random number for the third access attempt is determined using

$$z_3 = a \times z_2 \bmod m = 16,807 \times 282,475,249 \bmod 2,147,483,647$$
$$= 1,622,649,381$$

The access channel number for the third attempt is given by

$$k_3 = \left\lfloor \frac{N \times z_3}{m} \right\rfloor = \left\lfloor \frac{32 \times 1,622,649,381}{2,147,483,647} \right\rfloor = \lfloor 24.18 \rfloor = 24$$

5

Walsh Functions and CRC Codes

Walsh functions of "order 64" are extensively used in the IS-95 system [1]. In the forward link, the coded and interleaved traffic channel signal symbols are multiplied with distinct repeating Walsh sequences that are assigned to each channel for the duration of the call. The Walsh sequences (functions) are mutually orthogonal, ensuring that user signals are also orthogonal. Thus, it can be stated that, for the forward link, the IS-95 system is an orthogonal spread-spectrum system designed to limit or eliminate multiple access interference (MAI).

Theoretically, MAI can be reduced to zero if the signals in multiple access channels are mutually orthogonal. In practice, however, cochannel interference is still present because of multipath and signals from other cells that are not time-aligned with the desired signal. Delayed and attenuated replicas of the signals that arrive nonsynchronously are not orthogonal to the primary synchronously arriving components. These signals will cause interference. The signals from other base stations are received nonsynchronously with respect to the signal from the home cell and, not being orthogonal, also cause interference.

To reduce this interference, each forward link channel employs PN sequence modulation, each using the identical PN code sequence, which runs at the same clock rate as the orthogonal Walsh sequences. The elimination or rejection of the interference (the multipaths within the home cell and the signals from other cells) is due to the processing gain that the system provides in each channel. Thus, the multiple access scheme for the forward link is accomplished by the use of spreading orthogonal Walsh sequences (functions), whereas the purpose of the PN sequence application is to eliminate the interference to the extent the system's processing gains can provide for each channel.

The Walsh functions are used in the IS-95 system for the reverse link as well as the forward link. The ways in which the Walsh functions are used in each link are, however, quite different. For the reverse link, the Walsh sequences are employed as an orthogonal modulation code, which depends only on the data pattern, as explained in Section 4.3, forming a 64-ary orthogonal modulation system. When the Walsh functions (sequences) are used for multiplexing purposes in the forward link, their role is to distinguish the channel identities, whereas they are simply employed as orthogonal codewords for the reverse link.

So what are the Walsh functions? Aside from the orthogonality property, are there other properties of the Walsh functions that need to be understood? We now define the Walsh functions and study their properties. The materials presented in this chapter on the subject of Walsh functions are extensive, perhaps beyond their simple application to the IS-95 system. IS-95 does not specify how the Walsh functions for the reverse link are to be decoded. It is felt that, by understanding all the properties embedded in the Walsh functions—and there are many—one may envision several ways of generating or processing Walsh functions. This is the reason for discussing many forms of encoding and decoding the Walsh functions.

In the second half of this chapter, we discuss the theory and use of the cyclic redundancy check (CRC) codes in IS-95 for frame error detection. The approach to the topic is tutorial. Based on our experience in giving seminars on the IS-95 system to engineers, we found that many with even advanced degrees did not have the opportunity in their education to learn the basics of coding theory. As a consequence, they did not appreciate the coding schemes used in the IS-95 system, such as CRC frame error detection and forward error control coding. The materials presented in this chapter are intended to teach those engineers from the very beginning of coding theory, to the extent needed to understand the CRC coding in the IS-95 system. The development also forms a basis for Chapter 8, which deals with convolutional coding schemes and their use in IS-95. Therefore, the reader who is familiar with the fundamentals of coding theory may skip the second half of this chapter except for noting the description of the specific CRC error detection schemes employed in the IS-95 system.

5.1   Definition of the Walsh Functions

In 1923, J. L. Walsh published a paper entitled "A closed set of normal orthogonal functions" [2] in which he defined a system of orthogonal functions that is complete over the normalized interval (0, 1), each function taking only values of +1 or −1, except at a finite number of discontinuity points, where it takes the value zero.

Walsh stated that the functions (which we now call Walsh functions) are orthogonal, normalized, and complete. He meant by "orthogonal" that, if you multiply any two distinct functions and integrate them over the interval, the result is zero. By "normal," he meant that, if the two functions are one and the same, the integral of their product is unity. Finally, "complete" means something that is very mathematical and whose full discussion is beyond the purpose of this book; however, it means roughly that the set of orthogonal functions can be used to approximate any given function within the defined interval as a linear combination thereof, such that its mean square error in the interval tends to zero in the "limit in the mean" sense as the number of orthogonal functions increases without limit. For the purpose of this book, only the salient properties of the Walsh functions are summarized here as definitions.

We define the Walsh functions of order N as a set of N time functions, denoted $\{W_j(t); t \in (0, T), j = 0, 1, \ldots, N - 1\}$, such that

- $W_j(t)$ takes on the values $\{+1, -1\}$ except at the jumps, where it takes the value zero.

- $W_j(0) = 1$ for all j.

- $W_j(t)$ has precisely j sign changes (zero crossings) in the interval $(0, T)$.

- $\int_0^T W_j(t) W_k(t) \, dt = \begin{cases} 0, & \text{if } j \neq k \\ T, & \text{if } j = k \end{cases}$

- Each function $W_j(t)$ is either odd or even with respect to the midpoint of the interval.

A set of Walsh functions consists of N member functions, and they are ordered according to the number of zero crossings (sign changes). In the functional notation of

$$\{W_0(t), W_1(t), \ldots, W_j(t), \ldots, W_{N-1}(t)\}$$

the first function $W_0(t)$ has no zero crossings over the entire interval of $(0, T)$, whereas $W_1(t)$ has one zero crossing over the interval. Consider the functions shown in Figure 5.1, drawn on the interval $(0, T)$. They are clearly the Walsh functions of order 8 in view of the definitions given above:

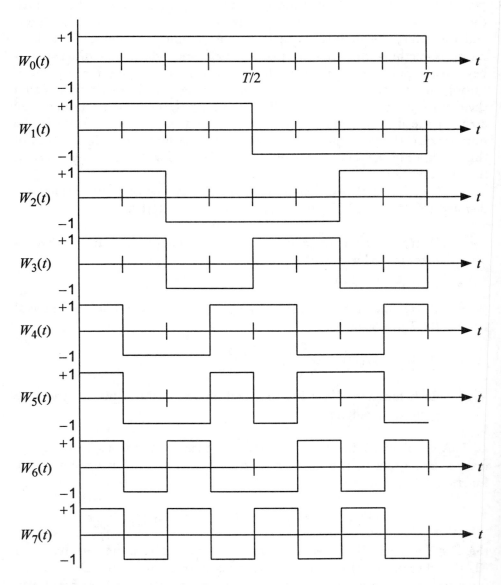

Figure 5.1 The Walsh functions of order 8.

- All eight functions take on the values $\{+1, -1\}$.

- Every function starts with the value $+1$; that is, $W_j(0) = 1$ for $j = 0$, $1, \ldots, 7$.

- The integer index of $W_j(t)$, $j = 0, 1, \ldots, 7$, accounts for the number of sign changes that each function has.

- We observe clearly, for any pair, the orthogonality of the functions:

$$\int_0^T W_j(t)\, W_k(t)\, dt = T\delta_{jk}, \quad j,\, k = 0, 1, \ldots, 7$$

where δ_{jk} is the Kronecker delta.

- The functions $W_0(t)$, $W_2(t)$, $W_4(t)$, and $W_6(t)$ are even functions with respect to the midpoint of the interval, $t = T/2$, whereas $W_1(t)$, $W_3(t)$, $W_5(t)$, and $W_7(t)$ are odd functions with respect to the midpoint $t = T/2$.

We have not yet explained, however, the ways in which these Walsh functions can be constructed. We have confirmed merely that the functions shown in Figure 5.1 are Walsh functions. Once we know how to generate these Walsh functions of any order N, we can use them in N-channel orthogonal multiplexing applications. We will see shortly that the generation of Walsh functions is very easily accomplished in many different ways.

The method of specifying the Walsh functions of arbitrary order $N = 2^K$, $K = 1, 2, \ldots$; that is, the computation of Walsh functions, had been a problem of considerable difficulty from the time Walsh published his paper until the year when Byrnes and Swick [3] published the paper entitled *Instant Walsh Functions*. Until Byrnes and Swick's method had come along, there were several ways of producing (or computing) the Walsh functions: they could be obtained from the Rademacher functions [4, 5] and from the solution of a certain difference equation [6, 7]. Whether from the Walsh's expression from the Rademacher functions (to be defined later) or from the difference equation, these conventional methods of computing the member functions of the Walsh function set of order $N = 2^K$ required knowledge of previously computed member functions in the set.

Byrne and Swick considered the inherent symmetry properties of the member functions of the Walsh functions and came up with a method of specifying each and every one of the member functions of any order at sight, without recourse to previously computed functions. Their method enables us to write down (or to generate or draw the waveform) "instantly" by looking at the K-tuple binary representation of the index integer "j" of the Walsh function $W_j(t)$ from the set of order $N = 2^K$. This method is described next, after which other methods are discussed.

5.2 Walsh Sequence Specifications (Instant Walsh Functions)

If we convert the ± 1 amplitudes of the Walsh functions to a binary logic $\{0, 1\}$ representation with the conversions $+1 \rightarrow$ "0" and $-1 \rightarrow$ "1" and use the binary representations for the entire set of index integers $j = 0, 1, \ldots, 7$ for the eight Walsh functions of order 8 in Figure 5.1, then we can write down the eight Walsh sequences as shown in Table 5.1; that is, $j \rightarrow W_j$, $j = 0, 1, \ldots, 7$.

When we consider a set of Walsh functions or sequences of order $N = 2^K$, we note that the sequences possess symmetry properties about K axes at $T/2$, $T/2^2, \ldots$, $T/2^K$, where T is the period (or interval) of the Walsh functions. Walsh functions have either even or odd symmetry about these axes. The axes are found in the order of $T/2^j$ for $j = K, K-1, \ldots, 1$, that is, $T/2^K$, $T/2^{K-1}, \ldots$, $T/2$. Consider an arbitrary Walsh sequence taken from the set of order $N = 2^4 = 16$:

$$W_{13} = 0\ 1\ 0\ 1\ 1\ 0\ 1\ 0\ 1\ 0\ 1\ 0\ 0\ 1\ 0\ 1$$

Table 5.1 The Walsh sequences of order 8

Index integer j	Index sequence	Walsh sequences of order $8 = 2^3$
0	0 0 0	$W_0 = 0\ 0\ 0\ 0\ 0\ 0\ 0\ 0$
1	0 0 1	$W_1 = 0\ 0\ 0\ 0\ 1\ 1\ 1\ 1$
2	0 1 0	$W_2 = 0\ 0\ 1\ 1\ 1\ 1\ 0\ 0$
3	0 1 1	$W_3 = 0\ 0\ 1\ 1\ 0\ 0\ 1\ 1$
4	1 0 0	$W_4 = 0\ 1\ 1\ 0\ 0\ 1\ 1\ 0$
5	1 0 1	$W_5 = 0\ 1\ 1\ 0\ 1\ 0\ 0\ 1$
6	1 1 0	$W_6 = 0\ 1\ 0\ 1\ 1\ 0\ 1\ 0$
7	1 1 1	$W_7 = 0\ 1\ 0\ 1\ 0\ 1\ 0\ 1$

The sequence has odd symmetry about the axis at $T/2^K = T/2^4 = T/16$:

$$\frac{T}{16}$$

| 0 | 1 | 0 1 1 0 1 0 1 0 1 0 0 1 0 1 |

The sequence has odd symmetry about the axis at $T/8$:

$$\frac{T}{8}$$

| 0 1 | 0 1 | 1 0 1 0 1 0 1 0 0 1 0 1 |

The sequence has even symmetry about the axis at $T/4$:

$$\frac{T}{4}$$

| 0 1 0 1 | 1 0 1 0 | 1 0 1 0 0 1 0 1 |

The sequence has odd symmetry about the axis at $T/2$:

$$\frac{T}{2}$$

| 0 1 0 1 1 0 1 0 | 1 0 1 0 0 1 0 1 |

The type of symmetry is indicated in the index sequence written as a binary K-tuple for the index integer j, that is, $j = (j_1, j_2, \ldots, j_K)$. If $j_k = 0$, the function has even symmetry about the axis at $T/2^{K-k+1}$, $k = 1, 2, \ldots, K$; if $j_k = 1$, the function has odd symmetry about that axis. The first entry in the sequence is always 0, as specified by the definition of the Walsh function; that is, $W_j(0) = 1$ or $W_{j0} = 0$ for all j.

The Walsh sequence W_{13}, which was observed above, can be analyzed as follows: In W_{13}, the index sequence is $(j_1, j_2, j_3, j_4) = (1, 1, 0, 1)$. Now:

$$j_1 = 1 \text{ implies that the symmetry is odd at axis } T/16$$

$$j_2 = 1 \text{ implies that the symmetry is odd at axis } T/8$$

$$j_3 = 0 \text{ implies that the symmetry is even at axis } T/4$$

$$j_4 = 1 \text{ implies that the symmetry is odd at axis } T/2$$

The sequence may now be written down, starting with 0, according to the above "symmetry command," as

$$W_{13} = 0 \ 1 \ 0 \ 1 \ 1 \ 0 \ 1 \ 0 \ 1 \ 0 \ 1 \ 0 \ 0 \ 1 \ 0 \ 1$$

which is the sequence that we started out with.

Another observation on the symmetry properties of the Walsh functions is in order, which is useful in decoding the "Walsh codes" used in the reverse link orthogonal modulation in the IS-95 system. We continue to use the Walsh functions of order $N = 2^K = 2^4 = 16$ in this illustration of the properties. Consider again the Walsh sequence W_{13}:

$$0\ 1\ 0\ 1\ 1\ 0\ 1\ 0\ 1\ 0\ 1\ 0\ 0\ 1\ 0\ 1 = W_{13}$$

As stated previously, the sequence has odd symmetry about the midpoint of the interval $(0, T)$, which is the axis at $T/2$. Call this axis $a_K = a_4$:

$$a_4 = \frac{T}{2}$$

| 0 | 1 | 0 | 1 | 1 | 0 | 1 | 0 | 1 | 0 | 1 | 0 | 0 | 1 | 0 | 1 |

The sequence has even symmetry about the midpoint of the subintervals defined above; namely, the midpoints of $(0, T/2)$ and $(T/2, T)$, which are the axes at $T/4$ and $3T/4$, respectively. Call these axes the set $a_{K-1} = a_3$:

$$a_3$$

$$\frac{T}{4} \qquad\qquad \frac{3T}{4}$$

| 0 | 1 | 0 | 1 | 1 | 0 | 1 | 0 | 1 | 0 | 1 | 0 | 0 | 1 | 0 | 1 |

The sequence has odd symmetry about the midpoints of the subintervals defined above; namely, the midpoints of $(0, T/4)$, $(T/4, T/2)$, $(T/2, 3T/4)$, and $(3T/4, T)$, which are the axes at $T/8$, $3T/8$, $5T/8$, and $7T/8$, respectively. Call these axes the set a_2:

$$a_2$$

$$\frac{T}{8} \qquad \frac{3T}{8} \qquad \frac{5T}{8} \qquad \frac{7T}{8}$$

| 0 | 1 | 0 | 1 | 1 | 0 | 1 | 0 | 1 | 0 | 1 | 0 | 0 | 1 | 0 | 1 |

The sequence has odd symmetry about the midpoints of the subintervals defined above; namely, the midpoints of $(0, T/8)$, $(T/8, T/4)$, $(T/4, 3T/8)$, $(3T/8, T/2)$, $(T/2, 5T/8)$, $(5T/8, 3T/4)$, $(3T/4, 7T/8)$, and $(7T/8, T)$, which are axes at $T/16$, $3T/16$, $5T/16$, $7T/16$, $9T/16$, $11T/16$, $13T/16$, and $15T/16$. Call these axes the set a_1:

$$a_1$$

$\frac{T}{16}$	$\frac{3T}{16}$	$\frac{5T}{16}$	$\frac{7T}{16}$	$\frac{9T}{16}$	$\frac{11T}{16}$	$\frac{13T}{16}$	$\frac{15T}{16}$

0	1	0	1	1	0	1	0	1	0	1	0	0	1	0	1

In general, the axes of symmetry can be found as follows:

- The Walsh functions have either even or odd symmetry about the midpoint of the interval $(0, T)$, which is the axis at $T/2$, called a_K.

- The Walsh functions have the same symmetry about the midpoints of the subintervals $(0, T/2)$ and $(T/2, T)$, which are the axes at $T/4$ and $3T/4$, called the set a_{K-1}.

- This process continues K times until the midpoints of the sub-intervals are $T/N, 3T/N, \ldots, (N-1)T/N$, where $N = 2^K$. This set of axes is called a_1, and the symmetry about these axes is the same.

The IS-95 system employs Walsh sequences (or functions) of order 64, and thus if we write down 64 sextuple index sequences, we can specify the entire set of 64 64-tuple Walsh sequences, based on instant Walsh function rules. Table 5.2 shows the Walsh sequences of order 64. The order in which the sequences are listed, in increasing order of the number of zero crossings, is different from that of the tables shown in the IS-95 system specification [1]. The sequences shown in the IS-95 documents are the Hadamard sequences, to be discussed shortly. The Walsh sequences and the Hadamard sequences are identical except that they are listed in a different order. We show how to convert the order of Hadamard sequence listings to Walsh sequence order in a later section of this chapter.

5.3   Walsh Function Generation

The Walsh functions can be generated (or computed) by many methods [8]. We can construct the Walsh functions by:

- Using Rademacher functions [9], as shown shortly;
- Using Hadamard matrices;
- Exploiting the symmetry properties of Walsh functions themselves.

Table 5.2 Walsh functions of order 64 (indexed by zero crossings)

W_0	0000000000000000 0000000000000000 0000000000000000 0000000000000000
W_1	0000000000000000 0000000000000000 1111111111111111 1111111111111111
W_2	0000000000000000 1111111111111111 1111111111111111 0000000000000000
W_3	0000000000000000 1111111111111111 0000000000000000 1111111111111111
W_4	0000000011111111 1111111100000000 0000000011111111 1111111100000000
W_5	0000000011111111 1111111100000000 1111111100000000 0000000011111111
W_6	0000000011111111 0000000011111111 1111111100000000 1111111100000000
W_7	0000000011111111 0000000011111111 0000000011111111 0000000011111111
W_8	0000111111110000 0000111111110000 0000111111110000 0000111111110000
W_9	0000111111110000 0000111111110000 1111000000001111 1111000000001111
W_{10}	0000111111110000 1111000000001111 1111000000001111 0000111111110000
W_{11}	0000111111110000 1111000000001111 0000111111110000 1111000000001111
W_{12}	0000111100001111 1111000011110000 0000111100001111 1111000011110000
W_{13}	0000111100001111 1111000011110000 1111000011110000 0000111100001111
W_{14}	0000111100001111 0000111100001111 1111000011110000 1111000011110000
W_{15}	0000111100001111 0000111100001111 0000111100001111 0000111100001111
W_{16}	0011110000111100 0011110000111100 0011110000111100 0011110000111100
W_{17}	0011110000111100 0011110000111100 1100001111000011 1100001111000011
W_{18}	0011110000111100 1100001111000011 1100001111000011 0011110000111100
W_{19}	0011110000111100 1100001111000011 0011110000111100 1100001111000011
W_{20}	0011110011000011 1100001100111100 0011110011000011 1100001100111100
W_{21}	0011110011000011 1100001100111100 1100001100111100 0011110011000011
W_{22}	0011110011000011 0011110011000011 1100001100111100 1100001100111100
W_{23}	0011110011000011 0011110011000011 0011110011000011 0011110011000011
W_{24}	0011001111001100 0011001111001100 0011001111001100 0011001111001100
W_{25}	0011001111001100 0011001111001100 1100110000110011 1100110000110011
W_{26}	0011001111001100 1100110000110011 1100110000110011 0011001111001100
W_{27}	0011001111001100 1100110000110011 0011001111001100 1100110000110011
W_{28}	0011001100110011 1100110011001100 0011001100110011 1100110011001100
W_{29}	0011001100110011 1100110011001100 1100110011001100 0011001100110011
W_{30}	0011001100110011 0011001100110011 1100110011001100 1100110011001100
W_{31}	0011001100110011 0011001100110011 0011001100110011 0011001100110011
W_{32}	0110011001100110 0110011001100110 0110011001100110 0110011001100110
W_{33}	0110011001100110 0110011001100110 1001100110011001 1001100110011001
W_{34}	0110011001100110 1001100110011001 1001100110011001 0110011001100110
W_{35}	0110011001100110 1001100110011001 0110011001100110 1001100110011001
W_{36}	0110011010011001 1001100101100110 0110011010011001 1001100101100110
W_{37}	0110011010011001 1001100101100110 1001100101100110 0110011010011001
W_{38}	0110011010011001 0110011010011001 1001100101100110 1001100101100110
W_{39}	0110011010011001 0110011010011001 0110011010011001 0110011010011001
W_{40}	0110100110010110 0110100110010110 0110100110010110 0110100110010110
W_{41}	0110100110010110 0110100110010110 1001011001101001 1001011001101001
W_{43}	0110100110010110 1001011001101001 0110100110010110 1001011001101001
W_{44}	0110100101101001 1001011010010110 0110100101101001 1001011010010110
W_{45}	0110100101101001 1001011010010110 1001011010010110 0110100101101001
W_{46}	0110100101101001 0110100101101001 1001011010010110 1001011010010110
W_{47}	0110100101101001 0110100101101001 0110100101101001 0110100101101001

Table 5.2 Walsh functions of order 64 (continued)

W_{48}	0101101001011010 0101101001011010 0101101001011010 0101101001011010
W_{49}	0101101001011010 0101101001011010 1010010110100101 1010010110100101
W_{50}	0101101001011010 1010010110100101 1010010110100101 0101101001011010
W_{51}	0101101001011010 1010010110100101 0101101001011010 1010010110100101
W_{52}	0101101010100101 1010010101011010 0101101010100101 1010010101011010
W_{53}	0101101010100101 1010010101011010 1010010101011010 0101101010100101
W_{54}	0101101010100101 0101101010100101 1010010101011010 1010010101011010
W_{55}	0101101010100101 0101101010100101 0101101010100101 0101101010100101
W_{56}	0101010110101010 0101010110101010 0101010110101010 0101010110101010
W_{57}	0101010110101010 0101010110101010 1010101001010101 1010101001010101
W_{58}	0101010110101010 1010101001010101 1010101001010101 0101010110101010
W_{59}	0101010110101010 1010101001010101 0101010110101010 1010101001010101
W_{60}	0101010101010101 1010101010101010 0101010101010101 1010101010101010
W_{61}	0101010101010101 1010101010101010 1010101010101010 0101010101010101
W_{62}	0101010101010101 0101010101010101 1010101010101010 1010101010101010
W_{63}	0101010101010101 0101010101010101 0101010101010101 0101010101010101

The Walsh functions are indexed according to the number of sign changes (zero crossings), which range from 0 to $N - 1$, and the index can be represented by a binary K-tuple, where $K = \log_2 N$. The binary representation for the entire set of index integers between 0 and $2^K - 1$ is expressed by the K-tuple

$$j = (j_1, j_2, \ldots, j_K) \tag{5.1}$$

such that

$$j = j_1 2^{K-1} + j_2 2^{K-2} + \cdots + j_K 2^0 \tag{5.2}$$

Let us now represent the K-tuple index-sequence of (5.1) as the K-component index sequence $\{X_j; j = 0, 1, \ldots, 2^K - 1\}$ as

$$X_j = (x_{j1}, x_{j2}, \ldots, x_{jK}) \tag{5.3}$$

and the corresponding Walsh sequence as

$$W_j = \left(w_{j0}, w_{j1}, \ldots, w_{j(N-1)} \right) \tag{5.4}$$

where $w_{j0} = 0$ for all j, and $w_{jn} = 0$ or 1 for $n = 1, 2, \ldots, N - 1$. In this notation, we can show a list of Walsh sequences of order 16 in Table 5.3.

We wish to distinguish the *functions* and the *sequences* in our presentations as we have done in the preceding paragraphs. Communications engineers have long been familiar with the concept of equivalence between logical operations using 0s and 1s and binary waveform modulation using

Table 5.3 Walsh sequences of order 16

Index sequences	Walsh sequences
$X_0 = 0\ 0\ 0\ 0$	$W_0 = 0\ 0\ 0\ 0\ 0\ 0\ 0\ 0\ 0\ 0\ 0\ 0\ 0\ 0\ 0\ 0$
$X_1 = 0\ 0\ 0\ 1$	$W_1 = 0\ 0\ 0\ 0\ 0\ 0\ 0\ 0\ 1\ 1\ 1\ 1\ 1\ 1\ 1\ 1$
$X_2 = 0\ 0\ 1\ 0$	$W_2 = 0\ 0\ 0\ 0\ 1\ 1\ 1\ 1\ 1\ 1\ 1\ 1\ 0\ 0\ 0\ 0$
$X_3 = 0\ 0\ 1\ 1$	$W_3 = 0\ 0\ 0\ 0\ 1\ 1\ 1\ 1\ 0\ 0\ 0\ 0\ 1\ 1\ 1\ 1$
$X_4 = 0\ 1\ 0\ 0$	$W_4 = 0\ 0\ 1\ 1\ 1\ 1\ 0\ 0\ 0\ 0\ 1\ 1\ 1\ 1\ 0\ 0$
$X_5 = 0\ 1\ 0\ 1$	$W_5 = 0\ 0\ 1\ 1\ 1\ 1\ 0\ 0\ 1\ 1\ 0\ 0\ 0\ 0\ 1\ 1$
$X_6 = 0\ 1\ 1\ 0$	$W_6 = 0\ 0\ 1\ 1\ 0\ 0\ 1\ 1\ 1\ 1\ 0\ 0\ 1\ 1\ 0\ 0$
$X_7 = 0\ 1\ 1\ 1$	$W_7 = 0\ 0\ 1\ 1\ 0\ 0\ 1\ 1\ 0\ 0\ 1\ 1\ 0\ 0\ 1\ 1$
$X_8 = 1\ 0\ 0\ 0$	$W_8 = 0\ 1\ 1\ 0\ 0\ 1\ 1\ 0\ 0\ 1\ 1\ 0\ 0\ 1\ 1\ 0$
$X_9 = 1\ 0\ 0\ 1$	$W_9 = 0\ 1\ 1\ 0\ 0\ 1\ 1\ 0\ 1\ 0\ 0\ 1\ 1\ 0\ 0\ 1$
$X_{10} = 1\ 0\ 1\ 0$	$W_{10} = 0\ 1\ 1\ 0\ 1\ 0\ 0\ 1\ 1\ 0\ 0\ 1\ 0\ 1\ 1\ 0$
$X_{11} = 1\ 0\ 1\ 1$	$W_{11} = 0\ 1\ 1\ 0\ 1\ 0\ 0\ 1\ 0\ 1\ 1\ 0\ 1\ 0\ 0\ 1$
$X_{12} = 1\ 1\ 0\ 0$	$W_{12} = 0\ 1\ 0\ 1\ 1\ 0\ 1\ 0\ 0\ 1\ 0\ 1\ 1\ 0\ 1\ 0$
$X_{13} = 1\ 1\ 0\ 1$	$W_{13} = 0\ 1\ 0\ 1\ 1\ 0\ 1\ 0\ 1\ 0\ 1\ 0\ 0\ 1\ 0\ 1$
$X_{14} = 1\ 1\ 1\ 0$	$W_{14} = 0\ 1\ 0\ 1\ 0\ 1\ 0\ 1\ 1\ 0\ 1\ 0\ 1\ 0\ 1\ 0$
$X_{15} = 1\ 1\ 1\ 1$	$W_{15} = 0\ 1\ 0\ 1\ 0\ 1\ 0\ 1\ 0\ 1\ 0\ 1\ 0\ 1\ 0\ 1$

"+1" and "−1," and between waveform multiplication, denoted by $\otimes$ and the modulo-2 addition, denoted by $\oplus$, corresponding to the same operations in system analysis. The distinction is really based on the law of combination between the elements within the set of functions or sequences that are defined under a suitable mathematical operation. To clarify this point and also for the purpose of later use in connection with the generation of Walsh functions or sequences, we will be a bit formal about the characterizations of the Walsh functions and Walsh sequences.

The Walsh functions, expressed in terms of ± 1 values, form a group under the multiplication operation (multiplicative group); whereas the Walsh sequences, expressed in terms of $(0, 1)$ values, form a group under modulo-2 addition (additive group). We review the definition of a group first, and then observe the properties of the Walsh functions and the Walsh sequences.

Definition. A group G is a set (ensemble) of objects, say $\{a, b, c, \ldots, \gamma\}$, for which a mathematical operation, which we will denote by "$*$" for the time being ("$*$" will be either "$+$" or "$\cdot$"), satisfies the following four properties:

- *Closure property.* G is closed under operation "$*$"; that is, if $a \in G$ and $b \in G$, then $a*b$ is also found in G ($a*b \in G$).

- *Associative law.* For any three elements in the set, the associative law is satisfied; that is, for any $a, b, c \in G$, the relation $(a*b)*c = a*(b*c)$ holds.

- *Identity element.* The set has an "identity element," such that for any element $a \in G$, the relation $a*(\text{identity element}) = a$ holds.

- *Inverse element.* Every element in the set has an "inverse element" in the set as well; that is, for every $a \in G$, there is an inverse element such that the relation $a*(\text{inverse element}) = (\text{identity element})$ holds.

We now need to specify the mathematical operation that we denoted as "$*$."

When we identify "$*$" to be the multiplicative operation, the Walsh functions, such as those shown previously in Figure 5.1 for the order $N = 8$, form a group under that multiplicative operation. The "closure property" is certainly satisfied, because when two Walsh functions are multiplied, say $W_3(t) \cdot W_5(t)$, we see the result of $W_6(t)$.

The satisfaction of the associative law by the Walsh functions is trivially evident. The identity function (element) is $W_0(t)$ for any order N, for when we multiply any Walsh function by $W_0(t)$, the result is that same function. Every function itself is its inverse element, since when we multiply a function by itself, we obtain $W_j^2(t) = W_0(t)$, the identity (element) function.

Now, if we define the operation "$*$" to be modulo-2 addition arithmetic, the Walsh sequences, not functions, form a group under that operation. Again, the defining four properties of the additive group are observed to be satisfied by Walsh sequences. The identity element is the all-zero sequence W_0, and because $W_j \oplus W_j = W_0$, every sequence has an inverse sequence (element), which is itself.

We see therefore that the Walsh functions $\{W_j(t)\}$ form a group under multiplication, while the Walsh sequences $\{W_j\}$ form a group under modulo-2 addition:

$$W_i(t) \cdot W_j(t) = W_r(t) \tag{5.5}$$

$$W_i \oplus W_j = W_r \tag{5.6}$$

The same statement is true of their corresponding index functions or sequences, so that $X_i(t) \cdot X_j(t) = X_r(t)$ and $\mathbf{X}_i \oplus \mathbf{X}_j = \mathbf{X}_r$.

Example 5.1 The Walsh sequences of order 8 form a group under modulo-2 addition. We observe the following example of the closure property:

$$\mathbf{W}_3 = 0 \ 0 \ 1 \ 1 \ 0 \ 0 \ 1 \ 1 \qquad\qquad \mathbf{X}_3 = 0 \ 1 \ 1$$

$$\mathbf{W}_5 = 0 \ 1 \ 1 \ 0 \ 1 \ 0 \ 0 \ 1 \qquad\qquad \mathbf{X}_5 = 1 \ 0 \ 1$$

$$\mathbf{W}_3 \oplus \mathbf{W}_5 = 0 \ 1 \ 0 \ 1 \ 1 \ 0 \ 1 \ 0 = \mathbf{W}_6 \qquad \mathbf{X}_3 \oplus \mathbf{X}_5 = 1 \ 1 \ 0 = \mathbf{X}_6$$

From here on we use the terms Walsh *functions* and Walsh *sequences* synonymously in the sense that they mean the same thing. We at times call the sequences *functions*, although the waveform functions are not called *sequences*.

5.3.1 Walsh Function Generation Using Rademacher Functions

Many ways of specifying the Walsh functions have been suggested over the years. As mentioned earlier, the Walsh functions as defined originally by Walsh [2] are ordered according to the number of sign changes or zero-crossings. Harmuth [6] showed a system of difference equations to specify the higher orders of Walsh functions in a given set of Walsh functions. In this section, we consider a method of generating Walsh functions using Rademacher functions. Other methods of generating Walsh functions are considered in later sections.

Rademacher functions $\{\mathcal{R}_n(t); t \in (0, T), n = 0, 1, \ldots, \log_2 N \}$ are a set of $1 + \log_2 N$ orthogonal functions consisting of $N = 2^K$ rectangular pulses that assume alternately the values $+1$ and -1 in an interval of $(0, T)$, except at jumps, where they take on the value 0 [9, 10]. The Rademacher functions with $N = 2^4 = 16$ pulses are shown in Figure 5.2, along with the sequence representations of the functions in the logical elements $\{0, 1\}$. The Rademacher functions $\{\mathcal{R}_n(t)\}$ can also be defined by the relation

$$\mathcal{R}_n(t) = sgn(\sin 2^n \pi t), \qquad t \in (0, T), n = 1, 2, \ldots, \log_2 N = K \qquad (5.7)$$

where $\mathcal{R}_0(t) \equiv 1$ and

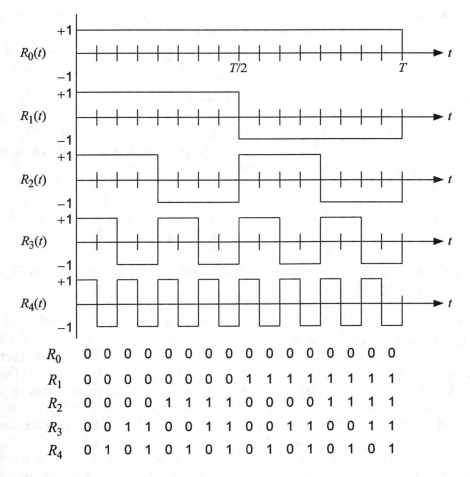

Figure 5.2 Rademacher functions with $N = 2^K = 2^4 = 16$ elements.

$$sgn(x) \triangleq \begin{cases} -1 & \text{for } x < 0 \\ 0 & \text{for } x = 0 \\ 1 & \text{for } x > 0 \end{cases} \tag{5.8}$$

The way Rademacher functions are constructed is as follows:

- Let $\mathcal{R}_0(t)$ be the function with the value 1 for the entire interval of duration T: $\mathcal{R}_0(t) = 1$.

- To obtain $\mathcal{R}_1(t)$, divide the interval $(0, T)$ in half, and let the value of $\mathcal{R}_1(t)$ in the first half of the interval be 1 and in the second half of the interval, -1:

$$\mathcal{R}_1(t) = \begin{cases} 1, & t \in (0, T/2) \\ -1, & t \in (T/2, T) \\ 0, & t = 0, T/2, T \end{cases}$$

- To obtain $\mathcal{R}_2(t)$, divide each of the intervals $(0, T/2)$ and $(T/2, T)$ in half and let the value of the function in the first half of each interval be 1 and in the second half of the interval, -1:

$$\mathcal{R}_2(t) = \begin{cases} 1, & t \in (0, T/4) \text{ and } t \in (T/2, 3T/4) \\ -1, & t \in (T/4, T/2) \text{ and } t \in (3T/4, T) \\ 0, & t = 0, T/4, T/2, 3T/4, T \end{cases}$$

- Repeat the previous steps $\log_2 N$ times until each subinterval is a single-pulse element.

The definition given in (5.7) for Rademacher functions says that a single-cycle square wave, following in the pattern of a sinewave, is drawn into each interval of length $T, T/2, T/2^2, \ldots, T/2^n, \ldots, T/2^K$. The Rademacher functions of $N = 2^K$ elements (pulses) account for $K + 1 = \log_2 N + 1$ of the N Walsh functions of order N, but the designation of Rademacher function $\mathcal{R}_n(t)$ has nothing to do with the relationship between the number of zero crossings and the index n. We now take up the subject of generating the Walsh functions using products of the Rademacher functions $\{\mathcal{R}_n(t)\}$.

The Walsh functions are developed as products of the Rademacher functions, based on the Gray code conversion of the Walsh function index sequence. The Gray code consists of N binary sequences, and each sequence is represented by $K = \log_2 N$ bits. The conversion of the index sequence $X_i = (x_{i1}, x_{i2}, \ldots, x_{iK})$ to the Gray code $G_i = (g_{i1}, g_{i2}, \ldots, g_{iK})$ is done as follows:

$$g_{i1} = x_{i1} \tag{5.9a}$$

$$g_{ij} = x_{i,j-1} \oplus x_{ij}, \; j = 2, 3, \ldots, K \tag{5.9b}$$

If we convert the index i, represented by the K-tuple sequence, to the corresponding Gray code by the rule of (5.9), the Gray code sequences show that the previous and following sequences (the adjacent sequences) differ in exactly one bit position.

Example 5.2 Gray code with 16 words (sequences). Find $G_{13} = (g_{13,1}, g_{13,2}, \ldots, g_{13,K})$ from $X_{13} = (x_{13,1}, x_{13,2}, x_{13,3}, x_{13,4}) = (1\ 1\ 0\ 1)$.

Solution:

$$g_{13,1} = x_{13,1} = 1$$

$$g_{13,2} = x_{13,1} \oplus x_{13,2} = 1 \oplus 1 = 0$$

$$g_{13,3} = x_{13,2} \oplus x_{13,3} = 1 \oplus 0 = 1 \qquad (5.10)$$

$$g_{13,4} = x_{13,3} \oplus x_{13,4} = 0 \oplus 1 = 1$$

and we obtain $G_{13} = (1\ 0\ 1\ 1)$. The remainder of the Gray code for $K = 4$ can be found by the rule, and the result is shown in Table 5.4.

Table 5.4 Gray code for $K = 4$

$X_i = x_{i1}\ x_{i2}\ x_{i3}\ x_{i4}$				$G_i = g_{i1}\ g_{i2}\ g_{i3}\ g_{i4}$			
$X_0 = 0$	0	0	0	$G_0 = 0$	0	0	0
$X_1 = 0$	0	0	1	$G_1 = 0$	0	0	1
$X_2 = 0$	0	1	0	$G_2 = 0$	0	1	1
$X_3 = 0$	0	1	1	$G_3 = 0$	0	1	0
$X_4 = 0$	1	0	0	$G_4 = 0$	1	1	0
$X_5 = 0$	1	0	1	$G_5 = 0$	1	1	1
$X_6 = 0$	1	1	0	$G_6 = 0$	1	0	1
$X_7 = 0$	1	1	1	$G_7 = 0$	1	0	0
$X_8 = 1$	0	0	0	$G_8 = 1$	1	0	0
$X_9 = 1$	0	0	1	$G_9 = 1$	1	0	1
$X_{10} = 1$	0	1	0	$G_{10} = 1$	1	1	1
$X_{11} = 1$	0	1	1	$G_{11} = 1$	1	1	0
$X_{12} = 1$	1	0	0	$G_{12} = 1$	0	1	0
$X_{13} = 1$	1	0	1	$G_{13} = 1$	0	1	1
$X_{14} = 1$	1	1	0	$G_{14} = 1$	0	0	1
$X_{15} = 1$	1	1	1	$G_{15} = 1$	0	0	0

To generate the set of Walsh functions of order $N = 2^K$ as products of the Rademacher functions in the set $\{\mathcal{R}_0, \mathcal{R}_1, \mathcal{R}_2, \ldots, \mathcal{R}_K\}$, perform the following steps:

1. Write the Walsh function index i of $\boldsymbol{W}_i$ in binary K-tuple form as the index sequence $\boldsymbol{X}_i = (x_{i1}, x_{i2}, \ldots, x_{iK})$, $x_{ij} \in \{0, 1\}$.

2. Convert $\boldsymbol{X}_i$ to the corresponding Gray code: $\boldsymbol{X}_i \to \boldsymbol{G}_i$.

3. For each Gray code element, associate a Rademacher sequence according to the rule, $g_{ij} \leftrightarrow \mathcal{R}_{K+1-j}$; that is, associate g_{i1} with $\mathcal{R}_K$, g_{i2} with $\mathcal{R}_{K-1}, \ldots, g_{iK}$ with $\mathcal{R}_1$.

4. The Walsh function $W_i(t)$ is formed as the product of $\mathcal{R}_0(t)$ and the Rademacher functions $\{\mathcal{R}_j(t)\}$ that are associated with nonzero elements of $\boldsymbol{G}_i$; that is:

$$W_i(t) = \mathcal{R}_0(t) \cdot \prod_{j:g_{ij}=1} \mathcal{R}_{K+1-j}(t) \tag{5.11a}$$

$$= \mathcal{R}_0(t) \cdot [\mathcal{R}_K(t) \text{ if } g_{i1} = 1] \cdot [\mathcal{R}_{K-1}(t) \text{ if } g_{i2} = 1] \times \cdots$$

$$\cdots \times [\mathcal{R}_1(t) \text{ if } g_{iK} = 1] \tag{5.11b}$$

The Walsh sequence $\boldsymbol{W}_i$ is formed as the modulo-2 sum of $\mathcal{R}_0$ and the Rademacher sequences $\{\mathcal{R}_j\}$ that are associated with nonzero elements of $\boldsymbol{G}_i$; that is:

$$\boldsymbol{W}_i = \mathcal{R}_0 \oplus \sum_{j:g_{ij}=1} \mathcal{R}_{K+1-j} \tag{5.12a}$$

$$= \mathcal{R}_0 \oplus [\mathcal{R}_K \text{ if } g_{i1} = 1] \oplus [\mathcal{R}_{K-1} \text{ if } g_{i2} = 1] \oplus \cdots$$

$$\cdots \oplus [\mathcal{R}_1 \text{ if } g_{iK} = 1] \tag{5.12b}$$

Example 5.3 Find the Walsh function $W_{13}(t)$ of order $N = 16$ as the product of Rademacher functions.

Solution: In the previous example, we found that $\boldsymbol{G}_{13} = (1 \ 0 \ 1 \ 1)$. Therefore, the Walsh function $W_{13}(t)$ is the product $W_{13}(t) = \mathcal{R}_0(t) \cdot \mathcal{R}_4(t) \cdot \mathcal{R}_2(t) \cdot \mathcal{R}_1(t)$ and the Walsh sequence $\boldsymbol{W}_{13}$ is the modulo-2 sum $\boldsymbol{W}_{13} = \mathcal{R}_0 \oplus \mathcal{R}_4 \oplus \mathcal{R}_2 \oplus \mathcal{R}_1$:

$$
\begin{array}{llcccccccccccccccc}
\mathcal{R}_0 &=& 0 & 0 & 0 & 0 & 0 & 0 & 0 & 0 & 0 & 0 & 0 & 0 & 0 & 0 & 0 & 0 \\
\mathcal{R}_4 &=& 0 & 1 & 0 & 1 & 0 & 1 & 0 & 1 & 0 & 1 & 0 & 1 & 0 & 1 & 0 & 1 \\
\mathcal{R}_2 &=& 0 & 0 & 0 & 0 & 1 & 1 & 1 & 1 & 0 & 0 & 0 & 0 & 1 & 1 & 1 & 1 \\
\mathcal{R}_1 &=& 0 & 0 & 0 & 0 & 0 & 0 & 0 & 0 & 1 & 1 & 1 & 1 & 1 & 1 & 1 & 1 \\
\hline
W_{13} &=& 0 & 1 & 0 & 1 & 1 & 0 & 1 & 0 & 1 & 0 & 1 & 0 & 0 & 1 & 0 & 1
\end{array}
$$

The entire set of Walsh sequences for $N = 16$ is shown in Table 5.5 as modulo-2 sums of Rademacher sequences.

5.3.2 Walsh Function Generation Using Hadamard Matrices

The Hadamard matrix [11, 12] is a square array of plus and minus ones, $\{+1, -1\}$, whose rows and columns are mutually orthogonal. If the first row and first column contain only plus ones, the matrix is said to be in normal form [11]. We can replace "+1" with "0" and "−1" with "1" to express the Hadamard matrix using the logic elements $\{0, 1\}$. The 2×2 Hadamard matrix of order 2 is

$$
\mathbf{H}_2 = \begin{bmatrix} 1 & 1 \\ 1 & -1 \end{bmatrix} \equiv \begin{bmatrix} 0 & 0 \\ 0 & 1 \end{bmatrix} \tag{5.13}
$$

If $\mathbf{H}_N$ is an $N \times N$ Hadamard matrix, then

$$
\mathbf{H}_N \mathbf{H}_N^{\mathsf{T}} = N \boldsymbol{I}_N \tag{5.14}
$$

where $\boldsymbol{I}_N$ is the $N \times N$ identity matrix. If $N \geq 1$ is the order of a Hadamard matrix, then $N = 1$, 2, or $4t$, where t is a positive integer. If $\mathbf{H}_a$ and $\mathbf{H}_b$ are Hadamard matrices of order a and b, respectively, then $\mathbf{H}_a \times \mathbf{H}_b = \mathbf{H}_{ab}$, the Hadamard matrix of order ab, when $\mathbf{H}_a \times \mathbf{H}_b$ is found by substituting $\mathbf{H}_b$ for $+1$ (or logic 0) in $\mathbf{H}_a$ and $-\mathbf{H}_b$ (or the complement of $\mathbf{H}_b$) for -1 (or logic 1) in $\mathbf{H}_a$. Consequently, if N is a power of two and it is understood that $\mathbf{H}_1 = [+1] \equiv [0]$, then $\mathbf{H}_{2N}$ can be found as follows:

$$
\mathbf{H}_{2N} = \begin{bmatrix} \mathbf{H}_N & \mathbf{H}_N \\ \mathbf{H}_N & \overline{\mathbf{H}}_N \end{bmatrix} \tag{5.15}
$$

Table 5.5 Rademacher sequence construction of Walsh sequences

$G_i = g_{i1}\ g_{i2}\ g_{i3}\ g_{i4}$	W_i
$G_0 = 0$ 0 0 0	$W_0 = \mathcal{R}_0$
$G_1 = 0$ 0 0 1	$W_1 = \mathcal{R}_0 \oplus \mathcal{R}_1 = \mathcal{R}_1$
$G_2 = 0$ 0 1 1	$W_2 = \mathcal{R}_0 \oplus \mathcal{R}_2 \oplus \mathcal{R}_1 = \mathcal{R}_2 \oplus \mathcal{R}_1$
$G_3 = 0$ 0 1 0	$W_3 = \mathcal{R}_0 \oplus \mathcal{R}_2 = \mathcal{R}_2$
$G_4 = 0$ 1 1 0	$W_4 = \mathcal{R}_0 \oplus \mathcal{R}_3 \oplus \mathcal{R}_2 = \mathcal{R}_3 \oplus \mathcal{R}_2$
$G_5 = 0$ 1 1 1	$W_5 = \mathcal{R}_0 \oplus \mathcal{R}_3 \oplus \mathcal{R}_2 \oplus \mathcal{R}_1 = \mathcal{R}_3 \oplus \mathcal{R}_2 \oplus \mathcal{R}_1$
$G_6 = 0$ 1 0 1	$W_6 = \mathcal{R}_0 \oplus \mathcal{R}_3 \oplus \mathcal{R}_1 = \mathcal{R}_3 \oplus \mathcal{R}_1$
$G_7 = 0$ 1 0 0	$W_7 = \mathcal{R}_0 \oplus \mathcal{R}_3 = \mathcal{R}_3$
$G_8 = 1$ 1 0 0	$W_8 = \mathcal{R}_0 \oplus \mathcal{R}_4 \oplus \mathcal{R}_3 = \mathcal{R}_4 \oplus \mathcal{R}_3$
$G_9 = 1$ 1 0 1	$W_9 = \mathcal{R}_0 \oplus \mathcal{R}_4 \oplus \mathcal{R}_3 \oplus \mathcal{R}_1 = \mathcal{R}_4 \oplus \mathcal{R}_3 \oplus \mathcal{R}_1$
$G_{10} = 1$ 1 1 1	$W_{10} = \mathcal{R}_0 \oplus \mathcal{R}_4 \oplus \mathcal{R}_3 \oplus \mathcal{R}_2 \oplus \mathcal{R}_1 = \mathcal{R}_4 \oplus \mathcal{R}_3 \oplus \mathcal{R}_2 \oplus \mathcal{R}_1$
$G_{11} = 1$ 1 1 0	$W_{11} = \mathcal{R}_0 \oplus \mathcal{R}_4 \oplus \mathcal{R}_3 \oplus \mathcal{R}_2 = \mathcal{R}_4 \oplus \mathcal{R}_3 \oplus \mathcal{R}_2$
$G_{12} = 1$ 0 1 0	$W_{12} = \mathcal{R}_0 \oplus \mathcal{R}_4 \oplus \mathcal{R}_2 = \mathcal{R}_4 \oplus \mathcal{R}_2$
$G_{13} = 1$ 0 1 1	$W_{13} = \mathcal{R}_0 \oplus \mathcal{R}_4 \oplus \mathcal{R}_2 \oplus \mathcal{R}_1 = \mathcal{R}_4 \oplus \mathcal{R}_2 \oplus \mathcal{R}_1$
$G_{14} = 1$ 0 0 1	$W_{14} = \mathcal{R}_0 \oplus \mathcal{R}_4 \oplus \mathcal{R}_1 = \mathcal{R}_4 \oplus \mathcal{R}_1$
$G_{15} = 1$ 0 0 0	$W_{15} = \mathcal{R}_0 \oplus \mathcal{R}_4 = \mathcal{R}_4$

where $\overline{\mathbf{H}}_N$ is the negative (complement) of $\mathbf{H}_N$. Hadamard matrices of order $N = 2^t$ can be formed by repeatedly multiplying the normal form of the $N = 2$ Hadamard matrix by itself.

Example 5.4 Construct the normal form of the order $N = 2^4 = 16$ Hadamard matrix.

Solution:

$$\mathbf{H}_2 = \begin{bmatrix} 1 & 1 \\ 1 & -1 \end{bmatrix} \qquad \mathbf{H}_4 = \mathbf{H}_2 \times \mathbf{H}_2 = \begin{bmatrix} 1 & 1 & 1 & 1 \\ 1 & -1 & 1 & -1 \\ 1 & 1 & -1 & -1 \\ 1 & -1 & -1 & 1 \end{bmatrix}$$

$$\mathbf{H_8 = H_2 \times H_4} = \begin{bmatrix} 1 & 1 & 1 & 1 & 1 & 1 & 1 & 1 \\ 1 & -1 & 1 & -1 & 1 & -1 & 1 & -1 \\ 1 & 1 & -1 & -1 & 1 & 1 & -1 & -1 \\ 1 & -1 & -1 & 1 & 1 & -1 & -1 & 1 \\ 1 & 1 & 1 & 1 & -1 & -1 & -1 & -1 \\ 1 & -1 & 1 & -1 & -1 & 1 & -1 & 1 \\ 1 & 1 & -1 & -1 & -1 & -1 & 1 & 1 \\ 1 & -1 & -1 & 1 & -1 & 1 & 1 & -1 \end{bmatrix}$$

$$\mathbf{H_{16} = H_2 \times H_8} =$$

$$\begin{bmatrix} 1 & 1 & 1 & 1 & 1 & 1 & 1 & 1 & 1 & 1 & 1 & 1 & 1 & 1 & 1 & 1 \\ 1 & -1 & 1 & -1 & 1 & -1 & 1 & -1 & 1 & -1 & 1 & -1 & 1 & -1 & 1 & -1 \\ 1 & 1 & -1 & -1 & 1 & 1 & -1 & -1 & 1 & 1 & -1 & -1 & 1 & 1 & -1 & -1 \\ 1 & -1 & -1 & 1 & 1 & -1 & -1 & 1 & 1 & -1 & -1 & 1 & 1 & -1 & -1 & 1 \\ 1 & 1 & 1 & 1 & -1 & -1 & -1 & -1 & 1 & 1 & 1 & 1 & -1 & -1 & -1 & -1 \\ 1 & -1 & 1 & -1 & -1 & 1 & -1 & 1 & 1 & -1 & 1 & -1 & -1 & 1 & -1 & 1 \\ 1 & 1 & -1 & -1 & -1 & -1 & 1 & 1 & 1 & 1 & -1 & -1 & -1 & -1 & 1 & 1 \\ 1 & -1 & -1 & 1 & -1 & 1 & 1 & -1 & 1 & -1 & -1 & 1 & -1 & 1 & 1 & -1 \\ 1 & 1 & 1 & 1 & 1 & 1 & 1 & 1 & -1 & -1 & -1 & -1 & -1 & -1 & -1 & -1 \\ 1 & -1 & 1 & -1 & 1 & -1 & 1 & -1 & -1 & 1 & -1 & 1 & -1 & 1 & -1 & 1 \\ 1 & 1 & -1 & -1 & 1 & 1 & -1 & -1 & -1 & -1 & 1 & 1 & -1 & -1 & 1 & 1 \\ 1 & -1 & -1 & 1 & 1 & -1 & -1 & 1 & -1 & 1 & 1 & -1 & -1 & 1 & 1 & -1 \\ 1 & 1 & 1 & 1 & -1 & -1 & -1 & -1 & -1 & -1 & -1 & -1 & 1 & 1 & 1 & 1 \\ 1 & -1 & 1 & -1 & -1 & 1 & -1 & 1 & -1 & 1 & -1 & 1 & 1 & -1 & 1 & -1 \\ 1 & 1 & -1 & -1 & -1 & -1 & 1 & 1 & -1 & -1 & 1 & 1 & 1 & 1 & -1 & -1 \\ 1 & -1 & -1 & 1 & -1 & 1 & 1 & -1 & -1 & 1 & 1 & -1 & 1 & -1 & -1 & 1 \end{bmatrix} \quad (5.16)$$

Let us express the Hadamard matrix of order 16 shown in (5.16) as binary sequences $\{H_i\}$ for its rows, with the row number i acting as the index, as shown in Table 5.6.

All the row and column sequences of Hadamard matrices are Walsh sequences if the order is $N = 2^t$. There is a difference, however, between the Walsh function representation and the Hadamard function representation in that neither the row nor the column index of Hadamard functions is related to the number of sign changes (zero crossings), as was the case for Walsh functions. Clearly the Walsh functions generated by the Hadamard matrix method are not indexed according to the number of sign changes. A method of converting their ordering is desirable.

The number of sign changes in Walsh function $W_i(t)$ is i, given in binary notation by the index sequence $X_i = (x_{i1}, x_{i2}, \ldots, x_{iK})$. The row of the Hadamard matrix that equals the Walsh sequence W_i may be indexed by the binary sequence $C_i = (c_{i1}, c_{i2}, \ldots, c_{iK})$. The relation between the elements of X_i and C_i is given by the following transformation [13]:

Table 5.6 Rows of the degree 16 Hadamard matrix

Rows of H_{16} as sequences
$H_0 = 0\ 0\ 0\ 0\ 0\ 0\ 0\ 0\ 0\ 0\ 0\ 0\ 0\ 0\ 0\ 0$
$H_1 = 0\ 1\ 0\ 1\ 0\ 1\ 0\ 1\ 0\ 1\ 0\ 1\ 0\ 1\ 0\ 1$
$H_2 = 0\ 0\ 1\ 1\ 0\ 0\ 1\ 1\ 0\ 0\ 1\ 1\ 0\ 0\ 1\ 1$
$H_3 = 0\ 1\ 1\ 0\ 0\ 1\ 1\ 0\ 0\ 1\ 1\ 0\ 0\ 1\ 1\ 0$
$H_4 = 0\ 0\ 0\ 0\ 1\ 1\ 1\ 1\ 0\ 0\ 0\ 0\ 1\ 1\ 1\ 1$
$H_5 = 0\ 1\ 0\ 1\ 1\ 0\ 1\ 0\ 0\ 1\ 0\ 1\ 1\ 0\ 1\ 0$
$H_6 = 0\ 0\ 1\ 1\ 1\ 1\ 0\ 0\ 0\ 0\ 1\ 1\ 1\ 1\ 0\ 0$
$H_7 = 0\ 1\ 1\ 0\ 1\ 0\ 0\ 1\ 0\ 1\ 1\ 0\ 1\ 0\ 0\ 1$
$H_8 = 0\ 0\ 0\ 0\ 0\ 0\ 0\ 0\ 1\ 1\ 1\ 1\ 1\ 1\ 1\ 1$
$H_9 = 0\ 1\ 0\ 1\ 0\ 1\ 0\ 1\ 1\ 0\ 1\ 0\ 1\ 0\ 1\ 0$
$H_{10} = 0\ 0\ 1\ 1\ 0\ 0\ 1\ 1\ 1\ 1\ 0\ 0\ 1\ 1\ 0\ 0$
$H_{11} = 0\ 1\ 1\ 0\ 0\ 1\ 1\ 0\ 1\ 0\ 0\ 1\ 1\ 0\ 0\ 1$
$H_{12} = 0\ 0\ 0\ 0\ 1\ 1\ 1\ 1\ 1\ 1\ 1\ 1\ 0\ 0\ 0\ 0$
$H_{13} = 0\ 1\ 0\ 1\ 1\ 0\ 1\ 0\ 1\ 0\ 1\ 0\ 0\ 1\ 0\ 1$
$H_{14} = 0\ 0\ 1\ 1\ 1\ 1\ 0\ 0\ 1\ 1\ 0\ 0\ 0\ 0\ 1\ 1$
$H_{15} = 0\ 1\ 1\ 0\ 1\ 0\ 0\ 1\ 1\ 0\ 0\ 1\ 0\ 1\ 1\ 0$

$$c_{i,K} = x_{i,1} \tag{5.17a}$$

$$c_{i,K-j} = x_{i,j} \oplus x_{i,j+1}, \quad j = 1, 2, \ldots, K-1 \tag{5.17b}$$

Example 5.5 Consider the Walsh sequences of order $N = 16 = 2^4$ and find the row of the Hadamard matrix H_{16} that corresponds to W_7.

Solution: Because $X_7 = (0\ \ 1\ \ 1\ \ 1)$, we have

$$c_{7,4} = x_{7,1} = 0$$

$$c_{7,3} = x_{7,1} \oplus x_{7,2} = 0 \oplus 1 = 1$$

$$c_{7,2} = x_{7,2} \oplus x_{7,3} = 1 \oplus 1 = 0$$

$$c_{7,1} = x_{7,3} \oplus x_{7,4} = 1 \oplus 1 = 0$$

Because $C_7 = (c_{7,1}, c_{7,2}, c_{7,3}, c_{7,4}) = (0\ \ 0\ \ 1\ \ 0)$, the second row of $\mathbf{H}_{16}$ is $\mathbf{W}_7$, that is, $\mathbf{H}_2 = \mathbf{W}_7$, as can be seen by comparing Table 5.6 with Table 5.3:

$$\mathbf{H}_2 = 0\ 0\ 1\ 1\ 0\ 0\ 1\ 1\ 0\ 0\ 1\ 1\ 0\ 0\ 1\ 1 = \mathbf{W}_7$$

The ordering for the remainder of the functions can be found similarly, and the result is shown in Table 5.7.

The table of "64-ary Walsh functions" given in IS-95 [1, p. 7-17], as pointed out previously, is actually a list of Hadamard sequences and the order of the sequences is related to that of Walsh functions as shown in Table 5.8.

The transformation of Walsh binary index sequences to find Hadamard matrix binary row number sequences, as described by (5.17), can be expressed in matrix notation as

$$\begin{bmatrix} c_{i1} \\ c_{i2} \\ c_{i3} \\ \vdots \\ c_{i,K-2} \\ c_{i,K-1} \\ c_{iK} \end{bmatrix} = \begin{bmatrix} 0 & 0 & 0 & \cdots & 0 & 1 & 1 \\ 0 & 0 & 0 & \cdots & 1 & 1 & 0 \\ 0 & 0 & 0 & \cdots & 1 & 0 & 0 \\ \vdots & \vdots & \vdots & & \vdots & \vdots & \vdots \\ 0 & 1 & 1 & \cdots & 0 & 0 & 0 \\ 1 & 1 & 0 & \cdots & 0 & 0 & 0 \\ 1 & 0 & 0 & \cdots & 0 & 0 & 0 \end{bmatrix} \begin{bmatrix} x_{i1} \\ x_{i2} \\ x_{i3} \\ \vdots \\ x_{i,K-2} \\ x_{i,K-1} \\ x_{iK} \end{bmatrix} \tag{5.18}$$

in which the summation involved in performing the matrix multiplication of the vector is understood to be modulo-2 summation.

We have presented the Walsh sequences of order $N = 2^K$ as binary N-tuples. By binary we meant that the components of the sequences assume a value from the finite field $\{0, 1\}$. What is a finite field? When we describe the algebraic structure of sequences or codes, we need some fundamental understanding of such terms as "finite field," "vector space," "subspace," and "linear dependence and independence." So, before we proceed further with the main topics of this section, let us pause here and review definitions of those terms for use in succeeding discussions.

5.3.3 Finite Fields

A finite field is a finite set of elements with two operations: *addition* and *multiplication*. A *finite field of elements* q is also called a *Galois field of elements* q, denoted by GF(q). Galois fields are named after the French math-

Table 5.7 Hadamard matrix rows corresponding to Walsh sequences

Walsh index	$x_1\ x_2\ x_3\ x_4$	$c_1\ c_2\ c_3\ c_4$	Hadamard row
0	0 0 0 0	0 0 0 0	0
1	0 0 0 1	1 0 0 0	8
2	0 0 1 0	1 1 0 0	12
3	0 0 1 1	0 1 0 0	4
4	0 1 0 0	0 1 1 0	6
5	0 1 0 1	1 1 1 0	14
6	0 1 1 0	1 0 1 0	10
7	0 1 1 1	0 0 1 0	2
8	1 0 0 0	0 0 1 1	3
9	1 0 0 1	1 0 1 1	11
10	1 0 1 0	1 1 1 1	15
11	1 0 1 1	0 1 1 1	7
12	1 1 0 0	0 1 1 0	6
13	1 1 0 1	1 1 0 1	13
14	1 1 1 0	1 0 0 1	9
15	1 1 1 1	0 0 0 1	1

ematician Ébariste Galois (1811–1832) who made major contributions to the theory of equations. He died at the age of 20, shot in a duel.

The operations performed on the inverse elements of fields imply two further operations, subtraction and division. Thus, the finite field $GF(q)$ consists of q elements that can be added, subtracted, multiplied, and divided almost like those employed in arithmetic operations with real and complex numbers. The rules of operations under addition "+" and multiplication " · " must satisfy the following conditions:

Table 5.8 Walsh functions of order 64, as indexed in IS-95 (W_i is the Walsh notation, and H_i is the Hadamard notation)

W_0	H_0	0000000000000000	0000000000000000	0000000000000000	0000000000000000
W_{63}	H_1	0101010101010101	0101010101010101	0101010101010101	0101010101010101
W_{31}	H_2	0011001100110011	0011001100110011	0011001100110011	0011001100110011
W_{32}	H_3	0110011001100110	0110011001100110	0110011001100110	0110011001100110
W_{15}	H_4	0000111100001111	0000111100001111	0000111100001111	0000111100001111
W_{48}	H_5	0101101001011010	0101101001011010	0101101001011010	0101101001011010
W_{16}	H_6	0011110000111100	0011110000111100	0011110000111100	0011110000111100
W_{47}	H_7	0110100101101001	0110100101101001	0110100101101001	0110100101101001
W_7	H_8	0000000011111111	0000000011111111	0000000011111111	0000000011111111
W_{56}	H_9	0101010110101010	0101010110101010	0101010110101010	0101010110101010
W_{24}	H_{10}	0011001111001100	0011001111001100	0011001111001100	0011001111001100
W_{39}	H_{11}	0110011010011001	0110011010011001	0110011010011001	0110011010011001
W_8	H_{12}	0000111111110000	0000111111110000	0000111111110000	0000111111110000
W_{55}	H_{13}	0101101010100101	0101101010100101	0101101010100101	0101101010100101
W_{23}	H_{14}	0011110011000011	0011110011000011	0011110011000011	0011110011000011
W_{40}	H_{15}	0110100110010110	0110100110010110	0110100110010110	0110100110010110
W_3	H_{16}	0000000000000000	1111111111111111	0000000000000000	1111111111111111
W_{60}	H_{17}	0101010101010101	1010101010101010	0101010101010101	1010101010101010
W_{28}	H_{18}	0011001100110011	1100110011001100	0011001100110011	1100110011001100
W_{35}	H_{19}	0110011001100110	1001100110011001	0110011001100110	1001100110011001
W_{12}	H_{20}	0000111100001111	1111000011110000	0000111100001111	1111000011110000
W_{51}	H_{21}	0101101001011010	1010010110100101	0101101001011010	1010010110100101
W_{19}	H_{22}	0011110000111100	1100001111000011	0011110000111100	1100001111000011
W_{44}	H_{23}	0110100101101001	1001011010010110	0110100101101001	1001011010010110
W_4	H_{24}	0000000011111111	1111111100000000	0000000011111111	1111111100000000
W_{59}	H_{25}	0101010110101010	1010101001010101	0101010110101010	1010101001010101
W_{27}	H_{26}	0011001111001100	1100110000110011	0011001111001100	1100110000110011
W_{36}	H_{27}	0110011010011001	1001100101100110	0110011010011001	1001100101100110
W_{11}	H_{28}	0000111111110000	1111000000001111	0000111111110000	1111000000001111
W_{52}	H_{29}	0101101010100101	1010010101011010	0101101010100101	1010010101011010
W_{20}	H_{30}	0011110011000011	1100001100111100	0011110011000011	1100001100111100
W_{43}	H_{31}	0110100110010110	1001011001101001	0110100110010110	1001011001101001
W_1	H_{32}	0000000000000000	0000000000000000	1111111111111111	1111111111111111
W_{62}	H_{33}	0101010101010101	0101010101010101	1010101010101010	1010101010101010
W_{30}	H_{34}	0011001100110011	0011001100110011	1100110011001100	1100110011001100
W_{33}	H_{35}	0110011001100110	0110011001100110	1001100110011001	1001100110011001
W_{14}	H_{36}	0000111100001111	0000111100001111	1111000011110000	1111000011110000
W_{49}	H_{37}	0101101001011010	0101101001011010	1010010110100101	1010010110100101
W_{17}	H_{38}	0011110000111100	0011110000111100	1100001111000011	1100001111000011
W_{46}	H_{39}	0110100101101001	0110100101101001	1001011010010110	1001011010010110
W_6	H_{40}	0000000011111111	0000000011111111	1111111100000000	1111111100000000
W_{57}	H_{41}	0101010110101010	0101010110101010	1010101001010101	1010101001010101
W_{25}	H_{42}	0011001111001100	0011001111001100	1100110000110011	1100110000110011
W_{38}	H_{43}	0110011010011001	0110011010011001	1001100101100110	1001100101100110
W_9	H_{44}	0000111111110000	0000111111110000	1111000000001111	1111000000001111
W_{54}	H_{45}	0101101010100101	0101101010100101	1010010101011010	1010010101011010
W_{22}	H_{46}	0011110011000011	0011110011000011	1100001100111100	1100001100111100
W_{41}	H_{47}	0110100110010110	0110100110010110	1001011001101001	1001011001101001

Table 5.8 Walsh functions of order 64, as indexed in IS-95 (continued)

W_2	H_{48}	0000000000000000 1111111111111111 1111111111111111 0000000000000000
W_{61}	H_{49}	0101010101010101 1010101010101010 1010101010101010 0101010101010101
W_{29}	H_{50}	0011001100110011 1100110011001100 1100110011001100 0011001100110011
W_{34}	H_{51}	0110011001100110 1001100110011001 1001100110011001 0110011001100110
W_{13}	H_{52}	0000111100001111 1111000011110000 1111000011110000 0000111100001111
W_{50}	H_{53}	0101101001011010 1010010110100101 1010010110100101 0101101001011010
W_{18}	H_{54}	0011110000111100 1100001111000011 1100001111000011 0011110000111100
W_{45}	H_{55}	0110100101101001 1001011010010110 1001011010010110 0110100101101001
W_5	H_{56}	0000000011111111 1111111100000000 1111111100000000 0000000011111111
W_{58}	H_{57}	0101010110101010 1010101001010101 1010101001010101 0101010110101010
W_{26}	H_{58}	0011001111001100 1100110000110011 1100110000110011 0011001111001100
W_{37}	H_{59}	0110011010011001 1001100101100110 1001100101100110 0110011010011001
W_{10}	H_{60}	0000111111110000 1111000000001111 1111000000001111 0000111111110000
W_{53}	H_{61}	0101101010100101 1010010101011010 1010010101011010 0101101010100101
W_{21}	H_{62}	0011110011000011 1100001100111100 1100001100111100 0011110011000011
W_{12}	H_{63}	0110100110010110 1001011001101001 1001011001101001 0110100110010110

1. The field is *closed*; that is, the sum or product of any two elements in $\mathrm{GF}(q)$ is also in $\mathrm{GF}(q)$.

2. The field always contains a unique *additive identity element* 0 and a unique *multiplicative identity element* 1, such that $\alpha + 0 = \alpha$ and $\alpha \cdot 1 = \alpha$ for any element $\alpha \in \mathrm{GF}(q)$.

3. For every element α, there is a unique *additive inverse element* $-\alpha$ such that $\alpha + (-\alpha) = 0$, and for $\alpha \neq 0$, there is a unique *multiplicative inverse element*, denoted α^{-1}, such that $\alpha \cdot \alpha^{-1} = 1$. Thus, the inverse operations, subtraction ($-$) and division ($\div$) are defined by

$$\alpha - \beta = \alpha + (-\beta) \quad \text{for any } \alpha, \beta \in \mathrm{GF}(q) \tag{5.19a}$$

$$\alpha \div \beta = \alpha \cdot (\beta^{-1}), \qquad \beta \neq 0, \tag{5.19b}$$

where $-\beta$ is the additive inverse element of β and β^{-1} is the multiplicative inverse element of β.

4. For each operation on elements $\alpha, \beta, \gamma \in \mathrm{GF}(q)$, the following rules apply:

$$\alpha + (\beta + \gamma) = (\alpha + \beta) + \gamma \qquad \text{(associative law)} \tag{5.20a}$$

$$\alpha \cdot (\beta \cdot \gamma) = (\alpha \cdot \beta) \cdot \gamma \qquad \text{(commutative law)} \tag{5.20b}$$

$$\alpha \cdot (\beta + \gamma) = \alpha \cdot \beta + \alpha \cdot \gamma \qquad \text{(distributive law)} \tag{5.20c}$$

From this definition of a field, we see that the set of all real numbers and the set of rational numbers under ordinary arithmetic operations are examples of fields. Their elements obey the rules of ordinary addition and multiplication, and conform to the four conditions of the field definition. The set of real numbers equal to or greater than zero, and the set of all integers under ordinary arithmetic, are not fields.

The number of elements in a field is called the *order* of the field, and the field GF(q) having a finite number of elements, q, is of interest to us. For example, GF(5) is a finite field containing the set of five elements {0, 1, 2, 3, 4}, which constitute a legitimate field under the rules of modulo-5 addition and multiplication operations. In GF(q), operations on integer elements are modulo-q addition and multiplication. The addition and multiplication tables of GF(5) are given in Table 5.9.

Observe that GF(5) = {0, 1, 2, 3, 4} contains the additive identity element 0 and the multliplicative identity element 1. The field is closed. For example, $2 + 3 = 5 \equiv 0 \bmod 5$ and $2 \cdot 3 = 6 \equiv 1 \bmod 5$. The unique additive inverse elements for {0, 1, 2, 3, 4} are observed from the addition table in Table 5.9 to be {0, 4, 3, 2, 1}, while the unique multiplicative inverse elements for the nonzero elements {1, 2, 3, 4} are observed from the multiplication table in Table 5.9 to be {1, 3, 2, 4}.

In a "prime field" GF(q), q must be a prime number. A finite field GF(p^m) exists for any p^m, where p is a prime number greater than 1 and m is an integer. In the prime field GF(p), the additive identity element 0 and the multiplicative identity element 1 are contained such that

$$\underbrace{1 + 1 + 1 + \cdots + 1}_{p \text{ times}} = 0 \bmod p \tag{5.21}$$

Table 5.9 Addition and multiplication tables for GF(5)

+	0	1	2	3	4
0	0	1	2	3	4
1	1	2	3	4	0
2	2	3	4	0	1
3	3	4	0	1	2
4	4	0	1	2	3

·	0	1	2	3	4
0	0	0	0	0	0
1	0	1	2	3	4
2	0	2	4	1	3
3	0	3	1	4	2
4	0	4	3	2	1

For example, in GF(5), $1 + 1 + 1 + 1 + 1 = 0 \bmod 5$, and $1 + 1 = 0 \bmod 2$ in GF(2). That is, the smallest number of times for which the multiplicative identity can be summed to give the additive identity is called the *characteristic* of the field. Thus, the characteristic of GF(5) is 5, and the characteristic of GF(2) is 2. We return to this topic later when we discuss the generation of pseudorandom sequences and make extensive use of the fact that the characteristic of GF(p^m) is p.

The simplest finite field of the form GF(p^m) with $m = 1$ is GF(2); it is called the prime field and is of much practical interest, as we will see. The addition and multiplication tables for GF(2) are given in Table 5.10, in which the special symbol $\oplus$ is used to denote modulo-2 addition.

If m is greater than 1, GF(p^m) is an extension field constructed as an extension of the prime field GF(p). For example, GF(25) = GF(5^2) and GF(8) = GF(2^3) are extension fields. We return to the subject of extension fields when we discuss pseudorandom sequence generation in Chapter 6 and coding theory in Section 5.9. Unless otherwise indicated, in our notations GF(q) and GF(p^m), we mean that $q = p^m$ and $q = p$ for $m = 1$, which is the case of a prime field.

5.3.4 Vector Spaces

In the mathematical description of sequences and codes, the concept of vector space plays an important role. The reader is familiar with the representation of vector v in two- or three-dimensional Euclidean space as the projections of v onto coordinate axes, and the vector is visualized as a line in the coordinate system. The familiar geometric vectors in two- or three-dimensional coordinate systems provide an intrinsic concept of the vector space that we define.

We interpret a vector in n-dimensional Euclidean space as a directed line and represent a vector v as an enumeration of its coordinates $(v_1, v_2, \ldots, v_n)$. The concept of a *vector space* V over a *field* F is now

Table 5.10 Addition and multiplication tables for GF(2)

$\oplus$	0	1		$\cdot$	0	1
0	0	1		0	0	0
1	1	0		1	0	1

defined as follows: a set V is called a *vector space* and its elements are called *vectors* $\{v_n\}$. The elements belonging to field F are called scalars. A vector space V over a field F produces a set of vectors and a set of scalars under the operations of addition and multiplication in a mathematical system of the family of geometric vectors in real numbers and in ordinary algebra. Vectors can be added (vector addition) and a scalar can multiply a vector (scalar multiplication). The addition of any two vectors u and v in V, that is, $u + v$, produces a vector that is also found in V. The multiplication of a scalar ν in F by a vector v in V, that is, $\nu \cdot v$, is also found in V. The results of both vector addition and scalar multiplication are always found in V because of the closure properties of groups and fields. We can state the definition of vector space more formally in what follows.

A *vector space* V *over a field* F is a set of elements, called vectors, which must satisfy the following properties:

- The vectors form a *commutative group* under vector addition.

- The *distributive law* applies, so that $\nu(u + v) = \nu u + \nu v$ and $(\nu + \gamma)u = \nu u + \gamma u$ for any $u, v \in$ V and $\nu, \gamma \in$ F.

- The *associative law* applies, such that $(\nu\gamma)u = \nu(\gamma u)$, for any $u \in$ V and $\nu, \gamma \in$ F.

- For the *multiplicative identity element*, $1 \in$ F, it follows that $1u = u$ for any $u \in$ V.

Note that there are two identity elements in vector space. The scalar field F has both the additive identity element 0, a scalar element, and a multiplicative identity element 1, also a scalar element. Because the vectors in V form a commutative group under the addition operation, there is an identity element, called the zero vector, that is 0, such that $u + 0 = 0 + u = u$ for any $u \in$ V. In other words, there are two identity elements, 0 and 0, a vector and a scalar, respectively.

Consider an ordered sequence of n symbols as a vector with n components:

$$C = (c_1, c_2, \ldots, c_n), \qquad c_i \in GF(2) \equiv \{0, 1\} \tag{5.22}$$

The sequence is an n-tuple over the field $GF(2)$. The set of all 2^n possible n-tuples is called a vector space over $GF(2)$, and we can denote it as V_n.

Example 5.6 For $n = 3$, the vector space V_3 over $GF(2)$ contains the following vectors:

$$V_3 = \{000, 001, 010, 011, 100, 101, 110, 111\} \qquad (5.23)$$

The addition and scalar multiplication operations of a binary n-tuple by a symbol over $GF(2)$ are defined as follows: Given the scalar ν, $C_1 = (c_{11}, c_{12}, \ldots, c_{1n})$, and $C_2 = (c_{21}, c_{22}, \ldots, c_{2n})$:

$$C_1 + C_2 = (c_{11} + c_{21}, c_{12} + c_{22}, \ldots, c_{1n} + c_{2n}) \qquad (5.24a)$$

$$\nu C_1 = (c_{11}, c_{12}, \ldots, c_{1n}) = (\nu c_{11}, \nu c_{12}, \ldots, \nu c_{1n}) \qquad (5.24b)$$

The concept of subspace needs to be defined. A set S of V_n is called a subspace if

- The all-zero vector $(0, 0, \ldots, 0)$ is in the set S, and

- The sum of any two vectors in S is also in S (closure property).

Example 5.7 Consider the following vectors constituting a subspace of V_4:

$$S = \{C_0, C_1, C_2, C_3\}$$

where

$$C_0 = (0\ 0\ 0\ 0), \quad C_1 = (0\ 1\ 0\ 1),$$
$$C_2 = (1\ 0\ 1\ 0), \quad C_3 = (1\ 1\ 1\ 1)$$

A simple check reveals that S satisfies the definition of subspace given above.

Consider the vectors $C_1, C_2, \ldots, C_K$ in V_n. A *linear combination* of these K vectors is another vector of the form

$$\nu_1 C_1 + \nu_2 C_2 + \cdots + \nu_K C_K, \quad \nu_i \in GF(2)$$

A set of vectors is said to be *linearly dependent* if and only if there are scalars $\nu_1, \nu_2, \ldots, \nu_K$, not all zero, such that

$$\nu_1 C_1 + \nu_2 C_2 + \cdots + \nu_K C_K = 0$$

The set of vectors $\{C_1, C_2, \ldots, C_K\}$ is said to be *linearly independent* if, whenever the above equation holds, it follows that

$$\nu_1 = \nu_2 = \cdots = \nu_K = 0$$

That is, if the vectors $C_1, C_2, \ldots, C_K$ are linearly independent, their linear combination cannot be identically zero unless the multiplying scalars are all chosen to be zero. A set of vectors is said to *span* (generate or determine) a vector space if every vector in the vector space is obtainable as a linear combination of the vectors in the set.

Example 5.8 Consider the following set of vectors:

$$C_1 = (1\ 0\ 0\ 0), C_2 = (0\ 1\ 0\ 0)$$
$$C_3 = (0\ 0\ 1\ 0), C_3 = (0\ 0\ 0\ 1)$$

We see that the vectors $\{C_1, C_2, C_3, C_4\}$ are linearly independent. Any vector in V_4 is a linear combination of this set of vectors, and hence the vectors C_1, C_2, C_3, C_4 span the vector space V_4.

In any vector space or subspace, there exists at least one set of linearly independent vectors that span the space. This set is called a *basis* of the vector space, and the number of vectors in the basis set is called the dimension of the vector space. Thus, the dimension of the vector space is the number of spanning vectors that can be used to generate the space—all vectors—by linear combinations thereof. For our applications here and in the sections to come, a matrix representation of a vector space is discussed briefly. In the example above, we showed four linearly independent vectors $\{C_1, C_2, C_3, C_4\}$ that could serve as the basis vectors to span the vector space comprising $2^4 = 16$ vectors as linear combinations thereof. We therefore need to specify all 16 possible combinations of scalars (a_1, a_2, a_3, a_4) from the field GF(2) to space V_4:

$$\boldsymbol{u} = a_1 C_1 + a_2 C_2 + a_3 C_3 + a_4 C_4$$
$$= a_1(1\ 0\ 0\ 0) + a_2(0\ 1\ 0\ 0) + a_3(0\ 0\ 1\ 0) + a_4(0\ 0\ 0\ 1)$$

$$= \begin{bmatrix} a_1 & a_2 & a_3 & a_4 \end{bmatrix} \begin{bmatrix} 1 & 0 & 0 & 0 \\ 0 & 1 & 0 & 0 \\ 0 & 0 & 1 & 0 \\ 0 & 0 & 0 & 1 \end{bmatrix}$$

We write the matrix form of u more compactly as

$$u = a\,\mathbf{G} \tag{5.25}$$

where the matrix multiplication of the 1×4 row vector a times the 4×4 matrix $\mathbf{G}$ yields a 1×4 row vector u. In this connection, we can now call the matrix, whose rows are the set of basis vectors for the vector space, a *generator matrix*. With a *generator matrix*, we can generate the entire set of 2^4 vectors in the vector space—all the quadruple vectors—by specifying all 2^4 quadruple combinations of the scalars (a_1, a_2, a_3, a_4), $a_i \in \mathrm{GF}(2)$, $i = 1, 2,$ 3, 4. In this example, we have chosen four linearly independent vectors as basis vectors, which happen to be in the form of a 4×4 unit (identity) matrix. Any four linearly independent vectors can serve as the basis vectors, and therefore as rows of the generator matrix.

5.3.5 Walsh Function Generation Using Basis Vectors

It is now clear that the Walsh sequences (functions) of order $N = 2^K$ form a K-dimensional vector space over $\mathrm{GF}(2)$ when we examine the Walsh sequences given in the tables and figures presented previously (that is, Tables 5.1 and 5.9, and Figure 5.1). In other words, all the N-tuple Walsh sequences of order $N = 2^K$ can be spanned by a set of K linearly independent basis sequences (vectors). Therefore, it is a matter of finding a suitable set of K N-tuple Walsh vectors to form a $K \times N$ generator matrix, so that we can generate (span) the entire set of the Walsh vectors (sequences) by specifying the 2^K K-tuple index vectors (sequences) in $\mathrm{GF}(2)$.

If the Walsh sequences are to be generated in the order according to the ascending numerical order of their binary index sequences, the generator matrix is easily specified. The Walsh sequences whose index vectors, $X_i = (x_{i1}, x_{i2}, \ldots, x_{iK})$, have a single 1, as in $(0, 0, 1, 0, \ldots, 0)$, will be the rows of the generator matrix. Such index vectors (sequences) form the rows of a $K \times K$ unit matrix, and the corresponding K row sequences of N-tuple Walsh vectors are linearly independent so that they can be used as basis vectors in the generator matrix. Index sequences that have a single 1 in the K-tuple correspond to indexes that are powers of 2.

Example 5.9 Walsh sequences of order $N = 2^4 = 16$. The generation of Walsh sequences can be accomplished in accordance with (5.25):

$$W_j = X_j G_W, \quad j = 0, 1, \ldots, 2^K = N \tag{5.26}$$

where X_j is the index sequence of W_j and G_W is a generator matrix. Because the index sequences X_8, X_4, X_2, and X_1 all have a single 1 in their K-tuples (each being a sequence of weight 1), we can specify the corresponding four Walsh sequences as the row vectors of the generator matrix G_W:

$$
\begin{aligned}
X_8 \rightarrow W_8 &= 0\ 1\ 1\ 0\ 0\ 1\ 1\ 0\ 0\ 1\ 1\ 0\ 0\ 1\ 1\ 0 \\
X_4 \rightarrow W_4 &= 0\ 0\ 1\ 1\ 1\ 1\ 0\ 0\ 0\ 0\ 1\ 1\ 1\ 1\ 0\ 0 \\
X_2 \rightarrow W_2 &= 0\ 0\ 0\ 0\ 1\ 1\ 1\ 1\ 1\ 1\ 1\ 1\ 0\ 0\ 0\ 0 \\
X_1 \rightarrow W_1 &= 0\ 0\ 0\ 0\ 0\ 0\ 0\ 0\ 1\ 1\ 1\ 1\ 1\ 1\ 1\ 1
\end{aligned}
\tag{5.27a}
$$

so that

$$
G_W =
\begin{bmatrix}
0 & 1 & 1 & 0 & 0 & 1 & 1 & 0 & 0 & 1 & 1 & 0 & 0 & 1 & 1 & 0 \\
0 & 0 & 1 & 1 & 1 & 1 & 0 & 0 & 0 & 0 & 1 & 1 & 1 & 1 & 0 & 0 \\
0 & 0 & 0 & 0 & 1 & 1 & 1 & 1 & 1 & 1 & 1 & 1 & 0 & 0 & 0 & 0 \\
0 & 0 & 0 & 0 & 0 & 0 & 0 & 0 & 1 & 1 & 1 & 1 & 1 & 1 & 1 & 1
\end{bmatrix}
\tag{5.27b}
$$

To find W_{13}, we simply compute the matrix multiplication

$$W_{13} = X_{13} G_W$$

$$
= \begin{bmatrix} 1 & 1 & 0 & 1 \end{bmatrix}
\begin{bmatrix}
0 & 1 & 1 & 0 & 0 & 1 & 1 & 0 & 0 & 1 & 1 & 0 & 0 & 1 & 1 & 0 \\
0 & 0 & 1 & 1 & 1 & 1 & 0 & 0 & 0 & 0 & 1 & 1 & 1 & 1 & 0 & 0 \\
0 & 0 & 0 & 0 & 1 & 1 & 1 & 1 & 1 & 1 & 1 & 1 & 0 & 0 & 0 & 0 \\
0 & 0 & 0 & 0 & 0 & 0 & 0 & 0 & 1 & 1 & 1 & 1 & 1 & 1 & 1 & 1
\end{bmatrix}
$$

$$
= \begin{bmatrix} 0 & 1 & 0 & 1 & 1 & 0 & 1 & 0 & 1 & 0 & 1 & 0 & 0 & 1 & 0 & 1 \end{bmatrix}
$$

$$\tag{5.27c}$$

The operation indicated by (5.27c) can be generalized with a hardware or software implementation to generate the Walsh sequence that is commanded by any index vector. To see this, let us denote an arbitrary index vector as

$$X = (x_1, x_2, x_3, x_4)$$

Then $\boldsymbol{W} = \boldsymbol{X}\,\mathbf{G}_W$

$$\begin{aligned}
&= (0,\, x_1,\, x_1 \oplus x_2,\, x_2,\, x_2 \oplus x_3,\, x_1 \oplus x_2 \oplus x_3,\, x_1 \oplus x_3,\, x_3,\, x_3 \oplus x_4, \\
&\qquad x_1 \oplus x_3 \oplus x_4,\, x_1 \oplus x_2 \oplus x_3 \oplus x_4,\, x_2 \oplus x_3 \oplus x_4,\, x_2 \oplus x_4, \\
&\qquad x_1 \oplus x_2 \oplus x_4,\, x_1 \oplus x_4,\, x_4) \qquad\qquad\qquad\qquad (5.27\text{d})
\end{aligned}$$

We can have a K-stage index sequence register that is connected to an N-stage sequence register as mappings of K-tuple index vectors $\{\boldsymbol{X}_j\}$ to N-tuple Walsh vectors $\{\boldsymbol{W}_j\}$. Figure 5.3 shows a scheme using shift registers for the generation of Walsh functions. This scheme is rooted to the discovery of the instant Walsh function [8]. The shift registers in Figure 5.3 are controlled by clock pulses; at each clock pulse of duration T seconds, the contents of the "index register" can be replaced with a new one (shifted in at K times the clock rate), while the contents of the "sequence register" can be read out at the speed of N times the clock rate [14].

We can also generate the Walsh sequences by using the Rademacher sequences as the rows of a generator matrix. If the index sequence is converted to a Gray code, the Walsh sequence indexing will be the same as before. Thus

$$\boldsymbol{W}_i = \boldsymbol{G}_i\,\mathbf{G}_R \qquad\qquad (5.28\text{a})$$

generates the Walsh sequences, where $\boldsymbol{G}_i$ is the Gray codeword for the index vector $\boldsymbol{X}_i$ and $\mathbf{G}_R$ is a $K \times 2^K$ generator matrix whose rows are the consecutive K Rademacher sequences:

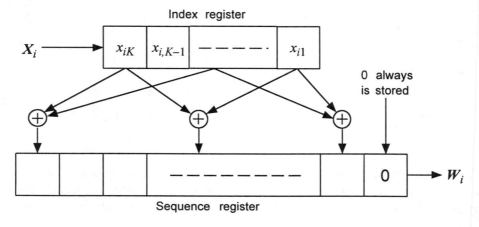

Figure 5.3 Walsh sequence generator using shift registers.

$$
\mathbf{G}_R = \begin{bmatrix} \mathcal{R}_K \\ \mathcal{R}_{K-1} \\ \vdots \\ \mathcal{R}_3 \\ \mathcal{R}_2 \\ \mathcal{R}_1 \end{bmatrix} \tag{5.28b}
$$

Example 5.10 Walsh sequences of order $N = 16 = 2^4 = 2^K$. The generator matrix using Rademacher sequences is

$$
\mathbf{G}_R = \begin{bmatrix}
0 & 1 & 0 & 1 & 0 & 1 & 0 & 1 & 0 & 1 & 0 & 1 & 0 & 1 & 0 & 1 \\
0 & 0 & 1 & 1 & 0 & 0 & 1 & 1 & 0 & 0 & 1 & 1 & 0 & 0 & 1 & 1 \\
0 & 0 & 0 & 0 & 1 & 1 & 1 & 1 & 0 & 0 & 0 & 0 & 1 & 1 & 1 & 1 \\
0 & 0 & 0 & 0 & 0 & 0 & 0 & 0 & 1 & 1 & 1 & 1 & 1 & 1 & 1 & 1
\end{bmatrix} \tag{5.28c}
$$

Find $\mathbf{W}_{13}$.

Solution: Because $\mathbf{X}_{13} = (1\ 1\ 0\ 1) \rightarrow \mathbf{G}_{13} = (1\ 0\ 1\ 1)$, and

$$
\mathbf{W}_{13} = \mathbf{G}_{13}\mathbf{G}_R
$$

$$
= \begin{bmatrix} 1 & 0 & 1 & 1 \end{bmatrix} \begin{bmatrix}
0 & 1 & 0 & 1 & 0 & 1 & 0 & 1 & 0 & 1 & 0 & 1 & 0 & 1 & 0 & 1 \\
0 & 0 & 1 & 1 & 0 & 0 & 1 & 1 & 0 & 0 & 1 & 1 & 0 & 0 & 1 & 1 \\
0 & 0 & 0 & 0 & 1 & 1 & 1 & 1 & 0 & 0 & 0 & 0 & 1 & 1 & 1 & 1 \\
0 & 0 & 0 & 0 & 0 & 0 & 0 & 0 & 1 & 1 & 1 & 1 & 1 & 1 & 1 & 1
\end{bmatrix}
$$

$$
= [0\ 1\ 0\ 1\ 1\ 0\ 1\ 0\ 1\ 0\ 1\ 0\ 0\ 1\ 0\ 1] \tag{5.28d}
$$

Another way of generating the Walsh sequences is to use a generator matrix formed from the set of Hadamard sequences. We noted previously that the Hadamard sequences are in fact Walsh sequences not listed in order of increasing numbers of zero crossings, but the index conversions were simply related to a Gray coding as shown in (5.18).

Consider the Walsh sequences of order $N = 16 = 2^4 = 2^K$ again, and let us construct the generator matrix $\mathbf{G}_H$ based on the basis vectors chosen from the Hadamard matrix. What we need to do is to generate the Hadamard sequences using a generator matrix formed from a selection of a set of basis vectors. The sequences generated this way are the Hadamard sequences, which we will call Walsh sequences with the understanding that the sequence designation with an index notation is different from the actual designation for the number of zero crossings. Now, let us pick the sequences from the Hadamard sequences of order $N = 16 = 2^4 = 2^K$ corresponding to the index vectors (j_1, j_2, j_3, j_4) with weight 1, to form a generator matrix $\mathbf{G}_H$.

The first row of $\mathbf{G}_H$ is $\mathbf{H}_8 = 0\ 0\ 0\ 0\ 0\ 0\ 0\ 0\ 1\ 1\ 1\ 1\ 1\ 1\ 1\ 1$

The second row of $\mathbf{G}_H$ is $\mathbf{H}_4 = 0\ 0\ 0\ 0\ 1\ 1\ 1\ 1\ 0\ 0\ 0\ 0\ 1\ 1\ 1\ 1$ (5.29a)

The third row of $\mathbf{G}_H$ is $\mathbf{H}_2 = 0\ 0\ 1\ 1\ 0\ 0\ 1\ 1\ 0\ 0\ 1\ 1\ 0\ 0\ 1\ 1$

The fourth row of $\mathbf{G}_H$ is $\mathbf{H}_1 = 0\ 1\ 0\ 1\ 0\ 1\ 0\ 1\ 0\ 1\ 0\ 1\ 0\ 1\ 0\ 1$

Note that these are the Rademacher sequences $\mathcal{R}_1, \mathcal{R}_2, \mathcal{R}_3$, and $\mathcal{R}_4$, respectively. Therefore, the generator matrix is

$$\mathbf{G}_H = \begin{bmatrix} 0 & 0 & 0 & 0 & 0 & 0 & 0 & 0 & 1 & 1 & 1 & 1 & 1 & 1 & 1 & 1 \\ 0 & 0 & 0 & 0 & 1 & 1 & 1 & 1 & 0 & 0 & 0 & 0 & 1 & 1 & 1 & 1 \\ 0 & 0 & 1 & 1 & 0 & 0 & 1 & 1 & 0 & 0 & 1 & 1 & 0 & 0 & 1 & 1 \\ 0 & 1 & 0 & 1 & 0 & 1 & 0 & 1 & 0 & 1 & 0 & 1 & 0 & 1 & 0 & 1 \end{bmatrix} \quad (5.29b)$$

From this matrix we can find any of the Hadamard sequences of order 16, say, $\mathbf{H}_9$. The index vector of $\mathbf{H}_9$ is $\mathbf{X}_{13} = (1\ 0\ 0\ 1)$, and thus

$$\mathbf{H}_9 = \mathbf{X}_9 \mathbf{G}_H$$

$$= \begin{bmatrix} 1 & 0 & 0 & 1 \end{bmatrix} \begin{bmatrix} 0 & 0 & 0 & 0 & 0 & 0 & 0 & 0 & 1 & 1 & 1 & 1 & 1 & 1 & 1 & 1 \\ 0 & 0 & 0 & 0 & 1 & 1 & 1 & 1 & 0 & 0 & 0 & 0 & 1 & 1 & 1 & 1 \\ 0 & 0 & 1 & 1 & 0 & 0 & 1 & 1 & 0 & 0 & 1 & 1 & 0 & 0 & 1 & 1 \\ 0 & 1 & 0 & 1 & 0 & 1 & 0 & 1 & 0 & 1 & 0 & 1 & 0 & 1 & 0 & 1 \end{bmatrix}$$

$$= \begin{bmatrix} 0 & 1 & 0 & 1 & 0 & 1 & 0 & 1 & 1 & 0 & 1 & 0 & 1 & 0 & 1 & 0 \end{bmatrix} \quad (5.29c)$$

We have shown several ways of computing or generating the Walsh sequences through the technique of generating matrices. The generating matrix is a set of row vectors that are linearly independent, which serve as basis vectors to generate (span) the rest of the sequences. The generator matrix was shown to consist of K row vectors in a system of $N = 2^K$ Walsh sequences (vectors), and the matrix can be constructed using K Walsh sequences, Rademacher functions, or Hadamard sequences.

The reason we are interested in the Walsh sequences is that the IS-95 system employs Walsh functions in the forward link for multiplexing the many different forward link channels, while Walsh sequences or Walsh codewords are used in the reverse link for power-efficient orthogonal modulation purposes. We now discuss the principles of orthogonal multiplexing, and then show some examples to appreciate clearly how each channel can recover the signal that was transmitted for each channel by the transmitter through the multiplexing scheme.

5.4 Orthogonal Walsh Functions for CDMA Applications

We have seen that the set of Walsh functions are mutually orthogonal over one period of the functions. Now we show how this property applies to orthogonal multiplexing applications [15], such as signaling on the forward link of the IS-95 cellular system [1].

5.4.1 Walsh Functions Used in the Forward Link

When we observe the CDMA forward link channel format, namely the pilot channel, sync channel, paging channels, and the traffic channels, we see that each channel's information symbol is modulo-2 added to a distinct, periodic Walsh sequence of order 64, before spreading by I- and Q-channel PN sequences (discussed in detail later). Modulo-2 additions of $(0, 1)$ binary sequences are, of course, equivalent to multiplications by $\pm$ waveforms (functions), as we have discussed previously. These Walsh functions are precisely synchronized with the information sequence of each channel.

Consider the traffic and paging channels, diagrammed in Figures 4.17 and 4.15, respectively. The coded symbols of the traffic and paging channels arrive at the modulo-2 adder at the rate of 19.2 kilosymbols per second (19.2 ksps), and each symbol's epoch is precisely synchronized with the beginning of the Walsh sequences, which run at 64 times the symbol rate, or 1.2288 Mcps. Thus, each coded symbol covers all of the 64 chips of one entire period of the Walsh sequence. The synchronization channel, diagrammed in Figure 4.13, differs in that the coded symbol rate is 4.8 ksps, and hence each symbol covers four periods of the Walsh sequences. In this sense, the Walsh functions, or sequences, are used as channel-distinct identification sequences.

The Walsh-covered channels are then further processed (spreading and filtering) before being combined with carrier modulation for transmission. At the receiving end, the mobile station, after despreading operations the intended channel symbol is recovered by correlation with the designated Walsh sequence.

Let us consider d_{ik}, the kth data symbol of the ith mobile user, and see how the intended signal can be recovered for each of the M mobile stations. Let

$$S_i(t) \triangleq \text{signal intended for mobile station } i = d_{ik} W_i(t)$$

$$= \pm \sqrt{E_s} W_i(t), \quad kT_s \leq t < (k+1)T_s \qquad (5.30a)$$

where E_s is the data symbol energy, T_s is the data symbol period (also the Walsh symbol period for the symbol rate of 19.2 ksps), and $W_i(t)$ denotes the ith Walsh function. In the absence of PN spreading and filtering, the total forward link baseband waveform $S_{tot}(t)$ then can be formulated as

$$S_{tot}(t) = \sum_{i=1}^{M} S_i(t) = S_1(t) + S_2(t) + \cdots + S_M(t)$$

$$= d_{1k}W_1(t) + d_{2k}W_2(t) + \cdots + d_{Mk}W_M(t), \quad kT_s \leq t < (k+1)T_s \quad (5.30b)$$

At the ith mobile station, the receiver is synchronized to correlate the incoming total base station transmitted signal $S_{tot}(t)$ with its assigned Walsh function $W_i(t)$, based on the principle that

$$\int_{kT_s}^{(k+1)T_x} S_j(t)W_i(t)\, dt = \int_{kT_s}^{(k+1)T_s} d_{jk}W_j(t)\, W_i(t)\, dt$$

$$= d_{jk}\, T_s\, \delta_{ji} \quad (5.30c)$$

Thus at the ith mobile station, as illustrated in Figure 5.4, all signals are rejected except the one intended for mobile i, because of the orthogonality of the Walsh functions:

$$\int_{kT_s}^{(k+1)T_x} S_{tot}(t)W_i(t)\, dt =$$

$$= \int_{kT_s}^{(k+1)T_s} d_{1k}W_1(t)\, W_i(t)\, dt + \int_{kT_s}^{(k+1)T_s} d_{2k}W_2(t)\, W_i(t)\, dt$$

$$+ \cdots + \int_{kT_s}^{(k+1)T_s} d_{ik}W_i(t)\, W_i(t)\, dt$$

$$+ \cdots + \int_{kT_s}^{(k+1)T_s} d_{Mk}W_M(t)\, W_i(t)\, dt$$

$$= 0 + 0 + \cdots + d_{ik}\, T_s + \cdots + 0$$

$$= \begin{cases} +\sqrt{E_s}\, T_s & \text{if the } k\text{th symbol is logic 0} \\ -\sqrt{E_s}\, T_s & \text{if the } k\text{th symbol is logic 1} \end{cases} \quad (5.30d)$$

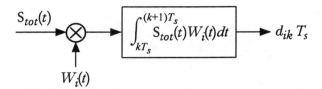

Figure 5.4 Principle of symbol recovery by Walsh function correlation.

Example 5.11 The IS-95 forward link uses orthogonal multiplexing of the various channel signals by exploiting the orthogonality of the set of Walsh functions of order 64. We now consider an example using Walsh functions of order 8 to illustrate the concept and show the principles involved. Let the eight ± 1 Walsh functions of order $N = 8$, $\{W_i(t), i = 0, 1, \ldots, 7\}$, be multiplied by different constants (symbol values), $\{d_{ik}, i = 0, 1, \ldots, 7;$ $k = 1, 2, \ldots\}$. That is, the values of the kth information symbol of the 8 individual channels are assumed to be as follows:

$$
\left.
\begin{aligned}
d_{0k}(t) &= 0.5 & d_{1k}(t) &= 1.0 \\
d_{2k}(t) &= 1.5 & d_{3k}(t) &= 0.8 \\
d_{4k}(t) &= 0.6 & d_{5k}(t) &= 2.0 \\
d_{6k}(t) &= 0.4 & d_{7k}(t) &= 0.3
\end{aligned}
\right\} \quad kT_s \le t < (k+1)T_s \quad (5.31)
$$

Let the duration of a Walsh chip be denoted T_c; each of the eight channel symbols of (5.31) is to be multiplied by a different Walsh function of order 8, such as those shown previously in Figure 5.1, and then summed (multiplexed) for transmission. The situation is depicted in Figure 5.5, and the waveform details for each channel are shown in Table 5.11. The total signal $S_{tot}(t)$ in Table 5.11 and Figure 5.5 denotes the sum of all eight active channel signals in sychronism, and it is plotted in Figure 5.6, along with a typical Walsh function, $W_7(t)$, "assigned" to channel number eight. In a real situation, the combined total signal $S_{tot}(t)$ is further processed, including modulation by a suitably chosen carrier waveform in the 800-MHz band for a cellular system or in the 2-GHz band for a PCS system. For illustration purposes, we omit carrier modulation.

At mobile station receiver i, a correlation operation is to be performed as illustrated previously in Figure 5.4. The objective of each mobile station is to recover its data symbol value d_{ik} through the indicated correlation

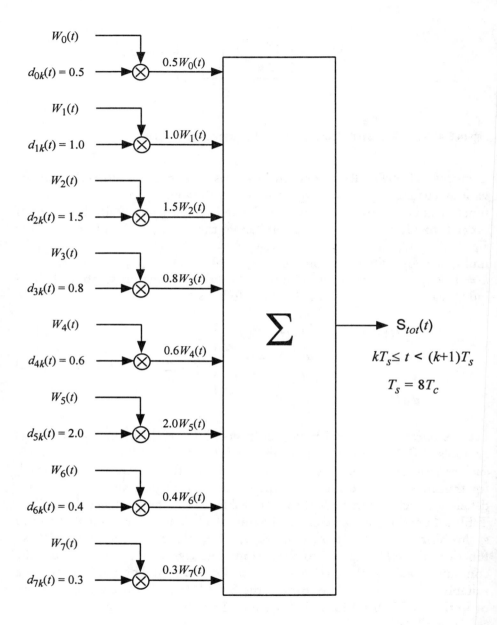

Figure 5.5 Example of Walsh function orthogonal multiplexing, $N = 8$.

Table 5.11 Details of Walsh multiplexing example of Figure 5.5

	$t=0$	$t=T_c$	$t=2T_c$	$t=3T_c$	$t=4T_c$	$t=5T_c$	$t=6T_c$	$t=7T_c$
$0.5W_0(t)$	0.5	0.5	0.5	0.5	0.5	0.5	0.5	0.5
$1.0W_1(t)$	1.0	1.0	1.0	1.0	−1.0	−1.0	−1.0	−1.0
$1.5W_2(t)$	1.5	1.5	−1.5	−1.5	−1.5	−1.5	1.5	1.5
$0.8W_3(t)$	0.8	0.8	−0.8	−0.8	0.8	0.8	−0.8	−0.8
$0.6W_4(t)$	0.6	−0.6	−0.6	0.6	0.6	−0.6	−0.6	0.6
$2.0W_5(t)$	2.0	−2.0	−2.0	2.0	−2.0	2.0	2.0	−2.0
$0.4W_6(t)$	0.4	−0.4	0.4	−0.4	−0.4	0.4	−0.4	0.4
$0.3W_7(t)$	0.3	−0.3	0.3	−0.3	0.3	−0.3	0.3	−0.3
$S_{tot}(t)$	7.1	0.5	−2.7	1.1	−2.7	0.3	1.5	−1.1

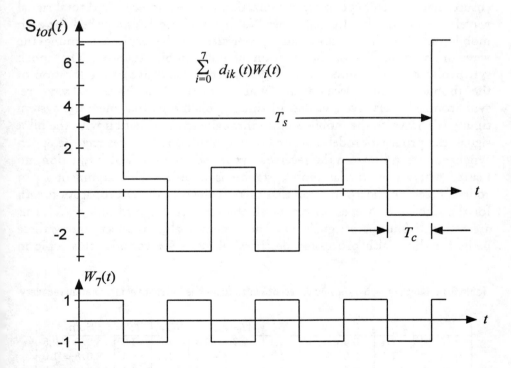

Figure 5.6 Total multiplexed signal for the $N=8$ example, compared with $W_7(t)$ for $t \in (0, T_s = 8T_c)$.

operation. In other words, each mobile station's receiver multiplies the received total signal $S_{tot}(t)$ by its assigned Walsh function $W_i(t)$ to recover the data transmitted by the base station transmitter. Mathematically, the correlation operation, as indicated by (5.30d), is performed to recover the channel symbol data. This operation for $N = 8$ example is depicted in Table 5.12, in which the first column denotes the multiplicative operation between the total received waveform $S_{tot}(t)$ and the Walsh function of each data channel, $W_i(t)$, $i = 0, 1, \ldots, 7$. Note that Table 5.12 shows clearly the recovered channel data $\{d_{ik}, i = 0, 1, \ldots, 7\}$ for the kth symbol; the results of the correlation given in the last column of the table are the transmitted data values assumed in Figure 5.5. This example illustrates the basic concept of orthogonal multiplexing.

Note that for the orthogonal multiplexing system to work, we must ensure that a perfect epoch synchronization between the received total signal waveform and the locally generated Walsh waveform is established for each mobile, so that the orthogonality properties can be exploited among the waveforms, regardless of the location of each mobile station. The epoch synchronization is ensured in the IS-95 system irrespective of the location of the mobile stations because the Walsh function epochs are always re-synchronized every even second of time. This means that once the system timing is known to the mobile station through synchronization with the pilot signal, the propagation delay related timing offset does not destroy the epoch synchronization between the received signal and the local Walsh function, no matter where a particular mobile station is located. This argument is, of course, based on the assumption that there is always a direct propagation path for the signal. However, in reality all the received forward link waveforms are reflected multipath signals, and the system is able to select among these paths for the Walsh processing described above. We consider this topic in

Table 5.12 Multiplying $S_{tot}(t)$ by different Walsh functions for channel information recovery

	$t = 0$	T_c	$2T_c$	$3T_c$	$4T_c$	$5T_c$	$6T_c$	$7T_c$	Sum/8
$\times\ W_0(t)$	7.1	0.5	−2.7	1.1	−2.7	0.3	1.5	−1.1	$4.0/8 = 0.5$
$\times\ W_1(t)$	7.1	0.5	−2.7	1.1	2.7	−0.3	−1.5	1.1	$8.0/8 = 1.0$
$\times\ W_2(t)$	7.1	0.5	2.7	−1.1	2.7	−0.3	1.5	−1.1	$12.0/8 = 1.5$
$\times\ W_3(t)$	7.1	0.5	2.7	−1.1	−2.7	0.3	−1.5	1.1	$6.4/8 = 0.8$
$\times\ W_4(t)$	7.1	−0.5	2.7	1.1	−2.7	−0.3	−1.5	−1.1	$4.8/8 = 0.6$
$\times\ W_5(t)$	7.1	−0.5	2.7	1.1	2.7	0.3	1.5	1.1	$16.0/8 = 2.0$
$\times\ W_6(t)$	7.1	−0.5	−2.7	−1.1	2.7	0.3	−1.5	−1.1	$3.2/8 = 0.4$
$\times\ W_7(t)$	7.1	−0.5	−2.7	−1.1	−2.7	−0.3	1.5	1.1	$2.4/8 = 0.3$

detail when we discuss path diversity, known as *Rake receiver combining*. Suffice it to say that, once the synchronization is established between the local Walsh function signal and the total received signal, be it a direct or reflected indirect path, the orthogonality properties are ensured between them as long as the transmitter side transmits the Walsh function mutliplexed signal at known time instants in synchronism with each channel, such as transmitting the Walsh function multiplexed signal using periodic Walsh functions that are resynchronized at every even second of the system time.

5.4.2 Walsh Functions Used in the Reverse Link

There are two channel categories in the reverse link of the IS-95 CDMA system: an access channel and a traffic channel. When we look at the block diagrams for these CDMA channel structures in Figures 4.21 and 4.22, respectively, we notice that they both employ a "64-ary orthogonal modulator," which follows the block interleaver, which outputs coded symbols at the rate of 28.8 ksps. The 64-ary orthogonal modulator is a block "channel encoder" of the (n, k) type, where the input to the encoder is k symbols (or bits if the input to the modulator is uncoded information bits) and the output is n symbols. In this sense, it is a $(64, 6)$ encoder.

The main reason for using this particular block coding scheme is to provide the base station receiver a measure of coherence over the duration of six coded symbols, which corresponds to the duration of two information bits because of the rate 1/3 convolutional coding. For the forward link, there is a pilot channel which provides a timing reference and a carrier phase reference for coherent demodulation. For the reverse link, there is no pilot channel that can provide a carrier reference for possible coherent demodulation. Therefore, the reverse link is necessarily a noncoherent channel, but a demodulation timing reference can be derived through demodulation of the coded 64 symbols.

Note that the $(64, 6)$ encoder for orthogonal modulation accepts the convolutionally coded symbols at a 28.8-ksps rate and puts out Walsh-encoded binary symbols at the rate of $28.8 \times 64/6 = 307.2$ ksps. The output of the $(64, 6)$ encoder is 64-ary Walsh coded symbols at the rate of $307.2/64 = 4.8$ ksps. The previously given tables 5.2 or 5.8 can be used as a "code book" for the encoder. The coded symbol selection rule given in IS-95 [1] is simply a means for specifying the index of the Hadamard sequences listed in Table 5.8. The sequence is chosen according to the rule

$$i = c_0 + 2c_1 + 4c_2 + 8c_3 + 16c_4 + 32c_5 \tag{5.32}$$

where i is the index number and the c_js are the coded symbol values for the index digits over $\{0, 1\}$. This method could be called a *table lookup scheme* that can be implemented in many possible ways in hardware.

Another possible way of encoding the reverse link is to use a shift register [15] based on the generator matrix method that we have discussed previously. According to this method, discussed in Section 5.3.5, the generator matrix **G** for the (64, 6) Walsh encoder can be chosen as follows:

$$\mathbf{G} = \begin{bmatrix} 0000000000000000000000000000000001111111111111111111111111111111 \\ 0000000000000001111111111111111000000000000000001111111111111111 \\ 0000000011111111000000001111111100000000111111110000000011111111 \\ 0000111100001111000011110000111100001111000011110000111100001111 \\ 0011001100110011001100110011001100110011001100110011001100110011 \\ 0101010101010101010101010101010101010101010101010101010101010101 \end{bmatrix}$$
$$\tag{5.33}$$

In this scheme, the encoder would operate exactly like the one shown previously in Figure 5.3, where the "index register" accepts convolutionally coded symbols six at a time at a 28.8-ksps rate and the "sequence register" reads out the 64 code symbols comprising one Walsh symbol at a 307.2-ksps rate (4.8 ksps for the Walsh symbols).

In the IS-95 modulation scheme, the Walsh coded chips generated at 307.2 ksps are scrambled by a long PN code sequence that is clocked at a rate four times faster (1.2288 Mcps) and thus the polarities of the Walsh coded chips are subject to random changes, four times per Walsh chip, according to the pseudorandom nature of the long PN code pattern.

The optimal receiver for the orthogonal Walsh code modulation, and its performance, is derived in Chapter 7.

5.5   Walsh Function Decoding

The Walsh encoded sequences used for orthogonal modulation on the reverse link, after further linear processing such as scrambling, I and Q short-PN-code modulation, and waveshaping by FIR filters, are carrier modulated and transmitted. The decoder at the base station receiver must decide as to which of the 64 possible Walsh symbols (64-chip sequences) was transmitted by the mobile station, and the method of formulating this symbol decision is based on a "comparison" technique. The comparison technique can take several different forms, all of which belong to the general method known as *correlation* to determine a measure of similarity between two functions. A high value of correlation means that the two functions are similar, and a low value of correlation means they are dissimilar.

When we compare two N-component sequences (N-tuples) A and B through correlation, we multiply the sequences, component by component, and the results are added. We denote this "multiply and sum" operation by $\langle A, B \rangle$, which is the scalar or inner product of the two sequences.

5.5.1 Correlation Decoding

Consider the Walsh functions of order N, such as those shown previously in Figure 5.1, where the functions have values ± 1 over an interval of length T: $\{W_j(t), t \in (0, T), j = 0, 1, 2, \ldots, N - 1\}$. The correlation between any pair of these functions, say $W_i(t)$ and $W_j(t)$, is defined as

$$\langle W_i(t), W_j(t) \rangle = \int_0^T W_i(t) W_j(t) \, dt \tag{5.34}$$

Functions with values ± 1 such as the Walsh functions can be treated like sequences with components in either $\{0, 1\}$ or $\{+1, -1\}$. Depending on which of these two representations of the binary components is selected, the correlation measure is calculated in two different ways. We explain with examples. Let the two sequences be given by

$$A = (a_1, a_2, \ldots, a_N), \qquad a_i \in \{+1, -1\}$$
$$B = (b_1, b_2, \ldots, b_N), \qquad b_i \in \{+1, -1\}$$

Then the correlation measure is calculated by adding up all the term-by-term or component-by-component products; that is:

$$\langle A, B \rangle = \sum_{i=1}^N a_i b_i = a_1 b_1 + a_2 b_2 + \cdots + a_N b_N \tag{5.35a}$$

Example 5.12 $\qquad A = (-1, -1, -1, +1, -1, +1, +1)$
$$B = (-1, +1, -1, -1, -1, +1, +1)$$

Their correlation is

$$\langle A, B \rangle = 1 + (-1) + 1 + (-1) + 1 + 1 + 1 = 3$$

Now, if the sequences A and B are given by

$$A = (a_1, a_2, \ldots, a_N), \qquad a_i \in \{0, 1\}$$
$$B = (b_1, b_2, \ldots, b_N), \qquad b_i \in \{0, 1\}$$

then the correlation is given by

$$\langle A, B \rangle = (\text{number of term-by-term agreements in } A \text{ and } B)$$
$$- (\text{number of term-by-term disagreements}) \qquad (5.35b)$$

Example 5.13 $A = (1, 1, 1, 0, 1, 0, 0)$, $B = (1, 0, 1, 1, 1, 0, 0)$. Their correlation is

$$\langle A, B \rangle = 5 - 2 = 3$$

The definition of correlation in (5.35b) can also be expressed in a different way. We observe that

$$(\text{number of agreements between } A \text{ and } B)$$
$$+ (\text{number of disagreements between } A \text{ and } B) = N$$

so that

$$(\text{number of agreements between } A \text{ and } B)$$
$$= N - (\text{number of disagreements between } A \text{ and } B)$$

If we define the "weight" of a sequence as the number of 1s in the sequence, then we can express (5.35b) as

$$\langle A, B \rangle = (\text{number of term-by-term agreements in } A \text{ and } B)$$
$$- (\text{number of term-by-term disagreements between } A \text{ and } B)$$
$$= [N - \text{weight}(A \oplus B)] - [\text{weight}(A \oplus B)]$$
$$= N - 2 \times \text{weight}(A \oplus B) \qquad (5.35c)$$

where $A \oplus B$ denotes the operation of term-by-term modulo-2 addition of the sequences A and B; that is:

$$A \oplus B = (a_1 \oplus b_1, \, a_2 \oplus b_2, \, \ldots, \, a_N \oplus b_N) \tag{5.36}$$

Example 5.14 $A = (1, 1, 1, 0, 1, 0, 0)$, $B = (1, 0, 1, 1, 1, 0, 0)$. Their correlation is

$$\langle A, B \rangle = 7 - 2 \cdot \text{weight}(0, 1, 0, 1, 0, 0, 0) = 7 - 2 \cdot 2 = 3$$

By definition, functions or sequences are uncorrelated (have zero correlation) if they are orthogonal. In a situation where there are N possible transmitted sequences, the decision as to which one of them is the most likely sequence is based on the highest correlation measure as a result of correlating the received sequence with stored versions of all N possible code sequences. If R is the received sequence and $\{W_i\}$ is the set of all possible sequences, the decision, denoted $\hat{W}$, is made such that

$$\hat{W} = W_j \quad \text{such that} \quad \langle R, W_j \rangle = \max_{i} \langle R, W_i \rangle \tag{5.37}$$

Let us consider an example using a set of Walsh sequences of order 16 as listed in Table 5.13. Assume that the received sequence is

$$R = 0 \; 1 \; 0 \; 0 \; 0 \; 0 \; 1 \; 0 \; 0 \; 0 \; 1 \; 0 \; 0 \; 0 \; 0 \tag{5.38a}$$

Note that this received sequence is not exactly equal to any of the sixteen Walsh functions—the sequence has been corrupted by noise or interference. To find the most likely transmitted sequence, we perform the correlation operation between R and each sequence of the set given in Table 5.13 and find the following results:

$$\langle R, W_0 \rangle = 10 \qquad\qquad \langle R, W_8 \rangle = -2$$
$$\langle R, W_1 \rangle = -2 \qquad\qquad \langle R, W_9 \rangle = 2$$
$$\langle R, W_2 \rangle = 2 \qquad\qquad \langle R, W_{10} \rangle = 6$$
$$\langle R, W_3 \rangle = -2 \qquad\qquad \langle R, W_{11} \rangle = 2$$
$$\langle R, W_4 \rangle = -2 \qquad\qquad \langle R, W_{12} \rangle = 2$$
$$\langle R, W_5 \rangle = -6 \qquad\qquad \langle R, W_{13} \rangle = -2 \tag{5.38b}$$
$$\langle R, W_6 \rangle = -2 \qquad\qquad \langle R, W_{14} \rangle = 2$$
$$\langle R, W_7 \rangle = 2 \qquad\qquad \langle R, W_{15} \rangle = 6$$

Table 5.13 Walsh sequences of order 16

$$
\begin{aligned}
W_0 &= 0\ 0\ 0\ 0\ 0\ 0\ 0\ 0\ 0\ 0\ 0\ 0\ 0\ 0\ 0\ 0 \\
W_1 &= 0\ 0\ 0\ 0\ 0\ 0\ 0\ 0\ 1\ 1\ 1\ 1\ 1\ 1\ 1\ 1 \\
W_2 &= 0\ 0\ 0\ 0\ 1\ 1\ 1\ 1\ 1\ 1\ 1\ 1\ 0\ 0\ 0\ 0 \\
W_3 &= 0\ 0\ 0\ 0\ 1\ 1\ 1\ 1\ 0\ 0\ 0\ 0\ 1\ 1\ 1\ 1 \\
W_4 &= 0\ 0\ 1\ 1\ 1\ 1\ 0\ 0\ 0\ 0\ 1\ 1\ 1\ 1\ 0\ 0 \\
W_5 &= 0\ 0\ 1\ 1\ 1\ 1\ 0\ 0\ 1\ 1\ 0\ 0\ 0\ 0\ 1\ 1 \\
W_6 &= 0\ 0\ 1\ 1\ 0\ 0\ 1\ 1\ 1\ 1\ 0\ 0\ 1\ 1\ 0\ 0 \\
W_7 &= 0\ 0\ 1\ 1\ 0\ 0\ 1\ 1\ 0\ 0\ 1\ 1\ 0\ 0\ 1\ 1 \\
W_8 &= 0\ 1\ 1\ 0\ 0\ 1\ 1\ 0\ 0\ 1\ 1\ 0\ 0\ 1\ 1\ 0 \\
W_9 &= 0\ 1\ 1\ 0\ 0\ 1\ 1\ 0\ 1\ 0\ 0\ 1\ 1\ 0\ 0\ 1 \\
W_{10} &= 0\ 1\ 1\ 0\ 1\ 0\ 0\ 1\ 1\ 0\ 0\ 1\ 0\ 1\ 1\ 0 \\
W_{11} &= 0\ 1\ 1\ 0\ 1\ 0\ 0\ 1\ 0\ 1\ 1\ 0\ 1\ 0\ 0\ 1 \\
W_{12} &= 0\ 1\ 0\ 1\ 1\ 0\ 1\ 0\ 0\ 1\ 0\ 1\ 1\ 0\ 1\ 0 \\
W_{13} &= 0\ 1\ 0\ 1\ 1\ 0\ 1\ 0\ 1\ 0\ 1\ 0\ 0\ 1\ 0\ 1 \\
W_{14} &= 0\ 1\ 0\ 1\ 0\ 1\ 0\ 1\ 1\ 0\ 1\ 0\ 1\ 0\ 1\ 0 \\
W_{15} &= 0\ 1\ 0\ 1\ 0\ 1\ 0\ 1\ 0\ 1\ 0\ 1\ 0\ 1\ 0\ 1 \\
W_0 &= 0\ 0\ 0\ 0\ 0\ 0\ 0\ 0\ 0\ 0\ 0\ 0\ 0\ 0\ 0\ 0
\end{aligned}
$$

We see that $\langle R, W_0 \rangle = 10 = \max_{i} \langle R, W_i \rangle$ and thus W_0 is the sequence that is most likely to have been transmitted.

There is yet another measure of comparison between two sequences, in terms of *distance*. We define this distance (also called *Hamming distance*, after its inventor) $D(A, B)$ between two N-tuple binary sequences A and B as the number of positions in which A and B disagree. For example, if

$$A = (1,\ 1,\ 1,\ 0,\ 1,\ 0,\ 0) \quad \text{and} \quad B = (0,\ 1,\ 0,\ 1,\ 0,\ 0,\ 0)$$

then
$$D(A, B) = 4$$

The fact that $\langle R, W_0 \rangle = 10$ in the example given above was the highest value of correlation meant that the received sequence R and the Walsh code sequence W_0 had the least number of disagreements. That is:

$$D(R, W_0) = 3 = \min_i D(R, W_i) \tag{5.39}$$

The "maximum correlation" decision criterion is seen to be equivalent to the "minimum distance" decision criterion. We need only to remember that the Hamming distance is a measure to be used in connection with sequences having components given in GF(2).

In the example just considered, we saw that the all-zero Walsh sequence (codeword) was determined to be the transmitted sequence when the received sequence had a weight 3. This means that the three errors were "corrected." This is no surprise when we recall a basic lemma in coding theory, that in a set of codewords W_0, W_1,..., W_N each of length N bits and with the Hamming distance $D(W_i, W_j) \geq 2t + 1$ for $i \neq j$, then all single, double, ..., t-tuple errors in transmission can be corrected; that is, all patterns of t or fewer errors are correctable, where

$$t = \left\lfloor \frac{d_{min} - 1}{2} \right\rfloor \tag{5.40}$$

where d_{min} is the minimum distance value of $D(W_i, W_j)$, $i \neq j$ and $\lfloor x \rfloor$ denotes the largest integer not greater than x. Thus, in the example we considered (i.e., the set of codewords consisting of Walsh sequences of order 16), the minimum distance is 8, and hence $t = \lfloor (8 - 1)/2 \rfloor = 3$ or fewer errors can be corrected. In general, the minimum distance of Walsh sequences of order $N = 2^K$ is 2^{K-1}, because all sequences, except the all-zero sequence, have the same weight, $N/2 = 2^{K-1}$. The Walsh sequences of order 64, therefore, can correct up to, and including, 15 errors in transmission.

To continue the example for Walsh codewords of order 16, suppose that the received sequence is

$$R = 0\ 1\ 0\ 1\ 0\ 0\ 0\ 1\ 0\ 0\ 0\ 1\ 0\ 0\ 0\ 0 \tag{5.41a}$$

The correlation results are:

$\langle R, W_0 \rangle = 8$	$\langle R, W_6 \rangle = 0$	$\langle R, W_{12} \rangle = 4$
$\langle R, W_1 \rangle = -4$	$\langle R, W_7 \rangle = 4$	$\langle R, W_{13} \rangle = 0$
$\langle R, W_2 \rangle = 0$	$\langle R, W_8 \rangle = -4$	$\langle R, W_{14} \rangle = 4$
$\langle R, W_3 \rangle = -4$	$\langle R, W_9 \rangle = 0$	$\langle R, W_{15} \rangle = 8$ (5.41b)
$\langle R, W_4 \rangle = 0$	$\langle R, W_{10} \rangle = 4$	
$\langle R, W_5 \rangle = -4$	$\langle R, W_{11} \rangle = 0$	

The correlations between R and W_0 and between R and W_{15} are both equal to 8, and they are maximum correlations. In this case, the most likely transmitted sequence is ambiguous because four errors cannot be corrected for the set of $N = 16$ Walsh sequence codewords. In terms of distance, we see that $D(R, W_0) = D(R, W_{15}) = 4$, while $D(R, W_i) > 4$ for $i \neq 0, 15$.

5.5.2 Fast Walsh Transform Decoding

A fast Walsh transform is possible using the interesting symmetry properties that the Walsh sequences possess. It was shown previously that the Walsh functions could be generated using symmetry properties. If no errors are present in the received Walsh codeword, the decoding process can be implemented by reversing the generation procedure. Recall that $W_i = (w_{i0}, w_{i1}, \ldots, w_{i,N-1})$ corresponds to the index vector $X_i = (x_{i1}, x_{i2}, \ldots, x_{iK})$, where $i = 0, 1, \ldots, N - 1$ and $N = 2^K$ is the order of the Walsh functions.

If $x_{ij} = 0$, the sequence has even symmetry about the set of axes called a_j, where $j = 1, 2, \ldots, K$. If $x_{ij} = 1$, the function has odd symmetry about that axis. To review these symmetry properties, consider the symmetry axes for a sequence taken from the set of Walsh functions of order 16, as shown in Figure 5.7. First, note that if symmetry is odd or even about an axis a, the bits reflected about the axis are identical. Next, observe the symmetries shown in Figure 5.8. In this example, the symmetry property about the axes

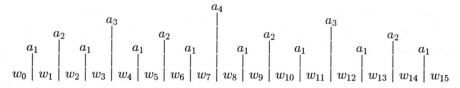

Figure 5.7 Walsh sequence symmetry axes for $N = 16$.

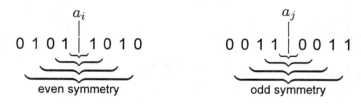

Figure 5.8 Example Walsh axis symmetries.

a_i and a_j involved four separate comparisons. In a real situation, some bits can be received in error, but an even or odd decision can be made using the majority decision rule. To put this method into practice, we assign a value of $+1$ to a symmetry measure for every pair of reflected bits that are identical about the axis a, and a value of -1 for every pair that differs.

Example 5.15 In the received sequence $R = 0\ 1\ 0\ 1\ 1\ 0\ 1\ 0$, the symmetry measure is $4(+1)$ about the central axis because all four reflected pairs are identical bits, whereas for the received sequence $R = 0\ 0\ 1\ 1\ 0\ 0\ 1\ 1$ gives a symmetry measure of $4(-1) = -4$ about the central axis because every pair of reflected bits is different.

Example 5.16 Consider the received sequence $R = 0\ 1\ 0\ 1\ 1\ 0\ 1\ 1$. We note that three reflected pairs with respect to the central axis are identical, while one pair is different. Therefore, the symmetry measure is $3(+1) + 1(-1) = 2$. By majority rule, the decision is even symmetry, and an error has been detected in the pair that differs (which of the two bits that is in error is not yet known). Using the additional information that each codeword except W_0 has the weight $N/2 = 4$, it can be decided that the last bit is in error, and it can be corrected.

Example 5.17 Consider the received sequence $R = 1\ 0\ 1\ 1\ 0\ 1\ 1\ 1$. The symmetry measure for this sequence is $2(+1) + 2(-1) = 0$. The symmetry measure equals zero because the number of reflected pairs having even symmetry equals the number having odd symmetry about the midpoint. Therefore, the symmetry in the interval cannot be classified as either odd or even.

We shall describe a generalized fast Walsh transform decoding rule. Decoding is based on the majority of even or odd symmetries about the axis in each set a_j, $j = 1, 2, \ldots, K$. The symmetry measure can be found by correlating the first half of a given interval with the reverse order of the second half, as shown in the following example:

Example 5.18 The correlation of the first half of the sequence given by

$$\overbrace{0\ 1\ 0\ 1}^{\text{1st half}}\ \underbrace{1\ 0\ 1\ 1}_{\text{2nd half}}$$

with the reverse of the second half is

$$\Big\langle \underbrace{0 \quad 1 \quad 0 \quad 1}_{\text{1st half}}, \quad \underbrace{1 \quad 0 \quad 1 \quad 1}_{\text{reverse of 2nd half}} \Big\rangle = 2$$

In this example, the received sequence is to be corrected by inverting the last bit, because it has been decided that the codeword transmitted has even symmetry, and the last bit both violates this symmetry and unbalances the number of zeros and ones.

Example 5.19 We consider an example using Walsh sequences of order $N = 16 = 2^4 = 2^K$. Let the received sequence R be given by

$$R = (r_1, r_2, \ldots, r_{16})$$

and let the decoded index sequence be given by

$$X = (x_1, x_2, x_3, x_4)$$

Assume that W_{13} was transmitted and corrupted by an error sequence E, resulting in the received sequence $R = W_{13} + E$:

$$W_{13} = 0 \ 1 \ 0 \ 1 \ 1 \ 0 \ 1 \ 0 \ 1 \ 0 \ 1 \ 0 \ 0 \ 1 \ 0 \ 1$$

$$E = 0 \ 1 \ 0 \ 0 \ 0 \ 0 \ 0 \ 0 \ 0 \ 0 \ 0 \ 0 \ 0 \ 0 \ 0 \ 0$$

$$R = 0 \ 0 \ 0 \ 1 \ 1 \ 0 \ 1 \ 0 \ 1 \ 0 \ 1 \ 0 \ 0 \ 1 \ 0 \ 1$$

First, find the sum of the symmetry measures about the set of a_1 axes:

$$
\begin{array}{cccccccccccccccc}
a_1 & & a_1 & & a_1 & & a_1 & & a_1 & & a_1 & & a_1 & & a_1 \\
| & & | & & | & & | & & | & & | & & | & & | \\
0 & | \ 0 & \ 0 & | \ 1 & \ 1 & | \ 0 & \ 1 & | \ 0 & \ 1 & | \ 0 & \ 1 & | \ 0 & \ 0 & | \ 1 & \ 0 & | \ 1 \\
r_1 & r_2 & r_3 & r_4 & r_5 & r_6 & r_7 & r_8 & r_9 & r_{10} & r_{11} & r_{12} & r_{13} & r_{14} & r_{15} & r_{16}
\end{array}
$$

$$\langle r_1, r_2 \rangle = \langle 0, 0 \rangle = +1 \qquad\qquad \langle r_3, r_4 \rangle = \langle 0, 1 \rangle = -1$$

$$\langle r_5, r_6 \rangle = \langle 1, 0 \rangle = -1 \qquad\qquad \langle r_7, r_8 \rangle = \langle 1, 0 \rangle = -1$$

$$\langle r_9, r_{10} \rangle = \langle 1, 0 \rangle = -1 \qquad\qquad \langle r_{11}, r_{12} \rangle = \langle 1, 0 \rangle = -1$$

$$\langle r_{13}, r_{14} \rangle = \langle 0, 1 \rangle = -1 \qquad\qquad \langle r_{15}, r_{16} \rangle = \langle 0, 1 \rangle = -1$$

$$\sum_{i=0}^{7} \langle r_{2i+1}, r_{2i+2} \rangle = -6 \implies \text{the majority of } a_1 \text{ symmetries are odd}$$

This finding implies that $x_1 = 1$, and thus $X = (1, x_2, x_3, x_4)$. Now, we find the sum of the symmetry measures about the a_2 set of axes.

$$
\begin{array}{cccccccccccccccc}
 & a_2 & & & & a_2 & & & & a_2 & & & & a_2 & & \\
0 & 0 & 0 & 1 & 1 & 0 & 1 & 0 & 1 & 0 & 1 & 0 & 0 & 1 & 0 & 1 \\
r_1 & r_2 & r_3 & r_4 & r_5 & r_6 & r_7 & r_8 & r_9 & r_{10} & r_{11} & r_{12} & r_{13} & r_{14} & r_{15} & r_{16}
\end{array}
$$

$$\langle (r_1, r_2), (r_4, r_3) \rangle = \langle (0, 0), (1, 0) \rangle = 0$$

$$\langle (r_5, r_6), (r_8, r_7) \rangle = \langle (1, 0), (0, 1) \rangle = -2$$

$$\langle (r_9, r_{10}), (r_{12}, r_{11}) \rangle = \langle (1, 0), (0, 1) \rangle = -2$$

$$\langle (r_{13}, r_{14}), (r_{16}, r_{15}) \rangle = \langle (0, 1), (1, 0) \rangle = -2$$

$$\sum_{i=0}^{3} \langle (r_{4i+1}, r_{4i+2}), (r_{4i+4}, r_{4i+3}) \rangle = -6$$

$$\implies \text{the majority of } a_2 \text{ symmetries are odd}$$

This finding implies that $x_2 = 1$, so that $X = (1, 1, x_3, x_4)$. The next step is to find the sum of the symmetry measures about the a_3 set of axes.

$$
\begin{array}{cccccccccccccccc}
 & & & a_3 & & & & & & & & a_3 & & & & \\
0 & 0 & 0 & 1 & 1 & 0 & 1 & 0 & 1 & 0 & 1 & 0 & 0 & 1 & 0 & 1 \\
r_1 & r_2 & r_3 & r_4 & r_5 & r_6 & r_7 & r_8 & r_9 & r_{10} & r_{11} & r_{12} & r_{13} & r_{14} & r_{15} & r_{16}
\end{array}
$$

$$\langle (r_1, r_2, r_3, r_4), (r_8, r_7, r_6, r_5) \rangle = \langle (0, 0, 0, 1), (0, 1, 0, 1) \rangle = 2$$

$$\langle (r_9, r_{10}, r_{11}, r_{12}), (r_{16}, r_{15}, r_{14}, r_{13}) \rangle = \langle (1, 0, 1, 0), (1, 0, 1, 0) \rangle = 4$$

$$\sum_{i=0}^{1} \langle (r_{8i+1}, r_{8i+2}, r_{8i+3}, r_{8i+4}), (r_{8i+8}, r_{8i+7}, r_{8i+6}, r_{8i+5}) \rangle = 6$$

$$\implies \text{the majority of } a_3 \text{ symmetries are even}$$

This finding implies that $x_3 = 0$, so that $X = (1, 1, 0, x_4)$. Finally, we calculate the symmetry measure about the a_1 axis.

$$a_1$$

$$
\begin{array}{cccccccc|cccccccc}
0 & 0 & 0 & 1 & 1 & 0 & 1 & 0 & 1 & 0 & 1 & 0 & 0 & 1 & 0 & 1 \\
r_1 & r_2 & r_3 & r_4 & r_5 & r_6 & r_7 & r_8 & r_9 & r_{10} & r_{11} & r_{12} & r_{13} & r_{14} & r_{15} & r_{16}
\end{array}
$$

$$\langle (r_1, r_2, r_3, r_4, r_5, r_6, r_7, r_8), (r_{16}, r_{15}, r_{14}, r_{13}, r_{12}, r_{11}, r_{10}, r_9) \rangle$$

$$= \langle (0,0,0,1,1,0,1,0), (1,0,1,0,0,1,0,1) \rangle = -6$$

Thus, $x_4 = 1$ and the decoded data sequence is now given as $X = (1, 1, 0, 1) = X_{13}$, which is the correct sequence.

The generalized decoding rule based on the fast Walsh transform can be stated as follows: In a set of Walsh sequences of order $N = 2^K$, correlate every other 2^{j-1}-tuple of symbols with the reverse order of the following 2^{j-1}-tuple. Take the sum of the correlation measures. If the sum is positive, the jth ($j = 1, 2, \ldots, K$) symbol of the index sequence is $x_j = 0$; if the sum is negative, then $x_j = 1$; and if the sum is zero, the index symbol is arbitrarily chosen; that is:

$$\sum_{i=0}^{N/2^j-1} \langle (r_{2^j \cdot i+1}, r_{2^j \cdot i+2}, \ldots, r_{2^j \cdot i+2^{j-1}}),$$

$$(r_{2^j \cdot i+2^j}, r_{2^j \cdot i+2^j-1}, \ldots, r_{2^j \cdot i+2^{j-1}+1}) \rangle$$

$$\begin{cases} > 0 \Rightarrow x_j = 0 \\ = 0 \Rightarrow \text{pick } x_j = 0 \text{ or } 1 \\ < 0 \Rightarrow x_j = 1 \end{cases} \tag{5.42}$$

In summary, we have considered two methods for decoding the Walsh sequences: correlation decoding and fast Walsh transform decoding. Another possible scheme that can be employed is *matched filter decoding*.

5.6 IS-95 Data Frames

In both cellular [1] and PCS [16] CDMA systems, the forward and reverse link signals are transmitted over the channel in frames or packets. The frame stuctures vary, depending upon the channel category and data rate, such as synchronization channel, paging channel, access channel, and traffic channel.

As an example, consider the forward traffic channel frame structures shown in Figure 5.9. The 9,600-bps, 4,800-bps, 2,400-bps, and 1,200-bps data rates for the traffic channel specify different frame structures. These data rates are the input data rates for the convolutional encoder, whose output

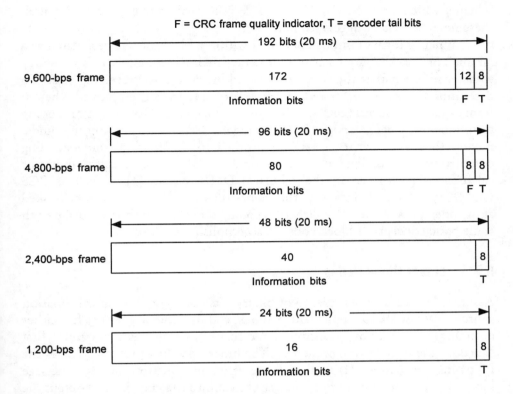

Figure 5.9 Forward traffic channel frame structures (from [1]).

is fed to the 20-ms delay block interleaver after symbol repetition to make the effective symbol rate 19.2 ksps for all four data rates, as shown previously in Figure 4.17.

For the 9,600-bps transmission rate, a total of 192 bits can be transmitted in a 20-ms frame duration. These 192 bits are composed of 172 "information" bits, followed by 12 *frame quality indicator bits* and 8 *encoder tail bits*. It is also observed from Figure 5.9 that the forward traffic frame for the 4,800-bps transmission rate consists of 96 bits that are composed of 80 information bits, 8 frame quality indicator bits, and 8 encoder tail bits. The forward traffic channel frames for the 2,400-bps and 1,200-bps transmission rates contain only information bits (40 and 16 bits, respectively) and 8 encoder tail bits each, without frame quality indicator bits. The frame quality indicator bits are "parity check" bits used in the system's error-detection scheme, which employs CRC codes [17]. We clarify and explain in detail all these new terms later in this section.

When information bits contained within a block of bits designated as a "frame" or "packet" are transmitted, as depicted in Figure 5.9, it is necessary to determine whether the frame is received in error at the receiving end. To determine the status of "error" or "no error," a scheme is employed wherein frame quality indicator bits are used in an automatic error detection coding technique using cyclic codes. In the following section, we develop the fundamental theory of cyclic codes and proceed to the level of understanding completely the design and operation of the frame quality indicator calculations performed in the cellular CDMA system [1] as well as PCS CDMA systems [16]. In a way, the frames shown in Figure 5.9 can be looked at as codewords in the error detection coding scheme, and thus we begin with some basic concepts of block codes to accomplish our objectives.

5.7 Linear Block Codes

Assume that we have a sequence of binary bits coming out of an information source. We wish to implement a block coding scheme [18–19] for either detecting or correcting transmission errors. In any case, we must first "encode" the information sequence. The block-encoding process involves the following procedure: (1) Collect k successive information bits as a message block; (2) feed the k bits of the message block into the encoder and obtain the coded sequence of n digits, where $n > k$, as diagrammed in Figure 5.10. The result of this procedure is an (n, k) linear code with rate k/n.

Because the message block consists of k information bits, 2^k distinct message blocks are possible, and the encoder output generates 2^k possible

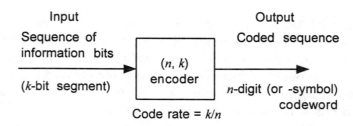

Figure 5.10 Concept of the encoding process.

coded sequences of length n digits, called *codewords*. We say that the block code consists of a set of 2^k codewords. The word "code" connotes an ensemble or a set, whereas the codewords are the elements in the set. Note that the n-digit codeword is an n-tuple, and thus it is a *code vector* in the vector space V_n of all n-tuples. The encoder is a machine (apparatus) or a mathematical rule that transforms the k information bits into an n-digit "coded" sequence.

Now, obviously the block code consisting of 2^k n-tuple codewords is a particular set of n-tuple sequences (vectors) chosen from the set of 2^n possible binary n-tuple vectors. This chosen set, based on a particular encoding rule, is the linear block code, and it is defined as a group:

Definition: A linear block code is a set of 2^k n-tuple vectors that form a subspace of the n-dimensional vector space V_n of all n-tuples.

From the definition of a linear code, the chosen code, being a subspace, must include the all-zero n-tuple, and the dimension of the subspace is k, as we defined these concepts previously in Section 5.3.4. Figure 5.11 depicts the concept of an (n, k) code in the vector space V_n of all n-tuples. The question is, how do we select 2^k n-tuple codewords from the set of all 2^n n-tuple vectors? This is the problem of encoding or designing a coding scheme.

Recall the Walsh sequences that we have defined in the first part of this chapter. The Walsh sequences of order 64 consist of $64 = 2^6$ 64-tuple vectors, and they are the codewords chosen in effect from all the 2^{64} possible 64-tuple sequences. It is in this sense that the Walsh sequences are codewords, and they form a subspace of the 64-dimensional subspace of all 64-tuples—the dimension of the subspace is only six, the number of digits in the index sequence. In fact, this is the orthogonal code that is used in the reverse link in CDMA systems [1, 16]. In Section 5.3.5, we treated the subject of generating Walsh functions using basis vectors. This is exactly the method we use in a block-encoding scheme.

The transformation of the message into a codeword is done in the encoder. The encoder's function for an (n, k) code can be fulfilled by specifying the *generator matrix* $\mathbf{G}$ of the code (see eq. (5.26)), which consists of k *linearly independent* n-tuple row vectors. Then all codewords can be generated by a *linear combination* of these row vectors. The coding therefore is to execute the operation of linear combining of the row vectors of the generator matrix in a manner dictated by the k-bit message sequence.

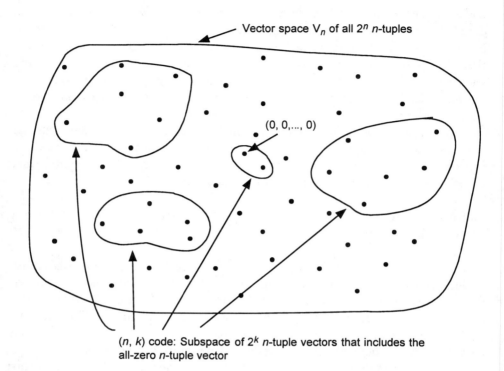

Figure 5.11 A code as a subspace of a vector space.

Let the generator matrix $\mathbf{G}$ be given as

$$\mathbf{G} = \begin{bmatrix} \boldsymbol{g}_0 \\ \boldsymbol{g}_1 \\ \vdots \\ \boldsymbol{g}_{k-1} \end{bmatrix} = \begin{bmatrix} g_{00} & g_{01} & \cdots & g_{0,n-1} \\ g_{10} & g_{11} & \cdots & g_{1,n-1} \\ \vdots & \vdots & & \vdots \\ g_{k-1,0} & g_{k-1,1} & \cdots & g_{k-1,n-1} \end{bmatrix} \qquad (5.43a)$$

and let the message sequence to be encoded be given by

$$\boldsymbol{m} = (m_0, m_1, \ldots, m_{k-1}), \quad m_i \in \mathrm{GF}(2) \ \text{for} \ 0 \le i \le k-1 \qquad (5.43b)$$

Then the codeword C of length n is determined as

$$C = m \cdot G = (m_0, m_1, \ldots, m_{k-1}) \begin{bmatrix} g_0 \\ g_1 \\ \vdots \\ g_{k-1} \end{bmatrix}$$

$$= m_0 g_0 + m_1 g_1 + \cdots + m_{k-1} g_{k-1} \qquad (5.43c)$$

Note that the encoder needs to store the row vectors of the generator matrix. How do we find row vectors that are linearly independent? This is really the question of "How do we discover a good code," and it is, like many other discoveries, not an easy question to answer. Trial and error is an answer! Fortunately, however, there are many good known codes, and we give examples as we proceed.

Example 5.20 Consider the generator matrix

$$G = \begin{bmatrix} 1 & 1 & 0 & 1 & 0 & 1 \\ 0 & 1 & 0 & 0 & 1 & 1 \\ 0 & 0 & 1 & 1 & 0 & 1 \end{bmatrix}$$

This matrix generates the $(n, k) = (6, 3)$ linear block code. The transformation of each of the $2^3 = 8$ possible 3-tuple message sequences into codewords of length 6 is shown in Table 5.14. In this example, the message encoding was accomplished using the equation

$$C = m \cdot G = (m_0, m_1, m_2) \begin{bmatrix} 1 & 1 & 0 & 1 & 0 & 1 \\ 0 & 1 & 0 & 0 & 1 & 1 \\ 0 & 0 & 1 & 1 & 0 & 1 \end{bmatrix}$$

$$= (m_0, m_0 + m_1, m_2, m_0 + m_2, m_1, m_0 + m_1 + m_2)$$

Note in the example that the message bits m_0, m_1, and m_2 appear in the codeword in the first, third, and fifth positions, and the $n - k = 3$ redundant digits occupy the second, fourth, and sixth positions. What is obvious here is that the original message bits, which appear in the codeword intact, are "scattered" around in the codeword. Why not make them appear in the first three positions or in the last three positions in codeword? Indeed we can do so! The way the original information bits appear in the example is a case of a nonsystematic code. We want to encode the message sequence into the codeword in such a way that the entire k-bit information sequence appears at

Table 5.14 (6, 3) Block encoding

3-bit message sequence $m_i = (m_{i0}, m_{i1}, m_{i2})$	Encoder transformation $m_i G$	Codewords of length 6 $C_i = (c_{i0}, c_{i1}, \ldots, c_{i5})$
$m_0 = 0\ 0\ 0$	$m_0 G$	$C_0 = 0\ 0\ 0\ 0\ 0\ 0$
$m_1 = 0\ 0\ 1$	$m_1 G$	$C_1 = 0\ 0\ 1\ 1\ 0\ 1$
$m_2 = 0\ 1\ 0$	$m_2 G$	$C_2 = 0\ 1\ 0\ 0\ 1\ 1$
$m_3 = 0\ 1\ 1$	$m_3 G$	$C_3 = 0\ 1\ 1\ 1\ 1\ 0$
$m_4 = 1\ 0\ 0$	$m_4 G$	$C_4 = 1\ 1\ 0\ 1\ 0\ 1$
$m_5 = 1\ 0\ 1$	$m_5 G$	$C_5 = 1\ 1\ 1\ 0\ 0\ 0$
$m_6 = 1\ 1\ 0$	$m_6 G$	$C_6 = 1\ 0\ 0\ 1\ 1\ 0$
$m_7 = 1\ 1\ 1$	$m_7 G$	$C_7 = 1\ 1\ 1\ 0\ 1\ 1$

the beginning or at the end of the codeword as suggested in Figure 5.12; when this order of the message bits is accomplished, the code is called a *systematic code*. A coding scheme is a powerful way of detecting and/or correcting any errors that have occurred during the transmission over the channel. The strength or the power of the code comes from the redundant digits in the codeword. Note, as illustrated in the example, that the redundant digits are computed as functions of the message (information) bits.[1] Although it is not clear at this point, the error detection and error correction capabilities of codes are due to these extra digits that are added to the message sequence and are called *redundant digits*. These redundant digits are known as *parity check digits* in coding theory, and as such, the codes are sometimes called *parity check codes*.

The term *parity check* goes back to the earliest parity-check coding scheme, in which simple error detection was accomplished using either an odd or even parity-check method: Given a code consisting of binary message sequences of length n, an extra digit is added to the codeword and its value

---

[1] The use of the words "bit" and "digit" is somewhat confusing. Usually the elements of the raw information sequence are called "bits," whereas those of the coded sequence are called "digits" or "symbols." However, any term can be used once the meaning is clear.

k-bit message	→	Message bits	Redundant digits
$\vert\leftarrow k \rightarrow\vert$		$\vert\leftarrow k \rightarrow\vert$	$\leftarrow n-k \rightarrow\vert$

or

k-bit message	→	Redundant digits	Message bits
$\vert\leftarrow k \rightarrow\vert$		$\vert\leftarrow n-k \rightarrow\vert$	$\leftarrow k \rightarrow\vert$

Figure 5.12 Systematic code formats.

chosen to make the total number of ones in the $n+1$-digit codeword either even (for even parity) or odd (for odd parity).

We have noted that the 9,600-bps and 4,800-bps frames of the IS-95 and PCS CDMA system forward traffic channels included reserved groups of 12 bits and 8 bits, respectively. These bits are parity check bits, which are called *frame quality indicators* [1 , 16] and are designed to protect the traffic channel message sequences in the sense that errors can be detected or corrected at the receiver, within the capability of the specific code.

Now, returning to the subject of *systematic* codes, to obtain the generator matrix G that gives a systematic code, we have only to perform elementary row or column manipulations on the given generator matrix (5.43a) to cause it to have one of the following forms:

$$\mathbf{G} = \left[\begin{array}{cccc|cccc} \multicolumn{4}{c}{\overbrace{\hspace{6cm}}^{k \times (n-k) \text{ matrix}}} & \multicolumn{4}{c}{\overbrace{\hspace{3cm}}^{k \times k \text{ identity matix}}} \\ q_{00} & q_{01} & \cdots & q_{0,n-k-1} & 1 & 0 & 0 \cdots & 0 \\ q_{10} & q_{11} & \cdots & q_{1,n-k-1} & 0 & 1 & 0 \cdots & 0 \\ \vdots & \vdots & & \vdots & \vdots & & & \vdots \\ q_{k-1,0} & q_{k-1,1} & \cdots & q_{k-1,n-k-1} & 0 & 0 & 0 \cdots & 1 \end{array}\right], \; q_{ij} = 0 \text{ or } 1 \tag{5.44a}$$

or

$$\mathbf{G} = \left[\begin{array}{cccc|cccc} \multicolumn{4}{c}{\overbrace{\hspace{3cm}}^{k \times k \text{ identity matix}}} & \multicolumn{4}{c}{\overbrace{\hspace{6cm}}^{k \times (n-k) \text{ matrix}}} \\ 1 & 0 & 0 \cdots & 0 & q_{00} & q_{01} & \cdots & q_{0,n-k-1} \\ 0 & 1 & 0 \cdots & 0 & q_{10} & q_{11} & \cdots & q_{1,n-k-1} \\ \vdots & & & \vdots & \vdots & \vdots & & \vdots \\ 0 & 0 & 0 \cdots & 1 & q_{k-1,0} & q_{k-1,1} & \cdots & q_{k-1,n-k-1} \end{array}\right], \; q_{ij} = 0 \text{ or } 1 \tag{5.44b}$$

The choice between (5.44a) and (5.44b) is whether the k information bits are to appear at the end of the codeword or at the beginning of the codeword, respectively, when written from left to right. Thus, the generator matrix of a systematic code can always be put into the forms shown above. That is, the systematic code generator matrix can be written as

$$G = [Q \quad I_k] \quad \text{or} \quad G = [I_k \quad Q] \tag{5.44c}$$

Let $m = (m_0, m_1, m_2, \ldots, m_{k-1})$ be the message sequence to be encoded. The encoder will then generate the codeword $C = (c_0, c_1, c_2, \ldots, c_{n-1})$ in a systematic form, by using (5.44a), as follows:

$$C = (m_0, m_1, \ldots, m_{k-1}) \begin{bmatrix} q_{00} & q_{01} & \cdots & q_{0,n-k-1} & 1 & 0 & 0 & \cdots & 0 \\ q_{10} & q_{11} & \cdots & q_{1,n-k-1} & 0 & 1 & 0 & \cdots & 0 \\ \vdots & \vdots & & \vdots & \vdots & & & & \vdots \\ q_{k-1,0} & q_{k-1,1} & \cdots & q_{k-1,n-k-1} & 0 & 0 & 0 & \cdots & 1 \end{bmatrix} \tag{5.45a}$$

which specifies the codeword digit computations given by

$$\left. \begin{aligned} c_0 &= m_0 q_{00} & + m_1 q_{10} & & + \cdots + m_{k-1} q_{k-1,0} \\ c_1 &= m_0 q_{01} & + m_1 q_{11} & & + \cdots + m_{k-1} q_{k-1,1} \\ &\ \vdots \\ c_{n-k-1} &= m_0 q_{0,n-k-1} + m_1 q_{1,n-k-1} + \cdots + m_{k-1} q_{k-1,n-k-1} \end{aligned} \right\} \begin{aligned} & n-k \\ & \text{parity check} \\ & \text{equations} \end{aligned} \tag{5.45b}$$

$$\left. \begin{aligned} c_{n-k} &= m_0 \\ c_{n-k+1} &= m_1 \\ &\ \vdots \\ c_{n-1} &= m_{k-1} \end{aligned} \right\} \text{original message sequence, at the end of the codeword}$$

Let us introduce an important mathematical fact concerning a generator matrix G. Suppose that G is a $k \times n$ rectangular matrix whose rows are k linearly independent n-tuple vectors as shown in (5.43a), which we repeat here for convenience:

$$G = \begin{bmatrix} g_{00} & g_{01} & \cdots & g_{0,n-1} \\ g_{10} & g_{11} & \cdots & g_{1,n-1} \\ \vdots & \vdots & & \vdots \\ g_{k-1,0} & g_{k-1,1} & \cdots & g_{k-1,n-1} \end{bmatrix} = \begin{bmatrix} g_0 \\ g_1 \\ \vdots \\ g_{k-1} \end{bmatrix} \tag{5.46a}$$

where

$$g_i \triangleq (g_{i0}, g_{i1}, \ldots, g_{i,n-1}), \quad g_{ij} \in \mathrm{GF}(2) \tag{5.46b}$$

Then all 2^k n-tuple vectors obtained as linear combinations of the row vectors of $\mathbf{G}$ form a k-dimensional subspace of the n-dimensional vector space V_n of all n-tuples. The subspace spanned by the row vectors of $\mathbf{G}$ is also called the *row space* of $\mathbf{G}$. Now, it is a mathematical fact that, for any $k \times n$ matrix $\mathbf{G}$ with linearly independent row vectors such as the generator matrix above, an $(n-k) \times n$ matrix $\mathbf{H}$ exists:

$$\mathbf{H} = \begin{bmatrix} h_{00} & h_{01} & \cdots & h_{0,n-1} \\ h_{10} & h_{11} & \cdots & h_{1,n-1} \\ \vdots & \vdots & & \vdots \\ h_{n-k-1,0} & h_{n-k-1,1} & \cdots & h_{n-k-1,n-1} \end{bmatrix} = \begin{bmatrix} \boldsymbol{h}_0 \\ \boldsymbol{h}_1 \\ \vdots \\ \boldsymbol{h}_{n-k-1} \end{bmatrix} \quad (5.46c)$$

where $$\boldsymbol{h}_j \triangleq (h_{j0}, h_{j1}, \ldots, h_{j,n-1}), \quad h_{jl} \in GF(2) \quad (5.46d)$$

such that the $n - k$ rows are linearly independent and such that any vector obtained as a linear combination of the row vectors of $\mathbf{G}$ (in the row space of $\mathbf{G}$) is orthogonal to all rows of $\mathbf{H}$. Also, any vector obtained as a linear combination of row vectors of $\mathbf{H}$ (in the row space of $\mathbf{H}$) is orthogonal to all rows of $\mathbf{G}$—any vector $\boldsymbol{X}$ in the row space of $\mathbf{G}$ and any vector $\boldsymbol{Y}$ in the row space of $\mathbf{H}$ are orthogonal. We say that the row space of $\mathbf{G}$ is the *null space* (or *dual space*) of $\mathbf{H}$, and the row space of $\mathbf{H}$ is the null space (or dual space) of $\mathbf{G}$. This is an important mathematical fact related to the matrix formulation of the coding problem, which we use in a later section when we "design" CRC codes. It is particularly easy to find the $\mathbf{H}$ matrix when the $\mathbf{G}$ matrix is given in a systematic code generator form. As we see in the following example, the effort of finding the $\mathbf{H}$ matrix involves only the transposition of the $\mathbf{G}$ matrix columns.

The 2^{n-k} vectors in the row space of $\mathbf{H}$ form an $(n, n-k)$ linear block code of length n and k parity check digits. This code is the null (dual) space of the (n, k) linear code generated by the matrix $\mathbf{G}$. The $(n, n-k)$ code is called the *dual code C_d*. Therefore, if $\boldsymbol{X} \in C$ and $\boldsymbol{Y} \in C_d$ then $\langle \boldsymbol{X}, \boldsymbol{Y} \rangle = 0$.

Example 5.21 Consider a $k \times n = 4 \times 7$ generator matrix given by

$$\mathbf{G} = \begin{bmatrix} 1 & 1 & 0 & \vdots & 1 & 0 & 0 & 0 \\ 0 & 1 & 1 & \vdots & 0 & 1 & 0 & 0 \\ 1 & 1 & 1 & \vdots & 0 & 0 & 1 & 0 \\ 1 & 0 & 1 & \vdots & 0 & 0 & 0 & 1 \end{bmatrix} \triangleq \begin{bmatrix} \boldsymbol{g}_1 \\ \boldsymbol{g}_2 \\ \boldsymbol{g}_3 \\ \boldsymbol{g}_4 \end{bmatrix}$$

If we specify the $(n - k) \times n = (7 - 4) \times 7$ matrix $\mathbf{H}$ by transposing the columns of $\mathbf{G}$ as shown below:

$$\mathbf{H} = \begin{bmatrix} 1 & 0 & 0 & \vdots & 1 & 0 & 1 & 1 \\ 0 & 1 & 0 & \vdots & 1 & 1 & 1 & 0 \\ 0 & 0 & 1 & \vdots & 0 & 1 & 1 & 1 \end{bmatrix} \triangleq \begin{bmatrix} \boldsymbol{h}_1 \\ \boldsymbol{h}_2 \\ \boldsymbol{h}_3 \end{bmatrix}$$

we can easily verify that the row space of $\mathbf{G}$ is the null space of $\mathbf{H}$ and vice versa. For example:

$$\langle \boldsymbol{g}_2, \boldsymbol{h}_3 \rangle = \boldsymbol{g}_2 \boldsymbol{h}_3^{\mathrm{T}} = \begin{bmatrix} 0 & 1 & 1 & 0 & 1 & 0 & 0 \end{bmatrix} \begin{bmatrix} 0 \\ 0 \\ 1 \\ 0 \\ 1 \\ 1 \\ 1 \end{bmatrix}$$

$$= 0 \cdot 0 + 1 \cdot 0 + 1 \cdot 1 + 0 \cdot 0 + 1 \cdot 1 + 0 \cdot 1 + 0 \cdot 1$$

$$= 1 + 1 = 0$$

Using other row vectors, we can show that the row space of $\mathbf{G}$ is indeed the null space of $\mathbf{H}$.

5.7.1 Parity Check Matrix

We have stated that any vector obtained as a linear combinatin of the row vectors of the generator matrix $\mathbf{G}$ is orthogonal to the row vectors of $\mathbf{H}$. The matrices $\mathbf{G}$ and $\mathbf{H}$ are given in (5.46a) and (5.46c), respectively. That is, if $C = (c_0, c_1, \ldots, c_{n-1})$ is a vector (codeword) in the row space of $\mathbf{G}$, then[2]

$$\langle C, \mathbf{H} \rangle = C \cdot \mathbf{H}^{\mathrm{T}} = 0 \tag{5.47a}$$

or

$$\mathbf{H} \cdot C^{\mathrm{T}} = 0 \tag{5.47b}$$

Using the form of the $\mathbf{H}$ matrix given in (5.46c), we have

---

[2] We used the notation $\langle \boldsymbol{a}, \boldsymbol{b} \rangle$ previously for the correlation of vectors $\boldsymbol{a}$ and $\boldsymbol{b}$. The correlation operation, dot product, and inner product all convey the same meaning, and $\langle \boldsymbol{a}, \boldsymbol{b} \rangle = \boldsymbol{a} \cdot \boldsymbol{b}^{\mathrm{T}}$ conveys the same meaning when $\boldsymbol{a}$ and $\boldsymbol{b}$ are row vectors and $\boldsymbol{b}^{\mathrm{T}}$ is the transpose of $\boldsymbol{b}$.

$$c_0 h_{00} + c_1 h_{01} + \cdots + c_{n-1} h_{0,n-1} = 0$$

$$c_0 h_{10} + c_1 h_{11} + \cdots + c_{n-1} h_{1,n-1} = 0$$

$$\vdots$$

$$c_0 h_{j0} + c_1 h_{j1} + \cdots + c_{n-1} h_{j,n-1} = 0 \qquad (5.47c)$$

$$\vdots$$

$$c_0 h_{n-k-1,0} + c_1 h_{n-k-1,1} + \cdots + c_{n-1} h_{n-k-1,n-1} = 0$$

This result gives alternative ways of characterizing the linear code generated by the generator matrix G; namely, that if any code is generated by G, it must satisfy equation (5.47a), (5.47b), or (5.47c).

The matrix H is called the *parity check matrix* of the code. The code C is a legitimate code if and only if it satisfies the parity check equation $C \cdot H^T = 0$. As we see later, this is the method or strategy employed at the receiving end to check if the received codeword satisfies the parity check equations and thus to verify whether it is received in error. Later, we develop ways of calculating the parity check equations using LFSR circuits.

The linear block code (n, k) can now be called a parity check code as well. It is also called a *group code* because of the fact that the code is a subspace of the vector space of all n-tuples. Thus, the term "(n, k) linear block code" can be referred to as a parity check code with codewords of length n and $n - k$ parity check digits, where k is the number of information digits.

It is instructive to revisit the set of parity check equations given in (5.47c) and express them as the parity check equations for an (n, k) code with codeword $C_i = (c_{i0}, c_{i1}, c_{i2}, \ldots, c_{ij}, \ldots, c_{i,n-1})$, $i = 0, 1, 2, \ldots, 2^k - 1$:

$$c_{i0} \begin{bmatrix} h_{00} \\ h_{10} \\ h_{20} \\ \vdots \\ h_{j0} \\ \vdots \\ h_{n-k-1,0} \end{bmatrix} + c_{i1} \begin{bmatrix} h_{01} \\ h_{11} \\ h_{21} \\ \vdots \\ h_{j1} \\ \vdots \\ h_{n-k-1,1} \end{bmatrix} + \cdots + c_{i,n-1} \begin{bmatrix} h_{0,n-1} \\ h_{1,n-1} \\ h_{2,n-1} \\ \vdots \\ h_{j,n-1} \\ \vdots \\ h_{n-k-1,n-1} \end{bmatrix} = \begin{bmatrix} 0 \\ 0 \\ 0 \\ \vdots \\ 0 \\ \vdots \\ 0 \end{bmatrix} \qquad (5.47d)$$

(5.47d) indicates that we can also define an (n, k) linear parity check code by first specifying a parity check matrix. In other words, we can specify H and then find the corresponding G matrix, or vice versa.

We now follow up on the statement made earlier that the row space of **G** and the row space of **H** are orthogonal. Consider the generator matrix of an (n, k) systematic code given by (5.46a). We repeat the equation here for convenience:

$$
\mathbf{G} = [\,\boldsymbol{Q} \quad \boldsymbol{I}_k\,] =
\overbrace{
\underbrace{\begin{bmatrix} q_{00} & q_{01} & \cdots & q_{0,\,n-k-1} \\ q_{10} & q_{11} & \cdots & q_{1,\,n-k-1} \\ \vdots & \vdots & & \vdots \\ q_{k-1,0} & q_{k-1,1} & \cdots & q_{k-1,n-k-1} \end{bmatrix}}_{k \times (n-k)}
\underbrace{\begin{bmatrix} 1 & 0 & 0 & \cdots & 0 \\ 0 & 1 & 0 & \cdots & 0 \\ & & & \vdots & \\ 0 & 0 & 0 & \cdots & 1 \end{bmatrix}}_{k \times k}}
\tag{5.48a}
$$

The parity check matrix **H** can then be obtained as

$$
\mathbf{H} = [\,\boldsymbol{I}_{n-k} \quad \boldsymbol{Q}^{\mathrm{T}}\,] =
\underbrace{\begin{bmatrix} 1 & 0 & 0 & \cdots & 0 \\ 0 & 1 & 0 & \cdots & 0 \\ & \vdots & & & \\ 0 & 0 & 0 & \cdots & 1 \end{bmatrix}}_{(n-k)\times(n-k)}
\underbrace{\begin{matrix} q_{00} & q_{10} & \cdots & q_{k-1,0} \\ q_{01} & q_{11} & \cdots & q_{k-1,1} \\ \vdots & & & \\ q_{0,n-k-1} & q_{1,n-k-1} & \cdots & q_{k-1,n-k-1} \end{matrix}}_{(n-k)\times k}
$$

$$
\tag{5.48b}
$$

We can easily verify that $\mathbf{G} \cdot \mathbf{H}^{\mathrm{T}} = \mathbf{0}$, as shown below:

$$
\mathbf{G} \cdot \mathbf{H}^{\mathrm{T}} = [\,\boldsymbol{Q} \quad \boldsymbol{I}_k\,] \begin{bmatrix} \boldsymbol{I}_{n-k} \\ \boldsymbol{Q} \end{bmatrix} = \left[\,\boldsymbol{Q}_{[k\times(n-k)]} \cdot \boldsymbol{I}_{n-k} + \boldsymbol{I}_k \cdot \boldsymbol{Q}_{[k\times(n-k)]}\,\right]
$$

$$
= \boldsymbol{Q}_{[k\times(n-k)]} + \boldsymbol{Q}_{[k\times(n-k)]}
$$

$$
= \begin{bmatrix} q_{00} \oplus q_{00} & q_{01} \oplus q_{01} & \cdots & q_{0,n-k-1} \oplus q_{0,n-k-1} \\ q_{10} \oplus q_{10} & q_{11} \oplus q_{11} & \cdots & q_{1,n-k-1} \oplus q_{1,n-k-1} \\ \vdots & & & \vdots \\ q_{k-1,0} \oplus q_{k-1,0} & q_{k-1,1} \oplus q_{k-1,1} & \cdots & q_{k-1,n-k-1} \oplus q_{k-1,n-k-1} \end{bmatrix} = \mathbf{0}
$$

$$
\tag{5.48c}
$$

Example 5.22 Consider the $(7, 4)$ linear code with the generator matrix **G** given in a systematic form as follows:

$$G = \begin{bmatrix} g_1 \\ g_2 \\ g_3 \\ g_4 \end{bmatrix} = \begin{bmatrix} 1 & 1 & 0 & 1 & 0 & 0 & 0 \\ 0 & 1 & 1 & 0 & 1 & 0 & 0 \\ 1 & 1 & 1 & 0 & 0 & 1 & 0 \\ 1 & 0 & 1 & 0 & 0 & 0 & 1 \end{bmatrix} = [\, Q_{[4\times 3]} \quad I_4 \,]$$

If the message is $m = (1\ 0\ 0\ 1)$, then the encoded codeword is obtained as follows:

$$C = m \cdot G = (1\ 0\ 0\ 1) \begin{bmatrix} g_1 \\ g_2 \\ g_3 \\ g_4 \end{bmatrix} = 1 \cdot g_1 + 0 \cdot g_2 + 0 \cdot g_3 + 1 \cdot g_4$$

$$= \left\{ \begin{array}{l} 1\ 1\ 0\ 1\ 0\ 0\ 0 \\ \oplus\ 1\ 0\ 1\ 0\ 0\ 0\ 1 \end{array} \right.$$

$$= \underbrace{0\ 1\ 1}_{\substack{\text{parity check} \\ \text{digits}}}\ \underbrace{1\ 0\ 0\ 1}_{\text{message}} \quad \Leftarrow \quad \text{codeword}$$

The corresponding parity check matrix is

$$H = \begin{bmatrix} I_{n-k} & Q^{T}_{[(n-k)\times k]} \end{bmatrix} = \begin{bmatrix} I_3 & Q^{T}_{[3\times 4]} \end{bmatrix} = \begin{bmatrix} 1 & 0 & 0 & 1 & 0 & 1 & 1 \\ 0 & 1 & 0 & 1 & 1 & 1 & 0 \\ 0 & 0 & 1 & 0 & 1 & 1 & 1 \end{bmatrix}$$

Note that the codeword $C = (0\ 1\ 1\ 1\ 0\ 0\ 1)$ satisfies the parity check equation:

$$H \cdot C^{T} = \begin{bmatrix} 1 & 0 & 0 & 1 & 0 & 1 & 1 \\ 0 & 1 & 0 & 1 & 1 & 1 & 0 \\ 0 & 0 & 1 & 0 & 1 & 1 & 1 \end{bmatrix} \begin{bmatrix} 0 \\ 1 \\ 1 \\ 1 \\ 0 \\ 0 \\ 1 \end{bmatrix} = \begin{bmatrix} 0 \\ 0 \\ 0 \end{bmatrix}$$

Example 5.23 A $(6, 3)$ linear code has a generator matrix G given by

$$G = \begin{bmatrix} 0 & 1 & 1 & 0 & 0 & 1 \\ 1 & 1 & 0 & 0 & 1 & 0 \\ 1 & 0 & 0 & 1 & 0 & 1 \end{bmatrix}$$

Find the parity check matrix for this code and list all $2^3 = 8$ codewords.

Solution: To solve this problem, we first need to recognize that the row vectors of this matrix are themselves codewords, and there are three check digits in each codeword that are determined as functions of the three information digits. Assuming that the codewords have the format $C = (c_1, c_2, c_3, m_1, m_2, m_3)$, where c_1, c_2, and c_3 are parity check digits and m_1, m_2, and m_3 are the message digits, and noting that the last three columns of $\mathbf{G}$ are linearly independent, we realize that each of the first three parity check digits must be obtained as linear combinations of the last three (information) digits. The generator matrix $\mathbf{G}$ will enable us to find scalar multipliers specifying the linear combinations of the information (message) digits [see (5.45b)]. Therefore, the simultaneous equations that need to be solved are

1st digits of C
$$\begin{bmatrix} 0 \\ 1 \\ 1 \end{bmatrix} = \alpha_1 \begin{bmatrix} 0 \\ 0 \\ 1 \end{bmatrix} + \alpha_2 \begin{bmatrix} 0 \\ 1 \\ 0 \end{bmatrix} + \alpha_3 \begin{bmatrix} 1 \\ 0 \\ 1 \end{bmatrix} \Rightarrow \alpha_1 = \alpha_2 = 1, \ \alpha_3 = 0$$

2nd digits of C
$$\begin{bmatrix} 1 \\ 1 \\ 0 \end{bmatrix} = \beta_1 \begin{bmatrix} 0 \\ 0 \\ 1 \end{bmatrix} + \beta_2 \begin{bmatrix} 0 \\ 1 \\ 0 \end{bmatrix} + \beta_3 \begin{bmatrix} 1 \\ 0 \\ 1 \end{bmatrix} \Rightarrow \beta_1 = \beta_2 = \beta_3 = 1$$

3rd digits of C
$$\begin{bmatrix} 1 \\ 0 \\ 0 \end{bmatrix} = \gamma_1 \begin{bmatrix} 0 \\ 0 \\ 1 \end{bmatrix} + \gamma_2 \begin{bmatrix} 0 \\ 1 \\ 0 \end{bmatrix} + \gamma_3 \begin{bmatrix} 1 \\ 0 \\ 1 \end{bmatrix} \Rightarrow \gamma_1 = \gamma_3 = 1, \ \gamma_2 = 0$$

For the assumed format of the codewords, these solutions of the simultaneous equations for the scalars determine that, in general

$$c_1 = \alpha_1 m_1 + \alpha_2 m_2 + \alpha_3 m_3 = m_1 + m_2$$

$$c_2 = \beta_1 m_1 + \beta_2 m_2 + \beta_3 m_3 = m_1 + m_2 + m_3$$

$$c_3 = \gamma_1 m_1 + \gamma_2 m_2 + \gamma_3 m_3 = m_1 + m_3$$

We thus obtain the parity check matrix $\mathbf{H}$:

$$\mathbf{H} = \begin{bmatrix} 1 & 0 & 0 & 1 & 1 & 0 \\ 0 & 1 & 0 & 1 & 1 & 1 \\ 0 & 0 & 1 & 1 & 0 & 1 \end{bmatrix}$$

A "formulaic"-oriented way of solving the problem is to perform elementary row operations on the given generator matrix $\mathbf{G}$ to bring it to the systematic form shown in (5.48a), and thus easily obtain $\mathbf{H}$ by transposing it to the parity check matrix form of (5.48b). So, let us obtain a systematic form, $\mathbf{G}'$, from $\mathbf{G}$ as

$$\mathbf{G}' = \begin{bmatrix} 1 & 1 & 1 & 1 & 0 & 0 \\ 1 & 1 & 0 & 0 & 1 & 0 \\ 0 & 1 & 1 & 0 & 0 & 1 \end{bmatrix} \begin{matrix} \leftarrow \text{row 1} + \text{row 3} \\ \leftarrow \text{row 2} \\ \leftarrow \text{row 3} \end{matrix}$$

which gives the parity check matrix shown above. The $(6,3)$ codewords are all obtained as linear combinations of row vectors of $\mathbf{G}$ and are listed as follows:

c_1	c_2	c_3	m_1	m_2	m_3
0	0	0	0	0	0
0	1	1	0	0	1
1	1	0	0	1	0
1	0	1	0	1	1
1	1	1	1	0	0
1	0	0	1	0	1
0	0	1	1	1	0
0	1	0	1	1	1

Example 5.24 A $(7,4)$ code is defined by the following parity check matrix:

$$\mathbf{H} = \begin{bmatrix} 0 & 0 & 0 & 1 & 1 & 1 & 1 \\ 0 & 1 & 1 & 0 & 0 & 1 & 1 \\ 1 & 0 & 1 & 0 & 1 & 0 & 1 \end{bmatrix}$$

Find the corresponding generator matrix $\mathbf{G}$ and list the codewords.

Solution: We begin with the construction of $\mathbf{H}'$ in systematic form from $\mathbf{H}$ by performing column operations (i.e., permuting the columns) to obtain

$$\mathbf{H}' = \begin{bmatrix} 1 & 0 & 0 & 0 & 1 & 1 & 1 \\ 0 & 1 & 0 & 1 & 0 & 1 & 1 \\ 0 & 0 & 1 & 1 & 1 & 0 & 1 \end{bmatrix}$$

The generator matrix is then given by

$$\mathbf{G} = \begin{bmatrix} 0 & 1 & 1 & 1 & 0 & 0 & 0 \\ 1 & 0 & 1 & 0 & 1 & 0 & 0 \\ 1 & 1 & 0 & 0 & 0 & 1 & 0 \\ 1 & 1 & 1 & 0 & 0 & 0 & 1 \end{bmatrix}$$

The $2^4 = 16$ codewords are listed below:

c_1	c_2	c_3	m_1	m_2	m_3	m_4
0	0	0	0	0	0	0
1	1	1	0	0	0	1
1	1	0	0	0	1	0
0	0	1	0	0	1	1
1	0	1	0	1	0	0
0	1	0	0	1	0	1
0	1	1	0	1	1	0
1	0	0	0	1	1	1
0	1	1	1	0	0	0
1	0	0	1	0	0	1
1	0	1	1	0	1	0
0	1	0	1	0	1	1
1	1	0	1	1	0	0
0	0	1	1	1	0	1
0	0	0	1	1	1	0
1	1	1	1	1	1	1

This example shows that a code can be defined with a parity check matrix without first defining a generator matrix for the code.

5.7.2   Concept of Syndrome and Error Detection

Consider an (n, k) code with the generator matrix $\mathbf{G}$ and the corresponding parity check matrix $\mathbf{H}$. The codewords are transmitted over a channel that introduces errors to the transmitted codewords, and the received sequence R is then the sum of the transmitted vector and an error pattern vector E as depicted in Figure 5.13. Let us specify the respective vectors as follows:

$$\text{Transmitted vector:} \qquad C = (c_0, c_1, c_2, \ldots, c_{n-1})$$
$$\text{Error pattern vector:} \qquad E = (e_0, e_1, e_2, \ldots, e_{n-1}) \qquad (5.49a)$$
$$\text{Received vector:} \qquad R = (r_0, r_1, r_2, \ldots, r_{n-1})$$

where c_i, r_i, and e_i all assume the values over GF(2). If $e_i = 1$, the received vector has an error at position i, and if $e_i = 0$, no error has occurred at that position. The received vector is then given by

$$R = C + E \qquad (5.49b)$$

The receiver needs to validate the vector R as a legitimate codeword. So, the receiver performs a "validation check" by subjecting the vector R to the parity check equations:

$$\mathbf{H} \cdot R^{\mathrm{T}} = \mathbf{H} \cdot (C + E)^{\mathrm{T}} = \mathbf{H} \cdot C^{\mathrm{T}} + \mathbf{H} \cdot E^{\mathrm{T}}$$
$$= \mathbf{H} \cdot E^{\mathrm{T}} \triangleq S^{\mathrm{T}} \qquad (5.49c)$$

Because C is a codeword, $\mathbf{H} \cdot C^{\mathrm{T}} = 0$, and in principle we only need to calculate

$$\mathbf{H} \cdot E^{\mathrm{T}} = S^{T} \quad \text{or} \quad E \cdot \mathbf{H}^{\mathrm{T}} = S \qquad (5.49d)$$

Because $\mathbf{H}$ is an $(n-k) \times n$ matrix, and E^{T} is an $(n \times 1)$ matrix, S^{T} is an $(n-k) \times 1$ matrix. Therefore, (5.49d) becomes

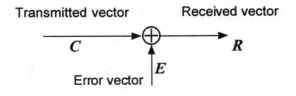

Figure 5.13 Received vector as the sum of the transmitted vector plus an error pattern vector.

$$
\boldsymbol{S}^{\mathrm{T}} = e_0 \begin{bmatrix} h_{00} \\ h_{10} \\ h_{20} \\ \vdots \\ h_{j0} \\ \vdots \\ h_{n-k-1,0} \end{bmatrix} + e_1 \begin{bmatrix} h_{01} \\ h_{11} \\ h_{21} \\ \vdots \\ h_{j1} \\ \vdots \\ h_{n-k-1,1} \end{bmatrix} + \cdots + e_{n-1} \begin{bmatrix} h_{0,n-1} \\ h_{1,n-1} \\ h_{2,n-1} \\ \vdots \\ h_{j,n-1} \\ \vdots \\ h_{n-k-1,n-1} \end{bmatrix} \qquad (5.49\mathrm{e})
$$

The $(n-k)$-tuple vector $\boldsymbol{S} = (s_0, s_1, s_2, \ldots, s_{n-k-1})$ is called the *syndrome* of the received vector $\boldsymbol{R}$.

Note in (5.49e) that the column vectors of the parity check matrix are summed after being multiplied by the vector components of the *error pattern vector* $\boldsymbol{E}$. As can be seen from (5.49e), if no error is introduced by the channel (i.e., $\boldsymbol{E} = (0, 0, \ldots, 0)$), then the syndrome vector $\boldsymbol{S}$ is a zero vector: $\boldsymbol{S} = (s_0, s_1, s_2, \ldots, s_{n-k-1}) = (0, 0, \ldots, 0)$. If $\boldsymbol{S} = \boldsymbol{0}$, the receiver accepts the received vector $\boldsymbol{R}$ as a legitimate codeword (error-free reception), and if $\boldsymbol{S} \neq \boldsymbol{0}$, the received vector is assumed to be in error. Thus, error detection is done by calculating the parity check equations to compute the *syndrome*.

It is possible that the errors that have actually occurred are not detectable; that is, the received vector $\boldsymbol{R}$ contains errors but $\boldsymbol{S}^{\mathrm{T}} = \mathbf{H} \cdot \boldsymbol{R}^{\mathrm{T}} = \boldsymbol{0}$. This situation could happen when the error-pattern vector $\boldsymbol{E}$ is identical to a nonzero codeword vector. That is, when $\boldsymbol{E} = \boldsymbol{C}_i$ for some $i \in \{0, 1, 2, \ldots, 2^k - 1\}$, $\mathbf{H} \cdot \boldsymbol{R}^{\mathrm{T}} = \mathbf{H} \cdot (\boldsymbol{C} + \boldsymbol{E})^{\mathrm{T}} = \mathbf{H} \cdot \boldsymbol{C}^{\mathrm{T}} + \mathbf{H} \cdot \boldsymbol{E}^{\mathrm{T}} = \boldsymbol{0} + \mathbf{H} \cdot \boldsymbol{C}_i^{\mathrm{T}} = \boldsymbol{0}$, or $\mathbf{H} \cdot \boldsymbol{R}^{\mathrm{T}} = \mathbf{H} \cdot (\boldsymbol{C} + \boldsymbol{E})^{\mathrm{T}} = \mathbf{H} \cdot (\boldsymbol{C} + \boldsymbol{C}_i)^{\mathrm{T}} = \mathbf{H} \cdot \boldsymbol{C}_j^{\mathrm{T}} = \boldsymbol{0}$, because the sum of two codewords is a codeword (closure property). Such error patterns are called *undetectable* error patterns, and when an undetectable error pattern occurs, the receiver makes an error in *error detection* and hence a *decoding error* also.

It would be instructive at this point to make a detailed observation of the way the syndrome is computed at the receiver. Let us again write the parity check matrix given in (5.48b) for an (n, k) systematic block code:

$$
\mathbf{H} = \begin{bmatrix} 1 & 0 & 0 & \cdots & 0 & q_{00} & q_{10} & \cdots & q_{k-1,0} \\ 0 & 1 & 0 & \cdots & 0 & q_{01} & q_{11} & \cdots & q_{k-1,1} \\ \vdots & & & & & \vdots & & & \\ 0 & 0 & 0 & \cdots & 1 & q_{0,n-k-1} & q_{1,n-k-1} & \cdots & q_{k-1,n-k-1} \end{bmatrix} \qquad (5.49\mathrm{f})
$$

The received vector is given by

$$\boldsymbol{R} = (\underbrace{r_0, r_1, \ldots, r_{n-k-1}}_{\substack{\text{received parity} \\ \text{check digits}}}, \underbrace{r_{n-k}, r_{n-k+1}, \ldots, r_{n-1}}_{\substack{\text{received} \\ \text{information digits}}}) \tag{5.49g}$$

Now, let us compute the syndrome of $\boldsymbol{R}$:

$$\boldsymbol{S}^{\mathrm{T}} = \mathbf{H} \cdot \boldsymbol{R}^{\mathrm{T}} = \begin{bmatrix} s_0 \\ s_1 \\ s_2 \\ \vdots \\ s_{n-k-1} \end{bmatrix} \tag{5.49h}$$

where the syndrome components are computed as follows:

$$s_0 = r_0 \quad + \overbrace{r_{n-k}\, q_{00} + r_{n-k+1}\, q_{10} + \cdots + r_{n-1}\, q_{k-1,0}}^{\text{receiver-computed parity check digit } c_0}$$

$$s_1 = r_1 \quad + \overbrace{r_{n-k}\, q_{01} + r_{n-k+1}\, q_{11} + \cdots + r_{n-1}\, q_{k-1,1}}^{\text{receiver-computed parity check digit } c_1}$$

$$\vdots \qquad\qquad \vdots \tag{5.49i}$$

$$s_{n-k-1} = r_{n-k-1} + \overbrace{r_{n-k}\, q_{0,n-k-1} + r_{n-k+1}\, q_{1,n-k-1} + \cdots + r_{n-1}\, q_{k-1,n-k-1}}^{\text{receiver-computed parity check digit } c_{n-k-1}}$$

In view of (5.45b), we can put the following interpretation on (5.49i): the syndrome is the vector sum of the received parity check digits and the parity check digits that the receiver computes using the received information digits $r_{n-k}, r_{n-k+1}, \ldots, r_{n-2}, r_{n-1}$ shown in (5.49g). The receiver performs the same computations that the transmitter did to find the parity check digits. When the received parity check digit r_i is identical to the computed parity check digit $r_{n-k}\, q_{0i} + r_{n-k+1}\, q_{1i} + \cdots + r_{n-1}\, q_{k-1,i}$ the ith syndrome vector component $s_i = 0$, and if not, $s_i = 1$ for $i = 0, 1, \ldots, n-k-1$.

Once the parity check matrix is known, the receiver can "design" the error-detection circuit, which is a syndrome calculating circuit. In the example above, we considered the $(7, 4)$ linear block code with a parity check matrix

$$\mathbf{H}' = \begin{bmatrix} 1 & 0 & 0 & 0 & 1 & 1 & 1 \\ 0 & 1 & 0 & 1 & 0 & 1 & 1 \\ 0 & 0 & 1 & 1 & 1 & 0 & 1 \end{bmatrix}$$

Let the received vector be $\boldsymbol{R} = (r_0, r_1, r_2, r_3, r_4, r_5, r_6)$. The syndrome is then given by

$$\boldsymbol{S}^{\mathrm{T}} = \begin{bmatrix} s_0 \\ s_1 \\ s_2 \end{bmatrix} = \begin{bmatrix} 1 & 0 & 0 & 0 & 1 & 1 & 1 \\ 0 & 1 & 0 & 1 & 0 & 1 & 1 \\ 0 & 0 & 1 & 1 & 1 & 0 & 1 \end{bmatrix} \begin{bmatrix} r_0 \\ r_1 \\ r_2 \\ r_3 \\ r_4 \\ r_5 \\ r_6 \end{bmatrix}$$

$$= \begin{bmatrix} r_0 + r_4 + r_5 + r_6 \\ r_1 + r_3 + r_5 + r_6 \\ r_2 + r_3 + r_4 + r_6 \end{bmatrix}$$

An error-detection circuit that can be used by the receiver is shown in Figure 5.14.

The syndrome of the received vector $\boldsymbol{R}$ actually depends on the error-pattern vector $\boldsymbol{E}$ as shown in (5.49d). Let us take a closer look at the relationship between the syndrome vector and the error-pattern vector by substituting the parity check matrix of (5.49f) in (5.49d):

$$\boldsymbol{S}^{\mathrm{T}} = \mathbf{H} \cdot \boldsymbol{E}^{\mathrm{T}}$$

$$= \begin{bmatrix} 1 & 0 & 0 & \cdots & 0 & q_{00} & q_{10} & \cdots & q_{k-1,0} \\ 0 & 1 & 0 & \cdots & 0 & q_{01} & q_{11} & \cdots & q_{k-1,1} \\ \vdots & & & & & \vdots & & & \\ 0 & 0 & 0 & \cdots & 1 & q_{0,n-k-1} & q_{1,n-k-1} & \cdots & q_{k-1,n-k-1} \end{bmatrix} \begin{bmatrix} e_0 \\ e_1 \\ \vdots \\ e_{n-k-1} \\ e_{n-k} \\ e_{n-k+1} \\ \vdots \\ e_{n-1} \end{bmatrix}$$

and we have

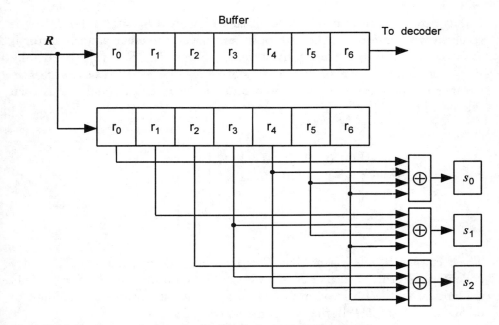

Figure 5.14 Syndrome calculator for $(7, 4)$ code (error-detection circuit).

$$s_0 = e_0 \quad + e_{n-k}\, q_{00} + e_{n-k+1}\, q_{10} + \cdots + e_{n-1}\, q_{k-1,0}$$
$$s_1 = e_1 \quad + e_{n-k}\, q_{01} + e_{n-k+1}\, q_{11} + \cdots + e_{n-1}\, q_{k-1,1}$$
$$\vdots \qquad\qquad \vdots \tag{5.49j}$$
$$s_{n-k-1} = e_{n-k-1} + e_{n-k}\, q_{0,n-k-1} + e_{n-k+1}\, q_{1,n-k-1} + \cdots + e_{n-1}\, q_{k-1,n-k-1}$$

Note that the components of the syndrome vector are linear combinations of the error digits, and thus they possess information on the error pattern. Any error-correction scheme that would be developed is based on the solution of the $n - k$ linear simultaneous equations given in (5.49j) to find the error pattern. Then, the correct decoding would simply be to find the transmitted code vector from $\mathbf{R} + \mathbf{E}$.

Example 5.25 Consider the parity check matrix $\mathbf{H}$ used in Example 5.24:

$$\mathbf{H} = \begin{bmatrix} 0 & 0 & 0 & 1 & 1 & 1 & 1 \\ 0 & 1 & 1 & 0 & 0 & 1 & 1 \\ 1 & 0 & 1 & 0 & 1 & 0 & 1 \end{bmatrix}$$

Let us now assume that the transmitted codeword $C_{15} = (1\ 1\ 1\ 1\ 1\ 1\ 1)$ is received with an error at the fifth position; that is, the error pattern vector is $E = (0\ 0\ 0\ 0\ 1\ 0\ 0)$. The received vector R is then, $R = (1\ 1\ 1\ 1\ 0\ 1\ 1)$. In a real situation, we of course do not know the error-pattern vector in advance of the reception. The syndrome of the received codeword is then calculated to be

$$
S^{\mathrm{T}} =
\begin{bmatrix}
0 & 0 & 0 & 1 & 1 & 1 & 1 \\
0 & 1 & 1 & 0 & 0 & 1 & 1 \\
1 & 0 & 1 & 0 & 1 & 0 & 1
\end{bmatrix}
\begin{bmatrix}
1 \\ 1 \\ 1 \\ 1 \\ 0 \\ 1 \\ 1
\end{bmatrix}
=
\begin{bmatrix}
1 \\ 0 \\ 1
\end{bmatrix}
$$

The fact that $S \neq 0$ means that R is received in error; that is, R is not a codeword. In this particular case, however, the syndrome vector $S = (1\ 0\ 1)$ is identical to the fifth column of $\mathbf{H}$, and that is the position of the digit in error. In view of (5.49f), the syndrome was calculated as

$$
S^{\mathrm{T}} = 0 \cdot \begin{bmatrix} 0 \\ 0 \\ 1 \end{bmatrix} + 0 \cdot \begin{bmatrix} 0 \\ 1 \\ 0 \end{bmatrix} + 0 \cdot \begin{bmatrix} 0 \\ 1 \\ 1 \end{bmatrix} + 0 \cdot \begin{bmatrix} 1 \\ 0 \\ 0 \end{bmatrix}
$$

$$
+ 1 \cdot \begin{bmatrix} 1 \\ 0 \\ 1 \end{bmatrix} + 0 \cdot \begin{bmatrix} 1 \\ 1 \\ 0 \end{bmatrix} + 0 \cdot \begin{bmatrix} 1 \\ 1 \\ 1 \end{bmatrix} = \begin{bmatrix} 1 \\ 0 \\ 1 \end{bmatrix}
$$

In many applications, the use of a block code is strictly for the purpose of error detection only, as in the case of IS-95 and PCS CDMA systems [1], [16], and no attempt is made to correct the errors using the block code.[3] The error-correction capabilities of block codes are discussed in many references [18–21].

The syndrome calculation in (5.49d) indicates that if the error-pattern vector is $E = (1, 0, 0, \ldots, 0)$, the syndrome vector is identical to the first column of the parity check matrix $\mathbf{H}$. If errors are made on the first two digits of the transmitted vector, then the syndrome is the sum of the first two columns of the parity check matrix. When a single error-detecting code is

---

[3] The errors that have occurred in the block (frame) are, of course, corrected by the Viterbi decoder.

used in a communications system, the error detection can easily be made, and, further, by seeing the specific syndrome vector, which is one of the columns of **H**, we can *locate* the error position. This is, in fact, the single-error-correcting *Hamming code*, which we briefly discuss before describing *cyclic codes*.

5.7.3 Hamming Codes

The single-error-correcting binary group codes were discovered in 1950 by Hamming [22], and were the first linear codes to be used for error control in digital data and communications systems. In fact, Hamming codes, which can correct only single errors, gave impetus to the discovery of t $(t > 1)$ error-correcting codes. There are many references that cover the subject of Hamming codes and error-control coding schemes [18–22]. Following definitions, we discuss the essential properties of Hamming codes for the reason that it brings out the basic mechanism of the error-detection scheme common to all linear block codes used for error detection, such as the schemes used in CDMA cellular and PCS systems.

There are many ways in which Hamming codes are defined and described. One way is to characterize the parity check matrices for the single-error-correcting Hamming codes; that is, if the code is capable of correcting all patterns of single errors, the columns of its parity check matrix must be all distinct and there should be no column having all zeros. In other words, if all columns of the parity check matrix **H** are distinct and nonzero, then a single error in any position results in a *syndrome* identical to the column vector in the same *location* of the error digit position in the received vector. This is clearly shown in (5.49f). This suggests that a Hamming code can be defined as a group code with the property that the ith column of the parity check matrix is the binary number i $(i \neq 0)$. This ensures that we can have all distinct nonzero columns of the parity check matrix. Note that here we define a code with a parity check matrix to begin with. One can then proceed to find the corresponding generator matrix to define or list the codewords, if necessary.

Recall that the parity check matrix **H** for an (n, k) linear code is such that the number of parity check digits, $n - k$, is equal to the number of linearly independent rows of the parity check matrix [see (5.46c) and (5.48b)]. We now need to know the *maximum* number of distinct nonzero columns corresponding to the required number of $n - k \triangleq m$ linearly independent rows of $\mathbf{H}$. The number is obviously $2^{n-k} - 1 = 2^m - 1$, a number that happens to be related to a PN generator using an m-stage LFSR. We follow up on this view later when we discuss the generation of cyclic codes. Therefore, we characterize the parity check matrix of a Hamming code (single-error-correcting code, of course) as one that consists of $2^{n-k} - 1 = 2^m - 1$ nonzero binary $(n - k)$-tuples arranged in any order.

With this reasoning, we can now present another definition [18] of Hamming codes: for any positive integer $m \geq 3$, there exists a Hamming code that has the following parameters:

Number of parity check symbols:	$m = n - k$
Codeword length:	$n = 2^m - 1$
Number of information symbols:	$k = (2^m - 1) - m$

Because Hamming codes are single-error-correcting codes, the minimal Hamming distance is 3; that is, $d_{min} = 3$ in view of (5.40). Hamming codes are perfect single-error-correcting binary group codes.

In coding theory, perfect codes are those that can correct all error vectors of t or fewer errors and no more than t errors. When we say that a given code is a t-error-correcting code, the code is usually meant to correct some error patterns of more than t errors, and such a code is not a perfect code. In an (n, k) t-error-correcting linear binary group code, let us consider the error pattern vector $\mathbf{E} = (e_0, e_1, \ldots, e_{n-1})$. Then there are $\binom{n}{0}$, $\binom{n}{1}$, $\binom{n}{2}, \ldots, \binom{n}{t}$ different numbers of n-tuple error patterns that can happen. Thus the number of possible t-tuple errors in an n-digit codeword is equal to the number of ways of choosing digit i out of n, namely $\binom{n}{i}$. The number of possible values is $2^{n-k} = 2^m$. Therefore, for a *perfect* t-error-correcting (n, k) code, we must have [18]

$$\binom{n}{0} + \binom{n}{1} + \cdots + \binom{n}{i} + \cdots + \binom{n}{t} = 2^{n-k} \tag{5.50a}$$

or

$$\left[1 + \binom{n}{1} + \cdots + \binom{n}{t} \right] 2^k = 2^n \tag{5.50b}$$

We can use (5.50b) as the sufficient condition for a *perfect code*. There are very few perfect codes [18], and the Hamming code is one of them.

For the $(7, 4)$ group code we have treated in an example above, the parity check matrix consists of all $2^{n-k} - 1 = 7$ distinct columns and hence it is the parity check matrix of a Hamming code. There are $\binom{7}{1} = 7$ different single-error patterns, and thus we see that $\left[1 + \binom{7}{1}\right]2^4 = 2^7$, which satisfies the condition (5.50b) for a perfect code.

There is another known binary perfect code, known as the $(23, 12)$ *Golay code* [19], which is widely used in communications systems for its multiple-error-correcting capability $(t = 3)$. In practice, the $(23, 12)$ Golay code is often lengthened to a $(24, 12)$ code to cause the code rate to be $1/2$, thereby extending its error-correction capabilities to include some 4-tuple error patterns. In almost all textbooks dealing with coding theory, the Golay code is discussed. Because $t = 3$, the minimum distance is $d_{min} = 7$, and for this code we have

$$\left[1 + \binom{23}{1} + \binom{23}{2} + \binom{23}{3}\right]2^{11} = 2^{23} \tag{5.50c}$$

which makes it a perfect code.

Though a trivial case, all $(n, 1)$ repetition codes of odd block length are perfect codes. Note that $k = 1$ and $t = [d_{min} - 1]/2 = (n - 1)/2$, because n is the minimal distance of a repetition code. From the relation

$$1 + \binom{n}{1} + \cdots + \binom{n}{n} = 2^n$$

for n odd we have half the terms:

$$1 + \binom{n}{1} + \cdots + \binom{n}{\frac{n-1}{2}} = \frac{2^n}{2}$$

Rearranging yields

$$\left[1 + \binom{n}{1} + \cdots + \binom{n}{\frac{n-1}{2}}\right]2^1 = 2^n$$

Substituting $k = 1$ and $t = (n - 1)/2$, we have

$$\left[1 + \binom{n}{1} + \cdots + \binom{n}{t}\right]2^k = 2^n$$

which meets the condition of a perfect code. Incidentally, the parity check matrix $\mathbf{H}$ for an $(n, 1)$ repetition code can be easily found from the generator matrix

$$\mathbf{G} = \underbrace{\begin{bmatrix} 1 & 1 & 1 & \cdots & 1 & 1 \end{bmatrix}}_{n} \tag{5.50d}$$

as

$$H = \begin{bmatrix} 1 & 0 & 0 & \cdots & 0 & 1 \\ 0 & 1 & 0 & \cdots & 0 & 1 \\ \vdots & & & & & \vdots \\ 0 & 0 & 0 & \cdots & 1 & 1 \end{bmatrix} \qquad (5.50e)$$

5.8 Cyclic Codes

We now present discussions on the generation and application of cyclic linear block codes. Cyclic codes, first discussed by Prange in 1957 [23], are widely used for error-detection and error-correction applications in many communications systems. The main reason for their wide use in practical systems is the simplicity of their implementation by means of LFSR circuits. Our treatment of cyclic codes, however, is limited to error-detection applications to the extent of understanding the CRC codes used as frame quality indicators in the IS-95 [1] and PCS CDMA [16] systems. The descriptions we give of the (n, k) linear block codes called *cyclic codes* are natural extensions of our general discussions of linear block codes using the algebra of polynomials. Our earlier discussions of PN sequence generation, in terms of an extended Galois field $\mathrm{GF}(2^m)$ generated by some irreducible binary polynomial of degree m, is also very helpful in our description of cyclic code implementation using LFSR circuits.

Definition. An (n, k) parity check code is called a cyclic code if it has the property that, whenever the n-tuple codeword $C = (c_0, c_1, \ldots, c_{n-1})$ is cyclically shifted (i.e., involving "end wraparound") one place to the right, the resulting n-tuple $C = (c_{n-1}, c_0, \ldots, c_{n-2})$ is also a codeword.

This definition implies that, if an n-tuple codeword C is cyclically shifted i places to the right, the resulting n-tuple $C^{(i)}$ is also a codeword. We may represent the codewords as *codeword polynomials*:

$$C = (c_0, c_1, \ldots, c_{n-1})$$
$$\leftrightarrow C(x) = c_0 + c_1 x + \cdots + c_{n-1} x^{n-1} \qquad (5.51a)$$

and

$$C^{(i)} = (c_{n-i}, c_{n-i+1}, \ldots, c_{n-1}, c_0, \ldots, c_{n-i-1})$$
$$\leftrightarrow C^{(i)}(x) = c_{n-i} + c_{n-i+1} x + \cdots$$
$$+ c_{n-1} x^{i-1} + c_0 x^i + c_1 x^{i+1} + \cdots + c_{n-1} x^{n-1} \qquad (5.51b)$$

To specify or generate such cyclic codes, let us revisit (5.44c); that is, let us express the ith codeword as

$$C_i = m_i \cdot G, \qquad i = 0, 1, \ldots, 2^k - 1 \tag{5.52a}$$

which indicates that a k-tuple message vector $m = (m_0, m_1, \ldots, m_{k-1})$ is encoded into an n-tuple codeword C through a $k \times n$ generator matrix G, whose rows are k linearly independent codewords of an (n, k) linear block code. The idea which we are promoting here is to represent the encoding process of (5.52a) in terms of polynomials. We simply replace (5.52a) with the following equation:

$$C^{(i)}(x) = m_i(x)\, g(x) \tag{5.52b}$$

where $C^{(i)}(x)$ is the *codeword polynomial*, $m_i(x)$ is the *message polynomial*, and $g(x)$ is a *generator polynomial*. Thus, if we specify a generator polynomial $g(x)$, we can list all the 2^k codeword polynomials corresponding to the entire set of n-tuple code vectors. The idea is to show the correspondence between the code vector C_i and the (cyclic) codeword polynomial $C^{(i)}(x)$, which corresponds to $C^{(i)}$.

Let us consider an example of a $(7, 4)$ linear block code. If we pick[4] the generator polynomial $g(x) = x^3 + x + 1$, the resulting codeword polynomials and the corresponding code vectors are those shown in Table 5.15. Note that the message vector m is given as a 4-tuple binary sequence representing 16 binary numbers. The codeword polynomials in Table 5.15 are not given in systematic form. We discuss the method of representing codeword polynomials in systematic form later in this section.

The analogy of (5.52b) to (5.52a) can be made explicit from the following example. Let us represent the polynomials $\{x^i g(x),\ i = 0, 1, 2, 3\}$ and their corresponding code vectors as follows:

$$
\begin{aligned}
g(x) &= 1 + x \quad\ + x^3 & &\leftrightarrow 1101000 \\
xg(x) &= \quad\ x + x^2 \quad\ + x^4 & &\leftrightarrow 0110100 \quad (5.52c) \\
x^2 g(x) &= \quad\quad\ x^2 + x^3 \quad\ + x^5 & &\leftrightarrow 0011010 \\
x^3 g(x) &= \quad\quad\quad\ x^3 + x^4 \quad + x^6 & &\leftrightarrow 0001101
\end{aligned}
$$

---

[4] We of course discuss how to pick a generator polynomial $g(x)$ in an (n, k) code, as the lowest degree polynomial that divides $x^n + 1$.

Table 5.15 Generation of a $(7,4)$ binary block code with $g(x) = x^3 + x + 1$

Message vector, m	$m(x)$	$C(x) = m(x)g(x)$	Code vector, C
0000	0	0	0000000
0001	1	$x^3 + x + 1$	0001011
0010	x	$x^4 + x^2 + x$	0010110
0011	$x + 1$	$x^4 + x^3 + x^2 + 1$	0011101
0100	x^2	$x^5 + x^3 + x^2$	0101100
0101	$x^2 + 1$	$x^5 + x^2 + x + 1$	0100111
0110	$x^2 + x$	$x^5 + x^4 + x^3 + x$	0111010
0111	$x^2 + x + 1$	$x^5 + x^4 + 1$	0110001
1000	x^3	$x^6 + x^4 + x^3$	1011000
1001	$x^3 + 1$	$x^6 + x^4 + x + 1$	1010011
1010	$x^3 + x$	$x^6 + x^3 + x^2 + x$	1001110
1011	$x^3 + x + 1$	$x^6 + x^2 + 1$	1000101
1100	$x^3 + x^2$	$x^6 + x^5 + x^4 + x^2$	1110100
1101	$x^3 + x^2 + 1$	$x^6 + x^5 + x^4 + x^3 + x^2 + x + 1$	1111111
1110	$x^3 + x^2 + x$	$x^6 + x^5 + x$	1100010
1111	$x^3 + x^2 + x + 1$	$x^6 + x^5 + x^3 + 1$	1101001

The four 7-tuple vectors forming a 4×7 matrix can serve as the generator matrix **G** in (5.52a) because these row vectors are linearly independent and, further, they are also codewords as can be checked out from Table 5.15.

It seems to us then that the generator polynomial $g(x) = x^3 + x + 1$ is capable of generating the entire set of $2^k = 2^4 = 16$ codeword polynomials. This is true. In (5.52c) we showed only four such polynomials. In $x^i g(x)$, where i is greater than 3 ($i \geq 4$), the resulting polynomials have degrees greater than 7. If $C(x)$ is a codeword polynomial, the definition of cyclic codes says that $x^i C(x)$ is also a codeword polynomial, because multiplication of a codeword polynomial by x^i is equivalent to a cyclic shift of the code vector. But if we multiply x^j by the code polynomial of degree $n - 1$, the resulting polynomial has the degree $n + j - 1$, which is greater than $n - 1$, the maximum degree of the codeword polynomial. Therefore, we need to reduce the degree $n + j - 1$ to a degree less than or equal to $n - 1$, so that the coefficients of the resulting polynomial are a code vector sequence.

The method of reducing a polynomial of degree greater than $n - 1$ to a polynomial of degree less than or equal to $n - 1$ is to reduce modulo some binary polynomial, say, $d(x)$, of degree n; that is, we divide by $d(x)$ and keep only the remainder. Thus, the situation we have is that we need to find $d(x)$ such that

$$x^i C(x) \bmod d(x) \equiv C^{(i)}(x) \tag{5.53}$$

where $C^{(i)}(x)$ is a cyclic codeword polynomial of degree $\leq n - 1$, and the symbol $\equiv$ is meant to be a "congruence"; that is, $x^i C(x) \bmod d(x)$ "is congruent to" $C^{(i)}(x)$. So, what is $d(x)$? The answer is that $d(x) = x^n + 1$. Explanations of this fact are found in the standard coding theory references cited earlier. Here, we give an explanation in terms of the development in Section 6.2 in connection with PN sequence generation.

From (6.20a) through (6.20f), we observe in Chapter 6 that the inverse (reciprocal), $1/f(x)$, of the characteristic polynomial of a shift register, $f(x)$, gives a sequence generated by the shift register in the form of an infinite series as the repetition of the (finite) "first period sequence" with period P, and the infinite sequence representation is given as a polynomial of degree $> P - 1$. To obtain the first period sequence polynomial with degree $P - 1$, (6.20c) indicates that the output sequence polynomial must be divided by $x^P + 1$ and use only the remainder, whose degree is less than or equal to $P - 1$.

Thus, in the generation of an (n, k) cyclic code, if $C(x) = c_0 + c_1 x + \cdots + c_{n-1} x^{n-1}$ is a codeword polynomial corresponding to the code vector $C = (c_0, c_1, \ldots, c_{n-1})$, then the codeword polynomial $C^{(i)}(x)$ corresponding to codeword vector

$$C^{(i)} = (c_{n-i}, c_{n-i+1}, \ldots, c_{n-1}, c_0, \ldots, c_{n-i-1})$$

is obtained as

$$C^{(i)}(x) = x^i C(x) \bmod (x^n + 1) \tag{5.54a}$$

$$= x^i [c_0 + c_1 x + \cdots + c_{n-1} x^{n-1}] \bmod (x^n + 1)$$

$$= [c_0 x^i + c_1 x^{i+1} + \cdots + c_{n-1} x^{n+i-1}] \bmod (x^n + 1) \tag{5.54b}$$

$$= [a(x)(x^n + 1) + C^{(i)}(x)] \bmod (x^n + 1) \tag{5.54c}$$

$$\equiv c_{n-i} + c_{n-i+1} x +$$
$$\cdots + c_{n-1} x^{i-1} + c_0 x^i + \cdots + c_{n-i-1} x^{n-1} \tag{5.54d}$$

Our discussions on (n, k) cyclic codes began with the analogy of an (n, k) linear block code, in that $C(x) = m(x)g(x)$ is equivalent to $C = m \cdot G$. We also have the parity check equations, $C \cdot H^T = 0$, indicating that there is a polynomial $h(x)$ called a *parity check polynomial* that is equivalent to the parity check matrix **H**.

If a generator polynomial of an (n, k) cyclic code $g(x)$ is picked, it must be one that divides $x^n + 1$; that is:

$$x^n + 1 = g(x)h(x) \tag{5.55}$$

This fact can be verified easily if we use (5.54a) and (5.54c) with (5.52b) substituted. From (5.54a) and (5.54c) we have

$$x^i C(x) = a(x)(x^n + 1) + C^{(i)}(x)$$

where $a(x)$ is the quotient. Substituting $C^{(i)}(x) = m_i(x)g(x)$ from (5.52b), we have

$$x^i g(x) = a(x)(x^n + 1) + m_i(x)g(x)$$

which gives $x^n + 1 = g(x)h(x)$, where

$$h(x) \triangleq \left[\frac{x^i + m_i(x)}{a(x)} \right]$$

Let us summarize the important facts on the generator polynomial and parity check polynomial of an (n, k) cyclic code: the code polynomials of an (n, k) cyclic code consist of all the multiples of a unique code-generating polynomial of degree $n - k$. The parity check polynomial of this code is given by $(x^n + 1)/g(x)$. Note that the degree of the generator polynomial is $n - k$, and the degree of the parity check polynomial is k; that is, $\deg[g(x)] + \deg[h(x)] = \deg[x^n + 1]$.

We discussed the *dual code* previously by saying that the code vectors obtained as linear combinations of the $n - k$ linearly independent row vectors of the parity check matrix **H** are the $(n, n - k)$ dual code. Thus, if $g(x)$ is the generator polynomial of an (n, k) cyclic code, the dual code of this (n, k) cyclic code is also cyclic and it is generated by the "reciprocal polynomial" $h^*(x) = x^k h(x^{-1})$, where $h(x) = (x^n + 1)/g(x)$. This is equivalent to the previously stated fact that a parity check matrix for an (n, k) code is a generator matrix for its dual code.

Example 5.26 Let us consider an example given in Section 6.2.1. There, an extended Galois field GF(2^3) is generated with a primitive polynomial

$x^3 + x + 1$. The distinct minimal polynomials are found to be $M_0(x) = 1 + x$, $M_1(x) = 1 + x^2 + x^3$, and $M_3(x) = 1 + x + x^3$, where

$$M_0(x) \cdot M_1(x) \cdot M_3(x) = (1 + x) \cdot (1 + x^2 + x^3) \cdot (1 + x + x^3)$$
$$= (1 + x + x^2 + x^4)(1 + x + x^3) = 1 + x^7$$

Now, suppose we wish to generate a $(7, 4)$ cyclic code with a generator polynomial $g(x) = M_3(x) = 1 + x + x^3$.

First, we see that $g(x)$ divides $x^7 + 1$, and thus it can be a generator polynomial. The parity check polynomial is then $h(x) = (x^7 + 1)/g(x) = 1 + x + x^2 + x^4$, which is the product of all the minimal polynomials except the one used as the generator polynomial. This example gives all the codeword polynomials shown in Table 5.15. Had we chosen the generator polynomial to be $g(x) = M_1(x) = 1 + x^2 + x^3$ instead, the parity check polynomial would have been $h(x) = (1 + x)(1 + x + x^3) = 1 + x^2 + x^3 + x^4$, indicating that the choice of $g(x)$ in the factorization of $x^n + 1$ is arbitrary as long as it is one of the lowest degree polynomials. But the choice is a matter of choosing a "good code" or a "bad (or not so good) code," from the point of view of given application purposes.

Example 5.27 In the previous example, the generator polynomial of a $(7, 4)$ cyclic code was chosen to be $g(x) = 1 + x + x^3$. The parity check polynomial was found to be $h(x) = (x^7 + 1)/g(x) = 1 + x + x^2 + x^4$. The generator polynomial for this code's $(7, 3)$ dual code is the reciprocal polynomial $h^*(x)$, and is given by

$$h^*(x) = x^4 h(x^{-1}) = 1 + x^2 + x^3 + x^4$$

An important fact we wish to point out here is that the $(n, k) = (7, 4)$ cyclic code generated by $g(x) = 1 + x + x^3$ is also completely specified by the parity check polynomial $h(x) = 1 + x + x^2 + x^4$, in the sense that encoding based on either $g(x) = 1 + x + x^3$ or $h(x) = 1 + x + x^2 + x^4$ produces the same codewords [18]. As we see when we consider shift register encoding circuits, LFSR configurations based on either $g(x)$ or $h^*(x)$ give identical (n, k) codes with length n and $n - k$ parity check digits.

5.8.1   Systematic Cyclic Codes

Throughout the discussions on (n, k) cyclic block codes, we represented a codeword of length n as a polynomial of degree $n - 1$ with coefficients from the codeword sequence. That is, $C(x) = c_0 + c_1 x + c_2 x^2 + \cdots + c_{n-1} x^{n-1}$ corresponding to the codeword sequence $C = (c_0, c_1, c_2, \ldots, c_{n-1})$, where the symbols $\{c_i\}$ are taken from GF(2). The systematic form of the codeword polynomial can, therefore, be given as follows:

$$C(x) = m_0 + m_1 x + \cdots + m_{k-1} x^{k-1} + p_0 x^k + \cdots + p_{n-k-1} x^{n-1} \quad (5.56a)$$

or

$$C(x) = p_0 + p_1 x + \cdots + p_{n-k-1} x^{n-k-1} + m_0 x^{n-k} + \cdots + m_{k-1} x^{n-1} \quad (5.56b)$$

where

$$m = (m_0, m_1, \ldots, m_{k-1})$$

and

$$p = (p_0, p_1, \ldots, p_{n-k-1})$$

are the k-tuple information digits and the $(n - k)$-tuple parity check digits, respectively. The representations in the form of (5.56a) and (5.56b) are equivalent. We use the form given in (5.56b), which means that the codeword polynomial corresponds to the encoded codeword that has all the k information digits intact in the end part of the code vector, as illustrated previously in Figure 5.12.

Since the message to be encoded is $(m_0, m_1, \ldots, m_{k-1})$ and the polynomial representing the message sequence is $m(x) = m_0 + m_1 x + \cdots + m_{k-1} x^{k-1}$, the systematic codeword polynomial in the form given in (5.56b) can be obtained if we multiply $m(x)$ by x^{n-k}. This is illustrated below:

$$m \rightarrow \boxed{m(x)} \rightarrow \boxed{x^{n-k} m(x)}$$
$$\downarrow$$
$$C(x) = \boxed{p_0 + p_1 x^1 + \cdots + p_{n-k-1} x^{n-k-1}} + \boxed{m_0 x^{n-k} + m_1 x^{n-k+1} + \cdots + m_{k-1} x^{n-1}}$$

$$(5.57a)$$

We now need to determine $p_0 + p_1 x + \cdots + p_{n-k-1} x^{n-k-1}$. The determination of this polynomial comes from the division operation of $x^{n-k} m(x)/g(x)$: because the degree $\deg[m(x)]$ is at most $k - 1$, $\deg[x^{n-k} m(x)] \leq n - 1$, and hence if we divide $x^{n-k} m(x)$ by g(x), whose degree is $n - k$, we have

$$x^{n-k}m(x) = a'(x)g(x) + r'(x) \qquad (5.57\text{b})$$

where $a'(x)$ and $r'(x)$ are the quotient and the remainder, respectively. We know that $\deg[r(x)] \leq n - k - 1$ because $\deg[g(x)] = n - k$. Therefore, we have a remainder polynomial of the form

$$r'(x) = r'_0 + r'_1 x + \cdots + r'_{n-k-1} x^{n-k-1} \qquad (5.57\text{c})$$

If we substitute (5.57c) in (5.57b), we have the result in the following form:

$$r'(x) + x^{n-k}m(x) = a'(x)g(x) \qquad (5.57\text{d})$$

or

$$r'_0 + r'_1 x + \cdots + r'_{n-k-1} x^{n-k-1} + m_0 x^{n-k} + \cdots + m_{k-1} x^{n-1}$$
$$= a'(x)g(x) \qquad (5.57\text{e})$$

The equation (5.57e) indicates that the left-hand side of the equal sign is a codeword polynomial of $n - 1$ in a systematic form and the right-hand side is a multiple of the generator polynomial $g(x)$. In other words, (5.57e) is identical to (5.57b), indicating that $r'_i = p_i$, $i = 0, 1, \ldots, n - k - 1$. We now have the procedure for encoding an (n, k) cyclic code in systematic form:

1. Select the generator polynomial of degree $n - k$.

2. From $(m_0, m_1, \ldots, m_{k-1})$, obtain
$$m(x) = m_0 + m_1 x + \cdots + m_{k-1} x^{k-1}.$$

3. Obtain $x^{n-k}m(x)$.

4. Obtain the remainder $r'(x)$ by dividing $x^{n-k}m(x)$ by the generator polynomial $g(x)$.

5. Obtain the codeword polynomial $C(x)$ as

$$C(x) = r'(x) + x^{n-k}m(x)$$
$$= r'_0 + r'_1 x + \cdots + r'_{n-k-1} x^{n-k-1} + m_0 x^{n-k} + \cdots + m_{k-1} x^{n-1}$$

6. The code vector is then obtained as

$$C = (r'_0, r'_1, \ldots, r'_{n-k-1}, m_0, m_1, \ldots, m_{k-1})$$
$$= (p_0, p_1, \ldots, p_{n-k-1}, m_0, m_1, \ldots, m_{k-1})$$

Example 5.28 Consider the $(7,4)$ cyclic code generated by $g(x) = 1 + x + x^3$. List all 16 codewords in systematic form. To solve this problem, we first represent the 4-tuple message sequences $m = (m_0, m_1, m_2, m_3)$ as polynomials of the form $m(x) = m_0 + m_1 x + m_2 x^2 + m_3 x^3$. Let us consider the case of $m_{10} = (0\,1\,0\,1)$, which gives $m_{10}(x) = x + x^3$. Thus

$$\frac{x^{n-k} m_{10}(x)}{g(x)} = \frac{x^3(x + x^3)}{1 + x + x^3} = \frac{x^4 + x^6}{1 + x + x^3} = \frac{(4,\,6)}{(0,\,1,\,3)}$$

where we use the notation $x^j = j$ for convenience. Carrying out the division:

$$
\begin{array}{r}
3,0 \qquad\; \to 1 + x^3 = a'_{10}(x) \\
3,1,0\,\overline{\big)\,6,4} \\
\underline{6,4,3} \\
3 \\
\underline{3,1,0} \\
1,0
\end{array}
$$

That is:

$$\frac{(4,\,6)}{(0,\,1,\,3)} \equiv (0,1) = 1 + x = r'(x)$$

Thus, the code polynomial in systematic form is

$$C_{10}(x) = r'(x) + x^3 m(x) = a'_{10}(x) g(x)$$
$$= 1 + x + x^4 + x^6 = (1 + x^3)(1 + x + x^3)$$

$$\underbrace{}_{\substack{\text{parity} \\ \text{check} \\ \text{digits}}} \quad \underbrace{}_{\substack{\text{original} \\ \text{message}}}$$

and $\qquad C_{10} = (\overbrace{1\ 1\ 0}\ \overbrace{0\ 1\ 0\ 1}).$

Take another example of $m_{14} = (0\ 1\ 1\ 1)$, and thus $m_{14}(x) = x + x^2 + x^3$:

$$\frac{x^3 m_{14}(x)}{g(x)} = \frac{x^4 + x^5 + x^6}{1 + x + x^3} = \frac{(4,5,6)}{(0,1,3)}$$

$$\begin{array}{r} 3,2 \\ \hline 3,1,0 \overline{\smash{\big)}\,6,5,4} \\ 6,4,3 \\ \hline 5,3 \\ 5,3,2 \\ \hline 2 \end{array} \quad \to x^2 + x^3 = a_{14}(x)$$

$$\to x^2 = r'(x)$$

Thus

$$C_{14}(x) = r'(x) + x^3 m_{14}(x) = a'_{14}(x)$$
$$= x^2 + x^4 + x^5 + x^6 = (x^2 + x^3)(1 + x + x^3)$$

and

$$\boldsymbol{C_{14}} = (0\ 0\ 1\ 0\ 1\ 1\ 1)$$

With the method used in these two example calculations, the entire set of 16 codewords can be obtained in systematic form, as shown in Table 5.16.

5.8.2 Encoders for Cyclic Codes

We have just seen how an (n, k) cyclic code in systematic form is encoded. The encoder performs the calculation to find (or determine) the parity check digits. The calculation was the division of $x^{n-k}m(x)$ by the generator polynomial $g(x)$ of degree $n - k$. Thus, we need a shift register circuit that divides $x^{n-k}m(x)$ by a polynomial of the form

$$g(x) = 1 + g_1 x + g_2 x^2 + \cdots + g_{n-k-1}x^{n-k-1} + x^{n-k}$$

In Chapter 6, it is shown that an LFSR of characteristic polynomial $g(x)$, mechanized as an MSRG or Galois configuration, is a polynomial division circuit. Therefore, the encoder for an (n, k) cyclic code with generator polynomial $g(x)$ of degree $n - k$ is basically an $(n - k)$-stage MSRG configuration, as shown in Figure 5.15.

The encoding is accomplished as follows:

1. Initially, all LFSR stages are set to logical zero, and the switches then set in their UP positions as shown in Figure 5.15.

Table 5.16 **Systematic** $(7,4)$ **cyclic code generated by** $g(x) = 1 + x + x^3$

Message	Codewords	Code polynomials
0 0 0 0	0 0 0 0 0 0 0	$C(x) = 0 = 0 \cdot g(x)$
1 0 0 0	1 1 0 1 0 0 0	$1 + x + x^3 = 1 \cdot g(x)$
0 1 0 0	0 1 1 0 1 0 0	$x + x^2 + x^4 = x \cdot g(x)$
1 1 0 0	1 0 1 1 1 0 0	$1 + x^2 + x^3 + x^4 = (1 + x) \cdot g(x)$
0 0 1 0	1 1 1 0 0 1 0	$1 + x + x^2 + x^5 = (1 + x^2) \cdot g(x)$
1 0 1 0	0 0 1 1 0 1 0	$x^2 + x^3 + x^5 = x^2 \cdot g(x)$
0 1 1 0	1 0 0 0 1 1 0	$1 + x^4 + x^5 = (1 + x + x^2) \cdot g(x)$
1 1 1 0	0 1 0 1 1 1 0	$x + x^3 + x^4 + x^5 = (x + x^2) \cdot g(x)$
0 0 0 1	1 0 1 0 0 0 1	$1 + x^2 + x^6 = (1 + x + x^3) \cdot g(x)$
1 0 0 1	0 1 1 1 0 0 1	$x + x^2 + x^3 + x^6 = (x + x^3) \cdot g(x)$
0 1 0 1	1 1 0 0 1 0 1	$1 + x + x^4 + x^6 = (1 + x^3) \cdot g(x)$
1 1 0 1	0 0 0 1 1 0 1	$x^3 + x^4 + x^6 = x^3 \cdot g(x)$
0 0 1 1	0 1 0 0 0 1 1	$x + x^5 + x^6 = (x + x^2 + x^3) \cdot g(x)$
1 0 1 1	1 0 0 1 0 1 1	$1 + x^3 + x^5 + x^6 = (1 + x + x^2 + x^3) \cdot g(x)$
0 1 1 1	0 0 1 0 1 1 1	$x^2 + x^4 + x^5 + x^6 = (x^2 + x^3) \cdot g(x)$
1 1 1 1	1 1 1 1 1 1 1	$1 + x + x^2 + x^3 + x^4 + x^5 + x^6$
		$= (1 + x^2 + x^3) \cdot g(x)$

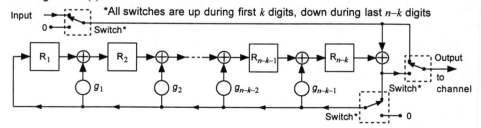

Figure 5.15 **Encoder for an** $(n,\ k)$ **cyclic code with generator polynomial**
$g(x) = 1 + g_1 x + g_2 x^2 + \cdots + g_{n-k-1} x^{n-k-1} + x^{n-k}.$

2. Then, the register is clocked k times to shift the k-bit information sequence $(m_0, m_1, \ldots, m_{k-1})$ into the LFSR, with m_{k-1} being the first to be shifted in, at the same time it is being sent to the channel [this corresponds to $x^{n-k}m(x)$].

3. The switches are set in their DOWN positions, and the register is clocked an additional $n - k$ times to read out the $n - k$ parity check digits to send to the channel.

4. The codeword $\boldsymbol{C} = (p_0, p_1, p_2, \ldots, p_{n-k-1}, m_0, m_1, \ldots, m_{k-1})$ was sent to the channel.

Example 5.29 Consider the $(7, 4)$ cyclic code generated by the generator polynomial $g(x) = 1 + x + x^3$. (1) Draw the encoder circuit for this generator polynomial; (2) explain the encoder operation using the matrix formulation for $\boldsymbol{m} = (0\,1\,1\,1)$, showing the contents of the shift register state; and (3) verify the encoder calculation of the parity check digits to be corrected by a manual calculation of the polynomial division.

(1) The encoder circuit is shown in Figure 5.16.

(2) To explain the operation of the encoder for the case of $\boldsymbol{m}_{14} = (0\,1\,1\,1)$, we use a modification of the matrix formulation for the MSRG configuration used in (6.17e) and (6.17f). That is, if $x_i(t)$ is the number stored in the register stage R_i of the encoder after clock pulse t and $x_i(t + 1)$ is the number stored in R_i after clock pulse $t + 1$, we have

$$\boldsymbol{X}(t + 1) = \boldsymbol{T}_M \boldsymbol{X}(t)$$

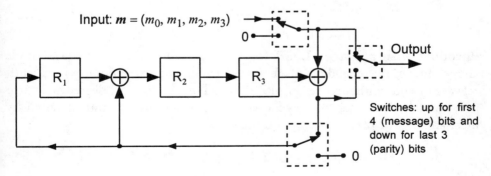

Figure 5.16 Encoder of $(7, 4)$ cyclic code with generator polynomial $g(x) = 1 + x + x^3$.

where

$$\boldsymbol{X}(t+1) = \begin{bmatrix} x_1(t+1) \\ x_2(t+1) \\ x_3(t+1) \end{bmatrix}, \quad \boldsymbol{X}(t) = \begin{bmatrix} x_1(t) \\ x_2(t) \\ x_3(t) \\ m(t) \end{bmatrix}, \quad \boldsymbol{T}_M = \begin{bmatrix} 0 & 0 & 1 & 1 \\ 1 & 0 & 1 & 1 \\ 0 & 1 & 0 & 0 \end{bmatrix}$$

Let $m(t)$ denote the message bit that enters the decoder at clock time t. With the message sequence of $\boldsymbol{m} = (0\ \underset{\leftarrow}{1}\ 1\ 1)$ shifting into the encoder in the order of 1, 1, 1, 0 (as the arrow indicates), we can observe the shift register contents (state vectors) as shown below:

t	$x_1(t), x_2(t), x_3(t)$			$m(t)$	$\leftarrow$	$\boldsymbol{X}^{\mathrm{T}}(t)$

t	$x_1(t)$	$x_2(t)$	$x_3(t)$	$m(t)$		
0	0	0	0	1		The message sequence is shifted into
1	1	1	0	1		the encoder in the order indicated.
2	1	0	1	1	$\downarrow$	
3	0	1	0	0		
4	0	0	1	$\rightarrow$	$\rightarrow$	0 0 1 | 0 1 1 1

$$\underset{\leftarrow}{} \qquad\qquad \text{Parity bits} \qquad \text{Message bits}$$

(3) The manual calculation for this case was done in a previous example; that is:

$$\frac{x^3 m_{14}(x)}{g(x)} = \frac{(4,5,6)}{(0,1,3)} \equiv 2 = x^2 \rightarrow r'(x) = 0 + 0 \cdot x + 1 \cdot x^2$$

which verifies the encoder operation to produce

$$\boldsymbol{C}_{14} = (p_0,\ p_1,\ p_2,\ m_0,\ m_1,\ m_2,\ m_3) = (0\ 0\ 1\ 0\ 1\ 1\ 1)$$

Encoding with a k-stage LFSR for (n, k) cyclic codes. As the example above shows, the encoding of an (n, k) cyclic code was done with a shift register circuit configured as an MSRG based on the $(n - k)$-degree generator polynomial. Encoding of a cyclic (n, k) code can also be implemented by mechanizing the parity check polynomial

$$h(x) = h_0 + h_1 x + h_2 x^2 + \cdots + h_{k-1} x^{k-1} + h_k x^k$$

This time, however, the mechanization of the parity check polynomial is done [18] in the "simple" shift register generator (SSRG) or Fibonacci configuration of the reciprocal polynomial

$$h^*(x) = x^k h(x^{-1}) = h_k + h_{k-1}x + \cdots + h_1 x^{k-1} + h_0 x^k \qquad (5.58)$$

as shown in Figure 5.17. This encoder circuit requires a k-stage LFSR, whereas the encoder based on the MSRG configuration requires $n - k$ stages, and hence the choice must be based on the economy of using fewer storage registers in the encoder circuit. The encoding is accomplished as follows:

1. Initially, all LFSR stages are set to logical zero, and the switch is set in its UP position.

2. The register is clocked k times to shift the k-bit information sequence $(m_0, m_1, \ldots, m_{k-1})$ into the LFSR, with m_{k-1} being the first to be shifted in, and simultaneously out, to the channel.

3. The switch is set to its DOWN position, and the register is clocked an additional $n - k$ times to read out the parity check digits to send to the channel.

Example 5.30 Assume that the generator polynomial for a $(15, 7)$ cyclic code is given by

$$g(x) = x^8 + x^7 + x^6 + x^4 + 1$$

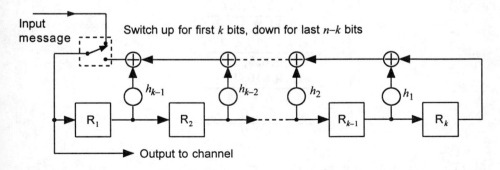

Figure 5.17 Encoder for an (n, k) cyclic code with parity check polynomial $h(x) = h_0 + h_1 x + \cdots + h_k x^k$.

This is the generator polynomial for a double-error-correcting cyclic code to be considered further in a later example.

(1) Determine the parity check digits corresponding to the message sequence $m = (1\ 0\ 1\ 0\ 1\ 0\ 1)$ by manual calculation.

(2) Draw the encoder circuit of the MSRG configuration of Figure 5.15, which is based on $g(x)$, and show that the codeword so generated is identical to that obtained in (1).

(3) Draw the encoder circuit of the SSRG configuration of Figure 5.17, which is based on $h(x)$, and show that the result is equivalent to the result obtained in (2).

Solution to (1):   Given $(n, k) = (15, 7)$, $m = (1\ 0\ 1\ 0\ 1\ 0\ 1)$, and $g(x) = x^8 + x^7 + x^6 + x^4 + 1$. The message polynomial is then $m(x) = x^6 + x^4 + x^2 + 1$, and $n - k = 8$:

$$\frac{x^{n-k}m(x)}{g(x)} = \frac{x^8(x^6 + x^4 + x^2 + 1)}{x^8 + x^7 + x^6 + x^4 + 1} = \frac{(14, 12, 10, 8)}{(8, 7, 6, 4, 0)}$$

$$\equiv (7, 6, 5, 2, 0) = r'(x)$$

$$
\begin{array}{r}
6, 5, 4, 2, 0 \\
\hline
8, 7, 6, 4, 0 \, \big|\, \overline{14, 12, 10, 8} \\
14, 13, 12, 10, 6 \\
\hline
13, 8, 6 \\
13, 12, 11, 9, 5 \\
\hline
12, 11, 9, 8, 6, 5 \\
12, 11, 10, 8, 4 \\
\hline
10, 9, 6, 5, 4 \\
10, 9, 8, 6, 2 \\
\hline
8, 5, 4, 2 \\
8, 7, 6, 4, 0 \\
\hline
7, 6, 5, 2, 0
\end{array}
$$

$$C(x) = r'(x) + x^8 m(x) = 1 + x^2 + x^5 + x^6 + x^7 + x^8 + x^{10} + x^{12} + x^{14}$$

$$C = (\underbrace{1\ 0\ 1\ 0\ 0\ 1\ 1\ 1}_{p}\ \underbrace{1\ 0\ 1\ 0\ 1\ 0\ 1}_{m})$$

$$\rightarrow r'(x) = 1 + x^2 + x^5 + x^6 + x^7$$

(2) The MSRG configuration of the encoder is shown in Figure 5.18. Now, the encoder operation can be explained as follows:

$$
\begin{bmatrix}
x_1(t+1) \\
x_2(t+1) \\
x_3(t+1) \\
x_4(t+1) \\
x_5(t+1) \\
x_6(t+1) \\
x_7(t+1) \\
x_8(t+1)
\end{bmatrix}
=
\begin{bmatrix}
0 & 0 & 0 & 0 & 0 & 0 & 0 & 1 & 1 \\
1 & 0 & 0 & 0 & 0 & 0 & 0 & 0 & 0 \\
0 & 1 & 0 & 0 & 0 & 0 & 0 & 0 & 0 \\
0 & 0 & 1 & 0 & 0 & 0 & 0 & 0 & 0 \\
0 & 0 & 0 & 1 & 0 & 0 & 0 & 1 & 1 \\
0 & 0 & 0 & 0 & 1 & 0 & 0 & 0 & 0 \\
0 & 0 & 0 & 0 & 0 & 1 & 0 & 1 & 1 \\
0 & 0 & 0 & 0 & 0 & 0 & 1 & 1 & 1
\end{bmatrix}
\begin{bmatrix}
x_1(t) \\
x_2(t) \\
x_3(t) \\
x_4(t) \\
x_5(t) \\
x_6(t) \\
x_7(t) \\
x_8(t) \\
m(t)
\end{bmatrix}
$$

t	x_1	x_2	x_3	x_4	x_5	x_6	x_7	x_8	m
0	0	0	0	0	0	0	0	0	1
1	1	0	0	0	1	0	1	1	0
2	1	1	0	0	1	1	1	0	1
3	1	1	1	0	1	1	0	0	0
4	0	1	1	1	0	1	1	0	1
5	1	0	1	1	0	0	0	0	0
6	0	1	0	1	1	0	0	0	1
7	1	0	1	0	0	1	1	1	↓

parity check digits   message digits

coded sequence = | 1 0 1 0 0 1 1 1 | 1 0 1 0 1 0 1 |

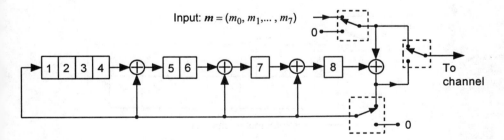

Figure 5.18 Encoder of $(15, 7)$ cyclic code with generator polynomial $g(x) = 1 + x^4 + x^6 + x^7 + x^8$.

(3) The encoder circuit for the $(15, 7)$ cyclic code in the SSRG is derived as follows:

$$\text{Given: } (n,\ k) = (15,\ 7),\ \boldsymbol{m} = (1\ 0\ 1\ 0\ 1\ 0\ 1),\ \text{and}$$

$$g(x) = x^8 + x^7 + x^6 + x^4 + 1$$

The parity check polynomial is obtained as

$$h(x) = \frac{x^{15}+1}{g(x)} = \frac{(15,0)}{(8,7,6,4,0)} = (7,6,4,0) = x^7 + x^6 + x^4 + 1$$

The reciprocal polynomial $h^*(x)$ is needed to "design" (draw) the SSRG configuration, and we have

$$h^*(x) = x^k h(x^{-1}) = x^8 h(x^{-1}) = x^7 + x^3 + x + 1$$

which is mechanized as shown in Figure 5.19. The operation of the circuit of Figure 5.19 is described as follows:

$$
\begin{bmatrix} x_1(t+1) \\ x_2(t+1) \\ x_3(t+1) \\ x_4(t+1) \\ x_5(t+1) \\ x_6(t+1) \\ x_7(t+1) \end{bmatrix}
=
\begin{bmatrix}
1 & 0 & 1 & 0 & 0 & 0 & 1 \\
1 & 0 & 0 & 0 & 0 & 0 & 0 \\
0 & 1 & 0 & 0 & 0 & 0 & 0 \\
0 & 0 & 1 & 0 & 0 & 0 & 0 \\
0 & 0 & 0 & 1 & 0 & 0 & 0 \\
0 & 0 & 0 & 0 & 1 & 0 & 0 \\
0 & 0 & 0 & 0 & 0 & 1 & 0
\end{bmatrix}
\begin{bmatrix} x_1(t) \\ x_2(t) \\ x_3(t) \\ x_4(t) \\ x_5(t) \\ x_6(t) \\ x_7(t) \end{bmatrix}
$$

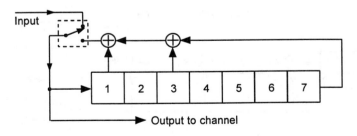

Figure 5.19 Encoder of $(15,\ 7)$ cyclic code with parity check polynomial $h(x) = x^7 + x^6 + x^4 + 1$.

t	x_1	x_2	x_3	x_4	x_5	x_6	x_7	
0	1	0	1	0	1	0	1	$\leftarrow m$
1	1	1	0	1	0	1	0	
2	1	1	1	0	1	0	1	
3	1	1	1	1	0	1	0	
4	0	1	1	1	1	0	1	
5	0	0	1	1	1	1	0	
6	1	0	0	1	1	1	1	
7	0	1	0	0	1	1	1	
8	1	0	1	0	0	1	1	

↑ parity check digits

Thus:

$$C = \underbrace{1\,0\,1\,0\,0\,1\,1\,1}_{p}\,\underbrace{1\,0\,1\,0\,1\,0\,1}_{m}$$

We observe that the codewords obtained by the encoder circuits of MSRG and SSRG configurations are identical.

5.8.3 Syndrome Calculation by Shift Register Circuits for Error Detection

We generate an (n, k) cyclic code by means of an LFSR circuit. Our purpose now is to design an automatic error detection scheme using an LFSR circuit that calculates the syndrome. Suppose that the received code vector of length n in the (n, k) cyclic code is given by

$$R = (r_0, r_1, r_2, \ldots, r_{n-1}) \tag{5.59a}$$

and its polynomial representation is given by

$$R(x) = r_0 + r_1 x + r_2 x^2 + \cdots + r_{n-1} x^{n-1} \tag{5.59b}$$

We seek a technique to process R to calculate the syndrome. In principle, if we multiply R by the parity check matrix, the syndrome results. Our aim is to accomplish this computation. In search of a method for calculating the syndrome by means of a shift register circuit, we first review the materials we covered earlier.

Suppose that the (n, k) cyclic code under discussion is a single-error correcting Hamming code. Then, the n columns of the parity check matrix $\mathbf{H}$ contain different (distinct) nonzero binary m-tuples, where $m \triangleq n - k$. For example, the parity check matrix for a single-error-correcting $(7, 4)$ Hamming code may be given as

$$\mathbf{H}' = \begin{matrix} 1 & 2 & 3 & 4 & 5 & 6 & 7 \\ \begin{bmatrix} 0 & 0 & 0 & 1 & 1 & 1 & 1 \\ 0 & 1 & 1 & 0 & 0 & 1 & 1 \\ 1 & 0 & 1 & 0 & 1 & 0 & 1 \end{bmatrix} \end{matrix} \tag{5.60a}$$

where the columns are the 3-tuple binary numbers 1 to 7. The generator polynomial $g(x)$ of this code has degree 3. Let us assume that the generator polynomial for this code was chosen to be

$$g(x) = 1 + x^2 + x^3$$

which is a primitive polynomial that generates a Galois field $\mathrm{GF}(2^3) = \mathrm{GF}(8)$, as we discuss in Section 6.2. Let us now specify the parity check matrix $\mathbf{H}$ whose columns are the nonzero elements of $\mathrm{GF}(2^3)$, arranged in such a way as to show successive powers of a primitive element $\alpha \in \mathrm{GF}(2^3)$:

$$\mathbf{H} = [\alpha^0, \alpha^1, \alpha^2, \alpha^3, \alpha^4, \alpha^5, \alpha^6] \tag{5.60b}$$

Note that the columns of $\mathbf{H}'$ in (5.60a) are also the nonzero elements of $\mathrm{GF}(2^3)$ in binary vector form, and hence the parity check matrices given in (5.60a) and (5.60b) both define a single-error-detecting (and also -correcting) Hamming code, although they do not exactly define the same codewords. To make them define identical codewords, we need to rearrange the columns of (5.60a) in accordance with $\alpha^3 = 1 + \alpha^2$ as follows:

$$\mathbf{H} = [\alpha^0, \alpha^1, \alpha^2, \alpha^3, \alpha^4, \alpha^5, \alpha^6]$$

$$= \begin{matrix} 1 & 2 & 4 & 5 & 7 & 3 & 6 \\ \begin{bmatrix} 1 & 0 & 0 & 1 & 1 & 1 & 0 \\ 0 & 1 & 0 & 0 & 1 & 1 & 1 \\ 0 & 0 & 1 & 1 & 1 & 0 & 1 \end{bmatrix} \end{matrix} \tag{5.60c}$$

In any way the columns are arranged, the code detects (and corrects) single errors. Now, the syndrome S can be calculated as

$$S^T = H \cdot R^T = [\alpha^0, \alpha^1, \alpha^2, \alpha^3, \alpha^4, \alpha^5, \alpha^6] \begin{bmatrix} r_0 \\ r_1 \\ r_2 \\ r_3 \\ r_4 \\ r_5 \\ r_6 \end{bmatrix}$$

$$= r_0\alpha^0 + r_1\alpha^1 + r_2\alpha^2 + r_3\alpha^3 + r_4\alpha^4 + r_5\alpha^5 + r_6\alpha^6 \tag{5.60d}$$

$$= r_0\begin{bmatrix}1\\0\\0\end{bmatrix} + r_1\begin{bmatrix}0\\1\\0\end{bmatrix} + r_2\begin{bmatrix}0\\0\\1\end{bmatrix} + r_3\begin{bmatrix}1\\0\\1\end{bmatrix} + r_4\begin{bmatrix}1\\1\\1\end{bmatrix} + r_5\begin{bmatrix}1\\1\\0\end{bmatrix} + r_6\begin{bmatrix}0\\1\\1\end{bmatrix}$$

$$= e_0\begin{bmatrix}1\\0\\0\end{bmatrix} + e_1\begin{bmatrix}0\\1\\0\end{bmatrix} + e_2\begin{bmatrix}0\\0\\1\end{bmatrix} + e_3\begin{bmatrix}1\\0\\1\end{bmatrix} + e_4\begin{bmatrix}1\\1\\1\end{bmatrix} + e_5\begin{bmatrix}1\\1\\0\end{bmatrix} + e_6\begin{bmatrix}0\\1\\1\end{bmatrix}$$

$$\tag{5.60e}$$

where we used the fact that $R = C + E$ and that $H \cdot C^T = 0$. In this particular case, the elements in the $GF(2^3)$ represent 3-tuple binary vectors as shown in (5.60e). Using the polynomial representation of the received code vector, the syndrome calculation can also be facilitated as follows:

$$S^T = H \cdot R^T = \sum_{i=0}^{n-1} r_i \alpha^i = R(\alpha) \tag{5.60f}$$

where $R(x)$ is the polynomial given in (5.59b). This indicates that the syndrome can be calculated as the received codeword polynomial evaluated at $x = \alpha$, which is the primitive root of $g(x)$. If we divide the received polynomial $R(x)$ by the generator polynomial $g(x)$, we have

$$R(x) = b'(x)g(x) + R'(x) \tag{5.60g}$$

where $b'(x)$ is the quotient and $R'(x)$ is the remainder. We are reminded, however, that the received codeword polynomial $R(x)$ can also be viewed as

$$R(x) = C(x) + E(x) \tag{5.60h}$$

The analogy of (5.60h) to (5.60g) is to be noted. By definition, every code-word polynomial is a multiple of the generator polynomial, and thus we have the following insights:

From (5.60g), $R(\alpha) = b'(\alpha)g(\alpha) + R'(\alpha)$

$$= R'(\alpha) \tag{5.61a}$$

From (5.60h), $R(\alpha) = C(\alpha) + E(\alpha)$

$$= E(\alpha) \tag{5.61b}$$

The syndrome can, therefore, be obtained as

$$\boldsymbol{S}^{\mathrm{T}} = R'(\alpha) = E(\alpha) \tag{5.61c}$$

An interpretation of (5.61c) is in order. It says that the syndrome vector can be evaluated from the remainder $R'(x)$ with α substituted for x, and also it can be evaluated from the error pattern polynomial $E(x)$ with $x = \alpha$. However, we do not know $E(x)$ in advance, whereas we do know (or can find) $R'(x)$. In fact, finding the error pattern from the knowledge of the syndrome is the key to the error-correction process, which we do not discuss in this book.

   Thus we now have found a way to calculate the syndrome; that is, to divide the received vector polynomial by the generator polynomial and obtain the remainder. The syndrome is the remainder evaluated at $x = \alpha$. This syndrome calculation can be accomplished by a shift register circuit that divides the received vector polynomial $R(x)$ by the same polynomial that generated the (n, k) cyclic code on the transmitter side. The syndrome calculator is shown in Figure 5.20.

   The syndrome calculator is the error detection circuit. Its operation is as follows:

1. Initially, the switch is in the UP position.

2. The received vector $\boldsymbol{R} = (r_0, r_1, \ldots, r_{n-1})$ is shifted into the register with all the stages initially set to zero.

3. As soon as the entire n digits of $\boldsymbol{R}$ have been shifted into the register, the contents in the $(n - k)$-stage register form the syndrome $\boldsymbol{S} = (s_1, s_2, \ldots, s_{n-k})$.

$\boldsymbol{R} = (r_0, r_1, ..., r_{n-1})$

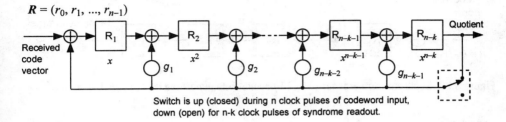

Switch is up (closed) during n clock pulses of codeword input,
down (open) for n-k clock pulses of syndrome readout.

Figure 5.20 Syndrome calculator circuit of an $(n,\ k)$ **cyclic code based on the generator polynomial** $g(x) = 1 + g_1 x + \cdots + g_{n-k} x^{n-k}$.

4. The syndrome digits are read out during $n - k$ clock pulses with the switch in the DOWN position.

Example 5.31 Let us consider the $(15, 11)$ binary cyclic code of length 15, generated by $g(x) = x^4 + x + 1$. We wish (a) to design an encoder and the syndrome calculator (error detection circuit), and (b) to calculate the syndrome for the received vector $\boldsymbol{R} = (1\ 0\ 1\ 1\ 1\ 0\ 1\ 0\ 0\ 0\ 1\ 1\ 0\ 1\ 0)$ using both manual and matrix methods.

(a) For the encoder design, this is a clear case of preferring an $n - k = 4$-stage LFSR circuit. An MSRG configuration based on $g(x) = x^4 + x + 1$, is shown in Figure 5.21. The syndrome calculator is the identical MSRG configuration based on the generator polynomial $g(x) = 1 + x + x^4$, except that it is used as a division circuit as shown in Figure 5.22.

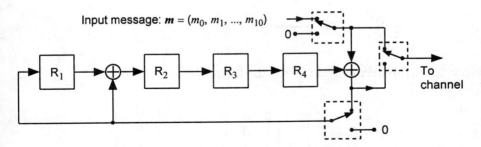

Figure 5.21 Encoder for $(15, 11)$ **cyclic code with generator polynomial** $g(x) = 1 + x + x^4$.

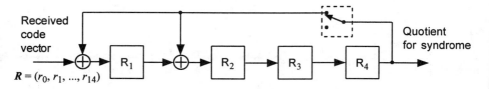

Figure 5.22 Syndrome detector for $(15, 11)$ **cyclic code generated by** $g(x) = 1 + x + x^4$.

(b) Given: $g(x) = x^4 + x + 1$, $\boldsymbol{R} = (1\ 0\ 1\ 1\ 1\ 0\ 1\ 0\ 0\ 0\ 1\ 1\ 0\ 1\ 0)$. By the manual method, we first express the polynomial for the received code vector as

$$R(x) = x^{14} + x^{12} + x^{11} + x^{10} + x^8 + x^4 + x^3 + x$$

Because the syndrome polynomial is given by $S(x) = R(x) \bmod g(x)$, we need to calculate $(14, 12, 11, 10, 8, 4, 3, 1)/(4, 1, 0)$:

$$
\begin{array}{r}
10, 8, 5, 2, 1,\ 0 \\
4, 1, 0\ \overline{\big)\ 14, 12, 11, 10, 8, 4, 3, 1} \\
14, 11, 10 \\
\hline
12, 8, 4, 3, 1 \\
12, 9, 8 \\
\hline
9, 4, 3, 1 \\
9, 6, 5 \\
\hline
6, 5, 4, 3, 1 \\
6, 3, 2 \\
\hline
5, 4, 2, 1 \\
5, 2, 1 \\
\hline
4 \\
4, 1, 0 \\
\hline
1, 0
\end{array}
$$

$\rightarrow R'(\alpha) = 1 + \alpha \rightarrow S = (1\,1\,0\,0)$

Now, by the matrix method, using the formulation

$$
\begin{bmatrix} s_1(t+1) \\ s_2(t+1) \\ s_3(t+1) \\ s_4(t+1) \end{bmatrix} =
\begin{bmatrix} 0 & 0 & 0 & 1 & 1 \\ 1 & 0 & 0 & 1 & 0 \\ 0 & 1 & 0 & 0 & 0 \\ 0 & 0 & 1 & 0 & 0 \end{bmatrix}
\begin{bmatrix} s_1(t) \\ s_2(t) \\ s_3(t) \\ s_4(t) \\ R(t) \end{bmatrix}
$$

The time history of the state of the shift register is shown in Table 5.17.

5.9 Binary BCH Codes

In the foregoing discussions of cyclic codes, we have not focused on the error-correcting capabilities of the codes, except when we described Hamming codes as being single-error-correcting codes. The reason for not having done so was intentional in that our main objective is to understand first the error detection, rather than correction, capabilities of the codes. Our remarks in Section 5.6 began with descriptions of the frame quality indicators used in the IS-95 and PCS CDMA systems. They are the schemes used in detecting possible errors in the data frames at both 4.8 kbps and 9.6 kbps, and our main purpose is to understand how they were designed and how error detections are accomplished in the CDMA systems.

We have come a long way since the beginning of the discussion of "What is a code?" and "How an (n, k) linear code is encoded." It was not our purpose to undertake a formal "coding theory" presentation all the while, but

Table 5.17 Shift register time history for example CRC encoder

t	s_1	s_2	s_3	s_4	R
0	0	0	0	0	1
1	1	0	0	0	0
2	0	1	0	0	1
3	1	0	1	0	1
4	1	1	0	1	1
5	0	0	1	0	0
6	0	0	0	1	1
7	0	1	0	0	0
8	0	0	1	0	0
9	0	0	0	1	0
10	1	1	0	0	1
11	1	1	1	0	1
12	1	1	1	1	0
13	1	0	1	1	1
14	0	0	0	1	0
15	1	1	0	0	$\rightarrow S$

We see that the syndrome is identically obtained to be $(1\,1\,0\,0)$ by both methods.

it was to help the reader to grasp the main ideas involved and be able to carry out the example problems pertaining to the subject under discussion.

We are now at a point where we can put the description of the cyclic codes developed thus far in a succinct general setting by introducing binary BCH codes, which are capable of correcting all patterns of t or fewer channel errors. We discuss t-error correcting BCH codes to the point of how they are generated and how the errors are detected.

The *Bose-Chaudhuri-Hocquenghem codes*, known as *BCH codes*, are a class of cyclic codes that are capable of multiple-error detection and correction [24, 25]. In this section, we restrict our discussion to binary BCH codes. Non-binary BCH codes are discussed in almost all books on coding theory [18, 20, 26]. The BCH codes can be looked at as a generalization of the Hamming codes for multiple-error detection and correction.

For any positive integers m ($m \geq 3$) and t ($t < 2^m - 1$), there exists a binary BCH code with block length $n = 2^m - 1$ and minimum distance $d_{min} \geq 2t + 1$ having no more than mt parity check digits. Thus, clearly the code is capable of detecting and correcting any pattern of t or fewer errors, and we call this code a t-error correcting BCH code. The t-error correcting binary BCH code of block length n is the cyclic code with a generator polynomial that is the product of the distinct minimal polynomials of the roots α, α^2, α^3, α^4, α^5, ..., α^{2t-1}, α^{2t}, which are from the elements of an extended Galois field $\mathrm{GF}(2^m)$ formed by a primitive irreducible binary polynomial of degree m.[5]

Another point that needs to be stressed in the definition of the t-error correcting BCH code of length $n = 2^m - 1$ is that the generator polynomial $g_t(x)$ is the *lowest degree polynomial* over $\mathrm{GF}(2)$, and that the degree of $g_t(x)$ is $n - k$, as is the case for any generator polynomial of the cyclic codes. We need also to note that the BCH codes are defined with the specification of the length of the code n, and the error correction parameter t, and that the parameter k, the information bit block length, is not known until after the generator polynomial $g_t(x)$ is specified according to the definition given above.

We now present examples of the construction of generator polynomial that defines a binary BCH code.

Example 5.32 As our first example, let us construct all binary BCH codes of block length 15. In Table 6.8 of Section 6.2.1, the conjugate classes of roots in $\mathrm{GF}(2^4) = \mathrm{GF}(16)$ generated by the primitive polynomial $f(x) = 1 + x +$

---

[5] The topic of extended Galois fields and their generation is treated in Section 6.2.

x^4 are shown, and the minimal polynomial for the element α^i is denoted as $M_i(x)$, which is also the minimal polynomial for all the elements belonging to the same conjugate class of which α^i is a member. For example, in Table 6.8, $M_1(x) = M_2(x) = M_4(x) = M_8(x)$ and $M_3(x) = M_6(x) = M_{12}(x) = M_9(x)$, and so forth.

Let us now denote by BCH(n, k, t) a t-error correcting BCH code with block length n with k information bits, and by $g_t(x)$ the generator polynomial of BCH(n, k, t). The definition of the generator polynomial $g_t(x)$ is

$$g_t(x) = \text{product of distinct minimal polynomials of } \{\alpha, \alpha^2, \alpha^3, \ldots, \alpha^{2t}\}$$

Now, for $t = 1$:

$$g_1(x) = \text{product of distinct minimal polynomials of } \{\alpha, \alpha^2\} = M_1(x)$$
$$= 1 + x + x^4$$

The degree of $g_1(x)$ is 4; that is, $n - k = 4$, and therefore $k = 11$. This code is BCH $(15, 11, 1)$, which denotes an $(n, k) = (15, 11)$ single-error-correcting Hamming code. For $t = 2$:

$$g_2(x) = \text{product of distinct minimal polynomials of } \{\alpha, \alpha^2, \alpha^3, \alpha^4\}$$
$$= M_1(x)M_3(x) = (1 + x + x^4)(1 + x + x^2 + x^3 + x^4)$$
$$= 1 + x^4 + x^6 + x^7 + x^8$$

The degree of $g_2(x)$ is 8; that is, $n - k = 8$, and therefore $k = 7$. This code is denoted as BCH $(15, 7, 2)$, which is also known as a double-error-correcting $(15, 7)$ cyclic code. For $t = 3$:

$$g_3(x) = \text{product of distinct minimal polynomials of } \{\alpha, \alpha^2, \alpha^3, \alpha^4, \alpha^5, \alpha^6\}$$
$$= M_1(x)M_3(x)M_5(x)$$
$$= (1 + x + x^4)(1 + x + x^2 + x^3 + x^4)(1 + x + x^2)$$
$$= 1 + x + x^2 + x^4 + x^5 + x^8 + x^{10}$$

We see that $\deg[g_3(x)] = n - k = 10 = $ number of parity check bits, and $k = 5$. This code is denoted BCH $(15, 5, 3)$.

The BCH codes are cyclic and the codewords have the elements α, α^2, ..., α^{2t} as roots and thus we can choose the generator polynomial $g_t(x)$ as the least common multiple (LCM) of the minimal polynomials of these roots; that is:

$$g_t(x) = \text{LCM}[M_1(x), M_2(x), \ldots, M_{2t-1}(x), M_{2t}(x)] \qquad (5.62a)$$

where $M_i(x)$ is the minimal polynomial of α^i. From the definition of the conjugate class in the extended Galois field $\text{GF}(2^m)$ generated by an irreducible primitive polynomial of degree m, we know that the even-power elements $\{\alpha^{2i}, 1 \leq i \leq 2^{m-2}\}$ of $\alpha^i \in \text{GF}(2^m)$ are the roots of the same minimal polynomial $M_i(x)$. Therefore, we can restate the definition of the generator polynomial of the code BCH (n, k, t) given in (5.62a) and write instead

$$g_t(x) = \text{LCM}[M_1(x), M_3(x), \ldots, M_{2t-1}(x)] \qquad (5.62b)$$

It is also obvious that the generator polynomial $g_t(x)$ of a t-error correcting BCH code, obtained from (5.62b), has a degree no greater than mt, because the minimal polynomial of an element in $\text{GF}(2^m)$ has a degree no greater than m. This fact implies that the number of parity check digits $(n - k)$ is $\leq mt$, as we stated at the beginning of this section.

Example 5.33 Let us construct all binary BCH (n, k, t) codes having $n = 31$. From Table 6.9 of Section 6.2.1 and the generator polynomial definition of (5.62b), we have the following results:

For $t = 1$, $\qquad g_1(x) = M_1(x) = 1 + x^2 + x^5 \rightarrow n - k = 5 \rightarrow k = 26$

This code is BCH $(31, 26, 1)$ with code rate $R_c = k/n = 26/31$. For $t = 2$:

$$\begin{aligned} g_2(x) &= \text{LCM}[M_1(x), M_3(x), M_5(x)] = M_1(x)M_3(x) \\ &= (1 + x^2 + x^5)(1 + x^2 + x^3 + x^4 + x^5) \\ &= 1 + x^3 + x^5 + x^6 + x^8 + x^9 + x^{10} \end{aligned}$$

The number of parity check digits is $n - k = 10 \rightarrow k = 21 \rightarrow$ BCH $(31, 21, 2)$ with $R_c = 21/31$. For $t = 3$:

$$g_3(x) = M_1(x)M_3(x)M_5(x)$$

$$= (0, 2, 5)(0, 2, 3, 4, 5)(0, 1, 2, 4, 5)$$

$$= (0, 1, 2, 3, 5, 7, 8, 9, 10, 11, 15)$$

$$= 1 + x + x^2 + x^3 + x^5 + x^7 + x^8 + x^9 + x^{10} + x^{11} + x^{15}$$

$n - k = 15 \rightarrow k = 16 \rightarrow \mathrm{BCH}(31, 16, 3)$ with $R_c = 16/31$. If we continue, we find that the four-error correcting BCH code and the five-error correcting BCH code have identical generator polynomials because $M_9(x) = M_5(x)$.

BCH codes can also be classified as being either primitive or non-primitive BCH codes. For a primitive BCH code, the block length n is $2^m - 1$, whereas for a nonprimitive BCH code, n may be any other odd number. Because α is a primitive element of $\mathrm{GF}(2^m)$, its order (see (6.8)) is $2^m - 1$, which is the block length of a BCH code. Thus, the above definition for a BCH code is that of a primitive BCH code. If, however, $2^m - 1$ is factorable, $2t$ consecutive powers of some nonprimitive element of $\mathrm{GF}(2^m)$ may instead be specified as roots of the codewords. The BCH code so generated is a nonprimitive BCH code and it has a block length that divides $2^m - 1$. In our discussions of the BCH codes, we confine our attention to primitive BCH codes.

Tables 5.18 through 5.20, adapted from [19], show the parameters of all primitive BCH (n, k, t) codes of length $n = 2^m - 1$ up to $n = 1023$. From Tables 5.18 through 5.20, we observe the following: for $t = 1$ and $t = 2$, $n - k = mt$; and for $t > 2$, $n - k < mt$, where $n = 2^m - 1$. For the single-error correcting BCH $(n, k, 1)$ codes, $n - k = m$; these codes are cyclic Hamming codes.

If our objective were the discussion of the decoding of t-error correcting BCH codes, we then would need to proceed with the subject of syndrome calculations for BCH codes. As we have stated at the outset of this chapter, our interest is limited to consideration of error detection of the codewords only. For this reason, we now move on the subject of frame quality indicators used in IS-95 and in CDMA PCS systems.

5.10 Frame and Message Structure Quality Indicators

We now return to the discussion of CRC codes. For example, the 9,600-bps and 4,800-bps frames of the forward link traffic channel, each including a frame quality indicator (FQI), were shown in Figure 5.9. These FQIs are based on CRC coding schemes that support two functions at the

Table 5.18 Binary BCH codes generated by primitive elements of order $n \leq 255$

n	k	t	n	k	t	n	k	t	n	k	t
7	4	1	63	10	13	127	8	31	255	115	21
15	11	1		7	15	255	247	1		107	22
	7	2	127	120	1		239	2		99	23
	5	3		113	2		231	3		91	25
31	26	1		106	3		223	4		87	26
	21	2		99	4		215	5		79	27
	16	3		92	5		207	6		71	29
	11	5		85	6		199	7		63	30
	6	7		78	7		191	8		55	31
63	57	1		71	9		187	9		47	42
	51	2		64	10		179	10		45	43
	45	3		57	11		171	11		37	45
	39	4		50	13		163	12		29	47
	36	5		43	14		155	13		21	55
	30	6		36	15		147	14		13	59
	24	6		29	21		139	15		9	63
	18	10		22	23		131	18			
	16	11		15	27		123	19			

Table 5.19 Binary BCH codes generated by primitive elements of order $n = 511$

k	t	k	t	k	t	k	t
502	1	367	16	238	37	103	61
493	2	358	18	229	38	94	62
484	3	349	19	220	39	85	63
475	4	340	20	211	41	76	85
466	5	331	21	202	42	67	87
457	6	322	22	193	43	58	91
448	7	313	23	184	45	49	93
439	8	304	25	175	46	40	95
430	9	295	26	166	47	31	109
421	10	286	27	157	51	28	111
412	11	277	28	148	53	19	119
403	12	268	29	139	54	10	127
394	13	259	30	130	55		
385	14	250	31	121	58		
376	15	241	36	112	59		

Table 5.20 Binary BCH codes generated by primitive elements of order $n = 1,023$

k	t	k	t	k	t	k	t	k	t
1013	1	808	22	598	46	393	79	193	118
1003	2	798	23	588	47	383	82	183	119
993	3	788	24	578	49	378	83	173	122
983	4	778	25	573	50	368	85	163	123
973	5	768	26	563	51	358	86	153	125
963	6	758	27	553	52	348	87	143	126
953	7	748	28	543	53	338	89	133	127
943	8	738	29	533	54	328	90	123	170
933	9	728	30	523	55	318	91	121	171
923	10	718	31	513	57	308	93	111	173
913	11	708	34	503	58	298	94	101	175
903	12	698	35	493	59	288	95	91	181
893	13	688	36	483	60	278	102	86	183
883	14	678	37	473	61	268	103	76	187
873	15	668	38	463	62	258	106	66	189
863	16	658	39	453	63	248	107	56	191
858	17	648	41	443	73	238	109	46	219
848	18	638	42	433	74	228	110	36	223
838	19	628	43	423	75	218	111	26	239
828	20	618	44	413	77	208	115	16	247
818	21	608	45	403	78	203	117	11	255

receiver. The first function is to determine whether the frame (packet) is received in error. The second function is to assist in the determination of the data rate of the received frame [1]; the final decision on the data rate is made at the decoder of the convolutional code. The FQI assists in the determination of the data rate in that the incoming symbols are fed into both 9,600-bps and 4,800-bps FQI detectors, so that an error-free frame can always be identified correctly. In the discussions to follow, we describe CRC computations used in both forward and reverse links in the IS-95 cellular system as well as CDMA PCS systems.

5.10.1 CRC Computations for the Forward Link Channels

The forward link traffic channel frame structures for 9,600 bps and 4,800 bps, shown previously in Figure 5.9, are redrawn in Figure 5.23.

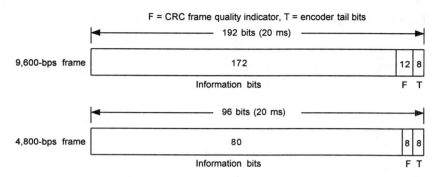

Figure 5.23 Forward traffic channel frame structures for 9,600-bps and 4,800-bps frames.

For both the 9,600-bps and 4,800-bps rates, the CRC bits are calculated on all bits within the frame, except the FQI itself and the convolutional encoder tail bits. The 9,600-bps transmission rate uses a 12-bit FQI, and the generator polynomial for this FQI is

$$g(x) = x^{12} + x^{11} + x^{10} + x^9 + x^8 + x^4 + x + 1 \qquad (5.63a)$$

and its encoder circuit is shown in Figure 5.24 (adapted from Figure 7.1.3.5.2-1 of [1]). Note from the figures and the descriptions given above that this is a linear cyclic code with $(n, k) = (172 + 12, 172) = (184, 12)$. Although the convolutional encoder tail bits are part of the 192-bit frame, they do not influence the FQI, and hence they are excluded from entering into the encoder circuit.

The 4,800-bps transmission rate uses an 8-bit FQI and its generator polynomial is

$$g(x) = x^8 + x^7 + x^4 + x^3 + x + 1 \qquad (5.63b)$$

The FQI's encoder circuit is shown in Figure 5.25 (adapted from Figure 7.1.3.5.2.1-2 of [1]).

Figures 5.24 and 5.25 are nothing more than examples of the (n, k) cyclic code in systematic form that we have described in Section 5.8.1, along with the encoder circuit shown in Figure 5.15 and its procedures. However, there is one difference between what was described in the encoding procedures of Section 5.8.1 and what is described in IS-95 concerning the initial loading of the elements in the shift registers. IS-95 gives the following procedure:

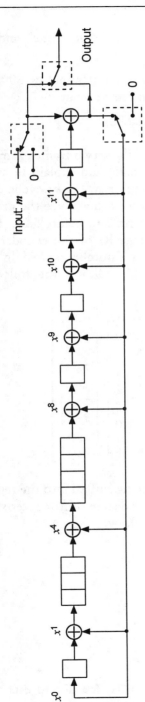

Figure 5.24 Forward traffic channel FQI calculations at the 9,600-bps rate.

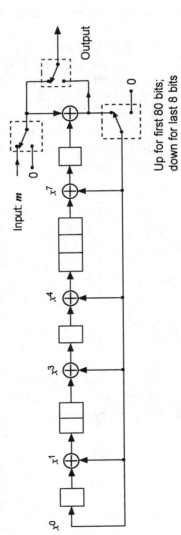

Figure 5.25 Forward traffic channel FQI calculations at the 4,800-bps rate.

- Initially, all shift register elements are set to logical one[6] and the switches shall are set in the UP position.

- The register is clocked 172 times (for the 192-bit frame) or 80 times (for the 96-bit frame) with the information bits as input.

- The bits are transmitted in the order calculated.

For a detailed analysis of the encoding processes of these two frame examples, we consider the case of 4,800-bps frames (Figure 5.25) and explain how this case fits into the subject considered in the encoding of an (n, k) cyclic code. This is an $(n, k) = (88, 80)$ cyclic code, and thus we can analyze the process of parity check digit calculation by means of the circuit in Figure 5.25. If we let $x_i(t)$ be the number stored in the register stage R_i of the encoder (see Figure 5.15) after clock pulse t and $x_i(t+1)$, the number stored in that register after clock pulse $t+1$, then Figure 5.25 indicates the following situation:

$$
\begin{bmatrix} x_1(t+1) \\ x_2(t+1) \\ x_3(t+1) \\ x_4(t+1) \\ x_5(t+1) \\ x_6(t+1) \\ x_7(t+1) \\ x_8(t+1) \end{bmatrix} = \begin{bmatrix} 0 & 0 & 0 & 0 & 0 & 0 & 0 & 1 & 1 \\ 1 & 0 & 0 & 0 & 0 & 0 & 0 & 0 & 0 \\ 0 & 1 & 0 & 0 & 0 & 0 & 0 & 0 & 0 \\ 0 & 0 & 1 & 0 & 0 & 0 & 0 & 1 & 1 \\ 0 & 0 & 0 & 1 & 0 & 0 & 0 & 1 & 1 \\ 0 & 0 & 0 & 0 & 1 & 0 & 0 & 0 & 0 \\ 0 & 0 & 0 & 0 & 0 & 1 & 0 & 0 & 0 \\ 0 & 0 & 0 & 0 & 0 & 0 & 1 & 1 & 1 \end{bmatrix} \begin{bmatrix} x_1(t) \\ x_2(t) \\ x_3(t) \\ x_4(t) \\ x_5(t) \\ x_6(t) \\ x_7(t) \\ x_8(t) \\ m(t) \end{bmatrix}
$$

where $m(t)$ is the message bit at time t, which is being shifted into the register circuit. IS-95 specifies that the initial loading of the 8-state shift register stages is all logical ones, and hence we have the situation shown.

---

[6] Initialization of the register to all ones causes the CRC for all-zero data to be nonzero.

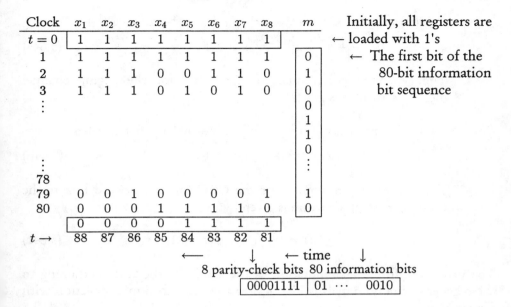

Clock	x_1	x_2	x_3	x_4	x_5	x_6	x_7	x_8	m	
$t = 0$	1	1	1	1	1	1	1	1		Initially, all registers are
1	1	1	1	1	1	1	1	1	0	← loaded with 1's
2	1	1	1	0	0	1	1	0	1	← The first bit of the
3	1	1	1	0	1	0	1	0	0	80-bit information
⋮									0	bit sequence
									1	
									1	
									0	
⋮									⋮	
78										
79	0	0	1	0	0	0	0	1	1	
80	0	0	0	1	1	1	1	0	0	
	0	0	0	0	1	1	1	1		
$t \rightarrow$	88	87	86	85	84	83	82	81		

←   ↓   ← time   ↓

8 parity-check bits 80 information bits

| 00001111 | 01 ⋯ 0010 |

The receiver has two syndrome calculators that are acting as frame error
(block error) detectors, one for 9,600 bps and the other for 4,800 bps. They
could be similar to those shown in Figure 5.26, which operate as follows:

- Initially, the switch is in the UP position and all the register stages are
  set to logical one.

- The received vector $\boldsymbol{R} = (r_0, r_1, \ldots, r_{N-1})$ is shifted into the
  register, where $N = 184$ for 9,600 bps and $N = 88$ for 4,800 bps.

- As soon as the entire N bits have been shifted into the register, the
  contents in the $(n-k)$-stage register form the syndrome, with
  $n - k = 12$ for 9,600 bps and $n - k = 8$ for 4,800 bps.

- The syndrome digits are read out either in a parallel or in a serial way
  to determine the error or no-error decision.

IS-95 describes varieties of CRC circuits employed for the forward link.
Now that we fully understand the mechanism for the CRC circuits, we only
list the generator polynomials for the CRCs used for *frame structures* as well
as *message structures*.

- The generator polynomial for the CRC for the sync channel message
  structure is given by

$$g(x) = x^{30} + x^{29} + x^{21} + x^{20} + x^{15} + x^{13} + x^{12} + x^{11} + x^8$$
$$+ x^7 + x^6 + x^2 + x + 1 \tag{5.64a}$$

- The generator polynomial for the CRC for the paging channel message structure is given by

$$g(x) = x^{30} + x^{29} + x^{21} + x^{20} + x^{15} + x^{13} + x^{12} + x^{11} + x^8$$
$$+ x^7 + x^6 + x^2 + x + 1 \tag{5.64b}$$

- The generator polynomial for the CRC for the forward link traffic channel message structure is given by[7]

$$g(x) = x^{16} + x^{12} + x^5 + 1 \tag{5.64c}$$

The procedure for the computation of the CRC for the cases pertaining to these generator polynomials is identical to that described in connection with Figure 5.26.

5.10.2 CRC Computations for the Reverse Link Channels

The reverse traffic channel frames sent at the 9,600-bps transmission rate consist of 192 bits, and the reverse traffic channel frames sent at the 4,800-bps transmission rate consist of 96 bits. These are exactly the same structure as for the forward link traffic channel frame structures shown in Figure 5.23. Thus, the reverse traffic channel FQIs for the 9,600-bps rate and the 4,800-bps rate are the same and, therefore, the generator polynomials for the reverse link traffic channel 9,600-bps and 4,800-bps rates are given as, respectively:

$$g(x) = x^{12} + x^{11} + x^{10} + x^9 + x^8 + x^4 + x + 1 \tag{5.65a}$$

and

$$g(x) = x^8 + x^7 + x^4 + x^3 + x + 1 \tag{5.65b}$$

The shift register circuits for these polynomials for the calculations of the respective CRCs are shown in Figures 5.24 and 5.25.

---

[7] This polynomial is the CRC-CCITT polynomial, which is the standard for European systems.

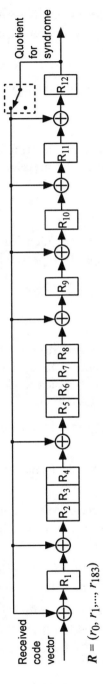

(a) 9,600-bps syndrome detector

$R = (r_0, r_1, ..., r_{183})$

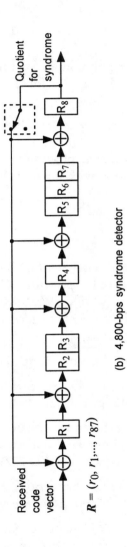

(b) 4,800-bps syndrome detector

$R = (r_0, r_1, ..., r_{87})$

Figure 5.26 Frame error-detection circuits for 9,600 bps and 4,800 bps.

As for the access channel message CRC, a 30-bit CRC is computed for each access channel signaling message. The generator polynomial is identical to that for the sync channel message structure, namely

$$g(x) = x^{30} + x^{29} + x^{21} + x^{20} + x^{15} + x^{13} + x^{12} + x^{11} + x^8$$
$$+ x^7 + x^6 + x^2 + x + 1 \qquad (5.65c)$$

The generator polynomial for the CRC for the reverse traffic channel message structure is the same as for the forward link traffic message structure, namely

$$g(x) = x^{16} + x^{12} + x^5 + 1 \qquad (5.65d)$$

Again, the computational procedures by the circuits are identical to those described in connection with Figure 5.26, and they are not repeated here.

References

[1] "Mobile Station-Base Station Compatibility Standard for Dual-Mode Wideband Spread Spectrum Cellular System," TIA/EIA Interim Standard 95 (IS-95), Washington: Telecommunications Industry Association, July 1993 (amended as IS-95A in May 1995).

[2] Walsh, J. L., "A Closed Set of Normal Orthogonal Functions," *American J. Mathematics*, Vol. 45, pp. 5–24, 1923.

[3] Byrnes, J. S., and D. A. Swick, "Instant Walsh Functions," *SIAM Review*, Vol. 12, p. 131, Jan. 1970.

[4] Corrington, M. S., and R. N. Adams, "Advanced Analytical and Signal Processing Techniques: Applications of Walsh Functions to Non-Linear Analysis," DTIC acc. no. AD-277942, Apr. 1962.

[5] Schreiber, H. H., "Bandwidth Requirements for Walsh Functions," *IEEE Trans. on Information Theory*, Vol. IT-16, pp. 491–492, July 1970.

[6] Harmuth, H., "A Generalized Concept of Frequency and Some Applications," *IEEE Trans. on Inform. Thy*, Vol. IT-14, pp. 375–382, May 1968.

[7]   Harmuth, H., *Transmission of Information by Orthogonal Functions*, New York: Springer-Verlag, pp. 5–26, 1969.

[8]   *Proc. 1970 Symposium on Applications of Walsh Functions*, Naval Research Laboratory, Washington, DC, March 31–April 3, 1970.

[9]   Alexits, G., *Convergence Problems of Orthogonal Series*, New York: Pergamon Press, 1961.

[10] Swick, D. A., "Walsh Function Generation," *IEEE Trans. on Information Theory* (Correspondence), Vol. IT-15, p. 167, Jan. 1969.

[11] Golomb, S. W., *Digital Communications With Space Applications*, Englewood Cliffs, NJ: Prentice-Hall, 1964.

[12] *Proc. 1971 Symposium on Applications of Walsh Functions*, held at Departmental Auditorium, Washington, DC, Apr. 13–15, 1971, DTIC acc. no. AD-727000.

[13] Ahmed, N., and K. Rao, "Walsh Functions and Hadamard Transform," *Proc. 1972 Symposium on Applications of Walsh Functions*, The Catholic Univ. of America, Washington, DC, pp. 8–13, Mar. 27–29, 1972 .

[14] *Proc. 1973 Symposium on Applications of Walsh Functions*, The Catholic Univ. of America, Washington, DC, Apr. 16–18, 1973.

[15] Lee, J. S., "Digital Generation of Walsh Functions for Orthogonal Multiplexing Applications," *Proc. 1973 Symposium on Applications of Walsh Functions*, Washington, DC, pp. 222–227, Apr. 1973. Also in DTIC acc. no. AD-763000.

[16] "Personal Station-Base Station Compatibility Requirements for 1.8 to 2.0 GHz Code Division Multiple Access (CDMA) Personal Communications Systems," ANSI J-STID-008, Washington: Telecommunications Industry Association, 1996.

[17] McNamara, J. E., *Technical Aspects of Data Communication*, Bedford, MA: Digital Equipment Corporation, 1977.

[18] Lin, S., and D. J. Costello, Jr., *Error Control Coding Fundamentals and Applications*, Englewood Cliffs, NJ: Prentice-Hall, 1983.

[19] Michelson, A. M., and A. H. Levesque, *Error-Control Techniques for Digital Communication*, New York: Wiley, 1985.

[20] Peterson, W. W., and E. J. Weldon, *Error Correcting Codes*, 2nd ed., Cambridge, MA: MIT Press, 1972.

[21]   Blahut, R. E., *Theory and Practice of Error Control Codes*, New York: Addison-Wesley, 1983.

[22]   Hamming, R. W., "Error Detecting and Error Correcting Codes," *Bell System Technical Journal*, Vol. 29, pp. 147–160, Apr. 1950.

[23]   Prange, E., "Cyclic Error-Correcting Codes in Two Symbols," AFCRC-TN-57, 1–3, Cambridge, MA: Air Force Cambridge Research Center, Sept. 1957.

[24]   Hocquenghem, A., "Codes Corecteurs d'Erreurs," *Chiffres*, Vol. 2, pp. 147–159, 1959.

[25]   Bose, R. C., and D. K. Ray-Chandhuri, "On a Class of Error Correcting Binary Group Codes," *Information and Control*, Vol. 3, pp. 68–79, Mar. 1960.

[26]   Berlekamp, E. R., *Algebraic Coding Theory*, New York: McGraw-Hill, 1968.

6

Theory and Application of Pseudonoise Sequences

6.1 Properties of Pseudonoise Sequences

Pseudorandom or pseudonoise (PN) sequences are used in data scrambling in the IS-95 system as well as for spread-spectrum modulation. Data scrambling is achieved by changing the data sequence "randomly" or in a noise-like fashion before transmission. At the receiver, the scrambled sequence is "changed back" to the original data sequence. The two concepts, "randomness" and "changing back," are the key ideas involved in understanding the CDMA system. If the generated sequence were completely random, the receiver would have no way to change back. On the other hand, if the receiver knows how to change back, the sequence cannot be completely random.

Consider the following sequences:

Data sequence	1 1 0 0 1 0 1 0 0 1 0 1 0 1 1...
Random sequence	1 0 1 0 0 0 0 1 0 1 1 0 1 0...
Transmitted sequence	0 1 1 0 1 0 1 1 0 0 1 1 1 1...

The transmitted sequence is a scrambled version of the data sequence obtained by the bit-by-bit modulo-2 addition of the data sequence and a random sequence. At the receiver, an identical "random" sequence is added to the received sequence, which in the absence of noise is the transmitted sequence:

Transmitted sequence	0 1 1 0 1 0 1 1 0 0 1 1 1 1...
Random sequence	1 0 1 0 0 0 0 1 0 1 1 0 1 0...
Data sequence	1 1 0 0 1 0 1 0 0 1 0 1 0 1 1...

This illustration reveals two fundamental requirements on the random sequence:

- It must be reproducible at the receiver;

- It must be reproduced in synchronism with the scrambling sequence at the transmitter.

These two requirements make it virtually impossible to use a completely random sequence and hence, in practice, we use a sequence that has sufficient randomness to be unrecognizable to unintended receivers and yet is deterministic to make it relatively easy to generate and to synchronize at the receiver.

The most important method of generating such binary sequences is by means of a linear feedback shift register (LFSR). For an LFSR sequence generator with n stages, the output sequence will always be periodic because, whatever the initial conditons of the shift register, after a finite number of clock pulses, the initial conditions must eventually be reproduced. Because the maximum number of different combinations of n binary digits is 2^n, the period cannot exceed 2^n. Because the all-zero condition, if reached, remains in the same state forever, it cannot appear in the shift register if the initial condition (initial loading or state) is not all zeros. Therefore, the maximum number of possible states is $2^n - 1$.

A shift register output sequence with the period $2^n - 1$ is called a "maximal length sequence" or "m-sequence" for short. M-sequences are also referred to as "pseudorandom sequences" or PN sequences. When PN sequences clocked at very high rates are modulated (multiplied) with data sequences in a communications system, such as the IS-95 system, it is a spread-spectrum system that provides $10 \log_{10}(R_N/R_b)$ dB of "processing gain," where R_N is the PN sequence rate and R_b is the data rate.

An example PN sequence generation with $n = 5$ (5-stage) is given in Figure 6.1 for the case of an initial state (or loading sequence) of 1 0 0 0 0. One period of the sequence S generated by the LFSR is

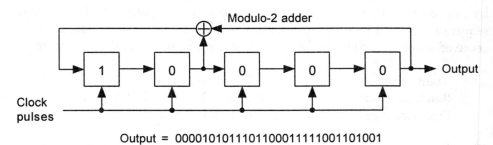

Output = 0000101011101100011111001101001

Figure 6.1 Example $n = 5$-stage LFSR sequence generator.

$$S = 0000101011101100011111001101001 \qquad (6.1)$$

The length of this sequence is $2^5 - 1 = 31$, so that it is an m-sequence. The sequences that qualify as m-sequences or PN sequences possess the following three properties [1]:

1. **The balance property.** In a complete period $P = 2^n - 1$ of a PN sequence, the number of 1s differs from the number of 0s by at most 1. This property is observed in the example sequence, as there are 16 1s and 15 0s.

2. **The run property.** There are $(2^n - 1 + 1)/2 = 2^{n-1}$ runs of consecutive 1s or 0s, and half of the runs are of length 1, $1/2^2$ of the runs are of length 2, $1/2^3$ of the runs are of length 3, etc. There is one run (of zeros) is of length $n - 1$, and 1 run (of 1s) is of length n. In the example sequence, let $P \triangleq 2^n - 1$, where $n = 5$. There are a total of $2^{5-1} = 16$ runs of consecutive 1s and 0s; there are $2^{5-2} = 8$ runs of length 1 (four runs of 1 and four runs of 0); there are $2^{5-3} = 4$ runs of length 2 (two runs of 11 and two runs of 00); there are $2^{5-4} = 2$ runs of length 3 (one run of 111 and one run of 000). There is one run of four consecutive 0s and one run of five consecutive 1s.

3. **The correlation property.** If a complete sequence is compared bit-by-bit with any shift of the sequence, the number of agreements minus the number of disagreements is always -1; that is, there is one more disagreement position than the number of agreement positions. Exploitation of this particular property of PN sequences makes it possible to design direct sequence (DS) spread-spectrum systems, such as CDMA systems, as we will have plenty of opportunities to see and analyze in this book.

Let us verify the correlation property with the example of the sequence S in (6.1). Suppose we shift the original sequence by four bits and compute the correlation. A four-bit is the new sequence $S^{(4)}$, starting from the fourth bit from the end of the sequence (four bits shifted to the right or wrapped around from the fourth bit from the end of the sequence). The comparison of S and $S^{(4)}$ reveals the following:

$$S = 00001010111011000111111001101001$$

$$S^{(4)} = 10101110110001111110011010010000 \qquad (6.2)$$

$$S \oplus S^{(4)} = 10100100001010111011100011111001$$

Thus the correlation is given, according to (5.35b), as

$$\langle S, S^{(4)} \rangle = (\text{number term-by-term agreements in } S \text{ and } S^{(4)})$$

$$- (\text{number of term-by-term disagreements})$$

$$= 31 - 2 \cdot \text{weight}\big(S \oplus S^{(4)}\big) = -1 \qquad (6.3)$$

The reader can verify that the correlation between S and any shift of this sequence is $\langle S, S^{(l)} \rangle = -1$ for $l = 1, 2, \ldots, 30$.

In the following sections, we will devote considerable effort to analyze these most interesting, mysterious, and useful linear sequences, so that we may fully understand and appreciate their use in CDMA cellular and PCS systems as a fundamental element of spread-spectrum modulation.

6.2 Extension Galois Fields and Primitive Polynomials

In Section 5.3.3, we introduced the concept of a finite field, called a Galois field, in connection with the discussion of Walsh sequences over a field of binary elements $\{0, 1\}$. In that section we also introduced the notion of extension field $\mathrm{GF}(q) = \mathrm{GF}(p^n)$, where $\mathrm{GF}(p)$ was called a prime field, which is a subfield of $\mathrm{GF}(p^n) \equiv \mathrm{GF}(q)$. In this section we present many examples of the generating extension fields and develop the relationship that exists between maximal-length periodic binary sequences (PN sequences) and extension fields.

The generation of an extension field requires consideration of an irreducible polynomial of degree n; that is, a degree-n polynomial that has no polynomial factors of degree less than $n > 0$. Let us now consider generating an extension field $\mathrm{GF}(2^n)$.

1. Select (pick) an irreducible polynomial of degree n, say, $f(x)$, given by

$$f(x) = x^n + a_{n-1} x^{n-1} + \cdots + a_2 x^2 + a_1 x^1 + x^0, \quad a_i \in \mathrm{GF}(2) \qquad (6.4)$$

2.   Select an abstract symbol, say, α, and assume that α is a root of the polynomial $f(x)$:

$$f(\alpha) = \alpha^n + a_{n-1}\alpha^{n-1} + \cdots + a_2\alpha^2 + a_1\alpha + 1 = 0 \qquad (6.5a)$$

so that

$$\alpha^n = a_{n-1}\alpha^n + a_{n-2}\alpha^{n-2} + \cdots + a_2\alpha + 1 \qquad (6.5b)$$

3.   Now, the 2^n elements of Galois field $GF(2^n)$ may be taken as the set of elements that are powers of α: $\{0, \alpha^0, \alpha^1, \alpha^2, \ldots, \alpha^{2^n-2}\}$.

Note that we have included a zero element as a unique additive identity element, without which the set cannot qualify as a field. Further, note that the elements in the set are based on the root α that satisfies the chosen irreducible polynomial, $f(x)$. Being a field, the two operations, addition and multiplication, are performed on the elements as follows:

$$\alpha^k \cdot \alpha^t = \alpha^{k+t} \qquad (6.6a)$$

and the addition of any two elements α^k and α^t can be carried out by two different methods. One method is to express α^k and α^t in the form of the polynomial (6.5b) for k, $t \geq n$, so that addition of the coefficients over $GF(2)$ of the corresponding powers of α in each of the polynomials may be done:

$$\alpha^k + \alpha^t = (c_{k0} + c_{k1}\alpha + \cdots + c_{k,n-1}\alpha^{n-1}) + (c_{t0} + c_{t1}\alpha + \cdots + c_{t,n-1}\alpha^{n-1})$$

$$= (c_{k0} \oplus c_{t0}) + (c_{k1} \oplus c_{t1})\alpha + \cdots + (c_{k,n-1} \oplus c_{t,n-1})\alpha^{n-1} \qquad (6.6b)$$

Example 6.1 Construct an extension field $GF(2^3) = GF(8)$.

Solution: (1) Select an irreducible polynomial of degree 3, say, $f(x) = 1 + x^2 + x^3$. (2) Assume that α is a root of $f(x)$. Then $f(\alpha) = 1 + \alpha^2 + \alpha^3 = 0 \Rightarrow \alpha^3 = \alpha^2 + 1$. (3) We can now construct the Galois field $GF(2^3)$ and represent the elements in powers of α, polynomials over $GF(2)$, and sequences (vectors) over $GF(2)$, as shown in Table 6.1.

In the sequence (vector) representations of the $GF(2^3)$ elements in Table 6.1, we used a fundamental idea in any sequence (or coding) theory. That is, the idea of representing a polynomial by a sequence, by reading the

Table 6.1 Representation of GF(2^3) generated from $f(x) = 1 + x^2 + x^3 : \alpha^3 = \alpha^2 + 1$

0 and Powers of α	Polynomials over GF(2)			Sequence over GF(2)
$0 =$		0		0 0 0
$\alpha^0 =$			1	0 0 1
$\alpha^1 =$		α		0 1 0
$\alpha^2 =$	α^2			1 0 0
$\alpha^3 =$	α^2		$+1$	1 0 1
$\alpha^4 =$	α^2	$+\alpha$	$+1$	1 1 1
$\alpha^5 =$		α	$+1$	0 1 1
$\alpha^6 =$	α^2	$+\alpha$		1 1 0

coefficients over GF(2) or vice versa. As we see later, this idea gave impetus to the development of modern sequence and coding theories. In the present example of GF(2^3), generated by the irreducible monic[1] binary polynomial $f(x) = x^3 + x^2 + 1$, additions and multiplications are performed in the ways suggested previously. From Table 6.1:

$$\alpha^3 + \alpha^5 = (\alpha^2 + 1) + (\alpha+1) = \alpha^2 + \alpha = \alpha^6$$
$$\alpha^5 + \alpha^6 = (\alpha+1) + (\alpha^2 + \alpha) = \alpha^2 + 1 = \alpha^3$$

Note that
$$\alpha^3 + \alpha^5 = (1 \ 0 \ 1) + (0 \ 1 \ 1) = (1 \ 1 \ 0) = \alpha^6$$
$$\alpha^5 + \alpha^6 = (0 \ 1 \ 1) + (1 \ 1 \ 0) = (1 \ 0 \ 1) = \alpha^3$$

For multiplications of the elements:

$$\alpha^3 \cdot \alpha^5 = \alpha^{8 \bmod 7} = \alpha$$
$$\alpha^2 \cdot \alpha^4 = \alpha^6$$

---

[1] A polynomial is called "monic" if the coefficient of the highest power is 1. In fact, all polynomials over GF(2) are monic polynomials.

We can quickly observe all the properties of the Galois field elements, such as additive and multiplicative identity elements, from the $GF(2^3)$ addition and multiplication tables that are shown in Table 6.2.

Another addition method, using the so-called Zech's log table, carries out the addition of two field elements given in powers of the root element α. Two elements α^i and α^j can be added using the rule

$$\alpha^i + \alpha^j = \alpha^i \left(1 + \alpha^{j-i \bmod (q-1)}\right)$$

$$= \alpha^i \alpha^{Z(j-i)} = \alpha^{i+Z(j-i)} \tag{6.7a}$$

Table 6.2 Addition and multiplication tables for $GF(2^3)$ generated by $f(x) = x^3 + x + 1$

$+$	0	1	α	α^2	α^3	α^4	α^5	α^6
0	0	1	α	α^2	α^3	α^4	α^5	α^6
1	1	0	α^5	α^3	α^2	α^6	α	α^4
α	α	α^5	0	α^6	α^4	α^3	1	α^2
α^2	α^2	α^3	α^6	0	1	α^5	α^4	α
α^3	α^3	α^2	α^4	1	0	α	α^6	α^5
α^4	α^4	α^6	α^3	α^5	α	0	α^2	1
α^5	α^5	α	1	α^4	α^6	α^2	0	α^3
α^6	α^6	α^4	α^2	α	α^5	1	α^3	0

$\times$	0	1	α	α^2	α^3	α^4	α^5	α^6
0	0	0	0	0	0	0	0	0
1	0	1	α	α^2	α^3	α^4	α^5	α^6
α	0	α	α^2	α^3	α^4	α^5	α^6	1
α^2	0	α^2	α^3	α^4	α^5	α^6	1	α
α^3	0	α^3	α^4	α^5	α^6	1	α	α^2
α^4	0	α^4	α^5	α^6	1	α	α^2	α^3
α^5	0	α^5	α^6	1	α	α^2	α^3	α^4
α^6	0	α^6	1	α	α^2	α^3	α^4	α^5

For example, in computing $\alpha^4 + \alpha^6 = \alpha^4(1 + \alpha^2) = \alpha^4 \alpha^{Z(2)}$, what is $Z(2)$? We can answer this question by constructing a table called "Zech's log table" for $Z(i)$, an integer, such that $1 + \alpha^i = \alpha^{Z(i)}$ for the field $\mathrm{GF}(q) = \{0, 1, \alpha, \ldots, \alpha^i, \ldots, \alpha^{q-2}\}$, $0 \le i \le q - 2$. An example illustrates the method of constructing Zech's log table.

Example 6.2 Consider $\mathrm{GF}(2^3) = \mathrm{GF}(8)$ generated by the irreducible polynomial $f(x) = x^3 + x^2 + 1$ with $\alpha^3 = \alpha^2 + 1$. We determine $Z(i)$, $0 \le i \le 6$, in an ad hoc manner as follows:

1. $1 + \alpha^2 = \alpha^3 \quad \Rightarrow \quad 1 + \alpha^2 = \alpha^{Z(2)} \quad \Rightarrow \quad Z(2) = 3.$

2. $1 + \alpha^0 = 0 \quad \Rightarrow \quad 1 + \alpha^0 = \alpha^{Z(0)} \quad \Rightarrow \quad Z(0) = -\infty.$

3. $1 + \alpha = \alpha^{Z(1)}.$

   Using $1 + \alpha^2 = \alpha^3$ and multiplying both sides by α^{-2},
   $$\alpha^{-2}(1 + \alpha^2) = 1 + \alpha^{-2} = \alpha^{-2}\alpha^3 = \alpha$$
   $$\Rightarrow 1 + \alpha = \alpha^{-2 \bmod 7} = \alpha^5 \quad \Rightarrow \quad Z(1) = 5.$$

4. $1 + \alpha^3 = \alpha^2 \quad \Rightarrow \quad 1 + \alpha^3 = \alpha^{Z(3)} = \alpha^2 \quad \Rightarrow \quad Z(3) = 2.$

5. $1 + \alpha^4 = (1 + \alpha^2)^2 = (\alpha^3)^2 = \alpha^6 \quad \Rightarrow \quad Z(4) = 6.$

6. $1 + \alpha^5 = \alpha$ from 3. $\quad \Rightarrow \quad 1 + \alpha^5 = \alpha^{Z(5)} = \alpha \quad \Rightarrow \quad Z(5) = 1.$

7. $1 + \alpha^6 = (1 + \alpha^3)^2 = (\alpha^2)^2 = \alpha^4 \quad \Rightarrow \quad Z(6) = 4.$

Thus, we construct Zech's log table for the specific case of $\alpha^3 = \alpha^2 + 1$ as shown below:

i	0	1	2	3	4	5	6
$Z(i)$	$-\infty$	5	3	2	6	1	4

(6.7b)

Using this table, we have

$$\alpha^5 + \alpha^6 = \alpha^5(1 + \alpha) = \alpha^{5 + Z(1)} = \alpha^{5+5} = \alpha^3$$

which is verified to be correct from Table 6.2.

Example 6.3 Construct $\mathrm{GF}(2^4)$ using the irreducible binary polynomial $f(x) = x^4 + x + 1$.

Solution: Assuming α to be a root of the polynomial, we have $\alpha^4 = \alpha + 1$. The reader can now construct Zech's log table and the addition and multiplication tables, respectively, as shown in Tables 6.3, 6.4, and 6.5.

The least positive integer P for which $\alpha^P = 1$ is called the order of element α. If the order of α is P, then the elements α^0, α^1, α^2, ..., α^{P-1} are all distinct, and thus the order of an element is the number of distinctive powers of that element. If $P = 2^n - 1$, where n is the degree of the irreducible polynomial that generated the GF(2^n), α is called a primitive

Table 6.3 Zech's log table for GF(2^4) generated by $f(x) = x^4 + x + 1$

i	0	1	2	3	4	5	6	7	8	9	10	11	12	13	14
$Z(i)$	$-\infty$	4	8	14	1	10	13	9	2	7	5	12	11	6	3

Table 6.4 Addition table for GF(2^4) generated by $f(x) = x^4 + x + 1$

+	0	1	α	α^2	α^3	α^4	α^5	α^6	α^7	α^8	α^9	α^{10}	α^{11}	α^{12}	α^{13}	α^{14}
0	0	1	α	α^2	α^3	α^4	α^5	α^6	α^7	α^8	α^9	α^{10}	α^{11}	α^{12}	α^{13}	α^{14}
1	1	0	α^4	α^8	α^{14}	α	α^{10}	α^{13}	α^9	α^2	α^7	α^5	α^{12}	α^{11}	α^6	α^3
α	α	α^4	0	α^5	α^9	1	α^2	α^{11}	α^{14}	α^{10}	α^3	α^8	α^6	α^{13}	α^{12}	α^7
α^2	α^2	α^8	α^5	0	α^6	α^{10}	α	α^3	α^{12}	1	α^{11}	α^4	α^9	α^7	α^{14}	α^{13}
α^3	α^3	α^{14}	α^9	α^6	0	α^7	α^{11}	α^2	α^4	α^{13}	α	α^{12}	α^5	α^{10}	α^8	1
α^4	α^4	α	1	α^{10}	α^7	0	α^8	α^{12}	α^3	α^5	α^{14}	α^2	α^{13}	α^6	α^{11}	α^9
α^5	α^5	α^{10}	α^2	α	α^{11}	α^8	0	α^9	α^{13}	α^4	α^6	1	α^3	α^{14}	α^7	α^{12}
α^6	α^6	α^{13}	α^{11}	α^3	α^2	α^{12}	α^9	0	α^{10}	α^{14}	α^5	α^7	α	α^4	1	α^8
α^7	α^7	α^9	α^{14}	α^{12}	α^4	α^3	α^{13}	α^{10}	0	α^{11}	1	α^6	α^8	α^2	α^5	α
α^8	α^8	α^2	α^{10}	1	α^{13}	α^5	α^4	α^{14}	α^{11}	0	α^{12}	α	α^7	α^9	α^3	α^6
α^9	α^9	α^7	α^3	α^{11}	α	α^{14}	α^6	α^5	1	α^{12}	0	α^{13}	α^2	α^8	α^{10}	α^4
α^{10}	α^{10}	α^5	α^8	α^4	α^{12}	α^2	1	α^7	α^6	α	α^{13}	0	α^{14}	α^3	α^9	α^{11}
α^{11}	α^{11}	α^{12}	α^6	α^9	α^5	α^{13}	α^3	α	α^8	α^7	α^2	α^{14}	0	1	α^4	α^{10}
α^{12}	α^{12}	α^{11}	α^{13}	α^7	α^{10}	α^6	α^{14}	α^4	α^2	α^9	α^8	α^3	1	0	α	α^5
α^{13}	α^{13}	α^6	α^{12}	α^{14}	α^8	α^{11}	α^7	1	α^5	α^3	α^{10}	α^9	α^4	α	0	α^2
α^{14}	α^{14}	α^3	α^7	α^{13}	1	α^9	α^{12}	α^8	α	α^6	α^4	α^{11}	α^{10}	α^5	α^2	0

Table 6.5 Multiplication table for GF(2^4) generated by $f(x) = x^4 + x + 1$

$\times$	0	1	α	α^2	α^3	α^4	α^5	α^6	α^7	α^8	α^9	α^{10}	α^{11}	α^{12}	α^{13}	α^{14}
0	0	0	0	0	0	0	0	0	0	0	0	0	0	0	0	0
1	0	1	α	α^2	α^3	α^4	α^5	α^6	α^7	α^8	α^9	α^{10}	α^{11}	α^{12}	α^{13}	α^{14}
α	0	α	α^2	α^3	α^4	α^5	α^6	α^7	α^8	α^9	α^{10}	α^{11}	α^{12}	α^{13}	α^{14}	1
α^2	0	α^2	α^3	α^4	α^5	α^6	α^7	α^8	α^9	α^{10}	α^{11}	α^{12}	α^{13}	α^{14}	1	α
α^3	0	α^3	α^4	α^5	α^6	α^7	α^8	α^9	α^{10}	α^{11}	α^{12}	α^{13}	α^{14}	1	α	α^2
α^4	0	α^4	α^5	α^6	α^7	α^8	α^9	α^{10}	α^{11}	α^{12}	α^{13}	α^{14}	1	α	α^2	α^3
α^5	0	α^5	α^6	α^7	α^8	α^9	α^{10}	α^{11}	α^{12}	α^{13}	α^{14}	1	α	α^2	α^3	α^4
α^6	0	α^6	α^7	α^8	α^9	α^{10}	α^{11}	α^{12}	α^{13}	α^{14}	1	α	α^2	α^3	α^4	α^5
α^7	0	α^7	α^8	α^9	α^{10}	α^{11}	α^{12}	α^{13}	α^{14}	1	α	α^2	α^3	α^4	α^5	α^6
α^8	0	α^8	α^9	α^{10}	α^{11}	α^{12}	α^{13}	α^{14}	1	α	α^2	α^3	α^4	α^5	α^6	α^7
α^9	0	α^9	α^{10}	α^{11}	α^{12}	α^{13}	α^{14}	1	α	α^2	α^3	α^4	α^5	α^6	α^7	α^8
α^{10}	0	α^{10}	α^{11}	α^{12}	α^{13}	α^{14}	1	α	α^2	α^3	α^4	α^5	α^6	α^7	α^8	α^9
α^{11}	0	α^{11}	α^{12}	α^{13}	α^{14}	1	α	α^2	α^3	α^4	α^5	α^6	α^7	α^8	α^9	α^{10}
α^{12}	0	α^{12}	α^{13}	α^{14}	1	α	α^2	α^3	α^4	α^5	α^6	α^7	α^8	α^9	α^{10}	α^{11}
α^{13}	0	α^{13}	α^{14}	1	α	α^2	α^3	α^4	α^5	α^6	α^7	α^8	α^9	α^{10}	α^{11}	α^{12}
α^{14}	0	α^{14}	1	α	α^2	α^3	α^4	α^5	α^6	α^7	α^8	α^9	α^{10}	α^{11}	α^{12}	α^{13}

element of the field GF(2^n). The order of any element $\alpha^k \in$ GF(2^n) can be calculated from the following formula [2]:

$$\text{Order of } \alpha^k = \frac{2^n - 1}{\text{GCD}(2^n - 1, k)} \tag{6.8}$$

where GCD(n, k) is the greatest common divisor of n and k. We should distinguish between the order of the field (number of elements in the field) and the order of elements in the field. The order of α^7 in the example extension field of Tables 6.3 to 6.5 can be calculated as

$$\text{Order of } \alpha^7 = \frac{2^4 - 1}{\text{GCD}(2^4 - 1, 7)} = \frac{15}{\text{GCD}(15, 7)} = \frac{15}{1} = 15$$

Any elements of $GF(2^n)$ whose powers generate all the nonzero elements of $GF(2^n)$ are called *primitive*. For example, the powers of α^4 of $GF(2^4) = GF(16)$ are:

$$(\alpha^4)^0 = 1 \qquad (\alpha^4)^1 = \alpha^4 \qquad (\alpha^4)^2 = \alpha^8$$
$$(\alpha^4)^3 = \alpha^{12} \qquad (\alpha^4)^4 = \alpha^{16} = \alpha \qquad (\alpha^4)^5 = \alpha^{20} = \alpha^5$$
$$(\alpha^4)^6 = \alpha^{24} = \alpha^9 \qquad (\alpha^4)^7 = \alpha^{28} = \alpha^{13} \qquad (\alpha^4)^8 = \alpha^{32} = \alpha^2$$
$$(\alpha^4)^9 = \alpha^{36} = \alpha^6 \qquad (\alpha^4)^{10} = \alpha^{40} = \alpha^{10} \qquad (\alpha^4)^{11} = \alpha^{44} = \alpha^{14}$$
$$(\alpha^4)^{12} = \alpha^{48} = \alpha^3 \qquad (\alpha^4)^{13} = \alpha^{52} = \alpha^7 \qquad (\alpha^4)^{14} = \alpha^{56} = \alpha^{11}$$

which generated $\{1, \alpha, \alpha^2, \ldots, \alpha^{13}, \alpha^{14}\}$, and thus α^4 is a primitive element of $GF(2^4)$. One can easily verify that α^3 is not a primitive element of $GF(2^4)$.

Example 6.4 Let us construct $GF(2^5) = GF(32)$ using an irreducible polynomial of degree 5: $f(x) = x^5 + x^2 + 1$. Again using α as a root of $f(x)$, we have $\alpha^5 = \alpha^2 + 1$ and Table 6.6 gives its representations.

A polynomial $f(x)$ of degree n that gives a complete set of 2^n distinctive elements, including 0, is called a primitive polynomial. Alternatively, an irreducible polynomial $f(x)$ of degree n is said to be primitive if $f(\alpha) = 0$ where α is a primitive element of $GF(2^n)$. We hasten to say here that an LFSR that is mechanized, based on a primitive polynomial, generates a maximal-length (PN) sequence. We present the development of this subject after some additional coverage of relevant materials in this and subsequent sections. It is not easy to recognize a primitive polynomial. However, there are tables of polynomials in which primitive polynomials are indicated. We also cover this subject later on.

6.2.1 Roots of Primitive Polynomials and Maximal-Length Binary Sequences

We considered several examples of Galois fields $GF(2^n)$ and gave representations of the fields for $n = 3, 4,$ and 5. The set of field elements in $GF(2^n)$ generated from an irreducible binary polynomial of degree n, say, $f(x)$, contains $2^n - 1$ nonzero elements and the zero element. We now

Table 6.6 A representation of $GF(2^5)$ generated from $x^5 + x^2 + 1$

0 and Powers of α	Polynomials over GF(2)	Sequence over GF(2)
$0 =$	0	0 0 0 0 0
$\alpha^0 =$	1	0 0 0 0 1
$\alpha^1 =$	α	0 0 0 1 0
$\alpha^2 =$	α^2	0 0 1 0 0
$\alpha^3 =$	α^3	0 1 0 0 0
$\alpha^4 =$	α^4	1 0 0 0 0
$\alpha^5 =$	$\alpha^2 + 1$	0 0 1 0 1
$\alpha^6 =$	$\alpha^3 + \alpha$	0 1 0 1 0
$\alpha^7 =$	$\alpha^4 + \alpha^2$	1 0 1 0 0
$\alpha^8 =$	$\alpha^3 + \alpha^2 + 1$	0 1 1 0 1
$\alpha^9 =$	$\alpha^4 + \alpha^3 + \alpha$	1 1 0 1 0
$\alpha^{10} =$	$\alpha^4 + 1$	1 0 0 0 1
$\alpha^{11} =$	$\alpha^2 + \alpha + 1$	0 0 1 1 1
$\alpha^{12} =$	$\alpha^3 + \alpha^2 + \alpha$	0 1 1 1 0
$\alpha^{13} =$	$\alpha^4 + \alpha^3 + \alpha^2$	1 1 1 0 0
$\alpha^{14} =$	$\alpha^4 + \alpha^3 + \alpha^2 + 1$	1 1 1 0 1
$\alpha^{15} =$	$\alpha^4 + \alpha^3 + \alpha^2 + \alpha + 1$	1 1 1 1 1
$\alpha^{16} =$	$\alpha^4 + \alpha^3 + \alpha + 1$	1 1 0 1 1
$\alpha^{17} =$	$\alpha^4 + \alpha + 1$	1 0 0 1 1
$\alpha^{18} =$	$\alpha + 1$	0 0 0 1 1
$\alpha^{19} =$	$\alpha^2 + \alpha$	0 0 1 1 0
$\alpha^{20} =$	$\alpha^3 + \alpha^2$	0 1 1 0 0
$\alpha^{21} =$	$\alpha^4 + \alpha^3$	1 1 0 0 0
$\alpha^{22} =$	$\alpha^4 + \alpha^2 + 1$	1 0 1 0 1
$\alpha^{23} =$	$\alpha^3 + \alpha^2 + \alpha + 1$	0 1 1 1 1
$\alpha^{24} =$	$\alpha^4 + \alpha^3 + \alpha^2 + \alpha$	1 1 1 1 0
$\alpha^{25} =$	$\alpha^4 + \alpha^3 + 1$	1 1 0 0 1
$\alpha^{26} =$	$\alpha^4 + \alpha^2 + \alpha + 1$	1 0 1 1 1
$\alpha^{27} =$	$\alpha^3 + \alpha + 1$	0 1 0 1 1
$\alpha^{28} =$	$\alpha^4 + \alpha^2 + \alpha$	1 0 1 1 0
$\alpha^{29} =$	$\alpha^3 + 1$	0 1 0 0 1
$\alpha^{30} =$	$\alpha^4 + \alpha$	1 0 0 1 0

see that n roots of any degree-n irreducible binary polynomial can be found in the Galois field $GF(2^n)$.

Let us first recognize, however, some fundamental facts about the roots of polynomials in a field of characteristic $p = 2$ [see (5.21)]; that is, over GF(2): If x_0 is a root of any polynomial over GF(2), then x_0^2, x_0^4, x_0^8, ..., $x_0^{2^{n-1}}$ are also the roots of the polynomial. To see this, let us consider

$$f(x) = a_0 + a_1 x + \cdots + a_{n-1} x^{n-1} + x^n \tag{6.9a}$$

Now, because x_0 is a root of $f(x)$, we have

$$f(x_0) = a_0 + a_1 x_0 + \cdots + a_{n-1} x_0^{n-1} + x_0^n = 0 \tag{6.9b}$$

To see that x_0^2 is also a root of $f(x)$, we need to calculate

$$f(x_0^2) = a_0 + a_1 x_0^2 + \cdots + a_{n-1} x_0^{2(n-1)} + x_0^{2n}$$
$$= a_0 + a_1(x_0^2) + \cdots + a_{n-1}(x_0^2)^{n-1} + (x_0^2)^n \tag{6.9c}$$

We say that (6.9c) is identical to

$$\left(a_0 + a_1 x_0 + \cdots + a_{n-1} x_0^{n-1} + x_0^n\right)^2 \tag{6.9d}$$

because, when we expand (6.9d), we obtain the terms equal to (6.9c) plus "cross terms" with coefficients of the form $2 \times (\cdot)$. But $2 = 1 + 1 = 0$ in a field of characteristic 2; that is, under modulo-2 addition. Thus, we have

$$f(x_0^2) = a_0 + a_1 x_0^2 + \cdots + a_{n-1} x_0^{2(n-1)} + x_0^{2n}$$
$$= \left(a_0 + a_1 x_0 + \cdots + a_{n-1} x_0^{n-1} + x_0^n\right)^2 = 0 \tag{6.9e}$$

We can continue this process to show that x_0^4, x_0^8, ..., $x_0^{2^{m-1}}$ are also roots of the polynomial $f(x)$.

Therefore, in the Galois field GF(2^3), if α is a root of $1 + x^2 + x^3$, then α^2 and α^4 are also roots of the polynomial. The fact that α, α^2, and α^4 are all roots of $1 + x^2 + x^3$ can be seen as follows:

$$1 + \alpha^2 + \alpha^3 = 0$$
$$1 + \left(\alpha^2\right)^2 + \left(\alpha^2\right)^3 = 1 + \alpha^4 + \alpha^6 = 1 + \alpha^4(1 + \alpha^2)$$
$$= 1 + \alpha^7 = 1 + \alpha^{7 \bmod 7} = 1 + \alpha^0 = 1 + 1 = 0$$

$$1 + \left(\alpha^4\right)^2 + \left(\alpha^4\right)^3 = 1 + \alpha + \alpha^5 = 1 + \alpha + \alpha^2(1 + \alpha^2)$$

$$= 1 + \alpha + \alpha^2 + \alpha(1 + \alpha^2) = 1 + \alpha^2 + \alpha^3 = 0$$

Now $\alpha^3 \in \mathrm{GF}(2^3)$ is a root of some irreducible polynomial, say, $f_3(x)$. By the results shown above, $f_3(\alpha^3) = f_3(\alpha^6) = f_3(\alpha^5) = 0$; that is, α^3, α^6, and $\alpha^{12} \equiv \alpha^5$ are the roots of $f_3(x)$. The polynomial $f_3(x)$ can be found by multiplying the roots:

$$f_3(x) = (x - \alpha^3)(x - \alpha^6)(x - \alpha^5) = x^3 + x + 1 \tag{6.10}$$

It can be verified easily that the order of these roots is 7, and hence $f_3(x)$ is also a primitive polynomial of degree 3.

If ω represents an element in $\mathrm{GF}(2^n) = \{0, 1, \alpha, \ldots, \alpha^{2^n-2}\}$, the elements $\{\omega^{2^0}, \omega^{2^1}, \omega^{2^2}, \ldots, \omega^{2^{n-1}}\}$ have the same order and these elements are the roots of the polynomial

$$M_\omega(x) = \prod_{i=0}^{n-1} \left(x - \omega^{2^i}\right) \tag{6.11}$$

and this polynomial is called the minimal polynomial of ω, and its degree n is called the degree of ω. The minimal polynomial is the lowest degree monic polynomial (one whose highest order term has the coefficient 1) having ω as a root. In the example just considered, if $\omega = \alpha$, the minimal polynomial of α, and hence α^2 and α^4, is the same as the primitive polynomial that generated the finite field. The polynomial $M_3(x)$ given in (6.11) is the minimal polynomial of the elements α^3, α^6, and $\alpha^{12} \equiv \alpha^5$. Because the elements ω, ω^2, ω^4, \ldots, $\omega^{2^{n-1}}$ must have the same minimal polynomial, these elements are called conjugates of ω, in the sense that the complex numbers "$+j$" and "$-j$" are called "conjugates" because of their being the roots of the same irreducible polynomial $x^2 + 1$.

We may thus group the elements of $\mathrm{GF}(2^n)$ into conjugate classes, as shown in the following examples.

Example 6.5 Construct a table showing the conjugate classes of roots in $\mathrm{GF}(2^3)$ generated by $f(x) = 1 + x^2 + x^3$: $\alpha^3 = \alpha^2 + 1$.

Solution: see Table 6.7. Note for future reference that the product of the minimal polynomials is given by

$$M_0(x)M_1(x)M_3(x) = (1 + x)(1 + x^2 + x^3)(1 + x + x^3) = 1 + x^7 \tag{6.12}$$

Table 6.7 Conjugate classes for GF(2^3) generated by $f(x) = 1 + x^2 + x^3$

Conjugate classes	Minimal polynomials $M_i(x)$	Order of roots
$\{0\}$		
$\{1\}$	$M_0(x) = 1 + x$	1
$\{\alpha, \alpha^2, \alpha^4\}$	$M_1(x) = 1 + x^2 + x^3$	7
$\{\alpha^3, \alpha^6, \alpha^5\}$	$M_3(x) = 1 + x + x^3$	7

Example 6.6 Construct a table showing the conjugate classes of roots in GF(2^4) generated by $f(x) = 1 + x + x^4$: $\alpha^4 = \alpha + 1$.

Solution: see Table 6.8.

Note that the order of the root α^3 is computed to be, by (6.8), $15/\text{GCD}(15, 3) = 15/3 = 5$, and the order of α^5 is 3. Again, the product of all the minimal polynomials is

$$(1 + x)(1 + x + x^4)(1 + x + x^2 + x^3 + x^4)(1 + x + x^2)(1 + x^3 + x^4)$$

$$= 1 + x^{15} \tag{6.13}$$

By looking at the order of the conjugate class, we immediately see whether the corresponding minimal polynomial is primitive. Note that $M_3(x)$ and $M_5(x)$ in the example are irreducible polynomials, but they are not primitive polynomials. That is, not every irreducible polynomial is primitive, whereas every primitive polynomial is irreducible. In the case of degree-4 polynomials

Table 6.8 Conjugate classes for GF(2^4) generated by $f(x) = 1 + x + x^4$

Conjugate classes	Minimal polynomials $M_i(x)$	Order of roots
$\{0\}$		
$\{1\}$	$M_0(x) = 1 + x$	1
$\{\alpha, \alpha^2, \alpha^4, \alpha^8\}$	$M_1(x) = 1 + x + x^4$	15
$\{\alpha^3, \alpha^6, \alpha^{12}, \alpha^9\}$	$M_3(x) = 1 + x + x^2 + x^3 + x^4$	5
$\{\alpha^5, \alpha^{10}\}$	$M_5(x) = 1 + x + x^2$	3
$\{\alpha^7, \alpha^{14}, \alpha^{13}, \alpha^{11}\}$	$M_7(x) = 1 + x^3 + x^4$	15

over GF(2), we note that there are only two primitive polynomials.

Example 6.7 Construct a table showing the conjugate classes of roots in GF(2^5) generated by $f(x) = 1 + x^2 + x^5$: $\alpha^5 = \alpha^2 + 1$.
Solution: see Table 6.9.

All of the degree-5 irreducible polynomials are primitive polynomials. Since the number $2^5 - 1 = 31$ is a prime number, it follows that GCD(31, k) = 1 for all k, which means that every element in GF(2^5) is primitive. This is an example of a Mersenne[2] prime-length code; that is, a prime of the form $2^n - 1$, where n is a prime. Not all such numbers are prime, however. For example, $2^{11} - 1$ is not a prime. Some Mersenne primes are listed in [1].
There is a simple formula that can be used to calculate $N_p(n)$, the number of primitive polynomials of degree n, and it is given by

$$N_p(n) = \frac{\phi(2^n - 1)}{n} \tag{6.14}$$

where $\phi(2^n - 1)$ is Euler's number, which is the number of positive integers including 1 that are relatively prime to the numbers less than $2^n - 1$. Using this formula, one can verify that $N_p(3) = \phi(7)/3 = 2$ and $N_p(5) = \phi(31)/5 = 6$, which checks out with the examples above. An example list of the number of possible primitive polynomials is given in Table 6.10.

Table 6.9 Conjugate classes for GF(2^5) generated by $f(x) = 1 + x^2 + x^5$

Conjugate classes	Minimal polynomials $M_i(x)$	Order of roots
$\{0\}$		
$\{1\}$	$M_0(x) = 1 + x$	1
$\{\alpha, \alpha^2, \alpha^4, \alpha^8, \alpha^{16}\}$	$M_1(x) = 1 + x^2 + x^5$	31
$\{\alpha^3, \alpha^6, \alpha^{12}, \alpha^{24}, \alpha^{17}\}$	$M_3(x) = 1 + x^2 + x^3 + x^4 + x^5$	31
$\{\alpha^5, \alpha^{10}, \alpha^{20}, \alpha^9, \alpha^{18}\}$	$M_5(x) = 1 + x + x^2 + x^4 + x^5$	31
$\{\alpha^7, \alpha^{14}, \alpha^{28}, \alpha^{25}, \alpha^{19}\}$	$M_7(x) = 1 + x + x^2 + x^3 + x^5$	31
$\{\alpha^{11}, \alpha^{22}, \alpha^{13}, \alpha^{26}, \alpha^{21}\}$	$M_{11}(x) = 1 + x + x^3 + x^4 + x^5$	31
$\{\alpha^{15}, \alpha^{30}, \alpha^{29}, \alpha^{27}, \alpha^{23}\}$	$M_{15}(x) = 1 + x^3 + x^5$	31

---

[2] Marin Mersenne (1588–1648), French monk, philosopher, and mathematician, who considered the numbers $2^p - 1$, where p is a prime, in an attempt to find a formula for prime numbers.

Table 6.10 Numbers of primitive polynomials of degree n $(2 \leq n \leq 20)$

Degree, n	Total number of primitive polynomials, $N_p(n) = \phi(2^n - 1)/n$
2	1
3	2
4	2
5	6
6	6
7	18
8	16
9	48
10	60
11	176
12	144
13	630
14	756
15	$1,800$
16	$2,048$
17	$7,710$
18	$7,716$
19	$27,594$
20	$24,000$

The field elements of $GF(2^n)$ can be generated by an LFSR mechanization of a binary primitive polynomial. The elements so generated account for every element of the Galois field except the zero element.

6.2.2 Reciprocal Polynomials and Tables of Irreducible Polynomials

Consider an irreducible polynomial of degree n given by

$$f(x) = 1 + c_1 x + c_2 x^2 + \cdots + c_{n-1} x^{n-1} + x^n \tag{6.15a}$$

Then its *reciprocal polynomial* $f^*(x)$ is given by

$$f^*(x) = x^n f(x^{-1})$$
$$= x^n(1 + c_1 x^{-1} + c_2 x^{-2} + \cdots + c_{n-1} x^{-n+1} + x^{-n})$$
$$= 1 + c_{n-1} x + c_{n-2} x^2 + \cdots + c_1 x^{n-1} + x^n \tag{6.15b}$$

If the degree-n polynomial $f(x)$ of (6.15a) is a representation of an $(n+1)$-tuple binary sequence

$$(1, c_1, c_2, \ldots, c_{n-1}, c_n), \qquad c_i \in \mathrm{GF}(2) \tag{6.15c}$$

the sequence represented by the reciprocal polynomial (6.15b) is

$$(1, c_{n-1}, c_{n-2}, \ldots, c_2, c_1, 1) \tag{6.15d}$$

In practice, therefore, we can easily write down the reciprocal polynomial by reading backward (denoted by $\Leftrightarrow$ below) the given sequence, which defined the polynomial $f(x)$. The following examples illustrate the method:

$$M_1(x) = x^5 + x^2 + 1 \rightarrow 1\,0\,0\,1\,0\,1$$
$$\Leftrightarrow 1\,0\,1\,0\,0\,1 \rightarrow M_1^*(x) = x^5 + x^3 + 1$$

$$M_2(x) = x^5 + x^4 + x^3 + x^2 + 1 \rightarrow 1\,1\,1\,1\,0\,1$$
$$\Leftrightarrow 1\,0\,1\,1\,1\,1 \rightarrow M_2^*(x) = x^5 + x^3 + x^2 + x + 1$$

$$M_3(x) = x^5 + x^4 + x^2 + x + 1 \rightarrow 1\,1\,0\,1\,1\,1$$
$$\Leftrightarrow 1\,1\,1\,0\,1\,1 \rightarrow M_3^*(x) = x^5 + x^4 + x^3 + x + 1$$

The task of finding the primitive polynomials of degree n is a difficult one, although we know, by (6.14), how many there are. Fortunately for practicing engineers, tables of irreducible polynomials are available in the literature. A well-known source, for example, is W. W. Peterson and E. J. Weldon's book [3]. An example entry from the table of Peterson and Weldon has the form

    Degree 5   1 45E   3 75G   5 67H

The numbers 45, 75, and 67 in this example represent the polynomials in octal notation:

$$45 \leftrightarrow 1\,0\,0\,1\,0\,1 \leftrightarrow x^5 + x^2 + 1$$
$$75 \leftrightarrow 1\,1\,1\,1\,0\,1 \leftrightarrow x^5 + x^4 + x^3 + x^2 + 1$$
$$67 \leftrightarrow 1\,1\,0\,1\,1\,1 \leftrightarrow x^5 + x^4 + x^2 + x + 1$$

The number 1 to the left of 45 means that if α^1 is assumed to be a root of the polynomial $45 = x^5 + x^2 + 1$, with $\alpha^5 = \alpha^2 + 1$, then the minimal polynomials of α^3 and α^5 are given by

$$\alpha^3 \rightarrow 75 \leftrightarrow f_3(x) = x^5 + x^4 + x^3 + x^2 + 1$$

$$\alpha^5 \rightarrow 67 \leftrightarrow f_5(x) = x^5 + x^4 + x^2 + x + 1$$

A letter E, F, G, or H following the polynomial code in the table indicates that the polynomial is primitive, whereas a letter A, B, C, or D indicates that the polynomial is not primitive. The number of primitive polynomials in degree-n polynomials is always an even number, and they are all reciprocal polynomial pairs. The table of Peterson and Weldon shows only half of the primitive polynomials, and we can always find their reciprocal polynomials by the methods we discussed previously. Some example polynomials were selected from Peterson and Weldon's table and are shown in Table 6.11.

Table 6.11 Some primitive polynomials over GF(2), from [3]

Deg.	Primitive polynomials	Deg.	Primitive polynomials
2	1 7H $\rightarrow x^2 + x + 1$	16	1 210013F $\rightarrow x^{16}+x^{12}+x^3+x+1$
3	1 13F $\rightarrow x^3 + x + 1$	17	1 400011E $\rightarrow x^{17} + x^3 + 1$
4	1 23F $\rightarrow x^4 + x + 1$	18	1 1000201E $\rightarrow x^{18} + x^7 + 1$
5	1 45F $\rightarrow x^5 + x^2 + 1$	19	1 2000047F $\rightarrow x^{19}+x^5+x^2+x+1$
6	1 103F $\rightarrow x^6 + x + 1$	20	1 4000011E $\rightarrow x^{20} + x^3 + 1$
7	1 211E $\rightarrow x^7 + x^3 + 1$	21	1 10000005E $\rightarrow x^{21} + x^2 + 1$
8	1 435E $\rightarrow x^8+x^4+x^3+x^2+1$	22	1 20000003F $\rightarrow x^{22} + x + 1$
9	1 1021E $\rightarrow x^9 + x^4 + 1$	23	1 40000041E $\rightarrow x^{23} + x^5 + 1$
10	1 2011E $\rightarrow x^{10} + x^3 + 1$	24	1 100000207F $\rightarrow x^{24}+x^7+x^2+x+1$
11	1 4005E $\rightarrow x^{11} + x^2 + 1$	25	1 200000011E $\rightarrow x^{25} + x^3 + 1$
12	1 10123F $\rightarrow x^{12}+x^6+x^4+x+1$	26	1 400000107F $\rightarrow x^{26}+x^6+x^2+x+1$
13	1 20033F $\rightarrow x^{13}+x^4+x^3+x+1$	27	1 1000000047E $\rightarrow x^{27}+x^5+x^2+x+1$
14	1 42103F $\rightarrow x^{14}+x^{10}+x^6+x+1$	28	1 2000000011E $\rightarrow x^{28} + x^3 + 1$
15	1 100003F $\rightarrow x^{15} + x + 1$	29	1 4000000005E $\rightarrow x^{29} + x^2 + 1$
		30	1 10040000007F $\rightarrow x^{30}+x^{23}+x^2+x+1$

6.2.3   Mechanization of Linear Feedback Shift Registers for Binary Irreducible Primitive Polynomials

Consider the general form of a monic binary irreducible primitive polynomial:

$$f(x) = 1 + c_1 x + c_2 x^2 + \cdots + c_i x^i + \cdots + c_{n-1} x^{n-1} + x^n, \quad c_i \in \mathrm{GF}(2)$$

$$(6.16)$$

If we generate $\mathrm{GF}(2^n)$ with this polynomial, using α as a root (primitive element), the set $\{\alpha^0, \alpha^1, \alpha^2, \ldots, \alpha^{2^n-2}\}$ accounts for the entire Galois field except the zero element. These elements can also be generated with "hardware" called LFSRs. There are two ways of "mechanizing" the binary irreducible (primitive) polynomial as an LFSR: Particular configurations are called the simple shift register generator (SSRG), as shown in Figure 6.2, and the modular shift register generator (MSRG), as shown in Figure 6.3.

The binary storage devices of the LFSR are designated in Figure 6.2 as R_1, R_2,..., R_n as each successive stage of the shift register. The coefficients

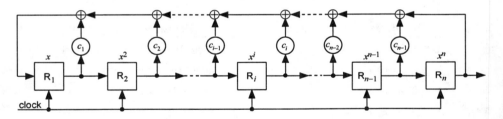

Figure 6.2 SSRG configuration of
$f(x) = 1 + c_1 x + c_2 x^2 + \cdots + c_i x^i + \cdots + c_{n-1} x^{n-1} + x^n.$

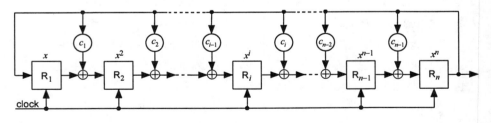

Figure 6.3 MSRG configuration of
$f(x) = 1 + c_1 x + c_2 x^2 + \cdots + c_i x^i + \cdots + c_{n-1} x^{n-1} + x^n.$

c_1, c_2,..., c_{n-1} of the polynomial play the role of "switches" in the mechanization. Note that the output of each stage R_i is connected to a modulo-2 adder through the switch c_i as a "tap" of the shift register. Therefore, if $c_i = 0$ for some stage i, there is no tap connection to the modulo-2 adder (and no adder), whereas if $c_i = 1$, the tap connection is made to the modulo-2 adder. Each register stage R_i is also identified with the power term x^i of the defining polynomial. When the proper tap connections are made in accordance with the given binary irreducible primitive polynomial, the combined (modulo-2 added) output of each stage is feedback to the first stage of the LFSR, R_1. To initiate the operation of the LFSR, we "load" an n-tuple sequence (vector) in the n-stage register and start clock pulses at an appropriate rate of design choice. As seen in the examples that follow, a nonzero n-tuple vector (loading vector) produces $2^n - 1$ distinct vectors in the register as the clock moves on. All the distinct $2^n - 1$ n-tuple vectors account for the sequence (or vector) representations of the Galois field elements of $GF(2^n)$, except the zero vector.

Another configuration for the mechanization of the primitive polynomial (6.16) is the MSRG shown in Figure 6.3. A distinct feature of the MSRG configuration is that the output of the last stage R_n is being fed back to each stage through the switches $\{c_i\}$ as shown in the figure. Note that the last stage is always fed back to the first stage R_1, and the output of stage R_i is modulo-2 added to the output of the last stage R_n, provided that $c_i \neq 0$, and fed to the next stage, R_{i+1}. Note that the configuration of the n-stage MSRG is to draw a modulo-2 adder at the output of each stage R_i if and only if the polynomial's coefficient c_i is not zero. If $c_i = 0$, the output of the stage R_i is directly shifted to the next stage. As in the case of the SSRG configuration, each register stage R_i is also identified with the power term x^i of the defining polynomial. When the n-stage register is loaded an n-tuple nonzero vector and starts shifting at the rate of the clock pulses, all of the nonzero elements in the Galois field $GF(2^n)$ are produced as the contents of the LFSR.

When the output is taken from the last stage of either the SSRG or the MSRG, the resulting sequence of length $2^n - 1$ is a pseudonoise sequence, which is one version (phase shift) out of $2^n - 1$ possible sequences. It is generated with a particular loading vector consisting of an n-tuple chosen from the $2^n - 1$ possible choices.

When the output is taken from the last stages of an SSRG and an MSRG, the resulting sequences are PN sequences. They are not the same PN sequences, however, in that they are not shifted versions of one another. In other words, they are distinct PN sequences. If the same PN sequences are desired from both an SSRG and an MSRG, we should mechanize the SSRG and MSRG in accordance with reciprocal polynomial pairs. Suppose we mechanize a given primitive polynomial $f(x)$ as an SSRG first, as shown in Figure 6.2, to generate a PN sequence. Then the same PN sequence can also be generated with an MSRG if we mechanize the reciprocal polynomial $f^*(x) = x^n f(x^{-1})$. The sequence so generated from the MSRG is the same sequence as generated from the SSRG, with only a possible relative phase difference. We consider several examples to show these relationships.

Another fact that needs to be pointed out is the sequence in which the Galois field elements are generated by an SSRG or an MSRG. Although both generators produce Galois fields GF(2^n), an MSRG generates the successive powers of α in GF(2^n), while an SSRG generates the elements in an arbitrary order. An example will illustrate this important fact below. Before taking up the examples that show many of these interesting properties related to the LFSRs, let us describe briefly the operations of the LFSRs.

If $x_i(t)$ is the number stored in R_i of the SSRG of Figure 6.2 after clock pulse t and $x_i(t+1)$ is the number stored in R_i after clock pulse $t+1$, for the polynomial shown in (6.16) we can write

$$x_n(t+1) = x_n(t)$$
$$x_{n-1}(t+1) = x_{n-1}(t)$$
$$\vdots \qquad\qquad\qquad (6.17a)$$
$$x_2(t+1) = x_2(t)$$
$$x_1(t+1) = x_n(t) + c_{n-1}x_{n-1}(t) + \cdots + c_2 x_2(t) + c_1 x_1(t)$$

where the addition is modulo-2. In matrix form,

$$\boldsymbol{X}(t+1) = \boldsymbol{T}_S\,\boldsymbol{X}(t) \qquad\qquad (6.17b)$$

where

$$\boldsymbol{X}(t) = \begin{bmatrix} x_n(t) \\ x_{n-1}(t) \\ \vdots \\ x_2(t) \\ x_1(t) \end{bmatrix}, \qquad \boldsymbol{X}(t+1) = \begin{bmatrix} x_n(t+1) \\ x_{n-1}(t+1) \\ \vdots \\ x_2(t+1) \\ x_1(t+1) \end{bmatrix} \qquad (6.17c)$$

and

$$T_S = \begin{bmatrix} 0 & 1 & 0 & 0 & \cdots & 0 & 0 \\ 0 & 0 & 1 & 0 & \cdots & 0 & 0 \\ 0 & 0 & 0 & 1 & \cdots & 0 & 0 \\ \vdots & & & & \vdots & & \vdots \\ 0 & 0 & 0 & 0 & \cdots & 0 & 1 \\ 1 & c_{n-1} & c_{n-2} & c_{n-3} & \cdots & c_2 & c_1 \end{bmatrix} \tag{6.17d}$$

The column vectors $X(t)$ and $X(t+1)$ are the "state" or content of the shift register after clock pulses t and $t+1$, respectively, and T_S is called the *characteristic matrix* of the SSRG. The primitive polynomial that generates the PN sequence is also called the *characteristic polynomial*, in the same sense that the matrix T_S is called the characteristic matrix.

For the MSRG of Figure 6.3, the transition equations are

$$x_n(t+1) = x_{n-1}(t) + c_{n-1}x_n(t)$$
$$x_{n-1}(t+1) = x_{n-2}(t) + c_{n-2}x_n(t)$$
$$\vdots \tag{6.17e}$$
$$x_2(t+1) = x_1(t) + c_1 x_n(t)$$
$$x_1(t+1) = x_n(t)$$

The matrix form of these equations is the same as (6.17a), except that the characteristic matrix for the MSRG is

$$T_M = \begin{bmatrix} c_{n-1} & 1 & 0 & 0 & \cdots & 0 & 0 \\ c_{n-2} & 0 & 1 & 0 & \cdots & 0 & 0 \\ c_{n-3} & 0 & 0 & 1 & \cdots & 0 & 0 \\ \vdots & & & & \vdots & & \vdots \\ c_1 & 0 & 0 & 0 & \cdots & 0 & 1 \\ 1 & 0 & 0 & 0 & \cdots & 0 & 0 \end{bmatrix} \tag{6.17f}$$

Example 6.8 Consider the irreducible primitive polynomial $f(x) = 1 + x^2 + x^3$. Mechanize an SSRG and an MSRG PN generator based on this polynomial, and list the state vectors as the clock pulses move on.

Solution: We use the characteristic matrix to compute the successive state vectors:

$$\begin{bmatrix} x_3(t+1) \\ x_2(t+1) \\ x_1(t+1) \end{bmatrix} = \begin{bmatrix} 0 & 1 & 0 \\ 0 & 0 & 1 \\ 1 & 1 & 0 \end{bmatrix} \begin{bmatrix} x_3(t) \\ x_2(t) \\ x_1(t) \end{bmatrix} \qquad \text{for an SSRG}$$

$$\begin{bmatrix} x_3(t+1) \\ x_2(t+1) \\ x_1(t+1) \end{bmatrix} = \begin{bmatrix} 1 & 1 & 0 \\ 0 & 0 & 1 \\ 1 & 0 & 0 \end{bmatrix} \begin{bmatrix} x_3(t) \\ x_2(t) \\ x_1(t) \end{bmatrix} \qquad \text{for an MSRG}$$

The two configurations are illustrated in Figure 6.4. If we assume that the initial state vector is $\boldsymbol{X}^{\mathrm{T}}(0) = [0\ 0\ 1]$, the succession of states for the two different LFSR configurations are as shown below in Table 6.12.

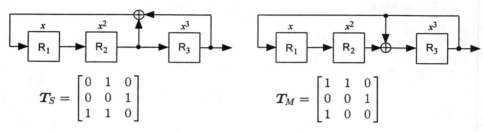

$$T_S = \begin{bmatrix} 0 & 1 & 0 \\ 0 & 0 & 1 \\ 1 & 1 & 0 \end{bmatrix} \qquad\qquad T_M = \begin{bmatrix} 1 & 1 & 0 \\ 0 & 0 & 1 \\ 1 & 0 & 0 \end{bmatrix}$$

Figure 6.4 SSRG and MSRG mechanizations of $f(x) = x^3 + x^2 + 1$.

Table 6.12 Succession of SSRG and MSRG states

Clock	SSRG		MSRG	
	$R_3\ R_2\ R_1$	State vectors in powers of α	$R_3\ R_2\ R_1$	State vectors in powers of α
$t = 0$	0 0 1	1	0 0 1	1
$t = 1$	0 1 0	α	0 1 0	α
$t = 2$	1 0 1	α^3	1 0 0	α^2
$t = 3$	0 1 1	α^5	1 0 1	α^3
$t = 4$	1 1 1	α^4	1 1 1	α^4
$t = 5$	1 1 0	α^6	0 1 1	α^5
$t = 6$	1 0 0	α^4	1 1 0	α^6
	Output from $R_3 = 0\ 0\ 1\ 0\ 1\ 1\ 1$		Output from $R_3 = 0\ 0\ 1\ 1\ 1\ 0\ 1$	
	Output from $R_2 = 0\ 1\ 0\ 1\ 1\ 1\ 0$		Output from $R_2 = 0\ 1\ 0\ 0\ 1\ 1\ 1$	
	Output from $R_1 = 1\ 0\ 1\ 1\ 1\ 0\ 0$		Output from $R_1 = 1\ 0\ 0\ 1\ 1\ 1\ 0$	

Observations on the example solutions in Table 6.12: The mechanizations of the primitive polynomial $f(x) = x^3 + x^2 + 1$ were done according to the rules stated. That is, for the SSRG configuration, the output of register R_2 is tapped (connected) to the modulo-2 adder feeding back to the input of the shift register because the coefficient $c_2 = 1$. For the MSRG, since $c_2 = 1$ there is a modulo-2 adder at the output of R_2 to combine this output with the feedback from the output of R_3, with the output of the adder shifted to R_3, as shown in the diagram in Figure 6.4. The SSRG operation and its state vectors were computed as follows:

$$
\begin{bmatrix} R_3(1) \\ R_2(1) \\ R_1(1) \end{bmatrix} = \begin{bmatrix} 0 & 1 & 0 \\ 0 & 0 & 1 \\ 1 & 1 & 0 \end{bmatrix} \begin{bmatrix} 0 \\ 0 \\ 1 \end{bmatrix} = \begin{bmatrix} 0 \\ 1 \\ 0 \end{bmatrix}, \quad
\begin{bmatrix} R_3(2) \\ R_2(2) \\ R_1(2) \end{bmatrix} = \begin{bmatrix} 0 & 1 & 0 \\ 0 & 0 & 1 \\ 1 & 1 & 0 \end{bmatrix} \begin{bmatrix} 0 \\ 1 \\ 0 \end{bmatrix} = \begin{bmatrix} 1 \\ 0 \\ 1 \end{bmatrix}
$$

$$
\begin{bmatrix} R_3(3) \\ R_2(3) \\ R_1(3) \end{bmatrix} = \begin{bmatrix} 0 & 1 & 0 \\ 0 & 0 & 1 \\ 1 & 1 & 0 \end{bmatrix} \begin{bmatrix} 1 \\ 0 \\ 1 \end{bmatrix} = \begin{bmatrix} 0 \\ 1 \\ 1 \end{bmatrix}, \quad
\begin{bmatrix} R_3(4) \\ R_2(4) \\ R_1(4) \end{bmatrix} = \begin{bmatrix} 0 & 1 & 0 \\ 0 & 0 & 1 \\ 1 & 1 & 0 \end{bmatrix} \begin{bmatrix} 0 \\ 1 \\ 1 \end{bmatrix} = \begin{bmatrix} 1 \\ 1 \\ 1 \end{bmatrix}
$$

$$
\begin{bmatrix} R_3(5) \\ R_2(5) \\ R_1(5) \end{bmatrix} = \begin{bmatrix} 0 & 1 & 0 \\ 0 & 0 & 1 \\ 1 & 1 & 0 \end{bmatrix} \begin{bmatrix} 1 \\ 1 \\ 1 \end{bmatrix} = \begin{bmatrix} 1 \\ 1 \\ 0 \end{bmatrix}, \quad
\begin{bmatrix} R_3(6) \\ R_2(6) \\ R_1(6) \end{bmatrix} = \begin{bmatrix} 0 & 1 & 0 \\ 0 & 0 & 1 \\ 1 & 1 & 0 \end{bmatrix} \begin{bmatrix} 1 \\ 1 \\ 0 \end{bmatrix} = \begin{bmatrix} 1 \\ 0 \\ 0 \end{bmatrix}
$$

From Table 6.1, which shows the representation of $GF(2^3)$ generated by $f(x) = 1 + x^2 + x^3$, in powers of α and polynomials over $GF(2)$, we can identify the state vectors in 3-tuples as "state vectors in powers of α" as shown in Table 6.12. Note that the generation of the SSRG state vectors is not in the order of successive powers of α in the field $GF(2^3)$. On the other hand, the state vectors in 3-tuples for the MSRG are given as the successive powers of α in the field $GF(2^3)$, according to the same table. The mechanization of irreducible binary polynomials in the MSRG configuration is an important practical application of the theory of PN sequence generation, as well as in cyclic code applications (discussed in Chapter 5).

Another distinct difference that can be seen in Table 6.12 between the two PN sequences generated by the SSRG and the MSRG is that, for the SSRG, the outputs taken from R_3 and R_2 are one-bit delayed versions of the previous stage's outputs. The same cannot be said about the MSRG, however. While the output of the SSRG taken from stage R_{n-i} $(i = 1, 2, \ldots,$ $n - 2)$ is a simple one-bit delayed version of the sequence from stage $R_{n-(i+1)}$, such a relationship does not exist in the case of the MSRG. The output from each stage of the MSRG configuration is more mathematically involved than the simple case of the SSRG. We derive detailed expressions for the MSRG register outputs in a later section. The SSRG output is based on a simple recursion relation of the form

$$a(k) = c_1 a(k - 1) + c_2 a(k - 2) + \cdots + c_n a(k - n), \quad k > n, \ c_i = 0 \text{ or } 1$$

(6.18)

For this reason, an LFSR configured as an SSRG is also known as the *Fibonacci configuration* (after the mathematician who studied linear recursion) [4]. An LFSR configured as an MSRG, on the other hand, is also known as the *Galois configuration* [4] because it is related to Galois field multiplication and division, as we have already seen. More is discussed on this topic in several sections to come in this chapter.

Note also from Table 6.12 that the sequence generated by the MSRG is the reverse of the sequence generated by the SSRG.

Example 6.9   Mechanize an MSRG configuration for the polynomial of Example 6.8 so that the PN sequences generated by it are the same sequences as given by the SSRG.

Solution: Recall that, given $f(x) = x^3 + x^2 + 1$, we need to obtain the reciprocal polynomial $f^*(x) = x^3 f(x^{-1})$ and mechanize $f^*(x)$ in an MSRG configuration to get the same PN sequence. Now,

$$f(x) = x^3 + x^2 + 1, \qquad f^*(x) = x^3 f(x^{-1}) = x^3 + x + 1$$

The mechanization of $f^*(x)$ is as shown in Figure 6.5, with the succession of states as shown in Table 6.13. Note that Table 6.1 cannot be used to identify the state vectors with powers of α, because a different degree-3 characteristic polynomial is involved. Note in Table 6.13 that the output sequence taken from the last stage, R_3, is the same as that for the SSRG in the previous example (Table 6.12). Had we started the generation of the sequence with the initial state vector 1 1 1 instead of 0 0 1, the output would have been 1 1 0 0 1 0 1, which is a 4-bit delayed version of the output sequence taken from R_3 of the SSRG in the Example 6.8.

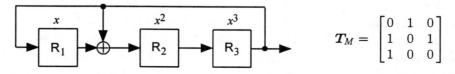

Figure 6.5 MSRG mechanization of $f^*(x) = x^3 + x + 1$.

Table 6.13 Succession of MSRG states for $f(x) = x^3 + x + 1$: $\alpha^3 = \alpha + 1$

Clock	R_3 R_2 R_1	State vectors in powers of α
$t = 0$	0 0 1	1
$t = 1$	0 1 0	α
$t = 2$	1 0 0	α^2
$t = 3$	0 1 1	α^3
$t = 4$	1 1 0	α^4
$t = 5$	1 1 1	α^5
$t = 6$	1 0 1	α^6
Output from $R_3 = 0$ 0 1 0 1 1 1		
Output from $R_2 = 0$ 1 0 1 1 1 0		
Output from $R_1 = 1$ 0 0 1 0 1 1		

Example 6.10 Construct the PN sequence generators for the primitive polynomial $f(x) = 1 + x^2 + x^5$ in both SSRG and MSRG configurations, and show the Galois field elements in vectors and in powers of the primitive element β.

Solution: see Figures 6.6 and 6.7, respectively, for the SSRG and MSRG.

Some important observations on the solutions of this example problem are in order: (1) When we compare the sequences taken from the last stage, R_5, of the two shift registers, we see that they are identical except for the point at which each sequence begins. In other words, these two sequences are identical up to a phase shift. (2) Although the output sequences are identical, the shift register vectors (Galois field elements) are not generated in the the identical order. (3) An MSRG (Galois) configuration is a multiplier of the shift register contents by the primitive element β. That is, if the initial vector is $\beta^2 = 0\ 0\ 1\ 0\ 0$, then, as the clock pulses move on, the register vectors (expressed in power form) become $\beta^2, \beta^3, \beta^4, \ldots, \beta^{30}, \beta^0, \beta^1$.

Having established the foundations for the theory of pseudorandom or pseudonoise sequence generation through the mechanizations of binary irreducible primitive polynomials of given degree choices, we are now ready to embark on keen examinations of the properties of such PN sequences in terms of practical generation of arbitrary phase shifts for engineering applications. There are some remarkable properties of the PN sequences, and also there are many efficient ways of generating the sequences for practical applications. The objective of the next few sections is to establish the basis for PN sequence generation by means of *masks*.

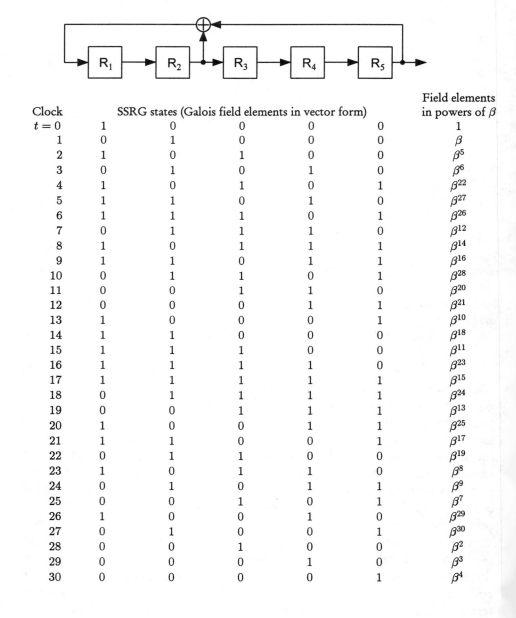

Clock	SSRG states (Galois field elements in vector form)					Field elements in powers of β
$t = 0$	1	0	0	0	0	1
1	0	1	0	0	0	β
2	1	0	1	0	0	β^5
3	0	1	0	1	0	β^6
4	1	0	1	0	1	β^{22}
5	1	1	0	1	0	β^{27}
6	1	1	1	0	1	β^{26}
7	0	1	1	1	0	β^{12}
8	1	0	1	1	1	β^{14}
9	1	1	0	1	1	β^{16}
10	0	1	1	0	1	β^{28}
11	0	0	1	1	0	β^{20}
12	0	0	0	1	1	β^{21}
13	1	0	0	0	1	β^{10}
14	1	1	0	0	0	β^{18}
15	1	1	1	0	0	β^{11}
16	1	1	1	1	0	β^{23}
17	1	1	1	1	1	β^{15}
18	0	1	1	1	1	β^{24}
19	0	0	1	1	1	β^{13}
20	1	0	0	1	1	β^{25}
21	1	1	0	0	1	β^{17}
22	0	1	1	0	0	β^{19}
23	1	0	1	1	0	β^8
24	0	1	0	1	1	β^9
25	0	0	1	0	1	β^7
26	1	0	0	1	0	β^{29}
27	0	1	0	0	1	β^{30}
28	0	0	1	0	0	β^2
29	0	0	0	1	0	β^3
30	0	0	0	0	1	β^4

Figure 6.6 SSRG (Fibonacci) configuration for $f(x) = 1 + x^2 + x^5$.

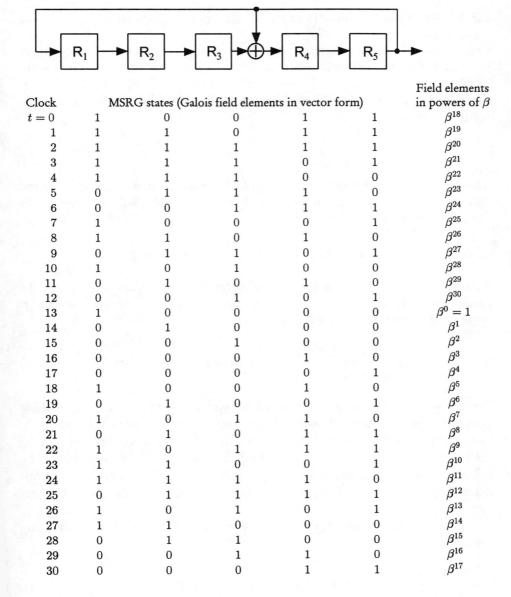

Clock	MSRG states (Galois field elements in vector form)					Field elements in powers of β
$t = 0$	1	0	0	1	1	β^{18}
1	1	1	0	1	1	β^{19}
2	1	1	1	1	1	β^{20}
3	1	1	1	0	1	β^{21}
4	1	1	1	0	0	β^{22}
5	0	1	1	1	0	β^{23}
6	0	0	1	1	1	β^{24}
7	1	0	0	0	1	β^{25}
8	1	1	0	1	0	β^{26}
9	0	1	1	0	1	β^{27}
10	1	0	1	0	0	β^{28}
11	0	1	0	1	0	β^{29}
12	0	0	1	0	1	β^{30}
13	1	0	0	0	0	$\beta^0 = 1$
14	0	1	0	0	0	β^1
15	0	0	1	0	0	β^2
16	0	0	0	1	0	β^3
17	0	0	0	0	1	β^4
18	1	0	0	1	0	β^5
19	0	1	0	0	1	β^6
20	1	0	1	1	0	β^7
21	0	1	0	1	1	β^8
22	1	0	1	1	1	β^9
23	1	1	0	0	1	β^{10}
24	1	1	1	1	0	β^{11}
25	0	1	1	1	1	β^{12}
26	1	0	1	0	1	β^{13}
27	1	1	0	0	0	β^{14}
28	0	1	1	0	0	β^{15}
29	0	0	1	1	0	β^{16}
30	0	0	0	1	1	β^{17}

Figure 6.7 MSRG (Galois) configuration for $f^*(x) = 1 + x^3 + x^5$.

6.2.4   State Vector Variations for PN Sequence Phase Shifts

Consider an SSRG mechanized from the primitive polynomial $f(x)$ of degree n. The LFSR for this situation is an n-stage generator with tap connections dictated by the given polynomial coefficients $\{c_i, i = 1, 2, \ldots, n\}$. A recursion equation of the form given by (6.18) applies for the Fibonacci type configuration, that is:

$$a_k = c_1 a_{k-1} + c_2 a_{k-2} + \cdots + c_n a_{k-n}, \qquad a_k, c_i \in \mathrm{GF}(2)$$

The output sequence from the SSRG can be denoted as $a_0, a_1, a_2, \ldots$, and this sequence can be represented as a polynomial, as we have done previously with the construction of Table 6.1:

$$a_0, a_1, a_2, \ldots \quad \Leftrightarrow \quad a(x) = a_0 + a_1 x + a_2 x^2 + \cdots$$

It can be shown that the power series for output sequence $a(x)$ can be calculated as the result of dividing a binary polynomial of degree less than n by the characteristic polynomial:

$$a(x) = \frac{g(x)}{f(x)}, \qquad g(x) = g_0 + g_1 x + \cdots + g_{n-1} x^{n-1} \tag{6.19}$$

Because $g_i = 0$ or 1, note that excluding $g(x) \equiv 0$ there are $2^n - 1$ possible numerator polynomials. For a PN sequence, which has period $P = 2^n - 1$, each of the possible numerator polynomials corresponds uniquely to one of the possible starting positions or "phase shifts" for the periodic sequence. The initial n terms of the sequence $a(x)$, which are called the initial conditions producing the particular phase shift of the sequence, can be calculated by long division, given $g(x)$.

For example, let $f(x) = 1 + x + x^3$ and $g(x) = 1 + x$. By long division we find that

$$\frac{1+x}{1+x+x^3} = 1 + x^3 + x^4 + x^5 + x^7 + x^{10} + x^{11} + x^{12} + \cdots$$

$$\overbrace{}^{\text{7-bit period}}$$

$$\rightarrow \quad \underbrace{1 \quad 0 \quad 0}_{\substack{\text{Initial} \\ \text{conditions}}} \quad 1 \quad 1 \quad 1 \quad 0 \quad 1 \quad 0 \quad 0 \quad 1 \quad 1 \quad 1 \quad 0 \ldots$$

Because the sequence is periodic with period P, its algebraic representation can be written

$$\frac{g(x)}{f(x)} = a(x) = \overbrace{a_0 + a_1 x + \cdots + a_{P-1}x^{P-1}}^{\text{Representation of first period}} + a_0 x^P + \cdots$$

$$= (a_0 + a_1 x + \cdots + a_{P-1}x^{P-1}) \cdot (1 + x^P + x^{2P} + x^{3P} + \cdots) \quad (6.20\text{a})$$

In modulo-2 arithmetic, addition and subtraction are the same operation, so that the geometric series in this equation can be expressed as

$$1 + x^P + x^{2P} + x^{3P} + \cdots = \frac{1}{1 - x^P} = \frac{1}{1 + x^P} \quad (6.20\text{b})$$

giving

$$\frac{g(x)}{f(x)} = a(x) = \frac{a_0 + a_1 x + \cdots + a_{P-1}x^{P-1}}{1 + x^P} \quad (6.20\text{c})$$

Let the first period of the sequence be denoted by the polynomial $b(x)$:

$$b(x) \triangleq a_0 + a_1 x + \cdots + a_{P-1}x^{P-1} \quad (6.20\text{d})$$

As written, $b(x)$ is a polynomial of degree $\leq P - 1$, since a_{P-1} can be zero. It is possible, however, to be more specific about the degree of $b(x)$. We observe that there is a definite relationship among the degrees of the polynomials $f(x)$, $g(x)$, and $b(x)$:

$$a(x) = \frac{\overbrace{g(x)}^{\text{degree} = D_g \leq n-1}}{\underbrace{f(x)}_{\text{degree} = n}} = \frac{\overbrace{b(x)}^{\text{degree} = D_b \leq P-1}}{\underbrace{1 + x^P}_{\text{degree} = P}} \quad (6.20\text{e})$$

$$\text{degree}\{f\} - \text{degree}\{g\} = P - \text{degree}\{b\} \Rightarrow D_b = D_g + P - n \quad (6.20\text{f})$$

The implication of the observation about the degree of the polynomial representing the first period of the PN sequence is that knowing the degree of $g(x)$, we can predict the number of 0s at the end of the first period; that is, the result given in (6.20f) implies that $b(x)$ ends with

$$(P - 1) - D_b = n - 1 - D_g \text{ zeros} \quad (6.20\text{g})$$

For example, if $g(x) = 1 + x$ (degree $D_g = 1$) and $f(x) = 1 + x + x^3$ ($n = 3$) we have seen previously that the period is $P = 7$. From this information alone, we can say that the number of zeros at the end of the first period is

$$n - 1 - D_g = 3 - 1 - 1 = 1$$

Indeed, the first period for this example is 1001110, with the polynomial representation $b(x) = 1 + x^3 + x^4 + x^5$.

There are $P = 2^n - 1$ different polynomials $b(x)$ for a given primitive polynomial $f(x)$, and each $b(x)$ is uniquely identified with a numerator polynomial $g(x)$. It is convenient to use a particular shift of the sequence as a reference, which involves reference cases of $g(x)$ and $b(x)$. The most common reference shift for PN sequences is the one for which $g(x) = 1$ (with degree $D_g = 0$). For this case, the degree of $b(x) \equiv b_0(x)$ is $D_b = D_g + P - n = P - n$. Let the subscript 0 denote the reference case. Then we can write

$$a_0(x) \triangleq \frac{1}{f(x)} = \frac{b_0(x)}{1 + x^P} \tag{6.21a}$$

Thus the starting bits in the sequence are understood to be those described by $b_0(x)$, and we know that the first period of the sequence $1/f(x)$ ends with $P - 1 - D_b = n - 1$ zeros. (This fact is very helpful later.)

Example 6.11 Consider the following sequence example showing the PN sequences corresponding to the polynomial $g(x)$ for the characteristic polynomial $f(x) = 1 + x + x^3$:

$$f(x) = 1 + x + x^3$$

Shift, k	First period, b_k	$g_k(x)$
0	1 1 1 0 1 0 0	1
1	0 1 1 1 0 1 0	x
2	0 0 1 1 1 0 1	x^2
3	1 0 0 1 1 1 0	$1 + x$
4	0 1 0 0 1 1 1	$x + x^2$
5	1 0 1 0 0 1 1	$1 + x + x^2$
6	1 1 0 1 0 0 1	$1 + x^2$

Corresponding to the "shift 1", we read the first period sequence 0 1 1 1 0 1 0. This sequence is the first period sequence corresponding to the "shift 0" with the entire sequence shifted *one* bit to the right and the last wrapped around to the beginning of the sequence to implement a one-bit *circular shift* to the right. Likewise, the last sequence, 1101001, is obtained from the first sequence by a *six*-bit circular shift to the right. We want to show the relationship between the number of bits shifted and the polynomial $g(x)$. For $g(x)$ to correspond to a circular shift of the bits in the first period of $1/f(x)$, it can be shown that

$$g(x) = x^k \text{ modulo } f(x) \triangleq g_k(x) \tag{6.21b}$$

That is, $g_k(x)$ must be taken as the remainder after dividing x^k by $f(x)$. Thus, in the example above,

$$x^0 = 1 + 0 \cdot (1 + x + x^3) \qquad \Rightarrow \qquad g_0(x) = 1$$
$$x^1 = x + 0 \cdot (1 + x + x^3) \qquad \Rightarrow \qquad g_1(x) = x$$
$$x^2 = x^2 + 0 \cdot (1 + x + x^3) \qquad \Rightarrow \qquad g_2(x) = x^2$$
$$x^3 = 1 + x + 1 \cdot (1 + x + x^3) \qquad \Rightarrow \qquad g_3(x) = 1 + x$$
$$x^4 = x + x^2 + x \cdot (1 + x + x^3) \qquad \Rightarrow \qquad g_4(x) = x + x^2$$
$$x^5 = 1 + x + x^2 + x^2 \cdot (1 + x + x^3) \qquad \Rightarrow \qquad g_5(x) = 1 + x + x^2$$
$$x^6 = 1 + x^2 + (1 + x + x^3)^2 \qquad \Rightarrow \qquad g_6(x) = 1 + x^2$$

What we have observed is that k, the number of bits being *circularly shifted*, is the same k in $g(x) = x^k$ in (6.21b). Thus, we can always relate the shifted PN sequence to numerator polynomial given in power form.

Note that when the numerator polynomial $g_k(x)$ has more than one nonzero term, the sequence shift $a_k(x)$ is the superposition of other shifts. For example, for the fourth shift,

$$a_4(x) = \frac{g_4(x)}{f(x)} = \frac{x + x^2}{f(x)} = \frac{x}{f(x)} + \frac{x^2}{f(x)} = a_1(x) + a_2(x)$$

b_1:		$0\ 1\ 1\ 1\ 0\ 1\ 0$	An illustration of the
b_2:	$\oplus$	$0\ 0\ 1\ 1\ 1\ 0\ 1$	closure property
b_4:		$0\ 1\ 0\ 0\ 1\ 1\ 1$	of PN sequences

Note that the sequence b_4 is obtained as the modulo-2 sum of sequences b_1 and b_2. This observation is also made in terms of $g_4(x)$, in that $g_4(x)$ is the sum of $g_1(x)$ and $g_2(x)$. We extend these observations in what follows.

In general, any shift of the reference sequence b_0 can be seen as the superposition of one or more of the first $n - 1$ shifts:

$$a_k(x) = \frac{g_k(x)}{f(x)} = \frac{g_{k0}+g_{k1}x+\cdots+g_{k,n-1}x^{n-1}}{f(x)}$$

$$= \sum_{i=0}^{n-1} g_{ki} \cdot \frac{x^i}{f(x)} = \sum_{i=0}^{n-1} g_{ki}\, a_i(x) \qquad (6.22)$$

In summary, there are unique relationships among the polynomials representing a particular shift of a PN sequence:

$$a_k(x),\ \text{shift } k \quad \Leftrightarrow \quad g_k(x) = x^k \text{ modulo } f(x),$$
$$\text{degree } D_g$$

$$\Updownarrow \qquad\qquad\qquad\qquad \Updownarrow$$

$$\overbrace{b_k(x),\ \text{first period, with } n - 1 - D_g \text{ zeros at end}}$$

6.3 Shift Register Implementation of PN Sequences

As illustrated in Figure 6.8, a straightforward implementation of a PN sequence is to use an n-stage shift register as a delay line and to cause the input to the shift register to be a function of its state:

$$f(x) = 1 + c_1 x + c_2 x^2 + \cdots + c_{n-1} x^{n-1} + x^n, \quad \text{where } c_i = 0 \text{ or } 1$$

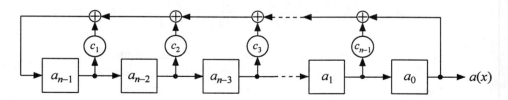

Figure 6.8 SSRG implementation with the initial loading vector $[a_{n-1}, a_{n-2}, \ldots, a_1, a_0]$.

Note that to generate the sequence $a(x) = a_0 + a_1 x + a_2 x^2 + \cdots$, the shift register must be initialized or "loaded" with the first n bits of the sequence. The loading or initial state can be calculated from a specified numerator polynomial, $g(x)$, by finding $(a_0, a_1, \ldots, a_{n-1})$ as the coefficients of the first n terms of the ratio of polynomials $g(x)/f(x)$. Conversely, given the loading[3] $(a_0, a_1, \ldots, a_{n-1})$, the numerator polynomial $g(x)$ can be found as

$$g(x) = \{f(x) \cdot (a_0 + a_1 x + \cdots + a_{n-1} x^{n-1})\}_{\text{terms of degree} < n} \qquad (6.23)$$

Example 6.12 Calculation of the loading vector from the numerator polynomial. Let $f(x) = 1 + x^3 + x^5$ and $g(x) = 1 + x$. The first 5 terms of $g(x)/f(x)$ are found by long division as follows:

$$
\begin{array}{l}
\;\; 1 + x + x^3 + x^4 + \cdots \quad \Rightarrow \qquad (a_0, a_1, a_2, a_3, a_4) \\
1 + x^3 + x^5 |\;\; 1 + x = (1, 1, 0, 1, 1)\\
\;\; 1 + x^3 + x^5 \\
\;\; x + x^3 + x^5 \\
\;\; x + x^4 + x^6 \\
\;\; x^3 + x^4 + x^5 + x^6 \\
\;\; x^3 + x^6 + x^8 \\
\;\; x^4 + x^5 + x^8
\end{array}
$$

Example 6.13 Calculation of the numerator polynomial from the loading vector. Let $f(x) = 1 + x^3 + x^5$ and the let the initial loading be $(1, 1, 0, 1, 1)$. The numerator polynomial is calculated as the degree $< n = 5$ terms of the product of $f(x)$ and the polynomial representation of the loading:

$$g(x) = \{f(x) \cdot (a_0 + a_1 x + a_2 x^2 + a_3 x^3 + a_4 x^4)\}_{\text{degree} < 5}$$

$$= \{(1 + x^3 + x^5) \cdot (1 + x + x^3 + x^4)\}_{\text{degree} < 5}$$

$$= \Big\{ \underbrace{1 + \cancel{x^3} + x^5}_{1 \cdot f(x)} + \underbrace{x + \cancel{x^4} + \cancel{x^6}}_{x \cdot f(x)} + \underbrace{\cancel{x^3} + \cancel{x^6} + x^8}_{x^3 \cdot f(x)} + \underbrace{\cancel{x^4} + x^7 + x^9}_{x^4 \cdot f(x)} \Big\}_{\text{degree} < 5}$$

$$= 1 + x$$

---

[3] Previously we identified the register stages with letters $R_1, R_2, \ldots, R_n$. We now designate the contents of the registers by the letters $a_0, a_1, a_2, \ldots, a_{n-1}$, with a_0 being the content in the last stage R_n, the first to be shifted out of the register.

In what follows, we examine the effects of the initial loading of the shift register upon the sequence at the output of the shift register generator. Special cases of the loading are used to establish certain properties of the shift register generators. Then certain general properties of the shift register generator are found, based on what we have established for the special cases.

6.3.1   Shift Register Generators With Special Loading Vectors

Let the loading of an SSRG be

$$\overbrace{\underbrace{1\ 0\ 0\cdots 0\ 0}_{n-1 \text{ zeros}}}$$

as illustrated in Figure 6.9. After the generator starts, the zeros that were loaded into the register appear at the output first, then the 1. We know that the 1 is the beginning of the reference sequence $a_0(x) = 1/f(x)$ because each period of $a_0(x)$ ends in $n-1$ zeros. Therefore, the output of the shift register, which is delayed from the first stage by $n-1$ bits, is $x^{n-1}/f(x)$. It follows also for this special case that the sequence appearing at the output of stage i is simply $a_{i-1}(x) = x^{i-1}/f(x)$, $i = 1, 2, \ldots, n$.

From this information, we want to determine the shift register output sequences when the initial loading is another special case: all zeros except for a 1 in the last stage:

$$\overbrace{\underbrace{0\ 0\cdots 0\ 0\ 1}_{n-1 \text{ zeros}}}$$

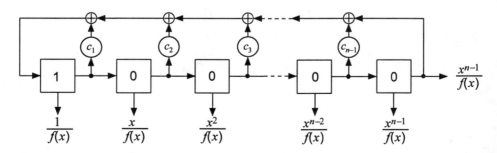

Figure 6.9   Special SSRG case of loading all zeros except the first stage.

Again, we use the fact that the period of the reference sequence $1/f(x)$ ends in $n - 1$ zeros. With the assumed loading, after the generator starts, the output of the shift register is a 1 followed by $n - 1$ zeros—the last n bits of the sequence $1/f(x)$. Thus, the output of the shift register is $1/f(x)$ rotated to the right by n bits, as shown in Figure 6.10:

$$\frac{x^n \text{ modulo } f(x)}{f(x)} \triangleq \frac{g_n(x)}{f(x)} \tag{6.24}$$

Because the sequence at the output (nth stage) of the shift register is an n-bit shift to the right of the reference sequence, it follows that the sequence at the ith stage is an i-bit shift to the right, or $x^i/f(x)$, $1 \leq i < n$. The situation can be understood from the following comparison of the $10\cdots0$ loading illustrated by Figure 6.9 and the $00\cdots01$ loading illustrated by Figure 6.10:

$$\text{Loading} = 1\ 0\ 0 \overset{\cdots}{\rightarrow} 0\ 0\ 0 \implies \text{output} = \overbrace{0\ 0\ 0 \cdots 0\ 0}^{\text{end of } 1/f(x)} \underset{\substack{\uparrow \\ \text{first stage} = 1/f(x) \\ = x^{n-1}/f(x)}}{1} \cdots$$

(last stage is at the left $\uparrow$)

$$\text{Loading} = 0\ 0\ 0 \overset{\cdots}{\rightarrow} 0\ 0\ 1 \implies \text{output} = \underset{\substack{\uparrow \\ \text{last stage} \\ = x^n/f(x)}}{1} \overbrace{0\ 0\cdots 0\ 0\ 0}^{\text{end of } 1/f(x)} \underset{\substack{\uparrow \\ \text{first stage} = x/f(x)}}{1} \cdots$$

Just as we did for SSRGs, we study the properties of MSRGs for some special initial states, then generalize the results that we find for the special

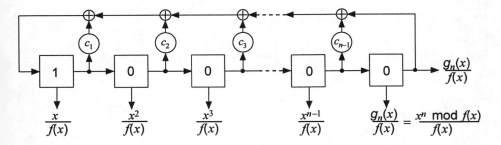

Figure 6.10 Special SSRG case of loading all zeros except the last stage.

cases. Let the initialization or loading of an MSRG be

$$\overbrace{1\ \underset{n-1\ \text{zeros}}{\underbrace{0\ 0\cdots 0\ 0}}}^{\longrightarrow}$$

After the generator starts, the zeros loaded into the register appear at the output first, then the 1, in the same order as for a SSRG for this special case, because as long as the last register has a zero in it, the feedback has no effect. Thus, the output sequence for this special case is

$$\text{MSRG special case: loading} = 100\cdots 0 \;\Rightarrow\; \text{output} = \frac{x^{n-1}}{f(x)} \qquad (6.25)$$

This case is illustrated in Figure 6.11. Unlike the SSRG for the same loading, it is not obvious what the sequence is for any of the other outputs of the MSRG stages, except for the first one.

Another special case: Let the initial loading of an MSRG be

$$\overbrace{\underset{n-1\ \text{zeros}}{\underbrace{0\ 0\cdots 0\ 0}}\ 1}^{\longrightarrow}$$

After the generator starts, the 1 loaded into the register appears at the output first, after which the output depends on the feedback connections, unlike the SSRG for this special case. However, because we know the succession of states, we can figure out the previous states.

This loading state is $s(\beta) = \beta^{n-1} = \beta^{n-1} \bmod f^*(\beta)$. Had the register been running, the previous $n - 1$ states would have been $s(\beta) = 1,\ \beta,\ \beta^2,\dots,$ β^{n-2}, for which the output would have been a series of $n - 1$ zeros, the end

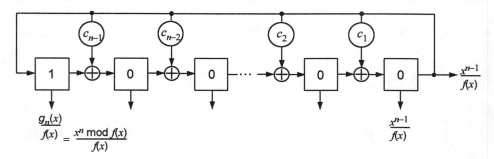

Figure 6.11 Special MSRG case of loading all zeros except the first stage.

of the sequence $1/f(x)$. Thus, the output sequence for this special case is

$$\text{MSRG special case: loading} = 00\cdots01 \Rightarrow \text{output} = \frac{1}{f(x)} \tag{6.26a}$$

This case is illustrated in Figure 6.12. Again, unlike the SSRG for the same loading, it is not obvious what the sequence is for any of the other outputs of the MSRG stages, except for the first one.

From the two special cases, we have observed the following results:

Initial state, $s(\beta)$	MSRG output	Numerator polynomial, $g(x)$
$10\cdots0 = \beta^0 = 1$	$x^{n-1}/f(x)$	x^{n-1}
$0\cdots01 = \beta^{n-1}$	$1/f(x)$	1

That is, when the initial state polynomial is

$$s(\beta) = r_1(0) + r_2(0)\,\beta + \cdots + r_n(0)\,\beta^{n-1} = \begin{cases} 1 & \text{or} \\ \beta^{n-1} \end{cases} \tag{6.26b}$$

then the numerator polynomial is

$$g(x) = g_0 + g_1\,x + \cdots + g_{n-1}\,x^{n-1} = \begin{cases} x^{n-1} & \text{or} \\ 1 \end{cases} \tag{6.26c}$$

The results for the special cases suggest a relationship between $g(x)$ and the reverse of $s(\beta)$, denoted $s^*(\beta)$, where

$$s^*(\beta) \triangleq \beta^{n-1}\, s(\beta^{-1}) = r_1(0)\,\beta^{n-1} + r_2(0)\,\beta^{n-2} + \cdots + r_{n-1}(0)\,\beta + r_n(0)$$

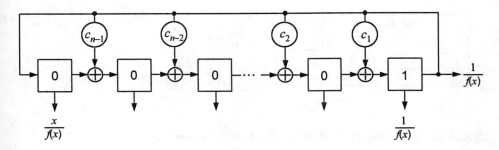

Figure 6.12 Special MSRG case of loading all zeros except the last stage.

It is shown in Appendix 6A that in general, the relationship between the initial state of the MSRG and the output sequence is

$$\text{Output} = \frac{g(x)}{f(x)} = \frac{x^{n-1}s(x^{-1})}{f(x)} = \frac{s^*(x)}{f(x)} \tag{6.27}$$

where $s^*(x)$ denotes the reverse of $s(\beta)$, with x substituted for β. In other words, the MSRG is a circuit that divides the polynomial representing the reverse of the initial state by the characteristic polynomial.

The relationship has been proved, namely that the numerator polynomial of the output sequence produced by an MSRG is the reverse of the polynomial representing the initial state of the MSRG, as illustrated in Figure 6.13.

When $s(x)$ is the initial state vector for the MSRG, the output sequence for the MSRG can always be represented by $s^*(x)/f(x)$. For example, let the characteristic polynomial and numerator polynomial of the desired sequence be $f(x) = 1 + x^2 + x^5$ and $g(x) = x^2 + x^3$. Then the sequence is

$$a(x) = \frac{x^2 + x^3}{1 + x^2 + x^5} = x^2 + x^3 + x^4 + x^5 + x^6 + x^9 + x^{10} + x^{12} + x^{15}$$
$$+ x^{20} + x^{22} + x^{24} + x^{25} + x^{26} + x^{28} + x^{29} + \cdots$$

$\Rightarrow$

$$\underset{\substack{0\ \ 1\ \ 2\ \ 3\ \ 4\ \ 5\ \ 6\ \ 7\ \ 8\ \ 9\ 10\ 11\ 12\ 13\ 14\ 15}}{0\ 0\ 1\ 1\ 1\ 1\ 1\ 0\ 0\ 1\ 1\ 0\ 1\ 0\ 0\ 1}\ \underset{\substack{16\ 17\ 18\ 19}}{0\ 0\ 0\ 0}\ \underset{\substack{20\ 21\ 22\ 23\ 24\ 25\ 26\ 27\ 28\ 29\ 30}}{1\ 0\ 1\ 0\ 1\ 1\ 1\ 0\ 1\ 1\ 0}\cdots$$
$$n - 1 = 4 \text{ zeros}$$

By inspection of the sequence, which is short enough to calculate in this example, we see that the output of the shift register generator is

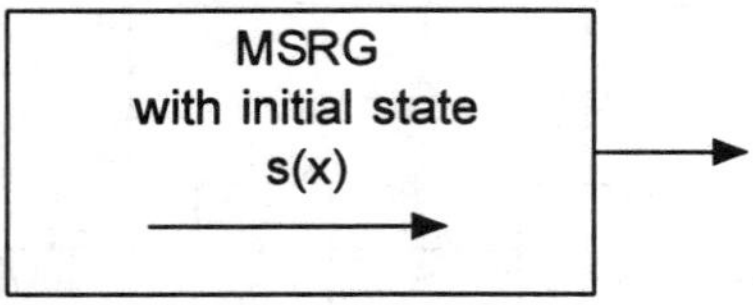

$$\frac{g(x)}{f(x)} = \frac{s^*(x)}{f(x)}$$

$$= \frac{r_n(0) + r_{n-1}(0)\,x + \cdots + r_1(0)\,x^{n-1}}{f(x)}$$

Figure 6.13 Relationship of MSRG initial state to output sequence.

$$\frac{g_{20}(x)}{f(x)} = \frac{x^{20} \text{ modulo } f(x)}{f(x)}$$

We observe that the shift of the sequence relative to $1/f(x)$ is the 20th shift to the right. The shift can be found algebraically by reversing the modulo $f(x)$ calculation (substitute $1 = 1 + f(x) = x^2 + x^5$ until a single term results):

$$
\begin{aligned}
x^k \text{ modulo } f(x) = g(x) &= x^2 + x^3 = x^2(1 + x) \\
&= x^2(x^2 + x^5 + x) = x^3(1 + x + x^4) \\
&= x^3(x^2 + x^5 + x + x^4) = x^4(1 + x + x^3 + x^4) \\
&= x^4(x^2 + x^5 + x + x^3 + x^4) = x^5(1 + x + x^2 + x^3 + x^4) \\
&= x^5(\cancel{x^2} + x^5 + x + \cancel{x^2} + x^3 + x^4) = x^6(1 + x^2 + x^3 + x^4) \\
&= x^6(\cancel{x^2} + x^5 + \cancel{x^2} + x^3 + x^4) = x^9(1 + x + x^2) \\
&= x^9(\cancel{x^2} + x^5 + x + \cancel{x^2}) = x^{10}(1 + x^4) \\
&= x^{10}(x^2 + x^5 + x^4) = x^{12}(1 + x^2 + x^3) \\
&= x^{12}(\cancel{x^2} + x^5 + \cancel{x^2} + x^3) = x^{15}(1 + x^2) \\
&= x^{15}(\cancel{x^2} + x^5 + \cancel{x^2}) = x^{20} \quad \rightarrow \quad \boxed{k = 20}
\end{aligned}
$$

For the same characteristic polynomial and numerator polynomial, the generation of the specified sequence using an SSRG is straightforward: Because $f(x)$ is given to be $1 + x^2 + x^5$, the feedback to the input stage consists of the modulo-2 addition of the second and fifth stages; also, the initial loading of the register is the first n bits of the sequence, $\underbrace{1\ 1\ 1\ 0\ 0}$. The SSRG implementation of the sequence is shown in Figure 6.14.

To produce the same sequence using an MSRG, we observe that the reciprocal polynomial is

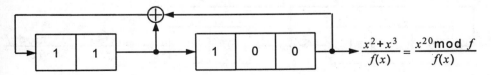

$$\frac{x^2 + x^3}{f(x)} = \frac{x^{20} \bmod f}{f(x)}$$

Figure 6.14 SSRG implementation of the sequence $\dfrac{x^2 + x^3}{1 + x^2 + x^5}$.

$$f^*(x) = x^5 \cdot f(x^{-1}) = x^5(1 + x^{-2} + x^{-5}) = x^5 + x^3 + 1 = 1 + x^3 + x^5$$

By observation of this polynomial, we know that the feedback is combined with the output of the third stage giving the MSRG implementation shown in Figure 6.15.

The initial register state of the MSRG is

$$s(x) = g^*(x) = x + x^2 \quad \Rightarrow \quad \overrightarrow{0\,1\,1\,0\,0}$$

We know that successive states of the MSRG may be represented as powers of β, modulo the reciprocal polynomial, $f^*(\beta) = 1 + \beta^3 + \beta^5$. Using the same algorithm we used previously to find k, we find that the power β^p that the initial state represents is

$$
\begin{aligned}
\beta^p \text{ modulo } f^*(\beta) = s(\beta) &= g^*(\beta) = \beta + \beta^2 \\
&= \beta(1 + \beta) = \beta(\beta^3 + \beta^5 + \beta) = \beta^2(1 + \beta^2 + \beta^4) \\
&= \beta^2(\beta^3 + \beta^5 + \beta^2 + \beta^4) = \beta^4(1 + \beta + \beta^2 + \beta^3) \\
&= \beta^4\left(\beta^3 + \beta^5 + \beta + \beta^2 + \beta^3\right) = \beta^5(1 + \beta + \beta^4) \\
&= \beta^5(\beta^3 + \beta^5 + \beta + \beta^4) = \beta^6(1 + \beta^2 + \beta^3 + \beta^4) \\
&= \beta^6\left(\beta^3 + \beta^5 + \beta^2 + \beta^3 + \beta^4\right) = \beta^8(1 + \beta^2 + \beta^3) \\
&= \beta^8\left(\beta^3 + \beta^5 + \beta^2 + \beta^3\right) = \beta^{10}(1 + \beta^3) \\
&= \beta^{10}\left(\beta^3 + \beta^5 + \beta^3\right) = \beta^{15} \quad \Rightarrow \quad \boxed{p = 15}
\end{aligned}
$$

The state, power of β modulo $f^*(\beta)$, and output sequence for this example are shown in Table 6.14.

6.3.2   Derivation of Sequences at the MSRG Outputs

We know that the output of the MSRG is the sequence $g(x)/f(x)$, where the numerator polynomial $g(x)$ is the reverse of the initial state of the shift

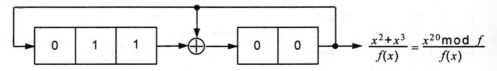

Figure 6.15 MSRG implementation of the sequence $\dfrac{x^2 + x^3}{1 + x^2 + x^5}$.

Table 6.14 States of the MSRG example of $f^*(x) = 1 + x^3 + x^5$

Time	State	Output
0	$0\,1\,\overrightarrow{1}\,0\,0 = \beta^{15}\mathrm{mod}\,f^* = \beta + \beta^2$	0
1	$0\,0\,1\,1\,0 = \beta^{16}\mathrm{mod}\,f^* = \beta^2 + \beta^3$	0
2	$0\,0\,0\,1\,1 = \beta^{17}\mathrm{mod}\,f^* = \beta^3 + \beta^4$	1
3	$1\,0\,0\,1\,1 = \beta^{18}\mathrm{mod}\,f^* = 1 + \beta^3 + \beta^4$	1
4	$1\,1\,0\,1\,1 = \beta^{19}\mathrm{mod}\,f^* = 1 + \beta + \beta^3 + \beta^4$	1
5	$1\,1\,1\,1\,1 = \beta^{20}\mathrm{mod}\,f^* = 1 + \beta + \beta^2 + \beta^3 + \beta^4$	1
6	$1\,1\,1\,0\,1 = \beta^{21}\mathrm{mod}\,f^* = 1 + \beta + \beta^2 + \beta^4$	1
7	$1\,1\,1\,0\,0 = \beta^{22}\mathrm{mod}\,f^* = 1 + \beta + \beta^2$	0
8	$0\,1\,1\,1\,0 = \beta^{23}\mathrm{mod}\,f^* = \beta + \beta^2 + \beta^3$	0
9	$0\,0\,1\,1\,1 = \beta^{24}\mathrm{mod}\,f^* = \beta^2 + \beta^3 + \beta^4$	1
10	$1\,0\,0\,0\,1 = \beta^{25}\mathrm{mod}\,f^* = 1 + \beta^4$	1
11	$1\,1\,0\,1\,0 = \beta^{26}\mathrm{mod}\,f^* = 1 + \beta + \beta^3$	0
12	$0\,1\,1\,0\,1 = \beta^{27}\mathrm{mod}\,f^* = \beta + \beta^2 + \beta^4$	1
13	$1\,0\,1\,0\,0 = \beta^{28}\mathrm{mod}\,f^* = 1 + \beta^2$	0
14	$0\,1\,0\,1\,0 = \beta^{29}\mathrm{mod}\,f^* = \beta + \beta^3$	0
15	$0\,0\,1\,0\,1 = \beta^{30}\mathrm{mod}\,f^* = \beta^2 + \beta^4$	1
16	$1\,0\,0\,0\,0 = \beta^{31}\mathrm{mod}\,f^* = \beta^0 = 1$	0
17	$0\,1\,0\,0\,0 = \beta$	0
18	$0\,0\,1\,0\,0 = \beta^2$	0
19	$0\,0\,0\,1\,0 = \beta^3$	0
20	$0\,0\,0\,0\,1 = \beta^4$	1
21	$1\,0\,0\,1\,0 = \beta^5\mathrm{mod}\,f^* = 1 + \beta^3$	0
22	$0\,1\,0\,0\,1 = \beta^6\mathrm{mod}\,f^* = \beta + \beta^4$	1
23	$1\,0\,1\,1\,0 = \beta^7\mathrm{mod}\,f^* = 1 + \beta^2 + \beta^3$	0
24	$0\,1\,0\,1\,1 = \beta^8\mathrm{mod}\,f^* = \beta + \beta^3 + \beta^4$	1
25	$1\,0\,1\,1\,1 = \beta^9\mathrm{mod}\,f^* = 1 + \beta^2 + \beta^3 + \beta^4$	1
26	$1\,1\,0\,0\,1 = \beta^{10}\mathrm{mod}\,f^* = 1 + \beta + \beta^4$	1
27	$1\,1\,1\,1\,0 = \beta^{11}\mathrm{mod}\,f^* = 1 + \beta + \beta^2 + \beta^3$	0
28	$0\,1\,1\,1\,1 = \beta^{12}\mathrm{mod}\,f^* = \beta + \beta^2 + \beta^3 + \beta^4$	1
29	$1\,0\,1\,0\,1 = \beta^{13}\mathrm{mod}\,f^* = 1 + \beta^2 + \beta^4$	1
30	$1\,1\,0\,0\,0 = \beta^{14}\mathrm{mod}\,f^* = 1 + \beta$	0

register when expressed as a polynomial, and that the contents for the registers R_n, $R_{n-1},\ldots$, R_2, R_1 at clock pulses t and $t+1$ for the MSRG configured for $f(x) = x^n + c_{n-1}x^{n-1} + \cdots + c_2x^2 + c_1x + 1$ are related by

$$r_n(t+1) = c_1 \, r_n(t) + r_{n-1}(t)$$
$$r_{n-1}(t+1) = c_2 \, r_n(t) + r_{n-2}(t)$$
$$\vdots \qquad \qquad \vdots$$
$$r_2(t+1) = c_{n-1} \, r_n(t) + r_1(t)$$
$$r_1(t+1) = r_n(t)$$

We derive algebraic expressions for the outputs of the individual registers of the MSRG. In matrix form, the transitions may be expressed by the equation

$$
\begin{bmatrix}
r_n(t+1) \\
r_{n-1}(t+1) \\
\vdots \\
r_2(t+1) \\
r_1(t+1)
\end{bmatrix}
=
\begin{bmatrix}
c_1 & 1 & 0 & \cdots & 0 & 0 & 0 \\
c_2 & 0 & 1 & \cdots & 0 & 0 & 0 \\
\vdots & & & \vdots & & & \vdots \\
c_{n-1} & 0 & 0 & \cdots & 0 & 0 & 1 \\
1 & 0 & 0 & \cdots & 0 & 0 & 0
\end{bmatrix}
\begin{bmatrix}
r_n(t) \\
r_{n-1}(t) \\
\vdots \\
r_2(t) \\
r_1(t)
\end{bmatrix}
\tag{6.28}
$$

We used the notation "x" in the polynomial expression to denote a 1-bit delay in a sequence, or shift to the right. Thus, if the time-domain notation for a sequence $a(x)$ is $A(t)$, then $A(t-1)$ corresponds to the sequence $x \cdot a(x)$. Similarly, the time-domain sequence notation $A(t+1)$ corresponds to the sequence $x^{-1} \cdot a(x) = a(x)/x$. From the relation

$$r_n(t+1) = c_1 \, r_n(t) + r_{n-1}(t)$$

we may deduce that

$$r_{n-1}(t) = c_1 \, r_n(t) + r_n(t+1)$$

where, in the modulo-2 arithmetic, addition and subtraction are the same. Thus, in the algebraic notation

$$r_{n-1}(x) = c_1 r_n(x) + \frac{r_n(x)}{x} = \frac{1 + c_1 x}{x} \cdot r_n(x) = \frac{1 + c_1 x}{x} \cdot \frac{g(x)}{f(x)}$$

where it is understood that the polynomial arithmetic is taken modulo $f(x)$, that is:

$$r_{n-1}(x) = \frac{\left[\dfrac{1 + c_1 x}{x} \cdot g(x) \right] \, \text{modulo} \, f(x)}{f(x)}$$

For example, suppose that a five-stage MSRG implements the sequence

$$\frac{g(x)}{f(x)} = \frac{1+x^2}{1+x^2+x^5} = r_5(x)$$

Note that $c_1 = 0$ for this example. Then the output of the fourth stage is

$$r_4(x) = \frac{\left[\dfrac{1+0\cdot x}{x}\cdot(1+x^2)\right] \text{ modulo } (1+x^2+x^5)}{1+x^2+x^5}$$

$$= \frac{\left[\dfrac{1+x^2}{x}\right] \text{ modulo } (1+x^2+x^5)}{1+x^2+x^5} = \frac{\left[\dfrac{x^5}{x}\right] \text{ modulo } (1+x^2+x^5)}{1+x^2+x^5}$$

$$= \frac{x^4}{1+x^2+x^5}$$

Having found $r_{n-1}(x)$, we proceed another step backward in the transition and note that

$$r_{n-1}(t+1) = c_2\,r_n(t) + r_{n-2}(t)$$

or

$$r_{n-2}(t) = c_2\,r_n(t) + r_{n-1}(t+1)$$

Thus, in the algebraic notation,

$$r_{n-2}(x) = c_2\,r_n(x) + \frac{r_{n-1}(x)}{x} = \left[c_2 + \frac{1+c_1x}{x^2}\right]\cdot r_n(x)$$

$$= \frac{1+c_1x+c_2x^2}{x^2}\cdot\frac{g(x)}{f(x)}$$

$$\equiv \frac{\left[\dfrac{1+c_1x+c_2x^2}{x^2}\cdot g(x)\right] \text{ modulo } f(x)}{f(x)}$$

Taking the same example of $f(x) = 1+x^2+x^5$ and $g(x) = 1+x^2$, we find that the output of the third register is

$$r_3(x) = \frac{\left[\dfrac{1 + 0 \cdot x + 1 \cdot x^2}{x^2} \cdot (1 + x^2)\right] \text{ modulo } (1 + x^2 + x^5)}{1 + x^2 + x^5}$$

$$= \frac{\left[\dfrac{1 + x^2}{x^2} \cdot (1 + x^2)\right] \text{ modulo } (1 + x^2 + x^5)}{1 + x^2 + x^5}$$

$$= \frac{\left[\dfrac{x^5}{x^2} \cdot x^5\right] \text{ modulo } (1 + x^2 + x^5)}{1 + x^2 + x^5} = \frac{x^8 \text{ modulo } (1 + x^2 + x^5)}{1 + x^2 + x^5}$$

$$= \frac{x^3(1 + x^2)}{1 + x^2 + x^5} = \frac{x^3 + x^5}{1 + x^2 + x^5} = \frac{1 + x^2 + x^3}{1 + x^2 + x^5}$$

Having seen how the algebraic expression for the MSRG stages is developed by working backward from the MSRG output (nth stage), we can postulate that the output of the $(n - i)$th stage is

$$r_{n-i}(x) = \frac{\left[\dfrac{1 + c_1 x + c_2 x^2 + \cdots + c_i x^i}{x^i} \cdot g(x)\right] \text{ modulo } f(x)}{f(x)} \tag{6.29}$$

As a check on this expression, we can apply it to finding the sequence at the first stage. Fortunately, we already know that the first stage receives direct feedback from the last stage and therefore may be written, $r_1(x) = x \cdot r_n(x)$.

To find the expression for the first stage, we set $n - i = 1$, or $i = n - 1$. This results in the formula

$$r_1(x) = \frac{\left[\dfrac{1 + c_1 x + c_2 x^2 + \cdots + c_{n-1} x^{n-1}}{x^{n-1}} \cdot g(x)\right] \text{ modulo } f(x)}{f(x)}$$

$$= \frac{\left[\dfrac{x^n \text{ modulo } f(x)}{x^{n-1}} \cdot g(x)\right] \text{ modulo } f(x)}{f(x)}$$

$$= \frac{\left[\dfrac{x^n}{x^{n-1}} \cdot g(x)\right] \text{ modulo } f(x)}{f(x)} = \frac{[x \cdot g(x)] \text{ modulo } f(x)}{f(x)}$$

This result confirms the correctness of the general formula.

To continue with the example of $f(x) = 1 + x^2 + x^5$ and $g(x) = 1 + x^2$, we find the remaining register outputs, having already found those of the third and fourth stages.

$$r_2(x) = \frac{\left[\dfrac{1+x^2}{x^3} \cdot (1+x^2)\right] \text{ modulo } (1+x^2+x^5)}{1+x^2+x^5}$$

$$= \frac{\left[\dfrac{x^5}{x^3} \cdot x^5\right] \text{ modulo } (1+x^2+x^5)}{1+x^2+x^5} = \frac{x^7 \text{ modulo } (1+x^2+x^5)}{1+x^2+x^5}$$

$$= \frac{x^2(1+x^2)}{1+x^2+x^5} = \frac{x^2+x^4}{1+x^2+x^5}$$

Finally, the first stage of the example is

$$r_1(x) = \frac{\left[\dfrac{1+x^2}{x^4} \cdot (1+x^2)\right] \text{ modulo } (1+x^2+x^5)}{1+x^2+x^5}$$

$$= \frac{\left[\dfrac{x^5}{x^4} \cdot x^5\right] \text{ modulo } (1+x^2+x^5)}{1+x^2+x^5} = \frac{x^6 \text{ modulo } (1+x^2+x^5)}{1+x^2+x^5}$$

$$= \frac{x(1+x^2)}{1+x^2+x^5} = \frac{x+x^3}{1+x^2+x^5}$$

6.3.3 The Use of Masks To Select a Sequence Phase Shift

For both the SSRG and the MSRG implementations of a maximal length sequence, we have shown how the particular phase or shift of the sequence can be controlled by selecting the initial loading of the shift register. Now we show how any desired shift of the sequence can be obtained without changing the loading, by combining certain outputs of the shift register stages. Which of the outputs are combined is specified by a mask vector. In this section, we show how this shift is accomplished for a special value of the shift register loading. In a following section, the technique is generalized to apply for any value of the loading.

Recall that the outputs of the stages of an SSRG are known shifts of the sequence when the initial loading is all zeros except the first stage, as shown in

Figure 6.16. The $P = 2^n - 1$ different shifts of the sequence correspond uniquely to the P different numerator polynomials $g(x)$ with degree $< n$.

Let a particular sequence be written

$$a(x) = \frac{g(x)}{f(x)} = \frac{g_0 + g_1 x + \cdots + g_{n-1} x^{n-1}}{f(x)}$$

$$= \frac{g_0}{f(x)} + \frac{g_1 x}{f(x)} + \cdots + \frac{g_{n-1} x^{n-1}}{f(x)}$$

$$= g_0 \cdot \frac{1}{f(x)} + g_1 \cdot \frac{x}{f(x)} + \cdots + g_{n-1} \cdot \frac{x^{n-1}}{f(x)} \tag{6.30}$$

It is easy to see how any shift of the sequence can be synthesized simply by combining the outputs of the SSRG for the loading $10\cdots00$: the mask consists of the coefficients of the powers in the polynomial $g(x)$, as illustrated in Figure 6.17. Using the polynomial $m(x) = m_0 + m_1 x + \cdots + m_{n-1} x^{n-1}$ to represent the mask vector $(m_0, m_1, \ldots, m_{n-1})$, the method for synthesizing the kth shift relative to $1/f(x)$ is to find $m(x) = x^k$ modulo $f(x) \equiv g(x)$.

For an MSRG, the specification of a mask for the special case of the initial loading is different from that of an SSRG, as is shown next. It is based on the well-known property that the successive states of the MSRG can be represented as successive powers of x, modulo the reciprocal polynomial $f^*(x)$. The MSRG initial loading of $10\cdots00$ has the polynomial representation $s_0(x) = 1$. The first n states for this loading are therefore those shown in Table 6.15. Note that each of these states has a weight of 1. Because each of the first n states has a weight of 1 when the loading is $10\cdots00$, the MSRG is acting like a simple delay line on which a 1 is "traveling." During the first n states for this special loading, the MSRG outputs may be thought of as nonoverlapping time pulses as illustrated in Figure 6.18.

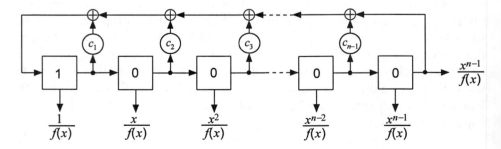

Figure 6.16 Outputs of the SSRG stages.

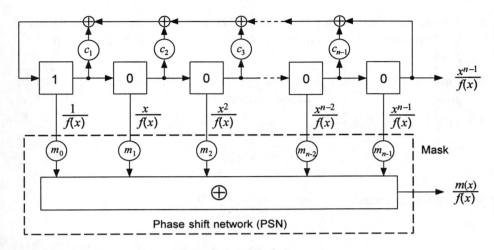

Figure 6.17 Sequence phase-shifting with an SSRG and loading $10\cdots0$.

The first n bits of the desired sequence can be programmed simply by selecting which of the MSRG output pulses are combined.

The observations made above lead to the following procedure for synthesizing an arbitrary shift of the sequence $1/f(x)$ for the special register initial loading $10\cdots0$:

Table 6.15 First n states for an MSRG with loading $10\cdots0$

Time	State polynomial	State
0	1	$100\cdots00$
1	x	$010\cdots00$
2	x^2	$001\cdots00$
$\vdots$	$\vdots$	$\vdots$
$n-1$	x^{n-1}	$000\cdots01$

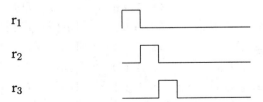

Figure 6.18 MSRG outputs in time for loading $10\cdots0$.

1. To obtain the kth shift to the right at the output of the MSRG, relative to $1/f(x)$, first find the required numerator polynomial $g_k(x) = x^k$ modulo $f(x)$.

2. Then, find the mask as the first n terms of the sequence specified by $g_k(x)/f(x)$, the division of the required numerator polynomial by $f(x)$.

3. That is, if the desired sequence is

$$\frac{g_k(x)}{f(x)} = \frac{x^k \text{ modulo } f(x)}{f(x)} = a_0 + a_1 x + \cdots + a_{n-1} x^{n-1} + \cdots$$

then set the "mask polynomial" equal to

$$m(x) = m_0 + m_1 x + \cdots + m_{n-1} x^{n-1}$$
$$= a_0 + a_1 x + \cdots + a_{n-1} x^{n-1} \qquad (6.31a)$$

This formula for the mask can also be written

$$m(x) = \left[\frac{x^k \text{modulo } f(x)}{f(x)} \right]_{\deg < n} \qquad (6.31b)$$

which denotes the terms of the ratio in the brackets having powers of x less than n.

Example 6.14 Comparison of SSRG and MSRG masks. Let $f(x) = 1 + x + x^3$, for which the reference sequence is 1110100.... For a shift register loading of $(r_1, r_2, r_3) = 100$, the SSRG and MSRG masks to achieve different delays are summarized in Table 6.16.

Table 6.16 SSRG and MSRG masks for $f(x) = 1 + x + x^3$ and loading 100

Delay, k	SSRG mask		MSRG mask
0	$g_0(x) = 1$	$\Rightarrow$ 1 0 0	$\{g_0(x)/f(x)\}_{3\,\text{bits}} = 1\ 1\ 1$
1	$g_1(x) = x$	$\Rightarrow$ 0 1 0	$\{g_0(x)/f(x)\}_{3\,\text{bits}} = 0\ 1\ 1$
2	$g_2(x) = x^2$	$\Rightarrow$ 0 0 1	$\{g_0(x)/f(x)\}_{3\,\text{bits}} = 0\ 0\ 1$
3	$g_3(x) = 1 + x$	$\Rightarrow$ 1 1 0	$\{g_0(x)/f(x)\}_{3\,\text{bits}} = 1\ 0\ 0$
4	$g_4(x) = x + x^2$	$\Rightarrow$ 0 1 1	$\{g_0(x)/f(x)\}_{3\,\text{bits}} = 0\ 1\ 0$
5	$g_5(x) = 1 + x + x^2$	$\Rightarrow$ 1 1 1	$\{g_0(x)/f(x)\}_{3\,\text{bits}} = 1\ 0\ 1$
6	$g_6(x) = 1 + x^2$	$\Rightarrow$ 1 0 1	$\{g_0(x)/f(x)\}_{3\,\text{bits}} = 1\ 1\ 0$

Thus, the mask for a shift register with the initial loading of $10\cdots00$ is

- The n bits describing the numerator polynomial, for an SSRG

- The first n bits of the desired shift of the sequence, for an MSRG.

We have shown how to generate an arbitrary shift of the sequence when the initial loading of the SSRG or MSRG is all zeros except for the first stage. This technique is useful for rapidly selecting different shifts of the sequence while maintaining a reference shift. It even can be used to generate several different shifts at the same time, by using multiple masks. Next, we show how to extend the synthesis procedure to an arbitrary initial loading, which allows us to specify the reference sequence and the desired shift independently.

6.3.4 Relationship Between the Mask and the Sequence Shift for Arbitrary Shift Register Loading

For the initial loading $10\cdots00$, the output of either the SSRG or the MSRG is known to be the shift of the sequence represented by $x^{n-1}/f(x)$. For an arbitrary initial loading, the output of the shift register will be some arbitrary shift, say the shift represented by $[x^q$ modulo $f(x)]/f(x)$—a different q for SSRG and MSRG for the same loading, of course. What must the mask be to obtain the kth sequence shift relative to $1/f(x)$? This question will be answered first for the SSRG, then for the MSRG.

When the output sequence of the SSRG is $x^{n-1}/f(x)$, we know from (6.30) that the output of the phase shift network must be

$$\frac{g_k(x)}{f(x)} = \frac{x^k \text{ modulo } f(x)}{f(x)} \quad \Rightarrow \quad m(x) = g_k(x) \tag{6.32}$$

We note in this case that the SSRG output is delayed (shifted to the right) with respect to the reference sequence $1/f(x)$ by $n-1$ bits, while the output of the phase shift network is delayed by k bits with respect to $1/f(x)$.

As time proceeds from the time of initialization, the output of the SSRG and the output of the phase-shift network (PSN) advance at the same rate. Therefore, they maintain the same relative delay that they had at initialization, which is

$$\text{Relative delay} \triangleq d_k = k - (n - 1) = k - n + 1 \qquad (6.33a)$$

This relative delay is a function of the mask; that is, it is a function of k. Further, if at the output of the SSRG a different initial loading produces the shift $[x^q \text{ modulo } f(x)]/f(x)$, while the mask continues to be $m(x) = g_k(x)$, the output of the PSN has the same relative delay d_k and is the shift $[x^{q+d_k} \text{ modulo } f(x)]/f(x)$, as illustrated by Figure 6.19.

For example, to show how the relative delay between the SSRG output and the PSN output is preserved, let the SSRG be connected to implement the sequence with characteristic polynomial $f(x) = 1 + x + x^4$, and let us consider two initial loadings:

- The special case 1000, which we know produces the sequence $x^3/f(x)$, and

- The arbitrary loading 0101.

First, we compare the sequence at the output of the SSRG with that out of a PSN with the mask $m(x) = g_4(x) = 1 + x$. That is, the mask combines

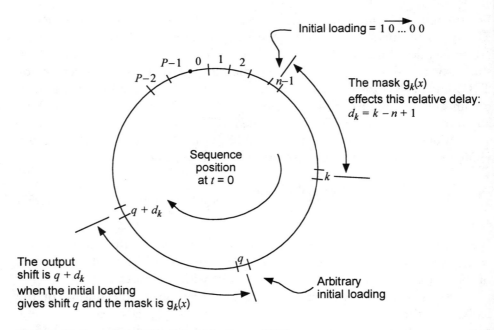

Figure 6.19 Preservation of relative delay for an SSRG.

the register outputs r_1 and r_2 to accomplish the sequence phase shift. The comparison is given in Table 6.17, showing that the mask 1100 produces a 1-bit relative delay regardless of the initial loading of the shift register.

The example demonstrates the following important fact for SSRGs with arbitrary mask polynomials and arbitrary loading:

> The arbitrary mask polynomial $m(x) = x^r$ modulo $f(x)$ (6.33b)
> implements the relative delay $d_r = r - n + 1$
> between the PSN output and the SSRG output,
> for any initial loading.

What use do we make of this fact? We find the mask polynomial x^r modulo $f(x) \equiv g_r(x)$ that produces the sequence $g_k(x)/f(x)$ at the output of the PSN. The steps to be followed are

1. Find the phase shift of the output of the PSN as a function of the initial loading and the mask polynomial.

Table 6.17 Example of the preservation of delay for an SSRG

Mask: $11\underset{\rightarrow}{0}0$

Initial loading: 1000		Initial loading: 0101	
Registers 1234	PSN Output	Registers 1234	PSN Output
1000	1	0101	1
1100	0	1010	1
1110	0	1101	0
1111 ← relative delay 0		0110	1
0111 = 1 → 1		0011	0
1011	1	1001	1
0101	1	0100	1
1010	1	0010	0
1101	0	0001	0
0110	1	1000	1
0011	0	1100	0
1001	1	1110	0
0100	1	1111 ← relative delay 0	
0010	0	0111 = 1 → 1	
0001	0	1011	1

2. Solve for the mask polynomial when the PSN output is the kth shift of the reference sequence $1/f(x)$.

Comment: originally, in the expression $d_k = k - n + 1$ for the relative delay in the special case, the expression for the delay was clearly seen to be based on

$$\text{Relative delay} = \text{PSN output shift } (k) - \text{SSRG output shift } (n-1) \quad (6.33c)$$

We learned that, in general, the correct interpretation of the expression for the relative delay is

$$\text{Relative delay} = \text{f}(\overbrace{\text{mask polynomial}}^{\text{parameter} = k}, \overbrace{\text{shift register length}}^{n}) \quad (6.33d)$$

With the objective of finding the mask polynomial that produces the kth shift of the reference sequence, we now formulate a general expression for the phase shift at the PSN output. Assume that the mask is $g_r(x) = x^r$ modulo $f(x)$, and that the initial loading of the shift register is such that the SSRG output is

$$\frac{g_q(x)}{f(x)} = \frac{x^q \text{ modulo } f(x)}{f(x)} \quad (6.33e)$$

It follows that the relative delay is $d_r = r - n + 1$, and the output of the PSN is

$$\frac{x^{q+d_r} \text{ modulo } f(x)}{f(x)} = \frac{x^k \text{ modulo } f(x)}{f(x)} = \frac{x^{\overbrace{q+r-n+1}^{\text{to be set to } k}} \text{ modulo } f(x)}{f(x)} \quad (6.33f)$$

A comparison for special and arbitrary loadings is shown in Figure 6.20. To make the output the kth shift, we must have r such that

$$r + q - n + 1 = k \quad (6.33g)$$

which implies that

$$\text{Mask polynomial} = m(x) = g_r(x) \quad (6.33h)$$

where

$$r = \text{desired shift } (k) - \text{SSRG shift } (q) + n - 1 \quad (6.33i)$$

The general procedure for finding the SSRG mask required to generate the kth shift of the sequence is the following:

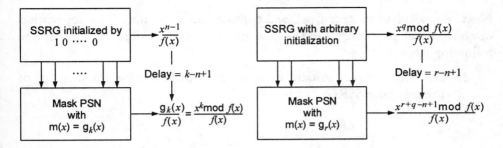

Figure 6.20 Comparison of SSRG masks for special and arbitrary loadings.

1. Find the shift of the reference SSRG from its numerator polynomial:

$$g_q(x) \;\Rightarrow\; x^q \;\Rightarrow\; q$$

Alternatively, if the initial state [first n terms of the sequence $a_q(x)$] is given, then find $g_q(x)$ first:

$$a_q(x) \;\Rightarrow\; g_q(x) = a_q(x) \cdot f(x) \quad \text{(terms of degree} < n).$$

2. Calculate the mask as

$$m(x) = x^{k-q+n-1} \text{ modulo } f(x)$$

Example 6.15 (SSRG mask) Suppose that it is desired to generate the 10th shift of the sequence $1/f(x)$, where $f(x) = 1 + x + x^4$. ($k = 10$ and $n = 4$.) The application of the general procedure for the initial loading 1000 takes the following steps:

1. For this special initialization we know that the output of the SSRG is shift $(n-1) = 3 = q$.

2. The mask is calculated as

$$m(x) = x^{k-q+n-1} \bmod f(x) = x^{10-3+4-1} \bmod f(x) = x^{10} \bmod f(x)$$
$$= x^2 \cdot \left(x^4\right)^2 = x^2 \cdot (1+x)^2 = x^2 \cdot (1+x^2) = x^2 + x^4$$
$$= x^2 + 1 + x = 1 + x + x^2 \;\Rightarrow\; 1110$$

Next, we calculate another mask and compare the two later. The application of the general procedure for the case of the initial loading 1001 takes the following steps:

1. For this arbitrary initialization, we have to determine the shift (q) of the reference SSRG:

$$\frac{g_q(x)}{1 + x + x^4} = a_q(x) = \underbrace{1 + x^3}_{\text{initialization}} + \cdots$$

$$\Rightarrow \quad g_q(x) = \{\text{degree} < n \text{ terms of } (1 + x^3) \cdot f(x)\}$$

$$= \left\{ \underbrace{1 + x + \not{x}^4}_{f(x)} + \underbrace{x^3 + \not{x}^4 + x^7}_{x^3 f(x)} \right\}_{\text{degree}<4} = 1 + x + x^3$$

2. To find q, we reverse the modular calculation of $g_q(x)$:

$$\Rightarrow \quad g_q(x) = 1 + x + x^3 = (\not{x} + x^4) + \not{x} + x^3 = x^3(1 + x)$$

$$= x^3[(\not{x} + x^4) + \not{x}] = x^7 \quad \Rightarrow \quad q = 7$$

3. Now, knowing q the mask for the initialization 1001 is calculated as

$$m(x) = x^{k-q+n-1} \bmod f(x) = x^{10-7+4-1} \bmod f(x) = x^6 \bmod f(x)$$

$$= x^2 \cdot x^4 = x^2 \cdot (1 + x) = x^2 + x^3 \quad \Rightarrow \quad 0\,0\,1\,1$$

Having calculated the two SSRG masks, one for the special initial loading of 1000 and one for the arbitrary initial loading of 1001, we compare the SSRG and PSN outputs to observe how the mask formulas applied in these two situations have enabled us to produce the same shift $(k = 10)$ at the output of the PSN. The comparison is shown in Figure 6.21.

Now we consider the MSRG shift register configuration. When the output of the MSRG is $x^{n-1}/f(x)$, we know that the output of the PSN is

$$a_k(x) = \frac{g_k(x)}{f(x)} = \frac{x^k \bmod f(x)}{f(x)}$$

when

$$m(x) = \{\text{first } n \text{ terms of } a_k(x)\} \qquad (6.34)$$

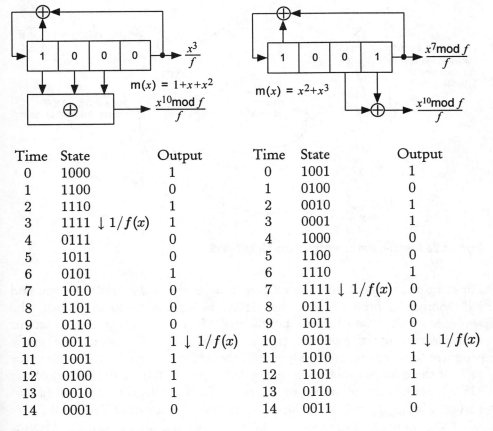

Time	State	Output		Time	State	Output
0	1000	1		0	1001	1
1	1100	0		1	0100	0
2	1110	1		2	0010	1
3	1111 ↓ $1/f(x)$	1		3	0001	1
4	0111	0		4	1000	0
5	1011	0		5	1100	0
6	0101	1		6	1110	1
7	1010	0		7	1111 ↓ $1/f(x)$	0
8	1101	0		8	0111	0
9	0110	0		9	1011	0
10	0011	1 ↓ $1/f(x)$		10	0101	1 ↓ $1/f(x)$
11	1001	1		11	1010	1
12	0100	1		12	1101	1
13	0010	1		13	0110	1
14	0001	0		14	0011	0

Figure 6.21 SSRG mask example.

As time proceeds, the output of the MSRG and the output of the phase-shift network maintain the same relative delay, which is

$$\text{Relative delay} \triangleq d_k = k - (n-1) = k - n + 1 \tag{6.35a}$$

This relative delay is a function of the mask; that is, k. Further, if at the output of the MSRG, a different initial loading produces the shift $[x^q \text{ modulo } f(x)]/f(x)$, while the mask continues to be $m(x) = \{g_k(x)/f(x)\}_{\deg<n}$, then the output of the phase-shift network will have the same relative delay d_k and be the sequence shift $[x^{q+d_k} \text{ modulo } f(x)]/f(x)$, as illustrated by Figure 6.22.

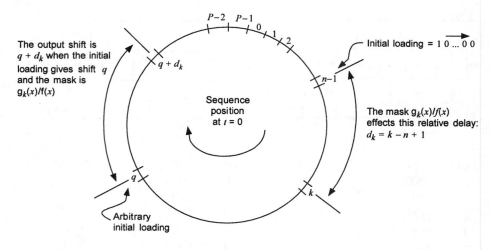

Figure 6.22 Preservation of relative delay for an MSRG.

For example, to show how the relative delay between the MSRG output and PSN output is preserved, let the MSRG be connected to implement the sequence with characteristic polynomial $f(x) = 1 + x + x^4$, and let us consider two initial loadings: (1) the special case 1000, which we know produces the sequence $x^3/f(x)$; and (2) the arbitrary loading 0101. Now for each of these loadings, let us compare the sequence shift at the output of the MSRG with that at the output of a PSN with the mask $m(x) = \{g_4(x)/f(x)\}_{\deg<4} = 1 \Rightarrow 1000$; the comparison is shown in Table 6.18.

The example demonstrates the following important fact for MSRGs with arbitrary mask polynomials and arbitrary loading:

> The mask polynomial $m(x) = \{g_k(x)/f(x)\}_{\deg<n}$ implements (6.35b)
> the relative delay $d_k = k - n + 1$ between the PSN output
> and the MSRG output, for any initial loading.

Therefore, if the mask is $\{g_r(x)/f(x) = [x^r \text{ modulo } f(x)]/f(x)\}_{\deg<n}$, and the MSRG output is $g_q(x)/f(x) = [x^q \text{ modulo } f(x)]/f(x)$, then the relative delay is $d_r = r - n + 1$ and the output of the PSN is

$$\frac{x^{q+d_r} \text{ modulo } f(x)}{f(x)} = \frac{x^k \text{ modulo } f(x)}{f(x)} = \frac{x^{q+r-n+1} \text{ modulo } f(x)}{f(x)} \qquad (6.35c)$$

Table 6.18 Example of the preservation of relative delay for an MSRG

Mask: $10\underset{\rightarrow}{0}0$

Initial loading: 1000			Initial loading: 0101		
Registers 1234		PSN Output	Registers 1234		PSN Output
1000		1	0101		0
0100		0	1011		1
0010		0	1100		1
0001 ← relative delay		0	0110		0
1001	= 1 →	1	0011		0
1101		1	1000		1
1111		1	0100		0
1110		1	0010		0
0111		0	0001 ← relative delay		0
1010		1	1001	= 1 →	1
0101		0	1101		1
1011		1	1111		1
1100		1	1110		1
0110		0	0111		0
0011		0	1010		1

The comparison of masks for special and arbitrary MSRG loadings is summarized in Figure 6.23.

To make the output the kth shift, we must have r such that

$$r + q - n + 1 = k \qquad (6.35d)$$

which implies that

$$\text{Mask polynomial} = m(x) = \left\{ \text{first } n \text{ terms of } \frac{g_r(x)}{f(x)} \right\} \qquad (6.35e)$$

where

$$k = \text{desired shift}, \quad q = \text{MSRG shift} \qquad (6.35f)$$

The general procedure for finding the MSRG mask required to generate the kth shift of the sequence is the following:

1. Find the shift of the reference MSRG from its numerator polynomial:

$$g_q(x) \;\Rightarrow\; x^q \;\Rightarrow\; q$$

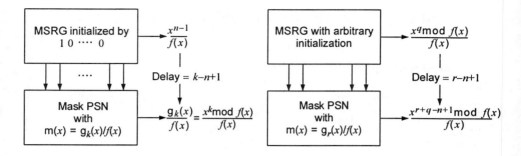

Figure 6.23 Comparison of MSRG masks for special and arbitrary loadings.

Alternatively, if the initial state $s(x)$ is given, then find $g_q(x)$ first:

$$s(x) \quad \Rightarrow \quad g_q(x) = s^*(x) \quad \text{(reverse of initial state)}$$

2. Calculate the MSRG mask as

$$m(x) = \left\{ \text{first } n \text{ terms of } \frac{x^{k-q+n-1} \text{ modulo } f(x)}{f(x)} \right\}$$

$$\equiv \left\{ \frac{x^{k-q+n-1} \text{ modulo } f(x)}{f(x)} \right\}_{\deg < n} \tag{6.35g}$$

Example 6.16 (MSRG mask) Suppose that it is desired to generate the 10th shift of the sequence $1/f(x)$, where $f(x) = 1 + x + x^4$. ($k = 10$ and $n = 4$.) We show how to meet this objective for two different MSRG initial loadings. For the mechanization of the MSRG, we need the reciprocal polynomial, which is $f^*(x) = 1 + x^3 + x^4$. The application of the general procedure for the initial loading 1000 follows:

1. For this special initialization, we know that the output of the MSRG is shift $(n - 1) = 3 = q$.

2. The parameter for the required mask is $r = k - q + n - 1 = 10$.

3. The MSRG mask is calculated as

$$m(x) = \left\{ \frac{x^{k-q+n-1} \text{ modulo } f(x)}{f(x)} \right\}_{\deg < n} = \left\{ \frac{x^{10} \text{ modulo } f(x)}{f(x)} \right\}_{\deg < 4}$$

The modular calculation gives

$$g_{10}(x) = x^{10} \text{ modulo}(1 + x + x^4) = x^2 \cdot (x^4)^2 = x^2(1 + x^2)$$
$$= x^2 + x^4 = x^2 + 1 + x = 1 + x + x^2$$

4. The mask giving shift 10 for the loading 1000 is then calculated as

$$m(x) = \left\{ \frac{1 + x + x^2}{1 + x + x^4} \right\}_{\deg < 4} = 1 + x^2 + x^3 \quad \Rightarrow \quad 1011$$

Next we calculate the mask required for different initial conditions and then compare the results of the two masks. The application of the general procedure for the initial loading 1001 follows:

1. For this arbitrary initialization, we have to determine the shift (q) of the reference MSRG:

$$g_q(x) = s^*(x) = \text{reverse of initial state polynomial} = 1 + x^3$$
$$\Rightarrow \qquad g_q(x) = 1 + x^3 = x + x^4 + x^3 = x(1 + x^2 + x^3)$$
$$= x(x + x^4 + x^2 + x^3) = x^2(1 + x + x^2 + x^3)$$
$$= x^2(\not{x} + x^4 + \not{x} + x^2 + x^3) = x^4(1 + x + x^2)$$
$$= x^4(\not{x} + x^4 + \not{x} + x^2) = x^6(1 + x^2) = x^6(1 + x)^2$$
$$= x^6(x^4)^2 = x^{14} \quad \Rightarrow \quad q = 14$$

2. The mask parameter in this case is $r = k - q + n - 1 = 10 - 14 + 4 - 1 = -1$, or $r = 15 - 1 = 14$, modulo the period, which is $P = 15$. The required mask is

$$m(x) = \left\{ \frac{x^{P-1} \text{ modulo } f(x)}{f(x)} \right\}_{\deg < 4} = \left\{ \frac{x^{14} \text{ modulo } f(x)}{f(x)} \right\}_{\deg < 4}$$

where

$$g_{14}(x) = x^{14} \text{ modulo}(1 + x + x^4)$$
$$= x^2 \cdot (x^4)^3 = x^2(1 + x + x^2 + x^3)$$
$$= x^2 + x^3 + x^4(1 + x) = x^2 + x^3 + (1 + x)^2$$
$$= \not{x}^2 + x^3 + 1 + \not{x}^2 = 1 + x^3$$

giving

$$m(x) = \left\{ \frac{1 + x^3}{1 + x + x^4} \right\}_{\deg < 4} = 1 + x + x^2 \;\; \Rightarrow \;\; 1\,1\,1\,0$$

The MSRG masks for the two different initial loadings are compared in Figure 6.24.

We have now completed the essential developments for the theory and practice of using masks for the generation of PN sequences with desired phase shifts by means of either SSRG (Fibonacci) or MSRG (Galois) configurations. For the case of MSRG configurations, we have derived formulas for the output sequence taken from any stage of the LFSR, and they serve for practicing engineers as a means to check or validate mask-based sequence generators, among possibly other uses.

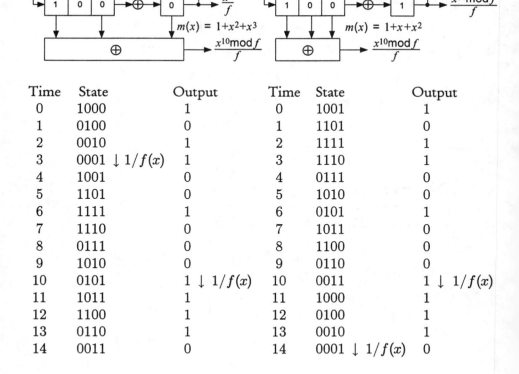

Time	State	Output	Time	State	Output
0	1000	1	0	1001	1
1	0100	0	1	1101	0
2	0010	1	2	1111	1
3	0001 $\downarrow 1/f(x)$	1	3	1110	1
4	1001	0	4	0111	0
5	1101	0	5	1010	0
6	1111	1	6	0101	1
7	1110	0	7	1011	0
8	0111	0	8	1100	0
9	1010	0	9	0110	0
10	0101	1 $\downarrow 1/f(x)$	10	0011	1 $\downarrow 1/f(x)$
11	1011	1	11	1000	1
12	1100	1	12	0100	1
13	0110	1	13	0010	1
14	0011	0	14	0001 $\downarrow 1/f(x)$	0

Figure 6.24 MSRG mask example.

As far as PN sequence generation is concerned, we now know just about everything that needs to be known in terms of generation techniques for specific PN code "offsets." In the subsections to follow, we consider examples to put these techniques into practice, and in addition, we make some observations on the PN sequences that are used in IS-95 systems.

On the following pages, we give various examples and illustrations of PN sequences and the use of masks to generate shifts of PN sequences, including a detailed five-stage MSRG example, discussion of the PN sequences specified in IS-95, discussion of the use of long-code offsets for IS-95, and derivation of masks for implementing the base station offsets for IS-95.

6.3.4.1 Five-Stage MSRG Example

Let the characteristic polynomial for the desired sequence be $f(x) = 1 + x^2 + x^5$. We calculate the MSRG mask for generating the sequence shift that is four chips delayed (shifted to the right) from the sequence at the MSRG output stage, for any initial loading. That is, for any shift q at the MSRG output relative to the reference sequence $1/f(x)$, we want to find the mask that produces the shift $q + 4$. For concreteness, we first use the example of the initial loading 11111, although the PSN is valid for any initial loading.

The output of the MSRG when the initial loading is 11111 is found as follows: The MSRG output sequence can be expressed algebraically as the division of $s^*(x)$, the reverse of the initial state polynomial, by $f(x)$. Thus, the initial loading given for the MSRG informs us directly that the output sequence is

$$a(x) = \frac{g(x)}{f(x)} \equiv \frac{s^*(x)}{f(x)} = \frac{1 + x + x^2 + x^3 + x^4}{1 + x^2 + x^5} \tag{6.36a}$$

which has the first period (output of the fifth stage)

$$\overbrace{1\ 1\ 0\ 0\ 1\ 1\ 0\ 1\ 0\ 0\ 1}\underbrace{\overbrace{0\ 0\ 0\ 0}^{n-1\ \text{zeros}}}_{\text{15-chip shift}}\ 1\ 0\ 1\ \ 0\ 1\ 1\ 1\ 0\ 1\ 1\ 0\ 0\ 0\ 1\ 1\ 1 \quad \rightarrow\ \frac{1}{f(x)}$$

Our objective is to find the mask that will give a 19-chip shift.

In this case, because the sequence is so short, we are able to see that the shift of the MSRG output sequence is $q = 15$ by inspection. Algebraically, the shift is determined by reversing the modulo-$f(x)$ calculation of $g(x)$:

$$
\begin{aligned}
g(x) = 1 + x + x^2 + x^3 + x^4 &= \overbrace{\cancel{x^2} + x^5}^{1+f(x)} + x + \cancel{x^2} + x^3 + x^4 \\
&= x(1 + x^2 + x^3 + x^4) = x(\cancel{x^2} + x^5 + \cancel{x^2} + x^3 + x^4) \\
&= x^4(1 + x + x^2) = x^4(\cancel{x^2} + x^5 + x + \cancel{x^2}) \\
&= x^5(1 + x^4) = x^5(x^2 + x^5 + x^4) \\
&= x^7(1 + x^2 + x^3) = x^7(\cancel{x^2} + x^5 + \cancel{x^2} + x^3) \\
&= x^{10}(1 + x^2) = x^{10}(\cancel{x^2} + x^5 + \cancel{x^2}) = x^{15} \quad \Rightarrow \quad q = 15
\end{aligned}
$$

Thus, as illustrated in Figure 6.25, the initial loading 11111 gives the output sequence $[x^{15} \text{ modulo } f(x)]/f(x)$.

Using the formula for the MSRG register outputs that is given in Section 6.3.4, with $c_1 = c_3 = c_4 = 0$ and $c_2 = 1$, and knowing that $g(x) = x^{15} \text{mod } f$, we can determine that the outputs of the first four stages are, respectively

$$
r_4(x) = r_{n-1}(x) = \frac{\left[\dfrac{1 + 0 \cdot x}{x} x^{15}\right] \text{mod } f}{f} = \frac{x^{14} \text{mod } f}{f} \tag{6.36b}
$$

$$
r_3(x) = r_{n-2}(x) = \frac{\left[\dfrac{1 + x^2}{x^2} x^{15}\right] \text{mod } f}{f} = \frac{[x^5 \cdot x^{13}] \text{mod } f}{f} = \frac{x^{18} \text{mod } f}{f}
$$

$$
\tag{6.36c}
$$

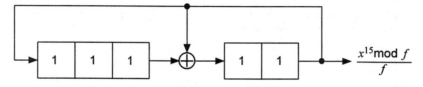

Figure 6.25 MSRG with initial loading 11111.

$$r_2(x) = r_{n-3}(x) = \frac{\left[\dfrac{1 + x^2 + 0 \cdot x^3}{x^3} x^{15}\right] \bmod f}{f} = \frac{x^{17} \bmod f}{f} \qquad (6.36d)$$

and

$$r_1(x) = r_{n-4}(x) = \frac{\left[\dfrac{1 + x^2 + 0 \cdot x^4}{x^4} x^{15}\right] \bmod f}{f} = \frac{x^{16} \bmod f}{f} \qquad (6.36e)$$

Thus, we deduce that the first periods of the sequences at the outputs of stages one through four are

First stage:

$$\underbrace{1\ 1\ 1\ 0\ 0\ 1\ 1\ 0\ 1\ 0\ 0\ 1}_{\text{16-chip shift}}\ \overbrace{0\ 0\ 0\ 0}^{n-1\ \text{zeros}}\ 1\ 0\ 1\ 0\ 1\ 1\ 1\ 0\ 1\ 1\ 0\ 0\ 0\ 1\ 1$$
$$\bigg|_{\longrightarrow}\ \tfrac{1}{f(x)}$$

Second stage:

$$\underbrace{1\ 1\ 1\ 1\ 0\ 0\ 1\ 1\ 0\ 1\ 0\ 0\ 1}_{\text{17-chip shift}}\ \overbrace{0\ 0\ 0\ 0}^{n-1\ \text{zeros}}\ 1\ 0\ 1\ 0\ 1\ 1\ 1\ 0\ 1\ 1\ 0\ 0\ 0\ 1$$
$$\bigg|_{\longrightarrow}\ \tfrac{1}{f(x)}$$

Third stage:

$$\underbrace{1\ 1\ 1\ 1\ 1\ 0\ 0\ 1\ 1\ 0\ 1\ 0\ 0\ 1}_{\text{18-chip shift}}\ \overbrace{0\ 0\ 0\ 0}^{n-1\ \text{zeros}}\ 1\ 0\ 1\ 0\ 1\ 1\ 1\ 0\ 1\ 1\ 0\ 0\ 0$$
$$\bigg|_{\longrightarrow}\ \tfrac{1}{f(x)}$$

Fourth stage:

$$\underbrace{1\ 0\ 0\ 1\ 1\ 0\ 1\ 0\ 0\ 1}_{\text{14-chip shift}}\ \overbrace{0\ 0\ 0\ 0}^{n-1\ \text{zeros}}\ 1\ 0\ 1\ 0\ 1\ 1\ 1\ 0\ 1\ 1\ 0\ 0\ 0\ 1\ 1\ 1\ 1$$
$$\bigg|_{\longrightarrow}\ \tfrac{1}{f(x)}$$

Having found the sequence shifts at the outputs of each of the five stages of the MSRG, we can use them later to check the solution for the MSRG mask that we are now going to find analytically. Recall that the analytical procedure for calculating the mask for an MSRG with arbitrary loading is the following:

1. Determine q, the shift of the MSRG output that is produced by the particular initial loading.

2. Given the desired shift k relative to $1/f(x)$, solve for the mask parameter $r = (k - q + n - 1) \bmod P$, where $P = 2^n - 1$ is the period of the sequence; or, given the desired relative delay d_r with respect to the sequence at the output of the MSRG, solve for $r = d_r + n - 1$.

3. Calculate the mask as the first n bits of $[x^r \bmod f(x)]/f(x)$.

As we have already seen, the shift of the output of the MSRG is $q = 15$, and the desired relative delay was stated to be $d_r = r - n + 1 = 4$, where r is the mask parameter to be found. Once found, the mask output sequence has this same day (shift to the left) relative to the MSRG output sequence for any initial loading. Solving for r gives

$$r = d_r + n - 1 = 4 + n - 1 = 4 + 5 - 1 = 8 \quad \Rightarrow \quad x^r = x^8$$

Because $r > n - 1 = 4$, x^r has to be reduced modulo $f(x)$ to find the numerator polynomial $g_r(x) = x^r \bmod f(x)$ that is needed in the calculation of the mask. The numerator polynomial $x^r \bmod f(x)$ for $r = 8$ that is needed to calculate the mask is determined as follows, using the fact that $f(x) = 1 + x^2 + x^5 = 0 \Rightarrow x^5 = 1 + x^2$:

$$x^8 = x^3 \cdot x^5 = x^3(1 + x^2) = x^3 + x^5 = 1 + x^2 + x^3$$

The mask is the first five bits of $g_8(x)/f(x)$, or

$$m(x) = \left\{ \frac{1 + x^2 + x^3}{1 + x^2 + x^5} \right\}_{\deg < 5} = \left\{ 1 + x^3 + x^8 + x^{10} + \cdots \right\}_{\deg < 5}$$

$$= 1 + x^3 \quad \Rightarrow \quad 1\,0\,0\,1\,0$$

This mask combines the outputs of the first and fourth stages of the MSRG to produce the desired four-chip relative delay.

Previously we found the sequences at all the MSRG stages when the initial loading is 11111. Now we are able to demonstrate that the combination of the sequences at the first and fourth stages does indeed produce the shift of the sequence that is four chips shifted to the right relative to that at the MSRG output (fifth stage). The modulo-2 combination of the first and fourth shift register stages is

First stage: 1 1 1 0 0 1 1 0 1 0 0 1 0 0 0 0 1 0 1 0 1 1 1 0 1 1 0 0 0 1 1
Fourth stage: 1 0 0 1 1 0 1 0 0 1 0 0 0 0 1 0 1 0 1 1 1 0 1 1 0 0 0 1 1 1 1
$$\overline{0\ 1\ 1\ 1\ 0\ 1\ 0\ 0\ 1\ 1\ 0\ 1\ 0\ 0\ 1\ 0\ 0\ 0\ 0}\ 1\ 0\ 1\ 0\ 1\ 1\ 1\ 0\ 1\ 1\ 0\ 0$$

19-chip shift

Because the MSRG output for this initial loading is the 15-chip shift, the relative delay of $19 - 15 = 4$ chips is achieved at the output of the PSN. A diagram of the PSN that achieves a relative delay of four chips for the MSRG with $f(x) = 1 + x^2 + x^5$ is in Figure 6.26.

Now we find the output sequences of the first, fourth, and fifth stages for some other, arbitrarily chosen initial loading of the MSRG. We do this to demonstrate that the relative delay of four chips is achieved by the PSN with the same mask 10010 that was just derived to achieve a four-chip delay for the loading 11111. Suppose that instead of 11111 the initial loading is $\overrightarrow{01011}$. This loading corresponds to the initial state polynomial $s(\beta) = \beta + \beta^3 + \beta^4$. For this initial state, we know that the numerator polynomial in the algebraic representation of the MSRG output sequence is

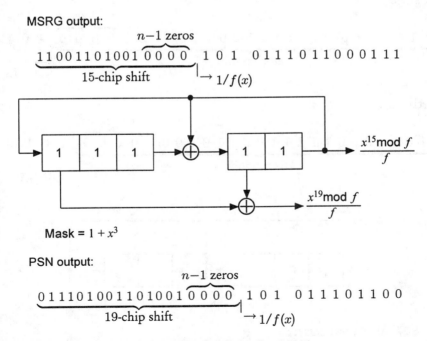

Figure 6.26 Five-stage MSRG mask example for the loading 11111.

$$g(x) = s^*(x) = x^4(x^{-4} + x^{-3} + x^{-1}) = 1 + x + x^3$$

Thus, the output of the MSRG is the sequence

$$a(x) = \frac{1 + x + x^3}{1 + x^2 + x^5} = 1 + x + x^2 + x^4 + x^5 + \cdots$$

which has the following as the first period:

Fifth stage:

$$\underbrace{1\ 1\ 1\ 0\ 1\ 1\ 0\ 0\ 0\ 1\ 1\ 1\ 1\ 0\ 0\ 1\ 1\ 0\ 1\ 0\ 0\ 1\ \overbrace{0\ 0\ 0\ 0}^{n-1\text{ zeros}}\ 1\ 0\ 1\ 0}_{27\text{-chip shift}} \quad \xrightarrow{} \quad \frac{1}{f(x)}$$

Knowing the structure of this particular MSRG, we observe from its diagram in Figure 6.27 that the sequence at the output of the first stage is one chip delayed from that of the fifth stage, which is delayed one chip from the output of the fourth stage. Thus we have

First stage:

$$\underbrace{0\ 1\ 1\ 1\ 0\ 1\ 1\ 0\ 0\ 0\ 1\ 1\ 1\ 1\ 0\ 0\ 1\ 1\ 0\ 1\ 0\ 0\ 1\ \overbrace{0\ 0\ 0\ 0}^{n-1\text{ zeros}}\ 1\ 0\ 1}_{28\text{-chip shift}} \quad \xrightarrow{} \quad \frac{1}{f(x)}$$

Fourth stage:

$$\underbrace{1\ 1\ 0\ 1\ 1\ 0\ 0\ 0\ 1\ 1\ 1\ 1\ 0\ 0\ 1\ 1\ 0\ 1\ 0\ 0\ 1\ \overbrace{0\ 0\ 0\ 0}^{n-1\text{ zeros}}\ 1\ 0\ 1\ 0\ 1}_{26\text{-chip shift}} \quad \xrightarrow{} \quad \frac{1}{f(x)}$$

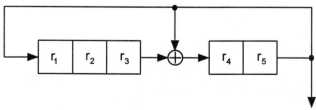

Figure 6.27 MSRG under study.

The sequence at the output of the PSN is the modulo-2 sum of the first and fourth stages, giving the sequence $1/f(x)$:

First stage:

0 1 1 1 0 1 1 0 0 0 1 1 1 1 1 0 0 1 1 0 1 0 0 1 0 0 0 0 1 0 1

$\downarrow$

$1/f(x)$

Fourth stage:

1 1 0 1 1 0 0 0 1 1 1 1 1 0 0 1 1 0 1 0 0 1 0 0 0 0 1 0 1 0 1

$\downarrow$

$1/f(x)$

PSN output:

$$\overbrace{}^{n-1\ \text{zeros}}$$

1 0 1 0 1 1 1 0 1 1 0 0 0 1 1 1 1 1 0 0 1 1 0 1 0 0 1 0 0 0 0 A diagram

$\downarrow \frac{1}{f(x)}$

showing that the four-chip delay of PSN relative to the MSRG output holds for the arbitrary initial conditions 01011 is given in Figure 6.28.

6.3.4.2 PN Sequences Specified in IS-95

Three PN sequences are specified in the CDMA digital cellular system common air interface standard known as IS-95:

- A "long" PN sequence ($n = 42$) that is used to scramble the user data with a different code shift for each user for multiple access purposes.

- Two "short" PN sequences ($n = 15$) that are used to spread the quadrature components of the forward and reverse link waveforms, with different code shifts for different cells.

We examine each of these sequences, particularly the masks that are employed to implement code shifts.

The IS-95 long PN code. The long PN sequence used in IS-95 to scramble user data has the following 42-degree characteristic polynomial:

$$f_S(x) = 1 + x^7 + x^9 + x^{11} + x^{15} + x^{16} + x^{17} + x^{20} + x^{21} + x^{23} + x^{24}$$

$$+ x^{25} + x^{26} + x^{32} + x^{35} + x^{36} + x^{37} + x^{39} + x^{40}$$

$$+ x^{41} + x^{42} \tag{6.37a}$$

MSRG output:

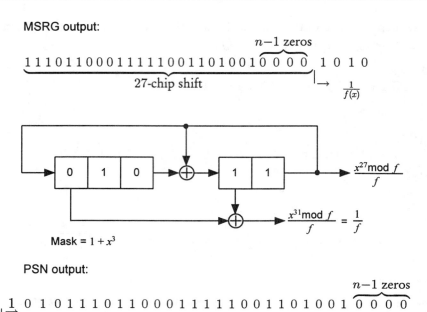

$$n-1 \text{ zeros}$$

$$\underbrace{1\ 1\ 1\ 0\ 1\ 1\ 0\ 0\ 0\ 1\ 1\ 1\ 1\ 1\ 0\ 0\ 1\ 1\ 0\ 1\ 0\ 0\ 1\ \overbrace{0\ 0\ 0\ 0}}_{\text{27-chip shift}}\ \underset{\longrightarrow\ \frac{1}{f(x)}}{1\ 0\ 1\ 0}$$

Mask = $1 + x^3$

PSN output:

$$n-1 \text{ zeros}$$

$$\underset{\overset{\longrightarrow}{\frac{1}{f(x)}}}{\underline{1}}\ 0\ 1\ 0\ 1\ 1\ 1\ 0\ 1\ 1\ 0\ 0\ 0\ 1\ 1\ 1\ 1\ 1\ 0\ 0\ 1\ 1\ 0\ 1\ 0\ 0\ 1\ \overbrace{0\ 0\ 0\ 0}$$

Figure 6.28 Five-stage MSRG mask example for the loading 01011.

The corresponding reciprocal polynomial is

$$f_S^*(x) = x^{42}\, f_S(x^{-1})$$
$$= 1 + x + x^2 + x^3 + x^5 + x^6 + x^7 + x^{10} + x^{16} + x^{17} + x^{18}$$
$$+ x^{19} + x^{21} + x^{22} + x^{25} + x^{26} + x^{27} + x^{31} + x^{33}$$
$$+ x^{35} + x^{42} \tag{6.37b}$$

Figure 6.29 shows the MSRG implementation of the long code given in IS-95.

System time for the CDMA cellular system is the same as that maintained by GPS. A reference phase shift of the long code, whose period lasts over 41 days, is synchronized to the CDMA system clock starting at a particular reference time. The reference code phase shift for the long code is specified in IS-95 to be the phase of the sequence that is produced when the MSRG output is the first logical one after 41 zeros, and the mask is all zeros except for a logical one in the most significant bit (MSB):

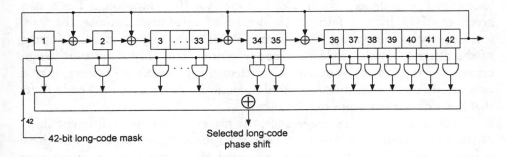

Figure 6.29 The IS-95 long PN code generator and mask.

$$\overbrace{\qquad\qquad}^{41\ zeros}$$

MSRG output: $\cdots 0\,0\,0 \cdots 0\,0\,0\ \underset{\uparrow}{1} \cdots$

Beginning of reference shift

Mask: $0\,0\,0 \cdots 0\,0\,0\,1$

From our knowledge of PN sequences and MSRG masks and from this information, we can state the following:

• The MSRG output for the reference phase is the shift described algebraically by $1/f_S(x)$, because this shift is uniquely identified by a 1 preceded by $n-1$ zeros. $\Rightarrow q = 0$.

• The mask that is specified is the first 42 bits of some shift of the sequence: $a_r(x) = 0 \cdot x^0 + 0 \cdot x^1 + \cdots + 0 \cdot x^{40} + 1 \cdot x^{41} + \cdots$. This shift is easily identified as

$$a_r(x) = \frac{x^{41}}{f_S(x)} = \frac{g_r(x)}{f_S(x)} \quad \Rightarrow \quad r = 41$$

• The relative delay between the MSRG output and that of the phase shifting network therefore is $d_r = r - (n-1) = 41 - 41 = 0$.

• Therefore the reference long-code phase shift is the sequence described by $1/f_S(x)$.

The masks specified for the long code on the forward link are shown in Figure 6.30. The 42-bit long-code masks "command" different shifts of the long-code PN sequence, which is $2^{42} - 1 \approx 4.4 \times 10^{12}$ chips long. The masks given in IS-95 have a fairly high degree of agreement because they are formatted with a certain prefix common to all users. This redundant format of the different masks is clearly a convenience. The only significant design criteria for the long-code masks, other than this convenience feature, is that two simultaneous transmissions should practically never have long-code shifts that are within a few chips of each other.

Although similar in appearance, the masks specify quite different shifts of the PN sequence. Because the code is so long, the probability that these segments will be the actual ones correlated at a given time is vanishingly small.

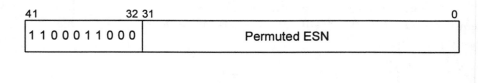

ACN: access channel number
PCN: paging channel number
BASE_ID: base station identification
PILOT_PN: PN offset for the forward CDMA channel
ESN: electronic serial number

Figure 6.30 Long-code masks used in IS-95.

Example 6.17 We now illustrate the issues involved in using MSRG masks with prefixes, similar to the IS-95 long-code masks shown in Figure 6.30. Consider the degree $n = 6$ PN generator with the characteristic polynomial $f(x) = 1 + x + x^6$. Also, to simplify the discussion, let the initial loading be $\overrightarrow{100000}$; for this loading, the mask is identical to the first six bits produced by the PSN. For example, with this loading, if the mask is $\overrightarrow{101100}$, then the first six bits of the sequence out of the PSN will be 101100.

One period of the 63-chip sequence $1/f(x)$ is the following:

111111010101100110111011010010011100010111100101000110000100000

There are 63 possible nonzero 6-bit masks, of the form

$$\text{Mask} = (m_1, m_2, m_3, m_4, m_5, m_6), \quad m_i = 0 \text{ or } 1$$

For the purposes of this discussion, let the first three bits of the mask be a prefix that is the same for all users, similar to the IS-95 mask format but on a much smaller scale. We show that the selection of such a prefix is not arbitrary: some prefixes are better than others.

For example, if the prefix is $(m_1, m_2, m_3) = 110$, the mask can be written 110xxx, where the last three bits ("xxx") accommodate eight users:

$$\text{Mask}(k) = \overrightarrow{110\text{xxx}}, \quad \text{xxx} = \text{binary of } k = 0, 1, \ldots, 7$$

The shifts of the 63-chip sequence that are produced by this example mask are the eight shifts that begin with 110. We may denote those shifts by pointing to their starting positions in the diagram in Figure 6.31. Using this kind of diagram, the starting points of the masks formatted with the prefixes 110 and 111 may be indicated as shown in Figure 6.32. Note how the starting points for the prefix 110 are more evenly spaced than those for the prefix 111. There is less likelihood that users with different propagation delays have the same code phase at the receiver if the prefix is 110 than there is if the prefix is 111. More example starting points for different prefixes are illustrated in Figure 6.33.

From this simple example, we observe that some mask prefixes are better than others. The best prefixes are those without long runs of ones or zeros, as the example illustrates.

xxx = 101   011 111 110 100   001   010   000

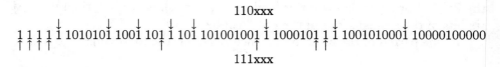

Figure 6.31 Location of code shifts selected by masks 110xxx.

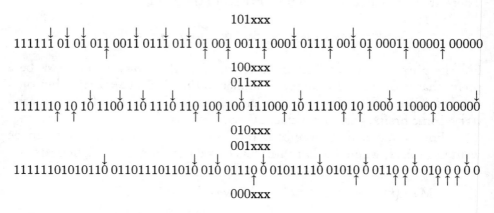

Figure 6.32 Location of code shifts selected by 110xxx and 111xxx.

Figure 6.33 Additional mask-selected starting positions.

Predicting the future state of the long PN code generator, as done in IS-95 synchronization channel messages, can be done using a transition matrix. An example algorithm for accomplishing this task is given in Appendix 6B.3.

The IS-95 short PN codes. Two "short" PN sequences used in IS-95 to spread the I and Q (quadrature) components of the forward and reverse link waveforms. Each implements a degree $n = 15$ characteristic polynomial. The I-channel short PN sequence has the characteristic polynomial

$$f_I(x) = 1 + x^2 + x^6 + x^7 + x^8 + x^{10} + x^{15} \tag{6.38a}$$

The Q-channel short PN sequence has the characteristic polynomial

$$f_Q(x) = 1 + x^3 + x^4 + x^5 + x^9 + x^{10} + x^{11} + x^{12} + x^{15} \tag{6.38b}$$

The period of the short PN codes is $2^{15} - 1 = 32{,}767$ chips. In the CDMA cellular system, to get a period of 2^{15}, the actual PN codes are these modified by the insertion of an additional zero to the 14 zeros at the end of each of the reference sequences $1/f_I(x)$ and $1/f_Q(x)$.

The MSRG implementation of the I- and Q-channel PN codes is illustrated in Figure 6.34. The reference or "zero offset" code phase shift for both of the short PN codes is specified in IS-95 to be the phase of the sequence that occurs when the shift register output is the first logical one after 15 zeros (including an extra zero), as illustrated in Figure 6.35.

The 15th zero that is inserted in these sequences occurs at the end of the run of 14 zeros that occurs normally. We know from our previous consideration of MSRGs that the normal sequence of $n - 1$ zeros at the output of an MSRG takes place for the first $n - 1$ consecutive states with weight 1:

$$100\cdots000, \quad 010\cdots000, \quad 001\cdots000, \ldots, \quad 000\cdots100, \quad 000\cdots010$$

An extra zero can be inserted in the sequence by forcing the MSRG to repeat any one of these states, either by reloading the shift register with that state just after it changes or by inhibiting the clock pulse input to the shift register for one chip duration. Another method is to count the chips—after 32,767 chips have been generated by the MSRG, the clock pulse input can be inhibited while a zero is substituted at the output as chip number 32,768.

IS-95 specifies that the short PN code sequences associated with different base stations are to be the same, except that they are to be offset in PN code phase by multiples of 64 chips. This gives a maximum of $2^{15}/64 = 512$ different starting positions for the short PN codes. IS-95 does not specify how the base station and mobile units are to generate the various short PN

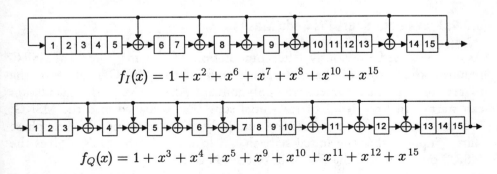

$$f_I(x) = 1 + x^2 + x^6 + x^7 + x^8 + x^{10} + x^{15}$$

$$f_Q(x) = 1 + x^3 + x^4 + x^5 + x^9 + x^{10} + x^{11} + x^{12} + x^{15}$$

Figure 6.34 MSRG implementation of IS-95 short PN codes.

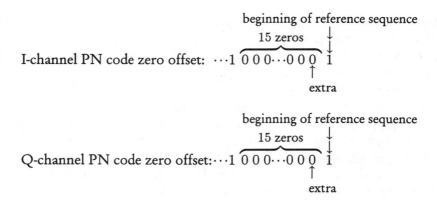

Figure 6.35 Short-code zero offset sequences.

code offsets.   Generation of shifts of zero-added sequences is discussed in Section 6.5.1.

The mobile unit needs the capability for generating any short PN code offset because it tracks the signals of different base stations, perhaps combining signals from several base stations at the same time.   There is also a requirement for the base station receiver to be able to switch back and forth between different PN code offsets during the several modes of reverse link transmission.

The foregoing considerations argue strongly for the use of masks to generate the short PN code offsets, as illustrated in Figure 6.36.   The $512 \times 2 = 1,024$ 15-bit masks for the two short PN code offsets can be precomputed and stored in the permanent memory of the CDMA equipment, using about 30 Kb of memory.

6.3.4.3 Example Short PN Code Masks

As we discussed previously, the computation of the mask for an MSRG requires knowledge of the code phase shift of the MSRG relative to the inverse of $f(x)$, the characteristic polynomial.   Alternatively, the calculation can start with knowledge of the initial state that is loaded into the MSRG, because that initial state corresponds uniquely to the resulting PN code phase shift.   Suppose that the initial state that is loaded into the MSRG gives the MSRG output that is the Kth shift of the PN code:

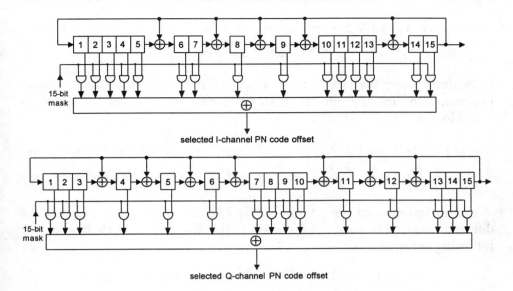

Figure 6.36 The I- and Q-channel MSRGs with masks to select the offset.

$$a_K(x) = \frac{x^K \text{ modulo } f(x)}{f(x)} \tag{6.38c}$$

Then the mask for the jth offset of 64 chips, having the mask parameter r_j, must produce a relative delay at the output of the phase shift network of

$$d_{r_j} = j\text{th desired offset phase} - \text{MSRG phase} = 64j - K \tag{6.38d}$$

The mask that produces this relative delay is the first $n = 15$ bits of the sequence with shift r_j, where

$$r_j = d_{r_j} + n - 1 = 64j - K + 14 \tag{6.38e}$$

For example, it is simple to initialize the MSRG's shift register by loading it with all 1s. Because the initial state of the shift register, when expressed as a polynomial, is the reverse of the output sequence numerator polynomial, we know this loading produces the MSRG output sequence

$$a_K(x) = \frac{g_K(x)}{f(x)}, \quad f(x) = f_I(x) \text{ or } f_Q(x)$$

where

$$g_K(x) = 1 + x + x^2 + x^3 + x^4 + x^5 + x^6 + x^7 + x^8 + x^9 + x^{10}$$
$$+ x^{11} + x^{12} + x^{13} + x^{14}$$

Recursively using the fact that $1 = 1 + f(x)$, as employed in the computer program listed in Appendix 6B.1, the numerator polynomial for the all-1s initial loading was found to be

$$g_K(x) = x^{10,719} \text{ modulo } f_I(x) \quad \Rightarrow \quad K = 10,719 \text{ for the I-channel short code}$$
$$= x^{3,601} \text{ modulo } f_Q(x) \quad \Rightarrow \quad K = 3,601 \text{ for the Q-channel short code}$$

With this information, for an all-1s MSRG initial loading, the masks for the different short PN code offsets need to be the first $n = 15$ bits of the following sequences:

$$a_{r_j}(x) = \frac{x^{r_j} \text{ modulo } f_I(x)}{f_I(x)} \quad \text{for the I channel, where}$$

$$r_j = 64j - 10,719 + 14 = 64j - 10,705$$

and

$$a_{r_j}(x) = \frac{x^{r_j} \text{ modulo } f_Q(x)}{f_Q(x)} \quad \text{for the Q channel, where}$$

$$r_j = 64j - 3,601 + 14 = 64j - 3,587$$

For the zero-offset base station pilot signal ($j = 0$), then modulo the period $P = 32,767$ the required mask parameters are

$$r_0 = 0 \cdot 64 - 10,705 = 32,767 - 10,705 = 22,062 \text{ for the I channel}$$

and

$$r_0 = 0 \cdot 64 - 3,587 = 32,767 - 3,587 = 29,180 \text{ for the Q channel}$$

Recursively applying the relation $x^{15} = x^{15} + f_I(x)$ results in finding that the I-channel numerator polynomial for a zero offset and an all-1s MSRG loading is

$$g_{I0}(x) = x^{22,062} \text{ modulo } f_I(x) = x^6 + x^8 + x^{10} + x^{11} + x^{13}$$

Recursively applying the relation $x^{15} = x^{15} + f_Q(x)$ results in finding that the Q-channel numerator polynomial for a zero offset and an all-1s MSRG loading is

$$g_{Q0}(x) = x^{29,180} \text{ modulo } f_I(x) = 1 + x + x^2 + x^4 + x^7 + x^8 + x^{11}$$

Appendix 6B.2 lists a computer program for these recursive computations.

Having found the required numerator polynomials, the masks are finally determined as the first 15 bits of the sequences given by the numerator polynomials:

$$m_{I0}(x) = \left\{ \frac{g_{I0}(x)}{f_I(x)} \right\}_{\deg < 15}$$

$$= \left\{ \frac{x^6 + x^8 + x^{10} + x^{11} + x^{13}}{1 + x^2 + x^6 + x^7 + x^8 + x^{10} + x^{15}} \right\}_{\deg < 15}$$

$$= x^6 + x^{10} + x^{11} + x^{13} + x^{14}$$

$$\Rightarrow \quad 000000100011011 = \text{I-channel mask}$$

and

$$m_{Q0}(x) = \left\{ \frac{g_{Q0}(x)}{f_Q(x)} \right\}_{\deg < 15}$$

$$= \left\{ \frac{1 + x + x^2 + x^4 + x^7 + x^8 + x^{11}}{1 + x^3 + x^4 + x^5 + x^9 + x^{10} + x^{11} + x^{12} + x^{15}} \right\}_{\deg < 15}$$

$$= 1 + x + x^2 + x^3 + x^4 + x^5 + x^6 + x^{11} + x^{14}$$

$$\Rightarrow \quad 111111100001001 = \text{Q-channel mask}$$

To show that these masks produce $1/f(x)$, we need first to find the leading bits in the reference PN sequences $1/f_I(x)$ and $1/f_Q(x)$. This is easily accomplished by long division:

$$\frac{1}{f_I(x)} = \frac{1}{1 + x^2 + x^6 + x^7 + x^8 + x^{10} + x^{15}}$$

$$= 1 + x^2 + x^4 + x^7 + x^{10} + x^{11} + x^{12} + x^{14} + \cdots$$

$$\Rightarrow 1\,0\,1\,0\,1\,0\,0\,1\,0\,0\,1\,1\,1\,0\,1\,0 \cdots$$

and

$$\frac{1}{f_Q(x)} = \frac{1}{1 + x^3 + x^4 + x^5 + x^9 + x^{10} + x^{11} + x^{12} + x^{15}}$$

$$= 1 + x^3 + x^4 + x^5 + x^6 + x^8 + x^{10} + x^{11} + x^{12} + x^{14} + \cdots$$

$$\Rightarrow 1\,0\,0\,1\,1\,1\,1\,0\,1\,0\,1\,1\,1\,0\,1\,0 \cdots$$

Verification that the I-channel mask produces $1/f_I(x)$ and that the Q-channel mask produces $1/f_Q(x)$ is shown in Tables 6.19 and 6.20. Figures 6.37 and 6.38 shows the corresponding PN generators and mask configurations.

Table 6.19 Output of I-channel mask

Mask: 0 0 0 0 0 0 1 0 0 0 1 1 0 1 1

MSRG state (1-15)	Mask output	$1/f_I(x)$
1 1 1 1 1 1 1 1 1 1 1 1 1 1 1	1	1
1 1 1 1 1 0 1 0 0 0 1 1 1 0 1	0	0
1 1 1 1 1 0 0 0 1 1 0 1 1 0 0	1	1
0 1 1 1 1 1 0 0 0 1 1 0 1 1 0	0	0
0 0 1 1 1 1 1 0 0 0 1 1 0 1 1	1	1
1 0 0 1 1 0 1 0 1 1 0 1 1 1 1	0	0
1 1 0 0 1 0 0 0 1 0 1 0 1 0 1	0	0
1 1 1 0 0 0 0 1 1 0 0 1 0 0 0	1	1
0 1 1 1 0 0 0 0 1 1 0 0 1 0 0	0	0
0 0 1 1 1 0 0 0 0 1 1 0 0 1 0	0	0
0 0 0 1 1 1 0 0 0 0 1 1 0 0 1	1	1
1 0 0 0 1 0 1 1 1 1 0 1 1 1 0	1	1
0 1 0 0 0 1 0 1 1 1 1 0 1 1 1	1	1
⋮	⋮	⋮

Table 6.20 Output of Q-channel mask

Mask: 1 1 1 1 1 1 1 0 0 0 0 1 0 0 1

MSRG state (1-15)	Mask output	$1/f_Q(x)$
1 1 1 1 1 1 1 1 1 1 1 1 1 1 1	1	1
1 1 1 0 0 0 0 1 1 1 0 0 0 1 1	0	0
1 1 1 0 1 1 1 0 1 1 0 1 1 0 1	0	0
1 1 1 0 1 0 0 1 0 1 0 1 0 1 0	1	1
0 1 1 1 0 1 0 0 1 0 1 0 1 0 1	1	1
1 0 1 0 0 1 0 0 0 1 1 0 1 1 0	1	1
0 1 0 1 0 0 1 0 0 0 1 1 0 1 1	1	1
1 0 1 1 0 1 1 1 0 0 1 0 0 0 1	0	0
1 1 0 0 0 1 0 1 1 0 1 0 1 0 0	1	1
0 1 1 0 0 0 1 0 1 1 0 1 0 1 0	0	0
0 0 1 1 0 0 0 1 0 1 1 0 1 0 1	1	1
1 0 0 0 0 1 1 0 1 0 0 0 1 1 0	1	1
0 1 0 0 0 0 1 1 0 1 0 0 0 1 1	1	1
⋮	⋮	⋮

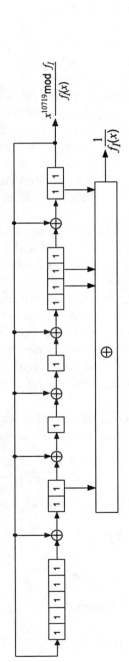

Figure 6.37 I-channel short-code generator with mask to produce zero offset.

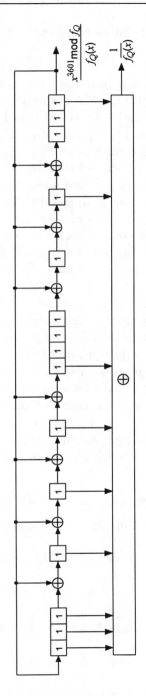

Figure 6.38 Q-channel short-code generator with mask to produce zero offset.

6.4 Autocorrelation and Cross-Correlation Properties of Binary Sequences

One of the most important reasons for the application of PN sequences in spread-spectrum CDMA systems is their *autocorrelation properties*. We have defined previously the correlation between two sequences as the number of bit-by-bit position agreements minus the number of disagreements. In Section 6.1, we stated that PN sequences must possess the correlation property that the number of term-by-term position agreements minus the number of term-by-term position disagreements is one in a full period of a PN sequence. In the examples of PN sequences that we have treated in this chapter, such as those of length $2^3 - 1 = 7$ and $2^4 - 1 = 15$, the correlation properties have indeed been observed to be validated.

When two sequences are correlated for given (fixed) phase shifts, a correlation value is obtained. When the two sequences are correlated for all phase shifts, we can plot the correlation values so obtained as a function of the phase shifts. Such a correlation plot is called a *correlation function*, in the sense that the correlation is a function of phase shift. Now, if the correlation function is computed for two sequences generated by the same PN generator, we call it an *autocorrelation function*. When the correlation function is obtained between two sequences that are generated by two different PN generators, it is called a *cross-correlation function*. We take up examples to illustrate these differences and some variations of these definitions in some cases.

Let us define the binary sequence notations to be used in this section. We consider two binary sequences a and b, each with period P as shown below:

$$a = (a_0, a_1, a_2, \ldots, a_{P-2}, a_{P-1}, a_0, \ldots) = \{a_i\}, \quad a_i \in \mathrm{GF}(2)$$

$$b = (b_0, b_1, b_2, \ldots, b_{P-2}, b_{P-1}, b_0, \ldots) = \{b_i\}, \quad b_i \in \mathrm{GF}(2)$$

Denote the sequence b shifted l positions to the right by

$$b^{(l)} = (b_{P-l}, b_{P-l+1}, \ldots, b_{P-2}, b_{P-1}, b_0, \ldots) = \{b_{i-l}\}$$

Denote the term-by-term modulo-2 addition of the sequences a and b by

$$a \oplus b = (a_0 \oplus b_0, a_1 \oplus b_1, \ldots) = \{a_i \oplus b_i\}$$

Denote the weight of a sequence a with period P by

$$W(a) = \text{\# of 1s in one period of } a$$

Given the binary sequence a and the shifted binary sequence $b^{(l)}$, both with period P, we define

$$A(a, b^{(l)}) \triangleq \text{\# of term-by-term agreements between } a \text{ and } b^{(l)}, \text{ and}$$

$$D(a, b^{(l)}) \triangleq \text{\# of term-by-term disagreements between } a \text{ and } b^{(l)}$$

The correlation $R_{a,b}(l)$ between a and $b^{(l)}$ is then given by

$$R_{a,b}(l) = A(a, b^{(l)}) - D(a, b^{(l)}) \tag{6.39a}$$

Note that

$$A(a, b^{(l)}) + D(a, b^{(l)}) = P \tag{6.39b}$$

Because under modulo-2 addition:

$$a_i \oplus b_{i-l} = \begin{cases} 0, & a_i = b_{i-l} \text{ (agreement)} \\ 1, & a_i \neq b_{i-l} \text{ (disagreement)} \end{cases}$$

it follows that

$$D(a, b^{(l)}) = W(a \oplus b^{(l)}) \tag{6.39c}$$

Therefore, the sequence cross-correlation function for a and b may be calculated as

$$\begin{aligned} R_{a,b}(l) &= A(a, b^{(l)}) - D(a, b^{(l)}) \\ &= [P - W(a \oplus b^{(l)})] - W(a \oplus b^{(l)}) \\ &= P - 2 \cdot W(a \oplus b^{(l)}) \end{aligned} \tag{6.39d}$$

Example 6.18 The computation of an example sequence cross-correlation function is shown below for the application of the formulas given above, and is plotted in Figure 6.39.

	$l = 0$		$l = 1$
a	1 1 1 0 1 0 0	a	1 1 1 0 1 0 0
b	1 0 1 1 1 0 0	$b^{(1)}$	0 1 0 1 1 1 0
$a \oplus b$	0 1 0 1 0 0 0	$a \oplus b^{(1)}$	1 0 1 1 0 1 0
$R_{a,b}(0) = 7 - 2(2) = 3$		$R_{a,b}(1) = 7 - 2(4) = -1$	

$$l = 2$$

a 1 1 1 0 1 0 0

$b^{(2)}$ 0 0 1 0 1 1 1

$a \oplus b^{(2)}$ 1 1 0 0 0 1 1

$R_{a,b}(2) = 7 - 2(4) = -1$

$$l = 3$$

a 1 1 1 0 1 0 0

$b^{(3)}$ 1 0 0 1 0 1 1

$a \oplus b^{(3)}$ 0 1 1 1 1 1 1

$R_{a,b}(3) = 7 - 2(6) = -5$

$$l = 4$$

a 1 1 1 0 1 0 0

$b^{(4)}$ 1 1 0 0 1 0 1

$a \oplus b^{(4)}$ 0 0 1 0 0 0 1

$R_{a,b}(4) = 7 - 2(2) = 3$

$$l = 5$$

a 1 1 1 0 1 0 0

$b^{(5)}$ 1 1 1 0 0 1 0

$a \oplus b^{(5)}$ 0 0 0 0 1 1 0

$R_{a,b}(5) = 7 - 2(2) = 3$

$$l = 6$$

a 1 1 1 0 1 0 0

$b^{(6)}$ 0 1 1 1 0 0 1

$a \oplus b^{(6)}$ 1 0 0 1 1 0 1

$R_{a,b}(6) = 7 - 2(4) = -1$

Sequence autocorrelation function. For $a = b$, the sequence cross-correlation function becomes the autocorrelation function

$$R_{a,a}(l) \equiv R_a(l) = P - 2 \cdot W\left(a \oplus a^{(l)}\right) \tag{6.40a}$$

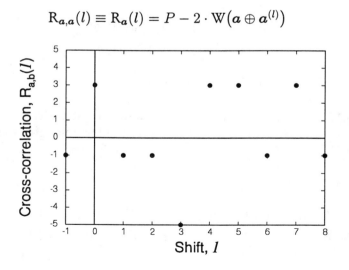

Figure 6.39 Sequence cross-correlation function example.

If a is a maximal-length sequence, the modulo-2 addition of a nonzero shift ($l \neq 0$) of the sequence $a^{(l)}$ to a produces the same sequence with a different shift, say, $a^{(l')}$, and therefore,

$$R_a(l \neq 0) = P - 2 \cdot W(a \oplus a^{(l)}) = P - 2 \cdot W(a^{(l')})$$
$$= P - 2 \cdot W(a) \qquad (6.40\text{b})$$

Because for a maximal-length sequence with $P = 2^n - 1$ the number of 1s is $(P+1)/2$ and the number of 0s is $(P-1)/2$, the autocorrelation function for a maximal-length sequence has the values

$$R_a(l) = \begin{cases} P, & l = 0, \pm P, \pm 2P, \dots \\ -1, & \text{otherwise} \end{cases} \qquad (6.40\text{c})$$

where n is the degree of the primitive polynomial.

Thus we have used the autocorrelation function formula (6.40b) to show that PN sequences over GF(2) give two-value ($P = 2^n - 1$ and -1) autocorrelation functions. Before we proceed further, however, we derive another formula for the correlation functions when the sequences are represented by "+1" and "−1" values instead of "0" and "1," which are really time functions having two-level values of $\{+1, -1\}$. We call them "waveforms" in a later discussion.

Let the original binary sequence of 1s and 0s be denoted by an italic *Roman* character, as we have done above, and let the corresponding ± 1 binary sequence be denoted by an uppercase ROMAN character:

Binary sequence over $\{0, 1\}$		Binary sequence over $\{+1, -1\}$
$1, 1, 0, 1, 0, 0, \dots$	$\rightarrow \boxed{\begin{array}{c} 1 \rightarrow -1 \\ 0 \rightarrow +1 \end{array}} \rightarrow$	$-1, -1, +1, -1, +1, +1, \dots$
$a_0, a_1, a_2, \dots$		$A_0, A_1, A_2, \dots$
$a_i = \dfrac{1 - A_i}{2}$		$A_i = 1 - 2a_i$

Given two ± 1 binary sequences with period P:

$$\mathbf{A} = (A_0, A_1, A_2, \dots, A_{P-1}, A_0, A_1, \dots) \quad \text{converted from } a$$

$$\mathbf{B} = (B_0, B_1, B_2, \dots, B_{P-1}, B_0, B_1, \dots) \quad \text{converted from } b$$

the cross-correlation function for **A** and **B** is defined as

$$R_{a,b}(l) = \sum_{i=0}^{P-1} A_i\, B_{i-l} \tag{6.41a}$$

$$= A_0\, B_{P-l} + A_1\, B_{P-l+1} + \cdots$$

$$+ A_{l-1}\, B_{P-1} + A_l\, B_0 + \cdots + A_{P-1}\, B_{P-l-1}$$

Note that $B_{-l} = B_{P-l}$.

The equivalence of the cross-correlation function formula given by (6.41a) with that given in (6.39d) is readily shown:

$$R_{a,b}(l) = \sum_{i=0}^{P-1} A_i\, B_{i-l}$$

$$= \sum_{i=0}^{P-1} (1 - 2a_i)(1 - 2b_{i-l}) = \sum_{i=0}^{P-1} [1 - 2(a_i + b_{i-l} - 2a_i\, b_{i-l})]$$

$$= \sum_{i=0}^{P-1} [1 - 2(a_i \oplus b_{i-l})] = P - 2 \cdot \mathrm{W}\big(a \oplus b^{(l)}\big) \tag{6.41b}$$

where we use $\oplus$ to denote modulo-2 arithmetic.

Example 6.19 An example sequence cross-correlation function is shown below. The values of $R_{a,b}(l)$ are computed using the formula in (6.41a).

	$l = 0$		$l = 1$
A	$-1\ -1\ -1\ +1\ -1\ +1\ +1$	**A**	$-1\ -1\ -1\ +1\ -1\ +1\ +1$
B	$-1\ +1\ -1\ -1\ -1\ +1\ +1$	$\mathbf{B}^{(1)}$	$+1\ -1\ +1\ -1\ -1\ -1\ +1$
$\times$	$+1\ -1\ +1\ -1\ +1\ +1\ +1$	$\times$	$-1\ +1\ -1\ -1\ +1\ -1\ +1$
	$R_{a,b}(0) = \text{sum} = 3$		$R_{a,b}(1) = \text{sum} = -1$
	$l = 2$		$l = 3$
A	$-1\ -1\ -1\ +1\ -1\ +1\ +1$	**A**	$-1\ -1\ -1\ +1\ -1\ +1\ +1$
$\mathbf{B}^{(2)}$	$+1\ +1\ -1\ +1\ -1\ -1\ -1$	$\mathbf{B}^{(3)}$	$-1\ +1\ +1\ -1\ +1\ -1\ -1$
$\times$	$-1\ -1\ +1\ +1\ +1\ -1\ -1$	$\times$	$+1\ -1\ -1\ -1\ -1\ -1\ -1$
	$R_{a,b}(2) = \text{sum} = -1$		$R_{a,b}(3) = \text{sum} = -5$

$$l = 4$$

A   −1 −1 −1 +1 −1 +1 +1

$B^{(4)}$ −1 −1 +1 +1 −1 +1 −1

×   +1 +1 −1 +1 +1 +1 −1

$R_{a,b}(4)$ = sum = 3

$$l = 5$$

A   −1 −1 −1 +1 −1 +1 +1

$B^{(5)}$ −1 −1 −1 +1 +1 −1 +1

×   +1 +1 +1 +1 −1 −1 +1

$R_{a,b}(5)$ = sum = 3

$$l = 6$$

A   −1 −1 −1 +1 −1 +1 +1

$B^{(6)}$ +1 −1 −1 −1 +1 +1 −1

×   −1 +1 +1 −1 −1 +1 −1

$R_{a,b}(6)$ = sum = −1

6.4.1 Correlation Function for Real-Time Signals

Let a *basic waveform* be denoted as $\varphi(t)$, which is defined on the interval $(0, T_c)$ as shown in Figure 6.40. Then the sequence over $\{+1, -1\}$ corresponding to the sequence a can be implemented by the real-time function

$$f_a(t) = \sum_{k=-\infty}^{\infty} A_k \varphi_a(t - kT_c), \qquad A_k \in \{+1, -1\}$$

in which $\varphi_a(t)$ is a basic waveform used for this particular sequence. The function $f_a(t)$ is a periodic function with period PT_c because

$$f_a(t + PT_c) = \sum_{k=-\infty}^{\infty} A_k \varphi_a(t + PT_c - kT_c)$$

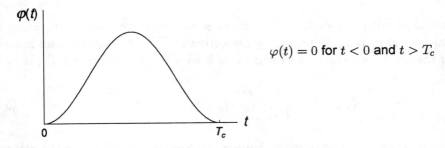

$\varphi(t) = 0$ for $t < 0$ and $t > T_c$

Figure 6.40 Basic waveform.

$$= \sum_{k=-\infty}^{\infty} A_k \, \varphi_a[t - (k - P)T_c] = \sum_{k=-\infty}^{\infty} A_{k+P} \, \varphi_a(t - kT_c)$$

$$= \sum_{k=-\infty}^{\infty} A_k \, \varphi_a(t - kT_c)$$

For nonperiodic waveforms $x(t)$ and $y(t)$ with finite energy, the cross-correlation function is defined as

$$R_{x,y}(\tau) = \int_{-\infty}^{\infty} x(t) \, y(t - \tau) \, dt \qquad (6.42a)$$

For periodic waveforms $x(t)$ and $y(t)$, with period T, the cross-correlation function is defined as

$$R_{x,y}(\tau) = \frac{1}{T} \int_{0}^{T} x(t) \, y(t - \tau) \, dt \qquad (6.42b)$$

The autocorrelation function for either type of waveform is defined as

$$R_{x,x}(\tau) \equiv R(\tau) \qquad R_{y,y}(\tau) \equiv R_y(\tau)$$

Now, let two binary waveforms be

$$f_a(t) = \sum_{k=-\infty}^{\infty} A_k \, \varphi_a(t - kT_c), \qquad A_k \in \{+1, -1\} \qquad (6.43a)$$

and

$$f_b(t) = \sum_{k=-\infty}^{\infty} B_k \, \varphi_b(t - kT_c), \qquad B_k \in \{+1, -1\} \qquad (6.43b)$$

associated with the periodic sequences a and b, respectively, and using the basic waveforms $\varphi_a(t)$ and $\varphi_b(t)$, respectively. The cross-correlation function for these two real-time waveforms, which have the same period PT_c, is

$$R_{f_a,f_b}(\tau) = \frac{1}{PT_c} \int_{0}^{PT_c} f_a(t) \, f_b(t - \tau) \, dt \qquad (6.44)$$

Our interest is to compute the correlation function (6.44) when the real-time functions $f_a(t)$ and $f_b(t)$ represent real-time PN signals based on given PN sequences. The calculation of (6.44) can be easily facilitated, however, when we use the following theorem:

Theorem: The cross-correlation function for $f_a(t)$ and $f_b(t)$ can be calculated from

$$R_{fa,fb}(\tau) = \frac{1}{PT_c} \sum_{k=-\infty}^{\infty} R_{a,b}(k)\, R_{\varphi_a,\varphi_b}(\tau - kT_c) \qquad (6.45a)$$

where $R_{a,b}(k)$ is the periodic sequence cross-correlation function and $R_{\varphi_a,\varphi_b}(\tau)$ is the nonperiodic cross-correlation function for the basic waveforms. The proof of (6.45a) is given in Appendix 6C.

The autocorrelation function for $a = b$ and $\varphi_a(t) = \varphi_b(t)$ is then of the form

$$R_{fa}(\tau) = \frac{1}{PT_c} \sum_{k=-\infty}^{\infty} R_a(k)\, R_{\varphi_a}(\tau - kT_c) \qquad (6.45b)$$

where $R_a(k)$ is the period sequence autocorrelation function and $R_{\varphi_a}(\tau)$ is nonperiodic autocorrelation function for the basic waveform.

Example 6.20 Let $\varphi(t)$, the *basic waveform*, be the unit rectangular pulse, as illustrated in Figure 6.41 and defined as

$$\varphi(t) = \text{Rect}\left(\frac{t}{T_c} - \frac{1}{2}\right) = \begin{cases} 1, & 0 \le t \le T_c \\ 0, & \text{otherwise} \end{cases} \qquad (6.46a)$$

The autocorrelation function for this basic waveform is

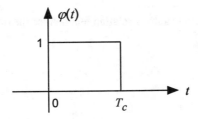

Figure 6.41 Rectangular pulse basic waveform.

$$R_\varphi(\tau) = \int_{-\infty}^{\infty} \varphi(t)\,\varphi(t-\tau)\,dt = \int_{L}^{U} 1 \cdot dt$$

where $(L, U) = (\tau, T_c)$ if $\tau > 0$ and $(L, U) = (0, T_c + \tau)$ if $\tau < 0$. Thus

$$R_\varphi(\tau) = \begin{cases} T_c - \tau, & 0 \le \tau \le T_c \\ T_c + \tau, & -T_c \le \tau \le 0 \\ 0, & \text{otherwise} \end{cases}$$

$$= \begin{cases} T_c\left(1 - \dfrac{|\tau|}{T_c}\right), & |\tau| \le T_c \\ 0, & \text{otherwise} \end{cases} \tag{6.46b}$$

which is the "triangular function" illustrated in Figure 6.42.

Example 6.21 We compute the autocorrelation function for the ± 1 sequence $\mathbf{A} = (-1, -1, -1, +1, -1, +1, +1)$, which corresponds to the $(0,1)$ sequence $\boldsymbol{a} = (1,1,1,0,1,0,0)$. Assume that the basic waveform is the rectangular pulse $\varphi(t)$ defined by $\varphi(t) = \text{Rect}(t/T_c - 1/2)$ as shown in Example 6.20. Now, from Examples 6.19 and 6.20, we have the sequence autocorrelation function $R_a(l)$ shown in Figure 6.43 and the autocorrelation function $R_\varphi(\tau)$ for the basic waveform shown in Figure 6.42.

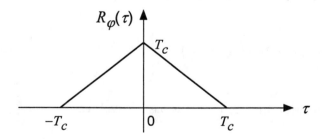

Figure 6.42 Triangular function: autocorrelation function for the rectangular pulse basic waveform.

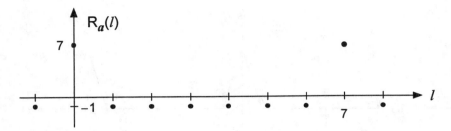

Figure 6.43 Sequence autocorrelation function for $a = (1, 1, 1, 0, 1, 0, 0)$.

The autocorrelation function for the real-time function $f_a(t)$ corresponding to A and using $\varphi(t) = \text{Rect}(t/T_c - 1/2)$ is given by (6.45b). Figure 6.44 gives a plot of the autocorrelation function before normalization by PT_c:

$$PT_c \cdot R_{fa}(\tau) = \sum_{k=-\infty}^{\infty} R_a(k) \, R_\varphi(\tau - kT_c) \qquad (6.46c)$$

Figure 6.44 shows that the autocorrelation function for the real-time signal, based on the sequence a or A, using the rectangular pulse basic waveform, is constructed as a sum (superposition) of different shifts of the basic waveform's autocorrelation function as dictated by the component amplitudes and "polarities" of the sequence A. This example is based on a PN sequence of length 7, and the periodic nature of the autocorrelation function $R_{fa}(\tau)$ (the solid line in Figure 6.44) is shown clearly in Figure 6.45 for the general case of a sequence period P.

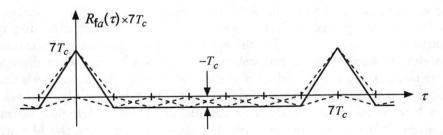

Figure 6.44 Autocorrelation function for real-time function.

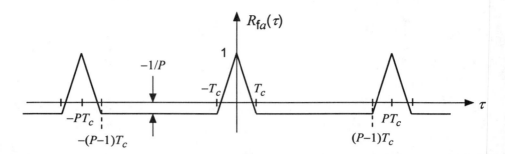

Figure 6.45 Normalized autocorrelation function for real time signal waveform.

An important observation needs to be made here: when the real-time waveform $f_a(t)$ is constructed using rectangular pulses, such as a rectangular wave PN signal waveform, its autocorrelation function $R_{fa}(\tau)$ is the same as the sequence autocorrelation function $R_a(l)$ normalized by P with straight lines connecting the points of the sequence autocorrelation function. Through this example, we have seen the evidence of the so-called "two-level autocorrelation function" of PN sequence waveforms.

In spread-spectrum communications systems employing PN sequences, the waveforms that are transmitted are not real-time signals based on rectangular pulse basic waveforms, in general. In the IS-95 system, the "basic waveform" is very different from the rectangular pulse shape for reasons having to do with the desire to minimize intersymbol interference and to confine most of the signal energy to a narrow bandwidth in order to reduce cochannel interference.

The two-level autocorrelation function property of PN sequences, as shown in Figure 6.45 for example, is the main reason that PN sequences are used in spread-spectrum CDMA systems. The autocorrelation function plotted in that figure shows that the signal level for a given code phase (corresponding to a particular mobile station, for example) under perfect alignment (synchronization) at the receiver is at its peak, whereas the signal levels under all other nonaligned conditions (corresponding to other mobile stations) are very low for sequences with large periods, P. This is a highly desirable situation. Thus, on the forward link of the IS-95 system, when the M distinct multiple-access (MA) system users are assigned separate PN code phases, the mobile receiver's ability to detect its own code phase is greatly enhanced under correlation procedures because of the two-level autocorrelation property of the PN sequence. It therefore can be stated that the autocorrelation property is used in the system for MA purposes, while for a waveform the orthogonality property can be used in the discrimination against (rejection of) signals from other base stations. Walsh functions are orthogonal, while PN sequences are not. The PN sequences have good autocorrelation properties, while the Walsh sequences do not.

It is of interest to see the distribution of signal energy of PN sequences in the frequency domain because thereby one can observe certain short-comings of their use as real-time signal waveforms in certain applications. Because the PN signals are periodic and belong to the class of infinite-energy signals, we derive the power spectral density of a PN signal, after discussing the partial correlation properties of the sequences.

6.4.2 Partial Correlation Functions of PN Sequences

We have seen that the autocorrelation function of a periodic PN sequence is two-valued, as shown for example in (6.40c). The two-valued autocorrelation function can be realized only when correlation over a full period of the PN sequence can take place; that is, the correlation integration time must be equal to the sequence period. In reality, however, this is not the case in most spread-spectrum systems. For example, in the IS-95 system, during the pilot signal acquisition process, a mobile station must search many possible I- and Q-PN code phases to lock in with the strongest signal, presumed to be the sequences transmitted by the nearest base station. The codes, generated by modified 15-stage PN generators, have $2^{15} = 32,768$ chips in a full period, and it would take 26.67 milliseconds to correlate an entire period of a pilot sequence for the determination of success or failure of the acquisition. This is too long a time to spend for testing one particular code phase, and a partial correlation must be used to make a decision.

In general, in spread-spectrum systems the correlation integration time is the information bit time duration. However, the duration of the PN sequence is much greater than that of a message bit or coded symbol, as in the case of the IS-95 reverse link operation as far as the long PN code is con-cerned. Thus, the subject of partial correlation is very important in any spread-spectrum system in which the duration of the sequence is much longer than the correlation time. Therefore, we need to examine the statistical performance measure of the partial correlation operation to answer such questions as "what happens" when correlation is performed over only a part of the PN sequence period. We will see that it is a reasonable thing to do because, if we correlate over M chips instead of P chips, the full length of the sequence, on average we expect to have the value of M, and a deviation from this value is not likely in most cases. This is what we intend to do in what follows.

Consider a typical PN sequence $\mathbf{A}$ over $\{+1, -1\}$, given by

$$\mathbf{A} = (A_0, A_1, A_2, \ldots, A_{P-1})$$

which is converted from the sequence a over GF(2). The full-length auto-correlation function of the sequence is

$$R_a(k) = \sum_{i=0}^{P-1} A_i A_{i-k}$$

$$= \begin{cases} \displaystyle\sum_{i=0}^{P-1} (A_i)^2 = \sum_{i=0}^{P-1} (1) = P, & k = 0, P, 2P, \ldots \\[4mm] \displaystyle\sum_{i=0}^{P-1} A_i A_{i-k} = \sum_{i=0}^{P-1} A_{i+k'} = -1, & \text{otherwise} \end{cases}$$

where we have used the fact that the sequence formed by the product $A_i A_{i-k}$ (corresponding the modulo-2 sum $a_i \oplus a_{i-k}$) is a different phase of the sequence, denoted $A_{i+k'}$, because of the closure property of PN sequences. Now, we define a partial correlation function, which integrates (sums) over $M < P$ chips:

$$R_M(k; m) \triangleq \sum_{i=0}^{M-1} A_{i+m} A_{i+m-k}$$

$$\text{for } M < P; \ k, m = 0, 1, \ldots, P-1 \tag{6.47}$$

Note that $M < P$, as opposed to the complete full-length correlation function, for which $M = P$. The first and second factors in the terms of the summation in (6.47) are (1) the sequence shifted by m chips (signifying an arbitrary starting point) and (2) the sequence shifted by $m - k$ chips ($-k$ chips relative to the first factor).

The partial correlation function definition, as given in (6.47), requires knowledge of the starting point m, and this requirement is not easily satisfied. When the sequence length is short, it may be possible to evaluate the partial correlation on a case-by-case basis for all possible values of m. For very long PN sequences, this approach is clearly not practical, and thus we may seek statistical expectations (or regularities) based on the assumption that the generated PN sequence is a sample function of a stationary random process. Therefore, the only thing we can do is ask: What can we expect the partial correlation to be in the statistical average sense?

This consideration motivates us to compute the expectation of the partial correlation measure, using the probability associated with the balance property of the sequence. Hence, we compute the mean value of the partial correlation function (6.47) as

$$
\begin{aligned}
E\{R_M(k; m)\} &= \sum_{i=0}^{M-1} \overline{A_{i+m} A_{i+m-k}} \\
&= \begin{cases} \displaystyle\sum_{i=0}^{M-1} \overline{(1)} = M, & k = 0, P, 2P, \ldots \\ \displaystyle\sum_{i=0}^{M-1} \overline{A_{i+m-k'}} = \sum_{i=0}^{M-1} \overline{\left(\frac{-1}{P}\right)} = \frac{-M}{P}, & k \neq 0, P, 2P, \ldots \end{cases}
\end{aligned}
\tag{6.48a}
$$

where $E\{\cdot\}$ and the overbar both indicate statistical averaging. Note that the starting position m is naturally averaged out. Note also that we used the identity, $A_{i+m}A_{i+m-k} = A_{i+m-k'}$, by invoking the closure property of PN sequences. The meaning of the result given in (6.48a) is very pleasing: If we integrate M chip positions out of P positions in the correlation operation, we can expect to have the result M when the sequences are lined up (synchronized); but if the sequences are not synchronized, we can expect the value $-M/P$. Note that if $M = P$, the result is completely identical to the two-valued, full-length correlation, that is:

$$
E\{R_{M=P}(k; m)\} = R_a(k) = \begin{cases} P, & k = 0, P, 2P, \ldots \\ -1, & k \neq 0, P, 2P, \ldots \end{cases}
\tag{6.48b}
$$

One more question needs to be asked: What would be the variation (or spread) of the partial correlation function from the average value of (6.48a)? To answer this question is to compute the variance measure of the partial correlation function. The variance of any random variable can be computed as the mean square value minus the square of the mean, and this is the procedure we follow. The mean square of the partial correlation function is formulated as follows:

$$
E\{R_M^2(k; m)\} = \sum_{i=0}^{M-1} \sum_{l=0}^{M-1} \overline{A_{i+m} A_{i+m-k} A_{l+m} A_{l+m-k}}
\tag{6.49a}
$$

We need to evaluate this expression under two separate assumptions:

(1) For $k = 0, P, 2P, \ldots$

$$E\{R_M^2(k; m)\} = \sum_{i=0}^{M-1} \sum_{l=0}^{M-1} \overline{(1) \cdot (1)} = M^2 \qquad (6.49b)$$

(2) For $k \neq 0, P, 2P, \ldots$

$$E\{R_M^2(k; m)\} = M + \sum_{\substack{i=0 \\ i \neq l}}^{M-1} \sum_{l=0}^{M-1} \overline{A_{i+m-k'} \, A_{l+m-k'}}$$

$$\text{(by the closure property)}$$

$$= M + \sum_{\substack{i=0 \\ i \neq l}}^{M-1} \sum_{l=0}^{M-1} \left(\frac{-1}{P}\right) \qquad \text{(closure property again)}$$

$$= M - \frac{M(M-1)}{P} \qquad (6.49c)$$

Thus, the variance of the partial correlation function is given by

$$\text{Var}\{R_M(k; m)\} = E\{R_M^2(k; m)\} - [E\{R_M(k; m)\}]^2$$

$$= \begin{cases} M^2 - (M)^2 = 0, & k = 0, P, 2P, \ldots \\ \left[M - \dfrac{M(M-1)}{P}\right] - \left(\dfrac{-M}{P}\right)^2 & \\ \quad = M\left(1 + \dfrac{1}{P}\right)\left(1 - \dfrac{M}{P}\right), & k \neq 0, P, 2P, \ldots \end{cases} \qquad (6.49d)$$

This result says the mean value of M is trustworthy if sequences are lined up (synchronized), because the variance is zero for that case, but if the sequences are not synchronized, the mean value of $-M/P$ can have a substantial variation. That is, under nonsynchronized conditions, the partial correlation values may differ drastically from the full-cycle correlation values.

Partial correlation functions for PN waveforms. Based on the correlation and partial correlation functions for PN sequences that we have discussed, we can give expressions for the partial correlation functions of PN waveforms.

The partial correlation function for a PN sequence real-time signal with chip (basic) waveform $\varphi(t)$ is given in exactly the same manner as (6.45b), where the correlation function of the full-length sequence is replaced by the partial correlation function of the sequence.

If we denote the partial correlation function for the waveform by $R_M(\tau)$, it is given by adapting (6.45b) to be

$$R_M(\tau) = \frac{1}{MT_c} \sum_{k=-\infty}^{\infty} R_M(k; m) R_\varphi(\tau - kT_c) \tag{6.50}$$

where $R_M(k; m)$ is the partial autocorrelation function for the PN sequence and $R_\varphi(\tau)$ is the autocorrelation function of the chip waveform. For the partial cross-correlation function of two PN waveforms, the theorem given in (6.45a) also applies, with the replacement of P with M and the replacement of the full-length cross-correlation function for the sequences with their partial cross-correlation function expression.

6.4.3 Spectral Properties of Binary Sequence Waveforms

Given the periodic sequence a over GF(2) with period P, and its corresponding sequence A over $\{+1, -1\}$, we are interested in the spectral properties of binary waveforms such as

$$f_a(t) = \sum_{k=-\infty}^{\infty} A_k \varphi_a(t - kT_c) \tag{6.51a}$$

where $\varphi_a(t)$ is a basic waveform that is zero outside the interval $(0, T_c)$. The Fourier transform of $f_a(t)$, denoted $\mathcal{F}\{f_a(t)\}$, is

$$\mathcal{F}\{f_a(t)\} = \sum_{k=-\infty}^{\infty} A_k \mathcal{F}\{\varphi_a(t - kT_c)\} = \mathcal{F}\{\varphi_a(t)\} \sum_{k=-\infty}^{\infty} A_k e^{-j\omega kT_c}$$

$$= \mathcal{F}\{\varphi_a(t)\} \mathcal{F}\left\{ \sum_{k=-\infty}^{\infty} A_k \delta(t - kT_c) \right\} \tag{6.51b}$$

Thus the spectrum of the binary waveform is the product of the spectrum of the basic waveform and that of a binary sequence of $\pm$ impulses. Another way to view this fact is to observe that $f_a(t)$ is the convolution of $\varphi_a(t)$ and the sequence of $\pm$ impulses:

$$f_a(t) = \{\varphi_a(t)\} * \left\{ \sum_{k=-\infty}^{\infty} A_k \, \delta(t - kT_c) \right\} \tag{6.51c}$$

where the asterisk ($*$) denotes the convolution operation. Note that if we take the Fourier transform of (6.51c), we have the equivalent expression in (6.51b).

Now let an "impulse train" $x(t)$ be given by

$$x(t) = \sum_{n=-\infty}^{\infty} \delta(t - nT_c) \tag{6.52a}$$

Because $x(t)$ is periodic with period T_c, it has the Fourier series

$$x(t) = \sum_{k=-\infty}^{\infty} c_k \, e^{j2\pi kt/T_c} \tag{6.52b}$$

where

$$c_k = \frac{1}{T_c} \int_{-T_c/2}^{T_c/2} x(t) \, dt = \frac{1}{T_c} \quad \text{for all } k$$

Thus

$$x(t) = \frac{1}{T_c} \sum_{k=-\infty}^{\infty} e^{j2\pi kt/T_c} = \frac{1}{T_c} \left[1 + 2 \sum_{k=1}^{\infty} \cos\left(\frac{2\pi kt}{T_c}\right) \right] \tag{6.52c}$$

A plot of a finite sum of cosines as a function of t/T_c is shown in Figure 6.46, illustrating how the sum converges to a train of impulses as the number of terms increases. As the sum of cosines becomes infinite, the train of impulses results:

$$\frac{1}{T_c} \sum_{k=-\infty}^{\infty} e^{j2\pi kt/T_c} = \frac{1}{T_c} \left[1 + 2 \sum_{k=1}^{\infty} \cos\left(\frac{2\pi kt}{T_c}\right) \right]$$

$$\rightarrow \sum_{n=-\infty}^{\infty} \delta(t - nT_c) \tag{6.52d}$$

This equivalence allows us to use the sum of cosines (Fourier series) form of the train of impulses to determine its Fourier transform:

$$\mathcal{F}\{x(t)\} = \frac{1}{T_c} \sum_{k=-\infty}^{\infty} \int_{-\infty}^{\infty} e^{-j2\pi ft} \cdot e^{j2\pi kt/T_c} = \frac{1}{T_c} \sum_{k=-\infty}^{\infty} \delta\left(f - \frac{k}{T_c}\right) \tag{6.52e}$$

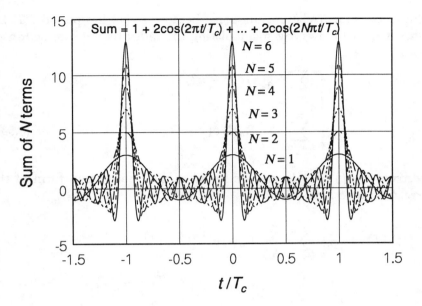

Figure 6.46 Plot of a finite sum of cosines as a function of t/T_c.

We therefore see that the Fourier transform of the time-domain train of impulses is a frequency-domain train of impulses.

Now consider a train of $\pm$ impulses. Let $x(t)$ be

$$x(t) = \sum_{k=-\infty}^{\infty} A_k\, \delta(t - kT_c) \tag{6.53a}$$

where $\{A_k\}$ is a ± 1 binary sequence with period P. Because $x(t)$ is periodic with period PT_c, the autocorrelation function of $x(t)$ is

$$R_x(\tau) = \frac{1}{PT_c} \int_0^{PT_c} x(t)\, x(t - \tau)\, dt = \frac{1}{PT_c} \sum_{k=-\infty}^{\infty} R_a(k)\, R_\delta(\tau - kT_c)$$

$$= \frac{1}{PT_c} \sum_{k=-\infty}^{\infty} R_a(k)\, \delta(\tau - kT_c) \tag{6.53b}$$

where

$$R_a(k) = \sum_{i=0}^{P-1} A_i\, A_{i-k}$$

is the sequence autocorrelation function. Because $R_x(\tau)$ has period PT_c, it can be written as follows, by manipulating the index of the summation to replace k by $kP + l$:

$$R_x(\tau) = \frac{1}{PT_c} \sum_{k=-\infty}^{\infty} R_a(k)\, \delta(\tau - kT_c)$$

$$= \frac{1}{PT_c} \sum_{k=-\infty}^{\infty} \sum_{l=0}^{P-1} R_a(l)\, \delta(\tau - lT_c - kPT_c) \qquad (6.53c)$$

The spectrum $S_x(f)$ of $x(t)$ may be found as the Fourier transform of the autocorrelation function:

$$S_x(f) = \frac{1}{PT_c} \sum_{k=-\infty}^{\infty} \sum_{l=0}^{P-1} R_a(l)\, e^{-j2\pi fT_c(l+kP)}$$

$$= \frac{1}{PT_c} \left(\sum_{l=0}^{P-1} R_a(l)\, e^{-j2\pi fT_c l} \right) \left(\sum_{k=-\infty}^{\infty} e^{-j2\pi fT_c kP} \right)$$

$$= \frac{1}{PT_c} \left(\sum_{l=0}^{P-1} R_a(l)\, e^{-j2\pi fT_c l} \right) \left[\frac{1}{PT_c} \sum_{k=-\infty}^{\infty} \delta\left(f - \frac{k}{PT_c} \right) \right]$$

$$= \frac{1}{(PT_c)^2} \sum_{k=-\infty}^{\infty} \left(\sum_{l=0}^{P-1} R_a(l)\, e^{-j2\pi fT_c l} \right) \delta\left(f - \frac{k}{PT_c} \right)$$

$$= \frac{1}{(PT_c)^2} \sum_{k=-\infty}^{\infty} \mathrm{DFT}_{Ra}(k)\, \delta\left(f - \frac{k}{PT_c} \right) \qquad (6.53d)$$

where $\mathrm{DFT}_{Ra}(k)$ is the discrete Fourier transform of the periodic sequence $R_a(l)$. Thus, the spectrum of the periodic binary sequence-modulated train of impulses in the time domain is a train of impulses in the frequency domain that is "modulated" by the periodic discrete Fourier transform (DFT) of the autocorrelation function of the binary sequence.

Properties of the DFT for a maximal-length sequence. For a maximal-length sequence (PN sequence),

$$R_a(0) = P \quad \text{and} \quad R_a(l) = -1 \text{ for } l \neq 0$$

Thus

$$\text{DFT}_{Ra}(k) = \sum_{l=0}^{P-1} \text{R}_a(l)\, e^{-j2\pi kl/P} \quad = P - \sum_{l=1}^{P-1} e^{-j2\pi kl/P}$$

$$= \begin{cases} P - (P-1) = 1, & \text{if } k \text{ is a multiple of } P \\ P - (0-1) = P+1, & \text{if } k \text{ is not a multiple of } P \end{cases} \quad (6.54)$$

In this development, we used the fact that

$$\sum_{l=0}^{P-1} e^{-j2\pi kl/P} = 0 \quad \text{for } k \text{ not a multiple of } P$$

We now can obtain the spectrum for a maximal-length real-time waveform. Because

$$f_a(t) = \sum_{k=-\infty}^{\infty} A_k\, \varphi_a(t - kT_c) = \varphi_a(t) * \sum_{k=-\infty}^{\infty} A_k\, \delta(t - kT_c)$$

using the Woodward notation [5] discussed in Chapter 1, the spectrum is

$$S_{fa}(f) = |\mathcal{F}\{\varphi_a(t)\}|^2 \cdot \frac{1}{(PT_c)^2} \sum_{k=-\infty}^{\infty} \text{DFT}_{Ra}(k)\, \delta\!\left(f - \frac{k}{PT_c}\right)$$

$$= \frac{1}{P^2} \text{sinc}^2(T_c f) \sum_{k=-\infty}^{\infty} \text{DFT}_{Ra}(k)\, \delta\!\left(f - \frac{k}{PT_c}\right) \quad (6.55)$$

for a rectangular pulse basic waveform (chip waveform). If we isolate the DC (zero frequency) term of the expression in (6.55) and substitute (6.54) for the DFT, the power spectrum of a PN real-time signal for a rectangular chip waveform becomes

$$S_{fa}(f) = \frac{1}{P^2} \delta(f) + \frac{P+1}{P^2} \sum_{\substack{k=-\infty \\ k\neq 0}}^{\infty} \text{sinc}^2(T_c f)\, \delta\!\left(f - \frac{k}{PT_c}\right)$$

$$= \frac{1}{P^2} \delta(f) + \sum_{\substack{k=-\infty \\ k\neq 0}}^{\infty} \left(\frac{P+1}{P^2}\right) \text{sinc}^2\!\left(\frac{k}{P}\right) \delta\!\left(f - \frac{k}{PT_c}\right) \quad (6.56)$$

which is plotted in Figure 6.47.

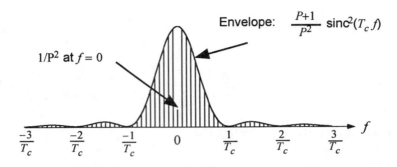

Figure 6.47 Power spectral density of a PN real-time signal for a rectangular chip waveform.

6.5 Operations on Maximal-Length Sequences

6.5.1 Orthogonalization

We have clearly seen that PN sequences are not orthogonal; that is, if two PN sequences, whether they are generated from the same primitive polynomial or from two distinct primitive polynomials, have nonzero correlation. In other words, neither the autocorrelation nor the cross-correlation of maximal-length sequence is zero. However, if we modify the PN sequences in certain ways, we can render them to be orthogonal to each other.

Consider the PN sequence generated by the primitive polynomial $f(x) = 1 + x^2 + x^5$ whose SSRG configuration is shown in Figure 6.48 with the initial loading of $1\,1\,1\,1\,1$. Let the 31-bit sequence generated from the initial vector 11111 be denoted $S(0)$. It is given by

$$S(0) = 1\,1\,1\,1\,1\,0\,0\,1\,1\,0\,1\,0\,0\,1\,0\,0\,0\,0\,1\,0\,1\,0\,1\,1\,1\,0\,1\,1\,0\,0\,0$$

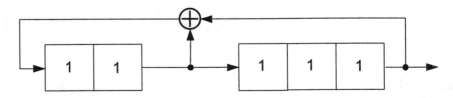

Figure 6.48 Five-stage, 31-bit PN sequence generator (Fibonacci configuration).

Let $S(2)$ denote this sequence shifted two bits to the right. Then

$$S(2) = 0\ 0\ 1\ 1\ 1\ 1\ 1\ 0\ 0\ 1\ 1\ 0\ 1\ 0\ 0\ 1\ 0\ 0\ 0\ 0\ 1\ 0\ 1\ 0\ 1\ 1\ 1\ 0\ 1\ 1\ 0$$

If we append a zero at the end of each of the above sequences (at the 32nd position), we have created two orthogonal sequences. In general, if we append a zero at the end of each PN sequence shift generated by an n-stage LFSR, we have a set of 2^{n-1} orthogonal sequences.

There is another approach to making orthogonal PN sequences generated by a given LFSR with n stages. Rather than adding a "0" at the end of each maximal-length sequence generated by a given state vector, we can insert a zero at the end of the longest "run length" of "0." There is only one such run of zeros, of length $n - 1$, for any PN sequence generator, due to the run property. This is the approach taken in IS-95 to bring about the orthogonality of certain pairs of PN sequences [6]. This method does not make all sequence shifts orthogonal, although the resulting sequences produce good cross-correlation and autocorrelation characteristics. Here we use the terms auto- and cross-correlation even though we mean only the sequences generated by the same PN generator.

The insertion of a zero at the end of $n - 1$ zeros in the IS-95 system is not strictly for the purpose of bringing about orthogonality between pairs of sequences that are assigned to different base stations. The insertion of an extra zero at the end of the run of $n - 1 = 14$ zeros in the two short PN codes in the IS-95 system makes the 15-stage PN generator periods $2^{15} = 32,768$ chips, which, at the PN clock rate of 1.2288 Mcps, run 75 cycles in every 2-sec interval. In the IS-95 system, the Walsh sequences also run at the clock rate of 1.2288 Mcps for the forward link, and the Walsh function generation is synchronized to every even-second period. Thus, the insertion of an extra zero in the PN sequences serves two purposes in this respect.

When a zero is added to a reference PN sequence by one of the methods discussed above, in effect the PN sequence generator is caused to repeat one of its states. It follows that the shifted sequence at the output of a mask-controlled PSN associated with that PN generator then repeats whatever output value is produced by that state, in whatever place in the output sequence the repetition occurs.

For example, consider the case illustrated in Figure 6.49. The insertion of a fourth (nth) zero at the end of the reference sequence $1/f(x)$ is accomplished by repeating the state $0010 = \beta^2 = \beta^{n-2}$. What effect does this insertion have on the output of a PSN? Because the zero insertion is

$$f(x) = 1 + x^3 + x^4, \quad f^*(\beta) = 1 + \beta + \beta^4$$

Time	State	Output
0	$1000 = 1 \bmod f^* = 1$	0
1	$0100 = \beta \bmod f^* = \beta$	0
2	$0010 = \beta^2 \bmod f^* = \beta^2$	0
2'	0010 (REPEAT)	0 (INSERTED ZERO)
3	$0001 = \beta^3 \bmod f^* = \beta^3$	1 ↓ $1/f(x)$
4	$1100 = \beta^4 \bmod f^* = 1 + \beta$	0
5	$0110 = \beta^5 \bmod f^* = \beta + \beta^2$	0
6	$0011 = \beta^6 \bmod f^* = \beta^2 + \beta^3$	1
7	$1101 = \beta^7 \bmod f^* = 1 + \beta + \beta^3$	1
8	$1010 = \beta^8 \bmod f^* = 1 + \beta^2$	0
9	$0101 = \beta^9 \bmod f^* = \beta + \beta^3$	1
10	$1110 = \beta^{10} \bmod f^* = 1 + \beta + \beta^2$	0
11	$0111 = \beta^{11} \bmod f^* = \beta + \beta^2 + \beta^3$	1
12	$1111 = \beta^{12} \bmod f^* = 1 + \beta + \beta^2 + \beta^3$	1
13	$1011 = \beta^{13} \bmod f^* = 1 + \beta^2 + \beta^3$	1
14	$1001 = \beta^{14} \bmod f^* = 1 + \beta^3$	1

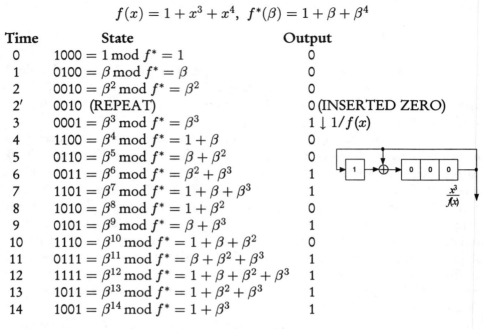

Figure 6.49 Example of MSRG with inserted zero.

equivalent to a "pause" in which one of the $2^n - 1$ bits in the reference PN sequence is repeated, the output of the PSN is subject to the pause at the same time, which in general is at a different position in the sequence with respect to the beginning of the $1/f(x)$ sequence. This effect is demonstrated for two different masks in Figure 6.50.

The example in Figure 6.50 involves a sequence with period $P = 2^4 - 1 = 15$ bits that is caused to have a period of 16 bits by the insertion of a logical 0 at the end of the reference sequence $1/f(x)$—in the sixteenth position in the modified sequence. We observe in Figure 6.50 (on the left-hand side of the figure) that the output of the mask that produces a relative delay of 7 bits has a logical 1 inserted at the ninth position in its (shifted) version of the reference sequence, and (on the right-hand side of the figure) that the output of the mask that produces a relative delay of 11 bits has a logical 1 inserted at the fifth position in its version of the reference sequence.

In general, the repeated bit in the shifted reference sequence is in the $(P + 1 - r)$th position, where r is the relative delay produced by the mask-controlled PSN. In IS-95, it is understood that the various offsets of the zero-

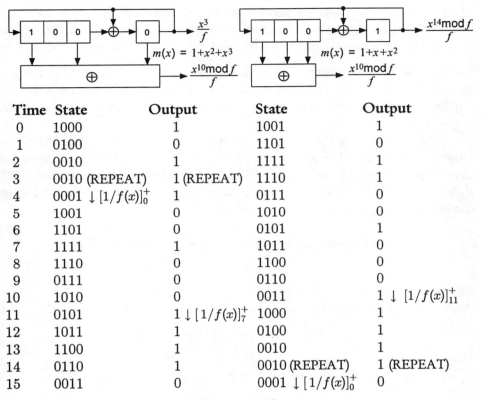

Time	State	Output	State	Output
0	1000	1	1001	1
1	0100	0	1101	0
2	0010	1	1111	1
3	0010 (REPEAT)	1 (REPEAT)	1110	1
4	0001 $\downarrow [1/f(x)]_0^+$	1	0111	0
5	1001	0	1010	0
6	1101	0	0101	1
7	1111	1	1011	0
8	1110	0	1100	0
9	0111	0	0110	0
10	1010	0	0011	1 $\downarrow [1/f(x)]_{11}^+$
11	0101	1 $\downarrow [1/f(x)]_7^+$	1000	1
12	1011	1	0100	1
13	1100	1	0010	1
14	0110	1	0010 (REPEAT)	1 (REPEAT)
15	0011	0	0001 $\downarrow [1/f(x)]_0^+$	0

Notation: Repeated chip for $[1/f(x)]_r^+$ is chip number $P + 1 - r$.

Figure 6.50 Examples of the effect of zero insertion on the PSN output.

added reference sequences for the two short PN codes should be identical except for the shift. Does this requirement eliminate the use of masks?

The great advantage to using masks to obtain sequence shifts can still be had while at the same time rectifying the effect we have seen by example in Figure 6.50. Without entering into the details of the implementation, we simply observe that a shifted version of the zero-added reference sequence is obtained using the following procedure, to be repeated for each period of the shifted sequence:

1. Use the mask output from the time of the beginning of the PN generator's reference sequence to the end of the run of $n - 1$ zeros in the mask output, and

2. Use a 1-chip-delayed version of the mask output from the time just after the run of $n-1$ consecutive zeros in the mask output to just before the beginning of the PN generator's reference sequence.

This technique is illustrated by the examples taken from Figure 6.50 that are shown in Figure 6.51, using the notation $[1/f(x)]_r^+$ to indicate the reference sequence with the repeated bit in the $P+1-r$ position of the P-bit sequence.

Finally, we note that the shifted sequences resulting from this technique have relative delays of one more bit than without a zero insertion. Thus, the masks for accomplishing relative delays of $64k$ bits in the zero-added IS-95 short PN codes should be computed so as to produce delays of $64k-1$ bits in the reference sequence without the added zero.

PN generator's reference sequence, with zero added:

1 1 1 1 0 1 0 1 1 0 0 1 0 0 0 0
$\rightarrow [1/f(x)]_0^+$

Case 1: Output of mask $1 + x^2 + x^3$ and its 1-bit delay:

1 0 0 1 0 0 0 1 1 1 1 0 1 0 1 1
$\rightarrow [1/f(x)]_7^+$

1 1 0 0 1 0 0 0 1 1 1 1 0 1 0 1
$\rightarrow [1/f(x)]_7^+$

Result of switching between mask, delayed mask as suggested:

1 0 0 1 0 0 0 0 1 1 1 1 0 1 0 1
8-bit shift $\rightarrow [1/f(x)]_0^+$

Case 2: Output of mask $1 + x + x^2$ and its 1-bit delay:

0 1 0 1 1 0 0 1 0 0 0 1 1 1 1 1
$\rightarrow [1/f(x)]_{11}^+$

1 0 1 0 1 1 0 0 1 0 0 0 1 1 1 1
$\rightarrow [1/f(x)]_{11}^+$

Result of switching between mask, delayed mask as suggested:

0 1 0 1 1 0 0 1 0 0 0 0 1 1 1 1
12-bit shift $\rightarrow [1/f(x)]_0^+$

Figure 6.51 Demonstration of zero-adding technique using a mask.

6.5.2 Decimation of PN Sequences

In the forward link of the IS-95 system, the paging channel and the traffic channels employ data-scrambling schemes by modulo-2 adding the 19.2-ksps data stream with a 19.2-ksps scrambling sequence, which is a sampled version of the long PN code sequence generated by a 42-stage LFSR. The long PN sequence, which runs at the clock rate of 1.2288 Mcps, is sampled once every 64-chip interval, thereby producing a 19.2-ksps sequence for scrambling purposes. The sampled version of a sequence is called a *decimation* of the original sequence [1]. Thus, we need to know some basic facts on decimations to characterize the properties of the sampled sequence. We treat this subject briefly in this section to analyze the relationship between the new, sampled sequence and the original sequence to answer such questions as, "Does the decimated sequence preserve the properties inherent to the pseudorandom sequence?"

For notational convenience in the following discussion, we call the original sequence the "A- sequence," denoted $a = (a_0, a_1, a_2, \ldots, a_{P-1}, a_0, \ldots)$, and the new sequence, which is the sampled version of the A sequence, will be called the "B-sequence" and be denoted $b = (b_0, b_1, b_2, \ldots)$. Consider a pseudorandom A-sequence of length $P = 2^n - 1$, which is generated by a primitive polynomial of degree n, denoted $f_A(x)$. Now, sample the A-sequence every k digits, where k is relatively prime to P, and form the B-sequence:

$$a \longrightarrow a_0, a_1, \ldots, \quad a_k, a_{k+1}, \ldots, \quad a_{2k}, a_{2k+1}, \ldots$$
$$b \longrightarrow b_0 \qquad\qquad b_1 \qquad\qquad b_2 \qquad \cdots \qquad (6.57a)$$

Because k and P are relatively prime to each other, the B-sequence does not repeat itself until every digit of the A-sequence has been sampled. Hence, the B-sequence is also a maximal-length sequence, and the length of the B-sequence is also P. In general, the decimation of a maximal-length sequence may or may not yield another maximal-length sequence. The B-sequence has period $P/\mathrm{GCD}(P, k)$. When the decimation yields a maximal-length sequence (i.e., when $\mathrm{GCD}(P, k) = 1$), the decimation is said to be a *proper* decimation; otherwise, it is an *improper* decimation.

Because the B-sequence is a maximal-length sequence, for a proper decimation there must be a primitive polynomial, $f_B(x)$, that will generate the B-sequence. We seek the polynomial $f_B(x)$. Let us briefly review some facts covered previously:

- If a primitive polynomial of degree n, $f(x)$, is mechanized as an SSRG and a PN sequence is obtained, the same sequence can also be realized from an MSRG that is mechanized from the reciprocal polynomial $f^*(x) = x^n f(x^{-1})$, except that the starting point of the sequence for each scheme may be different.

- An MSRG produces the field elements (state vectors) of $GF(2^n)$ in the order of consecutive powers of a primitive element. The digits of a PN sequence generated by a primitive polynomial $f(x)$ of degree n can be put into a one-to-one correspondence with the field elements of $GF(2^n)$, which are ordered as the consecutive powers of a root of the reciprocal polynomial $f^*(x)$.

For example, consider the sequence generated by the primitive polynomial $f(x) = x^5 + x^2 + 1$, mechanized in the Galois (MSRG) configuration with an initial loading of all zeros except the first stage, as shown in Figure 6.52. In the context of the present discussion, the A-sequence may be said to be powers of the root α^{-1} and the B-sequence, powers of the root β^{-1}:

$$
\begin{array}{cccccccc}
(\alpha^{-1})^0 = 1 & \alpha^{-1} & & \alpha^{-k} & \alpha^{-k+1} & & \alpha^{-2k} & \alpha^{-2k-1} \\
\updownarrow & \updownarrow & & \updownarrow & \updownarrow & & \updownarrow & \updownarrow \\
a_0 & , a_1 & ,\ldots, & a_k & , a_{k+1} & ,\ldots, & a_{2k} & , a_{2k+1} & ,\ldots \\
\downarrow & & & \downarrow & & & \downarrow & \\
b_0 & & & b_1 & & & b_2 & & \cdots \\
\updownarrow & & & \updownarrow & & & \updownarrow & \\
(\beta^{-1})^0 = 1 & & & \beta^{-1} & & & \beta^{-2} &
\end{array}
\qquad (6.57b)
$$

where α^{-1} is a root of $f_A^*(x)$ and β^{-1} is a root of $f_B^*(x)$ and $\beta^{-1} = (\alpha^{-1})^k$. In terms of field elements, the reciprocal polynomial has the following meaning:

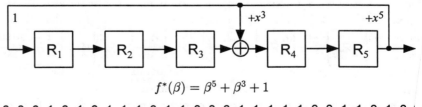

$$f^*(\beta) = \beta^5 + \beta^3 + 1$$

0 0 0 0 1 0 1 0 1 1 1 0 1 1 0 0 0 1 1 1 1 1 0 0 1 1 0 1 0 1 1

$\beta^0\ \beta^1\ \beta^2\ \beta^3\ \beta^4\ \beta^5\ \beta^6\ \beta^7\ \beta^8\ \beta^9\ \beta^{10}\ \beta^{11}\ \beta^{12}\ \beta^{13}\ \beta^{14}\ \beta^{15}\ \beta^{16}\ \beta^{17}\ \beta^{18}\ \beta^{19}\ \beta^{20}\ \beta^{21}\ \beta^{22}\ \beta^{23}\ \beta^{24}\ \beta^{25}\ \beta^{26}\ \beta^{27}\ \beta^{28}\ \beta^{29}\ \beta^{30}$

Figure 6.52 Correspondence between a PN sequence and powers of an element in a finite field.

if γ is a root of $f(x)$ (i.e., $f(\gamma) = 0$), then the inverse element γ^{-1} of γ is a root of the reciprocal polynomial $f^*(x)$ (i.e., $f^*(\gamma^{-1}) = 0$). An example will show this fact clearly.

Consider the finite field $GF(2^5)$ generated by the primitive polynomial $f(x) = x^5 + x^3 + 1$. The field element $\gamma = (0\,1\,0\,0\,0)$ is a root of $f(x)$; that is, $f(\gamma) = 0$. This can be verified as follows: from Figure 6.7, we observe that

$$\begin{aligned} \gamma^5 &= (1\,0\,0\,1\,0) \\ \gamma^3 &= (0\,0\,0\,1\,0) \\ 1 &= (1\,0\,0\,0\,0) \\ \hline \gamma^5 + \gamma^3 + 1 &= (0\,0\,0\,0\,0) \end{aligned}$$

Because the quintuple $(0\,0\,0\,0\,0)$ is the zero field element, we have $f(\gamma) = \gamma^5 + \gamma^3 + 1 = 0$. The reciprocal polynomial is $f^*(x) = x^5 + x^2 + 1$ and thus

$$\begin{aligned} \left(\gamma^{-1}\right)^5 &= \gamma^{-5} = \gamma^{26} = (1\,1\,0\,1\,0) \\ \left(\gamma^{-1}\right)^2 &= \gamma^{-2} = \gamma^{29} = (0\,1\,0\,1\,0) \\ 1 &= (1\,0\,0\,0\,0) \\ \hline \left(\gamma^{-1}\right)^5 + \left(\gamma^{-1}\right)^2 + 1 &= (0\,0\,0\,0\,0) \end{aligned}$$

Hence, $f^*(\gamma^{-1}) = (\gamma^{-1})^5 + (\gamma^{-1})^2 + 1 = 0$, which proves that γ^{-1} is a root of $f^*(x)$. Now from the results obtained above, $\beta^{-1} = (\alpha^{-1})^k$. Because each element has a unique inverse, it follows that $\beta = \alpha^k$. Further, the fact that α^{-1} is a root of $f_A^*(x)$ and β^{-1} is a root of $f_B^*(x)$ implies that α is a root of $f_A(x)$ and β is a root of $f_B(x)$, respectively. We can therefore conclude that if α is a root of a primitive polynomial $f_A(x)$ generating the A-sequence, and β is a root of the polynomial $f_B(x)$ generating the B-sequence, which is a sampled version of the A-sequence, sampled every k digits, then β is the kth power of α; that is, $\beta = \alpha^k$.

When $k = 2^m$, for $m \leq n - 1$, then $f_B(x) = f_A(x)$. This follows from the fact that in $GF(2^n)$ if α is a root of a polynomial $f(x)$, then $\{\alpha, \alpha^2, \alpha^4, \ldots, \alpha^{2^m}\}$ are also roots of $f(x)$. The notation $f_B(x) = f_A(x)$ means that the B-sequence is identical to the A-sequence, except for a possible difference in their starting points (phase shifts).

Example 6.22 Let $n = 3$, $P = 2^3 - 1 = 7$. Choose $f_A(x) = x^3 + x + 1$, with $1\,0\,1$ as the initial state. Then the generated A-sequence is

a_0	a_1	a_2	a_3	a_4	a_5	a_6
1	0	1	0	0	1	1

If $k = 2^2 = 4$, then

$$b_0 = a_0 = 1 \qquad\qquad b_4 = a_{16} = a_2 = 1$$
$$b_1 = a_4 = 0 \qquad\qquad b_5 = a_{20} = a_6 = 1$$
$$b_2 = a_8 = a_1 = 0 \qquad\qquad b_6 = a_{24} = a_3 = 0$$
$$b_3 = a_{12} = a_5 = 1.$$

The B-sequence for $k = 4$ is

b_0	b_1	b_2	b_3	b_4	b_5	b_6
1	0	0	1	1	1	0

Comparing these two sequences shows that the sampled sequence is the same as the original sequence except for the starting digits. If $k = 3$, then

$$b_0 = a_0 = 1 \qquad\qquad b_4 = a_{12} = a_5 = 1$$
$$b_1 = a_3 = 0 \qquad\qquad b_5 = a_{15} = a_1 = 0$$
$$b_2 = a_6 = 1 \qquad\qquad b_6 = a_{18} = a_4 = 0$$
$$b_3 = a_9 = a_2 = 1$$

The B-sequence for $k = 3$ is

b_0	b_1	b_2	b_3	b_4	b_5	b_6
1	0	1	1	1	0	0

This sequence is different from the original A-sequence. Because there are only two different pseudorandom sequences for $n = 3$, the polynomial $f_B(x)$ that generates the B-sequence must be the reciprocal polynomial of $f_A(x)$; hence, $f_B(x) = x^3 + x^2 + 1$.

Example 6.23 Let $n = 4$, $P = 2^4 - 1 = 15$. Choose $f_A(x) = x^4 + x + 1$, with $1\,1\,0\,1$ as the initial state. Then the A-sequence is

a_0	a_1	a_2	a_3	a_4	a_5	a_6	a_7	a_8	a_9	a_{10}	a_{11}	a_{12}	a_{13}	a_{14}
1	0	1	1	0	0	1	0	0	0	1	1	1	1	0

For $k = 2^2 = k$, the B-sequence is

b_0	b_1	b_2	b_3	b_4	b_5	b_6	b_7	b_8	b_9	b_{10}	b_{11}	b_{12}	b_{13}	b_{14}
1	0	0	1	0	0	0	1	1	1	1	0	1	0	1

which is a shifted version of the A-sequence. For $k = 7$, the B-sequence is

b_0	b_1	b_2	b_3	b_4	b_5	b_6	b_7	b_8	b_9	b_{10}	b_{11}	b_{12}	b_{13}	b_{14}
1	0	0	1	1	0	1	0	1	1	1	1	0	0	0

This sequence is different from the original A-sequence. Because there are only two distinct pseudorandom sequences for $n = 4$, the polynomial that generates the B-sequence must be $f_B(x) = f_A^*(x) = x^4 + x^3 + 1$.

Example 6.24 Let $n = 5$, $P = 2^5 - 1 = 31$. Choose $f_A(x) = x^5 + x^2 + 1$, with $1\,0\,0\,0\,0$ as the initial state. Then the A-sequence, the B-sequence obtained by sampling every fourth digit, and the B-sequence obtained by sampling every fifth digit are shown in Figure 6.53. Again, the B-sequence for $k = 2^2 = 4$ is identical with the A-sequence up to a phase shift and the B-sequence for $k = 5$ is different from the A-sequence. From the table of irreducible polynomials [3] we found the polynomial "5 67H." Thus, for $k = 5$, the generating polynomial for the B-sequence is $f_B(x) = x^5 + x^4 + x^2 + x + 1$.

We now summarize the results as follows. Let $f_A(x)$ be a primitive polynomial of degree n that generates a pseudorandom A-sequence of length $P = 2^n - 1$. Sample the A-sequence at every kth digit. When $\text{GCD}(P, k) = 1$, the sampled sequence (the B-sequence) is also pseudorandom and is generated by a primitive polynomial $f_B(x)$. If $k = 2^m$, for $m \le n - 1$, then $f_B(x) = f_A(x)$. If $k \ne 2^m$, then $f_B(x) \ne f_A(x)$. However, if α is a root of $f_A(x)$, then $\beta = \alpha^k$ is a root of $f_B(x)$. The polynomial $f_B(x)$ can be determined with aid of a table of irreducible polynomials.

```
   0 1 2 3 4 5 6 7 8 9 10 11 12 13 14 15 16 17 18 19 20 21 22 23 24 25 26 27 28 29 30
a  0 0 0 0 1 0 1 0 1 1  1  0  1  1  0  0  0  1  1  1  1  0  0  1  1  0  1  0  0  1
```

$$k = 4:$$

```
b  0 1 1 1 0 1 1 0 0 0 1 1 1 1 0 0 1 1 0 1 0 0 1 0 0 0 0 1 0 1
```

$$k = 5:$$

```
b  0 0 1 0 1 1 1 1 1 0 1 1 0 0 1 1 1 0 0 0 0 1 1 0 1 0 1 0 0 1 0
```

Figure 6.53 Pseudorandom sequence and its sampled sequences (decimations).

In the IS-95 system, the long code generated by

$$f_A(x) = x^{42} + x^{41} + x^{40} + x^{39} + x^{37} + x^{36} + x^{35} + x^{32} + x^{26} + x^{25}$$
$$+ x^{24} + x^{23} + x^{21} + x^{20} + x^{17} + x^{16} + x^{15}$$
$$+ x^{11} + x^9 + x^7 + 1 \tag{6.58}$$

is sampled every 64 digits (chips) to obtain a sequence modulo-2 added to the 19.2-ksps information digits. The sampled sequence (B-sequence) polynomial $f_B(x)$ is therefore identical to $f_A(x)$ because $\beta = \alpha^{64} = \alpha^{2^6}$ is a root of the minimal polynomial $f_A(x)$. In other words, the sampled (decimated) sequence is a pseudorandom sequence with all the desired attributes.

6.6 Gold Codes

The usefulness of the pseudorandom sequences in a spread-spectrum system depends in large part on their ideal autocorrelation properties. One of the randomness properties of the pseudorandom sequence is the correlation property; that is, if a complete sequence is compared, bit by bit, with any shift of itself, the number of agreements differs from the number of disagreements by at most one. From this property it is easy to see that the autocorrelation function of a pseudorandom sequence is of the two-valued shape described previously in Section 6.4.

The cross-correlation function between two different pseudorandom sequences of the same length is, however, an entirely different matter. It can have high peaks; and to make the matter worse, there is no simple method available to calculate the cross-correlation function between two pseudorandom sequences except by brute force calculation and simulation. For long sequences, this is not possible even with the fastest computers. To visualize the cross-correlation problem and how much is involved in the simulation, consider an example as follows:

The sequence generator $x^3 + x + 1$ with initial state $1\,0\,0$ generates

$$S_1 = 0\ 0\ 1\ 1\ 1\ 0\ 1$$

and the sequence generator $x^3 + x^2 + 1$ with initial state $1\,1\,1$ generates

$$S_2 = 1\ 1\ 1\ 0\ 0\ 1\ 0$$

The cross-correlation is computed by holding S_1 fixed, shifting S_2 one bit at a time, and comparing the two sequences to obtain the difference between the number of agreements and the number of disagreements, or by the formula given in (6.41b):

1. $\quad$ 0 0 1 1 1 0 1
    $\quad +$1 1 1 0 0 1 0
    $\quad \overline{\text{1 1 0 1 1 1 1}} \Rightarrow \quad R_{1,2}(0) = -5$

2. $\quad$ 0 0 1 1 1 0 1
    $\quad +$0 1 1 1 0 0 1
    $\quad \overline{\text{0 1 0 0 1 0 0}} \Rightarrow \quad R_{1,2}(1) = +3$

3. $\quad$ 0 0 1 1 1 0 1
    $\quad +$1 0 1 1 1 0 0
    $\quad \overline{\text{1 0 0 0 0 0 1}} \Rightarrow \quad R_{1,2}(2) = +3$

4. $\quad$ 0 0 1 1 1 0 1
    $\quad +$0 1 0 1 1 1 0
    $\quad \overline{\text{0 1 1 0 0 1 1}} \Rightarrow \quad R_{1,2}(3) = -1$

5. $\quad$ 0 0 1 1 1 0 1
    $\quad +$0 0 1 0 1 1 1
    $\quad \overline{\text{0 0 0 1 0 1 0}} \Rightarrow \quad R_{1,2}(4) = +3$

6. $\quad$ 0 0 1 1 1 0 1
    $\quad +$1 0 0 1 0 1 1
    $\quad \overline{\text{1 0 1 0 1 1 0}} \Rightarrow \quad R_{1,2}(5) = -1$

7. $\quad$ 0 0 1 1 1 0 1
    $\quad +$1 1 0 0 1 0 1
    $\quad \overline{\text{1 1 0 1 0 0 0}} \Rightarrow \quad R_{1,2}(6) = -1$

The cross-correlation function for the two PN sequence real-time signals, each with rectangular chip waveforms, is plotted in Figure 6.54.

In Section 6.2, we found all six PN generators having five stages (Table 6.9). Let us relabel the primitive polynomials as follows:

$$f_1(x) = x^5 + x^2 + 1$$
$$f_2(x) = x^5 + x^4 + x^3 + x^2 + 1$$
$$f_3(x) = x^5 + x^4 + x^2 + x + 1$$

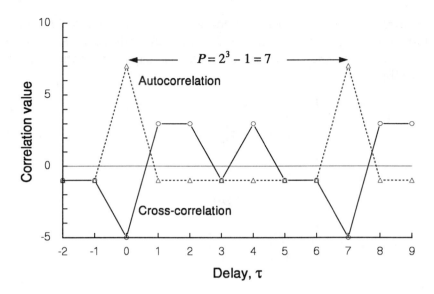

Figure 6.54 Autocorrelation and cross-correlation functions for two PN sequences generated by $x^3 + x + 1$ and $x^3 + x^2 + 1$.

$$f_4(x) = x^5 + x^3 + 1$$
$$f_5(x) = x^5 + x^3 + x^2 + x + 1$$
$$f_6(x) = x^5 + x^4 + x^3 + x + 1$$

The autocorrelation functions of the sequences generated by these polynomials and the cross-correlation functions between pairs of them are shown in Table 6.21; those generated by $f_1(x)$ and $f_4(x)$, and by $f_4(x)$ and $f_6(x)$, are plotted in Figure 6.55.

6.6.1 The Cross-Correlation Problem

The cross-correlation function between two distinct pseudorandom sequences is a very important consideration in MA communications systems where each user terminal (access terminal) is assigned a PN generator whose polynomial is distinct from all other user terminals. In fact, this is the type of CDMA spread-spectrum system used by the military. A distinction between the military type of CDMA system and the nonmilitary type such as IS-95 is that,

Table 6.21 Cross-correlations among maximal-length sequences generated by five-stage LFSRs

Delay	0	1	2	3	4	5	6	7	8	9	10	11	12	13	14	15	16	17	18	19	20	21	22	23	24	25	26	27	28	29	30
R_{11}	31	-1	-1	-1	-1	-1	-1	-1	-1	-1	-1	-1	-1	-1	-1	-1	-1	-1	-1	-1	-1	-1	-1	-1	-1	-1	-1	-1	-1	-1	-1
R_{12}	7	-1	-1	7	7	-1	7	-1	-1	-9	-1	-1	-1	-1	-9	-1	7	-1	-9	-9	7	-1	7	7	-1	-1	-1	-9	7	-9	7
R_{13}	7	7	7	-1	-1	-9	7	-1	-9	-1	-1	-9	-1	-1	-1	-1	7	-1	-9	-1	7	7	7	-1	7	-9	-1	-1	-1	7	-9
R_{14}	3	3	-5	-9	-5	3	-1	7	-9	3	3	3	-5	-5	-1	-9	3	-5	-1	3	-1	-1	11	7	7	-9	7	3	-9	3	7
R_{15}	-1	7	-1	-1	-1	-9	7	7	-1	-9	7	-9	-9	-9	-1	-1	7	-1	-1	7	7	-1	-1	-1	-1	7	-1	7	-9	-1	7
R_{16}	7	-1	-1	7	-1	-1	-1	7	-1	7	7	-9	-9	-9	-1	-9	7	-1	-1	-9	7	7	-1	-1	-1	-1	7	-9	-1	7	-1
R_{22}	31	-1	-1	-1	-1	-1	-1	-1	-1	-1	-1	-1	-1	-1	-1	-1	-1	-1	-1	-1	-1	-1	-1	-1	-1	-1	-1	-1	-1	-1	-1
R_{23}	7	7	-9	-1	-1	-9	-1	-1	-9	7	-1	-1	-1	-1	-9	-1	7	7	-9	-1	7	-1	-1	-1	7	-9	7	-1	7	-1	7
R_{24}	-1	7	-1	-1	-1	-9	7	7	-1	-9	7	-9	-9	-9	-1	-1	7	-1	-1	7	7	-1	-1	-1	-1	7	-1	7	-9	-1	7
R_{25}	3	-5	3	3	11	-9	-9	3	-9	-1	3	-5	-9	-1	-1	7	3	7	-5	3	-9	3	-1	-5	-1	7	7	3	3	-5	7
R_{26}	7	7	-9	7	7	-1	7	-9	-1	-1	7	-9	-9	-1	-1	-1	-1	7	7	-1	-9	-1	-9	-1	-1	7	-1	-1	-1	7	-1
R_{33}	31	-1	-1	-1	-1	-1	-1	-1	-1	-1	-1	-1	-1	-1	-1	-1	-1	-1	-1	-1	-1	-1	-1	-1	-1	-1	-1	-1	-1	-1	-1
R_{34}	7	-1	-1	7	-1	-1	-1	7	-1	7	7	-9	-9	-9	-1	-9	7	-1	-1	-9	7	7	-1	-1	-1	-1	7	-9	-1	7	-1
R_{35}	7	7	-9	7	7	-1	7	-9	-1	-1	7	-9	-9	-1	-1	-1	-1	7	7	-1	-9	-1	-9	-1	-1	7	-1	-1	-1	7	-1
R_{36}	11	3	3	-1	3	3	-1	7	3	3	3	-9	-1	-9	7	-5	3	-1	3	7	3	-9	-9	-5	-1	7	-9	-5	7	-5	-5
R_{44}	31	-1	-1	-1	-1	-1	-1	-1	-1	-1	-1	-1	-1	-1	-1	-1	-1	-1	-1	-1	-1	-1	-1	-1	-1	-1	-1	-1	-1	-1	-1
R_{45}	7	7	-9	7	-9	-1	-1	-1	7	7	-1	7	-9	-9	-1	7	-1	-9	-1	7	-1	-9	-1	-1	-1	7	-1	7	7	-1	-1
R_{46}	7	-9	7	-1	-1	-1	-9	7	-1	7	7	7	-1	-9	-1	7	-1	-1	-1	-1	-9	-1	-1	-9	-1	7	-9	-1	-1	7	7
R_{55}	31	-1	-1	-1	-1	-1	-1	-1	-1	-1	-1	-1	-1	-1	-1	-1	-1	-1	-1	-1	-1	-1	-1	-1	-1	-1	-1	-1	-1	-1	-1
R_{56}	7	7	7	-1	7	-9	7	-1	-1	-1	7	-1	-9	-1	7	7	-1	-9	-1	-1	-1	-1	7	-9	-1	-1	-9	-1	-1	-9	7
R_{66}	31	-1	-1	-1	-1	-1	-1	-1	-1	-1	-1	-1	-1	-1	-1	-1	-1	-1	-1	-1	-1	-1	-1	-1	-1	-1	-1	-1	-1	-1	-1

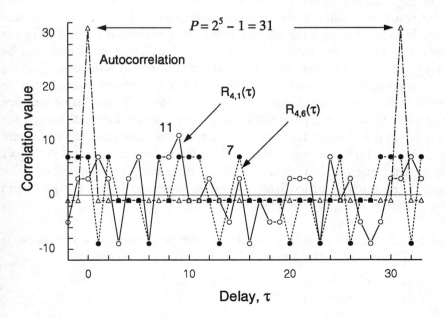

Figure 6.55 Correlation functions of PN sequence waveforms.

in the former, the communications channel condition does not permit a phase coherent PN code MA system, as opposed to the more controllable channel conditions of cellular or PCS applications in which mobililiy is not a particular concern. In a military or high-mobility environment, where the carrier phase tracking is of insurmountable difficulty, a CDMA system based on a single PN code generator, such as the IS-95 system, is not possible. The problem of assigning code generators with low cross-correlation peaks is an important consideration. Consider designing a MA spread-spectrum system using degree-5 primitive polynomials. As seen from Figure 6.55, the cross-correlation function $R_{4,1}(\tau)$ has a cross-correlation peak of 11, which is considered "high" in comparison with the autocorrelation peak of 31. If the user with the PN generator $f_1(x)$ is searching its code in an MA communications system, its receiver might be triggered by a false synchronization indication due to the relatively high degree of cross-correlation with PN generator $f_4(x)$.

To overcome the cross-correlation problem, Gold [7] considers the bit-by-bit modulo-2 sum of two pseudorandom sequences of the same length but generated by two distinct primitive polynomials, $f_1(x)$ and $f_2(x)$, as illustrated in Figure 6.56. If the length of the two PN sequences is $P = 2^n - 1$, then the resultant sequence also repeats itself after P bits. Further, if one sequence is kept fixed and the second sequence is shifted in time, a different resultant sequence is generated. In this way, P different sequences can be generated, one for each different time shift of the second sequence. Joining the two original PN sequences, altogether $2^n + 1$ different sequences can be generated with one pair of primitive polynomials. These sequences are referred to as *Gold sequences* or *Gold codes*; they are not maximal except for the two original PN sequences. Note that for an n-stage shift register, there are only $\phi(2^n - 1)/n$ PN sequences, as discussed in Section 6.2, and yet a pair of n-stage shift registers can generate $2^n + 1$ different Gold sequences. The increase in the number of available sequences is drastic. For example, from Table 6.10, for $n = 10$, $\phi(2^n - 1)/n = 60$, while $2^n + 1 = 1,025$.

Example 6.25 As an example of Gold sequence generation, consider the case of $n = 3$. There are only two primitive polynomials of degree 3:

$$f_1(x) = x^3 + x + 1 \qquad \text{generates} \qquad S_1 = 0\ 0\ 1\ 1\ 1\ 0\ 1$$
$$f_2(x) = x^3 + x^2 + 1 \qquad \text{generates} \qquad S_2 = 1\ 1\ 1\ 0\ 0\ 1\ 0$$

If we configure the Gold sequence generator as shown in Figure 6.56 using these polynomials, this pair generates nine Gold sequences:

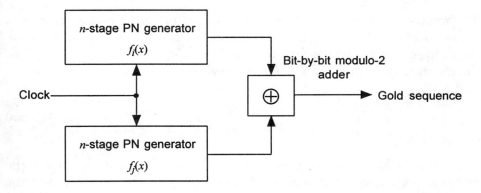

Figure 6.56 Gold sequence generator.

$$S_1 = 0\ 0\ 1\ 1\ 1\ 0\ 1 \qquad S_2 = 1\ 1\ 1\ 0\ 0\ 1\ 0$$
$$S_3 = 1\ 1\ 0\ 1\ 1\ 1\ 1 \qquad S_4 = 0\ 1\ 0\ 0\ 1\ 0\ 0$$
$$S_5 = 1\ 0\ 0\ 0\ 0\ 0\ 1 \qquad S_6 = 0\ 1\ 1\ 0\ 0\ 1\ 1$$
$$S_7 = 0\ 0\ 0\ 1\ 0\ 1\ 1 \qquad S_8 = 1\ 0\ 1\ 0\ 1\ 1\ 0$$
$$S_9 = 1\ 1\ 1\ 1\ 0\ 0\ 0$$

This is an illustration of having an increase in the number of available codes from

$$\left.\frac{\phi(2^n - 1)}{n}\right|_{n=3} = 2$$

to

$$(2^n + 1)|_{n=3} = 9$$

However, we have yet to address the problem of the cross-correlation functions of maximal-length sequences. Gold [7] addressed this question in a remarkable theorem. Observe Figure 6.55 and note that $R_{4,1}(\tau)$ has a peak cross-correlation value of 11, while that of $R_{4,6}(\tau)$ is 7, indicating that cross-correlation peaks for certain pairs of maximal-length sequences are higher than those of other pairs. In Gold's theorem, sequences generated by combining "preferred pairs" of maximal-length sequences give cross-correlation peaks that are no greater than the minimum possible cross-correlation peaks between any pair of maximal-length sequences of the same length. A method of finding the preferred pair is given in the following theorem [7].

Theorem. Let $f_1(x)$ be a primitive polynomial of degree n that generates a Galois field $GF(2^n) = \{0, 1, \alpha, \alpha^2, \ldots, \alpha^t, \ldots, \alpha^{2^n - 2}\}$, with α as a (primitive) root of the minimal polynomial $M_1(x) \equiv f_1(x)$. Let $M_t(x) \equiv f_t(x)$ be the minimal polynomial of $\alpha^t \in GF(2^n)$ such that

$$t = \begin{cases} 2^{\frac{n+1}{2}} + 1 & \text{if } n \text{ is odd} \\ 2^{\frac{n+2}{2}} + 1 & \text{if } n \text{ is even but } n \neq 0 \bmod 4 \end{cases} \tag{6.59a}$$

Then the Gold sequences generated by the configuration based on the minimal (primitive) polynomials $M_1(x) = f_1(x)$ and $M_t(x)$ satisfy the following cross-correlation function values:

$$|R_{1,t}(k)| \leq \begin{cases} 2^{\frac{n+1}{2}} + 1 & \text{if } n \text{ is odd} \\ 2^{\frac{n+2}{2}} + 1 & \text{if } n \text{ is even but } n \neq 0 \bmod 4 \end{cases} \tag{6.59b}$$

The polynomials $f_1(x) = M_1(x)$ and $M_t(x)$ form a preferred pair.

To illustrate the application of this theorem, we consider the case of $n = 11$. From Peterson's table of irreducible polynomials [3], we can choose the polynomial designated by 1 4005E as $f_1(x)$; that is:

$$f_1(x) = x^{11} + x^2 + 1$$

and

$$n = 11 \text{ is odd} \Rightarrow t = 2^{\frac{n+1}{2}} + 1 = 65$$

We want to find $M_{65}(x)$, which has α^{65} as a root. But the polynomial with 65 as its leading number is not listed in the table. In this case, we can do the following:

1. Calculate $(2^m \times 65) \bmod (2^n - 1) = (2^m \times 65) \bmod 2{,}047$ for $m = 1$, $2, \ldots, 11$. If $2^m \times 65$ is larger than 2,047, divide $2^m \times 65$ by 2,047 and replace $2^m \times 65$ with the remainder, and then continue the doubling process:

   $65, 130, 260, 520, 1{,}040, 2{,}080 \rightarrow 33, 66, 132, 264, 528, 1{,}056, 2{,}112 \rightarrow 65$

2. The doubling process ends at the eleventh time the number is doubled; at the twelfth time, the number goes back to the original number, 65.

3. $\alpha^{65}, \alpha^{130}, \ldots, \alpha^{1,056}$ are the roots of the same-numbered minimal polynomial, $M_{65}(x)$, for the conjugate class.

4. One of the elements in the conjugate class is α^{33}, and it has a polynomial listed in the table [3]: 33 7335G.

5. Therefore:

$$M_{65}(x) \equiv M_{33}(x)$$
$$= x^{11} + x^{10} + x^9 + x^7 + x^6 + x^4 + x^3 + x^2 + 1 \qquad (6.60)$$

There are $2^{11} + 1 = 2,049$ Gold sequences of length $2^{11} - 1 = 2,047$, and their cross-correlation peaks do not exceed the value of $t = 65$.

Another example, for the case of $n = 7$, for which $t = 17$. From [3], the polynomial listed for $f_1(x)$ is 1 211E, and the preferred pair is (1 211E, 9 277E), or

$$f_1(x) = x^7 + x^3 + 1 \qquad (6.61a)$$

and

$$f_t(x) = x^7 + x^5 + x^4 + x^3 + x^2 + x + 1 \qquad (6.61b)$$

This preferred pair gives the lowest cross-correlation peak, whose value is $t = 17$. There are pairs of primitive polynomials of degree 7 that produce sequences with cross-correlation peaks as high as 41; for example, the polynomial pairs (211, 367), (211, 313), and (211, 345), just to name a few. The Gold code for this example can be generated by the modulo-2 sum of the outputs of LFSRs mechanizing $f_1(x)$ and $f_t(x)$. It can also be generated by the LFSR mechanization of the product of the these polynomials, $f_1(x) \cdot f_t(x) = f(x)$. From (6.6-3a) and (6.6-3b), we have

$$f(x) = f_1(x) \cdot f_t(x)$$
$$= x^{14} + x^{12} + x^{11} + x^9 + x^7 + x^6 + x^2 + x + 1 \qquad (6.61c)$$

The SSRG (Fibonacci) configuration of the product polynomial $f(x)$ is shown in Figure 6.57, along with the sum-of-two-SSRG configuration.

The generation of sequences by the product of polynomials can be stated in general terms: if $f_1(x)$ and $f_2(x)$ are any polynomials, then the sequence that is obtained as the modulo-2 sum of the sequences generated by the characteristic polynomials $f_1(x)$ and $f_2(x)$ can be generated by the LFSR

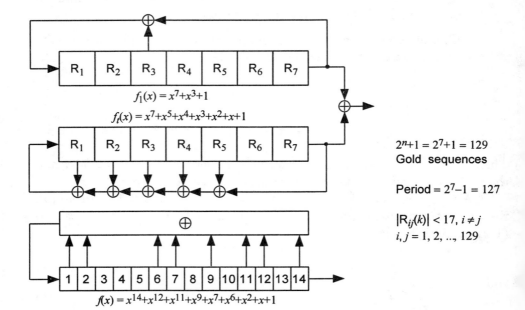

Figure 6.57 Equivalent single and double LFSR Gold sequence generators for $n = 7$.

with the characteristic polynomial $f(x) = f_1(x) \cdot f_2(x)$. A proof of this statement can be given as follows:

Let $a_1(x)$ and $a_2(x)$ be the sequences generated by the characteristic polynomials $f_1(x)$ and $f_2(x)$, respectively. Then, from our previous results, we have

$$a_1(x) = \frac{g_1(x)}{f_1(x)}, \quad \text{degree of } g_1(x) < \text{degree of } f_1(x) \tag{6.62a}$$

and

$$a_2(x) = \frac{g_2(x)}{f_2(x)}, \quad \text{degree of } g_2(x) < \text{degree of } f_2(x) \tag{6.62b}$$

for some numerator polynomials $g_1(x)$ and $g_2(x)$, where the degree of $g_i(x)$ is less than the degree of $f_i(x)$, $i = 1, 2$. Now:

$$a_1(x) + a_2(x) = \frac{g_1(x)}{f_1(x)} + \frac{g_2(x)}{f_2(x)} = \frac{f_2(x)g_1(x) + f_1(x)g_2(x)}{f_1(x)f_2(x)} \tag{6.62c}$$

which is of the form

$$a(x) = \frac{g(x)}{f(x)} \tag{6.62d}$$

with $a(x) = a_1(x) + a_2(x)$ and $f_2(x)g_1(x) + f_1(x)g_2(x)$. We therefore need only to verify that the degree of $g(x)$ is less than the degree of $f(x)$. Using $\deg[\,\cdot\,]$ to denote "degree of," we find that

$$
\begin{aligned}
\deg[g(x)] &= \deg[f_2(x)\,g_1(x) + f_1(x)\,g_2(x)] \\
&\leq \max\{\deg[f_2(x)\,g_1(x)],\, \deg[f_1(x)\,g_2(x)]\} \\
&= \max\{\deg[f_2(x)] + \deg[g_1(x)],\, \deg[f_1(x)] + \deg[g_2(x)]\} \\
&< \deg[f_1(x)] + \deg[f_2(x)] \\
&\Rightarrow \quad \deg[g(x)] < \deg[f(x)]
\end{aligned}
\tag{6.62e}
$$

Thus (6.62d) defines a sequence with characteristic polynomial $f(x) = f_1(x) \cdot f_2(x)$. If $f_1(x)$ and $f_2(x)$ are relatively prime, then the set of sum sequences can be shown to be equal to the set of sequences obtained from the product polynomial, which can be mechanized as a LFSR. Other varieties of Gold codes are discussed, along with examples, in [8].

6.6.2   Gold Codes and GPS Signal Structure

As stated in the description of the IS-95 system, the clock (time reference) is derived from the GPS, and the GPS signal plays an important role for bringing Universal Coordinated Time (UTC) to the base stations of the CDMA systems everywhere on a global basis. In this section, we describe briefly the signal structure of GPS to the extent of showing that the various satellite codes of the system are Gold codes.

The GPS signal consists of two components [9], Link 1 (L1), at a center frequency of 1,575.42 MHz and Link 2 (L2), at a center frequency of 1,227.6 MHz. Each of these center frequencies is a multiple of a 10.23-MHz clock rate used by a precision (P) code signal such that

$$
\begin{aligned}
L1 &= 1{,}575.42\,\text{MHz} = 154 \times 10.23\,\text{MHz} \\
L2 &= 1{,}227.60\,\text{MHz} = 120 \times 10.23\,\text{MHz}
\end{aligned}
\tag{6.63}
$$

and each of the L1 and L2 signals is modulated by either or both a 10.23-MHz clock rate P signal and/or by a 1.023-MHz clear/acquisition (C/A) signal. These two binary signals are formed, respectively, by a P code and a C/A code, each modulated by a data (D) sequence at 50 bps. The L1 in-phase (I) component of the carrier is modulated by the P-signal and the quadrature (Q) component is modulated by the C/A signal. The L2 signal is biphase-modulated by either the P code or the C/A code; normal operation provides for P-code modulation on the L2 signal.

The Link 1 signal from "satellite i" has the form

$$s_{1i}(t) = \underbrace{A_P\, P_i(t)\, D_i(t)\, \cos(\omega_1 t + \phi)}_{P} + \underbrace{A_C\, G_i(t)\, D_i(t)\, \sin(\omega_1 t + \phi)}_{C/A} \qquad (6.64a)$$

where ω_1 is the L1 angular frequency and ϕ is a small phase noise component. $P_i(t)$ is a ± 1 PN code of clock rate 10.23 MHz and has a period of exactly one week. $D_i(t)$ has a ± 1 amplitude at 50 bps, with 6-sec subframe and 30-sec frame time periods. The C/A code $G_i(t)$ is a unique Gold code of period 1,023 chips and has a clock rate of 1.023 MHz, giving a period of 1 ms. This code was selected to provide good MA properties for its period.

The C/A codes for the various satellites are Gold codes formed as products of two $2^{10} - 1 = 1{,}023$-chip PN sequences $G_1(t)$ and $G_2(t)$ over $\{+1, -1\}$. Thus the C/A code waveform from satellite i is

$$G_i(t) = G_1(t) \cdot G_2(t + N_i \cdot 10T) \qquad (6.64b)$$

where N_i determines the phase offset in chips between $G_1(t)$ and $G_2(t)$, and where $T = (10.23\,\text{MHz})^{-1}$ is the chip duration of the P-code signal, so that $10T$ is the C/A code chip duration. There are 1,023 different codes formed by the different offsets N_i (plus the original two sequences G_1 and G_2).

The $\mathbf{G_1}$ polynomial is given by

$$G_1(x) = x^{10} + x^3 + 1 \qquad (6.64c)$$

This "chosen polynomial" in octal notation is 2,011, and the polynomial $G_2(x)$ to form a preferred pair with $G_1(x)$ is the minimal polynomial for the conjugate class that includes the root α^t, where α is the primitive element of $G_1(x)$ and $t = 2^{(n+2)/2} + 1 = 65$. The conjugate class is

$$\{\alpha^{65},\ \alpha^{130},\ \alpha^{260},\ \alpha^{520},\ \alpha^{1040} \equiv \alpha^{17},\ \alpha^{34}, \dots\} \qquad (6.64d)$$

and the minimal polynomial that has α^{17} as a root is found in [3] to be listed as 3515G:

$$G_2(x) = x^{10} + x^9 + x^8 + x^6 + x^3 + x^2 + 1 \tag{6.64e}$$

The operational specification of the GPS [9] requires that the LFSRs for G_1 and G_2 be set to the all-ones states initially. To generate arbitrary phase shifts of the Gold codes, the initial state of the G_1 generator is fixed to the all-ones state as required, and the that of the G_2 generator is changed to provide for the phase shift.

The L2 signal is biphase modulated by either the P code or the C/A code as directed by ground command. The same 50-bps data stream that modulates the L1 carrier also modulates the L2 carrier. The reader is referred to [9] for further details on the GPS signal structure.

References

[1]   Golomb, S. W., *Shift Register Sequences*, Laguna Hills, CA: Aegean Park Press, 1982.

[2]   Berlekamp, E. R., *Algebraic Coding Theory*, New York: McGraw-Hill, 1968.

[3]   Peterson, W. W., and E. J. Weldon, *Error Correcting Codes*, 2nd ed., App. C, pp. 472–492, Cambridge, MA: MIT Press, 1972.

[4]   Simon, M. K., J. K. Omura, R. A. Scholtz, and B. K. Levitt, *Spread Spectrum Communications Handbook*, rev. ed., ch. 5, New York: McGraw-Hill, 1994.

[5]   Woodward, P. M., *Probability and Information Theory, with Applications to Radar*, New York: Pergamon Press, 1953.

[6]   "Mobile Station–Base Station Compatibility Standard for Dual-Mode Wideband Spread Spectrum Cellular System," TIA/EIA/IS-95, Washington: Telecommunications Industry Association, July 1993.

[7]   Gold, R., "Optimal Binary Sequences for Spread Spectrum Multiplexing," *IEEE Trans. on Information Theory*, Vol. IT-13, No. 4, pp. 619–621, Oct. 1967.

[8]   Holmes, J. K., *Coherent Spread Spectrum Systems*, ch. 11, New York: Wiley, 1982.

[9]   Spilker, J. J., Jr., "GPS Signal Structure and Performance Characteristics," *Journal of Navigation*, Vol. 1, pp. 29–53, Washington: Institute of Navigation, 1982.

Appendix 6A Inductive Proof of the Fact That $g(x) = s^*(x)$

Another way to state the relationship to be proved is the following:

$$\underbrace{x^{n-1-k} \text{ modulo } f(x)}_{g_{n-1-k}(x)} = \underbrace{\text{Rev}[s_k(x)]}_{s_k^*(x)} \tag{6A.1}$$

1. ($k = 0$): We have shown directly in the text that $g_{n-1}(x) = s_0^*(x)$. That is, the relationship to be proven is satisfied for $k = 0$.

2. ($k = 1$): For $k = 1$, the initial loading is $s_1(\beta) = \beta \Rightarrow \overrightarrow{010\cdots0}$. We know that for this particular loading, the output of the MSRG is 1 bit advanced (shifted to the left) relative to the output when the loading is $100\cdots0$. Therefore, $g(x) = g_{n-2}(x) = x^{n-2}$ when $s(\beta) = s_1(\beta) = \beta$, and the relationship is satisfied for $k = 1$.

3. For some value of k, we now assume that the relationship is true; that is:

$$\text{Rev}[s_k(x)] = g_{n-1-k}(x) \tag{6A.2a}$$

Given this assumption for the kth initial loading, if we can verify that relationship holds for $k + 1$, then the relationship is proved. Recall that $s(x)$ may be computed recursively:

$$s_{k+1}(x) = \begin{cases} x\,s_k(x) & \text{if } \deg[s_k(x)] < n - 1 \\ x\,s_k(x) + f^*(x) & \text{if } \deg[s_k(x)] = n - 1 \end{cases} \tag{6A.2b}$$

Because this recursion for $s(x)$ is conditional, we must consider two cases to verify that

$$\text{Rev}[s_{k+1}(x)] = g_{n-2-k}(x) \tag{6A.2c}$$

<u>Case 1:</u> $\deg[s_k(x)] < n - 1$; that is, using the coefficients $\{\alpha_i = 0 \text{ or } 1\}$:

$$s_k(x) = \alpha_0 + \alpha_1 x + \cdots + \alpha_{n-2} x^{n-2} + 0 \cdot x^{n-1} \tag{6A.3a}$$

and

$$g_{n-1-k}(x) = 0 + \alpha_{n-2} x + \cdots + \alpha_1 x^{n-2} + \alpha_0 x^{n-1} \tag{6A.3b}$$

In this case

$$s_{k+1}(x) = x \, s_k(x) = \alpha_0 x + \alpha_1 x^2 + \cdots + \alpha_{n-2} x^{n-1} \tag{6A.3c}$$

which has the reverse

$$\mathrm{Rev}[s_{k+1}(x)] = x^{n-1} s_{k+1}(x^{-1}) = \alpha_0 x^{n-2} + \alpha_1 x^{n-3} + \cdots + \alpha_{n-2}$$

$$= \frac{\alpha_0 x^{n-1} + \alpha_1 x^{n-2} + \cdots + \alpha_{n-2} x}{x} = \frac{g_{n-1-k}(x)}{x}$$

$$= g_{n-2-k}(x) \quad \Rightarrow \quad \boxed{\text{QED for case 1}} \tag{6A.3d}$$

<u>Case 2:</u> $\deg[s_k(x)] = n - 1$; that is, using the coefficients $\{\alpha_i = 0 \text{ or } 1\}$:

$$s_k(x) = \alpha_0 + \alpha_1 x + \cdots + \alpha_{n-2} x^{n-2} + x^{n-1} \tag{6A.4a}$$

and

$$g_{n-1-k}(x) = 1 + \alpha_{n-2} x + \cdots + \alpha_1 x^{n-2} + \alpha_0 x^{n-1} \tag{6A.4b}$$

In this case:

$$s_{k+1}(x) = x \, s_k(x) + f^*(x)$$

$$= 1 + \alpha_0 x + \alpha_1 x^2 + \cdots + \alpha_{n-2} x^{n-1} + x^n + f^*(x) \tag{6A.4c}$$

which has the reverse

$$\mathrm{Rev}[s_{k+1}(x)] = x^{n-1} s_{k+1}(x^{-1})$$

$$= x^{n-1} + \alpha_0 x^{n-2} + \alpha_1 x^{n-3} + \cdots + \alpha_{n-2} + \mathrm{Rev}[x^n + f^*(x)] \tag{6A.4d}$$

We observe that

$$\mathrm{Rev}[x^n + f^*(x)] = x^{n-1} \cdot [x^{-n} + f^*(x^{-1})]$$

$$= x^{n-1} \cdot [x^{-n} + x^{-n} f(x)] = \frac{1 + f(x)}{x} \tag{6A.5}$$

Thus, the reverse of the $k + 1$ state is

$$\text{Rev}[s_{k+1}(x)] = x^{n-1} + \alpha_0 x^{n-2} + \alpha_1 x^{n-3} + \cdots + \alpha_{n-2} + \frac{1 + f(x)}{x}$$

$$= \frac{\alpha_0 x^{n-1} + \alpha_1 x^{n-2} + \cdots + \alpha_{n-2} x + 1 + f(x)}{x}$$

$$= \frac{g_{n-1-k}(x) + f(x)}{x} \quad \left\{ \begin{array}{l} \text{It is necessary to add } f(x) \text{ to } g_{n-1-k}(x) \\ \text{before dividing by } x \text{ because the } x^0 \text{ term} \\ \text{of } g_{n-1-k}(x) \text{ is nonzero.} \end{array} \right.$$

$$= g_{n-2-k}(x) \quad \Rightarrow \quad \boxed{\text{QED for case 2}} \tag{6A.6}$$

Because the relationship was verified for both cases of step 3, the general validity of the relationship has been proved; namely that the numerator polynomial of the output sequence produced by an MSRG is the reverse of the polynomial representing the initial state of the MSRG.

Appendix 6B Computer Programs

6B.1 Program for Computing the Shift K

The BASIC program listed below, given a numerator polynomial $g_K(x)$ and a characteristic polynomial $f(x)$, reverses the computation of $g_K(x) = x^K$ modulo $f(x)$ to obtain the sequence shift K.

```
10  DIM G(16), F(16)
20  FOR I=1 TO 16 : READ F(I) : NEXT I
30  G(16)=1
40  FOR I=1 TO 15 : READ G(I) : NEXT I
50  K=0
60  GOSUB 200
70  IF SUM>14 GOTO 400                      stop if g(x) = 1
80  IF G(1)<0 GOTO 120
90  K=K+1
100 GOSUB 300
110 GOTO 60
120 FOR I=1 TO 16:G(I)=G(I)*P(I):NEXT I     g(x) ← g(x) + f(x)
130 GOTO 60
200 REM detect end
```

```
210 SUM=0
220 FOR I=2 TO 16 : SUM=SUM+G(I)  : NEXT I
230 RETURN
300 REM factor g(x)
310 FOR I=1 TO 15 : G(I)=G(I+1)  : NEXT I          g(x) ← g(x)/x
320 G(16)=1
330 RETURN
400 PRINT "K = "; K                     K = # times g(x) factored
410 STOP
500 DATA -1, 1, 1,-1,-1,-1, 1, 1       ← f(x) as binary word
510 DATA  1,-1,-1,-1,-1, 1, 1,-1
520 DATA -1,-1,-1,-1,-1,-1,-1,-1       ← gK(x) as binary word
530 DATA -1,-1,-1,-1,-1,-1,-1
```

The code annotations to the right read:

$g(x) \leftarrow g(x)/x$

$K = \#$ times $g(x)$ factored

$\leftarrow f(x)$ as binary word

$\leftarrow g_K(x)$ as binary word

6B.2 Program for Computing x^K Modulo $f(x)$

The BASIC program listed below, given a sequence shift K and a characteristic polynomial $f(x)$, performs the modular computation of $g_K(x) = x^K$ modulo $f(x)$ to obtain the numerator polynomial $g_K(x)$.

```
10  DIM G(16),F(16)
20  FOR I=1 TO 16:READ F(I):NEXT I
30  G(16)=-1
40  FOR I=1 TO 15:G(I)=1:NEXT I
50  INPUT "K = "; K
60  IF K<15 THEN STOP              program accepts K ≥ n = 15
70  D=K-15
80  GOSUB 200 ! ADD F TO REDUCE DEGREE OF G
90  IF D>0 THEN GOSUB 250
100 IF (D=0 AND G(16)>0) GOTO 400      stop when deg(g) < n
110 IF G(16)<0 GOTO 80
120 GOTO 90
200 REM add f to factored x^k
210 FOR I=1 TO 16:G(I)=G(I)*P(I): NEXT I
```

$$x^{l-n}(1+f+\cdots) \leftarrow x^{l-n}(1+\cdots)$$

```
220 RETURN
250 REM adjust intermediate g
270 FOR I=1 TO 15:G(17-I)=G(16-I):NEXT I
```

$$x^{l-n}(1+\cdots) \leftarrow x^l+\cdots$$

```
280 G(1)=1
290 D=D-1
```

```
300 RETURN
400 FOR I=1 TO 15: PRINT I, G(I): NEXT I     print {gᵢ} when done
500 DATA -1, 1,-1, 1, 1, 1,-1,-1     ← f(x) as a binary word
510 DATA -1, 1,-1, 1, 1, 1, 1,-1
```

Line 400: `print {g_i}$ when done`
Line 500: `← $f(x)$ as a binary word`

Instead of using this program to compute the modular reduction of x^K, the calculation can be based on the fact that the states of an MSRG are powers of x modulo $f(x)$ when the MSRG implements the recursion whose characteristic polynomial is $f^*(x)$. Therefore, the initial state $s_0(x) = x^0 = 1$ can be projected to the state $s_K(x) = x^K$ by multiplying the vector representing $s_0(x)$ by the Kth power of the MSRG's transition matrix to obtain the vector representing $s_K(x)$, whose components then can be interpreted as the coefficients of the terms of $g_K(x) = x^K \bmod f(x)$:

$$
s_K = \begin{bmatrix} g_{K,n-1} \\ g_{K,n-2} \\ g_{K,n-3} \\ \vdots \\ g_{K,1} \\ g_{K,0} \end{bmatrix} = \begin{bmatrix} c_{n-1} & 1 & 0 & \cdots & 0 \\ c_{n-2} & 0 & 1 & \cdots & 0 \\ c_{n-3} & 0 & 0 & \cdots & 0 \\ \vdots & & & & \vdots \\ c_1 & 0 & 0 & \cdots & 1 \\ 1 & 0 & 0 & \cdots & 0 \end{bmatrix}^K \begin{bmatrix} 0 \\ 0 \\ 0 \\ \vdots \\ 0 \\ 1 \end{bmatrix} \tag{6B.1}
$$

6B.3 Program for Computing Long PN Code Transition Matrix

The BASIC program listed below, given the IS-95 long PN code polynomial (6.36), computes the power P of the transition matrix for the MSRG implementation of the PN code. Note that in the program, P is double-precision in order to retain the needed accuracy. A good test of the program is to calculate the power $P = 2^{42} - 1 = 4398046511103$ and verify that the transition matrix raised to this power is an identity matrix. The program converts the power to a binary number (vector) and calculates the power as a product of the matrix raised to powers that are powers of 2.

```
10 DIM T(42,42),U(42,42),V(42,42),C(41)
20 GOSUB 1000                    initialize the transition matrix
30 INPUT "power = "; P#
35 LPRINT USING "#############"; P#
40 GOSUB 1300    !  CONVERT POWER TO VECTOR
45 GOSUB 2050    !  PRINT VECTOR
50 GOSUB 1400    !   MAKE V THE IDENTITY MATRIX
```

```
55 IF C(0)>.5 THEN GOSUB 1850    !   MAKE V=T
60 FOR R=1 TO IA
80 GOSUB 1100     !   U=T*T
90 GOSUB 1500     !   T=U
95 IF C(R)<.5 THEN GOTO 119
100 GOSUB 1600    !   U=T*V
110 GOSUB 1750    !   V=U
119 REM
120 NEXT R
930 FOR N=1 TO 42
940 FOR I=1 TO 42
950 LPRINT USING "#"; V(N,I);
960 NEXT I
970 LPRINT ""
980 NEXT N
999 STOP
1000 FOR N=1 TO 41
1010 T(N,N+1)=1
1020 NEXT N
1030 T(7,1)=1:T(9,1)=1:T(11,1)=1:T(15,1)=1:T(16,1)=1:T(17,1)=1
1040 T(20,1)=1:T(21,1)=1:T(23,1)=1:T(24,1)=1:T(25,1)=1
1050 T(26,1)=1:T(32,1)=1:T(35,1)=1:T(36,1)=1:T(37,1)=1
1060 T(39,1)=1:T(40,1)=1:T(41,1)=1:T(42,1)=1
1070 RETURN
1100 REM u=square of t
1130 FOR N=1 TO 42
1140 FOR I=1 TO 42
1150 U(N,I)=0
1160 FOR K=1 TO 42
1170 U(N,I)=U(N,I)+T(N,K)*T(K,I)
1175 IF U(N,I)>1 THEN U(N,I)=0
1180 NEXT K                        1370 IF I<42 GOTO 1330
1210 NEXT I                        1380 RETURN
1220 NEXT N                        1400 REM make V=I
1230 RETURN                        1410 FOR N=1 TO 42
1300 REM  P# --> vector            1420 V(N,N)=1
1320 I=0                           1430 NEXT N
1325 IA=41                         1440 RETURN
1330 Q#=P#/2                       1500 REM t=u
1350 P#=INT(Q#)                    1510 FOR N=1 TO 42
1360 C(I)=2*(Q#-P#)                1520 FOR I=1 TO 42
1362 IF P#=0 THEN IA=I             1530 T(N,I)=U(N,I)
1365 IF P#=0 THEN RETURN           1540 NEXT I
1366 I=I+1                         1550 NEXT N
```

```
1560 RETURN                              1860 FOR N=1 TO 42
1600 REM u=t*v                           1870 FOR I=1 TO 42
1610 FOR N=1 TO 42                       1880 V(N,I)=T(N,I)
1620 FOR I=1 TO 42                       1890 NEXT I
1630 U(N,I)=0                            1900 NEXT N
1640 FOR K=1 TO 42                       1910 RETURN
1650 U(N,I)=U(N,I)+T(N,K)*V(K,I)1950 REM   make u=v
1655 IF U(N,I)>1 THEN U(N,I)=0 1960 FOR N=1 TO 42
1660 NEXT K                              1970 FOR I=1 TO 42
1680 NEXT I                              1980 U(N,I)=V(N,I)
1690 NEXT N                              1990 NEXT I
1700 RETURN                              2000 NEXT N
1750 REM v=u                             2010 RETURN
1760 FOR N=1 TO 42                       2050 REM print c vector
1770 FOR I=1 TO 42                       2060 FOR N=0 TO 41
1780 V(N,I)=U(N,I)                       2070 LPRINT USING "#"; C(N);
1790 NEXT I                              2080 NEXT N
1800 NEXT N                              2090 LPRINT " "
1810 RETURN                              2095 LPRINT
1850 REM   make v=t                      2100 RETURN
```

Appendix 6C Proof of Correlation Function Theorem

The proof of the expression (6.45a) in the text is as follows:

$$R_{fa,fb}(\tau) = \frac{1}{T} \int_0^T f_a(t)\, f_b(t-\tau)\, dt$$

$$= \frac{1}{PT_c} \int_0^{PT_c} \overbrace{\sum_{j=-\infty}^{\infty} A_j\, \varphi_a(t-jT_c)}^{f_a(t)} \overbrace{\sum_{k=-\infty}^{\infty} B_k\, \varphi_b(t-\tau-kT_c)}^{f_a(t-\tau)} dt$$

$$= \frac{1}{PT_c} \sum_{j=-\infty}^{\infty} A_j \sum_{k=-\infty}^{\infty} B_k \int_0^{PT_c} \underbrace{\varphi_a(t-jT_c)}_{\substack{\text{nonzero in } (0,\, PT_c) \text{ only} \\ \text{for } jT_c \le t \le (j+1)T_c}} \varphi_b(t-\tau-kT_c)\, dt$$

$$= \frac{1}{PT_c} \sum_{j=0}^{P-1} A_j \sum_{k=-\infty}^{\infty} B_k \int_{jT_c}^{(j+1)T_c} \varphi_a(t-jT_c)\, \varphi_b(t-\tau-kT_c)\, dt$$

$$\downarrow \;\; \text{translation of integral's variable}$$

$$= \frac{1}{PT_c} \sum_{j=0}^{P-1} A_j \sum_{k=-\infty}^{\infty} B_k \int_0^{T_c} \varphi_a(t) \underbrace{\varphi_b(t + jT_c - \tau - kT_c)}_{\varphi_b[t - \tau - (k-j)T_c]} dt$$

$$= \frac{1}{PT_c} \sum_{j=0}^{P-1} A_j \sum_{k=-\infty}^{\infty} B_{k+j} \int_0^{T_c} \varphi_a(t) \varphi_b(t - \tau - kT_c) dt \qquad \begin{matrix} k \to k + j \\ \text{(translation)} \end{matrix}$$

Because $\varphi_a(t) = 0$ for $t < 0$ and $t > T_c$:

$$\int_0^{T_c} \varphi_a(t) \varphi_b(t - \tau - kT_c) dt = \int_{-\infty}^{\infty} \varphi_a(t) \varphi_b(t - \tau - kT_c) dt$$

$$= R_{\varphi_a, \varphi_b}(\tau + kT_c)$$

Thus, the cross-correlation function for $f_a(t)$ and $f_b(t)$ is

$$R_{fa, fb}(\tau) = \frac{1}{PT_c} \sum_{j=0}^{P-1} A_j \sum_{k=-\infty}^{\infty} B_{k+j} \int_0^{T_c} \varphi_a(t) \varphi_b(t - \tau - kT_c) dt$$

$$= \frac{1}{PT_c} \sum_{k=-\infty}^{\infty} \left(\sum_{j=0}^{P-1} A_j B_{k+j} \right) R_{\varphi_a, \varphi_b}(\tau + kT_c)$$

$$= \frac{1}{PT_c} \sum_{k=-\infty}^{\infty} R_{a,b}(-k) R_{\varphi_a, \varphi_b}(\tau + kT_c)$$

$$= \frac{1}{PT_c} \sum_{k=-\infty}^{\infty} R_{a,b}(k) R_{\varphi_a, \varphi_b}(\tau - kT_c) \qquad (k \text{ replaced by } -k)$$

Appendix 6D Extension of Correlation Theorem to Bandlimited Pulses

The derivation of (6.45a) that was presented in Appendix 6C applies to correlation and partial correlation functions for PN waveforms using baseband data pulses (basic waveforms) that are confined to a one-chip interval. In practical communications systems, it is desirable to shape the baseband data pulses to limit the bandwidth of the transmitted waveform. We now consider whether the correlation and partial correlation functions must be modified when filtered pulses are used in the PN waveform.

Suppose that the baseband data pulses are Dirac delta functions:

$$f_a(t) = \sum_{k=-\infty}^{\infty} A_k \, \delta(t - kT_c) \qquad (6D.1)$$

Because the pulses in this instance are confined to one-chip intervals, the derivation of Appendix 6C applies. Therefore the autocorrelation function in this case is

$$R_{fa}(\tau) = \frac{1}{PT_c} \sum_{k=-\infty}^{\infty} R_a(k) \, R_\delta(\tau - kT_c)$$

$$= \frac{1}{PT_c} \sum_{k=-\infty}^{\infty} R_a(k) \, \delta(\tau - kT_c) \qquad (6D.2)$$

Strictly speaking, a delta function cannot be implemented in hardware. However, in modern digital systems it is possible to synthesize filter impulse responses directly, producing the same effect as a filtered delta function.

From linear system theory, we know that the relationship between the spectrum at the output of a filter and the spectrum at the input of the filter is

$$\mathcal{S}_{out}(f) = |H(f)|^2 \cdot \mathcal{S}_{in}(f) \qquad (6D.3)$$

where $H(f)$ is the Fourier transform of the filter impulse response $h(t)$. Another fact of linear system theory is that the spectrum and the autocorrelation function are a Fourier transform pair. Taking the inverse Fourier transform of (6D.3) results in

$$\mathcal{F}^{-1}\{\mathcal{S}_{out}(f)\} = R_{out}(\tau) = R_h(\tau) * R_{in}(\tau) \qquad (6D.4a)$$

where the "filter autocorrelation function" is defined as

$$R_h(\tau) \triangleq \mathcal{F}^{-1}\{|H(f)|^2\} = \mathcal{F}^{-1}\{H(f) \cdot H^*(f)\}$$

$$= \int_{-\infty}^{\infty} e^{j2\pi f\tau} \left[\int_{-\infty}^{\infty} e^{-j2\pi ft} h(t) \, dt \right] \left[\int_{-\infty}^{\infty} e^{j2\pi fu} h^*(u) \, du \right] df$$

$$= \int_{-\infty}^{\infty} h(t) \int_{-\infty}^{\infty} h^*(u) \underbrace{\int_{-\infty}^{\infty} e^{j2\pi f\tau} e^{-j2\pi ft} e^{j2\pi fu} \, df}_{\delta(\tau - t + u)} \, du \, dt$$

$$= \int_{-\infty}^{\infty} h(t)\, h^*(t - \tau)\, dt = \int_{-\infty}^{\infty} h(t)\, h(t - \tau)\, dt$$

$$= \int_{-\infty}^{\infty} h(t + \tau)\, h(t)\, dt \tag{6D.4b}$$

Now, let the input to the filter be given by the train of impulses in (6D.1), whose autocorrelation function is given by (6D.2). According to (6D.4a), the autocorrelation function of $b(t)$, the filter output, is

$$R_b(\tau) = R_h(\tau) * \left[\frac{1}{PT_c} \sum_{k=-\infty}^{\infty} R_a(k)\, \delta(\tau - kT_c) \right]$$

$$= \frac{1}{PT_c} \sum_{k=-\infty}^{\infty} R_a(k)\, R_h(\tau - kT_c) \tag{6D.5}$$

It follows from this result that the expressions found for the correlation and partial correlation functions, which involve the pulse autocorrelation function $R_\varphi(\tau)$, can be applied to the case of bandlimited pulses. If the bandlimited pulses are produced by synthesis of a filter impulse response $h(t)$, then in the expressions $R_\varphi(\tau)$ may be replaced by $R_h(\tau)$, the filter autocorrelation function. For example, if the filter impulse response is the Nyquist pulse:

$$h(t) = \text{sinc}(t/T_c)$$

which has the autocorrelation function (see (7.116b))

$$R_h(\tau) = T_c\, \text{sinc}(\tau/T_c)$$

then the autocorrelation function for a PN signal out of the filter is

$$R_b(\tau) = \frac{T_c}{PT_c} \sum_{k=-\infty}^{\infty} R_a(k)\, \text{sinc}[(\tau - kT_c)/T_c]$$

$$= \frac{1}{P} \sum_{k=-\infty}^{\infty} R_a(k)\, \text{sinc}[(\tau - kT_c)/T_c] \tag{6D.6}$$

7

Modulation and Demodulation of IS-95 Spread-Spectrum Signals

We saw in Chapter 3 that the IS-95 digital cellular system employs spread-spectrum waveforms to implement a CDMA scheme and to provide spread-spectrum processing gain against interference. In this chapter, we review the principles of optimum receiver design and examine the modulation and demodulation of the IS-95 waveforms in the light of those principles.

7.1 Likelihood Function

With the statistical tools we developed in Chapter 1, we are in a position to derive receiver structures that are optimum in the sense that they are the *minimum probability of error receivers*. A *communication* requires a sender of the information (transmitter) and a receiver, and the entity between the two parties is the *communication channel*. The communication channel is characterized as a medium that disturbs the transmitted signal. The disturbing phenomenon is a random phenomenon that is normally described by some statistical distribution. The disturbing phenomenon can be attributed to many sources, and they are all called *noise*. The noise can be either atmospheric noise (see Section 1.5.10 on the discussion of lognormal distributions) or receiver-generated thermal noise (see Section 1.5.1). When we have the precise statistical description of the channel noise random processes, such as the Gaussian noise processes, we are able to derive optimal receivers that can be implemented in real-world situations. In reality, we are normally dealing with Gaussian noise channels, or additive white Gaussian noise (AWGN) channels.

A block diagram model of a communications system is depicted in Figure 7.1, in which $\{m_i\}$ denotes a set of M messages to be transmitted to

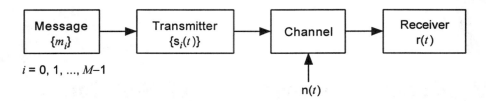

Figure 7.1 Block diagram description of a communications system over an AWGN channel.

the receiver by means of the signal set $\{s_i(t)\}$, with the mapping $m_i \rightarrow s_i(t)$. Assume that message m_i is transmitted by means of transmitter waveform $s_i(t)$. The received waveform is then given by the transmitted signal disturbed by the noise $n(t)$. If the channel is an AWGN channel, then

$$r(t) = s_i(t) + n(t) \tag{7.1}$$

where $n(t)$ is a sample function of a white Gaussian random process. For our purposes at this time, we can regard the RV $n(t)$, which is obtained from the Gaussian noise process $\{n(t)\}$ by sampling at time t, as a zero-mean Gaussian RV with variance $\sigma_n^2 = \frac{1}{2}\mathcal{N}_0$, as discussed in Section 1.5.1.4. Suppressing the time notation, $n(t)$ has the pdf

$$p_n(\alpha) = \frac{1}{\sqrt{2\pi}\sigma_n} \exp\left\{-\frac{\alpha^2}{2\sigma_n^2}\right\} \tag{7.2}$$

Now, because the transmitter sent the message m_i by means of the waveform $s_i(t)$, we can easily write down the pdf of the random variable $r(t)$, given that $s_i(t)$ is transmitted [see the discussion of conditional pdfs in connection with (1.48c)] as

$$p_r(\alpha \mid m_i) = \frac{1}{\sqrt{2\pi}\sigma_n} \exp\left\{-\frac{(\alpha - s_i)^2}{2\sigma_n^2}\right\} \tag{7.3}$$

The conditional pdf of the received waveform random variable, given that the message is transmitted, is called a *likelihood function* in view of the fact that the mode (most likely value) is the value s_i corresponding to the transmitted message.

7.1.1 Vector Representation of the Waveforms

A basic approach in the development of an optimal receiver design begins with the idea that the waveforms can be represented by N-dimensional vectors [1]. The analogy of this concept is similar to the representation of a continuous waveform by a set of Nyquist samples. The sampling strategy provides two options: one is to reconstruct the original continuous-time waveform by processing the sampled values through an interpolation filter; the other is to express the joint pdf of those sampled RVs. Our purpose for the vector representation of the time domain waveform is to have the option of the latter; namely, to be able to calculate the N-dimensional joint pdf. As will be shown below, with the joint pdf, we can then derive the optimal receiver design.

Thus, when the waveforms are replaced by vectors of the appropriate dimension, the *waveform* communications system becomes a *vector* communications system. The question to ask, then, is whether it is possible to replace all waveforms by finite-dimensional vectors in a communications system in which a set of M waveforms $\{s_i(t)\}$, $i = 0, 1, \ldots, M-1$ is used. The answer is yes, in that there are many ways in which a signal in the set $\{s_i(t)\}$ can be represented as an N-dimensional vector $s_i = (s_{i1}, s_{i2}, \ldots, s_{iN})$ by means of orthonormal (ON) waveforms $\{\varphi_j(t), 0 \le t \le T\}$, $j = 1, 2, \ldots, N$. That is:

$$s_i(t) = s_{i1}\,\varphi_1(t) + s_{i2}\,\varphi_2(t) + \cdots + s_{iN}\,\varphi_N(t) \qquad 0 \le t \le T \qquad (7.4a)$$

where the ON functions $\{\varphi_j(t)\}$ have the property[1]

$$\int_{-\infty}^{\infty} \varphi_j(t)\,\varphi_k(t)\,dt = \int_0^T \varphi_j(t)\,\varphi_k(t)\,dt = \delta_{jk} \triangleq \begin{cases} 1, & j = k \\ 0, & j \ne k \end{cases} \qquad (7.4b)$$

in which we use the Kronecker delta δ_{jk} as defined in (7.4b).

The representation of the waveforms $\{s_i(t)\}$ by the vectors $\{s_i\}$ can be done in two different ways. The first method may be termed the *waveform synthesis* approach, in that one chooses a set of ON functions $\{\varphi_j(t)\}$ and then synthesizes the waveforms by means of linear combinations of the ON functions of the form $s_i(t) = s_{i1}\,\varphi_1(t) + s_{i2}\,\varphi_2(t) + \cdots + s_{iN}\,\varphi_N(t)$ using the components of the vector $s_i = (s_{i1}, s_{i2}, \ldots, s_{iN})$ as coefficients. That is to say,

---

[1] The prefix "ortho" means "orthogonal," whereas the suffix "normal" means that the integral is unity for $j = k$.

given the vector $\mathbf{s}_i$ and *choosing*[2] any set of ON functions $\{\varphi_j(t)\}$, we obtain (synthesize) the transmitter waveform $s_i(t)$ by the procedure $\mathbf{s}_i \rightarrow s_i(t)$.

The second method may be called the *analysis* approach, in that one finds a set of ON functions $\{\varphi_j(t)\}$ from the *given* set of transmitter waveforms $\{s_i(t)\}$, that is, deriving the vectors $\{\mathbf{s}_i\}$, by using what is known as the Gram-Schmidt orthogonalization procedure [2, 3], which we explain in Appendix 7A and give an example to illustrate its application.

The two methods, the synthesis and analysis methods, are at our disposal and hence we can always represent the waveform communications system by the corresponding vector communications system, as illustrated in Figure 7.2.

The waveform representation method indicated in (7.4a) is analogous to the representation of the vectors $\{\mathbf{s}_i\}$ as

$$\mathbf{s}_i = s_{i1}\,\widehat{\boldsymbol{\varphi}}_1 + s_{i2}\,\widehat{\boldsymbol{\varphi}}_2 + \cdots + s_{iN}\,\widehat{\boldsymbol{\varphi}}_N \tag{7.4c}$$

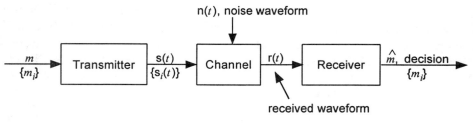

(a) Waveform communications system

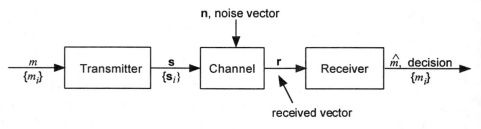

(b) Vector communications system

Figure 7.2 Representations of waveform and vector communications systems.

---

[2] The choice of the ON set $\{\varphi_j(t)\}$ involves the selection of transmitter waveforms, and hence it must be based on engineering considerations.

where $\widehat{\varphi}_j$ is viewed as the unit vector on the jth axis of the N mutually perpendicular axes $\varphi_1, \varphi_2, \ldots, \varphi_N$. Thus, the signal set $\{s_i(t)\}$, $i = 0, 1, \ldots, M - 1$, can now be represented as M vectors $\{s_i\}$ in an N-dimensional signal vector space. The concept of replacing $\{s_i(t)\}$ with $\{s_i\}$ is termed the *geometric representation of signals* or the *vector representation* of signal waveforms.

In (7.4a) the coefficients $\{s_{ij}\}$, $j = 1, 2, \ldots, N$ of the ON functions $\{\varphi_j(t)\}$, $j = 1, 2, \ldots, N$ are extracted from the following operation:

$$s_{ij} \triangleq \int_0^T s_i(t)\, \varphi_j(t)\, dt \qquad (7.4d)$$

since

$$\int_0^T s_i(t)\, \varphi_j(t)\, dt = \int_0^T \left[\sum_{k=1}^N s_{ik}\, \varphi_k(t) \right] \varphi_j(t)\, dt$$

$$= \sum_{k=1}^N s_{ik} \int_0^T \varphi_k(t)\, \varphi_j(t)\, dt = \sum_{k=1}^N s_{ik}\, \delta_{kj} = s_{ij}$$

The integral of the product of two functions is called a *correlation operation*, and thus the vector components $s_{i1}, s_{i1}, \ldots, s_{iN}$ are extracted from N correlators as shown in Figure 7.3.

Parseval's Theorem revisited [see (1.18g)]. Let $u(t)$ and $v(t)$ be two time functions with Fourier transforms $U(f)$ and $V(f)$, respectively. Assume that we have a set of ON functions $\{\varphi_j(t)\}$, $0 \le t \le T$, $j = 1, 2, \ldots, N$. Then

$$u(t) = \sum_{j=1}^N u_j\, \varphi_j(t), \qquad 0 \le t \le T \qquad (7.5a)$$

and

$$v(t) = \sum_{j=1}^N v_j\, \varphi_j(t), \qquad 0 \le t \le T \qquad (7.5b)$$

The vector representations of $u(t)$ and $v(t)$ are

$$\mathbf{u} = (u_1, u_2, \ldots, u_N) \qquad (7.5c)$$

$$\mathbf{v} = (v_1, v_2, \ldots, v_N) \qquad (7.5d)$$

Then

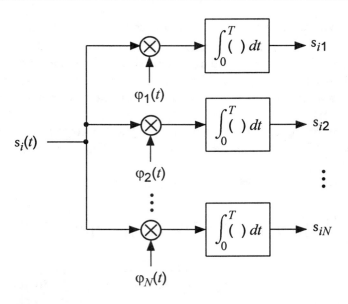

Figure 7.3 Extraction of vector components in $s_i = (s_{i1}, s_{i2}, \ldots, s_{iN})$ from the waveform $s_i(t)$ by correlating $s_i(t)$ with $\varphi_j(t), 0 \leq t \leq T, j = 1, 2, \ldots, N$.

$$\int_0^T u(t)\, v(t)\, dt = \int_0^T \left[\sum_{j=1}^N u_j\, \varphi_j(t)\right]\left[\sum_{k=1}^N v_k\, \varphi_k(t)\right] dt$$

$$= \sum_{j=1}^N u_j\, v_j \tag{7.5e}$$

and

$$\mathbf{u} \cdot \mathbf{v} = \mathbf{u}\, \mathbf{v}^{\mathrm{T}} = (u_1, u_2, \ldots, u_N) \begin{pmatrix} v_1 \\ v_2 \\ \vdots \\ v_N \end{pmatrix} = \sum_{j=1}^N u_j\, v_j \tag{7.5f}$$

From (7.5e) and (7.5f), we observe that the product of two time functions (signals) is equal to the scalar (dot) product of the two vectors representing the time functions. Let us use the following notations for the operations indicated in (7.5e) and (7.5f):

$$\langle u(t), v(t) \rangle \triangleq \int_{-\infty}^{\infty} u(t)\, v(t)\, dt = \int_0^T u(t)\, v(t)\, dt \tag{7.6a}$$

and

$$\langle \mathbf{u}, \mathbf{v} \rangle \triangleq \mathbf{u} \cdot \mathbf{v} = \mathbf{u} \, \mathbf{v}^{\mathrm{T}} \tag{7.6b}$$

Loosely speaking, (7.6a) is a scalar product (correlation operation) in "function space," whereas (7.6b) is a scalar product in vector space. Thus, the relations (7.5e) and (7.5f) imply that

$$\langle u(t), v(t) \rangle = \int_0^T u(t)\, v(t)\, dt = \langle \mathbf{u}, \mathbf{v} \rangle = \sum_{j=1}^N u_j v_j \tag{7.7a}$$

In Section 1.2, we have shown Parseval's Theorem (1.18g) as

$$\int_{-\infty}^{\infty} u(t)\, v^*(t)\, dt = \int_{-\infty}^{\infty} U(f)\, V^*(f)\, df$$

or

$$\int_{-\infty}^{\infty} u(t)\, v(t)\, dt = \int_{-\infty}^{\infty} U(f)\, V(-f)\, df$$

Thus we have for $u(t) = v(t)$,

$$\int_{-\infty}^{\infty} u^2(t)\, dt = \int_{-\infty}^{\infty} |U(f)|^2 df = |\mathbf{u}|^2 = \mathrm{E} \tag{7.7b}$$

where E is the energy[3] of the signal $u(t)$. For $u(t) = s(t)$, therefore:

$$|\mathbf{u}|^2 = |\mathbf{s}|^2 = \mathrm{E} = \langle \mathbf{s}, \mathbf{s} \rangle = \mathbf{s} \cdot \mathbf{s} = \mathbf{s} \, \mathbf{s}^{\mathrm{T}} \tag{7.7c}$$

is an expression that is prevalently used in discusssions of vector communications systems. For a vector $\mathbf{s}$, $|\mathbf{s}|^2$ is the square of the length of the vector, and hence (7.7b) and (7.7c) indicate that the signal energy is equal to the square of the length of the vector representing the signal waveform:

$$\mathrm{E} = \int_0^T s^2(t)\, dt = |\mathbf{s}|^2 \tag{7.7d}$$

Example 7.1 Let us consider the signal vectors s_0 and s_1 shown in the two-dimensional vector space in Figure 7.4. We wish to synthesize the wave-

---

[3] The implication is that the "voltage" $u(t)$ over a 1-ohm load is the basis of the signal "energy" definition.

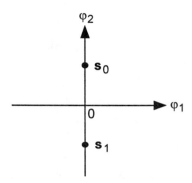

Figure 7.4 Two signal vectors in two-dimensional vector space.

forms corresponding to these vectors by choosing ON functions $\varphi_1(t)$ and $\varphi_2(t)$, $0 \leq t \leq T$. Because the choice of $\{\varphi_j(t)\}$ is the selection of transmitter waveforms, an engineering choice prompts us to consider the following waveforms:

$$\varphi_1(t) = \sqrt{\frac{2}{T}} \sin 2\pi f_0 t, \qquad 0 \leq t \leq T \tag{7.8a}$$

$$\varphi_2(t) = \sqrt{\frac{2}{T}} \cos 2\pi f_0 t, \qquad 0 \leq t \leq T \tag{7.8b}$$

where $f_0 = n/T$, n integer. Assuming $|s_0|^2 = |s_1|^2 = E$, the signal vectors are

$$\left. \begin{array}{l} s_0 = (s_{01}, s_{02}) = \left(0, \sqrt{E}\right) \\[2mm] s_1 = (s_{11}, s_{12}) = \left(0, -\sqrt{E}\right) \end{array} \right\} \tag{7.8c}$$

Because the waveforms $\{s_i(t)\}$, $i = 0, 1$ are given by

$$s_i(t) = s_{i1}\,\varphi_1(t) + s_{i2}\,\varphi_2(t) \tag{7.8d}$$

we have

$$s_0(t) = 0 \cdot \varphi_1(t) + \sqrt{E} \cdot \varphi_2(t) = \sqrt{\frac{2E}{T}} \cos 2\pi f_0 t, \qquad 0 \leq t \leq T \tag{7.8e}$$

and $\quad s_1(t) = 0 \cdot \varphi_1(t) - \sqrt{E} \cdot \varphi_2(t) = -\sqrt{\frac{2E}{T}} \cos 2\pi f_0 t, \; 0 \leq t \leq T \tag{7.8f}$

The signals $s_0(t)$ and $s_1(t)$ are of the form

$$s_i(t) = \sqrt{\frac{2E}{T}} \cos(2\pi f_0 t - i\pi) = \sqrt{\frac{2E}{T}} \cos(2\pi f_0 t - \theta_i), \quad i = 0, 1 \qquad (7.8g)$$

where $\theta_0 = 0$ and $\theta_1 = \pi$, and, thus, these waveforms are known as binary phase-shift keying (BPSK) signals or antipodal [1] signals, which we discuss in a later section dealing with modulation.

Example 7.2 Consider the four vectors shown in the two-dimensional ($N = 2$) signal vector space in Figure 7.5. Among several possible choices for $\varphi_1(t)$ and $\varphi_2(t)$, let us first consider the sinusoidal waveforms for the ON set as given previously in (7.8a) and (7.8b). Assuming that $|s_i|^2 = E$, $i = 0, 1, 2, 3$ the signal vectors are expressed as

$$s_0 = (0, \quad \sqrt{E}) \qquad\qquad s_1 = (-\sqrt{E}, 0)$$
$$s_2 = (0, -\sqrt{E}) \qquad\qquad s_3 = (\quad \sqrt{E}, 0) \qquad (7.9a)$$

Because the waveforms $\{s_i(t)\}$, $i = 0, 1, 2, 3$ are given by (7.8d), we have

$$s_0(t) = 0 \cdot \varphi_1(t) + \sqrt{E} \cdot \varphi_2(t) = \sqrt{\frac{2E}{T}} \cos 2\pi f_0 t, \quad 0 \le t \le T \qquad (7.9b)$$

$$s_1(t) = -\sqrt{E} \cdot \varphi_1(t) + 0 \cdot \varphi_2(t) = -\sqrt{\frac{2E}{T}} \sin 2\pi f_0 t, \quad 0 \le t \le T \qquad (7.9c)$$

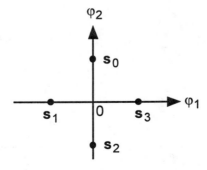

Figure 7.5 Four signals in a two-dimensional signal vector space.

$$s_2(t) = 0 \cdot \varphi_1(t) - \sqrt{E} \cdot \varphi_2(t) = -\sqrt{\frac{2E}{T}} \cos 2\pi f_0 t, \quad 0 \le t \le T \qquad (7.9d)$$

$$s_3(t) = \sqrt{E} \cdot \varphi_1(t) + 0 \cdot \varphi_2(t) = \sqrt{\frac{2E}{T}} \sin 2\pi f_0 t, \quad 0 \le t \le T \qquad (7.9e)$$

The waveforms $\{s_i(t)\}$ can be expressed as

$$s_i(t) = \sqrt{\frac{2E}{T}} \cos\left(2\pi f_0 t + i\frac{\pi}{2}\right), \quad i = 0, 1, 2, 3; \quad 0 \le t \le T \qquad (7.9f)$$

These waveforms are represented as four vectors in Figure 7.5, and they are two sets of BPSK signals, known as quadrature phase-shift keying (QPSK) signals, which we discuss in a later section dealing with modulation.

The two examples given above pertain to the case where the given signal vectors are transformed into transmitter waveforms based on sinusoidal ON functions as shown in (7.8a) and (7.8b). We now choose different ON functions and show that the resulting transmitter waveforms are totally different from those shown in the examples above for the same signal vectors shown in Figure 7.5.

Example 7.3 Now let us choose the ON functions $\{\varphi_1(t), \varphi_2(t)\}$ as shown in Figure 7.6, where the two ON waveforms are two nonoverlapping unit pulses, to synthesize the signal vectors $\{s_i\}$ shown in Figure 7.5. Because the signal vectors are given in (7.9a), the corresponding waveforms are those shown on the next page.

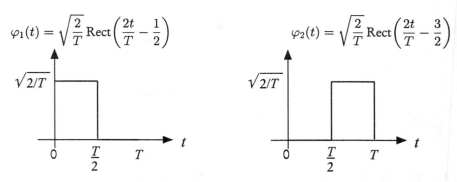

$$\varphi_1(t) = \sqrt{\frac{2}{T}} \, \text{Rect}\left(\frac{2t}{T} - \frac{1}{2}\right) \qquad\qquad \varphi_2(t) = \sqrt{\frac{2}{T}} \, \text{Rect}\left(\frac{2t}{T} - \frac{3}{2}\right)$$

Figure 7.6 Two nonoverlapping pulses as a set of ON functions for synthesis of signal waveforms.

$$s_0 = \left(0, \sqrt{E}\right) \quad \rightarrow s_0(t) = \sqrt{\frac{2E}{T}} \, \text{Rect}\left(\frac{2t}{T} - \frac{3}{2}\right), \quad 0 \le t \le T \qquad (7.10a)$$

$$s_1 = \left(-\sqrt{E}, 0\right) \rightarrow s_1(t) = -\sqrt{\frac{2E}{T}} \, \text{Rect}\left(\frac{2t}{T} - \frac{1}{2}\right), \quad 0 \le t \le T \qquad (7.10b)$$

$$s_2 = \left(0, -\sqrt{E}\right) \rightarrow s_2(t) = -\sqrt{\frac{2E}{T}} \, \text{Rect}\left(\frac{2t}{T} - \frac{3}{2}\right), \quad 0 \le t \le T \qquad (7.10c)$$

$$s_3 = \left(\sqrt{E}, 0\right) \quad \rightarrow s_3(t) = \sqrt{\frac{2E}{T}} \, \text{Rect}\left(\frac{2t}{T} - \frac{1}{2}\right), \quad 0 \le t \le T \qquad (7.10d)$$

These functions are illustrated in Figure 7.7.

7.1.2 Optimal Receiver Principles for Gaussian Channels

Consider Figure 7.8, which depicts an M-ary communications system in which a set of M messages $\{m_i\}$, $i = 0, 1, \ldots, M - 1$ is to be communicated to the receiving end by means of a set of M signal vectors $\{s_i\}$, $i = 0, 1, \ldots, M - 1$. That is, the system uses the mapping $\{m_i\} \rightarrow \{s_i\}$, where

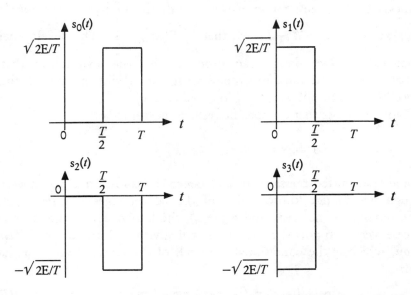

Figure 7.7 Transmitter waveforms for the signal vectors of Figure 7.5, synthesized with the ON waveforms shown in Figure 7.6.

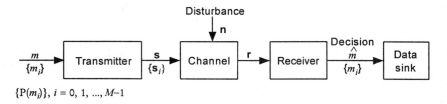

Figure 7.8 Vector communications system.

$$\mathbf{s}_i = (s_{i1}, s_{i2}, \ldots, s_{iN}), \quad i = 0, 1, \ldots, M - 1 \tag{7.11a}$$

The channel disturbs the transmitted vector components with the noise vector $\mathbf{n}$ given by

$$\mathbf{n} = (n_1, n_2, \ldots, n_N) \tag{7.11b}$$

and produces the vector $\mathbf{r}$ at the receiver:

$$\mathbf{r} = (r_1, r_2, \ldots, r_N) \tag{7.11c}$$

The messages $\{m_i\}$ are random events with the known *prior probabilities* $\{P(m_i)\}$, $i = 0, 1, \ldots, M - 1$ such that $\sum_{i=0}^{M-1} P(m_i) = 1$. The development of the optimal receiver design can proceed with the assumption that the disturbance is a noise random process with known distribution, so that we can specify the joint pdf $p_{n_1, n_2, \ldots, n_N}(\alpha_1, \alpha_2, \ldots, \alpha_N)$.

Now, suppose that the specific receiver vector is

$$\mathbf{r} = \boldsymbol{\gamma} = (\gamma_1, \gamma_2, \ldots, \gamma_N) \tag{7.12a}$$

for some transmitted message m, and hence the transmitted signal vector $\mathbf{s}$. Although we do not know which of the M messages was sent (i.e., the specific index i of the actual message m_i), the knowledge of the joint pdf of the noise vector $\mathbf{n} = (n_1, n_2, \ldots, n_N)$ will give the form of the *likelihood function*, which is the conditional joint pdf of the received vector $\mathbf{r}$, having the form

$$p_r(\gamma \mid s) = p_r(\gamma_1, \gamma_2, \ldots, \gamma_N \mid m) \tag{7.12b}$$

The optimal receiver, which we set out to specify, must be capable of determining which of the M possible messages $\{m_i\}$ is the transmitted message from the knowledge of the likelihood function (7.12b) and the prior probabilities $\{P(m_i)\}$. The optimal receiver design is based on the *maximum a posteriori probability (MAP)* receiver principle. Having received the vector $\mathbf{r} = \boldsymbol{\gamma}$, what is the probability that the reception of this particular vector value means the transmitted message is being m? In other words, we form the probability measure of the event m, given that $\mathbf{r} = \boldsymbol{\gamma}$, namely $P(m \mid \mathbf{r} = \boldsymbol{\gamma})$. In the literature, this probability is called the *a posteriori* probability, or *inverse probability*, to denote the obvious argument of the possible cause for the observed effect. One has to compute all M *a posteriori* probabilities $\{P(m \mid \mathbf{r} = \boldsymbol{\gamma})\}$, $i = 0, 1, \ldots, M - 1$ and choose the largest (maximum) value, and make the message decision, $\hat{m}$, corresponding to the maximum of $\{P(m \mid \mathbf{r} = \boldsymbol{\gamma})\}$.

The MAP receiver design criterion can further be elaborated as follows: Assuming that we have received the vector $\mathbf{r} = \boldsymbol{\gamma} = (\gamma_1, \gamma_2, \ldots, \gamma_N)$ the MAP criterion decides that the transmitted message is m_k if and only if the inverse probability (*a posteriori* probability) $P(m_k \mid \mathbf{r} = \boldsymbol{\gamma})$ is the maximum of the set $\{P(m \mid \mathbf{r} = \boldsymbol{\gamma})\}$. Using the notation $\hat{m}$ to mean the *message decision* at the receiver output, and "iff" to mean "if and only if," we state the MAP decision criterion like this:

$$\hat{m} = m_k \ \text{ iff } \ P(m_k \mid \mathbf{r} = \boldsymbol{\gamma}) = \max_i \ \{P(m_i \mid \mathbf{r} = \boldsymbol{\gamma})\} \qquad (7.12c)$$

But

$$P(m_i \mid \mathbf{r} = \boldsymbol{\gamma}) = \frac{p_r(\boldsymbol{\gamma}, m_i)}{p_r(\boldsymbol{\gamma})} = \frac{P(m_i)\, p_r(\boldsymbol{\gamma} \mid m = m_i)}{p_r(\boldsymbol{\gamma})} \qquad (7.12d)$$

where

$$p_r(\boldsymbol{\gamma}) = p_{r_1, r_2, \ldots, r_N}(\gamma_1, \gamma_2, \ldots, \gamma_N)$$

Assuming that the M possible messages are equiprobable, that is, their *a priori* message probabilities are all equal to $P(m_i) = 1/M$ for all i, and noting in (7.12d) that the factors independent of the index "i" do not affect the maximization process, we can further simplify (7.12c) as shown below:

$$\hat{m} = m_k \ \text{ iff } \ p_{r_1, r_2, \ldots, r_N}(\gamma_1, \gamma_2, \ldots, \gamma_N \mid \mathbf{s} = \mathbf{s}_k)$$

$$= \max_i \ \{p_{r_1, r_2, \ldots, r_N}(\gamma_1, \gamma_2, \ldots, \gamma_N \mid \mathbf{s} = \mathbf{s}_i)\} \qquad (7.12e)$$

Note that the right-hand side of (7.12e) is the set of M likelihood functions, and thus the MAP decision criterion is reduced to the maximum likelihood (ML) decision criterion when the input message probabilities are assumed equal. In the next section, we derive a practical receiver scheme that provides a sufficient condition that guarantees the ML decision rule for the practical AWGN channel. From here on we always mean the *ML decision rule* when referring to the optimal receiver decision criterion.

7.1.3 Correlation Receivers

Consider Figure 7.9, which describes an M-ary N-dimensional vector communications system over an additive Gaussian noise channel. We specify the signal vector, the noise vector and the received vector as follows:

$$m \leftrightarrow \mathbf{s} = (s_1, s_2, \ldots, s_N)$$
$$\mathbf{n} = (n_1, n_2, \ldots, n_N) \qquad (7.13a)$$
$$\mathbf{r} = \mathbf{s} + \mathbf{n} = (s_1 + n_1, s_2 + n_2, \ldots, s_N + n_N)$$

Assuming that $\mathbf{n}$ and $\mathbf{s}$ are statistically independent,[4] and that the N components of $\mathbf{n}$ are statistically independent, zero-mean Gaussian random variables, each with variance σ^2, we have the ML decision rule

$$\widehat{m} = m_k \text{ iff } \mathrm{p}_r(\gamma \mid \mathbf{s} = \mathbf{s}_k) = \max_i \{\mathrm{p}_r(\gamma \mid \mathbf{s} = \mathbf{s}_i)\}$$

$$= \max_i \{\mathrm{p}_n(\gamma - \mathbf{s}_i)\} \qquad (7.13b)$$

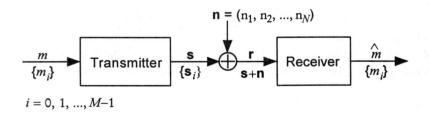

Figure 7.9 M-ary N-dimensional vector communications system.

---

[4] This condition means that $\mathrm{p}_r(\gamma \mid \mathbf{s}) = \mathrm{p}_n(\gamma - \mathbf{s} \mid \mathbf{s}) = \mathrm{p}_n(\gamma - \mathbf{s})$ or $\mathrm{p}_{n|s} = \mathrm{p}_n$.

where

$$p_n(\boldsymbol{\gamma} - \mathbf{s}_i) = p_{n_1, n_2, \ldots, n_N}(\gamma_1 - s_{i1}, \gamma_2 - s_{i2}, \ldots, \gamma_N - s_{iN}) \qquad (7.13\text{c})$$

Using the N independent and identically distributed Gaussian noise vector RVs:

$$p_n(\boldsymbol{\beta}) = p_{n_1, n_2, \ldots, n_N}(\beta_1, \beta_2, \ldots, \beta_N) = \frac{1}{(2\pi\sigma^2)^{N/2}} \exp\left(-\frac{1}{2\sigma^2} \sum_{j=1}^{N} \beta_j^2\right) \quad (7.13\text{d})$$

the ML decision criterion states

$$\hat{m} = m_k \text{ iff} \qquad p_r(\boldsymbol{\gamma} \mid \mathbf{s} = \mathbf{s}_k) = \max_i \{p_n(\boldsymbol{\gamma} - \mathbf{s}_i)\}$$

$$= \max_i \left\{\frac{1}{(2\pi\sigma^2)^{N/2}} \exp\left(-\frac{1}{2\sigma^2} \sum_{j=1}^{N} (\gamma_j - s_{ij})^2\right)\right\}$$

$$\Rightarrow \min_i \left\{\sum_{j=1}^{N} (\gamma_j - s_{ij})^2\right\} = \min_i \left\{|\boldsymbol{\gamma} - \mathbf{s}_i|^2\right\} \qquad (7.13\text{e})$$

in which the dot product of a vector with itself has been recognized as the square of its length. The implication of (7.13e) is that the ML decision criterion (optimal decision rule) is the *minimum distance* decision rule. Thus, conceptually speaking, when the received vector is $\boldsymbol{\gamma} = (\gamma_1, \gamma_2, \ldots, \gamma_N)$ the receiver displays it in an N-dimensional signal space in which the signal vectors $\{\mathbf{s}_i\}$ are already shown as points (known as the signal constellation); then the receiver picks the signal vector among the $\{\mathbf{s}_i\}$ which is the closest to the received vector. This is the concept implied in (7.13e), and we can most conveniently use this geometric interpretation for evaluating conditional error probabilities by drawing decision boundaries geometrically.

Let us take a further look at (7.13e). The ML decision rule can also be stated as follows:

$$\hat{m} = m_k \text{ iff} \qquad p_r(\boldsymbol{\gamma} \mid \mathbf{s} = \mathbf{s}_k) = \max_i \{p_n(\boldsymbol{\gamma} - \mathbf{s}_i)\}$$

$$\Rightarrow \min_i \left\{|\boldsymbol{\gamma} - \mathbf{s}_i|^2\right\}$$

$$= \min_i \left\{|\boldsymbol{\gamma}|^2 - 2\langle \boldsymbol{\gamma}, \mathbf{s}_i\rangle + |\mathbf{s}_i|^2\right\}$$

$$\Rightarrow \max_i \{\langle \boldsymbol{\gamma}, \mathbf{s}_i\rangle\} \qquad (7.13\text{f})$$

where we assume that each signal has the same energy

$$|\mathbf{s}_i|^2 = \int_0^T s_i^2(t)\, dt = E, \quad i = 0, 1, \ldots, M-1 \tag{7.13g}$$

and we note that the term $|\gamma|^2$ is independent of the decision index "i." We have come to observe that the maximum likelihood decision rule is *maximum correlation* decision rule as well as a *minimum distance* decision rule. The correlation operation is defined both in the vector space and in the function (waveform) space:

$$\langle r(t)\,,\, s_i(t)\rangle \triangleq \int_0^T r(t)\, s_i(t)\, dt \tag{7.13h}$$

$$\langle \mathbf{r}\,,\, \mathbf{s}\rangle \triangleq \sum_{j=1}^N r_j\, s_{ij} \tag{7.13i}$$

It is important to note that the vector communications system implements a correlation receiver, shown in Figure 7.10, based on the expression (7.13i), which is a sufficient condition for the ML decision rule.

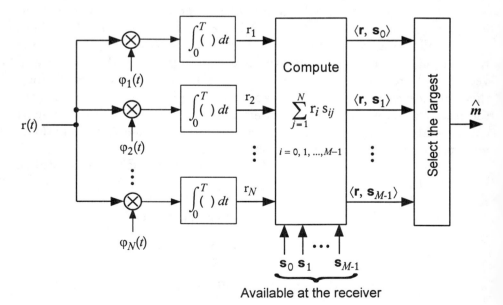

Figure 7.10 Correlation receiver that guarantees the ML decision rule for M-ary communications over N-dimensional vector channels in additive Gaussian noise.

The system in Figure 7.10 extracts the N received vector components $\{r_i\}$ for performing the M discrete-time correlations represented by (7.13j). We can still specify the correlation receiver for the waveform channels by mechanizing the M continuous-time correlations implied by (7.13i), as shown in Figure 7.11. Figures 7.10 and 7.11 are equivalent for the implementation of the ML decision rule. The distinctions between them pertain to their use in real situations and in the choice between *digital signal processing* or *analog signal processing* for the receiver implementation.

7.1.4 Matched Filter Receivers

The correlation receiver shown in Figure 7.10 can be replaced with one in which matched filters replace each of the multipler-integrator operations. Consider a linear filter with impulse response $h_j(t)$ and assume that the input and output are $r(t)$ and $y_j(t)$, respectively, as depicted in Figure 7.12. Then we have

$$y_j(t) = r(t) * h_j(t) = \int_0^t r(\alpha)\, h_j(t-\alpha)\, d\alpha \qquad (7.14a)$$

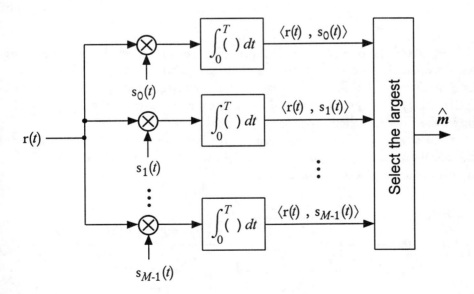

Figure 7.11 Correlation receiver that guarantees the ML decision rule for M-ary communications over the M-ary waveform channel in additive Gaussian noise.

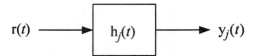

Figure 7.12 A linear time-invariant filter with impulse response $h_j(t)$.

Now, assume that the ON functions $\{\varphi_j(t)\}$ used in Figure 7.10 are identically zero outside the time interval $(0, T)$ and set

$$h_j(t) = \varphi_j(T - t) \tag{7.14b}$$

If we put (7.14b) in (7.14a), the filter output is

$$y_j(t) = \int_0^t r(\alpha)\, h_j(t - \alpha)\, d\alpha = \int_0^t r(\alpha)\, \varphi_j(T - t + \alpha)\, d\alpha \tag{7.14c}$$

and the output sampled at $t = T$ is

$$y_j(T) = \int_0^T r(\alpha)\, \varphi_j(\alpha)\, d\alpha \triangleq r_j \tag{7.14d}$$

That is, (7.14d) is exactly the operation being performed in the jth branch of the correlation receiver in Figure 7.10 to produce the received vector component r_j. The filter $h_j(t)$ that is a mirror image of $\varphi_j(t)$ with a delay of T; that is, $\varphi_j(T - t)$ as shown in (7.14b), is called a *matched filter*. Figure 7.13 shows the filter matched to $\varphi_j(t)$, $0 \le t \le T$.

An alternative implementation for the ML optimum receiver in lieu of the configuration of Figure 7.10 is shown in Figure 7.14, where the correlators are replaced by matched filters.

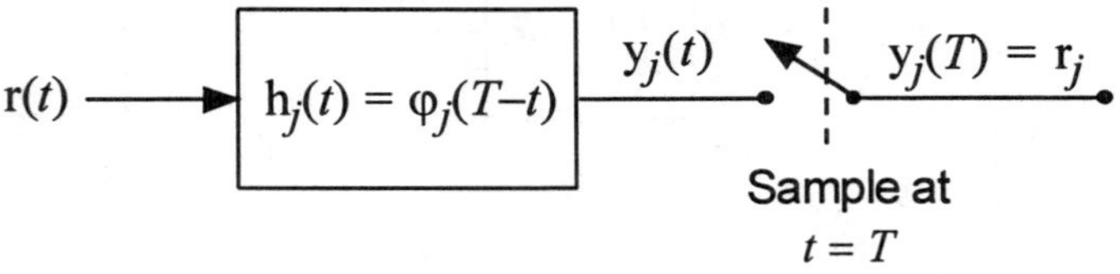

$$r(t) \longrightarrow \boxed{h_j(t) = \varphi_j(T{-}t)} \xrightarrow{\;y_j(t)\;} \diagdown \quad y_j(T) = r_j$$

Sample at
$t = T$

Figure 7.13 The matched filter for $\varphi_j(t)$, $0 \le t \le T$.

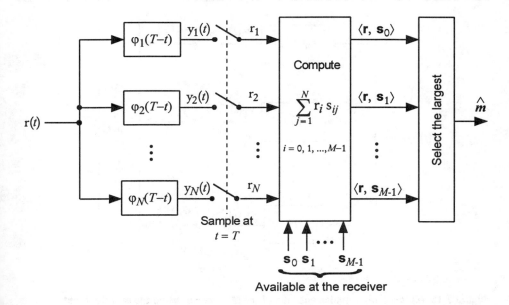

Figure 7.14 The matched filter receiver that guarantees the ML decision rule for M-ary communications over N-dimensional vector channels in additive Gaussian noise.

The choice between the correlation receiver configuration of Figure 7.10 and the matched filter receiver configuration of Figure 7.14 depends to a large extent on the choice between hardware designs for multipliers and filters. Under ideal conditions, both perform exactly the same way.

The correlation receiver using the signal waveforms $\{s_i(t)\}$, shown in Figure 7.11, can also be replaced with one using matched filters, as illustrated in Figure 7.15.

Let us examine the matched filter shown in Figure 7.16, which is a representation of a single branch of the receiver in Figure 7.15. We assume that the input $r(t)$ consists of the signal $s(t)$ and an AWGN component with two-sided spectral power density $\mathcal{N}_0/2$; that is:

$$r(t) = s(t) + n(t) \tag{7.15a}$$

with $s(t)$ defined on $0 \leq t \leq T$ and $\mathcal{S}_n(f) = \mathcal{N}_0/2$, $|f| < \infty$. Because

$$y(t) = r(t) * h(t) = [s(t) + n(t)] * h(t) = y_s(t) + y_n(t) \tag{7.15b}$$

we have

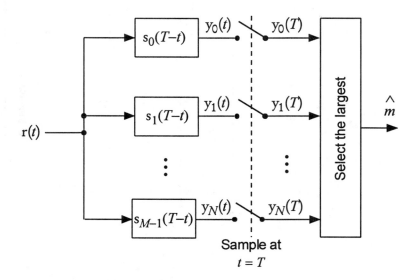

Figure 7.15 Matched filter implementation of the ML receiver using the waveform set $\{s_i(t), 0 \le t \le T\}, i = 0, 1, \ldots, M - 1$.

Figure 7.16 Matched filter receiver matched to the signal $s(t), 0 \le t \le T$.

$$y_s(t) \triangleq \int_{-\infty}^{\infty} s(\alpha)\, h(t - \alpha)\, d\alpha = \int_{-\infty}^{\infty} s(t - \alpha)\, h(\alpha)\, d\alpha \qquad (7.15c)$$

$$y_n(t) \triangleq \int_{-\infty}^{\infty} n(\alpha)\, h(t - \alpha)\, d\alpha = \int_{-\infty}^{\infty} n(t - \alpha)\, h(\alpha)\, d\alpha \qquad (7.15d)$$

The output SNR is

$$\left(\frac{S}{N}\right)_o \equiv (SNR)_o = \frac{E\{y_s^2(T)\}}{E\{y_n^2(T)\}} = \frac{y_s^2(T)}{y_n^2(T)} \qquad (7.15e)$$

From (7.15c) and (7.15d):

$$y_s^2(T) = \left[\int_{-\infty}^{\infty} s(t - \alpha)\, h(\alpha)\, d\alpha \right]^2 \tag{7.15f}$$

$$\overline{y_n^2(T)} = \int_{-\infty}^{\infty} \int_{-\infty}^{\infty} \overline{n(t - \alpha)\, n(t - \beta)}\, h(\alpha)\, h(\beta)\, d\alpha\, d\beta$$

$$= \int_{-\infty}^{\infty} \int_{-\infty}^{\infty} \tfrac{1}{2}\mathcal{N}_0\, \delta(\alpha - \beta)\, h(\alpha)\, h(\beta)\, d\alpha\, d\beta = \tfrac{1}{2}\mathcal{N}_0 \int_{-\infty}^{\infty} h^2(\alpha)\, d\alpha \tag{7.15g}$$

Thus, we have

$$\left(\frac{S}{N}\right)_o = \frac{y_s^2(T)}{E\{y_n^2(T)\}} = \frac{\left[\int_{-\infty}^{\infty} s(t - \alpha)\, h(\alpha)\, d\alpha \right]^2}{\tfrac{1}{2}\mathcal{N}_0 \int_{-\infty}^{\infty} h^2(\alpha)\, d\alpha} \tag{7.15h}$$

Using the *Schwarz inequality*, which is

$$\left[\int_{-\infty}^{\infty} a(t)\, b(t)\, dt \right]^2 \le \left[\int_{-\infty}^{\infty} a^2(t)\, dt \right] \left[\int_{-\infty}^{\infty} b^2(t)\, dt \right] \tag{7.15i}$$

with equality iff $a(t) = b(t)$, (7.15h) can now be expressed as

$$\left(\frac{S}{N}\right)_o \le \frac{\left[\int_{-\infty}^{\infty} s^2(t - \alpha)\, d\alpha \right]\left[\int_{-\infty}^{\infty} h^2(\alpha)\, d\alpha \right]}{\tfrac{1}{2}\mathcal{N}_0 \int_{-\infty}^{\infty} h^2(\alpha)\, d\alpha} = \frac{2}{\mathcal{N}_0} \int_{-\infty}^{\infty} s^2(\alpha)\, d\alpha \tag{7.15j}$$

Because $\int_{-\infty}^{\infty} s^2(\alpha)\, d\alpha = E$, we have

$$\left(\frac{S}{N}\right)_o \le \frac{2E}{\mathcal{N}_0} \tag{7.15k}$$

Note the equality condition of (7.15i); we see that when $h(t) = s(T - t)$ (i.e., the filter is matched to the signal), we have

$$\left(\frac{S}{N}\right)_{o,\max} = \frac{2E}{\mathcal{N}_0} \tag{7.16}$$

The matched filter, therefore, is not only equivalent to the correlator, but it is also optimum in the sense that the output gives the maximal output SNR, $(S/N)_{o,\,max}$.

Another remark on the matched filter is in order: If we take the Fourier transform of $h(t) = s(T - t)$, we obtain

$$H(f) = \mathcal{F}\{h(t)\} = \mathcal{F}\{s(T - t)\} = \int_{-\infty}^{\infty} s(T - t)\, e^{-j2\pi ft}\, dt$$

$$= \int_{-\infty}^{\infty} s(\alpha)\, e^{-j2\pi f(T-\alpha)}\, d\alpha = e^{-j2\pi fT} \int_{-\infty}^{\infty} s(\alpha)\, e^{j2\pi f\alpha}\, d\alpha$$

$$= S^*(f)\, e^{-j2\pi fT} \tag{7.17}$$

Note that the Fourier transform of the filter matched to the signal $s(t)$, $0 \le t \le T$, is the complex conjugate of the Fourier transform of the signal and, for this reason, the matched filter is also known as a *conjugate filter*.

Example 7.4 Let us consider that the input signal to the conjugate filter is $s(t) = \cos 2\pi f_0 t$, $0 \le t \le T$, with $f_0 = n/T$, n integer (T is an integer number of periods of f_0). Then $h(t) = s(T - t) = \cos[2\pi f_0(T - t)]$ and

$$y(t) = \int_{-\infty}^{\infty} s(\alpha)\, s(T - t + \alpha)\, d\alpha$$

$$= \begin{cases} \displaystyle\int_0^t \cos(2\pi f_0\alpha)\cos[2\pi f_0(T - t + \alpha)]\, d\alpha, & 0 \le t \le T \\[2ex] \displaystyle\int_{t-T}^T \cos(2\pi f_0\alpha)\cos[2\pi f_0(T - t + \alpha)]\, d\alpha, & T \le t \le 2T \end{cases}$$

$$= \begin{cases} \frac{1}{2}\left[t\cos(2\pi f_0 t) + \frac{1}{2\pi f_0}\sin(2\pi f_0 t)\right], & 0 \le t \le T \\[2ex] \frac{1}{2}\left[(2T - t)\cos(2\pi f_0 t) - \frac{1}{2\pi f_0}\sin(2\pi f_0 t)\right], & T \le t \le 2T \end{cases} \tag{7.18}$$

The output of the matched filter is shown in Figure 7.17 for the case of $f_0 T = 5$. Note that the output is maximum at the instant $t = T$.

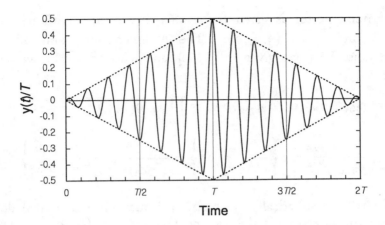

Figure 7.17 Output of filter matched to $\cos 2\pi f_0 t$, $0 \le t \le T$, with the maximum occurring at $t = T$.

7.1.5 Performance Evaluations for M-ary Communications in AWGN

We discussed ML receiver implementations in Section 7.1.3 and showed that there are two ways of accomplishing the optimal receiver signal processing at the receiver to render the optimal decision ($\hat{m}$) as to the transmitted message. They were the *minimum distance* decision rule and the *maximum correlation* decision rule, which could also be alternatively implemented by using matched filters. For digital communications, the performance measure of such optimal receivers is the *probability of error*. The decision rules are signal processing schemes implementing sufficient conditions that assure the optimality criteria, but they themselves do not reveal how well they perform in terms of the probability of error for a given M-ary communications system over an AWGN channel.

We have seen that the signals $\{s_i(t)\}$ are represented as M points $\{s_i\}$ in an N-dimensional vector space in an M-ary communications system. The minimum distance decision rule dictated the computation of the distance of the received vector $\mathbf{r} = \gamma$ from each of the signal vectors $\{s_i\}$, and the message decision is made on the basis of the minimum distance (shortest distance).

Decision regions and error probability calculations. Let us consider an M-ary communications system in an N-dimensional signal vector space: The M signal vectors are given as $\{s_i\}$, $i = 0, 1, \dots, M-1$ and the N-dimensional

Gaussian noise vector is given by $\mathbf{n} = (n_1, n_2, \ldots, n_N)$ with $\overline{n_j} = 0$, $j = 1, 2, \ldots, N$ and $\overline{n_j n_l} = \left(\frac{1}{2}\mathcal{N}_0\right)\delta_{jl}$. The N-dimensional joint pdf of the Gaussian noise RVs was given in (7.13d), and with $\sigma^2 = \frac{1}{2}\mathcal{N}_0$ we have the joint pdf

$$p_{\mathbf{n}}(\boldsymbol{\alpha}) = p_{n_1, n_2, \ldots, n_N}(\alpha_1, \alpha_2, \ldots, \alpha_N) = \prod_{j=1}^{N} p_{n_j}(\alpha_j) = \prod_{j=1}^{N} \frac{1}{\sqrt{\pi\mathcal{N}_0}} e^{-\alpha_j^2/\mathcal{N}_0}$$

$$= \frac{1}{(\pi\mathcal{N}_0)^{N/2}} \exp\left\{-\frac{1}{\mathcal{N}_0}\sum_{j=1}^{N}\alpha_j^2\right\} = \frac{1}{(\pi\mathcal{N}_0)^{N/2}} \exp\left\{-\frac{|\boldsymbol{\alpha}|^2}{\mathcal{N}_0}\right\} \quad (7.19a)$$

The pdf (7.19a) is spherically symmetric [1], meaning that $p_{\mathbf{n}}(\boldsymbol{\alpha})$ does not depend on the direction of the argument vector $\boldsymbol{\alpha} = (\alpha_1, \alpha_2, \ldots, \alpha_N)$, but only on the magnitude $|\boldsymbol{\alpha}| = \sqrt{\alpha_1^2 + \alpha_2^2 + \cdots + \alpha_N^2}$. Figure 7.18 shows an example of signal vectors $\{\mathbf{s}_i\}$ displayed as points in a two-dimensional (rectangular) vector space.

The signal vectors $\{\mathbf{s}_i\}$ shown in Figure 7.18 are known and available at the receiver, and, hence, when the received vector $\mathbf{r} = \boldsymbol{\gamma} = (\gamma_1, \gamma_2, \ldots, \gamma_N)$ is also displayed as a point in the vector space of that figure, the distance $|\boldsymbol{\gamma} - \mathbf{s}_i|$ is measured for all $i = 0, 1, \ldots, M - 1$. The performance evaluation for the system, however, must be based on the conditional probability of error or the conditional probability of correct decision. Let $P(C \mid m_i)$ denote the conditional probability of *correct decision*, given that message m_i (and hence $\mathbf{s}_i$) is transmitted. In Figure 7.18, we depict a situation where the correct decision

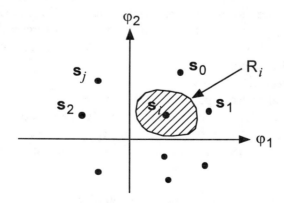

Figure 7.18 Signal vectors shown in two-dimensional signal vector space.

region R_i is shown as a shaded area for the signal vector s_i. That is, whenever the received vector $r = \gamma$ falls in the shaded region, the correct decision is made. Thus

$$P(C \mid m_i) = \text{Pr}\{r \text{ falls in region } R_i \mid m_i\} = \text{Pr}\{r \in R_i \mid m_i\} \qquad (7.19b)$$

Hence, the conditional probability of error $P(e \mid m_i)$ given that $m = m_i$ (message m_i is transmitted) is

$$P(e \mid m_i) = \text{Pr}\{r \text{ falls outside region } R_i \mid m_i\} = \text{Pr}\{r \notin R_i \mid m_i\} \qquad (7.19c)$$

Before we proceed further, we need to point out that the both conditional probability of error and the conditional probability of correct decision $\text{Pr}\{r = \gamma \in R_i \mid m_i\}$ are invariant to rotation and translation of vector coordinates [1], and this fact is illustrated in Figure 7.19. It should be noted, however, that the translation changes the distance of the vector from the origin and, hence, it affects (changes) the signal energy:

$$s_i \rightarrow s_i - a \qquad \text{means} \qquad |s_i|^2 \neq |s_i' - a|^2 \qquad (7.19d)$$

Assuming that we calculated the conditional probability of correct decision, given $m = m_i$, $i = 0, 1, \ldots, M - 1$, we then obtain the total unconditional probability of correct decision $P(C)$ by averaging over all message probabilities $\{P(m_i)\}$:

$$P(C) = \sum_{i=0}^{M-1} P(m_i) P(C \mid m_i) \qquad (7.20a)$$

Had we calculated the conditional probabilities of error $\{P(e \mid m_i)\}$, the total probability of error $P(e)$ is obtained by averaging them over all message probabilities:

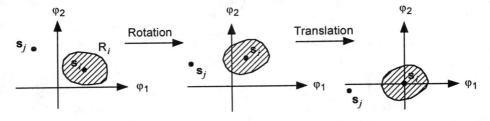

Figure 7.19 Illustration of conditional probability of correct decision being invariant to rotation and translation of coordinates.

$$P(e) = \sum_{i=0}^{M-1} P(m_i) P(e \mid m_i) \qquad (7.20b)$$

Because

$$P(C) + P(e) = 1 \qquad (7.20c)$$

we can also obtain the performance measure (i.e., the probability of error) by first obtaining the probability of correct decision:

$$P(e) = 1 - \sum_{i=0}^{M-1} P(m_i) P(C \mid m_i) \qquad (7.20d)$$

Example 7.5 Binary communication over an AWGN channel. Consider the two signal vectors shown in Figure 7.4, redrawn here in Figure 7.20(a). Note that two signal vectors are separated by the distance d, and each is distance $d/2$ from the origin, where

$$d \triangleq |s_0 - s_1| = \sqrt{\langle s_0 - s_1, s_0 - s_1 \rangle} = \left[\int_{-\infty}^{\infty} [s_0(t) - s_1(t)]^2 dt \right]^{1/2} \qquad (7.21a)$$

We wish to compute the probability of error for the binary communications system with each message probability equal to $\frac{1}{2}$. Suppose we rotate the signal set $90°$, so that the signal vectors lie on the φ_1 axis as shown in Figure 7.20(b). Now, recall from (7.13e) that the ML optimum decision rule is based on the minimum distance decoding criterion $\min_i |\gamma - s_i|^2$. First, let us consider the two vectors in Figure 7.20(a). Because the φ_1 axis is the locus of all points equidistant from s_0 and s_1, the decision rule is

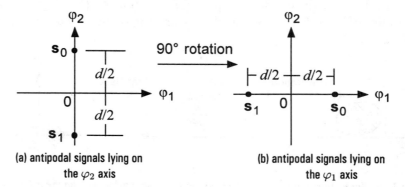

(a) antipodal signals lying on
the φ_2 axis

(b) antipodal signals lying on
the φ_1 axis

Figure 7.20 Binary signal vectors for equiprobable messages (examples of *antipodal* signals).

$$\widehat{m} = \begin{cases} m_0 \text{ iff } \mathbf{r} = \boldsymbol{\gamma} \text{ falls in the half-plane above the } \varphi_1 \text{ axis} \\ m_1 \text{ iff } \mathbf{r} = \boldsymbol{\gamma} \text{ falls in the half-plane below the } \varphi_1 \text{ axis} \end{cases}$$

Thus, the conditional error event, given m_0 is transmitted, is the received vector $\mathbf{r} = \boldsymbol{\gamma} = \mathbf{s}_0 + \mathbf{n}$ falling in the lower half-plane, below the φ_1 axis. The evaluation of $P(e \mid m_i)$ can be accomplished most conveniently if we assume that a translation is done in such a way that $\mathbf{s}_i$ is located at the origin $(0,0)$ of a new coordinate system. (This assumption is tantamount to drawing a new coordinate system with origin located at the signal position.) This is an application of the earlier statement that rotation and translation do not affect the error probability or the probability of correct decision. Therefore, the conditional probability of error $P(e \mid m_0)$ is given by

$$P(e \mid m_0) = \Pr\{\boldsymbol{\gamma} \in \text{ half-plane below } \varphi_1 \text{ axis} \mid \mathbf{s} = \mathbf{s}_0\}$$
$$= \Pr\{n_2 < -d/2\} = \Pr\{n_2 > d/2\} \qquad (7.21\text{b})$$

Similarly, when m_1 is sent, an error occurs if $\mathbf{r} = \boldsymbol{\gamma}$ is in the upper half-plane, with probability given by

$$P(e \mid m_1) = \Pr\{\boldsymbol{\gamma} \in \text{ half-plane above } \varphi_1 \text{ axis} \mid \mathbf{s} = \mathbf{s}_1\}$$
$$= \Pr\{n_2 > d/2\} \qquad (7.21\text{c})$$

But the pdf of n_i is given by

$$p_{n_i}(\alpha) = \frac{1}{\sqrt{2\pi}\sigma} e^{-\alpha^2/2\sigma^2} = \frac{1}{\sqrt{\pi\mathcal{N}_0}} e^{-\alpha^2/\mathcal{N}_0}, \quad i = 1, 2 \qquad (7.22\text{a})$$

Hence, we have

$$P(e \mid m_0) = P(e \mid m_1) = \Pr\{n_2 > d/2\} = \int_{d/2}^{\infty} p_{n_2}(\alpha)\, d\alpha$$

$$= \int_{d/2}^{\infty} \frac{1}{\sqrt{\pi\mathcal{N}_0}} e^{-\alpha^2/\mathcal{N}_0}\, d\alpha \triangleq Q\left(\frac{d}{\sqrt{2\mathcal{N}_0}}\right) = Q\left(\frac{d}{2\sigma}\right) \qquad (7.22\text{b})$$

where $Q(\cdot)$ is the Gaussian Q-function. For the equally likely message assumption, we have the total unconditional probability of error

$$P(e) = \sum_{i=0}^{1} P(m_i) P(e \mid m_i) = \tfrac{1}{2}[P(e \mid m_0) + P(e \mid m_1)]$$

$$= \frac{1}{2}\left[Q\left(\frac{d}{\sqrt{2\mathcal{N}_0}}\right) + Q\left(\frac{d}{\sqrt{2\mathcal{N}_0}}\right) \right] = Q\left(\frac{d}{\sqrt{2\mathcal{N}_0}}\right) \qquad (7.22c)$$

Had we considered the two signals shown in Figure 7.20(b), we would have calculated the identical conditional error probabilities

$$P(e \mid m_0) = \Pr\{n_1 < -d/2\} = \Pr\{n_1 > d/2\} = Q\left(\frac{d}{\sqrt{2\mathcal{N}_0}}\right) \qquad (7.22d)$$

and

$$P(e \mid m_1) = \Pr\{n_1 > d/2\} = Q\left(\frac{d}{\sqrt{2\mathcal{N}_0}}\right) \qquad (7.22e)$$

Thus, from (7.22a), the unconditional total probability of error equals the conditional probability of error, given by

$$P(e) = Q\left(\frac{d}{\sqrt{2\mathcal{N}_0}}\right) = Q\left(\frac{|s_0 - s_1|}{\sqrt{2\mathcal{N}_0}}\right) = Q\left(\sqrt{\frac{|s_0 - s_1|^2}{2\mathcal{N}_0}}\right) \qquad (7.23)$$

which is the probability of error for any binary (set of $M = 2$) signal vectors with mutual distance $d = |s_0 - s_1|$, no matter how they are located in the signal space. Further, the error probability shown in (7.23) is the minimum error probability because it is computed on the basis of the optimum (maximum likelihood) decision criterion. We previously called the antipodal signals BPSK signals and assumed that signal energy is $|s_i|^2 = E$, $i = 0, 1$ for each bit of the binary message. Thus, $d = 2\sqrt{E}$ and we have the well-known bit error probability or bit error rate (BER) expression

$$P_{BPSK}(e) = Q\left(\sqrt{\frac{2E}{\mathcal{N}_0}}\right) \qquad (7.24)$$

Example 7.6 Binary orthogonal signals. Figure 7.21 shows two orthogonal signal vectors with the distance

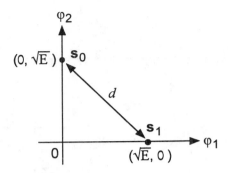

Figure 7.21 Binary orthogonal signals.

$$d = \left[\left(\sqrt{E}\right)^2 + \left(\sqrt{E}\right)^2 \right]^{1/2} = \sqrt{2E}$$

Using the "formula" given in (7.23), we have the probability of error for binary orthogonal (BO) signals

$$P_{BO}(e) = Q\left(\sqrt{\frac{|s_0 - s_1|^2}{2N_0}} \right) = Q\left(\sqrt{\frac{2E}{2N_0}} \right) = Q\left(\sqrt{\frac{E}{N_0}} \right) \qquad (7.25)$$

Note that, when compared with the antipodal signals of the BPSK system, the binary orthogonal system is 3 dB poorer than the BPSK system in BER performance. This comparison is shown in Figure 7.22.

7.1.6 Union Bound on the Probability of Error of M-ary Communications Systems

The exact computation of the probability of error for an M-ary communications system can be tedious and difficult at times when the signal constellation is such that the conditional probabilities of error for some signal vectors do not lend themselves to simple calculation. Hence, an upper bound on the probability of error can be an appropriate measure of system performance, just as the Chernoff bound and Central Limit Theorem are used in lieu of the exact calculation of required probability measures. The *union bound* on the

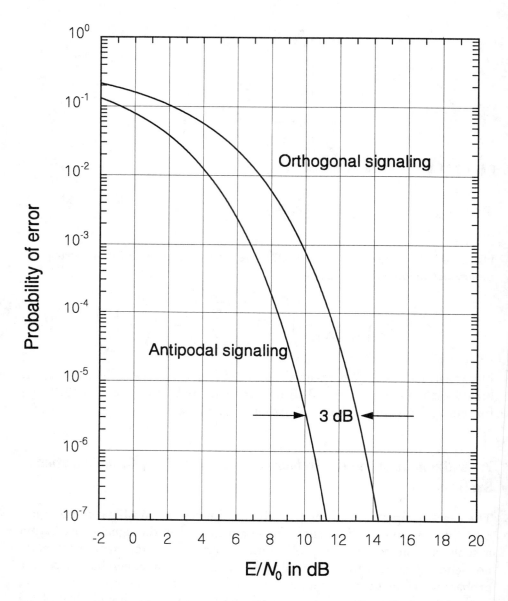

Figure 7.22 Probabilities of error for binary antipodal and binary orthogonal systems.

probability of error for an M-ary communications system is an important tool for performance analysis, and we derive this bound in this section.

Consider the signal constellation shown in Figure 7.23. Let us assume that the signal vector s_k is transmitted, and that the received vector is r. Then, according to the minimum distance decision rule, the receiver would make a wrong decision if the received vector r is closer to a signal vector other than s_k, the transmitted vector. To express this error event more succinctly, let us introduce a simple event notation as follows: Define

$$E_r(s_i, s_k) \triangleq \text{Event that the received vector } r \text{ is closer to the signal}$$
$$\text{vector } s_k \text{ than to the transmitted vector } s_i \quad (7.26a)$$

Clearly

$$E_r(s_i, s_k) \triangleq \text{Event}\{|r - s_k| < |r - s_i|\} \quad (7.26b)$$

Thus, given that the message m_i is transmitted (i.e., $m = m_i$), a (conditional) error event occurs whenever the received vector r is closer to at least one (or more) signal s_k, $k \neq i$ than it is to the transmitted signal vector s_k. The conditional probability of error $P(e \mid m_i)$ can then be written as the probability of the union of events:

$$P(e \mid m_i) = \text{Pr}\{E_r(s_i, s_0) \cup E_r(s_i, s_1) \cup \cdots \cup E_r(s_i, s_{i-1})$$
$$\cup E_r(s_i, s_{i+1}) \cup \cdots \cup E_r(s_i, s_{M-1})\} \quad (7.26c)$$

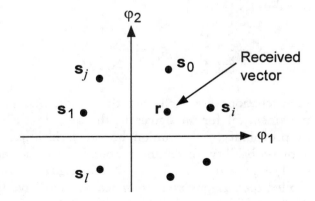

Figure 7.23 Signal constellation of $\{s_i\}$ for an M-ary communications system in a two-dimensional signal space.

The probability of the union of events A and B was given in (1.43d) as

$$P(A \cup B) = P(A) + P(B) - P(A \cap B) \leq P(A) + P(B)$$

where $A \cap B$ is the intersection of the events A and B. In the M-ary communications situation, if we ignore the intersection events of the events $\{E_{\mathbf{r}}(\mathbf{s}_i, \mathbf{s}_k)\}$ for all $k \neq i$, (7.26c) can be bounded above by the sum of the probabilities of the events $\{E_{\mathbf{r}}(\mathbf{s}_i, \mathbf{s}_k)\}$ for all $k \neq i$. Thus

$$P(e \mid m_i) \leq \sum_{\substack{k=0 \\ k \neq i}}^{M-1} P\big[E_{\mathbf{r}}(\mathbf{s}_i, \mathbf{s}_k)\big] = \sum_{\substack{k=0 \\ k \neq i}}^{M-1} \Pr\{|\mathbf{r} - \mathbf{s}_k| < |\mathbf{r} - \mathbf{s}_i| \mid m_i\} \qquad (7.26\text{d})$$

This is the union bound on the conditional probability of error. Note that each term depends only on the two signal vectors $\mathbf{s}_i$ and $\mathbf{s}_k$, and therefore we can use (7.23) to identify the probability of the kth term as

$$\Pr\{|\mathbf{r} - \mathbf{s}_k| < |\mathbf{r} - \mathbf{s}_i| \mid m_i\} = Q\left(\frac{|\mathbf{s}_i - \mathbf{s}_k|}{\sqrt{2\mathcal{N}_0}}\right) \qquad (7.27\text{a})$$

so that

$$P(e \mid m_i) \leq \sum_{\substack{k=0 \\ k \neq i}}^{M-1} Q\left(\frac{|\mathbf{s}_i - \mathbf{s}_k|}{\sqrt{2\mathcal{N}_0}}\right) \qquad (7.27\text{b})$$

The union bound on the total unconditional probability of error is obtained by averaging the conditional probability of error over all messages with their message prior probabilities $\{P(m_i)\}$. Thus

$$P(e) \leq \sum_{i=0}^{M-1} \sum_{\substack{k=0 \\ k \neq i}}^{M-1} P(m_i) \, Q\left(\frac{|\mathbf{s}_i - \mathbf{s}_k|}{\sqrt{2\mathcal{N}_0}}\right) \qquad (7.27\text{c})$$

Certain signal constellations are such that the conditional probability of correct decision is identical for all signals in the set. In such a case, the *conditional* error probability, or its union bound, is identical to the total *unconditional* error probability, or its union bound, for equal message probabilities $P(m_i) = 1/M$, $i = 0, 1, \ldots, M - 1$. The signal constellations of these types are called *completely symmetric signal sets* and have the property that "any relabeling of signal points can be undone by a rotation of coordinates, translation, and/or inversion of axis" [1].

Example 7.7 The five signal vectors shown in Figure 7.24 are to be used for a 5-ary communications system over an AWGN channel with two-sided power spectral density $N_0/2$. Find the probability of symbol error $P(e)$ assuming equal message probabilities.

Solution: To obtain the probability of error, we first consider calculating the conditional probability of correct decision $P(C \mid m_i)$ given that message m_i is transmitted. Thus, when the received vector is $\mathbf{r} = \boldsymbol{\gamma} = (\gamma_1, \gamma_2)$, we need to compute the probabilities

$$P(C \mid m_i) = \Pr\{\mathbf{r} = \boldsymbol{\gamma} \in R_i \mid m = m_i\} \tag{7.28a}$$

where R_i is the correct-decision region for signal s_i. We can see that the calculations for the probabilities (7.28a) are somewhat difficult because of the shape of the decision regions. Therefore, it is prudent to consider the union bound calculation for the probability of error measure of this 5-ary communications system. Using the geometry of Figure 7.25, we first compute the upper bound on $P(e \mid m_0)$ using the conditional union bound "formula" (7.27b):

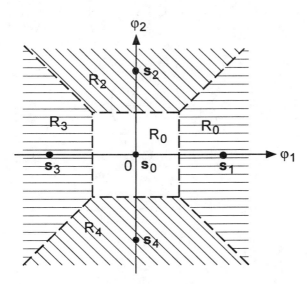

Figure 7.24 Five signal vectors $\{s_i\}$, $i = 0, 1, 2, 3, 4$ with decision regions $\{R_i\}$ separated by dotted lines.

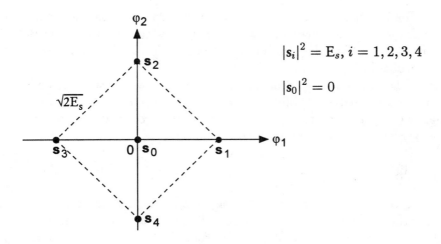

Figure 7.25 Five signal vectors for union-bound error probability calculation.

$$P(e \mid m_0) \leq \sum_{k=1}^{4} Q\left(\frac{|s_k - s_0|}{\sqrt{2\mathcal{N}_0}}\right) = \sum_{k=1}^{4} Q\left(\sqrt{\frac{|s_k|^2}{2\mathcal{N}_0}}\right) = 4Q\left(\sqrt{\frac{E_s}{2\mathcal{N}_0}}\right) \quad (7.28\text{b})$$

The upper bound for the case of $m = m_1$ is

$$P(e \mid m_1) \leq \sum_{\substack{k=0 \\ k \neq 1}}^{4} Q\left(\frac{|s_k - s_1|}{\sqrt{2\mathcal{N}_0}}\right) \quad (7.28\text{c})$$

where

$$|s_1 - s_0|^2 = |s_1|^2 = E_s$$

$$|s_1 - s_2|^2 = |s_1 - s_4|^2 = 2E_s$$

$$|s_1 - s_3|^2 = \left(2\sqrt{E_s}\right)^2 = 4E_s$$

Thus

$$P(e \mid m_1) \leq \sum_{\substack{k=0 \\ k \neq 1}}^{4} Q\left(\frac{|s_k - s_1|}{\sqrt{2\mathcal{N}_0}}\right)$$

$$= Q\left(\sqrt{\frac{E_s}{2\mathcal{N}_0}}\right) + 2Q\left(\sqrt{\frac{E_s}{\mathcal{N}_0}}\right) + Q\left(\sqrt{\frac{2E_s}{\mathcal{N}_0}}\right) \quad (7.28\text{d})$$

It is clear that the signal constellation $\{s_1, s_2, s_3, s_4\}$, not including s_0, is a *completely symmetric* constellation, and thus

$$P(e \mid m_i) = P(e \mid m_1), \quad i = 1, 2, 3, 4 \tag{7.28e}$$

The total unconditional union bound on the probability of error is therefore given by

$$P(e) \leq \sum_{i=0}^{4} \sum_{\substack{k=0 \\ k \neq i}}^{4} P(m_i) Q\left(\sqrt{\frac{|s_i - s_k|^2}{2\mathcal{N}_0}}\right)$$

$$= \frac{4}{5} Q\left(\sqrt{\frac{E_s}{2\mathcal{N}_0}}\right) + \frac{4}{5}\left[Q\left(\sqrt{\frac{E_s}{2\mathcal{N}_0}}\right) + 2Q\left(\sqrt{\frac{E_s}{\mathcal{N}_0}}\right) + Q\left(\sqrt{\frac{2E_s}{\mathcal{N}_0}}\right)\right]$$

$$= \frac{8}{5} Q\left(\sqrt{\frac{E_s}{2\mathcal{N}_0}}\right) + \frac{8}{5} Q\left(\sqrt{\frac{E_s}{\mathcal{N}_0}}\right) + \frac{4}{5} Q\left(\sqrt{\frac{2E_s}{\mathcal{N}_0}}\right) \tag{7.28f}$$

Example 7.8 Consider an M-ary communications system of equal-energy orthogonal signals $\{s_i\}$, $i = 0, 1, \ldots, M - 1$, to be used for communication over an AWGN channel with variance $\mathcal{N}_0/2$. All messages are equiprobable. The M signals can be represented as points, one each on the $N = M$ mutually orthogonal axes, as depicted in Figure 7.26 for the example of three orthogonal signals. For an $N = M$-dimensional signal space, the M-ary signals are represented as

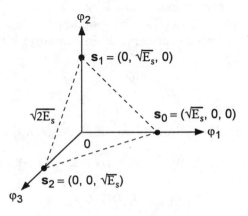

Figure 7.26 Three orthogonal signals in three-dimensional vector space.

$$\mathbf{s}_0 = \left(\sqrt{E_s},\, 0,\, 0,\, \dots,\, 0 \right)$$
$$\mathbf{s}_1 = \left(0,\, \sqrt{E_s},\, 0,\, \dots,\, 0 \right)$$
$$\vdots$$
$$\mathbf{s}_{M-1} = \left(0,\, 0,\, 0,\, \dots,\, \sqrt{E_s} \right) \tag{7.29a}$$

The exact calculation for the probability of error is shown in most of the standard textbooks on modulation [1]. A closed-form expression is not available. The signal set shown in Figure 7.26 is an example of a so-called completely symmetrical signal set, and, hence, for these M equally likely, equal-energy signals, the union bound is easily obtained to be

$$P(e) \le \sum_{i=0}^{M-1} \sum_{\substack{k=0 \\ k \ne i}}^{M-1} P(m_i)\, Q\left(\sqrt{\frac{|\mathbf{s}_i - \mathbf{s}_k|^2}{2\mathcal{N}_0}} \right) = \sum_{i=0}^{M-1} \sum_{\substack{k=0 \\ k \ne i}}^{M-1} \frac{1}{M}\, Q\left(\sqrt{\frac{E_s}{\mathcal{N}_0}} \right)$$

or

$$P(e) \le (M-1)\, Q\left(\sqrt{\frac{E_s}{\mathcal{N}_0}} \right) \tag{7.29b}$$

Using the approximation for the Q-function $Q(\alpha)$ in (1.49f), we have

$$P(e) < M\, e^{-E_s/2\mathcal{N}_0} \tag{7.29c}$$

In an M-ary communications system, each of the M messages $\{m_i\}$, $i = 0, 1, \dots, M-1$, can be represented by a K-tuple binary sequence with $M = 2^K$. The waveforms $\{s_i(t)\}$ corresponding to the signal vectors $\{\mathbf{s}_i\}$ are symbols, and each symbol represents K information bits. The bit rate, denoted R_b, is given by

$$R_b = \frac{K}{T} = \frac{\log_2 M}{T} \tag{7.30a}$$

where T is the symbol duration. If we substitute (7.30a) in (7.29c), we have

$$P(e) < M\, e^{-E_s/2\mathcal{N}_0} = 2^K e^{-E_s/2\mathcal{N}_0} = 2^{R_b T} e^{-E_s/2\mathcal{N}_0} \tag{7.30b}$$

Because

$$E_s = K E_b = T P_s \tag{7.30c}$$

where E_b is the energy per information bit and P_s is the average power, (7.30b) can be expressed as

$$P(e) < e^{K\ln 2}\, e^{-E_s/2\mathcal{N}_0} = \exp\left\{-K\left(\frac{E_b}{\mathcal{N}_0} - \ln 2\right)\right\} \qquad (7.30d)$$

or

$$P(e) < 2^{R_b T}\, e^{-TP_s/2\mathcal{N}_0} = e^{R_b T\ln 2 - TP_s/2\mathcal{N}_0}$$

$$= \exp\left\{-T\left(\frac{P_s}{2\mathcal{N}_0} - R_b\ln 2\right)\right\} \qquad (7.30e)$$

Let us now look at the conditions under which the probability of error $P(e)$ approaches zero. From (7.30d), we see that $P(e) \to 0$ as $K \to \infty$ under the condition of

$$\frac{E_b}{\mathcal{N}_0} > 2\ln 2 \simeq 1.4\,\mathrm{dB} \qquad (7.31a)$$

Also, (7.30e) states that $P(e) \to 0$ as $T \to \infty$ of the bit rate R_b meets the condition that

$$R_b < \frac{P_s}{\mathcal{N}_0}\left(\frac{1}{2\ln 2}\right) \simeq 0.7\,\frac{P_s}{\mathcal{N}_0} \qquad (7.31b)$$

The interpretations of (7.31a) and (7.31b) are important and significant in understanding the asymptotic behavior of orthogonal modulations. The condition (7.31a) has to do with a system with bit-by-bit transmission of the components of orthogonal signals, whereas (7.31b) pertains to schemes in which each message symbol uses an orthogonal signal waveform for transmission. The reader is referred to [1] and [4] for further discussion on this subject. We do, however, make reference to the conditions of (7.31a) and (7.31b) in Chapter 8 in connection with channel coding.

7.2   Modulation Schemes Used in the IS-95 System

In Chapter 4, "Overview of the IS-95 Standard," the forward and reverse links were described in block diagrams. With the exception of the pilot channel, which exists only in the forward link, we noted that every channel, in both forward and reverse links, convolutional channel coding is used. Also noted was the fact that every channel, including the pilot channel, employs a QPSK modulation scheme. The QPSK modulation used in the forward and reverse links has the characteristic feature that the identical information symbols are multiplied by distinct PN sequences in the in-phase (I) and quadrature (Q) branches of the transmitter. This modulation is different from conventional QPSK modulation, in which the I and Q branches take alternate information symbols [5–8].

The pilot channel allows the mobile stations to acquire the forward link carrier phase information, and, hence, forward link demodulation is accomplished coherently.

For the reverse link, however, no pilot channel is provided and, hence, coherent demodulation is not possible. The coded-and-interleaved information symbol sequence is transformed to a new sequence by a Walsh orthogonal block encoder, denoted $W(64, 6)$, in which each group of six input binary symbols is transformed to a block of 64 Walsh chips. The Walsh modulated symbols are multiplied by the mobile station-specific PN code sequence. This resultant sequence is offset QPSK (OQPSK) modulated for transmission to the base station. Because of the lack of phase information on the uplink (reverse link) carrier, the I- and Q-branch symbols cannot be coherently demodulated. The reverse link I- and Q-branch symbol decisions are not made immediately following carrier removal. Instead, the I- and Q-branch baseband values are passed on to the 64-ary Walsh orthogonal noncoherent demodulators. We analyze in detail the IS-95 CDMA reverse link orthogonal demodulation system in this chapter. By demodulating (decoding) the Walsh symbols in the phase incoherent demodulator, the corresponding six-symbol information sequence is extracted, thereby accomplishing demodulation without the benefit of having carrier phase information.

The performance measures of the demodulation schemes used in the forward and reverse links are analyzed for both the AWGN channel and Rayleigh fading channel environments. Included in this chapter is an analysis to show the reason behind the selection of QPSK modulation in the CDMA system instead of simple BPSK modulation.

7.2.1 Forward Link

All the forward link channels (pilot, sync, paging, and traffic) employ QPSK modulation for information symbol transmission. Figure 7.27 depicts the transmitter configuration for the nth channel, where n denotes any one of the forward link channels. When n denotes the pilot channel, the data waveform $D_n(t)$ is the constant 1, meaning that the pilot channel does not carry information. That is:

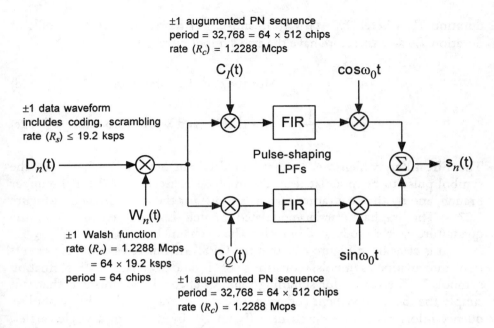

Figure 7.27 Transmitter configuration for the nth forward link channel.

$$D_{\text{pilot}}(t) \equiv 1 \qquad\qquad (7.32a)$$

In Figure 7.27, we see that the data waveform $D_n(t)$ is multiplied by the Walsh function $W_n(t)$, and then the resultant product $D_n(t) \cdot W_n(t)$ is fed to the upper branch as well as the lower branch. $C_I(t)$ represents the PN sequence for the upper branch and $C_Q(t)$, the PN sequence for the lower branch. Because of the fact that $W_n(t)$, $C_I(t)$, and $C_Q(t)$ represent respective sequences running at the same rate, the resultant sequences

$$I(t) = D_n(t) \times W_n(t) \times C_I(t) \qquad\qquad (7.32b)$$

and $\qquad$ $$Q(t) = D_n(t) \times W_n(t) \times C_Q(t) \qquad\qquad (7.32c)$$

are also random sequences with polarity ± 1 for each transmission symbol. Note that $D_n(t)$ represents coded and interleaved information symbols of duration T_s, whereas $W_n(t)$, $C_I(t)$, and $C_Q(t)$ represent "chip" symbols of duration T_c. We therefore have the two symbol rates[5]

$$R_s = \frac{1}{T_s} \qquad \text{for the coded-symbol rate} \qquad (7.32d)$$

$$R_c = \frac{1}{T_c} \qquad \text{for the PN and Walsh chip rates} \qquad (7.32e)$$

The FIR filter, which we discussed in detail in Section 1.4.2, shapes the symbol pulse to be modulated by the in-phase carrier ($\cos \omega_0 t$) for the upper branch, and by the quadrature carrier ($\sin \omega_0 t$) for the lower branch of Figure 7.27. The in-phase carrier-modulated branch is the I channel and the quadrature carrier-modulated branch is the Q channel.

It is clear from Figure 7.27 that the QPSK modulated symbol represents two information symbols, consisting of I- and Q-channel "information symbols." The reason for summing I and Q channel waveforms is that it is simply the most convenient and economic way of transmitting the symbol sequence information of each channel (I and Q), by means of a single carrier. To see this, let us ignore the effect of the FIR spectral shaping filters and express the QPSK modulated waveform as follows: Because $I(t)$ and $Q(t)$ assume the two possible values of ± 1, we may use indexes i and j to indicate that $I_i(t)$ and $Q_j(t)$ assume the value $+1$ for $i = j = 1$ and the value -1 for $i = j = -1$. We then have

$$s_n(t) = A_n[I_i(t)\cos \omega_0 t + Q_j(t) \sin \omega_0 t]$$

$$= A_n \sqrt{I_i^2(t) + Q_j^2(t)} \cos[\omega_o t - \phi_{ij}(t)], \quad 0 \le t \le T_c \qquad (7.33a)$$

where

$$\phi_{ij}(t) \triangleq \tan^{-1}\left[\frac{Q_j(t)}{I_i(t)}\right] \qquad (7.33b)$$

and A_n is an appropriate amplification factor for channel n. Therefore, there are four possible symbol pair patterns[6] $\{A_i, B_j\}$ or $\{a_i, b_j\}$ of

---

[5] Numerical relationships between R_s and R_c are shown in Tables 3.1 (sync channel), 3.2 (paging channel), and 3.3 (traffic channel) in Chapter 3.
[6] We use the notations A,B $\in \{+1, -1\}$ and $a, b \in \{0, 1\}$ as we did in Section 6.4 for denoting binary sequences over the respective sets.

$\{I_i(t),\ Q_j(t),\ 0 \le t \le T_c\}$, where $i, j \in \{+1, -1\}$ in (A_i, B_j) notation, whereas $i, j \in (0, 1)$ in $(a_i,\ b_j)$ notation. Thus

$$\{[I_i(t),\ Q_j(t)]\} = \{(A_i,\ B_j)\} = \{(1,1),\ (-1,1),\ (1,-1),\ (-1,-1)\} \qquad (7.34a)$$

or

$$\{[I_i(t),\ Q_j(t)]\} = \{(a_i,\ b_j)\} = \{(0,0),\ (1,0),\ (0,1),\ (1,1)\} \qquad (7.34b)$$

The corresponding phases of the carrier are

$$\{\phi_{ij}(t)\} \triangleq \{\tan^{-1}[Q_j(t)/I_i(t)]\} = \{45°,\ 135°,\ -45°,\ -135°\} \quad (7.34c)$$

The phase transitions corresponding to the symbol pair patterns are shown in Figure 7.28. We observe from Figure 7.28 that a given symbol pair can make a transition to all other possible pairs, including the same symbol pair pattern. For example, the symbol pair pattern $(I_i,\ Q_j) = (a_i,\ b_j) = (0,\ 0)$ can change to any of $\{(0,0),\ (0,1),\ (1,0),\ (1,1)\}$ at the next symbol interval. When a phase transition takes place by crossing the origin, the transmission wave shows a large variation in amplitude; other transitions may or may not involve a change in amplitude. Thus the QPSK waveform in general is a non-constant amplitude waveform. Nonconstant amplitude induces amplitude-to-phase conversion distortion if the transmission characteristic is not perfectly linear. For the forward link, the transmission characteristic can be designed to be linear, using a linear amplifier, thereby tolerating the amplitude variations due to the QPSK modulation.

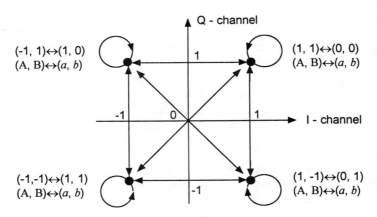

Figure 7.28 Signal constellation of QPSK modulation and carrier phase transitions with respect to symbol pair transitions.

On the other hand, for the reverse link, which is analyzed in Section 7.22, the IS-95 system employs OQPSK. In OQPSK modulation, the Q-channel symbol is one-half symbol delayed with respect to the I-channel data symbol, as illustrated in Figure 7.29. The I-channel in Figure 7.29 shows for both QPSK and OQPSK the symbol data sequence $1, -1, 1, 1, -1, 1$, while the Q-channel for QPSK shows the example sequence $-1, 1, 1, -1, 1, -1$. This is the case of the symbol pair pattern transitions $(1, -1) \rightarrow (-1, 1) \rightarrow (1, 1) \rightarrow (1, -1) \rightarrow (-1, 1) \rightarrow (1, -1)$. The corresponding phase transitions are

$$\text{(Reference, } -45°\text{)} \overset{180°}{\rightarrow} (135°) \overset{-90°}{\rightarrow} (45°) \overset{-90°}{\rightarrow} (-45°) \overset{180°}{\rightarrow} (135°) \overset{180°}{\rightarrow} (-45°)$$

(7.34d)

The Q-channel for OQPSK in Figure 7.29 shows the identical pattern as that for QPSK, but $T_s/2$ sec delayed, where T_s is the symbol duration. Let us observe the symbol pair patterns with respect to transitions of both the I-channel and the delayed Q-channel sequences. We then have

$$(1, -1) \rightarrow (-1, -1) \rightarrow (-1, 1) \rightarrow (1, 1) \rightarrow (1, 1) \rightarrow (1, 1) \rightarrow (1, -1)$$
$$\rightarrow (-1, -1) \rightarrow (-1, 1) \rightarrow (1, 1) \rightarrow (1, -1)$$

Now, regarding the first symbol pair $(1, -1)$ as the reference, these pair transitions will show the following phase transition pattern:

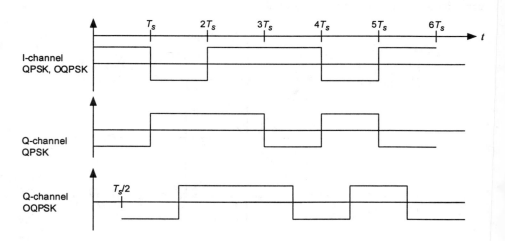

Figure 7.29 Symbol transition patterns in QPSK and OQPSK modulation.

$$(\text{Reference}, -45°) \xrightarrow{-90°} (-135°) \xrightarrow{-90°} (135°) \xrightarrow{-90°} (45°) \xrightarrow{0°} (45°) \xrightarrow{0°} (45°)$$

$$\xrightarrow{-90°} (-45°) \xrightarrow{-90°} (-135°) \xrightarrow{-90°} (135°) \xrightarrow{-90°} (45°) \xrightarrow{-90°} (-45°) \qquad (7.34e)$$

Note that in the case of OQPSK modulation, the phase transitions never exceed $\pm 90°$, and this fact is illustrated in Figure 7.30, in which the QPSK waveform shows $180°$ carrier phase changes, while the OQPSK waveform shows only $90°$ carrier phase changes.

A signal constellation for the OQPSK modulation is given in Figure 7.31, showing the symbol pair patterns along with the carrier phase transitions, which can be compared with those of QPSK shown in Figure 7.29.

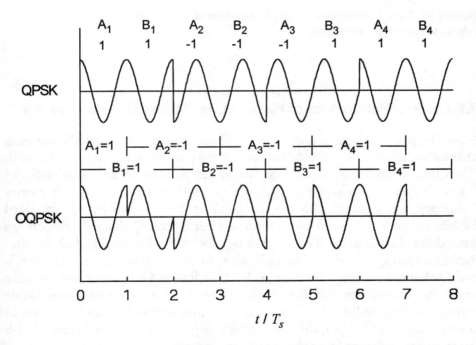

Figure 7.30 Illustration of the maximum carrier phase transitions in QPSK and OQPSK modulation systems.

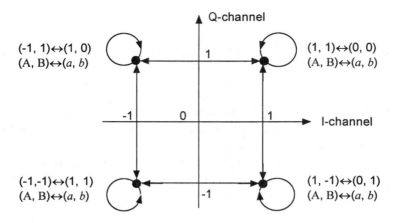

Figure 7.31 Signal constellation of OQPSK modulation and carrier phase transitions with respect to symbol pair transitions.

7.2.1.1 Error Performance of Forward Link Channel Symbols in AWGN

From Figure 7.27, we see that the modulation scheme is QPSK for each channel n, where n represents any one of 64 possible forward link channels. The QPSK modulation is made up of two independent BPSK modulations of I and Q carriers. The QPSK signal $s_n(t)$ is translated to a higher carrier frequency and radiated into the air for it to be received by the intended mobile station. At the mobile station receiver, I- and Q-channel symbols are demodulated separately. This is done by first extracting the I and Q decision variables through coherent multiplication of the received waveform with I- and Q-channel carriers. We first analyze the forward link error performance under the assumption that there is no multipath interference and no multiple access interference (MAI).[7] This is an ideal assumption because there are, in reality, multipaths from the transmitted signals that have different Walsh function multiplexing from that of the direct path.

Analysis under no multipath and no MAI due to signals from other cells. Consider Figure 7.32, which depicts the forward link Walsh function orthogonal multiplexing configuration consisting of $M = 64$ channels, each with the

---

[7] MAI due to own cell as well as adjacent cells is considered in the CDMA system performance analysis in Chapters 10 and 11.

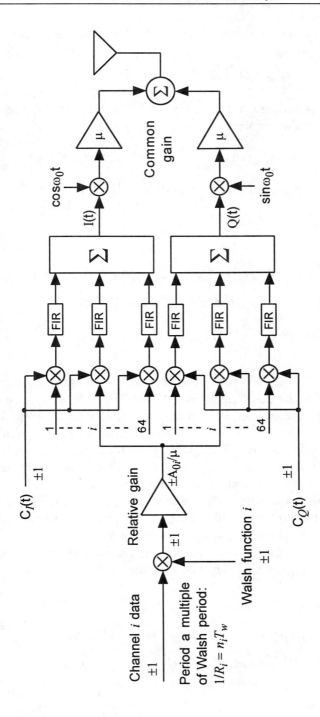

Figure 7.32 Forward link Walsh function orthogonal multiplexing configuration.

processing depicted previously in Figure 7.27. The transmitted forward link waveform can be written

$$s_{tr}(t) = I(t)\cos 2\pi f_0 t + Q(t)\sin 2\pi f_0 t \tag{7.35a}$$

where f_0 is the selected center frequency, and $I(t)$ and $Q(t)$ are the in-phase and quadrature-phase waveforms

$$
\begin{aligned}
I(t) &= \{C_I(t)[A_{00}W_0(t)D_0(t)]\}_{\text{shaped}} + \{C_I(t)[A_{01}W_1(t)D_1(t)]\}_{\text{shaped}} \\
&\quad + \cdots + \{C_I(t)[A_{0,63}W_{63}(t)D_{63}(t)]\}_{\text{shaped}} \\
&= \sum_{i=0}^{63}\{C_I(t)[A_{0i}W_i(t)D_i(t)]\}_{\text{shaped}}
\end{aligned} \tag{7.35b}
$$

and

$$
\begin{aligned}
Q(t) &= \{C_Q(t)[A_{00}W_0(t)D_0(t)]\}_{\text{shaped}} + \{C_Q(t)[A_{01}W_1(t)D_1(t)]\}_{\text{shaped}} \\
&\quad + \cdots + \{C_Q(t)[A_{0,63}W_{63}(t)D_{63}(t)]\}_{\text{shaped}} \\
&= \sum_{i=0}^{63}\{C_Q(t)[A_{0i}W_i(t)D_i(t)]\}_{\text{shaped}}
\end{aligned} \tag{7.35c}
$$

where $C_I(t)$ and $C_Q(t)$ are two distinct maximal length short ± 1 PN sequences clocked at the chip rate $R_c = 1.2288$ Mcps. The $\{A_{0i}\}$ are digitally controlled channel transmitter amplitudes, which can be different for each channel, depending on the signal power requirement of that channel at the receiving end (mobile station). The $\{W_i(t)\}$ are 64-chip ± 1 periodic Walsh functions, clocked at the same chip rate R_c. The $\{D_i(t)\}$ are ± 1 data symbols, possibly including coding and scrambling, and clocked at rates no greater than $R_c/64 = 19.2$ ksps. The pulse shaping controls the bandwidth of the waveform in the way explained in Section 1.4; we ignore the effects of this shaping in the error performance analysis to follow.

During reception of one data symbol, the received waveform $r(t)$ at the particular mobile using channel n can be expressed as

$$r(t) = s_{rec}(t) + n(t), \qquad 0 \le t \le T_w \tag{7.36a}$$

with

$$s_{rec}(t) = s_0(t) + s_1(t) + \cdots + s_{M-1}(t), \qquad M = 64$$

$$n(t) = n_c(t)\sqrt{2}\cos 2\pi f_0 t + n_s(t)\sqrt{2}\sin 2\pi f_0 t \tag{7.36b}$$

where

$$T_w = 64T_c = 64/R_c = \text{Walsh function period}$$

$$n(t) = G\left(0, \tfrac{1}{2}\mathcal{N}_0\right),\ n_c(t) = G\left(0, \tfrac{1}{2}\mathcal{N}_0\right),\ n_s(t) = G\left(0, \tfrac{1}{2}\mathcal{N}_0\right) \quad (7.36c)$$

and

$$s_i(t) = C_I(t)[A_i W_i(t) D_i(t)] \cos 2\pi f_0 t$$
$$+ C_Q(t)[A_i W_i(t) D_i(t)] \sin 2\pi f_0 t,\ i = 0, 1, \ldots, M-1 \quad (7.36d)$$

The $\{A_i\}$ denote the amplitudes of the received I- and Q-channel carriers, $i = 0, 1, \ldots, M-1$.

It is clear from observation of (7.36d) that the receiver needs to multiply the received waveform by both $C_I(t) \cos 2\pi f_0 t$ and $C_Q(t) \sin 2\pi f_0 t$ to obtain the information carrying Walsh function, $W_n(t)$. Hence, the receiver configuration for the nth channel receiver should be of the form shown in Figure 7.33. The quadrature components $r_c(t)$ and $r_s(t)$ developed by that receiver are

$$r_c(t) = \Big[r(t) \times \sqrt{2}\cos 2\pi f_0 t\Big]_{\text{lowpass}} \times C_I(t)$$

$$= \frac{1}{\sqrt{2}} \sum_{i=0}^{M-1} A_i C_I^2(t) D_i(t) W_i(t) + n_I(t)$$

$$= \frac{1}{\sqrt{2}} \sum_{i=0}^{M-1} A_i D_i(t) W_i(t) + n_I(t) \quad (7.37a)$$

and

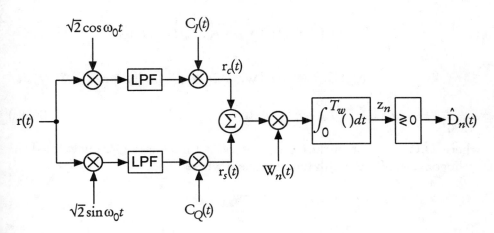

Figure 7.33 Optimal demodulator for nth forward link channel for symbol decision $\widehat{D}_n(t)$.

$$r_s(t) = \left[r(t) \times \sqrt{2}\sin 2\pi f_0 t \right]_{\text{lowpass}} \times C_Q(t)$$

$$= \frac{1}{\sqrt{2}} \sum_{i=0}^{M-1} A_i C_Q^2(t) D_i(t) W_i(t) + n_Q(t)$$

$$= \frac{1}{\sqrt{2}} \sum_{i=0}^{M-1} A_i D_i(t) W_i(t) + n_Q(t) \qquad (7.37b)$$

where

$$n_I(t) \triangleq n_c(t) C_I(t) \quad \text{and} \quad n_Q(t) \triangleq n_s(t) C_Q(t) \qquad (7.37c)$$

with $n_c(t)$ and $n_s(t)$ being uncorrelated noise waveforms with spectral density levels $\frac{1}{2}\mathcal{N}_0$ in the bandwidth of the lowpass filters. The sum of $r_c(t)$ and $r_s(t)$ is

$$r_c(t) + r_s(t) = \sqrt{2} \sum_{i=0}^{M-1} A_i D_i(t) W_i(t) + n_I(t) + n_Q(t) \qquad (7.37d)$$

Now, performing the correlation with the assigned mobile station Walsh function $W_n(t)$ to extract the decision $\widehat{D}_n(t)$ on the polarity of the data symbol $D_n(t)$, we have the decision statistic

$$z_n = \int_0^{T_w} [r_c(t) + r_s(t)] \times W_n(t)\, dt$$

$$= \sqrt{2} \int_0^{T_w} [A_0 D_0(t) W_0(t) + \cdots + A_n D_n(t) W_n(t) + \cdots$$

$$+ A_{63} D_{63}(t) W_{63}(t)] W_n(t)\, dt + \int_0^{T_w} [n_I(t) + n_Q(t)] W_n(t)\, dt$$

$$= \sqrt{2} A_n D_n(t) T_w + \eta \qquad (7.38a)$$

where $D_i(t)$ is constant during the interval of integration, we have used the orthogonality of the Walsh functions, and where

$$\eta \triangleq \int_0^{T_w} [n_I(t) + n_Q(t)] W_n(t)\, dt$$

$$= \int_0^{T_w} [n_c(t) C_I(t) + n_s(t) C_Q(t)] W_n(t)\, dt \qquad (7.38b)$$

The noise term η is a Gaussian RV with mean m_η and variance σ_η^2, denoted $\eta = G(m_\eta, \sigma_\eta^2)$, where the mean is

$$m_\eta = \int_0^{T_w} \overline{[n_c(t)\, C_I(t) + n_s(t)\, C_Q(t)]} W_n(t)\, dt = 0 \qquad (7.38c)$$

The variance of η is given by

$$\sigma_\eta^2 = \overline{\eta^2} = \int_0^{T_w} \int_0^{T_w} \overline{[n_I(t) + n_Q(t)][n_I(u) + n_Q(u)]} W_n(t) W_n(u)\, dt\, du$$

$$= \int_0^{T_w} \int_0^{T_w} \overline{[n_c(t)\, C_I(t) + n_s(t)\, C_Q(t)][n_c(u)\, C_I(u) + n_s(u)\, C_Q(u)]}$$

$$\times\, W_n(t) W_n(u)\, dt\, du$$

$$= \int_0^{T_w} \int_0^{T_w} \overline{n_c(t)\, n_c(u)} \cdot C_I(t)\, C_I(u) \cdot W_n(t) W_n(u)\, dt\, du$$

$$+ \int_0^{T_w} \int_0^{T_w} \overline{n_s(t)\, n_s(u)} \cdot C_Q(t)\, C_Q(u) \cdot W_n(t) W_n(u)\, dt\, du$$

$$= \int_0^{T_w} \int_0^{T_w} R_{n_c}(u - t) \cdot C_I(t)\, C_I(u) \cdot W_n(t) W_n(u)\, dt\, du$$

$$+ \int_0^{T_w} \int_0^{T_w} R_{n_s}(u - t) \cdot C_Q(t)\, C_Q(u) \cdot W_n(t) W_n(u)\, dt\, du \qquad (7.39a)$$

in which we have used the fact that

$$\overline{n_c(t)\, n_s(u)} \cdot C_I(t)\, C_Q(u) = 0 \qquad (7.39b)$$

It is clear that we need to calculate the noise correlation functions that appear in the integrand of (7.39a). Because the integral may be viewed as a lowpass filter, we can simplify the analysis by "lumping" all the filtering effects of the I and Q branches of the receiver into the integrals, as if the lowpass filters in Figure 7.33 were absent. Using this approach, the noise waveforms $n_c(t)$ and $n_s(t)$ can be treated as white noise with spectral density $\frac{1}{2}N_0$ [1] and with the corresponding correlation function $\frac{1}{2}N_0\, \delta(\tau)$, and the variance in (7.39a) becomes

$$\sigma_\eta^2 = \int_0^{T_w} \int_0^{T_w} \tfrac{1}{2}\mathcal{N}_0\,\delta(u-t) \cdot \mathrm{C}_I(t)\,\mathrm{C}_I(u) \cdot \mathrm{W}_n(t)\mathrm{W}_n(u)\,dt\,du$$

$$+ \int_0^{T_w} \int_0^{T_w} \tfrac{1}{2}\mathcal{N}_0\,\delta(u-t) \cdot \mathrm{C}_Q(t)\,\mathrm{C}_Q(u) \cdot \mathrm{W}_n(t)\mathrm{W}_n(u)\,dt\,du$$

$$= \tfrac{1}{2}\mathcal{N}_0 \int_0^{T_w} \mathrm{C}_I^2(t)\,\mathrm{W}_n^2(t)\,dt + \tfrac{1}{2}\mathcal{N}_0 \int_0^{T_w} \mathrm{C}_Q^2(t)\,\mathrm{W}_n^2(t)\,dt$$

$$= 2 \times \tfrac{1}{2}\mathcal{N}_0 T_w = \mathcal{N}_0 T_w \tag{7.40a}$$

because $\mathrm{C}_I^2(t) = \mathrm{C}_Q^2(t) = \mathrm{W}_n^2(t) = 1$. Therefore, the noise term defined in (7.38b) is given by

$$\eta = \mathrm{G}(0,\,\mathcal{N}_0 T_w) \tag{7.40b}$$

Because the decision variable is the Gaussian noise term η plus the data-dependent quantity $\sqrt{2}D_n A_n T_w = \pm\sqrt{2}A_n T_w$, the probability of error is identical to that of binary antipodal signals with distance $d = 2\sqrt{2}\,A_n T_w$, as illustrated in Figure 7.20. The data symbol error probability therefore is

$$\mathrm{P}_s(e) = \mathrm{Q}\!\left(\frac{d/2}{\sigma_\eta}\right) = \mathrm{Q}\!\left(\sqrt{\frac{\left(\sqrt{2}A_n T_w\right)^2}{\mathcal{N}_0 T_w}}\right) = \mathrm{Q}\!\left(\sqrt{\frac{2\mathrm{E}_w}{\mathcal{N}_0}}\right) \tag{7.41a}$$

using the fact that the energy of the signal over the duration of a Walsh period is $\mathrm{E}_w \triangleq A_n^2 T_w$. Note that the Walsh symbol, consisting of 64 Walsh chips, represents one coded information symbol; thus, using E_s to denote symbol energy, we have $\mathrm{E}_s = \mathrm{E}_w$ and

$$\mathrm{P}_s(e) = \mathrm{Q}\!\left(\sqrt{\frac{2\mathrm{E}_w}{\mathcal{N}_0}}\right) = \mathrm{Q}\!\left(\sqrt{\frac{2\mathrm{E}_s}{\mathcal{N}_0}}\right) \tag{7.41b}$$

In this analysis we used ideal assumptions[8] and represented the receiver as performing analog operations, such as integration, while the actual receiver employs digital operations.

---

[8] The ideal assumptions include perfect PN code synchronization. We will discuss PN code *acquisition* and *tracking* in a later section of this chapter.

7.2.1.2 Error Performance of Forward Link Channel Symbols in Rayleigh Fading

In this section, we obtain the error performance expression for forward link symbol modulation for the Rayleigh fading channel. Now, suppose that the transmitted QPSK signal, consisting of two independent BPSK modulations, is subjected to Rayleigh fading due to multipath effects. Then the received signal term for Walsh channel i included in (7.36a) can be expressed as

$$s_{i,rec}(t) = aI_i(t)\sqrt{\frac{2E_c}{T_c}}\cos 2\pi f_0 t + aQ_i(t)\sqrt{\frac{2E_c}{T_c}}\sin 2\pi f_0 t \qquad (7.42a)$$

where I_i and Q_i are the ± 1 parts of the quadrature modulations after factoring out the amplitude $A_i = \sqrt{2E_c/T_c}$, with E_c representing the chip energy in each quadrature channel without fading. In (7.42a), a is a Rayleigh RV (Rayleigh attenuation parameter) and its pdf is given by

$$p_a(\beta) = \frac{2\beta}{b}\,e^{-\beta^2/b}, \quad \beta \geq 0 \qquad (7.42b)$$

where $b = \overline{a^2}$ is the mean square value of the Rayleigh RV. In view of (7.42a) the Rayleigh fading has the effect of replacing the received symbol energy E_s with $a^2 E_s$, and thus the error probability of the BPSK system of quadrature PN chip-modulation symbol transmission in Rayleigh fading should be obtained by averaging the symbol error probability in (7.410b) over RV a:

$$P_s(e) = \overline{Q\left(\sqrt{\frac{2\beta^2 E_s}{\mathcal{N}_0}}\right)} \qquad (7.42c)$$

Using the pdf given in (7.42b), we then have

$$P_s(e) = \int_0^\infty \frac{2\beta}{b}\,e^{-\beta^2/b}\,Q\left(\sqrt{\frac{2\beta^2 E_s}{\mathcal{N}_0}}\right)\,d\beta \qquad (7.42d)$$

$$= \frac{1}{2}\left(1 - \sqrt{\frac{\overline{E_s}/\mathcal{N}_0}{1 + \overline{E_s}/\mathcal{N}_0}}\right), \qquad \overline{E_s} = \overline{a^2 E_s} = bE_s \qquad (7.42e)$$

The details of the calculation for the result of (7.42e) are carried out in Appendix 7B.

The antipodal signal AWGN channel error performance (7.41b) and Rayleigh fading channel error performance (7.42e) are plotted in Figure 7.34 with $P_s(e)$ replaced by $P_{BPSK}(e)$ and E_s replaced by E, so that the plots would lend themselves to a wide range of applications in which the energy refers to different parameters (bit, symbol, chip, etc.).

7.2.2 Reverse Link

As seen in the system description for the reverse link in Chapter 4, there are only two channel categories in the reverse link: access channel and traffic channel. There is no pilot channel as there is in the forward link. The signal processing for the reverse link channels at the base station, therefore, does not have the benefit of a pilot channel and, hence, the signal processing is noncoherent. The term "noncoherent" means that the receiver does not have knowledge of the random phase that is associated with the received carrier, and for that reason only the envelope of the received signal must be used for message decisions.

In what follows, we first discuss fundamentals related to the receiver structures for processing received signals with unknown random phase, and then move on to consider performance comparisons for the modulation systems used in the reverse link of the IS-95 system.

7.2.2.1 Noncoherent Signal Processing

Consider a communications system in which a set of M equally likely messages $\{m_i\}$, $i = 0, 1, \ldots, M - 1$ is to be transmitted to the receiving party by means of the baseband[9] waveforms $\{s_i(t), 0 \le t \le T\}$, $i = 0, 1, \ldots, M - 1$, with the mapping $m_i \to s_i(t)$. This system description is identical to that of the waveform communications system discussed in Section 1.5.2. We now assume that the transmitted signal $z_i(t)$ for message m_i is given by

$$z_i(t) = s_i(t) \sqrt{2} \cos(\omega_0 t - \theta'), \ 0 \le t \le T \tag{7.43a}$$

---

[9] By *baseband* signal we mean the signal before transmission by means of a carrier frequency, the signal occupying the spectrum in the lowpass region including $f = 0$.

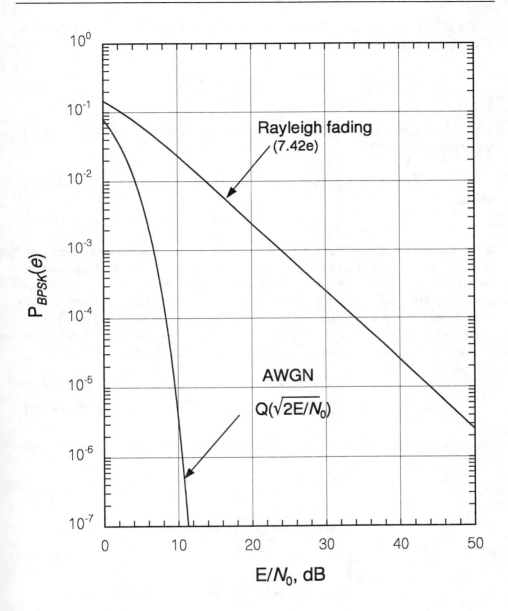

Figure 7.34 Performance of coherent BPSK modulation over AWGN and Rayleigh fading channels vs. $E_s / \mathcal{N}_0 = E / \mathcal{N}_0$.

where $s_i(t)$ is the chosen baseband waveform and θ' is a phase RV uniformly distributed over the interval $(0, 2\pi)$ with the pdf given by

$$p_{\theta'}(\alpha) = \begin{cases} 1/2\pi, & 0 \le \alpha \le 2\pi \\ 0, & \text{otherwise} \end{cases} \tag{7.43b}$$

The transmitted signal is corrupted by AWGN and the received waveform is given by

$$r(t) = s_i(t)\sqrt{2}\cos(\omega_0 t - \theta) + n(t), \ 0 \le t \le T \tag{7.43c}$$

where

$$n(t) = G(0, \mathcal{N}_0/2) \tag{7.43d}$$

We are dealing with a receiver that does not have the capability of estimating or knowing the carrier phase θ of the received signal. This situation, lack of knowledge of θ at the receiver, makes the signal processing noncoherent signal processing. The first step in implementing the optimal receiver in the face of unknown phase is to demodulate the received signal using quadrature carrier waveforms, as shown in Figure 7.35. The the in-phase channel output $r_c(t)$ and the quadrature channel output $r_s(t)$ are then given by

$$r_c(t) = s_i(t)\cos\theta + n_c(t) \tag{7.44a}$$
$$r_s(t) = s_i(t)\sin\theta + n_s(t) \tag{7.44b}$$

where $n_c(t)$ and $n_s(t)$ are independent zero-mean Gaussian RVs with equal

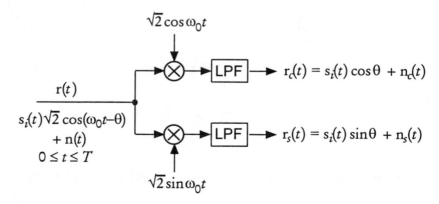

Figure 7.35 The first step in demodulating signals with random phase.

variances $\sigma_0^2 \triangleq \mathcal{N}_0 W$ when the lowpass filters' equivalent rectangular band-widths are from $-W$ to W. Note that the signal set $\{s_i(t)\}$ that is used at the transmitter is supposed to have been synthesized from a given set of signal vectors $\{\mathbf{s}_i\}$ based on the representation

$$\mathbf{s}_i = (s_{i1}, s_{i2}, \ldots, s_{iN}) \tag{7.45a}$$

and

$$s_i(t) = s_{i1}\varphi_1(t) + s_{i2}\varphi_2(t) + \cdots + s_{iN}\varphi_N(t) \tag{7.45b}$$

using the orthonormal functions $\{\varphi_j(t)\}$, $j = 1, 2, \ldots, N$ as described in Section 7.1.2. Recognition of this fact gives vector representations for (7.44a) and (7.44b) as

$$\mathbf{r}_c = \mathbf{s}_i \cos\theta + \mathbf{n}_c \tag{7.46a}$$

and

$$\mathbf{r}_s = \mathbf{s}_i \sin\theta + \mathbf{n}_s \tag{7.46b}$$

where $\mathbf{n}_c$ and $\mathbf{n}_s$ are N-dimensional zero-mean Gaussian random vectors independent of each other whose components are, respectively, the projections of $n_c(t)$ and $n_s(t)$ onto the orthonormal functions $\{\varphi_j(t)\}$. Because of the orthogonality of those functions, the components of $\mathbf{n}_c$ and $\mathbf{n}_s$ are also mutually independent with equal variances that may be considered to be [1] $\sigma^2 = \frac{1}{2}\mathcal{N}_0$, so that their joint pdf is given by

$$p_{\mathbf{n}_k}(\boldsymbol{\mu}) = p_{n_{k1}, n_{k2}, \ldots, n_{kN}}(\mu_1, \mu_2, \ldots, \mu_N), \quad k = c, s$$

$$= \frac{1}{(2\pi\sigma^2)^{N/2}} \exp\left\{ -\frac{1}{2\sigma^2} \sum_{m=1}^{N} \mu_m^2 \right\} = \frac{e^{-|\mu|^2/2\sigma^2}}{(2\pi\sigma^2)^{N/2}} \tag{7.46c}$$

The optimal decision rule is made on the basis of the received quadra-ture demodulator output vectors $\mathbf{r}_c$ and $\mathbf{r}_s$ [9] as given in (7.46a) and (7.46b). Now, assume that these vectors have the values $\mathbf{r}_c = \boldsymbol{\alpha} = (\alpha_1, \alpha_2, \ldots, \alpha_N)$ and $\mathbf{r}_s = \boldsymbol{\beta} = (\beta_1, \beta_2, \ldots, \beta_N)$. Note that the transmitted signal vector $\mathbf{s}_i$ appears in both of these vectors; this fact implies that we must consider the joint pdf of $\mathbf{r}_c$ and $\mathbf{r}_s$ in the optimum decision process. Thus, the likelihood function, given that the message m_i is transmitted, is the joint pdf of $\mathbf{r}_c$ and $\mathbf{r}_s$, given $m = m_i$ obtained by averaging the joint pdf, given $m = m_i$ and the carrier phase takes the value $\theta = \gamma$, over the possible values of γ. That is:

$$\text{Pr}_{r_c, r_s}(\boldsymbol{\alpha}, \boldsymbol{\beta} \mid m = m_i) = \overline{\text{Pr}_{r_c, r_s}(\boldsymbol{\alpha}, \boldsymbol{\beta} \mid m = m_i, \theta = \gamma)}$$

$$= \int_0^{2\pi} \text{Pr}_{r_c, r_s}(\boldsymbol{\alpha}, \boldsymbol{\beta} \mid m = m_i, \theta = \gamma)\, p_\theta(\gamma)\, d\gamma \quad (7.47a)$$

From (7.46a), (7.46b), and (7.46c) we have

$$\text{Pr}_{r_c, r_s}(\boldsymbol{\alpha}, \boldsymbol{\beta} \mid m = m_i, \theta = \gamma) = p_{n_c}(\boldsymbol{\alpha} - s_i \cos\gamma)\, p_{n_s}(\boldsymbol{\beta} - s_i \sin\gamma)$$

$$= \frac{1}{(2\pi\sigma^2)^{N/2}} \exp\left\{ -\frac{|\boldsymbol{\alpha} - s_i \cos\gamma|^2 + |\boldsymbol{\beta} - s_i \sin\gamma|^2}{2\sigma^2} \right\} \quad (7.47b)$$

The ML decision rule can now be stated as

$$\widehat{m} = m_k \text{ iff}$$

$$\text{Pr}_{r_c, r_s}(\boldsymbol{\alpha}, \boldsymbol{\beta} \mid m_k) = \max_i \left\{ \overline{\exp\left\{ -\frac{|\boldsymbol{\alpha} - s_i \cos\gamma|^2 + |\boldsymbol{\beta} - s_i \sin\gamma|^2}{2\sigma^2} \right\}} \right\}$$

$$\Rightarrow \max_i \left\{ \overline{\exp\left\{ \frac{\langle \boldsymbol{\alpha}, s_i \rangle \cos\gamma + \langle \boldsymbol{\beta}, s_i \rangle \sin\gamma}{\sigma^2} \right\}} \cdot \exp\left\{ -\frac{E_i}{2\sigma^2} \right\} \right\} \quad (7.47c)$$

where

$$E_i \triangleq |s_i|^2 = \int_0^T s_i^2(t)\, dt \quad (7.47d)$$

The optimal receiver structure is realized by mechanizing the sufficient conditions that satisfy the requirements of the right-hand side of (7.47c), and we derive these sufficient conditions. But first let us examine the expression in the exponent of the average in (7.47c). Because $\boldsymbol{\alpha}$ and $\boldsymbol{\beta}$ are particular vectors which r_c and r_s assumed, and γ is the particular value of the phase θ, the general expression for the exponent is

$$\langle r_c, s_i \rangle \cos\theta + \langle r_s, s_i \rangle \sin\theta = R_i \cos(\theta - \phi_i) \quad (7.48a)$$

where

$$R_i \triangleq \sqrt{(\langle r_c, s_i \rangle)^2 + (\langle r_s, s_i \rangle)^2} \quad (7.48b)$$

and

$$\phi_i \triangleq \tan^{-1} \frac{\langle r_s, s_i \rangle}{\langle r_c, s_i \rangle} \quad (7.48c)$$

The identity given by (7.48a) is easily understood with the aid of the transformation indicated in Figure 7.36. We also note that (see Section 7.1.1)

$$\langle \mathbf{r}_c , \mathbf{s}_i \rangle \equiv \int_0^T r_c(t)\, s_i(t)\, dt \equiv \langle r_c(t),\, s_i(t) \rangle \tag{7.48d}$$

and

$$\langle \mathbf{r}_s , \mathbf{s}_i \rangle \equiv \int_0^T r_s(t)\, s_i(t)\, dt \equiv \langle r_s(t),\, s_i(t) \rangle \tag{7.48e}$$

Using the identity of (7.48a), we have

$$\overline{\exp\left\{ \frac{\langle \mathbf{r}_c , \mathbf{s}_i \rangle \cos\theta + \langle \mathbf{r}_s , \mathbf{s}_i \rangle \sin\theta}{\sigma^2} \right\}} = \overline{\exp\left\{ \frac{R_i}{\sigma^2} \cos(\theta - \phi_i) \right\}}$$

$$= \int_0^{2\pi} e^{R_i \cos(\gamma - \phi_i)/\sigma^2}\, p_\theta(\gamma)\, d\gamma = \frac{1}{2\pi} \int_0^{2\pi} e^{R_i \cos(\gamma - \phi_i)/\sigma^2}\, d\gamma \tag{7.49a}$$

This integral is the modified Bessel function of order zero [see (1.88)], defined by

$$I_0(x) \triangleq \frac{1}{2\pi} \int_0^{2\pi} e^{x \cos\alpha}\, d\alpha = \frac{1}{2\pi} \int_0^{2\pi} e^{x \cos(\alpha - \phi)}\, d\alpha \tag{7.49b}$$

because of the periodicity of the cosine function. Thus, (7.49a) is

$$\frac{1}{2\pi} \int_0^{2\pi} e^{R_i \cos(\gamma - \phi_i)/\sigma^2}\, d\gamma = I_0\left(\frac{R_i}{\sigma^2} \right) \tag{7.49c}$$

and we now have the optimum decision rule

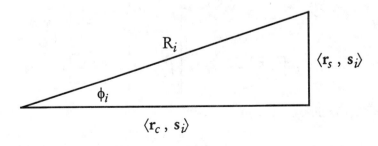

Figure 7.36 Transformation diagram for (7.48a).

$$\widehat{m} = m_k \quad \text{iff} \quad I_0\left(\frac{R_k}{\sigma^2}\right) e^{-E_k/2\sigma^2} = \max_i\left\{ I_0\left(\frac{R_i}{\sigma^2}\right) e^{-E_i/2\sigma^2} \right\} \tag{7.50a}$$

Let us impose some simplifying assumptions on this ML decision rule. First, we assume that $E_i = E_s$ for all $i = 0, 1, \ldots, M - 1$. Then (7.50a) becomes

$$\widehat{m} = m_k \quad \text{iff} \quad I_0\left(\frac{R_k}{\sigma^2}\right) = \max_i\left\{ I_0\left(\frac{R_i}{\sigma^2}\right) \right\} \tag{7.50b}$$

Next, since $I_0(x)$ is a monotonically increasing function of x (see Figure 1.72 in Chapter 1), we may write

$$\widehat{m} = m_k \quad \text{iff} \quad R_k = \max_i\{R_i\} \tag{7.50c}$$

or
$$\widehat{m} = m_k \quad \text{iff} \quad R_k^2 = \max_i\{R_i^2\} \tag{7.50d}$$

where R_i was defined in (7.48b), giving

$$R_i^2 = (\langle \mathbf{r}_c\,,\,\mathbf{s}_i \rangle)^2 + (\langle \mathbf{r}_s\,,\,\mathbf{s}_i \rangle)^2, \quad i = 0, 1, \ldots, M - 1 \tag{7.50e}$$

A mechanization of the decision variable R_i^2 for $s_i(t)$ is shown in Figure 7.37. Using the mathematical notation for the correlator operations, we can represent the decision variable generation in the alternative way shown in Figure 7.38. The correlations $\langle \mathbf{r}_c\,,\,\mathbf{s}_i \rangle$ and $\langle \mathbf{r}_s\,,\,\mathbf{s}_i \rangle$ are defined in (7.48d) and (7.48e), respectively. The entire configuration of the noncoherent correlation receiver for M-ary communications with random phase is shown in Figure 7.39.

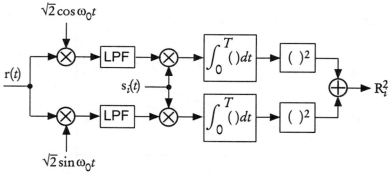

Figure 7.37 Noncoherent signal processing for generation of the decision variable R_i^2 corresponding to the signal $s_i(t)$.

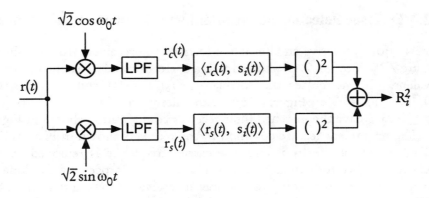

Figure 7.38 An alternative representation of Figure 7.37.

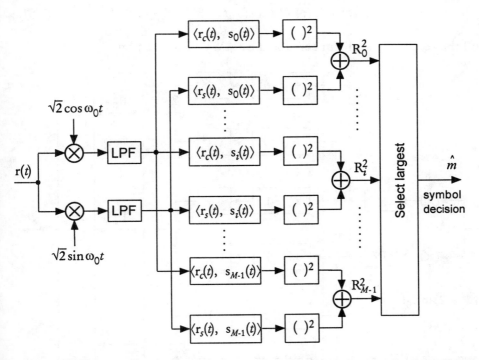

Figure 7.39 Noncoherent correlation receiver for equal-energy M-ary communications system with random phase.

7.2.2.2 Envelope Detection Receiver for M-ary Communications System

In this section we derive an alternative form to the receiver structure shown in Figure 7.39. Section 7.1.4 showed that a correlator and a matched filter are equivalent. If we replace the correlators $\langle r_c(t), s_i(t) \rangle$ and $\langle r_s(t), s_i(t) \rangle$, $i = 0, 1, \ldots, M-1$ in Figure 7.39 with filters matched to the $\{s_i(t)\}$, $0 \le t \le T$, we have the matched filter receiver configuration shown in Figure 7.40. The prominent feature common to the receivers shown in Figures 7.39 and 7.40 is that at the front end, the carrier frequency is removed by the action of the heterodyning, and hence the signal processing is done at baseband. We are now going to consider a receiver structure that processes the RF signal without removing the carrier frequency at the front end.

We assumed in (7.43a) that the transmitted signal is given by

$$z_i(t) = s_i(t) \sqrt{2} \cos(\omega_0 t - \theta), \quad 0 \le t \le T \tag{7.51a}$$

and the received waveform corrupted by AWGN is given by

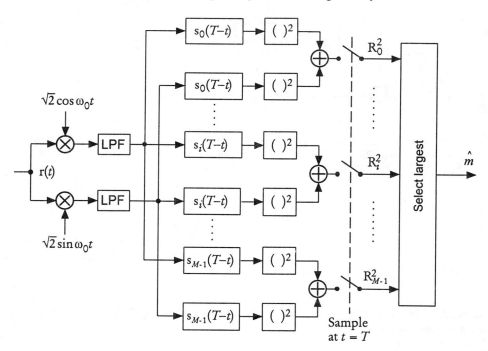

Figure 7.40 Noncoherent matched filter receiver for equal-energy M-ary communications system with random phase.

$$r(t) = z_i(t) + n(t), \quad 0 \leq t \leq T \tag{7.51b}$$

Now the idea that we promote here is to feed the received signal $r(t)$ to a bank of filters with impulse responses $\{h_i(t)\}$, $i = 0, 1, \ldots, M - 1$ matched to the bandpass signals $\{z_i(t)\}$, $i = 0, 1, \ldots, M - 1$ respectively. Note that

$$h_i(t) \triangleq z_i(T - t) = s_i(T - t) \sqrt{2} \cos(\omega_0 T - \omega_0 t - \theta)$$
$$= s_i(T - t) \sqrt{2} \cos(\omega_0 t - \theta') = s_i(T - t) \sqrt{2} \cos \omega_0 t \tag{7.51c}$$

In (7.51c) we assumed that $f_0 = n/T$, n an integer, and ignore the phase since it is inconsequential in the detection scheme to be presented.

Let us obtain the output of the matched filter, illustrated in Figure 7.41. The output may be written

$$y_i(t) = r(t) * h_i(t) = \int_0^t r(\alpha) \, h_i(t - \alpha) \, d\alpha$$

$$= \sqrt{2} \int_0^t r(\alpha) \, s_i(T - t + \alpha) \cos[\omega_0(t - \alpha)] \, d\alpha$$

$$= \sqrt{2} \cos \omega_0 t \int_0^t r(\alpha) \, s_i(T - t + \alpha) \cos \omega_0 \alpha \, d\alpha$$

$$+ \sqrt{2} \sin \omega_0 t \int_0^t r(\alpha) \, s_i(T - t + \alpha) \sin \omega_0 \alpha \, d\alpha$$

$$= y_{ci}(t) \cos \omega_0 t + y_{si}(t) \sin \omega_0 t \tag{7.51d}$$

where

$$y_{ci}(t) \triangleq \sqrt{2} \int_0^t r(\alpha) \, s_i(T - t + \alpha) \cos \omega_0 \alpha \, d\alpha \tag{7.52a}$$

and

$$y_{si}(t) \triangleq \sqrt{2} \int_0^t r(\alpha) \, s_i(T - t + \alpha) \sin \omega_0 \alpha \, d\alpha \tag{7.52b}$$

The matched filter output $y_i(t)$ given in (7.51d) can further be expressed as

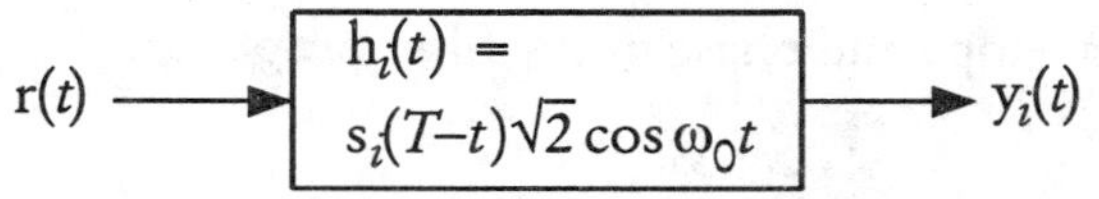

Figure 7.41 Filter matched to bandpass signal.

$$y_i(t) = y_{ci}(t) \cos \omega_0 t + y_{si}(t) \sin \omega_0 t = R_i(t) \cos[\omega_0 t - \phi_i(t)] \qquad (7.53a)$$

where

$$R_i(t) \triangleq \sqrt{y_{ci}^2(t) + y_{si}^2(t)} \qquad (7.53b)$$

is the envelope and

$$\phi_i(t) \triangleq \tan^{-1} \frac{y_{si}(t)}{y_{ci}(t)} \qquad (7.53c)$$

is the phase of the filter output $y_i(t)$.

Now, let us evaluate the expressions for $y_{ci}(t)$ and $y_{si}(t)$ given in (7.52a) and (7.52b), respectively, at the sample time $t = T$. Then we have

$$y_{ci}(T) \triangleq \sqrt{2} \int_0^T r(\alpha) \, s_i(\alpha) \cos \omega_0 \alpha \, d\alpha \qquad (7.54a)$$

and

$$y_{si}(T) \triangleq \sqrt{2} \int_0^T r(\alpha) \, s_i(\alpha) \sin \omega_0 \alpha \, d\alpha \qquad (7.54b)$$

In view of (7.48d) and (7.48e), the sampled quadrature components of the filter output are the quantities

$$y_{ci}(T) = \sqrt{2} \int_0^T r(\alpha) \, s_i(\alpha) \cos \omega_0 \alpha \, d\alpha = \int_0^T \underbrace{r(\alpha) \sqrt{2} \cos \omega_0 \alpha}_{r_c(\alpha)} \cdot s_i(\alpha) \, d\alpha$$

$$= \int_0^T r_c(\alpha) \, s_i(\alpha) \, d\alpha = \langle r_c(t), \, s_i(t) \rangle = \langle \mathbf{r}_c , \, \mathbf{s}_i \rangle \qquad (7.55a)$$

and similarly[10]

$$y_{si}(T) = \int_0^T \underbrace{r(\alpha) \sqrt{2} \sin \omega_0 \alpha}_{r_s(\alpha)} \cdot s_i(\alpha) \, d\alpha$$

$$= \int_0^T r_s(\alpha) \, s_i(\alpha) \, d\alpha = \langle r_s(t), \, s_i(t) \rangle = \langle \mathbf{r}_s , \, \mathbf{s}_i \rangle \qquad (7.55b)$$

and hence the envelope and phase of the filter output are

---

[10] The definitions of $r_c(t)$ and $r_s(t)$ in this development do not involve the post-heterodyning lowpass filters as in the previous sections, since local oscillators are not used.

$$R_i(T) = \sqrt{(\langle r_c(t),\, s_i(t) \rangle)^2 + (\langle r_s(t),\, s_i(t) \rangle)^2} \qquad (7.55c)$$

and

$$\phi_i(T) = \tan^{-1} \frac{\langle r_s(t),\, s_i(t) \rangle}{\langle r_c(t),\, s_i(t) \rangle} \qquad (7.55d)$$

The significance of these results is that the optimal receiver can also be configured using bandpass matched filters that are matched to the bandpass signals $\{z_i(t)\}$. The optimal decision then can be made on the basis of comparing the M matched filter outputs sampled at $t = T$, as indicated in Figure 7.42.

We have derived three different configurations for the optimal receiver for an M-ary equal-energy communications system: correlation receiver, matched filter receiver, and envelope detector receiver. All three deliver identical performance and are all called *noncoherent* receivers because they do not use knowledge of the received carrier phase, because of lack of a pilot signal or a means to estimate and track the phase at the receiver. All three receiver types extract decision variables that do not depend on the carrier's random phase. In the correlation receiver, the use of in-phase and quadrature channels eliminate the carrier phase from the decision variable by means of the simple trigonometric identity $\sin^2\theta + \cos^2\theta = 1$. The matched filter

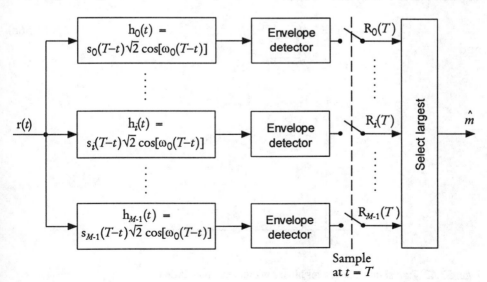

Figure 7.42 Envelope detector optimal receiver for equal-energy M-ary communications system with random phase.

receiver operates on the same principle, while the envelope detector receiver simply ignores the phase by sampling the slowly varying envelope of the matched filter output.

7.2.2.3 Noncoherent Binary Orthogonal System

A noncoherent binary orthogonal modulation system is one in which a set of two equally likely messages $\{m_i\}$, $i = 1, 2$ is to be communicated over an AWGN channel and the receiver employs noncoherent techniques for extraction of the decision variable, such as correlation, matched filtering, or envelope detection. The signal set $\{s_i(t), 0 \le t \le T\}$, $i = 1, 2$ can be represented in the signal vector space as shown in Figure 7.43.

Let us consider the following pair of ON functions:

$$\varphi_1(t) = \sqrt{2/T} \sin \omega_0 t, \quad 0 \le t \le T \tag{7.56a}$$

and

$$\varphi_2(t) = \sqrt{2/T} \cos \omega_0 t, \quad 0 \le t \le T \tag{7.56b}$$

Then, the signals that the transmitter uses for the signal vectors shown in Figure 7.43 can be synthesized as

$$s_1(t) = \sqrt{2E_s/T} \sin \omega_0 t, \quad 0 \le t \le T \tag{7.56c}$$

and

$$s_2(t) = \sqrt{2E_s/T} \cos \omega_0 t, \quad 0 \le t \le T \tag{7.56d}$$

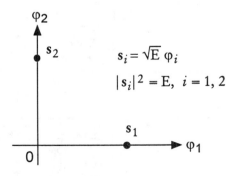

Figure 7.43 Two signal vectors for binary orthogonal modulation.

where the symbol energy is

$$E_s \triangleq \int_0^T s_i^2(t)\,dt = |s_i|^2, \ i = 1, 2 \tag{7.56e}$$

Another way of synthesizing the transmitter waveforms for the messages m_1 and m_2 corresponding to the signal vectors s_1 and s_2 is to consider two different ON functions, such as

$$\varphi_1'(t) = \sqrt{2/T}\cos\omega_1 t, \quad 0 \le t \le T \tag{7.57a}$$

and

$$\varphi_2'(t) = \sqrt{2/T}\cos\omega_2 t, \quad 0 \le t \le T \tag{7.57b}$$

where $\omega_i = 2\pi f_i$ with $f_i = n_i/T$ and n_1 and n_2 integer, such that

$$\int_0^T \varphi_1'(t)\,\varphi_2'(t)\,dt = 0 \tag{7.57c}$$

The corresponding transmitter waveforms are then given by

$$s_1(t) = \sqrt{2E/T}\cos 2\pi f_1 t, \quad 0 \le t \le T \tag{7.57d}$$

$$s_2(t) = \sqrt{2E/T}\cos 2\pi f_2 t, \quad 0 \le t \le T \tag{7.57e}$$

The waveforms $s_1(t)$ and $s_2(t)$ given in (7.57d) and (7.57e) are called BFSK waveforms, and the messages m_1 and m_2 are sometimes known as "mark" and "space."

When the receiver extracts the decision variables without using carrier phase information, the orthogonal system becomes the noncoherent binary orthogonal system or noncoherent BFSK system. The probability of error does not depend on the choice of any particular ON functions, and hence our analysis of the probability of error can be based on the decision variables we have dealt with in (7.50e). Thus, in this example, the optimal noncoherent receiver decision rule is

$$\hat{m} = m_1 \qquad \text{iff} \qquad R_1^2 > R_2^2 \tag{7.58a}$$

or

$$\hat{m} = m_1 \qquad \text{iff} \qquad R_1 > R_2 \tag{7.58b}$$

where the envelopes $\{R_i\}$ were defined in (7.50e) and transmitter signals $\{s_i(t)\}$ were defined to be either those given in (7.56c) and (7.56d) or those given in (7.57a) and (7.57b).

The probability of error for this noncoherent binary orthogonal system is identical for using square-law envelope detection based on (7.58a) or for using envelope detection based on (7.58b). However, the mathematical treatments of the two cases differ considerably. For our analysis, we use square-law envelope detection version based on (7.58a) and the transmitter waveforms we use are given in (7.57a) and (7.57b). Because

$$R_i^2 = [\langle r_c(t), s_i(t) \rangle]^2 + [\langle r_s(t), s_i(t) \rangle]^2 \tag{7.59a}$$

if we assume that m_1 is transmitted $(m = m_1)$, the decision rule is

$$\hat{m} = m_1 \text{ iff } \quad [\langle r_c(t), s_1(t) \rangle]^2 + [\langle r_s(t), s_1(t) \rangle]^2$$
$$> [\langle r_c(t), s_2(t) \rangle]^2 + [\langle r_s(t), s_2(t) \rangle]^2 \tag{7.59b}$$

Substituting for the quadrature components $r_c(t)$ and $r_s(t)$ from (7.44a) and (7.44b), respectively, we have

$$R_{c1} \triangleq \langle r_c(t), s_1(t) \rangle = \langle s_1(t) \cos \theta + n_c(t), s_1(t) \rangle$$
$$= \langle s_1(t), s_1(t) \rangle \cos \theta + \langle n_c(t), s_1(t) \rangle$$
$$= E_s \cos \theta + n_{c1} \tag{7.60a}$$

where

$$n_{c1} \triangleq \langle n_c(t), s_1(t) \rangle = \int_0^T n_c(t) \, s_1(t) \, dt \tag{7.60b}$$

and

$$R_{s1} \triangleq \langle r_s(t), s_1(t) \rangle = \langle s_1(t) \sin \theta + n_s(t), s_1(t) \rangle$$
$$= \langle s_1(t), s_1(t) \rangle \sin \theta + \langle n_s(t), s_1(t) \rangle$$
$$= E_s \sin \theta + n_{s1} \tag{7.61a}$$

where

$$n_{s1} \triangleq \langle n_s(t), s_1(t) \rangle = \int_0^T n_s(t) \, s_1(t) \, dt \tag{7.61b}$$

Also,

$$R_{c2} \triangleq \langle r_c(t), s_2(t) \rangle = \langle s_1(t) \cos \theta + n_c(t), s_2(t) \rangle$$
$$= \langle n_c(t), s_2(t) \rangle = n_{c2} \tag{7.62a}$$

where

$$n_{c2} \triangleq \langle n_c(t), s_2(t) \rangle = \int_0^T n_c(t) s_2(t) \, dt \qquad (7.62b)$$

and

$$R_{s2} \triangleq \langle r_s(t), s_2(t) \rangle = \langle s_1(t) \sin \theta + n_s(t), s_2(t) \rangle$$
$$= \langle n_s(t), s_2(t) \rangle = n_{s2} \qquad (7.63a)$$

where

$$n_{s2} \triangleq \langle n_s(t), s_2(t) \rangle = \int_0^T n_s(t) s_2(t) \, dt \qquad (7.63b)$$

It is easily shown that n_{c1}, n_{s1}, n_{c2}, and n_{s2} are all independent zero-mean Gaussian RVs with the same variance $\mathcal{N}_0 E_s / 2$, denoted

$$n_{ci} = G\left(0, \tfrac{1}{2}\mathcal{N}_0 E_s\right), \quad n_{si} = G\left(0, \tfrac{1}{2}\mathcal{N}_0 E_s\right), \quad i = 1, 2 \qquad (7.63c)$$

For example, the mean of n_{c2} is

$$\overline{n_{c2}} = \int_0^T \overline{n_c(t)} \, s_2(t) \, dt = 0 \qquad (7.63d)$$

and

$$\overline{n_{c2}^2} = \int_0^T \int_0^T \overline{n_c(t) n_c(u)} \cdot s_2(t) s_2(u) \, dt \, du$$
$$= \int_0^T \int_0^T \tfrac{1}{2}\mathcal{N}_0 \, \delta(t - u) \cdot s_2(t) s_2(u) \, dt \, du$$
$$= \tfrac{1}{2}\mathcal{N}_0 \int_0^T s_2^2(t) \, dt = \tfrac{1}{2}\mathcal{N}_0 \, E_s \qquad (7.63e)$$

Note that in (7.63e) we assumed that $\overline{n_c(t) n_c(u)} = \tfrac{1}{2}\mathcal{N}_0 \, \delta(t - u)$, the correlation function of white noise with spectral density $\tfrac{1}{2}\mathcal{N}_0$. Strictly speaking, because $n_c(t)$ is obtained as shown in Figure 7.35 with a lowpass filter, it is not white noise. However, its spectrum is flat over the bandwidth of interest, and for this reason $n_c(t)$ can be treated as white noise[1].

Likewise, $E\{n_{s2}\} = 0$ and $E\{n_{s2}^2\} = \tfrac{1}{2}\mathcal{N}_0 \, E_s$. Thus, we have the envelope decision variables given by

$$R_{c1} = E_s \cos \theta + n_{c1} = G\left(E_s \cos \theta, \tfrac{1}{2}\mathcal{N}_0 E_s\right) \qquad (7.64a)$$
$$R_{s1} = E_s \sin \theta + n_{s1} = G\left(E_s \sin \theta, \tfrac{1}{2}\mathcal{N}_0 E_s\right) \qquad (7.64b)$$

$$R_{c2} = n_{c2} = G\left(0, \tfrac{1}{2}\mathcal{N}_0 E_s\right) \tag{7.64c}$$

$$R_{s2} = n_{s2} = G\left(0, \tfrac{1}{2}\mathcal{N}_0 E_s\right) \tag{7.64d}$$

It is clear that R_1^2 and R_2^2 are scaled chi-squared RVs. Since normalization of a Gaussian RV to obtain a unit-variance is given by

$$G\left(\mu, \sigma^2\right) = \mu + G\left(0, \sigma^2\right) = \mu + \sigma G(0, 1) = \sigma G\left(\frac{\mu}{\sigma}, 1\right) \tag{7.65}$$

then the squared envelopes are

$$
\begin{aligned}
R_1^2 = R_{c1}^2 + R_{s1}^2 &= \left[G\left(E_s\cos\theta, \tfrac{1}{2}\mathcal{N}_0 E_s\right)\right]^2 + \left[G\left(E_s\cos\theta, \tfrac{1}{2}\mathcal{N}_0 E_s\right)\right]^2 \\
&= \left[\sqrt{\tfrac{1}{2}\mathcal{N}_0 E_s}\, G\left(\sqrt{\tfrac{2E_s}{\mathcal{N}_0}}\cos\theta, 1\right)\right]^2 + \left[\sqrt{\tfrac{1}{2}\mathcal{N}_0 E_s}\, G\left(\sqrt{\tfrac{2E_s}{\mathcal{N}_0}}\sin\theta, 1\right)\right]^2 \\
&= \tfrac{1}{2}\mathcal{N}_0 E_s\left\{\left[G\left(\sqrt{\tfrac{2E_s}{\mathcal{N}_0}}\cos\theta, 1\right)\right]^2 + \left[G\left(\sqrt{\tfrac{2E_s}{\mathcal{N}_0}}\sin\theta, 1\right)\right]^2\right\} \\
&= \sigma^2\chi^2(2;\lambda), \quad \sigma^2 = \tfrac{1}{2}\mathcal{N}_0 E_s
\end{aligned}
\tag{7.66a}
$$

and

$$
\begin{aligned}
R_2^2 = R_{c2}^2 + R_{s2}^2 &= \left[G\left(0, \tfrac{1}{2}\mathcal{N}_0 E_s\right)\right]^2 + \left[G\left(0, \tfrac{1}{2}\mathcal{N}_0 E_s\right)\right]^2 \\
&= \left[\sqrt{\tfrac{1}{2}\mathcal{N}_0 E_s}\, G(0, 1)\right]^2 + \left[\sqrt{\tfrac{1}{2}\mathcal{N}_0 E_s}\, G(0, 1)\right]^2 \\
&= \tfrac{1}{2}\mathcal{N}_0 E_s\left\{[G(0, 1)]^2 + [G(0, 1)]^2\right\} \\
&= \sigma^2\chi^2(2;0) \equiv \sigma^2\chi^2(2), \quad \sigma^2 = \tfrac{1}{2}\mathcal{N}_0 E_s
\end{aligned}
\tag{7.66b}
$$

These results are based on the fact (see Section 1.5.9) that the sum of the squares of two independent zero-mean, unit-variance Gaussian RVs is a central chi-squared RV with two degrees of freedom, and that such a sum involving nonzero mean Gaussian RVs is a noncentral chi-squared RV with two degrees of freedom and noncentrality parameter λ, given by

$$\lambda \triangleq \left[\sqrt{\frac{2E_s}{\mathcal{N}_0}}\cos\theta\right]^2 + \left[\sqrt{\frac{2E_s}{\mathcal{N}_0}}\sin\theta\right]^2 = \frac{2E_s}{\mathcal{N}_0} \tag{7.66c}$$

Because R_1 and R_2 have the same scale factor σ^2, we may perform the analysis using the normalized decision statistics

$$Z_1 \triangleq R_1/\sigma \quad \text{and} \quad Z_2 \triangleq R_2/\sigma \tag{7.66d}$$

whose pdfs are given by [see (1.107g) and (1.105f)]

$$p_{z_1}(\alpha) = \begin{cases} \frac{1}{2}e^{-(\alpha+\lambda)/2}I_0\left(\sqrt{\alpha\lambda}\right), & \alpha \geq 0 \\ 0, & \text{otherwise} \end{cases} \tag{7.67a}$$

and

$$p_{z_2}(\alpha) = \begin{cases} \frac{1}{2}e^{-\alpha/2}, & \alpha \geq 0 \\ 0, & \text{otherwise} \end{cases} \tag{7.67b}$$

To compute $P(e \mid m_1)$, the conditional probability of error, given that m_1 is transmitted, we first assume that the signal plus noise channel envelope takes a specific value β and then average the probability over the range of β. Thus

$$P(e \mid m_1, Z_1 = \beta) = \Pr\{Z_2 > \beta \mid Z_1 = \beta\} = \int_{\beta}^{\infty} p_{z_2}(\alpha)\,d\alpha$$

$$= \int_{\beta}^{\infty} \frac{1}{2}e^{-\alpha/2}\,d\alpha = e^{-\beta/2} \tag{7.68a}$$

Averaging over the RV Z_1 is accomplished as follows:

$$P(e \mid m_1) = E_{Z_1}\{P(e \mid m_1, Z_1 = \beta)\} = \int_0^{\infty} e^{-\beta/2} p_{z_1}(\beta)\,d\beta$$

$$= \int_0^{\infty} \frac{1}{2}e^{-\beta-\lambda/2}I_0\left(\sqrt{\beta\lambda}\right)\,d\beta \tag{7.68b}$$

Let the variable of integration β be transformed by $\beta = \frac{1}{2}y$. Then (7.68b) becomes

$$P(e \mid m_1) = \int_0^{\infty} \underbrace{\frac{1}{2}e^{-(y+\lambda)/2}I_0\left(\sqrt{y \cdot \frac{1}{2}\lambda}\right)}_{e^{-\lambda/4} \times \text{pdf for } \lambda' = \frac{1}{2}\lambda} \cdot \frac{1}{2}\,dy = \frac{1}{2}e^{-\lambda/4} \tag{7.68c}$$

Substituting for the noncentrality parameter λ, we obtain the error probability for the noncoherent binary orthogonal (NCBO) system as

$$P_{NCBO}(e) = \frac{1}{2}e^{-E_s/2N_0} \tag{7.69}$$

7.2.2.4 Noncoherent Binary Orthogonal System in Rayleigh Fading

For the case of the NCBO system in the Rayleigh fading channel, the error probability can be obtained by averaging the AWGN error probability, given that the energy is subject to fading. The conditional error probability is

$$P_{NCBO}(e \mid a) = \tfrac{1}{2} e^{-a^2 E_s / 2 N_0} \tag{7.70a}$$

given the Rayleigh attenuation factor a. The pdf of a Rayleigh RV was given in (7.42b) as

$$p_a(\alpha) = \frac{2\alpha}{b} e^{-\alpha^2 / b}, \quad \alpha \geq 0 \tag{7.70b}$$

where $b = \overline{a^2}$ is the mean square value of a. We then obtain the error probability by computing

$$P_{NCBO}(e) = \tfrac{1}{2} \overline{e^{-a^2 E_s / 2 N_0}} = \int_0^\infty \tfrac{1}{2} e^{-\alpha^2 E_s / 2 N_0} \frac{2\alpha}{b} e^{-\alpha^2 / b} \, d\alpha$$

$$= \frac{1}{b} \int_0^\infty \alpha \, e^{-k \alpha^2 / 2} \, d\alpha = \frac{1}{b} \int_0^\infty e^{-ku} \, du = \frac{1}{bk} \tag{7.70c}$$

where

$$k = \frac{2}{b} + \frac{E_s}{N_0} = \frac{1}{b}\left(2 + \frac{b E_s}{N_0} \right) = \frac{1}{b}\left(2 + \frac{\overline{E_s}}{N_0} \right) \tag{7.70d}$$

using $\overline{E_s} \triangleq b E_s = \overline{a^2} E_s$. Therefore, the error probability for the NCBO system in Rayleigh fading is

$$P_{NCBO}(e) = \frac{1}{2 + \overline{E_s}/N_0} \tag{7.71}$$

We plotted (7.69) and (7.71) in Figure 7.44 for comparison purposes.

7.2.2.5 IS-95 CDMA Reverse Link M-ary Orthogonal Modulation Scheme

From the system diagram of the reverse link channels shown in Chapter 4, we observe that after coding and interleaving the information sequence, the symbols are transmitted at the rate of $R_s = 28,800$ symbols/sec (sps). For

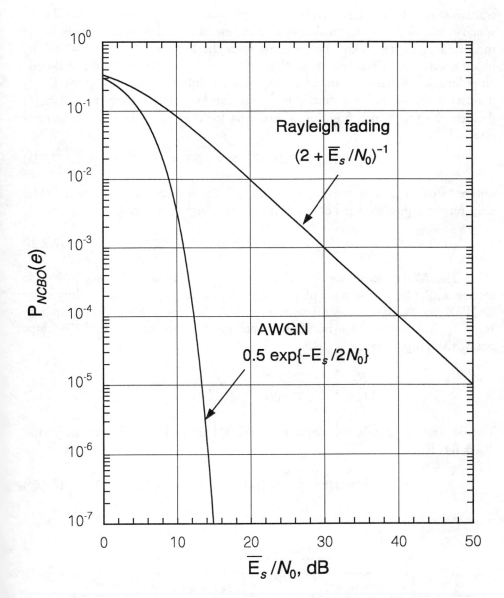

Figure 7.44 Probabilities of error for an NCBO system in AWGN and Rayleigh fading.

transmission, these data symbols are grouped into six-symbol blocks to specify one of $M = 64$ possible Walsh codewords (Walsh M-ary symbols), as indicated in Figure 7.45. The scheme is in the form of an $(n,\,k) = (64,\,6)$ block code, which we denote as $W(64,\,6)$ to indicate Walsh orthogonal-code chip length 64 with an input block of six symbols. In the figure, $D_k(t)$ denotes a six-symbol sequence (block) for the kth mobile user, and $W_{n(k)}(t)$ denotes the nth Walsh function of order 64 for that user. The Walsh symbol rate is

$$R_w = R_s/6 = 28{,}800/6 = 4{,}800 \, \text{wps} \tag{7.72a}$$

where "wps" means "words per second" or "Walsh symbols per second." The chip rate in chips/sec (cps) of the orthogonal Walsh sequence is

$$MR_w = MR_s/6 = 64 \times 4{,}800 = 307{,}200 \, \text{cps} \tag{7.72b}$$

The Walsh sequence $W_n(t)$ is then spread by a ± 1 long PN code sequence, $P_k(t)$, whose code phase is specific to the mobile user, followed by OQPSK modulation, as shown in Figure 7.46. Because the PN chip rate $R_c = 1/T_c$ is higher than the Walsh sequence chip rate, there are N PN chips per Walsh chip, where

$$N = \frac{R_c}{MR_w} = \frac{1{,}228{,}800}{307{,}200} = 4 \, \text{PN chips/Walsh chip} \tag{7.72c}$$

The 64 possible ± 1 Walsh sequences $\{W_n(t)\}$ are orthogonal over the period $T_w = 64/R_w$:

$$\langle W_n(t), W_q(t) \rangle = \int_0^{T_w} W_n(t) \, W_q(t) \, dt = T_w \delta_{nq} \tag{7.72d}$$

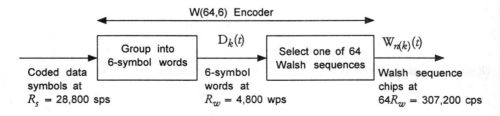

Figure 7.45 Reverse link Walsh orthogonal modulation concept.

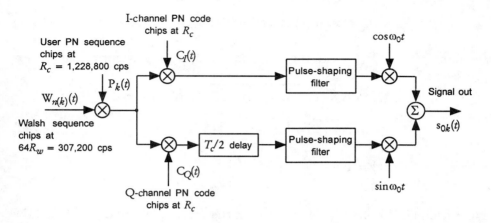

Figure 7.46 Diagram of reverse link spreading, pulse shaping, and OQPSK modulation.

Let us define $m_k(t)$ as the product of $W_n(t)$ and $P_k(t)$, that is:

$$m_k(t) \triangleq P_k(t)\, W_n(t) \tag{7.73a}$$

which carries the data for the kth user. Using these parameters and definitions for the reverse link M-ary orthogonal modulation, we specify the pertinent noncoherent receiver structure and evaluate its performance. We ignore the effects of the pulse shaping in our analysis.

From Figure 7.46, it is clear that the OQPSK modulated signal transmitted for the kth mobile user can be expressed as

$$s_{tr,k}(t) = m_k(t)\, C_I(t) \cos \omega_0 t + m_k\!\left(t - \tfrac{1}{2}T_c\right) C_Q\!\left(t - \tfrac{1}{2}T_c\right) \sin \omega_0 t \tag{7.73b}$$

where $\omega_0 = 2\pi f_0$ denotes the reverse link (uplink) carrier angular frequency, and $C_I(t)$ and $C_Q(t)$ are ± 1 short PN code sequences for the transmitter I and Q channels, respectively, which are clocked at the same chip rate as the long PN code sequence, $R_c = 1{,}228{,}800$ cps. At this point, we can drop the index "k" denoting "user k" and express the transmitted waveform as

$$s_{tr}(t) = A_0 \left[I(t) \cos \omega_0 t + Q\!\left(t - \tfrac{1}{2}T_c\right) \sin \omega_0 t \right] \tag{7.74a}$$

where

$$I(t) \triangleq m(t)\, C_I(t) = P(t)\, W_n(t)\, C_I(t) \tag{7.74b}$$

and

$$Q(t) \triangleq m(t) C_Q(t) = P(t) W_n(t) C_Q(t) \qquad (7.74c)$$

Because $I^2(t) = Q^2(t) = 1$, the OQPSK modulated waveform can be expressed as

$$s_{tr}(t) = A_0 \sqrt{I^2(t) + Q^2\left(t - \tfrac{1}{2}T_c\right)} \cos[\omega_0 t - \phi(t)]$$

$$= A_0 \sqrt{2} \cos[\omega_0 t - \phi(t)] \qquad (7.74d)$$

where
$$\phi(t) = \tan^{-1}\left[\frac{Q\left(t - \tfrac{1}{2}T_c\right)}{I(t)} \right] \qquad (7.74e)$$

Note from (7.74d) that the total power in the transmitted signal is A_0^2, and the powers in the I and Q channels are given by

$$P_{0,I} = \tfrac{1}{2}A_0^2 I^2(t) = \tfrac{1}{2}A_0^2 \quad \text{and} \quad P_{0,Q} = \tfrac{1}{2}A_0^2 Q^2\left(t - \tfrac{1}{2}T_c\right) = \tfrac{1}{2}A_0^2 \qquad (7.75a)$$

The chip energies for I and Q channels are

$$E_{c,I} \triangleq P_{0,I}T_c = \tfrac{1}{2}A_0^2 T_c \quad \text{and} \quad E_{c,I} \triangleq P_{0,I}T_c = \tfrac{1}{2}A_0^2 T_c \qquad (7.75b)$$

so that

$$A_0 = \sqrt{\frac{2E_{c,I}}{T_c}} = \sqrt{\frac{2E_{c,Q}}{T_c}} = \sqrt{\frac{2E_c}{T_c}} \qquad (7.75c)$$

where $E_c \triangleq E_{c,I} = E_{c,Q}$ is the chip energy in each of the I and Q channels. The total emitted OQPSK signal energy is thus given by

$$E_c' \triangleq A_0^2 T_c = 2E_c \qquad (7.75d)$$

7.2.2.6 Optimal Demodulation for IS-95 Reverse Link Waveforms

Let us assume that there is no multipath as well as no multiple access interference. Then the received reverse link waveform is given by

$$r(t) = s_{rec}(t) + n(t) \qquad (7.76a)$$

where $n(t)$ is assumed to be white Gaussian noise with two-sided power spectral density $\tfrac{1}{2}\mathcal{N}_0$. The signal term can be expressed as [see (7.33d)]

$$s_{rec}(t) = A\{m(t-\tau)\,C_I(t-\tau)\cos(\omega_1 t + \theta)$$
$$+ m(t - \tfrac{1}{2}T_c - \tau)\,C_Q(t - \tfrac{1}{2}T_c - \tau)\sin(\omega_1 t + \theta)\} \quad (7.76b)$$

where τ is the path delay and θ is a random phase of the carrier. The frequency notation in (7.76b) $f_1 = \omega_1/2\pi$ distinguishes the reverse link carrier frequency from that of the forward link, because the two are different. Because there is no pilot signal, nor any provision for carrier phase estimation at the base station for the reverse link, noncoherent signal processing (demodulation) is being performed.

Now, assuming that chip synchronization has been achieved,[11] so that we can set $\tau = 0$ in the received signal representation in (7.76b), we have for the received signal

$$s_{rec}(t) = A\big[I(t)\cos(\omega_1 t + \theta) + Q(t - \tfrac{1}{2}T_c)\sin(\omega_1 t + \theta)\big]$$
$$= A\big[I'(t)\cos\omega_1 t + Q'(t)\sin\omega_1 t\big] \quad (7.77a)$$

where

$$I'(t) \triangleq I(t)\cos\theta + Q(t - \tfrac{1}{2}T_c)\sin\theta \quad (7.77b)$$

and

$$Q'(t) \triangleq Q(t - \tfrac{1}{2}T_c)\cos\theta - I(t)\sin\theta \quad (7.77c)$$

Receiver signal processing at the front end. The first step for the signal processing is to obtain the quadrature components. This step is performed by multiplying the received waveform by I and Q carriers and then lowpass filtering to remove the $2f_1$ terms that result from the multiplication (heterodyning), as diagrammed in Figure 7.47. The received waveform, carrying the data in some Walsh function $W_n(t)$, is given by

$$r(t) = A\big[I'(t)\cos\omega_1 t + Q'(t)\sin\omega_1 t\big] + n(t) \quad (7.78a)$$

where $n(t)$ is white Gaussian noise with spectral density $\tfrac{1}{2}N_0$, given by

$$n(t) = n_c(t)\sqrt{2}\cos\omega_0 t + n_s(t)\sqrt{2}\sin\omega_0 t \quad (7.78b)$$

with $n_c(t)$ and $n_s(t)$ each being zero-mean Gaussian with variance $\tfrac{1}{2}N_0$. The received quadrature components shown in Figure 7.47 are

---

[11] Acquisition and synchronization are treated in Section 7.4.

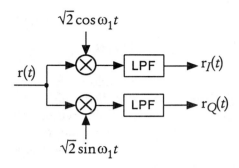

Figure 7.47 Extraction of quadrature components in the noncoherent receiver.

$$r_I(t) = \left[r(t) \times \sqrt{2}\cos\omega_1 t\right]_{\text{lowpass}} = \tfrac{1}{\sqrt{2}}A\,I'(t) + n_I(t) \qquad (7.78c)$$

and

$$r_Q(t) = \left[r(t) \times \sqrt{2}\sin\omega_1 t\right]_{\text{lowpass}} = \tfrac{1}{\sqrt{2}}A\,Q'(t) + n_Q(t) \qquad (7.78d)$$

in which $n_I(t) = n_c(t)$ and $n_Q(t) = n_c(t)$ are uncorrelated Gaussian noise terms, each with spectral density $\tfrac{1}{2}N_0$ in the bandwidth of its lowpass filter. Substituting for $I'(t)$ and $Q'(t)$ gives

$$\begin{aligned}
r_I(t) &= \tfrac{1}{\sqrt{2}}A\big[I(t)\cos\theta + Q\big(t - \tfrac{1}{2}T_c\big)\sin\theta\big] + n_I(t)\\
&= \tfrac{1}{\sqrt{2}}A\big[m(t)\,C_I(t)\cos\theta + m\big(t - \tfrac{1}{2}T_c\big)\,C_Q\big(t - \tfrac{1}{2}T_c\big)\sin\theta\big] + n_I(t)\\
&= \tfrac{1}{\sqrt{2}}A\big[P(t)\,W_n(t)\,C_I(t)\cos\theta\\
&\qquad + P\big(t - \tfrac{1}{2}T_c\big)\,W_n\big(t - \tfrac{1}{2}T_c\big)C_Q\big(t - \tfrac{1}{2}T_c\big)\sin\theta\big] + n_I(t) \qquad (7.79a)
\end{aligned}$$

and

$$\begin{aligned}
r_Q(t) &= \tfrac{1}{\sqrt{2}}A\big[Q\big(t - \tfrac{1}{2}T_c\big)\cos\theta - I(t)\sin\theta\big] + n_Q(t)\\
&= \tfrac{1}{\sqrt{2}}A\big[m\big(t - \tfrac{1}{2}T_c\big)\,C_Q\big(t - \tfrac{1}{2}T_c\big)\cos\theta - m(t)\,C_I(t)\sin\theta\big] + n_Q(t)\\
&= \tfrac{1}{\sqrt{2}}A\big[P\big(t - \tfrac{1}{2}T_c\big)\,W_n\big(t - \tfrac{1}{2}T_c\big)C_Q\big(t - \tfrac{1}{2}T_c\big)\cos\theta\\
&\qquad - P(t)\,W_n(t)\,C_I(t)\sin\theta\big] + n_Q(t) \qquad (7.79b)
\end{aligned}$$

The terms of (7.79a) and (7.79b) clearly indicate that each quadrature channel requires multiplications by $C_I(t) \times P(t)$ and $C_Q(t - \frac{1}{2}T_c) \times P(t - \frac{1}{2}T_c)$ to "despread" and "descramble" the signal terms, as depicted in Figure 7.48. The four output components in Figure 7.48 are obtained using the fact that

$$C_I^2(t) = P^2(t) = C_Q^2(t - \tfrac{1}{2}T_c) = P^2(t - \tfrac{1}{2}T_c) = 1$$

to develop the signal-plus-noise terms

$$d_{II}(t) \triangleq r_I(t) \times C_I(t) \times P(t) = \tfrac{1}{\sqrt{2}}AW_n(t)\cos\theta + n_{II}(t) \qquad (7.80a)$$

$$d_{IQ}(t) \triangleq r_I(t) \times C_Q(t - \tfrac{1}{2}T_c) \times P(t - \tfrac{1}{2}T_c)$$
$$= \tfrac{1}{\sqrt{2}}AW_n(t - \tfrac{1}{2}T_c)\sin\theta + n_{IQ}(t) \qquad (7.80b)$$

$$d_{QI}(t) \triangleq r_Q(t) \times C_I(t) \times P(t) = -\tfrac{1}{\sqrt{2}}AW_n(t)\sin\theta + n_{QI}(t) \quad (7.80c)$$

$$d_{QQ}(t) \triangleq r_Q(t) \times C_Q(t - \tfrac{1}{2}T_c) \times P(t - \tfrac{1}{2}T_c)$$
$$= \tfrac{1}{\sqrt{2}}AW_n(t - \tfrac{1}{2}T_c)\cos\theta + n_{QQ}(t) \qquad (7.80d)$$

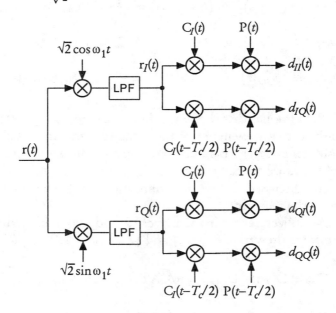

Figure 7.48 Despreading and descrambling operations on the quadrature components.

where the noise terms $n_{II}(t)$, $n_{IQ}(t)$, $n_{QI}(t)$, and $n_{QQ}(t)$ are specified later in the analysis of error performance of the demodulator.

Optimal receiver signal processing requires correlation of the components $\{d_{ij}(t)\}$ with stored replicas $\{W_q(t),\ W_q(t - \frac{1}{2}T_c)\ q = 1, 2, \ldots, M\}$ of the Walsh functions ($4M$ correlators). The correlation involving $d_{II}(t)$ produces

$$D_{II,q} \triangleq \langle d_{II}(t),\ W_q(t)\rangle = \langle \tfrac{1}{\sqrt{2}}AW_n(t)\cos\theta + n_{II}(t),\ W_q(t)\rangle$$

$$= \tfrac{1}{\sqrt{2}}A\cos\theta \int_0^{T_w} W_n(t)\,W_q(t)\,dt + \int_0^{T_w} n_{II}(t)\,W_q(t)\,dt$$

$$= \tfrac{1}{\sqrt{2}}AT_w\delta_{nq}\cos\theta + \xi_{II,q} \tag{7.81a}$$

using the orthogonality of the Walsh functions, expressed by

$$\int_0^{T_w} W_n(t)\,W_q(t)\,dt = T_w\delta_{nq} = \begin{cases} 1, & q = n \\ 0, & q \neq n \end{cases} \tag{7.81b}$$

Likewise, we have

$$D_{IQ,q} \triangleq \langle d_{IQ}(t),\ W_q(t - \tfrac{1}{2}T_c)\rangle = \tfrac{1}{\sqrt{2}}AT_w\delta_{nq}\sin\theta + \xi_{IQ,q} \tag{7.81c}$$

$$D_{QI,q} \triangleq \langle d_{QI}(t),\ W_q(t)\rangle = -\tfrac{1}{\sqrt{2}}AT_w\delta_{nq}\sin\theta + \xi_{QI,q} \tag{7.81d}$$

$$D_{QQ,q} \triangleq \langle d_{QQ}(t),\ W_q(t - \tfrac{1}{2}T_c)\rangle = \tfrac{1}{\sqrt{2}}AT_w\delta_{nq}\cos\theta + \xi_{QQ,q} \tag{7.81e}$$

Note that (7.81b) implies that the Walsh functions are ± 1 waveforms, just as the PN codes, so that the energy of the received signal in a Walsh period is A^2T_w without reference to parameters of any of the sequences used. The properties of the noise terms $\xi_{II,q}$, $\xi_{IQ,q}$, $\xi_{QI,q}$, and $\xi_{QQ,q}$ at the output of the correlators are discussed below in connection with the analysis of the performance of the receiver.

We now collect terms involving $\cos\theta$ and those involving $\sin\theta$, respectively, to form the variables $\{Z_{cq}\}$ and $\{Z_{sq}\}$, $q = 1, 2, \ldots, M$ as follows:

$$Z_{cq} \triangleq D_{II,q} + D_{QQ,q} = \sqrt{2}AT_w\delta_{nq}\cos\theta + n_{cq} \tag{7.82a}$$

and

$$Z_{sq} \triangleq D_{IQ,q} - D_{QI,q} = \sqrt{2}AT_w\delta_{nq}\sin\theta + n_{sq} \tag{7.82b}$$

where

$$n_{cq} \triangleq \xi_{II,q} + \xi_{QQ,q} \tag{7.82c}$$

and

$$n_{sq} \triangleq \xi_{IQ,q} - \xi_{QI,q} \tag{7.82d}$$

It is evident now that squared-envelope decision variables $\{R_q^2\}$ can be formed by summing the squares of Z_{cq} and Z_{sq}, thereby eliminating the dependence of the receiver decision on the random phase θ. The receiver therefore calculates the decision variables

$$R_q^2 = Z_{cq}^2 + Z_{sq}^2, \quad q = 1, 2, \ldots, M. \tag{7.82e}$$

The implementation of the optimal noncoherent receiver for M-ary Walsh function demodulation is diagrammed in Figure 7.49.

Analysis of the reverse link demodulator. We derived the optimal non-coherent demodulator for the reverse link. The demodulator forms the M squared-envelope decision variables $\{R_q^2, q = 1, 2, \ldots, M\}$, where $M = 64$ for the IS-95 system. We now determine the error performance of the demodulator for a single user without multipath interference.

Based on the optimal decision rule of (7.50d), we state that

$$\widehat{W}(t) = W_k(t) \quad \text{iff} \quad R_k^2 = \max_q\{R_q^2\} = \max_q\{Z_{cq}^2 + Z_{sq}^2\} \tag{7.83a}$$

where

$$Z_{cq} \triangleq \sqrt{2}AT_w\delta_{nq}\cos\theta + n_{cq} \tag{7.83b}$$

and

$$Z_{sq} \triangleq \sqrt{2}AT_w\delta_{nq}\sin\theta + n_{sq} \tag{7.83c}$$

The noise terms n_{cq} and n_{sq} are defined in (7.82c) and (7.82d), and it is shown in Appendix 7C that $\{(n_{cq}, n_{sq}) \, q = 1, 2, \ldots, M\}$ are independent, zero-mean Gaussian RVs, each with variance $\mathcal{N}_0 T_w$:

$$n_{cq} = G(0, \mathcal{N}_0 T_w), \quad n_{sq} = G(0, \mathcal{N}_0 T_w) \tag{7.83d}$$

Hence, the $\{(Z_{cq}, Z_{asq})\}$ are independent, nonzero-mean Gaussian RVs, each with variance $\mathcal{N}_0 T_w$:

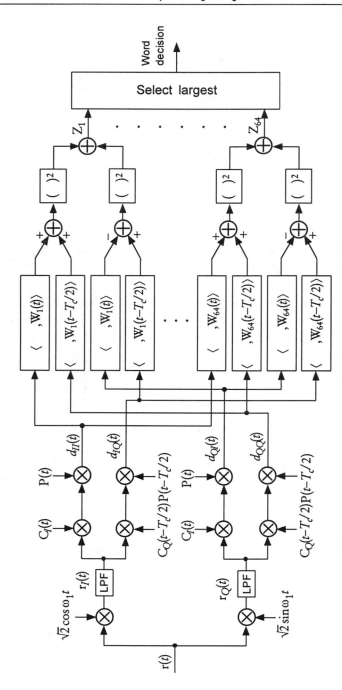

Figure 7.49 Optimal noncoherent receiver structure for IS-95 reverse link.

$$Z_{cq} = \begin{cases} G\left(\sqrt{2}AT_w\cos\theta,\ \mathcal{N}_0 T_w\right), & q = n \\ G(0,\ \mathcal{N}_0 T_w), & q \neq n \end{cases} \tag{7.84a}$$

and

$$Z_{sq} = \begin{cases} G\left(\sqrt{2}AT_w\sin\theta,\ \mathcal{N}_0 T_w\right), & q = n \\ G(0,\ \mathcal{N}_0 T_w), & q \neq n \end{cases} \tag{7.84b}$$

We therefore recognize that the decision variables $\{R_q^2,\ q = 1, 2, \ldots, M\}$ are scaled chi-squared RVs with two degrees of freedom:

$$R_q^2 = Z_{cq}^2 + Z_{sq}^2 = \begin{cases} \sigma^2 \chi^2(2;\lambda), & q = n \\ \sigma^2 \chi^2(2), & q \neq n \end{cases}, \ q = 1, 2, \ldots, M \tag{7.84c}$$

where $\sigma^2 = \mathcal{N}_0 T_w$ and the noncentrality parameter λ is given by

$$\lambda \triangleq \left[\frac{\sqrt{2}AT_w\cos\theta}{\sigma}\right]^2 + \left[\frac{\sqrt{2}AT_w\cos\theta}{\sigma}\right]^2 = \frac{2A^2 T_w}{\mathcal{N}_0} = \frac{2E_w}{\mathcal{N}_0} \tag{7.84d}$$

where

$$E_w \triangleq A^2 \int_0^{T_w} W_q^2(t)\, dt = A^2 T_w, \quad q = 1, 2, \ldots, M \tag{7.84e}$$

is the Walsh symbol energy and T_w is the Walsh symbol period.

Error probability analysis. Without loss of generality, suppose that the first Walsh word was sent. Then the decision variable R_1^2 is a scaled, *noncentral* chi-squared RV, given by

$$R_1^2 = Z_{c1}^2 + Z_{s1}^2 = \sigma^2 \chi^2(2;\lambda) \tag{7.85a}$$

and the remaining $M - 1$ decision variables are scaled, *central* chi-squared RVs, given by

$$R_q^2 = Z_{cq}^2 + Z_{sq}^2 = \sigma^2 \chi^2(2), \quad q \geq 2 \tag{7.85b}$$

Note that the decision variables are σ^2 times the standard forms of the chi-squared RVs, and hence we may normalize the decision variables as follows:

$$Z_1 \triangleq \frac{R_1^2}{\sigma^2} = \chi_1^2(2; \lambda) \quad \text{and} \quad Z_q \triangleq \frac{R_q^2}{\sigma^2} = \chi_q^2(2), \quad q \geq 2 \qquad (7.85c)$$

where $\chi_1^2(2; \lambda)$ is the notation for a noncentral chi-squared RV with two degrees of freedom and noncentrality parameter $\lambda = 2E_w/N_0$, and $\chi_q^2(2)$ denotes a central chi-squared RV with two degrees of freedom. The pdfs of Z_1 and Z_q, then, are given by [see (1.105f) and (1.107g)]

$$p_{z1}(\alpha; \lambda) = \begin{cases} \frac{1}{2} e^{-(\alpha+\lambda)/2} I_0\left(\sqrt{\alpha\lambda}\right), & \alpha \geq 0 \\ 0, & \text{otherwise} \end{cases} \qquad (7.86a)$$

$$\text{and} \qquad p_{zq}(\alpha) = \begin{cases} \frac{1}{2} e^{-\alpha/2}, & \alpha \geq 0, \\ 0, & \text{otherwise} \end{cases} \qquad q = 2, 3, \ldots, M \qquad (7.86b)$$

Let us denote the M Walsh functions as M "messages" $\{m_q\}$ corresponding to $\{W_q(t)\}$, $q = 1, 2, \ldots, M$. Then $P_w(e \mid m_1)$ for M equally likely Walsh symbols, given that $W_1(t)$ is transmitted, is equal to the total probability of error $P_w(e)$; that is, $P_w(e) = P_w(e \mid m_1)$. Because

$$P_w(e) = 1 - P_w(C) = 1 - P_w(C \mid m_1) \qquad (7.87a)$$

we first compute the probability of correct decision.

Assuming that the signal-plus-noise channel decision variable Z_1 has taken a value β, we have

$$P_w(C \mid m_1, Z_1 = \beta) = \Pr\{Z_2 < \beta, Z_3 < \beta, \ldots, Z_M < \beta \mid Z_1 = \beta\}$$

$$= \prod_{q=2}^{M} \Pr\{Z_q < \beta \mid Z_1 = \beta\} \qquad (7.87b)$$

$$= \prod_{q=2}^{M} \int_0^{\beta} \frac{1}{2} e^{-\alpha/2} \, d\alpha = \prod_{q=2}^{M} \left(1 - e^{-\beta/2}\right) = \left(1 - e^{-\beta/2}\right)^{M-1} \qquad (7.87c)$$

Averaged over Z_1, the conditional probability of correct decision becomes

$$P_w(C \mid m_1) = E_{Z_1}\{P_w(C \mid m_1, Z_1 = \beta)\}$$

$$= E_{Z_1}\left\{\left(1 - e^{-\beta/2}\right)^{M-1}\right\} \qquad (7.87d)$$

Substituting (7.87d) in (7.87a), we have

$$P_w(e) = 1 - E_{Z_1}\left\{\left(1 - e^{-\beta/2}\right)^{M-1}\right\} = 1 - E_{Z_1}\left\{\sum_{i=0}^{M-1}\binom{M-1}{i}(-1)^i e^{-i\beta/2}\right\}$$

$$= \sum_{i=1}^{M-1}\binom{M-1}{i}(-1)^{i+1}E_{Z_1}\left\{e^{-i\beta/2}\right\} \tag{7.87e}$$

where

$$
\begin{aligned}
E_{Z_1}\left\{e^{-i\beta/2}\right\} &= \int_0^\infty e^{-i\beta/2}\, p_{z1}(\beta)\, d\beta \\
&= \int_0^\infty e^{-i\beta/2}\, \tfrac{1}{2}\, e^{-(\beta+\lambda)/2}\, I_0\left(\sqrt{\beta\lambda}\right)\, d\beta \\
&= \int_0^\infty \tfrac{1}{2}\, e^{-[\beta(1+i)+\lambda]/2}\, I_0\left(\sqrt{\beta\lambda}\right)\, d\beta \\
&= \int_0^\infty \tfrac{1}{2}\, e^{-\frac{1}{2}\left[\beta(1+i)+\frac{\lambda}{1+i}+\lambda-\frac{\lambda}{1+i}\right]}\, I_0\left(\sqrt{(1+i)\beta\frac{\lambda}{1+i}}\right)\, d\beta \\
&= e^{-\frac{1}{2}\left(\lambda-\frac{\lambda}{1+i}\right)}\int_0^\infty \tfrac{1}{2}\, e^{-\frac{1}{2}[\beta(1+i)+\lambda']}\, I_0\left(\sqrt{(1+i)\beta\lambda'}\right)\, d\beta \\
&= \frac{1}{1+i}\, e^{-\frac{\lambda}{2}\left(\frac{i}{1+i}\right)}\, \underbrace{\int_0^\infty \tfrac{1}{2}\, e^{-\frac{1}{2}(x+\lambda')}\, I_0\left(\sqrt{x\lambda'}\right)\, dx}_{=\,1}
\end{aligned}
$$

Thus, the word error probability for the single-user case is

$$
\begin{aligned}
P_w(e) &= \sum_{i=1}^{M-1}\binom{M-1}{i}\frac{(-1)^{i+1}}{1+i}\exp\left\{-\frac{i}{1+i}\frac{\lambda}{2}\right\} \\
&= \sum_{i=1}^{M-1}\binom{M-1}{i}\frac{(-1)^{i+1}}{1+i}\exp\left\{-\frac{i}{1+i}\frac{E_w}{\mathcal{N}_0}\right\}
\end{aligned} \tag{7.87f}
$$

This expression has the same form as the symbol error probability for an M-ary noncoherent orthogonal (or M-ary noncoherent FSK (MFSK)) system [1, 7, 8], with the Walsh word energy E_w being replaced by the MFSK symbol energy. The symbol error probability expression (word error probability) of (7.87f) is plotted in Figure 7.50 as a function of word energy-to-noise spectral density with M as a parameter.

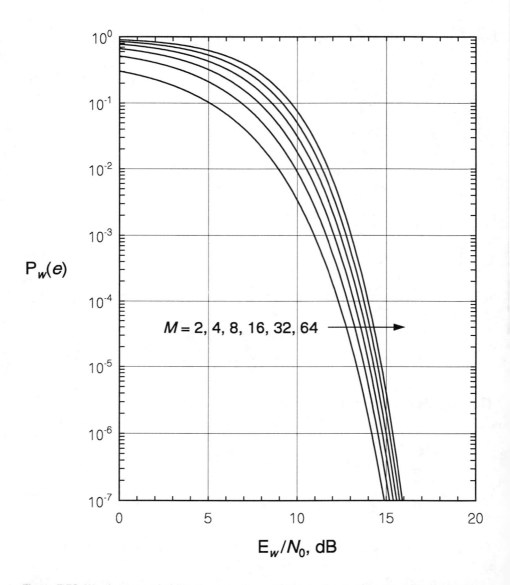

Figure 7.50 Word error probability for noncoherent M-ary orthogonal system. (The word energy E_w is sometimes called symbol energy, E_s.)

Bit error probability from word error probability. For M-ary orthogonal modulation systems, each word or M-ary symbol encodes $K = \log_2 M$ bits. The relationship between word energy and bit energy therefore is $E_w = KE_b$. Given the mathematical expression for the word error probability in (7.87f), the bit error probability, when M is a power of two, is obtained as

$$P_b\left(e; \frac{E_b}{N_0}\right) = \frac{M/2}{M-1} \cdot P_w\left(e; \frac{E_w}{N_0} = \frac{KE_b}{N_0}\right) \qquad (7.88)$$

The bit error probability for an M-ary orthogonal system is plotted in Figure 7.51 versus E_b/N_0, with M as a parameter. For the IS-95 reverse link, the $M = 64$ curve corresponds to the coded information symbol error rate, $P_s(e)$ versus E_s/N_0. That is, the "bits" in (7.57) are coded symbols running at 28.8 ksps, and "E_b/N_0" in Figure 7.51 should be read as "E_s/N_0," where E_s is the coded symbol energy, which equals $E_w/6$.

7.2.2.7 Reverse Link Performance in Rayleigh Fading

When the noncoherent M-ary orthogonal system's signal is subject to Rayleigh fading, the conditional word and bit error probabilities may be written

$$P_w(e \mid a) = \sum_{i=1}^{M-1} \binom{M-1}{i} \frac{(-1)^{i+1}}{1+i} \exp\left\{-\frac{i}{1+i} \frac{a^2 E_w}{N_0}\right\} \qquad (7.89a)$$

and

$$P_b(e \mid a) = \frac{M/2}{M-1} \sum_{i=1}^{M-1} \binom{M-1}{i} \frac{(-1)^{i+1}}{1+i} \exp\left\{-\frac{i}{1+i} \frac{Ka^2 E_b}{N_0}\right\} \qquad (7.89b)$$

in which $K = \log_2 M$ and a is a Rayleigh attenuation factor with the pdf given in (7.70b). We obtain the average word error probability by computing

$$P_w(e) = \sum_{i=1}^{M-1} \binom{M-1}{i} \frac{(-1)^{i+1}}{1+i} \overline{\exp\left\{-\frac{i}{1+i} \frac{a^2 E_w}{N_0}\right\}}$$

$$= \sum_{i=1}^{M-1} \binom{M-1}{i} \frac{(-1)^{i+1}}{1+i} \int_0^\infty \exp\left\{-\frac{i}{1+i} \frac{\alpha^2 E_w}{N_0}\right\} \frac{2\alpha}{b} e^{-\alpha^2/b} d\alpha$$

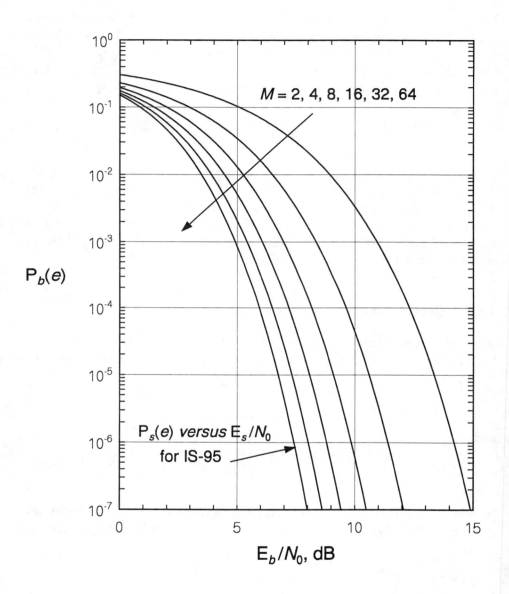

Figure 7.51 Bit error probability for noncoherent M-ary orthogonal system (for $M = 64$, $\mathbf{P}_b(e)$ should be read as $\mathbf{P}_s(e)$ versus $\mathbf{E}_s/\mathcal{N}_0 = \mathbf{E}_b/\mathcal{N}_0$ for the IS-95 system).

$$= \sum_{i=1}^{M-1} \binom{M-1}{i} \frac{(-1)^{i+1}}{1+i} \int_0^\infty \exp\left\{-\left(1 + \frac{i}{1+i}\frac{\overline{E_w}}{\mathcal{N}_0}\right)x\right\}dx$$

$$= \sum_{i=1}^{M-1} \binom{M-1}{i} \frac{(-1)^{i+1}}{1+i+i\cdot\overline{E_w}/\mathcal{N}_0} \qquad (7.89c)$$

using $\overline{E_w} = b E_w = b K E_b = K \overline{E_b}$. The average bit error probability is found by applying the formula (7.88) to the word error probability expression in (7.89c), and results in the graph shown in Figure 7.52. The IS-95 case of $M = 64$ is read from this figure by taking $P_b(e)$ to be $P_s(e)$, the average coded symbol error probability and $\overline{E_b} = \overline{E_s}$, the average coded symbol energy.

7.3 QPSK Versus BPSK

In the IS-95 CDMA digital cellular system, both forward and reverse links use forms of QPSK in which the same baseband data stream modulates both I and Q pseudorandom noise sequences. Conventional CDMA uses BPSK, in which the baseband data stream modulates a single PN sequence. In this section, a direct comparison is made of the two approaches to show the advantage of the IS-95 scheme. First, calculations are made assuming only BPSK spreading in a CDMA system, then the same quantities are calculated for QPSK spreading in a CDMA system for comparison. The development is similar to the method given in [10].

7.3.1 Analysis of a BPSK CDMA System

The *transmitted* waveform for the ith user in a BPSK CDMA system may be written

$$s_0^{(i)}(t) = A_0^{(i)}\cos(2\pi f_c t) \times \sum_n d_n^{(i)}\, c_n^{(i)}\, p_1(t - nT_c) \qquad (7.90a)$$

where $A_0^{(i)}$ is the transmitted amplitude, f_c is the carrier frequency, $d_n^{(i)}$ is the ± 1 data value and $c_n^{(i)}$ is the ± 1 PN code value during the nth chip interval T_c, and $p_1(t)$ is the impulse response of the pulse-shaping filter. Note that in the expression for the waveform, $d_n^{(i)}$ does not mean "the nth data symbol"

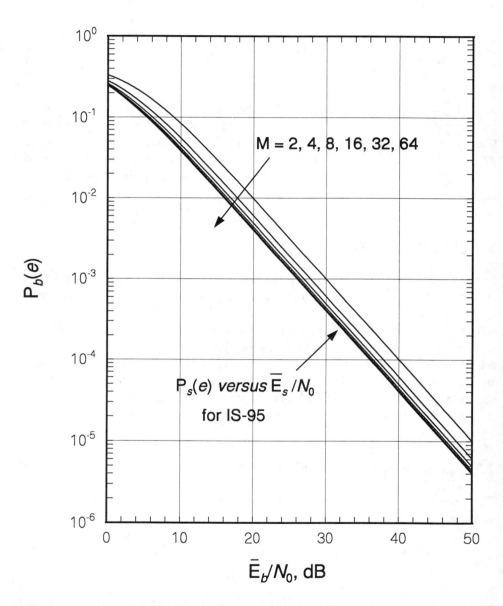

Figure 7.52 Bit error probability for noncoherent M-ary orthogonal system in Rayleigh fading (for $M = 64$, $P_b(e)$ should be read as $P_s(e)$ versus $\overline{E_s}/\mathcal{N}_0 = \overline{E_b}/\mathcal{N}_0$ for the IS-95 system).

but rather "the value of the data symbol at the time of the nth PN code chip." The *received* waveform at a particular location in the CDMA system can be written as noise plus the superposition of signals from multiple users:

$$r(t) = s(t) + n(t) = \sum_i s^{(i)}(t) + n(t) \qquad (7.90b)$$

in which the signal received from user i is modeled as

$$s^{(i)}(t) = A^{(i)} \cos\left(2\pi f_c t + \varphi_i\right) \times \sum_n d_n^{(i)} c_n^{(i)} \, p_1(t - nT_c - \tau_i) \qquad (7.90c)$$

where the received carrier amplitude and phase for user i at the common receiver location are $A^{(i)}$ and φ_i, and τ_i is the corresponding propagation time delay.

For the purpose of this analysis, we assume that the data value $d_n^{(i)}$ is equally likely to be $+1$ or -1 for all users, and that the PN codes are very long, so that $c_n^{(i)}$ is also equally likely to be $+1$ or -1 for all users. Further, we assume that the correlation and cross-correlation properties of the user PN codes are such that

$$E\{c_n^{(i)} c_m^{(j)}\} = \delta_{ij}\,\delta_{mn} = \begin{cases} 0, & i \neq j \\ 0, & i = j,\ m \neq n \\ 1, & i = j,\ m = n \end{cases} \qquad (7.91)$$

The kth CDMA receiver performs the following operations, as depicted in Figure 7.53:

1. Multiply the incoming waveform by the coherent oscillator reference for the kth signal, which is $\sqrt{2}\cos\left(2\pi f_c t + \varphi_k\right)$.

2. Lowpass filter the product, using the matched filter (*conjugate filter*) for the specially-shaped pulses, $p_1(t)$ [see (7.17)].

3. Sample the lowpass filter output to recover the chip stream.

4. Multiply the chip stream by the synchronous local PN sequence reference, which is $c^{(k)}(t)$, to recover the baseband data stream.

The statistics of the output of the BPSK CDMA receiver are studied in terms of first and second moments. Later, the same quantities for a QPSK CDMA system are derived and compared with those for BPSK.

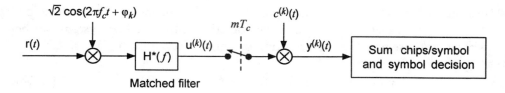

Figure 7.53 BPSK CDMA receiver processing.

The output of the matched lowpass filter at the kth BPSK CDMA receiver can be expressed by

$$u^{(k)}(t) = \left[r(t) \cdot \sqrt{2} \cos\left(2\pi f_c t + \varphi_k\right) \right]_{LP} \tag{7.92a}$$

$$= \frac{1}{\sqrt{2}} A^{(k)} \sum_n d_n^{(k)} c_n^{(k)} \, p_2(t - nT_c)$$

$$+ \frac{1}{\sqrt{2}} \cdot \sum_{i \neq k} A^{(i)} \cos\left(\varphi_i - \varphi_k\right) \sum_n d_n^{(i)} c_n^{(i)} \, p_2(t - nT_c - \tau_{ik})$$

$$+ n_{LP}(t) \tag{7.92b}$$

where $p_2(t)$ is the pulse shape that results from matched filtering of $p_1(t)$. The first term in (7.92b) is user k's synchronized signal, the second term is the sum of the unsynchronized signals of the other users, and the third term is Gaussian noise. We use the notation $\tau_{ik} = \tau_i - \tau_k$ to denote the time of arrival of the ith signal relative to that of user k.

It is assumed that $n(t)$, the noise component of $r(t)$, is white noise with two-sided spectral power density $\frac{1}{2} \mathcal{N}_0$. Because multiplication of $n(t)$ by the local oscillator waveform $\sqrt{2} \cos\left(2\pi f_c t + \varphi_k\right)$ shifts the white noise spectrum up and down by the amount f_c, and because the spectrum is flat to begin with, these operations do not affect the spectral shape—the noise is still white, with spectral power density level $\frac{1}{2} \mathcal{N}_0$. Therefore, $n_{LP}(t)$ is the result of passing white noise with spectral density level $\frac{1}{2} \mathcal{N}_0$ through a lowpass filter with frequency transfer function $H^*(f)$ that is matched to $H(f)$, the transmitter waveshaping filter's transfer function.

Since the transmitted shaped pulses are related to the pulse-shaping filter by the inverse Fourier transform

$$p_1(t) = \mathcal{F}^{-1}\{H(f)\} = \int_{-\infty}^{\infty} e^{j2\pi ft} H(f) \, df \tag{7.93a}$$

the pulse shape after matched filtering is related to the filter characteristic by

$$p_2(t) = \mathcal{F}^{-1}\{H(f) \cdot H^*(f)\} = \int_{-\infty}^{\infty} e^{j2\pi ft} |H(f)|^2 \, df \tag{7.93b}$$

This result expresses the fact that $p_2(t)$ is the impulse response of the cascading of the pulse-shaping filter in the transmitter and its matched filter in the receiver.

The output of the kth BPSK CDMA receiver at time $t = mT_c$ is the result of sampling the recovered chip stream at the lowpass filter output, then multiplying it by the local reference for the kth PN code:

$$y^{(k)}(mT_c) = u^{(k)}(mT_c) \cdot c^{(k)}(mT_c) \equiv u_m^{(k)} c_m^{(k)} = y_m^{(k)} \tag{7.94a}$$

where, using $P^{(i)} \triangleq [A^{(i)}]^2/2$ for the received power of the ith signal,

$$y_m^{(k)} = \sqrt{P^{(k)}} \sum_n d_n^{(k)} c_n^{(k)} c_m^{(k)} \, p_2(mT_c - nT_c)$$

$$+ \sum_{i \neq k} \sqrt{P^{(i)}} \cos(\varphi_i - \varphi_k) \sum_n d_n^{(i)} c_n^{(i)} c_m^{(k)} \, p_2(mT_c - nT_c - \tau_{ik})$$

$$+ n_{LP}(mT_c) c_m^{(k)} \tag{7.94b}$$

Each of these terms—the kth signal, the other signals, and the noise—is statistically independent of the other terms.

Given the data symbol value and the slowly varying relative carrier phases and relative delays, and using the assumptions in (7.91), the mean value of the kth BPSK CDMA receiver output sample is given by

$$E\{y_m^{(k)} \mid d_m^{(k)}, \{\varphi_i, \tau_{ik}\}\} = \sqrt{P^{(k)}} \sum_n E\{d_n^{(k)} \mid d_m^{(k)}\} E\{c_n^{(k)} c_m^{(k)}\}$$

$$\times p_2(mT_c - nT_c) \; + \sum_{i \neq k} \sqrt{P^{(i)}} \cos(\varphi_i - \varphi_k)$$

$$\times \sum_n E\{d_n^{(i)}\} E\{c_n^{(i)} c_m^{(k)}\} \, p_2(mT_c - nT_c - \tau_{ik})$$

$$+ E\{n_{LP}(mT_c)\} E\{c_m^{(k)}\}$$

$$= \sqrt{P^{(k)}} \, E\{d_m^{(k)} \mid d_m^{(k)}\} \, p_2(0) = \sqrt{P^{(k)}} \, d_m^{(k)} p_2(0) \tag{7.95}$$

because all other expectations equal zero. Note in view of this result that the components of user k's signal for chip pulses peaking at times other than $t = mT_c$ constitute PN chip intersymbol interference (ISI); these components may be nonzero because the pulse $p_2(t)$ is not confined to a one-chip interval.

Given the data symbol value and the relative phases and time delays, the variance of the kth BPSK CDMA receiver output sample, then, is the sum of the variances of each of the interference terms in (7.94b):

$$\mathrm{Var}\{y_m^{(k)} \mid d_m^{(k)}, \{\varphi_i, \tau_{ik}\}\} = \mathrm{Var}\left\{ \sqrt{P^{(k)}} \sum_{n \neq m} d_n^{(k)} c_n^{(k)} c_m^{(k)} \, p_2(mT_c - nT_c) \right\}$$

$$+ \sum_{i \neq k} \mathrm{Var}\left\{ \sqrt{P^{(i)}} \cos{(\varphi_i - \varphi_k)} \sum_n d_n^{(i)} c_n^{(i)} c_m^{(k)} \, p_2(mT_c - nT_c - \tau_{ik}) \right\}$$

$$+ \mathrm{Var}\left\{ n_{LP}(mT_c) \, c_m^{(k)} \right\} \tag{7.96a}$$

$$= \sigma_{ISI,k}^2 + \sigma_{MA,k}^2 + \sigma_N^2 \tag{7.96b}$$

and it is shown in Appendix 7D that the variance components are as follows:

$$\sigma_{ISI,k}^2 \triangleq P^{(k)} \sum_{\substack{n=-\infty \\ n \neq 0}}^{\infty} [p_2(nT_c)]^2 \tag{7.96c}$$

$$\sigma_{MA,k}^2 \triangleq \sum_{i \neq k} P^{(i)} \cos^2{(\varphi_i - \varphi_k)} \cdot \frac{1}{T_c} \int_{-\infty}^{\infty} |H(f)|^4 \, df \tag{7.96d}$$

$$\sigma_N^2 \triangleq \tfrac{1}{2} \mathcal{N}_0 \int_{-\infty}^{\infty} |H(f)|^2 \, df \tag{7.96e}$$

If the average of $\sigma_{MA,k}^2$ is taken with respect to the carrier phases, either over the ensemble of users, or over time, then (7.96d) becomes

$$\sigma_{MA,k}^2 = \sum_{i \neq k} P^{(i)} \cdot \frac{1}{2} \cdot \frac{1}{T_c} \int_{-\infty}^{\infty} |H(f)|^4 \, df \tag{7.96f}$$

7.3.2 Analysis of a QPSK CDMA System

Now we find the mean and variance of a receiver output sample when QPSK is used by the CDMA waveform. First, we need to introduce additional notation:

$c_{In}^{(i)}$ PN sequence, user i in-phase component

$c_{Qn}^{(i)}$ PN sequence, user i quadrature component

Using the adopted notation, the transmitted waveform for the ith user in the QPSK CDMA system may be written

$$s_0^{(i)}(t) = A_0^{(i)} \cos\left(2\pi f_c t\right) \times \sum_n d_n^{(i)} c_{In}^{(i)} \, \text{p}_1(t - nT_c)$$

$$+ A_0^{(i)} \sin\left(2\pi f_c t\right) \times \sum_n d_n^{(i)} c_{Qn}^{(i)} \, \text{p}_1(t - nT_c) \qquad (7.97a)$$

The signal received from user i is modeled as

$$s^{(i)}(t) = A^{(i)} \cos(2\pi f_c t + \varphi_i) \sum_n d_n^{(i)} c_{In}^{(i)} \, \text{p}_1(t - nT_c - \tau_i)$$

$$+ A^{(i)} \sin(2\pi f_c t + \varphi_i) \sum_n d_n^{(i)} c_{Qn}^{(i)} \, \text{p}_1(t - nT_c - \tau_i) \qquad (7.97b)$$

Note that the received signal power is $P^{(i)} = \left[A^{(i)}\right]^2$. The kth CDMA receiver performs the following operations, as diagrammed in Figure 7.54:

1. Multiply the incoming waveform by coherent I and Q oscillator references for the kth signal, which are $\cos\left(2\pi f_c t + \varphi_k\right)$ and $\sin\left(2\pi f_c t + \varphi_k\right)$. (Note the amplitudes of 1, not $\sqrt{2}$, to make the baseband power the same as in the BPSK case.)

2. Lowpass filter the two products, using the matched filter for the specially-shaped pulses, $\text{p}_1(t)$.

3. Sample the lowpass filter outputs to recover the I and Q chip streams.

4. Multiply the chip streams by their respective synchronous local PN sequence references, which are $c_I^{(i)}(t)$ and $c_Q^{(i)}(t)$, and add the results to recover the baseband data stream.

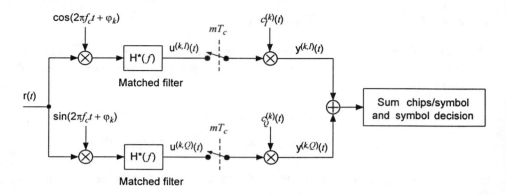

Figure 7.54 QPSK CDMA receiver processing.

The output of the I-channel matched lowpass filter at the kth QPSK CDMA receiver can be expressed by

$$u^{(k,I)}(t) \approx [r(t) \cdot \cos(2\pi f_c t + \varphi_k)]_{LP}$$

$$= \tfrac{1}{2} \cdot A^{(k)} \sum_n d_n^{(k)} c_{In}^{(k)} p_2(t - nT_c)$$

$$+ \tfrac{1}{2} \sum_{i \neq k} A^{(i)} \sum_n d_n^{(i)} \left[c_{In}^{(i)} \cos(\varphi_i - \varphi_k) + c_{Qn}^{(i)} \sin(\varphi_i - \varphi_k) \right] p_2(t - nT_c - \tau_{ik})$$

$$+ \tfrac{1}{\sqrt{2}} n_{LP}^{(I)}(t) \tag{7.98a}$$

where $p_2(t)$ is the pulse shape that results from matched filtering of $p_1(t)$. The output of the Q-channel matched lowpass filter at the kth QPSK CDMA receiver can be expressed by

$$u^{(k,Q)}(t) \approx [r(t) \cdot \sin(2\pi f_c t + \varphi_k)]_{LP}$$

$$= \tfrac{1}{2} \cdot A^{(k)} \sum_n d_n^{(k)} c_{Qn}^{(k)} p_2(t - nT_c)$$

$$+ \tfrac{1}{2} \sum_{i \neq k} A^{(i)} \sum_n d_n^{(i)} \left[c_{Qn}^{(i)} \cos(\varphi_i - \varphi_k) - c_{In}^{(i)} \sin(\varphi_i - \varphi_k) \right] p_2(t - nT_c - \tau_{ik})$$

$$+ \tfrac{1}{\sqrt{2}} n_{LP}^{(Q)}(t) \tag{7.98b}$$

As discussed previously for the BPSK system, the multiplication of white noise by $\sqrt{2}\cos(2\pi f_c t + \varphi_k)$ results in white noise with the same spectral density level. It follows that multiplication of white noise by $1 \cdot \cos(2\pi f_c t + \varphi_k)$ and by $1 \cdot \sin(2\pi f_c t + \varphi_k)$ results in two white noise terms, each of which has one-half the original spectral density level. At the same time instant, the two white noise terms are uncorrelated because of the orthogonality of the sine and cosine. For consistency, we use $n_{LP}^{(I)}(t)$ and $n_{LP}^{(Q)}(t)$ to denote the lowpass noise terms that result when oscillators with the amplitude $\sqrt{2}$ are used, just as for the BPSK case.

The output of the kth QPSK CDMA receiver is the result of sampling the recovered chip streams at the lowpass filter outputs, then multiplying them respectively by the local references for the kth I and Q PN codes and, finally, adding the products:

$$y^{(k)}(mT_c) = u^{(k,I)}(mT_c)\, c_{Im}^{(k)} + u^{(k,Q)}(mT_c)\, c_{Qm}^{(k)}$$

$$\equiv u_m^{(k,I)} c_{Im}^{(k)} + u_m^{(k,Q)} c_{Qm}^{(k)} = y_m^{(k)} \tag{7.99a}$$

where, using $P^{(i)} \triangleq \left[A^{(i)}\right]^2$,

$$y_m^{(k)} = \tfrac{1}{2}\sqrt{P^{(k)}} \sum_n d_n^{(k)} \left[c_{In}^{(k)} c_{Im}^{(k)} + c_{Qn}^{(k)} c_{Qm}^{(k)} \right] p_2(mT_c - nT_c)$$

$$+ \tfrac{1}{2}\sum_{i \neq k}\sqrt{P^{(i)}} \sum_n d_n^{(i)} \left\{ \left[c_{In}^{(i)} c_{Im}^{(k)} + c_{Qn}^{(i)} c_{Qm}^{(k)} \right] \cos(\varphi_i - \varphi_k) \right.$$

$$\left. + \left[c_{Qn}^{(i)} c_{Im}^{(k)} - c_{In}^{(i)} c_{Qm}^{(k)} \right] \sin(\varphi_i - \varphi_k) \right\} p_2(mT_c - nT_c - \tau_{ik})$$

$$+ \frac{1}{\sqrt{2}}\, n_{LP}^{(I)}(mT_c)\, c_{Im}^{(k)} + \frac{1}{\sqrt{2}}\, n_{LP}^{(Q)}(mT_c)\, c_{Qm}^{(k)} \tag{7.99b}$$

Note that the noise terms in (7.99b) have been scaled to reflect the different local oscillator amplitudes while retaining the same variances for the lowpass noise variables $n_{LP}^{(I)}$ and $n_{LP}^{(Q)}$ as for the noise term in (7.94b).

Given the data symbol value and the relative phases and time delays, and utilizing the assumptions in (7.91), the mean value of the mth sample of the output of the QPSK CDMA receiver for user k is

$$E\left\{y_m^{(k)} \mid d_m^{(k)}, \{\varphi_i, \tau_{ik}\}\right\} = \tfrac{1}{2}\sqrt{P^{(k)}} \sum_n E\left\{d_n^{(k)} \mid d_m^{(k)}\right\} p_2(mT_c - nT_c)$$

$$\times \left[E\left\{c_{In}^{(k)} c_{Im}^{(k)}\right\} + E\left\{c_{Qn}^{(k)} c_{Qm}^{(k)}\right\} \right]$$

$$+ \tfrac{1}{2} \sum_{i \neq k} \sqrt{P^{(i)}} \sum_n E\left\{d_n^{(i)}\right\} p_2(mT_c - nT_c - \tau_{ik})$$

$$\times \left\{ \left[E\left\{c_{In}^{(i)} c_{Im}^{(k)}\right\} + E\left\{c_{Qn}^{(i)} c_{Qm}^{(k)}\right\} \right] \cos(\varphi_i - \varphi_k) \right.$$

$$\left. + \left[E\left\{c_{Qn}^{(i)} c_{Im}^{(k)}\right\} - E\left\{c_{In}^{(i)} c_{Qm}^{(k)}\right\} \right] \sin(\varphi_i - \varphi_k) \right\}$$

$$+ \tfrac{1}{\sqrt{2}} E\left\{n_{LP}^{(I)}(mT_c)\right\} E\left\{c_{Im}^{(k)}\right\} + \tfrac{1}{\sqrt{2}} E\left\{n_{LP}^{(Q)}(mT_c)\right\} E\left\{c_{Qm}^{(k)}\right\}$$

$$= \tfrac{1}{2}\sqrt{P^{(k)}}\, E\left\{d_m^{(k)} \mid d_m^{(k)}\right\} \left[E\left\{\left[c_{Im}^{(k)}\right]^2\right\} + E\left\{\left[c_{Qm}^{(k)}\right]^2\right\} \right] p_2(0)$$

or
$$E\{y_m^{(k)} \mid d_m^{(k)}, \{\varphi_i, \tau_{ik}\}\} = \sqrt{P^{(k)}}\, d_m^{(k)} p_2(0) \qquad (7.100)$$

Note that this mean value is the same as (7.95), the mean for the BPSK case.

Given the data symbol value and the relative phases and delays, the variance of the kth QPSK CDMA receiver output sample is

$$\text{Var}\left\{y^{(k)}(mT_c) \mid x_m^{(k)}\right\}$$

$$= \text{Var}\left\{ \tfrac{1}{2}\sqrt{P^{(k)}} \sum_{n \neq m} d_n^{(k)} \left[c_{In}^{(k)} c_{Im}^{(k)} + c_{Qn}^{(k)} c_{Qm}^{(k)} \right] p_2(mT_c - nT_c) \right\}$$

$$+ \sum_{i \neq k} \text{Var}\left\{ \tfrac{1}{2}\sqrt{P^{(i)}} \sum_n d_n^{(i)} \left\{ \left[c_{In}^{(i)} c_{Im}^{(k)} + c_{Qn}^{(i)} c_{Qm}^{(k)} \right] \cos(\varphi_i - \varphi_k) \right. \right.$$

$$\left. \left. + \left[c_{Qn}^{(i)} c_{Im}^{(k)} - c_{In}^{(i)} c_{Qm}^{(k)} \right] \sin(\varphi_i - \varphi_k) \right\} p_2(mT_c - nT_c - \tau_{ik}) \right\}$$

$$+ \text{Var}\left\{ \tfrac{1}{\sqrt{2}} n_{LP}^{(I)}(mT_c)\, c_{Im}^{(k)} \right\} + \text{Var}\left\{ \tfrac{1}{\sqrt{2}} n_{LP}^{(Q)}(mT_c)\, c_{Qm}^{(k)} \right\} \qquad (7.101a)$$

$$= \sigma_{ISI,k}^2 + \sigma_{MA,k}^2 + \sigma_N^2 \qquad (7.101b)$$

where it is shown in Appendix 7D that the variance has the following components:

$$\sigma^2_{ISI,k} = \tfrac{1}{2} \, \mathrm{P}^{(k)} \sum_{\substack{n=-\infty \\ n\neq 0}}^{\infty} [\mathrm{p}_2(nT_c)]^2 \tag{7.101c}$$

$$\sigma^2_{MA,k} = \sum_{i\neq k} \mathrm{P}^{(i)} \cdot \frac{1}{2T_c} \int_{-\infty}^{\infty} |\mathrm{H}(f)|^4 \, df \tag{7.101d}$$

$$\sigma^2_{N} = \tfrac{1}{2} \, \mathcal{N}_0 \int_{-\infty}^{\infty} |\mathrm{H}(f)|^2 \, df \tag{7.101e}$$

7.3.3 Comparison of BPSK, QPSK Variances

The averaged variances computed for BPSK and QPSK data symbols are summarized in Table 7.1. We assumed that the relative carrier phases of the different multiple access users are slowly varying in comparison with the duration of a symbol, and the local oscillator powers in the two models were adjusted to make the mean values equal. The table shows that there is only one difference between the two systems, with respect to the averaged interference power (variance): the QPSK system has one-half the intersymbol interference power of the BPSK system.

On an instantaneous basis instead of an averaged basis, however, Table 7.1 reveals that there is a second difference: the multiple access interference variance (power) of BPSK is a function of the relative carrier phases, while that of QPSK is not. Because the QPSK multiple access interference power is

Table 7.1 Summary of average variances

	BPSK CDMA	QPSK CDMA
$\sigma^2_{ISI,k}$	$\mathrm{P}^{(k)} \sum\limits_{\substack{n=-\infty \\ n\neq 0}}^{\infty} [\mathrm{p}_2(nT_c)]^2$	$\tfrac{1}{2} \mathrm{P}^{(k)} \sum\limits_{\substack{n=-\infty \\ n\neq 0}}^{\infty} [\mathrm{p}_2(nT_c)]^2$
$\sigma^2_{MA,k}$	$\sum\limits_{i\neq k} \mathrm{P}^{(i)} \overline{\cos^2(\varphi_i - \varphi_k)} \frac{1}{T_c} \int_{-\infty}^{\infty} |\mathrm{H}(f)|^4 df$ $= \sum\limits_{i\neq k} \mathrm{P}^{(i)} \frac{1}{2T_c} \int_{-\infty}^{\infty} |\mathrm{H}(f)|^4 df$	$\sum\limits_{i\neq k} \mathrm{P}^{(i)} \frac{1}{2T_c} \int_{-\infty}^{\infty} |\mathrm{H}(f)|^4 df$
σ^2_{N}	$\tfrac{1}{2} \mathcal{N}_0 \int_{-\infty}^{\infty} |\mathrm{H}(f)|^2 df$	$\tfrac{1}{2} \mathcal{N}_0 \int_{-\infty}^{\infty} |\mathrm{H}(f)|^2 df$

not a function of the carrier phase differences, it is not subject to the fluctuation that can happen with BPSK CDMA, even though the two schemes have the same average MA interference power. Note that, for pulse shapes $p_1(t)$ either strictly time-limited to one chip period or strictly band-limited to $W = 1/T_c$, it can be shown that $p_2(nT_c) \approx 0$ for $n \neq 0$. Thus, it is possible to remove the ISI by pulse shaping, making BPSK CDMA and QPSK CDMA equivalent with respect to interchip ISI. Although it was shown in Chapter 1 that the IS-95 pulse shape fits neither category, it is true that values of the actual pulse shape at nonzero chip intervals is very small. Thus, the significant difference between BPSK CDMA and QPSK CDMA is that the instantaneous multiple access interference is not subject to fluctuation as the different carrier phases of the other users shift relative to that of the desired user.

7.4 PN Code Acquisition and Tracking

During all the previous discussions on the demodulation of spread-spectrum signals for forward and reverse links, we have always assumed that the receiver's PN sequence phase and the received signal's PN sequence were perfectly synchronized. In spread-spectrum communications systems, we simply cannot demodulate the data without despreading the spread-spectrum signal, which means that the synchronization is an absolutely essential requirement. Without the assumptions of perfect synchronization, therefore, we could not have proceeded with the analysis of the error performance as we have done in the previous sections.

We now discuss the issues related to PN sequence synchronization at the receiver, because we have developed most of the the essential "tools" and techniques for the pertinent analysis required. It is the purpose of this section to present the fundamentals related to PN code acquisition and tracking of the PN sequence in relation to IS-95 systems operation. In connection with PN sequence synchronization, the procedure consists of two parts:

- Initial determination of the PN sequence code phase, called *acquisition*, and

- Maintenance of the PN sequence code phase synchronization, called *tracking*.

Let us pose a question at this point: How does a mobile station acquire the initial synchronization? In the IS-95 CDMA system [11], there are two pre-designated (default) frequency allocations (FAs), called the *primary* CDMA channel and the *secondary* CDMA channel. At these frequencies, the system constantly transmits the pilot and synchronization signals, in addition to possible paging and traffic signals as well. The mobile station knows the frequency of the default FAs and, of course, knows the unique Walsh function that the transmitter uses for the pilot signal. That Walsh function is an all-zero sequence; thus the pilot channel waveform being transmitted is given by

$$s_{pilot}(t) = \left[\sum_n c_{In} \, w_n \, \mathrm{p}(t - nT_c) \right] \cdot A_0 \cos \omega_0 t$$

$$+ \left[\sum_n c_{Qn} \, w_n \, \mathrm{p}(t - nT_c) \right] \cdot A_0 \sin \omega_0 t, \ 0 \le t \le T_w \qquad (7.102a)$$

where

$c_{In} = $ nth chip polarity of the I-channel PN sequence

$w_n = $ nth chip polarity of the Walsh sequence $W_0 = 1$ for all n

$c_{Qn} = $ nth chip polarity of the Q-channel PN sequence

$A_0 = $ pilot channel signal amplitude

$\mathrm{p}(t) = $ shaped pulse for transmission

$T_w = $ period of Walsh functions $=$ data symbol period

For other forward link channels, the expression of (7.102a) is valid for one data symbol duration, the total forward link transmitted signal being of the form

$$s_{total}(t) = \mathrm{m}(t) \, s_{pilot}(t), \quad 0 \le t \le T_w \qquad (7.102b)$$

where

$$\mathrm{m}(t) \triangleq \sum_{i=0}^{63} g_i \, d_i \, W_i(t) = 1 + \sum_{i=1}^{63} g_i \, d_i \, W_i(t) \qquad (7.102c)$$

is the orthogonally multiplexed baseband combination of the forward link data streams, as explained in Section 4.2 and, by example, in Section 5.4.1. In (7.102c), $W_i(t)$ is the ith Walsh function, g_i denotes the amplitude of the nonpilot channel i relative to the pilot channel, and d_i represents the data symbol value in the interval. For example, for a traffic channel, the entire channel signal changes sign, depending on the sign of the data symbol d_i for "user i," where $d_i = +1$ or -1. The received forward link waveform is then given by

$$r(t) = s_{rec}(t) + n(t) \qquad (7.103a)$$

where $s_{rec}(t)$ and $n(t)$ are the received signal and noise terms, respectively, and the signal term is given as

$$s_{rec}(t) = m(t) \left\{ \left[\sum_n c_{In}\, p(t - nT_c) \right] \cdot A \cos\left(\omega_0 t + \phi_w\right) \right.$$

$$\left. + \left[\sum_n c_{Qn}\, p(t - nT_c) \right] \cdot A \sin\left(\omega_0 t + \phi_w\right) \right\} \qquad (7.103b)$$

where A is the received pilot signal amplitude and ϕ_w is the carrier phase, which is unknown but assumed to be constant for the entire duration of the symbol period, that is, for a period of 64 chips. We also assume that $n(t)$ represents both Gaussian noise and any other signals acting as interference, and it is modeled as AWGN of two-sided power spectral density $\frac{1}{2}\mathcal{N}_0$. The pilot signal chip energy of either the I or Q PN sequence, E_c, is given by

$$E_c = \tfrac{1}{2}A^2 T_c \qquad (7.103c)$$

so that the amplitude of the pilot signal in terms of quadrature chip energy is

$$A = \sqrt{2E_c/T_c} \qquad (7.103d)$$

For discussing sequence acquisition, we need not carry the shaped pulse expression $p(t)$ in the received signal representation. Thus, for simplicity

$$s_{rec}(t) = \sqrt{\frac{2E_c}{T_c}}\, m(t)[C_I(t)\cos(\omega_0 t + \phi_w) + C_Q(t)\sin(\omega_0 t + \phi_w)] \qquad (7.103e)$$

in which I and Q PN waveforms are denoted $C_I(t)$ and $C_Q(t)$, respectively.

Now, returning to the question posed earlier, the mobile station acquires coarse phase synchronization by performing a correlation operation between the received waveform and the mobile station's PN sequence, and this correlation is a serial search hypothesis test in that, if a given local PN code phase does not produce an acceptable autocorrelation peak, the system tries another PN code phase by "advancing" one chip of the $C_I(t)$ and $C_Q(t)$ codes. The decision as to the acquisition success or failure for a single trial is made over an interval equal to at least half the Walsh function period of 32 chips. The reason is that the synchronization acquisition is done while extracting the pilot channel from the total forward link signal in exactly the same manner as any other channel with an assigned Walsh function, say $W_i(t)$. As shown in Chapter 5, the only way we can receive the information transmitted specifically to a given receiver (mobile terminal) by the transmitter, along with other multiple access signals, is to multiply the *total* received signal by the assigned Walsh function and then integrate (or sum) the result over the Walsh function period. This is exactly the way it is done for the sync acquisition.

There are a couple of problems that need to be addressed at the outset here: One problem is that the mobile receiver does not know the carrier phase, and hence the demodulation must be done noncoherently, as is shown shortly. The other problem is a case when received nonpilot channel signals change polarity during the correlation period (integration period), because of the possible presence of data transitions in these other channels.

If the incoming received pilot signal is assumed to be maintaining the sequence polarities of both I-channel and Q-channel PN sequences, chances are that a synchronization is accomplished rather quickly. Otherwise, the serial correlation search process must continue until synchronization is acquired. With this background, let us review the fundamentals concerning the acquisition problem.

7.4.1 Review of Correlation Operations

Let **C** be a binary ± 1 sequence of period $P = 7$:

$$\mathbf{C} = (-1, -1, -1, +1, -1, +1, +1) \tag{7.104a}$$

If we compute the autocorrelation function of this sequence with "hardware," we may have a "system diagram" as shown in Figure 7.55. Or we may simply multiply the entire sequence **C** by a shifted version of **C** and the sum the result; that is, take their inner product. Because the sequence (7.104a) is a PN sequence, the autocorrelation has discrete values, as depicted in Figure 7.56.

$$\mathbf{C} = (c_0, c_1, c_2, ..., c_n, ..., c_{P-1})$$

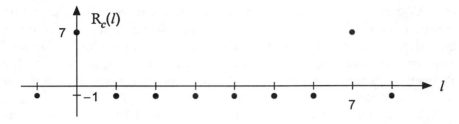

Figure 7.55 Sequence autocorrelation calculator.

Figure 7.56 Sequence autocorrelation for sequence C given in (7.104a).

For the waveform representation of the PN sequence C of (7.104a), we have

$$C(t) = \sum_{n=-\infty}^{\infty} c_n \varphi_c(t - nT_c) \qquad (7.104b)$$

where

$$\varphi_c(t) = \text{Rect}\left(\frac{t}{T_c} - \frac{1}{2}\right) \qquad (7.104c)$$

for the rectangular basic waveform (see Section 6.4.1) and

$$\varphi_c(t) = \text{sinc}(t/T_c) \qquad (7.104d)$$

for the sinc basic waveform. PN waveforms are shown in Figures 7.57 and 7.58 for the rectangular and sinc basic waveforms, respectively.

To obtain the autocorrelation function for the continuous-time waveform, we only need the autocorrelation functions for the basic waveforms given in (7.104c) and (7.104d). For $\varphi(t) = \text{Rect}(t/T_c - \frac{1}{2})$ [see (6.46b)]

$$R_\varphi(\tau) = \int_{-\infty}^{\infty} \varphi(t)\, \varphi(t - \tau)\, dt = \begin{cases} T_c(1 - |\tau|/T_c), & |\tau| \le T_c \\ 0, & \text{otherwise} \end{cases} \qquad (7.105a)$$

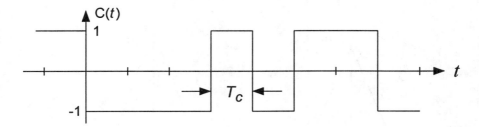

Figure 7.57 PN signal, using the rectangular basic waveform.

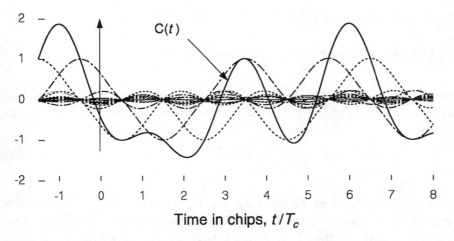

Time in chips, t/T_c

Figure 7.58 PN signal, using the sinc basic waveform.

and for $\varphi(t) = \mathrm{sinc}(t/T_c)$, using Parseval's theorem (1.18i) we have

$$R_\varphi(\tau) = \int_{-\infty}^{\infty} \varphi(t)\,\varphi(t-\tau)\,dt = \int_{-\infty}^{\infty} \mathcal{F}\{\varphi(t)\}\mathcal{F}^*\{\varphi(t-\tau)\}\,df$$

$$= \int_{-\infty}^{\infty} T_c\,\mathrm{Rect}(T_c f) \times T_c\,\mathrm{Rect}(T_c f)\,e^{-j2\pi f\tau}\,df$$

$$= T_c^2 \int_{-\infty}^{\infty} [\mathrm{Rect}(T_c f)]^2\,e^{-j2\pi f\tau}\,df = T_c^2 \int_{-\infty}^{\infty} \mathrm{Rect}(T_c f)\,e^{-j2\pi f\tau}\,df$$

$$= T_c^2 \cdot T_c^{-1}\mathrm{sinc}(\tau/T_c) = T_c\,\mathrm{sinc}(\tau/T_c) \qquad (7.105b)$$

These autocorrelation functions are plotted in Figure 7.59. Note that sinc basic waveform is a bandlimited waveform, such as the IS-95 FIR impulse

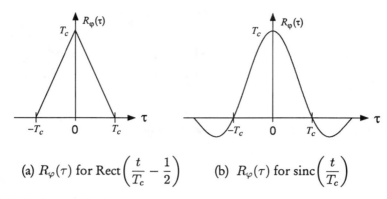

(a) $R_\varphi(\tau)$ for $\mathrm{Rect}\left(\dfrac{t}{T_c} - \dfrac{1}{2}\right)$ (b) $R_\varphi(\tau)$ for $\mathrm{sinc}\left(\dfrac{t}{T_c}\right)$

Figure 7.59 Autocorrelation functions of basic waveforms.

response in an approximate sense (see Section 1.4). To obtain the autocorrelation function for the sequence waveform, we use the result given in (6.45b):

$$\mathrm{R}_c(\tau) = \frac{1}{PT_c} \sum_{n=-\infty}^{\infty} \mathrm{R}_c(n)\, \mathrm{R}_\varphi(\tau - nT_c) \tag{7.106}$$

where $\mathrm{R}_c(n)$ is plotted in Figure 7.56 and $\mathrm{R}_\varphi(\tau)$ given in (7.105a) for the rectangular basic waveform and in (7.105b) for the sinc basic waveform. The corresponding periodic functions (7.106) for these basic waveforms are plotted in Figures 7.60 and 7.61.

Our next interest is to see how the autocorrelation function such as shown in Figures 7.60 and 7.61 can be obtained with hardware. To obtain the continuous-time function (waveform) of the autocorrelation function from the product of $C(t) \times C(t - \tau)$, one only needs to smooth out (or average) the product as shown in Figure 7.62. The average is done by a lowpass filter having an integration time much greater than the chip interval. Note that

$$\mathrm{R}_c(\tau) = \frac{1}{T} \int_0^T C(t)\, C(t - \tau)\, dt \tag{7.107a}$$

which is a "time average" autocorrelation function. The autocorrelation function $\mathrm{R}_c(\tau)$ can also be obtained by an ensemble average:

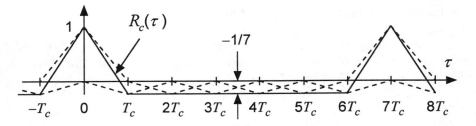

Figure 7.60 Autocorrelation function of the PN signal using the rectangular basic waveform.

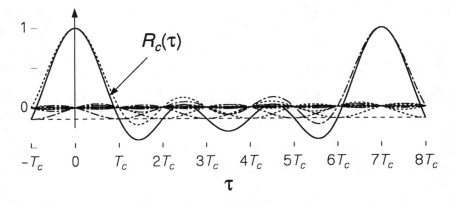

Figure 7.61 Autocorrelation function of the PN signal using the sinc basic waveform.

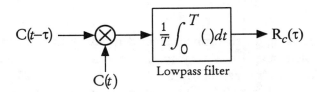

Figure 7.62 System that computes the autocorrelation function.

$$R_c(\tau) = E\{C(t)\,C(t-\tau)\} = \overline{C(t)\,C(t-\tau)} \qquad (7.107b)$$

If we assume that (7.107a) and (7.107b) give the same result, it is a case of the time-average autocorrelation function being equal to the ensemble-average autocorrelation function. The generation of the PN sequences being a random process, we call such a process an *ergodic* process [12].

7.4.2 Initial Sequence Phase Acquisition

As we have done in Section 7.2.2.6, the first step in the noncoherent demodulation process is to obtain the quadrature components from the received waveform given in (7.103a) and (7.103e):

$$r(t) = \sqrt{\frac{2\,E_c}{T_c}}\, m(t)[C_I(t)\cos(\omega_0 t + \phi_w) + C_Q(t)\sin(\omega_0 t + \phi_w)] + n(t)$$

$$= \sqrt{\frac{2\,E_c}{T_c}}\, m(t)\big[C_I'(t)\cos\omega_0 t + C_Q'(t)\sin\omega_0 t\big] + n(t) \qquad (7.108a)$$

where

$$C_I'(t) \triangleq C_I(t)\cos\phi_w + C_Q(t)\sin\phi_w \qquad (7.108b)$$

and

$$C_Q'(t) \triangleq C_Q(t)\cos\phi_w - C_I(t)\sin\phi_w \qquad (7.108c)$$

and where $C_I(t)$ and $C_Q(t)$ are the I- and Q-channel PN sequences, respectively, which assume ± 1 values. The total noise is assumed to be AWGN:

$$n(t) = \sqrt{2}\, n_c(t)\cos\omega_0 t - \sqrt{2}\, n_s(t)\sin\omega_0 t \qquad (7.109a)$$

with

$$\overline{n^2(t)} = \overline{n_c^2(t)} = \overline{n_s^2(t)} = \sigma^2 = \tfrac{1}{2}\mathcal{N}_0 \qquad (7.109b)$$

The signal processing for the noncoherent demodulation and sync acquisition decision variable extraction is very similar to that used for the noncoherent signal reception on the reverse link, shown in Figure 7.23. The system diagram is shown in Figure 7.63. The quadrature components are:

$$r_I(t) = \Big[r(t) \times \sqrt{2}\cos\omega_0 t\Big]_{\text{lowpass}} = \sqrt{\frac{E_c}{T_c}}\, m(t)\, C_I'(t) + n_I(t)$$

$$= \sqrt{\frac{E_c}{T_c}}\, m(t)\, C_I(t)\cos\phi_w + \sqrt{\frac{E_c}{T_c}}\, m(t)\, C_Q(t)\sin\phi_w + n_I(t) \quad (7.110a)$$

and

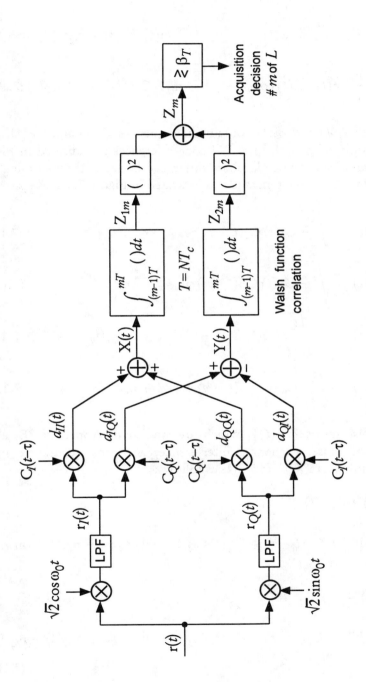

Figure 7.63 Diagram of acquisition processing for IS-95 signals.

$$r_Q(t) = \left[r(t) \times \sqrt{2} \sin \omega_0 t \right]_{\text{lowpass}} = \sqrt{\frac{E_c}{T_c}}\, m(t)\, C'_Q(t) + n_Q(t)$$

$$= \sqrt{\frac{E_c}{T_c}}\, m(t)\, C_Q(t) \cos \phi_w - \sqrt{\frac{E_c}{T_c}}\, m(t)\, C_I(t) \sin \phi_w + n_Q(t) \quad (7.110b)$$

Note that both $r_I(t)$ and $r_Q(t)$ contain the sequences $C_I(t)$ and $C_Q(t)$, and, thus, we multiply with both local PN code references as indicated in Figure 7.63. We assume that the code phase difference between the incoming PN sequences and the local references is τ as indicated in Figure 7.63. We see that

$$d_{II}(t) = \sqrt{\frac{E_c}{T_c}}\, C_I(t)\, C_I(t - \tau) \cos \phi_w + n_{II}(t) \qquad (7.111a)$$

$$d_{IQ}(t) = \sqrt{\frac{E_c}{T_c}}\, C_Q(t)\, C_Q(t - \tau) \sin \phi_w + n_{IQ}(t) \qquad (7.111b)$$

$$d_{QI}(t) = -\sqrt{\frac{E_c}{T_c}}\, C_I(t)\, C_I(t - \tau) \sin \phi_w + n_{QI}(t) \qquad (7.111c)$$

$$d_{QQ}(t) = \sqrt{\frac{E_c}{T_c}}\, C_Q(t)\, C_Q(t - \tau) \cos \phi_w + n_{QQ}(t) \qquad (7.111d)$$

where the noise terms $\{n_{ij}(t)\}$, which are studied in Appendix 7E, are a combination of noise, cochannel interference, and cross-quadrature interference. We now form $X(t)$ and $Y(t)$ as follows:

$$X(t) = d_{II}(t) + d_{QQ}(t)$$

$$= \sqrt{\frac{E_c}{T_c}}\, [C_I(t)\, C_I(t - \tau) + C_Q(t)\, C_Q(t - \tau)] \cos \phi_w + n_{II}(t) + n_{QQ}(t)$$

and $\hspace{9cm} (7.112a)$

$$Y(t) = d_{IQ}(t) - d_{QI}(t)$$

$$= \sqrt{\frac{E_c}{T_c}}\, [C_I(t)\, C_I(t - \tau) + C_Q(t)\, C_Q(t - \tau)] \sin \phi_w + n_{IQ}(t) - n_{QI}(t)$$

$$(7.112b)$$

Note that $X(t)$ and $Y(t)$ form the basis of a noncoherent detection procedure that eliminates the unknown carrier phase ϕ_w.

In Figure 7.63, the quantities $X(t)$ and $Y(t)$ are integrated over an observation interval $T = NT_c$ that is equivalent to some number of chips, N. As is shown, these integrations yield a measure of correlation between the phases of the incoming I and Q codes and those of the local reference PN codes. However, the integration also plays an important role in eliminating cochannel interference when the incoming and local reference PN codes agree.

From one viewpoint, when the PN codes are in agreement, the integration is in effect the partial correlation of the pilot Walsh function (a constant) with the Walsh functions of other active forward link channels that may be present during the acquisition process. Because the Walsh functions are orthogonal when correlated over an interval that is a multiple of 64 chips, N should be a multiple of 64 chips to effect such a correlation and to eliminate unwanted contributions to $X(t)$ and $Y(t)$ from channels other than the pilot channel.

From another viewpoint, the integration of $X(t)$ and $Y(t)$ is a determination of their average values over the interval. From the properties of the Walsh functions (see Section 5.2), we note that, for every Walsh function $W_i(t)$ for $i \geq 2$, there is an equal number of 0s and 1s ($+1$s and -1s) in the first and second halves of the Walsh function period. Hence, an integration over any multiple of one-half a Walsh function period (32 chips) can be effective in reducing the majority of forward link cochannel interference.

The acquisition processing consists of testing the current code phase of the local PN code references by observing the agreement of the incoming and local PN codes for one or more test intervals, after which a decision is made as to whether the codes agree (i.e., the code phases have been acquired). If the decision is that they do not agree, then another local PN code phase setting is selecting according to the search strategy. The integral of $X(t)$ over the mth test interval produces the statistic Z_{1m}, given by

$$
\begin{aligned}
Z_{1m} &= \int_{(m-1)T}^{mT} X(t)\,dt \\
&= \sqrt{\frac{E_c}{T_c}}\,\cos\phi_w \int_{(m-1)T}^{mT} [C_I(t)\,C_I(t-\tau) + C_Q(t)\,C_Q(t-\tau)]\,dt \\
&\quad + \int_{(m-1)T}^{mT} [n_{II}(t) + n_{QQ}(t)]\,dt
\end{aligned}
$$

$$= \sqrt{\frac{E_c}{T_c}} \cos \phi_w \times T[R_{cI}(\tau) + R_{cQ}(\tau)] + N_1 \tag{7.113a}$$

in which $R_{cI}(\tau)$ and $R_{cQ}(\tau)$ are the time-average autocorrelation functions of $C_I(t)$ and $C_Q(t)$, respectively. Similarly, the integration of $Y(t)$ over the mth test inteval produces the statistic Z_{2m}, given by

$$Z_{2m} = \int_{(m-1)T}^{mT} Y(t)\, dt$$

$$= \sqrt{\frac{E_c}{T_c}} \sin \phi_w \int_{(m-1)T}^{mT} [C_I(t)\, C_I(t-\tau) + C_Q(t)\, C_Q(t-\tau)]\, dt$$

$$+ \int_{(m-1)T}^{mT} [n_{IQ}(t) - n_{QI}(t)]\, dt$$

$$= \sqrt{\frac{E_c}{T_c}} \sin \phi_w \times T[R_{cI}(\tau) + R_{cQ}(\tau)] + N_2 \tag{7.113b}$$

in which the noise terms N_1 and N_2 are defined as the integrals of the noise terms in $X(t)$ and $Y(t)$, respectively. We assume that the autocorrelation functions $R_{cI}(\tau)$ and $R_{cQ}(\tau)$ are the same:

$$R_{cI}(\tau) = R_{cQ}(\tau) = R(\tau) \tag{7.114a}$$

Then we have

$$Z_{1m} = 2T\sqrt{\frac{E_c}{T_c}} R(\tau) \cos \phi_w + N_1 \tag{7.114b}$$

and

$$Z_{2m} = 2T\sqrt{\frac{E_c}{T_c}} R(\tau) \sin \phi_w + N_2 \tag{7.114c}$$

It is shown in Appendix 7E that

$$N_1 = G(0, \mathcal{N}_0'' T) \quad \text{and} \quad N_2 = G(0, \mathcal{N}_0'' T) \tag{7.114d}$$

where $\mathcal{N}_0' > \mathcal{N}_0$ is the effective noise spectral density that is due to both receiver noise and interference, which becomes receiver noise only ($\mathcal{N}_0' \approx \mathcal{N}_0$) when the acquisition procedure results in a zero phase difference ($\tau = 0$).

The decision variable is given by

$$Z_m = Z_{1m}^2 + Z_{2m}^2 \tag{7.115a}$$

where

$$
\begin{aligned}
Z_{1m} &= G\left[2T\sqrt{\frac{E_c}{T_c}}R(\tau)\cos\phi_w,\ \mathcal{N}_0'T\right] \\
&= \sqrt{\mathcal{N}_0'T}\cdot G\left[2\sqrt{\frac{T}{T_c}\frac{E_c}{\mathcal{N}_0'}}R(\tau)\cos\phi_w,\ 1\right]
\end{aligned}
\tag{7.115b}
$$

and

$$
\begin{aligned}
Z_{2m} &= G\left[2T\sqrt{\frac{E_c}{T_c}}R(\tau)\sin\phi_w,\ \mathcal{N}_0'T\right] \\
&= \sqrt{\mathcal{N}_0'T}\cdot G\left[2\sqrt{\frac{T}{T_c}\frac{E_c}{\mathcal{N}_0'}}R(\tau)\sin\phi_w,\ 1\right]
\end{aligned}
\tag{7.115c}
$$

Therefore, the decision variable Z is $\mathcal{N}_0'T$ times a noncentral chi-squared RV with two degrees of freedom, and the noncentrality parameter is

$$
\begin{aligned}
\lambda &= \left[2\sqrt{\frac{NE_c}{\mathcal{N}_0'}}R(\tau)\cos\phi_w\right]^2 + \left[2\sqrt{\frac{NE_c}{\mathcal{N}_0'}}R(\tau)\sin\phi_w\right]^2 \\
&= \left[2\sqrt{\frac{NE_c}{\mathcal{N}_0'}}R(\tau)\right]^2 = 4N\cdot R^2(\tau)\cdot\frac{E_c}{\mathcal{N}_0'}
\end{aligned}
\tag{7.115d}
$$

where $N = T/T_c$ and the functional form of the autocorrelation function $R(\tau)$ depends on the basic waveform used for the PN sequence. The pdf for Z_m then is

$$
p_{z_m}(\alpha) = \begin{cases} \dfrac{1}{2\sigma^2}\,e^{-\frac{1}{2}(\lambda+\alpha/\sigma^2)}\,I_0\left(\sqrt{\dfrac{\lambda\alpha}{\sigma^2}}\right), & \alpha \geq 0 \\[2ex] 0, & \text{otherwise} \end{cases}
\tag{7.116a}
$$

with $\sigma^2 = \mathcal{N}_0'T$. Let us view the synchronization test as a hypothesis test: H_1 is the hypothesis that the incoming and local PN signals are aligned within one PN chip [10], whereas H_0 denotes the alternative hypothesis. That is,

$$H_1: \ |\tau| \le T_c \ \rightarrow \ R(\tau) > 0, \ \mathcal{N}_0' \approx \mathcal{N}_0 \tag{7.116b}$$

$$H_0: \ |\tau| > T_c \ \rightarrow \ R(\tau) \approx 0, \ \mathcal{N}_0' > \mathcal{N}_0 \tag{7.116c}$$

The pdf (7.116a) conditioned on these hypotheses becomes

$$p_{z_m}(\alpha \mid H_0) = \frac{1}{2\mathcal{N}_0'T} \, e^{-\alpha/2\mathcal{N}_0'T} \tag{7.117a}$$

and

$$p_{z_m}(\alpha \mid H_1) = \frac{1}{2\mathcal{N}_0 T} \, e^{-\frac{1}{2}(\lambda + \alpha/\mathcal{N}_0 T)} \, I_0\left(\sqrt{\frac{\lambda\alpha}{\mathcal{N}_0 T}}\right) \tag{7.117b}$$

Using the pdfs of (7.117a) and (7.117b), we can evaluate the single-run acquisition detection and false alarm probabilities. To obtain the detection probability $P_D(m = 1)$ for a single test ($m = 1$), we need the false alarm threshold, which is chosen to achieve a specified false alarm probability $P_F(m = 1)$. The false alarm probability is given by

$$P_F(m = 1) = \Pr\{Z_m > \beta_T \mid H_0\}$$

$$= \int_{\beta_T}^{\infty} p_{z_m}(\alpha \mid H_0) \, d\alpha = \int_{\beta_T}^{\infty} \frac{1}{2\mathcal{N}_0'T} \, e^{-\alpha/2\mathcal{N}_0'T} \, d\alpha$$

$$= e^{-\beta_T/2\mathcal{N}_0'T} \tag{7.117c}$$

from which we obtain

$$\beta_T = -2\mathcal{N}_0'T \cdot \ln P_F(m = 1) \tag{7.117d}$$

Using this threshold, the detection probability is found to be

$$P_D(m = 1) = \Pr\{Z_m > \beta_T \mid H_1\} = \int_{\beta_T}^{\infty} p_{z_m}(\alpha \mid H_1) \, d\alpha$$

$$= \int_{\beta_T}^{\infty} \frac{1}{2\mathcal{N}_0 T} \, e^{-\frac{1}{2}(\lambda + \alpha/\mathcal{N}_0 T)} \, I_0\left(\sqrt{\frac{\lambda\alpha}{\mathcal{N}_0 T}}\right) d\alpha \tag{7.118a}$$

Let $\alpha = \mathcal{N}_0 T x^2$; then the integral in (7.118a) is transformed into

$$P_D(m = 1) = \int_{\sqrt{\beta_T/\mathcal{N}_0 T}}^{\infty} x\, e^{-\frac{1}{2}(\lambda + x^2)}\, I_0\left(\sqrt{\lambda}\, x\right) \tag{7.118b}$$

which is recognized as Marcum's Q-function [see (1.90a)] for $a = \sqrt{\lambda}$ and $\beta = \sqrt{\beta_T/\mathcal{N}_0 T}$, giving the result

$$P_D(m = 1) = Q\left(\sqrt{\lambda},\ \sqrt{\beta_T/\mathcal{N}_0 T}\right)$$

$$= Q\left(2\sqrt{\frac{NE_c}{\mathcal{N}_0}}\, R(\tau),\ \sqrt{-2\frac{\mathcal{N}_0'}{\mathcal{N}_0}\ln P_F(m = 1)}\right) \tag{7.118c}$$

where $R(\tau) \le R(0) = 1$. An upper bound on the detection probability is obtained, by assuming $\mathcal{N}_0' \approx \mathcal{N}_0$ and $R(\tau) \approx R(0) = 1$, as

$$P_D(m = 1) \le Q\left(2\sqrt{\frac{NE_c}{\mathcal{N}_0}},\ \sqrt{-2\ln P_F(m = 1)}\right) \tag{7.118d}$$

where

$$N = \frac{T}{T_c} = \text{number of PN chips integrated} \tag{7.118e}$$

Using the approximation to Marcum's Q-function by the Gaussian Q-function shown in (1.92b), known to be very accurate, the upper bound becomes

$$P_D(m = 1) \le Q\left(\sqrt{\beta_T/\mathcal{N}_0 T} - \sqrt{\lambda}\right)$$

$$= Q\left(\sqrt{-2\ln P_F(m = 1)} - 2\sqrt{\frac{NE_c}{\mathcal{N}_0}}\right) \tag{7.118f}$$

An illustrative case of the relation between acquisition false alarm and detection probability requirements is shown in Figure 7.64, which is a plot of the upper bound approximation (7.118d). For a given value of the received pilot chip energy-to-noise density ratio, the detection probability is proportional to N, the number of chips integrated at one time prior to the squaring and combining operations in Figure 7.63. For example, Figure 7.64 indicates that an integration time equivalent to 224 chips is required to

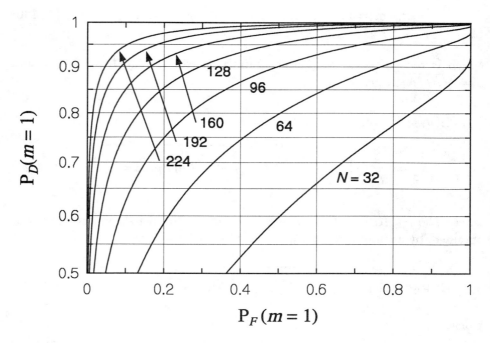

Figure 7.64 Upper bound on acquisition detection probability versus false alarm probability for the case of $E_c/\mathcal{N}_0 = -15$ dB.

achieve a detection probability of 95% while maintaining a false alarm probability of 10% when $E_c/\mathcal{N}_0$ equals -15 dB.

If L decision variables $\{Z_m\}$ are summed, the sum is a scaled noncentral chi-squared variable with $2L$ degrees of freedom and L times the noncentrality parameter given in (7.126d). For the same false alarm probability, the detection probability will increase if $L > 1$ is used. However, the total observation time is $LT = LNT_c$, and thus there is a tradeoff between the reliability of the acquisition decision and the time it takes to search the possible PN code phases, depending on the search strategy [10, 13, 14].

7.4.3 Code Tracking With a Delay-Lock Loop

When the initial serial search test results in a successful synchronization between the received and the local PN sequences, the timing reference is established as well. Once the sync is established to within a fraction of a chip duration T_c, the carrier frequency and phase are accurately determined continuously by conventional phase-lock loop (PLL) techniques [15–18], and further, the timing agreement between the incoming PN code phase and the local PN code phase, just established by the coarse acquisition method, must be tracked to bring the relative timing error to zero.

The purpose of this section is to present the fundamental concepts and techniques used in the code tracking, which is accomplished by what is known as *delay-lock loop (DLL)* tracking of PN signals [10, 19–22]. The original work on the DLL [19] as an optimal device for tracking the delay difference between the acquired sequence and the local sequence followed closely the scheme of a PLL that tracks the phase of a carrier, which is depicted in Figure 7.65. We discuss the fundamental principles of DLL tracking of PN sequences and its variations.

The code tracking loops perform either coherently or noncoherently: coherent loops use carrier phase information, whereas noncoherent loops do not require knowledge of carrier phase. In general, code tracking loops maintain synchronization of the receiver's replica of the spreading code by using two correlators called an *early correlator* and a *late correlator*. An early correlator uses a code reference waveform that is advanced in time by some fraction of a chip with respect to the currently estimated code phase. A late correlator uses a code reference waveform that is delayed by some fraction of a chip. The difference between early and late correlations is used to sense small deviations of the incoming spreading code's timing with respect to the early and late code timing.

The early and late correlation operations are depicted in Figure 7.66, in which the outputs of the lowpass filters are the autocorrelation functions

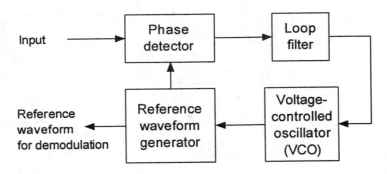

Figure 7.65 Essential features of a PLL.

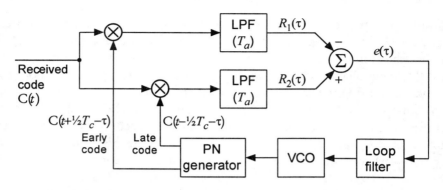

Figure 7.66 Essentials of a DLL for tracking PN code phase (T_c = PN chip period, T_a = integration time).

$R_1[\tau(t)]$ and $R_2[\tau(t)]$. The difference between these correlator outputs is taken to obtain an error signal $e(\tau)$, given by

$$e(\tau) = R_2(\tau) - R_1(\tau) \qquad (7.119)$$

These output signals are plotted in Figure 7.67. The output signal $e(\tau)$ plays the role of a correction signal, which is then used to drive a voltage-controlled oscillator (VCO). The VCO is a clock in the sense that it drives the PN generator such that, when the PN generator clock is lagging in phase in comparison with the incoming sequence phase, it drives the clock faster and vice versa.

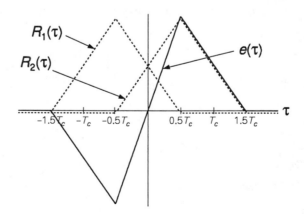

Figure 7.67 DLL error signal e(τ); early correlator output, $R_1(\tau)$; and late correlator output, $R_2(\tau)$.

From the DLL, the correct (on time) reference code is obtained for the use of despreading the received spread-spectrum signal, as the simplified diagram in Figure 7.68 illustrates. The correlator that uses the reference code is called the *punctual correlation* [23]. Code-tracking loops may be coherent or noncoherent. The illustration of Figure 7.68 is an example of coherent tracking; that is, we assumed that the receiver heterodyned the received waveform with a known carrier phase to produce the baseband signal $C(t - T_d)$. Thus, we observe from Figure 7.68 that the difference between the *early* and the *late* correlators is used to sense small deviations between the incoming code's timing and that of the nominal position of the *early-late* code to provide the reference waveform from the PN code generator (PNG) for despreading purposes.

Early-late tracking loops can also be classified into two types of configuration: (1) DLL, which we just discussed, and (2) the *tau-dither loop (TDL)*. The DLL is also known as the "full-time" early-late code tracking loop, whereas the TDL is called the "time-shared" early-late code tracking loop. There are also the variations of these basic DLL and TDL tracking loops: (3) *the double dither loop (DDL)* [24]; (4) the *product of sum and difference DLL (PSD DLL)* [25]; and (5) the *modified code tracking loop (MCTL)* [26]. Each of these tracking loops can be operated in a coherent or noncoherent implementation, depending on, as usual, the availability of carrier phase information during tracking. In a system such as IS-95, once timing has been acquired through the noncoherent acquisition scheme, phase and frequency

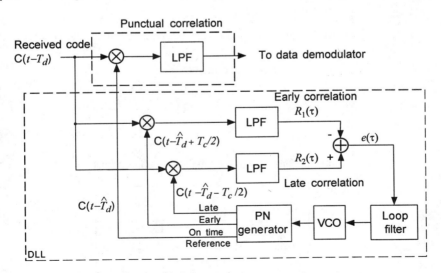

Figure 7.68 DLL and reference code tracking circuit [23].

(nominal value known *a priori*) can be accurately tracked by a PLL scheme, and hence coherent code tracking is always possible. We summarize well-known tracking loop configurations in Table 7.2. In what follows, we give the principles of operation for the DLL and TDL only.

7.4.3.1 Full-Time Noncoherent DLL Tracking

A noncoherent DLL configuration is given in Figure 7.69. This configuration can be a good candidate for BPSK signals modulated by a PN sequence; that is, the signal to be transmitted is of the form

$$s_t(t) = \sqrt{2P_t}\, C(t)\, m(t) \cos \omega_0 t \tag{7.120}$$

where $C(t)$ is the PN sequence waveform and $m(t)$ is the data sequence waveform, with P_t being the transmitted power. The received waveform is then of the form given by

$$r(t) = s_r(t) + n(t) \tag{7.121a}$$

with

$$s_r(t) = \sqrt{2P_r}\, C(t - T_d)\, m(t - T_d) \cos[\omega_0 t + \phi(t)] \tag{7.121b}$$

where

P_r = received signal power

T_d = propagation delay

$\phi(t)$ = random phase of the carrier, unknown to the receiver

and $n(t)$ is white Gaussian noise, represented as

$$n(t) = n_c(t)\sqrt{2}\cos[\omega_0 t + \phi(t)] - n_s(t)\sqrt{2}\sin[\omega_0 t + \phi(t)] \tag{7.121c}$$

Table 7.2 Summary of tracking loop configurations

Traditional early-late PN code tracking loops	1. DLL
	2. TDL
Variations on DLL	3. DDL
	4. PSD DLL
	5. MCTL

where $n_c(t)$ and $n_s(t)$ are lowpass noise waveforms that are statistically independent stationary Gaussian processes with two-sided power spectral density equal to $\frac{1}{2}\mathcal{N}_0$. Note that for this configuration, the error signal $e(\tau)$ is the difference of the squared envelopes of two signals, each multiplied by early and late PN code, respectively, as indicated in Figure 7.69.

Let us now consider the hypothetical case of a noncoherent tracking system for QPSK using the DLL of Figure 7.69. The transmitted signal is then assumed to be of the form

$$s_t(t) = \sqrt{P_t}\, m(t)\, C_I(t) \cos\omega_0 t + \sqrt{P_t}\, m(t)\, C_Q(t) \sin\omega_0 t \qquad (7.122a)$$

where P_t is the transmitted power and $C_I(t)$ and $C_Q(t)$ are two independent I and Q channel PN sequences running in complete synchronism at the same chip rate. This waveform mimics the IS-95 scheme, in which case $m(t)$ has the form of

$$m(t) = \sum_{i=0}^{M-1} g_i\, D_i(t)\, W_i(t) = 1 + \sum_{i=1}^{M-1} g_i\, D_i(t)\, W_i(t) \qquad (7.122b)$$

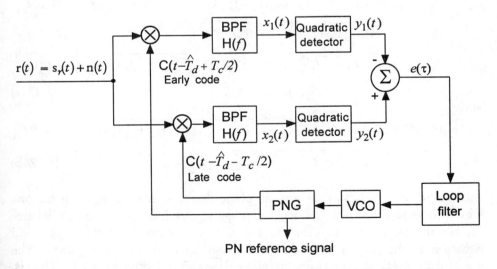

Figure 7.69 A noncoherent DLL for extraction of a PN reference signal for despreading purposes (T_c = chip duration and $\widehat{T}_d$ is the delay estimate).

where g_i is a channel gain relative to the $i = 0$ pilot channel, $D_i(t)$ is the data, and $W_i(t)$ is the Walsh function for channel (user) i. For our discussion of quadriphase tracking, however, these details are not of significance at this point. The received signal is then given by

$$r(t) = s_r(t) + n(t) \tag{7.123a}$$

where

$$s_r(t) = \sqrt{P_r}\, m(t - T_d)\left\{C_I(t - T_d)\cos[\omega_0 t + \phi(t)]\right.$$
$$\left. + C_Q(t - T_d)\sin[\omega_0 t + \phi(t)]\right\} \tag{7.123b}$$

where T_d, $\phi(t)$, and $n(t)$ have the same meanings as described above in connection with (7.121b). In a situation like this, with two PN sequences running in synchronism at the same chip rate, only one PN code needs to be tracked, say, $C_I(t)$. We then have

$$x_i(t) = \left[r(t) \times C_I\left(t - \widehat{T}_d \pm \tfrac{1}{2}T_c\right)\right]_{BPF}$$
$$= \sqrt{2P_r}\, K_i \overline{m(t - T_d)\, C_I(t - T_d)\, C_I\left(t - \widehat{T}_d \pm \tfrac{1}{2}T_c\right)}\cos[\omega_0 t + \theta(t)]$$
$$+ \sqrt{2P_r}\, K_i \overline{m(t - T_d)\, C_Q(t - T_d)\, C_I\left(t - \widehat{T}_d \pm \tfrac{1}{2}T_c\right)}\sin[\omega_0 t + \theta(t)]$$
$$+ \overline{n(t)\, C_I\left(t - \widehat{T}_d \pm \tfrac{1}{2}T_c\right)}$$
$$= \sqrt{2P_r}\, K_i \overline{m(t - T_d)\, C_I(t - T_d)\, C_I\left(t - \widehat{T}_d \pm \tfrac{1}{2}T_c\right)}\cos[\omega_0 t + \theta(t)]$$
$$+ N_i(t) \quad i = 1, 2 \tag{7.124a}$$

where

$$N_i(t) \triangleq \sqrt{2P_r}\, K_i \overline{m(t - T_d)\, C_Q(t - T_d)\, C_I\left(t - \widehat{T}_d \pm \tfrac{1}{2}T_c\right)}\sin[\omega_0 t + \theta(t)]$$
$$+ \overline{n(t)\, C_I\left(t - \widehat{T}_d \pm \tfrac{1}{2}T_c\right)} \quad i = 1, 2 \tag{7.124b}$$

and where the $\{K_i, i = 1, 2\}$ are the phase detector (multiplier) gain factors, which in general are not identical unless great care has been taken to balance the circuit design. The fact that K_1 and K_2 may not be identical in a practical situation is the reason why the DLL is so sensitive to these gains. The waveforms $x_1(t)$ and $x_2(t)$ are envelope-detected and then squared. That is, they are detected by quadratic detectors. These waveforms are of the form

$$x_i(t) = V_{i,s+n}(t)\cos[\omega_0 t + \xi_{i,s+n}(t)], \quad i = 1, 2 \tag{7.125a}$$

where $V_{i,s+n}(t)$ denotes the envelope of a signal plus noise waveform, and $\xi_{i,s+n}(t)$ denotes its phase. To see clearly the physical principles involved in the noncoherent DLL operation, a discussion on a *noiseless environment* is appropriate. Thus, we set $N_i(t) = 0$ in (7.124a) and, instead of (7.125a), have

$$x_i(t) = V_{i,s}(t)\cos[\omega_0 t + \xi_{i,s}(t)], \quad i = 1, 2 \tag{7.125b}$$

where $V_{i,s}(t)$ is the envelope of the original signal in (7.124a):

$$V_{i,s}(t) \triangleq \sqrt{2P_r}\, K_i\, \overline{m(t - T_d)}\, \overline{C_I(t - T_d)\, C_I\!\left(t - \widehat{T}_d \pm \tfrac{1}{2}T_c\right)} \tag{7.125c}$$

But

$$\overline{m(t - T_d)} = 1 + \overline{\sum_{i=1}^{M-1} g_i\, D_i(t)\, W_i(t)} = 1 \tag{7.125d}$$

and

$$\overline{C_I(t - T_d)\, C_I\!\left(t - \widehat{T}_d \pm \frac{1}{2}T_c\right)} = R_i\!\left(T_d - \widehat{T}_d \pm \tfrac{1}{2}T_c\right) \tag{7.125e}$$

giving

$$V_{i,s}(t) = \sqrt{2P_r}\, K_i \cdot 1 \cdot R_i\!\left(T_d - \widehat{T}_d \pm \tfrac{1}{2}T_c\right) \tag{7.125f}$$

and the difference of the quadratic detector outputs becomes

$$e(t) = y_2^2(t) - y_1^2(t) = V_{2,s}^2(t) - V_{1,s}^2(t)$$
$$= 2P_r\left[K_2^2 R_2^2\!\left(T_d - \widehat{T}_d - \tfrac{1}{2}T_c\right) - K_1^2 R_1^2\!\left(T_d - \widehat{T}_d + \tfrac{1}{2}T_c\right)\right] \tag{7.126a}$$

Note that we have maintained the phase detector gain factors K_1 and K_2 in the above expressions for the reason that the imbalance of these factors is responsible for performance errors of the DLL tracking system. With this aspect understood, let us consider the ideal case of of $K_1 = K_2$ and obtain the error signal term

$$e(t) = K\left[R_2^2\!\left(T_d - \widehat{T}_d - \tfrac{1}{2}T_c\right) - R_1^2\!\left(T_d - \widehat{T}_d + \tfrac{1}{2}T_c\right)\right] \tag{7.126b}$$

where

$$K \triangleq 2P_r\, K_i^2, \quad i = 1, 2 \tag{7.126c}$$

We started with the assumption that the early and late codes were separated in time by one chip period. For a general case of tighter or looser tracking, which depends on the system design requirement, we may define the following parameters:

$$\tau \triangleq T_d - \widehat{T}_d \tag{7.127a}$$

and

$$\Delta = \text{separation between the early and late codes} \tag{7.127b}$$

The parameter τ then is the delay estimator error and Δ can take the desired value such as $\Delta = T_c$, as we have assumed in our discussion, or $\Delta = 2T_c$, which corresponds to tapping the m and $m - 2$ stages of the PN generator for the early and late codes. With these new and general parameters, we can express the *normalized* error signal as

$$E(\tau, \Delta) \triangleq R_2^2\left(\tau - \tfrac{1}{2}\Delta\right) - R_1^2\left(\tau + \tfrac{1}{2}\Delta\right) \tag{7.128a}$$

The autocorrelation functions $R_1(\tau)$ and $R_2(\tau)$ should be identical for best operation of the DLL tracking system, and this is the assumption we make in our discussions that follow. Hence, we have

$$E(\tau, \Delta) = R^2\left(\tau - \tfrac{1}{2}\Delta\right) - R^2\left(\tau + \tfrac{1}{2}\Delta\right) \tag{7.128b}$$

As we have seen in Figure 7.59, the autocorrelation function of the PN code depends on the basic waveform $\varphi_c(t)$. If we assume $\varphi_c(t) = \text{Rect}\left(t/T_c - \tfrac{1}{2}\right)$ or $\varphi_c(t) = \text{sinc}(t/T_c)$ as we have done in (7.104c) and (7.104d) respectively, the normalized DLL error signal, called the *S-curve* or *discriminator characteristic* of the DLL, is as shown in Figures 7.70 and 7.71 for these cases.

We leave the subject of full-time noncoherent DLL tracking by saying that the noiseless analysis for the operation of the scheme clearly indicates that PN code tracking is accomplished by sensing the error signal over the range of $\tau \in \left(-\tfrac{1}{2}T_c, \tfrac{1}{2}T_c\right)$ for our choice of $\Delta = T_c$. The noise performance analysis of this system is tedious and rather involved. The reader interested in further study of this subject is referred to the cited references. The reference [22] in particular gives a succint summary of the noise performance analysis.

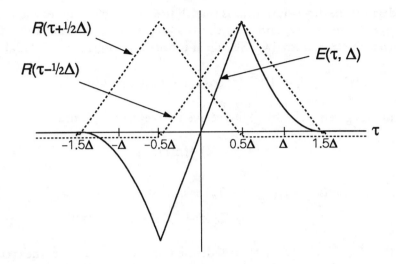

Figure 7.70 Error signal of full-time noncoherent DLL for chip basic waveform $\varphi_c(t) = \text{Rect}\left(t/T_c - \frac{1}{2}\right)$, drawn for $\Delta = T_c$.

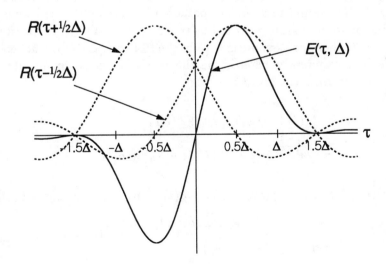

Figure 7.71 Error signal of full-time noncoherent DLL for chip basic waveform $\varphi_c(t) = \text{sinc}(t/T_c)$, drawn for $\Delta = T_c$.

7.4.3.2 Full-Time Coherent DLL Tracking

In this discussion, the assumption is that the receiver heterodynes the received QPSK signal coherently, and thus the tracking is done at baseband. Let us assume that the transmitted QPSK signal is the same as given in (7.122a)

$$s_t(t) = \sqrt{P_t}\, m(t)[C_I(t)\cos\omega_0 t + C_Q(t)\sin\omega_0 t] \tag{7.129a}$$

where $m(t)$ is given (7.122b). The received waveform is given by

$$r(t) = s_r(t) + n(t) \tag{7.129b}$$

where

$$s_r(t) = \sqrt{P_r}\, m(t - T_d)\{C_I(t - T_d)\cos[\omega_0 t + \theta(t)]$$
$$+ C_Q(t - T_d)\sin[\omega_0 t + \theta(t)]\} \tag{7.129c}$$

where P_r, T_d, $\theta(t)$, and $n(t)$ are defined identically as for the received waveform we had for the discussion of the noncoherent DLL tracking in the previous subsection. As we reasoned there, the I-channel PN code $C_I(t)$ and the Q-channel PN code $C_Q(t)$ are running at the same chip rate, although each is generated by a different tap connection PNG. Thus, tracking one code, say, $C_I(t)$, should be sufficient because $C_Q(t)$ is well synchronized with $C_I(t)$. If, for some reason, we need to track $C_I(t)$ and $C_Q(t)$ separately, we can do so. We therefore assume that only the PN code $C_I(t)$ is tracked.

We first coherently heterodyne the received waveform to obtain the baseband waveform $r_I(t)$, given by

$$r_I(t) = \left\{ r(t) \times \sqrt{2}\cos[\omega_0 t + \theta(t)] \right\}_{\text{lowpass}}$$
$$= \sqrt{P_r/2}\, m(t - T_d)\, C_I(t - T_d) + n_I(t) \tag{7.130a}$$

where

$$n_I(t) \triangleq \left\{ \left[n_c(t)\sqrt{2}\cos[\omega_0 t + \theta(t)] - n_s(t)\sqrt{2}\sin[\omega_0 t + \theta(t)] \right] \right.$$
$$\left. \times \sqrt{2}\cos[\omega_0 t + \theta(t)] \right\}_{\text{lowpass}}$$
$$= n_c(t) = G\left(0, \tfrac{1}{2}\mathcal{N}_0\right) \tag{7.130b}$$

The conceptual full-time coherent DLL tracking is depicted in Figure 7.72. The error signal $e'(t)$ is smoothed by the loop filter to give $e(t)$:

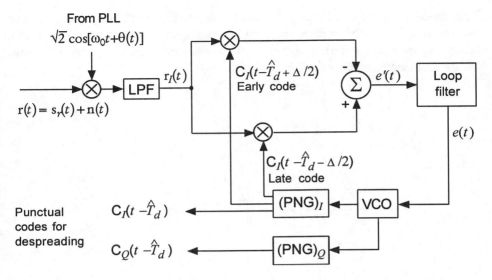

Figure 7.72 Conceptual scheme for full-time coherent DLL for code tracking using only I-channel PN code.

$$e(t) = \overline{e'(t)}$$

$$= \sqrt{\tfrac{1}{2}P_r}\,\overline{m(t - T_d)}\Big[K_2\,\overline{C_I(t - T_d)\,C_I\big(t - \widehat{T}_d - \tfrac{1}{2}\Delta\big)}$$

$$- K_1\,\overline{C_I(t - T_d)\,C_I\big(t - \widehat{T}_d + \tfrac{1}{2}\Delta\big)}\Big] + n_e(t)$$

$$= \sqrt{\tfrac{1}{2}P_r}\cdot 1\cdot\big[K_2 R\big(\tau - \tfrac{1}{2}\Delta\big) - K_1 R\big(\tau + \tfrac{1}{2}\Delta\big)\big] + n_e(t) \qquad (7.131a)$$

where

$$R\big(\tau - \tfrac{1}{2}\Delta\big) \triangleq \overline{C_I(t - T_d)\,C_I\big(t - \widehat{T}_d - \tfrac{1}{2}\Delta\big)} \qquad (7.131b)$$

$$R\big(\tau + \tfrac{1}{2}\Delta\big) \triangleq \overline{C_I(t - T_d)\,C_I\big(t - \widehat{T}_d + \tfrac{1}{2}\Delta\big)} \qquad (7.131c)$$

and $n_e(t)$ is the noise at the output of the loop filter. Again, we bring out the salient points in connection with the full-time coherent DLL tracking scheme without going through the noise performance analysis. The reader interested in further reading on noise analysis is referred to references [19–22, 27].

The normalized error signal in the noiseless environment is then simply given by

$$E(\tau, \Delta) \triangleq R\big(\tau - \tfrac{1}{2}\Delta\big) - R\big(\tau + \tfrac{1}{2}\Delta\big) \qquad (7.132)$$

This signal is plotted in Figures 7.73 and 7.74 for different assumptions about the shape of the PN code waveform. Note here that the autocorrelation functions are not squared, and, hence, the error signal is simply the difference of late and early autocorrelation functions.

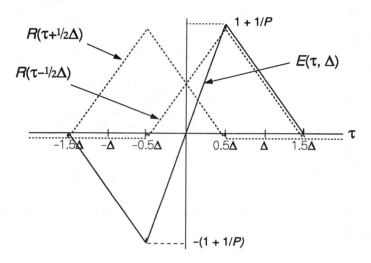

Figure 7.73 Error signal of full-time coherent DLL for chip basic waveform $\varphi_c = \text{Rect}\left(t/T_c - \frac{1}{2}\right)$ and for $\Delta = T_c$.

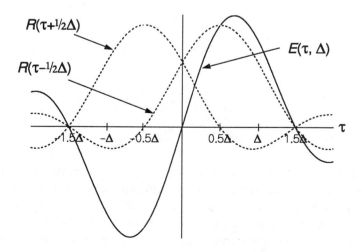

Figure 7.74 Error signal of full-time coherent DLL for chip basic waveform $\varphi_c = \text{sinc}(t/T_c)$ and for $\Delta = T_c$.

7.4.4 TDL Tracking

The full-time early-late DLL tracking system is widely used in spread-spectrum systems. There is one major problem associated with the system, however: the early and late channels must be precisely balanced. If they are not balanced, the DLL output discriminator is not zero even if the tracking error is zero. We have seen in earlier discussions, for example, that the phase detector gain factors K_i, $i = 1, 2$ are the possible cause of the imbalance, aside from other uneven component response characteristics. This problem can be overcome by time-sharing a single correlator by early and late codes, and this is the principle employed in the tau-dither early-late tracking loop. In a dithering loop, the incoming signal is correlated *alternately* with the early and the late codes [28].

Consider the noncoherent TDL illustrated in Figure 7.75. Note that the PNG phase is switched back and forth in accordance with the binary signal $q(t)$ as shown in the figure. Thus, the dithering loop can be looked upon as a DLL in which the early and late codes time-share the correlator. The advantage of the TDL over the DLL is that only a single correlator is used, thereby eliminating the potential cause of gain imbalance that is associated with a two-channel loop system such as the full-time DLL.

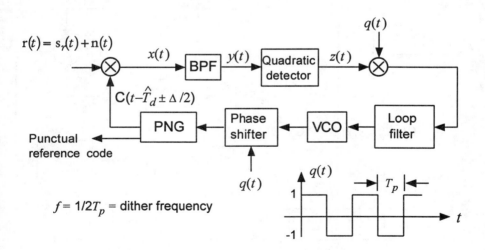

Figure 7.75 A noncoherent TDL.

To understand the principle of operation of the TDL more clearly, let us draw an equivalent loop model as illustrated in Figure 7.76. Let the received waveform be as given in the previous subsection, in (7.129b) and (7.129c). Again, we consider tracking only the PN code $C_I(t)$. Because the correlation and filtering operations will remove the nonpilot channels and the quadrature PN code $C_Q(t)$ from the tracking loop, we may model the received signal term as

$$s_r(t) = \sqrt{\tfrac{1}{2}P_r}\, C_I(t - T_d)\cos[\omega_0 t + \phi(t)] \tag{7.133a}$$

The switching signals $q_1(t)$ and $q_2(t)$ in Figure 7.76 are given by

$$q_1(t) = \tfrac{1}{2}[1 + q(t)], \qquad q_2(t) = \tfrac{1}{2}[1 - q(t)] \tag{7.133b}$$

The system diagram shown in Figure 7.76, using $q_1(t)$ and $q_2(t)$, is essentially a DLL. Hence, the noiseless analysis of its operation is the same as that given for the noncoherent DLL. The error signal is given by

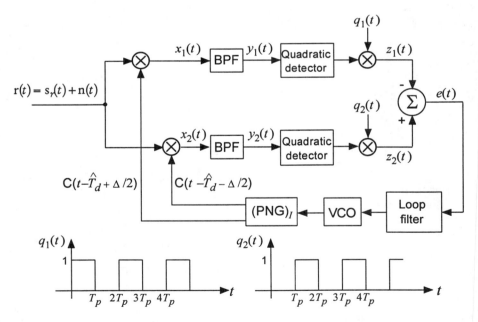

Figure 7.76 An equivalent loop model for a TDL.

$$e(t) = z_2(t) - z_1(t) \tag{7.134a}$$

where

$$z_1(t) = K \left[\sqrt{\tfrac{1}{2}P_r} \ \overline{C_I(t - T_d) \, C_I(t - \widehat{T}_d + \tfrac{1}{2}\Delta)} \right]^2 q_1(t)$$

$$= \tfrac{1}{2} K \, P_r \, R_I^2(T_d - \widehat{T}_d + \tfrac{1}{2}\Delta) \, q_1(t) \tag{7.134b}$$

and

$$z_2(t) = K \left[\sqrt{\tfrac{1}{2}P_r} \ \overline{C_I(t - T_d) \, C_I(t - \widehat{T}_d - \tfrac{1}{2}\Delta)} \right]^2 q_2(t)$$

$$= \tfrac{1}{2} K \, P_r \, R_I^2(T_d - \widehat{T}_d - \tfrac{1}{2}\Delta) \, q_2(t) \tag{7.134c}$$

By substituting (7.133b) in (7.134b) and (7.134c), we have for the error signal given in (7.134a)

$$
\begin{aligned}
e(t, \tau, \Delta) &= \tfrac{1}{2} K \, P_r \left[R_I^2(\tau - \tfrac{1}{2}\Delta) \, q_2(t) - R_I^2(\tau + \tfrac{1}{2}\Delta) \, q_1(t) \right] \\
&= \tfrac{1}{4} K \, P_r \left\{ R_I^2(\tau - \tfrac{1}{2}\Delta) \, [1 - q(t)] - R_I^2(\tau + \tfrac{1}{2}\Delta) \, [1 + q(t)] \right\} \\
&= \tfrac{1}{4} K \, P_r \left[R_I^2(\tau - \tfrac{1}{2}\Delta) - R_I^2(\tau + \tfrac{1}{2}\Delta) \right] \\
&\quad - q(t) \cdot \tfrac{1}{4} K \, P_r \left[R_I^2(\tau - \tfrac{1}{2}\Delta) + R_I^2(\tau + \tfrac{1}{2}\Delta) \right]
\end{aligned}
\tag{7.135a}
$$

where τ and Δ are as defined previously in (7.127a) and (7.127b), respectively. The normalized error function is then given by

$$
\begin{aligned}
E(t, \tau, \Delta) &= R_I^2(\tau - \tfrac{1}{2}\Delta) - R_I^2(\tau + \tfrac{1}{2}\Delta) \\
&\quad - q(t) \left[R_I^2(\tau - \tfrac{1}{2}\Delta) + R_I^2(\tau + \tfrac{1}{2}\Delta) \right]
\end{aligned}
\tag{7.135b}
$$

which consists of two parts: the first part is the familiar error function of the noncoherent full-time DLL, as plotted in Figures 7.70 and 7.71 for two different chip basic waveforms; and the second part is a sum of squares of the early and late autocorrelation functions, multiplied by the dither switching function $q(t)$. Therefore, if the dithering frequency is very high compared with the bandwidth of the loop filter, the second part of the error signal has virtually no influence on the VCO.

This brief analysis of the TDL in a noiseless environment shows that the TDL operation is the same as that of the full-time DLL. The reader interested in the noise performance analysis of the TDL is referred to references [22, 28, 29]. We do not discuss the variations on the basic early-late PN code tracking loops, such as the the DDL, the PSD DLL, and the MCTL. For these subjects, reference [22] is recommended for further reading.

7.5 Shaped Versus Unshaped PN Sequences for Despreading

It can be shown that it matters what shape of pulses are used in the local PN references used for despreading a direct sequence spread-spectrum signal [30]. In this section, an analysis in support of this statement is presented first, followed by a simulation that illustrates the relative significance of this issue.

7.5.1 Analysis of the Effect of Pulse Shape at the Receiver

We previously discussed the baseband pulse shaping specified for IS-95 transmissions and adopted the notations

$$p_0(t) = \text{rectangular (NRZ) data pulse}$$

$$p_1(t) = \text{IS-95 specified pulse shape}$$

In addition to the difference in shape of these pulses, it is pertinent to note the difference in their durations. Letting the center of the chip interval be the reference time, $p_0(t)$ has a duration of $\pm T_c/2$, while $p_1(t)$ (which is discussed in Section 1.4) extends to $\pm 6T_c$. The question we now address is what pulse shape should be used by the receiver to despread the incoming pulses that have the shape $p_1(t)$. First, we derive a quantitative measure of the effect of the pulse shape.

For simplicity, let us neglect filter and propagation delays and consider just the despreading of the IS-95 pilot signal's I channel, which is the shaped version of $C_I(t)$. Let us denote this waveform $C_1(t)$:

$$C_1(t) \triangleq [C_I(t)]_{\substack{\text{shaped} \\ \text{w/ } p_1(t)}} = \sum_k c_{Ik} \cdot p_1(t - kT_c) \tag{7.136}$$

using $\{c_{Ik}\}$ to denote the ± 1 PN sequence. Let the receiver's reference code waveform be in perfect synchronism, but having a different pulse shape, $p_2(t)$, let it be denoted $C_2(t)$:

$$C_2(t) \triangleq [C_I(t)]_{\substack{\text{shaped} \\ \text{w/ } p_2(t)}} = \sum_k c_{Ik} \cdot p_2(t - kT_c) \tag{7.137}$$

The despreading of the pilot signal is accomplished by integrating the product of $C_1(t)$ and $C_2(t)$, giving a measure of pilot energy:

$$E_p[T; p_2(t)] \triangleq \int_0^T C_1(t) \, C_2(t) \, dt \tag{7.138a}$$

Using $c_k \equiv c_{Ik}$ to simplify the notation, the integrand of (7.138a) can be written

$$C_1(t) \, C_2(t) = \left(\sum_k c_k \cdot p_1(t - kT_c) \right) \cdot \left(\sum_m c_m \cdot p_2(t - mT_c) \right)$$

$$= \sum_k c_k \, p_1(t - kT_c)$$

$$\times \{ c_k \, p_2(t - kT_c) + c_{k-1} \, p_2(t - kT_c + T_c)$$

$$+ c_{k+1} \, p_2(t - kT_c - T_c) + \cdots \}$$

$$= \sum_k c_k^2 \, p_1(t - kT_c) \, p_2(t - kT_c) + \text{interchip interference.} \tag{7.138b}$$

Thus, the pilot energy measure E_p has two parts: a part due to despreading of the chips in synchronism with each other, and a part due to the interchip interference (ICI) that occurs when the pulse shape has duration greater than T_c. This fact may be written

$$E_p[T; p_2(t)] = E_d[T; p_2(t)] + E_{ICI}[T; p_2(t)] \tag{7.138c}$$

Because $c_k^2 \equiv 1$, the portion of E_p due to despreading is

$$E_d[T; p_2(t)] = \int_0^T \sum_k p_1(t - kT_c) \, p_2(t - kT_c) \, dt$$

$$= \sum_k \int_0^T p_1(t - kT_c) \, p_2(t - kT_c) \, dt$$

$$= \sum_k \int_{-kT_c}^{T-kT_c} p_1(t) \, p_2(t) \, dt \tag{7.138d}$$

Pulse energy accumulated using a shaped reference PN code. For $p_2(t) = p_1(t)$, the same shape as the transmitted pulse, E_d becomes

$$E_d[T; p_1(t)] = \sum_k \int_{-kT_c}^{T-kT_c} p_1^2(t) \, dt = \sum_k \int_{L_1}^{U_1} p_1^2(t) \, dt \qquad (7.139a)$$

where the lower and upper limits of integration are modified to account for the duration of the pulse, taken to be $\pm 6T_c$:

$$\begin{array}{cc}
\text{lower limit of integral} & \text{upper limit of integral} \\
\downarrow & \downarrow \\
L_1 = \max\{-kT_c, -6T_c\} \text{ and } & U_1 = \min\{T - kT_c, 6T_c\} \\
\uparrow & \uparrow \\
\text{lower limit of integrand} & \text{upper limit of integrand}
\end{array} \qquad (7.139b)$$

That is, the effective lower limit of the integral is determined by the integrand when k has a relatively large positive value, and the effective upper limit is determined by the integrand when k has a relatively large negative value.

To evaluate (7.138a), it is useful to determine when the integral's limits are $\pm 6T_c$ and how many times; that is, how many complete double-sided pulse periods are contained in the integration interval. For $k \geq 6$, the lower limit of the integral is determined by the nonzero region of the integrand to be $L_1 = -6T_c$, and for $k \leq T/T_c - 6 \equiv N_c - 6$, the upper limit of the integral is determined by the integrand to be $U_1 = +6T_c$. Therefore, for $6 \leq k \leq N_c - 6$, where $N_c \triangleq T/T_c$ is the number of chips in the integration interval T, the integration interval is the same ($L_1 = -6T_c$ and $U_1 = 6T_c$), and all of the energy is captured for these pulses. The number of such pulses (number of k values) is

$$(U_1 - L_1)/T_c + 1 = (N_c - 6) - 6 + 1 = N_c - 11 \qquad (7.139c)$$

No pulse energy is captured for $k \leq -6$, when the lower limit of the integral is $+6T_c$ or greater, so that the integrand is zero for the whole interval of integration. Similarly, for $k \geq N_c + 6$, the upper limit of the integral is $-6T_c$, for which the integrand is zero for the whole interval of integration.

It follows that partial pulse energy is captured for the following values of k:

$$\underbrace{-6 < k < 6}_{\text{11 values of } k} \quad \text{and} \quad \underbrace{N_c - 6 < k < N_c + 6}_{\text{11 values of } k} \qquad (7.139d)$$

The number of pulses with partial energy, therefore, is

$$[(N_c + 6) - (N_c - 6) - 1] + [6 - (-6) - 1] = 11 + 11 = 22 \quad (7.139e)$$

So far, in evaluating the amount of pulse energy captured by the despreading receiver in the case of $p_2(t) = p_1(t)$, we have shown that the integration yields the full pulse energy for $N_c - 11$ values of k and partial pulse energy for 22 values of k. Allotting an average of one-half the pulse energy to the partial pulses, the total despreading energy E_d is approximately

$$E_d[T; p_1(t)] \approx \left[N_c - 11 + \tfrac{1}{2}(22) = N_c \right] \times \text{energy in one } p_1(t) \text{ pulse}$$

$$= N_c \int_{-6T_c}^{6T_c} p_1^2(t)\, dt \quad (7.139f)$$

The ICI energy $E_{ICI}[T; p_2(t)]$ can be expected to have an average integration value of zero because the PN code has not been removed in those terms.

Pulse energy accumulated using a nonshaped reference PN code. Now let the receiver pulse shape be $p_2(t) = p_0(t)$, the rectangular or NRZ (nonreturn to zero) pulse shape. The despreading energy becomes

$$E_d[T; p_0(t)] = \sum_k \int_{-kT_c}^{T-kT_c} p_1(t)\, p_0(t)\, dt = \sum_k \int_{L_0}^{U_0} p_1(t)\, dt \quad (7.140a)$$

because $p_0(t) = 1$ in the interval of integration, where the lower and upper limits of integration are modified to account for the $\pm T_c/2$ duration of the pulse $p_0(t)$:

$$L_0 = \max\{-kT_c, -T_c/2\} \quad \text{and} \quad U_0 = \min\{T - kT_c, T_c/2\} \quad (7.140b)$$

The chips for which the integration is over the maximal interval of $\pm T_c/2$ are those for which $k > \tfrac{1}{2}$ (the case for which L_0 equals $-T_c/2$) and $k < N_c - \tfrac{1}{2}$ (the case for which U_0 equals $+T_c/2$). The number of such chips is $N_c - 1$.

For $k > N_c - \tfrac{1}{2}$ and $k < \tfrac{1}{2}$, the integration is zero because the upper and lower limits converge. Thus, there are no chips for which there is a partial collection of pulse energy. The total despreading energy E_d for the case of an unshaped receiver PN code waveform is then

$$E_d[T; p_0(t)] = (N_c - 1) \int_{-T_c/2}^{T_c/2} p_1(t)\, dt \approx N_c \int_{-T_c/2}^{T_c/2} p_1(t)\, dt \qquad (7.140c)$$

The values of the pulse $p_1(t)$ at samples take at four times the chip rate were previously given in Table 1.3 as the discrete-time impulse response $h_0(i)$ of the FIR filter. Approximating the integrals in (7.139f) and (7.140c) by sums of samples gives

$$E_d[T; p_1(t)] \approx N_c \cdot \frac{T_c}{4} \cdot \sum_{i=0}^{47} h_0^2(i) = 3.940\, N_c \cdot (T_c/4) \qquad (7.141a)$$

and

$$E_d[T; p_0(t)] \approx N_c \cdot \frac{T_c}{4} \cdot \sum_{i=22}^{25} h_0(i) = 3.572\, N_c \cdot (T_c/4) \qquad (7.141b)$$

Thus, the advantage in using a shaped PN code waveform to despread the incoming PN signal is slight. This fact is next demonstrated by a numerical example.

7.5.2 Simulated Comparison of the Energies Accumulated

Simulated amples of the IS-95 shaped I and Q PN code baseband waveforms were generated at the rate of four samples per chip, for a time interval of 128 chips. The calculation of the despreading integral was done without the factor of $T_c/4$ that appears in (7.141a) and (7.141b). The results of the simulation are shown by the following figures:

- Figure 7.77: Side-by-side comparison of the IS-95 I-channel PN code waveform using $p_1(t)$ (shaped) and $p_0(t)$ (unshaped). The incoming PN waveform is shaped, and a comparison is to be made of the results of despreading using shaped and unshaped PN code waveforms.

- Figure 7.78: Side-by-side comparison of the IS-95 Q-channel PN code waveform using $p_1(t)$ (shaped) and $p_0(t)$ (unshaped).

- Figure 7.79: I-channel despreading energy for the two shaping cases.

- Figure 7.80: Q-channel despreading energy for the two shaping cases.

I channel PN code waveform

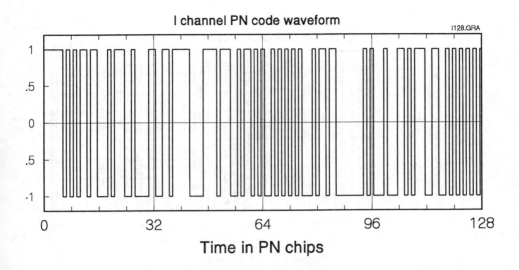

Time in PN chips

Shaped I channel PN code waveform

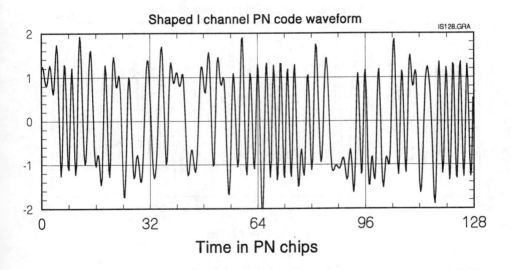

Time in PN chips

Figure 7.77 Candidates for the I-channel despreading waveform.

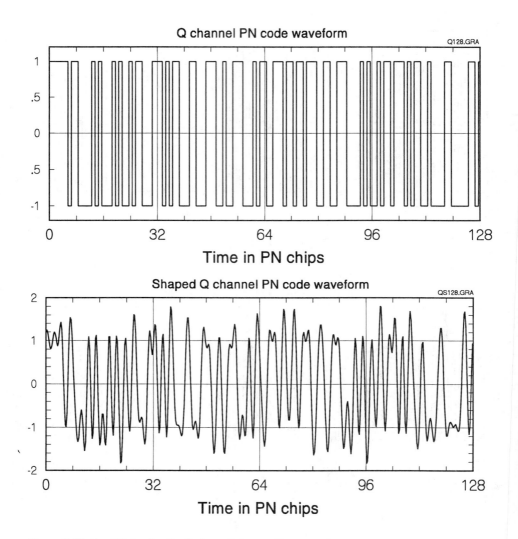

Figure 7.78 Candidates for the Q-channel despreading waveform.

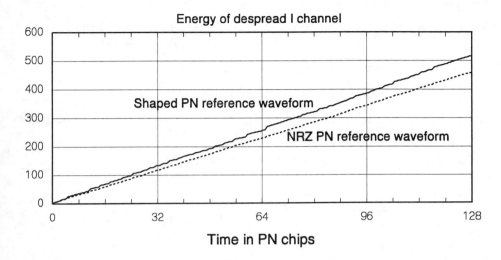

Figure 7.79 Comparison of I-channel despreading results for shaped and unshaped PN reference waveforms.

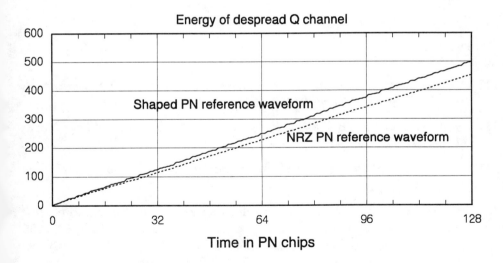

Figure 7.80 Comparison of Q-channel despreading results for shaped and unshaped PN reference waveforms.

We make the following observations on the numerical results for despreading:

- Given that the calculations (7.112a) and (7.112b) ignored the factor of $T_c/4$, the theory predicted despreading energy values of

  $3.940\,N_c$ for the shaped pulse, $p_1(t)$

  $3.572\,N_c$ for the unshaped pulse, $p_0(t)$

- The simulated integrations agree very well with the theory in that the integrations are practically linear functions of N_c.

- The slopes agree very well with the theory also:

Pulse	Predicted slope	Experimental values
$p_1(t)$	3.940	4.02 (I), 3.95 (Q)
$p_0(t)$	3.572	3.59 (I), 3.56 (Q)

We conclude from the analysis and the results that there is only a slight difference between using the optimal, shaped PN reference and one that is unshaped.

References

[1] Wozencraft, J. M., and I. M. Jacobs, *Principles of Communications Engineering*, New York: Wiley, 1965.

[2] Hildrebrand, F. B., *Methods of Applied Mathematics*, New York: Prentice-Hall, 1952.

[3] Friedland, B., *Principles and Techniques of Applied Mathematics*, New York: Wiley, 1956.

[4] Shannon, C. E., "A Mathematical Theory of Communication," *Bell System Technical Journal*, Vol. 27, pp. 379–433 (Part I) and pp. 623–656 (Part II). See also *Claude E. Shannon: Collected Papers*, edited by N. A. Sloane and A. D. Wyner, New York: IEEE Press, 1992.

[5] Taub, H., and D. L. Schilling, *Principles of Communication Systems*, 2nd ed., New York: McGraw-Hill, 1986.

[6] Anderson, J. B., T. Aulin, and C.-E. Sundberg, *Digital Phase Modulation*, New York: Plenum Press, 1986.

[7] Proakis, J. G., *Digital Communications*, 2nd ed., New York: McGraw-Hill, 1989.

[8] Simon, M. K., S. M. Hinedi, and W. C. Lindsey, *Digital Communication Techniques*, Englewood Cliffs, NJ: Prentice-Hall, 1995.

[9] Whalen, A. D., *Detection of Signals in Noise*, New York: Academic Press, 1971.

[10] Viterbi, A. J., *CDMA: Principles of Spread Spectrum Communication*, New York: Addison-Wesley, 1995.

[11] "Mobile Station-Base Station Compatibility Standard for Dual-Mode Wideband Spread Spectrum Cellular System," TIA/EIA Interim Standard 95 (IS-95), Washington: Telecommunications Industry Association, July 1993 (amended as IS-95-A in May 1995).

[12] Daveport, W. B., Jr., and W. L. Root, *An Introduction to the Theory of Random Signals and Noise*, New York: McGraw-Hill, 1958.

[13] Torrieri, D. J., *Principles of Secure Communication Systems*, Norwood, MA: Artech House, 1985.

[14] Miller, L. E., J. S. Lee, R. H. French, and D. J. Torrieri, "Analysis of an Antijam FH Acquisition Scheme," *IEEE Trans. on Communications*, Vol. 40, pp. 160–170, Jan. 1992.

[15] Gardner, F. M., *Phaselock Techniques*, New York: Wiley, 1968.

[16] Viterbi, A. J., *Principles of Coherent Communication*, New York: McGraw-Hill, 1966.

[17] Lindsey, W. C., and M. K. Simon, *Telecommunication Systems Engineering*, Englewood Cliffs, NJ: Prentice-Hall, 1973.

[18] Proakis, J. G., *Digital Communications* (2nd ed.), New York: McGraw-Hill, 1989.

[19] Spilker, J. J., Jr., and D. T. Magill, "The Delay-Lock Discriminator—An Optimum Tracking Device," *Proc. IRE*, Vol. 49, pp. 1403–1416, Sept. 1961.

[20] Spilker, J. J., Jr., "Delay-Lock Tracking of Binary Signals," *IRE Trans. on Space Electronics and Telemetry*, Vol. SET-9, pp. 1–8, Mar. 1963.

[21] Spilker, J. J., Jr., *Digital Communications by Satellite*, Englewood Cliffs, NJ: Prentice-Hall, 1977.

[22] Simon, M. K., J. K. Omura, R. H. Scholtz, and B. K. Levitt, *Spread Spectrum Communications Handbook* (rev. ed.), New York: McGraw-Hill, 1985.

[23] Spilker, J. J., Jr., "GPS Signal Structure and Performance Characteristics," *Journal of Navigation*, Vol. 1, Institute of Navigation, Washington, DC, pp. 29–53, 1982.

[24] Hopkins, P. M., "Double Dither Loop for Pseudonoise Code Tracking," *IEEE Trans. Aerospace and Electronics Syst.*, Vol. AES-13, No.6, pp. 644–650, Nov. 1977.

[25] LaFlame, D. T., "A Delay-Lock Loop Implementation Which Is Insensitive to Arm Gain Imbalance," *IEEE Trans. Communications*, Vol. COM-27, No. 10, pp. 1632–1633, Oct. 1979.

[26] Yost, R. A., and R. W. Boyd, "A Modified PN Code Tracking Loop: Its Performance Analysis and Comparative Evaluation," *IEEE Trans. Communications*, Vol. COM-30, No. 5, pp. 1027–1036, May 1982.

[27] Ziemer, R. E., and R. L. Peterson, *Digital Communications and Spread Spectrum Systems*, New York: Macmillan, 1985.

[28] Hartman, H. P., "Analysis of a Dithering Loop for PN Code Tracking," *IEEE Trans. Aerospace and Electronics Systems*, Vol. AES-10, No. 1, pp. 2–9, Jan. 1974.

[29] Simon, M. K., "Noncoherent Pseudonoise Code Tracking Performance of Spread Spectrum Receivers," *IEEE Trans. Communications*, Vol. COM-25, No. 3, pp. 327–345, Mar. 1977.

[30] Holmes, J. K., *Coherent Spread Spectrum Systems*, New York: Wiley, 1982.

Appendix 7A The Gram-Schmidt Orthogonalization Procedure

Given a signal set $\{s_i(t), 0 \leq t \leq T\}$, $i = 0, 1, \ldots, M-1$, we wish to obtain a set of orthonormal (ON) functions $\{\varphi_j(t), 0 \leq t \leq T\}$, $j = 1, 2, \ldots, N$ so that we can represent the waveforms $\{s_i(t)\}$ as a linear combination of the ON functions, like

$$s_i(t) = s_{i1}\,\varphi_1(t) + s_{i2}\,\varphi_2(t) + \cdots + s_{iN}\,\varphi_N(t) \tag{7A.1a}$$

With such a representation, the signal waveforms can be characterized as signal vectors $\{\mathbf{s}_i\}$, where

$$\mathbf{s}_i = (s_{i1}, s_{i2}, \ldots, s_{iN}) \tag{7A.1b}$$

This can be done by a method known as the *Gram-Schmidt orthogonalization procedure* [39–40]. If we start with a set of signal waveforms $s_0(t)$, $s_1(t)$,..., $s_{M-1}(t)$, the Gram-Schmidt procedure says that we can obtain a new set of orthogonal functions $\{y_i(t)\}$. If we assume that the original set of functions $\{s_0(t), s_1(t), \ldots, s_{M-1}(t)\}$ are linearly independent[12] functions, then a set of functions $\{y_1(t), y_2(t), \ldots, y_N(t)\}$ can be constructed such that the set $\{y_j(t)\}$, $j = 1, 2, \ldots, N$ is an orthogonal set. That is:

$$\langle y_i(t), y_j(t) \rangle \triangleq \int_0^T y_i(t) \, y_j(t) \, dt = 0 \quad \text{for} \quad i \neq j \tag{7A.1c}$$

and in this case $M = N$. If, however, the given set of signals $\{s_i(t)\}$ are not linearly independent, the number of orthogonal functions is always less than the number of given signal functions; that is, $N < M$ when $\{s_i(t)\}$ is a set of functions that are not linearly independent. As we see below, the orthogonal functions $\{y_j(t)\}$ are normalized to obtain the required ON set of functions $\{\varphi_j\}$, where

$$\varphi_j(t) = \frac{y_j(t)}{\sqrt{\langle y_j(t), y_j(t) \rangle}} = \frac{y_j(t)}{\left(\int_0^T y_i^2(t) \, dt \right)^{1/2}}, \, j = 0, 1, \ldots, N \tag{7A.1d}$$

Thus, the orthogonal functions are "interim" functions in view of the ultimate objective of obtaining the ON set $\{\varphi_j(t)\}$. The algorithm for constructing $y_j(t)$ is based on linear combinations of the original signal waveforms.

Without loss of generality, let us assume that the functions $s_0(t)$, $s_1(t)$, ..., $s_{M-1}(t)$ are linearly independent and proceed as follows:

---

[12] The reader is referred to Section 5.3.4 for formal definitions of *linear independence* and *linear dependence*.

1. We pick the first function $s_0(t)$ and set

$$y_1(t) = s_0(t), \ 0 \le t \le T \tag{7A.2a}$$

2. Using the second member of the set, $s_1(t)$, put

$$y_2(t) = s_1(t) + \gamma_1 y_1(t) \tag{7A.2b}$$

for some constant γ_1. Because $\langle y_2(t), y_1(t) \rangle = \langle s_1(t), y_1(t) \rangle + \gamma_1 \langle y_1(t), y_1(t) \rangle$, the desired orthogonality of $y_1(t)$ and $y_2(t)$, that is:

$$\langle y_1(t), y_2(t) \rangle = 0 \tag{7A.2c}$$

is assured by setting

$$\gamma_1 = -\frac{\langle s_1(t), y_1(t) \rangle}{\langle y_1(t), y_1(t) \rangle} \tag{7A.2d}$$

Thus
$$y_2(t) = s_0(t) - \frac{\langle s_1(t), y_1(t) \rangle}{\langle y_1(t), y_1(t) \rangle} y_1(t) \tag{7A.2e}$$

3. Next, we determine $y_3(t)$ that is orthogonal to $y_1(t)$ and $y_2(t)$ and is a linear combination of $\{s_2(t), y_1(t), y_2(t)\}$ and, hence, a linear combination of $\{s_0(t), s_1(t), s_2(t)\}$. So, let us put

$$y_3(t) = s_2(t) + \gamma_1 y_1(t) + \gamma_2 y_2(t) \tag{7A.3a}$$

From the orthogonality requirements, we have

$$\langle y_3(t), y_1(t) \rangle = \langle s_2(t), y_1(t) \rangle + \gamma_1 \langle y_1(t), y_1(t) \rangle$$
$$+ \gamma_2 \overbrace{\langle y_2(t), y_1(t) \rangle}^{0} = 0 \tag{7A.3b}$$

and

$$\langle y_3(t), y_2(t) \rangle = \langle s_2(t), y_2(t) \rangle + \gamma_1 \overbrace{\langle y_1(t), y_2(t) \rangle}^{0}$$
$$+ \gamma_2 \langle y_2(t), y_2(t) \rangle = 0 \tag{7A.3c}$$

from which we obtain the values for γ_1 and γ_2:

$$\gamma_1 = -\frac{\langle s_2(t), y_1(t) \rangle}{\langle y_1(t), y_1(t) \rangle} \qquad \gamma_2 = -\frac{\langle s_2(t), y_2(t) \rangle}{\langle y_1(t), y_1(t) \rangle} \tag{7A.3d}$$

Thus
$$y_3(t) = s_2(t) - \sum_{k=1}^{2} \frac{\langle s_2(t), y_k(t) \rangle}{\langle y_k(t), y_k(t) \rangle} y_k(t) \qquad (7A.3e)$$

4. The procedure continues in the manner described above, until the last member of the set $\{y_j(t)\}$ is obtained. The general expression can be given as

$$y_j(t) = s_{j-1}(t) - \sum_{k=1}^{j-1} \frac{\langle s_{j-1}(t), y_k(t) \rangle}{\langle y_k(t), y_k(t) \rangle} y_k(t) \qquad (7A.4)$$

for all $j = 1, 2, \ldots, N$.

5. The interim orthogonal functions $\{y_j(t)\}$ can now be normalized using (7.142d) to obtain the ON functions $\{\varphi_j(t)\}$.

If we had assumed that the given signal set $\{s_0(t), s_1(t), \ldots, s_{M-1}(t)\}$ is not a set of linearly independent functions, we would have obtained the interim functions $y_1(t), y_2(t), \ldots, y_N(t)$ where $N < M$. Consider (7.144c) above. Had we assumed that $\{s_0(t), s_1(t), s_2(t)\}$ are not linearly independent, $s_2(t)$ would then be expressible as a linear combination of $s_0(t)$ and $s_1(t)$ and, hence, $y_3(t) = 0$, meaning that the third function $y_3(t)$ would have been expressible as a linear combination of only two functions. We see this fact in the example below.

Example 7A.1 We consider a simple example to illustrate the method of applying the Gram-Schmidt orthogonalization procedure. Consider the four waveforms shown in Figure 7A.1. We obtain the vector representations for the waveforms by following the procedures discussed above.

1. Use the first waveform $s_0(t)$ and put $y_1(t) = s_0(t)$. Normalizing:

$$\varphi_1(t) = \frac{y_1(t)}{\sqrt{\langle y_1(t), y_1(t) \rangle}} = \frac{s_0(t)}{\left(\int_0^3 s_0^2(t)\, dt \right)^{1/2}} = \frac{1}{\sqrt{3}} s_0(t) \qquad (7A.5a)$$

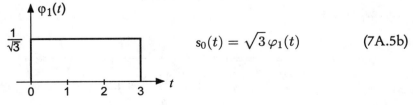

$$s_0(t) = \sqrt{3}\, \varphi_1(t) \qquad (7A.5b)$$

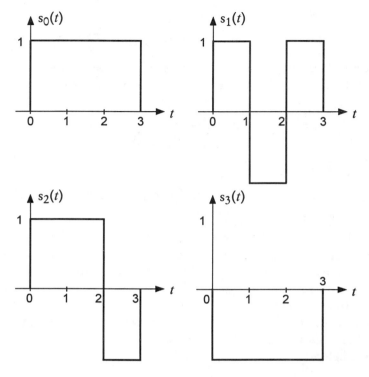

Figure 7A.1 Waveforms to be expressed as vectors.

2. Using the second waveform $s_1(t)$, we put

$$y_2(t) = s_1(t) - \frac{\langle s_1(t), y_1(t) \rangle}{\langle y_1(t), y_1(t) \rangle} y_1(t) \qquad (7A.6a)$$

where $\langle y_1(t), y_1(t) \rangle = 1$ and $\langle s_1(t), y_1(t) \rangle = \frac{1}{3}$. Thus

$$y_2(t) = -\tfrac{1}{3} s_0(t) + s_1(t) \qquad (7A.6b)$$

Normalizing $y_2(t)$:

$$\varphi_2(t) = \frac{y_2(t)}{\sqrt{\langle y_2(t), y_2(t) \rangle}} = \frac{y_2(t)}{\left(\int_0^3 y_2^2(t)\, dt \right)^{1/2}} = \sqrt{\frac{3}{8}}\, y_2(t) \qquad (7A.6c)$$

$y_2(t)$ and $\varphi_2(t)$ are illustrated in Figure 7A.2. From (7A.6b), we have

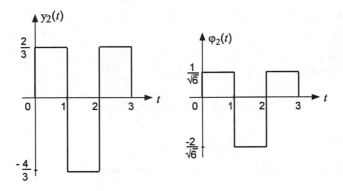

Figure 7A.2 The functions $y_2(t)$ and $\varphi_2(t)$.

$$s_1(t) = y_2(t) + \tfrac{1}{3} s_0(t) = \tfrac{2}{\sqrt{3}} \varphi_1(t) + 2\sqrt{\tfrac{2}{3}} \varphi_2(t) \qquad \text{(7A.6d)}$$

3. Using the third function:

$$y_3(t) = s_2(t) - \frac{\langle s_2(t), y_1(t) \rangle}{\langle y_1(t), y_1(t) \rangle} y_1(t) - \frac{\langle s_2(t), y_2(t) \rangle}{\langle y_2(t), y_2(t) \rangle} y_2(t) \qquad \text{(7A.7a)}$$

But $\langle y_1(t), y_1(t) \rangle = 3$, $\langle y_2(t), y_2(t) \rangle = \tfrac{8}{3}$, $\langle s_2(t), y_1(t) \rangle = 1$, and $\langle s_2(t), y_2(t) \rangle = -\tfrac{4}{3}$. Therefore:

$$y_3(t) = s_2(t) - \tfrac{1}{3} s_0(t) + \tfrac{4/3}{8/3} \left[s_1(t) - \tfrac{1}{3} s_0(t) \right]$$

$$= s_2(t) + \tfrac{1}{2} s_1(t) - \tfrac{1}{2} s_0(t) \equiv 0 \qquad \text{(7A.7b)}$$

Thus

$$s_2(t) = \tfrac{1}{2} s_0(t) - \tfrac{1}{2} s_1(t) = \tfrac{1}{2\sqrt{3}} \varphi_1(t) - \sqrt{\tfrac{2}{3}} \varphi_2(t) \qquad \text{(7A.7c)}$$

4. Using the last waveform:

$$y_4(t) = s_3(t) - \frac{\langle s_3(t), y_1(t) \rangle}{\langle y_1(t), y_1(t) \rangle} y_1(t) - \frac{\langle s_3(t), y_2(t) \rangle}{\langle y_2(t), y_2(t) \rangle} y_2(t)$$

$$- \frac{\langle s_3(t), y_3(t) \rangle}{\langle y_3(t), y_3(t) \rangle} \overbrace{y_3(t)}^{0} \qquad \text{(7A.8a)}$$

where $\langle y_1(t), y_1(t) \rangle = 3$, $\langle y_2(t), y_2(t) \rangle = \frac{8}{3}$, $\langle s_3(t), y_1(t) \rangle = -3$, and $\langle s_3(t), y_2(t) \rangle = 0$, giving

$$y_4(t) = s_3(t) + y_1(t) = s_3(t) + s_0(t) \equiv 0 \qquad (7A.8b)$$

Thus

$$s_3(t) = -s_0(t) = -\sqrt{3}\,\varphi_1(t) \qquad (7A.8c)$$

Summarizing:

$$
\begin{aligned}
s_0(t) &= \sqrt{3}\,\varphi_1(t) & &\rightarrow s_0 = \left(\sqrt{3},\, 0\right) \\
s_1(t) &= \tfrac{2}{\sqrt{3}}\,\varphi_1(t) + 2\sqrt{\tfrac{2}{3}}\,\varphi_2(t) & &\rightarrow s_1 = \left(\tfrac{2}{\sqrt{3}},\, 2\sqrt{\tfrac{2}{3}}\right) \\
s_2(t) &= \tfrac{1}{2\sqrt{3}}\,\varphi_1(t) - \sqrt{\tfrac{2}{3}}\,\varphi_2(t) & &\rightarrow s_2 = \left(\tfrac{1}{2\sqrt{3}},\, \sqrt{\tfrac{2}{3}}\right) \\
s_3(t) &= -\sqrt{3}\,\varphi_1(t) & &\rightarrow s_3 = \left(-\sqrt{3},\, 0\right)
\end{aligned}
\qquad (7A.9)
$$

Appendix 7B Average of BPSK Error Probability

The average symbol-error probability for a BPSK or QPSK system in Rayleigh fading is given by

$$P_s(e) = \int_0^\infty \frac{2\beta}{b}\, e^{-\beta^2/b}\, Q\left(\sqrt{\frac{2\beta^2 E_s}{N_0}}\right) d\beta \qquad (7B.1)$$

$$= \int_0^\infty e^{-x}\, Q\left(\sqrt{2\rho x}\right) dx, \qquad \rho = \frac{bE_s}{N_0} = \frac{\overline{E_s}}{N_0}$$

$$= \int_0^\infty e^{-x}\left[\int_{\sqrt{2\rho}}^\infty \frac{\sqrt{x}}{\sqrt{2\pi}}\, e^{-y^2 x/2}\, dy\right] dx$$

$$= \frac{1}{\sqrt{2\pi}} \int_{\sqrt{2\rho}}^\infty \left[\int_0^\infty x^{1/2}\, e^{-kx}\, dx\right] dy, \qquad k = 1 + \tfrac{1}{2}y^2$$

$$= \frac{1}{\sqrt{2\pi}} \int_{\sqrt{2\rho}}^\infty k^{-3/2}\, \Gamma\left(\tfrac{3}{2}\right) dy, \qquad \Gamma\left(\tfrac{3}{2}\right) = \tfrac{1}{2}\Gamma\left(\tfrac{1}{2}\right) = \tfrac{1}{2}\sqrt{\pi}$$

$$= \frac{1}{2\sqrt{2}} \int_{\sqrt{2\rho}}^\infty \left(1 + \tfrac{1}{2}y^2\right)^{-3/2} dy$$

$$= \frac{1}{2} \int_{\sqrt{\rho}}^{\infty} \left(1 + u^2\right)^{-3/2} du$$

$$= \frac{1}{2} \left[\sqrt{\frac{u^2}{1 + u^2}} \right]_{\sqrt{\rho}}^{\infty} = \frac{1}{2} \left(1 - \sqrt{\frac{\rho}{1 + \rho}}\right), \quad \rho = \frac{\overline{E_s}}{\mathcal{N}_0} \qquad (7B.2)$$

Appendix 7C Parameters of Integrated White Noise

In this appendix, we prove that $n_{cq} = G(0, \mathcal{N}_0 T_w)$ and $n_{sq} = G(0, \mathcal{N}_0 T_w)$ as stated in (7.83d). In (7.82c) and (7.82d) of the text we defined

$$n_{cq} \triangleq \xi_{II,q} + \xi_{QQ,q} \qquad (7C.1a)$$

and

$$n_{sq} \triangleq \xi_{IQ,q} - \xi_{QI,q} \qquad (7C.1b)$$

where, from (7.80a)–(7.80e), we further defined that

$$\xi_{II,q} \triangleq \langle n_{II}(t), W_q(t) \rangle = \int_0^{T_w} n_{II}(t) \, W_q(t) \, dt \qquad (7C.2a)$$

$$\xi_{IQ,q} \triangleq \langle n_{IQ}(t), W_q(t - \tfrac{1}{2}T_c) \rangle = \int_{\frac{1}{2}T_c}^{T_w + \frac{1}{2}T_c} n_{IQ}(t) \, W_q(t - \tfrac{1}{2}T_c) \, dt$$

$$\approx \int_0^{T_w} n_{IQ}(t) \, W_q(t - \tfrac{1}{2}T_c) \, dt \qquad (7C.2b)$$

$$\xi_{QI,q} \triangleq \langle n_{QI}(t), W_q(t) \rangle = \int_0^{T_w} n_{QI}(t) \, W_q(t) \, dt \qquad (7C.2c)$$

$$\xi_{QQ,q} \triangleq \langle n_{QQ}(t), W_q(t - \tfrac{1}{2}T_c) \rangle = \int_{\frac{1}{2}T_c}^{T_w + \frac{1}{2}T_c} n_{QQ}(t) \, W_q(t - \tfrac{1}{2}T_c) \, dt$$

$$\approx \int_0^{T_w} n_{QQ}(t) \, W_q(t - \tfrac{1}{2}T_c) \, dt \qquad (7C.2d)$$

Now, $n_{II}(t)$, $n_{IQ}(t)$, $n_{QI}(t)$, and $n_{QQ}(t)$, which are shown in (7.80a)–(7.80d) in the text, are defined as

$$n_{II}(t) \triangleq n_I(t) \, P(t) \, C_I(t)$$
$$+ \frac{1}{\sqrt{2}} A \, W_n\!\left(t - \tfrac{1}{2}T_c\right) P\!\left(t - \tfrac{1}{2}T_c\right) C_Q\!\left(t - \tfrac{1}{2}T_c\right) P(t) \, C_I(t) \sin\theta$$

$$(\text{7C.3a})$$

$$n_{IQ}(t) \triangleq n_I(t) \, P\!\left(t - \tfrac{1}{2}T_c\right) C_Q\!\left(t - \tfrac{1}{2}T_c\right)$$
$$+ \frac{1}{\sqrt{2}} A \, W_n(t) \, P(t) \, C_I(t) \, P\!\left(t - \tfrac{1}{2}T_c\right) C_Q\!\left(t - \tfrac{1}{2}T_c\right) \cos\theta \qquad (\text{7C.3b})$$

$$n_{QI}(t) \triangleq n_Q(t) \, P(t) \, C_I(t)$$
$$+ \frac{1}{\sqrt{2}} A \, W_n\!\left(t - \tfrac{1}{2}T_c\right) P\!\left(t - \tfrac{1}{2}T_c\right) C_Q\!\left(t - \tfrac{1}{2}T_c\right) P(t) \, C_I(t) \cos\theta$$

$$(\text{7C.3c})$$

$$n_{QQ}(t) \triangleq n_Q(t) \, P\!\left(t - \tfrac{1}{2}T_c\right) C_Q\!\left(t - \tfrac{1}{2}T_c\right)$$
$$- \frac{1}{\sqrt{2}} A \, W_n(t) \, P(t) \, C_I(t) \, P\!\left(t - \tfrac{1}{2}T_c\right) C_Q\!\left(t - \tfrac{1}{2}T_c\right) \sin\theta \qquad (\text{7C.3d})$$

where $n_I(t)$ and $n_Q(t)$, as defined in (7.78b)–(7.78d), are taken to be white Gaussian noise terms:

$$n_I(t) \equiv n_c(t) = G\!\left(0, \tfrac{1}{2}\mathcal{N}_0\right) \quad \text{and} \quad n_Q(t) \equiv n_s(t) = G\!\left(0, \tfrac{1}{2}\mathcal{N}_0\right) \qquad (\text{7C.4})$$

Now, returning to (7C.1a) above, we have

$$n_{cq} \triangleq \xi_{II,q} + \xi_{QQ,q} = \langle n_{II}(t), \, W_q(t) \rangle + \langle n_{QQ}(t), \, W_q\!\left(t - \tfrac{1}{2}T_c\right) \rangle \qquad (\text{7C.5a})$$

But from (7C.3a) and (7C.3d), we have

$$n_{II}(t) = \overbrace{n_I(t) \, P(t) \, C_I(t)}^{x_1(t)}$$

$$+ \overbrace{\frac{1}{\sqrt{2}} A \, W_n\!\left(t - \tfrac{1}{2}T_c\right) P\!\left(t - \tfrac{1}{2}T_c\right) C_Q\!\left(t - \tfrac{1}{2}T_c\right) P(t) \, C_I(t) \sin\theta}^{x_2(t)} \qquad (\text{7C.5b})$$

$$n_{QQ}(t) = \overbrace{n_Q(t) \, P\!\left(t - \tfrac{1}{2}T_c\right) C_Q\!\left(t - \tfrac{1}{2}T_c\right)}^{y_1(t)}$$

$$- \overbrace{\frac{1}{\sqrt{2}} A \, W_n(t) \, P(t) \, C_I(t) \, P\!\left(t - \tfrac{1}{2}T_c\right) C_Q\!\left(t - \tfrac{1}{2}T_c\right) \sin\theta}^{y_2(t)} \qquad (\text{7C.5c})$$

As indicated above, let us define

$$x_1(t) \triangleq n_I(t) \overbrace{P(t) C_I(t)}^{\pm 1 \text{ at } R_c} \tag{7C.5d}$$

$$x_2(t) \triangleq \frac{A}{\sqrt{2}} W_n\left(t - \tfrac{1}{2}T_c\right) \overbrace{P\left(t - \tfrac{1}{2}T_c\right) C_Q\left(t - \tfrac{1}{2}T_c\right) P(t) C_I(t)}^{\pm 1 \text{ at } 2R_c} \sin\theta \tag{7C.5e}$$

$$y_1(t) \triangleq n_Q(t) \overbrace{P\left(t - \tfrac{1}{2}T_c\right) C_Q\left(t - \tfrac{1}{2}T_c\right)}^{\pm 1 \text{ at } R_c} \tag{7C.5f}$$

$$y_2(t) \triangleq \frac{A}{\sqrt{2}} W_n(t) \overbrace{P(t) C_I(t) P\left(t - \tfrac{1}{2}T_c\right) C_Q\left(t - \tfrac{1}{2}T_c\right)}^{\pm 1 \text{ at } 2R_c} \sin\theta \tag{7C.5g}$$

Then in terms of x_1, x_2, y_1, and y_2, (7C.1a) is given by

$$
\begin{aligned}
n_{cq} &= \langle x_1(t) + x_2(t), W_q(t) \rangle + \langle y_1(t) - y_2(t), W_q\left(t - \tfrac{1}{2}T_c\right) \rangle \\
&= \left[\langle x_1(t), W_q(t) \rangle + \langle y_1(t), W_q\left(t - \tfrac{1}{2}T_c\right) \rangle \right] \\
&\quad + \left[\langle x_2(t), W_q(t) \rangle - \langle y_2(t), W_q\left(t - \tfrac{1}{2}T_c\right) \rangle \right]
\end{aligned} \tag{7C.6a}
$$

But it is easy to see that

$$
\begin{aligned}
&\langle x_2(t), W_q(t) \rangle - \langle y_2(t), W_q\left(t - \tfrac{1}{2}T_c\right) \rangle \\
&= \int_0^{T_w} \overbrace{P(t) C_I(t) P\left(t - \tfrac{1}{2}T_c\right) C_Q\left(t - \tfrac{1}{2}T_c\right)}^{a(t)} \\
&\quad \times \left[W_q(t) W_n\left(t - \tfrac{1}{2}T_c\right) - W_q\left(t - \tfrac{1}{2}T_c\right) W_n(t) \right] dt \\
&= \int_0^{T_w} a(t) \left[W_q(t) W_n\left(t - \tfrac{1}{2}T_c\right) - W_q\left(t - \tfrac{1}{2}T_c\right) W_n(t) \right] dt \\
&\approx 0
\end{aligned} \tag{7C.6b}
$$

ignoring the small effect of the half-chip delay. (The term is exactly equal to zero for $q = n$). Therefore, we can reduce (7C.5a) to

$$n_{cq} = \langle x_1(t), W_q(t) \rangle + \langle y_1(t), W_q\left(t - \tfrac{1}{2}T_c\right) \rangle = X_1 + Y_1 \tag{7C.6c}$$

where we defined

$$X_1 \triangleq \langle x_1(t), W_q(t) \rangle = \langle n_I(t) P(t) C_I(t), W_q(t) \rangle \qquad (7C.6d)$$

and

$$Y_1 \triangleq \langle y_1(t), W_q(t - \tfrac{1}{2}T_c) \rangle$$
$$= \langle n_Q(t) P(t - \tfrac{1}{2}T_c) C_Q(t - \tfrac{1}{2}T_c), W_q(t - \tfrac{1}{2}T_c) \rangle \qquad (7C.6e)$$

Clearly, X_1 and Y_1 are independent Gaussian RVs, and we need to specify their means and variances:

$$X_1 = G(m_x, \sigma_x^2) \qquad (7C.6f)$$

$$Y_1 = G(m_y, \sigma_y^2) \qquad (7C.6g)$$

where

$$m_x = \overline{X_1} = \overline{\langle x_1(t), W_q(t) \rangle} = \int_0^{T_w} \overline{n_I(t)} \, P(t) \, C_I(t) \, W_q(t) \, dt = 0 \quad (7C.7a)$$

$$\sigma_x^2 = \overline{\left(X_1 - \overline{X_1} \right)^2} = \overline{X_1^2} = \overline{\langle x_1(t), W_q(t) \rangle^2}$$

$$= \int_0^{T_w} \int_0^{T_w} \underbrace{\overline{n_I(t) \, n_I(u)}}_{\tfrac{1}{2}N_0 \, \delta(t-u)} P(t) \, P(u) \, C_I(t) \, C_I(u) \, W_q(t) \, W_q(u) \, dt \, du$$

$$= \tfrac{1}{2}N_0 \int_0^{T_w} P^2(t) \, C_I^2(t) \, W_q^2(t) \, dt = \tfrac{1}{2}N_0 T_w \qquad (7C.7b)$$

where $P^2(t) = C_I^2(t) = 1$ and

$$\int_0^{T_w} W_q^2(t) \, dt = T_w \qquad (7C.7c)$$

as defined in (7.72d). In the same manner, we can easily show that

$$m_y = \overline{Y_1} = \overline{\langle y_1(t), W_q(t - \tfrac{1}{2}T_c) \rangle} = 0 \qquad (7C.8a)$$

and

$$\sigma_y^2 = \overline{\left(Y_1 - \overline{Y_1} \right)^2} = \overline{Y_1^2} = \overline{\langle y_1(t), W_q(t - \tfrac{1}{2}T_c) \rangle^2} = \tfrac{1}{2}N_0 T_w \qquad (7C.8b)$$

Thus

$$X_1 = G\left(0, \tfrac{1}{2}N_0 T_w \right) \qquad (7C.8c)$$

and

$$Y_1 = G\left(0, \tfrac{1}{2}\mathcal{N}_0 T_w\right) \tag{7C.8d}$$

Therefore, $n_{cq} = X_1 + Y_1$ is a Gaussian RV with zero mean and variance $\sigma_x^2 + \sigma_y^2 = \mathcal{N}_0 T_w$; that is:

$$n_{cq} = G(0, \mathcal{N}_0 T_w) \tag{7C.8e}$$

Similar analysis for the RV n_{sq} shows that

$$n_{sq} = G(0, \mathcal{N}_0 T_w) \tag{7C.8f}$$

This proves the validity of (7.83d) in the text.

Appendix 7D Details of BPSK and QPSK Variances

The first variance in (7.96a) corresponds to interchip intersymbol interference within the kth user's signal:

$$\mathrm{Var}\left\{ \sqrt{P^{(k)}} \sum_{n \neq m} d_n^{(k)} c_n^{(k)} c_m^{(k)} \, p_2(mT_c - nT_c) \right\}$$

$$= P^{(k)} \, \mathrm{E}\left\{ \left[\sum_{n \neq m} d_n^{(k)} c_n^{(k)} c_m^{(k)} \, p_2(mT_c - nT_c) \right]^2 \right\}$$

$$= P^{(k)} \sum_{n \neq m} \sum_{l \neq m} \mathrm{E}\left\{ d_n^{(k)} d_l^{(k)} \right\} \underbrace{\mathrm{E}\left\{ c_n^{(k)} c_l^{(k)} \overbrace{\left[c_m^{(k)} \right]^2}^{1} \right\}}_{0 \text{ for } n \neq l}$$

$$\times p_2(mT_c - nT_c) \, p_2(mT_c - lT_c)$$

$$= P^{(k)} \sum_{n \neq m} [p_2(mT_c - nT_c)]^2 = P^{(k)} \sum_{\substack{n=-\infty \\ n \neq 0}}^{\infty} [p_2(nT_c)]^2 \triangleq \sigma_{ISI,k}^2 \tag{7D.1}$$

The second group of variances in (7.96a) corresponds to multiple access interference:

$$\sum_{i \neq k} \text{Var} \left\{ \sqrt{P^{(i)}} \cos(\varphi_i - \varphi_k) \sum_n d_n^{(i)} c_n^{(i)} c_m^{(k)} p_2(mT_c - nT_c - \tau_{ik}) \right\}$$

$$= \sum_{i \neq k} P^{(i)} \cos^2(\varphi_i - \varphi_k) \, \text{E} \left\{ \left[\sum_n d_n^{(i)} c_n^{(i)} c_m^{(k)} p_2(mT_c - nT_c - \tau_{ik}) \right]^2 \right\}$$

$$= \sum_{i \neq k} P^{(i)} \cos^2(\varphi_i - \varphi_k) \sum_n \sum_l \text{E} \left\{ d_n^{(i)} d_l^{(i)} \right\} \underbrace{\text{E} \left\{ c_n^{(i)} c_l^{(i)} \overbrace{\left[c_m^{(k)} \right]^2}^{1} \right\}}_{0 \text{ for } n \neq l}$$

$$\times p_2(mT_c - nT_c - \tau_{ik}) \, p_2(mT_c - lT_c - \tau_{ik})$$

$$= \sum_{i \neq k} P^{(i)} \cos^2(\varphi_i - \varphi_k) \sum_n \left[p_2(mT_c - nT_c - \tau_{ik}) \right]^2$$

$$= \sum_{i \neq k} P^{(i)} \cos^2(\varphi_i - \varphi_k) \sum_{n=-\infty}^{\infty} \left[p_2(nT_c - \tau_{ik}) \right]^2 \triangleq \sigma_{MA,k}^2 \qquad (7D.2a)$$

Note that the second summation in the (7D.2a) can be rewritten as

$$\sum_{n=-\infty}^{\infty} \left[p_2(nT_c - \tau_{ik}) \right]^2 = \sum_{n=-\infty}^{\infty} \left[\int_{-\infty}^{\infty} |H(f)|^2 \, e^{j2\pi f(nT_c - \tau_{ik})} \, df \right]^2$$

$$= \int_{-\infty}^{\infty} |H(f)|^2 \int_{-\infty}^{\infty} |H(\nu)|^2 e^{-j2\pi(f+\nu)\tau_{ik}} \sum_{n=-\infty}^{\infty} e^{j2\pi(f+\nu)nT_c} d\nu \, df$$

$$\qquad (7D.2b)$$

From (6.52d), and using the Fourier transform pair of Comb function $\leftrightarrow$ Rep function given in (1.23h), we recognize that

$$\sum_{n=-\infty}^{\infty} e^{j2\pi(f+\nu)nT_c} = \mathcal{F} \left\{ \sum_{n=-\infty}^{\infty} \delta(t - nT_c) \right\} \Bigg|_{f \to f+\nu}$$

$$= \mathcal{F} \{ \text{Comb}_{T_c}(1) \} |_{f+\nu} = \frac{1}{T_c} \text{Rep}_{1/T_c} \{ \delta(f + \nu) \}$$

$$= \frac{1}{T_c} \sum_{r=-\infty}^{\infty} \delta(f + \nu - r/T_c) \qquad (7D.2c)$$

Substituting (7D.2c) in (7D.2b) gives the following result:

$$\sum_{n=-\infty}^{\infty} [p_2(nT_c - \tau_{ik})]^2$$

$$= \sum_{r=-\infty}^{\infty} \frac{1}{T_c} \int_{-\infty}^{\infty} |H(f)|^2 \left[|H(\nu)|^2 e^{-j2\pi(f+\nu)\tau_{ik}} \right]_{\nu=-f+r/T_c} df$$

$$= \sum_{r=-\infty}^{\infty} \frac{1}{T_c} \int_{-\infty}^{\infty} |H(f)|^2 \cdot |H(-f + r/T_c)|^2 e^{-j2\pi(r/T_c)\tau_{ik}} df \qquad (7D.3)$$

Note that only the $r = 0$ term in (7D.3) is nonzero, because the filter transfer functions in the integrand are disjoint for $r \neq 0$:

$$= \frac{1}{T_c} \left\{ \int_{-\infty}^{\infty} |H(f)|^2 |H(-f)|^2 \, df \right.$$

$$+ \int_{-\infty}^{\infty} \underbrace{|H(f)|^2 \, |H(-f + 1/T_c)|^2}_{\approx \, 0 \text{ because spectra are disjoint}} e^{-j2\pi\tau_{ik}/T_c} \, df$$

$$\left. + \int_{-\infty}^{\infty} \underbrace{|H(f)|^2 \, |H(-f - 1/T_c)|^2}_{\approx \, 0 \text{ because spectra are disjoint}} e^{j2\pi\tau_{ik}/T_c} \, df + \cdots \right\}$$

$$\approx \frac{1}{T_c} \int_{-\infty}^{\infty} |H(f)|^4 \, df \quad \text{for } |H(f)|^2 = |H(-f)|^2 \qquad (7D.4)$$

assuming the bandwidth of $H(f)$ is no greater than $1/T_c$. Thus, $\sigma_{MA,k}^2$ may be written

$$\sigma_{MA,k}^2 = \sum_{i \neq k} P^{(i)} \cos^2(\varphi_i - \varphi_k) \cdot \frac{1}{T_c} \int_{-\infty}^{\infty} |H(f)|^4 \, df$$

Another way to calculate the multiple access interference variances:

$$E \left\{ \left[\sum_n d_n^{(i)} c_n^{(i)} c_m^{(k)} p_2(mT_c - nT_c - \tau_{ik}) \right]^2 \right\}$$

$$= E\left\{ \left| \sum_n d_n^{(i)} c_n^{(i)} c_m^{(k)} \int_{-\infty}^{\infty} |H(f)|^2 \, e^{j2\pi f [T_c(m-n) - \tau_{ik}]} \, df \right|^2 \right\}$$

$$= E\left\{ \left| c_m^{(k)} \int_{-\infty}^{\infty} |H(f)|^2 \, e^{j2\pi f (mT_c - \tau_{ik})} \underbrace{\sum_n d_n^{(i)} c_n^{(i)} e^{-j2\pi f n T_c}}_{\mathcal{F}\{\text{random impulse train}\}} \, df \right|^2 \right\}$$

Spectrally, the random impulse train can be viewed as white noise with spectral density $1/T_c$. Let $\eta(f)$ denote the Fourier transform of white noise with unit density. Then the expectation becomes

$$E\left\{ [c_m^{(k)}]^2 \frac{1}{T_c} \left| \int_{-\infty}^{\infty} |H(f)|^2 \, \eta(f) \, e^{j2\pi f (mT_c - \tau_{ik})} \, df \right|^2 \right\}$$

$$= \frac{1}{T_c} \int_{-\infty}^{\infty} |H(f)|^2 \int_{-\infty}^{\infty} |H(\nu)|^2 \underbrace{E\{\eta(f)\,\eta(\nu)\}}_{1 \cdot \delta(f - \nu)} e^{j2\pi(f-\nu)(mT_c - \tau_{ik})} d\nu \, df$$

$$= \frac{1}{T_c} \int_{-\infty}^{\infty} |H(f)|^4 \, df$$

The last variance in (7.96a) is that of the background noise:

$$\mathrm{Var}\{n_{LP}(mT_c)\, a_m^{(k)}\} = E\{n_{LP}^2(mT_c) \cdot 1\}$$

$$= \tfrac{1}{2} \mathcal{N}_0 \int_{-\infty}^{\infty} |H(f)|^2 \, df \triangleq \sigma_N^2 \qquad (7D.5)$$

For the QPSK case, evaluation of these variances proceeds just as for the case of BPSK. It is assumed that all PN sequences are uncorrelated with other sequences and uncorrelated with shifts of themselves:

$$\sigma_{ISI,k}^2 = \mathrm{Var}\left\{ \tfrac{1}{2} \sqrt{P^{(k)}} \sum_{n \neq m} d_n^{(k)} \left[c_{In}^{(k)} c_{Im}^{(k)} + c_{Qn}^{(k)} c_{Qm}^{(k)} \right] p_2(mT_c - nT_c) \right\}$$

$$= \tfrac{1}{4} P^{(k)} E\left\{ \left[\sum_{n \neq m} d_n^{(k)} \left[c_{In}^{(k)} c_{Im}^{(k)} + c_{Qn}^{(k)} c_{Qm}^{(k)} \right] p_2(mT_c - nT_c) \right]^2 \right\}$$

$$= \tfrac{1}{4} P^{(k)} \sum_{n \neq m} \sum_{l \neq m} E\left\{ d_n^{(k)} d_l^{(k)} \right\} p_2(mT_c - nT_c) \, p_2(mT_c - lT_c)$$

$$\times E\left\{ c_{In}^{(k)} c_{Il}^{(k)} \left[c_{Im}^{(k)} \right]^2 + c_{In}^{(k)} c_{Ql}^{(k)} \, c_{Im}^{(k)} c_{Qm}^{(k)} \right.$$

$$\left. + c_{Qn}^{(k)} c_{Il}^{(k)} \, c_{Qm}^{(k)} c_{Im}^{(k)} + c_{Qn}^{(k)} c_{Ql}^{(k)} \left[c_{Qm}^{(k)} \right]^2 \right\}$$

$$= \tfrac{1}{4} P^{(k)} \sum_{n \neq m} \sum_{l \neq m} E\left\{ d_n^{(k)} d_l^{(k)} \right\} p_2(mT_c - nT_c) \, p_2(mT_c - lT_c)$$

$$\times \left\{ E\left\{ c_{In}^{(k)} c_{Il}^{(k)} \right\} + E\left\{ c_{In}^{(k)} c_{Ql}^{(k)} \right\} E\left\{ c_{Im}^{(k)} c_{Qm}^{(k)} \right\} \right.$$

$$\left. + E\left\{ c_{Qn}^{(k)} c_{Il}^{(k)} \right\} E\left\{ c_{Qm}^{(k)} c_{Im}^{(k)} \right\} + E\left\{ c_{Qn}^{(k)} c_{Ql}^{(k)} \right\} \right\}$$

$$= \tfrac{1}{4} P^{(k)} \sum_{n \neq m} [p_2(mT_c - nT_c)]^2 \times (1 + 0 \cdot 0 + 0 \cdot 0 + 1)$$

$$= \tfrac{1}{2} P^{(k)} \sum_{\substack{n=-\infty \\ n \neq 0}}^{\infty} [p_2(nT_c)]^2 \tag{7D.6}$$

The multiple-access interference power due to the ith user is

$$\left[\sigma_{MA,k}^2\right]_i = \text{Var}\left\{ \tfrac{1}{2} \sqrt{P^{(i)}} \sum_n d_n^{(i)} \left\{ \left[c_{In}^{(i)} c_{Im}^{(k)} + c_{Qn}^{(i)} c_{Qm}^{(k)} \right] \cos(\varphi_i - \varphi_k) \right. \right.$$

$$\left. \left. + \left[c_{Qn}^{(i)} c_{Im}^{(k)} - c_{In}^{(i)} c_{Qm}^{(k)} \right] \sin(\varphi_i - \varphi_k) \right\} p_2(mT_c - nT_c - \tau_{ik}) \right\}$$

$$= \tfrac{1}{4} P^{(i)} \sum_n \sum_l E\left\{ d_n^{(i)} d_l^{(i)} \right\} p_2(mT_c - nT_c - \tau_{ik}) \, p_2(mT_c - lT_c - \tau_{ik})$$

$$\times \left\{ \cos^2(\varphi_i - \varphi_k) \, E\left\{ \left[c_{In}^{(i)} c_{Im}^{(k)} + c_{Qn}^{(i)} c_{Qm}^{(k)} \right] \left[c_{Il}^{(i)} c_{Im}^{(k)} + c_{Ql}^{(i)} c_{Qm}^{(k)} \right] \right\} \right.$$

$$+ \cos(\varphi_i - \varphi_k) \sin(\varphi_i - \varphi_k)$$

$$\times \mathrm{E}\left\{ \left[c_{In}^{(i)} c_{Im}^{(k)} + c_{Qn}^{(i)} c_{Qm}^{(k)} \right] \left[c_{Ql}^{(i)} c_{Im}^{(k)} - c_{Il}^{(i)} c_{Qm}^{(k)} \right] \right\}$$

$$+ \sin(\varphi_i - \varphi_k) \cos(\varphi_i - \varphi_k)$$

$$\times \mathrm{E}\left\{ \left[c_{Qn}^{(i)} c_{Im}^{(k)} - c_{In}^{(i)} c_{Qm}^{(k)} \right] \left[c_{Il}^{(i)} c_{Im}^{(k)} + c_{Ql}^{(i)} c_{Qm}^{(k)} \right] \right\}$$

$$+ \sin^2(\varphi_i - \varphi_k)$$

$$\times \mathrm{E}\left\{ \left[c_{Qn}^{(i)} c_{Im}^{(k)} - c_{In}^{(i)} c_{Qm}^{(k)} \right] \left[c_{Ql}^{(i)} c_{Im}^{(k)} - c_{Il}^{(i)} c_{Qm}^{(k)} \right] \right\} \right\}$$

$$= \tfrac{1}{4} \mathrm{P}^{(i)} \sum_n \sum_l \mathrm{E}\left\{ d_n^{(i)} d_l^{(i)} \right\} \mathrm{p}_2(mT_c - nT_c - \tau_{ik}) \, \mathrm{p}_2(mT_c - lT_c - \tau_{ik})$$

$$\times \left\{ \cos^2(\varphi_i - \varphi_k) \left[\mathrm{E}\left\{ c_{In}^{(i)} c_{Il}^{(i)} \left[c_{Im}^{(k)} \right]^2 \right\} + \mathrm{E}\left\{ c_{In}^{(i)} c_{Ql}^{(i)} \right\} \mathrm{E}\left\{ c_{Im}^{(k)} c_{Qm}^{(k)} \right\} \right. \right.$$

$$\left. + \mathrm{E}\left\{ c_{Qn}^{(i)} c_{Il}^{(i)} \right\} \mathrm{E}\left\{ c_{Qm}^{(k)} c_{Im}^{(k)} \right\} + \mathrm{E}\left\{ c_{Qn}^{(i)} c_{Ql}^{(i)} \left[c_{Qm}^{(k)} \right]^2 \right\} \right]$$

$$+ \cos(\varphi_i - \varphi_k) \sin(\varphi_i - \varphi_k) \left[\mathrm{E}\left\{ c_{In}^{(i)} c_{Ql}^{(i)} \left[c_{Im}^{(k)} \right]^2 \right\} - \mathrm{E}\left\{ c_{Qn}^{(i)} c_{Il}^{(i)} \left[c_{Qm}^{(k)} \right]^2 \right\} \right.$$

$$\left. - \mathrm{E}\left\{ c_{In}^{(i)} c_{Il}^{(i)} \right\} \mathrm{E}\left\{ c_{Im}^{(k)} c_{Qm}^{(k)} \right\} + \mathrm{E}\left\{ c_{Qn}^{(i)} c_{Ql}^{(i)} \right\} \mathrm{E}\left\{ c_{Qm}^{(k)} c_{Im}^{(k)} \right\} \right]$$

$$+ \sin(\varphi_i - \varphi_k) \cos(\varphi_i - \varphi_k) \left[\mathrm{E}\left\{ c_{Qn}^{(i)} c_{Il}^{(i)} \left[c_{Im}^{(k)} \right]^2 \right\} - \mathrm{E}\left\{ c_{In}^{(i)} c_{Ql}^{(i)} \left[c_{Qm}^{(k)} \right]^2 \right\} \right.$$

$$\left. - \mathrm{E}\left\{ c_{In}^{(i)} c_{Il}^{(i)} \right\} \mathrm{E}\left\{ c_{Qm}^{(k)} c_{Im}^{(k)} \right\} + \mathrm{E}\left\{ c_{Qn}^{(i)} c_{Ql}^{(i)} \right\} \mathrm{E}\left\{ c_{Im}^{(k)} c_{Qm}^{(k)} \right\} \right]$$

$$+ \sin^2(\varphi_i - \varphi_k) \left[\mathrm{E}\left\{ c_{Qn}^{(i)} c_{Ql}^{(i)} \left[c_{Im}^{(k)} \right]^2 \right\} - \mathrm{E}\left\{ c_{Qn}^{(i)} c_{Il}^{(i)} \right\} \mathrm{E}\left\{ c_{Im}^{(k)} c_{Qm}^{(k)} \right\} \right.$$

$$\left. \left. - \mathrm{E}\left\{ c_{In}^{(i)} c_{Ql}^{(i)} \right\} \mathrm{E}\left\{ c_{Qm}^{(k)} c_{Im}^{(k)} \right\} + \mathrm{E}\left\{ c_{In}^{(i)} c_{Il}^{(i)} \left[c_{Qm}^{(k)} \right]^2 \right\} \right] \right\}$$

$$= \tfrac{1}{4} \mathrm{P}^{(i)} \sum_n \left[\mathrm{p}_2(mT_c - nT_c - \tau_{ik}) \right]^2$$

$$\times \left\{ \cos^2(\varphi_i - \varphi_k) \left[\mathrm{E}\left\{ \left[c_{In}^{(i)} \right]^2 \cdot 1 \right\} + 2\,\mathrm{E}\left\{ c_{In}^{(i)} c_{Qn}^{(i)} \right\} \mathrm{E}\left\{ c_{Im}^{(k)} c_{Qm}^{(k)} \right\} \right. \right.$$

$$+ \mathrm{E}\left\{\left[c_{Qn}^{(i)}\right]^2 \cdot 1\right\}\right]$$

$$+ \cos(\varphi_i - \varphi_k)\sin(\varphi_i - \varphi_k)\left[\mathrm{E}\left\{c_{In}^{(i)} c_{Qn}^{(i)} \cdot 1 - c_{Qn}^{(i)} c_{In}^{(i)} \cdot 1\right\}\right.$$

$$\left. - \mathrm{E}\left\{\left[c_{In}^{(i)}\right]^2 - \left[c_{Qn}^{(i)}\right]^2\right\}\mathrm{E}\left\{c_{Im}^{(k)} c_{Qm}^{(k)}\right\}\right]$$

$$+ \sin(\varphi_i - \varphi_k)\cos(\varphi_i - \varphi_k)\left[\mathrm{E}\left\{c_{Qn}^{(i)} c_{In}^{(i)} \cdot 1 - c_{In}^{(i)} c_{Qn}^{(i)} \cdot 1\right\}\right.$$

$$\left. - \mathrm{E}\left\{\left[c_{In}^{(i)}\right]^2 - \left[c_{Qn}^{(i)}\right]^2\right\}\mathrm{E}\left\{c_{Qm}^{(k)} c_{Im}^{(k)}\right\}\right\}$$

$$+ \sin^2(\varphi_i - \varphi_k)\left[\mathrm{E}\left\{\left[c_{Qn}^{(i)}\right]^2 \cdot 1\right\} - 2\,\mathrm{E}\left\{c_{Qn}^{(i)} c_{In}^{(i)}\right\}\mathrm{E}\left\{c_{Im}^{(k)} c_{Qm}^{(k)}\right\}\right.$$

$$\left.\left. + \mathrm{E}\left\{\left[c_{In}^{(i)}\right]^2 \cdot 1\right\}\right]\right\} \tag{7D.7a}$$

$$= \tfrac{1}{4}\,\mathrm{P}^{(i)}\sum_n [\mathrm{p}_2(mT_c - nT_c - \tau_{ik})]^2$$

$$\times\{\cos^2(\varphi_i - \varphi_k)[1 + 2\cdot 0 + 1] + \cos(\varphi_i - \varphi_k)\sin(\varphi_i - \varphi_k)[0 - 0]$$

$$+ \sin(\varphi_i - \varphi_k)\cos(\varphi_i - \varphi_k)[0 - 0] + \sin^2(\varphi_i - \varphi_k)[1 - 2\cdot 0 + 1]\}$$

$$= \tfrac{1}{2}\,\mathrm{P}^{(i)}\sum_{n=-\infty}^{\infty}[\mathrm{p}_2(nT_c - \tau_{ik})]^2 \tag{7D.7b}$$

The fact that $\mathrm{E}\left\{d_n^{(i)} d_l^{(i)}\right\} = \delta_{nl}$ was used in (7D.7a). As shown above, (7D.7b) can be expressed as

$$[\sigma_{MA,k}^2]_i = \mathrm{P}^{(i)} \cdot \frac{1}{T_c}\int_{-\infty}^{\infty} df\,|\mathrm{H}(f)|^4 \tag{7D.8}$$

Finally, the total noise variance is

$$\mathrm{Var}\left\{\tfrac{1}{\sqrt{2}}\,\mathrm{n}_{LP}^{(I)}(mT_c)\,a_m^{(k,I)}\right\} + \mathrm{Var}\left\{\tfrac{1}{\sqrt{2}}\,\mathrm{n}_{LP}^{(Q)}(mT_c)\,a_m^{(k,Q)}\right\}$$

$$= \tfrac{1}{2}\mathrm{E}\left\{\left[\mathrm{n}_{LP}^{(I)}(mT_c)\right]^2\right\} + \tfrac{1}{2}\mathrm{E}\left\{\left[\mathrm{n}_{LP}^{(Q)}(mT_c)\right]^2\right\}$$

$$= \tfrac{1}{2}\left\{\tfrac{1}{2}\mathcal{N}_0\int_{-\infty}^{\infty}|\mathrm{H}(f)|^2\,df + \tfrac{1}{2}\mathcal{N}_0\int_{-\infty}^{\infty}|\mathrm{H}(f)|^2\,df\right\}$$

$$= \tfrac{1}{2} \mathcal{N}_0 \int_{-\infty}^{\infty} |\mathrm{H}(f)|^2 \, df \qquad (7\mathrm{D}.9)$$

Appendix 7E Acquisition Decision Noise Terms

The noise terms in (7.111a) to (7.111d), respectively, are

$$n_{II}(t) = \left[n_I(t) + \sqrt{\frac{E_c}{T_c}} \mathrm{m}(t)\, C_Q(t) \sin \phi_w \right.$$

$$\left. + \sqrt{\frac{E_c}{T_c}} [\mathrm{m}(t) - 1]\, C_I(t) \cos \phi_w \right] C_I(t - \tau) \qquad (7\mathrm{E}.1\mathrm{a})$$

$$n_{IQ}(t) = \left[n_I(t) + \sqrt{\frac{E_c}{T_c}} \mathrm{m}(t)\, C_I(t) \cos \phi_w \right.$$

$$\left. + \sqrt{\frac{E_c}{T_c}} [\mathrm{m}(t) - 1]\, C_Q(t) \sin \phi_w \right] C_Q(t - \tau) \qquad (7\mathrm{E}.1\mathrm{b})$$

$$n_{QI}(t) = \left[n_Q(t) + \sqrt{\frac{E_c}{T_c}} \mathrm{m}(t)\, C_Q(t) \cos \phi_w \right.$$

$$\left. - \sqrt{\frac{E_c}{T_c}} [\mathrm{m}(t) - 1]\, C_I(t) \sin \phi_w \right] C_I(t - \tau) \qquad (7\mathrm{E}.1\mathrm{c})$$

$$n_{QQ}(t) = \left[n_Q(t) - \sqrt{\frac{E_c}{T_c}} \mathrm{m}(t)\, C_I(t) \sin \phi_w \right.$$

$$\left. + \sqrt{\frac{E_c}{T_c}} [\mathrm{m}(t) - 1]\, C_Q(t) \cos \phi_w \right] C_Q(t - \tau) \qquad (7\mathrm{E}.1\mathrm{d})$$

The noise terms N_1 and N_2 in (7.113a) and (7.113b), respectively, are

$$N_1 = \int_0^T [n_{II}(t) + n_{QQ}(t)]\, dt$$

$$= \int_0^T n_I(t)\, C_I(t - \tau)\, dt + \int_0^T n_Q(t)\, C_Q(t - \tau)\, dt$$

$$+ \sqrt{\frac{E_c}{T_c}} \sin\phi_w \int_0^T m(t)\, [C_Q(t)\, C_I(t - \tau) - C_I(t)\, C_Q(t - \tau)]\, dt$$

$$+ \sqrt{\frac{E_c}{T_c}} \cos\phi_w \int_0^T [m(t) - 1]\, [C_I(t)\, C_I(t - \tau) + C_Q(t)\, C_Q(t - \tau)]\, dt$$

$$\tag{7E.2a}$$

and

$$N_2 = \int_0^T [n_{IQ}(t) - n_{QI}(t)]\, dt$$

$$= \int_0^T n_I(t)\, C_Q(t - \tau)\, dt - \int_0^T n_Q(t)\, C_I(t - \tau)\, dt$$

$$+ \sqrt{\frac{E_c}{T_c}} \cos\phi_w \int_0^T m(t)\, [C_I(t)\, C_Q(t - \tau) + C_Q(t)\, C_I(t - \tau)]\, dt$$

$$+ \sqrt{\frac{E_c}{T_c}} \sin\phi_w \int_0^T [m(t) - 1]\, [C_Q(t)\, C_Q(t - \tau) + C_I(t)\, C_I(t - \tau)]\, dt$$

$$\tag{7E.2b}$$

where from (7.102c)

$$m(t) = 1 + \sum_{i=1}^{63} g_i\, d_i\, W_i(t) \tag{7E.2c}$$

Clearly, the mean values of all these noise terms equal zero. The computation of their variances proceeds by first recognizing that the bandwidths of the lowpass filters in the I and Q branches that are shown in Figure 7.63 are much larger than the effective bandwidths of the integrators. Therefore, we may treat $n_I(t)$ and $n_Q(t)$ as from independent white noise processes with spectral density $\frac{1}{2}\mathcal{N}_0$ as if there were no filters [1].

Let $n_w(t)$ denote a white noise process, whose autocorrelation function is $\frac{1}{2}\mathcal{N}_0\, \delta(\tau)$. The variance of a noise term after multiplication by a unit-amplitude PN sequence and integration is

$$\text{Var}\left[\int_0^T n_w(t)\, C(t-\alpha)\, dt\right] = \int_0^T \int_0^T \overline{n_w(t)\, n_w(u)}\, C(t-\alpha)\, C(u-\alpha)\, dt\, du$$

$$= \int_0^T \int_0^T \tfrac{1}{2}\mathcal{N}_0\, \delta(t-u)\, C(t-\alpha)\, C(u-\alpha)\, dt\, du$$

$$= \tfrac{1}{2}\mathcal{N}_0 \int_0^T C^2(t-\alpha)\, dt = \tfrac{1}{2}\mathcal{N}_0 T \qquad (7E.3a)$$

because $C^2(t) = 1$. With this understanding, we realize that the variances of the Gaussian noise terms of N_1 and N_2 equal $2 \times \tfrac{1}{2}\mathcal{N}_0 T = \mathcal{N}_0 T$. We further note that these terms are uncorrelated, because

$$E\left\{\left[\int_0^T n_I(t)\, C_I(t-\tau)\, dt + \int_0^T n_Q(t)\, C_Q(t-\tau)\, dt\right]\right.$$

$$\left. \times \left[\int_0^T n_I(t)\, C_Q(t-\tau)\, dt - \int_0^T n_Q(t)\, C_I(t-\tau)\, dt\right]\right\}$$

$$= \int_0^T \int_0^T \overline{n_I(t)\, n_I(u)}\, C_I(t-\tau)\, C_Q(u-\tau)\, dt\, du$$

$$- \int_0^T \int_0^T \overline{n_Q(t)\, n_Q(u)}\, C_Q(t-\tau)\, C_I(u-\tau)\, dt\, du$$

$$= \tfrac{1}{2}\mathcal{N}_0 \int_0^T C_I(t-\tau)\, C_Q(t-\tau) - \tfrac{1}{2}\mathcal{N}_0 \int_0^T C_I(t-\tau)\, C_Q(t-\tau)$$

$$= 0 \qquad (7E.3b)$$

As for the cochannel and cross-quadrature interference terms in (7E.2a) through (7E.2b), we note that for the special case of $\tau = 0$, when the receiver PN codes are aligned with the incoming PN codes, we have

$$N_1 = G(0,\, \mathcal{N}_0 T) + \sqrt{\frac{E_c}{T_c}}\, \sin\phi_w \int_0^T m(t)\, \overbrace{[C_Q(t)\, C_I(t) - C_I(t)\, C_Q(t)]}^{0}\, dt$$

$$+ \sqrt{\frac{E_c}{T_c}}\, \cos\phi_w \int_0^T [m(t) - 1]\, \overbrace{[C_I^2(t) + C_Q^2(t)]}^{2}\, dt$$

$$= G(0, \mathcal{N}_0 T) + \sqrt{\frac{E_c}{T_c}} \cos \phi_w \sum_{i=1}^{63} g_i \, d_i \overbrace{\int_0^T W_i(t) \, dt}^{0}$$

$$= G(0, \mathcal{N}_0 T) \tag{7E.4a}$$

and

$$N_2 = G(0, \mathcal{N}_0 T) + \sqrt{\frac{E_c}{T_c}} \cos \phi_w \int_0^T m(t) \overbrace{[C_I(t) \, C_Q(t) - C_I(t) \, C_Q(t)]}^{0} \, dt$$

$$+ \sqrt{\frac{E_c}{T_c}} \sin \phi_w \int_0^T [m(t) - 1] \overbrace{[C_Q^2(t) + C_I^2(t)]}^{2} \, dt$$

$$= G(0, \mathcal{N}_0 T) + \sqrt{\frac{E_c}{T_c}} \sin \phi_w \sum_{i=1}^{63} g_i \, d_i \overbrace{\int_0^T W_i(t) \, dt}^{0}$$

$$= G(0, \mathcal{N}_0 T) \tag{7E.4b}$$

Thus, for $\tau = 0$, which pertains to the tracking situation, the cross-quadrature and the cochannel interference is zero. For the general case of $\tau \neq 0$, during acquisition, there is some residual interference. As a practical matter, we may simply treat N_1 and N_2 as having the variance $\mathcal{N}_0 T$ for $\tau = 0$ and $\mathcal{N}_0' T > \mathcal{N}_0 T$ during acquisition, until the code phase is accurately detected. Because of the fluctuation in the amount of traffic on the forward link, as well as fluctuations in propagation conditions, the false alarm threshold in the system diagram of Figure 7.63 must be determined empirically.

8

Convolutional Codes and Their Use in IS-95

8.1 Introduction

For both the forward link (downlink) and the reverse link (uplink) in the IS-95 system [1], convolutional coding schemes are used for forward error control (FEC) purposes as shown in Figures 8.1 and 8.2, which are reproductions of Figures 4.17 and 4.22 of Chapter 4. Achieving reliable communications and using the minimum possible signal power (and hence energy) are the objectives of communications system designers. Reliable information transmission and the minimal use of signal power are conflicting requirements to some extent. In a mobile cellular communications system, battery power is at a premium and, hence, the mobile station (handset terminal) can conserve the transmitter power resource longer if the most power-efficient modulation is employed.

When one has a limited transmitter power resource in a multiple access communications system, optimal or near-optimal modulation schemes must be considered, and indeed, in the IS-95 CDMA cellular system, this philosophy is practiced: the coherent BPSK modulation for the forward link and the M-ary orthogonal modulation used in the reverse link in the IS-95 system are both examples of power-efficient modulation.

If the signals that are transmitted deliver an energy level below a certain threshold value, the received signals are likely to produce errors at the receiver. The idea of using FEC is to help recover from errors that occur during the transmission over the channel. There are many types of FEC codes in terms of error-correction capabilities. FEC coding schemes are usually called *channel coding schemes* for the reasons just mentioned. The channel coding strategy is aimed at allowing the transmitter to use minimum possible signal energy in accomplishing the design objective of providing a specified error rate.

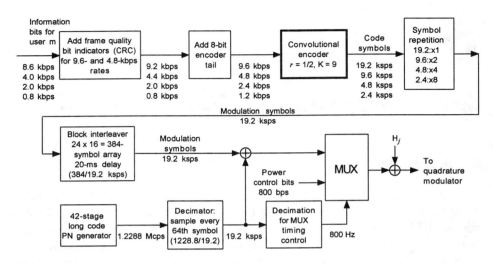

Figure 8.1 Forward link traffic channel structure.

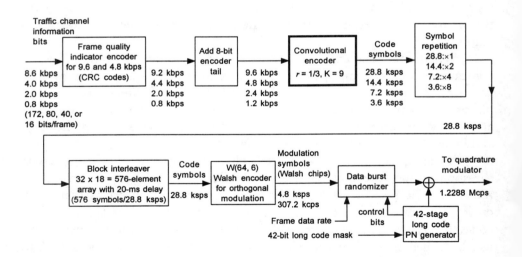

Figure 8.2 Reverse link traffic channel structure.

Another perspective on the use of channel coding in regard to the IS-95 CDMA spread-spectrum system can be observed: in Chapter 1, we saw that the capacity of a CDMA spread-spectrum system is given as

$$M = \frac{W/R_b}{E_b/\mathcal{N}_0} + 1 \tag{8.1}$$

where

M = number of users per CDMA bandwidth

W = spread-spectrum bandwidth

R_b = bit rate ($R_{b,max} = 9.6\,\text{kbps}$)

W/R_b is the processing gain and $E_b/\mathcal{N}_0$ is the required signal energy-to-noise density ratio. In digital communications, $E_b/\mathcal{N}_0$ is the figure of merit (or fidelity measure) for a specified error probability in a chosen modulation scheme.

Let S be the power at the receiver front end. The received signal power-to-noise density ratio is $S/\mathcal{N}_0$. The efficiency of a communications system is measured by the received $E_b/\mathcal{N}_0$, given by

$$\frac{E_b}{\mathcal{N}_0} = \frac{S}{\mathcal{N}_0} \cdot \frac{1}{R} \tag{8.2}$$

where R is the information bit rate in bits/sec. It is thus clear that the use of a good, efficient modulation technique reduces the required $E_b/\mathcal{N}_0$ for the specified bit-error rate (BER), and consequently we can increase the data rate, as seen in the equation

$$R = \frac{S/\mathcal{N}_0}{E_b/\mathcal{N}_0}$$

The remedy for reducing $E_b/\mathcal{N}_0$ can come from two sources: power-efficient modulation and coding, under the allowed bandwidth expansion constraint (if any). Let us return to (8.1). Because for a given processing gain the capacity is inversely proportional to the value of $E_b/\mathcal{N}_0$, lowering the value of $E_b/\mathcal{N}_0$ is a viable approach. But how low? For a given value of $E_b/\mathcal{N}_0$ that is required for a specified bit-error probability, lowering $E_b/\mathcal{N}_0$ is permitted, provided that the coding gain can restore the decrease. It is then plain that we need to consider a "powerful" coding scheme that provides high coding gain. In this chapter we show that the convolutional codes of "rate $\frac{1}{2}$ and constraint length 9" used for the forward link channels (see Figure 8.1) and of "rate $\frac{1}{3}$ and constraint length 9" used for the reverse link channels (see Figure 8.2) provide "powerful" coding gains. The capacity equation given in (8.1) is one that is valid under the assumption of perfect power control using an omnidirectional antenna. The equation implies that the way to increase (or enhance) the system capacity is to find a method for every communicator in the cell to use a minimal possible value of $E_b/\mathcal{N}_0$ and still achieve the required (specified) BER. The capacity enhancement realized by lowering the "required $E_b/\mathcal{N}_0$," $(E_b/\mathcal{N}_0)_{req}$, can be simply related to the *coding gain* as follows:

$$\text{Coding gain (dB)} \triangleq \left[\left(\frac{E_b}{\mathcal{N}_0}\right)dB\right]_{\text{uncoded}} - \left[\left(\frac{E_b}{\mathcal{N}_0}\right)dB\right]_{\text{coded}}$$

$$\equiv \Delta(E_b/\mathcal{N}_0) \tag{8.3}$$

In other words, coding reduces the $E_b/\mathcal{N}_0$ required to achieve a specified BER under a specified modulation scheme. This fact is illustrated in Figure 8.3, where (n, k) denotes a coding scheme of rate $r \triangleq k/n$, and $(n, k) = (1, 1)$ denotes an uncoded system. System designers seek a coding scheme that provides the largest possible coding gain of $\Delta(E_b/\mathcal{N}_0)$ under affordable complexity, cost, reliability, and so forth. Thus, by use of a channel coding scheme, we provide the needed coding gain to achieve the design objective. Clearly, the coding gain, $\Delta(E_b/\mathcal{N}_0)$, is the amount of improvement that is obtainable when a particular coding scheme, (n, k), is used for a specified BER. A practical way of determining coding gain is to plot BER as function of $E_b/\mathcal{N}_0$ for both coded (n, k) and uncoded $(1, 1)$ cases and to read the $\Delta(E_b/\mathcal{N}_0)$ at a specified BER rather than solving an equation. Thus, the amount of coding gain for a given coding scheme varies, depending on the specified BER. (See Section 8.5.2 for practical examples.) Figure 8.3 is a typical comparison of *coded* and *uncoded* system performance.

The "threshold" noted in Figure 8.3, $(E_b/\mathcal{N}_0)_{thr}$, is a phenomenon that exists in every coding scheme. It is the point of $E_b/\mathcal{N}_0$ at which the coded system begins to perform better than the uncoded system. For $E_b/\mathcal{N}_0 \leq (E_b/\mathcal{N}_0)_{thr}$, however, the coded system always performs worse than the uncoded system. An (n, k) coding scheme uses $n - k$ redundant (excess) symbols, and for values of $E_b/\mathcal{N}_0 < (E_b/\mathcal{N}_0)_{thr}$, the coding scheme simply operates before this investment of redundant symbols reaches the break-even point. All (n, k) codes display different values of the threshold point.

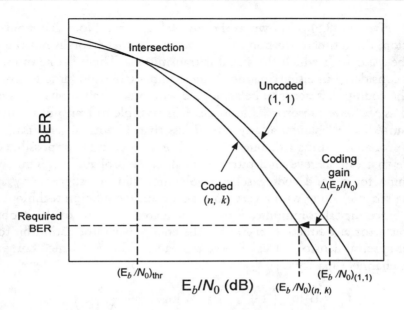

Figure 8.3 Illustration of coding gain in an (n, k) FEC coding scheme.

Let us ask a question: Is it possible to transmit a digital signal over a noisy communications channel without producing any error in the received signal at the output of the channel? The answer to this question was provided by C. E. Shannon in 1948 [2]. This was *yes*, under the following assumptions and conditions:

- The signal is transmitted over an AWGN channel, and

- The signal must be transmitted with a rate R bps that does not exceed a rate C bps, that is:

$$R \text{ (bps)} \leq C \text{ (bps)} \tag{8.4}$$

where C is given by

$$C = W \log_2\left(1 + \frac{S}{N}\right) \text{ bps} \tag{8.5}$$

and where W is the channel bandwidth and S/N is the SNR at the input of the receiver.

The C given in (8.5) is known as the *channel capacity*. Note that, contrary to intuition, the ultimate weapon in combatting channel errors is not the power but the *speed* with which the signal is transmitted. There is one more entity to be considered in citing Shannon's most remarkable and famous result, and it is the coding that needs to be associated with Shannon's limits. It is usually stated as *Shannon's theorem* as follows: It is possible to transmit information over an AWGN channel at any rate R less than C with an arbitrarily small error probability using a coding scheme. For $R > C$, it is not possible to find a code that can achieve an arbitrarily small error probability. Thus, we are searching for a good coding scheme (Shannon did not suggest any specific coding scheme) along with a very power-efficient modulation technique.

Since digital communications are evaluated in terms of the probability of a bit error as a function of $E_b/\mathcal{N}_0$, the energy per information bit-to-noise power spectral density, let us observe the value of $E_b/\mathcal{N}_0$, which corresponds to the channel capacity C:

$$C = W \log_2\left(1 + \frac{S}{N}\right) = W \log_2\left(1 + \frac{S}{\mathcal{N}_0 W}\right) \tag{8.6a}$$

with

$$N \triangleq \mathcal{N}_0 W \tag{8.6b}$$

in which we assumed an AWGN channel with noise power spectral density of $\frac{1}{2}\mathcal{N}_0$. Now, because

$$\left.\frac{E_b}{\mathcal{N}_0}\right|_{R=C} = \left.\frac{S}{\mathcal{N}_0 R}\right|_{R=C} = \frac{S}{\mathcal{N}_0 C} \tag{8.7a}$$

we have, from (8.6):

$$\left.\frac{E_b}{\mathcal{N}_0}\right|_{R=C} = \frac{S}{\mathcal{N}_0 W \log_2\left(1 + \dfrac{S}{\mathcal{N}_0 W}\right)} = \frac{S/N}{\log_2(1 + S/N)} \tag{8.7b}$$

The $E_b/\mathcal{N}_0$ resulting from the transmission of the bits at rate $R = C$ is plotted in Figure 8.4 as a function of S/N. We observe from Figure 8.4 that the minimum required $E_b/\mathcal{N}_0$ is $0.693 = -1.59\,\mathrm{dB} \cong -1.6\,\mathrm{dB}$ if the bit rate is the capacity rate C, provided, of course, that a coding scheme be found that can support the theoretical Shannon limit. Analytically, this result can be obtained as follows: from (8.6a) it is clear that, if the channel is noiseless,

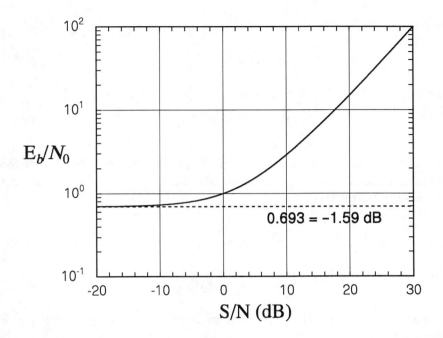

Figure 8.4 Required E_b/N_0 as a function of received S/N when the bits are transmitted at capacity rate C.

there is no noise power; that is, $N_0W \to 0$. This indicates that the capacity under such a noiseless channel would become infinite, as shown by

$$C|_{\substack{\text{noiseless} \\ \text{channel}}} = \lim_{N_0W \to 0} W \log_2 \left(1 + \frac{S}{N_0W}\right)$$

$$= \lim_{(S/N_0W) \to \infty} W \log_2 \left(1 + \frac{S}{N_0W}\right) \to \infty \qquad (8.8a)$$

On the other hand, if the bandwidth is increased without limit (i.e., $W \to \infty$), the concomitant increase in capacity, denoted C_∞, is obtained as

$$C_\infty = \lim_{W \to \infty} C = \lim_{W \to \infty} \left[\frac{S}{N_0} \frac{N_0W}{S} \log_2 \left(1 + \frac{S}{N_0W}\right)\right]$$

$$= \lim_{W \to \infty} \left[\frac{S}{\mathcal{N}_0} \left(1 + \frac{S}{\mathcal{N}_0 W} \right)^{1/(S/\mathcal{N}_0 W)} \right]$$

$$= \frac{S}{\mathcal{N}_0} \log_2 e = \frac{S}{\mathcal{N}_0} \frac{1}{ln2} \text{ bits/sec} \qquad (8.8b)$$

where we apply the well-known fact that

$$\lim_{x \to 0} (1 + x)^{1/x} = e \qquad (8.8c)$$

The capacity C_∞ under the assumption of infinite bandwidth is known as the *infinite bandwidth Shannon capacity limit*, and the required $E_b/\mathcal{N}_0$ under this capacity assumption is given by

$$\left. \frac{E_b}{\mathcal{N}_0} \right|_{R=C_\infty} = \frac{S}{\mathcal{N}_0 C_\infty} = \frac{S/\mathcal{N}_0}{(S/\mathcal{N}_0) \log_2 e} = ln2 = 0.693 \qquad (8.8d)$$

$$= -1.59 \, \text{dB} \cong -1.6 \, \text{dB} \qquad (8.8e)$$

Recall from (7.31a) that, as the number of bits/symbol $K \to \infty$, $P(e) \to 0$ provided that

$$\frac{E_b}{\mathcal{N}_0} > 2 \, ln2 \approx 1.4 \, \text{dB} \qquad (8.8f)$$

where $K \to \infty$ corresponds to the case of $W \to \infty$. What is observed here is that (8.8f) gives a value twice as large as the result of (8.8d). Remember that (8.8f) was obtained by using a union bound, and, thus, it is "3 dB off" from the exact result.

Let us again return to the capacity equation (8.1). It is interesting to compare the capacity measure of (8.1) with that attainable using the Shannon bound for an AWGN channel with an omnidirectional antenna. Using the ideal Shannon limit on the AWGN channel for error-free data transmission, the maximal possible capacity, M_0, under a given processing gain of W/R_b is given by [3, 4]

$$M_0 \cong M|_{(E_b/\mathcal{N}_0)=ln2} = \frac{W}{R_b} (\text{dB}) + 1.59 \, \text{dB} \qquad (8.9a)$$

If $W/R_b = 128$, as for the case of the IS-95 system, we have

$$M_0 = \frac{W/R_b}{E_b/\mathcal{N}_0} + 1 = \frac{128}{0.693} + 1 \cong 185 \qquad (8.9b)$$

Of course, the ideal maximal capacity of 185 can never be realized in the IS-95 system using 1.2288-MHz bandwidth for $R_b = 9.6$ kbps. It is nevertheless a theoretical limit, and hence we need to design a coding system that achieves a capacity as close to that limit as possible.

So, it is established that needs exist to employ a coding scheme with a large coding gain in order to maximize the user capacity of the CDMA cellular telephone system. The use of convolutional codes in the IS-95 system is based on this objective. We now introduce the basic concept of convolutional codes, define some fundamental terms, and then derive the performance measure of convolutional codes so that we can evaluate the actual convolutional codes that are used in the IS-95 system.

8.2 Convolutional Codes

In Chapter 5, in our treatment of CRC codes, we introduced the concept of a block code (n, k), where for every k information bits, n coded symbols are produced by the encoder, and we called n the block length and the ratio k/n was called the code rate r. We discussed the error-detection capability of the linear block code (n, k), which performs the error-detection function by syndrome calculation of the code that possesses $m = n - k$ redundant parity check digits. While in the (n, k) block code there is a clear code structure of block length n per k information bits, for the case of *convolutional codes* there is no parameter called *block length*, as will be seen shortly. The input stream of data bits is shifted into the convolutional encoder either one bit at a time or k bits at a time and, corresponding to that input, n symbols (coded digits) are transmitted into the channel. In a convolutional code, the one bit or the k bits being shifted into the encoder is not looked upon as a block of information bits to be encoded, but the input is looked upon as a continuous stream of data bits until all the message bits are encoded. For example, on the IS-95 forward link traffic channel (see Figure 8.1), the information bits at the rate of 9.6 kbps are encoded an entire 20-ms frame at a time, consisting of 9.6 kbps $\times$ 20 ms = 192 bits. For this long input information block, the coded output symbols are produced at the rate of 19.2 kbps for 20 ms, resulting in 384 coded symbols that are fed into the block interleaver, as shown in Figure 8.1. This encoding process repeats every 20 ms after "flushing out" the encoder by putting eight zeros at the end of the 192 bits of the information sequence.[1] In this sense, there is only one long input information block (of length 192 bits) and one corresponding encoded block length of 384 coded symbols. So there is no simple block structure that can be associated with a convolutional code and, as we see shortly, each of the coded symbols is linearly affected by a number of input information bits, much like the output of a discrete convolutional summation (see Section 1.3.4).

Historically, convolutional codes were first introduced by Elias [5] in 1954. Since the discovery of an efficient decoding algorithm in 1967 by A. J. Viterbi [6], which was demonstrated by J. Omura [7] in 1969 to be an ML decoding algorithm, the applications of convolutional codes with the Viterbi decoding algorithm have been almost indispensable channel coding schemes in modern communications systems, such as the IS-95 system. In this and the following sections, we introduce convolutional codes and the ways of specifying the codes in a *tutorial manner*, followed by the principles of the Viterbi algorithm. We then derive the classical performance measures for convolutional codes, which are union-bound error probability measures. The specific convolutional codes being used in the IS-95 system are then evaluated by the derived formulas.

8.2.1 Convolutional Encoders

In Chapter 5, we described an (n, k) block code generated by a matrix (generator matrix or parity check matrix) or polynomial (generator polynomial or parity check polynomial), where

$$k = \text{number of bits that form a block input to the block encoder} \tag{8.10a}$$

$$n = \text{number of bits (or symbols) out of the encoder associated with the } k\text{-bit input block} \tag{8.10b}$$

$$r \triangleq k/n = \text{code rate} \tag{8.10c}$$

A convolutional code is described by the three integers n, k, and K, where

$$k = \text{number of bits being shifted into the encoder at one time} \tag{8.11a}$$

---

[1] The "add 8-bit encoder tail" operation shown in the block diagram in Figure 8.1 is to supply eight bits, all zeros, for the clearing (flushing) job. More is said on this later.

n = number of encoder output symbols corresponding to the
    k information bits shifted into the encoder   (8.11b)

$r \triangleq k/n$ = code rate (same as block code)   (8.11c)

K = parameter known as the constraint length, which equals the
    number of k-tuple stages in the encoder shift register   (8.11d)

The encoder shift register consists of K k-bit stages and n output generators, as shown in Figure 8.5. On closer observation of the encoder diagram in Figure 8.5, we see that the first register is unnecessary, and therefore the required number of shift register stages is $Kk - 1$ rather Kk stages, and the elimination of the first stage is illustrated in Figure 8.6.

As evidenced in the encoder diagrams of Figures 8.5 and 8.6, the n-tuple "branch" word[2] depends not only on the k-tuple being shifted in, but also on the $K - 1$ k-tuples that have been previously shifted in. Thus, we say that convolutional encoders have "memory." The n coded symbols are produced

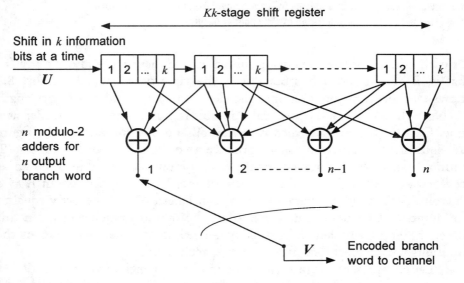

Figure 8.5 Convolutional encoder of rate k/n with constraint length K: $U = (u_1, u_2, \ldots, u_i, \ldots)$ is the input sequence shifted into the encoder k bits at a time, and for each k bits shifted in, an n-symbol (n-digit) "branch" word $V = (v_1, v_2, \ldots, v_n)$ is generated at the encoder output.

---

[2] It is made clear shortly as to why the n-tuple is called a "branch" word.

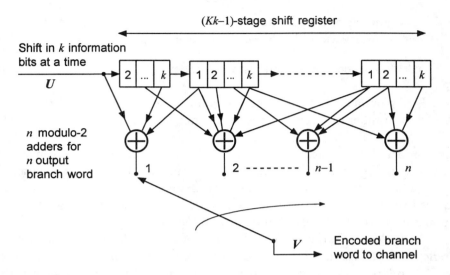

Figure 8.6 Convolutional encoder of rate k/n with constraint length K, using $(Kk-1)$-stage shift register: $U = (u_1, u_2, \ldots, u_i, \ldots)$ is the input sequence shifted into the encoder k bits at a time, and for each k bits shifted in, an n-symbol (n-digit) "branch" word $V = (v_1, v_2, \ldots, v_n)$ is generated at the encoder output.

for each k-tuple message bits being shifted in, and, hence, the rate is $r = k/n$; that is, k bits in and n bits out during the same time interval. Figure 8.7 shows two identical convolutional encoders of rate $1/2$ and constraint length 3. Note that $k = 1$ and, thus, only one bit is shifted into the encoder at a time. Clearly the encoders shown in the top and bottom parts of the figure are equivalent. These encoders produce two ($n = 2$) coded symbols (2-symbol branch word) for every single ($k = 1$) input bit shifted into the linear feedforward shift register and thus they are rate $1/2$ constraint length $K = 3$ convolutional encoders. Here the constraint length $K = 3$ is exactly equal to the number of encoding shifts that affect the output branch word. In this sense, the constraint length is directly related to the memory span of the input bits with respect to a given output branch word.

Let us represent the input information bit sequence by the vector U:

$$U = (u_1, u_2, \ldots, u_i, \ldots) \tag{8.12a}$$

and the encoder output branch word n-tuple corresponding to the input bit u_i by

$$V_i = (v_{i1}, v_{i2}, \ldots, v_{in}) \tag{8.12b}$$

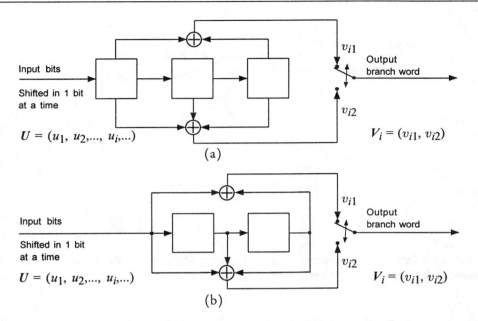

Figure 8.7 Rate $\frac{1}{2}$ convolutional encoders with constraint length $K = 3$. The (a) and (b) are identical encoders.

The term "constraint length K encoder" simply means that, if V_i is formed when the input u_i is shifted into the encoder, every component of the n-tuple word V_i is affected by K bits of the information sequence—$u_{i-K+1}, \ldots, u_{i-1},$ u_i—and each component of V_i is affected differently, depending on the encoder connections between the shift registers and the particular modulo-2 adder that forms the branch word component.

There are several ways of representing convolutional encoders, and each has its own merit in describing the properties and characteristics of convolutional codes. Some of the methods for representing convolutional codes are (1) encoder connection vectors, (2) encoder impulse response, (3) polynomial representations, and (4) state representations. We illustrate each with examples.

8.2.2 Encoder Connection Vector Representation

Consider Figure 8.8, which shows a convolutional encoder of rate $1/n$ and constraint length K. Define a K-tuple connection vector G_j:

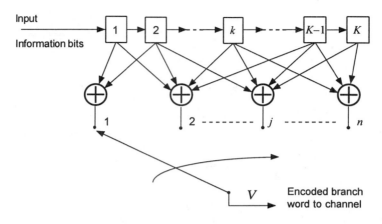

Figure 8.8 Rate $1/n$ convolutional encoder with constraint length K (the first stage can be removed and only $K - 1$ shift register stages are needed).

$$G_j = (g_{j1}, g_{j2}, \ldots, g_{jl}, \ldots, g_{jK}), \quad j = 1, 2, \ldots, n \tag{8.13a}$$

where

$$g_{jl} \triangleq \begin{cases} 0, & \text{if the } j\text{th modulo-2 adder is} \\ & \text{not connected to the } l\text{th stage} \\ & \text{of the encoder register} \\ 1, & \text{if connection is made between } j \text{ and } l \end{cases} \tag{8.13b}$$

The connection vector is a K-tuple vector, where K is the number of shift register stages, and there are n connection vectors, where n is the number of modulo-2 adders. It is very easy to draw an encoder diagram by looking at the connection vectors with a specification of the code rate $r = k/n$. The connection vectors are nothing but notations that indicate the way each modulo-2 adder is connected to the Kk shift register stages. Therefore, we can also use octal notation for the K-tuple connection vectors as illustrated in the following example.

Example 8.1 Draw the convolutional encoder of rate $\frac{1}{3}$ and constraint length 3, represented by the connection vectors

$$\left. \begin{array}{l} G_1 = (1 \ 0 \ 0) \ \rightarrow \ 4 \text{ in octal notation} \\ G_2 = (1 \ 0 \ 1) \ \rightarrow \ 5 \text{ in octal notation} \\ G_1 = (1 \ 1 \ 1) \ \rightarrow \ 7 \text{ in octal notation} \end{array} \right\} \tag{8.14}$$

Now, G_1 indicates that modulo-2 adder #1 is connected to the first stage, and G_2 indicates that modulo-2 adder #2 is connected to the first and third states, while G_3 implies that the third modulo-2 adder is connected to all three stages as shown in Figure 8.9.

Example 8.2 Draw the convolutional encoders for rate $\frac{2}{3}$ with constraint length 2 specified by the connection vectors $G_1 = (1\,0\,1\,1)$, $G_2 = (1\,1\,0\,1)$, and $G_3 = (1\,0\,1\,0)$. The encoders are shown in Figure 8.10.

8.2.3 Encoder Impulse Response Representation

When we discussed linear filters and convolutional integrals in Chapter 1, the idea that we exploited there was the principle of superposition in *predicting* the output of the *linear filter* by knowing the *impulse response* of the filter. Here, we are going to apply the same principle by looking at the convolutional encoder as a *linear system*. Given a convolutional encoder, the output sequence for the input 1, namely $(1\,0\,0\,0\cdots)$, is the impulse response of the encoder as the input 1 moves along the $\overrightarrow{K}$ stages of the encoder; that is, as

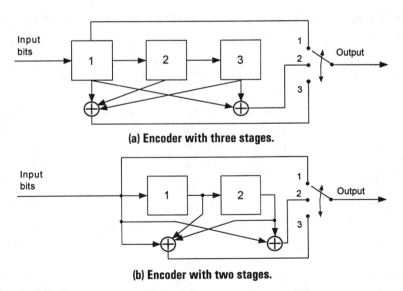

(a) Encoder with three stages.

(b) Encoder with two stages.

Figure 8.9 Convolutional encoder of rate $\frac{1}{3}$ with constraint length 3, specified by the connection vectors $G_1 = (1\,0\,0)$, $G_2 = (1\,0\,1)$, and $G_3 = (1\,1\,1)$. The (a) and (b) are the same encoder.

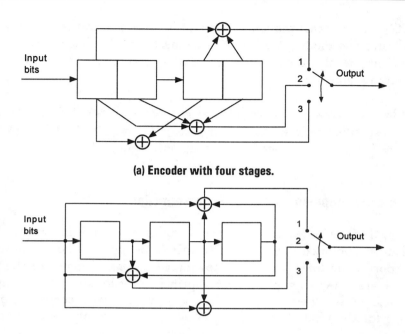

(a) Encoder with four stages.

(b) Encoder with three stages.

Figure 8.10 Convolutional coders of rate $\frac{2}{3}$ and constraint length 2 (two bits are shifted into the encoder at a time).

the K-tuple shift register vectors take the form given in (8.15a).

$$
K\text{-clock pulse period}
\begin{cases}
\begin{array}{ll}
K\text{-stage loading vectors} & \text{Response} \\
1\,0\,0\,0\cdots0\,0 & v_{11}\,v_{12}\cdots v_{1n} = \boldsymbol{V}_1 \\
0\,1\,0\,0\cdots0\,0 & v_{21}\,v_{22}\cdots v_{2n} = \boldsymbol{V}_2 \\
0\,0\,1\,0\cdots0\,0 & v_{31}\,v_{32}\cdots v_{3n} = \boldsymbol{V}_3 \\
\;\;\vdots \qquad\quad \vdots & \qquad\quad \vdots \\
0\,0\,0\,0\cdots1\,0 & \\
0\,0\,0\,0\cdots0\,1 & v_{K1}\,v_{K2}\cdots v_{Kn} = \boldsymbol{V}_K \\
0\,0\,0\,0\cdots0\,0 &
\end{array}
\end{cases}
\tag{8.15a}
$$

It is clear that, if the input data sequence is a K-tuple vector $\boldsymbol{U}_1 = (1\,0\,0\cdots0)$, then the response of the convolutional encoder is the branch word $\boldsymbol{V}_1 = (v_{11}, v_{12}, \ldots, v_{1n})$, and if the input sequence is $\boldsymbol{U}_{12} = (1\,1\,0\,0\cdots0)$, then the encoded sequence is the sum of the response to $(1\,0\,0\,0\cdots0)$ and $(0\,1\,0\,0\cdots0)$, that is:

$$V_{12} = V_1 + V_2 = (v_{11} + v_{21}, v_{12} + v_{22}, \ldots, v_{1n} + v_{2n}) \qquad (8.15b)$$

This is the superposition principle that was employed in deriving the convolution integral and convolution summation for the cases of continuous and discrete time systems, respectively. Let us consider the convolutional encoder of Figure 8.9, which is a rate $\frac{1}{3}$ encoder with constraint length 3. The details of the three-stage loading vectors and the corresponding branch outputs are shown in Table 8.1. Note that the first modulo-2 summer is connected only to the first stage, the second modulo-2 summer to the first and the third stages, and the third summer to all stages of the shift register. If we denote by x_i the content of the ith stage of the K-stage shift register, then the details of the encoder's loading vectors and the corresponding branch words are given as indicated in Table 8.1. The impulse response of this encoder is then given by

$$V_1, V_2, V_3 = 1\,1\,1, 0\,0\,1, 0\,1\,1 \qquad (8.15c)$$

It is clear that, once we have the impulse response of the convolutional encoder as shown in Table 8.1, we can predict the encoded symbol sequence corresponding to any combination of input information bits as a superposition of the responses of the corresponding impulse responses.

Example 8.3 Let the input data sequence for the rate $\frac{1}{3}$ encoder of Figure 8.9 be $(1\,1\,1\,0\,1)$. Note that the encoder's span is three bits (constraint length 3) and that the input sequence can be looked on as being composed of sequential impulses, as shown in Table 8.2. In that table, we use the notation $X_1(t_j)$ to denote the sequence $X_1(t_j) = (1\,0\,0)$ occurring at time t_j and Σ' to denote the modulo-2 sum of the sequences $\{X_1(t_j), j = 1, 2, \ldots\}$. Using the superposition principle, we can write down the encoded sequence corresponding to the input data sequence $(1\,1\,1\,0\,1)$ as shown in Table 8.3.

Table 8.1 Impulse response of the convolutional encoder of Figure 8.9

Three-stage loading vectors		Corresponding branch words $V_i = (v_{i1}, v_{i2}, v_{i3})$			
Vector X_i	Contents $x_{i1}\ x_{i2}\ x_{i3}$	Branch vector V_i	$v_{i1} = $ x_{i1}	$v_{i2} = $ $x_{i1} \oplus x_{i3}$	$v_{i3} = $ $x_{i1} \oplus x_{i2} \oplus x_{i3}$
X_1	1 0 0	V_1	1	1	1
X_2	0 1 0	V_2	0	0	1
X_3	0 0 1	V_3	0	1	1

Table 8.2 Decomposition of the data sequence as a sum of impulses

	t_1	t_2	t_3	t_4	t_5
$X_1(t_1)$	1	0	0		
$X_1(t_2)$		1	0	0	
$X_1(t_3)$			1	0	0
$X_1(t_4)$				0	0	0
$X_1(t_5)$					1	0	0
Input sequence $= \displaystyle\sum_{i=1}^{5}{}' X_1(t_i) = 1\,1\,1\,0\,1$					

Table 8.3 Prediction of encoded sequence by means of the impulse response of the encoder

			Input sequence to be encoded: 1 1 1 0 1
t_i	Input	Input vector	Output response
t_1	1	$X_1(t_1)$	$V_1(t_1) = 1\,1\,1\,0\,0\,1\,0\,1\,1$
t_2	1	$X_1(t_2)$	$V_1(t_2) = 1\,1\,1\,0\,0\,1\,0\,1\,1$
t_3	1	$X_1(t_3)$	$V_1(t_3) = 1\,1\,1\,0\,0\,1\,0\,1\,1$
t_4	0	$X_1(t_4)$	$V_1(t_4) = 0\,0\,0\,0\,0\,0\,0\,0\,0$
t_5	1	$X_1(t_5)$	$V_1(t_5) = 1\,1\,1\,0\,0\,1\,0\,1\,1$
t_6	0	0 (tail bits)	
Encoded output $= \displaystyle\sum_{j=1}^{5}{}' V_1(t_j) = 1\,1\,1\,1\,1\,0\,1\,0\,1\,0\,1\,0\,1\,0\,0\,0\,0\,1\,0\,1\,1$			

This suggests that a *generator matrix* G can be constructed from Table 8.3 as

$$
G = \begin{bmatrix}
1 & 1 & 1 & 0 & 0 & 1 & 0 & 1 & 1 & & & & & & \\
 & 1 & 1 & 1 & 0 & 0 & 1 & 0 & 1 & 1 & & & & & \\
 & & 1 & 1 & 1 & 0 & 0 & 1 & 0 & 1 & 1 & & & & \\
 & & & 1 & 1 & 1 & 0 & 0 & 1 & 0 & 1 & 1 & & & \\
 & & & & 1 & 1 & 1 & 0 & 0 & 1 & 0 & 1 & 1 & & \\
\vdots & & & & & & & & & & & & \vdots & & \vdots
\end{bmatrix}
$$

$$= \begin{bmatrix} 1 & 1 & 1 & 0 & 0 & 1 & 0 & 1 & 1 & 0 & 0 & 0 & 0 & 0 & 0 & 0 & 0 & 0 & \cdots \\ 0 & 0 & 0 & 1 & 1 & 1 & 0 & 0 & 1 & 0 & 1 & 1 & 0 & 0 & 0 & 0 & 0 & 0 & 0 & 0 & 0 \\ 0 & 0 & 0 & 0 & 0 & 0 & 1 & 1 & 1 & 0 & 0 & 1 & 0 & 1 & 1 & 0 & 0 & 0 & 0 & 0 & 0 \\ 0 & 0 & 0 & 0 & 0 & 0 & 0 & 0 & 0 & 1 & 1 & 1 & 0 & 0 & 1 & 0 & 1 & 1 & 0 & 0 & 0 \\ 0 & 0 & 0 & 0 & 0 & 0 & 0 & 0 & 0 & 0 & 0 & 0 & 1 & 1 & 1 & 0 & 0 & 1 & 0 & 1 & 1 \\ & \vdots & & & & & & & & & & & & \vdots & & & & \vdots & & & \end{bmatrix}$$

$$(8.15d)$$

The codeword corresponding to any input sequence U may be found by multiplying the input message vector by G, that is:

$$\text{Encoded sequence} = U\,G \qquad (8.15e)$$

where the operations are in modulo-2 arithmetic. We note that each row of the generator matrix G is identical to the preceding row but shifted $n = 3$ places to the right. Unlike block codes, convolutional codes do not have the information sequence in block form; that is, the codewords may have infinite length. The generator matrix shown in (8.15d) is a semi-infinite matrix. The encoded output sequence given in Table 8.3 can also be obtained based on the multiplication of (8.15e):

Encoded sequence $= U\,G$

$$= (1\,1\,1\,0\,1) \begin{bmatrix} 1 & 1 & 1 & 0 & 0 & 1 & 0 & 1 & 1 & & & & & & & & \\ & & & 1 & 1 & 1 & 0 & 0 & 1 & 0 & 1 & 1 & & & & & \\ & & & & & & 1 & 1 & 1 & 0 & 0 & 1 & 0 & 1 & 1 & & & \\ & & & & & & & & & 1 & 1 & 1 & 0 & 0 & 1 & 0 & 1 & 1 & & & \\ & & & & & & & & & & & & 1 & 1 & 1 & 0 & 0 & 1 & 0 & 1 & 1 \\ & \vdots & & & & & & & & & & & & \vdots & & & & \vdots & & & \end{bmatrix}$$

$$= 1\,1\,1\,1\,1\,0\,1\,0\,1\,0\,1\,0\,1\,0\,0\,0\,0\,1\,0\,1\,1 \qquad (8.15f)$$

which was obtained by adding the first, second, third, and fifth rows of the generator matrix G as dictated by the input sequence vector. Note that the encoded sequence in (8.15f) is identical to that in Table 8.3, as it should be.

In the discrete time system discussed in Section 1.3.4, we expressed the convolution summation as

$$y(n) = \sum_{k=0}^{n} h(k)\,x(n-k) \qquad (8.15g)$$

where $h(k)$ is the impulse response and $x(k)$ is the input. The name *convolutional encoder* is adapted from the application of the superposition principle in deriving the encoded sequence, much the same way as the convolution principle. Originally convolutional codes were called *recurrent codes* [5].

8.2.4 Polynomial Representation of the Encoder

We can also represent a convolutional encoder with a set of n *generator polynomials*, one for each of the n modulo-2 adders. This method is closely related to that of the *encoder connection vector* representation, in that we simply write a polynomial $G_j(x)$ based on the K-tuple encoder connection vector $\boldsymbol{G}_j$:

$$\boldsymbol{G}_j = (g_{j1}, g_{j2}, \ldots, g_{jl}, \ldots, g_{jK}) \qquad (8.16a)$$

$$G_j(x) = g_{j1} + g_{j2}x + \cdots + g_{jl}x^{l-1} + \cdots + g_{jK}x^{K-1} \qquad (8.16b)$$

This relationship is depicted in Figure 8.11, where we use the same encoder that we considered in Figure 8.9. In this method of representation, the message sequence is also represented as a message polynomial $U(x)$ based on the sequence vector $\boldsymbol{U} = (u_1, u_2, \ldots)$, that is:

$$U(x) = u_1 + u_2 x + \cdots + u_L x^{L-1} \qquad (8.16c)$$

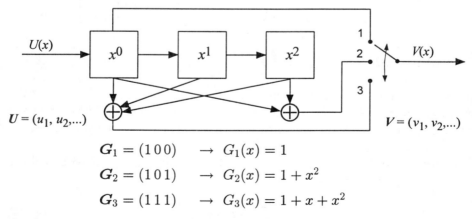

$$G_1 = (1\,0\,0) \quad \rightarrow \quad G_1(x) = 1$$
$$G_2 = (1\,0\,1) \quad \rightarrow \quad G_2(x) = 1 + x^2$$
$$G_3 = (1\,1\,1) \quad \rightarrow \quad G_3(x) = 1 + x + x^2$$

Figure 8.11 Polynomial representation of a convolutional encoder.

Example 8.4 Consider the message sequence $U = (1\,1\,1\,0\,1)$. Then

$$U(x) = 1 + x + x^2 + x^4 \tag{8.16d}$$

The encoded sequence polynomial is obtained by collecting the coefficients of the like power terms of the products $U(x)G_i(x)$, $i = 1, 2, \ldots, n$, and then the resultant coefficients are shown as the coded branch words in sequence. Let us explain this by the example of the encoder shown in Figure 8.11, where

$$G_1(x) = 1 \qquad G_2(x) = 1 + x^2 \qquad G_3(x) = 1 + x + x^2 \tag{8.16e}$$

Now, by multiplying the message polynomial $U(x)$ by each connection polynomial $G_i(x)$, $i = 1, 2, 3$, we have

$$V_1(x) = U(x)G_1(x) = (1 + x + x^2 + x^4)(1) = 1 + x + x^2 + x^4 \tag{8.16f}$$
$$V_2(x) = U(x)G_2(x) = (1 + x + x^2 + x^4)(1 + x^2) = 1 + x + x^3 + x^6$$
$$V_3(x) = U(x)G_3(x) = (1 + x + x^2 + x^4)(1 + x + x^2) = 1 + x^2 + x^5 + x^6$$

When we collect the coefficients of the like power terms of the polynomials $\{U(x)G_i(x)\}$ and represent them as column vector form as shown below, the column vector of the coefficient of x^i represents the encoded branch word:

$$[V(x)] = \begin{bmatrix} V_1(x) \\ V_2(x) \\ V_3(x) \end{bmatrix} = U(x) \begin{bmatrix} G_1(x) \\ G_2(x) \\ G_3(x) \end{bmatrix} = \begin{bmatrix} 1 \\ 1 \\ 1 \end{bmatrix} + \begin{bmatrix} 1 \\ 1 \\ 0 \end{bmatrix} x + \begin{bmatrix} 1 \\ 0 \\ 1 \end{bmatrix} x^2 + \begin{bmatrix} 0 \\ 1 \\ 0 \end{bmatrix} x^3$$

$$+ \begin{bmatrix} 1 \\ 0 \\ 0 \end{bmatrix} x^4 + \begin{bmatrix} 0 \\ 0 \\ 1 \end{bmatrix} x^5 + \begin{bmatrix} 0 \\ 1 \\ 1 \end{bmatrix} x^6 \tag{8.16g}$$

From (8.16g) we have the entire coded sequence of the rate $\frac{1}{3}$, $K = 3$ convolutional encoder to be

$$V = 111\ 110\ 101\ 010\ 100\ 001\ 011 \tag{8.16h}$$

which is identical to the sequence shown in Table 8.3, which was obtained on the basis of the impulse response representation of the encoder. The coded sequence of (8.16h) can be represented as a polynomial as follows: Because, from (8.16g), the output polynomials are

$$V_1(x) = U(x)G_1(x) = v_{10} + v_{11}x + v_{12}x^2 + \cdots$$
$$V_2(x) = U(x)G_2(x) = v_{20} + v_{21}x + v_{22}x^2 + \cdots$$
$$V_3(x) = U(x)G_3(x) = v_{30} + v_{31}x + v_{32}x^2 + \cdots$$

the encoded sequence can be represented as the polynomial

$$V(x) = V_1(x^3) + x\, V_2(x^3) + x^2 V_3(x^3) \tag{8.16i}$$

In (8.16i) the x is looked on as a delay element. The example we consider here specifies that

$$V_1\left(x^3\right) = 1 + x^3 + x^6 + x^{12}$$
$$x\, V_2\left(x^3\right) = x + x^4 + x^{10} + x^{19}$$
$$x^2 V_3\left(x^3\right) = x^2 + x^8 + x^{17} + x^{20}$$

The codeword, expressed as a polynomial, then becomes

$$V(x) = 1 + x + x^2 + x^3 + x^4 + x^6 + x^{10} + x^{12} + x^{17} + x^{20} \tag{8.16j}$$

which gives the sequence as shown in (8.16h).

8.2.5 State Representation of the Encoder

There is another way of characterizing the convolutional encoder, by which the coded sequence can be specified very effectively, which we discuss here as the state representation method. Let us again consider the encoder illustrated in Figure 8.9, using the circuit of part (b) of that figure, redrawn below in Figure 8.12.

The idea we exploit here is that if we know the *state* at the current clock pulse, we can always specify the encoded branch word when a new information bit is shifted into the encoder at the next clock pulse, at which time a *new state* is formed. Let us note the encoder under discussion in terms of the connection vector representation:

$$G_1 = (1\,0\,0) \qquad G_2 = (1\,0\,1) \qquad G_3 = (1\,1\,1) \tag{8.17}$$

Now suppose that the state is $(0\,0)$. If the input bit is 0 at the next clock pulse, the next state is $(0\,0)$ again; that is, $(00) \rightarrow (00)$. If, on the other hand, the next information bit is 1, the new state is (10), and the state transition is $(0\,0) \rightarrow (1\,0)$. Thus, according to the connection vectors[3] specified in (8.17), as the state transition is from $(0\,0)$ to $(0\,0)$ and from $(0\,0)$ to $(1\,0)$, the corresponding encoded branch words are $(0\,0\,0)$ and $(1\,1\,1)$, respectively.

Now suppose we are at state $(1\,0)$. Then, the next possible state is $(0\,1)$ or $(1\,1)$, depending upon the information bit being either a 0 or 1, respectively. The corresponding encoded branch word is, according to the encoding rules of (8.17), either $(0\,0\,1)$ or $(1\,1\,0)$, respectively. This encoding process using the state representation of the encoder shown in Figure 8.12 is depicted in Figure 8.13, along with the encoder and the connection vectors.

Example 8.5 Again let us look at the encoded sequence corresponding to the input information sequence $(1\,1\,1\,0\,1)$ for the rate $\frac{1}{3}$, $K = 3$ encoder. Starting from the state $a = (0\,0)$ in Figure 8.13, we can write down the encoded branch words corresponding to the state transitions as shown in Table 8.4.

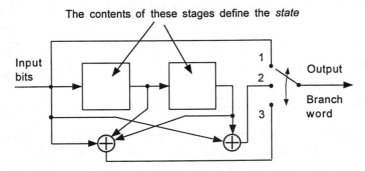

Figure 8.12 Rate $\frac{1}{3}$ constraint length 3 convolutional encoder. **The contents of the last $K - 1$ stages are defined as the _state_ of the encoder.**

---

[3] We are just using the connection vector representation here purely for convenience in lieu of directly observing the encoder circuit of Figure 8.12.

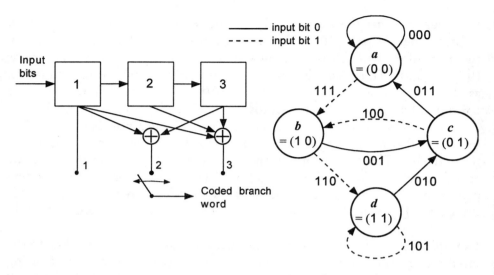

Figure 8.13 Encoder and state transition diagrams for the rate $\frac{1}{3}$, $K = 3$ encoder.

Table 8.4 Illustration of the use of the encoder state encoder state diagram of Figure 8.13

Input		1	→	1	→	1	→	0	→	1	→	0	→	0
States	00	→	10	→	11	→	11	→	01	→	10	→	01	→	00
	a	→	*b*	→	*d*	→	*d*	→	*c*	→	*b*	→	*c*	→	*a*
Brch.	↓		↓		↓		↓		↓		↓		↓		
wds.	111		110		101		010		100		001		011		

Example 8.6 Draw the state diagram for the convolutional encoder shown in Figure 8.14. Note that this is an example of rate k/n, where $k = 2$. Information bits to be encoded are shifted into the encoder two bits at a time. The two stages in the right side of the shift register constitute the current state and the *new state* is the incoming two information bits that are shifted into the two stages on the left side of the shift register. If the current state is assumed to be $a = (00)$, there are four different states to which the current state can move to, depending on the information bit pair patterns, namely (00), (01), (10), and (11). In fact, there are always these four state transition possibilities, no matter what the current state is. This situation is depicted in Figure 8.15 as a solution to this example.

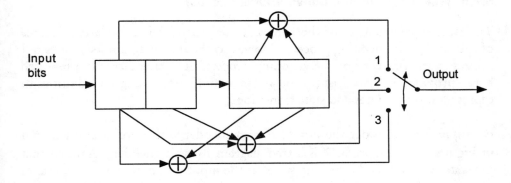

Figure 8.14 Rate $\frac{2}{3}$, $K = 2$ encoder.

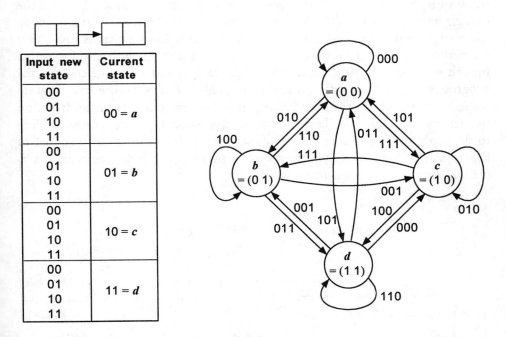

Figure 8.15 State diagram for the encoder of Figure 8.14.

8.2.6 Tree Diagram for a Convolutional Encoder

The state representation of the convolutional encoder is completely capable of specifying the encoding process. However, the state diagram itself does not provide a time history of the encoding process. The tree diagram, presented here, provides a time history of the encoding processing in addition to representation of states and state transitions.

Example 8.7 Consider the $r = \frac{1}{2}$, $K = 3$ encoder with state diagram shown in Figure 8.16. Figure 8.17 is a tree diagram for this encoder. Assume that the state $\boldsymbol{a} = (00)$ is the starting state. If the input is 0, a state transition takes place from $\boldsymbol{a}$ to $\boldsymbol{a}$, and if the input bit is 1, the state transition is from $\boldsymbol{a}$ to $\boldsymbol{b}$, where $\boldsymbol{b} = (10)$. The tree diagram shows these two state transitions that could occur at time $t = t_1$. When the input bit is 0, the branch word 00 is shown on the upper side of the tree, whereas when information bit is 1, the branch word is always shown on the lower side of the tree, with respect to the starting state position. At $t = t_2$, the four possible state transitions, two from the state $\boldsymbol{a}$ and two from the state $\boldsymbol{b}$, are shown with the respective branch word and the new state, and this process repeats along the tree as time progresses to $t = t_3$, $t = t_4, \dots$, and so on. Note that with the tree diagram we can always specify the encoded sequence corresponding to an information bit stream. For example, Table 8.5 shows the coded sequences corresponding to the input information sequences shown on the left side, for the encoder of Figure 8.16. The coded sequences are the *traces* of the code tree branches as dictated by the input sequence pattern.

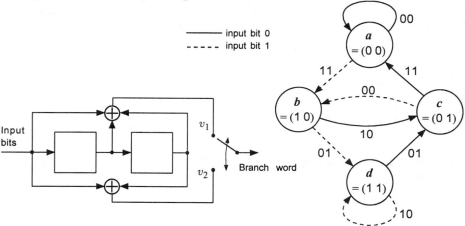

Figure 8.16 Encoder and state diagram for $r = \frac{1}{2}$, $K = 3$ convolutional code.

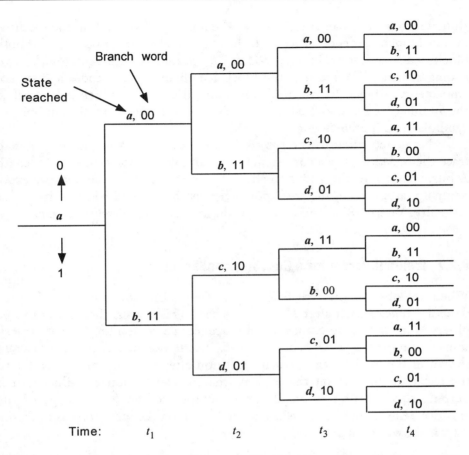

Figure 8.17 Tree diagram for the rate $\frac{1}{2}$, $K = 3$ convolutional encoder shown in Figure 8.16.

Table 8.5 Example of code tree-traced coded sequence for the encoder of Figure 8.16

Input sequence	Coded sequence
1 1 0 1 1 ··· →	1 1 0 1 0 1 0 0 0 1 ···
0 1 1 0 1 ··· →	0 0 1 1 0 1 0 1 0 0 ···
0 1 0 1 0 ··· →	0 0 1 1 1 0 0 0 1 0 ···

We have been using the term, coded "branch" word, and the basis for the use of this term is tree branch notion of representing the n-symbol coded sequences corresponding to the k input information bits. Thus, in a convolutional code, the code rate $r = k/n$ relates the number of information bits shifted into the encoder at one time to the number of symbols in the resulting output branch sequence. Although the code rate has the same form for both block codes and convolutional codes, the meaning is different for the two coding schemes. While n is the block length in a block code, it does not characterize a block length in a convolutional code, for which n is equal to the number of modulo-2 adders, which is in turn equal to the number of symbols in each branch word.

Now, returning to the subject of tree diagrams, if we are to use it to trace the coded sequence corresponding to an information sequence of length L bits, we have to show 2^L branches. The number of branches grows exponentially, and, hence, the tree approach must be avoided. Recognizing this difficulty, Viterbi [8] introduced a new diagram called the *trellis diagram*.

8.2.7 Trellis Diagram for a Convolutional Encoder

When we observe the tree diagram of Figure 8.17, we note that the tree structure repeats itself after $K = 3$ branches. This is a significant point to be observed in that some branch word sequences corresponding to state transitions, say $a\,b\,c\,a$ or $c\,a\,b\,c$, are periodic. In the example of the tree diagram shown in Figure 8.17, the branch repetition after the third branch is due to the fact that the first bit that entered the encoder no longer influences the output sequence after it has dropped out of the last (third) stage of the encoder. An example of a trellis diagram, based on the same code as in Figure 8.17, is shown in Figure 8.18.

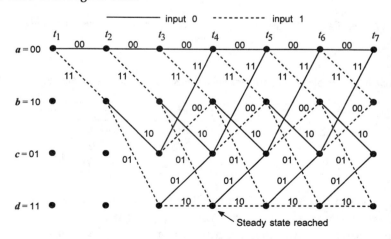

Figure 8.18 Trellis diagram for the rate $\frac{1}{2}$, $K = 3$ convolutional encoder of Figure 8.16.

Example 8.8 Consider the $r = \frac{1}{2}$, $K = 3$ convolutional encoder shown in Figure 8.19. Obtain (a) the connection vectors and code (connection) polynomials, (b) the state diagram, (c) the tree diagram, (d) the trellis diagram, and (e) the coded sequence for the information sequence 1010110101.

Solutions:

(a) $G_1 = (1\,1\,1)$ and $G_2 = (0\,1\,1)$

$G_1(x) = 1 + x + x^2, \quad G_2(x) = x + x^2$

(b) The state diagram is shown in Figure 8.20.

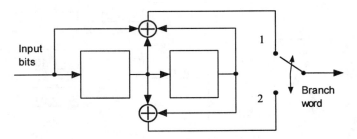

Figure 8.19 Rate $\frac{1}{2}$, $K = 3$ convolutional encoder.

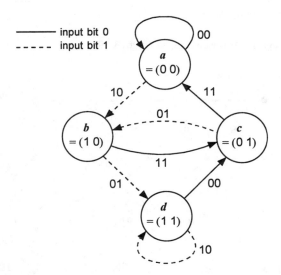

Figure 8.20 State diagram for the rate $\frac{1}{2}$, $K = 3$ convolutional encoder of Figure 8.19.

(c) By observing the state diagram, one can trace the coded tree branches as shown in Figure 8.21.

(d) The trellis diagram is shown in Figure 8.22.

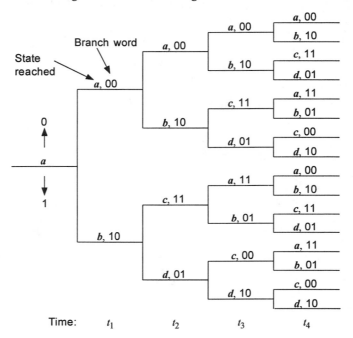

Figure 8.21 Tree diagram for the encoder of Figure 8.19.

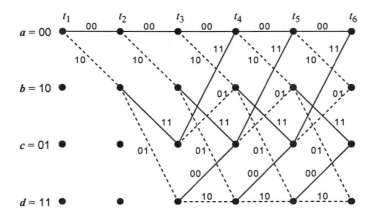

Figure 8.22 Trellis diagram for the encoder of Figure 8.19.

(e) The coded sequence is to be obtained for the information sequence

$$1\ 0\ 1\ 0\ 1\ 1\ 0\ 1\ 0\ 1\ \underbrace{0\ 0}$$
$$K - 1 = 2 \text{ zeros added to clear the register}$$

The coded sequence can be found from the encoder, connection vectors, state diagram, tree diagram, or trellis diagram, and it is given by

$$10\ 11\ 01\ 11\ 01\ 01\ 00\ 01\ 11\ 01\ \underbrace{11\ 11}$$
$$\text{to be discarded}$$

Example 8.9 Consider the rate $\frac{1}{2}$, $K = 4$ convolutional encoder shown in Figure 8.23. Obtain (a) the connection vectors and code polynomials, (b) the state diagram, (c) the trellis diagram, and (d) the coded sequence for the information sequence $1\ 0\ 1\ 0\ 1\ 1\ 0\ 1\ 0\ 1$.

Solutions:

(a) The connection vectors are $G_1 = (1111)$ and $G_2 = (1101)$. The code (connection) polynomials are $G_1(x) = 1 + x + x^2 + x^3$ and $G_2(x) = 1 + x + x^3$.

(b) As seen from the encoder shift register circuit, there are eight encoder states, which are given as 3-tuples of binary bits as follows:

$$a = \boxed{0\ |\ 0\ |\ 0} \quad b = \boxed{1\ |\ 0\ |\ 0} \quad c = \boxed{0\ |\ 1\ |\ 0} \quad d = \boxed{1\ |\ 1\ |\ 0}$$
$$e = \boxed{0\ |\ 0\ |\ 1} \quad f = \boxed{1\ |\ 0\ |\ 1} \quad g = \boxed{0\ |\ 1\ |\ 1} \quad h = \boxed{1\ |\ 1\ |\ 1}$$

The transitions between encoder states are shown in the state diagram of Figure 8.24.

(c) Because $K = 4$, there are $2^{K-1} = 2^3 = 8$ nodes, the same number as the number of states. The trellis diagram is shown in Figure 8.25.

(d) The coded sequence for the information sequence $1\ 0\ 1\ 0\ 1\ 1\ 0\ 1\ 0\ 1$ can be written by tracing the state diagram in the order indicated. Thus, the coded sequence is

$$11\ 11\ 01\ 00\ 01\ 11\ 01\ 10\ 00\ 01$$

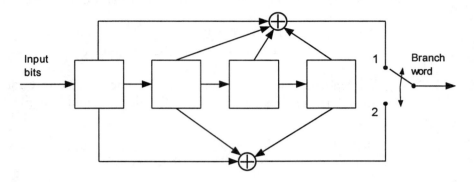

Figure 8.23 Rate $\frac{1}{2}$, $K = 4$ convolutional encoder.

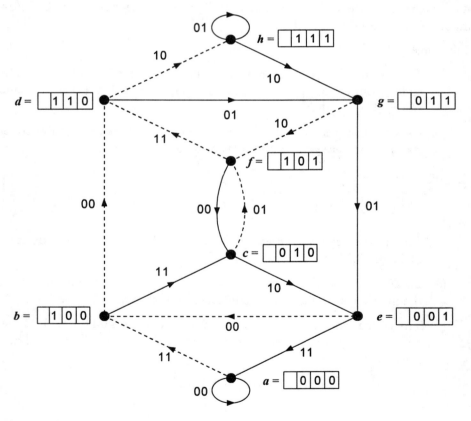

Figure 8.24 State diagram for the $r = \frac{1}{2}$, $K = 4$ encoder of Figure 8.23.

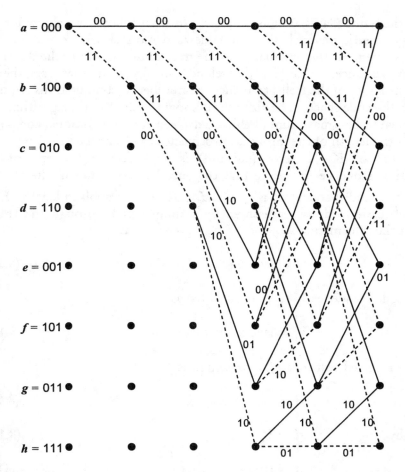

Figure 8.25 Trellis diagram for the $r = \frac{1}{2}$, $K = 4$ convolutional code defined by $G_1 = (1111)$ and $G_2 = (1101)$.

8.3 Maximum Likelihood Decoding of Convolutional Codes

8.3.1 Minimum Hamming Distance Decoding Rule

Maximum likelihood decoding involves searching the entire code space for the codeword which *most closely resembles* the received sequence. As we discussed in Chapter 7, the maximum likelihood decision rule is equivalent to the maximum correlation decision rule, as well as to the minimum distance rule for the white Gaussian noise channel. In what follows, we show, for the binary symmetric channel, that the codeword most closely resembling the received sequence is the one that gives the minimum distance in the metric of Hamming distance. Viterbi [8] developed an algorithm that uses the repetitive structure of the trellis to reduce the number of computations required to search the code space. The algorithm, known as the Viterbi algorithm, has been shown to be a maximum likelihood decoder for convolutional codes [6].

Let the length of the information sequence to be encoded be L bits. An encoder tail of $K-1$ zeros (tail bits) is appended to the end of the information sequence to ensure that the encoder terminates in the all-zero state. The encoded codeword is $N \triangleq n(L+K-1)$ symbols in length. Each of the 2^L codewords is represented by a unique path through the trellis. Now, let the information sequence U be represented as

$$U = (u_1, u_2, \ldots, u_L) \tag{8.18a}$$

and the coded sequence V to be transmitted as

$$V = (v_1, v_2, \ldots, v_N) \tag{8.18b}$$

Then the received sequence R, represented as

$$R = (r_1, r_2, \ldots, r_N) \tag{8.18c}$$

is given by

$$R = V + n \tag{8.18d}$$

where

$$n = (n_1, n_2, \ldots, n_N) \tag{8.18e}$$

is the independent, AWGN vector that disturbs the transmitted vector codeword. We then have a coded communications system where one of M equally likely messages $\{m_i, i = 0, 1, \ldots, 2^L - 1\}$ is transmitted over an AWGN channel by means of codewords $\{V_i, i = 0, 1, \ldots, 2^L - 1\}$. This is exactly the model we discussed in Section 7.1.2. The components of the independent noise vector n have the pdf given by

$$p_n(\alpha) = p_{n_1, n_2, \ldots n_N}(\alpha_1, \alpha_2, \ldots, \alpha_N) = \prod_{i=1}^{N} p_{n_i}(\alpha_i) \tag{8.18f}$$

With these assumptions, we now review the maximum likelihood decoding rule, which we discussed in Chapter 7, in the context of a coded system to derive the *sufficient condition* that satisfies the requirement of the maximum likelihood decision rule.

Let $m = m_i$ (i.e., codeword V_i is transmitted), and the received sequence is R. Then, the likelihood function of R given $V = V_i$ is the conditional probability

$$P(R \mid V_i) \tag{8.19a}$$

Then the MAP decision rule states that

$$\hat{m} = m_k \text{ iff } P(V_k \mid R) = \max_i \{P(V_i \mid R)\}$$

$$= \max_i \left\{ \frac{P(R \mid V_i)\, P(V_i)}{P(R)} \right\}$$

$$\Rightarrow \max_i \{P(R \mid V_i)\} \tag{8.19b}$$

which is the maximum likelihood decision rule, as expected. Let us digress for a moment and review the probability measures associated with the binary symmetric channel (BSC).

Binary symmetric channel. On the BSC, the probability that the channel causes a bit transition to its complement (bit error) is given by $p \leq \frac{1}{2}$, and the probability that the channel does not cause such a bit transition is then given by $1 - p$. This concept is depicted in Figure 8.26. When the coded symbols in a codeword $V_i = (v_{i1}, v_{i2}, \ldots, v_{iN})$ are subjected to the BSC, it is a case of N independent, repeated Bernoulli trials (see Example 1.17 in Section 1.5.3). Thus, the probability that k errors will occur in some combination in a block of N symbols transmitted over the BSC is given by

$$\binom{N}{k} p^k (1 - p)^{N-k} \tag{8.20a}$$

The probability that k or more errors (at least k errors) will occur is given by

$$\sum_{j=k}^{N} \binom{N}{j} p^j (1 - p)^{N-j} \tag{8.20b}$$

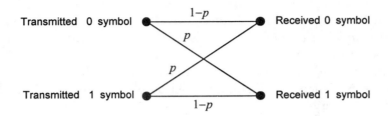

Figure 8.26 BSC model.

Define d_i to be the Hamming distance between the received sequence R and the codeword sequence V_i, that is:

$$d_i = \text{Hamming distance between } R \text{ and } V_i$$
$$\triangleq D(R, V_i) \tag{8.20c}$$

Then, on the BSC, the likelihood function can be expressed in terms of d_i as

$$P(R \mid V_i) = \Pr\{R \text{ differs from } V_i \text{ in } d_i \text{ symbol positions}\}$$
$$= \Pr\{R \text{ is received in error in } d_i \text{ symbol positions}\}$$
$$= p^{d_i}(1 - p)^{N - d_i} \tag{8.20d}$$

Let us now return to the subject of the maximum likelihood decision rule, given in (8.19b):

Maximum likelihood decision rule:
$$\hat{m} = m_k(V_k) \text{ iff } \quad P(R \mid V_k) = \max_i\{P(R \mid V_i)\}$$
$$= \max_i\left\{ p^{d_i}(1 - p)^{N - d_i} \right\} \tag{8.20e}$$

Because the logarithm is a monotonically increasing function of its argument, we can take the logarithm of both sides of (8.20e) to obtain

$$\max_i\{\log P(R \mid V_i)\} = \max_i\left\{ \log\left[p^{d_i}(1 - p)^{N - d_i} \right] \right\}$$
$$= \max_i\{d_i \log p + (N - d_i) \log(1 - p)\}$$

$$= \max_i \left\{ - d_i \log\left(\frac{1-p}{p}\right) + N\log(1-p) \right\} \qquad (8.20\text{f})$$

Because $p < \frac{1}{2}$, the quantity $(1-p)/p$ is always greater than unity, and $\mu \triangleq \log[(1-p)/p]$ is always positive. Similarly, if we define γ to be the index-independent term $N\log(1-p)$, that is, $\gamma \triangleq N\log(1-p)$, we have the maximum likelihood decoding rule that says

$$\widehat{V} = V_k \quad \text{iff} \quad \log P(R \mid V_k) = \max_i \{-\mu d_i + \gamma\}$$

or $\qquad \widehat{V} = V_k \quad \text{iff} \qquad\qquad d_k = \min_i \{d_i\} \qquad (8.20\text{g})$

This result is to the effect that the maximum likelihood decision rule on the BSC is the *minimum Hamming distance decoding rule*. In Chapter 7, we interpreted the maximum likelihood decision rule as the *minimum metric distance decision rule*.[4] In a later section of this chapter, we have occasion to evaluate coded system performance on both the Gaussian metric channel and Hamming's BSC.

8.3.2 Viterbi Decoding Algorithm

The Viterbi algorithm was discovered in 1967 by A. J. Viterbi [6], and it was proved in the well-known paper by J. K. Omura in 1969 [7] that the Viterbi algorithm is indeed a maximum likelihood decoding rule. The essence of the algorithm is the comparison between the received sequence at time $t = t_i$ and all the trellis paths arriving at the same state at $t = t_i$. When two paths enter the same state at $t = t_i$, the one having the maximum correlation (best metric) or minimum distance is chosen, and the chosen path is called the *surviving path*. The decoding process continues in this way deep into the trellis.

Example 8.10 Let us explain the principles involved in the algorithm with an example using the rate $\frac{1}{2}$, $K = 3$ encoder shown previously in Figure 8.16. The state diagram and trellis for this encoder are reproduced here in Figure 8.27. Suppose that the input sequence is

---

4 This criterion is also equivalent to the maximum correlation (best metric) decision rule.

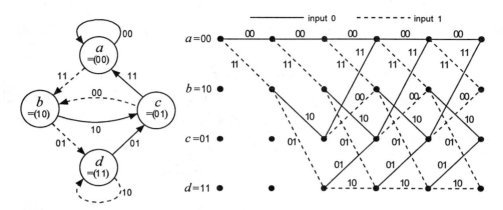

Figure 8.27 State diagram and trellis for the rate $\frac{1}{2}$, $K = 3$ convolutional encoder specified by $G_1 = (111)$ and $G_2 = (101)$.

$$U = (1\,0\,1\,0\,1\,1) \tag{8.21a}$$

Then the transmitted vector sequence (codeword) is given by

$$V = (11\;10\;00\;10\;00\;01) \tag{8.21b}$$

We now assume that the received sequence is given by

$$R = (11\;10\;10\;10\;00\;01) \tag{8.21c}$$

With these assumptions on the code vectors, we redraw the trellis along with the input sequence, transmitted sequence, and the received sequence, as shown in Figure 8.28. For each branch, the trellis diagram in Figure 8.28 shows the branch metric, given in Hamming distance between the branch word and the symbols of the received sequence R that were received during the branch interval. For example, the first two received symbols are $1\,1$, which have Hamming distances 2 and 0, respectively, from the branch words corresponding to the state transitions $a \to a$ and $a \to b$. Note that the trellis steady state is reached at time $t = t_4$, and it is known that the initial state of the encoder was $a = (00)$. If any two paths merge at a single state node, one of them always survives on the basis of the minimum Hamming distance decision rule and the other path is eliminated. At each time $t = t_i$ there are 2^{K-1} state nodes in the trellis, where K is the constraint length of the encoder. We observe the the decoding process as follows:

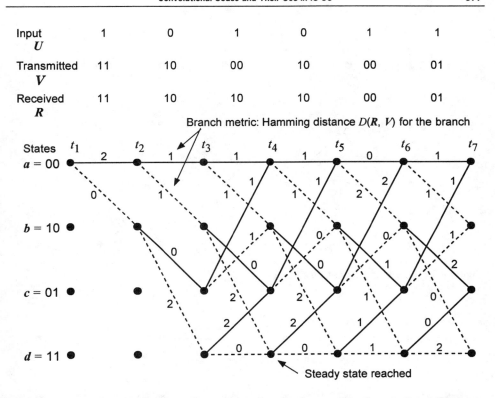

Input U	1	0	1	0	1	1
Transmitted V	11	10	00	10	00	01
Received R	11	10	10	10	00	01

Branch metric: Hamming distance $D(R, V)$ for the branch

Figure 8.28 Trellis with branch metric given by Hamming distance based on the received vector $R = (11\,10\,10\,10\,00\,01)$ for the encoder of Figure 8.27 $(r = \frac{1}{2}, K = 3)$.

(1) The surviving paths at each state at time $t = t_4$ are shown in Figure 8.29. Each path shown in this figure is the shorter of two possible paths in terms of Hamming distance. The surviving path at the state a at $t = t_4$ is the path that traversed the states $a \rightarrow b \rightarrow c \rightarrow a$ with total Hamming distance $0 + 0 + 1 = 1$, which was selected over the path that traversed the states $a \rightarrow a \rightarrow a \rightarrow a$ with a total Hamming distance $2 + 1 + 1 = 4$. At the state b (second node) at $t = t_4$, the survivor path $a \rightarrow b \rightarrow c \rightarrow b$ has Hamming distance 1; the defeated path $a \rightarrow a \rightarrow a \rightarrow b$ had distance 4. Note that the first input bit decision is still not made even at this delay, because there is more than one node at $t = t_2$ in a surviving path.

(2) At time $t = t_5$, the surviving path at each state is shown in Figure 8.30 along with the Hamming distances for each transition. Note that

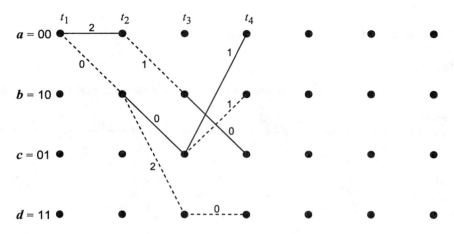

Figure 8.29 Surviving paths at time $t = t_4$.

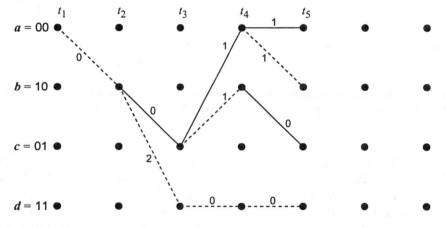

Figure 8.30 Surviving paths at time $t = t_5$.

for the first time, there is an input bit decision of 1 because all surviving paths have $a \rightarrow b$ (dotted line) as their first transition. It took a delay of five time units to make the first bit decision. This phenomenon is the *decoding delay*, and in a typical decoding process implementation, a decoding delay can be several times the constraint length, not to mention delays depending on algorithm efficiency.

(3) At time $t = t_6$, the surviving paths are as shown in Figure 8.31. In this particular example, the surviving path decisions at the nodes c and d were carried out on the basis of coin flipping, because the Hamming distances were identical between the contending pairs of paths at both states.

(4) At time $t = t_7$, the same processes were carried out, resulting in the surviving paths as shown in Figure 8.32. At this time, however, the

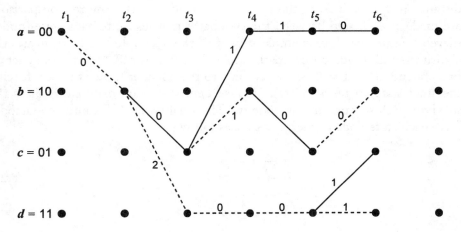

Figure 8.31 Surviving paths at time $t = t_6$.

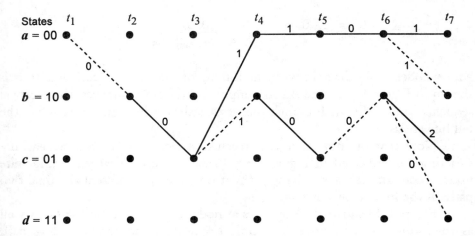

Figure 8.32 Surviving paths at time $t = t_7$.

second information bit decision is made to be 0, as seen in the figure from the fact that there is only one surviving transition in the second time interval. Note that the correct decision is made in view of the fact that the first two input bits that were encoded by the encoder were 10, as shown previouly in Figure 8.28.

We see from the example that the Viterbi algorithm requires storage of the minimum distance path at each node as well as its path metric, the minimum distance of the path from the received sequence. Furthermore, comparisons are made at each step between the two paths that lead into each state (node), thereby requiring four comparisons for the example. To truncate the algorithm and decide on one path rather than four (or 2^{K-1} paths in general), one appends $K - 1 = 2$ zeros (tail bits) to the encoder input, thereby forcing the final state to be $a = (00)$. Consequently, the ultimate survivor is the survivor path terminating at node a. The final two 00 bits into the encoder force each state node to merge with a as follows:

$$
\begin{aligned}
a &\rightarrow a \rightarrow a \\
b &\rightarrow c \rightarrow a \\
c &\rightarrow a \rightarrow a \\
d &\rightarrow c \rightarrow a
\end{aligned}
\tag{8.22}
$$

Let us consider this truncation process with the present example. We append the two dummy zeros at the end of the input data sequence to have

$$
U = (1 \ 0 \ 1 \ 0 \ 1 \ 1 \ \underbrace{0 \ 0}_{\text{dummy tail bits}} \ )
$$

The encoder trellis diagram with the forcing of the states to converge to a is shown in Figure 8.33. The encoder input and output sequences, the received sequence, and trellis with branch metrics are shown in Figure 8.34 when the tail bits are used.

Note that the final state a is reached at time $t = t_9$. Before reaching this time, as indicated in Figure 8.32, "battles" must be waged at the state nodes to determine the surviving paths at time $t = t_8$ to present the final two paths at the final node a at $t = t_9$.

Figure 8.35 shows the survivors at nodes a and c at time $t = t_8$, as well as their transitions to node a at time $t = t_9$, and the corresponding path metrics. The final selection at node a is clearly the lower path, and, hence, the decision (decoding) is 1 0 1 0 1 1 0 0, which is identical to the original

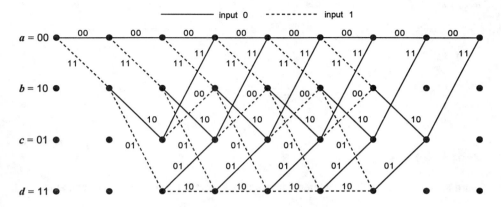

Figure 8.33 Trellis diagram showing the converge of the states to state a.

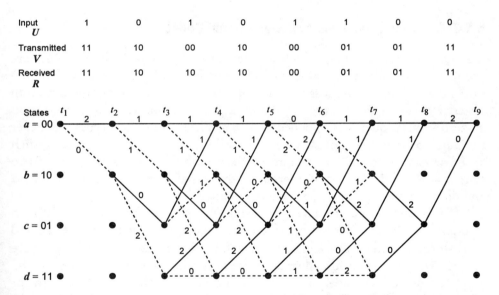

Figure 8.34 Trellis with branch metric given by Hamming distance based on the received vector $R = (11\,10\,10\,10\,00\,01\,01\,11)$ for the encoder of Figure 8.27 ($r = \frac{1}{2}$, $K = 3$).

uncoded sequence and tail bits. Note that the received sequence contained an error at the fifth symbol, but it was corrected in the process of decoding. We address the error-correcting capability of convolutional codes in a section to follow.

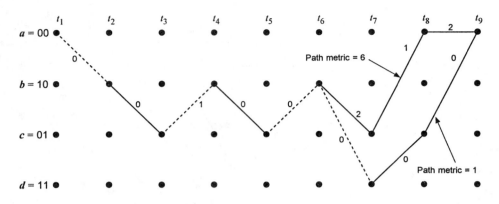

Figure 8.35 Path metric (accumulated Hamming distance) for the final two merging paths.

8.3.3 Distance Properties of Convolutional Codes

The convolutional codes, which are generated by linear shift registers, are group codes. Hence, the all-zero codeword is a codeword in any convolutional code. There is no loss in generality, therefore, in finding the minimum distance between the all-zero word and each codeword. If we assume that an all-zero input sequence is encoded and transmitted, an error occurs whenever the all-zero path does not survive. The minimum distance for making such an error can be found by examining every path from the state $a = (00)$ to state $a = (00)$. The paths of interest are those that start and end in the state a and do not return to state a anywhere in between.

We discuss the distance properties of convolutional codes with an example for the encoder we considered in Figure 8.16 for the rate $\frac{1}{2}$, $K = 3$ convolutional code. The trellis diagram for this encoder is shown in Figure 8.36 showing the Hamming distance metrics compared with the all-zero path.

Let us now consider all the paths that merge with the all-zero path for the first time. From Figure 8.36, it is seen that at times t_4, t_5, and t_6, paths begin to merge with the all-zero path. It is seen that there is just one merging path over (t_1, t_4) with distance 5 from the all-zero path $(a \xrightarrow{2} b \xrightarrow{1} c \xrightarrow{2} a)$. Also, there are two paths with distance 6: one over (t_1, t_5), $a \xrightarrow{2} b \xrightarrow{1} d \xrightarrow{1} c \xrightarrow{2} a$, and the other over (t_1, t_6), $a \xrightarrow{2} b \xrightarrow{1} c \xrightarrow{0} b \xrightarrow{1} c \xrightarrow{2} a$. For the path with distance 5, the input bits over (t_1, t_4) are $1\,0\,0$, and this sequence differs from the all-zero sequence $0\,0\,0$ in just one position (one bit error), whereas

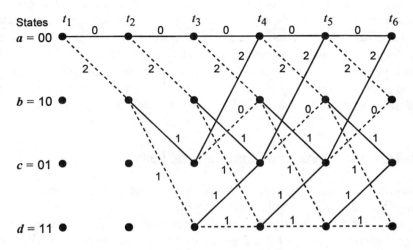

Figure 8.36 Trellis diagram with branches showing Hamming distance from the all-zero path.

for the distance-6 paths, the input bits are $1\,1\,0\,0$ over (t_1, t_5) and $1\,0\,1\,0\,0$ over (t_1, t_6). These input sequences differ in two positions (two bit errors) from the all-zero sequences $0\,0\,0\,0$ and $0\,0\,0\,0\,0$, respectively. Thus, we find in this situation that we are dealing with a set of codewords in which the minimum distance in the set of all arbitrarily long paths that diverge and remerge is 5.

From the theory of block codes, we know that the code with the minimum distance d_{min} can correct all t or less errors, where

$$t = \left\lfloor \frac{d_{min} - 1}{2} \right\rfloor \tag{8.23a}$$

and $\lfloor y \rfloor$ means the largest integer no greater than y. In the convolutional code, however, d_{min} is to be replaced by d_f, which is called the (*minimum*) *free distance*, so that the error-correcting capability of a convolutional code is

$$t = \left\lfloor \frac{d_f - 1}{2} \right\rfloor \tag{8.23b}$$

In the example under discussion, this means that any two channel errors can be corrected because such two errors will cause the received sequence to be at distance 2 from the correct transmitted sequence, while it is at least distance 3 from any other possible codewords. The characterization of a given convolutional code in terms of error-correcting capability appears to be based on the determination of the minimum free distance from the examination of the trellis diagram. Usually, a computer simulation method is used to find d_f for a convolutional code of large constraint length. Fortunately, an analytical method can be applied to obtain a closed-form expression whose expansion yields the free distance as well as all the distance information [8].

8.3.4 Transfer Functions of Convolutional Codes

As we just stated above, we need to know not only the minimum free distance but also all possible distance properties of the convolutional code under consideration to assess the error-correcting capabilities of the code in an average statistical sense. In this section, we show that we can find the distance properties of a convolutional code from its state diagram analytically.

Now consider the code characterized by the state diagram in Figure 8.27, which is the same encoder we discussed in the previous section. We redraw the state diagram by replacing every branch word with D^j, where j is the weight of the branch sequence. Because in this particular case, the branch words are two-symbol sequences, j is either 0, 1, or 2. This is shown in Figure 8.37. We then split open the node $a = (00)$ and display the result as shown in Figure 8.38. Because the node is split into two ends, we label one as the input state a and the other as the output state $e = (00)$. This is entirely consistent with the actual process of decoding the convolutional code, because the final survivor path is determined at the node a. The self-loop at the node a does not contribute any weight because it corresponds to an all-zero branch word.

In the previous section, we examined the trellis diagram in Figure 8.36 and observed that there is just one path over the time interval (t_1, t_4), corresponding to a node level 4, that had an accumulated Hamming distance 5. That path was $a \rightarrow b \rightarrow c \rightarrow a$. If we trace this same path in Figure 8.38, we have the total factor $D^2 \cdot D^1 \cdot D^2 = D^5$, and this indicates that the path $a \rightarrow b \rightarrow c \rightarrow a$ is at distance 5 from the correct all-zeros path. The paths $a \rightarrow b \rightarrow d \rightarrow c \rightarrow a$ and $a \rightarrow b \rightarrow c \rightarrow b \rightarrow c \rightarrow a$ correspond to the factors $D^2 \cdot D^1 \cdot D^1 \cdot D^2 = D^6$ and $D^2 \cdot D^1 \cdot D^0 \cdot D^1 \cdot D^2 = D^6$, respectively. This observation, concerning tracing the branches from the state a at time $t = t_1$ to the state a at time $t = t_j$ (or node level j), using the modified and relabeled state diagram of Figure 8.38, suggests that we can find all the paths

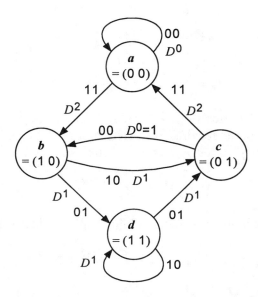

Figure 8.37 State diagram of the convolutional encoder of Figure 8.16 with branch words replaced with D^j, where j is the weight of the branch sequence.

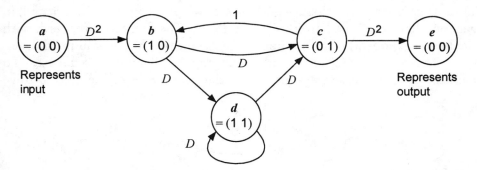

Figure 8.38 State diagram of Figure 8.37 split open at the node $a = (00)$.

merging with the all-zeros path at the node level j or time interval (t_1, t_j) by solving for the *generating function* or *transfer function* of the state diagram shown in Figure 8.38.

The generating function can be found by solving the following simultaneous state equations:

$$Z_b = D^2 Z_a + Z_c$$
$$Z_c = D Z_b + D Z_d$$
$$Z_d = D Z_b + D Z_d \qquad (8.24a)$$
$$Z_e = D^2 Z_c$$

where we use the state notations Z_a, Z_b, Z_c, Z_d, and Z_e to denote the sum of the products of all paths starting from state $a = (00)$ and ending at state b, c, d, and e, respectively. The generating function or transfer function, denoted $T(D)$, is defined as

$$T(D) \triangleq Z_e / Z_a \qquad (8.24b)$$

By solving the simultaneous equations in (8.24a), we obtain the result,

$$T(D) = \frac{Z_e}{Z_a} = \frac{D^5}{1 - 2D} = D^5 \left(1 + 2D + 4D^2 + 8D^3 + \cdots + 2^l D^l + \cdots \right)$$
$$= 1 \cdot D^5 + 2D^6 + 4D^7 + 8D^8 + \cdots + 2^l D^{l+5} + \cdots \qquad (8.24c)$$

The transfer function expanded in an infinite series form reveals the information we sought. The first term D^5 reveals that there is just one (coefficient of D^5 is 1) path with free distance 5, there are two paths with distance 6, and, in general, there are 2^l paths with distance $l + 5$. The first term (the term with the lowest exponent) is the minimum free distance of the convolutional code.

The result in (8.24c) is valid for an infinitely long code sequence. To truncate the series at some point, say, at the jth node level, we need to solve

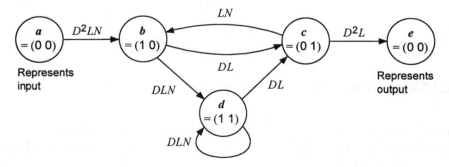

Figure 8.39 State diagram with branches labeled with Hamming weight D, length L, and N for branches caused by an input data 1.

for the transfer function for the modified state diagram of Figure 8.38, further modified as shown in Figure 8.39, in which we put the factor L on each branch as well as the factor N on those branches caused by the input data bit value 1. This is an efficient way of accounting for the length of a given path by means of the exponent of the L factor, because the exponent will increase by one every time a branch is traversed. Likewise, the exponent of the N term, which is included in the branch factor if that branch is caused by the input data 1 (dotted line branch in the original state diagram), delivers information on the number of bits in error in comparison with the all-zero data input data sequence. By forming simultaneous equations similar to those shown in (8.24a) for the state diagram of Figure 8.39, we have

$$Z_b = D^2 LN Z_a + LN Z_c$$
$$Z_c = DL Z_b + DL Z_d \tag{8.25a}$$
$$Z_d = DLN Z_b + DLN Z_d$$
$$Z_e = D^2 L Z_c$$

The transfer function $T(D, L, N)$ in this case is found to be

$$
\begin{aligned}
T(D, L, N) &= \frac{D^5 L^3 N}{1 - DL(1 + L)N} \\
&= D^5 L^3 N + D^6 L^4 (1 + L) N^2 + D^7 L^5 (1 + L)^2 N^3 + \cdots \\
&\quad + \cdots + D^{l+5} L^{l+3} (1 + L)^l N^{l+1} + \cdots
\end{aligned}
\tag{8.25b}
$$

Let us observe the implications of (8.25b):

(1) There is one path of free distance 5, which differs in one input bit position from the all-zeros path and has length 3;

(2) There are two paths of distance 6, one having length 4 and the other having length 5, and both differ in two input bits from the all-zeros path;

(3) If we are interested in the jth node level, we consider all terms up to and including the jth power (L^j) term.

With the determination of these properties of all paths, we can now evaluate error probabilities of the convolutional codes, which we consider in Section 8.4.

8.3.4.1 Systematic and Nonsystematic Convolutional Codes

A systematic convolutional code is one for which one of symbols of the branch word is just the data bit generating the branch word. An example of a systematic convolutional code is the one generated by the encoders shown in Figures 8.12 and 8.40.

We recall from the discussions on block codes in Chapter 7 that there is no difference between the systematic block code and nonsystematic block code from the performance standpoint. For convolutional codes, however, systematic codes and nonsystematic codes perform differently, in that their minimum free distances are different in general for the same constraint length [3, 8].

8.3.4.2 Catastrophic Error Propagation in Convolutional Codes

A catastrophic error is an event in which a finite number of channel symbol errors cause an infinite number of data bit errors to be decoded. A necessary and sufficient condition for a rate $1/n$ convolutional code to produce a catastrophic error is that all the code polynomials $G_i(x)$, $i = 1, 2, \ldots, n$ have a common factor [8]. For example, consider the encoder shown in Figure 8.41 and its code polynomials.

Indeed the code polynomials $G_1(x)$ and $G_2(x)$ have the common factor $1 + x$. When we draw the state diagram of the encoder as shown in Figure 8.42, we note that the state d has a closed loop with zero weight. In fact, catastrophic errors can occur if and only if any closed loop has a zero weight. To see this point more clearly, let us represent the state diagram as shown in Figure 8.43 using the D and N notation defined previously. If we assume that the all-zeros path is the correct path, any sequence different from an all-zero sequence is a sequence in error. Let us consider such an incorrect path, given

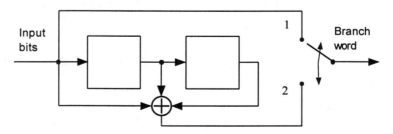

Figure 8.40 Rate $\frac{1}{2}$, $K = 3$ systematic convolutional encoder.

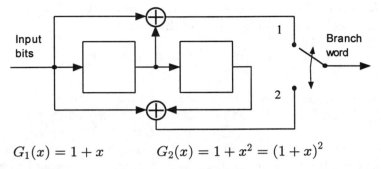

$$G_1(x) = 1 + x \qquad G_2(x) = 1 + x^2 = (1 + x)^2$$

Figure 8.41 Rate $\frac{1}{2}$, $K = 3$ convolutional encoder that may cause catastrophic error propagation.

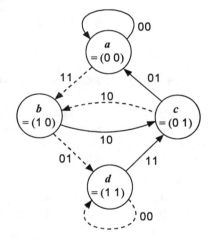

Figure 8.42 State diagram of the encoder in Figure 8.41.

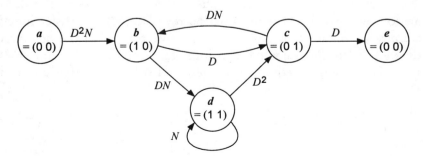

Figure 8.43 Split state diagram with branches labeled with D and N.

by $a \rightarrow b \rightarrow d \rightarrow d \rightarrow \cdots \rightarrow d \rightarrow c \rightarrow a$. From the state diagram in Figure 8.42, we can specify the code sequence of this path as

$$a \xrightarrow{01} b \xrightarrow{01} d \xrightarrow{00} d \xrightarrow{00} \cdots \xrightarrow{00} d \xrightarrow{01} c \xrightarrow{01} a$$

which has exactly six 1s, no matter how many times the zero-weight self-loop was used. From Figure 8.43, this observation is seen as equivalent to the expression of $(D^2 N)(DN)(N) \cdots (N)(D^2)(D) = D^6 N^n$, where D^6 is the weight of the entire sequence, and N simply means that the branch was caused by the data bit 1. Thus, for a BSC, four channel errors would induce the selection of this incorrect path, which may cause catastrophic error propagation. There are other necessary and sufficient conditions for catastrophic error propagation, and the reader is referred to Viterbi's classic paper [8] for further reading on this subject.

8.4 Performance Bounds for Viterbi Decoding of Convolutional Codes

When we discussed the trellis diagram of a typical convolutional encoder, we observed the distance properties of the code by means of the transfer function $T(D, L, N)$. We came away from that distance-property study with the observation that paths displayed not only a minimum distance but also the maximum distance for a typical node level j and many other different distances in between the minimum and the maximum. This fact implies that the probability of error calculation for the Viterbi decoding algorithm is not amenable to an exact analysis, and, in fact, for most of the practically employed convolutional codes, their performance measures are obtained by computer simulation. Analytically, however, we obtain performance bounds on the probability of an error in a code symbol as well as in the decoded information bits, based on the information revealed by the transfer function. The bounds are obtained as union bounds. We present in the next section bounds on the Viterbi decoding algorithm for both the BSC and AWGN channels. The analysis method to be presented is based on that given in [3] and [8].

8.4.1 Probability of Error Bounds for Hard Decision Decoding

We obtain the first-event probability, which is the probability that the correct path is rejected (not a survivor) for the first time at time $t = t_j$. We again consider the rate $\frac{1}{2}$, $K = 3$ convolutional encoder that was used as an example for the derivation of the transfer functions $T(D)$ and $T(D, L, N)$ given in (8.24b) and (8.25b), respectively. We have redrawn the encoder and its state diagram in Figure 8.44. The transfer function corresponding to this encoder was given in (8.25b) as

$$
T(D, L, N) = \frac{D^5 L^3 N}{1 - DL(1 + L)N}
$$

$$
= D^5 L^3 N + D^6 L^4 (1 + L)N^2 + D^7 L^5 (1 + L)^2 N^3 + \cdots
$$

$$
+ \cdots + D^{l+5} L^{l+3} (1 + L)^l N^{l+1} + \cdots \qquad (8.26a)
$$

The minimum free distance of this code is seen to be 5, and the sequence of *state transitions* that results in this *weight-5* excursion from the all-zeros path is seen to be

$$
a(t_1) \rightarrow b(t_2) \rightarrow c(t_3) \rightarrow d(t_4) \qquad (8.26b)
$$

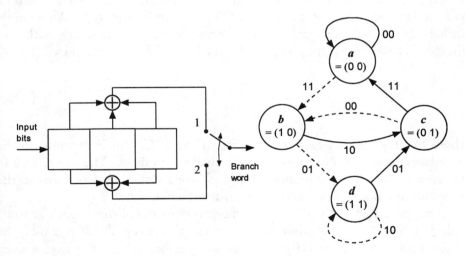

Figure 8.44 Rate $\frac{1}{2}$, $K = 3$ convolutional encoder.

The path sequence for the state transitions is $1\,1\,1\,0\,1\,1$ as seen from the trace of the state diagram. If the channel induces an error pattern such that the received sequence is favoring this sequence, $1\,1\,1\,0\,1\,1$, rather than the all-zeros sequence, $0\,0\,0\,0\,0\,0$, which is the correct sequence, then the all-zeros path will be eliminated from competition (will not survive) at node a at time $t = t_4$, denoted $a(t_4)$. Thus, in such a case, the maximum likelihood decoder will favor the incorrect path. At this point, the situation is equivalent to that of a binary communications system where m_0 and m_1 are two equally likely messages (to employ the terminology used in Section 7.1) to be communicated using codewords $C_0 = 0\,0\,0\,0\,0\,0$ and $C_1 = 1\,1\,1\,0\,1\,1$, respectively. The distance between C_0 and C_1 is

$$d_{min} = D(C_0, C_1) = 5 \qquad (8.26c)$$

The error-correcting capability of this two-codeword system is given by

$$t = \left\lfloor \frac{d_{min} - 1}{2} \right\rfloor = \left\lfloor \frac{5 - 1}{2} \right\rfloor = 2 \qquad (8.26d)$$

that is, as long as the channel induces one or two errors, the all-zeros sequence will be favored. If, however, three or more errors are made in the channel, then the correct all-zeros sequence will be rejected and the sequence $1\,1\,1\,0\,1\,1$ will be favored. Let us consider a BSC with p as the symbol crossover probability. Then the probability of error for a path sequence with free distance 5, denoted $P_2(e; 5)$, is given by the binomial distribution [see (1.59a)]

$$P_2(e; 5) = \sum_{k=3}^{5} \binom{5}{k} p^k (1 - p)^{5-k} \qquad (8.26e)$$

where p is the symbol error probability of the BSC, which depends on the modulation used for transmission of the channel symbols. The subscript 2 in the notation of (8.26e) indicates a two-codeword system that pertains to the convolutional decoder decision at state a for the first time.

Suppose that the path being compared with the all-zeros path at some node j at time $t = t_j$ has distance d from the all-zeros path. If d is odd, the all-zeros path will be correctly chosen if the number of errors in the received sequence is less than $(d + 1)/2$; otherwise, the incorrect path is chosen. Thus, the probability of selecting the incorrect path is

$$P_2(e; d) = \sum_{k=(d+1)/2}^{d} \binom{d}{k} p^k (1-p)^{d-k}, \quad d \text{ odd} \tag{8.27a}$$

If d is even, the incorrect path is selected when the number errors exceeds $\frac{1}{2}d$; that is, when $k \geq \frac{1}{2}d + 1$. If the number of errors equals $\frac{1}{2}d$, we must flip an honest coin and one of the two selected. Thus, in this situation, we have

$$P_2(e; d) = \sum_{k=\frac{1}{2}d+1}^{d} \binom{d}{k} p^k (1-p)^{d-k}$$

$$+ \frac{1}{2} \binom{d}{\frac{1}{2}d} p^{d/2} (1-p)^{d/2}, \quad d \text{ even} \tag{8.27b}$$

and the probability of incorrectly selecting a path with Hamming distance d is

$$P_2(e; d) = \begin{cases} \sum_{k=(d+1)/2}^{d} \binom{d}{k} p^k (1-p)^{d-k}, & d \text{ odd} \\ \\ \sum_{k=\frac{1}{2}d+1}^{d} \binom{d}{k} p^k (1-p)^{d-k} \\ \quad + \frac{1}{2} \binom{d}{\frac{1}{2}d} p^{d/2} (1-p)^{d/2}, & d \text{ even} \end{cases} \tag{8.27c}$$

We have seen from the transfer function $T(D)$ given in series form that there are many paths with different distances that merge with the all-zeros path at a given node, indicating thereby that there is no simple *exact* expression for the first-event error probability. Thus, we overbound the error probability using the union bound, which is the sum of the pairwise error probabilities $P_2(e; d)$ over all possible paths that merge with the all-zeros path at the given node. Now recall from (8.24c) and (8.25b) through (8.26a) that

$$T(D, L, N)|_{L=1, N=1} = T(D) = \sum_{k=d}^{\infty} a_k D^k \tag{8.28a}$$

where a_k is the number of nonzero paths with Hamming distance k, with respect to the all-zeros path, which merge with the all-zeros path at node j and diverge once from the all-zeros path at some previous node. For the case of the encoder that gave the transfer function (8.24c), the $\{a_k\}$ correspond to values $a_5 = 1$, $a_6 = 2$, $a_7 = 4, \ldots$, etc.

If there are a_k paths with weight k, and if we still use $P_2(e; k)$ to denote the pairwise error probability of paths with weight k, the union bound on the probability of decoding error is given by

$$P(e) < \sum_{k=d}^{\infty} a_k P_2(e; k) \tag{8.28b}$$

where $P_2(e; k)$ is given by (9.27c).

For the encoder characterized by the connection vectors $G_1 = (1\,1\,1)$ and $G_2 = (1\,0\,1)$, we had

$$T(D) = D^5 + 2D^6 + 4D^7 + \cdots + 2^k D^{k+5} + \cdots \tag{8.29a}$$

For this case, the first-event error probability is bounded by

$$P(e) < P_2(e; 5) + 2P_2(e; 6) + 4P_2(e; 7) + \cdots$$
$$+ 2^k P_2(e; k+5) + \cdots \tag{8.29b}$$

Viterbi [8] has also shown that the expression given in (8.27c) for the probability of incorrectly selecting a path with Hamming distance d can be bounded by

$$P_2(e; d) < [4p(1-p)]^{d/2} \tag{8.29c}$$

which is a more compact, albeit looser, bound than the exact first-event probability expression in (8.27c).[5] Thus, using (8.29b) we can have a looser upper bound on the first-event probability (decoding error)

$$P(e) < \sum_{k=d}^{\infty} a_k \left(\sqrt{4p(1-p)}\right)^k \tag{8.29d}$$

where p is the symbol crossover probability (symbol error probability) of the BSC. Comparing (8.29d) with (8.28a) suggests immediately that the first-

---

[5] We shall have occasion to observe the degree of "looseness" of (8.29b) as compared to (8.27c) in a plot of BER in Figure 8.47.

event error probability is overbounded by the transfer function $T(D)$ with $D = \sqrt{4p(1-p)}$; that is

$$P(e) < \sum_{k=d}^{\infty} a_k \left(\sqrt{4p(1-p)}\right)^k = T(D)\big|_{D=\sqrt{p(1-p)}} \qquad (8.29e)$$

8.4.2 Bit-Error Probability for the BSC

In the series expansion for the transfer function $T(D, N)$, the exponents of N indicate the number of nonzero information bits that are in error when an incorrect path is selected over the all-zeros path. Consider the transfer function given in (8.25b):

$$T(D, N) = \frac{D^5 N}{1 - 2DN} = D^5 N + 2D^6 N^2 + \cdots + 2^k D^{k+5} N^{k+1} + \cdots \qquad (8.30a)$$

By differentiating $T(D, N)$ with respect to N and setting $N = 1$, the exponents of N become multiplication factors of the corresponding $P_2(e; d)$. Because $T(D) = T(D, N)|_{N=1}$ is the upper bound on the first-event probability, each of the terms in $T(D)$ must be weighted by the exponent of N to obtain the bit-error probability $P_B(e)$. Now

$$\frac{\partial T(D, N)}{\partial N}\bigg|_{N=1} = D^5 + 2 \cdot 2D^6 + \cdots + (k+1) \cdot 2^k D^{k+5} + \cdots \qquad (8.30b)$$

From this expression, we obtain the bit-error probability $P_B(e)$ as we have obtained for the first-event error probability in (8.29d):

$$
\begin{array}{ccc}
\text{1 bit in error} & \text{2 bits in error} & \text{3 bits in error} \\
\downarrow & \downarrow & \downarrow
\end{array}
$$

$$P_B(e) < 1 \cdot P_2(e; 5) + 2 \cdot 2P_2(e; 6) + 3 \cdot 4P_2(e; 7) + \cdots$$

$$+ (k+1) \cdot 2^k P_2(e; k+5) + \cdots \qquad (8.30c)$$

$$
\begin{array}{c}
\uparrow \\
k+1 \text{ bits in error}
\end{array}
$$

where $P_2(e; d)$ was given by (8.27c) and (8.29e) as exact and bound expressions, respectively. If we use the bound in (8.29e), $P_2(e; d) < [4p(1-p)]^{d/2}$, we obtain a simple but weaker bound on the bit-error probability for the BSC:

$$P_B(e) < \left.\frac{\partial T(D, N)}{\partial N}\right|_{N=1,\, D=\sqrt{4p(1-p)}} \tag{8.30d}$$

8.4.3 Probability of Error Bounds for Soft-Decision Decoding

We now consider the error probability performance of Viterbi decoding in an AWGN channel with soft-decision decoding. We assume that antipodal PSK (BPSK) modulation is used for transmitted coded symbols over an AWGN channel with two-sided power spectral density $\frac{1}{2}\mathcal{N}_0$. The convolutional code is generated by an encoder of rate k/n, and the symbol energy E_s is related to the energy per information bit E_b by the expression

$$kE_b = nE_s \tag{8.31a}$$

so that

$$E_s = rE_b \tag{8.31b}$$

where

$$r \triangleq k/n \tag{8.31c}$$

is the code rate. For the infinite quantization soft-decision decoder, the probability of error in the pairwise comparison of two paths at a node that differ in d symbols is shown to be [3, 8]

$$P_2(e;\, d) = Q\left(\sqrt{\frac{2E_s}{\mathcal{N}_0} \cdot d}\right) = Q\left(\sqrt{\frac{2drE_b}{\mathcal{N}_0}}\right) \tag{8.31d}$$

where

$$Q(\alpha) = \int_\alpha^\infty \frac{1}{\sqrt{2\pi}}\, e^{-t^2/2}\, dt$$

This is the first-event error probability expression, which is the counterpart to the expression given in (8.27c), the probability of incorrectly selecting a path with Hamming distance d from the all-zeros sequence on a BSC. For the many other paths with different distances that merge with the all-zeros path, we have only to call upon the transfer function to provide the complete description:

$$T(D) = \sum_{k=d}^\infty a_k D^k \tag{8.31e}$$

where a_k denotes the number of nonzero paths of distance k from the all-zeros path, which merge with the all-zeros path for the first time. As we have done for the case of the BSC in (8.29c), the total first-event error probability for all pairs with distance $k \geq d$ is given in the form

$$P(e) < \sum_{k=d}^{\infty} a_k P_2(e;\, k) = \sum_{k=d}^{\infty} a_k Q\left(\sqrt{k \cdot \frac{2E_s}{\mathcal{N}_0}} \right) \qquad (8.31f)$$

To proceed, we invoke a useful inequality [8]:

$$Q(\sqrt{x + y}) \leq e^{-y/2} Q(\sqrt{x}) \qquad (8.32a)$$

which is proved in Appendix 8A. Thus, using this inequality for the Q-function, the first-event error probability for any pair of paths with distance $d + l$ is given by

$$P_2(e;\, d + l) = Q\left(\sqrt{(d + l)\frac{2E_s}{\mathcal{N}_0}} \right) \leq e^{-lE_s/\mathcal{N}_0} Q\left(\sqrt{d\,\frac{2E_s}{\mathcal{N}_0}} \right) \qquad (8.32b)$$

Now for $k \geq d$ in (8.31f), let $l = k - d$. Then

$$Q\left(\sqrt{k\frac{2E_s}{\mathcal{N}_0}} \right) = Q\left(\sqrt{(d + l)\frac{2E_s}{\mathcal{N}_0}} \right) \leq e^{-lE_s/\mathcal{N}_0} Q\left(\sqrt{d\,\frac{2E_s}{\mathcal{N}_0}} \right) \qquad (8.32c)$$

Hence the bound of (8.31f), using (8.32c), becomes

$$P(e) < \sum_{k=d}^{\infty} a_k P_2(e;\, k) = \sum_{k=d}^{\infty} a_k\, e^{-(k-d)E_s/\mathcal{N}_0} Q\left(\sqrt{d \cdot \frac{2E_s}{\mathcal{N}_0}} \right)$$

or

$$P(e) < Q\left(\sqrt{\frac{2dE_s}{\mathcal{N}_0}} \right) e^{dE_s/\mathcal{N}_0} \sum_{k=d}^{\infty} a_k \left[e^{-E_s/\mathcal{N}_0} \right]^k \qquad (8.32d)$$

In view of the transfer function of (8.31e), the bound (8.32d) can be expressed as

$$P(e) < Q\left(\sqrt{\frac{2dE_s}{N_0}}\right) e^{dE_s/N_0} T(D)|_{D=e^{-E_s/N_0}} \qquad (8.32e)$$

which is the total first-event error probability bound for soft-decision Viterbi decoding.

8.4.4 Bit-Error Probability Bounds for Soft-Decision Viterbi Decoding

The BER expression for soft-decision decoding is obtained the same way as for the case of the BSC in (8.30d) and is given by

$$P_B(e) < \sum_{k=d}^{\infty} b_k\, P_2(e;\, k) \qquad (8.33a)$$

where the $\{b_k\}$ are the coefficients of

$$\left.\frac{\partial T(D,\, N)}{\partial N}\right|_{N=1} = \sum_{k=d}^{\infty} b_k D^k \qquad (8.33b)$$

By using the same argument we employed in obtaining the $P_B(e)$ for the BSC in (8.30d), we now have a useful but loose upper bound expression for the bit error rate of soft-decision decoding of convolutional codes:

$$P(e) < Q\left(\sqrt{\frac{2dE_s}{N_0}}\right) e^{dE_s/N_0} \left.\frac{\partial T(D,\, N)}{\partial N}\right|_{N=1,D=e^{-E_s/N_0}} \qquad (8.33c)$$

This bound on the BER of soft-decision Viterbi decoding is for a code of rate $1/n$. For the codes with rate $r = k/n$, where $k > 1$, the expression (8.33c) is interpreted as $P_B(e;\, k = 1)$ and must be divided by k to obtain bounds on the average BER. That is:

$$P_B(e;\, k > 1) = \frac{1}{k} P_B(e;\, k = 1)$$

$$< \frac{1}{k} Q\left(\sqrt{\frac{2dE_s}{N_0}}\right) e^{dE_s/N_0} \left.\frac{\partial T(D,\, N)}{\partial N}\right|_{N=1,D=e^{-E_s/N_0}} \qquad (8.33d)$$

Let us summarize the error rate expressions we have derived thus far. For rate k/n convolutional codes with Viterbi decoding, we have

- BSC (hard-decision) with Hamming metric:

First-event error rate:

$$P(e) < T(D)|_{D=\sqrt{4p(1-p)}} \tag{8.34a}$$

Bit-error rate:

$$P_B(e) < \frac{1}{k} \frac{\partial T(D, N)}{\partial N}\bigg|_{N=1,\, D=\sqrt{4p(1-p)}} \tag{8.34b}$$

A tighter bound than (8.34b) can be computed using the binominal expressions of (8.27c) in (8.33a).

- AWGN channel (soft-decision) with correlation metric:

First-event error rate:

$$P(e) < Q\left(\sqrt{\frac{2dr\mathrm{E}_b}{\mathcal{N}_0}}\right) e^{dr\mathrm{E}_b/\mathcal{N}_0}\, T(D)|_{D=e^{-r\mathrm{E}_b/\mathcal{N}_0}} \tag{8.34c}$$

Bit-error rate:

$$P_B(e) < \frac{1}{k} Q\left(\sqrt{\frac{2dr\mathrm{E}_b}{\mathcal{N}_0}}\right) e^{dr\mathrm{E}_b/\mathcal{N}_0}\, \frac{\partial T(D, N)}{\partial N}\bigg|_{N=1,\, D=e^{-r\mathrm{E}_b/\mathcal{N}_0}} \tag{8.34d}$$

where $r = k/n$, p is the symbol error probability on the BSC, d is the minimum[6] free distance, and $\mathrm{E}_b/\mathcal{N}_0$ is the energy per bit-to-noise power spectral density ratio.

Example 8.11 Consider the rate $\frac{1}{3}$, $K = 3$ convolutional code characterized by the connection vectors $G_1 = (1\,0\,0)$, $G_2 = (1\,0\,1)$, and $G_3 = (1\,1\,1)$. (1) Draw the state diagram; (2) find the transfer function $T(D, N)$; (3) obtain the first-event error probability $P(e)$ for all pairs (total first-event error rate) and BER, assuming BPSK symbol modulation transmitted over an AWGN channel with one-sided power spectral density $\mathcal{N}_0$, for hard-decision Viterbi

---

[6] When *free distance* is mentioned without regard to either minimum or maximum, it usually means the *minimum free distance*.

decoding; (4) repeat (3) for soft-decision Viterbi decoding; and (5) plot the BER expressions in comparison with the uncoded system.

Solutions: (1) Based on the connection vectors, we have the encoder and state diagram as shown in Figure 8.45.

(2) The transfer function $T(D, N)$ can be obtained with the aid of the flow graph diagram for the state diagram as shown in Figure 8.46. From the flow graph, we set up the simultaneous equations:

$$X_b = D^3 N + DN X_c$$
$$X_c = DX_b + DX_d \qquad\qquad (8.35a)$$
$$X_d = D^2 N X_b + D^2 N X_d$$

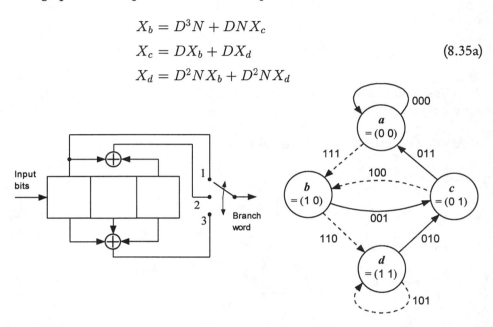

Figure 8.45 Encoder and state diagram for the rate $\frac{1}{3}$, $K = 3$ encoder of Example 8.11.

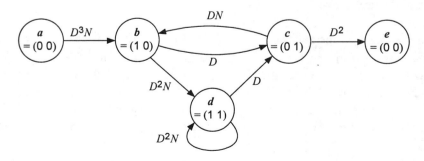

Figure 8.46 Flow graph diagram of the state diagram shown in Figure 8.45, with branches labeled with D and N.

and solve for the transfer function $T(D, N) = D^2 X_c$. Putting (8.35a) in matrix form:

$$\begin{bmatrix} 1 & -DN & 0 \\ -D & 1 & -D \\ -D^2N & 0 & 1-D^2N \end{bmatrix} \begin{bmatrix} X_b \\ X_c \\ X_d \end{bmatrix} = \begin{bmatrix} D^3N \\ 0 \\ 0 \end{bmatrix} \tag{8.35b}$$

we solve for X_c:

$$X_c = \frac{\begin{vmatrix} 1 & D^3N & 0 \\ -D & 0 & -D \\ -D^2N & 0 & 1-D^2N \end{vmatrix}}{\begin{vmatrix} 1 & -DN & 0 \\ -D & 1 & -D \\ -D^2N & 0 & 1-D^2N \end{vmatrix}} = \frac{D^3N(D - D^3N + D^3N)}{(1-D^2N) - DN(D-D^3N+D^3N)}$$

$$= \frac{D^4N}{1 - 2D^2N} \tag{8.35c}$$

so that

$$T(D, N) = D^2 X_c = \frac{D^6N}{1 - 2D^2N}$$

$$= D^6N + 2D^8N^2 + 4D^{10}N^3 + 8D^{12}N^4 + \cdots$$

$$+ 2^l D^{2l+6}N^{l+1} + \cdots \tag{8.35d}$$

from which we obtain

$$T(D) = T(D, N)|_{N=1} = \frac{D^6}{1 - 2D^2} \tag{8.35e}$$

(3) The symbol crossover probability (symbol error probability) of the BSC for binary antipodal modulation is given by

$$p = Q\left(\sqrt{\frac{2E_s}{N_0}}\right) = Q\left(\sqrt{\frac{2E_b}{3N_0}}\right) \tag{8.36a}$$

For the BSC, the first-event error probability is bounded by

$$P_{BSC}(e) < T(D)|_{D=\sqrt{4p(1-p)}} = \frac{[4p(1-p)]^3}{1 - 2[4p(1-p)]} \tag{8.36b}$$

where p is given by (8.36a). The BER bound on the BSC is given by

$$P_B(e) < \frac{\partial T(D, N)}{\partial N}\bigg|_{N=1, D=\sqrt{4p(1-p)}}$$

$$= \frac{\partial}{\partial N}\left(\frac{D^6 N}{1 - 2D^2 N}\right)\bigg|_{N=1, D=\sqrt{4p(1-p)}} = \frac{D^6}{(1 - 2D^2 N)^2}\bigg|_{N=1, D=\sqrt{4p(1-p)}}$$

Thus   (8.36c)

$$P_{B,BSC}(e) < \frac{[4p(1-p)]^3}{[1 - 8p(1-p)]^2} \tag{8.36d}$$

with p given by (8.36a).

(4)   For soft-decision Viterbi decoding, we have the first-event error probability bound

$$P_{AWGN}(e) < Q\left(\sqrt{\frac{2dr E_b}{\mathcal{N}_0}}\right) e^{dr E_b/\mathcal{N}_0} T(D)\big|_{D=e^{-r E_b/\mathcal{N}_0}} \tag{8.37a}$$

where $d = 6$, $r = 1/3$, and $T(D) = D^6/(1 - 2D^2)$. Hence

$$P_{AWGN}(e) < Q\left(\sqrt{\frac{4E_b}{\mathcal{N}_0}}\right) e^{2E_b/\mathcal{N}_0} \cdot \left[\frac{e^{-2E_b/\mathcal{N}_0}}{1 - 2e^{-2E_b/3\mathcal{N}_0}}\right]$$

or

$$P_{AWGN}(e) < Q\left(\sqrt{\frac{4E_b}{\mathcal{N}_0}}\right)\left(\frac{1}{1 - 2e^{-2E_b/3\mathcal{N}_0}}\right) \tag{8.37b}$$

The BER bound for soft-decision decoding is given by

$$P_{B,AWGN}(e) < Q\left(\sqrt{\frac{2dr E_b}{\mathcal{N}_0}}\right) e^{dr E_b/\mathcal{N}_0} \frac{\partial T(D, N)}{\partial N}\bigg|_{N=1, D=e^{-r E_b/\mathcal{N}_0}} \tag{8.37c}$$

where $d = 6$ and $r = 1/3$. Thus

$$P_{B,AWGN}(e) < Q\left(\sqrt{\frac{4E_b}{\mathcal{N}_0}}\right)\left(1 - 2e^{-2E_b/3\mathcal{N}_0}\right)^{-2} \tag{8.37d}$$

(5) The uncoded BPSK error probability is given by

$$P_{B,\text{uncoded}}(e) = Q\left(\sqrt{\frac{2E_b}{\mathcal{N}_0}}\right) \tag{8.38}$$

We have plotted the $P_{B,\text{uncoded}}(e)$, $P_{B,AWGN}(e)$, and $P_{B,BSC}(e)$ expressions in Figure 8.47 for comparison purposes. The looseness of the bound (8.29d) is evident in comparison with the binomial expressions of (8.27c) for the bit-error performance on the BSC. For this example, about 2-dB coding gain is observed using soft decisions, and less than 1 dB for hard decisions.

8.4.5 Estimates of Coding Gains of Convolutional Codes

It occurs at times that we need to have some idea of the amount of coding gain for a convolutional code, specified by rate and constraint length, and the assumed type of decoding (hard- or soft-decision) along with the minimum free distance of the code. To fill this need, we can base an estimate on simplifications of the error bounds that we have developed in the previous section.

For the BER bound on the BSC (8.36c) and from (8.33b) we have

$$P_{B,BSC}(e) < \left.\frac{\partial T(D, N)}{\partial N}\right|_{N=1, D=\sqrt{4p(1-p)}} = \left.\sum_{k=d}^{\infty} b_k D^k\right|_{D=\sqrt{4p(1-p)}}$$

$$= \left(b_d D^d + b_{d+1} D^{d+1} + \cdots\right)_{D=\sqrt{4p(1-p)}} \tag{8.39a}$$

where d is the minimum free distance and p is the symbol crossover probability for the BPSK modulation, given by (8.36a). We wish to simplify further the expression in (8.39a) by assuming that p is a very small number. Then the right-hand side of (8.39a) is dominated by the first term, and, hence, we have

$$P_{B,BSC}(e) < b_d D^d\big|_{D=\sqrt{4p(1-p)}} = b_d\left[\sqrt{4p(1-p)}\right]^d \approx b_d(4p)^{d/2} \tag{8.39b}$$

Replacing p by the Q-function upper bound of (1.45f), that is:

$$p = Q\left(\sqrt{\frac{2rE_b}{\mathcal{N}_0}}\right) < \tfrac{1}{2}e^{-rE_b/\mathcal{N}_0} \tag{8.39c}$$

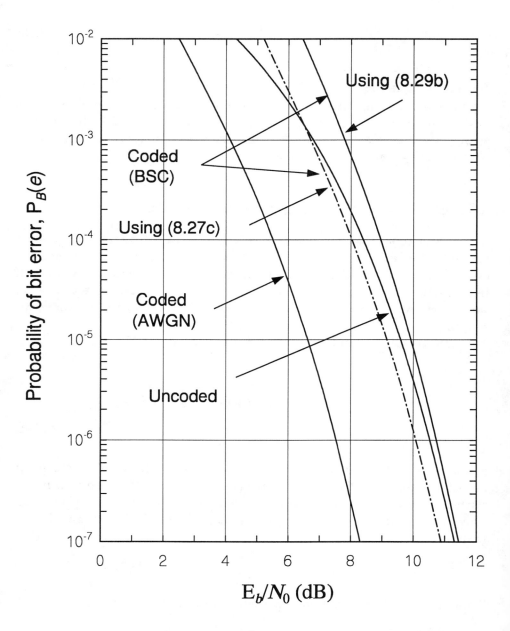

Figure 8.47 Error performance bounds of rate $\frac{1}{3}$, $K = 3$ convolutional code for BSC and AWGN channels.

we finally have the simple but loose upper bound of

$$P_{B,BSC}(e) < b_d \, 2^{d-1} e^{-rdE_b/2\mathcal{N}_0} \tag{8.39d}$$

The uncoded performance is given by

$$Q\left(\sqrt{\frac{2E_b}{\mathcal{N}_0}}\right) < \tfrac{1}{2} e^{-E_b/\mathcal{N}_0} \tag{8.39e}$$

By comparing the exponents of (8.39d) and (8.39e), we see that the coding has an advantage of $rd/2$ over the uncoded system in terms of $E_b/\mathcal{N}_0$ requirements. That is:

$$\Delta\left(\frac{E_b}{\mathcal{N}_0}\right) = \frac{(E_b/\mathcal{N}_0)_{\text{coded}}}{(E_b/\mathcal{N}_0)_{\text{uncoded}}} \approx rd/2 \tag{8.40a}$$

This can be taken as a lower bound on the coding gain [9] in view of the greater coding gain to be expected using soft decisions. Now, from the soft-decision Viterbi decoding bound given by (8.33d), we have

$$P_{B,AWGN}(e) < Q\left(\sqrt{\frac{2drE_b}{\mathcal{N}_0}}\right) e^{drE_b/\mathcal{N}_0} \left.\frac{\partial T(D,\,N)}{\partial N}\right|_{N=1,D=e^{-rE_b/\mathcal{N}_0}}$$

$$\approx Q\left(\sqrt{\frac{2drE_b}{\mathcal{N}_0}}\right) e^{drE_b/\mathcal{N}_0} \left\{ b_d \left[e^{-rE_b/\mathcal{N}_0} \right]^d \right\}$$

$$\approx \tfrac{1}{2} b_d e^{-rdE_b/\mathcal{N}_0} \tag{8.40b}$$

By comparing the exponents of (8.40b) and (8.39e), the coding gain advantage for soft-decision decoding is given by

$$\Delta\left(\frac{E_b}{\mathcal{N}_0}\right) \approx rd \tag{8.40c}$$

From (8.40a) and (8.40c), we draw an estimate of coding gain in terms of the bounds

$$\tfrac{1}{2}rd \le \left(\begin{array}{c} \text{Coding gain of convolutional code} \\ \text{with rate } r \text{ and free distance } d \end{array} \right) \le rd \qquad (8.40d)$$

The upper and lower bounds are separated by a 3-dB margin, which, to a good approximation, is a margin that in fact exists between hard- and soft-decision decoder implementations.

Some examples of coding gain upper bounds for convolutional codes of rate $\tfrac{1}{2}$ and rate $\tfrac{1}{3}$ are given in Table 8.6, where the upper bounds are calculated by the "formula" given in (8.40d) for the given free distance. It is plain that, once we know the free distance d of a given convolutional code of rate r, a coding gain upper bound can be estimated by using the "yardstick" of rd given in (8.40d).

Many researchers in the past discovered good (maximum free distance) convolutional codes of rate r with constraint length K. Their findings are summarized in Tables 8.7 and 8.8, taken from [10] and [11].

8.5 Convolutional Codes Used in the IS-95 CDMA System

From the IS-95 forward link traffic channel diagram in Figure 8.1, we see that the convolutional code used in the forward link is described as rate $\tfrac{1}{2}$ and constraint length $K = 9$. IS-95 [1] specifies the encoder for this code as shown in Figure 8.48. From the figure it is seen that the connection vectors for the encoder are

Table 8.6 Upper bounds on coding gain for some convolutional codes (from [9])

Rate 1/2 codes			Rate 1/3 codes		
K	d	Upper bound (dB)	K	d	Upper bound (dB)
3	5	3.97	3	8	4.26
4	6	4.76	4	10	5.23
5	7	5.43	5	12	6.02
6	8	6.00	6	13	6.37
7	10	6.99	7	15	6.99
8	10	6.99	8	16	7.27
9	12	7.78	9	18	7.78

Table 8.7 Distance properties of rate $\frac{1}{2}$ binary convolutional codes (from [10] and [11])

K	Connection vectors in octal		d
3	5	7	5
4	15	17	6
5	13	35	7
6	53	75	8
7	133	171	10
8	247	373	10
9	561	753	12
10	1167	1545	12
11	2335	3663	14
12	4335	5723	15
13	10533	17661	16
14	21765	27123	16

Table 8.8 Distance properties of rate $\frac{1}{3}$ binary convolutional codes (from [10] and [11])

K	Connection vectors in octal			d
3	5	7	7	8
4	13	15	17	10
5	25	33	37	12
6	47	53	75	13
7	133	145	175	15
8	225	331	367	16
9	557	663	711	18
10	1117	1365	1633	20
11	2355	2671	3175	22
12	4767	5723	6265	24
13	10533	10675	17661	24
14	21645	35661	37133	26

$$G_1 = (1\,1\,1\,1\,0\,1\,0\,1\,1) = (753)$$

and

$$G_2 = (1\,0\,1\,1\,1\,0\,0\,0\,1) = (561)$$

(8.41a)

The number of nodes (states) is $2^{K-1} = 2^8 = 256$ and, from Table 8.7, the octal notations for G_1 and G_2 indicate that the free distance of this particular convolutional code is

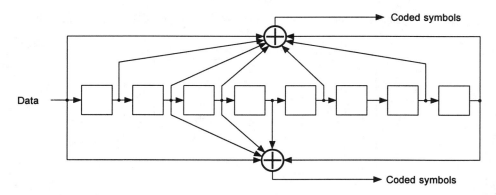

Figure 8.48 Convolutional encoder for the forward link: $r = \frac{1}{2}$, $K = 9$.

$$d_{\text{forward}} = 12 \qquad\qquad (8.41b)$$

which implies that a rough estimate of the coding gain is given by

$$\tfrac{1}{2}rd_{\text{forward}} = 4.8\,\text{dB} \leq \text{coding gain} \leq rd_{\text{forward}} = 7.8\,\text{dB} \qquad (8.41c)$$

For the IS-95 reverse link traffic channel diagram in Figure 8.2, we see that the convolutional code is rate $\frac{1}{3}$ and constraint length $K = 9$, and the encoder circuit given for this code in IS-95 is shown in Figure 8.49. The connection vectors are given by

$$\begin{aligned}
G_1 &= (1\,0\,1\,1\,0\,1\,1\,1\,1) = (557) \\
G_2 &= (1\,1\,0\,1\,1\,0\,0\,1\,1) = (663) \qquad\qquad (8.42a) \\
G_3 &= (1\,1\,1\,0\,0\,1\,0\,0\,1) = (711)
\end{aligned}$$

The number of nodes (states) in the trellis diagram is the same as that of the forward link encoder, $2^{K-1} = 2^8 = 256$. From Table 8.8, we find the octal notations for G_1, G_2, and G_3 listed for a free distance of

$$d_{\text{reverse}} = 18 \qquad\qquad (8.42b)$$

which implies that a loose estimate of the coding gain for this code is given by

$$\tfrac{1}{2}rd_{\text{forward}} = 4.8\,\text{dB} \leq \text{Coding Gain} \leq rd_{\text{forward}} = 7.8\,\text{dB} \qquad (8.42c)$$

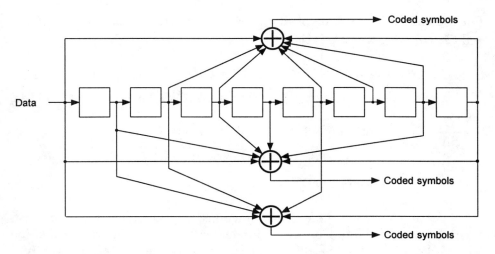

Figure 8.49 Convolutional encoder for the reverse link: $r = \frac{1}{3}$, $K = 9$.

Note that the coding gains estimated on the basis of the formula (8.40d) could not resolve the coding gain difference that really exists between the rate $\frac{1}{2}$ and rate $\frac{1}{3}$ convolutional codes of the same constraint length 9. The difference is noticed when we actually evaluate the performance bounds on the basis of the transfer functions, which we undertake in the next section.

8.5.1 Performance of the Convolutional Codes Used in the IS-95 System

For the convolutional code of $r = \frac{1}{2}$, $K = 9$ with $G_1 = (561)$ and $G_2 = (753)$, the coefficients of the generating function (transfer function) have been found through computer search and they are tabulated in the literature [12]:

$$T(D) = 11D^{12} + 50D^{14} + 286D^{16} + 1{,}630D^{18} + 9{,}639D^{20}$$
$$+ 55{,}152D^{22} + 320{,}782D^{24} + 1{,}859{,}184D^{26}$$
$$+ 10{,}777{,}264D^{28} + \cdots \tag{8.43a}$$

$$\left.\frac{\partial T(D,N)}{\partial N}\right|_{N=1} = 33D^{12} + 281D^{14} + 2{,}179D^{16} + 15{,}035D^{18} + 105{,}166D^{20}$$
$$+ 692{,}330D^{22} + 4{,}580{,}007D^{24} + 29{,}692{,}894D^{26}$$
$$+ 190{,}453{,}145D^{28} + \cdots \tag{8.43b}$$

For the convolutional code of rate $\frac{1}{3}$, $K = 9$ with $\boldsymbol{G_1} = (557)$, $\boldsymbol{G_2} = (663)$, and $\boldsymbol{G_3} = (711)$, we have [12]

$$T(D) = 5D^{18} + 7D^{20} + 36D^{22} + 85D^{24} + 204D^{26} + 636D^{28}$$
$$+ 1{,}927D^{30} + 5{,}416D^{32} + 15{,}769D^{34} + \cdots \tag{8.44a}$$

$$\left.\frac{\partial T(D,N)}{\partial N}\right|_{N=1} = 11D^{18} + 32D^{20} + 195D^{22} + 564D^{24} + 1{,}473D^{26}$$
$$+ 5{,}129D^{28} + 17{,}434D^{30} + 54{,}092D^{32}$$
$$+ 171{,}117D^{34} + \cdots \tag{8.44b}$$

Because we have the transfer functions, we can now obtain the first-event error probabilities or BERs for either hard-decision or soft-decision Viterbi decoding. For the BER bounds, we simply use (8.34b) for the BSC and (8.34d) for the AWGN channel to obtain

$$P_{B,BSC}(e) < \left.\frac{\partial T(D,N)}{\partial N}\right|_{N=1,D=\sqrt{4p(1-p)}} \tag{8.45a}$$

with
$$p = Q\left(\sqrt{\frac{2rE_b}{\mathcal{N}_0}}\right) \tag{8.45b}$$

and

$$P_{B,AWGN}(e) < Q\left(\sqrt{\frac{2drE_b}{\mathcal{N}_0}}\right) e^{drE_b/\mathcal{N}_0} \left.\frac{\partial T(D,N)}{\partial N}\right|_{N=1,D=e^{-rE_b/\mathcal{N}_0}} \tag{8.45c}$$

where the derivative of the transfer function is given in (8.43b) for the rate $\frac{1}{2}$ code and in (8.44b) for the rate $\frac{1}{3}$ code. BER hard- and soft-decision performance bounds are compared with each other and with the uncoded BER in Figures 8.50 and 8.51 for these two codes, respectively. Also shown in both figures (using a dashed line) is the tighter BSC bound calculated using the binomial expression of (8.33a) and (8.27c), which is seen to be about one-half dB tighter in this case. Note from Figures 8.50 and 8.51 that the rate $\frac{1}{3}$ code provides more coding gain as compared with the rate $\frac{1}{2}$ code. For example, at a BER of 10^{-7}, we read $(E_b/\mathcal{N}_0)_{r=1/3} \approx 4.4\,\text{dB}$ and $(E_b/\mathcal{N}_0)_{r=1/2} \approx 4.6\,\text{dB}$. The fractional coding gain advantage in favor of the rate $\frac{1}{3}$ code over the rate $\frac{1}{2}$ code was not revealed in the rough coding gain estimation measure given in (8.40d).

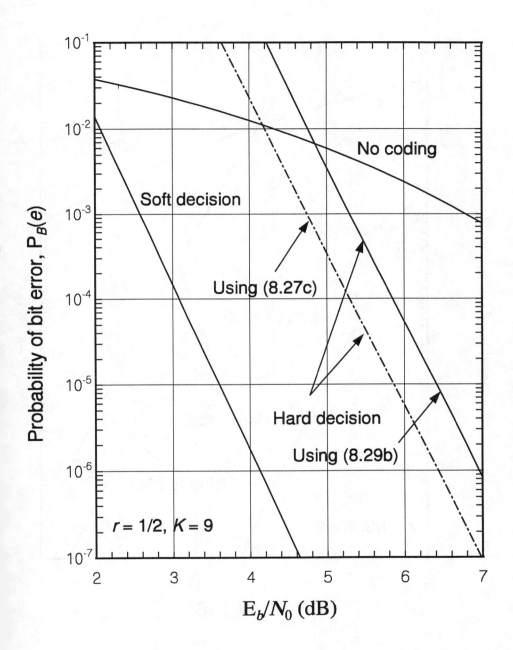

Figure 8.50 Performance for the forward link $r = \frac{1}{2}$**,** $K = 9$ **code.**

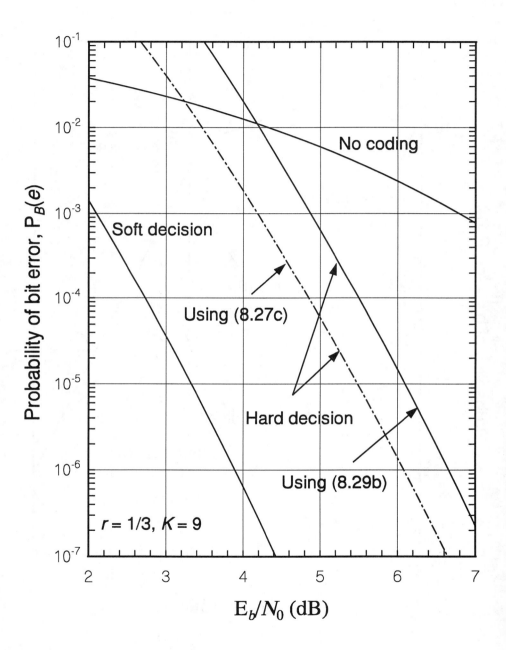

Figure 8.51 Performance for the reverse link $r = \frac{1}{3}$, $K = 9$ code.

8.5.2 Coding Gains Versus Constraint Length

In this chapter we have introduced the fundamentals of convolutional codes in terms of code generation, decoding concept of the Viterbi algorithm, and ways of evaluating error probability performance bounds. While the upper bounds for the BSC and the AWGN channel are reasonable performance estimates for large values of E_b/N_0, computer simulations must be used to obtain the results over the range of E_b/N_0 that is of interest. There are extensive simulation results published in the literature, and the work of Heller and Jacobs [13] is a frequently cited reference in this regard.

The convolutional codes used in the IS-95 system are noteworthy in that they are convolutional codes of large constraint length. Although large constraint-length convolutional codes increase complexity in terms of memory requirements in the Viterbi decoding process, as noted in our previous discussion, they provide higher coding gains for a given code rate.

Figures 8.52 and 8.53 show coding gains realizable at different error probabilities as functions of constraint length for rate $\frac{1}{2}$ and rate $\frac{1}{3}$ convolutional codes, respectively, for both soft-decision (solid lines) and hard-decision (dashed lines) Viterbi decoding [14–16].[7] The curves in each figure are parametric in the value of the error probability achieved: 10^{-3} down to 10^{-8} in decade steps. The most obvious feature of these figures is the fact that soft-decision decoding realizes about 2.3 dB or 2.4 dB more coding gain than hard-decision decoding. The next thing we notice, by comparing the two figures, is that generally the rate $\frac{1}{3}$ codes have higher gains than the rate $\frac{1}{2}$ codes (higher redundancy gives higher coding gain). Finally, we observe from the two figures that the coding gains are higher for lower error probabilities.

For cellular telephone applications in which digital voice is used, the curves in Figures 8.52 and 8.53 for higher values of error probability are of interest. If, as a rule of thumb, we pick a corrected digital voice BER of 10^{-3} as a target value, it is apparent that, with soft-decision decoding of the IS-95 convolutional codes, a coding gain on the order of 5 or 6 dB can be realized in practice. That is, using BPSK as a quick reference, instead of requiring an E_b/N_0 of about 7 dB (unfaded) or 24 dB (Rayleigh fading) to obtain the target BER (see Figure 7.34), an E_b/N_0 of 2 dB or 19 dB is required, respectively. The reduction in required E_b/N_0 directly affects the multiple access capacity of the system, as discussed in Section 1.1.3.

---

[7] The data in the figures are based on reading curves in [16] that are reprinted from [14]. For each error probability value, the coding gain is the difference between the E_b/N_0 required without coding and with coding.

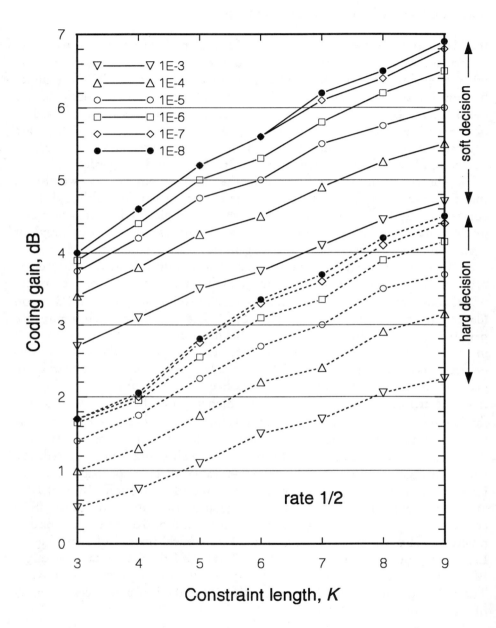

Figure 8.52 Dependence of realizable coding gains on constraint length for rate $\frac{1}{2}$ convolutional codes.

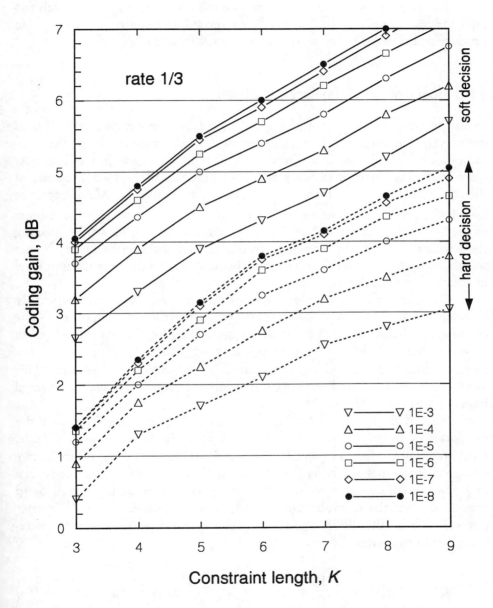

Figure 8.53 Dependence of realizable coding gains on constraint length for rate $\frac{1}{3}$ convolutional codes.

The required value of $E_b/\mathcal{N}_0$, despite the coding gain, is still much too high for case of fading. In Chapter 9, we show how the use of diversity can produce additional gain for reducing the required value of $E_b/\mathcal{N}_0$.

8.5.3 Quantization of the Received Signal

In our previous discussions, we have considered the two extreme cases of hard (two-level) and infinitely fine quantizations for the conversion of the continuous decision RV into a discrete RV for digital signal processing. As observed in the error probability curves of Figures 8.50 and 8.51, the use of the BSC results in the performance penalty of 2 to 3 dB of $E_b/\mathcal{N}_0$. Hence, the use of soft decisions (quantization level $\triangleq Q > 2$) is employed to regain most of the 2- to 3-dB loss due to hard ($Q = 2$) quantization [13, 16] while retaining the advantage of digital processing. In this section we describe briefly how a soft-decision ($Q > 2$) signal processing scheme relates to the quantization level in connection with the trellis diagram of a Viterbi decoder.

Consider a situation where the information sequence is convolutionally encoded by an encoder with rate $r = 1/n$ and constraint length K. The coded symbols are assumed to be BPSK modulated for transmission over an AWGN channel with a two-sided power spectral density of $\frac{1}{2}\mathcal{N}_0$. At the receiver, the demodulated signal is quantized for digital processing. This situation is depicted in Figure 8.54 in both waveform and vector channel notations, based on Figure 7.2.

Note that the set of "messages" $\{m_i\}$ in Figure 8.54 represents the categories of branch words. For example, in a rate $\frac{1}{2}$ convolutional code, $\{m_i\}$ represents the four words $\{00, 01, 10, 11\}$. For our example, we assume that the message m_i is one of 2^n possible messages (branch words) in a rate $1/n$ convolutional encoder. The binary symbol outputs (branch word sequence) from the convolutional encoder are modulated by an RF carrier $\cos\omega_0 t$. Thus, the BPSK modulated branch word waveform $s(t)$ corresponding to m is given by

$$s(t) = \sqrt{\frac{2E_s}{T_s}} \cos\omega_0 t \cdot \sum_{j=1}^{n} x_j \, p(t - jT_s) \tag{8.46a}$$

where x_j is ± 1 depending on whether the jth code symbol is 0 or 1. The function $p(t)$ is supposed to represent a unit energy "basic" waveform,

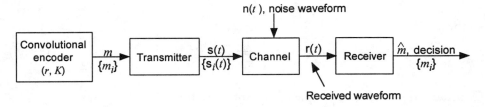

(a) Waveform communications system

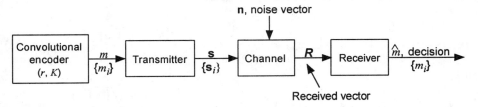

(b) Vector communications system

Figure 8.54 Representations of waveform and vector communications systems with coding.

$\omega_0 = 2\pi f_0$ is the angular carrier frequency, E_s is the code symbol energy, and T_s is the code symbol time duration.

The receiver in Figure 8.54 contains the demodulator (correlator) and quantizer as shown in Figure 8.55. It is clear that the RV r_j shown in Figure 8.55 is given by

$$r_j = x_j\sqrt{E_s} + n_j$$

$$= \begin{cases} \sqrt{E_s} + n_j & \text{if } x_j = \text{"0"} \\ -\sqrt{E_s} + n_j & \text{if } x_j = \text{"1"} \end{cases} \qquad (8.46b)$$

where

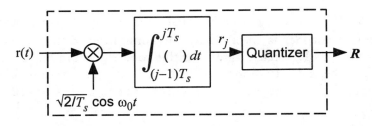

Figure 8.55 BPSK demodulator followed by quantizer.

$$n_j = G\left(0, \tfrac{1}{2}\mathcal{N}_0\right) \tag{8.46c}$$

so that

$$p_{n_j}(\alpha) = \frac{1}{\sqrt{\pi\mathcal{N}_0}}\, e^{-\alpha^2/\mathcal{N}_0} \tag{8.46d}$$

Consider the uniformly spaced quantizers for $Q = 2$, 4, and 8. The likelihood functions for the demodulator output RV r_j, given the transmitted code symbol is 0 and 1, respectively, are given by

$$p_{r_j}(\alpha \mid 0) = p_{r_j}\left(\alpha \mid s = \sqrt{E_s}\right) = \frac{1}{\sqrt{\pi\mathcal{N}_0}}\exp\left\{-\frac{\left(\alpha - \sqrt{E_s}\right)^2}{\mathcal{N}_0}\right\} \tag{8.47a}$$

and

$$p_{r_j}(\alpha \mid 1) = p_{r_j}\left(\alpha \mid s = -\sqrt{E_s}\right) = \frac{1}{\sqrt{\pi\mathcal{N}_0}}\exp\left\{-\frac{\left(\alpha + \sqrt{E_s}\right)^2}{\mathcal{N}_0}\right\} \tag{8.47b}$$

Therefore, we can relate the likelihood functions to the quantization levels in terms of uniformly spaced thresholds as depicted in Figure 8.56 or alternatively, in Figure 8.57 with metric assignments as indicated in Figure 8.56 for various quantization levels. In that figure, it is assumed that the receiver gain is adjusted so as to position the sampled received signal amplitude as shown. In Figure 8.56, the Q-level quantizer is shown as a partition of the r_j axis into Q segments, and the metrics (quantized values) are ordered so that the indication of zero being transmitted is associated with the smaller metric numbers, and the large numbers indicate that the likelihood of a zero is small. The metrics conditioned on the transmission of a logical one are simply the reverse of the values shown, as indicated in Figures 8.58 and 8.59.

We are now ready to explain the computation of the metric for a specific branch word in a convolutional code of rate $1/n$ under a soft-decision decoding scheme for $Q = 8$.

Example 8.12 Consider a convolutional code of rate $\frac{1}{2}$, constraint length $K = 3$, whose encoder is specified by the connection vectors $G_1 = (1\,0\,1)$ and $G_2 = (1\,1\,1)$. The state diagram for this code is given in Figure 8.60. We assume that the soft decision is based on 3-bit quantization (eight-level quantization). The decoder is now assumed to be at node $b = (10)$ and the

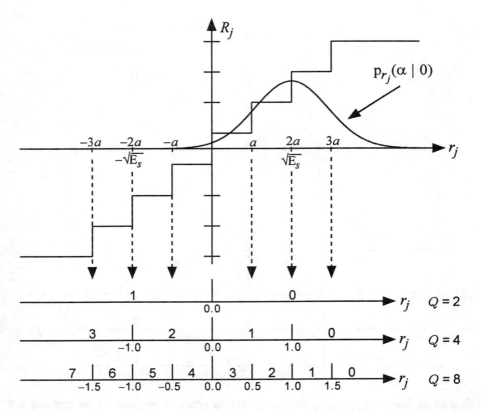

Figure 8.56 Uniformly spaced Q-level quantizer with metrics conditioned on the transmission of logic symbol O.

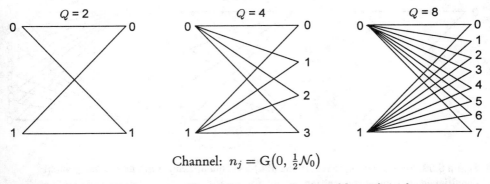

Channel: $n_j = G\left(0, \tfrac{1}{2}\mathcal{N}_0\right)$

Figure 8.57 Alternative representations of Q-level quantization with metric assignment conditioned on O being transmitted.

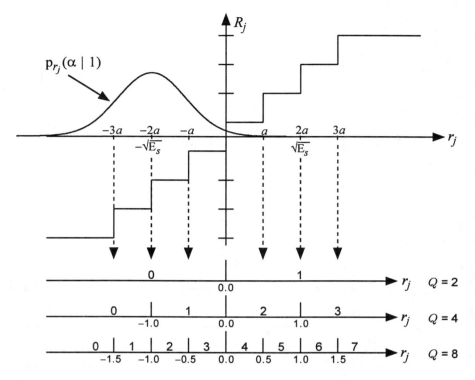

Figure 8.58 Uniformly spaced Q-level quantizer with metrics conditioned on the transmission of logic symbol 1.

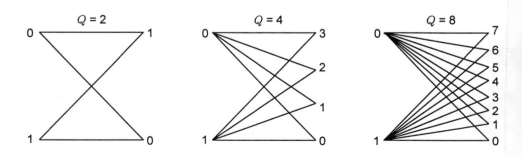

Figure 8.59 Alternative representations of Q-level quantization with metric assignment conditioned on 1 being transmitted.

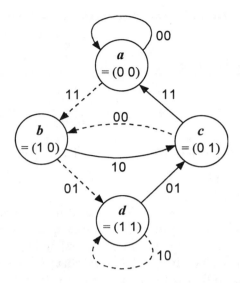

Figure 8.60 State diagram for rate $r = \frac{1}{2}$, constraint length $K = 3$ convolutional code used in Example 8.12.

two branch symbols are given by $r_j = 1.3$ and $r_{j+1} = 0.6$. What would be the next state the decoder will be arriving at, based on these values?

To answer this question, we need to examine the metric values of the two combined symbols under each symbol transmission assumption. Let M_0 be the branch metric associated with the assumption of the transmission of a 0 and let M_1 be the metric associated with the transmission of 1. From the state diagram, we see that the state $b = (10)$ can produce either branch words 01 or 10, the former for the next state $d = (01)$ and the latter for the next state $c = (11)$. Thus, the task is to determine which branch word is more likely to have been transmitted. Because the state transition is determined by the relationship between the information bit and the branch words as $0 \rightarrow 10$ and $1 \rightarrow 01$, we simply calculate the total metric values from Figures 8.56 and 8.57 as shown in Table 8.9. For example, the sample value $r_j = 1.3$ has a metric value of 1 conditioned on there being a 0 transmitted (Figure 8.56) and a metric value of 6 conditioned on there being a 1 transmitted (Figure 8.57). The reader can easily verify that, having reached the state $c = (01)$, for which the possible branch words are 11 and 00 (for bit values 0 and 1, respectively), if the next two quantizer outputs are assumed to be $r_{j+2} = -0.6$ and $r_{j+3} = 0.4$, we have

Table 8.9 Metric calculations for $Q = 8$ quantizer

Sample values	$r_j = 1.3,\ r_{j+1} = 0.6$	
Possible branch words	0 1	1 0
Next states	$d = (0\,1)$	$c = (1\,1)$
Metric sums	$M_0 = 1 + 5 = 6$	$M_1 = 6 + 2 = 8$
Decision	Because $M_0 < M_1$, 0 1 is more likely than 1 0	

$$M_0 = \text{metric}(-0.6 \mid 1) + \text{metric}(0.4 \mid 1) = 2 + 4 = 6$$
$$M_1 = \text{metric}(-0.6 \mid 0) + \text{metric}(0.4 \mid 0) = 5 + 3 = 8$$

so the next state would be $a = (0\,0)$ by deciding the branch word was 1 1.

As shown in this example, a simple quantizer can be employed in connection with the Viterbi decoding process, and such a scheme is called *soft-decision Viterbi decoding*. It is known [13] that eight-level quantization results in a loss of less than 0.25 dB as compared to infinitely fine quantization, and thus quantization to more than eight levels (3-bit quantization) produces little performance improvement.

References

[1]   "Mobile Station-Base Station Compatibility Standard for Dual-Mode Wideband Spread Spectrum Cellular System," TIA/EIA Interim Standard 95 (IS-95), Washington, D.C.: Telecommunications Industry Association, July 1993 (amended as IS-95-A in May 1995).

[2]   Shannon, C. E., "A Mathematical Theory of Communication," *Bell System Technical Journal*, Vol. 27, pp. 379–433 (Part I) and pp. 623–656 (Part II). See also *Claude E. Shannon: Collected Papers*, edited by N. A. Sloane and A. D. Wyner, IEEE Press, 1992.

[3]   Viterbi, A. J., and J. K. Omura, *Principles of Digital Communications and Coding*, New York: McGraw-Hill, 1979.

[4]   Lee, J. S., "Overview of the Technical Basis of Qualcomm's CDMA Cellular Telephone System Design–A View of North American TIA/EIA IS-95" (Invited paper), *Proc. 1994 IEEE International Conference on Communication Systems* (ICCS'94), pp. 353–358, Singapore: Nov. 14–18, 1994.

[5] Elias, P., "Error-Free Coding," *IRE Trans. on Information Theory*, Vol. IT-4, pp. 29–37, 1954.

[6] Viterbi, A. J., "Error Bounds for Convolutional Codes and an Asymptotically Optimum Decoding Algorithm," *IEEE Trans. on Information Theory*, Vol. IT-13, pp. 260–269, Apr. 1967.

[7] Omura, J. K., "On the Viterbi Decoding Algorithm," *IEEE Trans. on Information Theory*, Vol. IT-15, pp. 177–179, 1969.

[8] Viterbi, A. J., "Convolutional Codes and Their Performance in Communication Systems," *IEEE Trans. on Information Theory*, Vol. IT-17, pp. 751–772, Oct. 1971.

[9] Bhargava, V. K., D. Haccoun, R. Matyas, and P. Nuspl, *Digital Communications by Satellite*, New York: Wiley, 1981.

[10] Odenwalder, J. P., "Optimal Decoding of Convolutional Codes," Ph.D. Dissertation, Department of Systems Science, University of California, Los Angeles, 1970.

[11] Larsen, K. J., "Short Convolutional Codes with Maximum Free Distance for Rate 1/2, 1/3 and 1/4," *IEEE Trans. on Information Theory*, Vol. IT-19, pp. 371–372, May 1973.

[12] Conan, J., "The Weight Spectra of Some Short Low-Rate Convolutional Codes," *IEEE Trans. on Communications*, Vol. COM-32, pp. 1050–1053, Sept. 1984.

[13] Heller, J. A., and I. M. Jacobs, "Viterbi Decoding for Satellite and Space Communication," *IEEE Trans. on Communication Technology*, Vol. COM-19, pp. 835–848, Oct. 1971.

[14] Ziemer, R. E., and R. L. Peterson, *Introduction to Digital Communications*, New York: Macmillan, 1992.

[15] Michelson, A. M., and A. H. Levesque, *Error-Control Techniques for Digital Communication*, New York: Wiley-Interscience, 1985.

[16] Simon, M. K., S. M. Hinedi, and W. C. Lindsey, *Digital Communication Techniques: Signal Design and Detection*, Englewood Cliffs, NJ: Prentice Hall, 1995.

Selected Bibliography

Bahl, L. R., C. D. Cullum, W. D. Frazer, and F. Jelinek, "An Efficient Algorithm for Computing Free Distance," *IEEE Trans. on Information Theory*, Vol. IT-18, pp. 437–439, May 1972.

Begin, G., D. Haccoun, and C. Paquin, "Further Results on High-Rate Punctured Convolutional Codes for Viterbi and Sequential Decoding," *IEEE Trans. on Communication*, Vol. 38, pp. 1922–1926, Nov. 1990.

Bucher, E. A., and J. A. Heller, "Error Probability Bounds for Systematic Convolutional Codes," *IEEE Trans. on Information Theory*, Vol. IT-16, pp. 219–224, Mar. 1970.

Clark, G. C., Jr. and J. B. Cain, *Error-Correction Coding for Digital Communications*, New York: Plenum Press, 1981.

Costello, D. J., Jr., "Free Distance Bounds for Convolutional Codes," *IEEE Trans. on Information Theory*, Vol. IT-20, pp. 356–365, May 1974.

Forney, G. D., Jr., "Convolutional Codes I: Algebraic Structure," *IEEE Trans. on Information Theory*, Vol. IT-16, pp. 720–738, Nov. 1970.

Forney, G. D., Jr., "Convolutional Codes II: Maximum Likelihood Decoding," *Information and Control*, Vol. 25, pp. 222–266.

Forney, G. D., Jr., "Convolutional Codes III: Sequential Decoding," *Information and Control*, Vol. 25, pp. 267–297.

Heller, J. A., "Feedback Decoding of Convolutional Codes," *Advances in Communication Systems* (pp. 261–278), New York: Academic Press, 1975.

Hemmati, F., and D. J. Costello, Jr., "Truncation Error Probability in Viterbi Decoding," *IEEE Trans. on Communications*, Vol. COM-25, pp. 530–532, May 1977.

Johannesson, R., "Some Rate 1/3 and 1/4 Binary Codes With an Optimum Distance Profile," *IEEE Trans. on Information Theory*, Vol. IT-23, pp. 281–283, Mar. 1977.

Lin, S., and D. J. Costello, Jr., *Error Control Coding: Fundamentals and Applications*, Englewood Cliffs, NJ: Prentice Hall, 1983.

Peterson, W. W., and E. J. Weldon, Jr., *Error-Correcting Codes*, Cambridge, MA: MIT Press, 1991.

Post, K. A., "Explicit Evaluation of Viterbi's Union Bounds on Convolutional Code Performance for the Binary Symmetric Channel," *IEEE Trans. on Information Theory*, Vol. IT-23, pp. 403–404, May 1977.

Schalkwijk, J. P. M., K. A. Post, and J. P. J. C. Aarts, "On a Method of Calculating the Event Error Probability of Convolutional Codes With Maximum Likelihood Decoding," *IEEE Trans. on Information Theory*, Vol. IT-25, pp. 737–743, Nov. 1979.

van de Meeberg, L., "A Tightened Upper Bound on the Error Probability of Binary Convolutional Codes With Viterbi Decoding," *IEEE Trans. on Information Theory*, Vol. IT-20, pp. 389–391, May 1974.

Appendix 8A Proof of Q-Function Inequality

In the development of the upper bound for the decoded error rate of convolutional codes with soft-decision decoding, the following inequality is used:

$$Q(\sqrt{x+y}) \leq e^{-y/2} Q(\sqrt{x}) \text{ for } x \geq 0, \ y \geq 0 \qquad (8A.1)$$

This inequality is shown as follows:

$$Q(\sqrt{x+y}) \triangleq \int_{\sqrt{x+y}}^{\infty} \frac{1}{\sqrt{2\pi}} e^{-u^2/2} \, du \qquad (8A.2a)$$

$$= \int_{(x+y)/2}^{\infty} \frac{1}{2\sqrt{\pi t}} e^{-t} \, dt \quad \text{using } t = u^2/2$$

$$= \frac{1}{2\sqrt{\pi}} \int_{x/2}^{\infty} \frac{1}{\sqrt{\xi + y/2}} e^{-(\xi + y/2)} \, d\xi \quad \text{using } t = \xi + y/2$$

$$= \frac{e^{-y/2}}{\sqrt{2\pi}} \int_{x/2}^{\infty} \frac{1}{\sqrt{2\xi + y}} e^{-\xi} \, d\xi \leq \frac{e^{-y/2}}{\sqrt{2\pi}} \int_{x/2}^{\infty} \frac{1}{\sqrt{2\xi}} e^{-\xi} \, d\xi$$

$$= \frac{e^{-y/2}}{\sqrt{2\pi}} \int_{\sqrt{x}}^{\infty} e^{-u^2/2} \, du = e^{-y/2} Q(\sqrt{x}) \quad \text{using } \xi = \tfrac{1}{2}u^2 \qquad (8A.2b)$$

9

Diversity Techniques and Rake Processing

9.1 Introduction

In Chapter 2 we showed that signals in terrestrial communications systems, such as mobile cellular systems, experience multipath interference and fading. This fading can become quite severe in urban environments. Mobiles in motion also experience fading due to the Doppler spreading of various multipath components. If the fading causes many channel errors, methods to combat the fading must be used. When the fading is very slow, some type of feedback power control can be used to track the fades, as is done in the IS-95 system in the form of closed-loop power control (see Section 4.4.1). When the fading becomes fast, however, as with high mobile speeds, power control becomes ineffective.

When different multipath components fade independently, diversity reception of the signal is the method of choice. The rationale for this choice is that if p is the probability that a given path in a multipath environment is below a detection threshold, then the probability is p^L that all L paths in an L-path multipath situation are below the threshold, a number considerably smaller than p. The cost for diversity reception is additional receiver complexity, because of path tracking and the processing of additional signal components.

In this chapter, we describe and analyze several types of diversity techniques, including selection diversity (SD), equal gain combining (EGC), and maximal-ratio combining (MRC). We also discuss Rake diversity techniques in terms of its principle of operation and its application in the IS-95 system.

As discussed in detail in Section 4.4.1, power control is used in IS-95 primarily to ensure that all mobile signals are received by the base station antennas with the same power regardless of the differences of distance between the base station and the mobile stations, and this power control combats slow fading in the process. To be effective against fading, the power control scheme must be able to track the changes in the receiver power at the cell site. The desired value of average mobile transmit power on the reverse link can be determined at the mobile site by solving the link budget using the receiver forward link power at the mobile. This is the essence of the open-loop power control (see Section 4.4.1.1). However, because the forward and reverse links can fade independently, a closed-loop power control scheme must be used (see Section 4.4.1.2), wherein the cell receiver measures the power received from the mobile and instructs the mobile to raise or lower its transmit power.

9.2 Diversity Techniques

There are many different types of diversity that have been employed in various systems. Among these are

- *Frequency diversity*, transmitting or receiving the signal at different frequencies;

- *Time diversity*, transmitting or receiving the signal at different times;

- *Space diversity*, transmitting or receiving the signal at different locations;

- *Polarization diversity*, transmitting or receiving the signal with different polarizations.

Transmission of the same data on two frequencies spaced farther apart than the coherence bandwidth of the channel is one method of frequency diversity. This type of frequency diversity, however, is rarely used in conventional mobile systems because the coherence bandwidth is large. Frequency diversity is inherent in spread-spectrum systems, where the chip rate is greater than the coherence bandwidth.

Time diversity involves the transmission of redundant data at different times because fades are independent with respect to time. The simplest form of time diversity is repeating the data n times, which can be thought of as an $(n, 1)$ block code. A more efficient and practical method of time diversity for digital spread-spectrum systems is the use of interleaving (see Section 4.4.2) and error-correcting codes, as employed in the IS-95 system. The convolutional codes used in IS-95 are much superior to the repetition code. Fading tends to cause errors in bursts corresponding to the faded portion of the signal. Interleaving and deinterleaving are mechanisms by which a burst of errors can be randomized so that the random error-correcting capabilities of the convolutional codes can be exploited to the fullest extent.

Space diversity can be used in many different forms. In transmission space diversity, the signal can be emitted from multiple antennas at a single cell site or from several different cell sites as well. Space diversity can also take the form of tracking multipath components (path diversity), as explained in detail in Section 4.4.3 in connection with the IS-95 system's use of diversity in handoff situations.

In discussing the use of diversity methods in communications systems, it is useful to distinguish between the underlying physical principle of the diversity (time, frequency, space, etc.) and the receiver signal-processing techniques implemented to exploit the principle. In general, the physical principle is identified with communication resources (time, bandwidth, etc.). The signal-processing technique is designed to combine the diversity receptions in some way, and thus they can be referred to in general as *combining techniques*, as suggested in Figure 9.1. To a large extent, given that the diversity components of the signal have been received, the choice of combining technique may be any one of the possible techniques. However, there is often an engineering advantage to a particular combining technique in connection with a particular form of diversity.

It is not unusual for a communications system to employ multiple forms of diversity. As we stated in Chapter 4, the IS-95 system employs frequency, time, and space diversity schemes, all for the purpose of power control in the sense of minimizing the required amount of transmitter power to maintain a desired error rate. The use of polarization diversity, though not currently in use in IS-95, is certainly a future possibility to be considered for signal reception enhancement to combat fading in an application in which the orientation of the receiver antennas can be managed.

		Diversity type				
		A	B	C	D	···
Combining technique	X					
	Y					
	Z					
	⋮					

Figure 9.1 Matrix of diversity types and diversity combining techniques.

9.3 Diversity Selection and Combining Techniques

In a broad sense, there are three major generic diversity signal-processing techniques, and they are known as *selection diversity (SD)*, *equal gain combining (EGC)*, and *maximal ratio combining (MRC)*. The Rake technique of diversity reception, to be discussed in detail in Section 9.3.5, can be viewed as a form of either EGC or MRC, depending on its implementation. The version of Rake diversity reception employed in the IS-95 system is a form of MRC in that the signal multipath components selected by the "searcher receiver" and processed by separate receiver "fingers," are combined based on the maximal ratio combining principle. In this section, we go into the details of diversity selection and combining techniques, and methods for optimizing them.

9.3.1 Selection Diversity

Consider an L-fold diversity transmission system, which can use time diversity, frequency diversity, or space diversity. Regardless of the type of diversity, there must be L receivers in the sense that each of the L transmissions must be received separately. In the case of an L-fold time-diversity transmission, the receiver can be a single receiver used repeatedly L times. For frequency- and space-diversity transmissions, L separate receivers must be used.

In a selection diversity system, the signal diversity component (e.g., path) received with the highest SNR is chosen for signal demodulation, and the remaining $L - 1$ components are ignored. The selection of the signal component may be *predetection* or *postdetection*, meaning that the selection may be among the L RF receptions before receiver processing or among them after they have been demodulated to become baseband data streams. In the case of selection diversity, there is no difference in performance between predetection and postdetection processing because only one demodulated data stream is used, so that the degree of processing of each component before selection is a matter of engineering choice and expediency.[1] Some processing of the components is necessary to determine which one to select, such as determining their strengths.

---

[1] A possible diversity processing scenario is the selection of more than one signal component for processing when there are L components received.

The use of L-fold diversity is a time-honored method of combatting deep signal fades, and there are numerous classical references in this field, [1–4] just to cite a few. The problem of optimal diversity has also been treated in the literature for various types of diversity [4–7]. The applications of diversity schemes in mobile communication are also treated in the literature [8–9], and the IS-95 system has implemented several of these diversity strategies, including selection diversity. For example, selection diversity is used on the reverse link to select the strongest forward link signals from multiple cell sites.

In selection diversity, the diversity channel (the path) with the highest SNR is chosen as the signal to be demodulated. In order to describe this form of diversity, we can analyze selection diversity for any type of modulation we wish. We consider M-ary orthogonal (noncoherent) modulation, with BFSK as a special case, and other binary systems including BPSK, DPSK, and $\pi/4$-differentially encoded quadrature PSK ($\pi/4$-DQPSK). In all cases, we assume L-fold diversity transmissions.

Figure 9.2 depicts a selection diversity reception scheme for an L-fold diversity transmission system. Because the selection of the channel to be processed is based on the maximal SNR, we first need to find the pdf of the maximal SNR. To do this, we assume the following conditions:

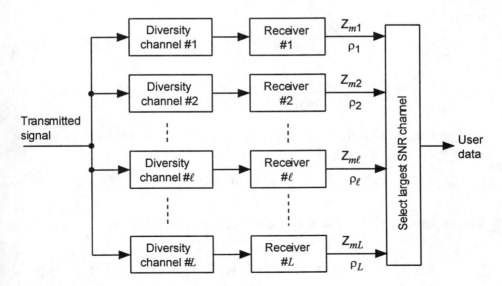

Figure 9.2 Postdetection selection-diversity receiver model.

- The random signal amplitude and carrier phase are constant during the period in which a signal is selected and demodulated.

- The noise in each diversity receiver is independent and has the same variance.

- The noise in each of the M-ary symbol channels is independent.

Let the SNR at the lth diversity channel receiver be denoted

$$\rho_l = \alpha_l^2 \frac{E_c}{\mathcal{N}_0} = \alpha_l^2 \frac{E_s}{L\mathcal{N}_0} \tag{9.1a}$$

where the $\{\alpha_l\}$ are independent Rayleigh attenuation factors with the pdf given by

$$p_{\alpha_l}(\beta) = \frac{2\beta}{\left(\overline{\alpha_l^2}\right)} e^{-\beta^2/\overline{\alpha_l^2}}, \quad l = 1, 2, \ldots, L \tag{9.1b}$$

The pdf of the $\{\rho_l\}$ is given by

$$p_{\rho_l}(\gamma) = \frac{1}{(\overline{\rho_l})} e^{-\gamma/\overline{\rho_l}}, \quad l = 1, 2, \ldots, L \tag{9.1c}$$

where

$$\overline{\rho_l} = E\{\rho_l\} = \frac{\overline{\alpha_l^2} E_s}{L\mathcal{N}_0} \tag{9.1d}$$

$$= \frac{\overline{E_s}}{L\mathcal{N}_0} = \overline{\rho_c} \tag{9.1e}$$

The probability that the SNR at any of the L receivers is below the value λ is given by

$$\Pr\{\rho_l < \lambda\} = \int_0^\lambda p_{\rho_l}(\gamma)\, d\gamma = 1 - e^{-\lambda/\overline{\rho_c}} \tag{9.2a}$$

The probability that SNRs at all L receivers are below λ is

$$\Pr\{\rho_1, \rho_2, \ldots, \rho_L < \lambda\} = \prod_{l=1}^{L} \Pr\{\rho_l < \lambda\} = \left(1 - e^{-\lambda/\overline{\rho_c}}\right)^L \tag{9.2b}$$

Now, the probability that all the SNRs are smaller than the given level is also the probability that the largest of them is smaller than the given level. Therefore, (9.2b) is the CDF for the signal with the largest SNR, ρ_{max}. The pdf of ρ_{max} can now be found by differentiating (9.2b) with respect to λ and evaluating at ρ_{max}:

$$\frac{d}{d\lambda}\left(1 - e^{-\lambda/\bar{\rho}_c}\right)^L \Bigg|_{\lambda=\rho_{max}} = L\left(1 - e^{-\lambda/\bar{\rho}_c}\right)^{L-1}\left(\frac{e^{-\lambda/\bar{\rho}_c}}{\bar{\rho}_c}\right)\Bigg|_{\lambda=\rho_{max}}$$

$$= \frac{L}{\bar{\rho}_c}\left(1 - e^{-\rho_{max}/\bar{\rho}_c}\right)^{L-1} e^{-\rho_{max}/\bar{\rho}_c} \tag{9.2c}$$

so that

$$\text{P}_{\rho_{max}}(\beta) = \frac{L}{\bar{\rho}_c}\left(1 - e^{-\beta/\bar{\rho}_c}\right)^{L-1} e^{-\beta/\bar{\rho}_c} \tag{9.2d}$$

This is the pdf of the maximal SNR over L diversity paths, originally derived by Pierce [3]. We can use this pdf to average the conditional probability of error, given that $\rho_{max} = \beta$, for any type of modulation under L-fold diversity with selection-diversity reception.

9.3.1.1 Noncoherent M-ary Frequency-Shift Keying (NCMFSK)

The modulation scheme to be analyzed under selection diversity reception is the MFSK system. For M-ary orthogonal signaling, the probability of error was derived in Chapter 7. Consistent with the receiver model of Figure 9.2, the conditional probability of symbol (word) error, given that the value of the maximal SNR of L diversity receptions is $\rho_{max} = \beta$, is given by

$$\text{P}_s(e \mid \rho_{max} = \beta) = \sum_{i=1}^{M-1}\binom{M-1}{i}\frac{(-1)^{i+1}}{i+1}\exp\left(-\frac{i}{i+1}\beta\right) \tag{9.3a}$$

where

$$\rho_{max} = \left(\alpha_l^2 \text{E}_c/\mathcal{N}_0\right)_{max} = \left(\alpha_l^2 \text{E}_s/L\mathcal{N}_0\right)_{max} = \beta \tag{9.3b}$$

The unconditional probability of error is obtained by averaging (9.3a) over the variable β, the given value of the RV whose pdf is given by (9.2d), and we have

$$P_s(e; L) = P_{NCMFSK}(e; L) = \int_0^\infty P_s(e \mid \rho_{max} = \beta) \, p_{\rho_{max}}(\beta) \, d\beta \qquad (9.3c)$$

$$= \sum_{i=1}^{M-1} \binom{M-1}{i} \frac{(-1)^{i+1}}{i+1} \prod_{l=1}^{L} \frac{l}{l + \bar{\rho}_c\left(\frac{i}{i+1}\right)} \qquad (9.3d)$$

which is also derived in [10] in a different manner. The derivation of (9.3d) from (9.3c) is given in Appendix 9A.1, in which it is also shown that an alternative form of (9.3d) is

$$P_s(e; L) = \sum_{i=1}^{M-1} \binom{M-1}{i} \frac{(-1)^{i+1}}{i+1} \cdot L \sum_{l=0}^{L-1} \binom{L-1}{l} \frac{(-1)^l}{l + 1 + \bar{\rho}_c\left(\frac{i}{i+1}\right)} \qquad (9.3e)$$

9.3.1.2 Noncoherent Binary Frequency-Shift Keying (NCBFSK)

This case corresponds to $M = 2$ in (9.3d) and (9.3e), leading respectively to the binary error probabilities given by

$$P_{NCBO}(e; L) = \frac{1}{2} \prod_{l=1}^{L} \frac{l}{l + \frac{1}{2}\bar{\rho}_c} = L \sum_{l=0}^{L-1} \binom{L-1}{l} \frac{(-1)^l}{2(l+1) + \bar{\rho}_c} \qquad (9.4)$$

The result in (9.4) can also be obtained directly by averaging the known probability of error for binary noncoherent orthogonal modulation (NCFSK or NCBO), which was shown in Chapter 7 to be

$$P_{NCBO}(e \mid \rho_{max} = \beta) = P_{NCFSK}(e \mid \rho_{max} = \beta) = \frac{1}{2} e^{-\beta/2} \qquad (9.5a)$$

Averaging (9.5a) over β, the value of the RV whose pdf is given by (9.2d), we have

$$P_{NCBO}(e; L) = \int_0^\infty P_{NCBO}(e \mid \rho_{max} = \beta) \, p_{\rho_{max}}(\beta) \, d\beta$$

$$= \int_0^\infty \frac{1}{2} e^{-\beta/2} \cdot \frac{L}{\bar{\rho}_c} \left(1 - e^{-\beta/\bar{\rho}_c}\right)^{L-1} e^{-\beta/\bar{\rho}_c} \, d\beta \qquad (9.5b)$$

Now, let $x = 1 - e^{-\beta/\bar{\rho}_c}$. Then (9.5b) becomes

$$P_{NCBO}(e; L) = \frac{L}{2} \int_0^1 x^{L-1} (1-x)^{\bar{\rho}_c/2} \, dx \qquad (9.5c)$$

Because

$$\int_0^1 x^{n-1}(1-x)^m dx = \frac{\Gamma(n)\,\Gamma(m+1)}{\Gamma(n+m+1)} \tag{9.5d}$$

$$\mathrm{P}_{NCBO}(e; L) = \mathrm{P}_{NCFSK}(e; L) = \frac{L}{2}\frac{\Gamma(L)\,\Gamma\!\left(\frac{1}{2}\overline{\rho}_c + 1\right)}{\Gamma\!\left(L + \frac{1}{2}\overline{\rho}_c + 1\right)}$$

$$= \frac{1}{2}\prod_{k=1}^{L}\frac{k}{k + \frac{1}{2}\overline{\rho}_c} = L\sum_{k=0}^{L-1}\binom{L-1}{k}\frac{(-1)^k}{\overline{\rho}_c + 2(1+k)} \tag{9.5e}$$

where, assuming a symbol is a bit,

$$\overline{\rho}_c \triangleq \overline{\mathrm{E}_s}/L\mathcal{N}_0 = \overline{\mathrm{E}_b}/L\mathcal{N}_0 \tag{9.5f}$$

The error probability of the BFSK system over L-fold diversity with selection-diversity reception is plotted in Figure 9.3. Note that the incremental diversity gain is highest for small orders of diversity.

9.3.1.3 BPSK Modulation

For a BPSK system, the conditonal error probability is given by

$$\mathrm{P}_{BPSK}(e) = \mathrm{Q}\!\left(\sqrt{\frac{2\mathrm{E}_s}{\mathcal{N}_0}}\right) = \mathrm{Q}\!\left(\sqrt{\frac{2\mathrm{E}_b}{\mathcal{N}_0}}\right) \tag{9.6a}$$

The error probability over L-fold diversity with selection-diversity reception is obtained from

$$\mathrm{P}_{BPSK}(e; L, \overline{\rho}_c) = \overline{\mathrm{P}_{BPSK}(e \mid \rho_{max} = \beta)} = \int_0^\infty \mathrm{Q}(\sqrt{2\beta})\,\mathrm{p}_{\rho_{max}}(\beta)\,d\beta$$

$$= \int_0^\infty \mathrm{Q}(\sqrt{2\beta})\frac{L}{\overline{\rho}_c}\left(1 - e^{-\beta/\overline{\rho}_c}\right)^{L-1} e^{-\beta/\overline{\rho}_c}\,d\beta \tag{9.6b}$$

$$= \frac{L}{2}\sum_{k=0}^{L-1}\binom{L-1}{k}\frac{(-1)^k}{k+1}\left[1 - \sqrt{\frac{\overline{\rho}_c}{k+1+\overline{\rho}_c}}\right] \tag{9.6c}$$

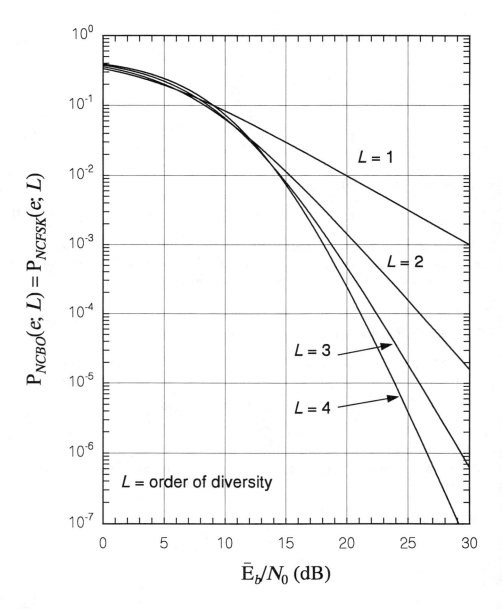

Figure 9.3 Probability of error for BFSK system over L-fold diversity with selection-diversity reception.

where, again assuming a symbol to be a bit,

$$\bar{\rho}_c = \overline{E}_s/L\mathcal{N}_0 = \overline{E}_b/L\mathcal{N}_0 \tag{9.6d}$$

The details of steps from (9.6b) to (9.6c) are shown in Appendix 9B.1. We plotted (9.6c) in Figure 9.4. We again note that the incremental diversity gain is larger for small orders of diversity.

9.3.1.4 $\pi/4$ DQPSK Modulation System With Differential Detection

The average bit-error rate for differentially detected $\pi/4$ DQPSK is given by [11]

$$P_{\pi/4}(e) = \frac{1}{2}\left[1 - Q\left(\sqrt{\frac{E_b}{\mathcal{N}_0}\left(2 + \sqrt{2}\right)}, \sqrt{\frac{E_b}{\mathcal{N}_0}\left(2 - \sqrt{2}\right)}\right)\right.$$

$$\left. + Q\left(\sqrt{\frac{E_b}{\mathcal{N}_0}\left(2 - \sqrt{2}\right)}, \sqrt{\frac{E_b}{\mathcal{N}_0}\left(2 + \sqrt{2}\right)}\right)\right] \tag{9.7a}$$

where $Q(\cdot, \cdot)$ is Marcum's Q-function. Using the close approximation to Marcum's Q-function discussed in Section 1.5.8.3, it is shown in [11] that this probability can be related to the Gaussian Q-function by the approximation

$$P_{\pi/4}(e) \approx Q\left(\sqrt{\frac{E_b}{\mathcal{N}_0}2\left(2 - \sqrt{2}\right)}\right) = Q\left(\sqrt{1.1716\frac{E_b}{\mathcal{N}_0}}\right)$$

which is very similar in form to $P_{BPSK}(e)$ given in (9.6a). In fact, as suggested in this approximation, $\pi/4$ DQPSK is 2.3 dB inferior to a BPSK system. In the case of Rayleigh fading, it is shown in [11] that the BER for $\pi/4$ DQPSK with no diversity is

$$P_{\pi/4}\left(e; 1, \frac{\overline{E}_b}{\mathcal{N}_0}\right) = \frac{1}{2}\left(1 - \frac{\overline{E}_b/\mathcal{N}_0}{\sqrt{\left(\overline{E}_b/\mathcal{N}_0\right)^2 + \frac{1}{4} + \frac{1}{2}\left(\overline{E}_b/\mathcal{N}_0\right)}}\right) \tag{9.7b}$$

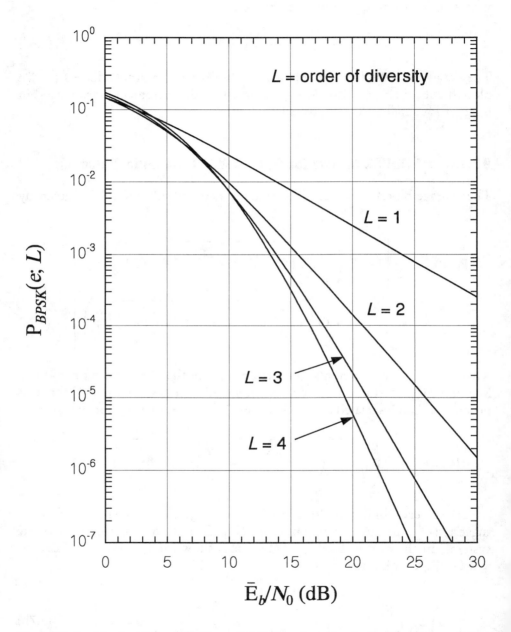

Figure 9.4 Error performance of BPSK over L-fold diversity with selection-diversity reception.

with L-fold selection diversity, it is shown in Appendix 9C.1 that the BER is

$$P_{\pi/4}(e;L) = L \sum_{k=0}^{L-1} \binom{L-1}{k} \frac{(-1)^k}{k+1} P_{\pi/4}\left(e;1,\frac{1}{k+1}\frac{\overline{E_b}}{LN_0}\right) \qquad (9.7c)$$

By analogy with (9.6c), however, we expect that the selection-diversity error probability for $\pi/4$ DQPSK is also shifted by 2.3 dB in terms of the input SNR with respect to the BPSK selection-diversity result:

$$P_{\pi/4}(e;L) \approx P_{BPSK}(e;L,\overline{\rho_c})\big|_{\overline{\rho_c} \to 0.5858\overline{\rho_c}}$$

$$= \frac{L}{2} \sum_{k=0}^{L-1} \binom{L-1}{k} \frac{(-1)^k}{k+1} \left[1 - \sqrt{\frac{0.5858\overline{\rho_c}}{k+1+0.5858\overline{\rho_c}}}\right] \qquad (9.7d)$$

Both (9.7c) and (9.7d) are plotted in Figure 9.5, showing that the approximation is well within 1 dB of the exact result. The approximation, however, is not as close as for the Gaussian channel. This result indicates that we should use (9.7c) rather than (9.7d) for Rayleigh fading.

We also observe that the BPSK selection-diversity result in (9.6c) has the same functional form as that for $\pi/4$ DQPSK in (9.7c). That is, in each case, the selection diversity result for $L > 1$ can be calculated using the error probability equation for $L = 1$ and appropriately scaling $\overline{E_b}/N_0$ as indicated. This suggests that the formula (9.7c) can be applied to other binary modulations, with $P_{\pi/4}(e;1)$ replaced by the appropriate Rayleigh fading error probability. This same type of "universal formula" will also be observed for the case of equal-gain diversity in Section 9.3.2.

9.3.2 Equal Gain Diversity Combining

9.3.2.1 M-ary Noncoherent Orthogonal Modulation System

We consider a situation in which an L-fold diversity scheme is implemented for an M-ary orthogonal system to combat the Rayleigh fading channel, and we first wish to evaluate its error performance when there is no fading. Of course, if the channel is assumed to be nonfading, one normally would not employ such a diversity scheme. Our ultimate objective is to obtain the error performance of the system using the diversity scheme in Rayleigh fading, and

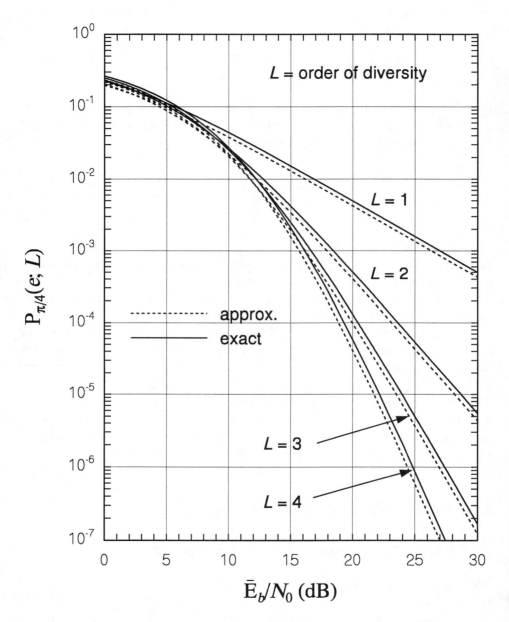

Figure 9.5 Error performance of $\pi/4$ DQPSK over L-fold diversity with selection-diversity reception.

the Gaussian channel performance serves as an interim result that leads us mathematically to the error performance under Rayleigh fading through the averaging operation.

Figure 9.6 depicts a typical postdetection, equal-gain diversity combining receiver for L-fold diversity. The symbol decision statistics for each branch are combined (summed) with equal gains to obtain overall decision statistics. We assume an M-ary noncoherent orthogonal system, with quadratic detection [10] as we discussed in detail in Chapter 7. Regardless of the type of diversity employed, for noncoherent quadratic detection and equal-gain combining, the analysis result is the same.

Suppose that the M symbol statistics from L independent receptions are summed, producing the M decision variables

$$Z_m = \sum_{l=1}^{L} Z_{ml}, \quad m = 1, \ldots, M \tag{9.8a}$$

where the RV Z_{ml} is a chi-squared random variable with two degrees of freedom, denoted

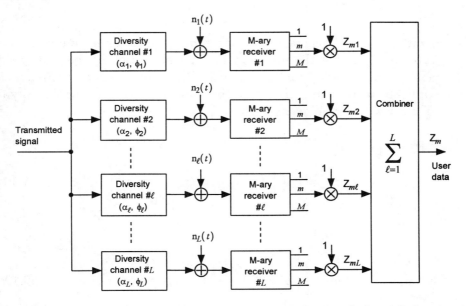

Figure 9.6 Postdetection equal-gain diversity-combining receiver. The $\{\alpha_l\}$ and $\{\phi_l\}$ are iid Rayleigh and uniform RVs, respectively, and the $\{n_l(t)\}$ are iid Gaussian RVs, $l = 1, 2, \ldots, L$.

$$Z_{ml} = \begin{cases} \sigma^2 \chi^2(2; \lambda_{ml}), & \text{symbol } m \text{ was sent} \\ \sigma^2 \chi^2(2), & \text{symbol } m \text{ was not sent} \end{cases} \tag{9.8b}$$

where σ^2 is the variance of the zero-mean Gaussian noise in diversity channel l for symbol m, assumed to be equal to that of the independent noise terms in all the other symbol and diversity channels. The noncentrality parameter λ_{ml} for the case in which the mth symbol was transmitted is equal to twice the SNR in diversity channel: $\lambda_{ml} = 2\rho$, assuming that this average quantity is equal for all diversity channels.

Because the sum of independent, equally weighted chi-squared RVs is also a chi-squared RV, for L-fold diversity and equal-gain combining we have the symbol decision statistics given by

$$Z_m = \begin{cases} \sigma^2 \chi^2(2L; \lambda_m), & \text{symbol } m \text{ was sent} \\ \sigma^2 \chi^2(2L), & \text{symbol } m \text{ was not sent} \end{cases} \tag{9.9a}$$

where

$$\lambda_m = \sum_{l=1}^{L} \lambda_{ml} = L \cdot \frac{2E_c}{\mathcal{N}_0} = 2L\rho = \frac{2E_s}{\mathcal{N}_0} \tag{9.9b}$$

Now, without loss of generality, let us assume that the first symbol is transmitted ($m = 1$), so that the first symbol channel in each of the L diversity receivers contains signal plus noise, while the other symbol channels are noise-only channels. Thus, given that the value of Z_1 is α, the probability of symbol error with L-fold diversity is

$$P_M(e; L \mid Z_1 = \alpha) = 1 - \Pr\{Z_2 < \alpha, Z_3 < \alpha, \dots, Z_M < \alpha\}$$

$$= 1 - \left[\int_0^\alpha p_{z_2}(\beta) \, d\beta \right]^{M-1} \tag{9.10a}$$

where

$$p_{z_m}(\beta) = p_{z_2}(\beta) = \frac{1}{\sigma^2} p_{\chi^2}\left(\frac{\beta}{\sigma^2}; 2L \right), \quad m \geq 2$$

$$= \frac{1}{2\sigma^2 \Gamma(L)} e^{-\beta/2\sigma^2} \left(\frac{\beta}{2\sigma^2} \right)^{L-1}, \quad \beta \geq 0 \tag{9.10b}$$

The unconditional symbol-error probability is obtained by averaging (9.10a) with the pdf of Z_1:

$$P_M(e; L) = 1 - \int_0^\infty p_{z_1}(\alpha) \left[\int_0^\alpha p_{z_2}(\beta) \, d\beta \right]^{M-1} d\alpha \tag{9.10c}$$

where

$$p_{z_1}(\alpha) = \frac{1}{\sigma^2} \, p_{\chi^2}\left(\frac{\alpha}{\sigma^2}; \, 2L, \, \lambda = 2L\rho \right)$$

$$= \frac{1}{2\sigma^2} \, e^{-L\rho - \alpha/2\sigma^2} \left(\frac{\alpha}{2\sigma^2 L\rho} \right)^{(L-1)/2} I_{L-1}\left(\sqrt{\frac{2L\rho\alpha}{\sigma^2}} \right), \, \alpha \geq 0 \tag{9.10d}$$

We show the details of the derivation in Appendix 9A.2. The final expression for the L-fold diversity error probability of an M-ary noncoherent orthogonal modulation in the AWGN channel with power spectral density $\frac{1}{2}\mathcal{N}_0$ is given by

$$P_M(e; L) = \sum_{q=1}^{M-1} \binom{M-1}{q} \frac{(-1)^{q+1}}{(q+1)^L} e^{-qL\rho/(q+1)}$$

$$\times \sum_{n=0}^{q(L-1)} \frac{c_n(q)}{(q+1)^n} \cdot \mathcal{L}_n^{(L-1)}\left(\frac{-L\rho}{q+1} \right) \tag{9.11a}$$

where

$$c_n(q) = \begin{cases} \dfrac{1}{n} \sum_{i=1}^{n} \binom{n}{i} [(q+1)i - n] \, c_{n-i}(q), & n \leq L - 1 \\[4mm] \dfrac{1}{n} \sum_{i=1}^{L-1} \binom{n}{i} [(q+1)i - n] \, c_{n-i}(q), & n \geq L - 1 \end{cases} \tag{9.11b}$$

with $c_0(q) = 1$, all q, and $\mathcal{L}_n^{(k)}(x)$ is the generalized Laguerre polynomial, defined by

$$\mathcal{L}_n^{(k)}(x) = \sum_{j=0}^{n} \binom{n+k}{n-j} \frac{(-x)^j}{j!}$$

$$= \binom{n+k}{n} - \binom{n+k}{n-1} x + \binom{n+k}{n-2} \frac{x^2}{2} + \cdots$$

$$\cdots + \binom{n+k}{1} \frac{(-x)^{n-1}}{(n-1)!} + \frac{(-x)^n}{n!} \tag{9.11c}$$

Example 9.1 The first three Laguerre polynomials are given by

$$\mathcal{L}_0^{(L-1)}\left(\frac{-L\rho}{q+1}\right) = \binom{L-1}{0} = 1$$

$$\mathcal{L}_1^{(L-1)}\left(\frac{-L\rho}{q+1}\right) = \binom{L}{1} + \binom{L}{0}\frac{L\rho}{q+1} = L + \frac{L\rho}{q+1}$$

$$\mathcal{L}_2^{(L-1)}\left(\frac{-L\rho}{q+1}\right) = \binom{L+1}{2} + \binom{L+1}{1}\frac{L\rho}{q+1} + \binom{L+1}{0}\frac{1}{2}\left(\frac{L\rho}{q+1}\right)^2$$

$$= \frac{L(L+1)}{2} + \frac{(L+1)L\rho}{q+1} + \frac{1}{2}\left(\frac{L\rho}{q+1}\right)^2$$

Example 9.2 The Laguerre polynomials can be computed recursively using the relation

$$\mathcal{L}_{n+1}^{(k)}(x) = \frac{1}{n+1}\left[(2n+k+1-x)\mathcal{L}_n^{(k)}(x) - (n+k)\mathcal{L}_{n-1}^{(k)}(x)\right]$$

with $\mathcal{L}_0^{(k)}(x) = 1$ and $\mathcal{L}_1^{(k)}(x) = k + 1 - x$. For example,

$$\mathcal{L}_2^{(L-1)}\left(\frac{-L\rho}{q+1}\right) = \frac{1}{2}\left[\left(2+L+\frac{L\rho}{q+1}\right)\mathcal{L}_1^{(L-1)}\left(\frac{-L\rho}{q+1}\right) - L\cdot\mathcal{L}_0^{(L-1)}\left(\frac{-L\rho}{q+1}\right)\right]$$

$$= \frac{1}{2}\left[\left(2+L+\frac{L\rho}{q+1}\right)\left(L+\frac{L\rho}{q+1}\right) - L\right]$$

$$= \frac{L(L+1)}{2} + \frac{(L+1)L\rho}{q+1} + \frac{1}{2}\left(\frac{L\rho}{q+1}\right)^2$$

Each M-ary symbol conveys $K = \log_2 M$ bits of information. Thus, the symbol energy-to-noise ratio is $L\rho = E_s/N_0 = KE_b/N_0$. Asymptotically, as ρ approaches zero, the symbol-error probability approaches $(M-1)/M$, the probability of error in a purely random selection of one out of M equally likely choices. Thus, we have

$$P_B(e; L) = \frac{1}{2}\cdot\left(\frac{M-1}{M}\right)^{-1}\cdot P_M(e; L) = \frac{M/2}{M-1}P_M(e; L) \qquad (9.12)$$

which is plotted in Figure 9.7 for $M = 2$ and $M = 64$. As the graphical results show, when diversity is used in combination with noncoherent M-ary

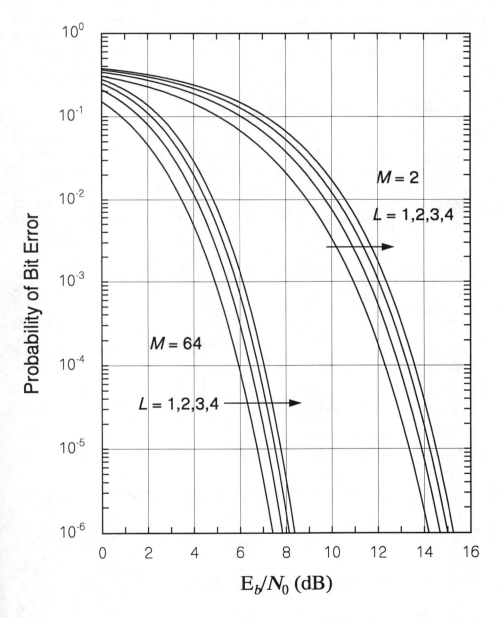

Figure 9.7 Bit-error performance over the AWGN channel for _L_-fold diversity combining of M-ary noncoherent orthogonal modulation.

orthogonal signaling on the Gaussian channel, the bit-error probability increases. This increase in the error probability is due to the fact that the amount of noise added by the diversity receptions is greater than the amount of signal power added, resulting in a net "noncoherent combining loss" [12].

9.3.2.2 MFSK With Rayleigh Fading

We saw in Section 7.2.2.7 that, in Rayleigh fading, the error performance of MFSK is proportional to $1/\rho_{avg}$. We now show that the error performance in Rayleigh fading can be improved by combining independent diversity receptions of the symbols by finding the average of the symbol error probability given in (9.12a) with respect to the pdf for the SNR, ρ. Starting at (9A.7a) in Appendix 9A.2, the average symbol error probability can be written

$$\overline{P_M(e;\,L)} = \sum_{q=1}^{M-1}\binom{M-1}{q}(-1)^{q+1}\sum_{n=0}^{q(L-1)}\frac{c_n(q)}{n!}$$

$$\times \int_0^\infty E_\rho\big\{2\sigma^2 p_{z_1}\big(2\sigma^2 x\big)\big\}\,x^n e^{-qx}\,dx \qquad (9.13a)$$

where, by the theorem of expectation (1.53c):

$$E_\rho\{f(\rho)\} = \int_0^\infty f(\alpha)\,p_\rho(\alpha)\,d\alpha = \int_0^\infty f(\alpha)\,\frac{1}{\rho_{avg}}\,e^{-\alpha/\rho_{avg}}\,d\alpha \qquad (9.13b)$$

The average of the pdf of Z_1 for $L=1$ is

$$E_\rho\big\{2\sigma^2 p_{z_1}\big(2\sigma^2 x\big)\big\} = \int_0^\infty e^{-x-\alpha}\,I_0\big(2\sqrt{\alpha x}\big)\,\frac{1}{\rho_{avg}}\,e^{-\alpha/\rho_{avg}}\,d\alpha \qquad (9.13c)$$

Let $u = \alpha(1 + 1/\rho_{avg}) = \alpha(1 + \rho_{avg})/\rho_{avg}$; the integral then becomes

$$E_\rho\big\{2\sigma^2 p_{z_1}\big(2\sigma^2 x\big)\big\} = \frac{e^{-x}}{1+\rho_{avg}}\int_0^\infty e^{-u}\,I_0\left(2\sqrt{u\cdot\frac{\rho_{avg}x}{1+\rho_{avg}}}\right)du$$

$$= \frac{e^{-x}}{1+\rho_{avg}}\int_0^\infty e^{\lambda/2}\,2\,p_{\chi^2}(2u;\,2,\lambda)\,du\,\bigg|_{\lambda=\frac{2\rho_{avg}x}{1+\rho_{avg}}}$$

$$= \frac{e^{-x}}{1+\rho_{avg}} \cdot \exp\left\{\frac{\rho_{avg}x}{1+\rho_{avg}}\right\} \cdot 1$$

$$= \frac{1}{1+\rho_{avg}} e^{-x/(1+\rho_{avg})}$$

so that

$$E_\rho\{p_{z_1}(x)\} = \frac{1}{2\sigma^2(1+\rho_{avg})} e^{-x/2\sigma^2(1+\rho_{avg})}$$

$$= \frac{1}{\sigma^2(1+\rho_{avg})} p_{\chi^2}\left(\frac{x}{\sigma^2(1+\rho_{avg})}; 2\right) \tag{9.13d}$$

In this development, we use the notations $p_{\chi^2}(\,\cdot\,;2,\lambda)$ and $p_{\chi^2}(\,\cdot\,;2)$ to indicate, respectively, the pdfs for a noncentral chi-squared RV with two degrees of freedom and noncentrality parameter λ and for a central chi-squared RV with two degrees of freedom, as defined in Section 1.5.9.1. Thus, $Z_1 = \sigma^2(1+\rho_{avg})\chi^2(2)$. This result implies that Z_1 in Rayleigh fading for $L > 1$ is $\sigma^2(1+\rho_{avg})$ times a $\chi^2(2L)$ RV, because the sum of equally weighted chi-squared RVs is also a chi-squared RV. Thus, the expectation of the pdf of Z_1 for $L > 1$ is

$$E\{p_{z_1}(x)\} = \frac{1}{\sigma^2(1+\rho_{avg})} p_{\chi^2}\left(\frac{x}{\sigma^2(1+\rho_{avg})}; 2L\right)$$

$$= \frac{1}{2\sigma^2(1+\rho_{avg})} \cdot \frac{1}{\Gamma(L)} \left[\frac{x}{2\sigma^2(1+\rho_{avg})}\right]^{L-1} e^{-x/2\sigma^2(1+\rho_{avg})}$$

and

$$E_\rho\{2\sigma^2 p_{z_1}(2\sigma^2 x)\} = \frac{2}{1+\rho_{avg}} p_{\chi^2}\left(\frac{2x}{1+\rho_{avg}}; 2L\right)$$

$$= \frac{1}{1+\rho_{avg}} \cdot \frac{1}{\Gamma(L)} \left[\frac{x}{1+\rho_{avg}}\right]^{L-1} e^{-x/(1+\rho_{avg})} \tag{9.13e}$$

Substituting this result in the error expression in (9.13a), we obtain

$$\overline{P_M(e;\,L)} = \sum_{q=1}^{M-1} \binom{M-1}{q}(-1)^{q+1} \sum_{n=0}^{q(L-1)} \frac{c_n(q)}{n!}$$

$$\times \int_0^\infty \frac{e^{-x/(1+\rho_{avg})}}{1+\rho_{avg}} \cdot \frac{1}{\Gamma(L)} \left[\frac{x}{1+\rho_{avg}}\right]^{L-1} x^n e^{-qx}\, dx \tag{9.14a}$$

The integral may be written

$$\frac{1}{\Gamma(L)} \cdot \frac{1}{(1+\rho_{avg})^L} \underbrace{\int_0^\infty x^{n+L-1} \exp\left\{-x\left(q + \frac{1}{1+\rho_{avg}}\right)\right\} dx}_{\left(\dfrac{1+\rho_{avg}}{1+q+q\rho_{avg}}\right)^{n+L} \times u^{n+L-1} e^{-u}\, du}$$

$$= \frac{\Gamma(L+n)}{\Gamma(L)} \cdot \frac{(1+\rho_{avg})^n}{(1+q+q\rho_{avg})^{n+L}} \qquad (9.14b)$$

Substituting this result in the error expression (9.14a) produces

$$\overline{P_M(e;\, L)} = \sum_{q=1}^{M-1} \binom{M-1}{q} \frac{(-1)^{q+1}}{(1+q+q\rho_{avg})^L}$$

$$\times \sum_{n=0}^{q(L-1)} c_n(q) \binom{L+n-1}{n} \left(\frac{1+\rho_{avg}}{1+q+q\rho_{avg}}\right)^n \qquad (9.14c)$$

Example 9.3: Binary case. For $M = 2$, the average bit-error probability is

$$\overline{P_B(e;\, L)} = \frac{1}{(2+\rho_{avg})^L} \sum_{n=0}^{L-1} \binom{L+n-1}{n} \left(\frac{1+\rho_{avg}}{2+\rho_{avg}}\right)^n \qquad (9.15a)$$

where

$$\rho_{avg} = \overline{E_b}/L\mathcal{N}_0$$

and (9.15a) can also be put into the form

$$\overline{P_B(e;\, L)} = \left(\overline{P_B(e;\, 1)}\right)^L \sum_{n=0}^{L-1} \binom{L+n-1}{n} \left(1 - \overline{P_B(e;\, 1)}\right)^n \qquad (9.15b)$$

Although we derived the result in (9.15b) from the M-ary noncoherent case, it holds as a formula for many other binary modulations with L-fold equal-gain diversity combining, such as BPSK, DPSK, and $\pi/4$ DQPSK [5, 8, 11].

Example 9.4: $M = 4$. For $M = 4$, the BER corresponding to (9.14c) becomes

$$\overline{P_B(e;\,L)} = \frac{2}{3}\sum_{q=1}^{3}\binom{3}{q}\frac{(-1)^{q+1}}{(1+q+q\rho_{avg})^L}$$

$$\times \sum_{n=0}^{q(L-1)} c_n(q)\binom{L+n-1}{n}\left(\frac{1+\rho_{avg}}{1+q+q\rho_{avg}}\right)^n \qquad (9.16)$$

where $\rho_{avg} = 2\overline{E_b}/L\mathcal{N}_0$. This expression cannot be made to have the simple relationship with the $L = 1$ expression that holds for the binary case.

The bit-error probability for M-ary orthogonal modulation in Rayleigh fading and L-fold diversity with equal-gain diversity combining is shown for $M = 2$, 4, and 64 in Figures 9.8 to 9.10, respectively. The realization of a diversity gain is clearly seen in these figures. For example, from Figure 9.8 we observe for $M = 2$ for an error probability of 10^{-5}, about $20\,\mathrm{dB}$ less SNR is required for $L = 2$ than for $L = 1$. The gain is about $25\,\mathrm{dB}$ for $L = 3$ and about $28\,\mathrm{dB}$ for $L = 4$. Note in these figures that $L = 1$ gives the lowest probability of error for low values of $E_b/\mathcal{N}_0$, but as $E_b/\mathcal{N}_0$ increases, the value of L giving the lowest BER increases.

The amount of diversity gain at a particular required value of the error probability as a function of L increases slightly with the value of M. This fact is illustrated in Figures 9.11 through 9.14, for $L = 1$ to 4, respectively; these figures show that the performance curves for $M = 2$, 4, 8, 16, 32, and 64 are increasingly spread apart as L is increased.

Note from Figures 9.11 to 9.14 that the error probability is proportional to ρ_{avg}^{-1} for high SNR, while at low SNR, the error probability increases with L. This behavior gives rise to an optimal value of diversity as a function of SNR or, alternatively, as a function of the desired value of error probability. As shown in Section 9.3.2.5, for BFSK ($M = 2$ NCBO system) and L-fold diversity combining, using the optimal value of diversity can restore the performance of the system in fading to within about $5\,\mathrm{dB}$ of its performance on the Gaussian channel.

9.3.2.3 BPSK Modulation Under L-fold Diversity With EGC Reception

We stated earlier that the BER formula for L-fold EGC diversity combining, given in (9.15b) for binary noncoherent orthogonal modulation, holds for many other binary modulations and, therefore, can be considered a "universal formula." That is, for the BPSK modulation we have only to use the formula given in (9.15b) to write

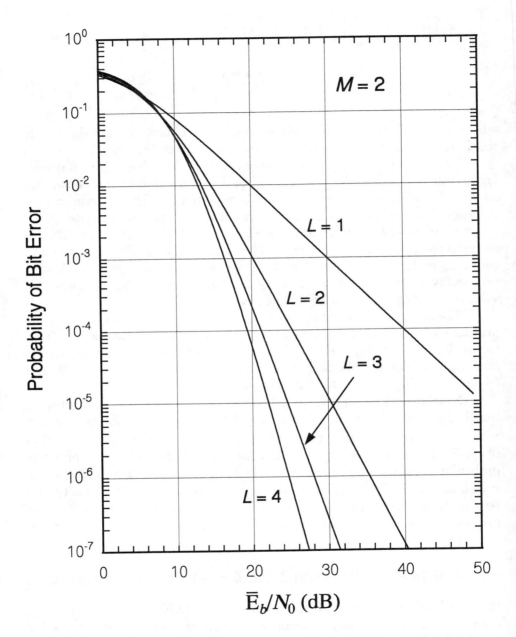

Figure 9.8 Bit-error performance for a binary orthogonal system with diversity in Rayleigh fading.

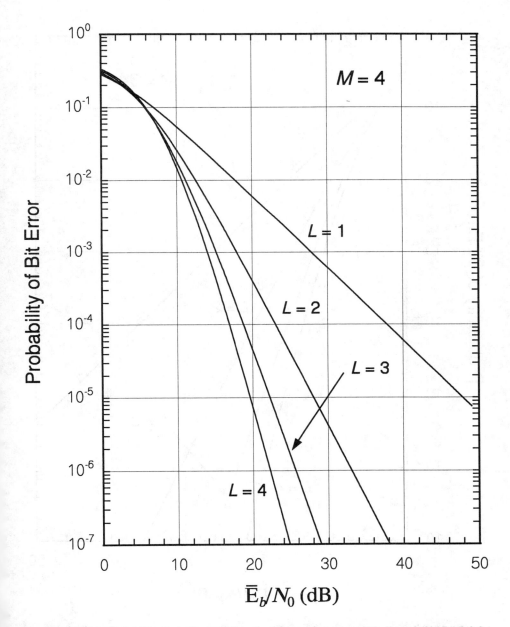

Figure 9.9 Bit-error performance for an $M = 4$ **orthogonal system with diversity in Rayleigh fading.**

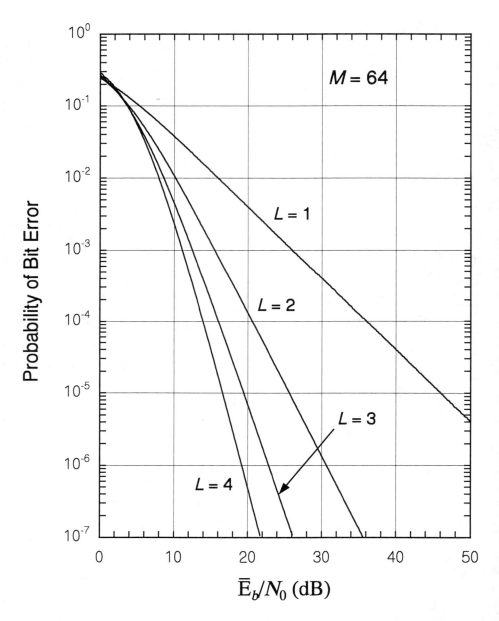

Figure 9.10 Bit-error performance for an $M = 64$ orthogonal system with diversity in Rayleigh fading.

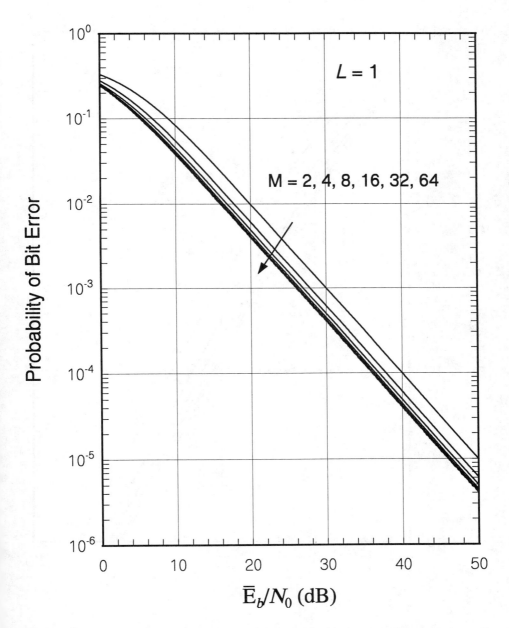

Figure 9.11 Bit-error performance M-ary orthogonal systems in Rayleigh fading for $L = 1$ (no diversity).

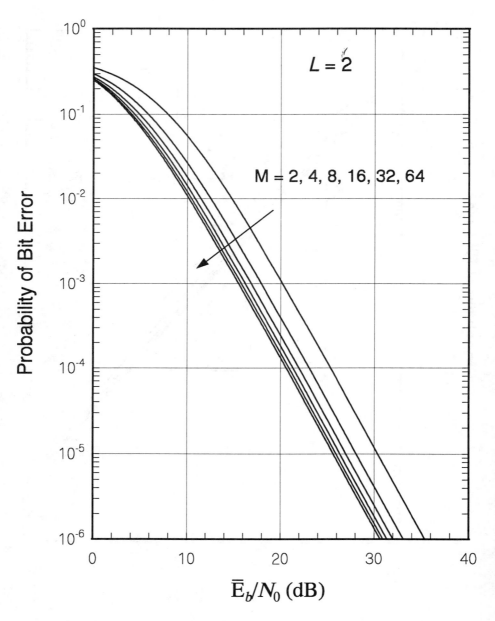

Figure 9.12 Bit-error performance M-ary orthogonal systems in Rayleigh fading for $L = 2$.

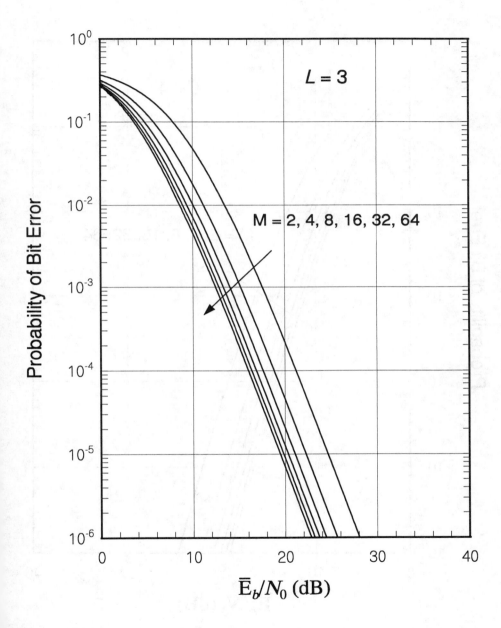

Figure 9.13 Bit-error performance M-ary orthogonal systems in Rayleigh fading for $L = 3$.

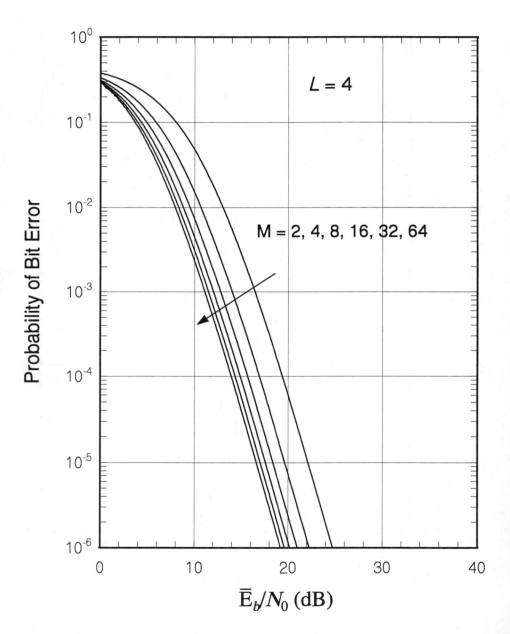

Figure 9.14 Bit-error performance M-ary orthogonal systems in Rayleigh fading for $L = 4$.

$$P_{BPSK}(e; L) = (\bar{p}_{BPSK})^L \sum_{l=0}^{L-1} \binom{L+l-1}{l}(1 - \bar{p}_{BPSK})^l \qquad (9.17a)$$

where $\bar{p}_{BPSK}$ is the BPSK error probability over the Rayleigh fading channel without diversity [see (7.42e)] given by

$$\bar{p}_{BPSK} \triangleq \frac{1}{2}\left(1 - \sqrt{\frac{\overline{E}_c/\mathcal{N}_0}{1 + \overline{E}_c/\mathcal{N}_0}}\right) = \frac{1}{2}\left(1 - \sqrt{\frac{\overline{E}_b/L\mathcal{N}_0}{1 + \overline{E}_b/L\mathcal{N}_0}}\right) \qquad (9.17b)$$

with $\overline{E}_b = L\overline{E}_c$. We now derive $P_{BPSK}(e; L)$ and show that the claim of universality of the formula is true at least for the case of BPSK modulation. For our derivation, we use the postdetection EGC combining receiver shown in Figure 9.15. We use the fact that, for equal-gain combining and independent noise in each diversity channel, the SNR at the output of the combiner is the sum of the channel SNRs.

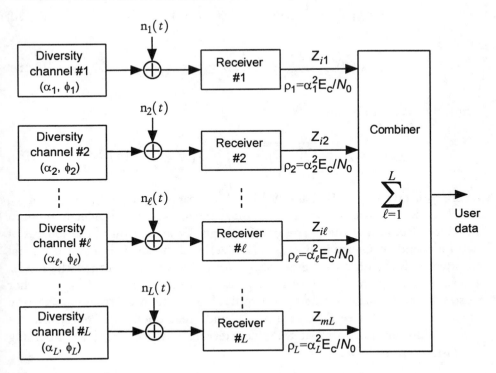

Figure 9.15 Postdetection EGC receiver for a BPSK L-fold diversity system.

The analysis is a slight extension of the method used in obtaining the fading channel error performance by averaging the AWGN channel error performance expression:

$$P_{BPSK}(e; L) = E_{\alpha_1, \alpha_2, \ldots, \alpha_L}\{P_{BPSK}(e; L \mid \alpha_1, \alpha_2, \ldots, \alpha_L)\}$$

$$= \overline{Q\left(\sqrt{\frac{2E_c}{\mathcal{N}_0}(\alpha_1^2 + \alpha_2^2 + \cdots + \alpha_L^2)}\right)} \qquad (9.18a)$$

Let

$$\gamma_c \triangleq \frac{E_c}{\mathcal{N}_0}(\alpha_1^2 + \alpha_2^2 + \cdots + \alpha_L^2) = \sum_{l=1}^{L} \rho_l \qquad (9.18b)$$

where

$$\rho_l \triangleq \alpha_l E_c/\mathcal{N}_0, \quad l = 1, 2, \ldots, L \qquad (9.18c)$$

with

$$p_{\rho_l}(\beta) = \frac{1}{\bar{\rho}_c} e^{-\beta/\bar{\rho}_c}, \quad \beta \geq 0, \; l = 1, 2, \ldots, L \qquad (9.18d)$$

and

$$p_{\alpha_l}(\beta) = \frac{2\beta}{\overline{(\alpha_l^2)}} e^{-\beta/\overline{\alpha_l^2}}, \quad \beta \geq 0, \; l = 1, 2, \ldots, L \qquad (9.18e)$$

Thus, the task is to obtain

$$P_{BPSK}(e; L) = \overline{Q(\sqrt{2\gamma_c})} = \int_0^\infty Q\left(\sqrt{2x}\right) p_{\gamma_c}(x)\, dx \qquad (9.19a)$$

where γ_c is given by (9.18b) and (9.18c). It appears that we need the pdf of γ_c, which is $E_c/\mathcal{N}_0$ times a sum of L independent squared Rayleigh random variables. Recall that the square of a Rayleigh RV is the sum of the squares of two independent Gaussian RVs [see (1.109e)] and, hence, it is a scaled central chi-squared RV with two degrees of freedom. This view identifies the RV γ_c as a scaled chi-squared RV with $2L$ degrees of freedom [see (1.133a)]. Another reasonable approach in finding the pdf of γ_c is to use "the characteristic function method" (see Section 1.5.1.2), and then obtain the pdf from the inverse Fourier transform of the characteristic function $C_{\gamma_c}(j\nu)$. Pursuing this latter approach, we have

$$C_{\gamma_c}(j\nu) = E\{e^{j\nu\gamma_c}\} = E\{e^{j\nu(\rho_1 + \rho_2 + \cdots + \rho_L)}\}$$

$$= \prod_{l=1}^{L} \overline{e^{j\nu\rho_l}} = \prod_{l=1}^{L} \int_0^\infty e^{j\nu\beta} \, p_{\rho_l}(\beta) \, d\beta$$

$$= \prod_{l=1}^{L} \int_0^\infty e^{j\nu\beta} \cdot \frac{1}{\overline{\rho}_c} e^{-\beta/\overline{\rho}_c} \, d\beta = \frac{1}{(1 - j\nu\overline{\rho}_c)^L} \qquad (9.19\text{b})$$

Indeed, we recognize (9.19b) to be the characteristic function of a scaled central chi-squared RV with $2L$ degrees of freedom [see (1.132e)]. That is:

$$\gamma_c = \tfrac{1}{2}\overline{\rho}_c \times \chi^2(2L) \qquad (9.19\text{c})$$

so that the pdf of γ_c is given by

$$p_{\gamma_c}(\beta) = \frac{\beta^{L-1}}{\left(\tfrac{1}{2}\overline{\rho}_c\right)^L 2^L \Gamma(L)} e^{-\beta/2(\tfrac{1}{2}\overline{\rho}_c)} = \frac{\beta^{L-1}}{(\overline{\rho}_c)^L \Gamma(L)} e^{-\beta/\overline{\rho}_c} \qquad (9.19\text{d})$$

Substituting (9.19d) in (9.19a) and carrying out the integration, we obtain

$$P_{BPSK}(e; L) = \overline{Q(\sqrt{2\gamma_c})} = \int_0^\infty Q\left(\sqrt{2x}\right) \frac{x^{L-1}}{(L-1)!(\overline{\rho}_c)^L} e^{-x/\overline{\rho}_c} \, dx \quad (9.19\text{e})$$

$$= (\overline{p}_{BPSK})^L \sum_{l=0}^{L-1} \binom{L+l-1}{l} (1 - \overline{p}_{BPSK})^l \qquad (9.19\text{f})$$

where

$$\overline{p}_{BPSK} \triangleq \frac{1}{2}\left(1 - \sqrt{\frac{\overline{\rho}_c}{1+\overline{\rho}_c}}\right) \qquad (9.19\text{g})$$

and

$$\overline{\rho}_c \triangleq \frac{\overline{\alpha^2 E_c}}{\mathcal{N}_0} = \frac{\overline{E}_c}{\mathcal{N}_0} = \frac{1}{L}\frac{\overline{\alpha^2 E_b}}{\mathcal{N}_0} = \frac{1}{L}\frac{\overline{E}_b}{\mathcal{N}_0} \qquad (9.19\text{h})$$

The steps from (9.19e) to (9.19f) are shown in Appendix 9B.2. The fact that (9.19f) is identical to (9.15b) suggests that the expression is a universal formula for L-fold diversity with EGC reception for any binary symbol (chip) modulation. (9.19f) is plotted in Figure 9.16 as a function of $\overline{E}_b/\mathcal{N}_0$ with the order of diversity as a parameter.

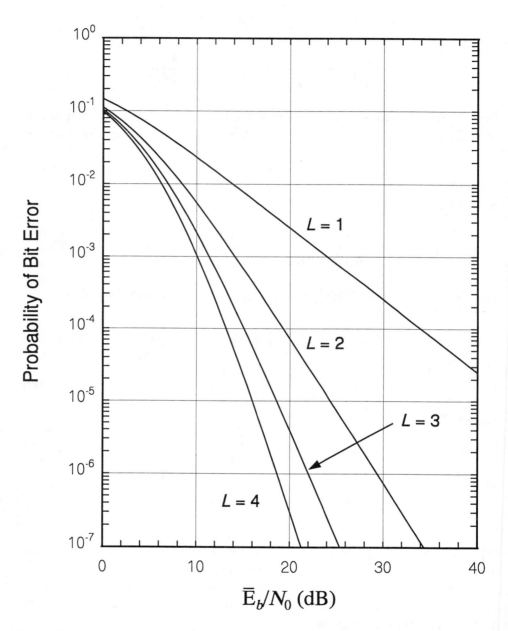

Figure 9.16 Equal-gain diversity reception of BPSK error performance with L-fold diversity over the Rayleigh fading channel.

9.3.2.4 $\pi/4$ DQPSK Modulation With Differential Detection Under L-fold Diversity With EGC Reception

For $\pi/4$ DQPSK modulation with differential detection in the Rayleigh fading channel, the symbol-error probability is given by [13]

$$\overline{P}_{\pi/4} = \frac{1}{2}\left(1 - \frac{\overline{E}_b/L\mathcal{N}_0}{\sqrt{\left(\overline{E}_b/L\mathcal{N}_0\right)^2 + \frac{1}{2} + 2\left(\overline{E}_b/L\mathcal{N}_0\right)}}\right) \qquad (9.20\text{a})$$

The derivation of this probability is also shown in Appendix 9C.1 as the case of $L = 1$. For L-fold diversity and EGC reception in Rayleigh fading, it is shown in Appendix 9C.2 that the BER for $\pi/4$ DQPSK is

$$\mathrm{P}_{\pi/4}(e; L) = \left(\overline{p}_{\pi/4}\right)^L \sum_{l=0}^{L-1} \binom{L+l-1}{l}\left(1 - \overline{p}_{\pi/4}\right)^l \qquad (9.20\text{b})$$

Note that this probability has the same form as (9.15b) and (9.19f) for BFSK and BPSK, respectively, in the way that the probability for $L > 1$ is related to the probability for $L = 1$. Calculations of (9.20b) are shown in Figure 9.17.

9.3.2.5 Noncoherent Binary Orthogonal System and Optimal Diversity

From the general M-ary noncoherent orthogonal system BER for L-fold diversity with EGC reception in Rayleigh fading, it was shown in (9.15a) to (9.15c) that the BER for the binary case of $M = 2$ is given by the formula

$$\mathrm{P}_{NCBO}(e; L) = \left(\overline{p}_{NCBO}\right)^L \sum_{l=0}^{L-1} \binom{L+l-1}{l}\left(1 - \overline{p}_{NCBO}\right)^l \qquad (9.21\text{a})$$

where

$$\overline{p}_{NCBO} = \frac{1}{2 + \overline{E}_c/\mathcal{N}_0} = \frac{1}{2 + \overline{E}_b/L\mathcal{N}_0} \qquad (9.21\text{b})$$

The BER is plotted in Figure 9.18 as a function of $\overline{E}_b/\mathcal{N}_0$ in dB, with the order of diversity as a parameter. Note that, as L increases, the equal-gain

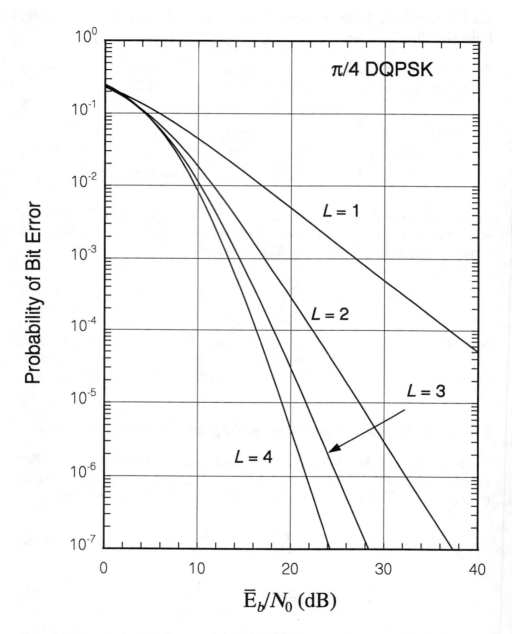

Figure 9.17 Equal-gain diversity reception of $\pi/4$ DQPSK error performance with L-fold diversity over the Rayleigh fading channel.

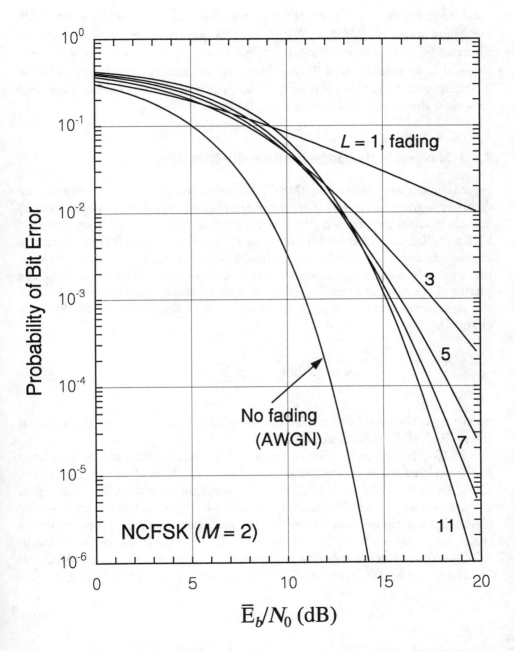

Figure 9.18 EGC reception of NCBO system under L-fold diversity over the Rayleigh fading channel.

combining results in a higher BER for low values of $\overline{E}_b/\mathcal{N}_0$ and a lower BER for high values of $\overline{E}_b/\mathcal{N}_0$. We chose this particular modulation case to illustrate how we can implement optimal diversity in the sense that there is value of L, as a function of $\overline{E}_b/\mathcal{N}_0$, that gives the minimum value of the BER in fading, and when this value of L is used, the probability of error is restored to within about 5 dB of the error probability for no fading.

9.3.3 Maximal Ratio Combining Diversity Reception

In maximal ratio combining (MRC) diversity reception, the phases of the diversity signals being added together are aligned (made mutually coherent) and their envelopes are weighted in proportion to the square roots of their SNRs. MRC, first discussed by Brennan [13], is the optimal form of diversity combining because it yields the maximal SNR achievable. As shown shortly, the optimal combining requires exact knowledge of SNRs as well as the phases of the diversity signals. MRC diversity combining achieves an SNR at the combiner output that is equal to the sum of the SNRs in the diversity channels:

$$\rho_T = (\text{SNR})_{o,max} = \sum_{j=1}^{L} \rho_j \qquad (9.22)$$

where ρ_T is the combined SNR and the $\{\rho_j;\ j = 1, 2, \ldots, L\}$ are the SNRs of the individual diversity channels.

MRC is used in the IS-95 system on both the forward and reverse links in the multipath diversity schemes, and, hence, the Rake diversity used in the CDMA cellular system is MRC diversity reception, as is discussed in detail in the next section. Basically, MRC is similar to EGC, except that each diversity channel is multiplied, not by an "equal gain," but by a gain proportional to the square root of the SNR of that channel. Therefore, the performance of EGC diversity reception behaves in a way that is similar to that of MRC and yields a lower bound, because MRC is the optimal form of combining.

9.3.3.1 Optimality Proof of MRC Diversity Reception

Here we prove that Brennan's MRC diversity reception does deliver the maximal possible SNR at the output of the diversity combiner. Consider Figure 9.19, which depicts the MRC scheme. For convenience we use complex envelope notation for the analysis. The received waveform for diversity channel l is given by

$$y_l(t) = s_l(t) + n_l(t) \tag{9.23a}$$

The conventional envelope and phase representation for the real-valued signal waveform is

$$s_l(t) = \alpha_l A_l \cos[\omega_0 t + \phi_l(t)]$$

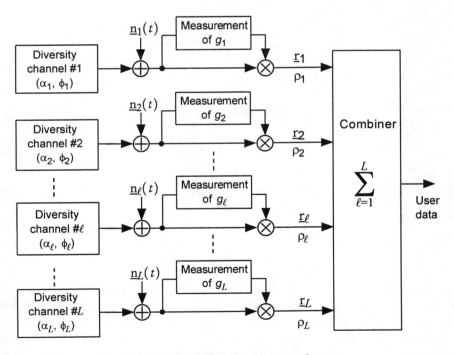

Figure 9.19 Complex envelope diagram of MRC diversity reception.

where α_l is the Rayleigh attenuation factor, A_l is the signal amplitude without fading, and $\phi_l(t)$ is the signal phase, for channel l. In complex envelope notation, we have

$$
\begin{aligned}
s_l(t) &= R_l(t)\cos[\omega_0 t + \phi_l(t)] \\
&= \text{Re}\big\{ R_l(t)\, e^{j\phi_l(t)} \cdot e^{j\omega_0 t} \big\} \\
&= \text{Re}\big\{ \underline{s}_l(t)\cdot e^{j\omega_0 t} \big\}
\end{aligned}
\tag{9.23b}
$$

where

$$
\begin{aligned}
\underline{s}_l(t) &\triangleq R_l(t)\, e^{j\phi_l(t)} = R_l(t)\cos\phi_l(t) + j\,R_l(t)\sin\phi_l(t) \\
&= s_{cl}(t) + j\, s_{sl}(t)
\end{aligned}
\tag{9.23c}
$$

Note that complex envelope notation is "lowpass" waveform notation. Likewise, for the noise term we have

$$
\begin{aligned}
n_l(t) &= n_{cl}(t)\cos\omega_0 t - n_{sl}(t)\sin\omega_0 t \\
&= \sqrt{n_{cl}^2(t) + n_{sl}^2(t)}\,\cos\left[\omega_0 t + \tan^{-1}\left(\frac{n_{sl}(t)}{n_{cl}(t)}\right)\right] \\
&= N_l(t)\cos[\omega_0 t + \theta_l(t)] = \text{Re}\big\{ \underline{n}_l(t)\cdot e^{j\omega_0 t} \big\}
\end{aligned}
\tag{9.23d}
$$

where $N_l(t)$ is the noise envelope, and the noise phase and complex envelope are

$$
\theta_l(t) \triangleq \tan^{-1}\left(\frac{n_{sl}(t)}{n_{cl}(t)}\right)
\tag{9.23e}
$$

$$
\underline{n}_l(t) \triangleq N_l(t)\, e^{j\theta_l(t)} = N_l(t)\cos\theta_l(t) + j\,N_l(t)\sin\theta_l(t)
\tag{9.23f}
$$

Thus, we have

$$
\underline{y}_l(t) = \underline{s}_l(t) + \underline{n}_l(t), \quad l = 1, 2, \ldots, L
\tag{9.24a}
$$

The principle of MRC reception is that we multiply the received waveform $y_l(t)$ by the gain g_l, which is complex in general, and then add the output of the diversity channels as indicated in Figure 9.19. Our objective is to find the specific form of g_l for $l = 1, 2, \ldots, L$. If we find g_l to be proportional to the square root of $(\text{SNR})_l$, the channel SNR, we have proven the principle of MRC and have shown that the total output SNR is the maximum.

From Figure 9.19, we note that the complex envelope of the lth waveform being combined is

$$\underline{r}_l(t) = g_l\, \underline{s}_l(t) + g_l\, \underline{n}_l(t), \quad l = 1, 2, \ldots, L \tag{9.24b}$$

Let us note that

$$\overline{|\underline{n}_l(t)|^2} = \overline{|n_{cl}(t) + j\, n_{sl}(t)|^2} = \overline{n_{cl}^2} + \overline{n_{sl}^2} = \sigma_l^2 + \sigma_l^2 = 2\sigma_l^2 \tag{9.24c}$$

where we assume that

$$\overline{n_{cl}\, n_{sl}} = \overline{n_{cl}} \cdot \overline{n_{sl}} = 0 \tag{9.24d}$$

To proceed with the development, we introduce the following complex variables:

$$u_l \triangleq g_l^* \sigma_l \tag{9.25a}$$

and

$$v_l \triangleq \frac{s_l}{\sigma_l} \tag{9.25b}$$

We then find that

$$u_l^* v_l = (g_l^* \sigma_l)^* \left(\frac{\underline{s}_l}{\sigma_l}\right) = g_l\, \underline{s}_l, \quad l = 1, 2, \ldots, L \tag{9.26a}$$

represents the signal term in the lth channel after the gain multiplication. Also, we see that

$$|v_l|^2 = \left|\frac{\underline{s}_l}{\sigma_l}\right|^2 = \frac{|\underline{s}_l|^2}{\sigma_l^2} = 2\rho_l \tag{9.26b}$$

where

$$\rho_l \triangleq \frac{|\underline{s}_l|^2}{2\sigma_l^2} \tag{9.26c}$$

is the SNR for channel l; that is, $(\text{SNR})_l$. The term

$$|u_l|^2 = (g_l^*\, \sigma_l)^* (g_l^*\, \sigma_l) = |g_l|^2 \sigma_l^2 \tag{9.26d}$$

is then the noise power at the lth channel after the gain multiplication. Because there are L inputs to the combiner, we form the vectors $\boldsymbol{U}$ and $\boldsymbol{V}$:

$$\boldsymbol{U} = (u_1, u_2, \ldots, u_L) \tag{9.27a}$$

$$\boldsymbol{V} = (v_1, v_2, \ldots, v_L) \tag{9.27b}$$

where $\{u_l\}$ and $\{v_l\}$ are defined in (9.25a) and (9.25b). The *Schwarz inequality* is given by

$$|\langle U, V \rangle|^2 \leq |U|^2 |V|^2 \tag{9.28a}$$

with equality if and only if

$$U = KV \tag{9.28b}$$

where K is a scalar. Thus, the Schwarz inequality in terms of the vectors of (9.25a) and (9.25b) may be written

$$\left| \sum_{l=1}^{L} u_l^* v_l \right|^2 \leq \left(\sum_{l=1}^{L} |u_l|^2 \right) \left(\sum_{l=1}^{L} |v_l|^2 \right) \tag{9.29a}$$

with equality if and only if

$$u_l = K v_l \tag{9.29b}$$

Let us examine each term in (9.29a):

$$\left| \sum_{l=1}^{L} u_l^* v_l \right|^2 = |u_1^* v_1 + u_2^* v_2 + \cdots + u_l^* v_l + \cdots + u_L^* v_L|^2$$

$$= |g_1 \underline{s}_1 + g_2 \underline{s}_2 + \cdots + g_l \underline{s}_l + \cdots + g_L \underline{s}_L|^2$$

$$= 2|S|^2 \tag{9.30a}$$

where $2|S|^2$ is the total complex signal power at the output of the combiner;

$$\sum_{l=1}^{L} |u_l|^2 = \sum_{l=1}^{L} |g_l^* \sigma_l|^2 = \sum_{l=1}^{L} |g_l|^2 \sigma_l^2 = 2\sigma_T^2 \tag{9.30b}$$

where $2\sigma_T^2$ is the total complex noise power;

$$\sum_{l=1}^{L} |v_l|^2 = \sum_{l=1}^{L} \left| \frac{\underline{s}_l}{\sigma_l} \right|^2 = \sum_{l=1}^{L} \frac{|\underline{s}_l|^2}{\sigma_l^2} = \sum_{l=1}^{L} 2\rho_l \tag{9.30c}$$

Substituting (9.30a), (9.30b), and (9.30c) into the Schwarz inequality in (9.20a), we have

$$2|S|^2 \leq 2\sigma_T^2 \cdot \sum_{l=1}^{L} 2\rho_l \qquad \text{or} \qquad \rho_T \equiv \frac{|S|^2}{2\sigma_T^2} \leq \sum_{l=1}^{L} \rho_l \qquad (9.31a)$$

if and only if

$$u_l = K v_l \qquad (9.31b)$$

The condition (9.31b) is

$$g_l^* \sigma_l = K \frac{s_l}{\sigma_l} \qquad \text{or} \qquad g_l = K \frac{s_l}{\sigma_l^2} \qquad (9.31c)$$

We thus conclude that

$$\rho_T = \frac{|S|^2}{2\sigma_T^2} = \sum_{l=1}^{L} \rho_l \qquad (9.32a)$$

if and only if the gain of each channel is proportional to the square root of the $(SNR)_l$ and inversely proportional to the rms value of the noise in that channel, with the same constant of proportionality (K) for all L diversity channels; that is:

$$g_l = K \frac{s_l}{\sigma_l^2} = K \frac{|s_l| e^{j\phi_l}}{\sigma_l^2} = K \frac{\sqrt{2\rho_l}}{\sigma_l} e^{j\phi_l} \qquad (9.32b)$$

Note also that the signal phase of the lth received complex envelope, $\phi_l(t)$, must be measured, where

$$\phi_l(t) = \tan^{-1} \frac{\text{Im}\{\underline{s}_l(t)\}}{\text{Re}\{\underline{s}_l(t)\}} \qquad (9.32c)$$

This proves the optimality of the MRC diversity reception scheme of Brennan [13]. To implement the MRC scheme, we need to have the means to measure the signal phase and SNR in each diversity channel to accomplish the coherent addition of all signals.

9.3.3.2 Example of MRC

Consider the dual-channel diversity combining depicted in Figure 9.20, where the signals in the two channels are coherently combined before detection based on the MRC principle. We assume that the carriers in the two channels are made coherent by some phase alignment process and that $n_1(t)$ and $n_2(t)$

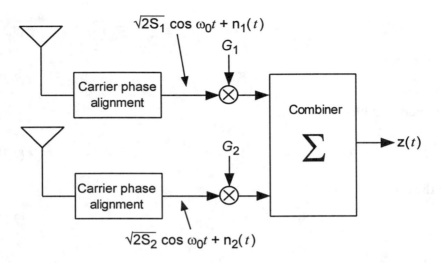

Figure 9.20 Coherent combining of two waveforms with independent noise components.

are statistically independent, zero-mean Gaussian noise waveforms with variances $\sigma_i^2 = N_i$, $i = 1, 2$. Note that the signal power in each channel is S_i, $i = 1, 2$. We wish to find the gains G_1 and G_2 that maximize the output SNR, denoted $(SNR)_o$, and to find its maximal value, $(SNR)_{o,max}$.

First, we need to express the output SNR as a function of G_1 and G_2, the parameters to be adjusted to maximize $(SNR)_o$. The output of the combiner is given by

$$z(t) = G_1\sqrt{2S_1}\cos\omega_0 t + G_2\sqrt{2S_2}\cos\omega_0 t + G_1 n_1(t) + G_2 n_2(t)$$
$$= \left(G_1\sqrt{2S_1} + G_2\sqrt{2S_2}\right)\cos\omega_0 t + G_1 n_1(t) + G_2 n_2(t) \qquad (9.33a)$$

The signal power P_s is given by

$$P_s = \tfrac{1}{2}\left(G_1\sqrt{2S_1} + G_2\sqrt{2S_2}\right)^2 = \left(G_1\sqrt{S_1} + G_2\sqrt{S_2}\right)^2 \qquad (9.33b)$$

and the noise power P_n is given by

$$P_n = \overline{[G_1 n_1(t) + G_2 n_2(t)]^2} = G_1^2\sigma_1^2 + G_2^2\sigma_2^2 = G_1^2 N_1 + G_2^2 N_2 \qquad (9.33c)$$

The output SNR is given by

$$(\text{SNR})_o = \frac{P_s}{P_n} = \frac{\left(G_1\sqrt{S_1} + G_2\sqrt{S_2}\right)^2}{G_1^2 N_1 + G_2^2 N_2} \qquad (9.34\text{a})$$

To the find the gains G_1 and G_2 that maximize $(\text{SNR})_o$, we take the partial derivatives with respect to G_1 and G_2 and set the results to zero:

$$\frac{\partial(\text{SNR})_o}{\partial G_1} = \frac{\partial}{\partial G_1}\left[\frac{\left(G_1\sqrt{S_1} + G_2\sqrt{S_2}\right)^2}{G_1^2 N_1 + G_2^2 N_2}\right] = 0 \qquad (9.34\text{b})$$

and

$$\frac{\partial(\text{SNR})_o}{\partial G_2} = \frac{\partial}{\partial G_2}\left[\frac{\left(G_1\sqrt{S_1} + G_2\sqrt{S_2}\right)^2}{G_1^2 N_1 + G_2^2 N_2}\right] = 0 \qquad (9.34\text{c})$$

From these partial derivatives, we obtain the simultaneous equations

$$\sqrt{S_1}N_2 G_2 - \sqrt{S_2}N_1 G_1 = 0 \qquad (9.34\text{d})$$

$$\sqrt{S_2}N_1 G_1 - \sqrt{S_1}N_2 G_2 = 0 \qquad (9.34\text{e})$$

Note that these equations are dependent and produce identical solutions, namely

$$\frac{G_1}{G_2} = \frac{N_2}{N_1}\sqrt{\frac{S_1}{S_2}} = \frac{\sqrt{S_1}/N_1}{\sqrt{S_2}/N_2} \qquad (9.34\text{f})$$

The implication of (9.34f) is that, while the gain G_i for channel i needs to be proportional to the rms signal power and inversely proportional to the noise power of each channel, it is the *relative* gain that matters; that is, had we set $G_2 = 1$, G_1 would have been found to be

$$G_1 = \frac{N_2}{N_1}\sqrt{\frac{S_1}{S_2}} \qquad (9.34\text{g})$$

Thus, the maximal $(\text{SNR})_o$ can be evaluated as

$$(\text{SNR})_{o,max} = \frac{\left(G_1\sqrt{S_1} + G_2\sqrt{S_2}\right)^2}{G_1^2 N_1 + G_2^2 N_2}\Bigg|_{\frac{G_1}{G_2} = \frac{N_2}{N_1}\sqrt{\frac{S_1}{S_2}}}$$

$$
= \frac{\left(\frac{G_1}{G_2}\sqrt{S_1} + \sqrt{S_2}\right)^2}{\left(\frac{G_1}{G_2}\right)^2 N_1 + N_2}\Bigg|_{\frac{G_1}{G_2} = \frac{N_2}{N_1}\sqrt{\frac{S_1}{S_2}}} = \frac{\left(\frac{N_1}{N_2} \cdot \frac{S_1}{\sqrt{S_2}} + \sqrt{S_2}\right)^2}{\frac{N_2^2}{N_1} \cdot \frac{S_1}{S_2} + N_2}
$$

$$
= \frac{\left(\frac{N_1}{N_2} \cdot \frac{S_1}{\sqrt{S_2}} + \sqrt{S_2}\right)^2}{\frac{N_2^2}{N_1} \cdot \frac{S_1}{S_2} + N_2} \cdot \frac{\left(\frac{S_2}{N_2^2}\right)}{\left(\frac{S_2}{N_2^2}\right)} = \frac{\left(\frac{S_1}{N_1} + \frac{S_2}{N_2}\right)^2}{\frac{S_1}{N_1} + \frac{S_2}{N_2}}
$$

$$
= \frac{S_1}{N_1} + \frac{S_2}{N_2} = (\text{SNR})_1 + (\text{SNR})_2 \tag{9.35}
$$

We therefore have the equivalent MRC diversity receiver as shown in Figure 9.21. This example illustrates the principle of MRC diversity reception that we developed in this section.

9.4 The Rake Receiver Concept

The Rake concept was introduced in 1958 by R. Price and P. E. Green through their paper [14] entitled, "A Communication Technique for Multipath Channels." They described an implementation of the Rake concept in equipment designed for operation in the HF band.

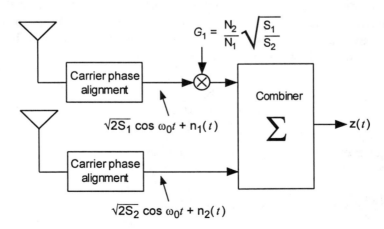

Figure 9.21 Alternative MRC diversity reception receiver to that of Figure 9.20.

For a fading multipath channel in which the maximal delay spread is T_m sec, the Rake concept uses a special wideband signal of bandwidth W Hz, where $W \gg 1/T_m$. The signal is structured to have a pseudorandom character, with an autocorrelation function whose width is on the order of $1/W$, that permits the isolation of individual multipath signal components for coherent combining. If a signal has the form of a PN sequence $C(t)$ modulating a carrier, we have the modulated waveform

$$m(t) = C(t)\cos\omega_0 t \qquad (9.36a)$$

where ω_0 is the carrier angular frequency. The signal's autocorrelation function is then given by

$$
\begin{aligned}
R_m(\tau) &= \overline{m(t)\,m(t+\tau)} \\
&= E\{C(t)\cos\omega_0 t \cdot C(t+\tau)\cos\left[\omega_0(t+\tau)\right]\} \\
&= E\{C(t)\,C(t+\tau)\} \cdot E\{\cos\omega_0 t \cos\left[\omega_0(t+\tau)\right]\} \\
&= R_c(\tau) \cdot \tfrac{1}{2}\cos\omega_0\tau \qquad (9.36b)
\end{aligned}
$$

where $R_c(\tau)$ has essentially zero correlation for $|\tau| > 1/W$, as shown in Figure 9.22. Thus, the autocorrelation function of the wideband PN sequence is approximately expressed as

$$R_c(\tau) \approx R_c(0)[1 - W|\tau|], \quad |\tau| < 1/W \qquad (9.36c)$$

$$= \begin{cases} R_c(0)(1 - W\tau), & 0 \le \tau < 1/W \\ R_c(0)(1 + W\tau), & -1/W < \tau < 0 \end{cases} \qquad (9.36d)$$

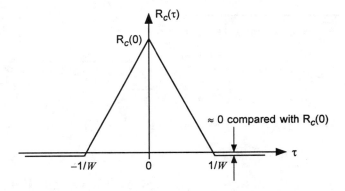

Figure 9.22 Autocorrelation function of a PN sequence with a rectangular basic waveform.

9.4.1 Basics of Rake Receiver Design

Recall that the output of a multipath channel can be modeled by the sum of attenuated and delayed versions of the signal (see Section 4.2.1.1):

$$Z(t) = \sum_n \alpha_n z(t - \tau_n) \tag{9.37a}$$

so that in complex envelope notation, we have

$$\underline{Z}(t) = \sum_n \alpha_n \underline{z}(t - \tau_n) e^{-j2\pi f_0 \tau_n} \tag{9.37b}$$

where the attenuation factors $\{\alpha_n\}$, possibly time varying, are real numbers that represent the propagation loss on individual paths, which are indexed by n. The delays $\{\tau_n\}$, also possibly time varying, represent path delays.

Suppose now that the signal is pseudorandom, with a correlation width of $1/W$. Then, the cross-correlation of the received signal with the local replica

$$\underline{z}_k = \underline{z}(t - k/W) \tag{9.37c}$$

selects the path with delay within $\pm 1/W$ of

$$\hat{\tau} = k/W \tag{9.37d}$$

if one exists, and produces an estimate of its amplitude and phase. The mechanism can be explained as follows:

$$\overline{\underline{Z}^* \underline{z}_k} = \overline{\sum_n \alpha_n \underline{z}^*(t - \tau_n) e^{j\omega_0 \tau_n} \underline{z}(t - \hat{\tau}_k)}$$

$$= \alpha_1 \overline{\underline{z}^*(t - \tau_1) e^{j\omega_0 \tau_1} \underline{z}(t - \hat{\tau}_k)} + \alpha_2 \overline{\underline{z}^*(t - \tau_2) e^{j\omega_0 \tau_2} \underline{z}(t - \hat{\tau}_k)}$$

$$+ \cdots + \alpha_n \overline{\underline{z}^*(t - \tau_n) e^{j\omega_0 \tau_n} \underline{z}(t - \hat{\tau}_k)} + \cdots$$

$$= \alpha_1 e^{j\omega_0 \tau_1} R_{\underline{z}}(\tau_1 - \hat{\tau}_k) + \alpha_2 e^{j\omega_0 \tau_2} R_{\underline{z}}(\tau_2 - \hat{\tau}_k) + \cdots$$

$$+ \alpha_n e^{j\omega_0 \tau_n} R_{\underline{z}}(\tau_n - \hat{\tau}_k) + \cdots$$

$$= \sum_n \alpha_n e^{j\omega_0 \tau_n} R_{\underline{z}}(\tau_n - \hat{\tau}_k) = \sum_n \alpha_n e^{j\omega_0 \tau_n} R_{\underline{z}}(\tau_n - k/W) \tag{9.37e}$$

From (9.36c) we know the form of the autocorrelation function to be

$$R_{\underline{z}}(\tau) = \begin{cases} R_{\underline{z}}(0)(1 - W|\tau|), & |\tau| < 1/W \\ 0, & |\tau| > 1/W \end{cases} \qquad (9.38a)$$

so that for some value of n (denoted n_1):

$$\overline{\underline{Z}^*\underline{z}_k} = \sum_n \alpha_n e^{j\omega_0\tau_n} R_{\underline{z}}(\tau_n - k/W) \qquad (9.38b)$$

$$\approx 0 + \cdots + 0 + \alpha_{n_1} e^{j\omega_0\tau_{n_1}} R_{\underline{z}}(0)(1 - W|\tau_{n_1} - k/W|) + 0 + \cdots \qquad (9.38c)$$

$$= 0 + \cdots + 0 + \alpha_{n_1} e^{j\omega_0\tau_{n_1}} R_{\underline{z}}(0)(1 - |W\tau_{n_1} - k|) + 0 + \cdots \qquad (9.38d)$$

if there is a $\tau_{n_1} \in \{\tau_n\}$ such that $|W\tau_{n_1} - k| < 1$. This selection process is illustrated in Figure 9.23.

The Rake receiver, as conceived by Price and Green, is made up of a bank of cross-correlators, each with successive delays of $1/W$. Each cross-correlator extracts from the total received signal only that portion corresponding to multipath contributions arriving at a particular delay. Because the multipath delay spread is assumed to be T_m, a total of

$$L = T_m \div (1/W) = T_m W \qquad (9.39)$$

versions of the transmitted signal can be isolated (resolved) and combined. Thus a Rake receiver uses a form of L-fold diversity, without requiring the replication of equipment (other than the cross-correlators) to ensure inde-

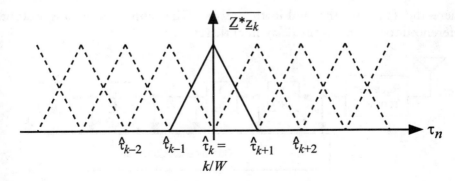

Figure 9.23 Mechanization of selecting the path with delay equal to $\widehat{\tau}_k$.

pendent diversity receptions. The basic Rake receiver structure is a tapped delay line that attempts to collect the signal energy from all the signal paths that fall within the span of the delay line and to combine them optimally, as depicted in Figure 9.24. The tapped delay line structure is reminiscent of an ordinary garden rake, so the nickname for the receiver used by Price and Green [14] is the so-called Rake receiver.

9.4.2 The Essence of Price and Green's Rake Concept

Price and Green used a PN sequence generator in combination with FSK to develop orthogonal wideband "mark" and "space" waveforms:

$$m_0(t) = PN(t) \cos 2\pi f_a t \quad \text{for mark waveform} \tag{9.40a}$$

and

$$m_1(t) = PN(t) \cos 2\pi f_b t \quad \text{for space waveform} \tag{9.40b}$$

where $PN(t)$ denotes a PN sequence time waveform, and f_a and f_b denote mark and space carrier frequencies, respectively. During a particular symbol interval, the received signal is either

$$s(t) = s_0(t) = \sum_n \alpha_n \, m_0(t - \tau_n) \tag{9.41a}$$

or

$$s(t) = s_1(t) = \sum_n \alpha_n \, m_1(t - \tau_n) \tag{9.41b}$$

where the $\{\tau_n\}$ are the multipath delays. The tapped delay line outputs (referenced to the *end* of the delay line) therefore are

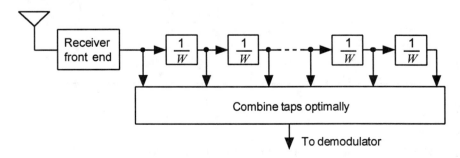

Figure 9.24 Tapped delay line combiner.

$$s\left(t + \frac{k}{W}\right) = s_0\left(t + \frac{k}{W}\right) = \sum_n \alpha_n\, m_0\left(t - \tau_n + \frac{k}{W}\right) \tag{9.42a}$$

or

$$s\left(t + \frac{k}{W}\right) = s_1\left(t + \frac{k}{W}\right) = \sum_n \alpha_n\, m_1\left(t - \tau_n + \frac{k}{W}\right) \tag{9.42b}$$

where W is the bandwidth of the wideband $PN(t)$ waveform, and the concept is depicted in Figure 9.25. The autocorrelation and cross-correlation functions of $m_0(t)$ and $m_1(t)$ are

$$
\begin{aligned}
R_{m_0}(\tau) &= \overline{m_0(t)\, m_0(t+\tau)} = \overline{PN(t)\, PN(t+\tau)}\,\overline{\cos(2\pi f_a t)\cos[2\pi f_a(t+\tau)]} \\
&= R_{PN}(\tau) \cdot \tfrac{1}{2}\cos(2\pi f_a \tau) \\
&\approx \begin{cases} R_{PN}(0)(1 - |W\tau|)\cdot\tfrac{1}{2}\cos(2\pi f_a\tau), & |W\tau| \le 1 \\ 0, & \text{otherwise} \end{cases}
\end{aligned} \tag{9.43a}
$$

$$
\begin{aligned}
R_{m_1}(\tau) &= \overline{m_1(t)\, m_1(t+\tau)} = \overline{PN(t)\, PN(t+\tau)}\,\overline{\cos(2\pi f_b t)\cos[2\pi f_b(t+\tau)]} \\
&= R_{PN}(\tau) \cdot \tfrac{1}{2}\cos(2\pi f_b \tau) \\
&\approx \begin{cases} R_{PN}(0)(1 - |W\tau|)\cdot\tfrac{1}{2}\cos(2\pi f_b\tau), & |W\tau| \le 1 \\ 0, & \text{otherwise} \end{cases}
\end{aligned} \tag{9.43b}
$$

and

$$
\begin{aligned}
R_{m_0 m_1}(\tau) &= \overline{m_0(t)\, m_1(t+\tau)} = \overline{PN(t)\, PN(t+\tau)}\,\overline{\cos(2\pi f_a t)\cos[2\pi f_b(t+\tau)]} \\
&\approx R_{PN}(\tau)\cdot 0 = 0
\end{aligned} \tag{9.43c}
$$

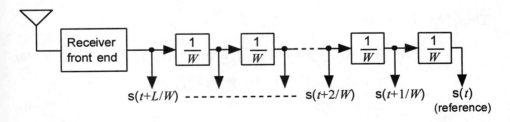

Figure 9.25 Tapped delay line Rake diversity reception concept used by Price and Green.

Price and Green used the correlation of the taps with the sum of $m_0(t)$ and $m_1(t)$ to estimate the multipath gains at the tap delays. Suppose that the mark symbol was transmitted. In principle, they used filtering to compute

$$X_k = \sum_n \overline{\alpha_n[m_0(t) + m_1(t)] m_0\left(t - \tau_n + \tfrac{k}{W}\right)}$$

$$= \sum_n \alpha_n \left[\overline{m_0(t) m_0\left(t - \tau_n + \tfrac{k}{W}\right)} + \overline{m_1(t) m_0\left(t - \tau_n + \tfrac{k}{W}\right)} \right]$$

$$(9.43\mathrm{d})$$

But

$$\overline{m_0(t) m_0\left(t - \tau_n + \tfrac{k}{W}\right)} = \overline{PN(t) PN\left(t - \tau_n + \tfrac{k}{W}\right)}$$

$$\times \overline{\cos(2\pi f_a t) \cos\left[2\pi f_a\left(t - \tau_n + \tfrac{k}{W}\right)\right]}$$

$$= R_{PN}\left(\tau_n - \tfrac{k}{W}\right) \cdot \tfrac{1}{2}\cos\left[2\pi f_a\left(\tau_n - \tfrac{k}{W}\right)\right]$$

$$\approx \begin{cases} R_{PN}(0)\left(1 - W\left|\tau_n - \tfrac{k}{W}\right|\right) \tfrac{1}{2}\cos\left[2\pi f_a\left(\tau_n - \tfrac{k}{W}\right)\right], & \left|\tau_n - \tfrac{k}{W}\right| < \tfrac{1}{W} \\ 0, & \text{otherwise} \end{cases}$$

$$= \begin{cases} R_{PN}(0)(1 - W|\epsilon_{nk}|) \tfrac{1}{2}\cos(2\pi f_a \epsilon_{nk}), & |\epsilon_{nk}| < \tfrac{1}{W} \\ 0, & \text{otherwise} \end{cases} \qquad (9.43\mathrm{e})$$

where

$$\epsilon_{nk} \triangleq \tau_n - k/W \approx 0 \qquad (9.43\mathrm{f})$$

and

$$\overline{m_1(t) m_0\left(t - \tau_n + \tfrac{k}{W}\right)} = 0 \qquad (9.43\mathrm{g})$$

Thus, (9.43d) becomes

$$X_k = \alpha_k R_{PN}(0)(1 - W|\epsilon_{nk}|) \tfrac{1}{2}\cos(2\pi f_a \epsilon_{nk}) \qquad (9.44\mathrm{a})$$

$$\approx \alpha_k R_{PN}(0)(1 - W \cdot 0) \tfrac{1}{2}\cos(2\pi f_a \cdot 0)$$

$$= \tfrac{1}{2}R_{PN}(0) \alpha_k \qquad (9.44\mathrm{b})$$

which gives the estimate $\widehat{\alpha}_k$ for the fading factor α_k for the mark symbol.

Similarly, for the case in which the space signal is being received, the cross-correlation gives

$$X_k \approx \tfrac{1}{2} R_{PN}(0)\, \alpha_k \qquad\qquad (9.44c)$$

Having the multipath gains $\{\widehat{\alpha}_k\}$ at the tap delays, the signals at the taps can be weighted in proportion to their gains and summed coherently. The summed result is then correlated with the possible message waveforms to form mark and space decision variables. Therefore, we see that the original Price and Green Rake receiver is a form of MRC diversity receiver, and, hence, the term Rake contains the connotation of MRC as a feature as well. This is depicted in the Price and Green diversity receiver configuration diagrammed in Figure 9.26.

Figure 9.26 is a simplified version of the Rake receiver used by Price and Green. In the original scheme, they used special heterodyning circuitry to refine the coherent combining. They also took advantage of the linearity of the various receiver operations to minimize the hardware. For example, the diagram shown in Figure 9.26 features correlations *after* the combining that both isolate the paths and form the symbol decision variables; that is, $m_0(t)$ and $m_1(t)$ are correlated with the sum of weighted and delayed

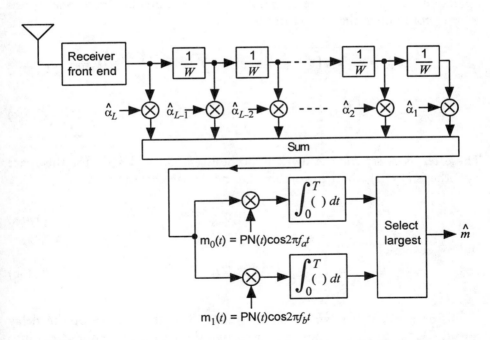

Figure 9.26 Essential features of the Rake receiver configuration invented by Price and Green [14]. (The $\{\widehat{\alpha}_k\}$ are fading parameter estimates.)

versions of the *total* input signal. The sum has the form

$$\sum_k \widehat{\alpha}_k \, s\!\left(t + \tfrac{k}{W}\right) = \sum_k \sum_n \widehat{\alpha}_k \alpha_n \, m_i\!\left(t - \tau_n + \tfrac{k}{W}\right), \quad i = 0, 1 \quad (9.45a)$$

and after the correlation with the signal waveforms $m_0(t) = \text{PN}(t) \cos 2\pi f_a t$ and $m_1(t) = \text{PN}(t) \cos 2\pi f_b t$, the multiple paths "line up" and are coherently combined:

$$\overline{m_j(t) \sum_k \widehat{\alpha}_k \, s\!\left(t + \tfrac{k}{W}\right)} = \sum_k \sum_n \widehat{\alpha}_k \alpha_n \, \overline{m_j(t) \, m_i\!\left(t - \tau_n + \tfrac{k}{W}\right)}$$

$$\approx \begin{cases} \displaystyle\sum_k \widehat{\alpha}_k \alpha_k \cdot \tfrac{1}{2} R_{\text{PN}}(0), & i = j \\[2mm] 0, & i \neq j \end{cases} \quad (9.45b)$$

where i and j take the values 0 and 1. An equivalent procedure, because the operations involved are linear, is to isolate the individual path waveforms *prior* to combining; that is, to form the terms

$$Z_{jk} = \overline{m_j(t) \, s\!\left(t + \tfrac{k}{W}\right)} = m_j(t) \sum_n \alpha_n m_i\!\left(t - \tau_n + \tfrac{k}{W}\right)$$

$$\approx \begin{cases} \alpha_k \cdot \tfrac{1}{2} R_{\text{PN}}(0), & j = i \\[2mm] 0, & j \neq i \end{cases} \quad (9.45c)$$

These terms, a set for the mark hypothesis ($j = 0$) and a set for the space hypothesis ($j = 1$), are then combined using the weights to produce

$$Z_0 = \sum_k \widehat{\alpha}_k Z_{0k} \quad (9.46a)$$

and

$$Z_1 = \sum_k \widehat{\alpha}_k Z_{1k} \quad (9.46b)$$

For a signal bandwidth much greater than the inverse of the delay spread ($W T_m \gg 1$), it is possible to resolve the multipath components into separate signals at delay intervals of

$$\Delta t = 1/W \qquad (9.47)$$

The relative amplitudes and phases of the existing multipath components are found by correlating the received waveform with delayed versions of the signal or vice versa. The energy in the multipaths can be recovered effectively by combining the (delay-compensated) multipaths in proportion to their strengths. This combining is a form of diversity and can help overcome fading. Multipaths with relative delays less than $\Delta t = 1/W$ cannot be resolved, and if existing, contribute to fading; in such cases forward error-correction coding and power control schemes play the dominant roles in mitigating the effects of fading.

9.4.3 The Use of the Rake Concept in IS-95

In CDMA cellular and PCS systems, the cell site-to-mobile (forward) link uses a "three-finger" Rake receiver, and the mobile-to-cell site (reverse) link uses a "four-finger" Rake receiver. The terminology, "n-finger" Rake, indicates that the number of paths that can be combined is n. Recall that Price and Green's original Rake receiver used L equally spaced delay-line taps for combining up to L paths.

The use of the term "Rake" in describing the IS-95 receiver design is somewhat different from the design originally put forward by Price and Green. The IS-95 Rake receivers detect and measure the parameters of multiple signals that may be used for diversity reception or for handoff purposes, and combine the multiple signal paths coherently (i.e., in symbol synchronization) after demodulation of each path signal (postdetection combining) based on the MRC principle.

In the IS-95 system, the detection and measurement of multipath parameters are performed by a "searcher receiver," which is programmed to compare incoming signals with portions of the I- and Q-channel PN codes. Multipath arrivals at the receiver unit manifest themselves as correlation peaks occurring at different times. A peak's magnitude is proportional to the envelope of the path signal, and the time of each peak, relative to the first arrival, gives a measurement of the path's delay. It is therefore clear that the IS-95 design takes the approach of determining what paths exist, rather than having a predetermined, fixed number of taps as in Price and Green's original Rake receiver.

The PN chip rate of 1.2288 MHz allows for resolution of multipaths at time intervals of $(1.2288 \times 10^6)^{-1} = 814 \times 10^9 = 0.814\,\mu s$. Because all the base stations use the same I and Q PN codes, differing only in code phase offset, not only multipaths but also other base stations will be eventually detected by correlation (in a different "search window" of arrival times) with the portion of the codes corresponding to the selected base stations. Thus, the searcher receiver can maintain a table of the stronger multipaths and/or base station signals for possible diversity combining or for handoff purposes. To be useful, the table must register time of arrival, signal strength, and the corresponding PN code offset.

On the reverse link, the base station's receiver that is assigned to track a particular mobile transmitter uses the zero-offset I- and Q-code times of arrival to confine the search to mobile signals from users affiliated with that base station. The searcher receiver at the base station can distinguish the desired mobile signal by means of its unique scrambling long PN code offset, acquired before voice or data transmission begins on the link, using a special preamble for that purpose. As the call proceeds, the searcher receiver is able to monitor the strengths of multipaths from the mobile unit to the base station and to use more than one path through diversity combining.

References

[1]   Montgomery, G. F., "Message Error in Diversity Frequency-Shift Reception," *Proc. of the IRE*, Vol. 42, pp. 1184–1187, July 1954.

[2]   Bond, F. E., and H. F. Meyer, "The Effect of Fading on Communication Circuits Subject to Interference," *Proc. of the IRE*, Vol. 45, pp. 636–642, May 1957.

[3]   Pierce, J. N., "Theoretical Diversity Improvement in Frequency-Shift Keying," *Proc. of the IRE*, Vol. 46, pp. 903–910, May 1958.

[4]   Hahn, P. M., "Theoretical Diversity Improvement in Multiple Frequency-Shift Keying," *IRE Trans. on Communication Syst.*, Vol. CS-10, pp. 117–184, June 1962.

[5]   Proakis, J. C., *Digital Communications*, New York: McGraw-Hill, 1983.

[6]   Leung, C., "Optimized Selection Diversity for Rayleigh Fading Channels," *IEEE Trans. on Communications*, Vol. COM-30, pp. 554–557, Mar. 1982.

[7] Stein, S., "Fading Channel Issues in System Engineering," *IEEE J. on Selected Areas in Communications*, Vol. SAC-5, pp. 68–89, Feb. 1987.

[8] Wozencraft, J. M., and I. M. Jacobs, *Principles of Communication Engineering*, New York: Wiley, 1965.

[9] Lee, W. C. Y., *Mobile Communication Engineering*, New York: McGraw-Hill, 1982.

[10] Kavehrad, M., and B. Ramamurthi, "Direct-Sequence Spread Spectrum With DPSK Modulation and Diversity for Indoor Wireless Communications," *IEEE Trans. on Communications*, Vol. COM-35, pp. 224–236, Feb. 1987.

[11] Miller, L. E., and J. S. Lee, "BER Expressions for Differentially Detected $\pi/4$ DQPSK Modulation," *IEEE Trans. on Communications*, Vol. 46, pp. 71–81, Jan. 1998.

[12] Lee, J. S., L. E. Miller, R. H. French, Y. K. Kim, and A. P. Kadrichu, "The Optimum Jamming Effects on Frequency-Hopping M-ary FSK Systems Under Certain ECCM Receiver Design Strategies," J. S. Lee Associates, Inc., final report to the Office of Naval Research under contract N00014-83-C-0312, Oct. 1984 (DTIC accession number AD-A147766).

[13] Brennan, D. G., "Linear Diversity Combining Techniques," *Proc. IRE*, Vol. 47, pp. 1075–1102, June 1959.

[14] Price, R., and P. E. Green, Jr., "A Communication Technique for Multipath Channels," *Proc. IRE*, Vol. 46, pp. 555–570, Mar. 1958.

[15] Gradshteyn, I. S., and I. M. Rhyzhik, *Table of Integrals, Series, and Products* (4th ed.), New York: Academic Press, 1965.

[16] Henrici, P., *Applied and Computational Complex Analysis*, Vol. 1, New York: Wiley, 1974.

[17] Abramowitz, M., and I. A. Stegun (eds.), *Handbook of Mathematical Functions*, National Bureau of Standards Applied Mathematics Series 55 (ninth printing), Washington, D.C.: Government Printing Office, 1970.

Selected Bibliography

Adachi, F., "Postdetection Optimal Diversity Combiner for DPSK Differential Detection," *IEEE Trans. on Vehicular Technology*, Vol. 42, pp. 326–337, Aug., 1993.

Adachi, F., and J. D. Parsons, "Unified Analysis of Postdetection Diversity for Binary Digital FM Mobile Radio," *IEEE Trans. on Vehicular Technology*, Vol. 37, pp. 189–198, Nov. 1988.

Adams, R. T., and B. M. Mindes, "Evaluation of IF and Baseband Diversity Combining Receivers," *IRE Trans. on Communications Syst.*, Vol. CS-6, pp. 8–13, June 1958.

Balaban, P., and J. Salz, "Optimum Diversity Combining and Equalization in Digital Data Transmission With Applications to Cellular Mobile Radio —Part I: Theoretical Considerations," *IEEE Trans. on Communications*, Vol. 40, pp. 885–907, May 1992.

Barrow, B. B., "Diversity Combination of Fading Signals With Unequal Mean Strengths," *IEEE Trans. on Communication Systems*, Vol. CS-11, pp. 73–78, Mar. 1963.

Beaulieu, N. C., and A. A. Abu-Dayya, "Analysis of Equal Gain Diversity on Nakagami Fading Channels," *IEEE Trans. on Communications*, Vol. 39, pp. 225–234, Feb. 1991.

Bello, P., and B. Nellin, "Predetection Diversity Combining With Selectively Fading Channels," *IRE Trans. on Communication Systems*, Vol. CS-12, pp. 34–42, Mar. 1962.

Chang, I., G. L. Stuber, and A. M. Bush, "Performance of Diversity Combining Techniques for DS/DPSK Signaling Over a Pulse Jammed Multipath-Fading Channel," *IEEE Trans. on Communications*, Vol. 38, pp. 1823–1834, Oct. 1990.

Chyi, G.-T., J. G. Proakis, and C. M. Keller, "On the Symbol Error Probability of Maximum-Selection Diversity Reception Schemes Over a Rayleigh Fading Channel," *IEEE Trans. on Communications*, Vol. 37, pp. 79–83, Jan. 1989.

Cox, D., "Antenna Diversity Performance in Mitigating the Effects of Portable Radiotelephone Orientation and Multipath Propagation," *IEEE Trans. on Communications*, Vol. COM-31, pp. 620–628, May 1983.

Gaarder, N. T., "Maximal-Ratio Diversity Combiners," *IEEE Trans. on Comm. Technology*, Vol. COM-15, pp. 492–500, Dec. 1967.

Gans, M. J., "The Effect of Gaussian Error in Maximal Ratio Combiners," *IEEE Trans. on Comm. Technology*, Vol. COM-19, pp. 492–500, Aug. 1971.

Gilbert, E. N., "Mobile Radio Diversity Reception," *Bell Systems Technical Journal*, pp. 2473–2492, Sept. 1969.

Glance, B., and L. J. Greenstein, "Frequency-Selective Fading Effects in Digital Mobile Radio with Diversity Combining," *IEEE Trans. on Communications*, Vol. COM-31, pp. 1085–1094, Sept. 1983.

Halpern, S. W., "The Theory of Operation of an Equal Gain Predetection Regenerative Diversity Combiner with Rayleigh Fading Channels," *IEEE Trans. on Communications*, Vol. COM-22, pp. 1099–1106, Aug. 1974.

Jakes, W. C., Jr., ed., *Microwave Mobile Communications*, New York: Wiley, 1974.

Kam, P. Y., "Adaptive Diversity Reception Over a Slow Nonselective Fading Channel," *IEEE Trans. on Communications*, Vol. COM-35, pp. 572–574, May 1987.

Kam, P. Y., "Optimal Detection of Digital Data Over the Nonselective Rayleigh Fading Channel with Divesity Reception," *IEEE Trans. on Communications*, Vol. 39, pp. 214–224, Feb. 1991.

Kchao, C., and G. L. Stuber, "Analysis of a Direct-Sequence Spread-Spectrum Cellular Radio System," *IEEE Trans. on Communications*, Vol. 41, pp. 1507–1516, Oct. 1993.

Komo, J. J., and A. Aridgides, "Diversity Cutoff Rate of the Rayleigh and Rician Fading Channels," *IEEE Trans. on Communications*, Vol. COM-35, pp. 762–764, July 1987.

Lee, W. C. Y., and Y. S. Yeh, "Polarization Diversity System for Mobile Radio," *IEEE Trans. on Communications*, Vol. COM-20, pp. 912–923, Oct. 1972.

Lee, W. C. Y., "Antenna Spacing Requirement for Mobile Radio Base-Station Diversity," *Bell System Technical Journal*, pp. 1859–1876, July-Aug. 1971.

Lehnert, J. S., and M. B. Pursley, "Multipath Diversity Reception of Spread-Spectrum Multiple-Access Communications," *IEEE Trans. on Communications*, Vol. COM-35, pp. 1189–1198, Nov. 1987.

Leung, C., "Optimized Selection Diversity for Rayleigh Fading Channels," *IEEE Trans. on Communications*, Vol. COM-30, pp. 554–557, Mar. 1982.

Lindsey, W. C., "Error Probability for Incoherent Diversity Reception," *IEEE Trans. on Information Theory*, Vol. IT-11, pp. 491–499, Oct. 1965.

Monsen, P., "Digital Transmission Performance on Fading Dispersive Diversity Channels," *IEEE Trans. on Communications*, Vol. COM-21, pp. 33–39, Jan. 1973.

Montgomery, G. F., "Message Error in Diversity Frequency-Shift Reception," *Proc. IRE*, Vol. 42, pp. 1184–1187, July 1954.

Packard, K. S., "Effect of Correlation on Combiner Diversity," *Proc. IRE*, Vol. 46, pp. 362–363, Jan. 1958.

Parsons, J. D., and J. G. Gardiner, *Mobile Communication Systems*, London: Blackie and Son Ltd., 1989.

Parsons, J. D., M. Henze, P. A. Ratliff, and M. J. Withers, "Diversity Techniques for Mobile Radio Reception," *The Radio and Electronic Engineer*, pp. 357–367, July 1975.

Parsons, J. D., and A. Pongsupaht, "Error-Rate Reduction in VHF Mobile Radio Data Systems Using Specific Diversity Reception Techniques," *IEE Proc. Part F*, pp. 475–484, Dec. 1980.

Pierce, J. N., and S. Stein, "Multiple Diversity with Nonindependent Fading," *Proc. IRE*, Vol. 48, pp. 89–104, Jan. 1960.

Schiff, L., "The Statistics of the In-Phase Component of the Branch Signal in a Pilot Derived Maximal Ratio Combiner," *IEEE Trans. on Communications*, Vol. COM-21, pp. 1319–1320, Nov. 1973.

Sesay, A. B., "Two-Stage Maximum Likelihood Estimation for Diversity Combining in Digital Mobile Radio," *IEEE Trans. on Communications*, Vol. 40, pp. 676–679, Apr. 1992.

Shortall, W. E., "A Switched Diversity Receiving System for Mobile Radio," *IEEE Trans. on Communications*, Vol. COM-21, pp. 1269–1275, Nov. 1973.

Sidwell, J. M., "A Diversity Combiner Giving Total Power Transfer," *IEE Proc.*, pp. 305–309, July 1962.

Staras, H., "Diversity Reception with Correlated Signals," *J. Applied Physics*, pp. 93–94, Jan. 1956.

Staras, H., "The Statistics of Combiner Diversity," *Proc. IRE*, Vol. 44, pp. 1057–1058, Aug. 1956.

Turin, G. L., "An Introduction to Spread-Spectrum Antimultipath Techniques and Their Application to Urban Digital Radio," *Proc. IEEE*, Vol. 68, pp. 328–353, Mar. 1980.

Turin, G. L., "On Optimal Diversity Reception," *IRE Trans. on Information Theory*, Vol. IT-7, pp. 154–166, July 1971.

Appendix 9A Derivation of M-ary Orthogonal Diversity Performances

9A.1 Selection Diversity

The integral to be solved is

$$\int_0^\infty P_s(e \mid \rho_{max} = \beta)\, p_{\rho_{max}}(\beta)\, d\beta$$

$$= \sum_{i=1}^{M-1} \binom{M-1}{i} \frac{(-1)^{i+1}}{i+1} \int_0^\infty \exp\left(-\frac{i}{i+1}\beta\right) p_{\rho_{max}}(\beta)\, d\beta \qquad (9A.1a)$$

where

$$p_{\rho_{max}}(\beta) = \frac{L}{\overline{\rho}_c}\left(1 - e^{-\beta/\overline{\rho}_c}\right)^{L-1} e^{-\beta/\overline{\rho}_c} \qquad (9A.1b)$$

Using integration formula 3.312.1 of [15], the integral in (9A.1a) can be put into the form

$$\frac{L}{\overline{\rho}_c}\int_0^\infty e^{-a\beta}\left(1 - e^{-\beta/\overline{\rho}_c}\right)^{L-1} d\beta$$

$$= \frac{L}{\overline{\rho}_c a}\int_0^\infty e^{-x}\left(1 - e^{-x/\overline{\rho}_c a}\right)^{L-1} dx \qquad (9A.2a)$$

$$= \frac{L}{\overline{\rho}_c a}\cdot \overline{\rho}_c a\, B(\overline{\rho}_c a, L) \qquad (9A.2b)$$

where

$$a = \frac{i}{i+1} + \frac{1}{\overline{\rho}_c} = \frac{1}{\overline{\rho}_c}\left(1 + \frac{i}{i+1}\overline{\rho}_c\right) \tag{9A.2c}$$

and the beta function is

$$\mathrm{B}(\mu, \nu) = \frac{\Gamma(\mu)\Gamma(\nu)}{\Gamma(\mu+\nu)} \tag{9A.2d}$$

Thus (9A.2b) can be written

$$
\begin{aligned}
L\mathrm{B}(\overline{\rho}_c a, L) &= \frac{L\Gamma(\overline{\rho}_c a)\Gamma(L)}{\Gamma(\overline{\rho}_c a + L)} = \frac{L!\,\Gamma\left(1 + \frac{i}{i+1}\overline{\rho}_c\right)}{\Gamma\left(L + 1 + \frac{i}{i+1}\overline{\rho}_c\right)} \\
&= \frac{L!}{\left(1 + \frac{i}{i+1}\overline{\rho}_c\right)\left(2 + \frac{i}{i+1}\overline{\rho}_c\right)\cdots\left(L + \frac{i}{i+1}\overline{\rho}_c\right)} \\
&= \prod_{l=1}^{L} \frac{l}{l + \frac{i}{i+1}\overline{\rho}_c} \tag{9A.2e}
\end{aligned}
$$

When substituted in (9A.1a), this result gives the error probability shown in (9.3d). Instead of this finite product form, (9A.2a) can be further transformed to yield a finite summation:

$$
\begin{aligned}
\frac{L}{\overline{\rho}_c a}\int_0^\infty e^{-x}\left(1 - e^{-x/\overline{\rho}_c a}\right)^{L-1} dx &= \frac{L}{\overline{\rho}_c a}\int_0^1 \left(1 - u^{1/\overline{\rho}_c a}\right)^{L-1} du \\
&= \frac{L}{\overline{\rho}_c a}\sum_{l=0}^{L-1}\binom{L-1}{l}(-1)^l \int_0^1 u^{l/\overline{\rho}_c a}\, du \\
&= L\sum_{l=0}^{L-1}\binom{L-1}{l}\frac{(-1)^l}{l + \overline{\rho}_c a} = L\sum_{l=0}^{L-1}\binom{L-1}{l}\frac{(-1)^l}{l + 1 + \frac{i}{i+1}\overline{\rho}_c} \tag{9A.3}
\end{aligned}
$$

When substituted in (9A.1a), this expression gives the error probability shown in (9.3e).

9A.2 EGC Diversity Reception

The inner integral in (9.10c) is determined to be

$$\int_0^{\alpha} p_{z_2}(\beta)\, d\beta = \int_0^{\alpha} \frac{1}{2\sigma^2 \Gamma(L)} e^{-\beta/2\sigma^2} \left(\frac{\beta}{2\sigma^2} \right)^{L-1} d\beta$$

$$= \int_0^{\alpha/2\sigma^2} \frac{1}{\Gamma(L)} x^{L-1} e^{-x}\, dx$$

$$= 1 - e^{-\alpha/2\sigma^2} \sum_{k=0}^{L-1} \frac{(\alpha/2\sigma^2)^k}{k!} \tag{9A.4a}$$

and raised to the $(M-1)$st power, it is

$$\left[\int_0^{\alpha} p_{z_2}(\beta)\, d\beta \right]^{M-1} = \left[1 - e^{-\alpha/2\sigma^2} \sum_{k=0}^{L-1} \frac{(\alpha/2\sigma^2)^k}{k!} \right]^{M-1}$$

$$= \sum_{q=0}^{M-1} \binom{M-1}{q} (-1)^q\, e^{-q\alpha/2\sigma^2} \left[\sum_{k=0}^{L-1} \frac{(\alpha/2\sigma^2)^k}{k!} \right]^q \tag{9A.4b}$$

The probability of symbol error then is

$$P_M(e; L) = 1 - \int_0^{\infty} p_{z_1}(\alpha) \cdot 1\, d\alpha$$

$$- \int_0^{\infty} p_{z_1}(\alpha) \cdot \sum_{q=1}^{M-1} \binom{M-1}{q} (-1)^q\, e^{-q\alpha/2\sigma^2} \left[\sum_{k=0}^{L-1} \frac{(\alpha/2\sigma^2)^k}{k!} \right]^q d\alpha$$

$$= \sum_{q=1}^{M-1} \binom{M-1}{q} (-1)^{q+1} \int_0^{\infty} p_{z_1}(\alpha)\, e^{-q\alpha/2\sigma^2} \left[\sum_{k=0}^{L-1} \frac{(\alpha/2\sigma^2)^k}{k!} \right]^q d\alpha$$

$$= \sum_{q=1}^{M-1} \binom{M-1}{q} (-1)^{q+1} \int_0^{\infty} 2\sigma^2 p_{z_1}(2\sigma^2 x)\, e^{-qx} \left[\sum_{k=0}^{L-1} \frac{x^k}{k!} \right]^q dx \tag{9A.4c}$$

To complete the derivation, it is necessary to expand the power of the summation in (9A.4c) and then carry out the indicated integration. Because the summation is finite, the power of the summation has the form

$$\left[\sum_{k=0}^{L-1} \frac{x^k}{k!} \right]^q = \sum_{n=0}^{q(L-1)} c_n(q) \cdot \frac{x^n}{n!} \tag{9A.5}$$

Using the J. C. P. Miller multinomial expansion [12, 16], the coefficients of the exansion, $\{c_n(q)\}$, can be shown to be

$$
c_n(q) = \begin{cases}
\dfrac{1}{n}\displaystyle\sum_{i=1}^{n} \binom{n}{i}[(q+1)i - n]\, c_{n-i}(q), & n \leq L-1 \\[3mm]
\dfrac{1}{n}\displaystyle\sum_{i=1}^{L-1} \binom{n}{i}[(q+1)i - n]\, c_{n-i}(q), & n \geq L-1
\end{cases}
\tag{9A.6}
$$

with $c_0(q) = 1$ for all q.

Example computation of the coefficients. Let $L = 3$ and $q = 4$. Then $q(L-1) = 9$ coefficients are needed, and are computed as follows:

$$c_0(4) = 1$$

$$c_1(4) = \tfrac{1}{1} \cdot \binom{1}{1}[5 \cdot 1 - 1]c_0(4) = 4c_0(4) = 4$$

$$c_2(4) = \tfrac{1}{2}\left\{\binom{2}{1}[5 \cdot 1 - 2]c_1(4) + \binom{2}{2}[5 \cdot 2 - 2]c_0(4)\right\}$$

$$\qquad = \tfrac{1}{2}\{2 \cdot 3 \cdot 4 - 1 \cdot 8 \cdot 1\} = 16$$

$$c_3(4) = \tfrac{1}{3}\{6c_2(4) + 21c_1(4)\} = 60$$

$$c_4(4) = \tfrac{1}{4}\{4c_3(4) + 36c_2(4)\} = 204$$

$$c_5(4) = \tfrac{1}{5}\{0c_4(4) + 50c_3(4)\} = 10c_3 = 600$$

$$c_6(4) = \tfrac{1}{6}\{-6c_5(4) + 60c_4(4)\} = 10c_4 - c_5 = 2{,}040 - 600 = 1{,}440$$

$$c_7(4) = \tfrac{1}{7}\{-14c_6(4) + 63c_5(4)\} = 9c_5 - 2c_6 = 5{,}400 - 2{,}880 = 2{,}520$$

$$c_8(4) = \tfrac{1}{8}\{-24c_7(4) + 56c_6(4)\} = 7c_6 - 3c_7 = 10{,}080 - 7{,}560 = 2{,}520$$

We may check these computations by direct calculation of the coefficients:

$$\left(1 + x + \tfrac{1}{2}x^2\right)^4 = \left(1 + 2x + 2x^2 + x^3 + \tfrac{1}{4}x^4\right)^2$$

$$= 1 + 4x + 8x^2 + 10x^3 + \tfrac{17}{2}x^4 + 5x^5 + 2x^6 + \tfrac{1}{2}x^7 + \tfrac{1}{16}x^8$$

$$= 1 + \tfrac{4}{1!}x + \tfrac{16}{2!}x^2 + \tfrac{60}{3!}x^3 + \tfrac{204}{4!}x^4 + \tfrac{600}{5!}x^5 + \tfrac{1{,}440}{6!}x^6$$

$$\qquad + \tfrac{2{,}520}{7!}x^7 + \tfrac{2{,}520}{8!}x^8$$

$$= 1 + \tfrac{c_1}{1!}x + \tfrac{c_2}{2!}x^2 + \tfrac{c_3}{3!}x^3 + \tfrac{c_4}{4!}x^4 + \tfrac{c_5}{5!}x^5 + \tfrac{c_6}{6!}x^6 + \tfrac{c_7}{7!}x^7 + \tfrac{c_8}{8!}x^8$$

Returning to the derivation, with the expansion of the power series, the error probability (9A.4c) becomes

$$P_M(e; L) = \sum_{q=1}^{M-1} \binom{M-1}{q} (-1)^{q+1} \sum_{n=0}^{q(L-1)} \frac{c_n(q)}{n!}$$

$$\times \int_0^\infty 2\sigma^2 p_{z_1}(2\sigma^2 x)\, x^n e^{-qx}\, dx \qquad (9A.7a)$$

where the integrand is

$$2\sigma^2 p_{z_1}(2\sigma^2 x)\, x^n e^{-qx} = e^{-(q+1)x - L\rho}\, x^n \left(\frac{x}{L\rho}\right)^{(L-1)/2} I_{L-1}\left(2\sqrt{L\rho x}\right)$$

$$(9A.7b)$$

From a table of integrals [15, formulas 6.643.2 and 9.220.2] we find that the integral in (9A.7a) is of the form

$$\int_0^\infty x^{\mu-1/2}\, e^{-\kappa x} I_{2\nu}\left(2\gamma\sqrt{x}\right) dx$$

$$= \frac{\Gamma\left(\mu+\nu+\tfrac{1}{2}\right)}{\Gamma(2\nu+1)} \cdot \frac{\gamma^{2\nu}}{\kappa^{\mu+\nu+1/2}}\, {}_1F_1\left(\mu+\nu+\tfrac{1}{2}; 2\nu+1; \frac{\gamma^2}{\kappa}\right) \qquad (9A.8a)$$

where

$$\mu = n + \tfrac{1}{2}L \qquad\qquad \nu = \tfrac{1}{2}(L-1)$$

$$\kappa = q+1 \qquad\qquad \gamma = \sqrt{L\rho}$$

and

$${}_1F_1(a; b; x) \triangleq \frac{\Gamma(b)}{\Gamma(a)} \sum_{i=0}^\infty \frac{x^i}{i!} \cdot \frac{\Gamma(a+i)}{\Gamma(b+i)} \qquad (9A.8b)$$

is the confluent hypergeometric function [17, Chapter 13]. Applying the solution for the integral results in the expression

$$P_M(e; L) = e^{-L\rho} \sum_{q=1}^{M-1} \binom{M-1}{q} \frac{(-1)^{q+1}}{(q+1)^L} \sum_{n=0}^{q(L-1)} \frac{c_n(q)}{n!\,(q+1)^n}$$

$$\times \frac{\Gamma(n+L)}{\Gamma(L)}\, {}_1F_1\left(n+L; L; \frac{L\rho}{q+1}\right) \qquad (9A.9a)$$

For computational purposes, this result can be simplified by using Kummer's transformation for the confluent hypergeometric function [17]

$$_1F_1(a; b; x) = e^x \, _1F_1(b - a; b; -x) \tag{9A.9b}$$

Applying Kummer's transformation to the confluent hypergeometric function in (9A.9a) results in a finite series:

$$_1F_1\left(n + L; L; \frac{L\rho}{q+1}\right) = e^{L\rho/(q+1)} \, _1F_1\left(-n; L; \frac{-L\rho}{q+1}\right)$$

$$= e^{L\rho/(q+1)} \sum_{i=0}^{\infty} \overbrace{\frac{\Gamma(L)\Gamma(-n+i)}{\Gamma(-n)\Gamma(L+i)}}^{\text{this factor} = 0 \text{ for } n > i} \frac{(-1)^i}{i!}\left(\frac{L\rho}{q+1}\right)^i$$

$$= e^{L\rho/(q+1)} \sum_{i=0}^{n} \frac{\Gamma(L)}{(L+i-1)!} \cdot \overbrace{\frac{(-1)^i n!/(n-i)!}{(-n)(-n+1)\cdots(-n+i-1)}}^{}$$

$$\times \frac{(-1)^i}{i!}\left(\frac{L\rho}{q+1}\right)^i$$

$$= \frac{n!\,\Gamma(L)}{(n+L-1)!} \, e^{L\rho/(q+1)} \overbrace{\sum_{i=0}^{n}\binom{n+L-1}{n-i}\frac{1}{i!}\left(\frac{L\rho}{q+1}\right)^i}^{\text{degree } n \text{ Laguerre polynomial}}$$

$$= \frac{n!\,\Gamma(L)}{(n+L-1)!} \, e^{L\rho/(q+1)} \, \mathcal{L}_n^{(L-1)}\left(\frac{-L\rho}{q+1}\right) \tag{9A.9c}$$

When this expression is substituted in (9A.9a), the error probability given in (9.11a) of the text results.

Appendix 9B Derivation of BPSK Diversity Performances

9B.1 Selection Diversity

The desired result is given by the integral

$$P(\overline{\rho}_c) = \int_0^\infty Q(\sqrt{2\beta}) \frac{L}{\overline{\rho}_c}(1 - e^{-\beta/\overline{\rho}_c})^{L-1} e^{-\beta/\overline{\rho}_c} \, d\beta \qquad (9B.1)$$

where $Q(\cdot)$ is the Gaussian Q-function. The solution proceeds as follows:

$$P(\overline{\rho}_c) = L \int_0^\infty Q(\sqrt{2\overline{\rho}_c x})(1 - e^{-x})^{L-1} e^{-x} \, dx$$

$$= L \sum_{k=0}^{L-1} \binom{L-1}{k} (-1)^k \int_0^\infty Q(\sqrt{2\overline{\rho}_c x}) \, e^{-(k+1)x} \, dx$$

$$= L \sum_{k=0}^{L-1} \binom{L-1}{k} \frac{(-1)^k}{k+1} \int_0^\infty Q\left(\sqrt{\frac{2\overline{\rho}_c y}{k+1}}\right) e^{-y} \, dy \qquad (9B.2)$$

Note that the integral in (9B.2) is the BPSK probability of error in Rayleigh fading and no diversity, as formulated in Appendix 7B in the line following (7B.1), with ρ replaced by $\overline{\rho}_c/(k+1)$. Therefore, with this substitution in (7B.2), we can write

$$\int_0^\infty Q\left(\sqrt{\frac{2\overline{\rho}_c y}{k+1}}\right) e^{-y} \, dy = \frac{1}{2}\left[1 - \sqrt{\frac{\overline{\rho}_c/(k+1)}{1 + \overline{\rho}_c/(k+1)}}\right]$$

$$= \frac{1}{2}\left[1 - \sqrt{\frac{\overline{\rho}_c}{k+1+\overline{\rho}_c}}\right] \qquad (9B.3)$$

and (9B.2) becomes

$$P(\overline{\rho}_c) = \frac{L}{2} \sum_{k=0}^{L-1} \binom{L-1}{k} \frac{(-1)^k}{k+1}\left[1 - \sqrt{\frac{\overline{\rho}_c}{k+1+\overline{\rho}_c}}\right] \qquad (9B.4)$$

9B.2 EGC Diversity Reception

The integral to be solved, (9.19e), can be put into the form

$$P(\overline{\rho}_c) = \int_0^\infty Q(\sqrt{2\overline{\rho}_c x}) \frac{x^{L-1}}{\Gamma(L)} e^{-x} \, dx \qquad (9B.5a)$$

$$= \frac{1}{\Gamma(L)\sqrt{2\pi}} \int_{\sqrt{2\overline{\rho}_c}}^{\infty} \left[\int_0^{\infty} x^{L-1/2}\, e^{-x(1+y^2/2)}\, dx \right] dy$$

$$= \frac{\Gamma(L+\frac{1}{2})}{\Gamma(L)\sqrt{\pi}} \int_{\sqrt{\overline{\rho}_c}}^{\infty} \frac{1}{(1+u^2)^{L+1/2}}\, du \tag{9B.5b}$$

in which

$$\frac{\Gamma(L+\frac{1}{2})}{\Gamma(L)\sqrt{\pi}} = \frac{\Gamma(L+\frac{1}{2})}{\Gamma(L)\Gamma(\frac{1}{2})} = \frac{(\frac{1}{2})(\frac{3}{2})\cdots(L-\frac{1}{2})}{\Gamma(L)}$$

$$= \frac{\Gamma(2L)}{2^{2L-1}\Gamma(L)\Gamma(L)} = \frac{1}{2^{2L-1}\,B(L,L)} \tag{9B.6}$$

where $B(\,\cdot\,,\,\cdot\,)$ is the beta function. Note that the lower limit of the integral in (9B.5b) is $\sqrt{\overline{\rho}_c}$, which causes the integration result to be a function of $\overline{\rho}_c$. We can make the result be a function of the binary probability of error p by making p be a limit of the integration by choosing the following transformation of variables. Let

$$y = \frac{1}{2}\left(1 - \sqrt{\frac{u^2}{1+u^2}}\right) \tag{9B.7a}$$

Then the integration limit $\sqrt{\overline{\rho}_c}$ in (9B.5b) becomes

$$p = \frac{1}{2}\left(1 - \sqrt{\frac{\overline{\rho}_c}{1+\overline{\rho}_c}}\right) \tag{9B.7b}$$

Solving for u as a function of y, we obtain

$$u = \sqrt{\frac{(1-2y)^2}{1-(1-2y)^2}} \quad \Rightarrow \quad \frac{du}{(1+u^2)^{L+1/2}} = 2^{2L-1}\, y^{L-1}(1-y)^{L-1}\, dy$$

so that (9B.5b) becomes

$$P(\overline{\rho}_c) = \frac{1}{B(L,L)} \int_0^p y^{L-1}(1-y)^{L-1}dy = I_p(L,L) \tag{9B.7c}$$

where $I_p(L,L)$ is the incomplete beta function [17, formula 6.61]. The integral is easily manipulated to solve for a computable expression:

$$P(\overline{p}_c) = \frac{1}{B(L,L)} \sum_{k=0}^{L-1} \binom{L-1}{k} (-1)^k \int_0^p y^{k+L-1} \, dy$$

$$= \frac{1}{B(L,L)} \sum_{k=0}^{L-1} \binom{L-1}{k} \frac{(-1)^k}{k+L} \, p^{k+L} \qquad (9B.7d)$$

Although this expression is easily computed, we want to show that it is equivalent to the "universal" binary error probability for L-fold diversity. This is accomplished by the following manipulation to convert the series in powers of p to a series in powers of $(1-p)$:

$$P(\overline{p}_c) = \frac{p^L}{B(L,L)} \sum_{k=0}^{L-1} \frac{(-1)^k \, \Gamma(L)}{k! \, \Gamma(L-k)} \frac{\Gamma(k+L)}{\Gamma(k+L+1)} [1-(1-p)]^k$$

$$= \frac{p^L}{B(L,L)} \sum_{k=0}^{L-1} \frac{(-1)^k \, \Gamma(L)}{k! \, \Gamma(L-k)} \frac{\Gamma(k+L)}{\Gamma(k+L+1)} \sum_{l=0}^{k} \binom{k}{l} (-1)^l (1-p)^l$$

$$= \frac{p^L}{B(L,L)} \sum_{k=l}^{L-1} \sum_{l=0}^{L-1} \frac{\Gamma(L)}{\Gamma(L-k)} \frac{\Gamma(k+L)}{\Gamma(k+L+1)} \frac{(-1)^{k+l}}{l! \, (k-l)!} (1-p)^l$$

$$= \frac{p^L}{B(L,L)} \sum_{l=0}^{L-1} \sum_{k=0}^{L-l-1} \frac{\Gamma(L)}{\Gamma(L-k-l)} \frac{\Gamma(k+l+L)}{\Gamma(k+l+L+1)}$$

$$\times \frac{(-1)^k}{l! \, k!} (1-p)^l$$

$$= \frac{p^L}{B(L,L)} \sum_{l=0}^{L-1} \frac{(-1)^l}{l!} (1-p)^l \cdot \Gamma(L)$$

$$\times \sum_{k=0}^{L-l-1} \frac{(-1)^k}{k!} \frac{\Gamma(k+l+L)}{\Gamma(L-k-l) \, \Gamma(k+l+L+1)} \qquad (9B.8a)$$

in which the order of summation has been exchanged. The sum over k may be recognized as related to the Gaussian hypergeometric function:

$$\sum_{k=0}^{\infty} \frac{(-1)^k}{k!} \frac{\Gamma(k+l+L)}{\Gamma(L-k-l) \, \Gamma(k+l+L+1)}$$

$$= \sum_{k=0}^{\infty} \frac{1}{k!} \frac{\Gamma(l+k-L+1)}{\Gamma(l-L+1)\Gamma(L-l)} \frac{\Gamma(k+l+L)}{\Gamma(k+l+L+1)}$$

$$= \frac{\Gamma(l+L)}{\Gamma(L-l)\Gamma(l+L+1)} \, {}_2F_1(l+L, l-L+1; l+L+1; 1) \qquad (9B.8b)$$

where we may use an infinite upper limit to the sum because the terms for $k > L - l - 1$ are zero, and the hypergeometric function is given by

$$\, {}_2F_1(a, b; c; x) \triangleq \sum_{k=0}^{\infty} \frac{x^k}{k!} \frac{\Gamma(k+a)}{\Gamma(a)} \frac{\Gamma(k+b)}{\Gamma(b)} \frac{\Gamma(c)}{\Gamma(k+c)} \qquad (9B.9a)$$

For the special case of a unity argument, we have [6, formula 15.1.20]

$$\, {}_2F_1(a, b; c; 1) = \frac{\Gamma(c)\Gamma(c-a-b)}{\Gamma(c-a)\Gamma(c-b)} \qquad (9B.9b)$$

Applying (9B.9b) to (9B.8b) results in

$$\sum_{k=0}^{\infty} \frac{(-1)^k}{k!} \frac{\Gamma(k+l+L)}{\Gamma(L-k+1)\Gamma(k+l+L+1)} = \frac{\Gamma(l+L)}{\Gamma(2L)}$$

Substituting this result in (9B.8a) gives

$$P(\bar{\rho}_c) = \frac{p^L \Gamma(L)}{B(L,L)} \sum_{l=0}^{L-1} \frac{(-1)^l (1-p)^l}{l!} \frac{\Gamma(l+L)}{\Gamma(2L)} = p^L \sum_{l=0}^{L-1} \frac{\Gamma(l+L)}{l! \, \Gamma(L)} (1-p)^l$$

$$= p^L \sum_{l=0}^{L-1} \binom{L-1+l}{l} (1-p)^l \qquad (9B.10)$$

which is the desired form, the "universal" formula.

Appendix 9C Derivation of $\pi/4$ DQPSK Diversity Performances

9C.1 Selection Diversity Performance

In [11] it is shown that for no noise correlation, the decision statistic for $\pi/4$ DQPSK is the difference of two independent, scaled noncentral χ^2 random variables with two degrees of freedom:

$$y = \tfrac{1}{4}\sigma^2\chi^2\left[2; \rho\left(2 + \sqrt{2}\right)\right] - \tfrac{1}{4}\sigma^2\chi^2\left[2; \rho\left(2 - \sqrt{2}\right)\right] \qquad (9C.1a)$$

where $\rho = 2E_b/N_0$. Given ρ, the characteristic function for this RV is

$$C_y(j\mu; \rho) = \frac{\exp\left\{\dfrac{j\mu\sigma^2\rho}{2} \cdot \dfrac{\sqrt{2} + j\mu\sigma^2}{(1 - j\mu\sigma^2/2)(1 + j\mu\sigma^2/2)}\right\}}{(1 - j\mu\sigma^2/2)(1 + j\mu\sigma^2/2)} \qquad (9C.1b)$$

Averaging this characteristic function over the pdf for ρ in the case of selection diversity gives

$$C_y(j\mu) = \int_0^\infty C_y(j\mu; \beta)\, p_\rho(\beta)\, d\beta$$

$$= \frac{L}{\overline{\rho}_c}\int_0^\infty C_y(j\mu; \beta)\, e^{-\beta/\overline{\rho}_c}\left(1 - e^{-\beta/\overline{\rho}_c}\right)^{L-1} d\beta$$

$$= L\int_0^\infty C_y(j\mu; \overline{\rho}_c x)\, e^{-x}(1 - e^{-x})^{L-1}\, dx$$

$$= L\sum_{l=0}^{L-1}\binom{L-1}{l}(-1)^l \int_0^\infty C_y(j\mu; \overline{\rho}_c x)\, e^{-(l+1)x}\, dx \qquad (9C.2a)$$

in which the integral is

$$\int_0^\infty C_y(j\mu; \overline{\rho}_c x)\, e^{-(l+1)x}\, dx = \frac{1}{(1 - j\mu\sigma^2/2)(1 + j\mu\sigma^2/2)}$$

$$\times \int_0^\infty \exp\left\{\frac{j\mu\sigma^2\overline{\rho}_c x}{2}\, \frac{\sqrt{2} + j\mu\sigma^2}{(1 - j\mu\sigma^2/2)(1 + j\mu\sigma^2/2)} - (l+1)x\right\} dx$$

$$= \frac{1}{(1 - j\mu\sigma^2/2)(1 + j\mu\sigma^2/2)} \cdot \frac{1}{l + 1 - \frac{j\mu\sigma^2\bar{\rho}_c x}{2} \frac{\sqrt{2}+j\mu\sigma^2}{(1-j\mu\sigma^2/2)(1+j\mu\sigma^2/2)}}$$

$$= \frac{1}{(l + 1)(1 - j\mu\sigma^2/2)(1 + j\mu\sigma^2/2) - \frac{j\mu\sigma^2\bar{\rho}_c x}{2}\left(\sqrt{2} + j\mu\sigma^2\right)}$$

$$= \frac{1}{l + 1}\left(\frac{A_l}{A_l + B_l} \cdot \frac{1}{1 - jA_l\mu} + \frac{B_l}{A_l + B_l} \cdot \frac{1}{1 + jB_l\mu}\right) \qquad \text{(9C.2b)}$$

where

$$A_l = \frac{\sigma^2}{2}\left\{\sqrt{1 + 2\frac{\bar{\rho}_c}{l+1} + \frac{1}{2}\left(\frac{\bar{\rho}_c}{l+1}\right)^2} + \frac{1}{\sqrt{2}}\frac{\bar{\rho}_c}{l+1}\right\} \qquad \text{(9C.2c)}$$

and

$$B_l = \frac{\sigma^2}{2}\left\{\sqrt{1 + 2\frac{\bar{\rho}_c}{l+1} + \frac{1}{2}\left(\frac{\bar{\rho}_c}{l+1}\right)^2} - \frac{1}{\sqrt{2}}\frac{\bar{\rho}_c}{l+1}\right\} \qquad \text{(9C.2d)}$$

Thus, the probability of error for Rayleigh fading and L-fold selection diversity is given by

$$P_{\pi/4}(e; L) = \Pr\{y < 0\}$$

$$= L\sum_{l=0}^{L-1}\binom{L-1}{l}\frac{(-1)^l}{l+1}\frac{B_l}{A_l + B_l}$$

$$= L\sum_{l=0}^{L-1}\binom{L-1}{l}\frac{(-1)^l}{l+1}\frac{1}{2}\left[1 - \frac{\frac{\overline{E}_b/\mathcal{N}_0}{l+1}}{\sqrt{\frac{1}{2} + 2\frac{\overline{E}_b/\mathcal{N}_0}{l+1} + \left(\frac{\overline{E}_b/\mathcal{N}_0}{l+1}\right)^2}}\right]$$

$$= L\sum_{l=0}^{L-1}\binom{L-1}{l}\frac{(-1)^l}{l+1} P_{\pi/4}\left(e; 1, \frac{\overline{E}_b/\mathcal{N}_0}{l+1}\right) \qquad \text{(9C.3)}$$

9C.2 EGC Diversity Reception

For $L = 1$, the characteristic function for the $\pi/4$ DQPSK in Rayleigh fading becomes the expression in (9C.2b) for $l = 0$:

$$C_y(j\mu; L = 1) = \frac{1}{(1 - jA\mu)(1 + jB\mu)} \tag{9C.4a}$$

where

$$A = \frac{\sigma^2}{2}\left\{ \sqrt{1 + 2\bar{\rho}_c + \frac{1}{2}(\bar{\rho}_c)^2} + \frac{\bar{\rho}_c}{\sqrt{2}} \right\} \tag{9C.4b}$$

and

$$B = \frac{\sigma^2}{2}\left\{ \sqrt{1 + 2\bar{\rho}_c + \frac{1}{2}(\bar{\rho}_c)^2} - \frac{\bar{\rho}_c}{\sqrt{2}} \right\} \tag{9C.4c}$$

For the case in which L independent diversity receptions of the signal are combined with equal gain, the characteristic function (9C.4a) becomes

$$C_y(j\mu; L > 1) = \frac{1}{(1 - jA\mu)^L(1 + jB\mu)^L} \tag{9C.4d}$$

which is the characteristic function for the difference between two independent chi-squared RVs, each with $2L$ degrees of freedom:

$$y = \frac{A}{2}\chi^2(2L) - \frac{B}{2}\chi^2(2L) \tag{9C.5a}$$

Therefore, the BER is

$$
\begin{aligned}
P_{\pi/4}(e; L) &= \Pr\{A\chi^2(2L) < B\chi^2(2L)\} \\
&= \Pr\left\{ \chi^2(2L) > \frac{A}{B}\chi^2(2L) \right\} \\
&= \int_0^\infty \frac{u^{L-1}}{2^L\Gamma(L)} e^{-u/2}\, du \int_{Au/B}^\infty \frac{v^{L-1}}{2^L\Gamma(L)} e^{-v/2}\, dv \\
&= \frac{1}{2^L} \sum_{l=0}^{L-1} \frac{[A/2B]^l}{l!\,\Gamma(L)} \int_0^\infty u^{l+L-1} e^{-u(1+A/B)/2}\, du
\end{aligned}
$$

$$= \sum_{l=0}^{L-1} \frac{[A/B]^l}{l!\,\Gamma(L)} \frac{\Gamma(l+L)}{(1+A/B)^{l+L}}$$

$$= \left(\frac{B}{A+B}\right)^L \sum_{l=0}^{L-1} \binom{L+l-1}{l} \left(\frac{A}{A+B}\right)^l$$

$$= \left[\mathrm{P}_{\pi/4}(e;1)\right]^L \sum_{l=0}^{L-1} \binom{L+l-1}{l} \left[1 - \mathrm{P}_{\pi/4}(e;1)\right]^l \qquad (9\mathrm{C}.5\mathrm{b})$$

10

CDMA Cellular System Design and Erlang Capacity

The basic concept of a cellular communications system is the division of the coverage area into cells in which the resource that provides the medium for communication is reused. As we saw in Chapter 3, in connection with analog cellular systems based on FDMA, the practical implementation of the cellular concept involves dealing with the reality of cochannel interference. This interference is due to the fact that the radio signals generated in a particular cell cannot be contained within that cell, but inevitably act as interference to receivers in neighboring cells to an extent that diminishes with distance. Therefore, an important aspect of analog cellular engineering is the tradeoff between spectral efficiency and the reuse distance that determines the user capacity of the system. As we see in this chapter, a digital cellular system based on CDMA must contend with not only cochannel interference from neighboring cells, but also cochannel interference *within* the cell, and the systems engineering design is based on managing this interference to achieve maximal capacity.

In explaining the basics of CDMA cellular systems engineering design in this chapter, we first introduce the parameters and quantities that characterize CDMA cells, most of which deal with cochannel interference. We then relate link margin to a measure of link reliability, derive the capacity of the system in terms of Erlangs of user communications traffic, and show the tradeoff between coverage and capacity for a CDMA system.

10.1 CDMA Cells

In this section, the basic cellular engineering issues for CDMA are discussed in terms of factors affecting cell size and capacity, including cochannel interference parameters and their effect on the forward and reverse link power budgets.

10.1.1 Forward Link Cochannel Interference

In assessing the cochannel interference on the IS-95 CDMA forward link [1], we first consider same-cell interference that is present even if there is only one cell. Then we consider other-cell interference that occurs for the general case of a multicell system.

10.1.1.1 Same-Cell Interference

Ideally, there is no same-cell forward link interference because the different CDMA forward link channels are orthogonal. Any forward link channel, including the pilot channel, can be selected by multiplying the despread chip stream by the appropriate Walsh function and summing (integrating) over 64 chips, the period of the Walsh functions. The result of this procedure is the baseband data stream for the selected channel, with no interference from other channels on the forward link. However, the mobile propagation channel introduces same-cell interference because of multipath receptions at the mobile unit's location.

Suppose that $\mathcal{I}_{0,f}$ is the total received spectral density at the mobile due to the home base station, including usable signal energy and interference due to multipath. Let the relative power of the kth multipath signal component be denoted by β_k, where $0 \leq \beta_k \leq 1$. That is, the portion of $\mathcal{I}_{0,f}$ due to the kth path is

$$\mathcal{I}_{0,k} \triangleq \text{spectral density for path } k = \beta_k \, \mathcal{I}_{0,f} \qquad (10.1a)$$

By definition, if there are K multipath components, then we have

$$\mathcal{I}_{0,f} = \sum_{k=1}^{K} \mathcal{I}_{0,k} = \mathcal{I}_{0,f} \sum_{k=1}^{K} \beta_k \quad \text{or} \quad \sum_{k=1}^{K} \beta_k = 1 \qquad (10.1b)$$

Now, suppose that at the mobile there are K receivers, each of which is set up to receive one of the K multipaths. If E_{b0} is the total bit energy received from the base station for a particular forward link channel, then the bit energy input to the jth receiver is

$$\mathrm{E}_{bj} \triangleq \text{bit energy for path } j = \beta_j \, \mathrm{E}_{b0} \qquad (10.1c)$$

The spectral density of the same-cell interference to the jth path, denoted $\mathcal{I}_{scj}$, is given by

$$\mathcal{I}_{scj} = \sum_{k \neq j} \mathcal{I}_{0,k} = \mathcal{I}_{0,f} \sum_{k \neq j} \beta_k = \mathcal{I}_{0,f}(1 - \beta_j) \qquad (10.1\text{d})$$

Thus, the effective bit energy-to-noise plus interference density ratio at a demodulator that is set up to receive the jth multipath is

$$\frac{E_{bj}}{(\mathcal{N}_{0,T})_j} = \frac{E_{b0}\,\beta_j}{\mathcal{N}_0 + \mathcal{I}_{0,oc} + \mathcal{I}_{0,f}(1 - \beta_j)} \qquad (10.2\text{a})$$

where

$\mathcal{N}_0$ = thermal noise spectral power density

$\mathcal{I}_{0,oc}$ = other-cell cochannel interference spectral power density (discussed in Section 10.1.1.2)

$\mathcal{I}_{0,f}$ = total same-cell received spectral power density

$(\mathcal{N}_{0,T})_j$ = total forward link noise-plus-interference density for path j

If maximal ratio combining of the K paths is used at the mobile unit, then the best overall bit energy ratio is the sum of the path bit energy ratios, as discussed in Chapter 9, which is given by

$$\frac{E_b}{\mathcal{N}_{0,T}} = \sum_{j=1}^{K} \frac{E_{bj}}{(\mathcal{N}_{0,T})_j} = \sum_{j=1}^{K} \frac{E_{b0}\,\beta_j}{\mathcal{N}_0 + \mathcal{I}_{0,oc} + \mathcal{I}_{0,f}(1 - \beta_j)} \qquad (10.2\text{b})$$

For mobiles close to the base station, the same-cell cochannel interference dominates, giving the approximate relation

$$\frac{E_b}{\mathcal{N}_{0,T}} \approx \sum_{j=1}^{K} \frac{E_{b0}\,\beta_j}{\mathcal{I}_{0,f}(1 - \beta_k)} = \frac{E_{b0}}{\mathcal{I}_{0,f}} \sum_{j=1}^{K} \frac{\beta_j}{1 - \beta_j} \triangleq \frac{E_{b0}}{\mathcal{I}_{0,sc}} \qquad (10.2\text{c})$$

where $\mathcal{I}_{0,sc}$ is the part of the total received same-cell density $\mathcal{I}_{0,f}$ that acts as interference. Thus the effective spectral power density for the same-cell cochannel interference can be defined as

$$\mathcal{I}_{0,sc} \triangleq \mathcal{I}_{0,f} \div \sum_{j=1}^{K} \frac{\beta_j}{1 - \beta_j} < \mathcal{I}_{0,f} \qquad (10.3)$$

Example 10.1 What is the same-cell forward link interference when there is no multipath, only a direct path? What is it when there are just two paths?

    Solution: For one path, $K = 1$ and $\alpha_1 = 1$, so that the forward link same-cell interference as defined in (10.3) becomes

$$\mathcal{I}_{0,sc} = \mathcal{I}_{0,f} \cdot \frac{1 - \beta_1}{\beta_1} = \mathcal{I}_{0,f} \cdot \frac{1 - 1}{1} = 0 \tag{10.4a}$$

There is no same-cell interference because the forward link channels are orthogonal. For two paths, $K = 2$ and $\beta_1 + \beta_2 = 1$, and the same-cell interference is given by

$$\mathcal{I}_{0,sc} = \mathcal{I}_{0,f} \cdot \frac{1}{\dfrac{\beta_1}{1 - \beta_1} + \dfrac{\beta_2}{1 - \beta_2}} = \mathcal{I}_{0,f} \cdot \frac{(1 - \beta_1)(1 - \beta_2)}{\beta_1(1 - \beta_2) + \beta_2(1 - \beta_1)}$$

$$= \mathcal{I}_{0,f} \cdot \frac{\beta_1\beta_2}{1 - 2\beta_1\beta_2} \quad \text{because} \quad \beta_1 + \beta_2 = 1 \tag{10.4b}$$

    A common assumption about same-cell forward link interference power is that it equals the power of the direct path [2, 3]. The implication of such an assumption is that

$$\sum_{k=2}^{K} \beta_k = \tfrac{1}{2} \tag{10.4c}$$

In terms of the relationship between $\mathcal{I}_{0,sc}$, $\mathcal{I}_{0,f}$, and the multipath power fractions $\{\beta_k\}$, the implication of (10.4c) is that the same-cell interference equals one-half of the total forward link spectral density received at the mobile receiver location.[1] That is:

$$\mathcal{I}_{0,sc} \approx \tfrac{1}{2}\mathcal{I}_{0,f} = \mathcal{I}_{0,f} \div \left[\frac{\tfrac{1}{2}}{1 - \tfrac{1}{2}} + \sum_{k=2}^{K} \frac{\beta_k}{1 - \beta_k} \right] \tag{10.5a}$$

In this equation, it is reasoned that, because the total multipath interference is one-half the total power, then the direct path must have one-half the total power (i.e., $\beta_1 = \tfrac{1}{2}$). Solving (10.5a) for the average value of the multipath

---

[1] Unless otherwise noted, "mobile location" in forward link calculations means a mobile at or near the edge of the cell's coverage area.

power fractions that correspond to the common assumption of $\mathcal{I}_{0,sc} = \frac{1}{2}\mathcal{I}_{0,f}$ leads to the solution

$$\overline{\left(\frac{\beta_k}{1-\beta_k}\right)_{k>1}} = \frac{1}{K-1} \tag{10.5b}$$

compared with the average power factor for the reflected paths of, from (10.4c),

$$\overline{(\beta_k)_{k>1}} = \frac{1}{2} \div (K-1) = \frac{1}{2(K-1)} \tag{10.5c}$$

10.1.1.2 Other-Cell Interference

Signals received from other CDMA sectors and other CDMA cells' base stations act as interference to a mobile receiver. The interference power from another cell tends to fluctuate and can be modeled as a lognormal random variable; that is, the interference power in decibels (dBW or dBm) is normal (Gaussian) random variable:

Interference power (dBm) = average (dBm) + zero-mean Gaussian RV

The average interference power can be predicted using a propagation power loss model, for example a loss proportional to the γth power of the distance. Mathematically, the forward link other-cell interference can be modeled as a lognormal random variable (see Section 1.5.10) by writing

Interference power (dB) $\triangleq 10 \log_{10} I_i$

$$= 10 \log_{10} \overline{I}_i + \sigma_{\mathrm{dB}} w_i, \quad w_i = \text{Gaussian RV}$$

or

$$I_i = \overline{I}_i \times 10^{\sigma_{\mathrm{dB}} w_i/10} = \text{Const} \times \frac{1}{r_i^\gamma} \times 10^{\sigma_{\mathrm{dB}} w_i/10} \tag{10.6}$$

where

$\overline{I}_i$ = median value of the interference power from base station i

r_i = distance from the mobile to the ith base station

γ = propagation power law

w_i = zero-mean, unit-variance Gaussian RV

and σ_{dB} is a standard deviation for the fluctuation in the range of 6 to 13 dB. For the purpose of analyzing other-cell interference, it is sufficient to deal with the median interference term $\overline{I}_i$ in (10.6) and to consider the influence of the spatial distribution of mobile stations; that is, the distribution of the $\{r_i\}$ that determines the values of the $\{\overline{I}_i\}$. In what follows, we omit the overbar and use the notations I_i and $\mathcal{I}_{0,i} = I_i/W$ to indicate a median interference power and its corresponding spectral power density.

To discuss the distribution of the mobile stations, let us first consider the geometry of other-cell interference to a mobile receiver at the position (r, θ_i) in the home cell, as illustrated in Figure 10.1. From that figure, we see that the distance of the ith interfering base station is

$$r_i(r, d_i, \theta_i) = \sqrt{r^2 + d_i^2 - 2r\,d_i\cos\theta_i} \tag{10.7a}$$

where d_i is the intercell distance; that is, the distance between the home base station and the interfering base station. A coordinate system suitable for calculating intercell distances, assuming a regular pattern of hexagonal cells, was shown in Figure 3.11. The coordinate system recognizes that there are "rings" of interfering cells around the home cell, as depicted in Figure 3.12, and the distances of these cells from the home cell are listed in Table 3.3 as multiples of either the center-to-corner cell radius, R_c, or the center-to-side cell radius, $R = \sqrt{3}\,R_c/2$. Combining the ideas represented in Figures 10.1 and 3.12 results in Figure 10.2, which illustrates the dependence of other-cell interference on the mobile position, (r, θ) and on the intercell distances $\{d_i\}$, which are the numbers appearing in each interfering cell.

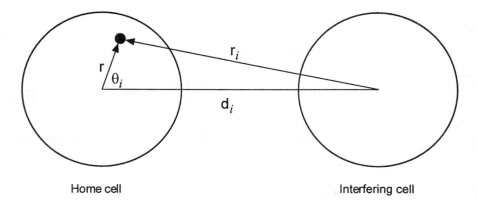

Home cell   Interfering cell

Figure 10.1 Geometry of forward link other-cell interference.

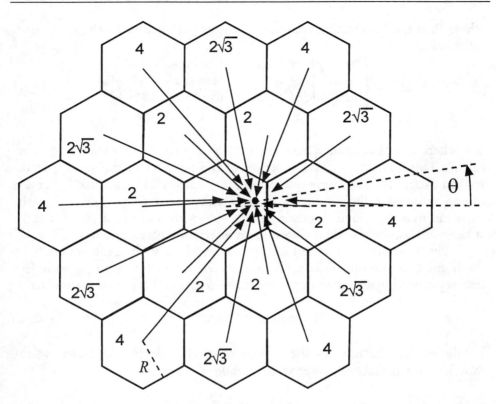

Figure 10.2 Forward link other-cell interference at mobile location (r, θ). **The number in each cell is the intercell distance** d_i **as a factor of** $R = \sqrt{3}\, R_c/2$.

We assume that propagation is proportional to the γ power; that is:

$$\text{Propagation loss, } L(r_i) = \frac{\text{interference power received } (I_i)}{\text{power transmitted } (P_t)} = \text{const.} \times r_i^{-\gamma}$$

where P_t is the base station transmitter power and r_i is the distance defined in (10.7a). Then the median value of the total other-cell interference power is the sum of the median values from neighboring cells, given by

$$I_{oc} = \sum_i I_i = \sum_i P_t \times L(r_i) = P_t \sum_i L(r_i)$$

$$= \text{Const} \times P_t \sum_i [r_i(r, d_i, \theta_i)]^{-\gamma} \qquad (10.7b)$$

where P_t is the base station transmitter power (assumed to be the same in all cells), where

$$\theta_i = \theta + \begin{cases} \text{Multiple of } 60°, & \text{first ring (6 cells)} \\ \text{Multiple of } 30°, & \text{second ring (12 cells)} \\ \text{Multiple of } \approx 20°, & \text{third ring (18 cells)} \\ \vdots \end{cases} \qquad (10.7c)$$

and where r_i is the distance from the mobile unit to the base station in other cell i. This quantity I_{oc} is a function of the position of the mobile as described by (r, θ) in the home cell, using the geometry of Figure 10.2. Because of the assumed cellular hexagonal symmetry, for a given value of r, the variation in θ is periodic, repeating every 60°, with maxima at $30° + n \cdot 60°$; at these angles, the mobile is closest to an interfering base station.

Let S denote the forward link power received at the mobile unit from the home base station in the absence of interference. For a propagation loss that is proportional to the γth power of the distance, this power is equal to

$$S = \text{const.} \times P_t \times r^{-\gamma} \qquad (10.7d)$$

It follows that the ratio of the average other-cell interference power at the mobile to the home cell power at the mobile is

$$\frac{I_{oc}}{S} = \sum_i \frac{r^\gamma}{[r_i(r, d_i, \theta_i)]^\gamma} = \sum_i \left[\frac{1}{\sqrt{1 + (d_i/r)^2 - 2(d_i/r)\cos\theta_i}} \right]^\gamma \qquad (10.7e)$$

$$= \sum_{i_1=1}^{6} \left[\frac{1}{\sqrt{1 + (d_{i_1}/r)^2 - 2(d_{i_1}/r)\cos\theta_{i_1}}} \right]^\gamma \quad \text{(first ring)}$$

$$+ \sum_{i_2=1}^{12} \left[\frac{1}{\sqrt{1 + (d_{i_2}/r)^2 - 2(d_{i_2}/r)\cos\theta_{i_2}}} \right]^\gamma \quad \text{(second ring)}$$

$$+ \sum_{i_3=1}^{18} \left[\frac{1}{\sqrt{1 + (d_{i_3}/r)^2 - 2(d_{i_3}/r)\cos\theta_{i_3}}} \right]^\gamma \quad \text{(third ring)}$$

$$+ \cdots$$

This ratio indicates the relative importance of the forward link interference due to other cells and may be calculated using the cellular ring intercell distance parameters shown in Figure 10.2 for each ring and the relative angles defined in (10.7a). For example, all six of the first-ring interfering cells are at distance $d_{i_1} = \sqrt{3}\,R_c = 2\,R$ from the home base station for all $\{\theta_{i_1}: \theta + 0°,$ $\theta + 60°, \theta + 120°, \theta + 180°, \theta + 240°, \theta + 300°\}$ and θ is the angular position of the mobile illustrated in Figure 10.2.

Example calculations of the interference-to-forward link power ratio using the first two rings of interfering cells—18 cells in all, as shown in Figure 10.2—are shown in Figures 10.3 and 10.4 for $\gamma = 4$. In those figures, the ratio in dB is shown as a function of θ for r $= 0.4\,R_c$ and $0.8\,R_c$, respectively. For each of these distances, we show only the first period of the ratio.

The peak value of the ratio I_{oc}/S (dB) is plotted in Figure 10.5 as a function of the normalized distance r/R_c and the power laws $\gamma = 3$, 3.5, and 4. Figure 10.3 shows that the maximum of the ratio I_{oc}/S is -16.27 dB for $\gamma = 4$, which occurs for $\theta = 0°$ and $60°$ when the mobile station is at the distance r $= 0.4\,R_c$. On the other hand, the maximal ratio of -0.93 dB is observed for a mobile station at distance r $= 0.8\,R_c$, as seen in Figure 10.4. In Figure 10.5, these peak values are plotted: for example, from the curve for $\gamma = 4$, we read the peak of the ratio I_{oc}/S to be about -16.3 dB at $r/R_c = 0.4$, whereas the peak value of -0.93 dB is seen for $r/R_c = 0.8$.

Figure 10.5 displays the extent of the interference power in comparison with the total forward link power for several propagation power laws. For a mobile near the edge of the cell ($r/R \approx 1$ or $r/R_c \approx 0.86$), the ratio $I_{oc}/S \approx$ 2.5 dB may be used as worst case value in forward link budget calculations. When the mobile is near the base station antenna, Figure 10.5 indicates that the other-cell forward link interference can be ignored (less than 30 dB down); this fact justifies the approximation used in going from (10.2b) to (10.2c).

Example 10.2 What is an approximate value of total forward link interference when the total received power from the home base station is S $= -95$ dBm? It is common practice to assume that the same-cell interference equals the total forward link power in the absence of interference.

Solution: Because the same-cell interference (due to multipath) is assumed to be equal to the forward link power S:

$$I_{sc} \approx S = -95 \text{ dBm} \tag{10.8a}$$

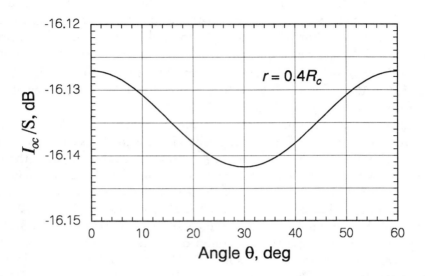

Figure 10.3 Interference-to-forward link power ratio for a mobile at distance $0.4R_c$ from the base station (fourth power propagation law).

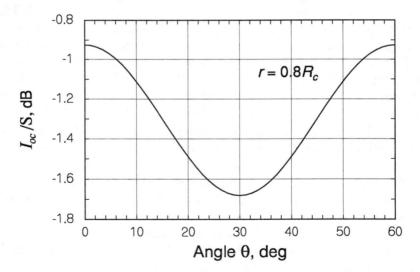

Figure 10.4 Interference-to-forward link power ratio for a mobile at distance $0.8R_c$ from the base station (fourth power propagation law).

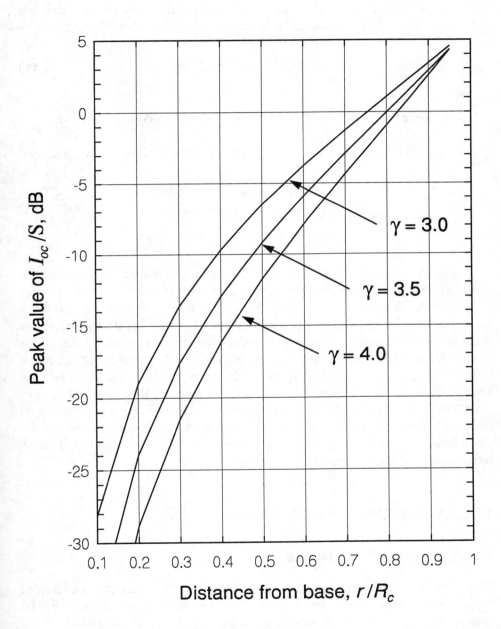

Figure 10.5 Peak value of ratio of other-cell interference to forward link power (parametric in propagation power law).

The other-cell interference for a mobile near the cell edge—which is the case of interest when determining the cell size—can be taken from Figure 10.5 to be

$$\frac{I_{oc}}{I_{sc}}(\text{dB}) \approx \frac{I_{oc}}{\text{S}}(\text{dB}) = 2.5\,\text{dB} \tag{10.8b}$$

so that

$$I_{oc}(\text{dBm}) \approx \text{S}(\text{dBm}) + 2.5\,\text{dB} = -95\,\text{dBm} + 2.5\,\text{dB} = -92.5\,\text{dBm}$$

Thus the total forward link interference power at the mobile, I_T, is given by

$$I_T \triangleq I_{sc} + I_{oc} = 10^{-9.5}\,\text{mW} + 10^{-9.25}\,\text{mW} \tag{10.8c}$$

$$= 8.79 \times 10^{-10}\,\text{mW} = 10^{-9.06}\,\text{mW}$$

$$= -90.6\,\text{dBm}$$

For systems analysis purposes, it has been assumed in the formulation of the same-cell interference term I_{sc} that it consists of multipath interference from the home cell only and the other-cell interference term I_{oc} has been treated as not having multipath components. In practice, it is possible for there to be interference from several paths originating from an adjacent base station transmitter, while experiencing no multipath interference from the home base station. Nevertheless, the model has a general application because both I_{sc} and I_{oc} have been referenced to S, the direct path (or earliest arrival) from the home base station. If measurement data indicate the presence of multipath from adjacent cells, this phenomenon can be accounted for by considering I_{sc} to be "multipath interference" rather than "same-cell interference" and adjusting the value accordingly.

10.1.2 Reverse Link Cochannel Interference

10.1.2.1 Same-Cell Interference

The same-cell interference on the reverse link consists of the superposition of signals from other mobile units at the base station receiver. For a CDMA system, the signals from all mobile users occupy the same bandwidth at the same time. The total interference is reasonably modeled for analysis purposes as bandlimited white noise.

Almost all the noise at the base station receiver is due to interfering mobile signals. The number of mobile users that can simultaneously engage in call transmissions (i.e., the system capacity) is maximized by making each signal's power the same at the base station, and as low as possible while achieving satisfactory link performance. Dynamic control of the mobile transmitter powers therefore is an essential part of the reverse link design. Because power control is used, for M mobile users in the cell, the amount of interference power at the base station receiver due to the same cell is assumed to be given by

$$I_{sc} = (M - 1) \cdot S \cdot \alpha_r \qquad (10.9a)$$

where

$$S = \text{power of each mobile at the receiver} \qquad (10.9b)$$

$$\alpha_r = \text{average reverse link voice activity factor} \qquad (10.9c)$$

10.1.2.2 Other-Cell Interference

Figure 10.6 depicts the geometry of other-cell interference on the reverse link. Assuming equal-sized cells using power control and a fourth-power propagation law, the transmit power of a mobile in an interfering cell with base station B is proportional to

Target power level at base station $B \times$ gain to offset propagation loss

$$= S \cdot r^4 \qquad (10.10a)$$

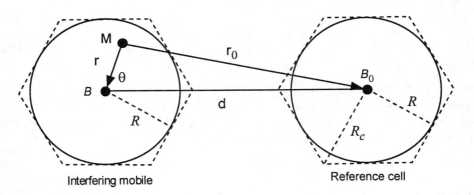

Interfering mobile   Reference cell

Figure 10.6 Geometry of reverse link interference (cell is represented by a circular area with radius R rather than $R_c = 2R/\sqrt{3}$).

where r is the distance of the mobile from its own base station B. A mobile in an interfering cell therefore presents to the reference base station B_0 (at distance r_0) the power

$$P_0 = \alpha_r S \cdot \left(\frac{r}{r_0}\right)^4 \tag{10.10b}$$

In terms of the angle θ between the line defining the intercell distance d and the line from the home cell to the mobile, the distance to the reference cell is given by

$$r_0 = \sqrt{r^2 + d^2 - 2r\,d\cos\theta} \tag{10.10c}$$

For analysis purposes, we assume that the M mobiles in the interfering cell are uniformly distributed in the cell and we approximate the hexagonal area of the cell by a circle of radius R, giving the user density[2]

$$\rho = \frac{\text{number of users}}{\text{area}} = \frac{M}{\pi R^2} \tag{10.11a}$$

The total power received at the reference base station due to users in the interfering cell can then be approximated by

$$P(d) = 2 \int_0^\pi d\theta \int_0^R r \cdot \frac{M}{\pi R^2} \cdot \alpha_r S \left(\frac{r}{r_0}\right)^4 dr \tag{10.11b}$$

For $d = \kappa R$, the integral for $P(d)$ can be solved to obtain [4]

$$P(d) = 2M\alpha_r S \left[2\kappa^2 \ln\left(\frac{\kappa^2}{\kappa^2 - 1}\right) - \frac{4\kappa^4 - 6\kappa^2 + 1}{2(\kappa^2 - 1)^2} \right] \tag{10.11c}$$

We can use the sector coordinate system (n, i) of Figure 3.11 to index the interfering cells[3] in a 60° sector by ring number ($n = 1, 2, \dots$) and the index of the sector's cells in the ring ($i = 0, 1, \dots, n - 1$). With this notation, the normalized distance for an interfering cell is

---

[2] We assume the cell is a hexagon *enclosing* a circle of radius R.
[3] Although the ring notation of Figure 3.11 pertains to hexagonal cells, we can still express the locations of circular interfering cells using the $d(n, i)$ coordinate system.

$$\kappa_{n,i} = \frac{d(n, i)}{R} = 2\sqrt{n^2 + i^2 - n\,i}, \qquad i < n \text{ for ring } n \qquad (10.11\text{d})$$

where $R = \sqrt{3}\,R_c/2$ is the center-to-edge cell radius. We also can normalize and rewrite (10.11c) as

$$\text{P}'[d(n,\,i)] \triangleq \frac{\text{P}[d(n,\,i)]}{M\alpha_r S} = 2\left[2\kappa^2 \ln\left(\frac{\kappa^2}{\kappa^2 - 1}\right) - \frac{4\kappa^4 - 6\kappa^2 + 1}{2\left(\kappa^2 - 1\right)^2}\right]_{\kappa=\kappa_{n,i}} \qquad (10.11\text{e})$$

A plot of (10.11e) is given in Figure 10.7. With this method for indexing the distances to interfering cells, the total interference power from other cells is [4]

$$I_{oc} = 6 \times \text{interference in one sector}$$

$$= 6\sum_{n=1}^{\infty}\sum_{i=1}^{n}\text{P}[d(n,i)] = (M\alpha_r S)\cdot 6\sum_{n=1}^{\infty}\sum_{i=1}^{n}\text{P}'[d(n,i)]$$

$$\approx (M\alpha_r S)\cdot 6\sum_{n=1}^{100}\sum_{i=1}^{n}\text{P}'[d(n,i)] \qquad (10.12\text{a})$$

$$= (M\alpha_r S)\cdot \xi \qquad (10.12\text{b})$$

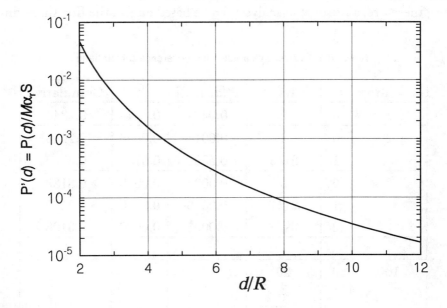

Figure 10.7 Other-cell interference function $\text{P}'(d)$ versus d/R.

where

$$\xi = 6 \sum_{n=1}^{100} \sum_{i=1}^{n} 2 \left[2\kappa^2 \, ln \left(\frac{\kappa^2}{\kappa^2 - 1} \right) - \frac{4\kappa^4 - 6\kappa^2 + 1}{2 \left(\kappa^2 - 1 \right)^2} \right]_{\kappa = \kappa_{n,i}} \tag{10.12c}$$

$$= \begin{cases} 0.33 & \text{for radius} = R \\ 0.42 & \text{for radius} = R_c \end{cases} \tag{10.12d}$$

is a *reuse fraction* and where $\kappa_{n,i}$ was defined in (10.11d); that is:

$$I_{oc} = \begin{cases} 0.33 \cdot M\alpha_r S = 0.33 \, (\text{total same-cell power}), & \text{radius} = R \\ 0.42 \cdot M\alpha_r S = 0.42 \, (\text{total same-cell power}), & \text{radius} = R_c \end{cases} \tag{10.12e}$$

Details of the sector interference calculations using R as the cell radius are shown in Table 10.1. As shown in the table, for ring 1 ($n = 1$), the intercell distance is $d = 2R$ and the interference power from each of the cells in the first ring is calculated to be $0.0474 \cdot M\alpha_r S$; this number can also be read from Figure 10.7 for $d/R = 2$. Thus, the total other-cell interference power in the first ring is $6 \times 0.0474 \cdot M\alpha_r S = 0.2844 \cdot M\alpha_r S$, or about 28% of the total reverse link power received from within the same cell. The total in the first three rings is $0.3198 \cdot M\alpha_r S$, which is close to the total for 100 rings; therefore, a reasonable analysis can be based on just the first three rings.

Table 10.1 Reverse link other-cell interference calculations

$n =$ Ring	i	$d(n, i)/R$	$P'[d(n, i)]$	$\times 6$	Cumulative
1	0	2	0.0474	0.2844	0.2844
2	0	4	0.0016	0.0096	0.2940
	1	$2\sqrt{3}$	0.0030	0.0180	0.3120
3	0	6	0.0003	0.0018	0.3138
	1	$2\sqrt{7}$	0.0005	0.0030	0.3168
	2	$2\sqrt{7}$	0.0005	0.0030	0.3198
$\vdots$					$\vdots$
100	0	20			
	$\vdots$	$\vdots$			$\vdots$
	99	$2\sqrt{9901}$			0.33

The estimate of total other-cell interference as being from 33 to 42% of the total power received from mobiles in the same cell, as indicated by (10.12e), is in good agreement with simulations [5].

10.1.2.3 CDMA Reuse Parameters

The total average interference power on the reverse link may be written

$$I_T = I_{sc} + I_{oc}$$

$$\equiv \text{same-cell interference power}$$

$$+ \text{other-cell interference power} \tag{10.13a}$$

where, in terms of spectral density:

$$I_{sc} = \mathcal{I}_{0,sc}W \quad \text{and} \quad I_{oc} = \mathcal{I}_{0,oc}W \tag{10.13b}$$

The same-cell interference power I_{sc} can be expressed as a multiple of the power for a single active reverse link transmitter, assuming that perfect power control causes all the reverse link transmissions to arrive at the base station receiver site with the same power S. Further, as shown previously, variable-rate digital voice information on IS-95 reverse link transmissions has less than a 100% duty factor, denoted here by α_r, so that the average same-cell interference power can be written

$$I_{sc} = (M - 1) \cdot \text{S} \cdot \alpha_r \tag{10.13c}$$

where M is the number of active reverse link users. We can express I_{oc}, the average other-cell reverse link interference power, as some fraction ξ of the total received same-cell power, as seen in (10.12e). We assume that each interfering cell has the same properties of the cell of interest, particularly the same number of active users on its reverse link M. The total same-cell reverse link power arriving at each base station is then $M\alpha_r S$, and at the base station of interest the other-cell interference can be written as some fraction of that amount of power:

$$I_{oc} = \xi \cdot M \cdot \alpha_r \cdot \text{S} \tag{10.14a}$$

in which the "reuse fraction" ξ has the interpretation

$$\xi = \frac{\text{total other-cell received power}}{\text{total same-cell received power}} = \text{reuse fraction} \quad (10.14b)$$

This parameter was calculated to be $\xi = 0.33$ in the previous subsection. It follows that the total average reverse link interference is given by

$$I_T = I_{sc} + I_{oc} = (M - 1)\alpha_r S + \xi M \alpha_r S$$

$$= [(1 + \xi)M - 1]\alpha_r S \quad (10.14c)$$

or

$$I_T = \left(\frac{M}{F_e} - 1\right)\alpha_r S \quad (10.14d)$$

where

$$F_e \triangleq \frac{1}{1 + \xi} \quad (10.14e)$$

is the "CDMA reuse efficiency," which equals the fraction of total received power that is due to emitters in the same cell and has the interpretation

$$F_e = \frac{1}{1 + \xi} = \frac{\text{total same-cell power}}{\text{total same- plus other-cell power}} < 1 \quad (10.14f)$$

Note that in a perfect cellular system, with no "leakage" of cochannel interference power from neighboring cells, the reuse efficiency would equal 1.

Another measure, F, called the "reuse factor," may be defined as the inverse of the reuse efficiency:

$$F \triangleq \frac{1}{F_e} = 1 + \xi \quad (10.14g)$$

All these CDMA reuse parameters are summarized in Table 10.2. In what follows, we relate these parameters to CDMA capacity.

10.1.2.4 CDMA Capacity Revisited

Consider a one-cell system. Neglecting thermal noise, the capacity M_c in terms of the number of active users is related to the system parameters by

$$\left(\frac{C}{I}\right)_{req} = \text{SNR}_{req} = \frac{\alpha_r S}{(M_c - 1)\alpha_r S} = \frac{1}{M_c - 1} \quad (10.15a)$$

Table 10.2 Reuse measures for CDMA cellular systems

Measure of interference	Expression
Reuse Factor	$F = 1 + \xi = \dfrac{1}{F_e}$
Reuse Efficiency	$F_e = \dfrac{1}{1 + \xi} = \dfrac{1}{F}$
Reuse Fraction	$\xi = F - 1 = \dfrac{1}{F_e} - 1$

Now, consider a multicell system. The capacity M of each cell is less than M_c because of other-cell interference. Again neglecting thermal noise, the carrier-to-interference ratio for a multicell system is related to the multicell capacity M by

$$\text{SNR}_{req} = \frac{\alpha_r S}{(M-1)\alpha_r S + \xi M \alpha_r S} = \frac{1}{(M-1) + \xi M} \qquad (10.15\text{b})$$

in view of the definition of the reuse fraction ξ in (10.12e). Setting the C/I value to the required value for both the single-cell case and the multicell case, we find that

$$\text{SNR}_{req} = \frac{1}{M_c - 1} = \frac{1}{(M-1) + \xi M}$$

or

$$M_c - 1 = M - 1 + \xi M \quad \rightarrow \quad M_c = M(1 + \xi) \qquad (10.15\text{c})$$

Thus, the multicell and single-cell capacities are related by

$$\frac{M_c}{M} = \frac{\text{single-cell capacity}}{\text{multicell capacity}} = 1 + \xi \triangleq F = \textit{reuse factor} \qquad (10.16\text{a})$$

Therefore:

$$M = \frac{M_c}{1 + \xi} = \frac{M_c}{F} = M_c F_e, \quad 0.33 \le \xi \le 0.42 \qquad (10.16\text{b})$$

The implication of (10.16b) is that the CDMA capacity under a multicell environment is equal to 70 to 75% of that under a single-cell environment.

The reuse factor for an AMPS cellular system is $F = 7$, so that the multicell capacity of the AMPS system is $1/F = 1/7 = 14\%$ of the single-cell capacity.

Example 10.3 (a) What are CDMA reuse factor and reuse efficiency when the reuse fraction is $\xi = 0.33$? (b) What are the values of CDMA reuse factor and reuse fraction when the reuse efficiency is $F_e = 0.66$?
  Solution (a) From Table 10.2, the reuse factor is calculated to be

$$F = 1 + \xi = 1.33$$

Again from Table 10.2, the reuse efficiency is calculated to be

$$F_e = \frac{1}{1+\xi} = \frac{1}{F} = \frac{1}{1.33} = 0.75 \tag{10.17a}$$

(b) From Table 10.2, we calculate for $F_e = 0.66$ the values

$$F = 1/F_e = 1.5 \quad \text{and} \quad \xi = F - 1 = 0.5. \tag{10.17b}$$

10.1.2.5 CDMA Cell Loading

CDMA cellular engineers use a concept of reverse link "cell loading" to monitor interference levels and to select the best operating parameters for a particular cell in the system. The cell loading is defined as

$$X = \text{cell loading} \triangleq \frac{\text{number of active users}}{\text{maximum allowable number of users}} \tag{10.18}$$

We now find X as a function of the relative powers of thermal noise and reverse link interference at the base station receiver.
  Suppose that the reverse link power control acts in such a way as to maintain a certain receiver desired SNR value that we denote ρ_0. The allowable number of users in the cell is found from using the relation for total interference in (10.13d) to write

$$\text{SNR} = \frac{S}{\sigma^2 + I_T} = \frac{S}{\sigma^2 + (M/F_e - 1)\alpha_r S} \overset{\text{set}}{=} \rho_0 \tag{10.19a}$$

from which we obtain

$$M = F_e \left(1 + \frac{1}{\rho_0 \alpha_r} - \frac{\sigma^2}{\alpha_r S} \right) \tag{10.19b}$$

An upper bound for this value of M is obtained by neglecting thermal noise to write

$$M_{max} \triangleq M|_{\sigma^2=0} = F_e \left(1 + \frac{1}{\rho_0 \alpha_r} \right) \tag{10.19c}$$

Solving now for X, the cell loading, by recognizing M/F_e as related to the total reverse link interference, we obtain

$$X \triangleq \frac{M}{M_{max}} = \frac{M}{M + \dfrac{\sigma^2}{S} \cdot \dfrac{F_e}{\alpha_r}} = \frac{(M/F_e) \cdot \alpha_r S}{\sigma^2 + (M/F_e) \cdot \alpha_r S}$$

$$\approx \frac{(M/F_e - 1)\alpha_r S}{\sigma^2 + (M/F_e - 1)\alpha_r S} = \frac{I_T}{\sigma^2 + I_T}, \quad \frac{M}{F_e} \gg 1 \tag{10.19d}$$

This relationship illustrates the fact that the system's capacity is self-limiting, because the amount of interference is proportional to the number of users in the same and other cells. The loading X is a convenient way to refer to the amount of potential capacity being used. Figure 10.8 gives the graphical relationship between interference and loading.

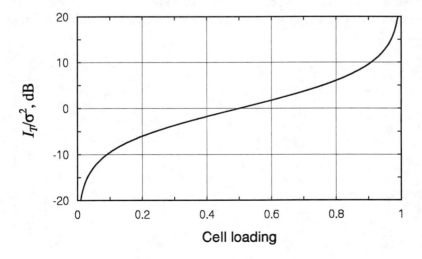

Figure 10.8 The ratio of interference to noise as a function of cell loading.

Example 10.4 What is the CDMA loading when the reverse link interference just equals the base station receiver's thermal noise? What is it when the interference is ten times the thermal noise?

Solution: When $I_T = \sigma^2$ is substituted in (10.19d), the loading is seen to be equal to $X = 0.5 = 50\%$; the same fact is observed in Figure 10.7 by noting that $I_T/\sigma^2 = 0$ dB when the abscissa is $X = 0.5$. When $I_T = 10\,\sigma^2$ or $I_T/\sigma^2 = 10$ dB, we find from Figure 10.7 that $X = 0.9$ or "90% loaded."

10.1.3 Cell Size

Transmissions in different cells of a CDMA cellular system are not divided into groups of frequency channels as in an FDMA cellular system, but are wideband and share a common spectral band. Rather than having the repeating frequency reuse plan of sets 1 to 7 of narrowband frequency channels shown on the left-hand side of Figure 10.9, a CDMA cellular system is designed to use the same set of frequencies in each cell, as indicated on the right-hand side of the figure by the 1 in each cell.

The effective size of a CDMA cell, expressed as the radius of a circle, is the maximal distance at which the forward or reverse link can operate reliably. This distance for a given radio link is equivalent to some maximal propagation loss that can be tolerated while receiving the signal with sufficient power to overcome noise and interference, as determined by the link's "power budget." Note that the effective CDMA cell radius can be different on the forward link than it is on the reverse link and, as discussed in

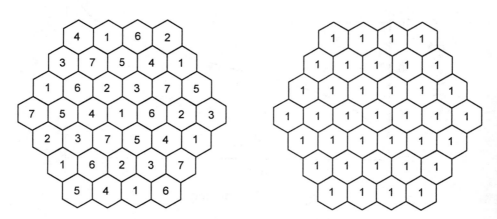

Figure 10.9 Comparison of frequency reuse patterns of FDMA and CDMA.

Section 2.1.6, may depend on the direction from or to the base station. In general, there is more interference on the reverse link, so that the forward link cell size tends to be greater than the reverse link cell size for the same amount of transmitted power for a particular channel. Therefore, as will be explained below, an effort is made to "balance" the forward and reverse links by adjusting power and other parameters to make their maximum tolerable propagation losses approximately equal.

10.1.3.1 Maximum Propagation Loss and the Cell Radius

Calculations of the link budget (forward or reverse) for a particular cell result in a value for the maximum tolerable propagation loss. Let that value be denoted L_{max}. Because propagation loss is proportional to link distance, this value of propagation loss implies a maximal distance for the link that is the effective radius of the cell or sector in a particular direction. In this section, the procedure for converting L_{max} to a value for cell radius is shown.

A general form for the propagation loss in dB as a function of distance—given the specification of other parameters such as frequency and antenna heights—is the following:

$$L(d_{km}) = L_1 + 10\gamma \log_{10} d_{km} \qquad (10.20a)$$

where

$$d_{km} = \text{link distance in kilometers} \qquad (10.20b)$$

$$L_1 = \text{intercept} \equiv \text{value of loss for } d_{km} = 1 \qquad (10.20c)$$

$$\gamma = \text{propagation power law} \qquad (10.20d)$$

At the cell radius, $d_{km} = R_{km}$ and the loss equals L_{max}. Therefore, an implicit equation for the radius in kilometers is

$$L_{max} = L(R_{km}) = L_1 + 10\gamma \log_{10} R_{km} \qquad (10.21a)$$

Solving the general equation (10.21a) for R_{km} gives

$$R_{km} = 10^{\frac{L_{max}-L_1}{10\gamma}} \qquad (10.21b)$$

Because $10 = e^{ln(10)} = e^{2.303}$, the expression for R_{km} can also be written

$$R_{km} = \exp\left\{ 2.303 \left(\frac{L_{max} - L_1}{10\gamma} \right) \right\} \qquad (10.21c)$$

Example 10.5 What is the formula for the cell radius for the CCIR propagation loss model? From Chapter 2, the parameters L_1 and 10γ in the general equation for the cell radius (10.21c) are

$$L_1 = 69.55 + 26.16 \log_{10} f_{MHz}$$
$$- [(1.1 \log_{10} f_{MHz} - 0.7)h_m - (1.56 \log_{10} f_{MHz} - 0.8)]$$
$$- 13.82 \log_{10} h_b - [30 - 25 \log_{10}(\%)]$$

$$= 38.75 + (27.72 - 1.1 h_m) \log_{10} f_{MHz} + 0.7 h_m$$
$$- 13.82 \log_{10} h_b + 25 \log_{10}(\%) \qquad (10.22a)$$

and

$$10\gamma = 44.9 - 6.55 \log_{10} h_b \qquad (10.22b)$$

where

$$h_m = \text{mobile antenna height in meters} \qquad (10.22c)$$
$$h_b = \text{base station antenna height in meters} \qquad (10.22d)$$
$$\% = \text{percentage of land covered by buildings} \qquad (10.22e)$$

Thus, a formula for the cell radius using the CCIR propagation model is

$$R_{km} = \exp\left\{ 2.303 \left(\frac{L_{max} - L_1}{10\gamma} \right) \right\}$$

$$= \exp\left\{ \frac{2.303}{44.9 - 6.55 \log_{10} h_b} [L_{max} - 38.75 - (27.72 - 1.1 h_m) \log_{10} f_{MHz} \right.$$

$$\left. - 0.7 h_m + 13.82 \log_{10} h_b - 25 \log_{10}(\%)] \right\}$$

$$= \exp\left\{ 2.303 \frac{L_{max} - 119.95 + 2.52 h_m + 13.82 \log_{10} h_b - 25 \log_{10}(\%)}{44.9 - 6.55 \log_{10} h_b} \right\}$$
$$\text{for } f_{MHz} = 850 \quad (10.22f)$$

$$= \exp\left\{ 2.303 \frac{L_{max} - 129.64 + 2.91 h_m + 13.82 \log_{10} h_b - 25 \log_{10}(\%)}{44.9 - 6.55 \log_{10} h_b} \right\}$$
$$\text{for } f_{MHz} = 1,900 \quad (10.22g)$$

The expressions in (10.22f) and (10.22g) are plotted in Figure 10.10 as functions of L_{max} and parameteric in the mobile antenna height, for the values of the other parameters indicated in the figure. The same information is presented in Figures 10.11 and 10.12, parametric in base station antenna height and percentage of buildings, respectively. Among the parameters varied in Figures 10.10 to 10.12, we observe that the cell radius is much more sensitive to the percentage of buildings than it is to the antenna heights, for practical values of the antenna heights.

Example 10.6 What is the formula for the cell radius for the Hata propagation loss model? From Chapter 2, the parameters L_1 and 10γ in the general equation (10.21c) are

$$L_1 = 69.55 + 26.16 \log_{10} f_{MHz}$$
$$- [(1.1 \log_{10} f_{MHz} - 0.7)h_m - (1.56 \log_{10} f_{MHz} - 0.8)]$$
$$- 13.82 \log_{10} h_b$$
$$- \begin{cases} 0, & \text{city}^4 \\ 2[\log_{10}(f_{MHz}/28)]^2 + 5.4, & \text{suburban} \\ 4.78(\log_{10} f_{MHz})^2 - 18.33 \log_{10} f_{MHz} + 40.94, & \text{open} \end{cases}$$

$$= \begin{cases} 68.75 + (27.72 - 1.1 h_m) \log_{10} f_{MHz} + 0.7 h_m - 13.82 \log_{10} h_b, & \text{city} \\ 59.16 + (33.51 - 1.1 h_m) \log_{10} f_{MHz} + 0.7 h_m - 13.82 \log_{10} h_b \\ \qquad - 2(\log_{10} f_{MHz})^2, & \text{suburban} \\ 27.81 + (46.05 - 1.1 h_m) \log_{10} f_{MHz} + 0.7 h_m - 13.82 \log_{10} h_b \\ \qquad - 4.78(\log_{10} f_{MHz})^2, & \text{open} \end{cases}$$
$$\tag{10.23a}$$

and
$$10\gamma = 44.9 - 6.55 \log_{10} h_b \tag{10.23b}$$

Thus, a formula for the cell radius at $f_{MHz} = 850$ using the Hata propagation model is

---

4 The difference between "large city" and "medium-small city" in the Hata model is negligible for the cellular band.

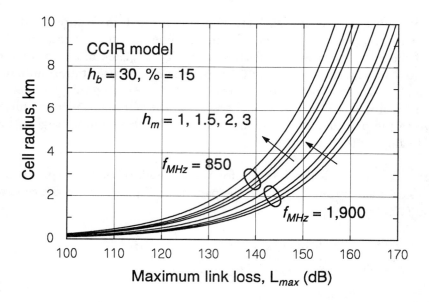

Figure 10.10 Cell radius versus maximum link loss using the CCIR propagation model, for 850 MHz and 1,900 MHz, with mobile antenna height varied.

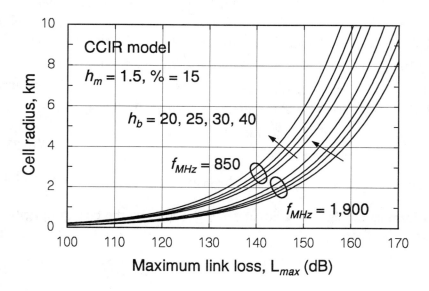

Figure 10.11 Cell radius versus maximum link loss using the CCIR propagation model, for 850 MHz and 1,900 MHz, with base station antenna height varied.

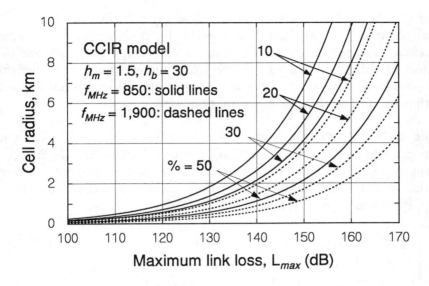

Figure 10.12 Cell radius versus maximum link loss using the CCIR propagation model, for 850 **MHz and** 1,900 **MHz, with building percentage varied.**

$$R_{km} = \begin{cases} \exp\left\{ 2.303 \dfrac{L_{max} - 149.95 + 2.52\,h_m + 13.82\log_{10}h_b}{44.9 - 6.55\log_{10}h_b} \right\}, & \text{city} \\[2ex] \exp\left\{ 2.303 \dfrac{L_{max} - 140.16 + 2.52\,h_m + 13.82\log_{10}h_b}{44.9 - 6.55\log_{10}h_b} \right\}, & \text{suburban} \\[2ex] \exp\left\{ 2.303 \dfrac{L_{max} - 121.69 + 2.52\,h_m + 13.82\log_{10}h_b}{44.9 - 6.55\log_{10}h_b} \right\}, & \text{open} \end{cases}$$

$$(10.23c)$$

and a formula for the cell radius at $f_{MHz} = 1,900$ using the Hata propagation model is

$$R_{km} = \begin{cases} \exp\left\{ 2.303 \dfrac{L_{max} - 159.64 + 2.91\,h_m + 13.82\log_{10}h_b}{44.9 - 6.55\log_{10}h_b} \right\}, & \text{city} \\[2ex] \exp\left\{ 2.303 \dfrac{L_{max} - 147.53 + 2.91\,h_m + 13.82\log_{10}h_b}{44.9 - 6.55\log_{10}h_b} \right\}, & \text{suburban} \\[2ex] \exp\left\{ 2.303 \dfrac{L_{max} - 127.41 + 2.91\,h_m + 13.82\log_{10}h_b}{44.9 - 6.55\log_{10}h_b} \right\}, & \text{open} \end{cases}$$

$$(10.23d)$$

Plots of (10.23c) and (10.23d) are shown in Figures 10.13 and 10.14 for a nominal suburban scenario and for, respectively, mobile and base station antenna heights varied. In Figure 10.15, the antenna heights are fixed and the scenario is varied. As in the case of the CCIR model, we observe that the scenario (building density) is potentially more influential in determining cell radius than the antenna heights in typical situations.

Example 10.7 What is the formula for the cell radius for the Walfisch-Ikegami propagation loss model (WIM)? From Section 2.1.4.2, in LOS situations and in NLOS situations where the base station antenna is either above the buildings ($\Delta h_b > 0$) or below them ($\Delta h_b \leq 0$), the parameters L_1 and 10γ in the general equation are[5]

$$
L_1 = \begin{cases}
42.64 + 20\log_{10} f_{MHz}, & \text{LOS situation} \\
\begin{aligned}69.55 + 26\log_{10}f_{MHz} - 10\log_{10}w - 9\log_{10}b \\ + 20\log_{10}\Delta h_m - 18\log_{10}(1 + \Delta h_b) + L_{ori},\end{aligned} & \text{NLOS, } \Delta h_b > 0 \\
\begin{aligned}69.55 + 26\log_{10}f_{MHz} - 10\log_{10}w - 9\log_{10}b \\ + 20\log_{10}\Delta h_m + 0.8\,|\Delta h_b| + L_{ori},\end{aligned} & \text{NLOS, } \Delta h_b \leq 0
\end{cases}
$$

and   (10.24a)

$$
10\gamma = \begin{cases}
26, & \text{LOS situation} \\
38, & \text{NLOS situation, } \Delta h_b > 0 \\
38 + 15|\Delta h_b|/h_B, & \text{NLOS situation, } \Delta h_b \leq 0
\end{cases} \qquad (10.24b)
$$

where

$$h_B = \text{building height in meters}$$

$$\Delta h_b = h_b - h_B$$

$$\Delta h_m = h_B - h_m$$

$$b = \text{building separation in meters (20--50m)} \qquad (10.24c)$$

$$w = \text{street width in meters } (\approx b/2)$$

$$L_{ori} = \text{mobile antenna orientation loss in dB}$$

---

[5] The adjustment for $d_{km} < 0.5$ for $\Delta h_b \leq 0$ is ignored here for simplicity.

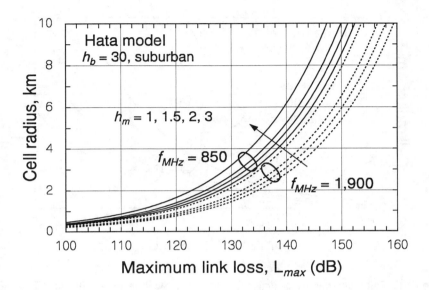

Figure 10.13 Cell radius versus maximum link loss using the Hata propagation model, for 850 MHz and 1,900 MHz, with mobile antenna height varied.

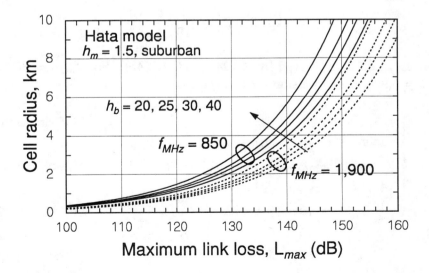

Figure 10.14 Cell radius versus maximum link loss using the Hata propagation model, for 850 MHz and 1,900 MHz, with base station antenna height varied.

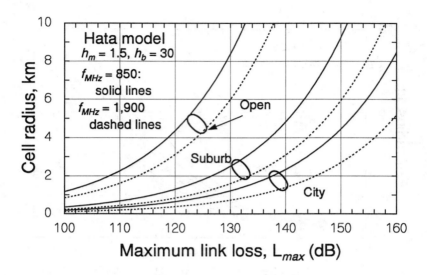

Figure 10.15 Cell radius versus maximum link loss using the Hata propagation model, for 850 **MHz and** 1,900 **MHz, with scenario varied.**

Thus, a formula for the cell radius using the WIM is

$$R_{km} = \exp\left\{2.303\left(\frac{L_{max} - L_1}{10\gamma}\right)\right\} \tag{10.24d}$$

where L_1 and 10γ are given by (10.24a) and (10.24b). For $f_{MHz} = 850$ and $f_{MHz} = 1,900$, the expression for L_1 becomes

$$L_1 = \begin{cases} 101.2, & \text{LOS situation} \\ 145.7 - 10\log_{10}w - 9\log_{10}b \\ \quad + 20\log_{10}\Delta h_m - 18\log_{10}(1 + \Delta h_b) + L_{ori}, & \text{NLOS, } \Delta h_b > 0 \\ 145.7 - 10\log_{10}w - 9\log_{10}b \\ \quad + 20\log_{10}\Delta h_m + 0.8\,|\Delta h_b| + L_{ori}, & \text{NLOS, } \Delta h_b \le 0 \\ \qquad\qquad\qquad \text{for } f_{MHz} = 850 \end{cases} \tag{10.25a}$$

and

$$
L_1 = \begin{cases}
108.2\,, & \text{LOS situation} \\[2mm]
\begin{aligned}
&154.8 - 10\log_{10}w - 9\log_{10}b \\
&\quad + 20\log_{10}\Delta h_m - 18\log_{10}(1+\Delta h_b) + L_{ori}\,,
\end{aligned} & \text{NLOS, } \Delta h_b > 0 \\[3mm]
\begin{aligned}
&154.8 - 10\log_{10}w - 9\log_{10}b \\
&\quad + 20\log_{10}\Delta h_m + 0.8\,|\Delta h_b| + L_{ori}\,,
\end{aligned} & \text{NLOS, } \Delta h_b \le 0
\end{cases}
$$

$$\text{for } f_{MHz} = 1{,}900 \qquad (10.25\text{b})$$

Parametric curves for the cell radius as a function of L_{max} for the NLOS WIM propagation loss model are given in Figures 10.16 to 10.19 for both $f_{MHz} = 850$ and $f_{MHz} = 1{,}900$. The parameters varied in these figures are, respectively, mobile antenna height, base station antenna height, building height, and building separation. In all the cases plotted in these figures, the height of the base station antenna is assumed to be above the level of building heights ($\Delta h_b > 0$), and for simplicity it is also assumed that the orientation loss is zero ($L_{ori} = 0$). We observe from these figures that mobile antenna height is not a sensitive parameter for determining cell radius using the WIM, but the base station antenna height, building height, and building separation are sensitive parameters for this propagation model.

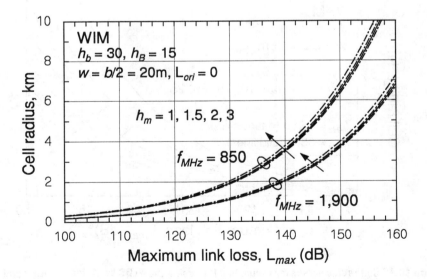

Figure 10.16 Cell radius versus maximum link loss using the NLOS WIM, for 850 MHz and 1,900 MHz, with mobile antenna height varied.

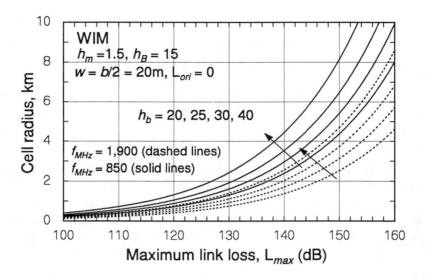

Figure 10.17 Cell radius versus maximum link loss using the NLOS WIM, for 850 MHz and 1,900 MHz, with base station antenna height varied.

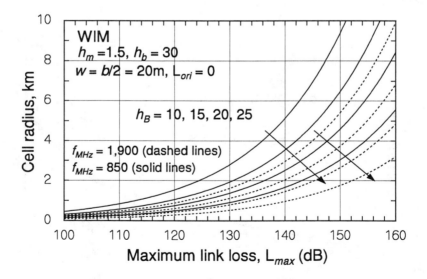

Figure 10.18 Cell radius versus maximum link loss using the NLOS WIM, for 850 MHz and 1,900 MHz, with building height varied.

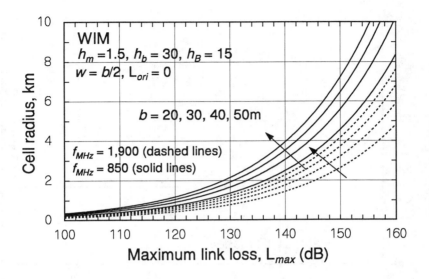

Figure 10.19 Cell radius versus maximum link loss using the NLOS WIM, for 850 MHz and 1,900 MHz, with building separation height varied.

Example 10.8 How does the propagation law parameter γ used in the cell radius calculation compare for the various propagation models?

Solution: A comparison of the propagation power γ for the CCIR, Hata, and WIM propagation loss models is presented in Figure 10.20. Note from (10.22b) and (10.23b) that γ is the same for the CCIR and Hata models, and that γ is not a function of frequency for either of these models or for the WIM, whose γ is given in (10.24b). The figure shows that the WIM power law generally has the same dependence on base station antenna height as the Hata model and is very similar to that of the Hata model for low building heights. As building heights increase beyond 15m, the WIM power law tends to be about 0.5 higher than the Hata power law for the whole range of base station antenna heights.

10.1.3.2 Forward Link Power Budget

For the forward link, the parameter that determines the cell size is $E_c/\mathcal{N}_{0,T}$, where $\mathcal{N}_{0,T}$ is the mobile's total input interference power spectral density, which can be expressed as

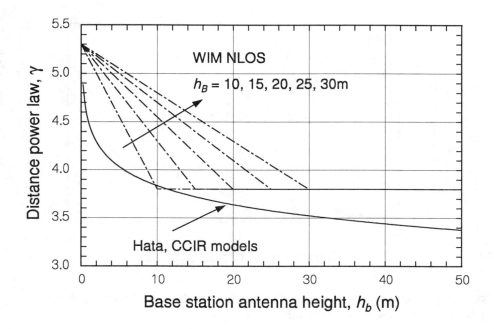

Figure 10.20 Comparison of propagation distance power laws for the CCIR, Hata, and WIM propagation loss models.

$$\frac{E_c}{\mathcal{N}_{0,T}} = \frac{[\zeta_p \cdot P_c \cdot G_c \cdot G_m \cdot 1/L]/W}{(\mathcal{N}_0)_m + \mathcal{I}_{0,sc} + \mathcal{I}_{0,oc}} = \frac{\zeta_p \cdot P_c \cdot G_c \cdot G_m \cdot 1/L}{(\mathcal{N}_0 W)_m + \mathcal{I}_{0,sc} W (1 + \mathcal{I}_{0,oc}/\mathcal{I}_{0,sc})}$$

(10.26)

where

ζ_p = fraction of cell transmit power allocated to the pilot channel

P_c = cell transmit power (watts)

W = spread-spectrum bandwidth

G_c , G_m = cell and mobile antenna gains, including cable losses

L = propagation loss on the forward link

$(\mathcal{N}_0)_m$ = thermal noise power spectral density at the mobile receiver

$\mathcal{I}_{0,sc}$ = interference spectral density due to same-cell transmissions

$\mathcal{I}_{0,oc}$ = interference spectral density due to other-cell transmissions

Using

$$L_T \triangleq \text{transmission power loss} = L/G_c G_m \tag{10.27a}$$

and assuming that the same-cell (multipath) interference is less than or equal to P_c/L_T; that is:

$$\text{Same-cell interference} = \mathcal{I}_{0,sc} W \leq \text{forward link received power} = P_c/L_T \tag{10.27b}$$

we replace $\mathcal{I}_{0,sc} W$ in (10.26) with P_c/L_T to obtain

$$\frac{E_c}{\mathcal{N}_{0,T}} \geq \left(\frac{E_c}{\mathcal{N}_{0,T}}\right)_{min} = \frac{\zeta_p P_c/L_T}{(\mathcal{N}_0 W)_m + (P_c/L_T)(1 + \mathcal{I}_{0,oc}/\mathcal{I}_{0,sc})} \tag{10.28a}$$

By assuming that the value of $E_c/\mathcal{N}_{0,T}$ on the right-hand side of (10.28a) is greater than or equal to $(E_c/\mathcal{N}_{0,T})_{req}$, the condition for satisfactory link operation, this equation may be solved for the maximum tolerable forward link transmission loss:

$$\left(\frac{E_c}{\mathcal{N}_{0,T}}\right)_{min} \geq \left(\frac{E_c}{\mathcal{N}_{0,T}}\right)_{req} \Rightarrow L_T \leq L_{Tmax} \tag{10.28b}$$

Solving for L_{Tmax} to determine the forward link cell size, we begin with the following inequality based on (10.28b):

$$\frac{\zeta_p P_c/L_T}{(\mathcal{N}_0 W)_m + (P_c/L_T)(1 + \mathcal{I}_{0,oc}/\mathcal{I}_{0,sc})} \geq \left(\frac{E_c}{\mathcal{N}_{0,T}}\right)_{req} \tag{10.28c}$$

Multiplying by the denominator of the left-hand side produces the inequality

$$\zeta_p P_c/L_T \geq \left(\frac{E_c}{\mathcal{N}_{0,T}}\right)_{req} [(\mathcal{N}_0 W)_m + (P_c/L_T)(1 + \mathcal{I}_{0,oc}/\mathcal{I}_{0,sc})]$$

which yields, after transposing terms containing L_T from the right to the left,

$$(P_c/L_T)\left[\zeta_p - \left(\frac{E_c}{\mathcal{N}_{0,T}}\right)_{req}(1 + \mathcal{I}_{0,oc}/\mathcal{I}_{0,sc})\right] \geq \left(\frac{E_c}{\mathcal{N}_{0,T}}\right)_{req}(\mathcal{N}_0 W)_m$$

Now solving for L_T, we obtain an inequality that defines the maximum tolerable transmission loss:

$$L_T \leq \frac{P_c \left[\zeta_p - \left(\dfrac{E_c}{\mathcal{N}_{0,T}} \right)_{req} \left(1 + \mathcal{I}_{0,oc}/\mathcal{I}_{0,sc} \right) \right]}{\left(\dfrac{E_c}{\mathcal{N}_{0,T}} \right)_{req} (\mathcal{N}_0 W)_m} \triangleq L_{Tmax} \qquad (10.29a)$$

Expressed in dB, the equation is

$$L_{Tmax}(\text{dB}) = P_c(\text{dBm}) + 10 \log_{10} \left[\zeta_p - \left(\frac{E_c}{\mathcal{N}_{0,T}} \right)_{req} \left(1 + \mathcal{I}_{0,oc}/\mathcal{I}_{0,sc} \right) \right]$$

$$- \left(\frac{E_c}{\mathcal{N}_{0,T}} \right)_{req} (\text{dB}) - (\mathcal{N}_0 W)_m(\text{dBm}) \qquad (10.29b)$$

Example 10.9 What is the maximum transmission loss on the forward link for a transmitter power of $P_c = 16$ W?

Solution: First, the given value for the base station's transmitter power is

$$P_c = 16\text{W} = 12\,\text{dBW} = 42\,\text{dBm} \qquad (10.30a)$$

of which we assume[6] that 15% is assigned to the CDMA pilot channel, or

$$\zeta_p = 0.15 \qquad (10.30b)$$

Because the pilot signal has no data modulation, the mobile receiver can integrate the signal for relatively long times, yielding a large processing gain for the pilot signal. A typical value for the mobile receiver's required pilot chip energy-to-noise spectral density ratio [2] is well below 0 dB:

$$(E_c/\mathcal{N}_{0,T})_{req} = -15\,\text{dB} \qquad (10.30c)$$

Using the noise figure approach to expressing the amount of receiver noise, as discussed in Section 3.3.1.2, the mobile receiver noise is

$$(\mathcal{N}_0 W)_m = (\mathcal{N}_0)_m \cdot W = kT_0 \cdot (\text{NF})_m \cdot W$$

---

[6] The optimal value of the pilot fraction ζ_p is studied in Chapter 11.

$$= -204\,\mathrm{dB(W/Hz)} + (\mathrm{NF})_m + 61\,\mathrm{dBHz}$$

$$= -143\,\mathrm{dBW} + (\mathrm{NF})_m = -113\,\mathrm{dBm} + (\mathrm{NF})_m \qquad (10.30\mathrm{d})$$

Typically, $(\mathrm{NF})_m = 8\,\mathrm{dB}$, giving $(\mathcal{N}_0 W)_m = -105\,\mathrm{dBm}$. Finally, we need a value for the ratio of other-cell interference to own-cell interference on the forward link, which is a quantity that fluctuates in real time. It was shown in Section 10.1.1 that a conservative value of the ratio of other-cell interference to same-cell interference is

$$\mathcal{I}_{0,oc}/\mathcal{I}_{0,sc} = 2.5\,\mathrm{dB} \qquad (10.30\mathrm{e})$$

This value of $\mathcal{I}_{0,oc}/\mathcal{I}_{0,sc}$ reflects the situation of a mobile near the cell edge.

Using the assumed typical values for the parameters, the maximum allowable forward link transmission loss for this example is found to be

$$L_{Tmax}(\mathrm{dB}) = P_c(\mathrm{dBm}) + 10\log_{10}\left[\varsigma_p - \left(\frac{E_c}{\mathcal{N}_{0,T}}\right)_{req}(1 + \mathcal{I}_{0,oc}/\mathcal{I}_{0,sc})\right]$$

$$- \left(\frac{E_c}{\mathcal{N}_{0,T}}\right)_{req}(\mathrm{dB}) - (\mathcal{N}_0 W)_m(\mathrm{dBm})$$

$$= 42 + 10\log_{10}\left[0.15 - 10^{-1.5}(1 + 10^{0.25})\right] - (-15) - (-105)$$

$$= 162 + 10\log_{10}\left[0.15 - .03162\,(1 + 1.78)\right]$$

$$= 162 - 12 = 150\,\mathrm{dB} \qquad (10.30\mathrm{f})$$

Example 10.10 What is the cell radius corresponding to the forward link transmission loss $L_{Tmax} = 150\,\mathrm{dB}$ if it is assumed that the mobile antenna gain is $G_m = 1 = 0\,\mathrm{dBi}$ and that the base station antenna gain is $G_c = 4 = 6\,\mathrm{dBi}$?

Solution: The example transmission loss is equivalent to a propagation loss of $L(\mathrm{dB}) = L_T(\mathrm{dB}) + 6 = 156\,\mathrm{dB}$. From Chapter 2, equation (2.28) and Table 2.2, a possible equation for the propagation loss of a cellular system based on the CCIR model in an urban area with 15% building density is

$$L(\mathrm{dB}) = [\alpha + \beta\log_{10}d_{km}]_{h_{base}=20m} = 126.93 + 36.4\log_{10}d_{km} \qquad (10.31\mathrm{a})$$

where d_{km} is the link distance in km. Given that $L = 156\,\mathrm{dB}$, (10.31a) can be solved for d_{km}, resulting in

$$d_{km} = 10^{(156-126.93)/36.4} = 10^{0.80} = 6.3\,\text{km} \approx 3.9\,\text{mi} \qquad (10.31\text{b})$$

This example shows, among other things, that 16W of forward link power is a relatively large amount of power, because a CDMA cell can be expected to service an area with a radius considerably less than 3.9 mi.

Using the values of the system parameters (transmitter power, noise figure, and the ratio $\mathcal{I}_{0,oc}/\mathcal{I}_{0,sc}$) assumed in examples 10.9 and 10.10, the formula for L_{Tmax} in (10.29b) becomes

$$L_{Tmax}(\text{dB}) = 147 + 10\log_{10}\left[\zeta_p - 2.78\left(\frac{E_c}{\mathcal{N}_{0,T}}\right)_{req}\right] - \left(\frac{E_c}{\mathcal{N}_{0,T}}\right)_{req} (\text{dB}) \quad (10.32)$$

This equation is plotted in Figure 10.21 as a function of the pilot fraction, with the required value of $E_c/\mathcal{N}_{0,T}$ as a parameter. The cell extends to those locations for which the forward link transmission loss does not exceed the amount plotted in Figure 10.21.

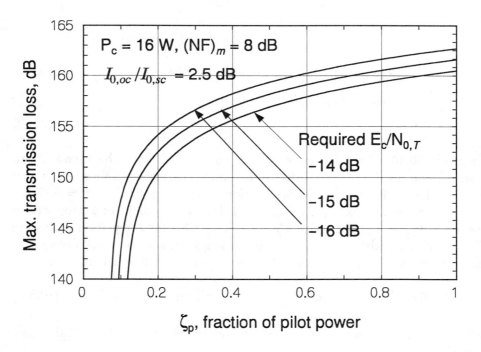

Figure 10.21 Example calculations of maximum tolerable forward link transmission loss.

Example 10.11 When the fraction of the base station's power that is assigned to the pilot signal is 20%, what is the maximum tolerable forward link transmission loss in dB if the required $E_c/\mathcal{N}_{0,T}$ at the mobile station is -15dB?

Solution: From the graph of transmission loss as a function of the pilot signal power fraction in Figure 10.21, we read for $\zeta_p = 0.2$ and $(E_c/\mathcal{N}_{0,T})_{req} = -15$ dB that the maximum tolerable loss is 154 dB.

Too much power on the forward link unbalances the forward and reverse links and results in unnecessary interference to mobiles in other cells. Link balancing is discussed below in Section 10.1.3.4 after the reverse link power budget is considered. A derivation of optimal forward link power allocation is presented in Chapter 11.

10.1.3.3 Reverse Link Power Budget

For the reverse link, the parameter that determines the cell size is the maximum value of the median link propagation loss, L_{med}, that may be experienced while maintaining a SNR at the base station receiver that is above a specified value. This SNR may be expressed by

$$\text{SNR} = \frac{P_m \cdot G_c \cdot G_m \cdot 1/L_{med}}{(\mathcal{N}_0 W)_c + \left(\frac{M}{F_e} - 1\right)\alpha_r \cdot P_m \cdot G_c \cdot G_m \cdot 1/L_{med}} \quad (10.33a)$$

where

$$\text{SNR} = \text{signal-to-noise power ratio at the base station}$$

$$P_m = \text{mobile's power amplifier output}$$

$$G_c = \text{cell antenna gain, including cable losses}$$

$$G_m = \text{mobile antenna gain, including cable losses}$$

$$L_{med} = \text{median reverse link path loss}$$

$$\alpha_r = \text{average voice activity factor}$$

$$F_e = \text{frequency reuse efficiency}$$

$$(\mathcal{N}_0 W)_c = \text{thermal noise at the cell receiver}$$

The denominator in (10.33a) is $\sigma^2 + I_T$, where

$$\sigma^2 \triangleq (\mathcal{N}_0 W)_c \tag{10.33b}$$

and

$$I_T \triangleq \left(\frac{M}{F_e} - 1 \right) \alpha_r \, P_m G_c G_m / L_{med} \tag{10.33c}$$

Because the cell loading, X, is defined as $X = I_T/(\sigma^2 + I_T)$, we may write

$$\sigma^2 + I_T = \frac{I_T}{X} = \frac{\sigma^2}{1 - X} \tag{10.33d}$$

Thus, the total noise and interference at the base station receiver is

$$(\mathcal{N}_0 W)_c + \left(\frac{M}{F_e} - 1 \right) \alpha_r \, P_m G_c G_m / L_{med} = \frac{(\mathcal{N}_0 W)_c}{1 - X} \tag{10.33e}$$

and, after substituting (10.33e) for the denominator of (10.33a), the maximum median reverse link path loss, denoted L_{max} (leading to the reverse link cell size) can be found from

$$SNR_{req} = \frac{P_m G_c G_m / L_{max}}{(\mathcal{N}_0 W)_c} (1 - X) \tag{10.34a}$$

or

$$L_{max} = \frac{P_m G_c G_m}{(\mathcal{N}_0 W)_c} \frac{(1 - X)}{SNR_{req}} = \frac{P_m G_c G_m}{(\mathcal{N}_0 W)_c} \frac{(1 - X)}{(E_b/\mathcal{N}_0)/(PG)} \tag{10.34b}$$

where we used the fact that $SNR_{req} = (E_b/\mathcal{N}_0)/(W/R_b) = (E_b/\mathcal{N}_0)/(PG)$. In dB units, (10.34b) can be written as

$$L_{max}(dB) = P_m(dBm) + G_c(dB) + G_m(dB) - SNR_{req}(dB)$$
$$- (\mathcal{N}_0 W)_c(dBm) + 10 \log_{10}(1 - X) \tag{10.34c}$$

Example 10.12 Calculation for L_{max}. The equation for the maximum allowable reverse link path loss is (10.34c). An example calculation is now given, including an explanation of each term in the calculation. First, the minimum required received carrier-to-noise power ratio on the reverse link is

$$SNR_{req} = \frac{E_b/\mathcal{N}_0}{W/R} = \frac{E_b/\mathcal{N}_0}{PG} \tag{10.35a}$$

where we assume the following numerical values:

$$E_b/\mathcal{N}_0 = \text{received bit energy-to-noise density ratio} = 7\,\text{dB}$$
$$\text{PG} = \text{processing gain} = W/R = 128 = 21\,\text{dB}$$

These assumptions give

$$\text{SNR}_{req} = \frac{E_b}{\mathcal{N}_0}\,(\text{dB}) - \frac{W}{R}\,(\text{dB}) = 7 - 21 = -14\,\text{dB} \qquad (10.35\text{b})$$

The term $(\mathcal{N}_0 W)_c \equiv (\mathcal{N}_0)_c \cdot W$ in the allowable loss calculation represents the receiver noise power at the cell site, as distinguished from interference. The noise spectral power density at the base station receiver may be written in terms of the *noise figure* NF, which equals

$$\text{NF} \triangleq 10\log_{10}\text{F} \qquad (10.35\text{c})$$

where F is the noise factor, a measure of the noise generated by a circuit, defined as the ratio of input SNR to output SNR, $\text{F} \triangleq (\text{SNR})_{in}/(\text{SNR})_{out}$. Then

$$\mathcal{N}_0 = kT\text{F} = 1.38 \times 10^{-23} \times 293 \times \text{F} = 4.04 \times 10^{-21} \times \text{F} \text{ in W/Hz}$$
$$= -204\,\text{dB(W/Hz)} + \text{NF(dB)} = -174\,\text{dB(mW/Hz)} + \text{NF(dB)}$$

A typical value of the receiver noise figure at the base station is $5\,\text{dB}$ [3, 4], giving

$$(\mathcal{N}_0)_c = -204 + 5 = -199\,\text{dB(W/Hz)} = -169\,\text{dB(mW/Hz)}$$

Because the receiver noise bandwidth in dB units is $W = 1.2288\,\text{MHz} = 61.1$ dBHz, the noise power at the base station receiver is found to be

$$(\mathcal{N}_0 W)_c = -299 + 61.1 = -138.1\,\text{dBW} = -108.1\,\text{dBm} \qquad (10.35\text{d})$$

Typical values of the remaining parameters in the equation for the allowable reverse link path loss are

- $P_m = 200\,\text{mW} = 23\,\text{dBm}$. This is the maximal power that can be delivered by the mobile unit's power amplifier. Because the system has power control, most of the time the mobile unit actually emits much less power. Here we are interested in how far away the mobile can be, and so we consider the maximum value of its power.

- $G_c = 6 \, \text{dBi}$, $G_m = 0 \, \text{dBi}$. These typical values of antenna gain, relative to an isotropic antenna gain (less cable losses), correspond to a directive (sector) antenna at the base station, and an omnidirectional antenna at the mobile unit. For the purposes of the allowable loss calculation, relatively conservative (low) values for these parameters should be used.

- $X = 90\% \rightarrow 10 \log_{10}(1 - X) = -10 \, \text{dB}$. This is a conservative value because a high value of loading represents more interference.

Using the typical values for the parameters, the maximum allowable reverse link path loss is found to be

$$L_{max}(\text{dB}) = P_m(\text{dB}) + G_c(\text{dB}) + G_m(\text{dB}) - \text{SNR}_{req}(\text{dB})$$
$$- (\mathcal{N}_0 W)_c(\text{dB}) + 10 \log_{10}(1 - X)$$

or

$$L_{max}(\text{dB}) = 23 + 6 + 0 - (-14) - (-108) - 10 = 141 \, \text{dB} \qquad (10.35\text{e})$$

Any mobile within the contour defined by this value of path loss can operate on the reverse link without exceeding its maximal power.

For parametric studies, the following slightly more general version of (10.34c), the equation for the maximum path loss, is useful:

$$L_{max}(\text{dB}) = P_m(\text{dBm}) + G_c(\text{dB}) + G_m(\text{dB}) - \left(\frac{E_b}{\mathcal{N}_0}\right)_{req}(\text{dB})$$
$$+ \text{PG}(\text{dB}) - (\mathcal{N}_0 W)_c(\text{dBm}) + 10 \log_{10}(1 - X)(10.36\text{a})$$

Using the same assumed values of the fixed system parameters (antenna gains, transmitter power, noise figure, bandwidth, and processing gain), the formula in (10.36a) becomes

$$L_{max}(\text{dB}) = 23 + 6 + 0 - \left(\frac{E_b}{\mathcal{N}_0}\right)_{req}(\text{dB}) + 21 - (-108) + 10 \log_{10}(1 - X)$$
$$= 158 - \left(\frac{E_b}{\mathcal{N}_0}\right)_{req} \text{dB} + 10 \log_{10}(1 - X) \qquad (10.36\text{b})$$

A plot of the reverse link maximum path loss versus cell loading is given in Figure 10.22 for different values of E_b/N_0 and the parameters given previously.

Example 10.13 When the reverse link loading is 60%, what is the maximum tolerable reverse link loss in dB if the required E_b/N_0 at the base station receiver is 6 dB?

    Solution: From the graph of maximum path loss as a function of the loading in Figure 10.22, we read for $X = 60\% = 0.6$ and $E_b/N_0 = 6$ dB that the maximum tolerable loss is 148 dB. Using a propagation formula as before for the forward link, the maximum tolerable reverse link loss can be converted to a distance, the reverse link cell radius or cell size.

10.1.3.4 Link Balancing

Too much power on the forward link results in unnecessary interference to mobiles in other cells. Too much power in the reverse link decreases the effective cell capacity. It is desirable to *balance* the system so that the coverage area of the forward link coincides with the boundary of the loca-

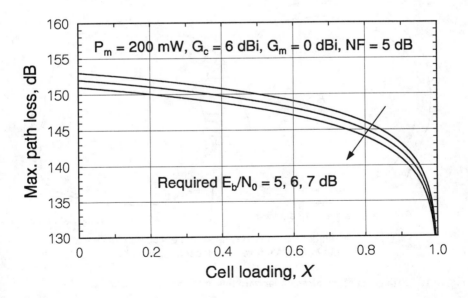

Figure 10.22 Maximum reverse link path loss versus cell loading, parametric in the required value of SNR.

tions for which the reverse link path loss is within the allowable range; that is, to the extent possible it is good to make the forward link cell size and the reverse link cell size equal.

A balanced system design makes handoff smooth and reduces interference. But too strong a forward link or too weak a reverse link, as depicted in Figure 10.23, gives rise to problems as mobile receivers in the handoff area experience pilot signals as interference, not handoff opportunities. On the other hand, for too weak a forward link or too strong a reverse link, the situation illustrated in Figure 10.24, mobiles in the area between the cells may lose contact with any base station because the reverse link interference is too strong. To keep the forward and reverse links' boundaries the same (or nearly the same), the system is designed to minimize a balance factor, B, defined as

$$B \triangleq L_{\text{forward}}(\text{dB}) - L_{\text{reverse}}(\text{dB}) \tag{10.37a}$$

where previously we found

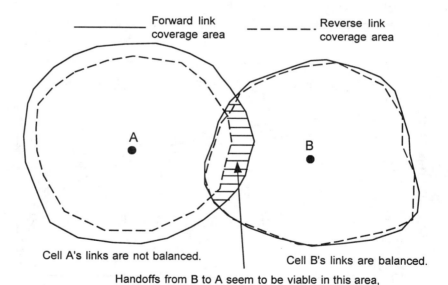

Figure 10.23 Concept of too strong a forward link or too weak a reverse link.

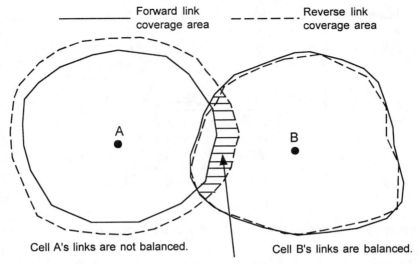

Forward link coverage area ——— Reverse link coverage area – – – –

A

B

Cell A's links are not balanced.

Cell B's links are balanced.

Handoffs from B to A do not occur in this area, so the mobiles in cell A generate more interference to cell B than necessary

Figure 10.24 Concept of too weak a forward link or too strong a reverse link.

$$L_{\text{reverse}}(\text{dB}) = L_{max}(\text{dB})$$

$$= P_m(\text{dBm}) + G_c(\text{dB}) + G_m(\text{dB}) - \text{SNR}_{req}(\text{dB})$$

$$- (\mathcal{N}_0 W)_c(\text{dBm}) + 10\log_{10}(1 - X) \quad (10.37\text{b})$$

and

$$L_{\text{forward}}(\text{dB}) = L_{Tmax}(\text{dB}) + G_c(\text{dB}) + G_m(\text{dB})$$

$$= P_c(\text{dBm}) + G_c(\text{dB}) + G_m(\text{dB})$$

$$+ 10\log_{10}\left[\zeta_p - \left(\frac{E_c}{\mathcal{N}_{0,T}}\right)_{req}\left(1 + \frac{\mathcal{I}_{0,oc}}{\mathcal{I}_{0,sc}}\right)\right]$$

$$- \left(\frac{E_c}{\mathcal{N}_{tot}}\right)_{req}(\text{dB}) - (\mathcal{N}_0 W)_m(\text{dBm}) \quad (10.37\text{c})$$

Using these expressions for forward and reverse propagation losses, we can write the balance factor as

$$\mathcal{B} = L_{forward}(dB) - L_{reverse}(dB)$$

$$= P_c(dBm) + G_c(dB) + G_m(dB) + 10\log_{10}\left[\zeta_p - \left(\frac{E_c}{\mathcal{N}_{0,T}}\right)_{req}\left(1 + \frac{\mathcal{I}_{0,oc}}{\mathcal{I}_{0,sc}}\right)\right]$$

$$- \left(\frac{E_c}{\mathcal{N}_{0,T}}\right)_{req}(dB) - \underbrace{(\mathcal{N}_0 W)_m(dBm)}_{kT_0 W(dBm)+(NF)_m} - P_m(dBm) - G_c(dB)$$

$$- G_m(dB) + SNR_{req}(dB) + \underbrace{(\mathcal{N}_0 W)_c(dBm)}_{kT_0 W(dBm)+(NF)_c} - 10\log_{10}(1 - X)$$

$$\text{(10.37d)}$$

The antenna gain terms and the $kT_0 W$ terms cancel, leaving

$$\mathcal{B} = P_c(dBm) + 10\log_{10}\left[\zeta_p - \left(\frac{E_c}{\mathcal{N}_{0,T}}\right)_{req}\left(1 + \frac{\mathcal{I}_{0,oc}}{\mathcal{I}_{0,sc}}\right)\right] - \left(\frac{E_c}{\mathcal{N}_{0,T}}\right)_{req}(dB)$$

$$- (NF)_m(dB) - P_m(dBm) + SNR_{req}(dB) + (NF)_c(dB)$$
$$- 10\log_{10}(1 - X) \qquad\qquad\qquad\qquad\qquad\qquad \text{(10.37e)}$$

Grouping similar terms, we have

$$\mathcal{B} = [SNR_{req}(dB) - (E_c/\mathcal{N}_{tot})_{req}(dB)] + [(NF)_c(dB) - (NF)_m(dB)]$$

$$+ [P_c(dBm) - P_m(dBm)] + 10\log_{10}\left[\frac{\zeta_p - (E_c/\mathcal{N}_{tot})_{req}(1 + \mathcal{I}_{oc}/\mathcal{I}_{sc})}{1 - X}\right]$$

$$\text{(10.37f)}$$

The balance in the system may be expressed in terms of whether the system is *reverse link limited* (the reverse link radius is relatively too small) or *forward link limited* (the forward link radius is relatively too small). The decision can be made using the rule

$$\begin{cases} \mathcal{B} < -\delta: & \text{The system is reverse link limited.} \\ |\mathcal{B}| \le \delta: & \text{The system is balanced.} \\ \mathcal{B} > \delta: & \text{The system is forward link limited.} \end{cases} \qquad \text{(10.37g)}$$

The parameter $\delta > 0$ (at most 1 or 2 dB) is used to take into account a tolerance in all the factors involved in calculating $\mathcal{B}$.

Example 10.14 (Balanced design.) Given the following forward and reverse link parameters, find the value of the forward link pilot fraction, ζ_p, required to balance the system.

Forward link	Reverse link
$P_c = 10\,\text{W} = 40\,\text{dBm}$	$P_m = 200\,\text{mW} = 23\,\text{dBm}$
$G_c = 6\,\text{dBi}$	$G_m = 0\,\text{dBi}$
$\text{SNR}_{req} = 7 - 21 = -14\,\text{dB}$	$\left(\dfrac{E_c}{\mathcal{N}_{0,T}}\right)_{min} = -15\,\text{dB}$
$\mathcal{I}_{0,oc}/\mathcal{I}_{0,sc} = 2.5\,\text{dB}$	$\text{Loading} = X = 50\%$
$(\text{NF})_m = 8\,\text{dB}$	$(\text{NF})_c = 5\,\text{dB}$
	$\text{Voice activity}: \alpha_r = 0.4$

Solution: To make $\mathcal{B} = 0$, we write

$$0 = [\text{SNR}_{req} - (E_c/\mathcal{N}_{tot})_{min}] + [(\text{NF})_c - (\text{NF})_m] + [P_c - P_m]$$

$$+ 10\log_{10}\left[\frac{\zeta_p - (E_c/\mathcal{N}_{0,T})_{req}(1 + \mathcal{I}_{0,oc}/\mathcal{I}_{0,sc})}{1 - X}\right]$$

$$= [-14 + 15] + [5 - 8] + [40 - 23]$$

$$+ 10\log_{10}\left[\frac{\zeta_p - 10^{-1.5}(1 + 10^{0.25})}{1 - 0.5}\right]$$

or

$$\zeta_p = 10^{-1.5}(1 + 10^{0.25}) + 0.5 \times 10^{-1.5}$$

$$= (1.5 + 10^{0.25})10^{-1.5} = 0.104 = 10.4\%$$

To balance the links, the designer in this example situation should allocate 10.4% of the cell transmitter power to the pilot signal. It is shown in Chapter 11 that the optimal fraction of forward link power for the pilot (or any other forward link channel) is "dynamic" in that it changes as, for example, the numbers of users in the same cell and in other cells changes.

10.2 CDMA Link Reliability and Erlang Capacity

In Section 3.3.2, we discussed coverage and capacity issues in connection with the basics of cellular engineering and introduced a measure of link reliability for a cellular system in terms of a margin that compensates for the variation in mobile signal level that may occur through fading or shadowing. In this section, we extend the consideration of reliability measures to a CDMA system to obtain the margins required for the specified reliability in diversity reception of soft handoff. In a CDMA cellular system, there is a positive handoff gain as opposed to the negative handoff gain that was noted for the case of the hard handoff used in the AMPS FDMA cellular system. It is shown that the positive handoff gain of a CDMA cellular system significantly reduces the margin required to provide a specified level of link reliability. A comparison is also made in this section between AMPS and CDMA cellular systems in terms of margins required for various levels of reliability.

We then consider the subject of the Erlang capacity of a CDMA system, which is a subject of current research. We analyze CDMA Erlang capacity using both central limit theorem (Gaussian) and lognormal approximations to the sum of lognormal interference powers. The purpose of this analysis is to obtain Erlang traffic measures for a CDMA system, which can be used as the basis of system design in terms of area coverage and total area capacity. The section concludes with example calculations of base station density as a function of Erlang requirements and cell loading.

10.2.1 Link Reliability and Link Margin

In the mobile radio propagation environment, the received signal power level is subject to shadowing, which causes the received signal level to be an RV. The randomness of the signal level can be attributed to a variation in the amount of propagation loss, which in dB units is very well represented as a Gaussian RV whose median[7] is a function of the link distance d:

$$L(\text{dB}) = G(L_{med}, \sigma_{dB}) = L_{med} + \sigma_{dB} G(0, 1) \tag{10.38a}$$

where $G(\mu, \sigma^2)$ denotes a Gaussian RV with mean μ and variance σ^2. Because the received signal power level is random, it is appropriate to speak of

---

[7] Note that the median and mean of a Gaussian RV are equal.

the probability that the received SNR exceeds system requirements, expressed as a required value of SNR, denoted SNR_{req}. Formulated as a measure of "link reliability," that probability may be written

$$P_{rel} \triangleq \Pr\{SNR > SNR_{req}\} \tag{10.38b}$$

and an objective of the system design is to achieve a certain value of link reliability, say $P_{rel} = 0.9$ (SNR requirement exceeded 90% of the time).

To overcome the variation in propagation loss to some degree, and to achieve a desired value of link reliability, the reverse link budget presented in Section 10.1.3.3 must allow for more transmitter power than is necessary to meet the receiver SNR requirements in the absence of shadowing.

10.2.1.1 Link Margin for No Interference

A generic equation for a link budget (in dB) in the absence of multiple-access interference may be written

$$SNR_{med} = P_m + G_m - L_{med}(d) + G_c - L_c - N_c + PG$$

$$= SNR_{req} + E_{dB} \tag{10.39a}$$

$$\geq SNR_{req} + M_{dB} \tag{10.39b}$$

where

$$P_m = \text{mobile transmitter power, in dBm}$$

$$G_m, G_c = \text{mobile, base station antenna gains in dBi}$$

$$L_{med}(d) = \text{median propagation loss}$$

$$L_c = \text{base station cabling losses}$$

$$N_c = \text{noise power at the receiver, in dBm}$$

$$PG = \text{processing gain}$$

In (10.39a) E_{dB} is the excess of the received SNR over the required SNR in dB, and in (10.39b) M_{dB} is an operating margin allowing for unpredictable SNR variation. Thus, a link budget that includes margin in effect stipulates the higher system SNR requirement of $SNR_{req} + M_{dB}$ to be met by the median received SNR.

Rearranging (10.39b) to solve for $L_{med}(d)$ results in the expression

$$L_{med}(d) \leq P_m + G_m + G_c - L_c - N_c + PG - SNR_{req} - M_{dB}$$

or

$$L_{med}(d) \leq L_{max} = L_0 - M_{dB} \qquad (10.39c)$$

where

$$L_0 = P_m + G_m + G_c - L_c - N_c + PG - SNR_{req} \qquad (10.39d)$$

which indicates that the amount of tolerable propagation loss for the link is reduced by the amount of the margin, which in turn implies a reduction in the cell size due to the margin.

If the link budget is modified by substituting the actual propagation loss L(dB) given in (10.38a) in place of the median propagation loss, the link reliability can be written as

$$
\begin{aligned}
P_{rel} &= Pr\{SNR \geq SNR_{req}\} \\
&= Pr\{P_m + G_m - L(dB) + G_c - L_c - N_c + PG \geq SNR_{req}\} \\
&= Pr\{P_m + G_m - L_{med}(d) - \sigma_{dB}\, G(0,1) \\
&\qquad\qquad + G_c - L_c - N_c + PG \geq SNR_{req}\} \\
&= Pr\{SNR_{med} - \sigma_{dB}\, G(0,1) \geq SNR_{req}\} \qquad (10.40a)
\end{aligned}
$$

$$= Pr\left\{ G(0,1) \leq \frac{SNR_{med} - SNR_{req}}{\sigma_{dB}} \right\} \qquad (10.40b)$$

$$= Pr\left\{ G(0,1) \leq \frac{M_{dB}}{\sigma_{dB}} \right\} \qquad (10.40c)$$

Viewed another way, the reliability can also be written as

$$P_{rel} = Pr\{L(dB) \leq P_m + G_m + G_c - L_c - N_c + PG - SNR_{req}\} \qquad (10.40d)$$

or

$$P_{rel} = Pr\{L(dB) \leq L_0\} \qquad (10.40e)$$

where

$$L_0 \triangleq P_m + G_m + G_c - L_c - N_c + PG - SNR_{req} \qquad (10.40f)$$

Thus, the SNR reliability requirement is seen to be a propagation loss reliability requirement, to be satisfied by controlling the median propagation loss through the link distance d, as indicated in (10.39c). Substituting (10.38a) for L(dB), in view of (10.39c) we obtain

$$P_{rel} = \Pr\{L_{med} + \sigma_{dB}\, G(0, 1) \leq L_0\}$$

$$= \Pr\left\{ G(0, 1) \leq \frac{L_0 - L_{med}}{\sigma_{dB}} \right\} \tag{10.41a}$$

$$= \Pr\{G(0, 1) \leq M_{dB}/\sigma_{dB}\} \tag{10.41b}$$

Example 10.15 What margin is required to achieve 90% reliability?

Solution: If the requirement is 90% link reliability, then we have the equation

$$0.90 = \Pr\{G(0, 1) \leq M_{dB}/\sigma_{dB}\} \quad \rightarrow \quad M_{dB} = 1.28155\, \sigma_{dB} \tag{10.42}$$

using the method discussed in Section 3.3.2.1. The amount of margin required is proportional not only to the reliability but also to the degree of variability of the propagation loss expressed in terms of σ_{dB}, the standard deviation of the propagation loss in dB.

10.2.1.2 Link Margin and Power Control

A typical value of the standard deviation for the propagation loss variation is $\sigma_{dB} = 8$ dB [6, 7] (see Section 2.1.5.1), which becomes the standard deviation of the SNR if there is no power control. With power control, the mobile transmitter power can be adjusted to adapt to the variations in the propagation loss, and this adaptation has the effect of reducing the variations. A typical value for the standard deviation of the SNR in that case is $\sigma_{dB} = 2.5$ dB [8]. Thus, from (10.42) the single-cell margin needed for 90% link reliability without power control is

$$M_{dB} = 1.28155\, \sigma_{dB} = 1.28155 \times 8\,\text{dB} = 10.25\,\text{dB} \tag{10.43a}$$

With power control, the single-cell margin needed for 90% link reliability is

$$M_{dB} = 1.28155\, \sigma_{dB} = 1.28155 \times 2.5\,\text{dB} = 3.20\,\text{dB} \tag{10.43b}$$

Figure 10.25 illustrates the fact that the median SNR must be higher for $\sigma_{dB} = 8$ dB to achieve the same link reliability as for $\sigma_{dB} = 2.5$ dB. Values of

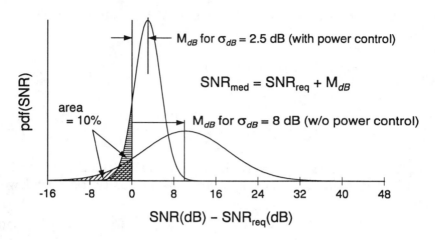

Figure 10.25 Comparison of reverse link SNR distributions with and without power control.

the no-interference margin for several different values of the link reliability
are given in Table 10.3.

10.2.1.3 Margin Required With Interference

Having found the margin when there is no interference, we now consider
what margin is required when same- and other-cell interference is present.
The median CDMA reverse link SNR may be expressed by

$$\text{SNR} = \frac{P_m \cdot G_c \cdot G_m \cdot 1/L_{med}}{(\mathcal{N}_0 W)_c + \left(\frac{M}{F_e} - 1\right)\alpha_r \cdot P_m \cdot G_c \cdot G_m \cdot 1/L_{med}} \tag{10.44}$$

Table 10.3 Required link margins with and without power control

P_{rel}	M_{dB}	M_{dB} for $\sigma_{dB} = 8$ dB	M_{dB} for $\sigma_{dB} = 2.5$ dB
0.7	$0.52240\,\sigma_{dB}$	4.18 dB	1.31 dB
0.75	$0.67449\,\sigma_{dB}$	5.40 dB	1.69 dB
0.8	$0.84162\,\sigma_{dB}$	6.73 dB	2.10 dB
0.85	$1.03643\,\sigma_{dB}$	8.29 dB	2.59 dB
0.9	$1.28155\,\sigma_{dB}$	10.25 dB	3.20 dB
0.95	$1.64485\,\sigma_{dB}$	13.16 dB	4.11 dB

Note that the median propagation loss L_{med} appears in both the numerator and denominator of (10.44) because both the interference power and the desired signal power are affected by the propagation losses. However, each of the interfering users experiences a different amount of propagation loss at a given time. Before, to analyze the margin, we substituted the random loss $L(dB)$ for L_{med} and found the P_{rel} as a function of the margin. It would not be correct to substitute the same random propagation loss $L(dB)$ in place of L_{med} in both numerator and denominator. The loss to the desired signal in the numerator and the losses to the interfering signals in the denominator should be treated as having different random components, as analyzed in [9] and [10]. However, as discussed in Section 10.1.3.3, it is valid and expedient to take the view that the amount of interference is controlled to maintain a certain value of the loading parameter X, giving the SNR expression

$$\text{SNR} = \frac{P_m \cdot G_c \cdot G_m \cdot 1/L_{med}}{(\mathcal{N}_0 W)_c}(1 - X) \tag{10.45}$$

Note that (10.45) has L_{med} only in the numerator and, in fact, the SNR expression differs from the no-interference case only by the factor $(1 - X)$. Therefore, it is reasonable to substitute a random propagation loss for L_{med} in (10.45) to formulate the link reliability. It is clear that following this procedure leads to the same mobile transmitter power margin in dB as for the no-interference case. What makes a difference in margin, however, is the provision of the CDMA system for diversity reception and for soft handoff.

10.2.1.4 Margin for Diversity Reception and Soft Handoff

In a multicell CDMA system based on IS-95, it is possible for multiple paths of a mobile user's reverse link signal to be processed by one or more base station receivers. When the mobile is near the border between two cells or two sectors, its waveform can be processed by the two base station receivers. Data from the mobile are forwarded by each base station to the MTSO, where the best source is selected on a frame-by-frame (20-ms) basis.

In this situation, a form of selection diversity[8] is operative, described as "soft handoff." As analyzed in Section 3.3.2.2, the equation for the margin M_{dB} is given by

---

[8] On the forward link, it is possible to use optimal (maximal ratio) diversity combining instead of selection diversity. See Section 4.4.2.

$$L\left(\frac{M_{dB}}{\sigma_{dB}}, \ 0, \ -\sqrt{\frac{1-\rho}{2}}\right) = \frac{1 - P_{rel}}{2} \tag{10.46a}$$

where ρ is the correlation coefficient between the Gaussian propagation losses in dB of the two paths from the mobile to the two base stations, and $L(a, b, c)$ is the bivariate Gaussian probability integral [11]:

$$L(a, b, c) \triangleq \int_a^\infty \int_b^\infty \frac{1}{2\pi(1 - c^2)} \exp\left\{-\frac{x^2 - 2cxy + y^2}{2(1 - c^2)}\right\} dy\, dx \tag{10.46b}$$

A plot of the relative margin M_{dB}/σ_{dB} is given in Figure 10.26 as a function of ρ for link reliability equal to 70%, 80%, and 90%. The data for these plots were obtained from curves in Figures 26.2 to 26.4 of [11].

Example 10.16 For soft handoff using selection diversity, what is the relative margin required to achieve 90% link reliability?
    Solution: From Figure 10.26, we observe that the value of M_{dB}/σ_{dB} depends on the value of the correlation coefficient ρ between the Gaussian propagation losses (in dB) from the mobile to the two base stations. For example, if the two losses are uncorrelated ($\rho = 0$), the value of M_{dB}/σ_{dB} giving $P_{rel} = 0.9$ is $M_{dB}/\sigma_{dB} = 0.48$. With respect to the relative margin found for no diversity, which in Table 10.3 is given as 1.28155, the soft handoff feature of the CDMA system provides a gain of $1.28155\,\sigma_{dB} - 0.48\,\sigma_{dB} = 0.80\,\sigma_{dB}$ or 6.4 dB when $\sigma_{dB} = 8$ dB.

    We note from Figure 10.26 that the required margin is low for negative correlation, when it is likely that the other link is good when one of them is bad, and the required margin is high for positive correlation, when the links tend to be either good or bad at the same time. A value of $\rho = 0.5$ is perhaps typical [12]. For $P_{rel} = 0.9$ and a correlation of $\rho = 0.5$ between propagation losses in adjacent cells, from the figure we find that the margin required at the cell boundary (distance R from either base station) is $M_{dB}(R) = 0.77\,\sigma_{dB}$, which equals 6.16 dB for $\sigma_{dB} = 8$ dB, and 1.93 dB for $\sigma_{dB} = 2.5$ dB. Thus, relative to the single cell (no handoff) case, for which the required margin for 90% reliability is $M_{dB}(R) = 10.25$ dB for $\sigma_{dB} = 8$ and $M_{dB}(R) = 3.20$ dB for $\sigma_{dB} = 8$, there are gains of 4.09 and 1.27 dB, respectively, when soft handoff is used[13].

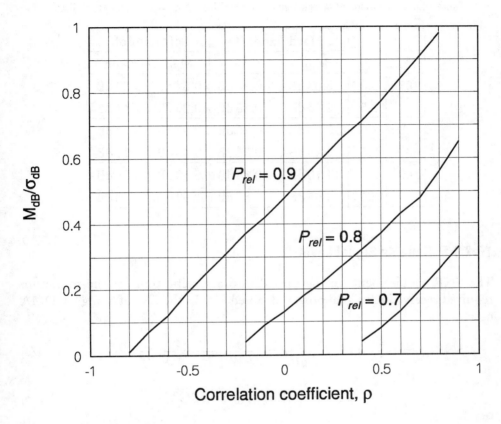

Figure 10.26 Relative margin in dB versus correlation.

In Section 3.3.2.2, it was shown that a hard handoff system must use additional margin to prevent excessive handoffs near the cell edge, and that increasing the required SNR by 5% ($\kappa = 1.05$ as described in Table 3.13) for this purpose results in a handoff "gain" of $-0.85\,$dB for a fourth-power propagation law. In Table 10.4, the handoff gains of hard and soft handoff systems are compared for different values of the link reliability and for values of σ_{dB} with and without power control.

Table 10.4 Comparison of margins and handoff gains for $\rho = 0.5$ and $\kappa = 1.05$

P_{rel}	σ_{dB}	Hard handoff		Soft handoff	
		M_{dB}	Gain	M_{dB}	Gain
0.7	8.0	5.03 dB	−0.85 dB	0.64 dB	3.54 dB
	2.5	2.16 dB	−0.85 dB	0.20 dB	1.11 dB
0.8	8.0	7.58 dB	−0.85 dB	2.96 dB	3.77 dB
	2.5	2.95 dB	−0.85 dB	0.93 dB	1.18 dB
0.9	8.0	11.10 dB	−0.85 dB	6.16 dB	4.09 dB
	2.5	4.05 dB	−0.85 dB	1.92 dB	1.28 dB

10.2.1.5 Reliable Signal Level

The SNR requirement can be turned into a specification for the minimum required received RF signal power, denoted P'_S. In the case of a single CDMA user:

$$\text{SNR}_{req} = \left(\frac{E_b}{\mathcal{N}_0}\right)_{req} = \frac{P'_S}{\mathcal{N}_0 R_b} = \frac{P'_S}{\mathcal{N}_0 W} \cdot \frac{W}{R_b} = \frac{P'_S}{\mathcal{N}_0 W} \cdot \text{PG} \qquad (10.47\text{a})$$

$$\rightarrow P'_S = \text{SNR}_{req} \cdot \mathcal{N}_0 W \div \text{PG} \qquad (10.47\text{b})$$

or

$$P'_S(\text{dBm}) = \text{SNR}_{req}(\text{dB}) + \mathcal{N}_0(\text{dBm/Hz}) + W(\text{dBHz}) - \text{PG}(\text{dB}) \qquad (10.47\text{c})$$

Adding the margin to the minimum required signal results in the *reliable signal level (RSL)*, given by

$$\text{RSL}(\text{dBm}) = P'_S(\text{dBm}) + M_{dB} \qquad (10.47\text{d})$$

$$= P'_S(\text{dBm}) + M_{dB}|_{\text{single cell}} - \text{handoff gain (dB)} \qquad (10.47\text{e})$$

Example 10.17 What are the minimum required signal level and the reliable signal level for a single CDMA user assuming typical values of the system parameters, with and without power control?

Solution: (a) Minimum required signal. The minimum required signal is calculated from (10.47c) as follows:

Required $E_b/\mathcal{N}_0$ in dB	7.0 dB
+ $\mathcal{N}_0$ for 5 dB noise figure	−169.0 dBm/Hz
+ Bandwidth in dBHz	60.9 dBHz
− Processing gain	21.1 dB
Minimum required signal P'_S	−122.2 dBm

(b) RSL under no power control. Using (10.47e) and assuming $\rho = 0.5$ and $\sigma_{db} = 8$, then the required signal power at the receiver for no power control is found as follows:

Minimum required signal P'_S	−122.20 dBm
+ One-cell margin for 90% reliability	10.25 dB
− Handoff gain	4.09 dB
RSL	−116.04 dBm

(c) RSL under power control. Using (10.47e) and assuming $\rho = 0.5$ and $\sigma_{db} = 2.5$, then the required signal power at the receiver for power control is found as follows:

Minimum required signal P'_S	−122.20 dBm
+ One-cell margin for 90% reliability	3.20 dB
− Handoff gain	1.28 dB
RSL	−120.28 dBm

10.2.2 Erlang Capacity

As discussed in Section 3.2, the Erlang B model can be applied directly to the reverse links of FDMA and TDMA cellular telephone systems, where the channels available for assignment to mobile calls are based on frequency division multiplexing. The application of the Erlang B theory to FDMA and TDMA cellular systems was straightforward because the number of available channels for those systems is fixed. In practice, the number of channels in a CDMA cellular system is not fixed, but fluctuates because the reverse link is interference-limited. The mechanism for blocking in a CDMA cellular system needs to be examined before an application of the Erlang B theory can be made.

In a CDMA cellular system, there is no fixed number of channels as in an FDMA or TDMA system, because the capacity (allowable number of users) depends on the degree of interference.[9] Blocking occurs when the reverse link multiple access interference power reaches a predetermined level that is set to maintain acceptable signal quality. If the total user interference at a base station receiver exceeds some threshold, the system blocks (denies access) to the next user who attempts to place a call. The number of users for which the CDMA blocking probability, denoted B_{CDMA}, equals a certain value (usually 1 or 2%) is defined to be the *Erlang capacity* of the system and is related to an equivalent number of channels in an FDMA or TDMA cellular system. Thus, the calculation of the CDMA blocking probability is based on an analysis of the other-user interference.

Because the number of users (call traffic) at a given time is random, and the interference power from a user is an RV, the probability of blocking leads to an estimate of the average number of active users that is termed the Erlang capacity of the CDMA cell or sector. The determination of Erlang capacity depends on the assumptions about the probability distributions of the call traffic and user interference. In this section, a comparison is made between the CDMA Erlang capacities determined using two different assumed distributions for the total user interference power: Gaussian and lognormal.

10.2.2.1 Formulation of the Blocking Probability

First we consider a single, isolated CDMA cell with M active users. The total reverse link signal-plus-noise power received at the base station can be written as

$$\underbrace{\alpha_{r1} P_1 + \alpha_{r2} P_2 + \cdots + \alpha_{rM} P_M}_{M \text{ reverse link signals}} + \underbrace{(\mathcal{N}_0 W)_c}_{\text{noise power}} \qquad (10.48a)$$

where

- The $\{\alpha_{ri}\}$ are random variables representing the reverse link voice activity, which have the experimental values given by [3]

$$E\{\alpha_{ri}\} = \overline{\alpha}_r = 0.4 \quad \text{and} \quad E\{\alpha_{ri}^2\} = \overline{\alpha_r^2} = 0.31 \qquad (10.48b)$$

- The $\{P_i\}$ are the random signal powers for the M active users.

- The number of signals M is itself an RV, assumed to have a Poisson distribution, so that $E\{M\} = \overline{M} = \text{Var}\{M\}$.

---

[9] Unless otherwise noted, we discuss CDMA capacity in terms of the requirements for the interference-limited reverse link.

To a potential $(M + 1)$st reverse link user, the total power for the M active users and the thermal noise is interference power. Thus, we may write

$$I' \equiv \mathcal{I}_0' W = \alpha_{r1} P_1 + \alpha_{r2} P_2 + \cdots + \alpha_{rM} P_M + (\mathcal{N}_0 W)_c \qquad (10.48c)$$

where

$$\mathcal{I}_0' = \frac{I'}{W} = \frac{\alpha_{r1} P_1 + \alpha_{r2} P_2 + \cdots + \alpha_{rM} P_M}{W} + \mathcal{N}_0 \qquad (10.48d)$$

is the power spectral density level for the total received interference power. Normalized by $\mathcal{I}_0' R_b$, where R_b is the data bit rate, the total interference is characterized by the quantity

$$\frac{I'}{\mathcal{I}_0' R_b} = \frac{W}{R_b} = \alpha_{r1} \frac{E_{b1}}{\mathcal{I}_0'} + \alpha_{r2} \frac{E_{b2}}{\mathcal{I}_0'} + \cdots + \alpha_{rM} \frac{E_{bM}}{\mathcal{I}_0'} + \frac{\mathcal{N}_0}{\mathcal{I}_0'} \cdot \frac{W}{R_b}$$

$$= Z + \frac{\mathcal{N}_0}{\mathcal{I}_0'} \cdot \frac{W}{R_b} = Z + \eta \cdot \frac{W}{R_b} \qquad (10.49a)$$

where

$$Z \triangleq \sum_{i=1}^{M} \alpha_{ri} \, \rho_i = \tfrac{W}{R_b}(1 - \eta) \qquad (10.49b)$$

$$\rho_i \triangleq E_{bi}/\mathcal{I}_0' \qquad (10.49c)$$

and

$$\eta \triangleq \mathcal{N}_0/\mathcal{I}_0' \qquad (10.49d)$$

is a parameter indicating the loading of the CDMA system and W/R_b is the spread-spectrum processing gain. Given the value of η, the quality of the channel that is available to the $(M + 1)$st mobile user is characterized by the value of the random variable Z; if Z exceeds some threshold value, then the channel is effectively unavailable (blocked) to the $(M + 1)$st user. In terms of the distribution of the random variable Z, the probability that the $(M + 1)$st mobile CDMA user will be blocked is the probability that Z exceeds some threshold value Z_0, as a function of a threshold value of the interference parameter η_0:

$$B_{CDMA} = \Pr\left\{ Z > Z_0 = \frac{W}{R_b}(1 - \eta_0) \right\} \qquad (10.50a)$$

$$= \Pr\left\{ \sum_{i=1}^{M} \alpha_{ri} \, \rho_i > \frac{W}{R_b}(1 - \eta_0) \right\} \qquad (10.50b)$$

If a probability density function $p_z(x)$ is known or assumed for Z, then the evaluation of B_{CDMA} is simply a matter of integrating that pdf over the region defined by $Z > Z_0$:

$$B_{CDMA} = \int_{Z_0}^{\infty} dx \, p_z(x) \tag{10.51a}$$

The exact pdf of Z is not known, however, so an approximation is needed to compute B_{CDMA}.

The CDMA blocking probability can be manipulated to the form

$$B_{CDMA} = \Pr\{Z > Z_0\} = \Pr\left\{ \frac{Z - E\{Z\}}{\sqrt{\mathrm{Var}\{Z\}}} > \frac{Z_0 - E\{Z\}}{\sqrt{\mathrm{Var}\{Z\}}} \right\}$$

$$= Q_Z\left(\frac{Z_0 - E\{Z\}}{\sqrt{\mathrm{Var}\{Z\}}} \right) \tag{10.51b}$$

where $Q_Z(\cdot)$ is a notation for the complementary cumulative distribution function of the standardized (zero-mean, unit variance) version of the RV Z.

The approximation methods to be considered in what follows are

- Gaussian approximation: Based on the fact that Z is a sum (Central Limit Theorem) we can write

$$Q_Z(x) \approx Q(x) = \int_x^{\infty} \frac{1}{\sqrt{2\pi}} e^{-t^2/2} \, dt \tag{10.51c}$$

that is, the blocking probability can be calculated using

$$B_{CDMA} = Q\left(\frac{Z_0 - E\{Z\}}{\sqrt{\mathrm{Var}\{Z\}}} \right) \tag{10.51d}$$

- Lognormal approximation: Based on the fact the SNRs in the sum are lognormal, Z itself can be approximately characterized as a lognormal variable.

These approximation methods are based on identifying the actual mean and variance of Z with the mean and variance of a Gaussian RV and the mean and variance of a lognormal RV, respectively. Next, we find the mean and variance of Z and specify its relation to Gaussian and lognormal RVs.

10.2.2.2 Mean and Variance of Z

The form of the interference statistic Z is the weighted sum of the M RVs $\{\rho_i, \ i = 1, 2, \ldots, M\}$:

$$
\begin{aligned}
Z &= \alpha_{r1} \frac{E_{b1}}{\mathcal{I}_0'} + \alpha_{r2} \frac{E_{b2}}{\mathcal{I}_0'} + \cdots + \alpha_{rM} \frac{E_{bM}}{\mathcal{I}_0'} \\
&= \alpha_{r1} \rho_1 + \alpha_{r2} \rho_2 + \cdots + \alpha_{rM} \rho_M
\end{aligned}
\tag{10.52a}
$$

Propagation measurements in general indicate that received signal powers, when expressed in dB units, are nearly Gaussian. Therefore, the $\{\rho_i\}$ in (10.52a) also, when measured in dB units, are close to having a Gaussian probability distribution with median m_{dB} and standard deviation σ_{dB}:

$$
\rho_i(\text{dB}) = 10 \log_{10} \rho_i = m_{dB} + \sigma_{dB} G_i, \qquad G_i = \mathrm{G}(0, 1)
\tag{10.52b}
$$

Therefore, the RV ρ_i is lognormal and can be written

$$
\begin{aligned}
\rho_i &= 10^{(m_{dB} + \sigma_{dB} G_i)/10} = \left(e^{ln 10} \right)^{(m_{dB} + \sigma_{dB} G_i)/10} \\
&= e^{\beta(m_{dB} + \sigma_{dB} G_i)}, \ \text{using } \beta = (ln 10)/10
\end{aligned}
\tag{10.52c}
$$

The median, mean, and mean square of ρ_i are assumed to be same for all i and are obtained as follows:

- Median:

$$
\begin{aligned}
\tfrac{1}{2} &\triangleq \Pr\{\rho_i \leq \rho_{med}\} = \Pr\left\{ e^{\beta(m_{dB} + \sigma_{dB} G)} \leq \rho_{med} \right\} \\
&= \Pr\left\{ G \leq \frac{\tfrac{1}{\beta} ln \, \rho_{med} - m_{dB}}{\sigma_{dB}} = G_{med} = 0 \right\}
\end{aligned}
$$

Thus $\qquad\qquad\qquad \rho_{med} = e^{\beta m_{dB}}$ $\qquad\qquad$ (10.53a)

- Mean:

$$E\{\rho_i\} = E\left\{e^{\beta(m_{dB}+\sigma_{dB}G)}\right\} = e^{\beta m_{dB}} E\left\{e^{\beta\sigma_{dB}G}\right\}$$

$$= \rho_{med} E\left\{e^{vG}\right\}\Big|_{v=\beta\sigma_{dB}}$$

where

$$E\left\{e^{vG}\right\} = M_G(v) = e^{v^2/2} \quad \text{(MGF, see Section 1.5.4)}$$

Thus

$$E\{\rho_i\} = \rho_{med} M_G(\beta\sigma_{dB}) = \rho_{med} e^{\frac{1}{2}(\beta\sigma_{dB})^2} \tag{10.53b}$$

- Mean square:

$$E\{\rho_i^2\} = E\left\{\left[e^{\beta(m_{dB}+\sigma_{dB}G)}\right]^2\right\} = E\left\{e^{2\beta(m_{dB}+\sigma_{dB}G)}\right\}$$

$$= e^{2\beta m_{dB}} E\left\{e^{2\beta\sigma_{dB}G}\right\} = \rho_{med}^2 M_G(2\beta\sigma_{dB})$$

$$= \rho_{med}^2 e^{\frac{1}{2}(2\beta\sigma_{dB})^2} = \rho_{med}^2 e^{2(\beta\sigma_{dB})^2} \tag{10.53c}$$

Because M is analogous to the number of calls in progress through a switch, which has a Poisson distribution, it is reasonable to postulate that the mean and variance of M are equal, as in the case of a Poisson RV. The mean, mean square, and variance of Z therefore are

$$E\{Z\} = E_M\{E\{Z \mid M\}\} = E_M\left\{\sum_{i=1}^{M} E\{\alpha_{ri}\,\rho_i\}\right\}$$

$$= E_M\{M\,E\{\alpha_{ri}\,\rho_i\}\} = E\{M\}\,E\{\alpha_{ri}\,\rho_i\}$$

$$= \overline{M}\,E\{\alpha_{ri}\,\rho_i\} = \overline{M}\,\overline{\alpha}_r\,e^{\beta m_{dB}+\frac{1}{2}\beta^2\sigma_{dB}^2} \tag{10.54a}$$

$$E\{Z^2\} = E_M\{E\{Z^2 \mid M\}\} = E_M\left\{\sum_{i=1}^{M}\sum_{j=1}^{M} E\{\alpha_{ri}\,\alpha_{rj}\,\rho_i\,\rho_j\}\right\}$$

$$= E_M\Big\{\underbrace{M\,E\{\alpha_{ri}^2\,\rho_i^2\}}_{i=j\ \text{terms}} + \underbrace{M(M-1)[E\{\alpha_{ri}\,\rho_i\}]^2}_{i\neq j\ \text{terms}}\Big\}$$

$$= E_M\big\{M\,\{E\{\alpha_{ri}^2\,\rho_i^2\} - [E\{\alpha_{ri}\,\rho_i\}]^2\} + M^2[E\{\alpha_{ri}\,\rho_i\}]^2\big\}$$

$$= \overline{M}\,\mathrm{Var}\{\alpha_{ri}\,\rho_i\} + \overline{M^2}\,[E\{\alpha_{ri}\,\rho_i\}]^2 \tag{10.54b}$$

$$\text{Var}\{Z\} = \text{E}\{Z^2\} - [\text{E}\{Z\}]^2$$

$$= \overline{M} \, \text{Var}\{\alpha_{ri} \, \rho_i\} + \left(\overline{M^2} - \overline{M}^2\right) [\text{E}\{\alpha_{ri} \, \rho_i\}]^2$$

$$= \overline{M} \, \text{Var}\{\alpha_{ri} \, \rho_i\} + \text{Var}\{M\} [\text{E}\{\alpha_{ri} \, \rho_i\}]^2$$

$$= \overline{M} \left\{ \text{Var}\{\alpha_{ri} \, \rho_i\} + [\text{E}\{\alpha_{ri} \, \rho_i\}]^2 \right\}, \quad \text{because } \text{Var}\{M\} = \overline{M}$$

$$= \overline{M} \, \text{E}\{\alpha_{ri}^2 \, \rho_i^2\} = \overline{M} \, \overline{\alpha_r^2} \, e^{2\beta m_{dB} + 2\beta^2 \sigma_{dB}^2} \qquad (10.54c)$$

The interference due to mobiles in other cells can be accounted for by using first- and second-order *frequency reuse factors* $F = 1 + \xi$ and $F' = 1 + \xi'$, respectively, where (see Section 10.1.2.3)

$$\xi = \frac{\text{Total other-cell received (median) power}}{\text{Total same-cell received (median) power}} \qquad (10.55a)$$

and $\qquad \xi' = \dfrac{\text{Total other-cell mean square received power}}{\text{Total same-cell mean square received power}} \qquad (10.55b)$

In Appendix 10A, an analysis is presented that leads to a second-order reuse fraction value of $\xi' = 0.086$. In [3] and [8], ξ' was assumed to be equal to ξ as an upper bound, and the values $\xi = \xi' = 0.55$ were used. With this method of accounting for interference from other cells, the mean and variance for Z become

$$\text{E}\{Z\} = \overline{M} \, \overline{\alpha}_r \, \rho_{med} e^{\frac{1}{2}\beta^2 \sigma_{dB}^2} \cdot (1 + \xi) \qquad (10.56a)$$

and

$$\text{Var}\{Z\} = \overline{M} \, \overline{\alpha_r^2} \, \rho_{med}^2 e^{2\beta^2 \sigma_{dB}^2} \cdot (1 + \xi') \qquad (10.56b)$$

Approximations for the probability distribution of Z. Because the M RVs $\{\rho_i, i = 1, 2, \ldots, M\}$ are lognormal, the interference statistic Z is the weighted sum of lognormal RVs. One approximation for the distribution of Z is based on assuming that the summing of variables to produce Z causes its distribution to converge to a Gaussian distribution according to the CLT. Another approach is to assume that the lognormal character of the $\{\rho_i\}$ makes Z have an approximately lognormal distribution. Thus:

$$Z = \alpha_{r1} \, \rho_1 + \alpha_{r2} \, \rho_2 + \cdots + \alpha_{rM} \, \rho_M \qquad (10.57a)$$

$$= \alpha_{r1}e^{\beta(m_{dB}+\sigma_{dB}G_1)} + \alpha_{r2}e^{\beta(m_{dB}+\sigma_{dB}G_2)} + \cdots$$

$$\cdots + \alpha_{rM}e^{\beta(m_{dB}+\sigma_{dB}G_M)} \qquad (10.57b)$$

$$\approx \begin{cases} m_M + \sigma_M G, & \text{Gaussian approximation} \\ e^{m_M+\sigma_M G}, & \text{Lognormal approximation} \end{cases} \qquad (10.57c)$$

10.2.2.3 CDMA Blocking Probability Formula for Gaussian Assumptions

Under the Gaussian assumption, the mean and variance of Z are identified as the mean and variance of a Gaussian RV, $m_M + \sigma_M G$:

$$B_{CDMA} = \Pr\{Z > Z_0\}$$

$$\approx \Pr\{m_M + \sigma_M G > Z_0\}$$

$$= \Pr\left\{G > \frac{Z_0 - m_M}{\sigma_M}\right\} = Q\left(\frac{Z_0 - \mathrm{E}\{Z\}}{\sqrt{\mathrm{Var}\{Z\}}}\right)$$

Substituting the expressions for the mean and variance of Z, we obtain a general expression for the CDMA blocking probability under the Gaussian approximation for the interference statistic, given by

$$B_{CMDA} = Q\left(\frac{\frac{W}{R_b}(1-\eta_0) - \overline{M}\,\overline{\alpha_r}\,\rho_{med}\,e^{\frac{1}{2}\beta^2\sigma_{dB}^2}(1+\xi)}{\sqrt{\overline{M}\,\overline{\alpha_r^2}\,\rho_{med}^2\,e^{2\beta^2\sigma_{dB}^2}\,(1+\xi')}}\right) \qquad (10.58a)$$

in which the Erlang capacity is $\overline{M}$. Because the interference parameter η is related to the cell loading discussed in Section 10.1.2.5 by

$$\eta = 1 - X \qquad (10.58b)$$

we may convert the threshold η_0 into a loading threshold $X_0 = 1 - \eta_0$ and write (10.58a) as

$$B_{CMDA} = Q\left(\frac{\frac{W}{R_b}X_0 - \overline{M}\,\overline{\alpha_r}\,\rho_{med}\,e^{\frac{1}{2}\beta^2\sigma_{dB}^2}(1+\xi)}{\sqrt{\overline{M}\,\overline{\alpha_r^2}\,\rho_{med}^2\,e^{2\beta^2\sigma_{dB}^2}\,(1+\xi')}}\right) \qquad (10.58c)$$

Plots of B_{CDMA} versus $\overline{M}$ are shown in Figure 10.27 for a single cell ($\xi = \xi' = 0$) and for multiple cells, using $\xi = \xi' = 0.55$ [3] and the parameter values $\sigma_{dB} = 2.5\,\text{dB}$, $W = 1.2288\,\text{MHz}$, $R_b = 9.6\,\text{kbps}$, $X_0 = 0.9$, $\overline{\alpha_r} = 0.4$, and $\overline{\alpha_r^2} = 0.31$. The plots are parametric in $m_{dB} = E_b/\mathcal{N}_0$, which takes the values 5, 6, and 7 dB. We observe that the value of $E_b/\mathcal{N}_0$ needed for link operations profoundly affects the average number of users that can be accommodated at a given level of blocking; raising the $E_b/\mathcal{N}_0$ requirement increases the blocking probability for the same value of $\overline{M}$ or decreases the capacity for the same probability. For example, when the blocking probability is chosen to be $B_{CDMA} = 10^{-2} = 0.01 = 1\%$, Figure 10.27 shows for multiple cells that the corresponding value of the Erlang capacity $\overline{M}$ is 18 for $m_{dB} = 7\,\text{dB}$, 24 for $m_{dB} = 6\,\text{dB}$, and 33 for $m_{dB} = 5\,\text{dB}$.

Let us denote the Erlang capacity for a single cell by $\overline{M}_c$ and the capacity for multiple cells by $\overline{M}$, similar to the notation used in (10.16a). For $B_{CDMA} = 1.0\%$ and $E_b/\mathcal{N}_0 = 6\,\text{dB}$, we then read from Figure 10.27 that $\overline{M}_c = 37.5$ and $\overline{M} = 24.1$. The ratio of these values is $37.5/24.1 = 1.56$, a value that is approximately equal to the assumed value of the reuse factor, $F = 1 + \xi = 1.55$. This is consistent with the definition of the reuse factor in (10.16a).

Assuming that $m_{dB} = E_b/\mathcal{N}_0 = 6\,\text{dB}$, the effect of varying the loading threshold X_0 on B_{CDMA} and $\overline{M}$ is illustrated in Figure 10.28, in which X_0 takes the values $X_0 = 0.66$, 0.75, and 0.9. These values correspond to the multiple access interference power being twice, three times, and nine times as strong as the thermal noise. Raising the loading threshold has the effect of relaxing the system requirements, and is seen in Figure 10.28 to result in either a decrease in the blocking probability for the same value of $\overline{M}$, or an increase in $\overline{M}$ for the same value of B_{CDMA}.

Example 10.18 What is the form of the blocking probability as a function of the Erlang capacity?

Solution: If we substitute specific numerical parameter values into the general expression (10.58c), such as $\sigma_{dB} = 2.5\,\text{dB}$, $m_{dB} = 7\,\text{dB}$, $W = 1.2288$ MHz, $R_b = 9.6\,\text{kbps}$, $X_0 = 0.9$, $\overline{\alpha_r} = 0.4$, $\overline{\alpha_r^2} = 0.31$, we obtain

$$B_{CDMA} = Q\left(\frac{115.2 - 2.37(1 + \xi)\overline{M}}{3.89\sqrt{(1 + \xi')\overline{M}}} \right) \tag{10.59a}$$

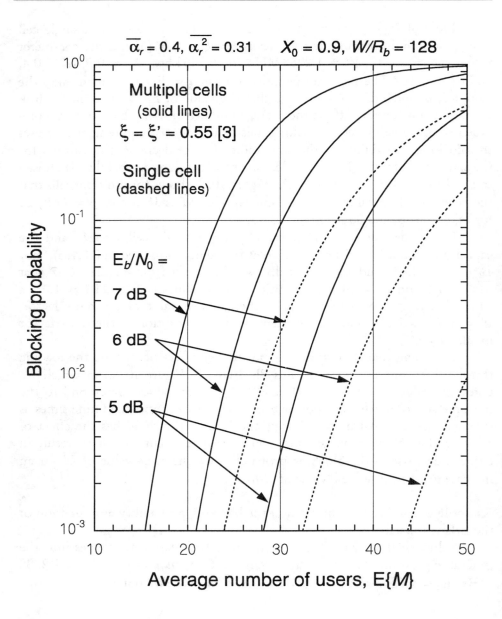

Figure 10.27 CDMA blocking probability (Gaussian approximation) versus average number of mobile users, SNR requirement varied.

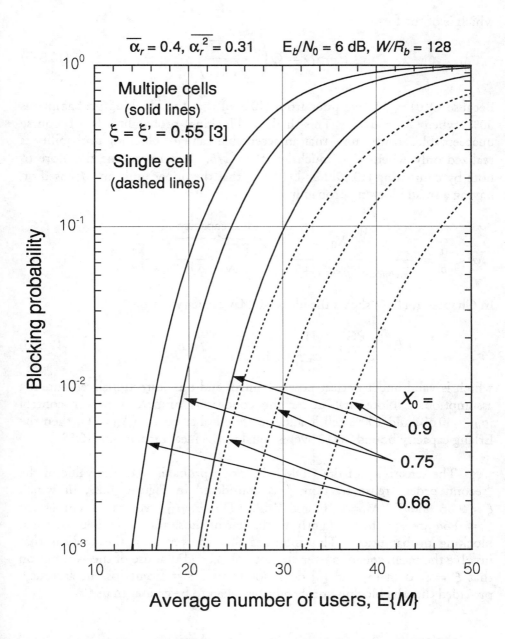

Figure 10.28 CDMA blocking probability (Gaussian approximation) versus average number of mobile users, loading threshold varied.

which is of the form

$$B_{CDMA} = Q\left(\frac{a - b\overline{M}}{\sqrt{c\overline{M}}}\right) \qquad (10.59b)$$

Because $Q(0) = 0.5$, we infer from (10.59b) that the blocking probability is 50% when $\overline{M} = a/b$. This high a blocking probability is of course unacceptable, so we know that an acceptable value of blocking probability is realized only when $\overline{M}$ is much less than a/b. It is interesting therefore to note by comparing (10.59b) and (10.58c) that the upper limit on $\overline{M}$ based on having a small blocking probability is

$$\overline{M} < \frac{a}{b} = \frac{\dfrac{W}{R_b} X_0}{\overline{\alpha}_r \, \rho_{med} \, e^{\frac{1}{2}\beta^2 \sigma_{dB}^2}(1 + \xi)} = \overbrace{\frac{\mathrm{PG}}{E_b/\mathcal{N}_0} \cdot \frac{1}{\overline{\alpha}_r \, F}}^{\text{ideal capacity}} \cdot \frac{X_0}{e^{\frac{1}{2}\beta^2 \sigma_{dB}^2}} \qquad (10.59c)$$

In Chapter 1, (1.10) shows the ideal CDMA capacity as

$$M = \frac{\mathrm{PG}}{E_b/\mathcal{N}_0} \cdot \frac{1}{\alpha_r \, F} = \frac{\mathrm{PG}}{E_b/\mathcal{N}_0} \cdot \frac{1}{\alpha_r} \cdot F_e \qquad (10.59d)$$

which is valid under perfect power control and omnidirectional cell antenna assumptions. Note that under the conditions of perfect power control ($\sigma_{dB} = 0\,\mathrm{dB}$) and 100% cell loading in the ideal situation ($X_0 = 1$), then the Erlang capacity bound in (10.59c) is equal to the ideal capacity in (10.59d).

The sensitivity of the CDMA blocking probability to the value of the "second order" reuse fraction ξ' is considered in Figure 10.29, in which $\xi = 0.55$ and $E_b/\mathcal{N}_0 = 5$, 6, and 7 dB. The B_{CDMA} versus $\overline{M}$ curves for $\xi' = 0.55$ are seen to be slightly to the left of those for $\xi' = 0.086$ for small blocking probabilities. This indicates that the terms in the analysis that involve the mean square power are not critical and that use of the assumption that $\xi' = \xi$ as in [3] and [8] does not incur a significant loss in accuracy, provided that the blocking probability is taken to be greater than 1%.

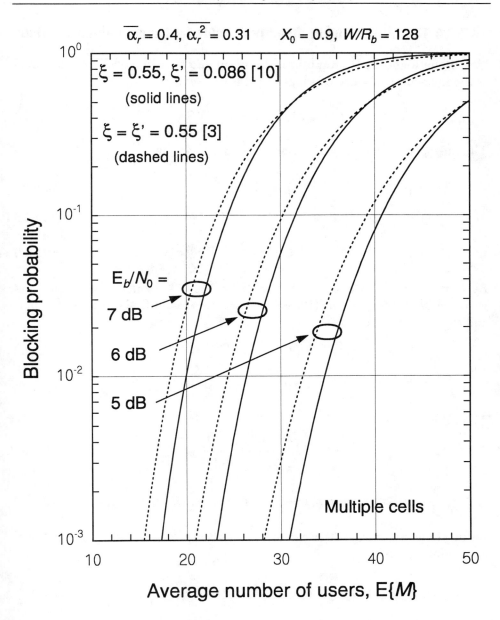

Figure 10.29 CDMA blocking probability (Gaussian approximation) versus average number of mobile users, reuse fractions varied.

10.2.2.4 CDMA Blocking Probability Formula for Lognormal Assumptions

Under the lognormal assumption, the mean and variance of Z are identified as the mean and variance of a lognormal RV, ζ, where

$$\zeta = e^{m_M + \sigma_M G} \tag{10.60a}$$

The mean, mean square, and variance of ζ are given by

$$E\{\zeta\} = e^{m_M} E\{e^{\sigma_M G}\} = e^{m_M + \frac{1}{2}\sigma_M^2} \tag{10.60b}$$

$$E\{\zeta^2\} = e^{2m_M} E\{e^{2\sigma_M G}\} = e^{2m_M + 2\sigma_M^2} \tag{10.60c}$$

$$Var\{\zeta\} = E\{\zeta^2\} - [E\{\zeta\}]^2 = e^{2m_M + \sigma_M^2} \left[e^{\sigma_M^2} - 1 \right] \tag{10.60d}$$

By setting $E\{Z\} = E\{\zeta\}$ and $Var\{Z\} = Var\{\zeta\}$, where the mean and variance of Z are given in (10.56a) and (10.56b), as in [14] we solve for m_M and σ_M^2:

$$\overline{M}\,\overline{\alpha}_r\, e^{\beta m_{dB} + \frac{1}{2}\beta^2 \sigma_{dB}^2} (1 + \xi) = e^{m_M + \frac{1}{2}\sigma_M^2} \tag{10.61a}$$

and

$$\overline{M}\,\overline{\alpha_r^2}\, e^{2\beta m_{dB} + 2\beta^2 \sigma_{dB}^2} (1 + \xi') = e^{2m_M + \sigma_M^2} \left[e^{\sigma_M^2} - 1 \right] \tag{10.61b}$$

The solution is

$$\sigma_M^2 = ln\left[\frac{\overline{\alpha_r^2}\,(1 + \xi')e^{\beta^2 \sigma_{dB}^2}}{\overline{M}\,(\overline{\alpha}_r)^2 (1 + \xi)^2} + 1 \right] \tag{10.61c}$$

and

$$m_M = ln\left[\overline{M}\,\overline{\alpha}_r\,(1 + \xi)\right] + \beta m_{dB} + \tfrac{1}{2}(\beta^2 \sigma_{dB}^2 - \sigma_M^2) \tag{10.61d}$$

Using these parameters, the blocking probability formula for the lognormal approximation is

$$B_{CDMA} = Pr\{Z > Z_0\} \approx Pr\{e^{m_M + \sigma_M G} > Z_0\} = Q\left(\frac{ln Z_0 - m_M}{\sigma_M} \right)$$

Substituting the expressions for m_M and σ_M, we obtain general expressions for the CDMA blocking probability under the lognormal approximation for the interference statistic, given by

$$B_{CDMA} = Q\left(\frac{ln\left[\frac{W}{R_b}(1-\eta_0)\right]-ln\left[\overline{M}\,\overline{\alpha}_r\,(1+\xi)\right]-\beta m_{dB}}{\sqrt{ln\left[\frac{\overline{\alpha_r^2}\,(1+\xi')e^{\beta^2\sigma_{dB}^2}}{\overline{M}\,(\overline{\alpha}_r)^2(1+\xi)^2}+1\right]}}\right.$$

$$\left.-\frac{\frac{1}{2}\left\{\beta^2\sigma_{dB}^2-ln\left[\frac{\overline{\alpha_r^2}\,(1+\xi')e^{\beta^2\sigma_{dB}^2}}{\overline{M}\,(\overline{\alpha}_r)^2(1+\xi)^2}+1\right]\right\}}{\sqrt{ln\left[\frac{\overline{\alpha_r^2}\,(1+\xi')e^{\beta^2\sigma_{dB}^2}}{\overline{M}\,(\overline{\alpha}_r)^2(1+\xi)^2}+1\right]}}\right) \quad (10.62a)$$

in which the Erlang capacity is $\overline{M}$. Because the interference parameter η is $\eta = 1 - X$ as given by (10.58b), we may convert the threshold η_0 into a loading threshold $X_0 = 1 - \eta_0$ and write (10.62a) as

$$B_{CDMA} = Q\left(\frac{ln\left[\frac{W}{R_b}X_0\right]-ln\left[\overline{M}\,\overline{\alpha}_r\,(1+\xi)\right]-\beta m_{dB}}{\sqrt{ln\left[\frac{\overline{\alpha_r^2}\,(1+\xi')e^{\beta^2\sigma_{dB}^2}}{\overline{M}\,(\overline{\alpha}_r)^2(1+\xi)^2}+1\right]}}\right.$$

$$\left.-\frac{\frac{1}{2}\left\{\beta^2\sigma_{dB}^2-ln\left[\frac{\overline{\alpha_r^2}\,(1+\xi')e^{\beta^2\sigma_{dB}^2}}{\overline{M}\,(\overline{\alpha}_r)^2(1+\xi)^2}+1\right]\right\}}{\sqrt{ln\left[\frac{\overline{\alpha_r^2}\,(1+\xi')e^{\beta^2\sigma_{dB}^2}}{\overline{M}\,(\overline{\alpha}_r)^2(1+\xi)^2}+1\right]}}\right) \quad (10.62b)$$

Plots of (10.62b) for B_{CDMA} versus $\overline{M}$ are shown in Figure 10.30 for a single cell ($\xi = \xi' = 0$) and for multiple cells, using $\xi = \xi' = 0.55$ [3] and the parameter values $\sigma_{dB} = 2.5\,dB$, $W = 1.2288\,MHz$, $R_b = 9.6\,kbps$, $X_0 = 0.9$, $\overline{\alpha}_r = 0.4$, and $\overline{\alpha_r^2} = 0.31$. The plots are parametric in $m_{dB} = E_b/\mathcal{N}_0$, which takes the values 5, 6, and 7 dB. As for the Gaussian approximation in Figure 10.27, we observe that the value of $E_b/\mathcal{N}_0$ needed for link operations greatly affects the value of $\overline{M}$ for a given value B_{CDMA}; the higher the $E_b/\mathcal{N}_0$, the lower the $\overline{M}$ value, for given blocking probability. For example, when $B_{CDMA} = 10^{-2} = 0.01 = 1\%$, Figure 10.30 shows for multiple cells that the corresponding (integer) value of the Erlang capacity $\overline{M}$ is 16 for $E_b/\mathcal{N}_0 = 7$ dB, 22 for $E_b/\mathcal{N}_0 = 6\,dB$, and 29 for $E_b/\mathcal{N}_0 = 5\,dB$.

In Figure 10.31, the cell loading threshold X_0 is varied for the lognormal approximation and the case of $E_b/\mathcal{N}_0 = 6\,dB$. Again we see that raising or lowering X_0 has a significant effect on the Erlang capacity $\overline{M}$ for a

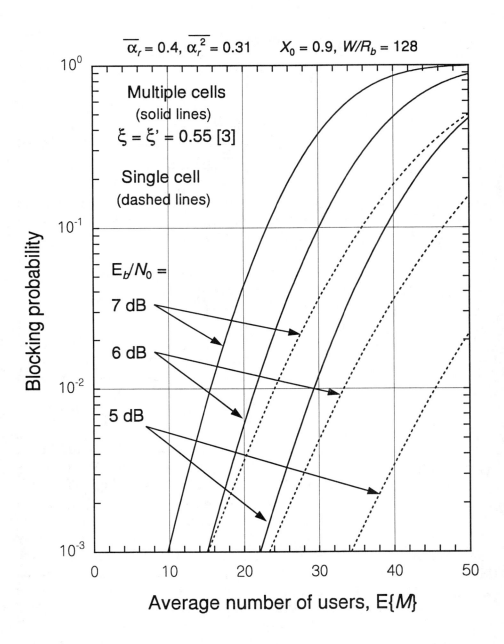

Figure 10.30 CDMA blocking probability (lognormal approximation) versus average number mobile users, SNR requirement varied.

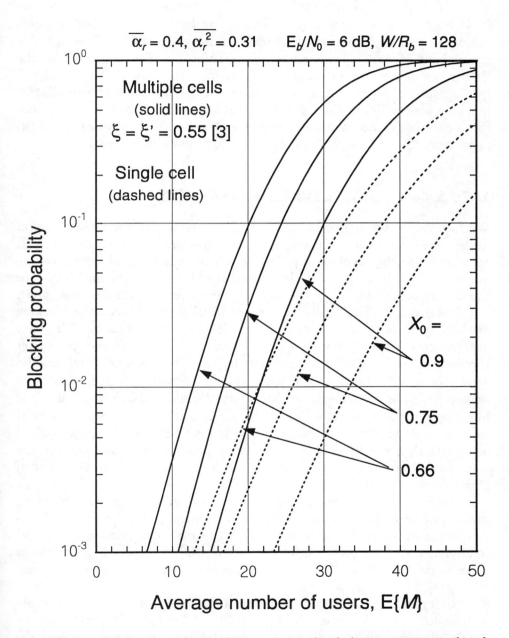

Figure 10.31 CDMA blocking probability (lognormal approximation) versus average number of mobile users, loading threshold varied.

given value of the blocking probability. The amount of increase or decrease in $\overline{M}$ is greater than the amount of increase or decrease in X_0. For example, raising X_0 by 20% from 0.75 to 0.9 increases the value of $\overline{M}$ at $B_{CDMA} = 1\%$ from about 16 to 22, or about 35%, indicating the high sensitivity of the Erlang capacity to the cell loading. Therefore, the threshold value of cell loading should be chosen very carefully. The sensitivity of the CDMA blocking probability to the value of the second-order reuse fraction ξ' under the lognormal approximation is shown in Figure 10.32, in which the first-order reuse fraction $\xi = 0.55$ and $E_b/\mathcal{N}_0 = 5$, 6, and 7 dB are used.

10.2.2.5 Comparison of CDMA Blocking Probabilities

We have derived Erlang capacity "formulas" for the CDMA cellular system under two separate approximations for the interference statistic: a Gaussian approximation, by invoking the CLT, and lognormal approximation, on the assumption that the sum of M lognormal RVs is also a lognormal RV. For the Gaussian approximation, the blocking probability expressions were given in (10.58a) and (10.58c), while for the lognormal approximation the blocking probability expressions were given in (10.62a) and (10.62b). We have expressed the blocking probability as a function of the interference parameter threshold η_0 and the cell loading threshold X_0. These blocking probabilities were plotted as a function of Erlang capacity $\overline{M}$ with the median value of $m_{dB} = E_b/\mathcal{N}_0$ as a parameter for three different values of 5, 6, and 7 dB. The blocking probabilities plotted in the figures also show two separate cases of using second-order reuse fractions of $\xi' = 0.55$ and $\xi' = 0.086$. Figures 10.29 and 10.32 clearly indicate that the blocking probability is insensitive to the value of the second-order reuse fraction ξ' as demonstrated for the cases of $\xi' = 0.55$ and $\xi' = 0.086$.

In Figure 10.33 a comparison of the Gaussian and lognormal blocking probability approximations is made for $\xi = 0.55$ and $\xi' = 0.086$. Clearly, it is demonstrated that the Gaussian and lognormal approximations to the interference statistic give similar results for CDMA blocking probabilities greater than 1%. Therefore, we may choose the simpler Gaussian expression.

Sensitivity of B_{CDMA} to ξ. In Figure 10.34, we plot the blocking probabilities for two different values of $\xi = 0.33$ and $\xi = 0.55$. For each case, we use the identical value $\xi' = 0.086$. We also plot, for comparison purposes, the case of $(\xi, \xi') = (0.55, 0.086)$ and the single-cell blocking probability. Note that the effect of changing the value of ξ is significant, in contrast to what we

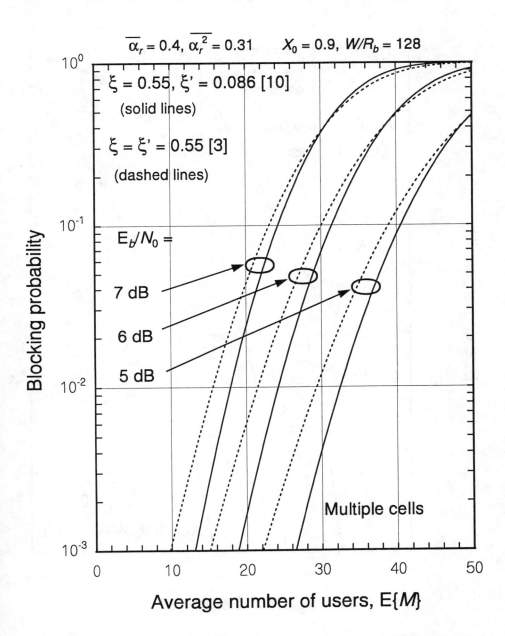

Figure 10.32 CDMA blocking probability (lognormal approximation) versus average number of mobile users, reuse fractions varied.

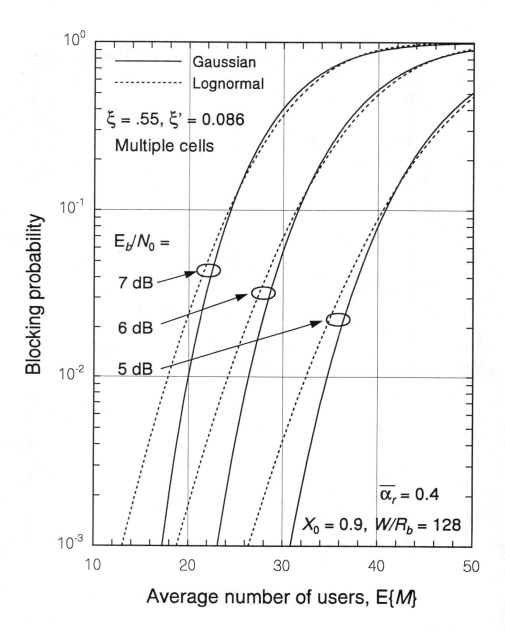

Figure 10.33 Comparison of Gaussian and lognormal blocking probability approximations for $\xi = 0.55$ **and** $\xi' = 0.086$.

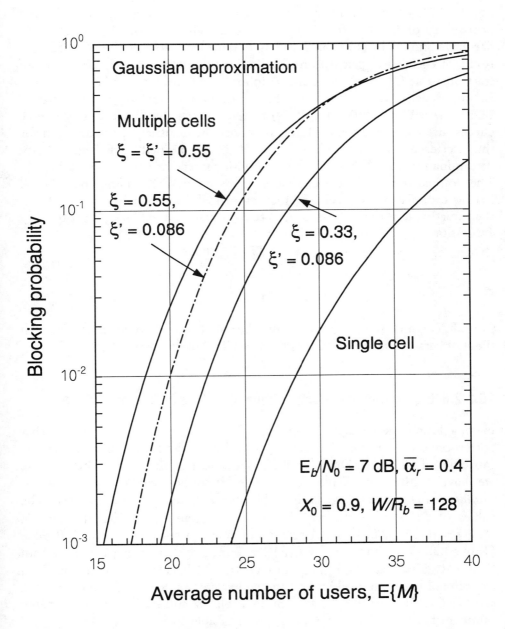

Figure 10.34 Comparison of CDMA blocking probabilities for different reuse fraction values.

observe about the sensitivity of B_{CDMA} to ξ'. For example, for a blocking probability of 1%, $\xi = 0.33$ gives $\overline{M} = 22.5$, while $\xi = 0.55$ gives $\overline{M} = 20$. We have shown in (10.16b) that ξ is bounded between 0.33 and 0.42 as a result of theoretical calculations, while $\xi = 0.55$ is a simulation-based reuse fraction value [5]. In the sections to follow, we consider calculations for the Erlang capacity of a CDMA telephone system and compare it with that of FDMA (AMPS) and TDMA (IS-54) systems, using the blocking probability curves that we have developed thus far under the Gaussian approximation for the interference statistic. Thus, Figure 10.29 can be used for Erlang capacity determination for all cases of interest with respect to the values of E_b/N_0. The comparison of Erlang capacities is based on first reading the CDMA Erlang capacity for a specified blocking probability from Figure 10.29 and then treating it as the offered load, so that we can use the Erlang B probability expression

$$B_{CDMA} = \frac{\left(\overline{M}\right)^N / N!}{\sum\limits_{n=0}^{N} \left(\overline{M}\right)^n / n!} \tag{10.63}$$

to find N, which is an "equivalent number of channels" to be compared with the numbers of channels in the FDMA and TDMA systems.

10.2.2.6 Erlang Capacity Comparisons of CDMA, FDMA, and TDMA

Having found the equivalent number of channels in a CDMA system that corresponds to a value of blocking probability, we now can compare the capacities of CDMA, FDMA, and TDMA cellular systems. For this purpose we have constructed Table 10.5. The Erlang capacities and equivalent numbers of channels (N) given in Table 10.5 were developed as follows: The values of $\overline{M}$ were read from the curves in Figure 10.29, for the case of $(\xi, \xi') = (0.55, 0.086)$, and the values of N are based on the Erlang B table (Table 3.2). For example, for $E_b/N_0 = 7\,\mathrm{dB}$ and $B = B_{CDMA} = 1\%$, from Figure 10.29 we read the Erlang capacity $\overline{M} = 20\,\mathrm{Erlangs}$. Now, for $\overline{M} = A$, the offered load, we need to find the equivalent number of channels N that satisfies (10.63). For this N, we look in Table 3.2 and read the corresponding number of channels as $N = 30$, as shown in Table 10.5.

The number of channels in Table 10.5 is for a CDMA cellular system in one 1.25-MHz band (one FA). Assuming a noncontiguous 12.5-MHz cellular band, the CDMA system can use as many as nine 1.25-MHz FAs.

Table 10.5 Erlang capacities and corresponding numbers of channels, based on Figure 10.29, for $\xi = 0.55$ and $\xi' = 0.086$

B_{CDMA}	$E_b/\mathcal{N}_0 = 5\,\mathrm{dB}$	$E_b/\mathcal{N}_0 = 6\,\mathrm{dB}$	$E_b/\mathcal{N}_0 = 7\,\mathrm{dB}$
1%	$\overline{M} = 35$	$\overline{M} = 26$	$\overline{M} = 20$
	$N = 46$	$N = 37$	$N = 30$
2%	$\overline{M} = 36$	$\overline{M} = 28$	$\overline{M} = 21$
	$N = 46$	$N = 36$	$N = 29$

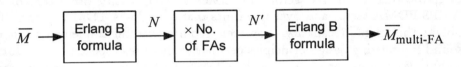

Figure 10.35 Procedure for determining multi-FA CDMA Erlang capacity.

Thus, to compare the effective number of CDMA channels available in a particular sector with the number of channels available in the AMPS and TDMA (IS-54) [15] cellular systems (19 and 57, respectively), the numbers for N in Table 10.5 must be multiplied by the number of FAs, which is nine. After this multiplication is done, the corresponding Erlang capacity can be obtained from an Erlang B table or calculation by identifying the offered load with a multi-FA CDMA Erlang capacity denoted by $\overline{M}_{\text{multi-FA}}$. As diagrammed in Figure 10.35, this procedure of finding the offered load corresponding the aggregate number of channels (nine times N in Table 10.5) assumes that all those $N' = 9N$ channels are available for assignment and that the mobile is capable of accessing any of them.

Let us consider a numerical example. For $B_{CDMA} = 1\%$, $\xi = 0.55$, $\xi' = 0.086$, and $E_b/\mathcal{N}_0 = 7\,\mathrm{dB}$, we have $N = 30$ from Table 10.5. Thus, the effective number of CDMA channels in a 9-FA system is $N' = 30 \times 9 = 270$. We now need to find the offered load $A = \overline{M}_{\text{multi-FA}}$ corresponding to $N' = 270$ and 1% blocking. The offered load giving this numerical value according to the Erlang B formula of (10.63), with $\overline{M}$ replaced by $\overline{M}_{\text{multi-FA}}$, is calculated to be 248 Erlangs. The offered load for an AMPS sector, which supports 19 channels, gives 12.3 Erlangs for a blocking probability of 2%, as shown in Table 3.2. For a single sector of an IS-54 TDMA system, which supports $19 \times 3 = 57$ channels, gives 46.8 Erlangs for the identical blocking probability of 1%. Thus, using the notation of $M_{CDMA} \triangleq \overline{M}_{\text{multi-FA}}$, the ratios of Erlang capacities are computed to be

$$\frac{M_{CDMA}}{A_{AMPS}} = \frac{248}{12.3} = 20.2 \text{ (CDMA advantage over AMPS)} \qquad (10.64a)$$

$$\frac{M_{CDMA}}{A_{TDMA}} = \frac{248}{46.8} = 5.3 \text{ (CDMA advantage over IS-54)} \qquad (10.64b)$$

Therefore, over a 12.5-MHz band with nine FAs, the total number of Erlangs of traffic that a CDMA sector can service is over twenty times that of an AMPS FDMA sector and over five times that of an IS-54 TDMA sector. To obtain this gain in capacity, however, requires careful implementation of sound engineering design principles peculiar to the spread-spectrum waveform techniques that are employed by the CDMA system, as is discussed in detail in Chapter 11.

10.2.2.7 Number of Subscribers During the Busy Hour

We are interested in computing not only the number of active users at a given time, but also the number of subscribers that may be supported by the CDMA system in any given cell. In terms of traffic theory, consider a telephone switch and its trunk of N lines. For system-planning purposes, the traffic during the "busy hour" of the day is used. Typically, a user is likely to be on the telephone at any given time during the busy hour with the probability of 0.02 to 0.03; that is, each subscriber is considered to offer $A_0 = 0.02$ to 0.03 Erlangs of traffic. The number of subscribers (M_S) that can be supported by the trunk during the busy hour for a specified blocking probability that results in the total load A then is given by the formula

$$M_S = A/A_0 \qquad (10.65a)$$

Example 10.19 Consider that eight FAs are available from the B' band of 10 MHz. If 20 milliErlangs per subscriber is assumed to be the individual offered load during the busy hour, what would be the supportable number of subscribers per CDMA sector for $B_{CDMA} = 2\%$, $R_b = 9.6$ kbps, $\xi = 0.55$, $\xi' = 0.086$, and $E_b/N_0 = 6$ dB?

Solution: From Table 10.5, we read that the Erlang capacity and equivalent number of channels for a single FA are $\overline{M} = 28$ and $N = 36$ for $B_{CDMA} = 2\%$ and $E_b/\mathcal{N}_0 = 6\,\text{dB}$. For eight FAs, the number of channels is $N' = 8 \times 36 = 288$, which corresponds to an offered load calculated to be 274 Erlangs. Thus $M_{CDMA} = 274\,\text{E}$ and the number of subscribers for $A_0 = 0.02\,\text{E}$ is given by

$$M_S = \frac{M_{CDMA}}{A_0} = \frac{274}{0.02} = 13{,}700 \text{ subscribers per sector} \qquad (10.65b)$$

This example indicates that a CDMA system operator can assume the number of supportable subscribers for a single CDMA sector, based on a 2% blocking probability, to be 13, 700 subscribers. Assuming three sectors per CDMA cell, this example indicates that a total number $13{,}700 \times 3 = 41{,}100$ subscribers per cell.

An AMPS system for 2% blocking would support

$$M_S(\text{AMPS}) = \frac{A_{AMPS}}{A_0} = \frac{12.3}{0.02} = 615 \text{ subscribers per sector} \qquad (10.65c)$$

which indicates a total of $3 \times 615 = 1{,}845$ subscribers in one cell. In this case, the advantage of CDMA over AMPS in terms of the ratio of subscribers is

$$\frac{41{,}100 \text{ subscribers per cell}}{1{,}845 \text{ subscribers per cell}} = 22.2 \qquad (10.65d)$$

10.2.3 CDMA Area Coverage Analysis

The area coverage of a base station sector in a cellular system is the geographical area in which a mobile unit may communicate with the base station with sufficient signal level at the base station receiver to satisfy the service requirements. The coverage of individual cellular base stations determines how many base stations are required to provide service to a particular area. In a CDMA system, the coverage is limited by the multiple access interference on the reverse link rather than by the sensitivity of the base station receiver. In this section, we show the relation between coverage and capacity for a CDMA cellular system [16] by considering the following analysis steps:

- Find the required received signal level as a function of the traffic level in Erlangs.

- Use a propagation loss model to find the radius and area of cells to achieve the minimum signal as a function of traffic in Erlangs.

- Interpret the cell area in terms of base station density as a function of traffic.

10.2.3.1 Required Received Signal Level as a Function of Loading

The relation of CDMA coverage to capacity begins with determining the effect of cellular traffic on the median ratio of signal power (P_S) to total interference power (I) plus noise power (σ^2).[10]

When loaded to capacity, by definition, the received ratio $P_S/(\sigma^2 + I)$ is just sufficient to meet the link requirements, but does not meet requirements if another user is added to the reverse link. The maximum allowable number of users depends upon the amount of tolerable interference. To maintain the same SNR at the base station receiver when there is interference as when there is no interference requires that the mobile transmitter power be increased when there is interference. Let P_S be the median signal power required at the receiver when there is interference, and let P'_S be the median required signal power at the receiver when there is no interference; then assuming that the reverse link power control perfectly adjusts the median mobile transmitter powers to adapt to the interference, the system maintains the constant signal-to-interference-plus-noise power ratio at the receiver that is given by

$$\frac{P_S}{\sigma^2 + I} = \frac{P'_S}{\sigma^2} \quad \text{or} \quad \frac{P_S}{P'_S} = \frac{\sigma^2 + I}{\sigma^2} \tag{10.66a}$$

But from the definition of loading, we have from (10.19c)

$$X = \frac{M}{M_{max}} = \frac{I}{\sigma^2 + I} \tag{10.66b}$$

---

[10] Unless otherwise indicated, for convenience we use I to denote the total reverse link interference I_T.

where M_{max} is the maximum number of mobile users that can be accommodated while meeting the SNR requirement at the receiver. Because our interest is to investigate required signal power level as the number of users varies from one to some maximum allowable number in a given FA, we need to identify that "maximum number of users" to proceed with the relative loading, X. We investigate two cases: one for M_{max} is that of the capacity equation that we derived earlier under the perfect power control assumption (ideal capacity equation), and the other one is that of the maximum realizable Erlang capacity that we derived in the previous section.

Ideal capacity equation. When we derive the ideal capacity equation, we have neglected thermal noise, and perfect power control is assumed. The expression was for the maximum number of users was given in (10.59d) as

$$M_{max}(\mathrm{E}_b/\mathcal{N}_0) = \frac{\mathrm{PG}}{(\mathrm{E}_b/\mathcal{N}_0)_{req}\, F\, \overline{\alpha}_r} \qquad (10.66c)$$

where F is the CDMA reuse factor and $\overline{\alpha}_r$ is average reverse link voice activity factor. Substituting (10.66b) in (10.66a) gives

$$\frac{\mathrm{P}_S}{\mathrm{P}'_S} = \frac{\sigma^2 + I}{\sigma^2} = \frac{1}{\left(\frac{\sigma^2}{\sigma^2 + I}\right)} = \frac{1}{1 - X}$$

and it follows that

$$\mathrm{P}_S = \frac{\mathrm{P}'_S}{1 - X} = \frac{\mathrm{P}'_S}{1 - M/M_{max}} \qquad (10.66d)$$

where the value of P'_S, the signal power that is required at the receiver in the absence of interference, is given in (10.47c) of Section 10.2.1.5. According to the perfect power control model, then, the required median value of the received signal level in dBm units must increase as a function of the loading as follows:

$$\mathrm{P}_S(\mathrm{dBm}) = \mathrm{P}'_S(\mathrm{dBm}) - 10\log_{10}(1 - X) \qquad (10.67a)$$

$$= \mathrm{P}'_S(\mathrm{dBm}) - 10\log_{10}(1 - M/M_{max}) \qquad (10.67b)$$

where $\mathrm{P}'_S(\mathrm{dBm})$ is the single-user minimum receiver power given by (10.47c):

$$\mathrm{P}'_S(\mathrm{dBm}) = (\mathrm{E}_b/\mathcal{N}_0)_{req}(\mathrm{dB}) + \mathcal{N}_0(\mathrm{dB/Hz}) + W(\mathrm{dBHz}) - \mathrm{PG}(\mathrm{dB})$$

$$= (E_b/\mathcal{N}_0)_{req}(dB) - 169.0\,dBm/Hz + 60.9\,dBHz - 21.1\,dB$$

$$= (E_b/\mathcal{N}_0)_{req}(dB) - 129.2\,dBm \tag{10.67c}$$

Assume that we wish to increase link reliability. For this purpose, the target value of SNR in dB is not $(E_b/\mathcal{N}_0)_{req}(dB)$ but $(E_b/\mathcal{N}_0)_{req}(dB) + M_{dB}$, where M_{dB} is the margin in dB. Table 10.6 gives the link margin for several values of link reliability and of σ_{dB}, which is reproduced from Table 10.4. Because each user is using more power to achieve the margin, a smaller number of users can be accommodated, so the value of $M_{max}(E_b/\mathcal{N}_0)$ in (10.66c) must be replaced by $M_{max}(E_b/\mathcal{N}_0; M_{dB})$:

$$M_{max}(E_b/\mathcal{N}_0; M_{dB}) \triangleq M_{max}(E_b/\mathcal{N}_0) \div 10^{M_{dB}/10} \tag{10.68a}$$

$$= \frac{PG}{(E_b/\mathcal{N}_0)_{req}\, F\, \overline{\alpha}_r\, 10^{M_{dB}/10}}$$

$$= \frac{PG}{(E_b/\mathcal{N}_0)_{req}\, F\, \overline{\alpha}_r} \cdot \frac{1}{e^{\beta M_{dB}}}, \quad \beta \triangleq \frac{\ln 10}{10} \tag{10.68b}$$

Note also that the signal power without interference P'_S must be increased by the margin in order to achieve a given link reliability, as discussed in Section 10.2.1.5. The signal power required at the base station receiver to the number of users is related to the number of users is then

$$P_S(dBm) = P'_S(dBm) + M_{dB} - 10\log_{10}(1 - M/M_{max}) \tag{10.68c}$$

where $P'_S(dBm)$ is given in (10.67c)

Example 10.20 Find the maximum number of mobile users in a CDMA sector for a single FA with and without a link reliability margin for the following assumptions:

Table 10.6 Link margins for soft handoff

P_{rel}	M_{dB}	
	$\sigma_{dB} = 8$	$\sigma_{dB} = 2.5$
0.70	0.64 dB	0.20 dB
0.80	2.96 dB	0.93 dB
0.90	6.16 dB	1.92 dB

$P_{rel} = 90\%$, $\sigma_{dB} = 2.5\,\text{dB}$, $F = 1.55$, $\overline{\alpha}_r = 0.4$, $(E_b/\mathcal{N}_0)_{req} = 7\,\text{dB}$, and $PG = 128$.

Solution: Assuming a 90% link reliability requirement, from Table 10.6 we have the margin $M_{dB} = 1.92\,\text{dB}$ for $\sigma_{dB} = 2.5$, and the values of M_{max} for the assumptions are

$$M_{max}(7\,\text{dB}; 0\,\text{dB}) = \frac{128}{10^{0.7} \cdot 1.55 \cdot 0.4} = 41.2 \qquad (10.68\text{d})$$

and

$$M_{max}(7\,\text{dB}; 1.92\,\text{dB}) = \frac{128}{10^{0.7} \cdot 1.55 \cdot 0.4} \cdot \frac{1}{10^{0.192}} = 26.5 \qquad (10.68\text{e})$$

The adoption of a 1.92-dB margin cuts the maximum number of users by about 1.92 dB, or a factor of 1.556.

Equation (10.68c) is plotted in Figure 10.36 for a link margin of 0.93 dB when $E_b/\mathcal{N}_0$ is varied, and for the parameter values $F = 1.55$, $\overline{\alpha}_r = 0.4$, $PG = 128$. The margin corresponds to a link reliability of $P_{rel} = 0.8$ and $\sigma_{dB} = 2.5\,\text{dB}$. Note in the figure that the required signal power rises as the number of users increases. Near the maximum number of users, the power requirement rises to infinity asymptotically at some number of users, M_{max}, that varies with the assumed values of the parameters. From (10.68b), the maximum values are

$$M_{max}(7\,\text{dB}; 0.93\,\text{dB}) = 33.3$$

$$M_{max}(6\,\text{dB}; 0.93\,\text{dB}) = 41.9$$

$$M_{max}(5\,\text{dB}; 0.93\,\text{dB}) = 52.7$$

These numbers increase as $E_b/\mathcal{N}_0$ decreases because the power requirement is less. The physical meaning of this mathematical behavior is that, beyond this value of M, it is impossible for the power control of the system, even a perfect power control, to cause the median received SNRs of all the users to exceed the desired amount.

The effect on required signal power of both reliability requirements and the degree of power control are shown in Figure 10.37, in which the same values of F, $\overline{\alpha}_r$, and PG are used as in Figure 10.36. A 7-dB $E_b/\mathcal{N}_0$ value is assumed and P_{rel}, the link reliability, and σ_{dB}, the standard deviation of the received power in dBm, are varied as indicated in the figure. It is obvious in the figure that the required power and the maximum number of users are

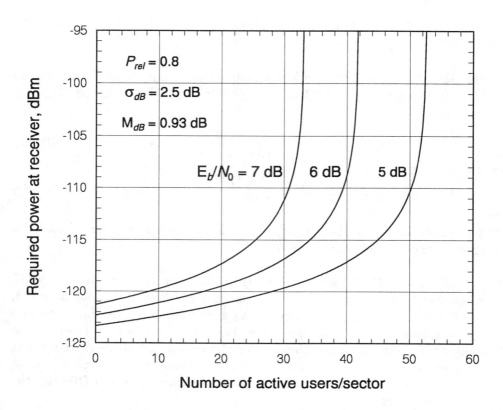

Figure 10.36 CDMA required signal power at receiver versus number of users, $E_b/\mathcal{N}_0$ varied.

reduced for a lower reliability requirement and for more efficient power control (smaller σ_{dB}).

Erlang capacity equation. In this section, we shall examine required power at the receiver for the case of the Erlang capacity-derived number of users for a specified blocking probability. We consider the Erlang capacity obtained under the specified blocking probability as a maximum, and then examine the power requirement when the number of active users varies from 1 to M_{max}, based on (10.67b):

$$P_S(\text{dBm}) = P'_S(\text{dBm}) - 10\log_{10}\left(1 - M/\overline{M}\right) \qquad (10.69a)$$

$$= P'_S(\text{dBm}) - 10\log_{10}(1 - M/M_{max}) \qquad (10.69b)$$

where $\qquad M_{max} = M_{max}(E_b/\mathcal{N}_0; B_{CDMA})$ $\qquad\qquad$ (10.69c)

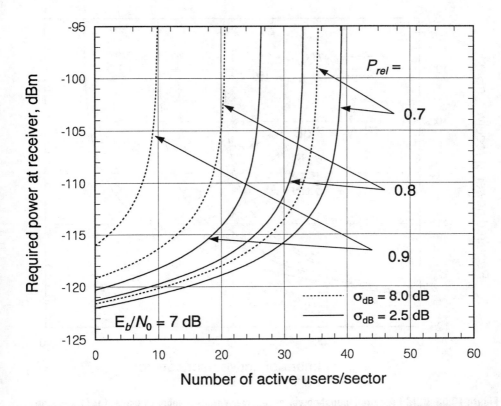

Figure 10.37 CDMA required signal power at receiver versus number of users, reliability and power control parameters varied.

and P'_S(dBm) is given by (10.67c). Using this approach, P_S(dBm) can be plotted as a function of the user traffic, as shown in Figure 10.38, based on the blocking probabilities and values of $\overline{M}$ in Table 10.5. Figure 10.38 depicts the rise in required signal power at the receiver, modeled by (10.69b), as the number of users increases.

Erlang capacity for multi-FA CDMA system. As in Section 10.2.2.6, let us consider a numerical example: for $B_{CDMA} = 2\%$, $\xi = 0.55$, and $\xi' = 0.086$, we have the values of N for different E_b/N_0 assumptions from Table 10.5. The effective number of CDMA channels with $N = 30$ in a 9-FA system is $N' = 30 \times 9 = 270$. Thus, the 9-FA values of N' are

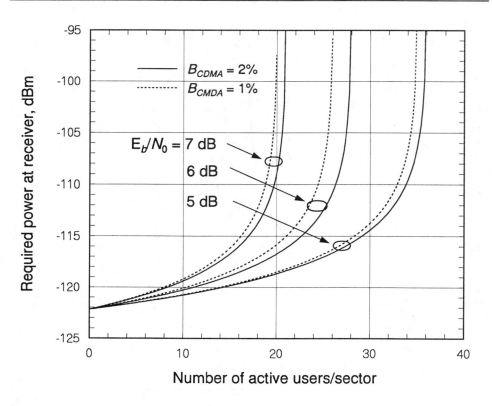

Figure 10.38 CDMA required signal power at receiver versus number of users, for $E_b/\mathcal{N}_0$ and blocking probability varied.

$$N' = 9 \times 46 = 414 \quad \text{for} \quad E_b/\mathcal{N}_0 = 5\,\text{dB}$$

$$N' = 9 \times 36 = 324 \quad \text{for} \quad E_b/\mathcal{N}_0 = 6\,\text{dB}$$

$$N' = 9 \times 29 = 261 \quad \text{for} \quad E_b/\mathcal{N}_0 = 7\,\text{dB}$$

We now need to find the offered loads $A = \overline{M}_{\text{multi-FA}}$ corresponding to these N' values and 2% blocking. According to the procedure diagrammed in Figure 10.35, we find the following offered loads giving these numerical values according to the Erlang B formula of (10.63), with $\overline{M}$ replaced by $\overline{M}_{\text{multi-FA}}$. We use the notation $\overline{M}_{\text{multi-FA}}(E_b/\mathcal{N}_0; B_{CDMA})$ as the maximum in (10.69b). These numbers are calculated to be as follows:

$$\overline{M}_{\text{multi-FA}}(5\,\text{dB}; 2\%) \rightarrow M_{max} = 400$$

$$\overline{M}_{\text{multi-FA}}(6\,\text{dB}; 2\%) \rightarrow M_{max} = 310$$

$$\overline{M}_{\text{multi-FA}}(7\,\text{dB}; 2\%) \rightarrow M_{max} = 247$$

Figure 10.39 shows the required signal power at the base station receiver for a nine-FA CDMA system as a function of the number of users. This model projects that each mobile user's signal power must be around $-120\,\text{dBm}$ at the base station receiver when there are $M = 120$ users in the system, evenly spread over the nine FAs.

10.2.3.2 Cell Radius as a Function of Cell Loading

Having found the value of signal power needed at the base station receiver by an IS-95 CDMA cellular system as a function of the user traffic, we next convert the signal powers into cell radii through use of a propagation loss model and relate the radii to the traffic load. We also expand the assessment with consideration of cable losses, antenna gains, and an allowance for the

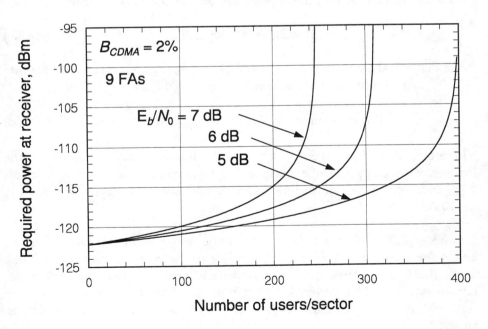

Figure 10.39 CDMA required signal power at receiver versus number of users for a nine-FA system.

additional signal attenuation that may occur when the mobile user is inside a building. The cell radius is obtained by finding the distance at which the propagation loss, when used in the link budget, results in the received signal level being equal to its required value, as a function of the traffic load.

In Section 10.1.3.1, the cell radius was found by setting the maximum tolerable reverse link propagation loss equal to the propagation loss from a mobile at the cell edge to the base station (when the link distance is the cell radius in kilometers, R_{km}). This definition of the cell radius provides an equation for R_{km}, as follows: from (10.20a) we have

$$L_{max} \triangleq L(R_{km}) = L_1 + 10\,\gamma\log_{10} R_{km} \qquad (10.70a)$$

or

$$R_{km}(\text{dBkm}) = 10\log_{10} R_{km} = \frac{L_{max} - L_1}{\gamma} \qquad (10.70b)$$

where $R_{km}(\text{dBkm})$ is the radius expressed in dB relative to 1 km, and L_1 and $10\,\gamma$ are the effective "intercept" and "slope" of the propagation loss equation on a semi-logarithmic plot of propagation loss in dB versus distance:

$$L(d_{km}) = L_1 + 10\,\gamma\log_{10} d_{km} \qquad (10.70c)$$

Note that γ is the distance "power law" of the propagation loss model because (10.70b) can be written in absolute units (i. e., not dB) as

$$\text{Propagation loss} = 10^{L_1/10} \times d_{km}^{\gamma} \propto d_{km}^{\gamma} \qquad (10.70d)$$

Therefore, to find a relation for the cell radius in terms of the cell traffic loading, we need to find an expression for the maximum link propagation loss L_{max} in terms of the cell loading and then substitute this expression for L_{max} in (10.70b).

Example 10.21 What are the values of the propagation loss parameters L_1 and γ for a center frequency in the middle of the reverse link cellular band, using the CCIR propagation loss model and assuming that the area under consideration is 10% covered with buildings? Assume also that the base station and mobile antenna heights are $h_b = 30$m and $h_m = 1.5$m, respectively.

Solution: The reverse link cellular band covers the frequencies 824 to 849 MHz, so a frequency in the middle of this band (representative of the possible reverse link frequencies) is $f = 835$ MHz. The CCIR propagation loss formula (see Section 2.1.4.1) for a link center frequency of $f = 835$ MHz, base station antenna height $h_b = 30$m, and mobile antenna height $h_m = 1.5$m becomes

$$L(d_{km}) = 95.56 + 35.22 \log_{10} d_{km} + 25 \log_{10}(\%) \qquad (10.71a)$$

$$= 120.56 + 35.22 \log_{10} d_{km}, \qquad 10\% \text{ buildings} \qquad (10.71b)$$

where d_{km} is the link distance expressed in kilometers and "%" denotes the percentage of the area that is covered by buildings. Thus, by equating (10.71b) with (10.70a), we find the solutions for L_1 and γ to be $L_1 = 120.56$ dB and $\gamma = 35.22/10 = 3.522$.

Our task now is to find an expression for the maximum link propagation loss in terms of the cell loading. We found in the previous subsection that the minimum signal power required at the base station receiver can be written

$$P_S(\text{dBm}) = P'_S(\text{dBm}) + M_{dB} - 10 \log_{10}(1 - M/M_{max}) \qquad (10.72a)$$

where P'_S was defined in (10.47c) and (10.67c) as the power required for no interference when no link reliability margin is used (i.e., when $M_{dB} = 0$):[11]

$$P'_S(\text{dBm}) = (E_b/N_0)_{req}(\text{dB}) + (N_0W)_c(\text{dBm}) - (PG)(\text{dB}) \qquad (10.72b)$$

$$= (E_b/N_0)_{req}(\text{dB}) - 129.2 \text{ dBm} \qquad (10.72c)$$

and the ideal maximum number of users is given by, from (10.68b):

$$M_{max}(E_b/N_0; M_{dB}) = \frac{PG}{(E_b/N_0)_{req} \cdot F \cdot \overline{\alpha}_r} \cdot \frac{1}{10^{M_{dB}/10}} \qquad (10.72d)$$

It follows that the maximum tolerable propagation loss is the loss that, given the maximal mobile transmitter power and the various gains and non-propagation losses in the reverse link power budget, results in P_S being

---

[11] We assume that $PG = 128 = 21.1$ dB and that the base station receiver has a noise figure of 5 dB, resulting in $(N_0W)_c = -108.1$ dBm, as discussed in Section 10.1.3.3.

delivered to the base station receiver. An equation expressing this condition is the following:

$$P_S(\text{dBm}) = \text{transmitter power} + \text{gains} - \text{losses} - L_{max}$$

$$= P_R(\text{dBm})|_{\text{no loss}} - L_{max} \qquad (10.73a)$$

where

$$P_R(\text{dBm})|_{\text{no loss}} \triangleq \text{transmitter power} + \text{gains} - \text{losses}$$

$$= P_m - L_m + G_m - L_p - L_b + G_c - L_c \qquad (10.73b)$$

$P_R(\text{dBm})|_{\text{no loss}}$ denotes the mobile unit's power that would be delivered to the base station receiver if there were no propagation loss. Thus:

$$L_{max} = P_R(\text{dBm})|_{\text{no loss}} - P_S(\text{dBm}) \qquad (10.73c)$$

Typical values of the link budget parameters itemized in (10.73b) are given in Table 10.7 [16]. When these parameter values are substituted in (10.73b), then

$$P_R(\text{dBm})|_{\text{no loss}} = 23\,\text{dBm} - 0\,\text{dB} + 2.1\,\text{dBi} - 3\,\text{dB}$$

$$- 10\,\text{dB} + 14.1\,\text{dBi} - 2\,\text{dB} \quad = 24.2\,\text{dBm} \quad (10.73d)$$

In Section 10.1.3.3, (10.34c) was given for the maximum loss as a function of the cell loading parameter X. For simplicity, that equation involves just the antenna gains. If the additional detailed losses included in (10.73c) are added, and the margin considered in (10.72a) is introduced, then (10.34c) agrees with formulation here, which, using (10.72a), can be expressed as

Table 10.7 CDMA reverse link power budget parameters

Parameter	Symbol	Value
Mobile transmitter power	P_m	23 dBm
Mobile cable losses	L_m	0 dB
Mobile antenna gain	G_m	2.1 dBi
Mobile antenna orientation loss	L_p	3 dB
Allowance for building penetration	L_b	10 dB
Base station antenna gain	G_c	14.1 dBi
Base station cable losses	L_c	2 dB

$$L_{max}(dB) = P_R(dBm)|_{no\,loss} - P_S(dBm)$$

$$= P_R(dBm)|_{no\,loss} - P'_S(dBm) - M_{dB}$$

$$+ 10\log_{10}(1 - M/M_{max}) \qquad (10.73e)$$

Now we substitute (10.73e) for L_{max} in (10.70b) to obtain the desired expression for the cell radius as a function of the cell loading:

$$R_{km}(dBkm) = 10\log_{10} R_{km} = \frac{L_{max} - L_1}{\gamma}$$

$$= \frac{P_R(dBm)|_{no\,loss} - L_1 - P'_S(dBm) - M_{dB} + 10\log_{10}(1 - M/M_{max})}{\gamma}$$

$$(10.74)$$

This expression gives the maximal cell radius that can be obtained for the mobile transmitter power assumed in the calculation of $P_R(dBm)|_{no\,loss}$.

Example 10.22 Find a numerical expression for the cell radius based on (10.74) when the CCIR model of Example 10.21 and the link budget parameters of Table 10.7 are used.

Solution: If the parameters of Table 10.7 are used, the received power without a propagation loss is $P_R(dBm)|_{no\,loss} = 24.2\,dBm$. From (10.72c), the required power for interference and no margin is $P'_S(dBm) = (E_b/\mathcal{N}_0)_{req}(dB) - 129.2\,dBm$. And, from Example 10.21, the values of L_1 and γ are $L_1 = 120.56$ and $\gamma = 3.522$. Substituting all these into (10.74), we get an expression that is parametric in $E_b/\mathcal{N}_0$, M_{dB}, M, and M_{max}:

$$R_{km}(dBkm) = \frac{1}{3.522}\left[24.2 - 120.56 - \left(\frac{E_b}{\mathcal{N}_0}(dB) - 129.2\right)\right.$$

$$\left. - M_{dB} + 10\log_{10}\left(1 - \frac{M}{M_{max}}\right)\right]$$

$$= \frac{1}{3.522}\left[32.84 - \frac{E_b}{\mathcal{N}_0}(dB) - M_{dB} + 10\log_{10}\left(1 - \frac{M}{M_{max}}\right)\right]$$

$$(10.75a)$$

To illustrate the dependence of the cell radius relationship to M on assumed values of $E_b/\mathcal{N}_0$ and margin, we use (10.75a) in the example above to write

$$R_{km} = 10^{R_{km}(\text{dBkm})/10}$$

$$= 10^{32.84/35.22}\left(1 - \frac{M}{M_{max}}\right)^{10/35.22}\left[\frac{E_b}{\mathcal{N}_0} \cdot 10^{M_{dB}/10}\right]^{-1/3.522}$$

$$= 8.559\left(1 - \frac{M}{M_{max}}\right)^{0.284}\left[\frac{E_b}{\mathcal{N}_0} \cdot 10^{M_{dB}/10}\right]^{-0.284} \tag{10.75b}$$

Using the ideal capacity expression for M_{max} that is given in (10.72d) the cell radius expression (10.75b) is plotted in Figure 10.40 for different values of the combination $(M_{dB} + E_b/\mathcal{N}_0)$ in dB. The figure shows clearly that the margin and/or $E_b/\mathcal{N}_0$ requirement for the reverse link greatly affects the cell size and the rate at which the cell size decreases when there are M users. The number of users in Figure 10.40 pertains to a CDMA system with one FA. To get the variation in cell size as a function of the number of users in a multi-FA CDMA system, simply multiply the number of users plotted in Figure 10.40 by the number of FAs.

If the values of M_{max} based on the analysis of Erlang capacity are used instead of those based on ideal capacity, then the dependence of cell radius on the number of users is modeled by the curves plotted in Figure 10.41, in which it was assumed that the margin is zero. The curves in the figure pertain to a CDMA system with a single FA, and the values of M_{max} were taken from Table 10.5. To extend the result to a multi-FA system, the procedure illustrated in Figure 10.35 must be followed to ascertain the value of M_{max} that represents the Erlang capacity for a multi-FA system.

10.2.3.3 Base Station Density

The maximal cell radius expression shown in (10.75b) can be extended to give the cell area as a function of the number of active CDMA users, and this area can be compared with some unit area, to render the minimal base station density per unit area. The expression for cell radius can also be manipulated to obtain a model for the density of active users.

The area of a hexagonal cell with the center-to-corner radius R_c is

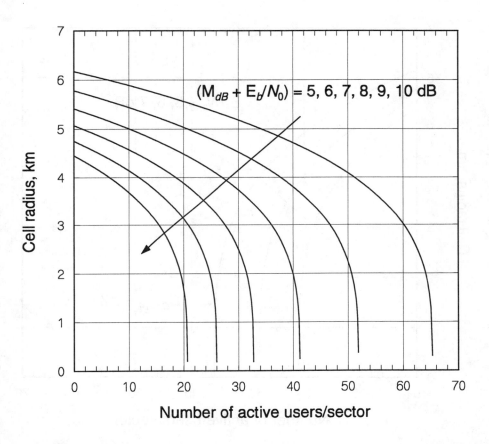

Figure 10.40 CDMA cell radius versus number of active users, margin and $E_b/\mathcal{N}_0$ requirement varied.

$$a_{cell} = \left(3\sqrt{3}/2\right) R_c^2 \tag{10.76a}$$

Let the quantity y be defined as the base station density as follows:

$$y \triangleq \text{base station density} = \text{number of base stations per unit area}$$

$$= \frac{\text{unit area}}{\text{area of cell}} = \frac{a_0}{a_{cell}} = \left. \frac{2a_0}{3\sqrt{3}\,R_c^2} \right|_{R_c = R_{km}} = \frac{0.3849\,a_0}{R_{km}^2} \tag{10.76b}$$

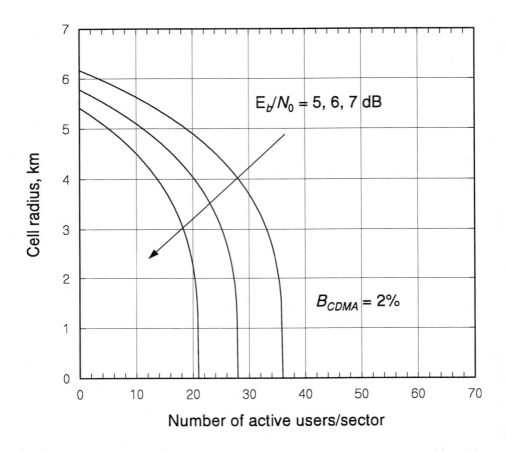

Figure 10.41 CDMA cell radius versus number of active users, M_{max} **based on Erlang capacity analysis.**

using a_0 to denote the unit area expressed in square kilometers, such as 10 km^2, and so forth. The units of the area of the cell are the units of the cell radius. Taking the radius of the cell in units of kilometers, we may substitute R_{km} in (10.75b) for R_c in (10.76b) to obtain the minimal base station density as a function of the number of users per sector:

$$y = 0.3849\, a_0 \div \left[8.559 \left(1 - \frac{M}{M_{max}} \right)^{0.284} \left(\frac{E_b}{N_0} \cdot 10^{M_{dB}/10} \right)^{-0.284} \right]^2$$

$$= \left(5.254 \times 10^{-3}\right) a_0 \left[\frac{(E_b/\mathcal{N}_0) \cdot 10^{M_{dB}/10}}{1 - M/M_{max}} \right]^{0.568} \tag{10.76c}$$

This expression pertains to the CCIR model of propagation loss for an area that is 10% covered by buildings and for $f = 835\,\text{MHz}$, as discussed in examples 10.21 and 10.22. It can be extended to other models by substituting numerical values for L_1 and γ in the cell radius given in (10.74).

Example 10.23 Find the minimal base station density as a function of the number of users per sector when the propagation loss is calculated using the NLOS WIM with frequency $f = 835\,\text{MHz}$, base station antenna height $h_b = 30\text{m}$, mobile antenna height $h_m = 1.5\text{m}$, building height $h_B = 20\text{m}$, building separation $b = 40\text{m}$, and width of street $w = 20\text{m}$. Assume an orientation loss of $0\,\text{dB}$.

Solution: Because $\Delta h_b = h_b - h_B = 30 - 20 = 10\,\text{m} > 0$, we substitute the given parameters in the NLOS WIM formula given in (2.34a) and obtain

$$L(d_{km}) = 69.55 + 38\log_{10} d_{km} + 26\log_{10} 835 - 10\log_{10} 20$$

$$- 9\log_{10} 40 + 20\log_{10}(20 - 1.5) - 18\log_{10}(1 + 10) + 0$$

$$= 124.7 + 38\log_{10} d_{km} \tag{10.77}$$

By equating (10.77) with (10.70a), we find that $L_1 = 124.7\,\text{dB}$ and $\gamma = 3.8$. Substituting these values in (10.74) results in the maximal cell radius formula

$$R_{km}(\text{dBkm}) = \frac{1}{3.8}\left[24.2 - 124.7 - \left(\frac{E_b}{\mathcal{N}_0}(\text{dB}) - 129.2 \right) \right.$$

$$\left. - M_{dB} + 10\log_{10}\left(1 - \frac{M}{M_{max}} \right) \right]$$

$$= \frac{1}{3.8}\left[28.7 - \frac{E_b}{\mathcal{N}_0}(\text{dB}) - M_{dB} + 10\log_{10}\left(1 - \frac{M}{M_{max}} \right) \right]$$

$$\tag{10.78b}$$

or

$$R_{km} = 5.692 \left(1 - \frac{M}{M_{max}} \right)^{0.263} \left[\frac{E_b}{\mathcal{N}_0} \cdot 10^{M_{dB}/10} \right]^{-0.263} \tag{10.78c}$$

Substituting this cell radius in the formula for base station density given in (10.76b) results in the following expression for the minimal base station density using the NLOS WIM:

$$y = .01188\, a_0 \left[\frac{(E_b/\mathcal{N}_0) \cdot 10^{M_{dB}/10}}{1 - M/M_{max}} \right]^{0.526} \tag{10.78c}$$

Figure 10.42 shows the base station density per $a_0 = 10\,\text{km}^2$ as a function of cell loading, $X = M/M_{max}$, for the CCIR propagation model result (10.76c) and for the NLOS WIM result (10.78c). As indicated in the figure, the parameters used in the CCIR example correspond to a suburban environment, while those in the WIM example correspond to an urban environment. The base station densities are seen in the figure to be about 2.5 times as high for the urban situation as for the suburban situation, because of higher propagation loss. Figure 10.42 can be interpreted in terms of a number of users per sector, cell, or multi-FA system by reading $M = X M_{max}$ using the appropriate value of M_{max}.

Example 10.24 An estimate of the number of base stations is to be made by using an 80% loading factor ($X = 0.8$) for areas of 10, 20,..., 90, $100\,\text{km}^2$. Construct a table for the number of base stations (#BS), for CCIR propagation model (suburban) and WIM propagation model (urban), assuming an $E_b/\mathcal{N}_0$ requirement of 7 dB and the system parameter values used in Figure 10.42.

Solution: From Figure 10.42, at $X = 0.8$ the value of base station density for $E_b/\mathcal{N}_0 = 7\,\text{dB}$ is 0.33 base stations per $10\,\text{km}^2$ for the CCIR propagation model, which pertains to a suburban situation, and 0.65 base stations per $10\,\text{km}^2$ for the WIM, which pertains to an urban situation. Thus, the numbers of base stations (rounded up to whole numbers) for the various area sizes are as shown in Table 10.8. For example, if the area size is $50\,\text{km}^2$, then $5 \times 0.33 = 1.65 \rightarrow 2$ base stations would be needed in an suburban environment, while $5 \times 0.65 = 3.25 \rightarrow 4$ base stations would be required in an urban environment.

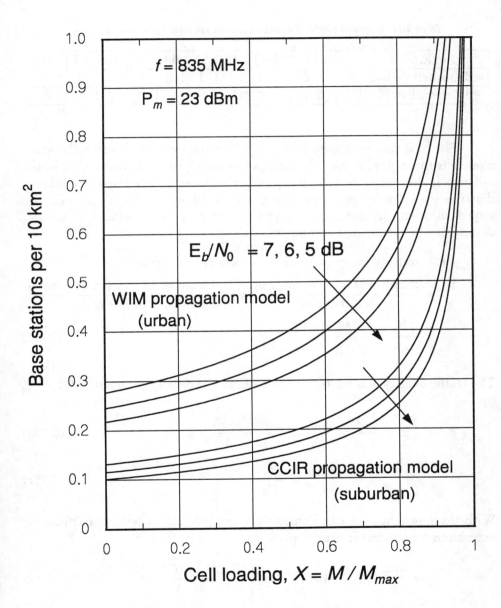

Figure 10.42 Base station density versus cell loading, with $E_b/\mathcal{N}_0$ as a parameter.

Table 10.8 Required numbers of base stations estimated in Example 10.24

Area in km^2	10	20	30	40	50	60	70	80	90	100
#BS, urban (WIM)	1	2	2	3	4	4	5	6	6	7
#BS, suburban (CCIR)	1	1	1	2	2	2	3	3	4	4

The base station density increases as loading increases because the interference is increasing, having the effect of reducing the cell radius; that is, the cell size is decreasing because the number of users is increasing. It follows that the user density per unit area is increasing more than if the cell size stayed constant. Assuming three sectors per cell, an equation for the user density can be formulated as follows:

$$x \triangleq \text{traffic density} = \text{number of users per unit area}$$

$$= \frac{\text{number of users in one cell}}{\text{area of one cell}} = \frac{3 \times \text{number of users in one sector}}{\text{area of one cell}}$$

$$= \frac{3M}{a_{cell}} = \frac{2M}{\sqrt{3}\, R_c^2} \tag{10.79a}$$

The traffic density x is related to the base station density y, given in (10.76b), by

$$x = \frac{3M}{a_{cell}} = \frac{3M}{a_0} \times \frac{a_0}{a_{cell}} = \frac{3M}{a_0} \times y \tag{10.79b}$$

or

$$y = \frac{a_0}{3M} x \tag{10.79c}$$

With these insights, we can immediately use (10.76c) to write the following expression for the traffic density per km^2:

$$x = \frac{3M}{a_0} \times (5.254 \times 10^{-3}) a_0 \left[\frac{(E_b/\mathcal{N}_0) \cdot 10^{M_{dB}/10}}{1 - M/M_{max}} \right]^{0.568}$$

$$= 0.0158\, M \left[\frac{(E_b/\mathcal{N}_0) \cdot 10^{M_{dB}/10}}{1 - M/M_{max}} \right]^{0.568} \tag{10.79d}$$

Figure 10.43 shows the traffic density per $10\,\mathrm{km}^2$, obtained by multiplying (10.79d) by 10, as a function of the number of users for the case of $E_b/\mathcal{N}_0 = 7$ dB and no margin. The figure pertains to a CDMA system with one FA, but it is easily adapted to a multi-FA system by simply multiplying the traffic density times the number of FAs. Shown in Figure 10.43 is the fact that the two propagation models compared in Figure 10.42 yield different traffic density results because they produce different values for the maximal cell radius.

Example 10.25 Suppose that it has been estimated that the potential number of cellular customers is 100 per km^2 in a suburban area and 500 per km^2 in an urban area. Construct a table showing the number of FAs required (#FAs) for both the urban area and the suburban area to accommodate these customers, as a function of cell loading, for $X = 0.2, 0.3, \ldots, 0.8$. Assume that each customer offers $A_0 = 0.02\,\mathrm{Erlangs}$ of traffic and that the $E_b/\mathcal{N}_0$ requirement is 7 dB.

Solution: For the urban area, the desired Erlang capacity per km^2 is found by dividing the number of customers per km^2 by A_0 to get $500 \times 0.02 = 10\,\mathrm{E}$ per km^2, or $100\,\mathrm{E}/10\,\mathrm{km}^2$. The desired Erlang capacity per km^2 for the suburban area is $100 \times A_0 = 2\,\mathrm{E/km}^2$ or $20\,\mathrm{E}/10\,\mathrm{km}^2$. From Figure 10.43, we read the Erlang capacity per $10\,\mathrm{km}^2$ for 1 FA and determine how many FAs are required. For example, for 30% loading ($X = 0.3$), Figure 10.43 shows that the urban system can accommodate about $16\,\mathrm{E}/10\,\mathrm{km}^2$, so that the number of FAs required in this instance is $(100\,\mathrm{E}/10\,\mathrm{km}^2) \div (16\,\mathrm{E}/10\,\mathrm{km}^2) = 6.25 \rightarrow 7$ FAs. Table 10.9 shows the numbers of FAs required for different loadings. Note that when the number of FAs exceeds the number that can be fit into the bandwidth allocated to the CDMA system, the requirement is satisfied by using smaller cells.

Table 10.9 Numbers of FAs required

Cell loading	0.2	0.3	0.4	0.5	0.6	0.7	0.8
#FAs, urban (WIM)	11	7	5	4	3	2	1
#FAs, suburban (CCIR)	5	3	2	2	1	1	1

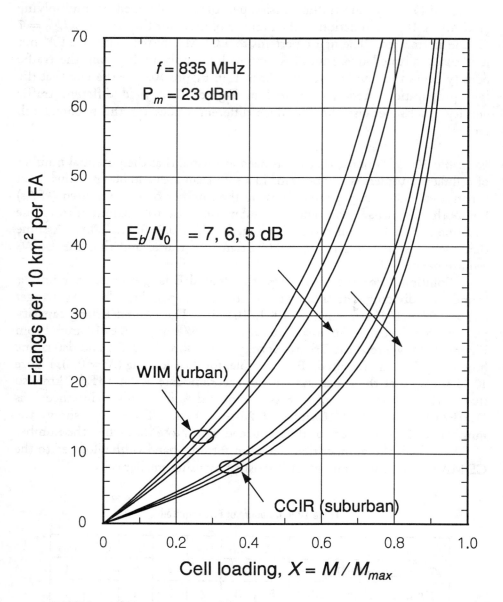

Figure 10.43 Traffic density in Erlangs per unit area of $a_0 = 10\ \mathrm{km}^2$ versus cell loading.

Now let us observe the relationship between the base station density (y) and the traffic density (x). Because we obtained both of these quantities as functions of the loading $X = M/M_{max}$, in the form $x = x(X)$ in (10.79d) and $y = y(X)$ in (10.76c), we can plot them against one another to observe the relationship between these quantities. Figures 10.44 and 10.45 give these plots for the CCIR propagation model and the WIM, respectively. In each figure, the value of the loading common to both the base station density and the traffic density is indicated. Note that for each propagation model, these figures provide the minimum number of base stations for any unit area that may be specified in a given requirement, as a function of traffic density, with loading and $E_b/\mathcal{N}_0$ as parameters. Although the figures are plotted for zero margin, we can always plot the required curves by replacing $E_b/\mathcal{N}_0$ with $M_{dB} + E_b/\mathcal{N}_0$ in (10.76c) and (10.79d).

Example 10.26 Find the subscriber capacity of a system covering an area of $36\,\text{km}^2$ with a minimal CDMA installation using a loading of no greater than 60%. Assume that $A_0 = 0.02$ and $E_b/\mathcal{N}_0 = 7\,\text{dB}$, and find the capacity for the urban area we assumed in the previous examples.

Solution: At 60% loading, we read from Figure 10.45 (for the WIM) that the urban system requires a minimum of 0.45 base stations per $10\,\text{km}^2$. For the postulated $36\,\text{km}^2$ area, then, the minimum number of base stations is equal to $0.45 \times 3.6 = 1.62 \rightarrow 2$ base stations. From Figure 10.45 again, we read that a sector with 1 FA at 60% loading can provide about 33 E of capacity per $10\,\text{km}^2$ of area, or 118.8 E in $36\,\text{km}^2$ at the same time that the theoretical value of 1.62 base stations is obtained. Because we have rounded up to two base stations, the number of Erlangs for the urban area is $118.8 \times 2/1.62 = 147$ E. Assuming $A_0 = 0.02$, the two base stations can service up to $147/0.02 = 7,350$ subscribers.

References

[1] "Mobile Station-Base Station Compatibility Standard for Dual-Mode Wideband Spread Spectrum Cellular System," TIA/EIA Interim Standard 95 (IS-95), Washington, DC: Telecommunications Industry Association, July 1993 (amended as IS-95-A in May 1995).

[2] Lee, J. S., and L. E. Miller, "Dynamic Allocation of CDMA Forward Link Power for PCS and Cellular Systems" (invited paper), *Proc. 2nd CDMA Internat'l Conf.*, pp. 95–99, Oct. 21–24, 1997, Seoul, Korea.

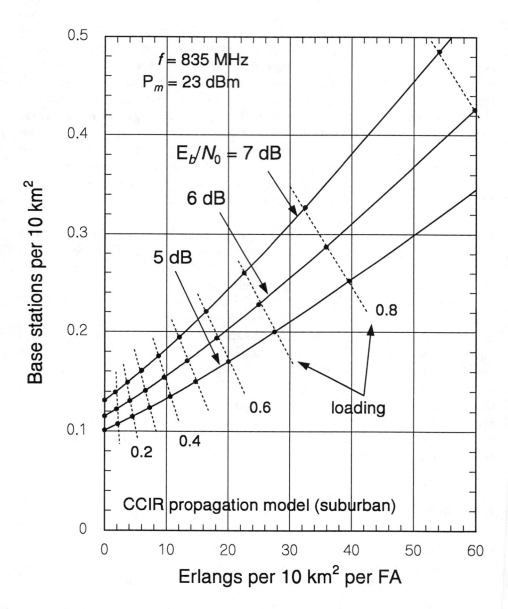

Figure 10.44 Base station density versus traffic density for CCIR model.

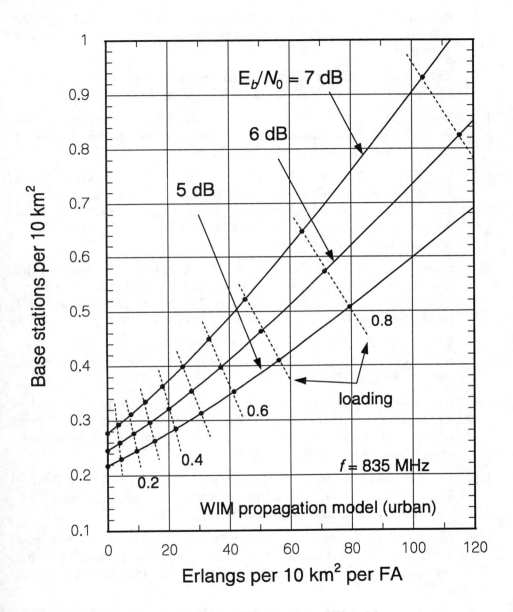

Figure 10.45 Base station density versus traffic density for WIM model.

[3]   Padovani, R., "Reverse Link Performance of IS-95 Based Cellular
      Systems," *IEEE Personal Communications Magazine*, pp. 28–34, 3rd
      quarter 1994.

[4]   Kim, K. I., "CDMA Cellular Engineering Issues," *IEEE Trans. on
      Vehicular Technology*, Vol. 42, pp. 345–350, Aug. 1993.

[5]   Ross, A. H. M., and K. S. Gilhousen, "CDMA Technology and the IS-
      95 North American Standard," *The Mobile Communications Handbook*
      (J. D. Gibson, ed.), CRC Press/IEEE Press, 1996.

[6]   Longley, A. G., and P. L. Rice, "Prediction of Tropospheric Radio
      Transmission Loss Over Irregular Terrain: A Computer Method—
      1968," U.S. Dept. of Commerce, Environmental Science Services
      Admin. (ESSA) technical report ERL 79-ITS 67, July 1968 (DTIC
      accession number AD-676874).

[7]   Hufford, G. A., A. G. Longley, and W. A. Kissick, "A Guide to the
      Use of the ITS Irregular Terrain Model in the Area Prediction Mode,"
      Natl. Telecom. and Info. Admin. (NTIA) report 82-100, Apr. 1982
      (available from NTIS as document number PB82-217977).

[8]   Viterbi, A. M., and A. J. Viterbi, "Erlang Capacity of a Power Con-
      trolled CDMA System," *IEEE J. on Selected Areas in Communications*,
      Vol. 11, pp. 892–900, Aug. 1993.

[9]   Jansen, M. G., and R. Prasad, "Capacity, Throughput, and Delay
      Analysis of a Cellular DS CDMA System with Imperfect Power
      Control and Imperfect Sectorization," *IEEE Trans. on Vehicular Tech-
      nology*, Vol. 44, pp. 67–75, Feb. 1995.

[10] Lee, J. S., and L. E. Miller, "On the Erlang Capacity of CDMA Cellular
      Systems," *Proc. 1995 IEEE Global Telecommunications Conference*, pp.
      1877–1883 (paper 54.2), Singapore, Nov.13–17, 1995.

[11] Abramowitz, M., and I. A. Stegun (eds.), *Handbook of Mathematical
      Functions*, National Bureau of Standards Applied Mathematics Series 55,
      Washington, DC: Government Printing Office, 1964 (ninth printing,
      1970).

[12] Viterbi, A. J., A. M. Viterbi, K. S. Gilhousen, and E. Zehavi, "Soft
      Handoff Extends CDMA Cell Coverage and Increases Reverse Link
      Capacity," *IEEE J. on Selected Areas in Communications*, Vol. 12, pp.
      1281–1288, Oct. 1994.

[13] Lee, J. S., and L. E. Miller, "Comparison of Handoff Gains for CDMA (IS-95) and FDMA Cellular Systems," *Proc. 2nd International Workshop on Multi-Dimensional Mobile Communications (MDMC '96)*, pp. 353–357, July 18–20, 1996, Seoul, Korea.

[14] Fenton, L. F., "The Sum of a Log-Normal Probability Distribution in Scattered Transmission Systems," *IRE Trans. on Communication Systems*, Vol. CS-8, pp. 57–67, Mar. 1960.

[15] "Dual-Mode Subscriber Equipment-Network Equipment Compatibility Specifications," TIA/EIA Interim Standard 54 (IS-54)," Washington, DC: Telecommunications Industry Association, 1989.

[16] Wheatley, C., "Trading Coverage for Capacity in Cellular Systems: A System Perspective," *Microwave Journal*, pp. 62–79, July 1995.

[17] Gradshteyn, I. S., and I. M. Rhyzhik, *Table of Integrals, Series, and Products* (4th ed.), New York: Academic Press, 1965.

Appendix 10A Analysis of Second-Order Reuse Fraction

The blocking probability for a CDMA cellular system is formulated in terms of the reuse fractions ξ and ξ' for the received mobile interference power and squared power, respectively. An estimate of ξ was derived in Section 10.1.2.2, based on the methodology in [4]. In this appendix, we extend that methodology to derive an estimate for ξ', which may be termed the "second order reuse fraction."

Assuming a density of M users uniformly distributed over each cell area πR^2, the average total interference power from a cell at distance d was formulated in Section 10.1.2.2 as

$$P(d) = \frac{2MP_0}{\pi R^2} \int_0^R dr\, r^5 \int_0^\pi d\theta \, \frac{1}{\left(r^2 + d^2 + 2r\, d \cos\theta\right)^2} \tag{10A.1a}$$

Using the parameter $\kappa = d/R$, the integrals may be solved to give

$$P(d = \kappa R) = 2MP_0 \left[2\kappa^2 \ln\left(\frac{\kappa^2}{\kappa^2 - 1}\right) - \frac{4\kappa^4 - 6\kappa^2 + 1}{2(\kappa^2 - 1)^2} \right] \tag{10A.1b}$$

Because the total received mobile power in a given cell is MP_0, the ratio of total received power from all cells to that for one cell is the reuse factor

$$\text{Reuse factor} \triangleq 1 + \xi = \frac{MP_0 + \sum_\kappa P(d = \kappa R)}{MP_0} \tag{10A.2a}$$

where the reuse fraction ξ is

$$\xi = \frac{\sum_\kappa P(\kappa R)}{MP_0} = 2\sum_\kappa \left[2\kappa^2 \ln\left(\frac{\kappa^2}{\kappa^2 - 1}\right) - \frac{4\kappa^4 - 6\kappa^2 + 1}{2(\kappa^2 - 1)^2} \right] \tag{10A.2b}$$

$$= \frac{\dfrac{1}{M\mathcal{I}_0 R_b} \sum_\kappa P(d = \kappa R)}{\dfrac{1}{M\mathcal{I}_0 R_b} MP_0} \approx \frac{\rho_{med}\left(\substack{\text{other} \\ \text{cells}}\right)}{\rho_{med}\left(\substack{\text{same} \\ \text{cell}}\right)} \tag{10A.2c}$$

The approximation involved is in assuming that the ratio of the averages equals the ratio of the medians of the interference powers.

For a regular cellular pattern with R equal to the nonoverlapping cell radius, there are six adjacent cells each having $\kappa = 2$; a second ring with six cells having $\kappa = 4$ and six having $\kappa = 2\sqrt{3}$; a third ring with six cells having $\kappa = 6$ and twelve having $\kappa = 2\sqrt{7}$; and so forth. As discussed in Section 10.1.2.2, the sum over the values of κ gives $F = 0.3198$ for the first three rings and $F \approx 0.33$ for a large number of terms [4]. Tests and simulations [3], [8] give a higher value (≈ 0.55) attributable to imperfect power control and a non-fourth power propagation law in particular situations, among other factors differing from this analysis.

Intuitively, because for small values $x^2 \ll x$, we expect that the ratio of other-cell squared power to same-cell squared power will be much less than the ratio of powers. Thus, the authors of [8] recognized that their assumption of $\xi' = \xi$ would produce an upper bound on the total squared power of the interference, and, hence, a lower bound on the system capacity. Extending the method in [4], the square of the power received from a mobile in another cell can be written as

$$P_{other}^2 = P_0^2 \left(r/r_0\right)^8 \tag{10A.3a}$$

giving the total squared power

$$P_2(d) = \frac{2MP_0^2}{\pi R^2} \int_0^R dr\, r^9 \int_0^\pi d\theta\, \frac{1}{(r^2 + d^2 + 2r\,d\cos\theta)^4} \qquad (10A.3b)$$

The inner integral involved in the calculation of the total squared interference power from another cell is

$$C(r) = \frac{1}{\pi} \int_0^\pi d\theta\, \frac{1}{(r^2 + d^2 + 2d\,r\cos\theta)^4} = \frac{d^6 + 9d^4 r^2 + 9d^2 r^4 + r^6}{(d^2 - r^2)^7}$$

found by applying integration formulas 2.554.1 and 2.553.3 in [17]. Thus, the outer integral becomes

$$f_2(R, d) = \int_0^R dr\, r^9\, C(r) = \int_0^R dr\, \frac{d^6 r^9 + 9d^4 r^{11} + 9d^2 r^{13} + r^{15}}{(d^2 - r^2)^7}$$

Repeated integration by parts and the use of integration formula 2.152.1 in [17] give the result

$$f_2(R, d) = \frac{R^8(d^6 + 9d^4 R^2 + 9d^2 R^4 + R^6)}{12\,(d^2 - R^2)^6} - \frac{R^6(4d^6 + 45d^4 R^2 + 54d^2 R^4 + 7R^6)}{60\,(d^2 - R^2)^5}$$

$$+ \frac{R^4(2d^6 + 30d^4 R^2 + 45d^2 R^4 + 7R^6)}{40\,(d^2 - R^2)^4} - \frac{R^2(4d^6 + 90d^4 R^2 + 180d^2 R^4 + 35R^6)}{120\,(d^2 - R^2)^3}$$

$$+ \frac{d^6 + 45d^4 R^2 + 135d^2 R^4 + 35R^6}{60\,(d^2 - R^2)^2} - \frac{3d^4 + 18d^2 R^2 + 7R^4}{4\,(d^2 - R^2)}$$

$$+ \frac{11d^2}{15} - \frac{7}{2}R^2 + 8d^2\ln\left(\frac{d^2}{d^2 - R^2}\right) \qquad (10A.3c)$$

Using the new integration results, the new quantity $P_2(d)$ can be written

$$P_2(d = \kappa R) = \frac{2MP_0^2}{R^2} \cdot f_2(R, d)\big|_{d=\kappa R}$$

$$= 2M\mathrm{P}_0^2 \left\{ \frac{\kappa^6 + 9\kappa^4 + 9\kappa^2 + 1}{12\left(\kappa^2 - 1\right)^6} - \frac{4\kappa^6 + 45\kappa^4 + 54\kappa^2 + 7}{60\left(\kappa^2 - 1\right)^5} \right.$$

$$+ \frac{2\kappa^6 + 30\kappa^4 + 45\kappa^2 + 7}{40\left(\kappa^2 - 1\right)^4} - \frac{4\kappa^6 + 90\kappa^4 + 180\kappa^2 + 35}{120\left(\kappa^2 - 1\right)^3}$$

$$+ \frac{\kappa^6 + 45\kappa^4 + 135\kappa^2 + 35}{60\left(\kappa^2 - 1\right)^2} - \frac{3\kappa^4 + 18\kappa^2 + 7}{4\left(\kappa^2 - 1\right)}$$

$$\left. + \frac{11\kappa^2}{15} - \frac{7}{2} + 8\kappa^2 \ln\left(\frac{\kappa^2}{\kappa^2 - 1}\right) \right\} \tag{10A.3d}$$

Considerations similar to those discussed above lead to the definition of the squared-power reuse fraction

$$\xi' = \frac{1}{M\mathrm{P}_0^2} \sum_\kappa \mathrm{P}_2(d = \kappa R) \tag{10A.4a}$$

$$= \frac{\dfrac{1}{M(\mathcal{I}_0 R_b)^2} \displaystyle\sum_\kappa \mathrm{P}_2(d = \kappa R)}{\dfrac{1}{M(\mathcal{I}_0 R_b)^2} M\mathrm{P}_0^2} \approx \frac{\rho_{med}^2\left(\begin{smallmatrix}\text{other}\\\text{cells}\end{smallmatrix}\right)}{\rho_{med}^2\left(\begin{smallmatrix}\text{same}\\\text{cell}\end{smallmatrix}\right)} \tag{10A.4b}$$

The approximation involved is in equating the ratio of mean squares to the ratio of squared medians of the interference powers. Summed over the first two rings of adjacent cells, this quantity is calculated to be

$$\xi' = 0.086$$

with negligible contributions to the sum from third and farther rings of cells.

11

CDMA Optimization Issues

We saw in Chapters 1 through 10 that a significant increase in user capacity is promised by second-generation CDMA digital cellular and PCS systems designed according to the IS-95 air interface [1]. However, the benefit of an increase in capacity is obtained at the cost of an increase in the complexity of the waveform modulation and in the system control and network management segments of the system. We have many system and waveform parameters to select that specify the operation of the system and that, consequently, determine system performance. The task of optimizing a CDMA cellular or PCS system can be viewed as the determination of the best values of those many parameters for a particular system in a particular environment on a long- or short-term basis, perhaps even in real time.

For example, before a CDMA cellular system is deployed, we need to assign PN code offsets to the different base station/sectors in the system. There are as many as 512 different offsets to choose from before reusing any of them. Can these offsets be selected arbitrarily, or are there constraints to guide in the selection for a particular installation? Another example: The forward link waveform is an orthogonally multiplexed combination of pilot, synchronization, paging, and traffic channel digitally modulated waveforms; all told, there can be 30 to 40 channels being transmitted simultaneously on the forward link carrier. What considerations govern the allocation of power among these channels? In particular, how much of the forward link power should be allocated to the pilot signal? Can the allocation be based on a one-time determination of the conditions in the cell, or must it be continually adjusted to achieve good CDMA system performance?

In this final chapter, we discuss selected aspects of CDMA system optimization that are most critical to the operation of the system at, or near, its best performance. We introduce the subject of CDMA system optimization by considering the factors affecting the selection and reuse of PN pilot offsets and show that the selection of the offsets for neighboring cells and sectors is not arbitrary but requires careful consideration. Next, we analyze the requirements for forward link channel transmitter powers and present a simultaneous solution for these powers that takes into account the dynamics of forward link power control and the different processing gains of each type of channel. We show that there is a definite relationship among the forward link powers that satisfies each channel's requirements, including fade margins. Using parametric variations on the power solutions, we show how the different parameters under control of the system operator affect the distribution of power among the channels, including the choice of fade margins and the realizability of fade margins. These results lead to studies of means for balancing the capacity of forward and reverse links to maintain good system performance and of possible implementations of dynamic forward link power allocation. The emphasis on the forward link in these topics reflects the fact that the task of CDMA system optimization is centered on parameter selections that are made concerning the forward link [2, 3].

11.1 Selection of Pilot PN Code Offsets

In the IS-95 CDMA cellular system, each base station or sector transmits a pilot PN-coded signal that is generated by quadrature modulating the sinusoidal carrier by "short" I and Q PN codes, without any baseband data modulation. Each base station uses the same PN codes. The form of the pilot signal as transmitted by base station i is

$$s_i(t) = A[C_I(t; \phi_i) \cos(2\pi f_c t + \theta) + C_Q(t; \phi_i) \sin(2\pi f_c t + \theta)] \qquad (11.1)$$

where C_I and C_Q are the in-phase and cross-quadrature $\pm$ PN code-sequence waveforms, respectively, ϕ_i is a code-phase offset that uniquely identifies the base station, and θ is a random initial carrier phase. As illustrated in Figure 11.1, the CDMA system has as many as $M = 512$ offsets (short code starting positions) spaced at $N = 64$-chip intervals in the short code period of $MN = 2^{15} = 32,768$ chips. This period is accomplished for each short code by appending a zero to the maximal length sequence generated by a 15-stage shift register PN generator.[1]

The chip rate for the short PN codes is 1.2288 MHz, giving a chip duration of $T_c = \frac{1}{1.2288} = 0.814\,\mu s$ and a signal bandwidth of approximately 1.25 MHz. The period of the short codes in time is

---

[1] See Section 6.3.4.2 for a detailed description of the short PN codes in IS-95.

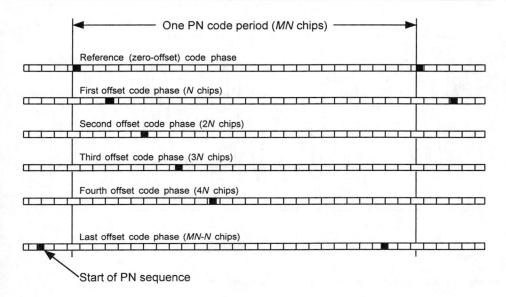

Figure 11.1 CDMA pilot PN code offsets.

$$32{,}768 T_c = \frac{32{,}768}{1{,}228{,}800}\,\sec = \frac{80}{3}\,\text{ms} = 26.66\,\text{ms} \qquad (11.2)$$

To control the out-of-band spectrum of the pilot signal, the quadrature PN sequence waveforms are generated using lowpass FIR filtering that gives a smooth shape to the $\pm$ pulses (negative polarity $\Leftrightarrow$ logical "1" and positive polarity $\Leftrightarrow$ logical "0") and extends the time over which they are nonzero, as explained in Section 1.4.2. A section of of the shaped in-phase PN code baseband waveform is shown in Figure 11.2.

11.1.1 The Role of PN Offsets in System Operation

When a mobile CDMA cellular unit is turned on, it acquires (locks onto) the strongest base station pilot signal and continually tracks that pilot signal in frequency and in PN code phase. Because all the CDMA base stations use the same PN codes in their pilot signals, except with different offsets, the mobile station has only to make a single search through all code phases to find the strongest base station signal. The mobile uses the acquired pilot signal to establish synchronism with system timing, which then enables the mobile to

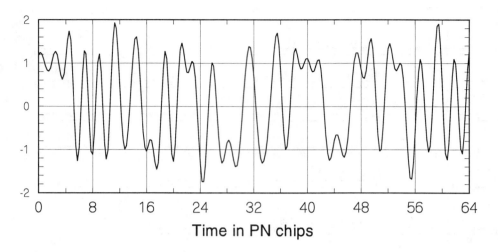

Time in PN chips

Figure 11.2 Shaped I-channel PN code baseband waveform.

obtain system control information from messages transmitted by the base station on the forward link's synchronization and paging channels. Data on the paging channels are scrambled by a long PN code, whose synchronization is known after the mobile reads the (unscrambled) data on the synchronization channel.

The synchronization channel for a particular base station has the same PN code phase offset as the pilot signal and therefore can be demodulated by the mobile after it is locked onto the pilot. The synchronization channel repeatedly broadcasts a message starting at the beginning of one of the periods of the short code PN sequences, and containing identification and timing parameters that include

- Base station and cellular system identification numbers;

- Base station PN code offset as a multiple of $N = 64$ chips (offset index);

- System time and long PN code state at a predetermined time after the completion of the message (320 ms, minus the offset) to allow the receiver time to be set up for this operation.

By sending the mobile the base station's long PN code state that is valid at a known time, the sync message enables the mobile to synchronize its long PN code generator. Subsequently, the mobile can unscramble paging and traffic channel data that have been scrambled using the long PN code.

The mobile learns the base station's identity from the sync message. After acquiring the strongest pilot signal, which is likely to be that of the nearest base station, the mobile is affiliated with that particular station and is under its control. The mobile does not "know" which base station it has locked onto until it reads the sync message, which tells the mobile the identification number of the base station and its short PN code offset index, which we denote by

$$m \in \left\{ 0, 1, ..., M - 1; \ M = \frac{32{,}768}{N} = \frac{32{,}768}{64} = 512 \right\} \tag{11.3}$$

The short PN code offset for the base station, in PN code chips, is $Nm = 64m$. Knowing both the offset of the affiliated base station and the system timing permits the mobile to synchronize its own reference (zero offset) short PN code and long PN code generators with the same timing convention used at all the other base stations and active mobile stations.

The mobile can measure the strength of other base station pilots. The mobile has several digital receivers at its command, all contained in the unit's modem chip. Several of the mobile's receivers are designed to lock onto and track individual base station multipath components and to demodulate them for Rake (diversity) combining. One of the mobile's receivers is specially designed to search for base station pilot signals and their multipaths and to store information about their strengths in a table in its memory, as suggested in Table 11.1.

The pilot offset measurements play a crucial role in handoff operations. As the mobile station changes position, it continually becomes farther away from some base stations and closer to other base stations, and it may move across cell boundaries. To maintain the integrity of a call while the mobile moves across cell boundaries, a cellular system is designed to hand off calls

Table 11.1 Format of data on PN offsets measured by the mobile

Base station ID	Path offset in chips	Relative strength

from one base station to another base station with a stronger signal. The CDMA cellular system has an additional feature called "soft handoff" that permits the mobile to process signals from two different base stations at the same time, using a form of space diversity. Soft handoff is possible because the mobile station has multiple digital receivers for processing calls. In anticipation of possible handoffs or possible space diversity opportunities, the mobile maintains lists of base station signals that its search receiver has found:

- The *active set* is the list of pilot signals that are being used for the current telephone call.

- The *candidate set* is the list of pilot signals not presently assigned to handle calls, but strong enough to be used successfully.

- The *neighbor set* is the list of pilot signals that have been found by the search receiver, but are not strong enough for good demodulation results.

- The *remaining set* is the set of all pilot offsets that are possible, but have yet to be found by the search receiver, based on the fact that the eligible values of pilot offsets are increments of $N = 64$ PN chips.

11.1.2 Pilot Offset Search Parameters

The search for pilot signals is based on distance considerations. The CDMA base station specifies, for each of the four types of pilot signal sets, the size (in PN chips) of a "window" in which to search for pilot signal paths. The window size, expressible as $\pm K_m$ chips with respect to the nominal offset value of $64m$ chips for the pilot with offset index m, can also be thought of as a window on the distance that the mobile is from another base station.

To understand this concept, first consider the fact that physical distance can be expressed in terms of PN chips, as illustrated in Figure 11.3. The correspondence between physical distance d and the path distance in chips D is

$$d = \text{physical distance} = \text{speed of light} \times \text{path delay} = c\tau$$

$$= cT_c \times \frac{\tau}{T_c} = \text{distance for 1 chip} \times \text{distance in chips}$$

$$= cT_c \times D \tag{11.4a}$$

Thus

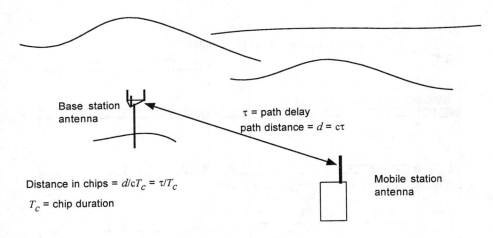

Figure 11.3 Delay in pilot signal arrival due to distance.

$$D = \frac{\tau}{T_c} = \frac{d}{cT_c}$$

$$= \frac{d_{km}}{(3 \times 10^5 \text{km/sec})(1/1228800\,\text{Hz})} = 4.1\,d_{km}\,\text{chips} \qquad (11.4b)$$

That is, the distance in chips is 4.1 times the distance in kilometers, d_{km}. It must be kept in mind that the path between the base station and the mobile is not necessarily a direct, straight-line path.

At the mobile station, the time reference—which is aligned with the beginning of the zero-offset short PN codes—is delayed by propagation from that of the base station. Assuming that the mobile has used the sync message information from the base station with which it is currently affiliated, the relative timing of that base station, the mobile, and another base station is as illustrated in Figure 11.4. There is a chip delay, say, D_1, between the system time that is kept at the affiliated base station and the system time that is kept at the mobile, as a function of the mobile's distance from the base station. Therefore, although each base station transmits its PN code using precisely the same timing, in general, the propagation delays between the different base stations and a particular mobile cause the apparent timing of the direct path pilot signals from each base station to be different. For this reason, and because it is possible for the mobile to receive multipaths, the search for other pilot signals must take into account these possibilities by searching in a "window" about the nominal code phases that are possible for the base stations, as illustrated in Figure 11.5.

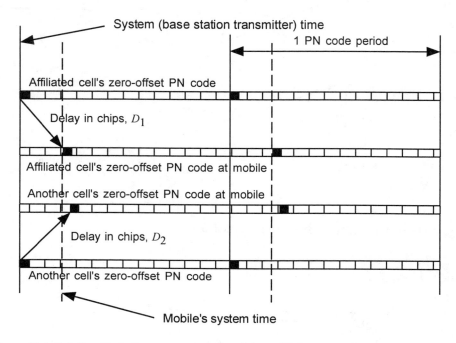

Figure 11.4 Relationship between system time and the mobile's system time.

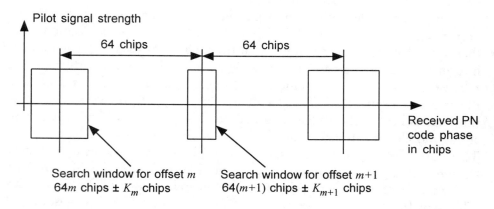

Figure 11.5 Pilot offset search window.

Figure 11.5 depicts the fact that, although the mobile receiver knows that the possible pilot offset chip positions are spaced 64 chips apart, the geometry of the relative locations of the mobile and the existing base stations may cause the actual or "perceived" chip position of a pilot signal to be "off" by one or more chips to the "left" (early) or to the right "late." The figure also indicates that, in principle, the amount of variation in perceived pilot offset is different for each potential offset position: offset position m in Figure 11.5 is shown as having a uncertainty window of $\pm K_m$ chips, while offset position $m + 1$ is shown as having a (smaller) uncertainty window of $\pm K_{m+1}$ chips, where K_m and K_{m+1} may be different. However, even if the relative locations of the mobiles and the various base stations were known and available for the determination of pilot offset search windows, there are other sources of uncertainty that contribute to the consideration of search window size, as we discuss after considering an example problem.

Example 11.1 Let a mobile receiver be located at the relative distances of 0.75 km, 1.5 km, and 3 km from base stations A, B, and C, respectively, as depicted in Figure 11.6. What are the relative pilot offset delays (in chips) and the "perceived" delays at the mobile receiver for these base stations?

    Solution: In view of (11.4b), the delays in chips are $D_A = 0.75$ km $\times 4.1$ chips/km $= 3.1$ chips for base station A, $D_B = 1.5 \times 4.1 = 6.2$ chips for base station B, and $D_C = 3 \times 4.1 = 12.3$ chips for base station C. Presumably, the mobile is affiliated with base station A, so that the mobile's time reference is aligned with that of the signal arriving from A; in that case, the pilot offset from B appears to be $D_B - D_A = 6.2 - 3.1 = 3.1$ chips late and that from C appears to be $D_C - D_A = 12.3 - 3.1 = 9.2$ chips late.

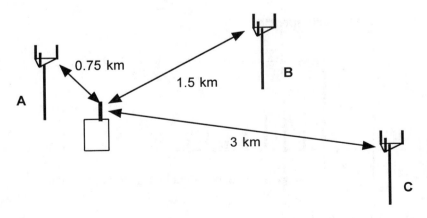

Figure 11.6 Example relative distances of a mobile and several base stations.

11.1.2.1 Effect of Multipath on Search Window

It is possible for a mobile receiver to establish its system timing reference by synchronizing to a reflected path signal from the affiliated base station, rather than to a direct path. This can happen because the strength of a reflected path can, for a short period of time, be greater than that of the direct path (or shortest path, if there is no direct path), even though on the average the strengths of reflected paths may decrease with delay as sketched in Figure 11.7. Therefore, in searching for multipath signals from the affiliated base station, the mobile's search window should allow for an uncertainty on the order of the delay spread, T_m. That is, a search window size of $\pm K$ chips that is based on multipath considerations would use an assumed value of delay spread to obtain

$$K = T_m/T_c \qquad (11.5a)$$

This same consideration would apply to the sizes of the search windows at the other pilot offset chip positions, because the mobile does not know the multipath delay status of the pilot energy detected during a given search interval.

Example 11.2 What would be the size in chips of a pilot offset search window based on (11.5a) when the delay spread is assumed to be 5 μs?

Solution: K in (11.5a) is simply the number of chips in 5 μs, which is calculated to be

$$K = 5\,\mu s/0.814\mu s = 6.1\,\text{chips} \qquad (11.5b)$$

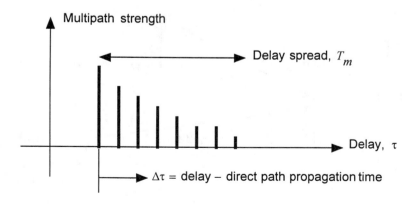

Figure 11.7 Multipath delay spread.

11.1.2.2 Bounds on Relative Delays

Assume that a mobile receiver has affiliated (synchronized its timing) using a pilot signal received from the nearest base station, which we designate as base station A. As we have discussed previously, the mobile's timing is delayed from that at A by D_A chips. The offset of a more distant base station (B) appears to be late by the chip delay of the signal from B relative to D_A; that is, by $D_B - D_A$. It follows that the window size for the purpose of measuring the strengths of nonaffiliated base stations should allow for the possible values of $D_B - D_A$.

If we assume that the mobile is synchronized to the nearest base station, then A is closer than B and, therefore, $D_B - D_A > 0$. If we cannot assume that the mobile has synchronized to the nearest station, then the window for B paths must allow for the more general possibilities, and the search window would be centered on the nominal offset position of B, which we may denote by ϕ_B, and vary ϕ according to the bounds given by

$$\min\{D_B - D_A\} \leq \phi - \phi_B \leq \max\{D_B - D_A\} \qquad (11.6a)$$

That is, we use ϕ to denote the values of the offset in chips to be searched. The maximum and minimum values of $D_B - D_A$ can be estimated, based on the spacing of the base stations in the cellular system and the sensitivity of the mobile receiver.

Assuming that the forward link power is adjusted throughout the cellular system to minimize interference to adjacent cells, the cell radius is approximately the maximal distance to the mobile from which any base station's signal appears to be strong enough for demodulation. In practice, this distance is made slightly larger than the physical cell radius in order to provide for soft handoff (base station diversity) at mobile positions that are near the border between two cells. Thus bounds can be put on $D_B - D_A$ based on searching for signals in cells that are adjacent to the mobile's "home" cell. As indicated in Figure 11.8, the minimal relative chip distance is zero and occurs when the mobile is equidistant from A and B; the maximal chip distance occurs when the mobile is as far away from B as possible while still being in A's cell. The maximal relative distance is $2R = \sqrt{3}\,R_c$. In chips, the maximal relative delay of an adjacent cell therefore is

$$\max\{D_B - D_A\} = \sqrt{3} \times 4.1\,R_{km} = 7.1\,R_{km} \qquad (11.6b)$$

where we use R_{km} to express the cell radius R_c in kilometers.

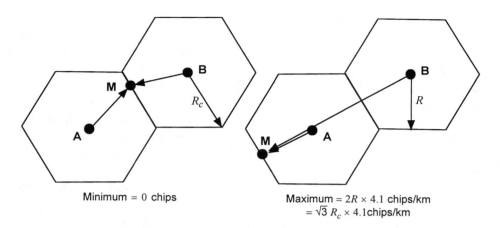

Figure 11.8 Minimal and maximal relative chip distances.

11.1.2.3 IS-95 Search Window Parameters

We saw that the setting of a search window size $\pm K_m$ for pilot signal offset index m must take into account the multipath delay spread and the relative positions of base station m and the base station upon which the mobile's system timing is based. If the base station software keeps track of the position of the mobile, or in some other fashion estimates the distances between the mobile and all the base stations in the system, then the base station that is controlling the mobile (base station A in the example) can instruct the mobile to use a different value of the search window size $\pm K_m$ for each base station. If this detailed relative delay information is not developed by the base station, the search window sizes must be somewhat larger than necessary to account for the uncertainties in the mobile's position and in the values of path delays.

    The IS-95 CDMA system does, in fact, provide for different search window sizes to be sent to the mobile in the *Systems Parameter Message* on the paging channel—not for individual base stations, but for the categories of detected pilot signals that were mentioned previously. The search window parameters and their possible values are listed in Table 11.2 [1]. The mapping of these 4-bit parameters into a window size is shown graphically in Figure 11.9. Note that the mapping of the search window parameter permits the specification of search window sizes from ± 1 PN chip to ± 226 PN chips.

Table 11.2 IS-95 search window parameters

Pilot set	Search window parameter	Range of values	Window size $(2K_m)$
Active set	SRCH_WIN_A	0–15 (4 bits)	2–452
Candidate set			
Neighbor set	SRCH_WIN_N		
Remaining set	SRCH_WIN_R		

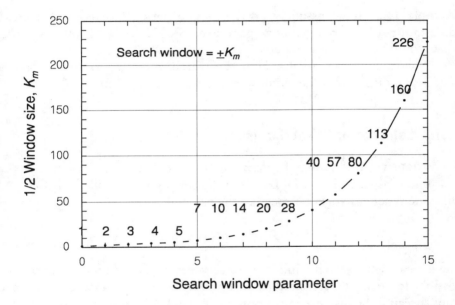

Figure 11.9 Mapping of IS-95 search window parameter values.

The range of the delay uncertainty corresponding to these window sizes is

$$\pm T_c = \pm 0.814\,\mu s \quad \text{to} \quad \pm 226T_c = \pm 184\,\mu s$$

The range of the differential distance (e.g., $d_B - d_A$) corresponding to these window sizes is

$$\pm 1cT_c = \pm 244\text{m} \quad \text{to} \quad \pm 226cT_c = \pm 55.1\,\text{km}$$

Given that the pilot PN code offset spacing is $N = 64$ chips, the size of the search window that is selected by the base station for directing the pilot search at the mobile usually is something much less than ± 64 chips. However, telling the mobile to search using a certain set of possible offsets using a search window size greater than ± 64 chips is a way for the base station to direct the mobile to "scan" the possible PN code offsets for measurement purposes. The selection of search window sizes obviously is affected by the PN code offset spacing N. The factors to be considered in choosing N are discussed next.

Example 11.3 If we assume that the mobile station has received from the base station the search window parameter SRCH_WIN_A value of 1001, what would be the search window size?

Solution: From Figure 11.9, the search window parameter value of $1001 = 9$ indicates that the window size is $K_m = \pm 28$ chips.

11.1.3 Selection of Offset Spacing

For various reasons, it may be desirable to use an offset spacing greater than $N = 64$ chips; the IS-95 CDMA system has a parameter called PILOT_INC (pilot increment factor) that specifies the value of N as

$$N = 64 \times \text{PILOT\_INC} = \text{offset index increment}$$

The issue of what search window sizes to select is affected by the fact that it is possible for distant base stations to have differential delay greater than 64 chips, because the distance corresponding to a 64-chip delay is only $64/4.1 = 15.6$ km. Thus, if the mobile is affiliated with a base station having offset index m, the pilot signal from another base station having offset index $m - 1$ appears in the search window for offset m, if the other base station is about 15 or 16 km farther away.

This ambiguous situation is illustrated by Figure 11.10, in which a mobile is assumed to be at a distance $d_A = 0.75$ km from its affiliated base station A with pilot PN code offset m, and at a distance $d_B = 16.2$ km from another base station (B) with pilot PN code offset $m - 1$. The differential delay in this example is $(16.2 \text{ km} - 0.75 \text{ km}) \times 4.1 \text{ chips/km} = 63.3$ chips. If the offset spacing is $N = 64$ chips, the pilot signal from base station B appears in the search window of base station A's pilot as if it were a multipath

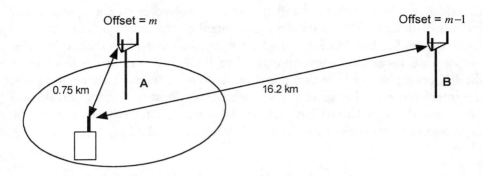

Offset = m Offset = $m-1$

0.75 km   A   16.2 km   B

Figure 11.10 Situation of PN offset ambiguity.

component of base station A. This situation obviously has the potential for false identification of B signals as belonging to A, and there is also the possibility that a pilot signal from B can "jam" a pilot signal from A that has the same apparent PN code offset. The ambiguous situation illustrated in Figure 11.10 can be avoided if the offsets assigned to the different base stations are selected carefully, or the offset index increment (spacing) is increased ($N > 64$), or both.

By increasing the value of PILOT_INC, the system operator can make the value of the offset spacing, N, be any multiple of 64 chips. It has to be remembered, though, that the number of offsets is $M = 2^{15}/N$; the number of different PN code offsets is inversely proportional to N. Thus, the selection of N for the system may be constrained by the desired number of offsets that are available for assignment to base stations or base station sectors.

In addition to ambiguity, as we have noted, the situation that was depicted in Figure 11.10 involves potential interference. An analysis of the required value of N can be based on consideration of the interference [4]. If it is assumed that offset index ambiguity is unavoidable because of the need to have many different PN code offsets available for the base stations and sectors, then a reasonable approach is to find a way to bound the interference caused by the distant base station or sector because of the ambiguity.

To begin the analysis, let the following notation be used:

D_i = distance in chips between the mobile and base station i

$D_{ij} = D_i - D_j$ = differential distance in chips

m_i = offset index of base station i

R_{km} = cell radius in kilometers

Consider two base stations (i and j) that are involved in an ambiguous offset index situation, with base station i in control of the mobile. The PN code phase offsets for base stations i and j are $m_i N$ and $m_j N$, respectively. For propagation losses that are proportional to the μth power of the distance, if the PN code phases of the two pilots, as received at the mobile, are different, the interference to the signal from base station i (power S_i after correlation) caused by the signal from base station j (power I_j after correlation) can be expressed in terms of the ratio (not considering thermal noise)

$$\frac{S_i}{I_j}(\text{dB}) = 10 \log_{10}\left[\left(\frac{D_j}{D_i}\right)^{\mu}\right] + \text{processing gain in dB}$$

$$= 10\mu \log_{10}\left(\frac{D_j}{D_i}\right) + \text{PG} \tag{11.7a}$$

where we assumed equal transmitter power from each base station, and μ is the propagation power law. The amount of processing gain depends on how many chips of pilot signal energy are accumulated in the pilot tracking part of the receiver. For example, using the distances in the Figure 11.10 and $\mu = 3$ gives

$$D_i = 0.75 \,\text{km} \times 4.1 \,\text{chips/km} = 3.1 \,\text{chips}$$

$$D_j = 16.2 \,\text{km} \times 4.1 \,\text{chips/km} = 66.4 \,\text{chips}$$

$$\frac{S_i}{I_j}(\text{dB}) = 10\mu \log_{10}\left(\frac{D_j}{D_i}\right) + \text{PG}$$

$$= 30 \log_{10}\left(\frac{66.4 \,\text{chips}}{3.1 \,\text{chips}}\right) + \text{PG}$$

$$= 39.9 \,\text{dB} + \text{PG}$$

For this example, the amount of interference presented by the ambiguous situation is tolerable, since PG can be 15 dB (32 chips) or more (see Section 7.4). If the PN code phases of the two pilots are the same, as received at the mobile, for propagation losses that are proportional to the μth power of the distance, the interference to the signal from base station i caused by the signal from base station j can be expressed in terms of the ratio

$$\frac{S_i}{I_j}(\text{dB}) = 10 \log_{10}\left[\left(\frac{D_j}{D_i}\right)^{\mu}\right] = 10\mu \log_{10}\left(\frac{D_j}{D_i}\right) \tag{11.7b}$$

assuming equal transmitter powers. That is, the interference ratio in this situation is a function of the ratio of the distances and is not improved by the spread-spectrum processing gain. In the numerical example just considered, for the same apparent code phases, the ratio is

$$\frac{S_i}{I_j}(\text{dB}) = 10\mu \log_{10}\left(\frac{D_j}{D_i}\right) = 30\log_{10}\left(\frac{66.4 \text{ chips}}{3.1 \text{ chips}}\right)$$
$$= 39.9 \text{ dB}$$

which is still an acceptable value.

Let base station i be in control of the mobile; then base station j interferes because of offset ambiguity with no processing gain improvement, as expressed by (11.7b), when the differential distance is such that the apparent code phase received by the mobile from the two base stations is the same (within one chip). A mathematical statement of this condition of interference for the case of the nearest possible interfering base station giving this kind of interference is

$$D_j - D_i = D_{ji} = \text{code separation between } j \text{ and } i = 1 \times N = N$$

where the value of N is to be found such that the interference is acceptably small. A mathematical statement of the condition that the interference is acceptable is

$$\frac{S_i}{I_j}(\text{dB}) = 10\mu \log_{10}\left(\frac{D_j}{D_i}\right) > \text{threshold} = \rho_0 \qquad (11.8\text{a})$$

where ρ_0 denotes a baseband SNR threshold. The equation for the condition $S_i/I_j > \rho_0$ can be rearranged by the following development:

$$10\mu \log_{10}\left(\frac{D_j}{D_i}\right) > \rho_0 \quad \Rightarrow \quad \log_{10}\left(\frac{D_j}{D_i}\right) > \frac{\rho_0}{10\mu}$$

$$\Rightarrow \quad \frac{D_j}{D_i} > 10^{\rho_0/10\mu} \qquad (11.8\text{b})$$

Substituting for $D_j = D_i + N$ in the preceding expression gives

$$\frac{D_i + N}{D_i} > 10^{\rho_0/10\mu} \quad \Rightarrow \quad \frac{N}{D_i} > 10^{\rho_0/10\mu} - 1 \qquad (11.8\text{c})$$

Thus, we have the following lower bound for the offset spacing N:

$$N > D_i \left[10^{\rho_0/10\mu} - 1\right] \tag{11.8d}$$

In this manner a consideration of the degree of interference posed by another base station with an adjacent offset led to a lower bound on the offset increment N. For example, if the acceptable correlated interference threshold is $\rho_0 = 24\,\text{dB}$ [4] and the propagation power law is $\mu = 3$ (a conservative value in terms of interference), then the lower bound is

$$N > D_i \left[10^{24/30} - 1\right] = D_i[10^{0.8} - 1] = D_i[6.31 - 1]$$

or

$$N > 5.31 D_i \tag{11.9a}$$

This example shows that, given the values of ρ_0 and μ, the lower bound on the offset increment is a function of the distance (in PN code chips) of the mobile from its affiliated base station. Because the distance in chips is $D_i = 4.1d_i$, where d_i is the distance in kilometers, the lower bound can be written

$$N > 5.31 \times 4.1d_i = 21.77d_i \tag{11.9b}$$

Because by definition, the distance of a mobile from its affiliated base station is limited by $d_i \leq R_{km}$, where R_{km} is the cell radius in kilometers, then to cover all situations from (11.9b), we must require that

$$N > 21.77 R_{km} \tag{11.9c}$$

if the correlated base station interference ratio threshold is $\rho_0 = 24\,\text{dB}$ and $\mu = 3$, assuming that it is possible for adjacent cells to have adjacent offsets.

In the IS-95 CDMA cellular system, the value of the offset increment N is restricted to multiples of 64 chips. The lower bound on N then must be rounded up to the nearest multiple of 64. For example, the rounded lower bound equals 64 chips for $21.77 R_{km} \leq 64$, or the cell radius less than or equal to $64/21.77 = 2.94$ kilometers $\approx 3\,\text{km}$. This fact indicates that, according to the lower bound analysis, the minimal offset separation between base stations should increase by 64 chips for each 3 km of cell radius when $\rho_0 = 24\,\text{dB}$ and $\mu = 3$. The values of the offset increment, N, and the corresponding cell radius, R_{km}, are shown in Table 11.3, along with the number of offsets, $M = \lfloor 32768/N \rfloor$.

Table 11.3 Offset spacing and number of offsets for different cell sizes

N	R_{km}	M
64	$\leq 3\,\text{km}$	512
128	$3\text{--}6\,\text{km}$	256
192	$6\text{--}9\,\text{km}$	170
256	$9\text{--}12\,\text{km}$	128
320	$12\text{--}15\,\text{km}$	102
384	$15\text{--}18\,\text{km}$	85
448	$18\text{--}21\,\text{km}$	73
512	$21\text{--}24\,\text{km}$	64

This table shows that the larger the cells, the smaller the number of offset indexes that are available for assignment to base stations, according to the lower bound analysis.

The lower bound analysis is useful for gaining a preliminary estimate of the offset increment N and the number of offset indexes. In a given practical CDMA engineering situation, the use of directional sector antennas and careful selection of PN code offsets for the base station sites can control the PN code interference and thereby make it possible to use a smaller value of N to increase M, the number of offsets, or to reuse offset indexes without changing N or M as calculated by the lower bound, in much the same way as frequency channels are reused in FDMA cellular systems.

11.2 Optimal Allocation of CDMA Forward Link Power

The forward link in the IS-95 CDMA cellular system features four different kinds of channel:

- A continuously transmitted CDMA *pilot channel* that provides a PN code and signal strength reference for mobile terminals seeking a base station with which to affiliate;

- A continuously transmitted *sync channel* that provides base station identification and a system timing reference to the mobile terminals;

- Up to 7 *paging channels* that inform mobiles of incoming calls and other call-related information and instructions;

- As many as 55 *traffic channels* over which digital voice and other data are transmitted during calls.

These channels are simultaneously transmitted by the base station using Walsh function orthogonal multiplexing. Because the forward link channels are simultaneously transmitted on the same PN code carrier, they share the same link power budget gain and loss parameters. However, the different channel categories have different baseband data rates and different SNR or E_b/N_0 requirements, and the base station transmits each type of channel at a different power level to meet those requirements.

11.2.1 Forward Link Channel SNR Requirements

11.2.1.1 Pilot Channel

The *pilot channel* is unmodulated, so that its effective data rate is the PN code chip rate, $R_b = R_c = 1.2288\,\text{MHz}$, and its SNR requirement is expressed in terms of the received pilot channel $E_c/N_{0,T}$, where E_c is the pilot channel chip energy and $N_{0,T} = N_0 + \mathcal{I}_{0,T}$ is the effective noise-plus-interference spectral power density. Assuming the noise bandwidth to be equal to the chip rate; i.e., $W = R_c$, the received pilot channel's $E_c/N_{0,T}$ is numerically equal to its SNR:

$$\left(\frac{E_c}{N_{0,T}}\right)_{\text{pilot}} = \frac{S_{\text{pil}}\,T_c}{N_0 + \mathcal{I}_{0,T}} = \frac{S_{\text{pil}}}{(N_0 + \mathcal{I}_{0,T})W} = \frac{S_{\text{pil}}}{N_m + I_T} \qquad (11.10a)$$

where S_{pil} is the received pilot channel power, $T_c = 1/W$, $N_m = N_0 W$ is the mobile receiver's thermal noise power, and $I_T = \mathcal{I}_{0,T}W$ is the effective noise power of the received forward link interference. Note that the processing gain for the pilot channel is given by

$$(\text{PG})_{\text{pil}} = \left(\frac{W}{R_b}\right)_{\text{pil}} = \frac{W}{R_c} = 1 \qquad (11.10b)$$

A typical threshold or required value of $E_c/N_{0,T}$ for the pilot channel is -15 dB (see Figure 10.21). This low value is feasible because the pilot channel is unmodulated and energy can accumulate for a relatively long observation time.

11.2.1.2 Sync Channel

The *sync channel* has a baseband data rate of $R_b = 1,200\,\text{bps}$, and its SNR requirement is expressed as a threshold for the received sync channel $E_b/\mathcal{N}_{0,T}$. The received sync channel $E_b/\mathcal{N}_{0,T}$ can be related to its SNR as follows:

$$
\left(\frac{E_b}{\mathcal{N}_{0,T}}\right)_{\text{sync}} = \frac{S_{\text{sync}}\,T_b}{\mathcal{N}_0 + \mathcal{I}_{0,T}} = \frac{S_{\text{sync}}}{(\mathcal{N}_0 + \mathcal{I}_{0,T})R_b} \cdot \frac{W}{W} = \frac{S_{\text{sync}}}{N_m + I_T} \cdot \frac{W}{R_b}
$$

$$
= \frac{S_{\text{sync}}}{N_m + I_T} \cdot (\text{PG})_{\text{sync}} = \frac{1,024\,S_{\text{sync}}}{N_m + I_T} \tag{11.11a}
$$

where S_{sync} is the received sync channel power and $T_b = 1/R_b$. Note that the sync channel has a processing gain of

$$
(\text{PG})_{\text{sync}} = \left(\frac{W}{R_b}\right)_{\text{sync}} = \frac{1.2288\,\text{MHz}}{1,200\,\text{bps}} = 1,024 \tag{11.11b}
$$

A typical threshold or required value of $E_b/\mathcal{N}_{0,T}$ for the sync channel is 6 dB. Note that the sync channel is transmitted with error-control coding (see Figure 4.13).

11.2.1.3 Paging Channels

Each *paging channel* has a baseband data rate of $R_b = 4,800\,\text{bps}$ or $R_b = 9,600\,\text{bps}$, and its SNR requirement is expressed as a threshold for the received paging channel $E_b/\mathcal{N}_{0,T}$. For $R_b = 4,800\,\text{bps}$, the received sync channel $E_b/\mathcal{N}_{0,T}$ can be related to its SNR as follows:

$$
\left(\frac{E_b}{\mathcal{N}_{0,T}}\right)_{\text{paging}} = \frac{S_{\text{pag}}\,T_b}{\mathcal{N}_0 + \mathcal{I}_{0,T}} = \frac{S_{\text{pag}}}{(\mathcal{N}_0 + \mathcal{I}_{0,T})R_b} \cdot \frac{W}{W} = \frac{S_{\text{pag}}}{N_m + I_T} \cdot \frac{W}{R_b}
$$

$$
= \frac{S_{\text{pag}}}{N_m + I_T} \cdot (\text{PG})_{\text{pag}} = \frac{256\,S_{\text{pag}}}{N_m + I_T} \tag{11.12a}
$$

where S_{pag} is the received paging channel power. Note that the paging channels have a processing gain of

$$(PG)_{pag} = \left(\frac{W}{R_b}\right)_{pag} = \frac{1.2288\,\text{MHz}}{4,800\,\text{bps}} = 256 \qquad (11.12b)$$

A typical threshold or required value of $E_b/\mathcal{N}_{0,T}$ for the paging channel is 6 dB. Note that the paging channel is transmitted with error-control coding (see Figure 4.15).

11.2.1.4 Traffic Channels

Each *traffic channel* has a variable baseband data rate of up to $R_b = 9,600\,\text{bps}$, and its SNR requirement is expressed as a threshold for the received traffic channel $E_b/\mathcal{N}_{0,T}$. The received traffic channel $E_b/\mathcal{N}_{0,T}$ can be related to its SNR as follows:

$$\left(\frac{E_b}{\mathcal{N}_{0,T}}\right)_{traffic} = \frac{S_{traf}\,T_b}{\mathcal{N}_0 + \mathcal{I}_T} = \frac{S_{traf}}{(\mathcal{N}_0 + \mathcal{I}_T)R_b} \cdot \frac{W}{W} = \frac{S_{traf}}{N_m + I_T} \cdot \frac{W}{R_b}$$

$$= \frac{S_{traf}}{N_m + I_T} \cdot (PG)_{traf} = \frac{128\,S_{traf}}{N_m + I_T} \qquad (11.13a)$$

where S_{traf} is the received traffic channel power. Assuming $R_b = 9,600$, the traffic channel has a processing gain of

$$(PG)_{traf} = \left(\frac{W}{R_b}\right)_{traf} = \frac{1.2288\,\text{MHz}}{9,600\,\text{bps}} = 128 \qquad (11.13b)$$

A typical threshold or required value of $E_b/\mathcal{N}_{0,T}$ for the traffic channel is 7 dB. Note that the traffic channel is transmitted with error-control coding (see Figure 4.17).

11.2.1.5 Interference and Noise Terms

The total interference power on the forward link I_T is a component of the SNRs of all the channels and may be broken down into same-cell interference power I_{sc} and other-cell interference power I_{oc}.

The forward link channels are transmitted orthogonally, so that ideally, there is no same-cell multiple access interference on the forward link. However, due to multipath delays, there is same-cell forward link interference. Let the total *received* forward link power in the absence of interference be denoted S as defined in (10.7d); then, as discussed in Section 10.1.1, the same-cell interference power on the forward link can be written as a factor times S:

$$I_{sc} = K_{same} S \qquad (11.14a)$$

In the absence of specific multipath information, it is common to use $K_{same} \approx 1$ as an estimate (see Example 10.2).

The forward link power received from other base stations at a particular mobile's location acts as interference. This other-cell interference power can also be written as a factor times S, as implied by the graphs in Figure 10.5:

$$I_{oc} = K_{other} S \qquad (11.14b)$$

The strength of the other-cell interference depends on the mobile's location within the cell. A reasonable estimate of other-cell interference power is obtained by using the value $K_{other} \approx 2.5\,dB = 1.778$, as shown in (10.8b).

The total forward link interference power is denoted $I_T = I_{sc} + I_{oc}$ as in (10.12a), and we have

$$I_T = I_{sc} + I_{oc} = (K_{same} + K_{other})S = K_f S \qquad (11.14c)$$

where

$$K_f \triangleq K_{same} + K_{other} \qquad (11.14d)$$

At the mobile receiver, the effective thermal noise power is calculated as $N_m = (N_0 W)_m$. For a typical mobile noise figure of 8 dB, the receiver noise power is (see Example 10.9)

$$N_m = (N_0 W)_m = -105.0\,dBm \quad \text{for} \quad (NF)_m = 8\,dB \qquad (11.14e)$$

11.2.2 Total Forward Link Power

Having related the forward link interference power to the received forward link signal power, it is now appropriate to develop the details of this received signal power. In doing so, we find it convenient to define a quantity that we call the *forward link traffic channel power control factor*, denoted K_{traf}, and we characterize the net link losses that this power will experience.

At the base station transmitter, the total power delivered by the power amplifier can be written as

$$P_{total} = P_{pil} + P_{sync} + N_p P_{pag} + K_{traf} M \, \alpha_f P_{traf} \qquad (11.15)$$

where

$$P_{pil} = \text{transmitter pilot channel power}$$

$$P_{sync} = \text{transmitter sync channel power}$$

$$P_{pag} = \text{transmitter paging channel power}$$

$$P_{traf} = \text{transmitter traffic channel power}$$

$$N_p = \text{number of active paging channels}$$

$$M = \text{number of active traffic channels}$$

$$\alpha_f = \text{forward link voice activity factor}$$

$$K_{traf} = \text{forward link power control factor}$$

For convenience, the traffic channel transmitter power P_{traf} used in this formulation is taken to be the power transmitted for a mobile user at the cell edge. Note that in (11.15) the total traffic channel transmitted power is given by the term $K_{traf} M \, \alpha_f P_{traf}$. Although there are assumed to be M active traffic channels, the total power in these channels is not $M P_{traf}$ because the (average) power for any channel is reduced by the average voice activity factor, $\alpha_f \leq 1$. In the same manner, the factor $K_{traf} < 1$ in (11.15) is a parameter that takes into account the fact that most of the mobile users are not located at the cell edge, and assumes the use of some degree of forward link power control by the base station.

11.2.2.1 Forward Link Power Control Factor

The base station has a rough idea of how much power is needed to support a traffic channel, because the base station is engaged in power control on the reverse link. If it is assumed that the forward link power for a particular traffic channel is adjusted to deliver the required median value of power to the mobile's location (and no more, to reduce interference), then it follows that the traffic channel power $P_{traf}(r)$ required for a mobile at a distance r from the base station is inversely proportional to propagation loss within the cell, and may be related to the power required for a mobile at the cell edge (distance R) by

$$P_{\text{traf}}(r) = P_{\text{traf}}(R) \cdot \left(\frac{r}{R}\right)^{\gamma}, \quad \gamma = \text{power law}, \, R = \text{cell radius} \quad (11.16a)$$

We assume that mobiles are equally likely to be at any distance r from the base station between the values $r = 0$ and $r = R$ (at the edge). Then a pdf for the distance of mobiles from the base station can be written as (see Section 1.5)

$$p_r(\alpha) = \begin{cases} 1/R, & 0 \leq \alpha \leq R \\ 0, & \text{otherwise} \end{cases} \quad (11.16b)$$

The average required forward link traffic channel power is

$$\overline{P}_{\text{traf}} = \int_0^R P_{\text{traf}}(\alpha) \, p_r(\alpha) \, d\alpha = \int_0^R P_{\text{traf}}(R) \left(\frac{\alpha}{R}\right)^{\gamma} \cdot \frac{1}{R} \, d\alpha$$

$$= P_{\text{traf}}(R) \int_0^1 x^{\gamma} \, dx = P_{\text{traf}}(R) \cdot \frac{1}{\gamma + 1}$$

That is:

$$\overline{P}_{\text{traf}} = K_{\text{traf}} \cdot P_{\text{traf}}(R) \quad (11.16c)$$

where

$$K_{\text{traf}} \triangleq \frac{1}{\gamma + 1}, \quad \text{mobiles uniformly distributed over distance} \quad (11.16d)$$

According to this analysis, the average forward link traffic channel power is about $1/(\gamma + 1)$ times the traffic channel power needed to serve a mobile at the edge of the cell. Because the power law is commonly between $\gamma = 3$ and $\gamma = 4$, a value of K_{traf} on the order of 0.2 or 0.3 would be reasonable for the assumptions, and $K_{\text{traf}} = 0.5$ would be a conservative value.

If, instead of being uniformly distributed in distance from the base station, the mobiles are uniformly distributed in area within the sector, then a pdf for the locations (r, θ) of mobiles within the sector is

$$f_{r,\theta}(\alpha, \beta) = \begin{cases} \dfrac{2}{R^2 \, \theta_s}, & 0 \leq \alpha \leq R, \; 0 \leq \beta \leq \theta_s \\ 0, & \text{otherwise} \end{cases} \quad (11.16e)$$

where θ_s is the sector angle. This density function is illustrated in Figure 11.11. The average forward link traffic channel power is

$$\overline{P}_{traf} = \int_0^{\theta_s} \int_0^R \alpha\, P_{traf}(\alpha)\, f_{r,\theta}(\alpha,\beta)\, d\alpha\, d\beta = \int_0^R \alpha\, P_{traf}(R) \left(\frac{\alpha}{R}\right)^\gamma \cdot \frac{2}{R^2}\, d\alpha$$

$$= P_{traf}(R) \int_0^1 2x^{\gamma+1}\, dx = P_{traf}(R) \cdot \frac{2}{\gamma+2}$$

That is:

$$\overline{P}_{traf} = K_{traf} \cdot P_{traf}(R) \tag{11.16f}$$

where

$$K_{traf} \triangleq \frac{2}{\gamma+2} = \frac{1}{\frac{1}{2}\gamma+1}, \quad \text{mobiles uniformly distributed over area} \tag{11.16g}$$

According to this analysis, the average forward link traffic channel power is about $1/(0.5\gamma + 1)$ times the traffic channel power needed to serve a mobile at the edge of the cell. For a power law between $\gamma = 3$ and $\gamma = 4$, a value of K_{traf} on the order of 0.3 or 0.4 would be reasonable for the assumptions, and $K_{traf} = 0.5$ would still be a conservative value.

11.2.2.2 Net Losses on the Forward Link

In Section 10.1.3.2, we defined the forward link transmission loss L_T as the net loss on the link. Generalizing that definition to include not only antenna

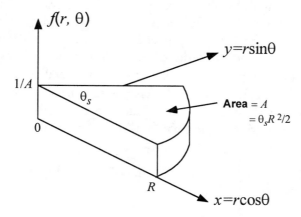

Figure 11.11 Uniform distribution of mobiles in a sector area.

gains but the other loss factors that we considered in Table 10.7 for the reverse link, assumed to apply also to the forward link, we can write

$$L_T(d) \triangleq \frac{\text{propagation loss} \times \text{other losses}}{\text{gains}} = \frac{L(d) \cdot L_{rm} \cdot L_{tc}}{G_m \cdot G_c} \qquad (11.17a)$$

where

$L(d)$ = median link propagation loss at distance d

L_{rm} = mobile receiver losses (cable, etc.)

L_{tc} = base station transmitter losses (cable, etc.)

G_m = mobile antenna gain

G_c = base station antenna gain

Given the total power delivered by the base station transmitter's power amplifier, the received forward link power at the location of a mobile on the cell edge, in the absence of interference, is

$$P_r = P_{\text{total}}/L_T(R) \qquad (11.17b)$$

Because the losses and gains are common to all the forward link channels, the received channel powers to be used in calculating the channel SNRs are

$$S_{\text{pil}} = P_{\text{pil}}/L_T(R)$$
$$S_{\text{sync}} = P_{\text{sync}}/L_T(R)$$
$$S_{\text{pag}} = P_{\text{pag}}/L_T(R) \qquad (11.17c)$$
$$S_{\text{traf}} = P_{\text{traf}}/L_T(R)$$

11.2.3 Solution for Forward Link Powers

With the preceding background and definitions, we are now in a position to solve for the forward link powers P_{pil}, P_{sync}, P_{pag}, and P_{traf} that will satisfy the constraints placed on the total transmitter power and the link budget for each forward link channel, as expressed in (11.10a), (11.11a), (11.12a), and (11.13a). Our formulation for this problem starts with the zero-margin case, then is extended to the general case of nonzero margin. The zero-margin constraints are the following [5]:

$$P_{total} = P_{pil} + P_{sync} + N_p P_{pag} + K_{traf} M \alpha_f P_{traf} \leq P_{max} \qquad (11.18a)$$

$$\left(\frac{E_c}{\mathcal{N}_{0,T}}\right)_{pilot} = \frac{S_{pil}}{N_m + I_T} = \frac{P_{pil}/L_T(R)}{N_m + K_f P_{total}/L_T(R)}$$

$$= \frac{P_{pil}}{N_m L_T(R) + K_f P_{total}} \geq \rho_{pil} \qquad (11.18b)$$

$$\left(\frac{E_b}{\mathcal{N}_{0,T}}\right)_{sync} = \frac{1{,}024\, S_{sync}}{N_m + I_T} = \frac{1{,}024\, P_{sync}/L_T(R)}{N_m + K_f P_{total}/L_T(R)}$$

$$= \frac{1{,}024\, P_{sync}}{N_m L_T(R) + K_f P_{total}} \geq \rho_{sync} \qquad (11.18c)$$

$$\left(\frac{E_b}{\mathcal{N}_{0,T}}\right)_{pag} = \frac{256\, S_{pag}}{N_m + I_T} = \frac{256\, P_{pag}/L_T(R)}{N_m + K_f P_{total}/L_T(R)}$$

$$= \frac{256\, P_{pag}}{N_m L_T + K_f P_{total}} \geq \rho_{pag} \qquad (11.18d)$$

$$\left(\frac{E_b}{\mathcal{N}_{0,T}}\right)_{traf} = \frac{128\, S_{traf}}{N_m + I_T} = \frac{128\, P_{traf}/L_T(R)}{N_m + K_f P_{total}/L_T(R)}$$

$$= \frac{128\, P_{traf}}{N_m L_T(R) + K_f P_{total}} \geq \rho_{traf} \qquad (11.18e)$$

where ρ_{pil}, ρ_{sync}, ρ_{pag}, and ρ_{traf} are the required values of, respectively, $(E_c/\mathcal{N}_{0,T})_{pil}$, $(E_b/\mathcal{N}_{0,T})_{sync}$, $(E_b/\mathcal{N}_{0,T})_{pag}$, and $(E_b/\mathcal{N}_{0,T})_{traf}$. Nominal values for the IS-95 system are

$$\rho_{pil} = -15\,dB, \qquad \rho_{sync} = \rho_{pag} = 6\,dB, \quad and \quad \rho_{traf} = 7\,dB \qquad (11.18f)$$

These equations can be put in the form of a system of simultaneous equations to be solved for the channel powers under operational constraints. For example, the condition for satisfying the requirements of the pilot channel can be manipulated as follows: from (11.18b) with (11.15) substituted, we have

$$\frac{P_{pil}}{N_m L_T(R) + K_f\big(P_{pil} + P_{sync} + N_p P_{pag} + K_{traf} M \alpha_f P_{traf}\big)} \geq \rho_{pil}$$

$$\Rightarrow$$

$$P_{pil} \geq \rho_{pil}\big[N_m L_T(R) + K_f\big(P_{pil} + P_{sync} + N_p P_{pag} + K_{traf} M \alpha_f P_{traf}\big)\big]$$

which results in the linear equation

$$P_{pil}(a_1 - 1) - P_{sync} - N_p P_{pag} - K_{traf} M \alpha_f P_{traf} \geq b \qquad (11.19a)$$

where

$$a_1 \triangleq 1/\rho_{pil} K_f \qquad (11.19b)$$

and

$$b \triangleq N_m L_T(R)/K_f > 0 \qquad (11.19c)$$

Similarly, the conditions for satisfying the requirements of the sync, paging, and traffic channels as given in (11.18c), (11.18d), and (11.18e) can be manipulated into the following linear equation forms:

$$-P_{pil} + P_{sync}(a_2 - 1) - N_p P_{pag} - K_{traf} M \alpha_f P_{traf} \geq b > 0 \qquad (11.19d)$$

where

$$a_2 \triangleq (PG)_{sync}/\rho_{sync} K_f = 1024/\rho_{sync} K_f \qquad (11.19e)$$

$$-P_{pil} - P_{sync} + P_{pag}(a_3 - N_p) - K_{traf} M \alpha_f P_{traf} \geq b > 0 \qquad (11.19f)$$

where

$$a_3 \triangleq (PG)_{pag}/\rho_{pag} K_f = 256/\rho_{pag} K_f \qquad (11.19g)$$

and

$$-P_{pil} - P_{sync} - N_p P_{pag} + (a_4 - K_{traf} M \alpha_f) P_{traf} \geq b > 0 \qquad (11.19h)$$

where

$$a_4 \triangleq (PG)_{traf}/\rho_{traf} K_f = 128/\rho_{traf} K_f \qquad (11.19i)$$

We now consider a modification of these equations to include fade margins, M_j(dB), $j = 1$ (pilot), 2 (sync), 3 (paging), and 4 (traffic), to obtain more general equations. This modification can be performed simply as follows: a fade margin may be included in the analysis by stipulating that the SNRs exceed their "required" values by some amount, resulting in the new SNR requirements given by

$$\rho_j' \triangleq \rho_j \cdot 10^{M_j(dB)/10}, \qquad j = 1, 2, 3, 4 \qquad (11.20a)$$

Thus, the quantities a_1, a_2, a_3, and a_4 defined above now become, for the general case of nonzero margins:

$$a'_j = \frac{(PG)_j}{10^{M_j(dB)/10}\rho_j K_f} = \frac{(PG)_j}{\rho'_j K_f} \tag{11.20b}$$

Having formulated the required simultaneous constraint equations as *linear equations* as shown above, if the equality conditions are assumed to be satisfied at the cell edge, then the problem is a system of linear equations with the following matrix form:

$$\begin{bmatrix} a'_1 - 1 & -1 & -N_p & -K_{traf} M \alpha_f \\ -1 & a'_2 - 1 & -N_p & -K_{traf} M \alpha_f \\ -1 & -1 & a'_3 - N_p & -K_{traf} M \alpha_f \\ -1 & -1 & -N_p & a'_4 - K_{traf} M \alpha_f \end{bmatrix} \begin{bmatrix} P_{pil} \\ P_{sync} \\ P_{pag} \\ P_{traf} \end{bmatrix} = \begin{bmatrix} b \\ b \\ b \\ b \end{bmatrix} \tag{11.21a}$$

The determinant of the matrix is

$$\Delta = \left[1 - \frac{1}{a'_1} - \frac{1}{a'_2} - N_p \frac{1}{a'_3} - K_{traf} M \alpha_f \frac{1}{a'_4} \right] a'_1 a'_2 a'_3 a'_4$$

$$= \left[1 - K_f \left(\rho'_1 + \frac{\rho'_2}{(PG)_{sync}} + N_p \frac{\rho'_3}{(PG)_{pag}} + K_{traf} M \alpha_f \frac{\rho'_4}{(PG)_{traf}} \right) \right]$$

$$\times \frac{(PG)_{sync} \cdot (PG)_{pag} \cdot (PG)_{traf}}{\rho'_1 \rho'_2 \rho'_3 \rho'_4 \cdot K_f^4} \tag{11.21b}$$

From this expression we observe that a constraint on the solution (since powers must be positive numbers) is

$$K_f \left(\rho'_1 + \frac{\rho'_2}{(PG)_{sync}} + N_p \frac{\rho'_3}{(PG)_{pag}} + K_{traf} M \alpha_f \frac{\rho'_4}{(PG)_{traf}} \right) < 1 \tag{11.21c}$$

Later, we use this expression to derive a bound on M, the number of active users.

Substituting for $L_T(R)$, the solutions for the four forward link channel transmitter powers are

$$P_{pil} = \cfrac{N_m\, \rho'_{pil}\, L(R)\, L_{tc}\, L_{rm}/G_c G_m}{1 - K_f\left(\rho'_{pil} + \cfrac{\rho'_{sync}}{(PG)_{sync}} + N_p \cfrac{\rho'_{pag}}{(PG)_{pag}} + K_{traf}\, M\alpha_f \cfrac{\rho'_{traf}}{(PG)_{traf}}\right)}$$

$$(11.22a)$$

$$P_{sync} = \cfrac{N_m\, \rho'_{sync}\, L(R)\, L_{tc}\, L_{rm}/(PG)_{sync} G_c G_m}{1 - K_f\left(\rho'_{pil} + \cfrac{\rho'_{sync}}{(PG)_{sync}} + N_p \cfrac{\rho'_{pag}}{(PG)_{pag}} + K_{traf}\, M\alpha_f \cfrac{\rho'_{traf}}{(PG)_{traf}}\right)}$$

$$(11.22b)$$

$$P_{pag} = \cfrac{N_m\, \rho'_{pag}\, L(R)\, L_{tc}\, L_{rm}/(PG)_{pag} G_c G_m}{1 - K_f\left(\rho'_{pil} + \cfrac{\rho'_{sync}}{(PG)_{sync}} + N_p \cfrac{\rho'_{pag}}{(PG)_{pag}} + K_{traf}\, M\alpha_f \cfrac{\rho'_{traf}}{(PG)_{traf}}\right)}$$

$$(11.22c)$$

$$P_{traf} = \cfrac{N_m\, \rho'_{traf}\, L(R)\, L_{tc}\, L_{rm}/(PG)_{traf} G_c G_m}{1 - K_f\left(\rho'_{pil} + \cfrac{\rho'_{sync}}{(PG)_{sync}} + N_p \cfrac{\rho'_{pag}}{(PG)_{pag}} + K_{traf}\, M\alpha_f \cfrac{\rho'_{traf}}{(PG)_{traf}}\right)}$$

$$(11.22d)$$

Note from these equations that the solution for each channel's required power is a function of not only the SNR requirement for all the channels, including its own, but also the data rates of all the forward link channels as shown in terms of the processing gain of each channel. This means that the channel power requirements are interdependent. Note also that each channel power is a function of the number of active users M, which implies that ideally the forward link powers must be controlled in real time (dynamic power allocation) by using feedback involving all the parameters indicated in the denominators of the equations. In a later section, we discuss means for implementing these optimal power settings.

Example 11.4 Using the forward link transmitter power solutions in (11.22a) to (11.22d), calculate the required transmitter power of each channel and the total power for the case of zero-valued fade margins in all channels for the following parameter values:

$$N_m = -105\,\text{dBm}, \quad L_T(R) = 130\,\text{dB}, \quad K_f = 2.778,$$

$$N_p = 1, \quad M = 25, \quad K_{\text{traf}} = 0.5, \quad \text{and} \quad \alpha_f = 0.4$$

(11.23a)

Solution: Using the SNR requirements for no margin ($\rho_j = \rho'_j$) given in (11.18f), and substituting the channel processing gains given in (11.11b), (11.12b), and (11.13b), and substituting the parameter values in (11.23a), the expression in parentheses in the denominators of the equations for the powers is

$$\left(\rho'_{\text{pil}} + \frac{\rho'_{\text{sync}}}{(\text{PG})_{\text{sync}}} + N_p \frac{\rho'_{\text{pag}}}{(\text{PG})_{\text{pag}}} + K_{\text{traf}}\, M \alpha_f \frac{\rho'_{\text{traf}}}{(\text{PG})_{\text{traf}}} \right) \Bigg|_{M_j(\text{dB})=0,\ \text{all}\ j}$$

$$= 10^{-1.5} + \frac{10^{0.6}}{1024} + N_p \cdot \frac{10^{0.6}}{256} + 0.5 \cdot M \cdot 0.4 \cdot \frac{10^{0.7}}{128}$$

$$= 0.0355 + 0.0156\, N_p + 0.00783\, M \tag{11.23b}$$

$$= 0.2469 = -6.075\,\text{dB} \tag{11.23c}$$

Thus, the denominators are all equal to $1 - 0.2469\, K_f = 0.3143 = -5.029\,\text{dB}$. The numerator factor common to all the powers is

$$N_m L_T(R) = -105\,\text{dBm} + 130\,\text{dB} = 25\,\text{dBm}$$

Using this value, we calculate the required forward link powers as follows:

$$P_{\text{pil}}(\text{dBm}) = 25\,\text{dBm} + (-15\,\text{dB}) - (-5.029\,\text{dB})$$

$$= 15.029\,\text{dBm} = 31.83\,\text{mW}$$

$$P_{\text{sync}}(\text{dBm}) = 25\,\text{dBm} + 6\,\text{dB} - 10\log_{10}(1{,}024) - (-5.029\,\text{dB})$$

$$= 5.926\,\text{dBm} = 3.91\,\text{mW}$$

$$P_{\text{pag}}(\text{dBm}) = 25\,\text{dBm} + 6\,\text{dB} - 10\log_{10}(256) - (-5.029\,\text{dB})$$

$$= 11.947\,\text{dBm} = 15.66\,\text{mW}$$

$$P_{traf}(dBm) = 25\,dBm + 7\,dB - 10\log_{10}(128) - (-5.029\,dB)$$

$$= 15.957\,dBm = 39.42\,mW$$

Thus, the total forward link power is

$$P_{total} = P_{pil} + P_{sync} + N_p\,P_{pag} + K_{traf}\,M\,\alpha_f\,P_{traf}$$

$$= 31.83 + 3.91 + 1 \cdot 15.66 + 0.5 \cdot 25 \cdot 0.4 \cdot 39.42\,mW$$

$$= 248.4\,mW = 24.0\,dBm$$

Comments on Example 11.4: If the antenna gains and various losses listed in Table 10.7 for the reverse link are valid for the forward link, then the postulated transmission loss in (11.23a), $L_T(R) = 130\,dB$, corresponds to a propagation loss to the cell edge of about 129 dB. From Figure 10.12, we observe that this propagation loss is typical of a cellular application with the cell radius on the order of 1 to 2 km. In view of this cell size, it is perhaps not surprising that the total forward link power requirement for Example 11.4 is only about one-quarter of a Watt. Of that power, the fraction allocated to the pilot channel is $31.83/248.4 = 12.8\%$ of the total power. What may seem surprising is that the total power of a one-quarter Watt is for a cell loaded with 25 active users, about 60% of reverse link capacity with no margin, according to the "ideal capacity" analysis in Section 10.2.3. It would be natural for a cellular operator to want to use more power than this required amount of power in an effort to increase system reliability. However, we show later that increasing power arbitrarily does not necessarily improve systems performance.

Example 11.5 Calculate the forward link powers for the same parameter values as in Example 11.4, except that each channel has a 1-dB fade margin.

Solution: By assumption, we have $\rho'_{pil} = -14\,dB$, $\rho'_{sync} = \rho'_{pag} = 7\,dB$, and $\rho'_{traf} = 8\,dB$. Because the same value of margin is used for each type of channel, we may factor the denominator term calculated in (11.23c) as follows:

$$\left(\rho'_{pil} + \frac{\rho'_{sync}}{(PG)_{sync}} + N_p \frac{\rho'_{pag}}{(PG)_{pag}} + K_{traf}\,M\,\alpha_f \frac{\rho'_{traf}}{(PG)_{traf}}\right)\Bigg|_{M_j(dB)=M(dB),\ all\ j}$$

$$= \left(\rho_{\text{pil}} + \frac{\rho_{\text{sync}}}{(\text{PG})_{\text{sync}}} + N_p \frac{\rho_{\text{pag}}}{(\text{PG})_{\text{pag}}} + \text{K}_{\text{traf}} \, M \alpha_{\text{f}} \frac{\rho_{\text{traf}}}{(\text{PG})_{\text{traf}}} \right) \cdot 10^{M(\text{dB})/10}$$

$$= \left(\rho'_{\text{pil}} + \frac{\rho'_{\text{sync}}}{(\text{PG})_{\text{sync}}} + N_p \frac{\rho'_{\text{pag}}}{(\text{PG})_{\text{pag}}} + \text{K}_{\text{traf}} \, M \alpha_{\text{f}} \frac{\rho'_{\text{traf}}}{(\text{PG})_{\text{traf}}} \right) \Bigg|_{\text{in dB}} + \text{M(dB)}$$

Thus, this term is increased from $-6.075\,\text{dB}$ in example 11.4 to -5.075 dB $= 0.3108$ in this example, and the denominators are all equal to $1 - 0.3108\,\text{K}_{\text{f}} = 0.1366 = -8.647\,\text{dB}$. Using this value, we calculate the required forward link powers as follows:

$$P_{\text{pil}}(\text{dBm}) = 25\,\text{dBm} + (-14\,\text{dB}) - (-8.647\,\text{dB})$$
$$= 19.647\,\text{dBm} = 92.19\,\text{mW}$$

$$P_{\text{sync}}(\text{dBm}) = 25\,\text{dBm} + 7\,\text{dB} - 10\log_{10}(1024) - (-8.647\,\text{dB})$$
$$= 10.544\,\text{dBm} = 11.33\,\text{mW}$$

$$P_{\text{pag}}(\text{dBm}) = 25\,\text{dBm} + 7\,\text{dB} - 10\log_{10}(256) - (-8.647\,\text{dB})$$
$$= 16.565\,\text{dBm} = 45.34\,\text{mW}$$

$$P_{\text{traf}}(\text{dBm}) = 25\,\text{dBm} + 8\,\text{dB} - 10\log_{10}(128) - (-8.647\,\text{dB})$$
$$= 20.575\,\text{dBm} = 114.15\,\text{mW}$$

Thus the total forward link power is

$$P_{\text{total}} = P_{\text{pil}} + P_{\text{sync}} + N_p \, P_{\text{pag}} + \text{K}_{\text{traf}} \, M \alpha_{\text{f}} \, P_{\text{traf}}$$
$$= 92.19 + 11.33 + 1 \cdot 45.34 + 0.5 \cdot 25 \cdot 0.4 \cdot 114.15\,\text{mW}$$
$$= 719.6\,\text{mW} = 28.6\,\text{dBm}$$

We note that the 1 dB margin added in example 11.5 caused the amount of required power to increase from 24 dBm in the zero-margin case of example 11.4 to 28.6 dBm, a power increase of 4.6 dB that is considerably higher than the increase in margin. These examples reveal a high sensitivity of the forward link power solutions to the fade margin. Later, we demonstrate the sensitivity of the optimal forward link power solutions to the parameters.

11.2.3.1 Allocated Channel Power as a Fraction of Total Power

The fraction of the forward link power allocated to the pilot channel in Example 11.5, with a 1-dB margin in all channels, is $92.19/719.6 = 0.128$. In Example 11.4, with no margin, the fraction of the forward link power allocated to the pilot channel was calculated to be $31.83/284.4 = 0.128$. Thus, the pilot fraction is the same.

This result, that the power fraction for each channel is independent of margin when all the channel have same margin, can be explained as follows: (11.22a) to (11.22d) are of the form

$$P_{\text{pil}} = \rho'_{\text{pil}} \times \nu\left(\rho'_{\text{pil}}, \rho'_{\text{sync}}, \rho'_{\text{pag}}, \rho'_{\text{traf}}\right) \tag{11.24a}$$

$$P_{\text{sync}} = \frac{\rho'_{\text{sync}}}{(\text{PG})_{\text{sync}}} \times \nu\left(\rho'_{\text{pil}}, \rho'_{\text{sync}}, \rho'_{\text{pag}}, \rho'_{\text{traf}}\right) \tag{11.24b}$$

$$P_{\text{pag}} = \frac{\rho'_{\text{pag}}}{(\text{PG})_{\text{pag}}} \times \nu\left(\rho'_{\text{pil}}, \rho'_{\text{sync}}, \rho'_{\text{pag}}, \rho'_{\text{traf}}\right) \tag{11.24c}$$

$$P_{\text{traf}} = \frac{\rho'_{\text{traf}}}{(\text{PG})_{\text{traf}}} \times \nu\left(\rho'_{\text{pil}}, \rho'_{\text{sync}}, \rho'_{\text{pag}}, \rho'_{\text{traf}}\right) \tag{11.24d}$$

where the common factor is

$$\nu\left(\rho'_{\text{pil}}, \rho'_{\text{sync}}, \rho'_{\text{pag}}, \rho'_{\text{traf}}\right) \tag{11.24e}$$

$$\triangleq \frac{N_m L_T(R)}{1 - K_f\left(\rho'_{\text{pil}} + \dfrac{\rho'_{\text{sync}}}{(\text{PG})_{\text{sync}}} + N_p \dfrac{\rho'_{\text{pag}}}{(\text{PG})_{\text{pag}}} + K_{\text{traf}} M \alpha_f \dfrac{\rho'_{\text{traf}}}{(\text{PG})_{\text{traf}}}\right)}$$

Therefore, the fraction, ζ_j, for channel type j, is given by

$$\zeta_j \triangleq \frac{P_j}{P_{\text{total}}} = \frac{\rho'_j/(\text{PG})_j}{\rho'_{\text{pil}} + \dfrac{\rho'_{\text{sync}}}{(\text{PG})_{\text{sync}}} + N_p \dfrac{\rho'_{\text{pag}}}{(\text{PG})_{\text{pag}}} + K_{\text{traf}} M \alpha_f \dfrac{\rho'_{\text{traf}}}{(\text{PG})_{\text{traf}}}} \tag{11.25a}$$

When the margins are equal in all the channels (i.e., $M_j(\text{dB}) = M(\text{dB})$ for all j), then (11.25a) becomes

$$\zeta_j = \frac{\rho_j/(PG)_j}{\rho_{pil} + \dfrac{\rho_{sync}}{(PG)_{sync}} + N_p \dfrac{\rho_{pag}}{(PG)_{pag}} + K_{traf}\, M\, \alpha_f \dfrac{\rho_{traf}}{(PG)_{traf}}} \qquad (11.25b)$$

which is independent of the margin. This explains why the pilot fraction is the same in Examples 11.4 and 11.5. Also note that the fractions do not depend on the interference factor K_f or on the transmission loss $L_T(R)$, although the amount is very sensitive to these parameters, as we see in the next section. For the parameter values used in Examples 11.4 and 11.5, and assuming each channel has the same amount of margin, the fractions for the channels are as given in Table 11.4. In the case of a traffic channel, Table 11.4 gives the fraction for a mobile at the cell edge. It is clear from the expressions for the fractions in Table 11.4 that the fraction is primarily a function of the number of active users M, although it is also a function of the number of paging channels N_p. For example, for $N_p = 1$ paging channel, the pilot fraction ζ_{pil} may be written

$$\zeta_{pil} = \frac{0.0316}{0.0511 + 0.00783\, M} = \frac{0.6184}{1 + 0.1532\, M} \qquad (11.26)$$

Note that the pilot channel fraction is inversely proportional to the number of active users. The same statement can be made about all the channels, as indicated in Table 11.4.

Table 11.4 Fractions of the total power allocated to the channels in the equal-margin case

Channel	Fraction	
Pilot	$\zeta_1 =$	$\dfrac{0.0316}{0.0355 + 0.0156\, N_p + 0.00783\, M}$
Sync	$\zeta_2 =$	$\dfrac{0.00389}{0.0355 + 0.0156\, N_p + 0.00783\, M}$
Paging	$\zeta_3 =$	$\dfrac{0.0156}{0.0355 + 0.0156\, N_p + 0.00783\, M}$
Traffic	$\zeta_4 =$	$\dfrac{0.0392}{0.0355 + 0.0156\, N_p + 0.00783\, M}$

11.2.3.2 Parametric Variations in the Power Solutions

Because the forward link transmitter power for each of the four channel types varies by the same proportion for fixed SNR requirements, a study of the effects of the several parameters determining the value of $\nu\left(\rho'_{\text{pil}}, \rho'_{\text{sync}}, \rho'_{\text{pag}}, \rho'_{\text{traf}}\right)$ can be made using calculations of just the pilot channel transmitter power P_{pil} and its fraction of the total transmitter power. In addition to margin, the parameters to be studied are L_T, K_f, and M. For convenience, we assume that the same fade margin is used on each channel when there is a fade margin. That is, we assume that $M_j(\text{dB})$ in (11.20c) equals $M(\text{dB})$ for all the channels.

Because the transmitter powers are each directly proportional to the receiver noise power N_m, and to the net link loss L_T, plots of P_{pil} in dBm versus L_T in dB are straight lines with slopes equal to 1, as shown in Figure 11.12. In Figure 11.12 we show the dependence of P_{pil} on L_T parametric in M, the number of users and for the typical parameter values given in (11.23a). Note from the figure that the increase in power with M, which is needed to overcome the same- and other-cell interference, is not linear: the amount of increment in P_{pil} is not identical for the uniform increments of M shown.

Let us now observe the functional behavior of P_{pil} versus M. Assuming $K_{\text{same}} = 1$, the total interference factor is $K_f = 1 + K_{\text{other}}$. In Figure 11.13, we show the dependence of pilot power on M with K_{other} varied [see (11.14b)]. In Figure 11.13, we see that there is a value of M at which the required pilot power goes to infinity; this value can be interpreted as the user capacity of the forward link. More is said about this concept later. The increase in P_{pil} with an increase in K_{other} for a given value of M is expected in order to overcome the increased interference. For a system that is not heavily loaded ($M < 25$), the sensitivity of P_{pil} to K_{other} can be characterized as follows: a half-dB increase in K_{other} requires at most a half-dB increase in P_{pil}; that is, the required increase in pilot power in this instance is less than an increase in the other-cell interference factor.

Figure 11.14 shows the required pilot power as a function of the margin with the active number of users M as a parameter, assuming that all the channels have the same value of fade margin, $M(\text{dB})$. It is important to note from the figure that the amount of margin that can be achieved by the system is limited to some upper value as a function of M because the transmitter power, no matter how large, cannot produce a greater margin. For example, the margin can be no greater than 1 dB when there are $M = 30$ users.

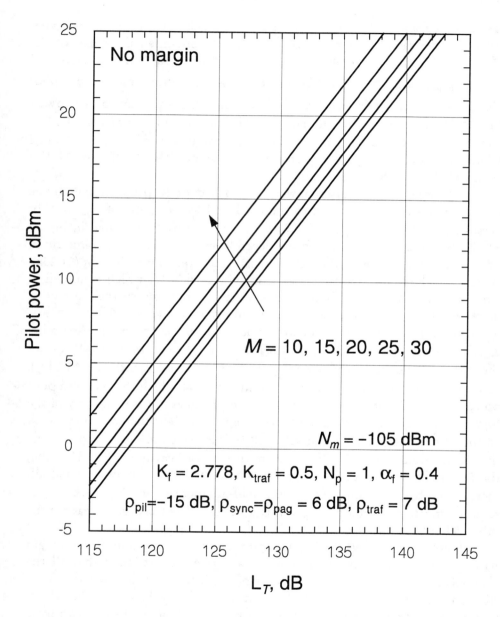

Figure 11.12 Effect of transmission loss on required pilot channel power.

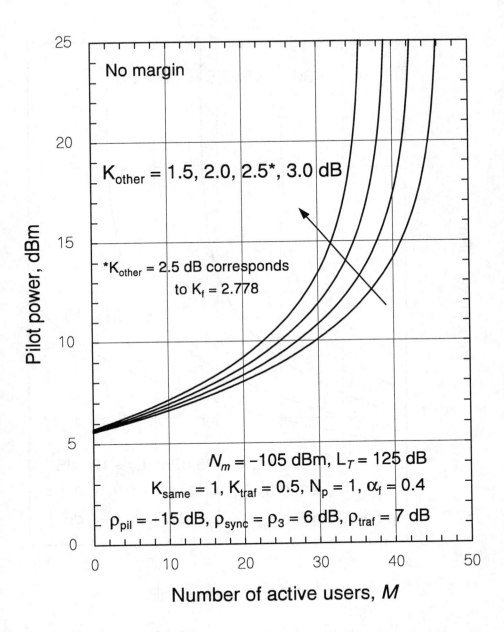

Figure 11.13 Effect of other-cell interference factor on P_{pil} as a function of M.

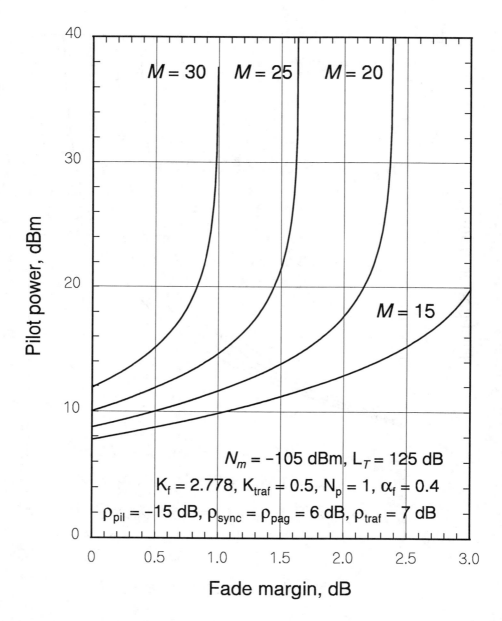

Figure 11.14 Required pilot channel transmitter power as a function of fade margin.

The situation is that, according to our assumptions, the amount of interference from other cells is proportional to the total power received at the edge of the cell; therefore, increasing transmitter power also increases interference. When the interference power is small compared with the mobile receiver's thermal noise, increasing the base station power improves the SNR at the receiver, but as the interference power becomes greater than the thermal noise, there is no more improvement in the SNR to be had by increasing transmitter power. In Section 11.3, we discuss the issue of margin more thoroughly, and in Section 11.4, we consider the implications of the limitation on SNR improvement exhibited in Figure 11.14 for the assessment of forward link user capacity.

To emphasize the fact that all the required channel transmitter powers have the same behavior as the pilot channel with respect to its sensitivity to the parameters, we show in Figure 11.15 the required power of each channel and the corresponding required total power as functions of M. The parameter values used in the calculations are shown in the figure. The curve labeled "one traffic channel" is a plot of P_{traf}, while the curve labeled "all traffic channels" is a plot of $K_{traf} M \alpha_f P_{traf}$. In this figure, it is evident that the conditions for meeting the requirements for all the channels cannot be met simultaneously when the number of active users exceeds $M = 35$. This is an example of the CDMA "soft capacity" that is a function of many system parameters rather than a "hard" or fixed maximum number of users. We observe also that the total power in this example is $1\,\mathrm{W} = 0\,\mathrm{dBW}$ for $M = 26$ users. This amount of power is relatively small, because of the processing gain of the CDMA modulation and also from the assumed conditions; the assumed 145-dB median propagation loss corresponds to the relatively small cellular radius of about 3 to 4 km at cellular frequencies and about 1 to 3 km at PCS frequencies.

11.3 Selection of Forward Link Fade Margins

In this section, we review the basic principles involved in delivering a forward link signal to the mobile location to meet a specified SNR objective, where the SNR value is chosen to implement a link reliability objective for system optimization. We show that the amount of fade margin at the mobile receiver is limited by the multiple access interference, so that the desired margin may not be achievable for the number of active CDMA users that is desired. We also interpret the form of the solutions for the forward link powers in a way that explains the fundamental principles that are involved.

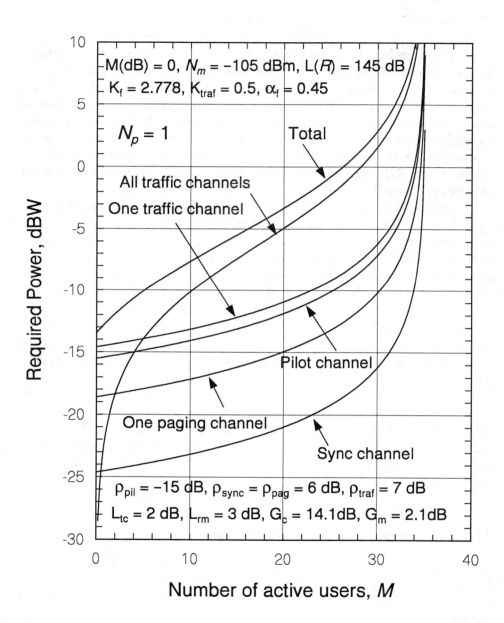

Figure 11.15 Required channel powers in dBW versus M for $N_p = 1$.

11.3.1 Limits on Receiver Margin

As we have seen, both analytically and in the numerical examples presented in Figures 11.11 to 11.15, the effective SNR at the mobile receiver for a given type of channel is proportional to P_t, the transmitted power for the channel:

$$SNR_{rec} \propto P_t \qquad (11.27a)$$

Other factors being equal (path loss, etc.), the transmitted power for the forward link Walsh channel is required to be proportional to the desired SNR at the mobile receiver:

$$P_t \propto SNR_{req} \qquad (11.27b)$$

The ratio of median receiver SNR for a given Walsh channel to its required SNR is the receiver margin; expressed in dB, the margin is defined by

$$SNR_{med}(dB) - SNR_{req}(dB) \triangleq M(dB) = \text{receiver margin}^2 \qquad (11.28a)$$

Let the margin at the receiver for Walsh channel j in absolute units (not dB) be denoted by $\Gamma_{r,j}$, where

$$\Gamma_{r,j} \triangleq 10^{M_j(dB)/10} \qquad (11.28b)$$

In general, the amount of margin experienced at the receiver for a given amount of forward link transmitter power is a function of multiple-access and multipath interference arising chiefly from forward link transmissions from, respectively, other cells and the same cell. Thus, we may write

$$\Gamma_{r,j} \equiv \Gamma_{r,j}(K_f) \qquad (11.28c)$$

where K_f was defined in (11.14d) as the ratio of total interference power received by a mobile at the cell edge to the total direct path forward link signal power received at that location.

   The margin experienced at the receiver is due to transmitting more power than required to obtain S_{req} at the receiver. In general, the power that is required at the receiver is a function of the interference.

---

2 In this section, the notation $M(dB)$ is equivalent to M_{dB}: $M(dB) \equiv M_{dB}$.

11.3.1.1 Receiver and Transmitter Powers Under No Interference

If there is no interference ($K_f = 0$), the required receiver power for Walsh channel j is

$$S_{0,j} \triangleq S_{req,j}(0) = SNR_{req,j} \times \text{noise power at mobile}$$

$$= SNR_{req,j} \times N_m \tag{11.29a}$$

$S_{0,j}$ is obtained from a required amount of transmitter power in the absence of interference:

$$P_{t0,j} \triangleq S_{0,j} \times \text{losses} = SNR_{req,j} \times N_m \times \text{losses} \tag{11.29b}$$

At the forward link transmitter, we may express the amount of power transmitted in terms of a "transmitter margin" or *excess power ratio*, which is not a function of the interference:

$$(\text{excess power ratio})_j \triangleq \Gamma_{t,j} = \frac{P_{t,j}}{P_{t0,j}} \tag{11.29c}$$

In dB units, we denote this transmitter margin by $M_{t,j}(\text{dB})$:

$$M_{t,j}(\text{dB}) \triangleq 10 \log_{10} \Gamma_{t,j} \tag{11.29d}$$

Although $P_{t,j}$, $\Gamma_{t,j}$, and $M_{t,j}(\text{dB})$ are not functions of the interference, we may use the notations $P_{t,j}(K_f)$, $\Gamma_{t,j}(K_f)$, and $M_{t,j}(\text{dB}; K_f)$ to indicate the interference conditions considered in a particular case, either $K_f = 0$ or K_f equal to some nonzero value.

At the mobile receiver, we may express the median SNR in terms of the receiver margin, which, in general, is a function of the interference:

$$(\text{receiver margin})_j \triangleq \Gamma_{r,j}(K_f) = \frac{SNR_{med,j}(K_f)}{SNR_{req,j}} \tag{11.30a}$$

For no interference ($K_f = 0$), the margin at the receiver is the same as the transmitter margin:

$$\Gamma_{r,j}(0) = \frac{\text{SNR}_{med,j}(0)}{\text{SNR}_{req,j}} = \frac{S_j \div N_m}{S_{0,j} \div N_m} = \frac{S_j}{S_{0,j}} \qquad (11.30\text{b})$$

This fact implies that, with or without interference, the received power S_j is the minimum required power times the receiver margin that would be experienced when there is no interference:

$$S_j = \Gamma_{r,j}(0)\, S_{0,j} \qquad (11.30\text{c})$$

The receiver margin in the absence of interference can also be written

$$\Gamma_{r,j}(0) = \frac{S_j \times \text{Losses}}{S_{0,j} \times \text{Losses}} = \frac{P_{t,j}}{P_{t0,j}} = \Gamma_{t,j} \qquad (11.30\text{d})$$

which shows that the receiver and transmitter margins are equal when there is no interference.

Example 11.6 Let the mobile receiver noise power be $N_m = -105\,\text{dBm}$. What receiver power $S_{0,j}$ is needed for Walsh channel j in the absence of interference to achieve an SNR of $-10\,\text{dB}$? If the net losses on the forward link are $120\,\text{dB}$, what transmitter power $P_{t0,j}$ is required in the absence of interference? What should $P_{t,j}$ be in the absence of interference if a receiver margin of $M(\text{dB}) = 2\,\text{dB}$ is to be implemented?

Solution: The required receiver power is given by

$$S_{0,j} = \text{SNR}_{req,j} \times N_m = -10\,\text{dB} + (-105\,\text{dBm}) = -115\,\text{dBm}$$

The power required at the transmitter is given by

$$P_{t0,j} = S_{0,j} \times \text{Losses} = -115\,\text{dBm} + 120\,\text{dB} = 5\,\text{dBm} = 3.16\,\text{mW}$$

In the absence of interference the transmitter margin is equal to the receiver margin, so the power required to implement a 2 dB margin at the receiver is

$$P_{t,j} = 5\,\text{dBm} + 2\,\text{dB} = 7\,\text{dBm} = 5.01\,\text{mW}$$

11.3.1.2 Receiver and Transmitter Powers When There Is Interference

The total interference I_T was defined in (11.14c) to be

$$I_T = K_f S \tag{11.31a}$$

where S denotes the total transmitted power received at a particular mobile station, and hence

$$S = \sum_j S_j \tag{11.31b}$$

where

$$S_j = \text{Received power on Walsh channel } j \tag{11.31c}$$

Therefore,

$$I_T = K_f S = K_f \cdot \sum_j S_j = K_f \cdot \sum_j P_{t,j}/\text{Losses} \tag{11.31d}$$

where $P_{t,j}$ denotes the power transmitted for forward link Walsh channel j. We show shortly that the receiver margin and the transmitter margin are quite different under interference conditions.

Including interference from the same and other cells, we use (11.30c) to show that the median receiver SNR for Walsh channel j is

$$SNR_{med,j}(K_f) = \frac{S_j}{\text{noise} + \text{interference}} = \frac{S_j}{N_m + K_f \cdot \sum_j S_j}$$

$$= \frac{\Gamma_{r,j}(0) S_{0,j}}{N_m + K_f \sum_j \Gamma_{r,j}(0) S_{0,j}} = \frac{\Gamma_{t,j} S_{0,j}}{N_m + K_f \sum_j \Gamma_{t,j} S_{0,j}} \tag{11.31e}$$

But from (11.29a), $S_{0,j} = SNR_{req,j} \times N_m$, so that we may write (11.31e) as

$$SNR_{med,j}(K_f) = \frac{\Gamma_{t,j} N_m \times SNR_{req,j}}{N_m + K_f \sum_j \Gamma_{t,j} N_m \times SNR_{req,j}}$$

$$= \frac{\Gamma_{t,j} SNR_{req,j}}{1 + K_f \sum_j \Gamma_{t,j} SNR_{req,j}} \tag{11.31f}$$

Because by definition, $\text{SNR}_{med,j}(K_f) = \Gamma_{r,j}(K_f)\,\text{SNR}_{req,j}$, the receiver margin is found by dividing (11.31e) by $\text{SNR}_{req,j}$ to obtain an expression for the received margin as a function of the transmitter margins:

$$\Gamma_{r,j}(K_f) = \frac{\text{SNR}_{med,j}(K_f)}{\text{SNR}_{req,j}} = \frac{\Gamma_{t,j}}{1 + K_f \sum_j \Gamma_{t,j}\,\text{SNR}_{req,j}} \qquad (11.32a)$$

If all the Walsh channels have the same margins at the transmitter, then they have the same receiver margins, and we may write (11.32a) as

$$\Gamma_{r,j}(K_f) = \Gamma_r(K_f) = \frac{\Gamma_t}{1 + K_f \Gamma_t \sum_j \text{SNR}_{req,j}} \qquad (11.32b)$$

From (11.32b), it is clear that the receiver margin is less than the transmitter margin when $K_f \neq 0$; that is, when there is interference. The presence of interference requires more transmitter margin to achieve the same amount of receiver margin. This fact is illustrated in Figure 11.16.

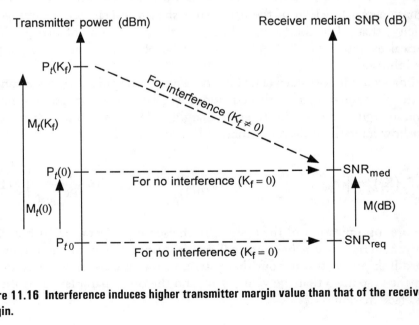

Figure 11.16 Interference induces higher transmitter margin value than that of the receiver margin.

Using the notation of Section 11.2 for the required SNRs for the different forward link Walsh channels, we can write the receiver margin (11.32b) as

$$\Gamma_r(K_f) = \frac{\Gamma_t}{1 + K_f\,\Gamma_t\left(\rho_{\text{pil}} + \dfrac{\rho_{\text{sync}}}{(PG)_{\text{sync}}} + N_p\dfrac{\rho_{\text{pag}}}{(PG)_{\text{pag}}} + K_{\text{traf}}\,M\,\alpha_f\dfrac{\rho_{\text{traf}}}{(PG)_{\text{traf}}}\right)}$$

$$(11.32c)$$

If we substitute the nominal values for the IS-95 system as given in (11.18a) through (11.18f), the receiver margin is given by

$$\Gamma_r(K_f) = \frac{\Gamma_t}{1 + K_f\,\Gamma_t\left(0.0355 + 0.0156\,N_p + 0.0392\,K_{\text{traf}}M\,\alpha_f\right)} \qquad (11.32d)$$

11.3.2 Numerical Examples of CDMA Margin

In what follows, we present numerical examples for the achievable CDMA link margin as a function of the parameters shown in (11.32c), assuming, for simplicity, that all of the forward links use the same value of margin. The numerical examples of CDMA margin can be observed from the equations derived thus far.

For example, consider (11.32c). Among several things that we can see from the functional form of this equation, we observe an aymptotic limit on receiver margin, so that the receiver margin is less than some constant no matter how large the transmitter margin Γ_t becomes. That is:

$$\lim_{\Gamma_t \to \infty} \Gamma_r(K_f) = \left(\rho_{\text{pil}} + \frac{\rho_{\text{sync}}}{(PG)_{\text{sync}}} + N_p\frac{\rho_{\text{pag}}}{(PG)_{\text{pag}}} + K_{\text{traf}}\,M\,\alpha_f\frac{\rho_{\text{traf}}}{(PG)_{\text{traf}}}\right)^{-1} \quad (11.33)$$

Therefore, optimization of the system (in the sense of operating with as high a receiver margin as can be obtained) requires using the minimal amount of forward link power throughout the system, to minimize interference, and to design the system so that the value of M and the other parameters afford the desired fade margin.

11.3.2.1 Receiver Margin Versus Transmitter Margin

Calculations of (11.32c) are shown in Figure 11.17, in which the receiver margin is plotted as a function of the transmitter margin, with the number of users as a parameter and with the assumptions stated in the figure. The limitation on achievable margin that we discussed is clearly seen; the asymptotic amount of achievable margin is inversely proportional to the number of active users. For example, when there are $M = 20$ users, the maximum attainable receiver margin is less than 2 dB. This limiting behavior is shown in Figure 11.17 in contrast to the no-interference relationship between the receiver and transmitter margins. Another observation that can be made from Figure 11.18 is the value of transmitter margin that is necessary just to meet the receiver SNR requirements with a zero-dB receiver margin. Even for as small a number of users as $M = 5$, about 1.3-dB transmitter margin is required to break even. For $M = 20$ the transmitter margin must be at least 4.3 dB to break even and for $M = 30$ it must be at least 9 dB.

In many instances it is more natural to ask: what transmitter margin is needed to achieve a specified receiver margin? The answer to this question is the replotting of Figure 11.17 to obtain Figure 11.18. The plotting of these curves was facilitated by noting that (11.32c) can be inverted to obtain the transmitter margin as a function of receiver margin:

$$\Gamma_t = \frac{\Gamma_r\left(K_f\right)}{1 - K_f\,\Gamma_r\left(K_f\right)\left(\rho_{\text{pil}} + \dfrac{\rho_{\text{sync}}}{(\text{PG})_{\text{sync}}} + N_p\dfrac{\rho_{\text{pag}}}{(\text{PG})_{\text{pag}}} + K_{\text{traf}}\,M\,\alpha_f\dfrac{\rho_{\text{traf}}}{(\text{PG})_{\text{traf}}}\right)}$$

(11.34a)

Thus, in the case of equal margins for all channels the transmitted power for the jth forward link channel can be written as follows:

$$P_{t,j} = \Gamma_t\,P_{t0,j}$$

$$= \frac{\Gamma_r\left(K_f\right)P_{t0,j}}{1 - K_f\,\Gamma_r\left(K_f\right)\left(\rho_{\text{pil}} + \dfrac{\rho_{\text{sync}}}{(\text{PG})_{\text{sync}}} + N_p\dfrac{\rho_{\text{pag}}}{(\text{PG})_{\text{pag}}} + K_{\text{traf}}\,M\,\alpha_f\dfrac{\rho_{\text{traf}}}{(\text{PG})_{\text{traf}}}\right)}$$

(11.34b)

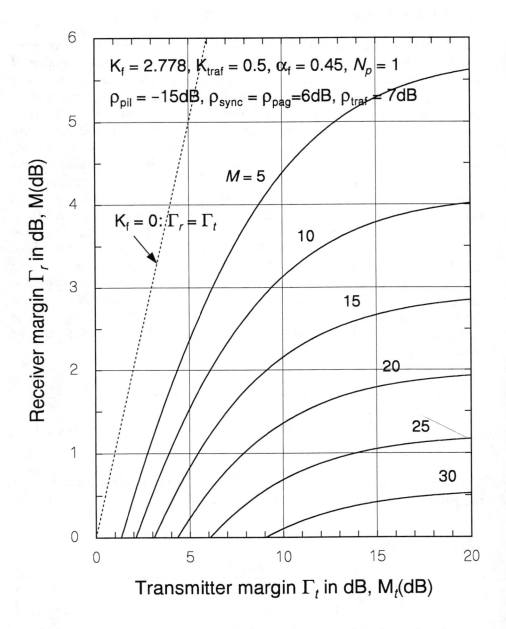

Figure 11.17 Receiver margin versus transmitter margin, parametric in the number of users.

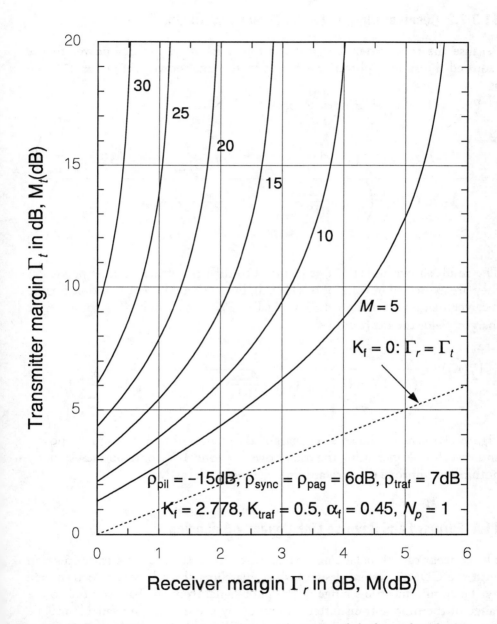

Figure 11.18 Transmitter margin versus receiver margin, parametric in the number of users.

11.3.2.2 Receiver Margin Versus Total Forward Link Power

Because the transmitter margin Γ_t is the ratio of transmitter power to the required transmitter power, we can write the total power defined in (11.18a) as

$$P_{total} = \Gamma_t \times \text{receiver noise} \times \text{losses} \times \sum_j \text{SNR}_{req,j}$$

$$= \frac{\Gamma_r\left(K_f\right) N_m L_T(R) \left(\rho_{pil} + \dfrac{\rho_{sync}}{(PG)_{sync}} + N_p \dfrac{\rho_{pag}}{(PG)_{pag}} + K_{traf}\, M\alpha_f \dfrac{\rho_{traf}}{(PG)_{traf}}\right)}{1 - K_f \Gamma_r\left(K_f\right) \left(\rho_{pil} + \dfrac{\rho_{sync}}{(PG)_{sync}} + N_p \dfrac{\rho_{pag}}{(PG)_{pag}} + K_{traf}\, M\alpha_f \dfrac{\rho_{traf}}{(PG)_{traf}}\right)}$$

$$(11.35a)$$

The result shown in (11.35a) is identical to the sum of the channel powers as they are expressed in (11.22a) to (11.22d), but with a different notation for the receiver margin. Setting (11.35a) equal to P_{total} and solving for the receiver margin yields the expression

$$\Gamma_r\left(K_f\right) = \frac{P_{total}}{\left(N_m\, L_T + K_f\, P_{total}\right)\left(\rho_{pil} + \dfrac{\rho_{sync}}{1024} + N_p \dfrac{\rho_{pag}}{256} + K_{traf}\, M\alpha_f \dfrac{\rho_{traf}}{128}\right)}$$

$$(11.35b)$$

Figure 11.19 shows the receiver margin as function of total transmitter power and reveals that increasing the total power beyond 1 W contributes absolutely nothing to improve the forward link channel margins.[3]

11.4 Forward and Reverse Link Capacity Balancing

The concept of link balancing was discussed in Section 10.1.3.4 in connection with the CDMA system design task of matching the cell size, as determined by the communications range of the pilot signal from the base station, to the range of the mobile transmitter, as limited by the maximum available mobile

---

[3] We find it worthwhile to make this point rather forcefully because we have observed the tendency of engineers to try to apply their experience in non-cellular RF situations to the cellular situation.

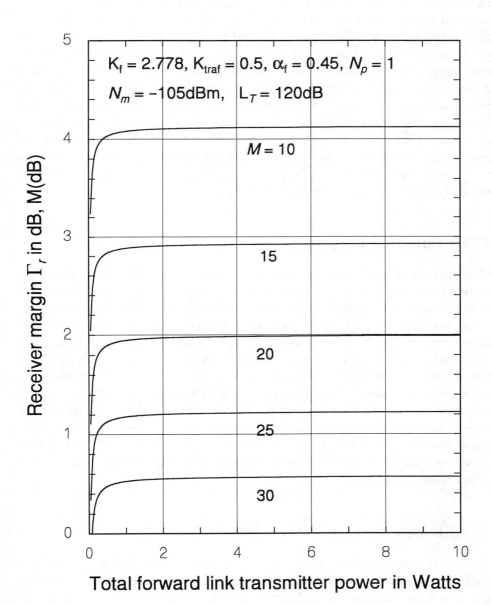

Figure 11.19 Receiver margin versus total forward link power, parametric in the number of users.

transmitter power. In the analysis of link balancing in Chapter 10, we assumed a given amount of total forward link power and showed how the forward and reverse links can be balanced by selecting the "right" amount of pilot channel power through the value of the pilot fraction, ζ_{pil} [4]. We saw in this chapter from an analysis of the required amount of forward link channel powers that the total forward link power, if optimally selected as its minimum required value, is not fixed but varies as the number of active users changes, and that this power is very sensitive to the assumed or actual number of users. Therefore, in this section, we revisit the subject of forward and reverse link balancing in the system optimization setting. We introduce in the process a new concept of "capacity balancing" that is useful for selecting optimal CDMA system parameter values.

11.4.1 Forward Link Capacity

One upper limit to the number of traffic channels in the IS-95 CDMA cellular system is the number of orthogonally multiplexed forward link channels that is available for assignment to user traffic. Nominally, 55 channels are designated as traffic channels. However, we have seen for the reverse link that 40 active users is a relatively high number of users, and the reverse link capacity is limited by interference to a number of users that is considerably less than the number of available forward link channels. In our presentation of numerical examples of the optimal forward link powers in Section 11.2, we observed that on the forward link the interference also limits the number of active users that can be accommodated while satisfying link SNR requirements. In what follows, we return to the forward link power solutions and derive expressions for forward link capacity as a function of interference parameters.

11.4.1.1 Asymptotic Forward Link Capacity

We saw that the required amount of forward link power "blows up" at some value of the number of users, M. For example, in Figure 11.13, we observe that the required pilot channel power goes to infinity as M approaches the value of $M = 39$ when $K_{other} = 2.5\,dB$. This behavior can be interpreted in

---

[4] In Chapter 10, the symbol ζ_p was used; here a slightly different notation is used to distinguish the pilot fraction from the paging channel fraction.

terms of a limit to the forward link user capacity. Mathematically, the observed behavior of the forward link powers can be traced to the fact that each channel's power is of the form given in (11.24a) to (11.24e), which has a singularity as M approaches some value. That is, for the power to be positive and finite, the following restriction must be applied to the denominator of the factor $\nu\left(\rho'_{\text{pil}}, \rho'_{\text{sync}}, \rho'_{\text{pag}}, \rho'_{\text{traf}}\right)$ that is common to all the forward link power solutions:

$$1 - K_f \left(\rho'_{\text{pil}} + \frac{\rho'_{\text{sync}}}{(\text{PG})_{\text{sync}}} + N_p \frac{\rho'_{\text{pag}}}{(\text{PG})_{\text{pag}}} + K_{\text{traf}} M \alpha_f \frac{\rho'_{\text{traf}}}{(\text{PG})_{\text{traf}}} \right) > 0 \tag{11.36a}$$

which can be manipulated to obtain the inequality

$$M < \frac{(\text{PG})_{\text{traf}}}{K_{\text{traf}} \alpha_f \rho'_{\text{traf}}} \left[\frac{1}{K_f} - \rho'_{\text{pil}} - \frac{\rho'_{\text{sync}}}{(\text{PG})_{\text{sync}}} - N_p \frac{\rho'_{\text{pag}}}{(\text{PG})_{\text{pag}}} \right] \tag{11.36b}$$

$$= \frac{(\text{PG})_{\text{traf}}}{K_{\text{traf}} \alpha_f \rho_{\text{traf}}} \left[\frac{10^{-M(\text{dB})/10}}{K_f} - \rho_{\text{pil}} - \frac{\rho_{\text{sync}}}{(\text{PG})_{\text{sync}}} - N_p \frac{\rho_{\text{pag}}}{(\text{PG})_{\text{pag}}} \right] \tag{11.36c}$$

In going from (11.36b) to (11.36c), we assumed that all the forward link Walsh channels have the same specified margin, M(dB). To gain some physical meaning of "asymptotic forward link capacity," we plot the upper bound of M given in (11.36c) with specific parameter assumptions: $(\text{PG})_{\text{traf}} = 128$, $(\text{PG})_{\text{sync}} = 1,024$, and $(\text{PG})_{\text{pag}} = 256$. We will call this value the *forward link asymptotic capacity*, denoted by M_∞. That is:

$$M_\infty \triangleq \frac{128}{K_{\text{traf}} \alpha_f \rho_{\text{traf}}} \left[\frac{10^{-M(\text{dB})/10}}{K_f} - \rho_{\text{pil}} - \frac{\rho_{\text{sync}}}{1024} - N_p \frac{\rho_{\text{pag}}}{256} \right] \tag{11.37a}$$

Plots of (11.37a) are given in Figure 11.20 as a function of the margin in dB. It should be recognized that this asymptotic capacity is a value that can never be reached because the amount of transmitter power available will be used up before M reaches this value. Note that the asymptotic capacity of Figure 11.20 is for the specific cases of the parameters indicated, and, as such, it is also parameter-dependent. The figure also shows that M_∞ is inversely proportional to the forward link power control factor K_{traf}.

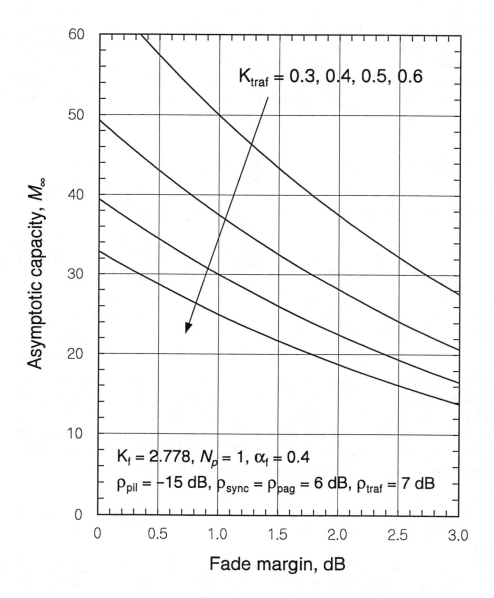

Figure 11.20 Forward link asymptotic capacity versus fade margin, parametric in K_{traf}.

11.4.1.2 Power-Limited Forward Link Capacity

The user capacity of the forward link can be said to have been reached when the total transmitter power equals the maximal power constraint given in (11.18a); that is, when the total forward link power equals P_{max}. This condition is expressed when we write (11.18a) with equality by substituting (11.22a) through (11.22d):

$$P_{total} =$$

$$\frac{10^{M(dB)/10} N_m L_T(R) \left(\rho_{pil} + \dfrac{\rho_{sync}}{(PG)_{sync}} + N_p \dfrac{\rho_{pag}}{(PG)_{pag}} + K_{traf} M \alpha_f \dfrac{\rho_{traf}}{(PG)_{traf}} \right)}{1 - 10^{M(dB)/10} K_f \left(\rho_{pil} + \dfrac{\rho_{sync}}{(PG)_{sync}} + N_p \dfrac{\rho_{pag}}{(PG)_{pag}} + K_{traf} M \alpha_f \dfrac{\rho_{traf}}{(PG)_{traf}} \right)}$$

$$\overset{set}{=} P_{max} \tag{11.37b}$$

where

$$L_T(R) \overset{\triangle}{=} L(R) L_{tc} L_{rm} / G_c G_m \tag{11.37c}$$

Solving for M in this equation, which assumes that all the channels have the same margin, results in an expression for what can be called the *power-limited capacity* of the forwark link, denoted $M(P_{max})$:

$$M(P_{max}) = \frac{(PG)_{traf}}{K_{traf} \, \alpha_f \, \rho_{traf}} \left[\frac{10^{-M(dB)/10} P_{max}}{N_m L_T(R) + K_f P_{max}} - \rho_{pil} - \frac{\rho_{sync}}{(PG)_{sync}} - N_p \frac{\rho_{pag}}{(PG)_{pag}} \right] \tag{11.37d}$$

$$= \frac{128}{K_{traf} \, \alpha_f \, \rho_{traf}} \left[\frac{10^{-M(dB)/10} P_{max}}{N_m L_T(R) + K_f P_{max}} - \rho_{pil} - \frac{\rho_{sync}}{1024} - N_p \frac{\rho_{pag}}{256} \right] \tag{11.37e}$$

Calculations of (11.37d) are shown in Figure 11.21 and 11.22 for two values of P_{max}: 1 W and 10 W. In Figure 11.21, we assume that the transmission loss is 125 dB; while in Figure 11.22 we assume that it is 135 dB, corresponding to a larger cell. In each figure, K_{traf} is varied as a parameter. Figure 11.21 shows that when $L_T(R) = 125$ dB, there is little difference in power-limited capacity between the cases of $P_{max} = 1$ W and $P_{max} = 10$ W. That is, using ten times

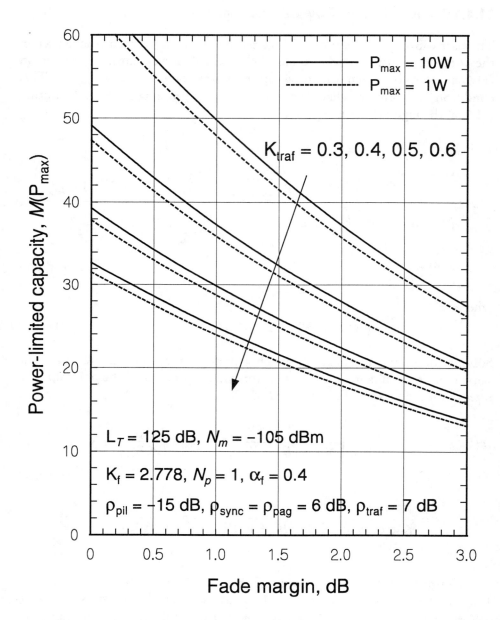

Figure 11.21 Power-limited capacity versus fade margin for $L_T = 125$ dB and parametric in K_{traf}.

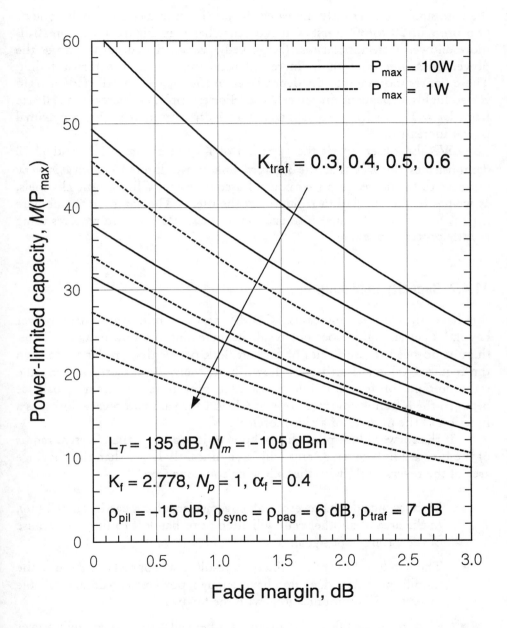

Figure 11.22 Power-limited capacity versus fade margin for $L_T = 135$ **dB and parametric in** K_{traf}.

the base station power only increased the possible number of users by one or two users in 20 to 30 users! On the other hand, in Figure 11.22 for 10 dB more allowed transmission loss (larger cell), we see that using ten times the power does indeed result in increased user capacity on the forward link. These figures indicate that the dependence of the capacity on the fade margin is also highly dependent on the cell size. For example, an increase in cell size from $L_T = 125$ to $L_T = 135$ showed a 4 : 1 capacity increase for a maximal power increase of 10 : 1.

We also observe that the variation of K_{traf} in Figures 11.21 and 11.22 demonstrates the fact that the forward link power-limited capacity is very sensitive to the degree of power control used on forward link traffic channels, as well as to the physical distribution of the users. The common theme here is that, by whatever means, optimization of the CDMA system requires using as little power as possible.

11.4.2 Capacity Balancing

We have shown that increasing CDMA forward link transmitter power beyond a certain value does not result in increasing the link margin. Thus, there is no need to use values of forward link power that are far more than the required values. Further, as shown here, there is a reason for using the right amount of forward link transmitter power—to maintain a balance between the effective coverage areas of CDMA forward and reverse links, as a function of the number of active mobile users.

If the forward link pilot channel power is too large, the effective radius of the forward link coverage area will be significantly larger than the coverage area of the reverse link, with the following consequences:

- Mobiles outside the reverse link coverage area for a particular cell (i.e., in the area of another cell) will attempt to handoff to the cell because of the strong pilot signal.

- The handoff, however, is likely to result in a dropped call because the mobile probably does not have enough power to maintain reliable reverse link communications with the far-away cell.

- Also, if the mobile in the area of another cell does have enough power and does establish a call with the cell, it will produce excess interference on the reverse link of that other cell.

If the forward link pilot channel power is too small, the effective radius of the forward link coverage area is significantly smaller than the coverage area of the reverse link, with the following consequences:

- Mobiles inside the cell's reverse link coverage area experience poor forward link quality at the edge of that area.

- Mobiles will have erratic handoff behavior because (assuming all the cells use the same forward link power allocation policy) the pilot signals of adjacent cells will be weak in the regions between cells.

The radius of the forward link coverage area and that of the reverse link coverage area are approximately equal if the propagation losses on forward and reverse links are equal. Using this concept, a link-balancing parameter $\mathcal{B}$, as defined in (10.37a), is given by

$$\mathcal{B} \triangleq L_{reverse} - L_{forward} \quad (\text{dB}) \tag{11.38}$$

and maintained close to a zero value, $\pm \delta$, where δ is a tolerance that acknowledges the uncertainty in the parameters used to calculate $\mathcal{B}$.

The forward and reverse link propagation losses can be expressed in terms of the parameters of each link's power budget. The median received SNR for the forward link pilot channel for a mobile at the cell edge can be expressed as follows:

$$\text{SNR}_{med,pil} = \frac{P_{pil}/L_T(R)}{N_m + K_f P_{total}/L_T(R)} = \frac{P_{pil}}{N_m L_T(R) + K_f P_{total}}$$

$$= \frac{\zeta_{pil} P_{total}}{N_m L_T(R) + K_f P_{total}} \tag{11.39a}$$

Solving the forward link SNR equation for $L_T(R)$ results in the equation

$$L_T = \frac{1}{N_m} \left(\frac{\zeta_{pil} P_{total}}{\text{SNR}_{med,pil}} - K_f P_{total} \right) = \frac{P_{total}}{N_m} \left(\frac{\zeta_{pil}}{\text{SNR}_{med,pil}} - K_f \right) \tag{11.39b}$$

where the pilot fraction has been shown to be[5]

---

[5] Note that the balancing analysis here now departs from that in Chapter 10 by expressing the pilot fraction as a function of M.

$$\zeta_{\text{pil}} = \frac{\rho_{\text{pil}}}{\rho_{\text{pil}} + \dfrac{\rho_{\text{sync}}}{(\text{PG})_{\text{sync}}} + N_p \cdot \dfrac{\rho_{\text{pag}}}{(\text{PG})_{\text{pag}}} + K_{\text{traf}} M \alpha_{\text{f}} \cdot \dfrac{\rho_{\text{traf}}}{(\text{PG})_{\text{traf}}}} \qquad (11.39c)$$

In dB, the solution for $L_T(R)$ on the forward link is

$$L_T(\text{dB}) = P_{\text{total}}(\text{dBm}) - N_m(\text{dBm}) + 10\log_{10}\left(\frac{\zeta_{\text{pil}}}{\text{SNR}_{med,\text{pil}}} - K_{\text{f}}\right) \qquad (11.39d)$$

We assume that the intended value of $\text{SNR}_{med,\text{pil}}$ in (11.39d) is the required pilot SNR plus a margin:

$$\text{SNR}_{med,\text{pil}} = \rho_{\text{pil}} \cdot \Gamma_r = \rho_{\text{pil}}(\text{dB}) + M(\text{dB}) \qquad (11.39e)$$

The median received SNR for the reverse link traffic channel for a mobile at the cell edge can be expressed as follows:

$$\begin{aligned}
\text{SNR}_{med} &= \frac{P_m/L_T(R)}{N_c + \left(\dfrac{M}{F_e} - 1\right)\alpha_r P_m/L_T(R)} \\[2mm]
&= \frac{P_m}{N_c L_T(R) + \left(\dfrac{M}{F_e} - 1\right)\alpha_r P_m} \qquad (11.40a)
\end{aligned}$$

where P_m is the maximal mobile transmitter power[6], N_c is the base station receiver noise power, α_r is the reverse link voice activity factor, and F_e is the reuse efficiency of the reverse link. Solving the reverse link SNR equation for $L_T(R)$ results in the equation

$$L_T(R) = \frac{P_m}{N_c}\left[\frac{1}{\text{SNR}_{med}} - \left(\frac{M}{F_e} - 1\right)\alpha_r\right] \qquad (11.40b)$$

In dB, the solution for $L_T(R)$ on the reverse link is

---

[6] Note that P_m is assumed to be the maximal transmitter power here because the transmission loss is specified for a distance equal to the cell radius R. For a different distance $r < R$, P_m should be interpreted simply as the value of mobile transmitter power.

$$L_T(dB) = P_m(dBm) - N_c(dBm) + 10\log_{10}\left[\frac{1}{SNR_{med}} - \left(\frac{M}{F_e} - 1\right)\alpha_r\right]$$

$$(11.40c)$$

We assume that the intended value of SNR_{med} is the required traffic SNR plus a margin:

$$SNR_{med} = \frac{\Gamma_{rev}(E_b/\mathcal{N}_0)_{rev}}{(PG)_{rev}}$$

$$= \left(\frac{E_b}{\mathcal{N}_0}\right)_{rev}(dB) - (PG)_{rev}(dB) + M_{rev}(dB) \qquad (11.40d)$$

where $(E_b/\mathcal{N}_0)_{rev}$ is the reverse link traffic channel $E_b/\mathcal{N}_0$ requirement, $(PG)_{rev}$ is the reverse link processing gain, and Γ_{rev} is the reverse link margin, and

$$M_{rev}(dB) \triangleq 10\log_{10}\Gamma_{rev} \qquad (11.40e)$$

Given that the antenna gains and implementation losses are the same for forward and reverse links, we can express the balancing parameter in terms of the net losses on the two links:

$$\mathcal{B} = (L_T)_{reverse} - (L_T)_{forward} \quad \text{in dB} \qquad (11.41a)$$

Substituting for the two transmission losses, we find that the balancing parameter $\mathcal{B}$ is given by

$$\mathcal{B} = \left\{P_m(dBm) - N_c(dBm) + 10\log_{10}\left[\frac{1}{SNR_{med}} - \left(\frac{M}{F_e} - 1\right)\alpha_r\right]\right\}$$

$$- \left\{P_{total}(dBm) - N_m(dBm) + 10\log_{10}\left(\frac{\zeta_{pil}}{SNR_{med,pil}} - K_f\right)\right\} \quad (11.41b)$$

If we set $\mathcal{B}$ equal to zero, the target value for balancing, then the following expression for forward link total power is found:

$$P_{total}(dBm) = P_m(dBm) - N_c(dBm)$$

$$+ 10 \log_{10} \left[\frac{1}{SNR_{med}} - \left(\frac{M}{F_e} - 1 \right) \alpha_r \right]$$

$$+ N_m(dBm) - 10 \log_{10} \left(\frac{\zeta_{pil}}{SNR_{med,pil}} - K_f \right) \qquad (11.42a)$$

which in absolute units becomes

$$P_{total} = P_m \cdot \frac{N_m}{N_c} \cdot \frac{\dfrac{1}{SNR_{med}} - \left(\dfrac{M}{F_e} - 1 \right) \alpha_r}{\dfrac{\zeta_{pil}}{SNR_{med,pil}} - K_f} \qquad (11.42b)$$

Because the pilot power is simply $P_{pil} = \zeta_{pil} \cdot P_{total}$, we can also write

$$P_{pil} = P_m \cdot \zeta_{pil} \cdot \frac{N_m}{N_c} \cdot \frac{\dfrac{1}{SNR_{med}} - \left(\dfrac{M}{F_e} - 1 \right) \alpha_r}{\dfrac{\zeta_{pil}}{SNR_{med,pil}} - K_f} \qquad (11.42c)$$

$$= P_m \cdot \zeta_{pil} \cdot \frac{N_m}{N_c} \cdot \frac{SNR_{med,pil}}{SNR_{med}} \cdot \frac{1 - \left(\dfrac{M}{F_e} - 1 \right) \alpha_r \, SNR_{med}}{\zeta_{pil} - K_f \, SNR_{med,pil}} \qquad (11.42d)$$

We now observe the relationship between the pilot channel power and the mobile transmitter power under capacity balance conditions by plotting (11.42d) as a function of M with other system parameters. Typically, the mobile's receiver noise figure is 3 dB more than the noise of the base station receiver, and, hence, we assume that mobile receiver noise is twice that at the base station, so that $N_m/N_c = 2$ is a reasonable assumption. The respective median SNRs to use in (11.42d) are

$$SNR_{med,pil} = \rho_{pil} \, \Gamma_r = 10^{-15/10} \, \Gamma_r = 0.0316 \, \Gamma_r \qquad (11.42e)$$

and

$$SNR_{med} = \frac{\rho_{rev}}{(PG)_{rev}} \Gamma_{rev} = \frac{10^{7/10}}{128} \, \Gamma_{rev} = 0.0392 \, \Gamma_{rev} \qquad (11.42f)$$

Note that Γ_r is the forward link receiver margin, whereas Γ_{rev} is the reverse link base station receiver margin. Using the expression for ζ_{pil} given in (11.39c) and substituting assumed numerical values in (11.42d) that have been introduced for the parameters [see (11.18f) and (11.23a)], including $F_e = 0.75$ as in (10.17a), (11.42d) becomes

$$
\mathrm{P_{pil}} = \mathrm{P}_m \times \overbrace{\frac{.0316}{.0316 + .00389 + .0156 N_p + .00881 M}}^{\zeta_{\mathrm{pil}}} \times 2 \times \frac{.0316\,\Gamma_r}{.0392\,\Gamma_{\mathrm{rev}}}
$$

$$
\times \frac{1 - (1.33M - 1) \times .4 \times .0392\,\Gamma_{\mathrm{rev}}}{\dfrac{.0316}{.0316 + .00389 + .0156 N_p + .00881 M} - 2.778 \times .0316\,\Gamma_r}
$$

$$
= \mathrm{P}_m \times \frac{\Gamma_r}{\Gamma_{\mathrm{rev}}} \times 1.615 \times \frac{1 - .0157\,\Gamma_{\mathrm{rev}}(1.33M - 1)}{1 - 2.778\,\Gamma_r(.0355 + .0156 N_p + .00881 M)}
$$

$$
\tag{11.42g}
$$

Calculations of (11.42g) are shown in Figures 11.23 and 11.24 as a function of the number of users in the form of the ratio of the pilot channel transmitter power $\mathrm{P_{pil}}$, normalized by the mobile station transmit power P_m. In Figure 11.23, the forward and reverse link margins are varied as a parameter, while in Figure 11.24 the forward link power control factor $\mathrm{K_{traf}}$ is varied as a parameter. These figures display the "correct" amount of pilot channel transmitter power as the number of traffic channels varies. We see in Figure 11.24 that the pilot power starts at about twice the value of P_{max} and increases as M increases. This figure also reveals the correct relationship between the transmitter mobile power and the pilot power under the capacity-balanced condition, which can also be exploited as means for open-loop power control.

Note in Figure 11.24 that the value of $\mathrm{K_{traf}}$ determines whether the ratio $\mathrm{P_{pil}}/\mathrm{P}_m$ decreases or increases as a function of M. When pilot power increases more than mobile power, it is because there is a higher effective value of the interference factor on the forward link than on the reverse link.

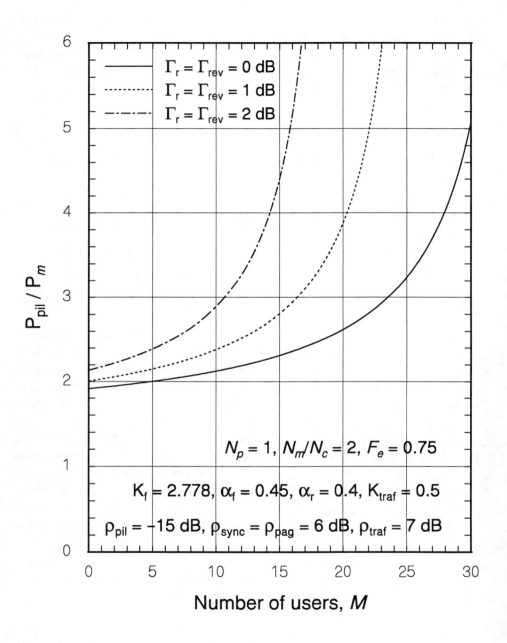

Figure 11.23 P_{pil}/P_m as a function of the number of users.

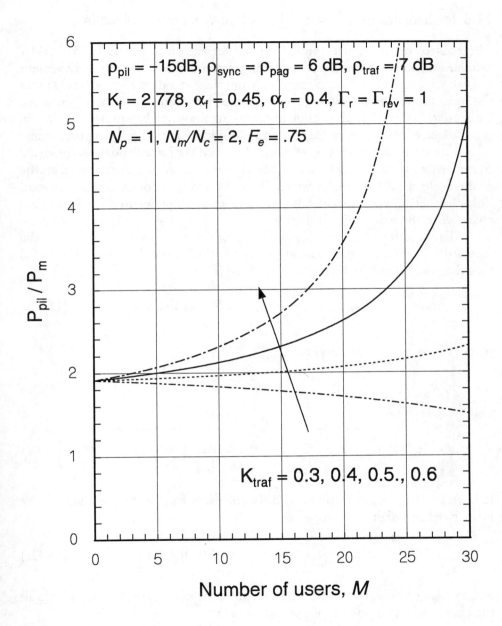

Figure 11.24 P_{pil}/P_m as a function of the number of users, parametric in K_{traf}.

11.5 Implementation of Forward Link Dynamic Power Allocation

The control of transmitter power on the forward link of the IS-95 CDMA cellular system has been shown to be of great importance for the achievement of high user capacity. The common air interface standard IS-95 states [1] that the base station "may" enable forward link power control for traffic channels. This power control is an option that may or may not be employed. When this power control is enabled, the mobile station periodically reports frame error rate statistics to the base station. The base station may use the reported frame error rate statistics to adjust the transmit power of the forward traffic channel. In what follows, an approach is described for controlling all forward link channel powers through the effective gains implemented for individual channels in the base station equipment.

The solutions we have obtained for the optimal power allocation of the forward link Walsh channels were given in (11.24a) to (11.24d). We noted that the power for Walsh channel j is of the form

$$P_j = \frac{\rho'_j}{(PG)_j}\, \nu, \quad \text{for Walsh channel } j \qquad (11.43a)$$

where ν was given in (11.24e) to be

$$\nu\left(\rho'_{\text{pil}},\, \rho'_{\text{sync}},\, \rho'_{\text{pag}},\, \rho'_{\text{traf}}\right) \qquad (11.43b)$$

$$\triangleq \frac{N_m L_T(R)}{1 - K_f\left(\rho'_{\text{pil}} + \dfrac{\rho'_{\text{sync}}}{(PG)_{\text{sync}}} + N_p \dfrac{\rho'_{\text{pag}}}{(PG)_{\text{pag}}} + K_{\text{traf}}\, M \alpha_f \dfrac{\rho'_{\text{traf}}}{(PG)_{\text{traf}}}\right)}$$

It is clear that we can implement different gains for the Walsh channels by using baseband digital voltage gains

$$d_j = \sqrt{\rho'_j/(PG)_j}, \quad \text{Walsh channel } j \qquad (11.43c)$$

and by using a power amplifier to implement the common gain that affects all the channels with a power gain proportional to

$$\text{Common power gain} = \nu\left(\rho'_{\text{pil}},\, \rho'_{\text{sync}},\, \rho'_{\text{pag}},\, \rho'_{\text{traf}}\right) \qquad (11.43d)$$

We now consider a numerical example to illustrate the principle of implementing optimal forward link channel powers. Let the system parameter values shown in Table 11.5 be assumed. The relative powers of the different channels may be controlled by giving each channel a different gain. This is most easily implemented at baseband, as the conceptual diagram in Figure 11.25 illustrates. The following Walsh channel relative digital gains are calculated for the nominal IS-95 parameter values as shown in Table 11.5:

$$d_1 \triangleq \sqrt{\rho_1'/(\mathrm{PG})_1} = \sqrt{.0316/1} \qquad = 0.1778 \text{ for the pilot channel}$$

$$d_2 \triangleq \sqrt{\rho_2'/(\mathrm{PG})_2} = \sqrt{3.98/1024} \qquad = 0.0624 \text{ for the sync channel}$$

$$d_3 \triangleq \sqrt{\rho_3'/(\mathrm{PG})_3} = \sqrt{3.98/256} \qquad = 0.1247 \text{ for each paging channel}$$

$$d_4 \triangleq \sqrt{\rho_j'/(\mathrm{PG})_j} = \sqrt{5.01/128} \qquad = 0.1979 \text{ for each traffic channel}$$

$$(11.43e)$$

where we assumed zero margin for simplicity ($\rho_j' = \rho_j$).

The diagram in Figure 11.25 shows how different gains for the Walsh channels achieves the objective of causing the channels to have different relative powers. To achieve the desired transmitter output power, there needs to be additional gain in the CDMA transmitter, denoted μ, at the RF power amplifier. Thus, there are the relative gains that are different for each channel, and there is a common gain that affects all the channels in the same way and causes the correct amount of output power to be delivered to the transmitter antenna.

Let R_{out} be the output load seen by the power amplifier, and let the power output for a particular Walsh channel be denoted P_j. The amount of common voltage gain that is needed is

Table 11.5 Assumed parameter values

Parameter	Value	Parameter	Value
Pilot $(E_c/N_0)_{req}$	-15 dB	Receiver noise power	-105 dBm
Sync $(E_b/N_0)_{req}$	6 dB	Base xmission losses	2 dB
Paging $(E_b/N_0)_{req}$	6 dB	Mobile reception losses	3 dB
Traffic $(E_b/N_0)_{req}$	7 dB	Voice activity factor	0.4
Base antenna gain	14.1 dBi	Power control factor	0.5
Mobile antenna gain	2.1 dBi	Interference factor	2.778

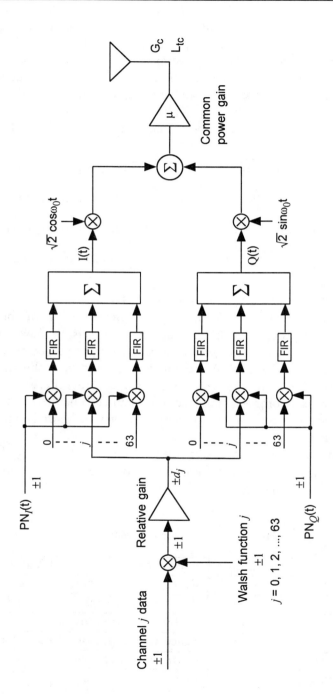

Figure 11.25 Dynamic forward link channel power control.

$$\text{Common voltage gain} = \sqrt{\mu} = \frac{\text{Desired channel output amplitude}}{\text{Relative channel amplitude} \times V} \quad (11.44a)$$

where $\pm V$ is the voltage corresponding to the baseband ± 1 logic values of the data. Since Power $= (\text{Amplitude})^2/\text{Load}$, then

$$\text{Amplitude} = \sqrt{\text{Power} \times \text{Load}} \quad (11.44b)$$

and the desired channel output amplitude is

$$\text{Amplitude} = \sqrt{(P_I + P_Q) \times R_{out}} = \sqrt{P_j \times R_{out}} \quad (11.44c)$$

where $P_I = \frac{1}{2} P_j$ and $P_Q = \frac{1}{2} P_j$ are the I-channel and Q-channel output powers and R_{out} is the amplifier load in ohms. For example, the ± 1 data logic may correspond to ± 3 volts ($V = 3$), and the impedance seen by the transmitter power amplifier output may be 50 ohms.

Thus, using Γ to denote the margin, assumed to be the same in each channel, and selecting any channel j to determine the needed gain, the needed output power gain is

$$
\begin{aligned}
\text{Output power gain} = \mu &\triangleq \frac{P_j}{(d_j \times V)^2/R_{out}} \\[2mm]
&= \frac{\rho_j \times \nu}{d_j \times V/R_{out}} \\[2mm]
&= \frac{\Gamma \cdot \nu\left(\rho'_{pil}, \rho'_{sync}, \rho'_{pag}, \rho'_{traf}\right)}{V^2/R_{out}} \quad (11.44d)
\end{aligned}
$$

In terms of the detailed parameters, the output power gain is given by

$$
\begin{aligned}
\mu &= \frac{\Gamma\, N_m L_T}{1 - \Gamma K_f\left(\rho_{traf} + \frac{\rho_{sync}}{1024} + N_p \frac{\rho_{pag}}{256} + K_{traf} M \alpha_f \frac{\rho_{traf}}{128}\right)} \cdot \frac{1}{V^2/R_{out}} \\[4mm]
&= \frac{\Gamma\, N_m\, L \cdot L_{tc} \cdot L_{rm}/G_c G_m (V^2/R_{out})}{1 - \Gamma K_f\left(\rho_{traf} + \frac{\rho_{sync}}{1024} + N_p \frac{\rho_{pag}}{256} + K_{traf} M \alpha_f \frac{\rho_{traf}}{128}\right)} \quad (11.44e)
\end{aligned}
$$

For N_p and M varying, for $V^2/R_{out} = (3)^2/50 = 0.18\,\text{W} = 180\,\text{mW}$, and for the nominal system parameters presented previously:

$$\mu = \sqrt{\frac{\Gamma \cdot 10^{(-10.5+L(\text{dB})/10+.2+.3-1.41-.21)}/180}{1 - \Gamma \cdot 2.778(0.0355 + N_p \cdot 0.0156 + 0.5 \cdot M \cdot 0.45 \cdot 0.0392)}}$$

$$= 10^{(L(\text{dB})-138.8\,\text{dB})/20}\sqrt{\frac{\Gamma}{1 - 2.778\,\Gamma(0.0355 + 0.01567 N_p + 0.00881 M)}}$$

$$(11.44\text{f})$$

Properties of the multiplexed waveform. Recall from the orthogonal multiplexing example that was shown in Figures 5.5 and 5.6 that the Walsh function-multiplexed waveform has periodic peaks due to the agreement of the Walsh functions with each other in certain chip positions. To continue the numerical example used above, we show waveforms obtained from the superposition of differently weighted Walsh channels, and from observations of the waveforms we comment on requirements for CDMA infrastructure and test equipment. Suppose that the gains given in (11.43e) are used to simulate an IS-95 baseband data waveform comprised of the sum of 18 Walsh channels, as follows:

- Pilot signal with voltage gain 0.1778 and Walsh function H_0[7];

- Sync channel with voltage gain 0.0624 and Walsh function H_{32};

- Two active paging channels with voltage gains 0.1247 and Walsh functions H_1 and H_2;

- Two active traffic channels to mobiles at the edge, with voltage gains 0.1979 and Walsh functions H_8 and H_9;

- Four active traffic channels to mobiles at distances requiring half the power for a mobile at the cell edge, with voltage gains $0.1979/\sqrt{2} = 0.14$ and Walsh functions H_{10} to H_{13};

- Eight active traffic channels to mobiles at distances requiring one-fourth the power for a mobile at the cell edge, with voltage gains $0.1979/2 = 0.099$ and Walsh functions H_{14} to H_{21}.

---

[7] Here we use the H_i notation for the 64-chip Walsh functions, indexed as in IS-95. See Table 5.8 for a listing of these functions.

In addition, for each pair of paging and traffic channels, we assume that one has an input data value of +1 (logic 0) and the other has an input data value of −1 (logic 1). Figure 11.26 shows the superposition of these weighted Walsh channels prior to combining with I- and Q-channel PN codes, FIR filtering, and modulation by sinusoidal carriers for forward link transmission. In anticipation of comparing this figure with the results of FIR filtering, a six-chip delay is included, as was included in Figures 1.50 and 1.53 to simulate the IS-95 FIR filter delay. Recall that the first chip in each 64-chip Walsh sequence has the same value. Therefore, since the pairs of paging and traffic channels cancel when their Walsh function values agree, the first (delayed) Walsh chip value is the sum of pilot and sync channel amplitudes, so that the first (delayed) value is $0.1778 + 0.0624 = 0.2402$. Thereafter, the total amplitude of the signal depends on the particular combination of Walsh function agreements and disagreements. If all the data-modulated Walsh chips had the same sign for some chip time, the total amplitude would equal

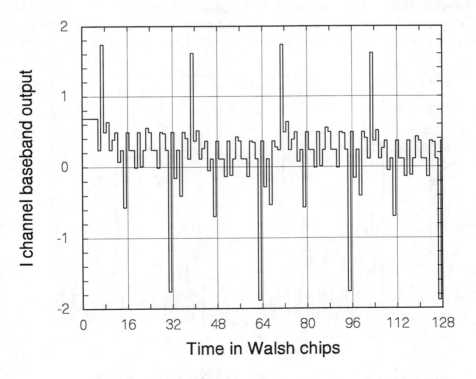

Figure 11.26 Simulated Walsh-multiplexed combination of pilot, sync, paging, and traffic channels with different voltage gains.

the sum of the gains of all the simulated channels, or ± 2.2374; we see from Figure 11.26 that the total amplitude for the particular Walsh functions that are used in this example varies between −1.85 and +1.75, and the amplitude waveform, in addition to having a period of 64 chips, is quite "peaky."

After the waveform of Figure 11.26 is separately combined with I- and Q-channel PN codes, and then is filtered and used to modulate cosine and sine carriers, the resulting envelope is the waveform shown in Figure 11.27. Note that the envelope waveform is characterized by relatively large peaks that occur periodically if the channel input data are held constant, as in this simulation. There are two practical implications of this characteristic behavior of the IS-95 forward link waveform. First, the forward link transmitter's power amplifier must have very good linearity in order to deliver the Walsh-multiplexed waveform without significant distortion. Second, it is obvious that the testing of such amplifiers using a channel simulator cannot be performed using bandlimited noise, even though the signal spectrum resembles that of bandlimited noise, because the envelope of the actual IS-95 time waveform does not resemble that of a noise waveform but has distinctive pulse-like features with a significantly large dynamic range.

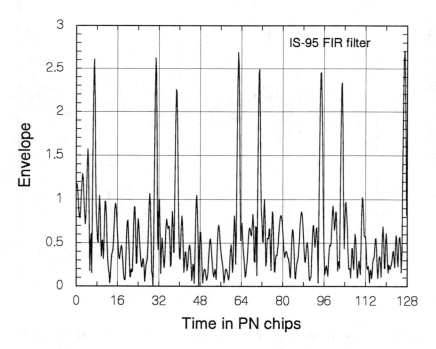

Figure 11.27 Envelope of simulated IS-95 signal.

Implementation of power allocation using measurements. The solution for forward link powers presented in Section 11.2 is based on modeling the total forward link power using the quantity K_{traf}, the forward link power control factor, to estimate the total traffic channel power in terms of the power needed to transmit to a mobile user at the edge of the cell. An alternative solution can be based on using a measurement of the actual total traffic power instead of that estimate. Under this approach, the equations to be solved, equations (11.18a) to (11.18e), are replaced by the following:

$$P_{total} = P_{pil} + P_{sync} + N_p P_{pag} + P_{tt} \leq P_{max} \tag{11.45a}$$

$$\left(\frac{E_c}{\mathcal{N}_{0,T}}\right)_{pilot} = \frac{(PG)_{pil} P_{pil}}{N_m L_T(R) + K_f P_{total}} \geq \rho_{pil} \tag{11.45b}$$

$$\left(\frac{E_b}{\mathcal{N}_{0,T}}\right)_{sync} = \frac{(PG)_{sync} P_{sync}}{N_m L_T(R) + K_f P_{total}} \geq \rho_{sync} \tag{11.45c}$$

$$\left(\frac{E_b}{\mathcal{N}_{0,T}}\right)_{pag} = \frac{(PG)_{pag} P_{pag}}{N_m L_T + K_f P_{total}} \geq \rho_{pag} \tag{11.45d}$$

where P_{tt} denotes the actual total traffic channel power on the forward link, assuming that this power is being controlled by the CDMA system on a per-user basis. Taking the case of equality for each of these equations, the corresponding joint solutions for the signaling (non-traffic) channels are

$$P_{pil} = \frac{\left(N_m L_T + K_f P_{tt}\right) \rho_{pil}/(PG)_{pil}}{1 - K_f\left(\rho_{pil} + \dfrac{\rho_{sync}}{(PG)_{sync}} + N_p\dfrac{\rho_{pag}}{(PG)_{pag}}\right)} \tag{11.46a}$$

$$P_{sync} = \frac{\left(N_m L_T + K_f P_{tt}\right) \rho_{sync}/(PG)_{sync}}{1 - K_f\left(\rho_{pil} + \dfrac{\rho_{sync}}{(PG)_{sync}} + N_p\dfrac{\rho_{pag}}{(PG)_{pag}}\right)} \tag{11.46b}$$

$$P_{pag} = \frac{\left(N_m L_T + K_f P_{tt}\right) \rho_{pag}/(PG)_{pag}}{1 - K_f\left(\rho_{pil} + \dfrac{\rho_{sync}}{(PG)_{sync}} + N_p\dfrac{\rho_{pag}}{(PG)_{pag}}\right)} \tag{11.46c}$$

In this solution, the SNR requirements with margin (ρ'_{pil}, ρ'_{sync}, and ρ'_{pag}) may be used in place of the requirements with zero-margin. Note that the formulation does not result in a solution for traffic channel power, since it is assumed that the forward link power control is operative. If desired, however, we can calculate a value of traffic channel power for initializing the forward traffic power control loop by replacing ρ_{pag} and $(PG)_{pag}$ in the numerator of (11.46c) with ρ_{traf} and $(PG)_{traf}$, respectively.

Also, instead of estimating other-cell interference as K_{other} times the received forward link power at the cell edge, we can find the signaling channel powers using measurements or some other estimate of the term I_{other}, resulting in the pilot channel power solution (for example) given by

$$
P_{pil} = \frac{\left[(N_m + I_{other})L_T + K_{same}P_{tt}\right]\rho_{pil}}{1 - K_{same}\left(\rho_{pil} + \dfrac{\rho_{sync}}{(PG)_{sync}} + N_p\dfrac{\rho_{pag}}{(PG)_{pag}}\right)}
\tag{11.47}
$$

Note that, with or without these measurements, the solution for pilot power results in a value of the pilot power fraction ζ_{pil} that is not a fixed value but adapts to the amount of traffic and interference.

References

[1] "Mobile Station-Base Station Compatibility Standard for Dual-Mode Wideband Spread Spectrum Cellular System," TIA/EIA Interim Standard 95 (IS-95), Washington, DC: Telecommunications Industry Association, July 1993 (amended as IS-95-A in May 1995).

[2] Singer, A., "Improving System Performance," *Wireless Review*, pp. 74–76, Feb. 1, 1998.

[3] Owens, D., "The Big Picture," *CDMA Spectrum*, pp. 36–39, Dec. 1997.

[4] Qualcomm, Inc., *CDMA System Engineering Training Handbook*, draft version X1, 1993.

[5] Lee, J. S., and L. E. Miller, "Dynamic Allocation of CDMA Forward Link Power for PCS and Cellular Systems" (invited paper), *Proc. 2nd CDMA Internat'l Conf.*, pp. 95–99, Oct. 21–24, 1997, Seoul, Korea.

About the Authors

Jhong Sam Lee received his B.S.E.E. from The University of Oklahoma in 1959, his M.S.E. from The George Washington University in 1961, and his D.Sc. from The George Washington University in 1967, all in electrical engineering. He is founder and President of J. S. Lee Associates, Inc. (JSLAI), and has been responsible for all program initiations and management in the areas of special communications, electronic warfare (EW) systems, anti-submarine warfare (ASW) signal processing, and spread-spectrum systems. Dr. Lee is a Fellow of the IEEE.

He was an Assistant Professor at The George Washington University (1965 to 1968) and an Associate Professor at The Catholic University of America (1969 to 1973), teaching digital communications and detection and coding theories. Dr. Lee has authored or co-authored more than 70 papers in IEEE publications.

From 1973 to 1976, Dr. Lee was Associate Director at Magnavox Advanced Systems Analysis Office. From 1968 to 1969, he was Advisory Engineer at IBM. He was a consultant with the U.S. Naval Research Laboratory (1965 to 1973) and with COMSAT Laboratories (1965, 1971).

Leonard E. Miller received his B.E.E. from Rensselaer Polytechnic Institute in 1964, his M.S.E.E. from Purdue University in 1966, and his Ph.D. in electrical engineering from The Catholic University of America in 1973.

Dr. Miller is Vice President for Research with JSLAI. Since joining JSLAI in 1978, he has performed detailed analysis of the design and performance of military tactical communications and EW systems, as well as commercial communications systems. He developed analytical models for independent evaluation of the survivability of tactical communications networks, including MSE, JTIDS, and EPLRS.

From 1964 to 1978, Dr. Miller was an electrical engineer at the Naval Surface Weapons Center in Maryland, where he developed and evaluated underwater detection and tracking algorithms and designed instrumentation for submarines and underwater weapons.

In addition to many technical reports and memoranda, Dr. Miller has more than 40 papers in IEEE publications. He serves as a reviewer for several IEEE transactions and for *Electronic Letters*. He is a Senior Member of the IEEE.

Index

Recent Titles in the Artech House
Mobile Communications Series

John Walker, Series Editor

Wireless Communications in Developing Countries: Cellular and Satellite Systems, Rachael E. Schwartz

Wireless Technician's Handbook, Andrew Miceli

For further information on these and other Artech House titles, including previously considered out-of-print books now available through our In-Print-Forever® (IPF®) program, contact:

Artech House
685 Canton Street
Norwood, MA 02062
Phone: 781-769-9750
Fax: 781-769-6334
e-mail: artech@artechhouse.com

Artech House
46 Gillingham Street
London SW1V 1AH UK
Phone: +44 (0)20 7596-8750
Fax: +44 (0)20 7630-0166
e-mail: artech-uk@artechhouse.com

Find us on the World Wide Web at:
www.artechhouse.com